Geophysical Monograph Series

Geophysical Monograph Series

Geophysical Monograph 286

Salt in the Earth Sciences
Evaporite Rocks and Salt Deposition

Webster Mohriak

This Work is a co-publication of the American Geophysical Union and John Wiley and Sons, Inc.

AGU
ADVANCING EARTH
AND SPACE SCIENCES

WILEY

This edition first published 2025

Published by John Wiley & Sons, Inc., Hoboken, New Jersey

Published under the aegis of the AGU Publications Committee

Matthew Giampoala, Vice President, Publications
Steven A. Hauck, II, Chair, Publications Committee
For details about the American Geophysical Union visit us at www.agu.org.

Registered Office
John Wiley & Sons, Inc., 111 River Street, Hoboken, NJ 07030, USA

Editorial Office
111 River Street, Hoboken, NJ 07030, USA
For details of our global editorial offices, customer services, and more information about Wiley products visit us at www.wiley.com.

Wiley also publishes its books in a variety of electronic formats and by print-on-demand. Some content that appears in standard print versions of this book may not be available in other formats. For more information about Wiley products, visit our web site at www.wiley.com.

The manufacturer's authorized representative according to the EU General Product Safety Regulation is Wiley-VCH GmbH, Boschstr. 12, 69469 Weinheim, Germany, e-mail: Product_Safety@wiley.com.

Library of Congress Cataloging-in-Publication Data Applied for:
Hardback ISBN 9781119405214

Cover Design: Wiley
Cover Image: Mount Sedom, Dead Sea; courtesy of Assaf Tsabar

Set in 10/12pt Times New Roman by Straive, Pondicherry, India

CONTENTS

PREFACE

Salt is formed when water collects on the Earth's surface and then evaporates, leaving behind various mineral deposits called evaporites. Over thousands of years and in many parts of the world, salt deposits have formed in sedimentary basins in extensional, convergent, and strike-slip settings. Over time, pressure and tectonics resulted in salt flow and changes in the structure and shape of salt sequences. Understanding the distribution and characteristics of salt deposits is of geological and socioeconomic interest, as salt serves various purposes in human society and evaporite basins house many of the world's hydrocarbon reserves.

Geological models for salt structures are based on observations of surface outcrops or interpretations of the subsurface structures derived from indirect geophysical methods and exploratory drilling. Recent advances in geophysical technologies, such as seismic imaging, have improved the interpretation of subsurface evaporite structures. Studies based on 2D and 3D seismic datasets, physical and numerical models, and the interpretation of deep seismic reflection profiles help develop novel concepts about basin evolution. This knowledge has significant applications in basin analysis, applied geology, and the mineral and petroleum industries. In the past century, the study of salt tectonics has contributed to global basin analysis projects, resulting in major hydrocarbon discoveries in giant salt basins.

This two-volume set, *Salt in the Earth Sciences*, presents a comprehensive, global overview of salt deposition and deformation in sedimentary basins, bringing together data analysis, observations, theories, and modeling. The first volume, *Salt in the Earth Sciences: Evaporite Rocks and Salt Deposition*, covers the basic principles and properties of evaporite rocks (Part I) and then moves to more complex examples of salt deposition in diverse sedimentary environments (Part II). The second volume, *Salt in the Earth Sciences: Basin Analysis and Salt Tectonics*, examines giant salt basins around the world and presents more advanced discussions of salt tectonic events and their global implications. Each volume functions as a standalone book, but readers can explore both volumes together to enhance their understanding of the respective topics. The volumes are linked by their global perspectives of selected salt basins.

The first chapter in *Salt in the Earth Sciences: Evaporite Rocks and Salt Deposition* presents a brief overview of salt use by humanity since prehistoric times, discussing its influence on past civilizations and its impact on the modern industrial world. Chapters 2–7 cover the basic chemical and physical principles of salt deposition in sedimentary basins, with key concepts presented in plain language. These chapters also discuss the effects of gravity and tectonic forces on rock salt deformation, as well as the development of intriguing salt structures exposed in orogenic belts or hidden in the subsurface of many sedimentary basins. The basic principles of seismic methods applied to the identification of salt structures in the subsurface are briefly summarized, and several seismic examples are included in the discussion of classical salt basins.

Chapters 8–12 discuss the key sedimentological models for salt deposition, comparing them with modern environments and ancient basins. These chapters compare modern rift lakes in Africa, North and South America, and Chinese salt lakes in the Tibetan Plateau to giant salt basins of the ancient past. Several chapters discuss the Dead Sea and Mediterranean Sea genetic models as analogs for ancient salt sequences accumulated in deep and shallow water settings.

Each chapter includes an extensive bibliography with key references. Readers will also benefit from descriptions of natural events, photographs of natural outcrops, satellite images, and geological maps throughout the book.

The content should be useful to undergraduate and graduate students studying Earth sciences. Since the book presents an introduction to the importance of salt for basin analysis projects in the petroleum and mining industries, it will also be of use to professionals in the fields of petroleum exploration, salt mining, and engineering. Students from other scientific fields, including archaeology, anthropology, biology, physics, and chemistry, will also find material of interest in this book. Readers need only a basic understanding of geological principles applied to sedimentary basin analysis to grasp the content explored in this book.

Webster Mohriak
Rio de Janeiro State University, Brazil

ACKNOWLEDGMENTS

This book presents a comprehensive overview of salt deposition principles, fundamental geological concepts, and salt tectonics methods applied to basin analysis, a field that has been extensively studied and researched by various experts since the last century. Some chapters are expanded from topics discussed in the book *Sal: Geologia e Tectônica* published in 2006 by Editora Beca in Brazil. The book was organized by Webster Mohriak, Peter Szatmari, and Sylvia M. Couto Anjos, and most chapters included the participation of several Petrobras explorationists and researchers. In addition, the present book benefits from decades of technological advancements and studies conducted by renowned specialists working in international salt basins since the publication of the second edition of Petrobras's book in 2009.

I would like to express my special gratitude to Martin Jackson, a lifelong colleague who coordinated the Salt Tectonics project at the University of Texas's Applied Geodynamics Laboratory (AGL) for more than two decades and participated in many technical discussions involving the South Atlantic basins, with the help of many of AGL's staff, particularly B. Vendeville, G. Guglielmo, M. Hudec, and T. Dooley. The book also includes Rio de Janeiro State University (UERJ) training materials that have been used in several MS and PhD courses since 2012.

Many chapters in the book include results from academic projects and technical works from several MS and PhD projects conducted at UERJ, which utilized numerous seismic profiles of salt tectonics from the Brazilian margin, mainly contributed by the Brazilian National Agency of Petroleum, Natural Gas and Biofuels (ANP). The book also includes several fresh, unpublished seismic profiles from numerous basins worldwide, courtesy of several seismic vendors. Multiple NASA satellite images and public databases were utilized to identify several geographical features. The University of Bern in Switzerland and the Nagra project provided high-resolution images of salt rocks in the Alps. I am grateful to CGG Multi-Client & New Ventures for providing high-quality seismic images, particularly Sara Pink-Zerling and Madhurima Bhattacharya, and for selecting didactic seismic profiles from many salt basins in the South Atlantic, Caspian Sea, and North Sea. PGS is thanked for supplying high-resolution seismic data images from the Santos Basin, particularly Stephane Dezaunay and Hermann Lebit. I would also like to acknowledge the support of Petrobras, ANP, the National Research Council, FAPERJ, and other research agencies during the preparation of several studies that are synthesized in the present book.

Rituparna Bose, the Wiley Books Editor, originated this book project when she invited me to submit a proposal following the 2016 International Geological Congress in South Africa. I collaborated with numerous colleagues and friends from the petroleum industry and academia during the writing process. Preliminary chapters were written from 2017 to 2020. Stephanie Wischer critically read the early chapters from 2020 to 2021, and I am grateful for her efforts in providing some guidelines and helping to simplify complicated subjects. Ana Kruger, Gabor Tari, Ian Davison, and Simon Stewart reviewed the final volume. I am grateful to many Wiley staff, particularly Mandy Collison, Layla Harden, Noel McGlinchey, Keerthana Govindarajan, Nathanael Mcgavin, Neena Ganjoo, Summers Scholl, Frank Weinreich, and Ranjith Kumar Thanigasalam, for their help throughout the writing, review, and revision process. Thanks also to Estella Atekwana from the AGU Books Editorial Board and Jenny Lunn and Dara Liling from AGU Publications for steering my book through its journey. Devipriya Somasundaram, Permissions Specialist and Content Operations leader for John Wiley & Sons, obtained approval for all the figures adapted from previous publications. During the copyediting process, Rathi Aravind and Aravind Kannankara thoroughly checked all the revised chapters, resulting in a clean, corrected final product. The PDFs for all the chapters were prepared by Avinash Makarla.

I am grateful to numerous lifetime colleagues who provided valuable suggestions and technical material for the chapters, particularly Al Danforth, Bruno Vendeville, Carl Fiduk, Christian Heine, Christopher Jackson, Christopher Scotese, David Brown, David Quirk, Evando Bartholazzi, Frank Peel, Gabor Tari, Hemin Koyi, Ian Alsop, Ian Davison, Ian Norton, Ian West, James Pindell, Janos Urai, Joseph Cartwright, Juan Soto, Katherine Giles, Magdalena Scheck-Wenderoth, Mark Rowan, Martin Jackson, Michael Hudec, Paul Post, Paul Weimer, Peter Kukla, Peter Scholle, Piotr Krzywiec, Sverre Planke, Thomas Smith, and Tim Dooley. Several researchers contributed technical material to the chapters, including A. Savin, Amos Frumkin, Barbara Cavalazzi, Christof Leitner, Daniel Traber, Eszter Banffy, Herman Lebit, Hoora Sni, Irina Filina, Jane Selvestone, Janos Urai, Javier García-Veigas, Josep Antoni Muñoz, Jūrg Alean,

Justyna Nowińska, Katherine Giles, Kelly Kimberly, Khalil Sarkarinejad, Krzysztof Bukowski, Mahdi Najafi, Marco Roveri, Marcos Roberto Fetter Lopes, Mark Wilson, Martin Mazurek, Michael Lazar, Nico Augustin, Oliver Weller, Oscar Fernandez Bellon, Philippe Crochet, Riadh Ahmadi, Simon Kemp, Simon Stewart, Stan Celestian, Stanislaw Burliga, Tony Waltham, Vinicio Manzi, Xavier Fort, Yohann Poprawski, and many others. Many Brazilian colleagues also contributed to the improvement of the didactic content of several chapters, particularly Andres C. Gordon, David Castro, Fabia Emanuela Rafaloski Bobco, Francisco R. Abrantes Jr., Gabriela Salomão Martins, Gisele Nery Camargo, Leonardo Borghi, Leonardo Muniz Pichel, Luciana Brelaz, Marcelo Bassetto, Marcio Rocha Mello, Marcos Fetter, Monica Heilbron, Paulo Palagi, Peter Szatmari, Stephanie Wischer, Sylvia dos Anjos, and Thiago Rodrigues de Mattos.

Several anonymous reviewers provided helpful feedback on chapters and insightful suggestions that helped to improve the scientific content. I am grateful for their dedication and advice.

QuillBot AI software was used when preparing this manuscript to check for typographic errors and make grammatical corrections.

1
A Brief History of Salt and Its Influence on Civilization

ABSTRACT

Humans have used salt, a chemical compound that forms evaporite minerals in sedimentary basins, since prehistoric times, and it is essential to modern societies. Salt production and trading contributed to the wealth of nations in medieval and industrial times. In the industrialized world, salt exploitation is an economic asset contributing to the food, agricultural, fertilizer, and chemical industries. Evaporite basins are often associated with petroleum provinces, as salt layers deform and create hydrocarbon accumulation traps.

1.1. INTRODUCTION

Salt is a naturally occurring chemical compound that forms various types of evaporite minerals, which are precipitated in aqueous environments such as shallow seas and lakes. The most common salt in nature is the mineral halite, composed of sodium chloride (NaCl). This chemical precipitate is characterized by a weak ionic bond, forming a substance that easily dissolves in water. Salt is a vital component of life, particularly potassium salt, which is widely used in the agricultural industry as an essential chemical compound in fertilizers.

Salt has the power to change the physical properties of liquids, making it useful in various industries. For example, when mixed with water, salt can significantly reduce freezing or melting points, making it essential for the manufacture of explosives, chlorine gas, soda products, soft drinks and cleaners, plastics, pharmaceuticals, and several chemical compounds used for medical purposes.

Salt minerals can be obtained by harvesting the precipitated minerals from evaporated salt pans. This process dates back to prehistoric times. During the Neolithic period, nomadic tribes sourced salt from raw meat, fish, and seafood. As agriculture developed, fireplaces were used to cook grains and meat, necessitating more salt for food preparation (Salt Association, 2020). Salt has also been used as a preservative for meat and fish conservation, making it a staple in diets in northern Africa, Southern Europe, Southern China, Scandinavia, and other regions. This preservative enabled Europeans to embark on sea exploration during the Middle Ages, sailing around the African continent to find new routes to India.

Salt was commonly obtained from natural evaporative ponds in arid regions like the Saharan region in northern Africa and many other places worldwide. In many regions, such as the Danakil Depression or Assal Lake in Djibouti, rainfall is scarce and the solar heat causes halite to precipitate when seawater has concentrated into brine with about 10 times the salinity of normal seawater.

Salt can also be extracted from sedimentary basins as brine from deep aquifers or reservoirs enriched with salt minerals. Salt minerals are mined from ancient evaporite beds that occur in the subsurface or outcrop as huge mountainous features known as "salt diapirs." In the modern world, less than 10% of the total salt produced is used in food preparation, while more than 90% goes to industrial applications, particularly agriculture and chemical plants, and mainly salting roads in wintertime.

Due to the fact that some evaporites are valuable economic resources, surface pits and subsurface salt mines frequently exploit bedded salt and salt diapirs. Conventional mining uses open pits to excavate and extract the salt minerals, while solution mining uses wells that inject water to dissolve the salt layers in the subsurface. Concentrated brines are then pumped back to the surface, where natural or artificial solar ponds are employed to speed up the evaporation process to obtain pure salt crystals and other economically important compounds.

Salt in the Earth Sciences: Evaporite Rocks and Salt Deposition, Geophysical Monograph 286, First Edition. Webster Mohriak.

DOI: 10.1002/9781119405245.ch01

These direct uses of salt minerals are important for the mining industry, as many mineral deposits are associated with evaporite sediments. Igneous intrusions into salt structures are relatively frequent. As a result, magmatism contributes to mineralizing brines and salt bodies provide seals for the flow of hydrothermal fluids. Ore deposits in northeastern Mexico's Sabinas Basin were formed due to the Cenozoic Laramide tectono-magmatic event, which mobilized basinal brines and resulted in the enrichment of valuable metals. These minerals were precipitated through hydrothermal fluid interaction with Mesozoic carbonate and evaporite sediments (González-Sánchez et al., 2009).

Evaporites are directly or indirectly associated with many petroleum accumulations in the most prolific petroleum provinces worldwide, such as the Middle East, the North Sea, the South Atlantic, and the Gulf of Mexico. In this century, giant hydrocarbon accumulations in the South Atlantic salt basins have become one of the main targets for the petroleum industry (see Chapter 3 in Mohriak, 2025).

1.2. HISTORICAL AND ARCHAEOLOGICAL ASPECTS OF SALT USAGE

Salt has a rich history in modern society, with various branches of science focusing on its minerals, including geology, chemistry, physics, engineering, biology, archaeology, and anthropology. The use of salt has been linked to life, death, health, religion, economics, rebellions, war, piracy, and magical powers. Ancient philosophers considered salt to be the fifth element, complementing the other four vital elements (earth, water, wind, and fire). Homer considered salt a divine substance, and Plato described it as a magical substance. Pythagoras (580 BCE–500 BCE, where BCE corresponds to Before Common Era) wrote, "Salt is born of the purest of parents: the sun and the sea." Chinese proverbs also considered salt one of the "seven necessities of life."

Several civilizations in different parts of the planet have considered salt a symbol of fertility and used it in religious ceremonies and covenants with their gods. Pliny the Elder, a renowned naturalist and philosopher who lived in the 1st century of the Common Era (CE), wrote *Naturalis Historia*, one of the first encyclopedic works in the world (Jones, 1938). Pliny is frequently quoted for his famous saying "A civilized life is impossible without salt". Hippocrates, the father of medicine, encouraged his fellow doctors in ancient Greece to use salt water to heal various ailments and diseases. Salt is known as a cheap and easily available antibiotic in medicine and has been used to sterilize wounds since prehistoric times.

During the pharaohs' times in the ancient Egyptian Empire (between the 16th and 11th centuries BCE), salt had the property of delaying the body's tissues from putrefying and decomposing after death. A mixture of halite (sodium chloride) and trona (sodium carbonate), known as "natron," was commonly employed in the mummification process. Natron, a drying substance that absorbs water and destroys grease and fat in the human body, hinders the rotting and decaying of body tissues. Archaeologists have found evidence indicating that natron has been mined from lakes and quarries in several localities in North Africa for thousands of years. Large amounts of natron are found in the sedimentary beds of several Egyptian lakes in the region known as the Natron Valley (or Wadi El Natrun) in northern Egypt, near the Nile Valley. Priests in ancient Egypt used aggregates based on natron during medical procedures, believing that the chemical substances present in the mud of the salty lakes contained preservative substances that deterred the decomposition of human and animal flesh after death.

In modern biology, microorganisms are thought to be the cause of flesh decay. The antimicrobial properties of salt solutions are linked to their ability to lower water activity, which in turn lowers the amount of bacteria that get into organic tissues. The ancient Egyptians considered the substances in salty mud to act miraculously and started using natron to anoint the corpses of their dead pharaohs, hoping its magical powers might halt nature's destructive processes after death. The pharaohs' corpses were embalmed in the process of creating mummies, which were buried together with all necessary material goods for life after resurrection, including treasures, food, and furnishing. Some of the mummies were also impregnated with natural asphalt or bitumen collected along the Dead Sea shores (Wijnker et al., 2006).

Archaeologists have found evidence indicating that natron has been mined from lakes and quarries in several localities in North Africa for thousands of years. Large amounts of natron are found in the sedimentary beds of several Egyptian lakes in the region known as the Natron Valley (or Wadi El Natrun) in northern Egypt, near the Nile Valley (Lucas, 1932). The ancient Egyptians believed salty mud substances to be miraculous and began using natron to anoint the corpses of their dead pharaohs. Mummies were buried with necessary material goods for life after resurrection, including treasures, food, and furnishings. The Egyptians correctly inferred that the chemical substances present in the mud of the salty lakes contained preservative substances that deterred the decomposition of human and animal flesh after death.

Some mummies were impregnated with natural asphalt or bitumen collected along the Dead Sea shores. The Dead Sea, located to the east of the Sinai Peninsula, was part of the Egyptian Empire at that time. This tradition continued even after the New Kingdom's last periods (circa 750 BCE) and during Roman colonization (Clark et al., 2016). Modern biological sciences interpret flesh decay as caused by microorganisms, with anaerobic bacteria and fungi attacking organic tissues and releasing rotten-egg odors. Salt solutions' antimicrobial properties reduce water activity, reducing bacterial contamination in organic tissues (Wijnker et al., 2006).

In conclusion, salt has a rich history in modern society, with multiple branches of science and historical accounts highlighting its various applications in various fields. Several books have been dedicated to describing the salt odyssey through time. For example, the books by Kurlansky (2002) and Warren (2006, 2016) are rich sources and reference publications for many tales and facts about the impact of salt on civilizations, as will be summarized in this chapter.

1.3. PHYSIOLOGICAL ASPECTS OF SALT

Salt is as essential to life as oxygen, water, vitamins, carbohydrates, proteins, and fats. All mammal embryos develop in a salty amniotic fluid produced in the mother's womb, which is similar to the origin of life on Earth. This liquid environment is reminiscent of the origin of life on this planet, considering that primordial unicellular organisms lived in seawater, as postulated in the classical work by Darwin (1859) and discussed in many papers and textbooks thereafter (e.g., Alexander et al., 2011; McClendon, 1999; Sverdrup et al., 1942). Life started in a "primitive soup" containing inorganic salts and various organic substances (Oparin, 1968).

The human body has distinct fluid systems, including blood plasma, lymphatic fluid, and extracellular fluids, all salty liquids with specific chemistry to balance vital processes. Salt is an essential part of human metabolism, and every person needs a daily intake of salt to remain alive and function properly. Some specialists suggest that a normal person needs at least 2 g of salt to replace the amounts lost through sweat or excreted in urine. The amount of salt required depends on physical activity, climate, and other social and economic factors. The total quantity of salt included in the typical U.S. diet amounts to 9.6 g/d (Cordain et al., 2005). However, the British National Health Service (NHS) recommends a diet not exceeding 6 g/day for adults and less for children.

Salt helps balance and replenish all the body's electrolytes, maintaining intracellular and extracellular fluids in the right balance to equilibrate metabolic processes. Chemical compounds derived from salts, such as potassium and sodium ions, are present in the blood of healthy people, carrying nutrients into cells, stabilizing stomach acidity, acting on digestive fermentation processes, and controlling blood pressure. Natural salt has been used as a preservative and condiment to cook meat and vegetables since the Stone Age. It stimulates salivation to enhance digestive processes, and some salts, particularly those rich in magnesium, have been used to remedy digestive problems, accelerate cell metabolism, and act as an anti-allergic agent. Some foods are naturally high in salt content, such as eggs, nuts, and sea life. Salt is also present in fresh cow's milk, and some salty cheeses are especially tasty. Since prehistoric times, milk from a mother's breast has been considered an extremely healthy source of salty nutrients for newborns.

1.4. SALT IN THE BIBLE

Salt is a common metaphor in the Bible, often used as a symbol of purification and positive values. In Kings 2:21, it is mentioned that the Lord healed the waters with salt, and covenants between men and God were sealed with salt. In Leviticus 2:13, it is written: "And every oblation of thy meat offering shalt thou season with salt; neither shalt thou suffer the salt of the covenant of thy God to be lacking from thy meat offering; with all thine offerings thou shalt offer salt."

In the New Testament (Matthew 5:13), Jesus said to his disciples, "Ye are the salt of the earth: but if the salt has lost his savor, wherewith shall it be salted? it is thenceforth good for nothing, but to be cast out, and to be trodden under foot of men." This sentence is both praise and a warning, indicating that the disciples were valuable as long as they kept up with their vows to serve God and preserve their faith. The positive meaning of the metaphor is again emphasized in the next verse from Matthew 5, when Jesus refers to his disciples as the "light of the world."

Salt was also used in ancient times in military warfare, with the strategy of punishing the conquered cities. Salting the earth destroyed the soil's capability of producing food for the defeated armies and population. Salt as a weapon or sign of punishment is mentioned several times in the Holy Bible. In the Book of Judges (9:45), it is written, "And Abimelech fought against the city all that day; and he took the city, and sowed it with salt."

The destruction of Sodom and Gomorrah as described in the Bible is also associated with tales involving salt (Neev & Emery, 1995). According to the Book of Genesis, if Abraham's family fled without looking back at the fire that would burn the cities down, they would avoid God' wrath and save themselves. Abraham's nephew, Lot, obeyed God's order, but his wife looked back to see the flames and was transformed into a salt pillar.

The Siege of Carthage, a three-year siege between 149 and 146 BCE, was the last of the Punic Wars between the Phoenicians and the Roman Republic. The war resulted in the annihilation of one of the world's wealthiest cities (Kiernan, 2004). According to some reports, the military actions involved salting the soil of the conquered lands to ensure permanent infertility, but historians have questioned this claim (Ridley, 1986; Stevens, 1988).

Many stories in the Bible refer to global destruction, plagues, and floods, possibly corresponding to natural hazards that affected several cultures. The transformation of Lot's wife into a salt pillar is related to geology. In the Dead Sea region, particularly along the Mount Sedom shores, several salt columns stand upright, resembling the shapes of human bodies. These pillars are the remains of evaporite rocks that collapsed from ancient caves in the area. Mount Sedom, which appears on the cover of this book, is an active salt diapir, with halite and gypsum rising from great depths and extruding onto the surface with vertical layers of bedded salt (see Chapter 3). One of these columns that broke away from the salt diapir displays an intriguing outline that is reminiscent of a woman's body. At present, it is a famous tourist attraction in Israel, attracting many international visitors.

1.5. FOLKLORIC ASPECTS OF SALT

Salt is a highly valued and sought-after commodity since the beginning of agriculture until around the 19th century. It is essential to life and has been used in various ways, such as a dietary food, a metaphor for energy, and a source of salt for many animals. The roots of many words in several languages around the world often use salt as a pleasant taste for the senses, such as "sauce," "sausage," "salami," and "salad."

The Roman word for salt (*sal*) is derived from the goddess's name. Salus was the daughter of the god of medicine, Asclepius, indicating some kind of relationship between salt and medicine. Salt has given its name to an endless number of cities, towns, and localities. For example, Salzburg in Austria is associated with the castle and fortress that protected the salt mines, and Hallstatt is a German name for a place with salt.

In ancient Latium, a man in love was called *salax*, meaning "in a salted state," which is the origin of the word "salacious." In Spanish, pretty and smiling girls are associated with the word "salerosidad," implying a joyfulness and appreciation of good things in life. In medieval Europe, some wives had the custom of brushing their husbands' skin and salting the sore parts, hoping it would make them healthier and more virile.

The Assale Ring, an intriguing circular salt structure in northern Ethiopia, is considered sacred by the tribes that live in the Afar region. The evaporites that outcrop at the surface are used as a traditional medicine to heal some diseases, even in modern times (Cavalazzi et al., 2019).

Some prehistoric people in Europe and Asia actively produced and traded salt as a valuable commodity, as indicated by several rudimentary salt mines (Harding, 2013, 2014). In the Middle Ages, salt was considered as precious as gold, with caravans crossing the African deserts exchanging salt for precious metals, ivory, goat skins, and slaves. Historical accounts report that salt could be exchanged at an ounce-for-ounce rate with gold dust.

Some authors suggest that the word "salary" originated from the Latin word for salt because the Roman Legion soldiers were paid in salt. The origin of this story is related to a famous quote by Pliny, the Elder, in his book "Natural History" (Jones, 1938). In book XXXI, it is written that in Roman times salt was used "on service abroad, from which comes the term 'salary' (salt money); it had great importance among the men of old, as is clear from the name of the Salarian Way, since by it, according to agreement, salt was imported to the Sabine." However, there is no proof that salt was routinely used as money in Roman times, although the soldier's monthly allowance was certainly used to buy salt (Kurlansky, 2002).

Some fables mention that early trade in ancient Greece involved the exchange of salt for slaves, leading to the depreciating expression that those who did not work well were "not worth their salt." According to this folkloric myth, some free men who worked as professionals were also paid in salt for their services, and the ones who excelled in their jobs were "worth their salt."

However, some historians and writers contest these stories as mythic. Marco Polo's travels throughout Asia in the 13th century are accepted as true stories, as they mention the use of salt coins by the Chinese. These inventive people boiled saltwater in small pans until the brine was transformed into a sticky mixture that was pressed into circular cakes. When government officials stamped these flat disks with the Grand Khan's image, they were transformed into coins that could be legally used as a trading currency or exchanged for other goods via bartering. According to Marco Polo's tales, some of these salt coins had a prodigious value in Tibet (Kurlansky, 2002).

Salt is a common theme in poetic literature, particularly in the works of Yehuda Amichai, a German-born poet who lived in Israel and wrote about contemporary themes. His poems often revolve around the saga of the Hebrew people and their land, highlighting human conflicts. Amichai's famous poem, which repeatedly uses the word salt, suggests that while some dramatic events should be forgotten, others should be remembered by those affected by warfare and destruction. Amichai experienced severe emotional distress as a result of her childhood friend's bicycle accident, which required her to amputate a leg. When the Nazi repression started several years later, she was unable to escape and was killed in the Holocaust.

> White salt,
> Salt of mercy,
> Salt that makes you forget,
> Salt that makes you remember.
>
> (Yehuda Amichai)

Pablo Neruda, a renowned South American writer, is known for his famous poem "Ode to Salt," which vividly depicts his memories of visiting salt mines in Chile. The poem personifies salt as a living element with sound and voice, a feeling most people experience when visiting the salt mountains in Iran. The poem's terse structure conveys the sound of salt and the importance of salt landscapes as an inspirational theme. Neruda's poem hints that salt, under its simple disguise, conceals a hidden natural wonder that is mysterious and powerful beyond imagination.

> This salt
> in the salt cellar
> I once saw in the salt mines.
> I know
> you won't
> believe me
> but
> it sings
> salt sings, the skin
> of the salt mines
> sings
> with a mouth smothered
> by the earth.
> I shivered in those
> solitudes
> when I heard
> the voice
> of
> the salt
> in the desert.
>
> (Pablo Neruda)

Salt is an intriguing component of sedimentary basins, and Earth scientists have scrutinized many aspects of salt formation and mapped complex geological structures outcropping in some regions of the world. In recent decades, geophysical methods have revealed aspects hidden from sight, imaging the subsurface of many onshore and offshore petroleum provinces. The salt deposits in onshore and offshore basins will be the focus of this book.

1.6. PREHISTORICAL SALT USAGE AND CULTURAL REVOLUTIONS

The next sections will briefly overview salt deposition through geological time and its influence on the development of civilizations and cultural revolutions from prehistoric times to modern-day societies.

1.6.1. Salt Deposition Through Geological Time

The development of life on Earth began in the Archean period when simple life forms started to adapt to a changing world. During the geological eras, only some species survived after catastrophic events that changed the face of the planet and destroyed most existing life in several extinction episodes throughout the Phanerozoic (Glikson & Groves, 2016). Some of these episodes are related to climate changes that dramatically reduced the amount of water that covered shallow seas, gulfs, and lakes (e.g., Hansen & Sato, 2012; Warren, 2006, 2008, 2010, 2014).

Some researchers interpret that some giant basins were evaporated during periods of extreme aridity, leaving only salt deposits as a testament to desiccation episodes. The majority of life forms cannot survive in these hypersaline environments, and the sedimentary layers deposited during these evaporitic intervals are notable for having few fossils that have survived. Some biological events on Earth are intrinsically related to geology, such as mass extinction episodes that coincide with cataclysmic events, increased volcanism, or asteroid impacts (see Chapter 9 in Mohriak, 2025). Geological events such as sea-level fluctuations are associated with glaciation epochs and variations of carbon dioxide (CO_2) in the atmosphere (Glikson, 2005; Morgan et al., 2004; Royer et al., 2004).

Salt deposition episodes in giant salt basins are considered extraordinary events throughout geological time (Figure 1.1). Some events were catastrophic and coincided with extinctions and volcanic activity, resulting in rapid changes in the environment of deposition and drastically affecting biological systems. Some of these intervals coincide with the deposition of evaporites in giant salt basins, such as the Hormuz salt in Arabia, the Late Devonian salt in Europe, and the Permo-Triassic salt deposits in many European basins. During the Messinian Salinity Crisis, when evaporite deposits covered the Mediterranean Sea, most of the marine life was annihilated when a deep ocean was desiccated, according to some hypotheses (Cita, 1983; Hsü, 1972).

The Phanerozoic chronostratigraphic chart (Figure 1.1) displays major glaciations, volcanic events, extinctions, and epochs of large salt basin development. Several researchers (e.g., Glikson, 2005; Sepkoski Jr., 1996; Wignall, 2001) have noted that continental flood basalts and large igneous provinces (LIPs) events and mass extinctions coincide with radiometric and paleontological ages at various intervals. These coincident events occur at the Early Cambrian (513 Ma), Late Devonian (374–359 Ma), Permo-Triassic boundary (251 Ma), Late Triassic (201 Ma), and Cretaceous-Tertiary boundary (66 Ma). Some of these intervals coincide with the deposition of evaporites in giant salt basins. The Hormuz salt in Arabia is Neoproterozoic-Cambrian (see Chapter 3). Late Devonian salt is mainly observed in eastern Europe (Ukraine and Russia). Middle-Devonian evaporites are found in North America's Rocky Mountain foreland basins (see Chapter 8 in Mohriak, 2025). The Permo-Triassic salt deposits are widespread in European basins, such as the North German Basin, the North Sea, offshore Iberia, and the northwest African margins (e.g., Soto et al., 2017).

Evaporite basins are linked to magmatic events and climatic changes that occurred after global glaciations.

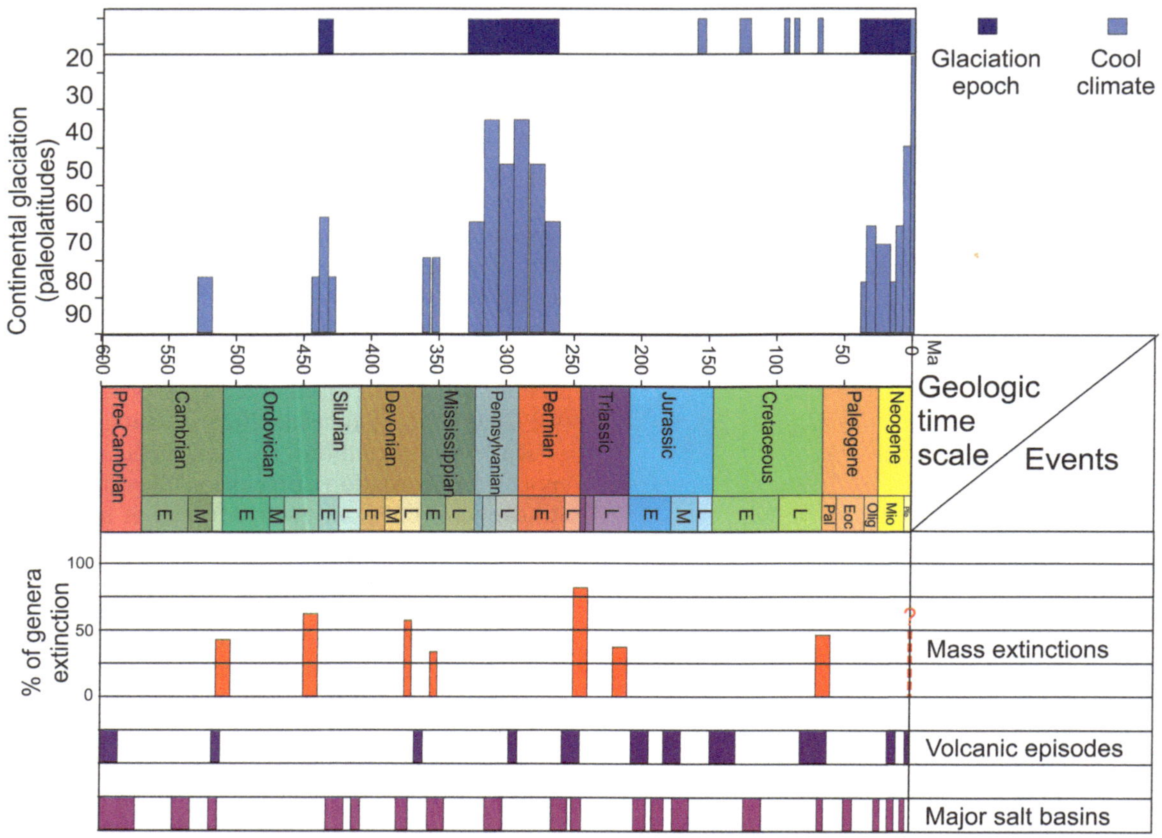

Figure 1.1 The stratigraphic chart displays the Phanerozoic divisions of the Paleozoic, Mesozoic, and Cenozoic eras, highlighting key geological events. It shows the epochs of glaciations, mass extinction episodes, volcanic episodes, and the development of major evaporitic basins. The upper graph shows intervals when temperatures were below average. The lower graph shows (1) mass extinction episodes, with the height of red bars representing the percentages of species that were extinct; (2) volcanic episodes (purple bars), which are associated with large igneous provinces; and (3) the major evaporitic basins (magenta color). Major salt deposition events are recorded in various eras, including the Neoproterozoic, Silurian, Carboniferous, Late Permian to Early Triassic, Early Cretaceous, and Middle to Late Miocene.

Examples include the Neoproterozoic-Cambrian salt basins in Arabia and Iran, which formed at the end of the Neoproterozoic glaciation (Gradstein et al., 2005, 2012; Ogg et al., 2016; Warren, 2006, 2008, 2010, 2014). The Zechstein salt basin in north-central Europe formed after a glacial episode in the Carboniferous, which coincided with a period of rather low CO_2 content in the atmosphere (Royer et al., 2004). The snow belt likely reached low latitudes on the central European continent and even South America, as indicated by Permian glacial sediments in onshore cratonic basins (see Chapter 8 in Mohriak, 2025). The Hettangian salt of the Central and North Atlantic is related to one of the largest igneous provinces in the world, corresponding to the CAMP event (Davison, 2005; Soto et al., 2017).

The geological time scale is correlated with some major evolutionary events that affected life on Earth and led to the development of the human species (Figure 1.2). Some geological intervals coincide with the ages of major ancient salt basins, such as the Late Permian, Jurassic, Early Cretaceous, Middle Miocene, and Late Miocene. Climate factors involving glaciation and extreme aridity leading to salt deposition seem to be linked with mass extinction episodes and the development of new forms of life (e.g., Warren, 2006, 2008, 2010, 2014, 2016). However, plate tectonic events are the most important factors in controlling basin evolution and salt deposition. For example, rifting and the formation of oceanic crust were critical in establishing the depocenter for evaporite deposition in the Gulf of Mexico in the Middle Jurassic (see Chapter 4 in Mohriak, 2025). The Messinian Salinity Crisis resulted in the massive deposition of halite on the bottom of the Mediterranean Sea, partly because of the isolation of the basin by closing the

Era	Period	Epoch	Mapb	Major evolutionary events	Major salt basins
Cenozoic (Age of mammals)	Quaternary	Holocene	0.01	Beginning of agriculture and farming	Localized salt deposits; last glacial episode about 70,000 years BCE.
		Pleistocene	2.6	Various hominin species at beginning; several hominins disappear but *Homo sapiens* evolves and spreads	Localized salt deposits, Eastern and Northern Africa
	Neogene	Pliocene	5.3	Old World monkeys adaptive radiation; Apes decrease; first bipeds and small hominoids species radiate from Africa, first stone tools at 2.5 Ma	Localized salt deposits, Northern Africa, Dead Sea, Anatolia, Central Europe
		Miocene	23	First Old World monkeys; Adaptive radiation of Hominoids (Age of the Apes)	Major salt basin in the Red Sea (15 Ma), Mediterranean (5.5 Ma)
	Paleogene	Oligocene	34	Radiation of anthropoids; First New World monkeys (27 Ma)	Localized salt deposits Rhine Graben, Iberia
		Eocene	56	Adaptive radiation of Prosimians, but most are extinct by end of Eocene; First anthropoids appear on earth, e.g., *Eosimias* (45 Ma) and *Baliina* (40 Ma)	Localized salt deposits Rhine Graben, Iberia
		Paleocene	66	Adaptive radiation of placental mammals; first Primates (prosimians)	
Mesozoic (Age of dinosaurs)	Cretaceous		145	Flowering plants; mammals and birds spread **(76% of all species become extinct at end of Mesozoic about 66 Ma)**	Major salt basin in the South Atlantic (115 Ma)
	Jurassic		201	Many dinosaurs; primitive mammals spread; toothed birds	Major salt basin in the Gulf of Mexico (169 Ma) and Central Europe
	Triassic		251	First dinosaurs; first egg laying mammals	Major salt basin in the Central Atlantic (201 Ma); sat deposits in rift basins
Paleozoic	Permian		299	Spreading of reptiles, insects, first mammal-like reptiles **(96% of all species become extinct at end of Paleozoic)**	Major salt basin in Europe (256 Ma); Permain salt basins in North America (Southwest U.S.) and South America (Parnaiba)
	Carboniferous		359	Reptiles, modern insects **(82% of species become extinct about 380 Ma)**	Salt basins in North America (central U.S.) and South America (Amazon)
	Devonian		416	First amphibians; forests appear	Salt basin in central – Eastern Europe
	Silurian		444	First land plants	Major salt basin in North America
	Ordovician		488	Invertebrates dominate the world; first vertebrates **(85% of species become extinct about 440 Ma)**	
	Cambrian		542	Invertebrates	Major salt basin in the Middle East
Proterozoic (earliest life forms)	Precambrian		1,000	Protozoa, sponges, algae	Salt basins in Australia, Africa, Middle East
			2,500	First evidence of life (bacteria)	
Azoic (no life)	Archean		4,550	Origin of Earth	

Figure 1.2 The simplified geologic time scale shows the major evolutionary events and the distribution of major salt basins on Earth. Credit: Adapted from Walker et al. (2018).

Gibraltar Strait that connected with the Atlantic Ocean. Salt deposition ended when this connection was reestablished in the Pliocene (see Chapter 5 in Mohriak, 2025).

According to Walker et al. (2018), the Holocene (the last 10,000 years) was characterized by a biological catastrophe that saw the abrupt extinction of numerous species at a rate up to 1,000 times that of background extinctions (see, e.g., Cowie et al., 2022). These extinctions are not related to natural phenomena but are mainly linked to anthropogenic factors, such as the colonization by Europeans of some continents previously occupied by native animal and vegetal species.

1.6.2. Early Humans and Salt Lakes

Early human settlements in the East African rift system were not accidental, as humans began using salt for food preservation even in nomadic cultures. By the end of the Paleolithic Epoch, some hunter-gatherers in coastal areas dipped food in seawater or evaporated seawater to obtain dry salt, similar to the habits of some Polynesian cultures during their first contact with European societies (Cordain et al., 2005). However, there is no evidence of widespread salt extraction from seawater or mining of inland salt deposits during the Paleolithic or Mesolithic (about 100 ka, where ka means a thousand years) periods. Only during the Neolithic and Holocene (last 10 ka), industrial salt exploitation was recorded in some regions where salt occurred at the surface, such as the Cardona region of Spain (Cordain et al., 2005).

Early humans, belonging to the *Homo erectus* species, lived around 600,000 years ago (600 ka) and traveled around areas with caves for shelter during cold weather. They quickly learned how to control fire and cook meat, a turning point for human evolution. This allowed humans to prepare a more varied diet that was not exclusively dependent on agriculture or picking seasonal fruits and vegetables. The use of salt to prepare food in cold climates allowed populations to migrate from Africa and spread across northern continents, such as Europe and Asia. Modern humans evolved from a single branch of mammals. Apes, chimpanzees, and modern humans share a common ancestor. These species formed a new lineage between ~9.3 million and ~6.5 million years ago (Ma) that developed bipedalism, manipulation, and later, cognition (Almécija et al., 2021; McClendon, 1999; Potts & Sloan, 2010). Hominoid, a broad term that refers to several species of the superfamily *Hominoidea*, includes apes and humans (Owen-Smith, 2021; Stringer & McKie, 2015).

Many other species preceding the genus *Homo* lived on Earth from that early beginning, such as *Australopithecus*, which walked on Earth about 2.5 million years ago (Figure 1.3). About 2.0 million years ago, these hominoids left their savannah homeland to venture across rivers, mountains, and deserts to establish nomadic tribes in North Africa, Europe, and Asia. The closest species that evolved from apes into a human-like animal was *Homo rudolfensis*, which lived around the lakes of the East African rift system, and *H. erectus*, a species that lived in Asia (e.g., Harari, 2014; Sponheimer et al., 2013). These species are the direct ancestors of modern humans and are called *hominins*, which belong to the family *Hominidae*, almost all of which are extinct; only *Homo sapiens* have survived (Figure 1.4). While the "hominoid" term also includes the ancestors of apes, the term "hominin" applies only to the ancestors of modern human beings (Almécija et al., 2021; Higham, 2021).

Hominins lived around the lakes of the East African rift system and used stone or bone artifacts (e.g., Cuthbert & Ashley, 2014; Owen-Smith, 2021; Stringer & McKie, 2015). During the Paleolithic period, hominoids developed skills indicating increasing brain sizes (Glikson & Groves, 2016; Potts & Sloan, 2010). *Homo neanderthalensis* lived in Europe and the Middle East about 500 ka ago, while *Homo soloensis* thrived in Indonesia. *H. sapiens* began dominating the world around 125 ka, coexisting with remnants of *H. neanderthalensis* that inhabited Europe and Asia until around 40 ka. In the Neolithic period (around 10 ka), major advancements such as agriculture and farming led to the exponential growth of the

Prehistoric Ages

Chronology	Achievements
Paleolithic Age (c. 2 Ma to 100,000 years BCE)	Corresponds to the Stone Age, characterized by the first appearance of hominids in Africa in late Pliocene to Pleistocene, evolving to *Homo neanderthalensis and Homo sapiens* at circa 500,000 years BCE. First use of fire and use of rudimentary artifacts made by stones and bones. Some species of hominids such as *Homo neanderthalensis and Homo sapiens* co-existed at circa 100,000 years BCE.
Mesolithic Age (transitional period between the Stone Age and the Neolithic Age)	*Homo neanderthalensis* became extinct at circa 40,000 to 10,000 years BCE, and *Homo sapiens*reigned absolute on Earth after the disappearance of all other hominin species. Humankind adapted to changing environments after the demise of the last glacial and survived by hunting edible animals, fishing and collecting food around lakeside and seashore environments. Some tribes started to adventure across forests, mountains and seas to explore distant lands across Europe, Asia and the Americas.
Neolithic Age (c. 10,000–8,000 years BCE) Bronze Age: 2300 to 700 BCE Iron Age: 700 BCE to 1 BCE	Development of agriculture, domestication of animals, manufacture of pottery, texture, and fabrication of some essential commodities such as salt

Figure 1.3 Prehistoric chronology of humankind with main cultural and economic achievements. Credit: Adapted from Britannica/www.britannica.com/last accessed March 13, 2024.

Hominin Species

Species	Locations (Age)
Australopithecus	East Africa (2.5 Ma)
Homo rudolfensis	East African lakes (2 Ma)
Homo erectus	Africa, Arabia, East Asia (1.8–1.6 Ma)
Homo soloensis	Indonesia (500,000–50,000 years BCE)
Homo neanderthalensis	Europe and Middle East (500,000–40,000 years BCE)
Homo sapiens	Africa, Middle East, Europe, Asia (dominating the planet since c. 125,000 year BCE)

Figure 1.4 Chronological evolution of the main hominin species and the main locations where fossils have been found by archaeologists. Credit: Adapted from Britannica/www.britannica.com/last accessed March 13, 2024.

human population. This event coincided with the demise of the last glacial age on Earth (Clark et al., 2009).

H. sapiens began dominating the world around 125 ka after the other hominin species started to disappear in the Mesolithic period. In the Neolithic period, major advancements, such as the development of agriculture and farming, resulted in the exponential growth of the human population, which coincided with the demise of the last glacial age on Earth. *H. sapiens* emerged during the last phases of the glacial period, associated with the Eemian interglacial epoch (circa 124 ka), when temperatures rose by about 5 °C. The use of fire allowed humans to migrate from the East African lakes to colder regions in the Middle East, southern and southeastern Asia, and Southern Europe. Glikson and Groves (2016) emphasized that Darwin considered language and fire the two most significant achievements of humanity, directly responsible for the technological advances conquered by our species.

The evolution of the first hominins is linked to climate changes in the East African rift system (Maslin et al., 2014). During the Pleistocene, short episodes of humidity and extreme aridity affected water levels in East African lakes, leading to a long-term drying trend. Hominins had to adapt to new habitats where freshwater could only be obtained from groundwater springs, which were sourced from underground aquifers (Cuthbert & Ashley, 2014). These springs continued to flow fresh water while the rivers dried up and the lake waters became increasingly saline. Eventually, these lakes dried out and accumulated a salt layer.

Lake Natron in the East African rift system, near the border between Tanzania and Kenya (Figure 1.5A), provides a didactic example of such a catastrophic change. Lake Natron is the world's most caustic body of water due to its unique color derived from halophilic microorganisms, particularly the blue-green algae known as Spirulina, which is characterized by red pigments. Millions of birds come to the lake every year to use it as a breeding ground for their young. The flamingoes that feed on these algae also acquire a pinkish color. The lake sediments that form on the lake's shores and accumulate in the center also exhibit pink salt encrustations, resulting from impurities in the mineral halite.

Lake Natron occupies a large depression in the middle of a rift system that is controlled by active normal faults and is associated with several volcanic centers, such as the Ngorongoro volcano (Figure 1.5A). The East African rift depression diverges toward the south, forming two separated rift valleys with Lake Eyasi to the west and Lake Manyara to the east of the Ngorongoro volcanic highlands. The detailed satellite image (Figure 1.5B) shows a bright red color that distinguishes this lake from other lakes in the rift system.

Evaporite sediments accumulate in the shallow lake center, less than 3 m deep, and dry up completely during dry periods (Figure 1.5B). The lake was once rich in chemical compounds associated with evaporite minerals, such as sodium carbonate (natron and trona) and sodium chloride (halite). Solar evaporation increased salinity, resulting in concentrated brines that were too inhospitable for hominins and other animals. The active groundwater system in the Olduvai Basin provided potable water during dry periods, supporting vegetation, watering holes, and freshwater lakes. Early hominins migrated to cooler climates in the northern hemisphere due to the desiccation of East African lakes during extreme aridity (Cuthbert & Ashley, 2014). Archaeologists have recovered remnants of the early hominins in the Olduvai Basin, which lies to the west of the Ngorongoro volcanic centers (Figure 1.5A).

One of the reasons that can be associated with the decision of the early hominins to migrate to cooler climates in the northern hemisphere was the desiccation of the East African lakes during periods of extreme aridity. When the glacial period waned 100 ka ago, *H. sapiens* began migrating from the African savannahs

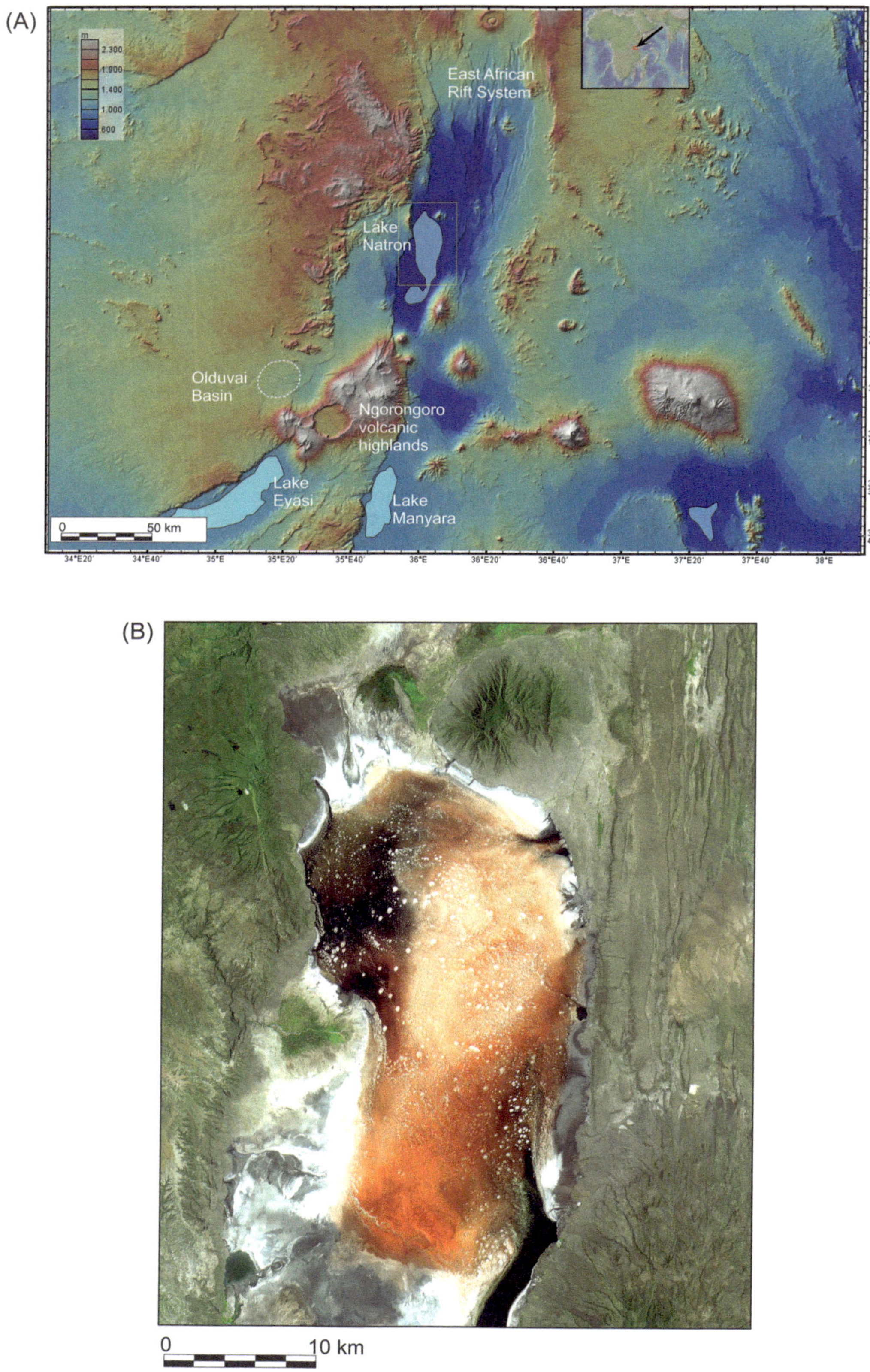

Figure 1.5 (A) The East African rift system, located at the border between Tanzania and Kenya, features a graben depression and rift border uplifts. The zoom image, centered on Lake Natron, shows the accumulated evaporite sediments on the lake's floor. The lake is north of the Ngorongoro volcanic highlands, while the Olduvai Basin is west of the highlands. (B) A satellite image of Lake Natron shows the evaporite sediments, remnant lake waters, and a green circular structure corresponding to a volcano. The area has higher topography and more vegetation than the barren salt flats at the rift depression. Credit: NASA/Public domain.

to the northern hemisphere. The favorable climate and abundance of water bodies attracted a growing population that began adapting to new ways of life through the agricultural revolution (Owen-Smith, 2021; Stringer & McKie, 2015).

1.6.3. Cultural Revolutions from Prehistorical to Historical Times

The turning point in humankind's evolution was the mastery of fire, which was most necessary for survival during intermittent glacial epochs in the late Pleistocene (e.g., Glikson & Groves, 2016). Hominins dominated the learning of how to ignite and control fire thousands of years before the agricultural revolution when they learned fire could heat caves. Archaeological evidence indicates that tribes of *H. sapiens* moved out of Africa around 125 ka and quickly expanded throughout Arabia, Europe, and Asia, replacing the existing populations of *H. erectus* and *H. neanderthalensis* (Figure 1.4). The radical changes in human culture that have allowed this phenomenal occupation of a planet by a single species are linked with the development of humans' brain capacity to think logically and communicate ideas to other members of the species (Harari, 2014).

The Agricultural Revolution in Neolithic times (circa 10 ka) marked a drastic change for humankind, with a wide-scale transition from ancient cultures that survived by hunting wild animals and gathering food as nomad tribes to a more evolved culture, with tribes fixed around agricultural sites. In these regions, the population settled around a central farming area and thrived to survive in hostile environments through the development of planned agriculture and the fabrication of natural products essential for food preparation, such as salt. The first settlements were established around the lakes in the East African rift system or along the shores of large rivers such as the Tigris, Euphrates, and Nile. Some communities were established in flatlands near desertic areas around the Red Sea, the Gulf of Aqaba, and the Dead Sea. The proximity of these settlements to arid regions such as the Sahara Desert, south of the Mediterranean Sea, resulted in the first exploitation of salt deposits by harvesting the surface of desiccated water bodies. Within a short time, humankind learned how to use fire and produce salt in temperate climates by heating brine in pottery artifacts.

1.6.4. Archaeological Evidence of Salt Production in Prehistoric Times

Salt has been a significant contributor to the foundations of civilizations since Neolithic times, as it helped eliminate dependency on fresh food and enabled long-distance transportation of salted meat and vegetables. From the Stone Age to the Bronze and Iron Ages, salt was a highly valued commodity, with some ancient civilizations considering it a form of currency used in trading (Kurlansky, 2002). Some cultures were actively engaged in producing and trading salt from solar evaporation, brine boiling, and mining salt rock from outcrops in Europe and Asia (Banffy, 2015; Weller, 2015).

Animals also depend on salt, and they can travel many miles to find salt springs where they lick the salted crust. Tribesmen who hunted wildlife in search of food followed these trails, and their families established their housing along the way. Many roads connecting cities within Europe and Africa were initially built as trails to transport salt. These roads connected producing regions, that were economically viable due to climate aridity to regions where salt could not be easily produced due to the rainy seasons (Kurlansky, 2002).

Prehistoric saltworks have been discovered in many places in Africa, Arabia, Asia, and Europe, particularly in regions where salt layers occur at the surface. These sites include the Afar Province in northern Africa, the Dead Sea, and some regions close to mountain ranges, such as the Alps, Pyrenees, Andes, Atlas, Zagros, and the Himalayas (see Chapter 3). Archaeological evidence indicates that salt was produced in rudimentary saltworks or salterns around desiccated lakes or dried rivers that abound around the Sahara Desert in Africa, a region characterized by climate aridity (Figure 1.6). In addition, some salt was also produced in some cooler regions in Europe, such as the salt ponds in western France (Harding, 2013, 2014; Weller, 2015).

In most arid environments, solar evaporation is sufficient to concentrate the saltwater in inland lakes or coastal lagoons, forming natural brines that fill lakes, lagoons, and shallow seas (see Chapter 8). However, in temperate climates such as those in Central Europe and Asia, prehistoric civilizations used coarse ceramic materials, bronze, or iron pans to boil the water, which evaporated and precipitated salt crystals. The salt residue accumulated at the bottom of the pans and was collected with primitive spoons. Archaeologists use these ceramic artifacts as crucial evidence in their interpretation of prehistoric salt production (Brigand & Weller, 2015; Harding, 2013, 2014; Weller, 2015).

Several Neolithic civilizations developed rudimentary techniques to produce salt using pottery vessels. These apparatuses were heated over a fire and frequently broke, leaving fragments that now constitute an important archaeological object. These broken pieces of primitive equipment used to produce salt are known as "briquetage," which mainly correspond to coarse ceramic tools used for brine boiling. The term briquetage also includes the pieces of equipment used in cutting the salt blocks (Weller, 2015).

Briquetage and other evidence indicate that some salt has been produced since 2500 BCE in many places, including Europe, Africa, Asia, China, and the Americas

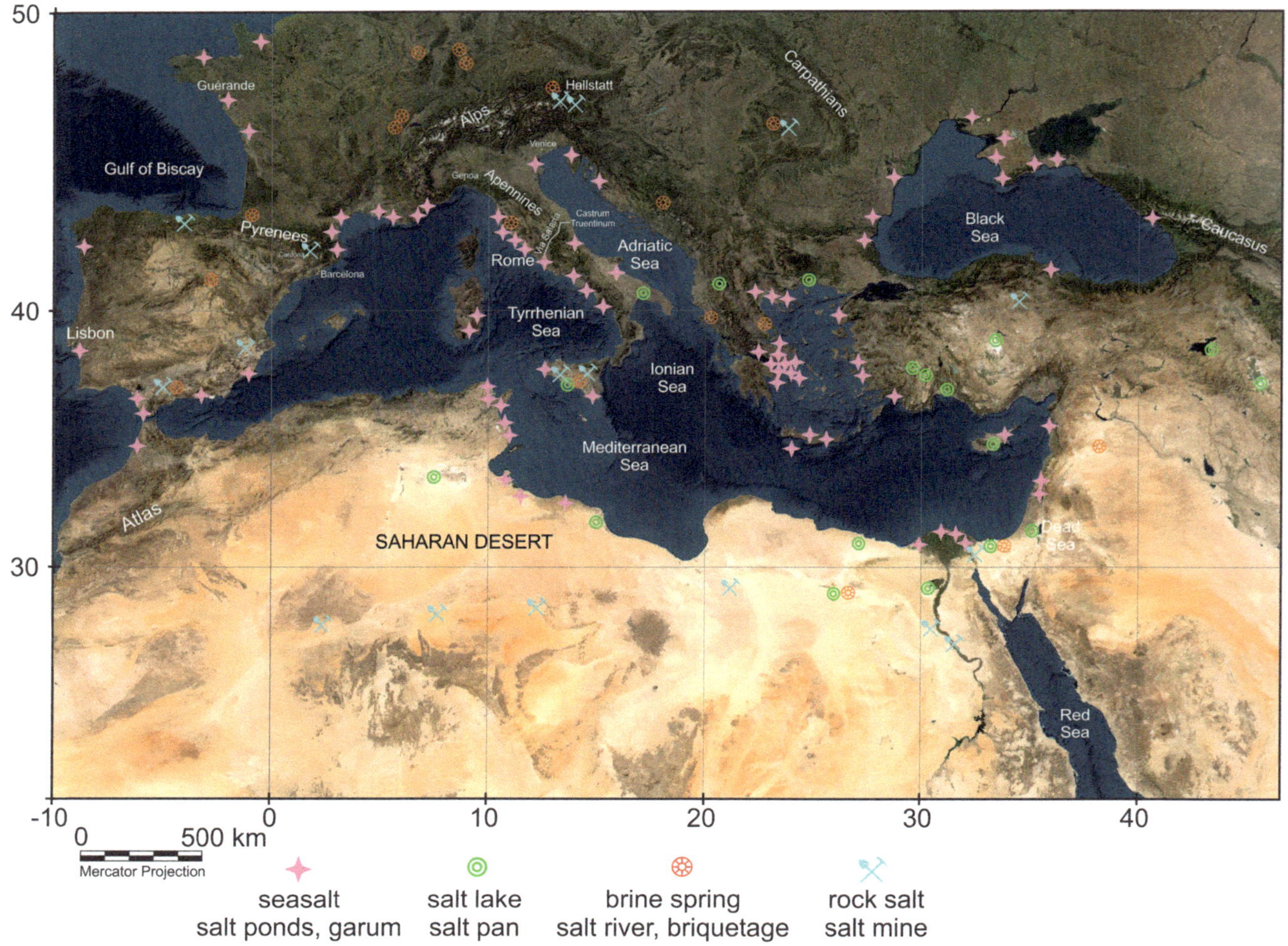

Figure 1.6 The Sahara Desert in northern Africa and the main mountain ranges on the European mainland have archaeological evidence of Neolithic salt production. Salt works were produced by evaporating seawater, obtained from salt lakes and salt pans, or extracted from salt mines. Methods of producing salt works included evaporating seawater, obtaining salt from dried salt lakes and pans, or extracting salt from salt mines. Some brine springs produce salt, as indicated by briquetage. The Via Salaria was an important route during the Roman Empire, connecting the Tyrrhenian Sea saltworks near Rome to the ancient city of Castrum Truentinum. Genoa and Venice were economic and cultural centers in Italy, famous for wealthy merchants trading salt and other commodities during the Medieval Ages. The Cardona salt mine, located south of the Pyrenees, is the only salt diapir in Western Europe that has been mined since prehistoric times. Other salt mines exist in the Alps, Carpathians, Caucasus, and Arabian Peninsula. A significant salt production center, situated on the western Atlantic coast of France near Guérande, utilizes the solar evaporation of shallow salt pans. Credit: Adapted from Saile (2015).

(Weller, 2002, 2015). Given that the earliest remnants of salt production may have vanished over time, there is some evidence that humans produced salt for unrecorded millennia before the first archaeological indications recorded its production.

Salt occurrences are prevalent in Europe, particularly in flatlands and mountain ranges in Iberia, Central Europe, and Romania (Figure 1.6). These saltworks were primarily located near favorable arid areas in the Sahara Desert, Iberia, the Alpine region, and along the Carpathian Mountains, where salt layers occur at the surface or at shallow depths underground (Banffy, 2015). Salt was also produced from evaporating water from salt springs, which occur in several regions across Europe, the Middle East, and Asia. Salt basins formed during distinct geological times have been exploited by humankind since prehistoric times due to salt spring occurrences (Figure 1.7). These springs are associated with

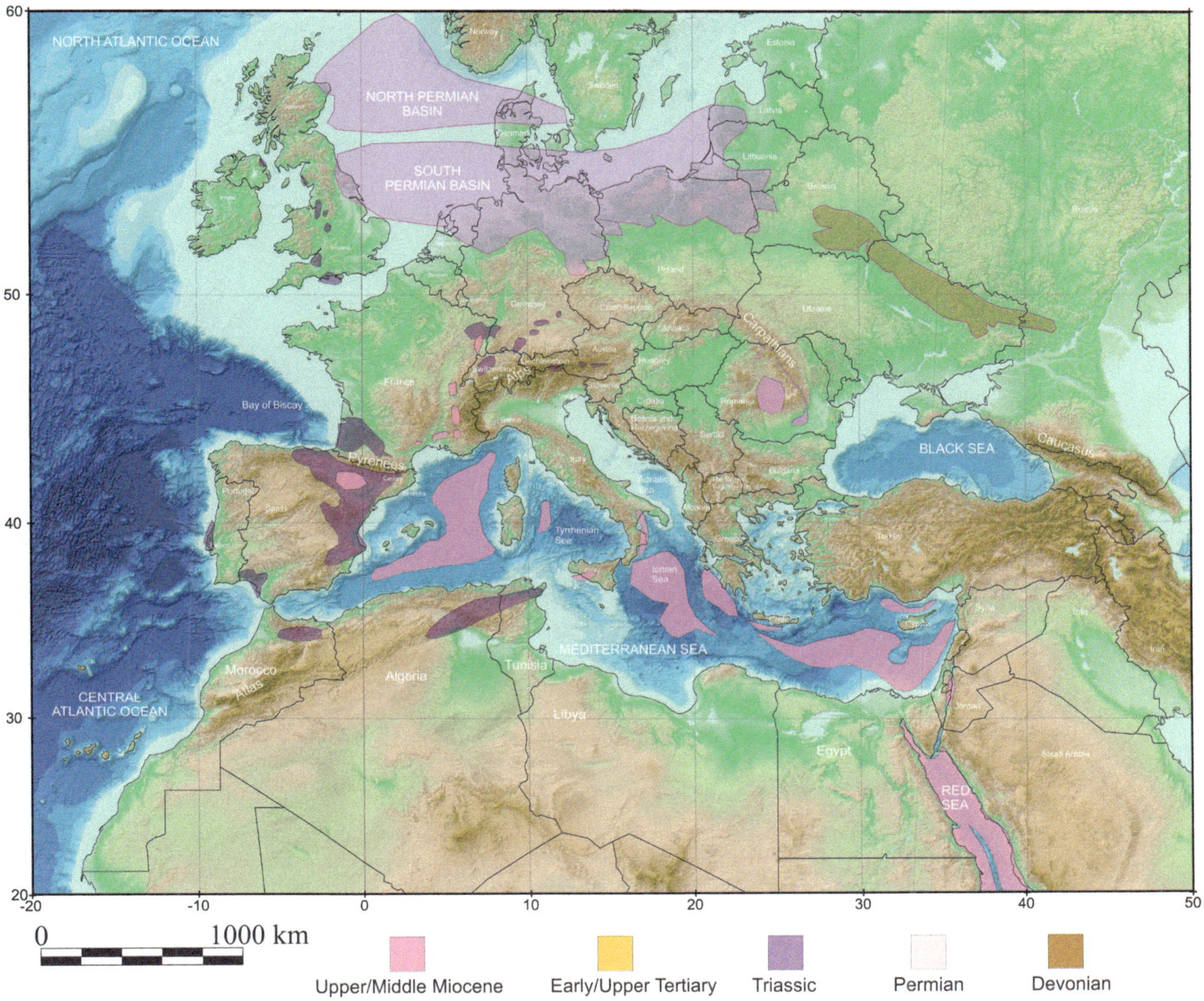

Figure 1.7 The North African and European sector of the global bathymetric chart (based on GEBCO, 2019) shows the main salt basins in Europe, the Mediterranean Sea, and the Arabian Peninsula. The chart outlines evaporite occurrences from various geological ages, including Devonian, Permian, Triassic, Tertiary, and Miocene, spanning from the Dnieper-Donets Basin to the Carpathian Mountains region, the Mediterranean Sea, the Red Sea, and the Dead Sea in the Arabian Peninsula. Credit: Adapted from Harding (2014).

evaporite layers that outcrop on land as bedded salt or form diapiric structures, such as the magnificent salt mountains in the Alps, Carpathians, or Zagros region of Iran (see Chapter 3).

According to some authors (e.g., Banffy, 2015), some of these salt occurrences were the focal points for the first human incursions into the European mainland north of the Alps (Figure 1.8). Some of these salt structures have been known to the inhabitants of these regions for millennia, such as the Cardona Mountain in Spain, which hosts a salt mine that has been exploited since prehistoric times (Weller, 2002).

In the Late Neolithic times (around the 5th millennium BCE), Lengyel culture tribes occupied the Carpathian region, expanding toward Germany, France, Poland, Romania, and Ukraine (Figure 1.8). These tribes are associated with the Linear Pottery Culture (LBK), which were the first inhabitants of this European region to leave nomadic life and adapt to a life as farmers (Banffy, 2015). The LBK corresponds to a major archaeological horizon of the European Neolithic period, associated with the earliest farming cultures that flourished in Europe between c. 5500 and 4500 BCE (e.g., Nikitin et al., 2019).

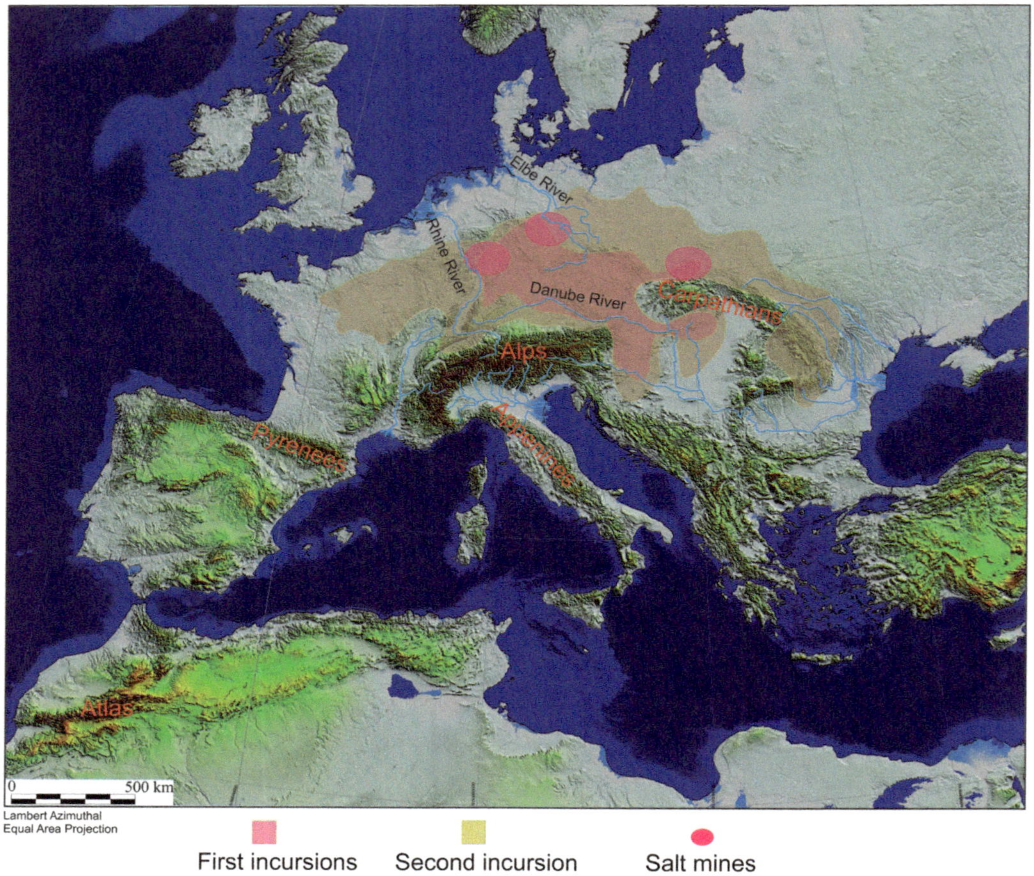

Figure 1.8 The map depicts the distribution of the first farmers in Central Europe during Neolithic times, with three major salt regions marked in red. The first prehistoric expansion sites are pink ellipses, with tribes crossing valleys along the Danube River between the Alps and Carpathians. The second expansion routes, to the east of the Elbe River and west of the Rhine River in Central Europe, are highlighted in beige, centered on salt mines north of the Alps and Carpathians.

The communities established a new way of life, including domesticating animals and cultivating land for vegetables and cereals. They spread to the northern regions, crossing valleys between the Alps and the Carpathians and following the Danube River. In the first phase, they reached southern Poland, where salt was mined in the Krakow salt mine. They advanced to the Halle/Saale region east of the Rhine River and dominated the region west of the Elbe River (Figure 1.3). After 150 years, the second wave of expansion throughout Europe was achieved, advancing across the Rhine River and expanding toward the Paris Basin in the west. New routes advanced across the Elbe River and reached Ukraine in the eastern region, near the Carpathians (Figure 1.8). These regions were close to salt-producing sites. The Wieliczka salt mine near Krakow was a significant point in the first phase of human cultural and economic expansion in Europe and Asia, with archaeological evidence suggesting that the incursion routes were directed to regions with more readily available salt sources (Banffy, 2015).

Salt production in Europe dates back to prehistoric Neolithic times, with briquetage being the earliest examples found in buried sites or abandoned salt mines across Europe (Weller, 2015). Salt-making was established at least 6,000 years BCE, and fragments of primitive apparatus employed in the process of heating brine in rudimentary pans have been found in excavations of buried sites or within abandoned salt mines across Europe, from Iberia in the west to the region east of the Carpathians (Figure 1.9).

Salt pans were initially made of baked clay or ceramics (Figure 1.10), but later more sophisticated articles were made from different metals, such as bronze, lead, and iron. The technological change from ceramics to bronze, then to lead, and then to iron was accompanied by improvements in heating techniques. Initially, the

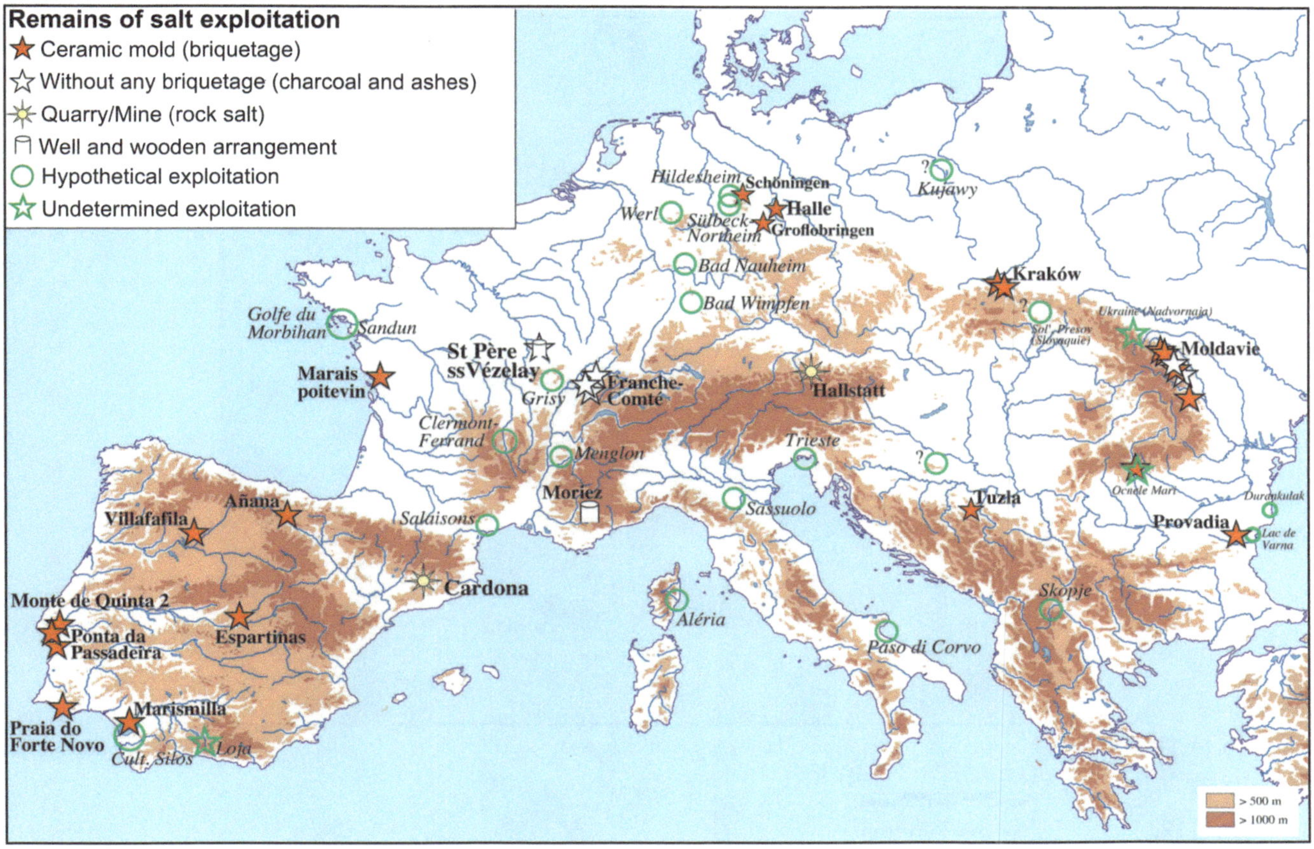

Figure 1.9 The map shows the location of major salt exploitation remains in Europe, including the oldest salt mines in Austria and Cardona in northern Spain. These regions have been exploiting rock salt since prehistoric times, with briquetage found in Portugal, Spain, France, Germany, Austria, Poland, Italy, and the Carpathian Mountains region. Rock salt has been exploited in these regions since prehistoric times. Briquetage has been found throughout Europe, particularly in Portugal (near Lisbon), Spain, France, Germany, Austria, Poland, Italy, and the Carpathian Mountains region. Credit: Weller (2015)/University of Ljubljana Press/Public domain.

furnaces used in salt production burned wood, but with technological advances, they started to burn coal. Brines obtained from salt springs or extracted from surface and underground salt layers were pumped into large pans, which boiled the water away. Workers in these factories accompanied the formation of salt minerals, mainly halite crystals, which accumulated at the bottom of the pans. The residues were scraped out or transferred to a smaller pan that collected the salt, and then more brine was added to the large pan to continue the concentration process on a large scale (e.g., Weller, 2015).

The Hallstatt salt mine in Austria was one of the most important salt mines in the Alpine region, hosting several other salt mines during the Neolithic, such as the Bad Reichenhall prehistoric salt mine in Germany, whose production rivaled that of Hallstatt (Figure 1.9). The Cardona salt mine is a unique diapiric structure south of the Pyrenees Mountains in Spain (Figure 1.11). The salt mountain displays a surface height of about 100 m above the ground (Miralles et al., 2000). The salt diapir extends for at least 1 km below the surface, as indicated by salt exploration boreholes. Archaeological findings indicate that primitive tribes mined salt from the Cardona Mountains as far back as 3500 BCE (Harding, 2013).

The Romans, who believed that rock salt from the mines was of higher quality than salt scooped up from the sea or obtained through evaporation in the North African saltworks, left the first written record of salt production in Cardona. Salt from the Cardona region has been extracted from mines that operated from the Neolithic through medieval times until the Industrial Revolution. In the 20th century, the salt industry in Cardona aimed at producing sylvite rather than halite (Harding, 2013, 2014; Weller, 2002).

Many archaeological sites in Southern Europe have provided evidence of salt production in prehistoric times,

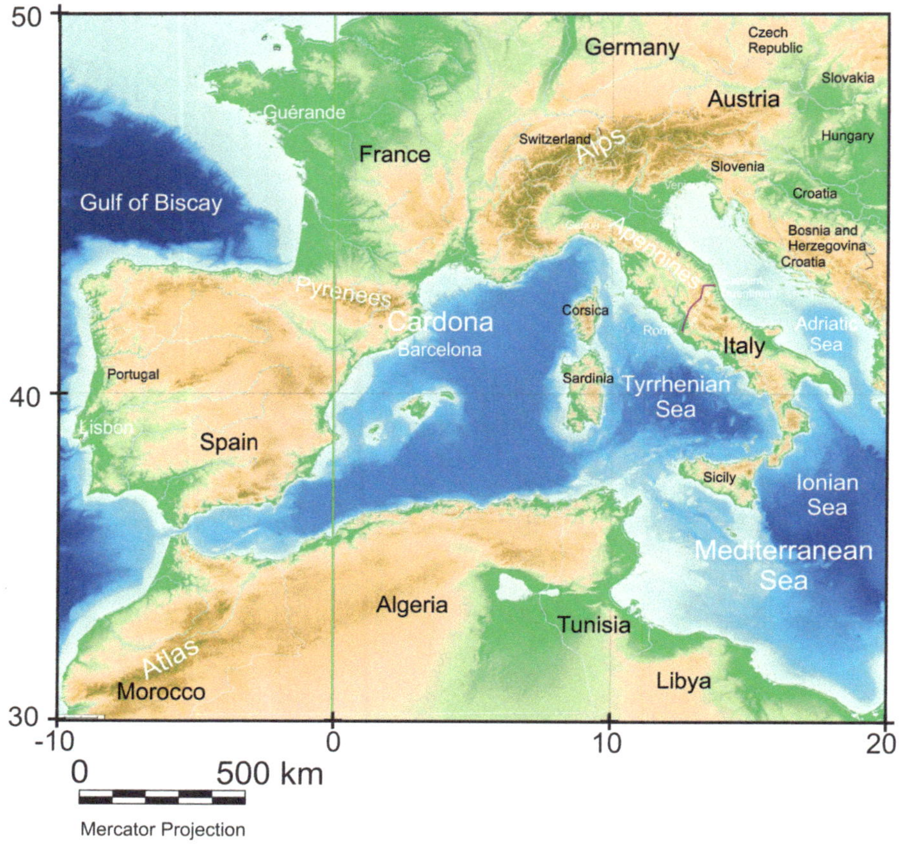

Figure 1.10 The topobathymetric map of Southern Europe shows the Pyrenees and Alps, where archaeological evidence for prehistoric salt production can be found. The Cardona salt mine in northern Spain and the Hallstatt salt mine in Austria are examples of salt production centers. The Via Salaria connected the Tyrrhenian Sea saltworks near Rome to the Adriatic Sea saltworks. Salt was also extracted from mines in Sicily and saltpans across the Saharan desert, from Morocco to Tunisia. Credit: Topography from GEBCO (2019).

Figure 1.11 The Cardona salt mine in Spain, located near Barcelona, features gypsum layers folded by salt flow. At the top of the hill, Tertiary sedimentary rocks and soil cover the evaporites, with trees visible as scales. The white to light gray colors represent gypsum layers.

as indicated by briquetage, mining tools, and blocks of salt cut from salt mines (Figure 1.12). The relationship of humankind with salt is expressed not only by prehistoric archaeological sites but also by the modern mining and tourism industries. Some salt mines from medieval Europe and South America have been converted to sanctuaries displaying cathedrals carved out of the salt rock, such as those near Salzburg, Austria, or the Praid salt mine in Romania. Some salt mines in Romania are touted as health destinations to alleviate respiratory diseases and allergic ailments (Goldan et al., 2010).

1.6.5. Salt Production in Ancient China

Salt production in ancient China has a long history, with evidence of significant activity in the city of Zhongba during the 2nd millennium BCE (Flad et al., 2005). The Qui country in ancient Eastern China became prosperous due to salt trading and salt taxes. Salt was obtained from salterns at the salt marshes of northern Jiangsu and near

Figure 1.12 Archaeological evidence of salt production in Southern Europe during the Neolithic period includes briquetage studies and salt blocks obtained from evaporite rocks. (A) Mining tools for the extraction of salt. (B) Salt blocks cut by the prehistoric miners. Credit: Courtesy of Weller (2015).

Bohay Bay in eastern China, which has become one of the largest salt-producing areas in China since the Ming dynasty.

The Sichuan Basin in southwest China is renowned for its technological breakthroughs in the salt production industry, making it the largest inland salt production district in China (Chiang, 1976). Many technological advances in salt mining and borehole drilling are linked to the Chinese (e.g., Suihong, 2016). Chinese engineers constructed the first deep boreholes to be drilled in the world in search of underground brine and salt deposits. By the 3rd century BCE, bamboo was used as drilling equipment and pipelines to connect salt springs with salt pans (Suihong, 2016). Between 100 and 400 CE, the Chinese produced salt from the dissolution of underground evaporite layers in the Sichuan Province. They discovered a black substance that impregnated reservoir rocks, which burned easily together with the gas present in the reservoirs.

Chinese engineers developed a way to connect oil and gas wells with salt springs and salt pans using bamboo pipes. They invented a pioneering technique of using hydrocarbon fuels to produce the heat necessary to boil brines and produce salt. This was the first time humans used petroleum from a sedimentary basin as fuel to produce salt from brines in reservoirs of the same basin. The Chinese also discovered that by mixing other salt compounds, such as potassium nitrate, with sulfur and carbon, they could manufacture a powder that was easily ignited, producing a powerful explosion. This discovery led to the invention of gunpowder, which was later used in warfare by many armies worldwide.

1.7. SALT TRADING AND THE WEALTH OF NATIONS

The Scientific Revolution, which began around 1500 CE, led to the development of tools to control the natural world. Salt trading, a commodity, significantly impacted economic development and created an industry that generated significant profits for empires and governments. However, salt taxes have led to rebellions and conflicts until the Industrial Revolution. Salt has been exploited globally, not only in Europe and Asia but also in pre-Columbian civilizations in the New World before European colonization.

1.7.1. Salt in Europe

Salt in Southern Europe has been produced since prehistoric times (Figure 1.9). A brisk salt-mining industry and salt-shipping trade centered in Southern Europe from around 1000 BCE onward (Harding, 2013, 2014; Saile, 2015; Weller, 2002, 2015). Ancient cities were founded near saltworks, where salt could be produced on a large scale by evaporating seawater. Rome's origins began on the hills behind ancient saltworks built by tribes at the mouth of the Tiber River, which runs from the Apennines to the western coast of Italy at the Tyrrhenian Sea.

During the Bronze Age, the Romans built the main road in Europe to transport salt to other cities across the Apennines. By the 2nd millennium BCE, horsemen were using the Via Salaria as a trail. During the Roman Empire, the Via Salaria and many other roads were rebuilt to connect the capital Rome with distant conquered provinces, many of which were used to transport food and goods, including salt, from the colonies in northern Africa. The Via Salaria became the main commercial route that crossed the Italian Peninsula, linking Rome to the shores of the Adriatic Sea near the city of Castrum Truentinum (Figure 1.6). The Via Salaria turned out to be one of the keystones in the creation of wealth for the Roman Empire (Marzano, 2013; Saile, 2015).

As the Romans were eager to have saltworks closer to their domains, they developed salt production poles in coastal cities using solar ponds. Venice, renowned for its lowlands, was one of the cities where solar evaporation pools were built using water from the Adriatic Sea. In the Medieval Age, Italian merchants established their commerce in Venice and Genoa, which became economic, cultural, and trading centers for several commodities, including salt produced in the nearby ponds and from other coastal cities along the Adriatic Sea (Marzano, 2013).

In regions far from the seashore, salt was produced from salt mines that exploited salt diapirs that outcropped in mountain ranges, such as Austria (Hallstatt salt diapir) and Spain (Cardona salt diapir). One of the most important salt structures is located in the Salzkammergut region of Austria, where Roman and Celtic remains have been found near a salt mine considered one of the oldest salt exploitations in the world, producing salt for almost 7,000 years (Figure 1.8).

The Hallstatt salt mine, located in Austria, has been in operation since the Bronze Age (3300–1200 BCE) and became one of Europe's main salt production sites by the Iron Age (1200 BCE to 1000 CE). By 1000 BCE, the near-surface salt deposits were nearly exhausted, but some salt was inferred to occur underground. The underground salt mine produced large quantities of rock salt, which was transported north and south of the Alps. The mine was so important due to the salt trade and archaeological finds that an Early Iron Age culture has been named after the region.

During the Iron Age, the miners created a distinctive heart-shaped trademark consisting of blocks of salt, each about 40 kg in mass. However, a catastrophic landslide in the middle of the Iron Age caused a massive influx of sediment into the galleries, killing many salt workers and interrupting mining activities by 400 BCE. Roman occupation and access to Adriatic salt decreased the relevance of the mine in Southern Europe, but some operations persisted until the Middle Ages (Festi et al., 2021; Grabner et al., 2021; Harding, 2013; Reschreiter & Kowarik, 2019, and references therein).

In the late 19th century, archaeologists excavated over 2,000 graves near the Hallstatt salt mine, discovering the corpses of prehistoric Celtic people. The bodies were encased, and healed back into rock salt, and later buried in Christian cemeteries. The archaeological finds included mining tools, remnants of equipment, clothing, and food residues in their feces (Gamsjäger et al., 2008; Maixner et al., 2021; Reschreiter & Kowarik, 2019). The Hallstatt salt mine is now a World Cultural Heritage Center and an iconic archaeological site studied by international scientists.

1.7.2. Salt in North America

The Yucatan Peninsula, which is now part of Mexico, Belize, and Guatemala, has been a source of salt for at least 2,000 years before the present (McKillop, 2005). Archaeological evidence suggests that ancient Mayan cultures traded salt as an important commodity with other cultures in the region (Andrews, 1983; McKillop, 2019).

Salt-making from evaporating pans was an activity that Indigenous Americans had undertaken around the same time as Africans, Chinese, and Europeans. The Mayans also learned to extract small quantities of salt from some species of plants, particularly the halophytes, which grow in waters of high salinity in semi-deserts, mangrove swamps, and marshes (Robinson & McKillop, 2014).

However, the rudimentary method of obtaining salt by burning plants cannot compete with more traditional means of solar evaporation in brine pools or collecting salt precipitated in shallow lakes in arid deserts. Most briquetage evidence in the Yucatan Peninsula points to salt production by solar evaporation (Andrews, 1983; McKillop, 2019). The ancient Maya in southern Belize boiled brine in salt kitchens made of poles and thatch, preserved below the sea floor in mangrove peat. There are thousands of wooden building posts that have been mapped at several archaeological sites (McKillop, 1995, 2019).

There are archaeological indications that native inhabitants in pre-Columbus times obtained salt from salt springs in some hills near the Gulf of Mexico coastal region (Brown, 1999, 2015). In southern Louisiana State, there is a place originally called "Ile Petite Anse," which means "little harbor" in French, and is nowadays known as Avery Island, named after the landowners in the 18th century. Avery Island corresponds to a raised area within a swamp on the Gulf Coast, forming almost an island surrounded by dark waterways associated with fluvial tributaries and bays (Figure 1.13).

In the 18th century, fragments of Indian pottery and broken artifacts were accidentally found near one of the salt springs on Petite Anse. Since then, archaeologists have discovered many other mounds interpreted as piles of briquetage associated with ancient salt production in the coastal plains. The briquetage indicated that some sites in the southern United States corresponded to prehistoric saltworks, similar to the ones developed by the early Romans in Europe and North Africa.

Both ancient pre-Columbus Indian cultures and European colonizers were unaware that the sources of the salt springs were hidden underground. The relationship between salt springs and evaporite layers in the subsurface was recognized only during the American Civil War. This discovery indicated that the Petite Anse salt springs were associated with the dissolution of massive salt layers in the subsurface of the marshlands. Farmers and geologists living in the region by the end of the 19th century were also unaware of the implications of this shallow salt

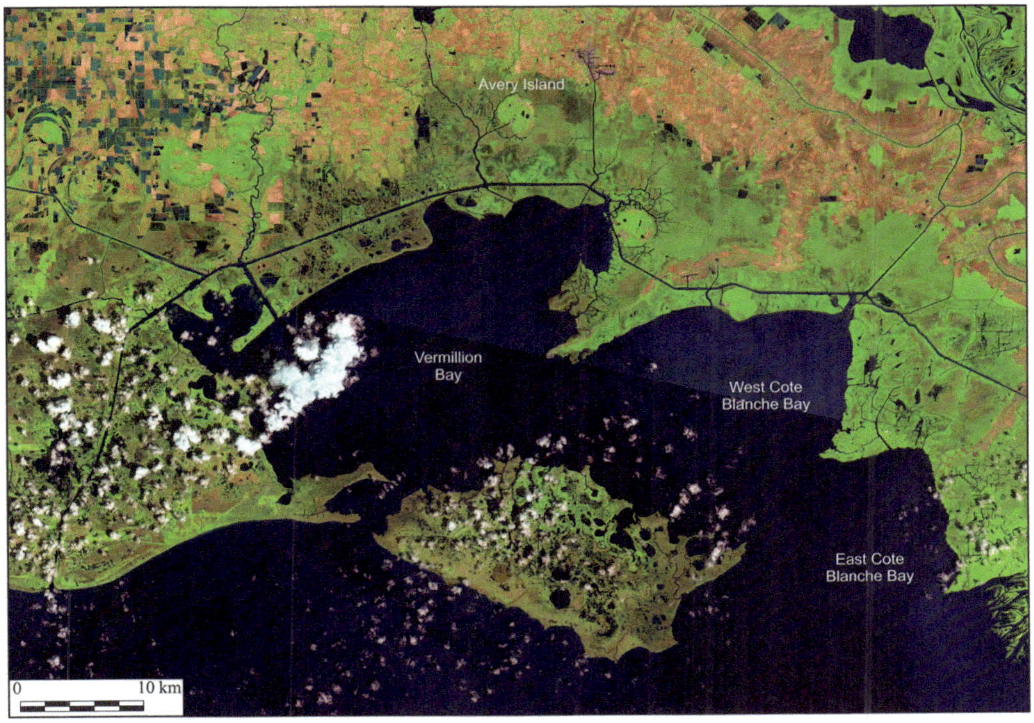

Figure 1.13 A NASA satellite image of the northern Gulf of Mexico region near Marsh Island in Vermilion Bay shows the circular outline of salt diapirs on land. Archaeological studies have discovered briquetage from ancient salt works exploited by North American native tribes. Salt has been produced from these diapirs, both onshore and offshore, and many of these structures are associated with hydrocarbon accumulations, which changed the petroleum industry in the early 20th century.

occurrence on Avery Island for their understanding of Gulf of Mexico geology.

Ancient pre-Columbus Indian cultures and European colonizers were unaware of the hidden sources of salt springs in the subsurface. The relationship between salt springs and evaporite layers in the subsurface was only recognized during the American Civil War. In 1862, a black slave, working on a sugar plantation, on a sugar plantation, found a white layer of solid rock salt, mainly composed of gypsum and halite, at a depth of about 5 m. This discovery indicated that the Petite Anse salt springs were associated with the dissolution of massive salt layers in the subsurface of the marshlands. The discovery also hinted at the presence of numerous salt structures in the subsurface (e.g., Taylor, 1938). The discovery of petroleum associated with a salt structure in the Spindletop deep well in Texas in 1901 changed the Gulf of Mexico story, revealing the region's rich history of salt structures (see Chapter 4 in Mohriak, 2025).

1.7.3. Salt in Central America

The Aztecs, Mayans, and Incas were three advanced civilizations in the Americas that coexisted with European conquerors for a while. However, Spanish soldiers eventually overthrew them and took control of their cities, leaving only a few scattered descendants of these tribes in Latin America. During their golden age, these civilizations discovered salt springs in their mountainous territories. This discovery was an essential component of the colonization process conducted by the Europeans. Salt, gems, and precious metals influenced the foundation of many cities in the New World after Christopher Columbus' discovery of the Americas in the 15th century. The search for and exploitation of these precious items led to violent colonization by European cultures and the almost annihilation of native civilizations, including the Aztec and Mayan empires in Mexico and Central America, after fierce fighting that lasted for decades (e.g., Craig et al., 2011).

Pre-Columbus cultures in the Americas were renowned for their use of precious metals (silver and gold) and salt in their religious ceremonies, as in the fable of El Dorado in Colombia. Salt sources were known since prehistoric times by Indigenous tribes in Central America and the Andean Mountains, and salt was exploited in coastal areas and desert lakes. Salt deposits were precious assets, and many tribes aiming to control them fought bloody wars with competitors.

The Aztecs, Mayans, and Incas employed salt preparations in medicine and as religious offerings to their gods. The Mayan and Aztec economies flourished in the Middle Ages by trading agricultural products and salt with other tribes living on the Yucatan Peninsula and nearby Caribbean islands in Central America (McKillop, 2005; Robinson & McKillop, 2014). However, the reason that favored the rich Mayan culture, a dry climate that was essential for salt precipitation, also resulted in the fall of their empire. Extremely dry climatic episodes affected this inhospitable land from time to time, with drought events occurring repeatedly between ca. 770 and 1100 CE (Hodell et al., 2005).

The Yucatan Peninsula was the largest salt producer in pre-Columbian America and remained a prominent producer for a few centuries into the modern world. The arrival of Spanish, Portuguese, and Dutch conquerors in the Americas led to a new culture obsessed with the search for salt, pepper, wood, and precious metals such as gold and silver. Due to the increased use of salt in processing food for New World settlers and the introduction of cattle by Europeans across North and South America, salt demand increased in the metalworking industry during the period of European colonization.

1.7.4. Salt in the Americas After the Arrival of the Europeans

The mid-16th century saw the invention of a silver mining process in Mexico by the Spaniards, which improved the efficiency of producing precious metals. This technique involved separating silver from its ore using salt to extract impurities during a rudimentary melting process. Europeans built large-scale saltworks adjacent to silver mines in many places across the New World, including North America. Gold mining in the Sierra Nevada region of the Western United States was the catalyst for the rapid growth and development of California in the 19th century. Production of silver and gold was accompanied by increasing salt output from areas where salt could be easily exploited, such as the Great Salt Lake in Utah (Clark, 1984).

The Chibcha Indians were part of an important tribe living in the Andean highlands of what is now Colombia, in northern South America. The Chibcha culture was renowned for its advanced knowledge of salt extraction techniques from the mountainous Andean region. Pre-Chibcha civilization already used salt from around 150 CE in the area of Zipaquirá, and the Chibchas themselves exploited salt from the 9th century CE (Glassner, 1970). Some South American cultures produced salt by evaporating brine extracted from saltpans, which abound in the Andean highlands, for example, in the Uyuni Salar in Bolivia, and from the dry lakes that occur in high-altitude deserts, such as the Atacama Salar in Chile (see Chapter 10).

The El Dorado legend from Colombia deeply impressed European explorers who were searching for precious metals in the Andean region in the 16th century (Glassner, 1970). The story links salt with the search for gold, as the Spanish conquistadores demanded knowledge from their prisoners, Indigenous tribes from the Muisca confederation, about the real source of their riches. The Spaniards did not understand when they were taken to the salt mines exploited by the Chibchas for centuries instead of the gold and silver mines they dreamed about. Similar to other civilizations, the Indians considered salt and gold the most precious minerals that their gods provided on Earth. It is also possible that the Indians from Colombia had some clues that the origin of the famous emeralds coveted by the conquerors was associated with salt deposits. Scientists now recognize that hydrothermal fluids coming from granitic intrusions form these precious gems at the interface between carbonate and salt layers (see Chapter 6).

The legend of El Dorado originally described a man impregnated with gold powder and revered during religious ceremonies. After the initiation ritual, he was bathed in a saline lake to remove these particles and clean his skin in a purifying rite. After the arrival of the Europeans, many explorers tried to find this lake, whose bottom would have been collecting layers of gold dust for centuries. Today, scientists assume that there is a geological reason why some lakes are salty and deep, such as those found in the Andean foreland basins. These lakes are located on top of a subsurface salt structure that was affected by dissolution, many meters below the lake. Geological maps indicate that many salt diapirs occur in the Andean foreland basins, as observed in places such as Zipaquirá, a city near Bogotá. The Zipaquirá mountain hosts an old salt mine that has been transformed into a famous salt cathedral and a museum. Most of the columns, ornaments, and architectural details in the Zipaquirá Salt Cathedral are hand-carved from pure halite.

1.8. SALT PRODUCTION IN THE PREINDUSTRIAL WORLD

Salt production in Europe, Africa, and Asia was a significant economic factor from premedieval times to the Middle Ages. It established productive regions for trade and industrialization. The relationship between salt occurrences and petroleum accumulations in sedimentary basins was established in Europe and North America only during the Industrial Revolution in the 19th century (see Chapter 6 in Mohriak, 2025).

Many centuries before the Industrial Revolution, salt production was a significant factor in wealth creation for regions like the Roman Mediterranean Empire and the Western Saharan region (McDougall, 1990; Marzano, 2013). Throughout human history, autocratic rulers and governments have adopted the practice of imposing salt monopolies and taxing salt consumption, which has often led to violent popular rebellions against authorities. These conflicts date from the Middle Ages to the modern world.

1.8.1. Salt Production and Trading Expansion to Northern Africa

From pre-biblical times to the Middle Ages, salt was a valuable commodity for various ancient civilizations. The Roman Empire experienced exponential growth in political and economic power during the 1st century of the Common Era, leading to the expansion and rebuilding of roads to transport salt from producing regions to European capital cities. The Via Salaria, an example of these early roads, connected Rome to other cities across the Apennines at the Adriatic Sea (Figure 1.1). For the next few centuries, Rome would be considered the center of the world after the demise of the Greek Empire (e.g., Luttwak, 2016).

During the late Roman Empire and throughout the Middle Ages, salt was transported along salt roads into Europe, where it was consumed by Germanic and Celtic tribes. Caravans carrying thousands of camels traversed the Sahara Desert, bringing salt and other precious goods to inland markets in North Africa (McDougall, 1990). The Romans were forced to establish extensions of the Via Salaria to distant and forsaken places such as the Dead Sea and North African and Asian salt mines. These regions actively produced salt from desert lakes such as the Dead Sea, Tuz Gölü in Turkey, the Afar region in northwest Africa, and the Assal Lake in Djibouti (Erdoğu et al., 2003; Nissenbaum, 1993; Woldekiros, 2019). These inland hypersaline lakes became salt havens for the European civilization, whose nearshore saltworks along the Adriatic Sea had been declining since the collapse of the Roman Empire (Marzano, 2013).

Salt-making was well-established in many cities along the Western European coasts, such as the Guérande region of western France, but was mainly produced on the North African and Arabian coasts (Harding, 2013). Saltworks were established mainly across the Saharan desert belt, where Miocene and younger salt deposits occur along the Mediterranean and Red Sea coastlines (Figure 1.2). These saltworks obtained halite by exploiting sedimentary layers outcropping on the surface or at shallow depths underground, as in the Afar region of northern Ethiopia (Woldekiros, 2019).

Salt crusts accumulated on dried-up salt flats were locally dissolved during wet periods, forming brines in ephemeral lakes and evaporite pans. Many salt pans and artificial saltworks were built at the northern, southern, and eastern margins of the Mediterranean Sea. These saltworks have provided plenty of salt since prehistoric times and up to the present, such as the salt pans located in the eastern part of Tunisia or the western coast of Israel (e.g., Galili & Arenson, 2017).

During the early centuries of the Middle Ages, salt production was an extremely profitable (and sometimes illegal) economic activity in Europe. Salt mines in mountainous regions and saltworks in coastal cities were mainly established in Central Europe, aiming to produce salt for a growing population. However, the scale of salt production in the Alpine and Carpathian regions was relatively small compared to other salt provinces in northern Africa, where salt could be easily obtained due to climate aridity (McDougall, 1990). Some of these solar ponds have been actively operating for the last 500 years in many countries, such as Tunisia, Ethiopia, and Djibouti (Cartwright, 2019; McDougall, 1990).

Salt played a key role in the history of the ancient city of Timbuktu in northwestern Africa, particularly during the great trading empire of Mali from the 13th through the 16th centuries. Merchants exchanged goods from North and West Africa, as well as from Mediterranean cities across the Sahara Desert, leading to a flourishing market in Timbuktu. The wealth from trading gold and salt for other commodities resulted in one of the richest cities in Africa during the Middle Ages, developing some economic and cultural centers that rivaled European cities (Mingren, 2019 and references therein).

The decline of the northwestern African empires started when the Portuguese, Spanish, and Dutch navigators ventured to explore the Atlantic Ocean and their trading ships discovered more economic routes to bring salt and other goods to Europe. After the Europeans discovered and colonized the Americas in the 15th and 16th centuries, the African continent started to decline. The slave trade and the establishment of European colonies in Africa resulted in an economic depression for the African population, which was accentuated in the 19th and 20th centuries.

1.8.2. Salt Mining in Medieval Europe

Slanic Prahova, a small town in Romania's sub-Carpathian landscape, is home to one of the largest salt mines in medieval Europe. The city, situated at an altitude of about 400 m, has been exploiting salt deposits since the Middle Ages (Dordea et al., 2013). Other important salt mines in medieval Europe include Hallstatt in Austria and Wieliczka in Poland. Some authors believe that the Wieliczka salt mine is the oldest continuously operating industrial venture in Poland (e.g., Garlicki, 2008), responsible for one-third of the total country's income in the

Medieval Ages, when salt exploration was a royal monopoly (Hallett, 2002).

During the Middle Ages, salt was intensively produced in Western Europe, either by brine evaporation in solar ponds or by mining salt diapirs like the Cardona salt mine in northern Spain (Figure 1.10). The evaporite rocks in the Cardona Hills outcrop at the surface as gypsum, a white mineral composed of calcium sulfate. The Cardona mine is considered the only rock salt mountain in Western Europe that has been continuously exploited since Neolithic times.

However, Italian cities like Genoa still dominated the salt market in Europe around the 12th century CE. Salt merchants considered it cheaper to transport salt to Barcelona using ships from Genoa than to transport salt from the nearby Cardona mine. Land transport was too expensive and uneconomic in medieval Europe. During the Middle Ages, better roads were built, and the salt produced in the Cardona mine was transported to the port in Barcelona and exported across the Mediterranean to many European regions. Salt became one of the main Spanish products marketed throughout the Middle Ages, with Barcelona becoming the cultural and economic capital of the Catalonia region.

1.8.3. Salt Mining Techniques and Scientific Methods in the Preindustrial World

During the Middle Ages, salt production in Southern Europe relied on various techniques, including salt diapirs, cutting blocks of salt from halite crusts in salt pans, obtaining salt by evaporating seawater in solar saltworks, or collecting salty water from springs and evaporating the bulk of the water in ceramic pans. These traditional methods were similar to prehistoric techniques developed by ancient civilizations. The Romans were the first to develop a more advanced technique of salt production on a large scale on the Mediterranean shores, gathering seawater in large, shallow basins to obtain large amounts of salt by solar evaporation. The salt workers scraped the bottom of these saltworks and piled up the salt to be transported to other places. These traditional ways of salt production were described by Pliny the Elder in the book "Natural History," as discussed by Jones (1938).

Leaching, a method that was likely first used by the Chinese or Europeans, was a significant innovation in medieval salt production. The water absorbed salt from the surrounding walls and formed a brine that could be later boiled on a conventional briquetage. The salt dissolution technique was probably the most important technique developed by salt miners during the Middle Ages. The dissolution process started by drilling a borehole to the top of the salt layer and injecting water into the formation. The Chinese probably applied the technique using bamboo pipes to obtain brine and gas from the subsurface in petroleum-rich basins (Suihong, 2016). In Europe, salt was mined by dissolution methods even before the 17th century, as indicated by the industrial pipelines built to transport brine from the mountainous regions in Germany and Austria into the saltpans in the valleys (Harding, 2013, 2014).

One of the major advances that revolutionized salt mining and petroleum production was the use of drilling technology. In ancient times, Chinese engineers pioneered percussion-drilling techniques to obtain brines from the subsurface, and many boreholes were drilled to depths of a few hundred meters in the Sichuan Basin. Around the 19th century, the percussion-drilling technique was applied in saltworks across Europe and the United States. However, this technique was superseded by the modern rotary drill, which has been employed by the salt mining and petroleum industries since the beginning of the 20th century.

Georgius Agricola, a German naturalist, introduced observation-based practices to the study of the earth, publishing "De Re Metallica" in 1556, which is considered the first scientific publication in medieval Europe to discuss physical aspects of the natural world (Barton, 2016). This book constituted a rich source of practical knowledge for earth science professionals, presenting basic principles and focusing on the practical applications of geology for mining, refining, and smelting metals and salt.

A description of practical methods to obtain salt by boiling brines included very didactic drawings showing the wooden apparatuses used by ancient salt workers (Harding, 2013). Because of the postmortem publication of these important works on mining techniques in the Middle Ages, Georgius Agricola is considered an outstanding historical figure responsible for the development of the earth sciences and is rightfully called the "Forefather of Geology."

James Hutton, a Scottish naturalist, is often referred to as the "Father of Geology." He and Charles Lyell were pioneers in European geological studies, advocating for natural causes rather than divine intervention. Hutton (1788, 1795) postulated that natural phenomena cannot be fully observed by humans, but their effects should be interpreted using the present as a key to the past. This scientific methodology was crucial in the development of earth sciences during the Industrial Revolution, contributing to the explanation of ancient salt basins by analyzing modern environments where evaporite minerals form through chemical precipitation (see Chapter 8).

1.8.4. Salt and Wars in North America

Following the discovery of the Americas in 1492, Europeans began colonizing the New World in the 1500s, with Great Britain establishing colonies in North America by the late 16th century. Salt was in short supply due to the inadequate climate for salt production from evaporation ponds, and most salt used by the fish industry in the United States and Eastern Canada was imported from England. Some salt was produced in the American colonies, particularly from mining shallow salt in the Michigan Basin, especially in the region of Onondaga County in New York (Kappel, 2000). Some small Caribbean islands produced large quantities of salt by solar evaporation in coastal ponds, particularly the Turks and Caicos Islands (see the classical overview of these islands by de Booy, 1918).

In the 18th century, salt production in the Caribbean Sea declined steadily, leading the American colonies to import salt and other goods from other European countries, including Portugal, Spain, and France. American merchants began selling salted cod not only to the British colonies but also to the Spanish colonies in North and South America.

By the middle of the 18th century, the British government imposed heavy taxes on the U.S. colonies to penalize them for trading with other nations instead of England. This deteriorated the political and economic situation in many of the 13 North American colonies. The first outcries for independence started in New England by the second half of the 18th century. By 1775, the British government decided to declare a naval blockade to stop illegal trading with other nations. This caused an immediate salt shortage, which affected the entire population of the New England colonies. Following years of fierce fighting between the British and Americans, an armistice was signed in Paris in 1783. The British government recognized the sovereignty of the United States, formally ending a war that lasted for 8 years (Crout, 1983).

1.8.5. Salt and the U.S. Civil War

The Southern States Confederacy's defeat during the American Civil War may have been due to the salt shortage that affected their soldiers (e.g., Smith, 2011; Snodgrass, 2020). In the second half of the 19th century, the northern states were responsible for 85% of the U.S. salt production, mainly from shallow underground salt layers in the Paleozoic basins extending toward Canada (Kappel, 2000). As the salt production in the southern states was negligible (Bowman, 1969), these colonies relied heavily on salt importation from Europe or the Caribbean islands.

Consequently, the Northern States Union had a well-fed army whose daily diet was supplied with plenty of meat, milk, and vegetables. The Confederate soldiers from the southern states had a poorer diet, mainly based on fish and meat. During the Civil War, salt production was greatly reduced, and a salt shortage resulted in fewer healthy foods. One of the first acts US President Abraham Lincoln conducted in 1861, at the beginning of the American Civil War, was to order a naval blockade of all salt-importing ships to the southern ports and ban imported salt distribution to supply farmers' needs across the southern states. This act lasted until the end of the war in 1865.

The poor nutrition affected not only the soldiers but also the general population in the southern states. The Northern Union generals ordered the destruction of saltworks as a strategy to defeat the enemy. The Union Army considered giving salt or food to the enemy a crime or high treason, leading to emergency regulations restricting salt use and even regulating food and diet. By 1864–1865, the situation had become critical in the southern states, leading farmers to divert slaves from sugar and cotton plantations to operate the reduced number of saltworks. The government established emergency regulations restricting the use of salt by the population and even regulating their food and diet. These enforcing acts were more intolerant and tyrannical than the legislation the British government imposed on the inhabitants during the colonial period (Hilldrup, 1945).

The fate of the Southern Confederate Army might have been different if they knew that the Gulf Coast of the southern states concealed massive salt deposits, with reserves reaching millions of tons. The discovery of salt in Petite Anse occurred during the last stages of the Civil War, too late to have an effect on the United States history. The Confederate Army was defeated by guns and famine, partly due to the lack of salt, which was more abundant in the northern states.

Salt deposits on the Gulf Coast occur only a few meters below the surface in some salt diapirs, but they have only been properly characterized by earth scientists in the 20th century. Borehole drilling operations conducted by petroleum explorationists indicated a huge quantity of salt underground in the southern part of the United States, from Texas to Florida. More importantly, it became clear that there was an association between salt structures and petroleum accumulations, which contributed to the dawn of the Industrial Revolution (see Chapter 6 in Mohriak, 2025).

1.8.6. Salt Rebellion in India

Salt played a significant role in the fight for independence in British colonies, particularly in India, where the 1857 national uprising was a revolt against British rule (e.g., Engler & Engler, 2014). The 18th and 19th centuries saw numerous independence movements in Asia, including India, which culminated in the Indian Independence Act of 1947, ending British rule. The British colonizers' imposition of salt taxes was one of the main causes of the fight for independence in India. Mahatma Gandhi used salt as a symbol to leverage his nationalistic project during India's independence movement. He organized the Salt Satyagraha protest to demonstrate against the unjust salt tax and strict laws prohibiting private salt production.

In 1930, Gandhi's "march to the sea" was a famous example of nonviolent resistance against the British Empire. Thousands of people followed Gandhi's 24-day journey to Dandi, where they collected salt from marshes, which was later claimed to be illegal by British officials. The Salt March attracted thousands of followers for nonviolent resistance against the British Empire. Gandhi's nationalistic movement defied the law, triggering a political process that eventually earned India's independence from Great Britain (Engler & Engler, 2014).

Mahatma Gandhi's nationalistic movement openly defied the law with this nonviolent civil disobedience, triggering a political process that eventually earned India its independence from Great Britain. India's independence was formally accepted by the British with the signing of the Indian Independence Act in 1947. However, the roots of the anti-imperialist riots and rebellions date back to the early years of the 19th century, when the British rulers imposed a salt tax and monopolized salt production.

1.9. SALT USAGE FOLLOWING THE INDUSTRIAL REVOLUTION

The 19th-century Industrial Revolution led to a significant increase in the use of evaporite minerals in modern societies. This was influenced by the understanding of geological concepts using scientific methods to interpret the past, as proposed by renowned naturalists (e.g., Hutton, 1788, 1795; Lyell, 1830, 1865).

Two global regions, China and the Transylvanian Basin in eastern Europe, demonstrated significant advancements in salt production and applications. China's drilling technologies in the eastern world paved the way for drilling deeper targets in sedimentary basins. The petroleum exploration in the Transylvanian Basin in Romania, and the Gulf of Mexico in North America, highlighted the importance of salt as an indicator of hydrocarbon accumulations. The salt sciences, involving chemistry, physics, and geology, played a crucial role in the development of technologies for finding salt and hydrocarbon deposits in sedimentary basins, with applications in agriculture, medicine, and other scientific fields.

1.9.1. Salt Production in Modern China

The Chinese have been a pioneer in deep-drilling technologies since the mid-11th century CE, using bamboo to reach depths hundreds of meters below the surface (Suihong, 2016). The borehole drilling technology in China, which was well ahead of the world in the Middle Ages, continued to improve in the 17th and 18th centuries. The Sichuan Province, which produced about one-fifth of China's salt during the early years of the Industrial Revolution, is considered one of the oldest salt-producing provinces in the world. Salt and hydrocarbons have been exploited in Sichuan for over 2,000 years (Flad et al., 2005; Suihong, 2016).

In 1835, a well drilled in Zigong to obtain brine from deep layers in the Sichuan Basin was deepened to about 800 m when it struck a reservoir rock with natural gas. At depths of about 900 m, the borehole penetrated porous rocks filled with natural brines. Drilling continued for at least another 100 m, establishing a world record by drilling an exploratory well that reached a total depth of more than 1,000 m. This discovery provided one of the first hints worldwide about the intriguing relationship between salt and hydrocarbons. This evidence would be confirmed only in the early years of the 20th century by drilling the salt diapirs in the onshore portion of the Gulf Coast in the southern United States (see Chapter 6 in Mohriak, 2025).

Technological advances in China changed the way scientists studied rocks in sedimentary basins across the world. By drilling deep wells, earth scientists investigated not only the rocks near the surface but also the deep layers underground, hidden from sight. This new science was called *geology*, and it rapidly advanced when English naturalists such as Hutton and Lyell started recording their observations about the processes that shape the natural world.

After discovering new accumulations of brine and gas in the deeper portions of the Sichuan Basin at the beginning of the Industrial Revolution, the Chinese drilled hundreds of brine wells in the first few decades of the 20th century. Some authors consider the exploratory well Ba-1, drilled in the Sichuan Basin in 1937, to be the first natural gas well drilled by a rotary drilling rig, establishing a technological breakthrough in the petroleum industry. China also conducted the first borehole logging using geophysical techniques (Xinhua, 2017).

However, the Sichuan salt industry has been slowly declining since the 20th century. By the 1960s, Zigong was only a provincial town with 300,000 people living among medieval brine derricks. In the 21st century, Sichuan salt producers in Zigong are established in a modern industrial city and operate modern rotary drills, applying modern methods for rock salt mining. Active exploratory drilling now aims to search for petroleum accumulations at deeper levels, below the salt layers underground. Recent technical innovations as well as a better understanding of shale gas exploration methods have led to the discovery of several giant gas fields in the Sichuan Basin. The basin ranks as the second-largest gas-producing region in China, with the Tarim Basin still holding the largest amount of gas in place (Dai et al., 2021; Xinhua, 2017).

Total salt production in the world reached about 290 million metric tons in 2022, according to statistical reports (U.S. Geological Survey, 2023). China is the largest salt producer (Figure 1.14), responsible for 22% of the total global production. India is the second-largest salt producer, with about 45 million tons of salt produced in 2022. The United States is the third-largest salt producer, with about 14% of the total world production, corresponding to 42 million metric tons, but is also the world's largest consumer of salt (U.S. Geological Survey, 2023). Germany, Canada, and Australia are also important global producers, with more than 10 million tons of salt produced in 2022 by each country. Chile, Mexico, Brazil, Turkey, the Netherlands, and Russia are

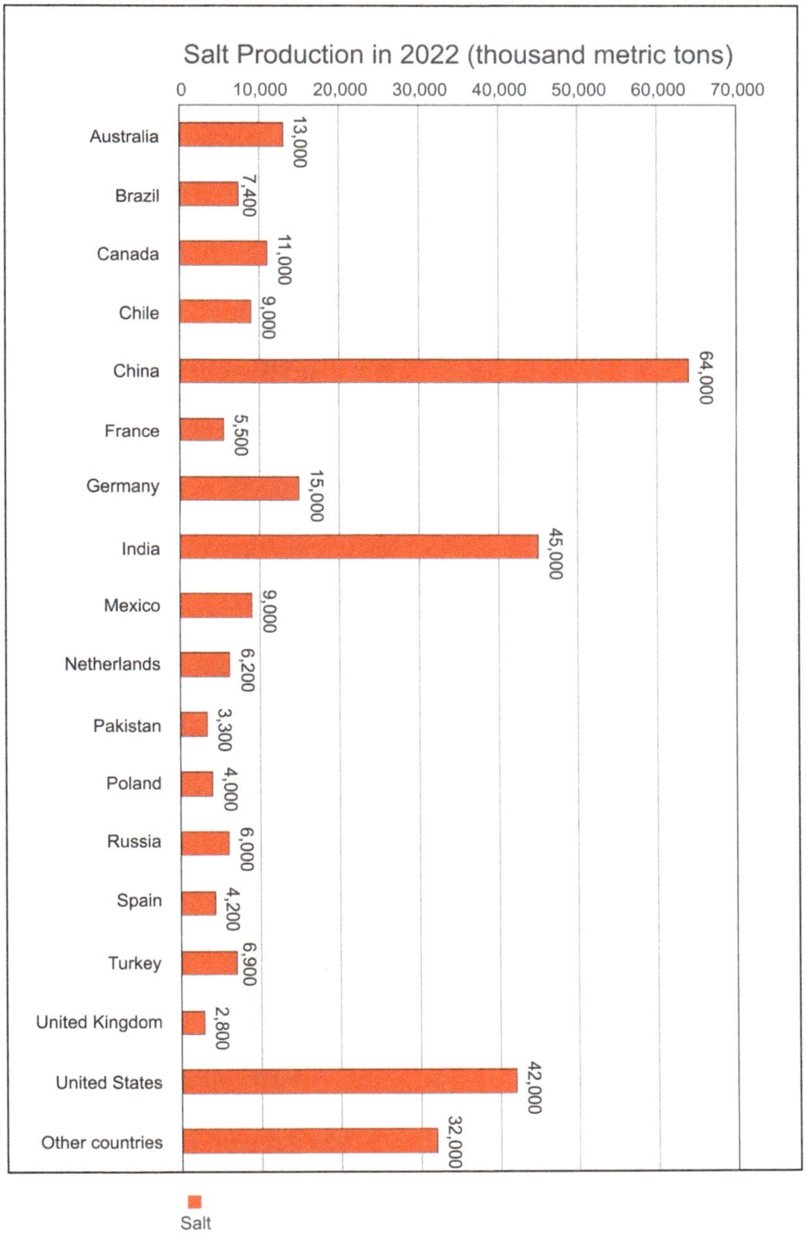

Figure 1.14 The bar graph shows the countries with the largest salt production in 2022 when total global salt production was about 290 million tons. China was the largest salt producer in the world, with 64 million tons.

next in rank, producing between 9.0 and 6.0 million tons of salt in 2022 (U.S. Geological Survey, 2023).

1.9.2. Salt Usage in the New World

Salt production in the Caribbean Sea has been primarily obtained from the islands of Caicos and Turks, where solar salt ponds concentrated marine seawater into brines that precipitated evaporite minerals (De Booy, 1918). The salt crystals were raked by hand and piled up along the salinas before being exported to North America and other places. The Industrial Revolution in the 19th and 20th centuries required massive production of salt, which ultimately doomed small-scale saltworks in Central America and the Caribbean Sea islands.

Onondaga County, New York, is famous for its saltworks, which have been in operation since the first Jesuit missionaries arrived in 1654. Syracuse became known as "Salt City" for many decades during the 18th and 19th centuries (Kappel, 2000). Salt production declined during the Civil War, with a peak in 1862. Salt mines established in the Syracuse area exploited underground salt for over two centuries, but most of the salt was produced by solution mining (Kappel, 2000).

The New York State government controlled all salt rights surrounding Onondaga Lake between 1797 and 1908. However, salt production declined steadily during the first decades of the 20th century due to diminishing brine concentrations in springs and brine depletion in subsurface reservoirs. Most of the production facilities fell into the hands of small, privately owned companies, which continued to produce salt at a reduced scale until the First World War. In the 1920s, storms ruined most of the saltworks, and the salt industry in Onondaga County came to an end (Kappel, 2000).

However, some traditional saltworks continue to operate as small-scale salt industries worldwide, such as harvesting salt pans and the mining of mining shallow salt deposits in places like the Danakil region of Ethiopia. Pink salt is salt mined in Pakistan mountain ranges, some saltworks still operate at Guérande in France, and there are some salt farms in the Incheon and Sinan regions of South Korea. At present, the largest operating saltworks on Earth are located in Laguna Ojo de Liebre in Baja California, on the Pacific coast of Mexico (see Chapter 10).

1.9.3. Petroleum and Mineral Exploration in Salt Basins

The Industrial Revolution in the late 18th and early 19th centuries led to the development of modern scientific technologies for petroleum exploration in sedimentary basins. Colonel Edwin Drake, who had studied drilling techniques for salt exploration, drilled an exploratory well near Titusville, Pennsylvania, in 1859. The borehole found hydrocarbons near the surface at a depth of less than 20 m. However, subsequent drilling with his new technology failed to find significant petroleum reserves. The company that hired Colonel Drake to drill the wells abandoned him, and he died some years later as an impoverished old man. Colonel Drake was unaware that he had ushered in a new era of petroleum exploration in sedimentary basins. Many entrepreneurs became fabulously rich by producing petroleum using Colonel Drake's methods for drilling deep wells (Garvin-Donohue and Birghenthal, 1979).

A few years later, Canadian investors decided to drill for oil in Ontario, near the shores of Lake Huron. Since the War of Independence, Michigan's Lower Peninsula and Southwestern Ontario have become targets for salt exploitation from the vast shallow evaporite deposits in the Michigan Basin. The Ontario salt fields have become one of the most productive saltworks in the modern world, producing salt from the underground layers of the Michigan Basin (Kieta, 2018). The Sifto salt mine is the largest in the world, producing salt from galleries more than 500 m below Lake Huron. Huge reserves of salt deposits are also found in the Williston Basin, on the western side of the United States, near the border with Canada. Paleozoic salt layers occur east of the Rocky Mountains, forming the largest occurrences of bedded potassium salt in the world. The Canadian province of Saskatchewan has become the largest global producer of sylvite (potassium chloride), derived from Devonian evaporite deposits (see Chapter 8 in Mohriak, 2025).

In 1900, Pattillo Higgins and Anthony Lucas, two American entrepreneurs searching for oil in the southern states near the Gulf Coast, decided to drill a topographically elevated hill related to a salt dome near the Texas coastline. The rising of salt masses, forming diapiric structures with the top near the surface, controlled the formation of both hills. The spot they initially chose to drill was called "Sour Hill Mound" because of its unpleasant sulfur smell, which they guessed was related to hydrocarbons. Despite the advice of experienced geologists, chemists, and drilling engineers, the enterprise was doomed to fail due to the widespread belief that there should be no oil in Texas. Anthony Lucas' perseverance in drilling other hills led to the discovery of the Spindletop oil well in 1901, which was considered a turning point for the oil industry in the United States (see Chapter 6 in Mohriak, 2025).

The successful well proved the Gulf Coast had oil deep underground, and some wells could produce more than a hundred thousand barrels of oil per day (Barton & Paxson, 1925; Eby & Halbouty, 1937). In its first 65 years of operation, Spindletop produced 145 million barrels of oil and is considered by many researchers to be the stepping stone for petroleum exploration in the Gulf of Mexico. The offshore province between the United States and Mexico, extending from the continental shelf to the ultradeep water region, holds one of the largest salt basins in the world and is extremely rich in petroleum (see Chapter 4 in Mohriak, 2025).

Until the beginning of the 20th century, Russia was the largest oil producer in the world, with several discoveries in the Baku region. The United States surpassed the oil production of all other countries due to Spindletop and other discoveries in Texas in the early 1900s. Oil production from the onshore and offshore Gulf of Mexico increased in the second half of the 20th century, but it steadily declined a few decades after the major discoveries were made. Other players in the petroleum business appeared on the scene, including the Middle East countries, particularly the region of the Persian Gulf (see Chapter 12).

Although there are some hydrocarbon accumulations on land in Southern Europe, the exploration of the Zechstein salt basin in the offshore region resulted in major petroleum discoveries in the North Sea, starting in the 1960s and continuing up to the present (see Chapter 2 in Mohriak, 2025). In the 1970s, several basins in the South Atlantic started to produce large quantities of petroleum, particularly Brazil, Angola, and Nigeria (see Chapter 3 in Mohriak, 2025).

Salt has impacted human societies in multiple aspects, with an intriguing relationship with the origin of life on Earth and multiple connections with culture. Salt has profoundly affected the social and economic development of humankind and civilizations. The salt sciences have advanced extraordinarily from the first hypotheses proposed to explain the genesis of evaporite basins on Earth. Many early hypotheses related salt deposition to a global flood, magmatic activity, and extreme climate changes resulting in solar evaporation and desiccation of deep oceans.

The evaporite sequences that occur in the subsurface of the Paleozoic basins in the northern United States, and the Jurassic salt diapirs that outcrop on the northern Gulf Coast, belong to sedimentary basins formed in different geological times and different tectonic settings. Giant salt basins occur in many regions of the world, such as the Mediterranean Sea, Central Europe, the Gulf of Mexico, the South Atlantic, and Asia. Earth scientists still debate how thousands of meters of halite precipitated rapidly in ancient sedimentary environments that have no counterparts in the modern world. There is growing evidence that some planets in the Solar System accumulated salt in ancient seas, and some sedimentary environments were similar to the extremely arid environments presently recognized on Earth (see Chapter 9 in Mohriak, 2025).

REFERENCES

Agricola, G. (1556). *De Re Metallica*. Libri XII, original version in Latin, trans. H. C. Hoover & L. H. Hoover. New York: Dover Publications, 585 pp., reprint, 1950.

Alexander, L., Budziak, A., Desonie, D., Kelchner, C., Lambert, J., Menzel, T., et al. (2011). *Life on an ocean planet*. Current Publishing Corp., 824 pp.

Almécija, S., Hammond, A. S., Thompson, N. E., Pugh, K. D., Moyà-Solà, S., & Alba, D. M. (2021). Fossil apes and human evolution. *Science*, *372*(6542): eabb4363. doi: 10.1126/science.abb4363

Andrews, A. P. (1983). *Maya salt production and trade*. US: University of Arizona Press, 173 pp.

Banffy, E. (2015). The beginnings of salt exploitation in the Carpathian basin (6th–5th millennium BC). *Documenta Praehistorica, XLII*, 197–209.

Barton, D. C., & Paxson, R. B. (1925). The Spindletop salt dome and oil field, Jefferson County, Texas. *American Association of Petroleum Geologists Bulletin*, *9*(3), 594–612.

Barton, I. F. (2016). Georgius Agricola's De Re Metallica in early modern scholarship. *Earth Sciences History*, *35*(2), 265–282.

Bowman, L. G. (1969). The scarcity of salt in Virginia during the American Revolution. *The Virginia Magazine of History and Biography*, *77*(4), 464–472. Virginia Historical Society. https://www.jstor.org/stable/4247510

Brigand, R., & Weller, O. (Eds.) (2015). *Archaeology of salt: Approaching an invisible past*. Leiden: Sidestone Press, 227 pp.

Brown, I. W. (1999). Salt manufacture and trade from the perspective of Avery Island, *Louisiana. Midcontinental Journal of Archaeology*, *24*(2), 113–151.

Brown, I. W. (2015). The Petite Anse Project: Archaeological investigations along the West-central coast of Louisiana 1978–1979. *Lousiana Archaeological Society*, Special Publication No. 4, Borgo Publishing, 442 pp.

Cartwright, M. (2019). The salt trade of ancient West Africa. *World History Encyclopedia*. Retrieved from https://www.worldhistory.org/article/1342/the-salt-trade-of-ancient-west-africa/

Cavalazzi, B., Barbieri, R., Gómez, F., Capaccioni, B., Olsson-Francis, K., Pondrelli, M., et al. (2019). The Dallol Geothermal Area, Northern Afar (Ethiopia)—An exceptional planetary field analog on Earth. *Astrobiology*, *19*(4), 553–578. https://doi.org/10.1089/ast.2018.1926

Chiang, T.-C. (1976). The production of salt in China, 1644-1911. *Annals of the Association of American Geographers*, *66*(4), 516–530.

Cita, M. B. (1983). The Messinian salinity crisis in the Mediterranean: A review. In H. Berckhemer & K. J. Hsü (Eds.), *Alpine Mediterranean Geodynamics, Geodynamics Series* (Vol. 7, pp. 113–140). Washington, DC: American Geophysical Union. doi:10.1029/GD007

Clark, J. L. (1984). The Mormon Church and Utah Salt Manufacturing 1847-1918. *Journal of the Southwest*, *26*(3), 225–242. https://www.jstor.org/stable/40169303

Clark, K. A., Ikram, S., & Evershed, R. P. (2016). The significance of petroleum bitumen in ancient Egyptian mummies.

Philosophical Transactions of the Royal Society A: Mathematical, Physical and Engineering Sciences, *374*(2079), 20160229.

Clark, P. U., Dyke, A. S., Shakun, J. D., Carlson, A. E., Clark, J., Wohlfarth, B., et al. (2009). The last glacial maximum. *Science*, *325*(5941), 710–714.

Cordain, L., Eaton, S. B., Sebastian, A., Mann, N., Lindeberg, S., Watkins, B. A., et al. (2005). Origins and evolution of the Western diet: health implications for the 21st century. *The American Journal of Clinical Nutrition*, *81*(2), 341–354. https://doi.org/10.1093/ajcn.81.2.341

Cowie, R. H., Bouchet, P., & Fontaine, B. (2022). The sixth mass extinction: Fact, fiction or speculation? *Biological Reviews*, *97*(2), 640–663. https://doi.org/10.1111/brv.12816

Craig, A. M., Graham, W. A., Kagan, D., Ozment, S., & Turner, F. M. (2011). *The Heritage of World Civilizations: Brief Edition* (Combined Volume, 5th ed.). UK: Pearson Prentice Hall, 976 pp.

Crout, R. R. (1983). In search of a "Just and lasting peace": The treaty of 1783, Louis XVI, Vergennes, and the regeneration of the realm. *The International History Review*, *5*(3), 364–398. https://www.jstor.org/stable/40105315

Cuthbert, M. O., & Ashley, G. M. (2014). A spring forward for hominin evolution in East Africa. *PLoS ONE*, *9*(9), e107358, 1–9. doi:10.1371/journal.pone.0107358

Dai, J., Ni, Y., Liu, Q., Wu, X., Gong, D., Hong, F., et al. (2021). Sichuan super gas basin in southwest China. *Petroleum Exploration and Development*, *48*(6), 1251–1259. https://doi.org/10.1016/S1876-3804(21)60284-7

Darwin, C. (1859). *On the origin of species by means of natural selection*. London: John Murray. Republished by ElecBook, London, 1997, 421 pp.

Davison, I. (2005). Central Atlantic margin basins of North West Africa: Geology and hydrocarbon potential (Morocco to Guinea). *Journal of African Earth Sciences*, *43*, 254–274.

de Booy, T. (1918). The Turks and Caicos islands, British West Indies. *Geographical Review*, *6*(1), 37–51. https://www.jstor.org/stable/207448

Dordea, D., Nedelcu, M., Gorie, J., & Necsoiu, I. (2013). *The salt mountain. GeoExpro*, *13*(6), 80–84.

Eby, J. B., & Halbouty, M. T. (1937). Spindletop oil field, Jefferson County, Texas, *American Association of Petroleum Geologists Bulletin*, *21*(4), 475–490.

Engler, M., & Engler, P. (2014). How did Gandhi win? Lessons from the Salt March. *Dissent Magazine, October 10*, 2014. https://www.dissentmagazine.org/blog/gandhi-win-lessons-salt-march-social-movements

Erdoğu, B., Özbaşaran, M., Erdoğu, R., & Chapman, J. (2003). Prehistoric salt exploitation in Tuz Gölü, Central Anatolia: Preliminary investigations. *Anatolia Antiqua*, *11*, 11–19. https://doi.org/10.3406/anata.2003.992

Festi, D., Brandner, D., Grabner, M., Knierzinger, W., Reschreiter, H., & Kowari, K. (2021). 3500 years of environmental sustainability in the large-scale alpine mining district of Hallstatt, *Austria. Journal of Archaeological Science Reports*, *35*, 102670.

Flad, R., Zhu, J., Wang, C., Chen, P., von Falkenhausen, L., Sun, Z., & Li, S. (2005). Archaeological and chemical evidence for early salt production in China. *Proceedings of the National Academy of Sciences of the United States of America*, *102*(35), 12618–12622.

Galili, E., & Arenson, S. (2017). *The ancient and modern salt industry on the Mediterranean coast of Israel*. Salt of the Earth Company Ltd., Haifa, 108 pp.

Gamsjäger, R., Kern, A., Kowarik, K., Maix, S., Reschreiter, H., & Schwab, A. (2008). *Hallstatt 7000*. Edition Lammerhuber, 456 pp.

Garlicki, A. (2008). Salt mines at Bochnia and Wieliczka. *Przeglad Geologiczny*, *56*(8/1), 663–669.

Garvin-Donohue, C., & Birghenthal, J. (1979). The Drake oil well. *The American Society of Mechanical Engineers*, *16* pp.

GEBCO (2019). General Bathymetric Chart of the Oceans. British Oceanographic Data Centre, U.K., website www.gebco.net.

Glassner, M. (1970). The Chibchas: A history and re-evaluation. *The Americas*, *26*(3), 302–327. http://www.jstor.org/stable/980080

Glikson, A. (2005). Asteroid/comet impact clusters, flood basalts and mass extinctions: Significance of isotopic age overlaps. *Earth and Planetary Science Letters*, *236*, 933– 937. doi:10.1016/j.epsl.2005.05.007

Glikson, A., & Groves, C. (2016). *Climate, fire and human evolution: The deep time dimensions of the Anthropocene*. Switzerland: Springer International Publishing, 227 pp.

Goldan, T., Danciu, C., & Nistor, C. (2010). Touristic and medical facilities in Romanian salt mine cavities. *Research Journal of Agricultural Science*, *42*(3), 2010579.

González-Sánchez, F., Camprubí, A., González-Partida, E., Puente-Solís, R., Canet, C., Centeno-García, E., & Atudorei, V. (2009). Regional stratigraphy and distribution of epigenetic stratabound celestine, fluorite, barite and Pb–Zn deposits in the MVT province of northeastern Mexico. *Miner Deposita*, *44*, 343–361. doi 10.1007/s00126-008-0212-4

Grabner, M., Wachter, E. Nicolussi, K., Bolka, M., Sormaz, T., Steier, P., et al. (2021). Prehistoric salt mining in Hallstatt, Austria. New chronologies out of small wooden fragments. *Dendrochronologia*, *66*, 125814.

Gradstein, F. M. Ogg, J. G. Schmitz, M. D., & Ogg, G. (2012). *The geologic time scale 2012*. Amsterdam: Elsevier, 1144 pp.

Gradstein, F. M., Ogg, J. G., & Smith, A. G. (2005). *A geologic time scale 2004*. Cambridge, U.K.: Cambridge University Press, 589 pp.

Hallett, D. (2002). The Wieliczka salt mine. *Geology Today*, *18*(5), 182–185.

Hansen, J. E., & Sato, M. (2012). Paleoclimate implications for human-made climate change. In A. Berger, F. Mesinger, & D. Sijacki (Eds.), *Climate change inferences from paleoclimate and regional aspects* (pp. 21–47). Wien: Springer.

Harari, Y. N. (2014). *Sapiens: A brief history of humankind*. New York: Harper, 464 pp.

Harding, A. (2013). *Salt in prehistoric Europe*. Leiden: Sidestone Press, 162 pp.

Harding, A. (2014). The prehistoric exploitation of salt in Europe. *Geological Quarterly*, *58*(3), 591–596. doi: http://dx.doi.org/10.7306/gq.1164

Higham, T. (2021). *The world before us: How science is revealing a new story of our human origins*. London: Penguin Books Limited, 296 pp.

Hilldrup, R. L. (1945). The salt supply of North Carolina during the American Revolution. *The North Carolina Historical Review*, *22*(4), 393–417. https://www.jstor.org/stable/23526672

Hodell, D. A., Brenner, M., & Curtis, J. H. (2005). Terminal Classic drought in the northern Maya lowlands inferred from multiple sediment cores in Lake Chichancanab (Mexico), *Quaternary Science Reviews*, *24*, 12–13, 1413–1427.

Hsü, K. J. (1972). Origin of saline giants: A critical review after the discovery of the Mediterranean evaporite. *Earth-Science Reviews*, *8*, 371–396.

Hutton, J. (1788, 1795). *Theory of the Earth*. U.K.: Global Grey, 447 pp.

Jones, W. H. S. (1938). *Pliny Natural history* (Vol. *VIII*). London: William Heinemann Ltd., 596 pp. https://archive.org/details/naturalhistory08plinuoft/page/432/mode/2up

Kappel, W. M. (2000). Salt production in Syracuse, New York ("The Salt City") and the hydrogeology of the Onondaga Creek Valley. *U.S. Geological Survey*, Fact Sheet FS 139-00, 8 pp.

Kiernan, B. (2004). The first genocide: Carthage, 146 BC. *Diogenes*, *51*(3), 27–39.

Kieta, H. M. (2018). *A comparison of the development of the salt industries in Michigan and Ontario. Honors College, Theses*, *41*. https://digitalcommons.wayne.edu/honorstheses/41

Kurlansky, M. (2002). *Salt: A World History*. New York: Walker & Company. ISBN: 978-0-802-77944-1.

Lucas, A. (1932). The occurrence of natron in ancient Egypt. *The Journal of Egyptian Archaeology*, *18*(1/2), 62–66.

Luttwak, E. N. (2016). *The grand strategy of the Roman Empire: From the first century A.D. to the third*. Baltimore: Johns Hopkins University Press, 296 pp. https://doi.org/10.56021/9781421419442

Lyell, C. (1830). *Principles of geology, being an attempt to explain the former changes of the Earth's surface, by reference to causes now in operation*. London: Murray, 385 pp.

Lyell, C. (1865). *Elements of geology or the ancient changes of the Earth and its inhabitants as illustrated by geological monuments*. London: Murray, 794 pp.

Maixner, F., Sarhan, M. S., Huang, K. D., Tett, A., Schoenafinger, A., Zink, A., et al. (2021). Hallstatt miners consumed blue cheese and beer during the Iron Age and retained a non-Westernized gut microbiome until the Baroque period. *Current Biology*, *31*, 5149–5162. https://doi.org/10.1016/j.cub.2021.09.031

Marzano, A. (2013). Salt. In A. Marzano (Ed.), *Harvesting the Sea: The exploitation of marine resources in the Roman Mediterranean* (pp. 123–141). https://doi.org/10.1093/acprof:oso/9780199675623.003.0005

Maslin, M. A., Brierley, C. M., Milner, A. M., Shultz, S., Trauth, M. H., & Wilson, K. E. (2014). East African climate pulses and early human evolution. *Quaternary Science Reviews*, *101*, 1–17.

McClendon, J. H. (1999). The origin of life. *Earth-Science Reviews*, *47*(1–2), 71–93. https://doi.org/10.1016/S0012-8252(99)00015-X

McDougall, E. A. (1990). Salts of the Western Sahara: Myths, mysteries, and historical significance. *The International Journal of African Historical Studies*, *23*(2), 231–257. https://doi.org/10.2307/219336

McKillop, H. (1995). Underwater archaeology, salt production, and coastal Maya trade at Stingray Lagoon, *Belize. Latin American Antiquity*, *6*(3), 214–228.

McKillop, H. (2005). Finds in Belize document Late Classic Maya salt making and canoe transport. *Procedures of the National Academy of Sciences*, *102*(15), 5630–5634.

McKillop, H. (2019). *Maya salt works*. Florida: University Press of Florida, 324 pp.

Mingren, W. (2019). When salt was traded for gold: The salt trade of West Africa that built kingdoms and spread culture. *Ancient Origins*, updated 28 April, 2019. https://www.ancient-origins.net/history-ancient-traditions/salt-trade-0011802

Miralles, L., Sans, M., Pueyo, J. J., & Santanach, P. (2000). Recrystallization salt fabric in a shear zone (Cardona diapir, southern Pyrenees, Spain). In B. Vendeville, Y. Mart, & J. L. Vigneresse (Eds.), *Salt, shale and igneous diapirs in and around Europe* (Vol. 174, pp. 149–167). Geological Society, London, Special Publications. https://doi.org/10.1144/GSL.SP.1999.174.01.09

Mohriak, W. (2025). *Salt in the Earth sciences: Basin analysis and salt tectonics*. Washington, DC: American Geophysical Union. doi:10.1002/9781394307067

Morgan, J. P., Reston, T. J., & Ranero, C. R. (2004). Contemporaneous mass extinctions, continental flood basalts, and "impact signals": Are mantle plume-induced lithospheric gas explosions the causal link?, *Earth Planetary Science Letters*, *217*, 263–284. doi:10.1016/S0012-821X(03)00602-2

Neev, D., & Emery, K. O. (1995). *The destruction of Sodom, Gomorrah, and Jericho: Geological, climatological and archaeological background*. Oxford: Oxford University Press, 188 pp.

Nikitin, A. G., Stadler, P., Kotova, N., Teschler-Nicola, M., Price, T. D., Hoover, J., et al. (2019). Interactions between earliest Linearbandkeramik farmers and central European hunter gatherers at the dawn of European Neolithization. *Nature Scientific Reports*, *9*, 19544. https://doi.org/10.1038/s41598-019-56029-2

Nissenbaum, A. (1993). The Dead Sea—An economic resource for 10 000 years. *Hydrobiologia*, *267*(1–3), 127–141. https://doi.org/10.1007/BF00018795

Ogg, J. G., Ogg, G., & Gradstein, F. M. (2016). *A concise geologic time scale*. Amsterdam: Elsevier, 234 pp.

Oparin, A. I. (1968). *Genesis and evolutionary development of life*. London: Academic Press, 203 pp.

Owen-Smith, N. (2021). *Only in Africa—The ecology of human evolution*. Cambridge: Cambridge University Press, 376 pp.

Potts, R., & Sloan, C. (2010). *What does it mean to be human?* Smithsonian National Museum of Natural History, 176 pp.

Reschreiter, H., & Kowarik, K. (2019). Bronze Age mining in Hallstatt. A new picture of everyday life in the salt mines and beyond. *Archaeologia Austriaca*, Band *103*/2019, 99–136. doi: 10.1553/archaeologia103s99

Robinson, M., & McKillop, H. (2014). Fuelling the ancient Maya salt industry. *Economic Botany*, *68*, 96–108. https://doi.org/10.1007/s12231-014-9263-x

Royer, D. L., Berner, R. A., Montañez, I. P., Tabor, N. J., & Beerling, D. J. (2004). CO_2 as a primary driver of phanerozoic climate. *Geological Society of America, GSA Today*, *14*(3), 4–10.

Saile, T. (2015). Competing on unequal terms: Saltworks at the turn of the Christian era. In R. Brigand & O. Weller (Eds.), *Archaeology of Salt—Approaching an invisible past* (pp. 199–209). Leiden: Sidestone Press.

Salt Association (2020). Ancient times. https://saltassociation.co.uk/education/salt-history/early-history/

Sepkoski, J. J., Jr. (1996). Patterns of Phanerozoic extinction: A perspective from global data bases. In O. H. Walliser (Ed.), *Global events and event stratigraphy in the phanerozoic* (pp. 35–51). Berlin Heidelberg: Springer-Verlag.

Smith, A. F. (2011). *Starving the South: How the North won the Civil War*. New York: St. Martin's Press, 295 pp.

Snodgrass, M. E. (2020). *The Civil War era and reconstruction*. New York: Taylor & Francis Group, 3958 pp.

Soto, J. I., Flinch, J. F., & Tari, G. (2017). Permo-Triassic basins and tectonics in Europe, North Africa and the Atlantic Margins: A synthesis. In J. I. Soto, J. F. Flinch, & G. Tari (Eds.), *Permo-triassic salt provinces of Europe, North Africa and the Atlantic Margins: Tectonics and hydrocarbon potential* (pp. 3–41). Amsterdam: Elsevier.

Sponheimer, M., Lee-Thorp, J. A., Reed, K. E., & Ungar, P. S. (2013). *Early hominin paleoecology*. Boulder: University Press of Colorado, 470 pp.

Stringer, C., & McKie, R. (2015). *African exodus: The origins of modern humanity*. New York: Henry Holt and Company, Inc., 304 pp.

Suihong, S. (2016). A perfect play of salt, bamboo and gas: The first petroleum well (breakthrough) in the World. *American Association of Petroleum Geologists, Search and Discovery Article* #70220 (2016).

Sverdrup, H. U., Johnson, M. W., & Fleming, R. H. (1942). *The oceans, Their physics, chemistry, and general biology*. New York: Prentice-Hall, 1087 pp. http://ark.cdlib.org/ark:/13030/kt167nb66r/

Taylor, R. E. (1938). *Origin of the Cap Rock of Louisiana Salt Domes*. PhD thesis, Louisiana State University, LSU Historical Dissertations and Theses. 7793. https://digitalcommons.lsu.edu/gradschool_disstheses/7793.

U.S. Geological Survey (2023). Mineral Commodity Summaries. 2023. *U.S. Geological Survey*, 210 p. https://pubs.er.usgs.gov/publication/mcs2023

Walker, J. D., Geissman, J. W., Bowring, S. A., & Babcock, L. E. (2018), (Compilers). *Geologic time scale, v. 5.0, Geological Society of America*. https://doi.org/10.1130/2018.CTS005R3C.

Warren, J. K. (2006). *Evaporites: Sediments, resources and hydrocarbons*. Berlin: Springer Verlag, 1035p.

Warren, J. K. (2008). Salt as sediment in the Central European Basin system as seen from a deep time perspective. In R. Littke, U. Bayer, D. Gajewski, & S. Nelskamp (Eds.), *Dynamics of complex intracontinental basins: The central European basin* (Chapter 5.1, pp. 249–276). Berlin: Springer.

Warren, J. K. (2010). Evaporites through time: Tectonic, climatic and eustatic controls in marine and nonmarine deposits. *Earth and Science Review*, *98*, 217–268.

Warren, J. K. (2014). *Evaporite sedimentology: Importance in hydrocarbon accumulation*. Prentice Hall Advanced Reference Series, 285 pp.

Warren, J. K. (2016). *Evaporites—A geological compendium*. Berlin: Springer, 1807 pp. ISBN 978-3-319-13511-3.

Weller, O. (2002). The earliest rock salt exploitation in Europe: A salt mountain in the Spanish Neolithic. *Antiquity*, *76*, 317–318.

Weller, O. (2015). First salt making in Europe: An overview from Neolithic times. *Documenta Praehistorica*, *XLII*, 185–196.

Wignall, P. B. (2001). Large igneous provinces and mass extinctions. *Earth-Science Reviews*, *53*, 1–33.

Wijnker, J. J., Koop, G., & Lipman, L. J. A. (2006). Antimicrobial properties of salt (NaCl) used for the preservation of natural casings. *Food Microbiology*, *23*(7), 657–662. doi: 10.1016/j.fm.2005.11.004

Woldekiros, H. S. (2019). The route most traveled: The Afar salt trail, Northern Ethiopia. *Chungara Revista de Antropología Chilena*, *51*(1), 95–110.

Xinhua, M. (2017). A golden era for natural gas development in the Sichuan Basin. *Natural Gas Industry*, *B4*, 163–173.

Part I

Chemical and Physical Properties of Evaporite Rocks

2

Basic Scientific Concepts for the Interpretation of Evaporites

ABSTRACT

The first scientific salt deposition models were based on analogies between modern and ancient sedimentary environments, where evaporite minerals precipitated and accumulated, forming thick layers. Factors controlling evaporite mineral deposition include climate, water budget, and ions dissolved in brines. Salt layers occur in continental, lacustrine, gulfs, inland seas, and marginal marine basins. Most modern salt basins are associated with shallow water environments, like the sabkhas in northern Africa and the Arabian Peninsula. The Dead Sea is a deep lake with halite precipitation accumulating on its bottom, constituting the only modern analog for salt deposition in deepwater settings.

This chapter examines early theories for the formation of evaporite rocks, which were unconstrained by scientific methods until the Industrial Revolution and the emergence of geological sciences in the 18th century. Chemical principles were applied in the 19th century, and sedimentological models were proposed in the 20th century. The chapter discusses salt deposition in lacustrine and marine environments and hydrological and limnological concepts related to continental evaporitic environments. The Dead Sea Lake is a classic modern analog for salt deposition in lacustrine environments. The chapter briefly discusses recent models that have been proposed for halite precipitation in the Dead Sea deepwater settings.

2.1. EARLY CONCEPTS FOR THE DEPOSITION OF EVAPORITES

Evaporites are chemical precipitates formed from the concentration of seawater and brines in water bodies, particularly shallow seas, gulfs, and isolated, hypersaline lakes. Brine concentration results from a water budget deficit in the hydrological cycle. The water lost by solar evaporation exceeds the water input into the system, either from atmospheric precipitation, fluvial, hydrothermal springs, glacial melt, and/or groundwater (Kendall, 2010; Warren, 2006, 2016).

There are hundreds of specialized publications dealing with the interpretation of evaporite basins in the geological literature (e.g., Schreiber & Hsü, 1980; Sonnenfeld, 2003; Warren, 2021). Many works focus not only on the sedimentological aspects of evaporite deposition but also scrutinize the chemical and physical properties of brine and salt minerals (Babel & Schreiber, 2014; Talbot, 2005). Early theories for the formation of evaporite rocks were based on religious concepts, such as the Dead Sea and other salt lakes formed as a result of torrential rains during a global flood (Magruder, 2009; Rudwick, 2009). However, many observations refute this hypothesis. The Dead Sea evaporite rocks are deeply buried under 8 km and more of sediments (e.g., Al-Zoubi & ten Brink, 2001; Ben-Avraham, 1997). Extensional tectonics and the weight of the Pliocene to Quaternary sediments on these evaporite layers resulted in salt deformation and upward flow, forming salt diapirs on land, such as Mount Sedom salt diapir, and below the lake's surface, such as the En Gedi salt dome observed in the northern lake (see Chapter 11). In addition, the lake is still depositing evaporite sediments at present (e.g., Steinhorn et al., 1979; Steinhorn, 1983).

Many other controversial hypotheses about the origin of evaporite sediments have been proposed and abandoned since the Industrial Revolution when modern geological theories were established in the earth sciences.

Salt in the Earth Sciences: Evaporite Rocks and Salt Deposition, Geophysical Monograph 286, First Edition. Webster Mohriak.

DOI: 10.1002/9781119405245.ch02

Two main schools of thought were prevalent in Europe in the 18th century, based on the neptunism and plutonism theories. The neptunists, led by German mineralogist Abraham Gottlob Werner (1749–1817), believed that the source for igneous rocks was a common ancient sea and that all minerals observed in crystalline rocks were precipitated from seawater. The advocates of these ideas postulated that evaporite minerals are chemical compounds that precipitate from seawater brines concentrated by evaporation. Consequently, the halite crystals in evaporite layers were considered equivalent to the quartz crystals in granites.

The neptunistic school of geology postulated that early Earth had a primordial ocean with chemical elements dissolved in seawater, leading to the formation of evaporite rocks. These rocks were formed by the crystallization and precipitation of these compounds, resulting in solid minerals. According to the neptunists, the evaporite rocks were formed in a very specific sequence in nature. First, the oldest and hardest rocks, which constituted the granites, gneisses, and schists, then basalts; and finally, limestones and other common chemical and sedimentary rocks, which were formed by the erosion of previous rocks.

James Hutton, the founder of modern geology, made the case that natural laws can explain geological phenomena, establishing a set of guidelines that would later come to be known as the "uniformitarianism principle" (Lyell, 1830). The scientific methodology that underlies modern geology and other sciences is based on the principle of uniformitarianism, which establishes that "the present is the key to the past." The concept has been modified and adapted to the new knowledge acquired in the 19th and 20th centuries (Romano, 2015).

Hutton (1788, 1795) applied this principle to the genesis of evaporite rocks, advocating a magmatic origin. According to Hutton's arguments, magma can be observed in volcanic eruptions both on land and under oceanic waters, and the formation of granite and basalt is associated with the magma temperature dropping below the melting point. According to the followers of the plutonist hypothesis, igneous rocks were formed by the consolidation of molten rock masses, as observed in active volcanoes that erupted lava and formed basalt layers. The igneous rocks consolidated and hardened as the minerals crystallized with a reduction in temperature, following the principles established by experimental chemistry. According to the plutonists, evaporites are chemical rocks that should also follow a similar origin by precipitation from melts (Hutton, 1788, 1795).

The debate over the origin of evaporitic rocks continued during the 20th century. Although at present most researchers interpret salt deposition results from brine concentration due to solar heating (e.g., Talbot, 2008), several recent works have discussed alternative hypotheses to explain the formation of giant salt basins by other mechanisms instead of solar evaporation. These include supercritical water associated with igneous activity and fluids emanating from serpentinized mantle rocks (e.g., Debure et al., 2019; Hovland et al., 2006, 2008, 2018a, 2018b).

The American Association of Petroleum Geologists held the first international conference to discuss the origin of salt domes in the Gulf of Mexico and other regions. Some researchers suggested a primordial ocean as the explanation (Voitesti, 1926), but this hypothesis was contested by other geoscientists. Van Waterschoot Van der Gracht (1926) argued that salt layers occurred at different ages across different sedimentary basins. Some salt structures were not related to the earliest sedimentary beds above the Precambrian rocks but were formed much later in geological time. DeGolyer (1926) scrutinized speculative genetic interpretations for salt deposits in the Gulf of Mexico. His classic paper highlights that "Throughout all of these periods, like the chorus in a Greek tragedy, run constantly recurring theories of dome-origin as a result, direct or indirect, of volcanic activity...because structurally an igneous plug is the only geological feature known to us that even remotely resembles a salt dome."

Following the intense debate on the origin of igneous rocks during the 18th century, neptunism was eventually rejected by earth scientists in the 19th century. Chemistry studies were essential in establishing the relationship between evaporite minerals and seawater. These experiments demonstrated that it is possible to precipitate the majority of salt compounds found in nature by evaporating seawater and increasing brine concentrations above the solubility of the minerals (Usiglio, 1849a, 1849b). One of the reasons for the conflicting interpretations of salt genesis is that uniformitarianism cannot be applied to the formation of giant salt basins recorded in the ancient past. There are no modern examples in the world of salt deposition in deep basins that can be considered counterparts of the giant saline basins (Hsü, 1972; Warren, 2006, 2010, 2016). Many authors (e.g., Hsü, 1972; Montaron & Tapponnier, 2012; Sonnenfeld & Perthuisot, 1989) have advocated that the most important and deepest salt basins formed during catastrophic events, but evaporite deposition is a complex process that may take hundreds of thousands to millions of years to accumulate a large thickness of salt (see Chapter 1).

Geological studies have revealed that many regions have underground salt deposits that are not visible on the Earth's surface. These deposits are often hidden in sedimentary basins, such as the Amazon Basin in northern Brazil, where thick halite sequences were deposited during the late Carboniferous and early Permian. These evaporite rocks were only discovered when the petroleum industry started to drill below the younger sedimentary sequences in the past decades (Mohriak et al., 2012). The Central European Basin was believed to be underlain by an enormous mass of late Paleozoic and early Mesozoic evaporite sediments, which underlie parts of Germany, Poland, and Britain.

The Alps, Pyrenees, Apennines, and Mediterranean Sea basins contain the majority of salt deposits exploited in southern Europe since prehistoric times (see Chapter 1). These deposits have different ages, and deformation is linked to geodynamic processes related to plate tectonics and gravity forces. The Late Paleozoic sedimentary basins that extended through the North-Central European region formed one of the largest and most important salt basins in the world. Many evaporite basins in Europe, Africa, and the Americas were formed during the Pangea breakup and accumulated salt in the Permian and Triassic rift basins (Soto et al., 2017).

Salt was also interpreted in the cores of several mountain ranges in Africa and Europe, such as the Atlas, Alps, and Pyrenees (see Chapter 3). These deposits are much younger than the salt layers identified in the Zagros Mountains in Iran, which accumulated salt within the rift basins formed during the Late Precambrian and also in several other geological intervals in the Mesozoic and Cenozoic (see Chapter 9 in Mohriak, 2025).

The South Atlantic salt basins, consisting of Late Aptian evaporite deposits, were once a single salt province but were separated by oceanic spreading centers. At present, these basins are located on different plates (see Chapter 3 in Mohriak, 2025). The Red Sea formed a single salt basin in the Miocene, but the separation of the salt provinces by incipient oceanic ridges has already started (see Chapter 1 in Mohriak, 2025). The Mediterranean Sea accumulated evaporites in the Late Miocene, forming the last giant salt basin on Earth (see Chapter 5 in Mohriak, 2025).

Scientific principles and plate movement reconstructions from the Precambrian to the present day help understand how salt basins formed in ancient geological environments, using basic principles and climate changes to reconstruct their formation (Kendall et al., 2009; Warren, 2010). Basin analysis applied to salt basins involves interpreting sedimentary environments associated with evaporite deposition, the structural control of salt layers, and deformation by tectono-magmatic episodes. Geological analysis conducted by exploration companies integrates surface sedimentary occurrences with subsurface interpretation using geophysical methods (see Chapter 7).

2.2. SALT DEPOSITION AND SEDIMENTARY ENVIRONMENTS

Evaporite sediments are formed by the evaporation of saline water in shallow seas and lakes in arid environments (Sonnenfeld & Perthuisot, 1989; Warren, 2006, 2016). The name "evaporite" comes from the Latin word "evaporare" , which means to change a liquid into vapor, and the Greek suffix "litho", which means stone. Evaporite rocks are observed in nature due to the restriction of shallow inland seas in desertic climates, leading to brine concentration and the precipitation of chemical compounds.

The Persian Gulf and the Red Sea are examples of restricted basins that are modern counterparts for the large areas that accumulated thick evaporite sequences in the ancient past. However, these two modern gulfs are very different in terms of tectonic settings and sedimentary environments. The Persian Gulf is a shallow, foredeep basin associated with the convergence of the Arabian and Eurasian plates. The Red Sea is a deep basin related to the divergence of the African and Arabian plates. Both basins are located close to the Sahara Desert and exhibit a deficit in the hydrological budget. These observations are key to understanding how giant salt basins are formed on Earth, either by closing old oceans or opening new ones through plate tectonics (Kendall et al., 2009; Warren, 2010).

Most ancient evaporites in the world's sedimentary basins are composed of minerals formed by the concentration of seawater-derived brine (e.g., Hsü, 1972; Warren, 2016). When the inland seas in arid regions are isolated from the world's oceans, they can expose large quantities of seawater to evaporation, concentrating the brine with salt ions to the point of precipitation of chemical compounds (e.g., Bischof, 1854; Ochsenius, 1877, 1888; Usiglio, 1849a, 1849b). The relative proportions of global water distribution indicate that most of the water on Earth is saline and concentrated in the oceans (Figure 2.1). More than 96% of the water on Earth comes from oceans, seas, lakes, and bays. As a result, the majority of the water is salty. Less than 2% of the total water capacity is made up of fresh and saline groundwater, while only 0.13% is made up of fresh and saline lakes. Only a small percentage of the global water supply is potable freshwater with a salinity of less than 1,000 ppm (Shiklomanov, 1993; U.S. Geological Survey, 2018).

All the giant salt basins are related to the evaporation of seawater. However, there are many examples of important evaporite basins with continental, nonmarine brine as the only source for the salt deposits. For example, the modern evaporite deposits recorded in the Tibetan Plateau hypersaline lakes (see Chapter 10), and some hypersaline lakes in the United States, such as the Great Salt Lake in Utah and the Death Valley in California (see Chapter 9), are sourced from continental brines (Hardie, 1984 1991; Sonnenfeld, 1985; Smoot & Lowenstein, 1991).

Climate, physical factors, and water geochemistry all have an impact on salt deposits, which are a component

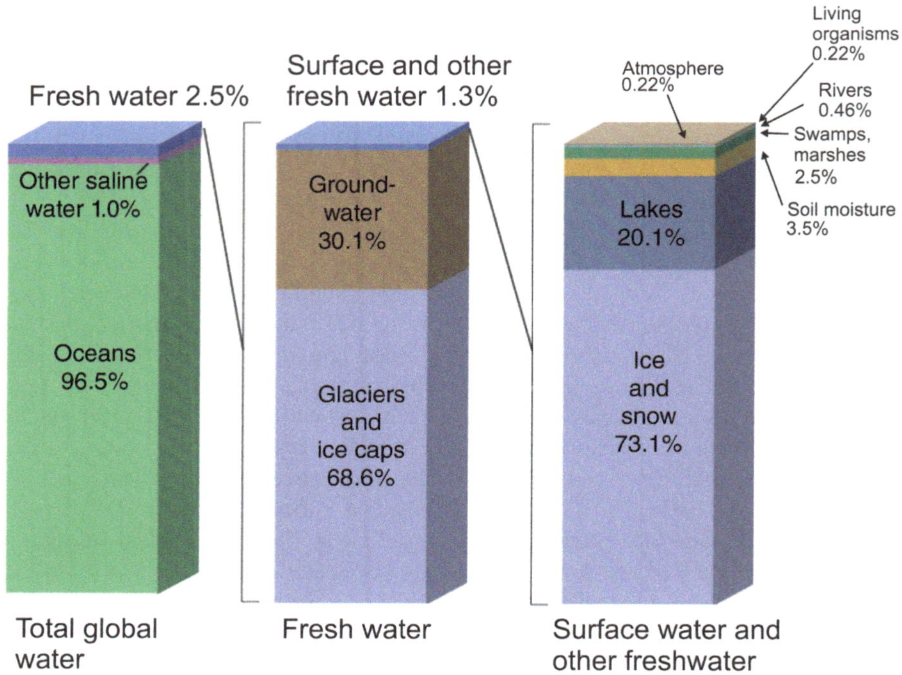

Figure 2.1 The diagram shows that only 2.5% of global water is freshwater, with 68.6% stored in ice caps and 30% underground. Most potable freshwater is associated with glaciers and ice caps, with less than one-third in subsurface or accumulated lakes and rivers. Credit: Adapted from USGS (2022) after Shiklomanov (1993).

of the hydrological cycle (e.g., Babel & Schreiber, 2014; Talbot, 2005). Water vapor from oceans, inland seas, lakes, and rivers evaporates, condensing and precipitating. Part of the water that infiltrates the soil is stored in aquifers in the subsurface, forming groundwater. When these aquifers are in contact with salt layers, the formation water becomes more saline and may reach high concentrations of dissolved solids, resulting in concentrated brines. Groundwater can pick up other minerals as it comes into contact with different rocks (such as Na-rich feldspars), becoming more saline without direct contact with a salt layer. The pores in the subsurface reservoirs are filled with fluids, mainly water and hydrocarbons. Most basin analysis projects analyze the petrophysical characteristics of the subsurface reservoirs and aquifers, including the origin of the formation waters (see Chapter 5). The formation waters are classified according to their salinity content (e.g., parts per thousand).

Salinity is the salt concentration of water in subsurface reservoirs and surface pools, expressed as the total concentration of dissolved salts in the water (e.g., Tundisi & Tundisi, 2011; Webb, 2021). Salinity can be easily measured in laboratories by evaporating a small sample and determining the weight of the residues (salts, minerals, and other solid residues). Total dissolved solids (TDS) can be expressed in percentages (%), milligrams per liter (mg/l), or parts per million (ppm). Normal seawater is characterized by an almost constant salinity value of 35,000 mg/l (35,000 ppm, or 3.5%). Freshwater and brackish water have reduced amounts of these solid compounds (less than 5,000 ppm). Saline water shows a salinity of up to 100,000 ppm, resulting in the precipitation of carbonate minerals such as aragonite (see Chapter 4). Sulfate minerals (e.g., gypsum) are precipitated when seawater is concentrated by four to five times the original concentration (Braitsch, 1971; Sonnenfeld, 2003). Hypersaline water and brines are characterized by higher salinities of up to 300,000 ppm. Halite crystals start to precipitate from a hypersaline solution when the brine concentration exceeds a threshold value of about 10 times the original seawater concentration (Braitsch, 1971; Sonnenfeld, 2003).

Several authors (e.g., Babel & Schreiber, 2014) recommend determining the salinity of seawater or groundwater brines by measuring their conductivity to electricity, resulting in a dimensionless unit known as the "practical salinity unit" (PSU). The standard salinity of seawater is the mean global salinity, corresponding to about 35‰ (parts per thousand), or 36.5 PSU. However, depending on geographic location, depth, and climatic factors, ocean salinity at specific sites may vary from the global average. Waters with higher salinity are denser and underlie the less saline water layers, resulting in vertical and horizontal variations in the global pattern of ocean salinity (Webb, 2021).

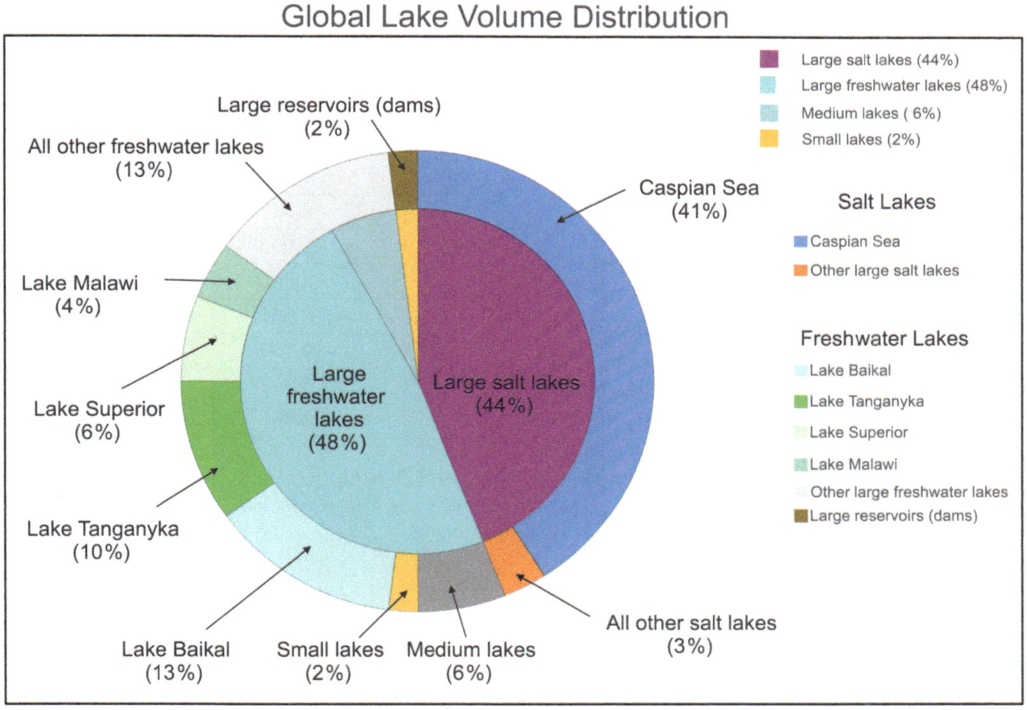

Figure 2.2 The pie chart shows the global volume distribution of water stored in freshwater and saline lakes as of 2016. Freshwater lakes account for 54% of the world's lakes, and the largest ones are Lake Baikal in Siberia and Lake Tanganyika in Africa. Large salt lakes account for 44% of the lacustrine global water distribution. The Caspian Sea is the largest salt lake, but its salinity is only one-third of normal seawater. Several small salt lakes, such as the Dead Sea, display extremely high salinities, and the brines can precipitate evaporite minerals. Credit: Adapted from Messager et al. (2016).

The world's distribution of lakes can be split into freshwater and saltwater lakes (Figure 2.2). Lake Baikal is the largest freshwater lake in the world, with 13% of all freshwater available on Earth. The Caspian Sea comprises about 41% of all lacustrine water (saline and fresh) in the world, corresponding to the largest salt lake by water volume. Hypersaline lakes are the only present-day analogs for salt deposition in ancient evaporite basins.

Modern-day evaporite deposits are often found in salt lakes in arid regions, with some lakes having slightly above-normal salinities, such as the Great Salt Lake in the United States (see Chapter 10). Although all saline and hypersaline lakes make up only 3% of the total water volume (Figure 2.2), hypersaline lakes are the only present-day analogs for salt deposition in ancient evaporite basins (Warren, 2006, 2016). Hypersaline lakes are characterized by extreme brine concentration and the precipitation of highly soluble minerals like natron and halite. The Kara Bogaz Gol, a hypersaline shallow lake in Turkmenistan, is filled with saline water from the Caspian Sea and has been the site of massive halite precipitation events in the past century (see Chapter 10). Most salt lakes where evaporites precipitate correspond to shallow, small depressions, such as the salt lakes in the Death Valley of California, southwestern United States (see Chapter 9). However, some are relatively large, such as the hypersaline lakes in the Qarhan salt flats of the Qaidam Basin in China (see Chapter 10). The only hypersaline, deep lake in the world is the Dead Sea, located between Israel and Jordan (Neev & Emery, 1967; Warren, 2006, 2016).

2.2.1. Chemical Composition of Evaporites

When the brine concentration exceeds the solubility of mineral compounds, elements in the brine dissolve in water as ions and precipitate as salts, which affects the chemical composition of evaporites. The original brine may be marine, nonmarine, or a hybrid mixture of continental and seawater (e.g., Babel & Schreiber, 2014; Warren, 2006, 2016). Locally, the brine includes hydrothermal spring contributions and water input from underground aquifers or seepage through barriers (Hardie, 1991; Kendall, 2010).

The aridity of the environment controls the water balance in both marine and nonmarine settings. Arid deserts have a mean annual precipitation of less than 250 mm/km^2 of water, while semiarid deserts have a mean annual rainfall between 250 and 500 mm (Warren, 2010, 2016). Arid regions experience a greater loss of water from the hydrological system than the inflow of freshwater, resulting in a negative budget in the water cycle (Babel & Schreiber, 2014; Sonnenfeld & Perthuisot, 1989). Many factors control the water cycle, including precipitation, evapotranspiration, condensation, infiltration, surface runoff, and subsurface water flow (e.g., Cui et al., 2018).

To understand the chemical composition of evaporite minerals, it is essential to analyze the Earth's composition in terms of the abundance of chemical elements in the crust and the globe. Oxygen, silicon, aluminum, iron, and magnesium are the most abundant chemical elements in the crust, with percentages decreasing from about 46% to less than 5% (Press et al., 2003). Iron and magnesium are much more abundant on the globe than in the crust, suggesting a larger proportion of these elements occurs in the Earth's mantle and core. Magnesium, calcium, potassium, and sodium are relatively abundant elements in evaporite rocks, mainly concentrated in igneous rocks. Most chemical elements in evaporite minerals appear in small percentages in the Earth's crust, while only sodium and chlorine occur in large percentages in seawater. Only two elements (sodium and chlorine) occur in large percentages in seawater. Since the 19th century, several authors have used this fact to support a genetic link between seawater and the formation of halite deposits (e.g., Bischof, 1854; Braitsch, 1971; Ochsenius, 1877, 1888; Sonnenfeld & Perthuisot, 1989; Usiglio, 1849a, 1849b).

2.2.2. Chemical Components of Normal Marine Seawater

The proportion of major ions in normal seawater is almost constant in all oceans, but some regions show slight variations due to climatic zones on Earth. Temperature, aridity, fluvial input, and glacial melting are a few factors that affect these anomalies (Webb, 2021). For example, the Red Sea, an elongated gulf between Africa and Arabia, is slightly more saline than the Atlantic Ocean due to its location north of the equator and the narrow Hormuz Strait at its southern end (see Chapter 1 in Mohriak, 2025). In satellite images, differences in color represent these salinity differences. Slight differences in salinity are also observed at the southernmost tip of South America, creating an almost vertical barrier between the Pacific and South Atlantic oceans. Plumes of less saline water associated with river input are also distinguished in some satellite images (Vinogradova et al., 2019).

Normal seawater contains a variety of dissolved ions. These ions, in descending order of abundance by mass, include Cl^-, Na^+, SO_4^{2-}, Mg^{2+}, Ca^{2+}, K^+, HCO_3^-, and Br^- (Figure 2.3). These eight ions account for over 99% of the dissolved ions in seawater. The remaining 1% can be ignored in the analysis of common evaporite minerals.

Ions	Concentration (weight in parts per thousand, ‰)	Salts	Percentage by weight (%)
Na^+	10,560	NaCl	78.03
Mg^{++}	1,270	$MgCl_2$	9.21
Ca^{++}	0.400	$MgSO_4$	6.53
K^+	0.380	$MgSO_4$	3.84
Sr^{++}	0.008	KCl	2.11
Cl^-	19,000	$CaCO_3$	0.33
SO_4^{--}	2,650	$MgBr_2$	0.25
HCO_3^-	0.140	$SrSO_4$	0.05
Br^-	0.065		
F^-	0.001		

Figure 2.3 Mean contents of ions and salts in seawater with normal salinity. Bromine and fluorine do not form salts in seawater.

However, there are other economically important ions, such as lithium, present in seawater and continental brines (see Chapter 9 in Mohriak, 2025).

Seawater evaporated by solar heat primarily consists of sodium chloride (NaCl). A bottle filled with 1 liter of seawater contains several ions dissolved in the aqueous solution, but only a few are significant. If weighed, the liquid and solid content in this bottle would correspond to a total weight of about 1 kg. The dissolved salts include 19 g of chlorine (Cl), 11 g of sodium (Na), 3 g of sulfates (SO_4^{-2}), 1.5 g of magnesium (Mg), 0.35 g of calcium (Ca), and 0.35 g of potassium (K). All other elements in the solution weigh less than 0.01 g (Braitsch, 1971; Babel & Schreiber, 2014).

2.2.3. Chemical Components of Continental Lacustrine Water

Continental waters in lakes have a distinct chemical composition compared to seawater. Lakes are primarily sourced by fluvial or meteoric waters and typically contain less than 1 g/l of dissolved ions (Babel & Schreiber, 2014; Renaut & Gierlowski-Kordesch, 2010). However, some lakes can reach higher salinities due to extreme evaporation or enrichment by hydrothermal springs or seeps (e.g., Lowenstein & Risacher, 2009).

Based on published scientific reports and textbooks (e.g., Braitsch, 1971; Warren, 2006, 2010, 2016; Webb, 2021), Figure 2.4 shows the mean estimated ranges of salinities for several lakes and oceans worldwide. Most hypersaline lakes with salt deposition are located in arid to hyperarid deserts, with less than 25 mm of annual precipitation and no seasonal rainfall. Hyperarid

Salinity (g/kg, ‰)	Lake	Type	Region
403	Don Juan Pond	Salt lake	Antarctica
400	Lake Retba	Salt lake	Senegal
350	Lake Vanda	Salt lake	Antarctica
350	Karabogaz	Lagoon	Turkmenistan
348	Lake Assal	Salt lake	Djibouti
337	Dead Sea	Salt lake	Israel, Jordan
85–280	Lake Urmia	Salt lake	Iran
50–270	Great Salt Lake	Salt lake	United States
180	Little Manitou Lake	Salt lake	Canada
50–99	Mono Lake	Salt lake	United States
44	Salton Sea	Salt lake	United States
38	Mediterranean Sea	Inland sea	Southern Europe to North Africa
36–41	Red Sea	Inland sea	Middle East
30–40	Lake Natron	Salt lake	Tanzania
35	Pacific Ocean	Major ocean	Western and Eastern Hemisphere
35	Atlantic Ocean	Major ocean	Western and Eastern Hemisphere
12	Caspian Sea	Inland sea	Central Asia

Figure 2.4 Comparison of salinities in various lakes and oceans worldwide, expressed in g/kg or parts per thousand. Credit: Adapted from Warren (2006).

environments are characterized by less than 25 mm of annual precipitation and generally exhibit no seasonal rhythm of rainfall (e.g., Laity, 2008).

Some lakes concentrate brines in cold temperatures, such as the saltiest lake in the world, the Don Juan Pond in Antarctica, with a salinity between 300 and 400 g/kg of water (Figure 2.4). Despite temperatures reaching −50 °C, the lake never freezes due to its high salinity, dominated by calcium chloride (Siegel et al., 1979). Several lakes or lagoons are characterized by brines that concentrate to the point of precipitating soluble evaporite minerals, such as the Dead Sea, where halite is precipitated several times a year (Sirota, 2020; Zak, 1974, 1997). This phenomenon has been investigated by earth scientists from various specializations, including chemistry, hydrology, and limnology, with key questions being: 1) why halite is deposited on the Dead Sea's lake bottom during certain months and not in others; and 2) why the accumulated sediments exhibit different textures. These topics will be addressed in subsequent sections of this chapter.

Most evaporite salts are ionic compounds that contain the major ions present in seawater (such as Na, Ca, Mg, K, Cl, SO_4^{-2}, and CO_3^{-2}) in varying proportions. However, there are also minor ionic constituents (such as B, Ba, Sr, Br, and I) and varying amounts of structural water in the molecules of some evaporite minerals. Some continental lakes are characterized by a chemical composition that differs significantly from simply evaporating normal seawater. Although the chemistry of the Dead Sea brine varies through time as a consequence of climate, water budget, and decreasing water levels in the past decades, the chemical analysis indicates a consistent concentration of some ions when compared to the chemistry of normal seawater (e.g., Beyth et al., 1998; Warren, 2016). Figure 2.5 shows the concentration of the main ions in normal seawater and a comparison with the concentrated brines of the Dead Sea. Observe the enrichment of the element bromine in the Dead Sea brines (more than 5 g/l) as compared to the very small quantity (less than 0.01 g/l) present in normal seawater. The anomalous bromine concentration in the Dead Sea brine is one of the highest in the world (Wisniak, 2002).

2.2.4. Salt Precipitation from Normal Seawater

Since the 19th century, the majority of researchers have agreed that the majority of the evaporite minerals in ancient salt basins are the result of normal seawater evaporation, with some small salt basins genetically linked to nonmarine water (e.g., Braitsch, 1971; Hardie, 1984; Smoot & Lowenstein, 1991). Usiglio (1849a, 1849b) was one of the first chemists to reproduce the sequence of mineral precipitation by heating a container of seawater

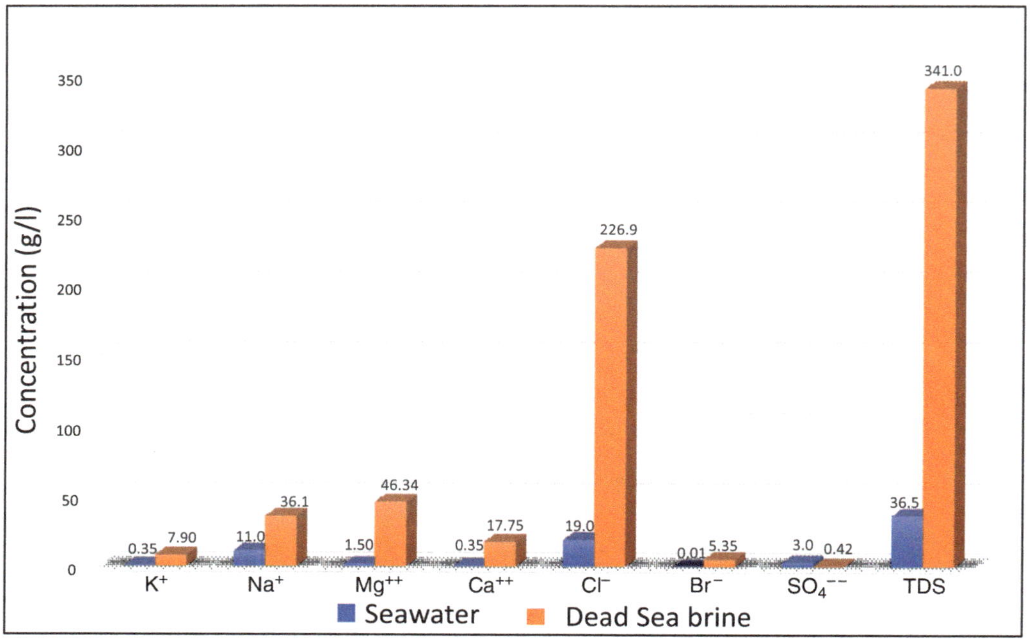

Figure 2.5 Concentration (g/l) of main ions in standard seawater from the world's oceans compared with the Dead Sea brine. The amount of total dissolved solids in the Dead Sea brine is almost 10 times higher than in normal seawater. Credit: Data from Beyth et al. (1998).

until the water evaporated completely. After drying up the last drops of water, all that was left was a salt residue at the bottom of the container. The chemical analysis of this residue showed that the compounds are similar to those observed in evaporite rocks (Figure 2.6). As evaporation progressed and water was removed, sediments and mineral crystals precipitated in a specific order from concentrated saline water. Initially, clays were deposited due to the Mediterranean seawater carrying fine-grained sediments from fluvial and deltaic siliciclastic input. As seawater concentration increased, the siliciclastic sediments were followed by carbonates, sulfates, and chlorides, particularly the mineral halite, which formed a large percentage of the total solids accumulated at the bottom of the container (Braitsch, 1971; Usiglio, 1849a, 1849b).

When normal seawater evaporates, it precipitates a predictable amount of salts in a sequence that depends on the mineral's solubility (e.g., Braitsch, 1971). The first mineral to form is calcite or aragonite ($CaCO_3$), which precipitates when the solution reaches about twice the normal seawater concentration (60 per mille). When the brine has been concentrated five times the normal concentration of seawater (150 per mille), the next mineral to precipitate is gypsum ($CaSO_4.2H_2O$). The mineral halite (NaCl) precipitates when the brine is concentrated to more than 10 times the original seawater concentration

Mineral Component	Concentration	Water loss (%)	Brine density (g/cc)
Seawater	1 ×	0	1.04
Calcite	2–3 ×	50	1.10
Gypsum	5 ×	80	1.126
Halite	11 ×	90	1.214
K, Mg salts	63 ×	98.7	1.290

Figure 2.6 Usiglio's experiment results display the sequence of mineral deposition following the evaporation of Mediterranean seawater. Credit: Based on Usiglio (1849).

(340 to 360 per mille). The more soluble salts (with K and Mg) only precipitate when the brine has reached a concentration greater than 60 times that of the original seawater salinity (Braitsch, 1971; Sonnenfeld, 2003).

G. Bischof, a famous German chemist, was probably the first researcher to discuss scientifically the precipitation of chemical compounds in sedimentary environments (Hsü, 1972). In a seminal publication, Bischof (1854) suggested that evaporites are precipitated by the evaporation of saline water in a lagoon partially isolated from the sea by sandbanks. This hypothesis was extended in the late 19th century by Carl Ochsenius to the famous "bar theory," which invokes sandbars as sills or barriers that might partially isolate the lagoons from the open seas. These lagoons would then evaporate under a warm and arid climate, forming thick salt deposits with different chemical compositions depending on their solubility (Braitsch, 1971; Ochsenius, 1877, 1888).

Most researchers involved in several branches of science, including chemists, sedimentologists, and biologists, admit a genetic association of common evaporite minerals with seawater. Some theories associate saline environments with the origin of life (e.g., Javor, 1989; Webb, 2021). There is evidence for life even under extreme environmental conditions. Many researchers are involved in the study of extremophiles, the organisms that can survive in highly stressful arid environments (see Chapter 9 in Mohriak, 2025). However, the primordial origin of the chemical elements that form the evaporite minerals is still a matter of debate. The chemical enrichment of certain ions required in these crystallization processes is a challenging theme addressed by many investigations (e.g., Braitsch, 1971; Hardie, 1991; Sonnenfeld, 2003).

2.2.5. Origin of Chemical Elements in Brines and Evaporites

The origin of the chemical elements in brines and evaporite minerals in giant salt basins is an issue frequently discussed in the geological literature. In many basins, there is disagreement on the source of the ions that formed the evaporite sequences. Salt deposits can derive from normal seawater, hydrothermal water from seeps, continental brine, or a mixture of these sources (Hardie, 1984, 1991; Sonnenfeld & Perthuisot, 1989; Sonnenfeld, 2003; Warren, 2021).

The origin of salinity in the oceans has been debated by chemists and earth scientists since Usiglio's classical experiments. One hypothesis suggests that the water that initially formed on the planet was freshwater, resulting from rainfall that accumulated on the Earth's surface, either in continental lakes or gulfs between continental landmasses. During the Earth's evolution, the water in the seas became saline, as indicated by droplets of seawater recovered from fluid inclusions in quartz crystals associated with rocks dated as Archean (Marty et al., 2018; Spear et al., 2014).

The main question concerns the source of the ions that enriched the primordial freshwater. It is often assumed that some of these ions are related to volcanism. The element chlorine (Cl) comes from volcanic eruptions; sodium (Na^+) comes from weathering of minerals that are commonly present in igneous and metamorphic rocks, mainly feldspars; and potassium (K^+) comes from feldspar and is subsequently largely captured by clay minerals in the soil because of its large ion size (Press et al., 2003). Clay minerals can easily absorb water molecules in their crystalline structure, similar to some evaporite minerals such as carnallite (see Chapter 4).

Sulfur (S), a yellow crust found in fumaroles, is a common element derived from volcanic eruptions. Carbon dioxide (CO_2), originating from the atmosphere, fills reservoir rocks in sedimentary basins due to magmatic activity interactions with carbonate rocks. Some authors suggest an origin in volcanic arcs, mid-ocean ridges, and continental rifts (e.g., Wong et al., 2019). According to other authors, a remote source of CO_2 might be located in the mantle, resulting from the metamorphism of carbonate sediments in subduction zones, or be related to volcanic activity in rift zones (e.g., Hovland et al., 2018a, 2018b; Muirhead et al., 2020).

2.2.6. Dissolution and Preservation of Evaporite Deposits

Erosion and metamorphism are destructive factors that affect sedimentary basins, including evaporite deposits. Some salt minerals are extremely soluble and rarely preserved in the sedimentary record, while some carbonate and sulfate minerals can be preserved from dissolution for millions of years (Warren, 1991). Evaporite deposits have been present on the planet since at least the Neoproterozoic (between 1,000 and 542 Ma), with large evaporitic basins in Iran, South America, and Australia (e.g., Schmid, 2017; Warren, 2006, 2010, 2016). Outcropping evaporite layers from Proterozoic sequences have been preserved for hundreds of millions of years, as seen in the Katanga Copperbelt of Central Africa (Jackson et al., 2003).

In north-central Brazil, some authors suggest that a Neoproterozoic sedimentary sequence in the São Francisco Basin might contain evaporite deposits in the deepest parts (e.g., Misi et al., 2015). There is an intimate association between rocks hosting lead and zinc mineralizations and evaporitic environments, as indicated by sedimentological analysis of boreholes drilled to investigate carbonate sedimentary sequences in many basins worldwide (Misi et al., 2015).

Climate plays a vital role in the deposition of salt minerals, and the preservation of salt minerals in outcrops depends on the appropriate climatic conditions. The chemical composition of the evaporite mineral is a key element in its preservation potential throughout the geological record. Salt dissolution is observed within and below the depositional surface, along with features resulting from leaching and dissolution associated with ancient karstification processes (see Chapter 8 in Mohriak, 2025).

Salt dissolution is recognized in many regions worldwide, such as Europe, Asia, Africa, South America, and Arabia. The salt layers in the Neoproterozoic basins of South America and Africa are easily dissolved when evaporite deposits occur near the surface. In tropical rainy regions, only remnants of less soluble evaporites, such as aragonite, calcite, and gypsum, are preserved in sedimentary layers exposed to weathering (see Chapter 9 in Mohriak, 2025). The Zagros Mountains are located in an extremely arid desert, which can preserve salt rocks from dissolution, resulting in outcrops of halite and gypsum (see Chapter 3).

Karstification of the evaporite layers that comprise the Lisan Peninsula and other regions around the Dead Sea indicates both the natural and anthropogenic effects of the dissolution caused by rainfall and groundwater circulation (Closson, 2007; Frumkin, 1994, 1996). Subaqueous salt diapirs and Pliocene salt deposits have been dissolving in the Dead Sea since the Quaternary, due to water dilution following the last glacial episodes on Earth (e.g., Levy et al., 2018) and freshwater runoff that dissolved halite units in the subsurface (Kiro et al., 2016). During the interglacial periods, the lake levels dropped, resulting in the precipitation of evaporite minerals (Kiro et al., 2016). Modern halite layers accumulate on the lake's bottom when the brine is oversaturated with respect to halite, particularly in cold seasons (Steinhorn, 1983; Stein et al., 2010; Sirota et al., 2016, 2018; Sirota, 2020).

Salt dissolution and concomitant precipitation are recorded in many modern salt basins worldwide. Marine incursions causing halite dissolution by less concentrated brines in subaqueous environments are recorded in the coastal depressions in Tunisia (Babel & Schreiber, 2014; Sonnenfeld & Perthuisot, 1989; Warren, 2006, 2016). Salt dissolution features have been recognized in the subsurface of several giant salt basins on land, such as the Michigan Basin of the Northeastern United States or the Saskatchewan region in western Canada. Examples of salt dissolution features in the offshore region include the Gulf of Mexico in the United States, the Santos Basin in the South Atlantic, and the Levant Basin in the Eastern Mediterranean (see Chapter 8 in Mohriak, 2025).

2.2.7. Climatic Controls in Evaporitic Environments

Evaporite environments are primarily found in low-latitude desert areas, where shallow lakes dry up periodically and precipitate salt (Figure 2.7). These regions are controlled by their geographic location, typically in tropical belts between the Equator and 30° N and 30° S latitudes (e.g., Press et al., 2003; Webb, 2021). Solar energy has its maximum intensity near the equator, with rays perpendicular to the surface, and is much reduced at the poles, where the rays are oblique to the surface. This results in atmospheric circulation transporting energy from the equator toward the poles, reducing the equator-to-pole temperature gradient. The thermal equator on Earth runs from southern North America, northern South America, central Africa, southern Arabia and India, and northern Australia, resulting in several deserts in adjacent regions (Figure 2.7).

Deserts in western South America are associated with a lack of water vapor due to the rain shadow formed from the Andean mountain ranges (Figure 2.7). These highlands block the transfer of humidity from the Pacific Ocean and coastal regions to the interior of the continent. High mountains, such as the Sierra Nevada in North America, form barriers for air masses that form at the seaside of the continental landmass near the thermal equator. The driest deserts in South America, such as the Atacama desert in Chile or the Uyuni desert in Bolivia, are located in the rain shadow of the Coastal Ranges and Andes Mountains (see Chapter 10). The continental depressions behind these mountains are characterized by air with extremely low humidity after the water is removed by the rains near the coast (Garreaud, 2009). The Atacama region receives only 1 mm of rain every 5–10 years, making it one of the driest deserts in the world, although its latitude and altitude result in relatively cool weather. In North America, large deserts in Nevada, California, New Mexico, and Utah are located in the rain shadow of the Sierra Nevada (Figure 2.7).

The Danakil Mountains near the Red Sea coast control the desert salt flats in the Afar depression. Topographic elevations exert a strong influence on evaporite deposition in the Danakil Depression, with altitudes below sea level. Hovland et al. (2018a, 2018b) pointed out that the evaporite layers that have built up on the Danakil Depression, which are hundreds of meters thick and made of Quaternary halite, are linked to an active volcanic area in the Afar region. Talbot (2008) dismissed this association and postulated that the salt flats with thick halite deposits presently exposed on the surface are remnants of previous lagoons and lakes that dried up in the Quaternary (see Chapter 9).

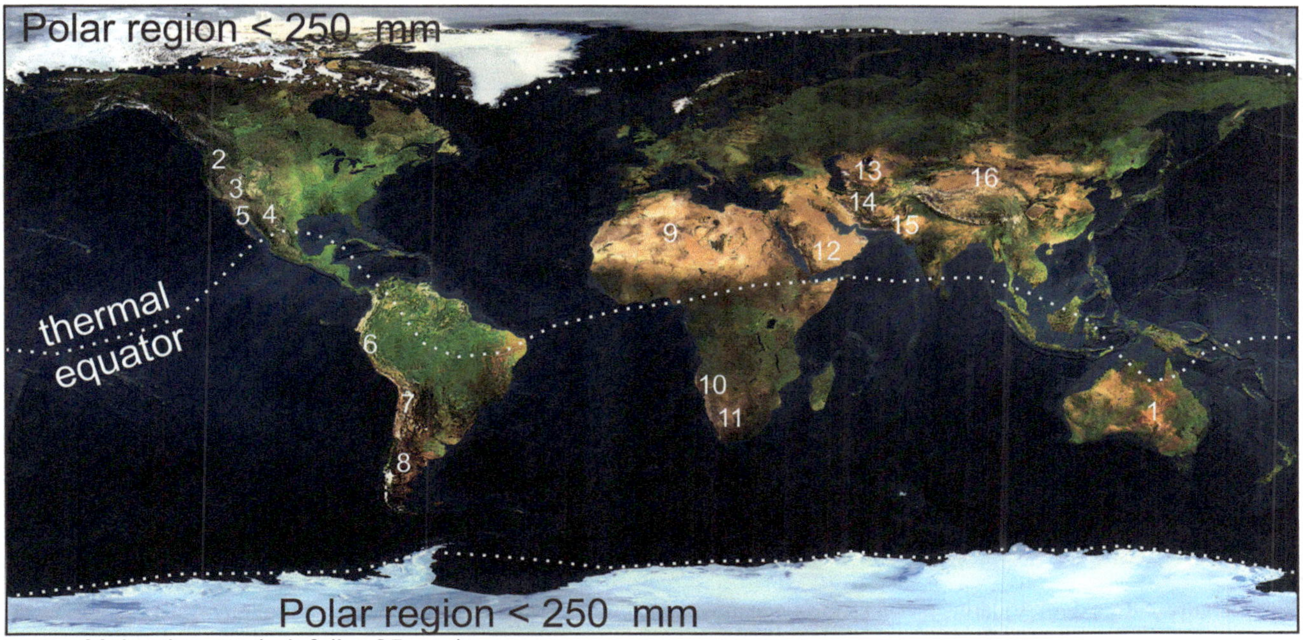

Figure 2.7 A global map shows some modern deserts where rainfall is less than 25 mm per year. These arid regions are mainly located in parts of North America, South America, Africa, the Middle East, and Asia. The major deserts (as numbered by Warren, 2006) are located close to the thermal equator, in the tropical belts between latitudes 30° N and 30° S. The thermal equator is characterized by ascending atmospheric currents, which descend in the tropical belts to the north and south. The deserts in western South America are formed in the rain shadow of the Andes Mountains. The polar regions are characterized by reduced rainfall—less than 250 mm per year. Credit: Warren (2006)/with permission of Springer Nature.

2.2.8. Climatic Controls on Ocean Salinity

The largest concentration of dissolved salt on Earth is in the oceans, and factors affecting global salinity patterns include water temperature, oceanic circulation, climate belts, and water cycle parameters like pluvial and fluvial input and evaporation (Press et al., 2003; Webb, 2021). The Atlantic Ocean has higher salinity due to the overturning water circulation around the Americas, Europe, and Africa. The Mediterranean, Red Sea, and Arabian Gulf also have high salinity due to their proximity to the Sahara Desert. Higher than normal salinities are also observed in the Central Atlantic Ocean and along the eastern Brazilian margins (Figure 2.8).

Large river-influenced regions, like the Amazon, have freshwater sources that diminish seawater salinity. Near the poles, regions with lower temperatures and glacial meltwater have the lowest global salinity values. Some Caribbean Sea islands and their coastal regions have warmer temperatures and form ocean belts with higher salinity. Seawater density and salinity are increased by evaporation in shallow seas, lagoons, and hypersaline coastal lakes, forming brines that precipitate evaporites in salt ponds (Sonnenfeld & Perthuisot, 1989; Smoot & Lowenstein, 1991; Warren, 1991).

The global distribution of salinity (Figure 2.8) reveals lower salinities in the northern hemisphere, particularly in the Pacific Ocean region from Asia to North America (NASA Aquarius Mission, 2017). Lower temperatures reduce seawater's capacity to dissolve salt minerals, which also controls the capacity of salt lakes to dissolve salt ions within a water mass. This principle also applies to oceanic waters, which are saltier near the surface due to intense evaporation (Dodson, 2004; Webb, 2021).

Salinity variations in the global oceans are associated with various mechanisms, including evaporation during arid periods, episodic rainfall or precipitation, surface runoff and inflow by rivers, subsurface inflow by water table fluctuations, and ice freezing and snow melting (Webb, 2021).

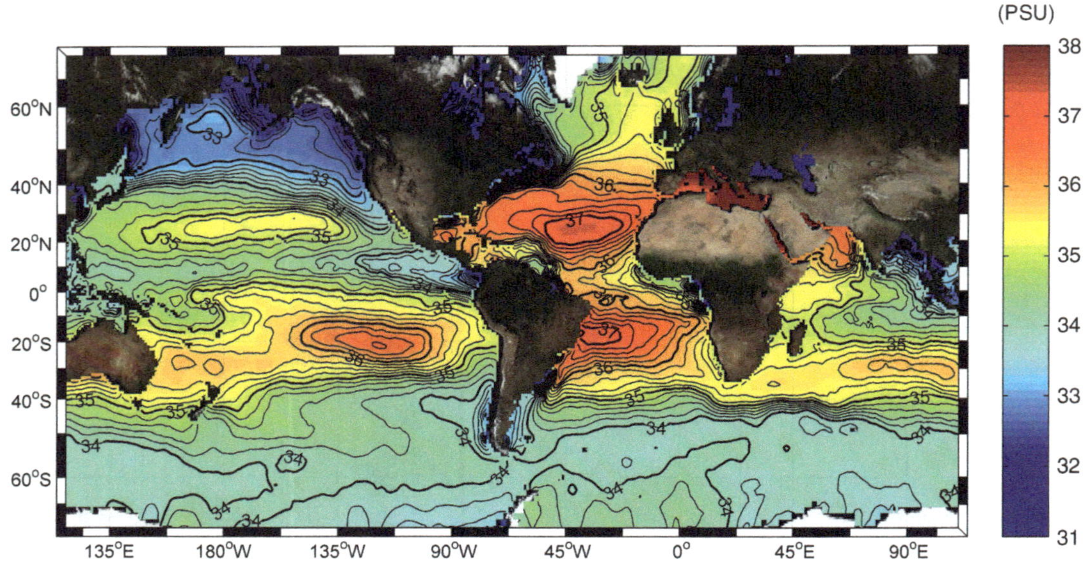

Figure 2.8 NASA's Aquarius project shows ocean water salinity distribution worldwide. The salinity is expressed in practical salinity units (PSU), equivalent to about 1 g/l of total dissolved solids in the seawater. The standard salinity of seawater is the standard global salinity, corresponding to about 36.5 PSU. More saline waters are found in belts near the thermal equator, between latitudes 40° N and 40° S. Credit: NASA/Public domain.

The physical properties of the oceanic water column change with increasing depth, with deeper layers being cooler than the upper layers and less saline. There is a gradual increase in water density from the surface to the ocean bottom. For example, in the central segment of the South Atlantic Ocean, salinity reaches 37–38 PSU at depths less than a few hundred meters but drops to around 35 PSU at greater depths (Figure 2.9). The seawater density increases slowly from 1025 kg/m^3 at the surface to about 1050 kg/m^3 at the ocean bottom due to the open ocean and submarine currents circulating the waters, resulting in homogenization of salinity (Figure 2.8).

2.3. SALT DEPOSITION IN LACUSTRINE AND SHALLOW MARINE ENVIRONMENTS

In this section, the basic principles of lacustrine environments will be succinctly presented as a background for the hydrological and sedimentological discussions presented subsequently, with a focus on Dead Sea Lake. Although the lake has been traditionally divided into northern (deep water) and southern (shallow water) basins, in the present century only the northern Dead Sea can be considered a natural lake. Artificial evaporation ponds constructed by the salt industry are now occupying the southern basin, which has dried up. The geological literature on the deposition of evaporite minerals in lacustrine environments, such as the Dead Sea and many other hypersaline lakes worldwide, is vast due to its scientific and economic importance for the mining and petroleum industries (e.g., Katz, 2001; Renaut & Gierlowski-Kordesch, 2010; Warren, 2006, 2016).

Modern examples of shallow seas and intracontinental depressions are periodically flooded, forming ephemeral lakes where salt is deposited after basin desiccation. Basin desiccation means almost total evaporation of lake water by solar heat, forming an extremely concentrated brine that precipitates evaporite minerals on the basin floor. Desiccated lakes form a crust of halite and other evaporite minerals that are exposed on the surface, resulting in a salt flat. Salt deposition may occur in lagoons, which are depressions filled with a shallow body of water separated from a larger water mass by narrow strips of land, sandy bars, reefs, or barrier islands. Some lagoons are located close to the seashores of major oceans or marine gulfs and are periodically flooded with seawater. Lagoonal transitional environments are characterized by the mixing of marine and nonmarine waters, as there is ample connection between the lagoon and the oceanic waters (Warren, 2006, 2016).

During the transition from the upper Miocene to the early Pliocene, an elongated lagoon formed in the region of the Dead Sea–Kinneret rift valley. The lagoon was connected with the eastern Mediterranean Sea by a narrow channel in the region of the Yisre'el Valley (e.g., Agnon, 2005; Stein, 2014). The restricted Sedom Lagoon

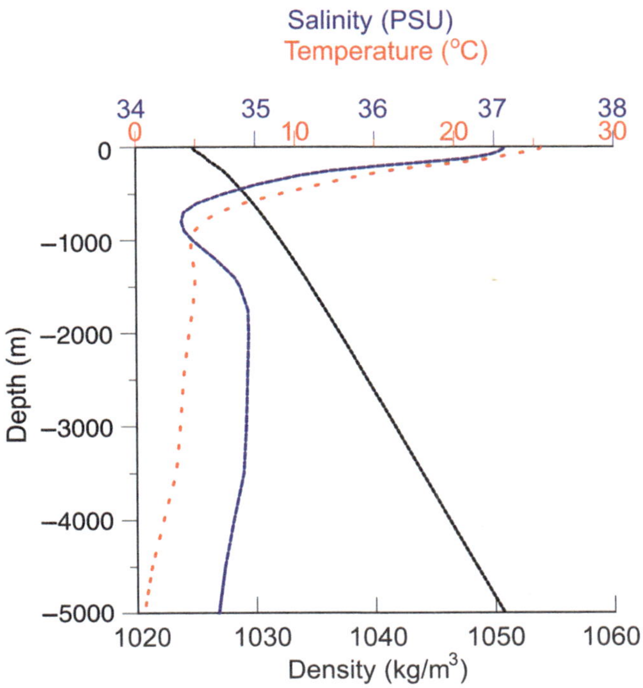

Figure 2.9 Salinity, temperature, and density variation for the oceanic water column in the South Atlantic Brazilian margin (30° W, 20° S). Note the rapid change in temperature and salinity between 100 and 900 m below sea level. The temperature is nearly constant after 1,000 m of depth, which corresponds to a temperature of about 4 °C. There is a slight decrease to 0 °C at the bottom of the ocean, around 5,000 m deep. The salinity reaches a minimum at a depth of about 800–900 m, then increases to about 35 PSU at a depth of about 1,500 m, and slowly decreases to the maximum ocean depth. The density of the seawater increases from 1,025 kg/m^3 near the sea surface to a maximum of 1,050 kg/m^3 at the maximum ocean depth. Credit: Aquarius Mission (2017)/NASA/Public domain.

was evaporated, depositing salt layers that accumulated on the basin's floor. The evaporite sediments are overlain by continental lacustrine sediments deposited around 3 Ma, as the basin was isolated from the Mediterranean Sea (e.g., Torfstein, 2017).

Salinas are depressions that may occur close to the seashore but have only sporadic communication between the lake and the oceanic waters, resulting in the concentration of brine and salt precipitation. Saline lakes may occur far away or close to the sea, but in a salina, the isolation of the lake is enhanced by water evaporation in arid climates, resulting in a water surface below the global sea level. This phenomenon is observed in the Danakil Depression in northern Africa, which was flooded by the Red Sea waters in the Quaternary, and Lake MacLeod in Northwest Australia, which was periodically flooded by the Indian Ocean (see Chapter 9).

After the desiccation of large saline lakes, the evaporite deposits result in extensive salt flats. The remnant, ephemeral ponds of hypersaline water that seasonally occupy these depressions are known as *playa lakes*. Most playa lakes are rather small, such as the Badwater Pond in Death Valley, while some large playa lakes, such as the hypersaline lakes in the Qaidam Basin of Tibet, are very large. Some dry salt flats may exceed 10,000 km^2 (see Chapter 10). Earth scientists have studied several present-day continental lakes and shallow marine environments as analogs for ancient giant saline basins (see Chapter 8).

Modern salt basins are located in diverse regions and geological settings, such as the salt lakes in Africa, South America, North America, Asia, Australia, and India. Most of these regions have been the subject of limnological studies involving the hydrological analysis of the lake system, the quantification of the water cycle volumes, the determination of brine salinity variations through time, and the mineralogy of the evaporite sequences (e.g., Dodson, 2004; Wetzel, 2001; Warren, 2006, 2010, 2016).

Evaporite sediments accumulated in large salt basins display different depositional facies (see Chapter 8). Evaporite deposits occur as subaerial efflorescent crusts or in subaqueous lacustrine and marine environments. The subaqueous accumulation of surface-nucleated crystals corresponds to crystal cumulates. The subaqueous bottom precipitates are formed at the bottom of the basin, resulting in crystal bottom growth and the accumulation of bottom crusts. Intra-sediment precipitates are formed in sabkha supra-tidal environments, displacing or incorporating host sediments in the evaporite matrix (see Chapter 8). Clastic accumulations result from the erosion of evaporite particles and their transport to the basin depocenter (Kendall, 2010).

2.3.1. Basic Characteristics of Lake Systems

The science that focuses on lake systems is called *limnology*, a combined word derived from the Greek root *límnē*, which means a pool or marsh, and *logia*, which means study. Limnology is an interdisciplinary science that focuses on lake systems, analyzing the geographic and geological components of these systems to understand natural phenomena (Dodson, 2004; Tundisi & Tundisi, 2011). It involves concepts from geography, geology, chemistry, physics, and biology, relating the earth's climate system with a quantitative analysis of the water cycle. Reeves (1968) published a classic introductory book on the applications of several scientific methods to the study of ancient lacustrine environments. The geological literature is extremely rich in works involving lacustrine sedimentation, lake hydrology, and the geochemistry of lacustrine brine and sediments (e.g., Boehrer and Schultze, 2008; Dodson, 2004; Schulz, 2001; Tundisi & Tundisi, 2011; Wetzel, 2001).

Several factors control sedimentation in lacustrine systems, including climate, geology, time, hydrology, and biology. Deep lakes have a range of water columns from a few meters to several hundred meters, while shallow lakes can have a cover of a few meters in wet periods. Extremely shallow lakes typically exhibit only a few centimeters of water during dry weather and desiccate completely during extended drought periods (Renaut & Gierlowski-Kordesch, 2010).

Lakes are classified according to their physical and chemical characteristics, such as size, volume, salinity, and maximum depth to the lake's bottom (Figure 2.2). Deep lakes display water columns ranging from a few tens of meters to several hundred meters. Extremely shallow lakes normally exhibit only a few centimeters of water during dry weather and eventually desiccate completely during extended drought periods (Dodson, 2004; Tundrisi & Tundisi, 2011).

Limnologists also subdivide lake components according to their thermal structure, which is based on the temperature of the water layers (Figure 2.10). Most lakes are divided into three main layers: epilimnion, metalimnion, and hypolimnion. The superficial layer is called the *epilimnion*, while the transition zone from the superficial to deeper portions is called the *metalimnion*. The intermediate layer corresponds to the thermocline, where temperature varies rapidly with depth. The deeper portions of the water bodies correspond to the hypolimnion, with *hypo* derived from the Greek root meaning deep, thus referring to the deeper portions of the water column (Figure 2.10).

The thermocline (which is another word with Greek roots, with *thermo* meaning temperature and *cline* indicating variation or inclination) is a temperature gradient that occurs in both oceans and lakes, affecting the salinity and density of water. It occurs when the warmer layer above (epilimnion) receives solar radiation during the day, which is more heated in tropical regions than in high latitudes (Dodson, 2004; Tundisi & Tundisi, 2011; Wetzel, 2001).

The hypolimnion, below the thermocline, underlies the photic zone and receives little heat from radiation. The maximum density of water (1 g/cc) is reached when the temperature is about 4 ° C and the dense water is trapped at the hypolimnion (Wetzel, 2001). The lack of circulation between the epilimnion and hypolimnion affects the lake's chemistry and biology. Since the waters do not mix, there is increasingly less oxygen below the thermocline, and organisms in the water (particularly bacteria) deplete the available oxygen, resulting in an anoxic environment (Dodson, 2004; Reeves, 1968; Tundisi & Tundisi, 2011).

Most continental lakes are shallow, with a total depth reaching only tens or a few hundred meters. In oceans, the physical parameters display fewer variations because the water column can reach thousands of meters. When under the influence of gravity, the colder, denser, superficial water moves down to the deeper portions of the lake, while the lighter, deeper water layer rises to the surface, resulting in a phenomenon called *lake overturn*.

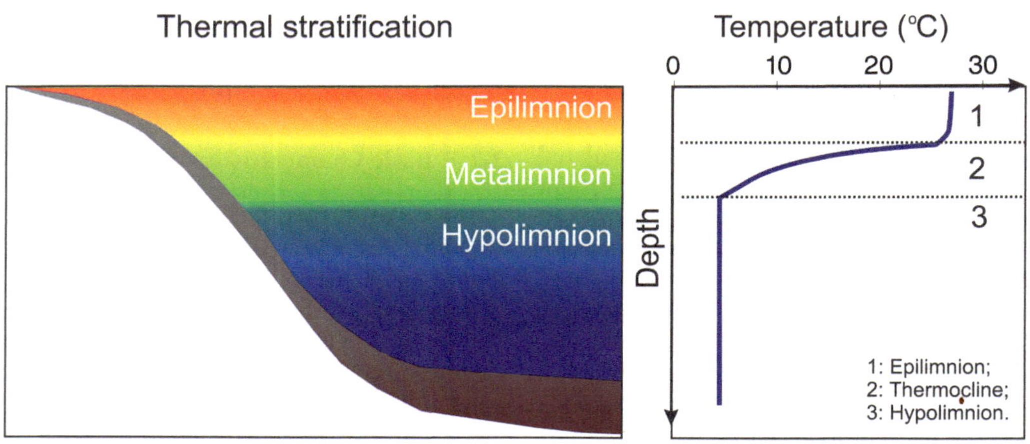

Figure 2.10 Schematic diagram showing the thermal classification of water layers in lakes. 1. epilimnion, corresponding to the upper mixed layer, is characterized by warmer, less dense water; 2. metalimnion, corresponding to the middle layer, where temperature changes more rapidly with depth; thus, the thermocline is defined as the region where dT/dz is maximum; 3. hypolimnion, corresponding to the lower layer, is characterized by cooler, denser water. Credit: Adapted from Dodson (2004).

This event results in the mixing of the upper and deeper water layers, and a uniform temperature is established across the whole of the water column. Deep waters are more enriched in nutrients than superficial waters, particularly in polar regions (Dodson, 2004; Reeves, 1968; Tundisi & Tundisi, 2011).

Modern lakes and shallow seas are typically layered according to the physical properties of their waters, particularly temperature, brine salinity, and density. In most lakes, the denser, more saline waters are located at the base of the water column, in the hypolimnion. These deeper layers tend to be hydrologically more stable than the more superficial waters, which are subject to rapid salinity fluctuations due to the input of freshwater by runoff and precipitation. Some lakes are characterized by harsh winters when the temperature drops near the freezing point and the superficial water layer density becomes greater than the deeper water layer in the hypolimnion. During these episodes, the hypolimnion can be warmer and less dense than the epilimnion waters (Tundisi & Tundisi, 2011; Wetzel, 2001).

When under the influence of gravity, the colder, denser, superficial water moves down to the deeper portions of the lake. The lighter, deeper water layer rises to the surface, resulting in lake overturn (Dodson, 2004; Reeves, 1968; Tundisi & Tundisi, 2011). This event results in the mixing of the upper and deeper water layers, and a uniform temperature is established across the whole of the water column (Dodson, 2004). Usually, deep waters are more enriched in nutrients than the superficial waters, particularly in polar regions. The overturn of cold, nutrient-rich waters mixing with warm, superficial waters often results in an environment rich in oxygen, causing a bloom of phytoplankton. Living organisms that feed on phytoplankton, such as fish, thrive in these regions. That is why Arctic and Antarctic waters that border Nordic countries are extremely productive and constitute a rich resource for the fishing industry.

Lakes can be categorized into several types, including freshwater lakes like Lake Baikal, which exhibit low salinity from top to bottom (Figure 2.11-top), and hypersaline lakes like the Dead Sea, which undergo multiple episodes of stratification and non-stratification (Figure 2.11-center). Chemoclines are layers where chemistry variations are more rapid than in the layer above or below, similar to temperature variations in the thermocline. In the case of salt content, this *cline* is termed the *halocline* (halo meaning salt and cline, a word derived from the Greek root *klinein*, meaning slope or variation). The halocline is a layer with rapid variations in salt content, while the metalimnion corresponds to a transition zone associated with rapid variations in brine density (Figure 2.10). Since the quantity of salt also controls the density of the water, the metalimnion corresponds to a transition zone associated with rapid variations in brine density, and this layer is called the *pycnocline* (a combining word derived from Greek, with *pycno* meaning density).

Holomictic lakes have mixed upper, middle, and lower layers, resulting in a homogeneous body of non-stratified water (Figure 2.11). This period is generally characterized by stable stratification during the summer season, with winter overturning when the bottom layers rise to the surface (Dodson, 2004; Gertman & Hecht, 2002; Gertman, 2012). The opposite term, *meromictic* (from the Greek root *meros* meaning part), means that the upper and lower layers never mix totally throughout the year, with no overturn during winter. Meromictic lakes never totally mix their water layers because denser, more stagnant bottom waters in the hypolimnion are isolated from the epilimnion by the thermocline (Figure 2.11). However, the thermocline is controlled by seasonal variations, and some lakes mix their waters at least once a year. During these episodes, the density at the epilimnion is equal to the density of the hypolimnion (Dodson, 2004; Gertman & Hecht, 2002; Gertman, 2012).

Large lakes in North America are classified as monomictic lakes, such as Lake Huron, Lake Michigan, Lake Ontario, and Lake Superior, because their water layers mix twice a year, during winter and spring. During summer, their waters are warmer (reaching a mean temperature of about 20° C), and so they have a circulation pattern independent of the denser water layers at the bottom, which remain at a temperature of about 4° C (Dodson, 2004). In winter, when temperatures approach 0 °C, the upper and lower layers of these lakes mix completely, and the bottom layers come to the surface in the overturning process (Dodson, 2004; Gertman & Hecht, 2002; Gertman, 2012). When there is a complete mixing of the water column in a lake, it is called *non-stratified*, and there are no differences in salinity from top to bottom (Figure 2.11). Hypersaline lakes, such as the Dead Sea, experience episodes of lake overturn, resulting in salt mineral precipitation throughout the water column (Steinhorn et al., 1979; Steinhorn, 1983, 1985).

2.3.2. Thermal Stratification in Lakes

Thermal stratification in lacustrine systems is a significant area of interest in limnology and salt sciences due to its impact on chemical and water density stratifications (Dodson, 2004; Gertman & Hecht, 2002; Gertman, 2012; Tundisi & Tundisi, 2011). This stratification can control the cyclicity associated with salinity fluctuations over time, resulting in the precipitation of different minerals. Some lakes are characterized by freshwater, while most saline lakes exhibit stratification and the development of a characteristic thermocline and halocline. Hypersaline lakes do not show this stratification,

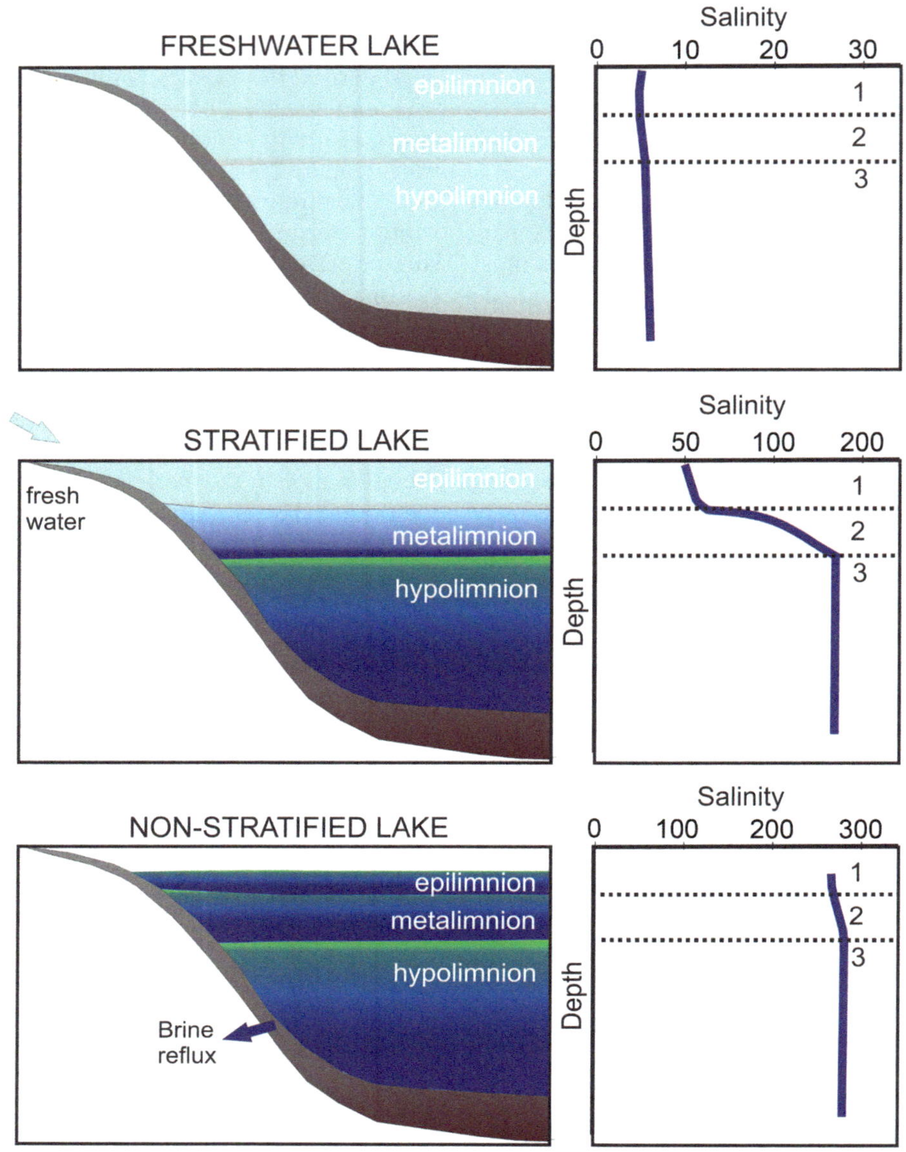

Figure 2.11 Top: salinity profiles for freshwater lakes (e.g., the Baikal Lake). Center: stratified lakes, characterized by a rapid increase in salinity at the metalimnion (e.g., the Dead Sea). Bottom: non-stratified lakes, where the salinity is constant from top to bottom (e.g., the Dead Sea during overturn periods). Credit: Adapted from Dodson (2004).

and the water column is characterized by a homogeneous temperature and brine concentration from the lake surface to the bottom during certain overturn episodes. These non-stratified hypersaline lakes are of great interest to the salt sciences, particularly when their brines reach the point of precipitating halite and other salts (Figure 2.11).

Thermal stratification in lakes is also of great interest to biology because it controls the distribution of essential chemical elements, such as oxygen. The solubility of oxygen in water is highly dependent on temperature, with a 50% decrease within a temperature range common in high-latitude lakes. The solubility of gases in water is also dependent on pressure, with the bottom layers of deep lakes or oceans holding much less oxygen than the upper layers (Figure 2.12).

Carbon dioxide (CO_2) is 200 times more soluble in water than oxygen, and in stratified lakes, the water at the bottom is under greater pressure. Consequently, it can dissolve a much greater quantity of CO_2 than the upper layers of the epilimnion. This can lead to an explosion of

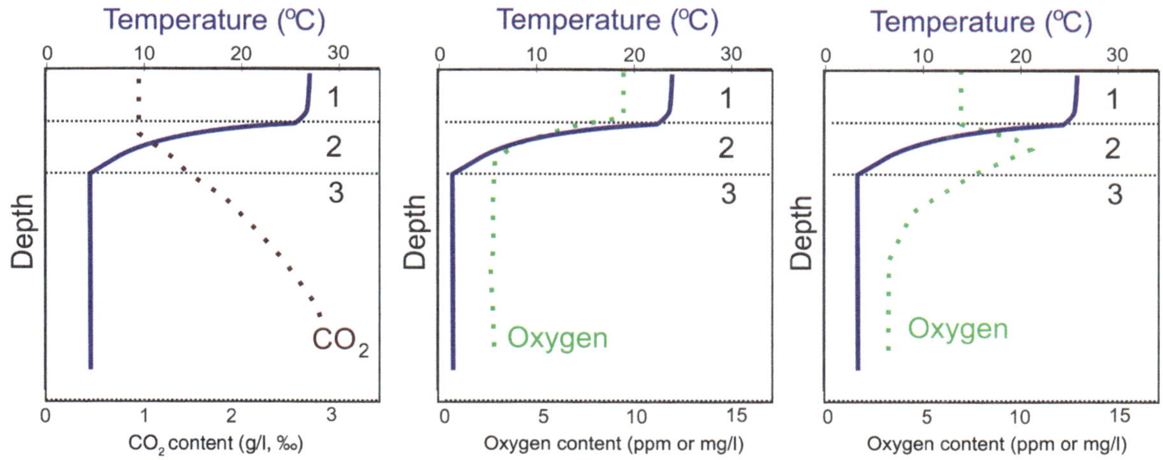

Figure 2.12 Carbonic gas (CO_2) and oxygen (O_2) distribution in stratified lakes. Layer 1 corresponds to the epilimnion, Layer 2 to the metalimnion, and Layer 3 to the hypolimnion. The thermocline in stratified lakes shows a rapid decrease in temperature with depth. In the left-column graph, the CO_2 shows a marked increase in content from the surface to the bottom of the lake, due to the increase in pressure. The center graph shows that the oxygen content decreases from the surface (which is warmer) to the bottom of the lake (which is cooler). There is a rapid change in oxygen content in the thermocline, with a rapid decrease and then stabilization of the oxygen levels when the temperature stabilizes around 4 °C. The right-column graph shows that in some lakes, the thermocline is characterized by an increase in oxygen content at the intermediate layer. Credit: Adapted from Dodson (2004).

gases when the bottom waters rise to the lake's surface, releasing large amounts of gas due to the decrease in pressure. This can be a deadly hazard when the dissolved gases are poisonous, such as in bottom waters that contain H_2S. Even CO_2, when released in large quantities, can kill living organisms that depend on oxygen. When the deep water layers of Lake Nyos in Africa came to the surface in 1986, this ominous lake overturning resulted in the disastrous displacement of air on the lake's surface. The outburst formed a gas cloud (mainly composed of CO_2) that killed, by suffocation, a large number of people who lived in nearby villages (Tanyileke et al., 2019).

The salinity of the brines also affects the solubility of gases, including oxygen and CO_2. The TDS increases in water with increasing temperature, resulting in increased salinity, while the solubility of gases decreases exponentially (Dodson, 2004). Hypersaline lakes are characterized by minor amounts of oxygen in the hypolimnion, resulting in extremely anoxic environments where only a few types of organic lifeforms can survive (Figure 2.13). Only specialized forms of life can adapt to such stressful environments, and several works have been published focusing on the microbiology of lacustrine systems in extremely arid environments (e.g., Javor, 1989).

2.3.3. Lacustrine Settings and Organic Productivity

Salt deposition is primarily associated with continental environments, particularly lakes in arid regions or near shallow seas and gulfs. Large and deep oceans like the South Atlantic or Pacific Ocean do not exhibit extensive salt deposition. The deepest lakes on Earth, such as Lake Baikal in Asia and Lake Tanganyika in Africa, are characterized by their extreme purity and high transparency, resulting in crystalline waters (Figure 2.2).

In terms of water depth, the deepest lake in the world at present is Lake Baikal, located in Siberia, north of the border with Mongolia. Its maximum depth is estimated at 1642 m below the lake level, corresponding to a lake bottom at 1187 m below global sea level, as the lake surface is located at an elevation of 455 m above mean sea level. Lake Baikal is a tourist attraction not only because of its great depth but also because it shows an annual cycle of freezing and melting its surface. In summer, the ice melts and the freshwaters are characterized by extreme purity due to extremely low mineralization degrees and high transparency, resulting in crystalline waters. Lake Baikal is an exorheic basin, meaning that drainage flows away from the water body. It is highly influenced by the melting of glaciers that replenish the water that flows outward with pure freshwater. Lake Baikal is characterized by an extremely rich biota (Dodson, 2004; Morgunova et al., 2022).

Lake Tanganyika in Africa, one of the deepest lakes in the world, is also characterized by rich biota and organic productivity. Many authors have discussed the source rock deposition on the lake's bottom, which is characterized by an anoxic zone (e.g., Cohen, 1990). Figure 2.13

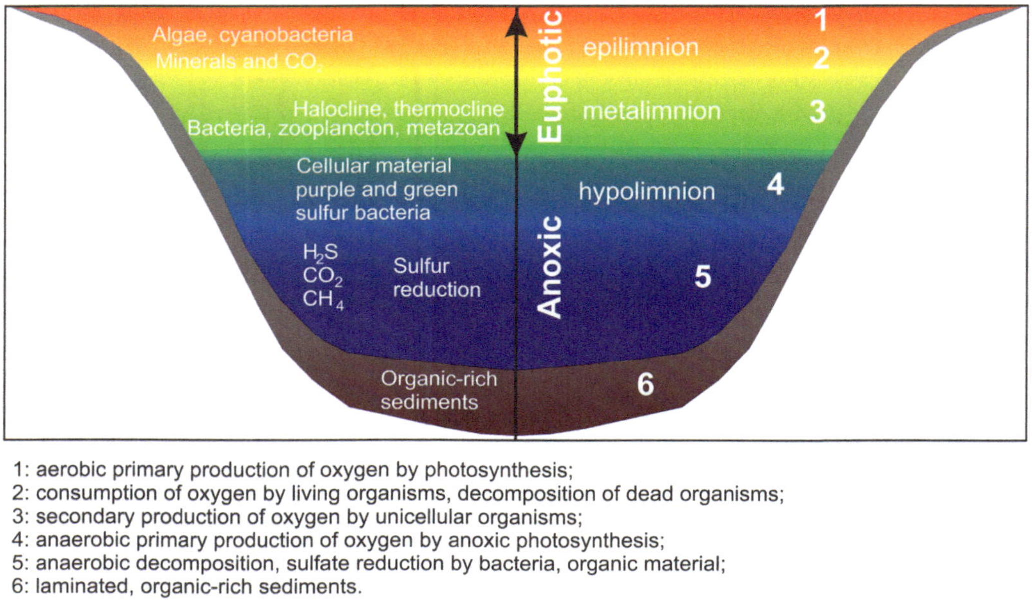

Figure 2.13 Life productivity in stratified lakes. The epilimnion is characterized by algae and cyanobacteria, whereas the metalimnion is characterized by bacteria and zooplankton. The hypolimnion is deficient in oxygen, and the bacteria live by absorbing sulfur into their organisms, and the metabolism results in sulfur reduction. The bottom of anoxic lakes is characterized by black, organic-rich sediments that eventually constitute rich source rocks for hydrocarbon generation.

shows the schematic diagram of a deep lake with euphotic and anoxic zones. The upper water layer of a deep lake contains euphotic and anoxic zones, which control the distribution of algae, phytoplankton, zooplankton, and microbial communities. Lake Tanganyka's hypolimnion is characterized by the anaerobic production of oxygen, the decomposition of dead organisms, sulfur reduction by bacteria, and the accumulation of organic-rich sediments (Katz, 2001; Warren, 2006, 2016).

Only a few organic species can survive when salinities in the upper layers of salt lakes reach the range of gypsum precipitation (>140‰), corresponding to a brine concentration more than five times that of normal seawater (Figure 2.6). However, when this superficial layer is freshened by the addition of less saline water due to river inflow or rain in wet climate episodes, salinity can drop into the carbonate domain (70 to 140‰), causing cyanobacteria and green algae blooms and photosynthetic microbes to color moderately saline waters pink or red. These phenomena can be seen in many saline lakes in the world, from the saline lakes in California to the Natron Lake in Africa (Dodson, 2004; Renaut & Gierlowski-Kordesch, 2010; Tundusi & Tundisi, 2011). Because the cyanobacteria and algae that flourish in mesohaline conditions yield sapropelic organic matter, source rocks associated with evaporitic basins tend to be oil-prone or type I, as observed in the lacustrine pre-salt sediments in the South Atlantic continental margins, where most of the oil is associated with saline lacustrine environments (Malek-Aslani, 1980; Mello & Maxwell, 1991).

2.4. THE DEAD SEA: THE PARADIGM FOR DEEP LAKES WITH EVAPORITE DEPOSITION

The Dead Sea, located in a depression extending from the Arava to the Jordan valleys, receives limited fluvial input, mainly from the Jordan River. The lake has an endorheic drainage system, with most water evaporating due to its arid desert location. The Dead Sea Works, an Israeli company, pumps out water to the southern basin for industrial use. Despite the belief that Lake Baikal's pure freshwaters favor life, the Dead Sea is renowned for the scarcity of life in its waters. The lake's biblical name is the "Sea of Salt," and the lake and its adjacent salt mountain, Mount Sedom, are mentioned several times in the Bible (see Chapter 1).

Asphalt blocks were collected from the lake's shores since the Early Bronze Age (about 4,000 years ago) to glue flint implements to wooden handles used in agriculture as a biocidal agent in agricultural practices (Nissenbaum, 1979, 1980, 1993, 1994). During the Roman period, asphalt was one of the substances used in Egypt in mummification processes (Nissenbaum & Buckley, 2013). From the first millennium CE until the 19th century, the Dead Sea was almost unexplored. Since the beginning of the 20th century, many studies have focused on the lake's physiography, chemistry, and geology (e.g., Cady, 1904; Quennell, 1958).

Dead Sea Lake is one of the most studied hypersaline lakes in the world due to its unique hydrological, biological, and geological characteristics. The extreme salinity in the Dead Sea results in a scarcity of living organisms, with scant bottom mud and a layer of gypsum and halite covering much of the lake floor (Anati et al., 1987; Ben-Avraham, 1997; Garfunkel, 1997; Neev & Emery, 1967; Sirota, 2020; Warren, 2016). However, the top 6 m of a scientific borehole drilled in the basin's depocenter revealed a large proportion of siliciclastic, fine-grained detrital rocks, mainly composed of silt and marl (Goldstein et al., 2020; Neugebauer et al., 2014).

Some researchers have highlighted that the Dead Sea is not devoid of vestiges of life, as there is evidence for the presence of halophilic bacteria in the water column (hypersaline brine) and some microorganisms that survive on the anoxic sediments below the lake's bottom (Nissenbaum, 1979, 1980).

The Dead Sea, a hypersaline lake, is home to various microbial forms of life that have adapted to its harsh conditions. These include bacteria, algae, and fungi, with the indigenous biota in the Dead Sea brine being specialized organisms that have adapted to the most extreme environment on Earth (e.g., Ionescu et al., 2012; Javor, 1989; Nissenbaum, 1975). The main organisms in the Dead Sea biota are the unicellular green alga Dunaliella and several red halophilic Archaea (Gavrieli & Oren, 2004). Some underwater springs in the Dead Sea harbor dense archaeal microbial communities, partly unknown in other hypersaline environments (Ionescu et al., 2012).

The Dead Sea is one of the global sites investigated as an analog for the habitability of extraterrestrial worlds (e.g., Preston & Dartnell, 2014). Astrobiology is a growing field of research involving the study of microorganisms that have adapted to extreme environments, providing clues for similar environments on other planets of the solar system (see Chapter 9 in Mohriak, 2025).

Compared to Lake Baikal or Lake Tanganyika, the Dead Sea is a rather small lake that separates Israel from Jordan, but it corresponds to the deepest hypersaline lake in the world (e.g., Anati et al., 1987; Neev & Emery, 1967; Warren, 2006, 2016). It measures only 50 km in length and about 15 km in width at its widest point, covering an area of about 630 m^2 (Sirota et al., 2016). The Dead Sea lake surface was measured at 437 m below the global mean sea level (bmsl) in 2022, corresponding to the lowest elevation on Earth. As the measured water level was 430 m bmsl in 2016, the lake level has been falling at a rate of about 1 m/year in the 21st century (Sirota et al., 2016, 2018; Sirota, 2020).

The Dead Sea Graben is characterized by two sub-basins with rift sediments and modern water masses that have become disconnected in the past decades. The northern basin hosts a natural, deep lake that is shrinking due to natural causes and the intensive use of water for agricultural and industrial purposes (e.g., Gertman & Hecht, 2002; Gertman, 2012). The southern basin is an artificial, industrial pond that is almost dry due to the intense evaporation in the saltworks. The salt industry has exploited these artificial salt ponds for more than 50 years, and some of the chemical compounds are exported to other countries, including potash and bromine (Wisniak, 2002).

The Dead Sea's sedimentary basins have different rift structures. The northern sub-basin has a sedimentary column that is about 6 to 8 km thick, while the southern basin has a much thicker sedimentary column that is almost 14 km thick from the Holocene to the basement (Garfunkel & Ben-Avraham, 1996; Ben-Avraham & Schubert, 2006; Ben-Avraham et al., 2008). The northern sub-basin hosts a deep, natural lake with a maximum measured bathymetry of circa 290 m, resulting in a lake bottom of about 730 m bmsl (Sirota et al., 2016; Golan et al., 2017; Sirota, 2020). The southern basin, located to the south of the Lisan Peninsula, is an artificial, much shallower pond, less than 3 m deep on average, corresponding to a salt pan that has been desiccated several times in the Holocene (Bookman et al., 2006).

The Dead Sea brine salinity is about 337 g/kg, corresponding to 34.2%, which is about 10 times the normal seawater salinity (see Figure 2.5). Golan et al. (2017) measured the weight of TDS in the deep water layer (50–250 m) and obtained a mean value of 348 kg/m^3. This concentration results in a very high brine density, up to 1.24 kg/l, which makes it easy for people to swim or even float in the water (Moses et al., 2006). The brine displays a chemical composition that includes several ions that commonly occur in seawater, such as Na^+, Mg^{++}, Ca^{++}, and Cl^- (Figure 2.5). In addition, several other ions appear in minor amounts in seawater but exhibit an extremely high concentration in the Dead Sea brine, such as the bromide (Br^-) ion (Zak, 1974; Beyth et al., 1998). The bromide content in the Dead Sea (5.2 g/l) is several orders of magnitude higher than the content in seawater (0.065 mg/l), resulting in the richest brine concentration worldwide (Salameh et al., 2016).

The high mineral content of the Dead Sea brine has been reputed to have therapeutic properties, as some chemical elements in the water penetrate the skin and contribute to a healing effect (Moses et al., 2006). The extreme aridity of the Dead Sea results in lower levels of pollen and allergens in the atmosphere, and the lake's surface, located hundreds of meters below global sea level, helps in the oxygenation of body cells. Some European Union countries subsidize medical treatment in the Dead Sea for mild skin diseases like psoriasis, as climatotherapy can be used without risking side effects associated with hospitalization and traditional medications (Moses et al., 2006).

2.4.1. Brief Discussion of the Geology and Hydrology of the Dead Sea Basin

The Dead Sea Lake is part of a deep depression in the Earth's crust, forming the lowest land surface along the Dead Sea Transform Fault (DSTF), according to global topobathymetric datasets (e.g., Amante & Eakins, 2009; GEBCO, 2019). The DSTF zone extends from the northern Red Sea, forming the Gulf of Aqaba, Arava Valley, Sea of Galilee, and the Jordan and Hula valleys (Figure 2.14). The deformation zone reaches the Palmyride fold-thrust belt in Syria and the East Anatolian Fault in Turkey, near the Taurus Mountains, a region affected by transpressional tectonics (Ben-Avraham et al., 2012; Fedorik et al., 2022; Garfunkel, 1981, 1997, 2014; Garfunkel & Ben-Avraham, 1996). The transcurrent fault zone is associated with the northward movement of the Arabian Plate relative to the Sinai sub-plate, forming a restraining segment in Lebanon that links with the Palmyride deformation zone in Syria. Plate velocity ranges from 4.4 mm/yr in the Dead Sea Lake region to 6.2 mm/yr north of the Palmyra fold-thrust belt in Syria (Gomez et al., 2007).

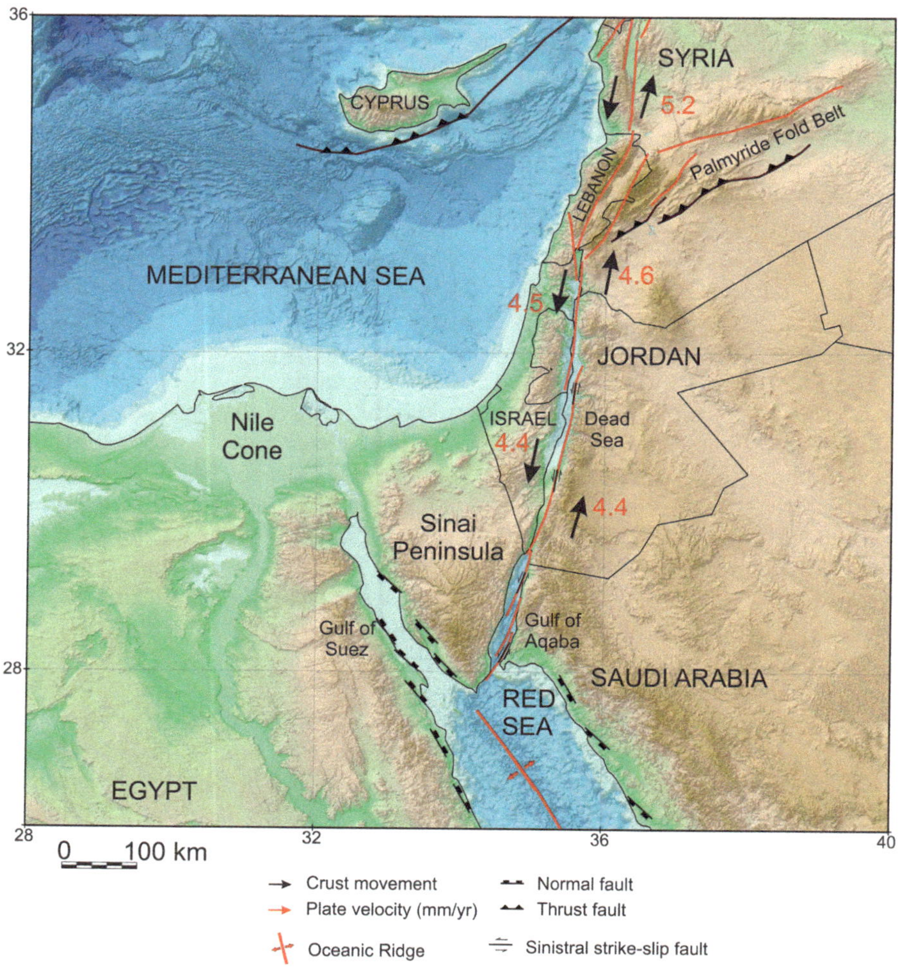

Figure 2.14 The topographic map of Eastern Egypt shows the Nile Cone in the eastern Mediterranean Sea, the Sinai Peninsula, and the Arabian Peninsula. The Dead Sea Transform Fault Zone (DSTF) extends from the Red Sea spreading center to Syria and displays left-lateral motion. The rift depressions along the fault zones are characterized by altitudes below the global sea level, as indicated by the light blue color along the valleys. The DSTF is characterized by left-lateral motion, with a velocity of about 4.4 mm/yr between the Sinai Peninsula and the Jordan segment of the Arabian Peninsula.

The Dead Sea transform fault is a tectonic trough that controls the depocenters of several transtensional rift troughs associated with the movement of the Arabian plate relative to the Sinai sub-plate, attached to the African plate (Figure 2.14). The Dead Sea Lake, a remnant of the larger Lisan paleolake, occupies part of the depression. The southern basin has been desiccated but exhibits greenish to light blue reflectivity in satellite images due to artificial ponds built by the Dead Sea Works. The northern basin is a natural, deep lake characterized by a dark blue color (Figure 2.15).

The Jordan River is the only major source of water flowing into the Dead Sea depression, resulting in an endorheic basin with an elongated valley and remnant lakes like Dead Sea Lake, Lake Kinneret, and Lake Hula (Figure 2.14). The Dead Sea Basin is a tectonic trough formed by transtension, resulting in deep, narrow pull-apart basins occupied by fluvial and lacustrine systems since the Pliocene (Ben-Avraham, 1997; Ben-Avraham et al., 2012; Garfunkel, 2014). The Dead Sea modern lake was preceded by other lakes in the geological past, including Lisan Lake in the Late Pleistocene, Amora

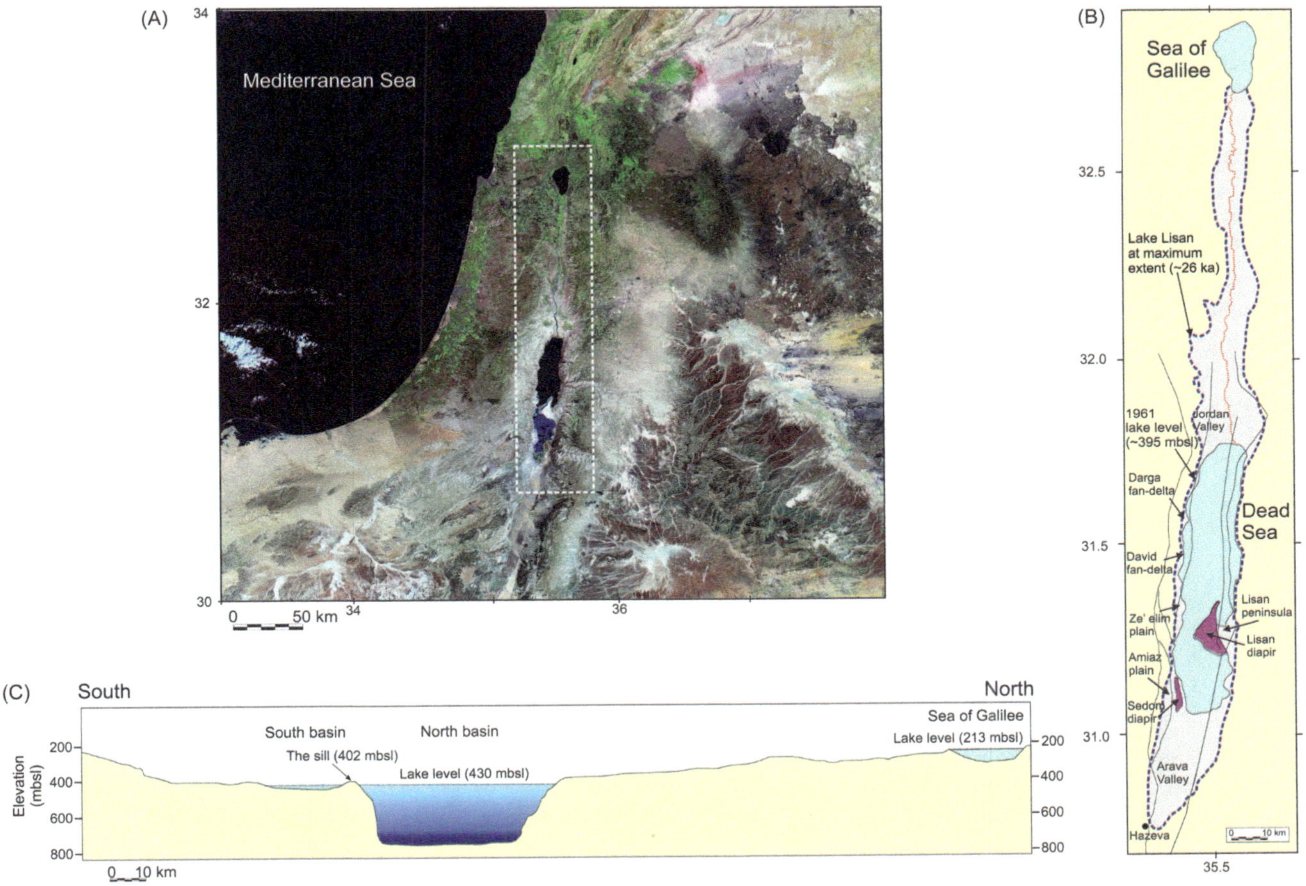

Figure 2.15 (A) Landsat image of the Levant Basin margin in the Eastern Mediterranean Sea, showing the depression of the River Jordan between the Sea of Galilee and the Dead Sea. The dashed white line marks the outline of the detailed map to the right. Credit: Courtesy of NASA. (B) Reconstruction of the lacustrine sediments associated with the ancient Lake Lisan (shown in gray color). During the last glacial maximum lake level (around 26 ka), the lake shorelines extended from the Arava Valley in the south to the Sea of Galilee in the north. The Dead Sea Lake occupies a relatively small area in the central part of the previous Lisan paleolake. The magenta color outlines show that there are two salt diapir outcrops in the vicinity of the Dead Sea Lake. The Lisan diapir is located between the northern lake and the artificial evaporation ponds of the Dead Sea Works. The Sedom diapir is located onshore, near the salt pans. Credit: Adapted from Bookman et al. (2006). (C) Schematic geological cross-section showing a longitudinal profile across the Dead Sea Basin. The salt pans in the southern basin are artificial ponds fed with water from the northern basin. The Dead Sea Lake is a remnant water body that exhibits a water column of about 300 m. The Sea of Galilee, located in the northern part of the Dead Sea Basin, is a rather shallow lake with freshwater sourced from the Jordan River. Credit: Adapted from Bookman et al. (2006).

Lake in the Pliocene to Late Pleistocene, and Sedom Lake in the Early Pliocene (Bartov et al., 2002; Torfstein, 2017).

The Dead Sea lacustrine system's origins can be traced back to the Neogene ingression of marine waters from the Mediterranean Sea into the valley depression, which occurred between 6 and 3 million years ago (Zak, 1974, 1997). This event formed the Sedom Lagoon, which repeatedly evaporated and precipitated evaporite sediments, resulting in a thick salt layer known as *Sedom Fm.* (Neev & Emery, 1967; Stein, 2001, 2014). These evaporite layers were covered by other lacustrine layers that formed when the climate changed, which caused the paleolake level to rise and fall over time. These fluctuations in lake level have been correlated to long-scale climate changes, particularly epochs of glaciation during the Pleistocene (Bookman et al., 2006; Stein, 2014).

Lake Lisan formed around 70 ka, and during the Late Pleistocene, a large water body occupied the depression from the Sea of Galilee to the Arava Valley (Figure 2.15). After the last glaciation, around 15 ka, the paleolake started to shrink, evolving into the present-day lake systems associated with the hypersaline Dead Sea and the freshwater Sea of Galilee (e.g., Bartov et al., 2002).

During its maximum extent, between 26 and 23 ka, Lisan Lake reached its maximum elevation of about 164 m bmsl and probably reached a total length of about 200 km and a much larger width, as indicated by the Late Pleistocene lacustrine sediments that surround the depression (Abu Ghazleh & Kempe, 2009; Bartov et al., 2002). Lake Lisan has significantly decreased since the last glaciation, with a mean lake level of around 400 m bmsl during most of the Holocene (Bartov et al., 2002). Late Pleistocene lacustrine sediments are exposed in the valleys of the Amiaz Plain, located west of Mount Sedom (e.g., Bartov et al., 2006).

Currently, only the northern Dead Sea is a deep lake, while the southern basin has been desiccated and the Sea of Galilee is covered by shallow water (Figure 2.15). The Sea of Galilee (Lake Kinneret) is characterized by a higher lake level (−213 m) and a maximum water depth of about 50 m (Eppelbaum et al., 2007). At present (2022), the Dead Sea Lake is characterized by a surface level of 437 m bmsl, and due to intense evaporation and agricultural use, the lake level has been falling at a rate of 1 m/yr in the past decades (Bookman et al., 2004, 2006; Bartov et al., 2006).

The Dead Sea has experienced dramatic changes in the 20th century due to the increased use of Jordan River water for irrigation systems. Until the 1970s, there were two water bodies connected via a narrow passage west of the Lisan Peninsula. A photographic comparison of three snapshots of the Dead Sea in recent decades shows the shrinking water surface and disappearance of the connection (Figure 2.16). In the 2000s, the northern and southern basins were separated by a sill 28 m higher than the northern lake surface (Figure 2.15). However, in the

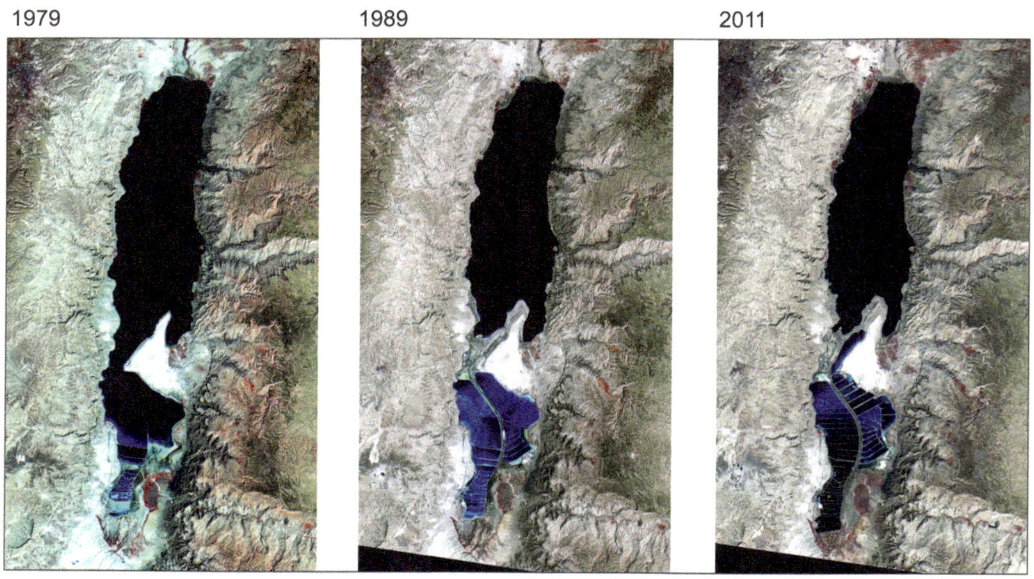

Figure 2.16 Sequential satellite images of the Dead Sea region show the changes in the lake's geomorphology from 1979 to 2011. Observe that since the 1970s, the water communication between the northern (shown with deep blue colors) and southern (lighter blue colors) basins has been greatly reduced. This is a consequence of the lake level fall, which at present is declining at a rate of about 1 m per year. Credit: Robert Simmon/NASA/Public domain.

present decade, the difference in altitude between the sill and lake level is about 35 m. The lowering water level has exposed the lakebed on the Lisan Peninsula. At present, the peninsula corresponds to a strip of dry land that, in satellite images, forms a white patch with a triangular shape (Figure 2.16).

2.4.2. Evaporite Sedimentation in the Dead Sea

The Dead Sea northern lake is a modern analog for halite precipitation in deep water-deep basin settings (Kiro et al., 2016; Sirota et al., 2016, 2017, 2018; Torfstein, 2008; Warren, 2006, 2016). The deepest parts of the lake accumulated thick deposits of halite minerals, which precipitated both at the air-water interface and on the lake's bottom due to the supersaturation of the brine column with sodium chloride (Sirota et al., 2016). The first direct observation of halite minerals precipitating on the Dead Sea floor in the late 1970s indicated that the crystals were controlled by seasonal temperature variations, thermohaline stratification, and halite saturation, resulting in deposits of layered halite with different textures (Steinhorn, 1983, 1985; Sirota et al., 2016, 2018).

Due to variations in lake level, the Late Pleistocene to Holocene sedimentary deposits in the lake center are thickened by the dissolution of the halite deposits in the shallow parts of the lake (e.g., Kiro et al., 2017). The cyclicity of the evaporite deposits also reflects seasonal variations, as the salt layers are thinned in the shallow basin margins by intensive dissolution during the summer, when the warmer brine dissolves a larger amount of sodium chloride in the epilimnion (e.g., Sirota, 2020).

The Late Pleistocene to Holocene salt deposits recorded in scientific boreholes drilled in the Northern Lake correspond to a fraction of the original evaporite layer that accumulated in the Dead Sea Basin. The salt deposits formed during the late Miocene to early Pliocene marine incursion that connected the Mediterranean Sea to the Dead Sea Basin (Torfstein, 2008; Zak, 1967, 1974, 1997). The rapid episodes of salt deposition resulted from the disconnection of the lagoon from the Mediterranean Sea, which resulted in a hypersaline lacustrine system. The desiccation episodes accumulated thick halite beds on the basin's floor, forming evaporitic cycles associated with the Sedom formation (Figure 2.17).

Based on the restoration of the salt thickness in the Sedom salt diapir, Ben-Avraham and Katsman (2015) estimated the original salt layer covering the basin floor to reach a thickness of about 2.3 km. According to seismic interpretation and exploratory boreholes (e.g., Ben-Avraham & Katsman, 2015; Kashai & Croker, 1987), the original salt deposits in the basin depocenter reached a thickness of about 2 km. Other authors, assuming that evaporite deposition might be five times faster than the accumulation of siliciclastic sediments, postulated that a 2 km halite layer could be deposited within less than 0.2 Ma (Garfunkel, 1997).

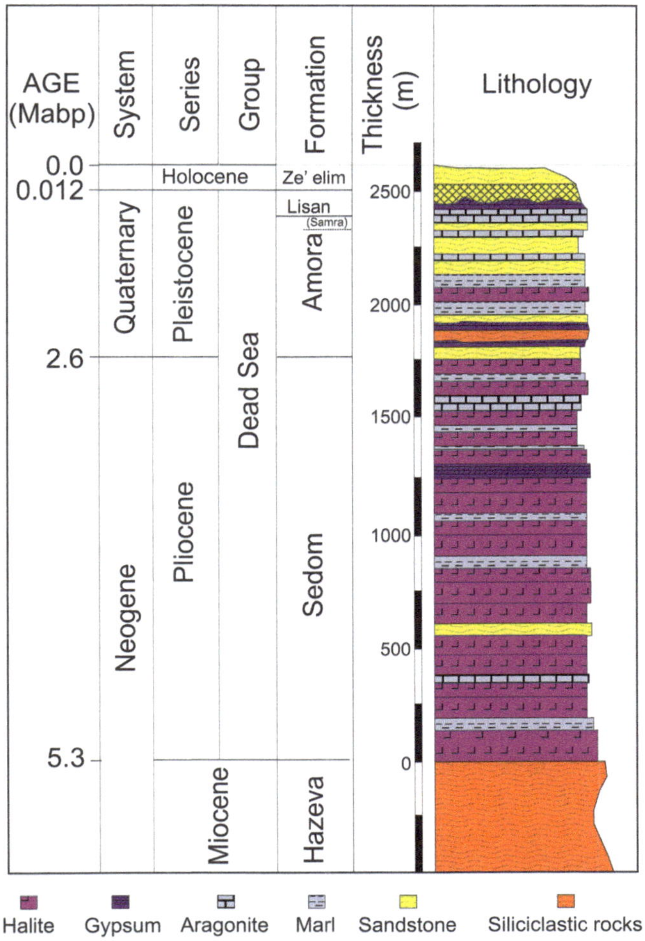

Figure 2.17 Simplified lithostratigraphic column of the upper Neogene and Quaternary sedimentary successions in the Dead Sea. The Pliocene is characterized by a thick sequence of evaporite rocks, predominantly halite, and minor intercalations of siliciclastic deposits. Lacustrine deposits are mainly siliciclastic rocks (marl), with intercalations of chemical precipitates (aragonite, gypsum, and halite) resulting from brine concentration during intervals of lake-level decline. Credit: Adapted from Zak (1967).

The Pliocene Sedom Fm. evaporite cycles are formed by halite layers intercalated with carbonate and siliciclastic sequences (Sirota, 2020; Zak, 1967, 1974, 1997). However, during the Late Pleistocene and Holocene, most of the lacustrine deposits corresponded to carbonate and siliciclastic sediments intercalated with rather thin layers of evaporite sediments, up to about 10 m in thickness, as indicated by the scientific borehole drilled in the northern lake (Coianiz et al., 2019, 2020; Garber et al., 1987; Neugebauer et al., 2014). Boreholes drilled to investigate the salt layer precipitated around 11–10 ka on the Dead Sea paleolake's floor revealed a maximum thickness of 23 m (Baer et al., 2023).

The generalized stratigraphic column of the Dead Sea region (Figure 2.17) is based on geological information from outcrops near the Mount Sedom salt diapir (Torfstein, 2008; Zak, 1967, 1974, 1997). The pre-salt deposits from the Miocene are made up of siliciclastic rocks from the Hazeva Group, with a maximum sedimentary thickness estimated at about 2,500 m (Calvo & Bartov, 2001). The upper part of the Hazeva Group was deposited during tectonic activity in the basin, with Miocene sediments overlying unconformably the Cretaceous to Eocene rock units that are exposed today on land (Calvo & Bartov, 2001).

The syn-rift sediments were abruptly overlain by the evaporite rocks. The Pliocene salt sequences (Sedom Fm.) are overlain by Pleistocene lacustrine sediments associated with the Amora Fm., which spans an interval of almost 3 million years and reaches a thickness of several kilometers (Torfstein, 2017). This paleogeography is indicated by the lacustrine sediments deposited to the west of the Sedom salt diapir, located close to the present-day salt evaporation ponds in the southern basin (Figure 2.18).

Loading of the Sedom Fm. salt deposits with the Amora and Lisan Fm. sedimentary sequences resulted in the deformation of the original mother salt layer (Alsop et al., 2015, 2016; Zak & Freund, 1980, 1981). Several salt structures are observed at the surface near the lake's margins and identified in the subsurface by geophysical methods (e.g., Al-Zoubi & ten Brink, 2001; Al-Zoubi et al., 2002; Kashai & Croker, 1987). The salt column known as "Lot's wife" is a remnant outcrop of the original Sedom Fm. salt deposits (see Chapter 1). Based on the displacements measured in the salt mirrors observed in the region, there is evidence that the Sedom salt diapir has been significantly uplifted during the Quaternary (Frumkin, 1996; Weinberger et al., 2006).

2.4.3. Dead Sea Lake Level Fluctuations

Lake Lisan, the largest and most recent paleolake in the Dead Sea depression, formed after the end of the last ice age during the transition from the Pleistocene to the Holocene (Stein, 2001, 2014). The Lisan paleolake extended across the Arava and Jordan River valleys and reached the Sea of Galilee in the north (Figure 2.15). From 70 to 12 ka, Lisan Lake's level was more than 100 m above the current Dead Sea level

Figure 2.18 A detailed image of the Dead Sea's southern basin shows the saltworks and the Mount Sedom salt diapir, which runs parallel to the southwestern border of the artificial ponds. The view is toward the southwest. Beyond the topographic high, the Amiaz Plain valleys are characterized by outcrops of the lacustrine sediments of the precursory Amora and Lisan lakes, deposited during the Pleistocene. Credit: Courtesy of A. Tsabar.

(Bartov et al., 2002; Bookman et al., 2006; Waldmann et al., 2009).

Limnological studies show that the water bodies expanded and contracted according to the basin's physiography and changes in water balance (e.g., Lensky et al., 2013; Stein, 2001; Torfstein, 2008). These changes depend on factors such as tectonics and global climate variations, particularly those affecting the Northern Hemisphere (Torfstein et al., 2013; Waldmann et al., 2009).

Lisan Lake expanded during glacial episodes, resulting in more wet climates (Bartov et al., 2002; Stein et al., 2010). The paleolake reached its maximum extent at around 26 ka, when the water level was around 160–170 m below the global mean sea level (bmsl), corresponding to more than 250 m above the present-day lake surface (Figure 2.19). During interglacial periods, which were drier, the lake contracted to smaller water bodies (Bartov et al., 2002; Bookman et al., 2006; Stein, 2001; Stein et al., 2010; Torfstein et al., 2013, 2015). During the interval 60–14 ka, at least 17 lake-level maxima are observed, coinciding with Greenland warming events (Torfstein et al., 2013).

The Dead Sea Lake's water level fluctuated dramatically in the Late Pleistocene (Figure 2.19) due to climate warming, melting glaciers in the northern hemisphere, and increasing global sea levels (Bookman et al., 2006; Müller et al., 2022; Stein et al., 2010; Torfstein et al., 2013). The lake reached a maximum of about 160 m bmsl between 30 and 20 ka, approximately 240 m above typical Holocene levels of about 400 m bmsl. Lacustrine sediments associated with the precursory Lake Lisan can be seen today outcropping in valleys of the Amiaz Plain, several tens of meters above the modern Dead Sea lake level (Abu Ghazleh & Kempe, 2009).

A large drop in lake level is observed between 15 and 10 ka, likely coincident with the time humankind started to populate the lake's shores. There is archaeological evidence for cereal cultivation and goat herding between 8,000 and 6,000 years BCE. (Before Common Era), particularly in the region surrounding Jericho (Neumann & Zwickel, 2019). The beginning of farming activity coincided with more humid climatic conditions and high lake levels (Figure 2.19).

During the Bronze and Iron Ages (ca. 5500–2500 BCE), the occupation of the Dead Sea region was associated with abrupt variations in the lake level. During the Middle Bronze Age, the lake reached an elevation of ~370 m bmsl, but the lake surface declined to about 414 m bmsl during the Late Bronze Age (Kagan et al., 2015). The climate became increasingly arid, and the lake level reached a minimum around 1400 BCE. (Neumann & Zwickel, 2019). The periods of maximum climate aridity in the Late Bronze Age coincided with the collapse of some Eastern Mediterranean civilizations, leading to drought intervals that affected the Middle East and southern Europe regions for several decades (Kaniewski et al., 2015).

During the transition from the Late Bronze to the Early Iron Age, the Dead Sea Lake recovered a water level of about 408 m bmsl, which affected freshwater activity in

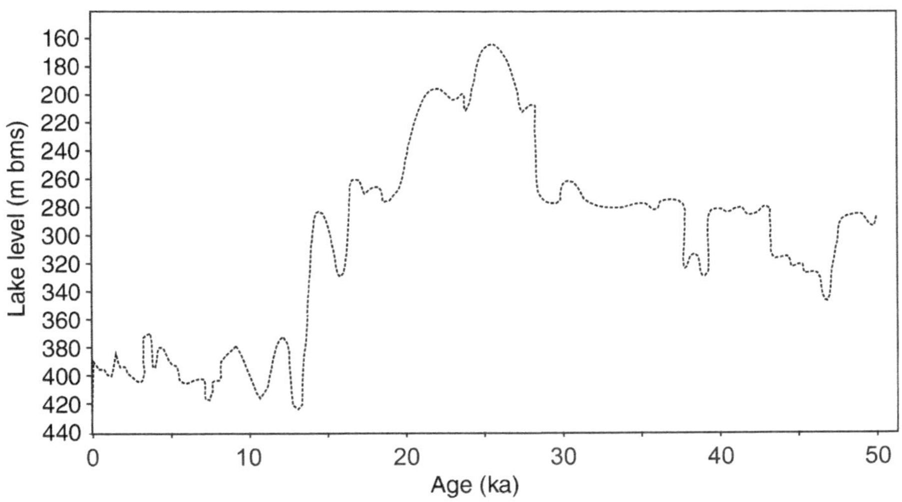

Figure 2.19 The Dead Sea lake level fluctuated over the past 50 years, reaching 400 m below mean sea level in the Holocene. The highest level was 160 m between 30 and 20 ka, during the largest lake extension in the late Pleistocene. During the maximum expansion of the lake, the water level was more than 200 m higher than it is today, depositing lacustrine sediments on the Amiaz Plain. The Holocene began after the Younger Dryas, around 12 ka, resulting in the formation of the Dead Sea Lake, which has shrunk to its present water level of circa 437 m bmsl. Credit: Modified from Torfstein (2008) and Stein et al. (2010).

the Judean Hills, especially in the oases of En-Gedi and Jericho. During the Holocene, favorable hydrological and environmental conditions facilitated a wave of Iron Age settlements in the region around the Dead Sea (Kagan et al., 2015). The lake-level fluctuations in the Late Pleistocene and Holocene have been analyzed by researchers using diverse methods in hydrological investigations (e.g., Lisker et al., 2007; Lu et al., 2021; Torfstein, 2013, 2015; Waldmann et al., 2009). Some authors base their interpretation on biblical records that have been scientifically calibrated by radiocarbon-dated geological and archaeological evidence (Frumkin & Elitzur, 2002).

Based on the biblical descriptions of the Dead Sea region, there are three historically documented phases of the Dead Sea lake levels. The first corresponds to the low-lake levels in the interval 2000–1500 BCE; the second to the high-lake levels between ca. 1000 and 700 BCE; and the third to the low-lake levels between ca. 1000 and 700 BCE (Frumkin & Elitzur, 2002). The historical records indicate that during the driest intervals, the southern basin was completely desiccated (Stein et al., 2010).

The trend of falling lake levels continued until the 1970s, when a narrow passage between the northern and southern lakes, to the west of the Lisan Peninsula, could still be seen in satellite images (Figure 2.16). In the last two decades of the 20th century, the Dead Sea Lake started to shrink rapidly due to climatic causes and anthropogenic intervention (Gavrieli & Oren, 2004). In 1980, the water level was measured at 400 m bmsl, and the lake surface area was only 680 km^2. Currently, the lake level is falling at a rate of about 1 m/year (Lensky et al., 2005).

In 2021, the lake's surface was at −436 m, and the surface area was reduced to about 600 km^2, shrinking by more than 10% in four decades. Although climate changes and overexploitation of natural resources threaten the Dead Sea, some researchers postulate that the water level will fall substantially in the next decades but the lake will stabilize at a lower level (Gertman, 2012; Yechieliet et al., 1998), or a controversial conveyance project will be implemented in the future, aiming to recover the lake level by linking the Red Sea with the Dead Sea (e.g., Abu-Qubu et al., 2022).

2.4.4. Salinity Variations in the Dead Sea

The northern Dead Sea Lake's brines are characterized by a unique Ca-Chloride chemical composition and a salinity of about 340 g/l (~276 g/kg), which corresponds to a concentration almost nine times the standard ocean salinity (Gavrieli & Stein, 2006). The salinity of the Dead Sea has changed dramatically since the last century, with changes in evaporite deposition associated with lake stratification (Figure 2.20).

In the 1950s and until the late 1970s, the brine was not as concentrated as it is today, and the brine density at the epilimnion was about 1.205 g/cc in 1959 (Figure 2.20A). Halite crystals were not seen precipitating from the Dead Sea water until the 1970s, with only gypsum accumulating on the lake's bottom due to lower brine concentrations (Warren, 2006, 2016). Neev and Emery (1967), in a comprehensive study of the Dead Sea water chemistry, described the precipitation of gypsum crystals during the 1960s as a rain-like mass of white snow-like flakes that slowly descended from the epilimnion to accumulate at the lake bottom. These crystals formed a thin white blanket that covered the black carbonate mud in deep water. During the late 1970s, the falling of the lake level and the rapid increase in brine concentration resulted in the precipitation of halite crystals on the lake's bottom for the first time in hundreds of years (Gertman, 2012; Steinhorn, 1983, 1985).

At present, in the 21st century, halite can be seen accumulating on the lake's bottom following a seasonal cycle (Sirota et al., 2016, 2017, 2018). Until the late 1970s, the Dead Sea was considered one of the best examples of thermally stratified lakes in the world. The waters were meromictic, or "permanently" stratified, and had been so since the 16th century. The topmost 35 m of the water column (epilimnion) had an average density of 1.205 g/cc. Below the thermocline and pycnocline, the hypolimnion temperature was a constant 21.5 °C and the brine density was 1.232 g/cc (Figure 2.20A). The lake was stratified by dilution, resulting in a meromictic lake with density and thermal stratification, with a less salty upper layer exhibiting a lower density than the deeper layers (Anati et al., 1987).

The lake hydrology changed substantially in the late 1970s. Following a rapid lowering of the lake surface level by almost 3.5–4 m in early 1979, a long period of stratification ended with the overturn of the lake water during the winter. The upper and bottom water layers were mixed and replaced by homogeneous water with almost no difference in salinity, resulting in a holomictic lake (Steinhorn et al., 1979; Steinhorn, 1985). The upper and bottom water layers were mixed and replaced by homogeneous water with almost no difference in salinity, resulting in a holomictic lake. The abrupt changes in lake conditions were mainly caused by human activity, such as diverting the Jordan River to irrigate farms in Jordan, Israel, and Syria, and the exploitation of salt minerals by the Dead Sea Works (Gertman & Hecht, 2002; Gertman, 2012; Lensky et al., 2005; Levy & Gvirtzman, 2021; Sirota, 2020; Torfstein, 2008).

The explanation for the lake overturn phenomenon in the Dead Sea is based on limnological studies (e.g., Gertman, 2012; Warren, 2016). In 1975, due to increased evaporation and the extreme brine concentration of the epilimnion, the salinity of the superficial layer increased

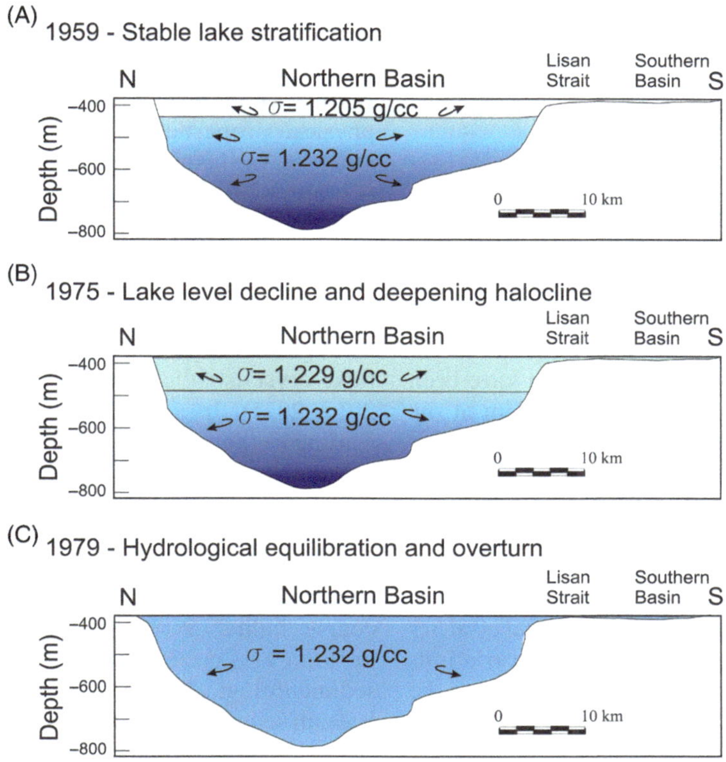

Figure 2.20 Dead Sea brine densities in the second half of the 20th century (1959, 1975, and 1979). The lake was a stratified water mass for centuries, but in 1979, an overturning episode was recorded in the lake. A very cold winter cooled the epilimnion waters, increasing the density of the upper layer and deepening the halocline. The overturning of the bottom waters resulted in mixing with the upper layers, releasing gases from the anoxic lake bottom. The change in water stratification resulted in the formation of a holomictic lake and the precipitation of halite throughout the water column. Credit: Adapted from Warren (2006).

to about 1.229 g/cc and the pycnocline deepened to about 100 m below the lake surface (Figure 2.20B). However, the temperature in the superficial water was strongly reduced by a very cold winter in 1979, causing the upper water layer to become denser and move toward the bottom of the lake, mixing with the lower water mass in the hypolimnion (Figure 2.20C).

The overturn event mixed the bottom waters with the superficial waters. As a result, the densities of the upper and lower water masses equalized, with a mean density of around 1.232 g/cc. For the first time after many centuries, a lake overturn was recorded and witnessed by Earth scientists working in the Dead Sea (compare with Figure 2.11). In 1979, halite crystals started to precipitate on the lake's surface and accumulate on the lake floor, which for centuries had predominantly black muddy sediments deposited in an anoxic environment. As the bottom waters came to the surface during the overturn, the gases brought large quantities of H_2S that were previously entrapped in the sediments. When these gases were continuously released into the atmosphere, a strong smell of rotten eggs persisted for several weeks around the shores of the Dead Sea's northern lake (Gertman, 2012; Steinhorn, 1983, 1985; Warren, 2016).

As the Dead Sea became a holomictic lake with a homogeneous body of water with no thermal or chemical stratification, several consequences resulted from the phenomenon. Measurements of the brine salinity revealed a total salinity of about 276 g/kg from the surface to the bottom of the water column. The lake was holomictic only during the three months of winter, while during the summer, the superficial layers were heated and developed a thermal stratification (Steinhorn, 1985).

Since the overturn episode in 1979, there have been several other episodes of mixing the denser bottom waters with the less dense waters of the upper layer. As a consequence, the Dead Sea in the past decades has continually changed states from a meromictic to a mostly holomictic lake during the cold season, indicating hydrological instability and resulting in widespread deposition of halite minerals (Arnon et al., 2019; Gertman & Hecht, 2002; Gertman, 2012; Lensky et al., 2005; Sirota, 2020).

2.4.5. Modern Observations of Halite Precipitation in the Dead Sea

The Dead Sea Lake was famous for its "white rain" of microscopic flakes of aragonite and gypsum (Garber et al., 1987; Neev & Emery, 1967; Warren, 2006, 2016). In recent years, the lake has been studied extensively to understand the factors controlling halite precipitation and accumulation (e.g., Lazar & Ben-Avraham, 2002; Sirota, 2020). Since the 1980s, halite minerals have been observed to precipitate at the margins and bottom of the northern lake, indicating changes in the hydrological systems and evaporite deposition cycles (Gertman, 2012; Sirota et al., 2016, 2017, 2021; Warren, 2016).

The brine chemistry in the Dead Sea Lake has been extensively studied to determine the factors controlling halite precipitation and accumulation (e.g., Arnon et al., 2016, 2019; Lensky et al., 2013). An annual cyclicity has been observed in the lake, as expressed by variations in physical properties measured in the lake brine over the past decades (Figure 2.21).

The largest variations in water temperature, salinity, and density occur in autumn and summer, when the Dead Sea is a holomictic lake but thermally stratified. During winter, the lake is a holomictic lake with uniform physical properties (Figure 2.22). Sediments accumulating on the lake's bottom also exhibit seasonal cyclicity (Figure 2.23), with climatic control on the morphology of halite crystals resulting in sedimentary deposits with different textures (e.g., Arnon et al., 2016, 2019; Sirota et al., 2016, 2017, 2018).

Halite accumulation in the Dead Sea is not uniform throughout the year, with winter resulting in small snowflake crystals called *cumulates* (e.g., Sirota et al., 2016, 2017, 2021). In summer, the water column is warmer, leading to the undersaturation of brine in the epilimnion, causing the dissolution of salt deposits and concurrent hypolimnetic precipitation (Arnon et al., 2016, 2019; Sirota et al., 2016, 2017; Sirota, 2021). The process of halite dissolution involves the dissolution of halite deposits, which are transported basinward, forming coarse halite crystals that accumulate on the lake's bottom. This cyclicity in halite deposits is reflected in the sedimentological analysis of cores recovered from lake-bottom sediments (Sirota, 2020; Sirota et al., 2021).

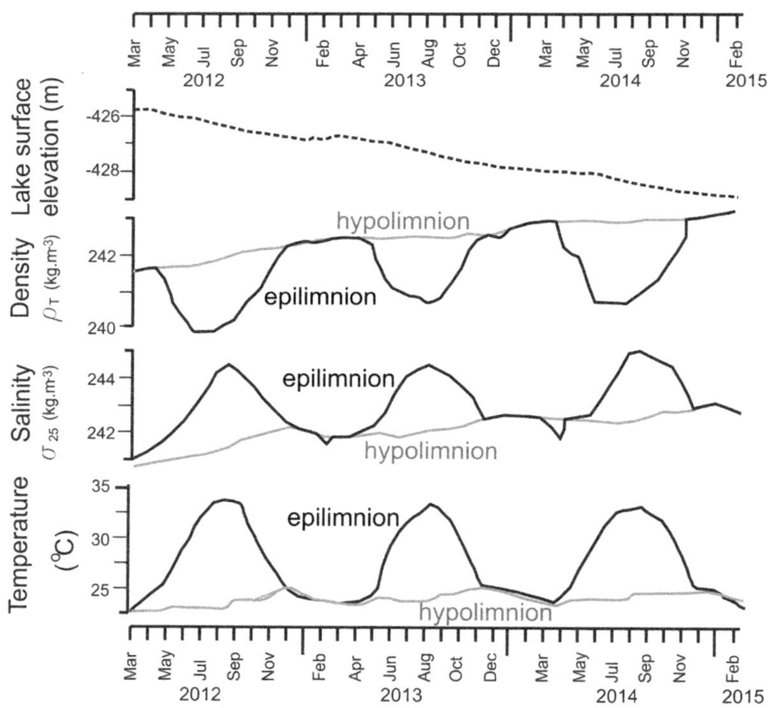

Figure 2.21 Seasonal variations of Dead Sea Lake surface, density, salinity, and temperature. Summer has high temperatures and high salinity, while winter has lower temperatures, decreased salinity, and increased density. Hypolimnion variations are smaller, as the bottom layers are less affected by variations in the superficial water layers. Credit: Adapted from Arnon et al. (2016).

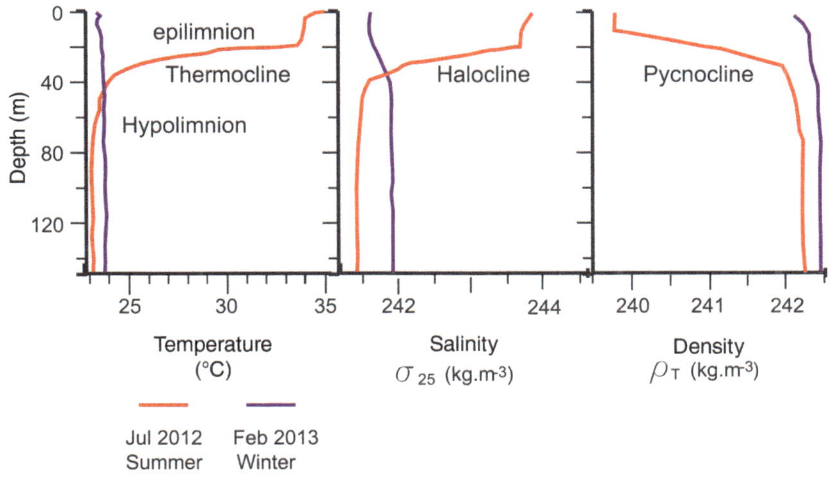

Figure 2.22 The Dead Sea northern lake exhibits contrasting temperature, salinity, and density profiles during summer and winter. There are small variations in the physical properties of the water column during winter, as the lake is holomictic. During summer, the lake is meromictic and characterized by a well-developed thermocline, halocline, and pycnocline. Credit: Adapted from Arnon et al. (2016).

Arnon et al. (2016, 2019) discussed the influence of annual variations in the lacustrine environment, including solar radiation, which influences lake level, water temperature, salinity, and density. The Dead Sea brines are saturated in autumn and winter when temperatures reach supersaturation, depositing halite on the lake's bottom (Figure 2.22). Conversely, when temperatures increase from April to October, the waters are undersaturated for halite, resulting in the dissolution of previously deposited layers in the epilimnion and halite crystal growth (Figure 2.23). The sizes of halite crystals deposited in winter are very small, up to 1 mm, whereas in summer, the crystal sizes vary from 3 to 10 mm (Sirota et al., 2017).

Sirota et al. (2017) provided one of the few accounts of direct, in situ observations of halite precipitation and dissolution in both shallow- and deepwater settings of the Dead Sea using underwater cameras. The investigation combined hydrologic and limnologic observations and textural characterization of chemical precipitates, mainly represented by halite deposits. Although minor amounts of sulfate minerals are expected to precipitate in the lake, gypsum precipitation has not been documented so far (Sirota et al., 2017, 2021; Sirota, 2020).

According to the observations, halite deposition in the northern lake is associated with seasonal alternations of climate and brine saturation, leading to distinct crystallization mechanisms (Sirota et al., 2017, 2018; 2021; Sirota, 2020). The temperature variations from winter, spring, summer, and autumn control the salinity concentrations in the lake brines, resulting in the sequential development of salt precipitation and dissolution. Figure 2.24 is a summary of these observations, showing the seasonal variations of halite precipitation as compared with Figures 2.24, 2.25, and 2.26 (e.g., Arnon et al., 2016, 2019).

In summer, when temperatures in the epilimnion are higher than 30° C, the degree of halite saturation is reduced as warmer brines can dissolve a larger number of ions. This process results in the dissolution of halite layers in the epilimnion. This process is transferred to the hypolimnion via double diffusion salt fingering, resulting in the precipitation of halite crystals (Arnon et al., 2016, 2019). The hypolimnion, on the other hand, experiences lower temperatures and has a higher brine concentration, leading to the precipitation of coarse (1–10 mm) and well-developed cubic halite crystals. These crystals form metric-scale mounds with an increasingly rough and undulating surface on the lake floor. The halite crystals are formed in the hypolimnion under a slight supersaturation, with an average degree of halite saturation of about 0.3 kg/m^3 (Sirota et al., 2016, 2017; Sirota, 2020).

During winter, the temperature in the epilimnion approaches the hypolimnion (less than 25° C), causing a greater degree of halite saturation (ranging from 0.45 to 0.65 kg/m^3) in the cooled lake brine. This is due to the reduced capacity of the colder brine to maintain the dissolved halite ions. Halite precipitates as aggregates of fine-grained crystals (< 0.1 mm), resembling a rain of white snowflakes falling on the lake floor. These crystals settle in a holomictic water column, forming a smooth depositional surface under high supersaturation of the lake brine (Sirota et al., 2016, 2017, 2018; Sirota, 2020).

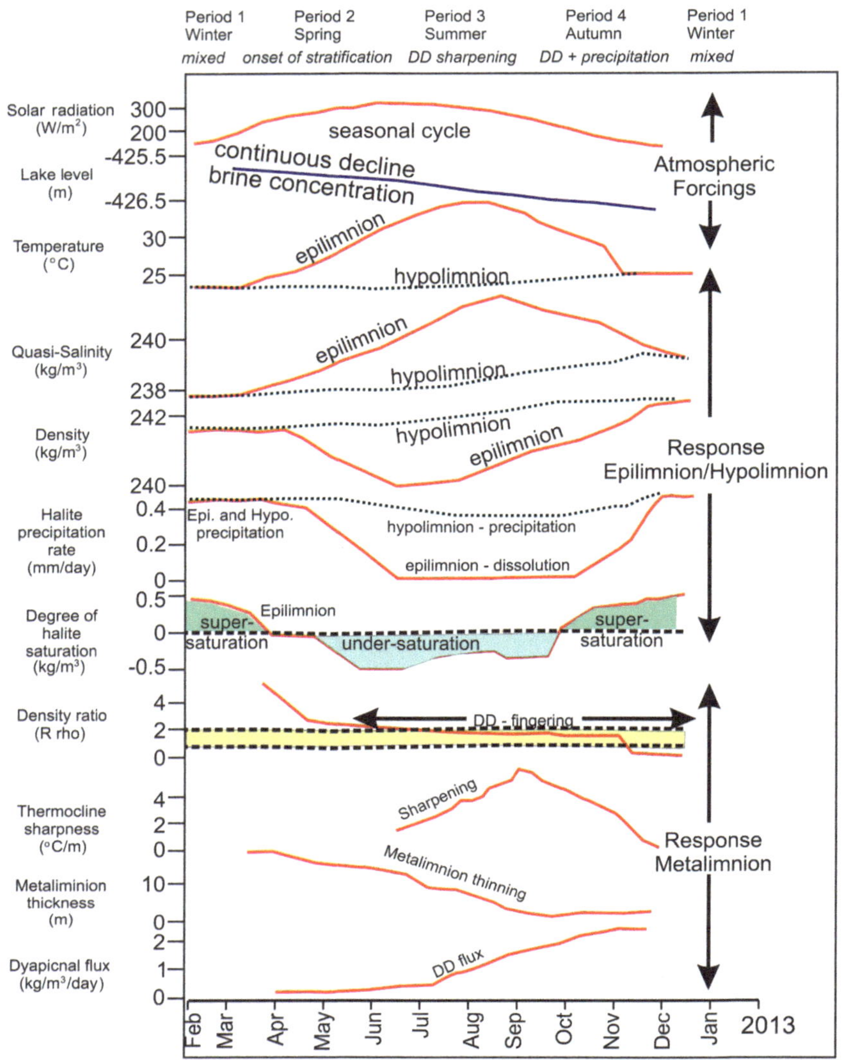

Figure 2.23 Seasonal salinity variations in the Dead Sea cause halite deposition cyclicity. Winter accumulates salt layers, while summer temperatures undersaturate the lake. Supersaturation occurs between winter and spring, causing massive precipitation of small crystals and forming a white carpet on the Dead Sea floor. This annual cyclicity is observed in sediment cores that recovered thin layers of evaporite sediments that accumulated on the lake's bottom. Credit: Arnon et al. (2016)/with permission of John Wiley & Sons.

Based on the direct observations of halite deposition in the Dead Sea and numerical models, several authors (e.g., Arnon et al., 2016; Ouillon et al., 2019) have proposed a mechanism for precipitating halite. The process involves the mixing of epilimnion and hypolimnion brines in a process called "double-diffusive salt fingers." The upper layer of the lake is warmer and saltier than the lower layers due to intense evaporation and its greater capacity for retaining halite ions in the solution. This layer overlies the colder, less salty brine in the hypolimnion, which is also saturated with halite. These more saturated upper layers form fingers of overconcentrated brines that cross the interface between the lake's layers, resulting in the mixing of less dense brines with denser brines.

A unique asymmetry of brine flow is observed in the Dead Sea. The descending, cooling fingers become supersaturated and precipitate halite, while ascending warming fingers are undersaturated in the warmer layer of a stratified lake (Arnon et al., 2016). The double-diffusive instability originates in the metalimnion, the interface between the two layers (Figure 2.24). This results in a constant and strong downward salinity flux, with the upper layers becoming less salty and the hypolimnion becoming oversaturated and precipitating halite. The halite focusing

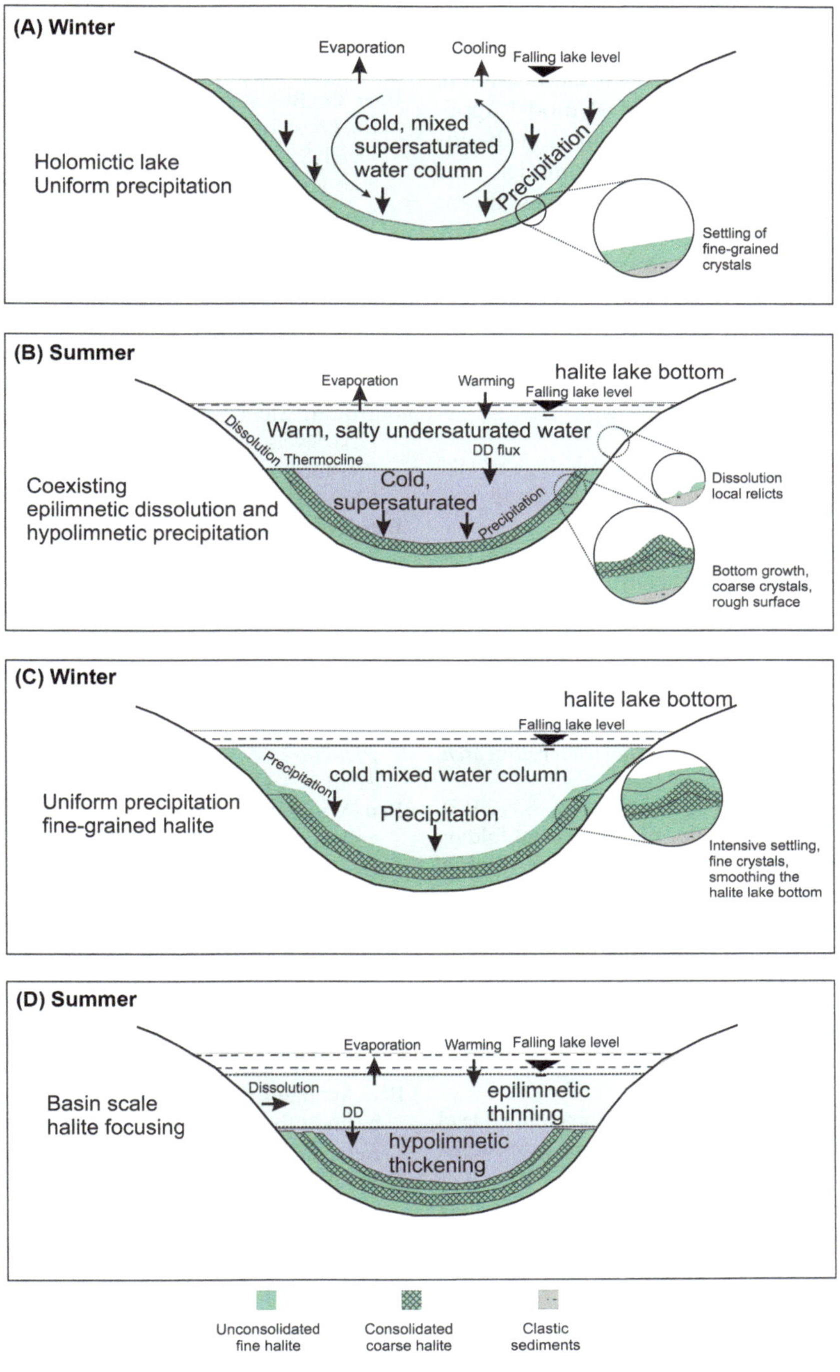

Figure 2.24 The Dead Sea experiences seasonal salt precipitation due to climatic changes. Wintertime sees supersaturation of halite crystals, while summertime sees undersaturation and dissolution of previously deposited evaporites. Large halite crystals that have thickened due to hypolimnetic processes precipitate at the lake's bottom. Credit: Arnon et al. (2016)/with permission of John Wiley & Sons.

mechanism by double diffusion results in the hypolimnion thickening, forming thick salt layers that accumulate on the lake's bottom (Ouillon et al., 2019).

The warm, undersaturated epilimnion is characterized by halite dissolution, resulting in the epilimnetic thinning at the lake's margin (Figure 2.24D). The net result is a constant and strong downward salinity flux, with the upper layers becoming less salty and the hypolimnion becoming oversaturated and precipitating halite. The halite focusing mechanism by double diffusion results in hypolimnion thickening, forming thick salt layers that accumulate on the lake's bottom (Figure 2.24D).

This explanation for salt deposition and accumulation in the Dead Sea is the only sedimentological model applicable to thick halite deposits recorded in giant saline basins worldwide. The basin desiccation and sabkha models were initially applied to explain salt deposits in the largest basins worldwide. However, these models have been increasingly challenged by observations and numerical simulations conducted by earth scientists in the present century (see Chapter 8).

REFERENCES

Abu Ghazleh, S., & Kempe, S. (2009). Geomorphology of lake Lisan terraces along the eastern coast of the Dead Sea, *Jordan. Geomorphology*, *108*(3–4), 246–263.

Abu-Qubu, J., Rimawi, O., Anbar, A., Alebous, T., & Abu-Hamatteh, Z. (2022). Hydrochemistry of the mixed Dead Sea–Red Sea water under different impoundment scenarios as a time dependent state. *Journal of Geoscience and Environment Protection*, *10*, 94–108. doi: 10.4236/gep.2022.101007

Agnon, A. (2005). The late quaternary limnological history of lake Kinneret (Sea of Galilee), Israel. *Quaternary Research*, *63*, 60–77. doi:10.1016/j.yqres.2004.09.004

Alsop, G.I., Weinberger, R., Levi, T., & Marco, S. (2015). Deformation within an exposed salt wall: Recumbent folding and extrusion of evaporites in the Dead Sea Basin. *Journal of Structural Geology*, *70*, 95–118.

Alsop, G. I., Weinberger, R., Levi, T., & Marco, S. (2016). Cycles of passive versus active diapirism recorded along an exposed salt wall. *Journal of Structural Geology*, *84*, 47–67.

Al-Zoubi, A., Shulman, H., & Ben-Avraham, Z. (2002). Seismic reflection profiles across the southern Dead Sea basin. *Tectonophysics*, *346*, 61–69.

Al-Zoubi, A., & ten Brink, U. S. (2001). Salt diapirs in the Dead Sea basin and their relationship to Quaternary extensional tectonics. *Marine and Petroleum Geology*, *18*, 779–797.

Amante, C., & Eakins, B. W. (2009). *ETOPO1 1 arc-minute global relief model: procedures, data sources and analysis*. NOAA Technical Memorandum NESDIS NGDC-24, National Geophysical Data Center Marine Geology and Geophysics Division, Boulder, Colorado, March 2009, 19 pp.

Anati, D. A. Stiller, M., Shasha, S., & Gat, J. R. (1987). Changes in the thermo-haline structure of the Dead Sea: 1979-1984. *Earth and Planetary Science Letters*, *84*, 109–121.

Arnon, A., Brenner, S., Selker, J. S., Gertman, I., & Lensky, N. G. (2019). Seasonal dynamics of internal waves governed by stratification stability and wind: Analysis of high-resolution observations from the Dead Sea. *Limnology and Oceanography*, *64*(5), 1864–1882. doi: 10.1002/lno.11156

Arnon, A., Selker, J. S., & Lensky, N. G. (2016). Thermohaline stratification and double diffusion diapycnal fluxes in the hypersaline Dead Sea. *Limnology and Oceanography*, *61*(4), 1214–1231, doi: 10.1002/lno.10285

Babel, M., & Schreiber, B. C. (2014). Geochemistry of evaporites and evolution of seawater. *Treatise on Geochemistry* (2nd ed., Chapter 9.17, pp. 483–560). Amsterdam: Elsevier.

Baer, G., Bernstein, M., Yechieli, Y. et al. (2023). Elevation and thickness of the 11–10 Kyr old 'Sinkholes Salt' layer in the Dead Sea: Clues to past limnology, paleo-bathymetry and lake levels. *Journal of Paleolimnology*, *70*, 159–173. https://doi.org/10.1007/s10933-023-00286-9

Bartov, Y., Bookman, R., & Enzel, Y. (2006). Current depositional environments at the Dead Sea margins as indicators of past lake levels. In Y. Enzel, A. Agnon, & M. Stein (Eds.), *New Frontiers in Dead Sea Paleoenvironmental research* (Vol. *401*, pp. 127–140). Geological Society of America, Special Paper. doi:10.1130/2006.2401(08)

Bartov, Y., Stein, M., Enzel, Y., Agnon, A., & Reches, Z. (2002). Lake levels and sequence stratigraphy of Lake Lisan, the late Pleistocene precursor of the Dead Sea, *Quaternary Research*, *57*, 9–21.

Ben-Avraham, Z. (1997). Geophysical framework of the Dead Sea: Structure and tectonic. In T. M. Niemi, Z. Ben-Avraham, & J. R. Gat (Eds.), *The Dead Sea: The lake and its settings, Oxford Monographs on Geology and Geophysics* (Vol. *36*, pp. 22–35). New York: Oxford University Press.

Ben-Avraham, Z., Garfunkel, Z., & Lazar, M. (2008). Geology and evolution of the southern Dead Sea fault with emphasis on subsurface structure. *Annual Review of Earth and Planetary Sciences*, *36*, 357–387. doi:10.1146/annurev.earth.36.031207.124201

Ben-Avraham, Z., & Katsman, R. (2015). The formation of graben morphology inthe Dead Sea Fault, and its implications. *Geophysical Research Letters*, *42*, 6989–6996. doi:10.1002/2015GL065111

Ben-Avraham, Z., Lazar, M., Garfunkel, Z., Reshef, M., Ginzburg, A., Rotstein, Y., et al. (2012), Structural styles along the Dead Sea fault. In D. G. Roberts & A. W. Bally (Eds.), *Regional Geology and Tectonics: Phanerozoic rift systems and sedimentary basins* (Vol. 1B, pp. 617–633). doi:10.1016/B978-0-444-56357-6.00016-0

Ben-Avraham, Z., & Schubert, G. (2006). Deep "drop down" basin in the southern Dead Sea. *Earth and Planetary Science Letters*, *251*, 254–263. doi:10.1016/j.epsl.2006.09.008

Beyth, M., Gertman, I., Weinstein, R., & Gavrieli, I. (1998). End-brine mixing in the Dead Sea. *Geological Survey of Israel*, Report GSI/30/98, 23pp.

Bischof, G. (1854). *Elements of chemical and physical geology* (Vol. *1*), tr. B. H. Paul & J. Drummond. London: The Cavendish Society, Harrison & Sons, St. Martin's Lane, 455pp.

Boehrer, B., & Schultze, M. (2008). Stratification of lake. *Reviews of Geophysics*, *46*(2), 1–27. https://doi.org/10.1029/2006RG000210

Bookman, R., Bartov, Y., Enzel, Y., & Stein, M. (2006). Quaternary lake levels in the Dead Sea basin: Two centuries of research. In Y. Enzel, A. Agnon, & M. Stein (Eds.), *New frontiers in Dead Sea paleoenvironmental research* (Vol. *401*, pp. 155–170). Geological Society of America, Special Paper. USA. doi: 10.1130/2006.2401(10).

Bookman, R., Enzel, Y., Agnon, A., & Stein, M. (2004). Late Holocene lake levels of the Dead Sea. *Geological Society of America Bulletin*, *116*(5/6), 555–571. doi: 10.1130/B25286.1

Braitsch, O. (1971). *Salt deposits: Their origin and composition*. New York: Springer-Verlag, 279 pp.

Cady, P. (1904). The historical and physical geography of the Dead Sea region. *Bulletin of the American Geographical Society*, *36*(10), 577–589. https://www.jstor.org/stable/197977

Calvo, R., & Bartov, Y. (2001). Hazeva Group, southern Israel: New observations, and their implications for its stratigraphy, paleogeography, and tectono-sedimentary regime. *Israel Journal of Earth Sciences*, *50*, 71–99. doi: 10.1560/B02L-6K04-UFQL-KUE350: 71–99.

Closson, D., LaMoreaux, P. E., Karaki, N. A., & al-Fugha, H. (2007). Karst system developed in salt layers of the Lisan Peninsula, Dead Sea, Jordan. *Environmental Geology*, *52*, 155–172. doi 10.1007/s00254-006-0469-9

Cohen, A. S. (1990). Tectono-stratigraphic model for sedimentation in Lake Tanganyika, Africa. In B. J. Katz (Ed.), *Lacustrine basin exploration: Case studies and modern analogs*, *AAPG Memoir* (Vol. 50, 137–150).

Coianiz, L., Bialik, O. M., Ben-Avraham, Z., & Lazar, M. (2019). Late Quaternary lacustrine deposits of the Dead Sea basin: High resolution sequence stratigraphy from downhole logging data. *Quaternary Science Reviews*, *210*, 175–189.

Coianiz, L., Schattner, U., Lang, G., Ben-Avraham, Z., & Lazar, M. (2020). Between plate and salt tectonics—New stratigraphic constraints on the architecture and timing of the Dead Sea basin during the Late Quaternary. *Basin Research*, *32*(4), 636–651.

Cui, Y., Chen, X., Gao, J., Yan, B., Tang, G., & Hong, Y. (2018). Global water cycle and remote sensing big data: Overview, challenge, and opportunities. *Big Earth Data*, *2*(3), 282–297. doi:10.1080/20964471.2018.1548052

Debure, M., Lassin, A., Marty, N. C., Claret, F., Virgone, A., Calassou, S., & Gaucher, E. C. (2019). Thermodynamic evidence of giant salt deposit formation by serpentinization: An alternative mechanism to solar evaporation. *Nature Science Reports*, *9*, 11720. https://doi.org/10.1038/s41598-019-48138-9

DeGolyer, E. (1926). Origin of North American salt domes. In R. C. Moore, W. E. Pratt, D. C. Barton, A. Deussen, & J. P. D. Hull (Eds.), *Geology of salt dome oil fields* (Vol. *1*, pp. 1–44). AAPG Special Publication.

Dodson, S. I. (2004). *Introduction to limnology* (1st ed.). New York: McGraw-Hill, 400 pp.

Eppelbaum, L., Ben-Avraham, Z., & Katz, Y. (2007). Structure of the Sea of Galilee and Kinarot valley derived from combined geological-geophysical analysis. *First Break*, *25*, 43–50.

Fedorik, J., Maesano, F. E., & Afifi, A. M. (2022). A validated geomechanical model for the strike-slip restraining bend in Lebanon. *Nature Scientific Reports*, *12*, 20071. https://doi.org/10.1038/s41598-022-24718-0

Frumkin, A. (1994). Hydrology and denudation rates of halite karst. *Journal of Hydrology*, *162*, 171–189.

Frumkin, A. (1996). Uplift rate relative to base-levels of a salt diapir (Dead Sea Basin, Israel) as indicated by cave levels. In G. I. Alsop, D. J. Blundell, & I. Davison (Eds.), *Salt tectonics* (Vol. *100*, pp. 41–47). Geological Society, Special Publication.

Frumkin, A., & Elitzur, Y. (2002). Historic Dead Sea level fluctuations calibrated with geological and archaeological evidence. *Quaternary Research*, *57*(03), 334–342. doi:10.1006/qres.2002.2330

Garber, R. A., Levy, Y., & Friedman, G. M. (1987). The sedimentology of the Dead Sea. *Carbonates and Evaporites*, *2*, 43–57. https://doi.org/10.1007/BF03174303

Garfunkel, Z. (1981). Internal structure of the Dead Sea leaky transform (rift) in relation to plate kinematics. *Tectonophysics*, *80*, 81–108.

Garfunkel, Z. (1997). The history and formation of the Dead Sea basin. In T. M. Niemi, Z. Ben-Avraham, & J. R. Gat (Eds.), *The Dead Sea; The lake and its settings, Oxford Monograph on Geology and Geophysics* (Vol. *36*, pp. 36–56). Oxford: Oxford University Press.

Garfunkel, Z. (2014). Lateral motion and deformation along the Dead Sea transform. In Z. Garfunkel, Z. Ben-Avraham, & E. Kagan (Eds.), *Dead Sea transform fault system: Reviews* (pp. 109–150). Dordrecht: Springer.

Garfunkel, Z., & Ben-Avraham, Z. (1996). The structure of the Dead Sea basin. *Tectonophysics*, *226*, 155–176.

Garreaud, R. D. (2009). The Andes climate and weather. *Advances in Geosciences*, *22*, 3–11. www.adv-geosci.net/22/3/2009/

Gavrieli, I., & Oren, A. (2004). The Dead Sea as a dying lake. In J. C. J. Nihoul, P. O. Zavialov, & P. P. Micklin (Eds.), *Dying and Dead Seas climatic versus anthropic causes* (pp. 287–305). Netherlands: Kluwer Academic Publishers. doi:10.1007/978-94-007-0967-6_11

Gavrieli, I., & Stein, M. (2006). On the origin and fate of the brines in the Dead Sea basin. In Y. Enzel, A. Agnon, & M. Stein (Eds.), *New frontiers in Dead Sea paleoenvironmental research* (Vol. *401*, 183–194), Geological Society of America, USA. Special Paper 401. doi: 10.1130/2006.2401(12).

GEBCO (2019). General Bathymetric Chart of the Oceans. British Oceanographic Data Centre, U.K., website www.gebco.net.

Gertman, I. (2012). Dead Sea. In L. Bengtsson, R. W. Herschy, & R. W. Fairbridge (Eds.), *Encyclopedia of Lakes and Reservoirs, Encyclopedia of Earth Sciences Series* (pp. 212–216). Dordrecht: Springer. https://doi.org/10.1007/978-1-4020-4410-6_211

Gertman, I., & Hecht, A. (2002). The Dead Sea hydrography from 1992 to 2000. *Journal of Marine Systems*, *35*, 169–181.

Golan, R., Lazar, B., Wurgaft, E., Lensky, N., Ganor, J., & Gavrieli, I. (2017). Continuous CO_2 escape from the hypersaline Dead Sea caused by aragonite precipitation. *Geochimica et Cosmochimica Acta*, *207*, 43–56. http://dx.doi.org/10.1016/j.gca.2017.02.031

Goldstein, S. L., Kiro, Y., Torfstein, A., Kitagawa, H., Tierney, J., & Stein, M. (2020). Revised chronology of the ICDP Dead Sea deep drill core relates drier-wetter-drier climate cycles to insolation over the past 220 kyr. *Quaternary Science Reviews*, *244*, 106460. https://doi.org/10.1016/j.quascirev.2020.106460

Gomez, F., Karam, G., Khawlie, M., McClusky S., Vernant P., Reilinger R., et al. (2007). Global Positioning System measurements of strain accumulation and slip transfer through the restraining bend along the Dead Sea fault system in Lebanon. *Geophysical Journal International*, *168*(3), 1021–1028. doi:10.1111/j.1365-246X.2006.03328.x

Hardie, L. A. (1984). Evaporites: Marine or non-marine. *American Journal of Science*, *283*, 193–240.

Hardie, L. A. (1991). On the significance of evaporites. *Annual Review Earth and Planetary Science*, *290*, 43–106.

Hovland, M., Rueslaatten H., & Johnsen, H. K. (2008). Hydrothermal salt—But how much? Reply to Christopher Talbot on his comments to our articles. *Marine and Petroleum Geology*, *25*, 203–204.

Hovland, M., Rueslåtten, H., & Johnsen, H. K. (2018a). Large salt accumulations as a consequence of hydrothermal processes associated with "Wilson cycles": A review, Part 1: Towards a new understanding. *Marine and Petroleum Geology*, *92*, 987–1009.

Hovland, M., Rueslåtten, H., & Johnsen, H. K. (2018b). Large salt accumulations as a consequence of hydrothermal processes associated with "Wilson cycles": A review, Part 2: Application of a new salt-forming model on selected cases. *Marine and Petroleum Geology*, *92*, 128–148.

Hovland, M., Rueslåtten, H. G., Johnsen, H. K., Kvamme, B., & Kuznetsova, T. (2006). Salt formation associated with sub-surface boiling and supercritical water. *Marine and Petroleum Geology*, *23*, 855–869.

Hsü, K. J. (1972). Origin of saline giants: A critical review after the discovery of the Mediterranean Evaporite. *Earth-Science Reviews*, *8*, 371–396.

Hutton, J. (1788, 1795). *Theory of the Earth*. Global Grey, 447 pp.

Ionescu, D., Siebert, C., Polerecky, L., Munwes, Y. Y., Lott, C., Häusler, S., et al. (2012) Microbial and chemical characterization of underwater fresh water springs in the Dead Sea. *PLoS ONE*, *7*(6): e38319. https://doi.org/10.1371/journal.pone.0038319

Jackson, M. P. A., Warin, O. N., Woad, G. M., & Hudec, M. R. (2003). Neoproterozoic allochthonous salt tectonics during the Lufilian orogeny in the Katangan Copperbelt, central Africa, *Geological Society of America Bulletin*, *115*(3), 314–330.

Javor, B. (1989). *Hypersaline environments: Microbiology and biogeochemistry*. Berlin Heidelberg: Springer-Verlag, 328 pp.

Kagan, E. J., Langgut, D., Boaretto, E., Herald Neumann, F., & Stein, M. (2015). Dead Sea levels during the Bronze and Iron ages. *Radiocarbon*, *57*(02), 237–252. doi:10.2458/azu_rc.57.18560

Kaniewsk, D., Guiot, J., & Van Campo, E. (2015). Drought and societal collapse 3200 years ago in the Eastern Mediterranean: A review. *Climate Change*, *6*(4), 369–382. doi:10.1002/wcc.345

Kashai, E. L., & Croker, P. F. (1987). Structural geometry and evolution of the Dead Sea-Jordan rift system as deduced from new subsurface data. *Tectonophysics*, *141*, 33–60.

Katz, B. J. (2001). Lacustrine basin hydrocarbon exploration—Current thoughts. *Journal of Paleolimnology*, *26*, 161–179. https://doi.org/10.1023/A:1011173805661

Kendall, A. C. (2010). Marine evaporites. In N. P. James, & R. W. Dalrymple (Eds.), *Facies Models* (Vol. *4*, pp. 505–539). Geotext 6, Geological Association of Canada.

Kendall, C. G. St. C., Weber, L. J., & Alsharhan, A. S. (2009). The Giant Oil Field Evaporite Association: A function of the Wilson cycle, climate, basin position and sea level. *AAPG Search and Discovery, Article* #40471.

Kiro, Y., Goldstein, S. L., Garcia-Veigas, J., Levy, E., Kushnir, Y., Stein, M., & Lazar, B. (2017). Relationships between lake-level changes and water and salt budgets in the Dead Sea during extreme aridities in the Eastern Mediterranean. *Earth and Planetary Science Letters*, *464*, 211–226. http://dx.doi.org/10.1016/j.epsl.2017.01.043

Kiro, Y., Goldstein, S. L., Lazar, B., & Stein, M. (2016). Environmental implications of salt facies in the Dead Sea. *Geological Society of American Bulletin*, *128*(5/6), 824–841. doi: 10.1130/B31357.1

Laity, J. J. (2008). *Deserts and desert environments. Environmental systems and global change series 2*. Chichester, U.K.: Wiley-Blackwell, 342 pp.

Lazar, M., & Ben-Avraham, Z. (2002). First images from the bottom of the Dead Sea—Indications of recent tectonic activity. *Israel Journal of Earth Sciences*, *51*(3–4), 211–218. doi: 10.1560/5MPK-UJDN-L6CV-024K

Lensky, N. G., Dvorkin, Y., Lyakhovsky, V., Gertman, I., & Gavrieli, I. (2005). Water, salt, and energy balances of the Dead Sea. *Water Resources Research*, *41*, 1–13. doi:10.1029/2005WR004084

Lensky, N. G., Gertman, I., Arnon, A., Ozer, T., Biton, E., Gavrieli, I., et al. (2013). *Currents and hydrography of the Dead Sea: A study for the Salt Recovery Project, 2nd Report—August 2012 Survey*. Geological Survey of Israel/Israel Oceanographic and Limnological Research Report, 30 pp.

Levy, E. J., Yechieli, Y., Gavrieli, I., Lazard, B., Kiro, Y., Stein, M., & Sivan, O. (2018). Salt precipitation and dissolution in the late Quaternary Dead Sea: Evidence from chemical and δ^{37}Cl composition of pore fluids and halites. *Earth and Planetary Science Letters*, *487*, 127–137. https://doi.org/10.1016/j.epsl.2018.02.003

Levy, Y., & Gvirtzman, H. (2021). Industry-driven versus natural groundwater flow regime at the Dead Sea Coastal Aquifer. *Water*, *13*(4), 498. https://doi.org/10.3390/w13040498

Lisker, S., Porat, R., Davidovich, U., Eshel, H., Lauritzen, S.-E., & Frumkin, A. (2007). Late Quaternary environmental and human events at En Gedi, reflected by the geology and archaeology of the Moringa Cave (Dead Sea area, Israel). *Quaternary Research*, *68*(02), 203–212. doi:10.1016/j.yqres.2007.03.01

Lowenstein, T. K., & Risacher, F. (2009). Closed basin brine evolution and the influence of Ca–Cl inflow waters: Death valley and Bristol dry lake California, Qaidam Basin, China, and Salar de Atacama, Chile. *Aquatic Geochemistry*, *15*, 71–94.

Lu, Y., Moernaut, J., Waldmann, N., Bookman, R., Ian Alsop, G., Hubert-Ferrari, A., et al. (2021). Orbital- and millennial-scale changes in lake-levels facilitate earthquake-triggered mass failures in the Dead Sea Basin. *Geophysical Research Letters*, *48*(4), e2021GL093391. https://doi.org/10.1029/2021GL093391

Lyell, C. (1830). *Principles of geology, being an attempt to explain the former changes of the earth's surface, by reference to causes now in operation*. London: Murray, 385 pp.

Magruder, K. V. (2009). *The idiom of a six day creation and global depictions in Theories of the Earth*. In M. Kolbl-Ebert (Ed.), *Geology and Religion: A history of harmony and hostility* (Vol. 310, pp. 49–66). The Geological Society, London, Special Publications.

Malek-Aslani, M. (1980). Environmental and diagenetic controls of carbonate and evaporite source rocks. *Gulf Coast Association of Geological Societies Transactions*, *30*, 445–456.

Marty, B., Avice, G., Bekaert, D. V., & Broadley, M. W. (2018). Salinity of the Archaean oceans from analysis of fluid inclusions in quartz. *Comptes Rendus Geosciences*, *350*, 154–163.

Mello, M. R., & Maxwell, J. R. (1991). Organic geochemical and biological marker characterization of source rocks and oils derived from lacustrine environments in the Brazilian continental margin. In B. J. Katz (Ed.), *Lacustrine basin exploration, American Association of Petroleum Geologists Memoir* (Vol. 50, pp. 77–98).

Messager, M., Lehner, B., Grill, G., Nedeva, I., & Schmitt, O. (2016). Estimating the volume and age of water stored in global lakes using a geo-statistical approach. Nature Communications 7, 13603. https://doi.org/10.1038/ncomms13603

Misi, A., Iyer, S. S. S., Coelho, C. E. S., Tassinari, C. C. G., Franca-Rocha, W. J. S., & Cunha, I. A. (2015). Sediment hosted lead–zinc deposits of the Neoproterozoic Bambuı´ Group and correlative sequences, São Francisco Craton, Brazil: A review and a possible metallogenic evolution model. *Ore Geology Reviews*, *26*, 263–304.

Mohriak, W. (2025). *Salt in the Earth sciences: Basin analysis and salt tectonics*. Washington, DC: American Geophysical Union. doi:10.1002/9781394307067

Mohriak, W. U., Szatmari, P., & Anjos, S. (2012). Salt: Geology and tectonics of selected Brazilian basins in their global context. In G. I. Alsop, S. G. Archer, A. J. Hartley, N. T. Grant, & R. Hodgkinson (Eds.), Salt tectonics, sediments and prospectivity (Vol. 363, pp. 131–158). Geological Society, London, Special Publications. doi:10.1144/SP363.7

Montaron, B., & Tapponnier, P. (2012). *The making of giant salt basins: An introduction to tectonics and evaporite deposition,* SPE ATW, Amman, Jordan, June 10–12, 2012, *Abstracts*. doi: 10.3997/2214-4609.20132079

Morgunova, I., Semenov, P., Kursheva, A., Litvinenko, I., Malyshev, S., Bukin, S., et al. (2022). Molecular indicators of sources and biodegradation of organic matter in sediments of fluid discharge zones of lake Baikal. *Geosciences*, *12*(2), 72. https://doi.org/10.3390/geosciences12020072

Moses, S. W., David, M., Goldhammer, E., Tal, A., & Sukenik, S. (2006). The Dead Sea, A unique natural health resort. *The Israel Medical Association Journal (IMAJ)*, *8*, 483–488.

Müller, D., Neugebauer, I., Ben Dor, Y., Enzel, Y., Schwab, M. J., Tjallingii, R., & Brauer, A. (2022). Phases of stability during major hydroclimate change ending the Last Glacial in the Levant. *Nature Scientific Reports*, *12*, 60252. doi: 10.1038/s41598-022-10217-9

Muirhead, J. D., Fischer, T. P., Oliva, S. J., Lalzer, A., van Wijk, J., Currie, C. A., et al. (2020). Displaced cratonic mantle concentrates deep carbon during continental rifting. *Nature*, *582*, 67–72. https://doi.org/10.1038/s41586-020-2328-3

NASA Aquarius Mission (2017). *Climatology*. https://aquarius.oceansciences.org/cgi/gal_climatology.htm

Neev, D., & Emery, K. (1967). The Dead Sea: Depositional processes and environments of evaporites. *Geological Survey of Israel Bulletin*, *41*, 1–147.

Neev, D., & Hall, J. K. (1979). Geophysical investigations in the Dead Sea. *Sedimentary Geology*, *23*, 209–238. https://doi.org/10.1016/0037-0738(79)90015-0

Neugebauer, I., Brauer, A., Schwab, M. J., Waldmann, N. D., Enzel, Y., Kitagawa, H., et al. (2014). Lithology of the long sediment record recovered by the ICDP Dead Sea Deep Drilling Project (DSDDP), *Quaternary Science Reviews*, *102*, 149–165.

Neumann, F. H., & Zwickel, W. (2019). Settlements, climate and vegetation at the Dead Sea from the Neolithic until the Crusader period. In M. Peilstöcker & S. Wolfram (Eds.), *Life at the Dead Sea*. Proceedings of the International Conference, State Museum of Archaeology Chemnitz, February 21–24, 2018, Chemnitz, Zaphon, pp. 75–94.

Nissenbaum, A. (1975). The microbiology and biogeochemistry of the Dead Sea. *Microbial Ecology*, *2*(2), 139–161. https://www.jstor.org/stable/4250452

Nissenbaum, A. (1979). Life in a Dead Sea: Fables, allegories, and scientific search. *BioScience*, *29*(3), 153–157. doi: 10.2307/1307795

Nissenbaum, A. (1980). Hypersaline brines and evaporitic environments. *Developments in sedimentology* (Vol. 28). Amsterdam: Elsevier, 270 pp.

Nissenbaum, A. (1993). The Dead Sea—An economic resource for 10 000 years. *Hydrobiologia*, *267*, 127–141. https://doi.org/10.1007/BF00018795

Nissenbaum, A. (1994). Sodom, Gomorrah and the other lost cities of the plain—A climatic perspective. *Climatic Change*, *26*(4), 435–446. https://doi.org/10.1007/BF01094406

Nissenbaum, A., & Buckley, S. (2013). Dead Sea Asphalt in ancient Egyptian mummies—Why? *Archaeometry*, *55*(3), 563–568. https://doi.org/10.1111/j.1475-4754.2012.00713.x

Ochsenius, K. (1877). *Die Bilding der Steinsalzlager und ihrer mutter augensalze*. C.E.M. Pfeffer, Halle, 172 pp.

Ochsenius, K. (1888). On the formation of rock salt beds and mother liquor salts. *Philadelphia Academic Society Proceedings, Pl 2*, 181–187.

Quennel, A. M. (1958). The structural and geomorphic evolution of the Dead Sea rift. *Quarterly Journal of the Geological Society*, *114*(1–4), 1–24. doi:10.1144/gsjgs.114.1.0001

Ouillon, R., Lensky, N. G., Lyakhovsky, V., Arnon, A., & Meiburg, E. (2019). Halite precipitation from double-diffusive salt fingers in the Dead Sea: Numerical simulations. *Water Resources Research*, *55*, 4252–4265.

Press, F., Siever, R., Grotzinger, J., & Jordan, T. H. (2003). *Understanding Earth* (4th ed.). New York: W. H. Freeman, 568 pp.

Preston, L. J., & Dartnell, L. R. (2014). Planetary habitability: Lessons learned from terrestrial analogues. *International Journal of Astrobiology*, *13*(1), 81–98. doi:10.1017/S1473550413000396

Reeves, C. C., Jr. (1968). *Introduction to paleolimnology: Developments in sedimentology* (Vol. *11*), 228 pp. Elsevier.

Renaut, R. W., & Gierlowski-Kordesch, E. H. (2010). Lakes. In N. P. James & R. W. Dalrymple (Eds.), *Facies Models* (Vol. 4, pp. 541–575). Geological Association of Canada, Geotext 6.

Romano, M. (2015). Reviewing the term uniformitarianism in modern Earth sciences. *Earth-Science Reviews*, *148*, 65–76.

Rudwick, M. J. S. (2009). *Biblical flood and geological deluge: The amicable dissociation of geology and Genesis*. In M. Kolbl-Ebert (Ed.), *Theories of the Earth*. In *Geology and Religion: A history of harmony and hostility* (Vol. 310, pp. 103–110). The Geological Society, London, Special Publications.

Salameh, E., Tarawneh, A., & Al-Raggad, M. (2016). Origin of high bromide concentration in the water sources in Jordan and in the Dead Sea water. *Arabian Journal of Geosciences*, *9*(5), *414*, 1–20. doi 10.1007/s12517-016-2431-9

Schmid, S. (2017). Neoproterozoic evaporites and their role in carbon isotope chemostratigraphy (Amadeus Basin, Australia). *Precambrian Research*, *290*, 16–31. doi:10.1016/j.precamres.2016.12.004

Schreiber, B. C., & Hsü, K. J. (1980). Evaporites. In G. D. Hobson (Ed.), *Developments in petroleum geology* (Vol. *2*, pp. 87–138). London: Applied Science Publishers.

Schulz, K. L. (2001). Limnology. https://www.esf.edu/efb/schulz/Limnology/Limnology2.htm

Shiklomanov, I. A. (1993). World fresh water resources. In P. H. Gleick (Ed.), *Water in crisis: A guide to the World's fresh water resources* (Chapter 2, pp. 13–24). New York: Oxford University Press.

Siegel, B. Z., McMurth, G., Siegel, S. M., Chen, J., & Larock, P. (1979). Life in the calcium chloride environment of Don Juan Pond, *Antarctica. Nature*, *280*(5725), 828–829. doi:10.1038/280828a0

Sirota, I. (2020). *The dynamics of Halite deposition in the Dead Sea and its implications to Thick Halite Sequences in the Geological Record*, 141 pp. Geological Survey of Israel, Report, GSI/30/2020.

Sirota, I., Arnon, A., & Lensky, N. G. (2016). Seasonal variations of halite saturation in the Dead Sea. *Water Resources Research*, *52*, 7151–7162. doi:10.1002/2016WR018974.

Sirota, I., Enzel, Y., & Lensky, N. G. (2017). Temperature seasonality control on modern halite layers in the Dead Sea: In situ observations, *Geological Society of America Bulletin*, *129*(9/10), 1181–1194. doi: 10.1130/B31661.1.

Sirota, I., Enzel, Y., & Lensky, N. G. (2018). Halite focusing and amplification of salt layer thickness: From the Dead Sea to deep hypersaline basins. *Geology*, *46*(10), 851–854.

Sirota, I., Enzel, Y., Mor, Z., Ben Moshe, L., Eyal, H., Lowenstein, T. K., & Lensky, N. G. (2021). Sedimentology and stratigraphy of a modern halite sequence formed under Dead Sea level fall. *Sedimentology. 68*(3), 1069–1090. https://doi.org/10.1111/sed.12814

Smoot, J. P., & Lowenstein, T. K. (1991). Depositional environments of non-marine evaporites. In J. L. Melvin (Ed.), *Evaporites, petroleum and mineral resources: Developments in sedimentology* (Vol. *50*, Chapter 3, pp. 189–347). Amsterdam: Elsevier.

Sonnenfeld, P. (1985). Evaporites; marine or non-marine? *American Journal of Science*, *285*, 661–672.

Sonnenfeld, P. (2003). Evaporites. In Meyers, R. A. (Ed.), *Encyclopedia of physical science and technology* (pp. 653–671). doi:10.1016/b0-12-227410-5/00234-9

Sonnenfeld, P., & Perthuisot, J. -P. (1989). *Brines and evaporites. Short Course in Geology, Volume 3, 28th International Geological Congress*. American Geophysical Union, Washington, DC, 126 pp.

Soto, J. I., Flinch, J. F., & Tari, G. (2017). Permo-Triassic Basins and Tectonics in Europe, North Africa and the Atlantic margins: A synthesis. In J. I. Soto, J. F. Flinch, & G. Tari (Eds.), *Permo-Triassic salt provinces of Europe, North Africa and the Atlantic margins tectonics and hydrocarbon potential* (pp. 3–41). Amsterdam: Elsevier.

Spear, N., Holland, H. D., Garcia-Veígas, J., Lowenstein, T. K., Giegengack, R., & Peters, H. (2014). Analyses of fluid inclusions in Neoproterozoic marine halite provide oldest measurement of seawater chemistry. *Geology*, *42*, 103–106. doi:10.1130/G34913.1

Stein, M. (2001). The sedimentary and geochemical record of Neogene-Quaternary water bodies in the Dead Sea basin-inferences for the regional paleoclimatic history. *Journal of Paleolimnology*, *26*(3), 271–282. doi:10.1023/a:1017529228186

Stein, M. (2014). The evolution of Neogene-Quaternary water-bodies in the Dead Sea Rift Valley. In Z. Garfunkel, Z. Ben-Avraham, & E. Kagan (Eds.), *Dead Sea Transform fault system: Reviews, Modern approaches in solid Earth sciences* (Vol. *6*, pp. 279–316). Dordrecht: Springer. doi 10.1007/978-94-017-8872-4_10

Stein, M., Torfstein, A., Gavrieli, I., & Yechieli, I. (2010). Abrupt aridities and salt deposition in the post-glacial Dead Sea and their North Atlantic connection. *Quaternary Science Reviews*, *29*(3), 567–575.

Steinhorn, I. (1983). In situ salt precipitation at the Dead Sea. *Limnology Oceanography*, *28*(3), 580–583.

Steinhorn, I. (1985). The disappearance of the long term meromictic stratification of the Dead Sea. *Limnology and Oceanography*, *30*(3), 451–472.

Steinhorn, I. Assaf, G., Gat, J. R., Nishry, A., Nissenbaum, A., Stiller, M., et al. (1979). The Dead Sea: Deepening of the mixolimnion signifies the overture to overturn of the water column. *Science*, *206*(4414), 55–57. doi:10.1126/science.206.4414.55

Talbot, C. J. (2008). Hydrothermal salt—But how much? *Marine and Petroleum Geology*, *25*, 191–202.

Talbot, M. R. (2005). Lake processes and deposits. In R. C. Selley, L. R. M. Cocks, & I. R. Plimer (Eds.), *Encyclopedia of Geology* (pp. 550–561). doi:10.1016/b0-12-369396-9/00493-7

Tanyileke, G., Ntchantcho, R., Fantong, W. Y., Aka, F. T., & Hell, J. V. (2019). 30 years of the Lakes Nyos and Monoun gas disasters: A scientific, technological, institutional and social adventure. *Journal of African Earth Sciences*, *150*, 415–424.

Torfstein, A. (2008). Brine—Freshwater interplay and effects on the evolution of saline lakes: The Dead Sea rift terminal lakes, 290 pp. Geological Survey of Israel, Report GSI /20/08.

Torfstein, A. (2017). The Amora formation, Dead Sea basin. In Y. Enzel & O. Bar (Eds.), *Quaternary of the Levant-Environments, Climate change, and humans* (pp. 91–98). Cambridge: Cambridge University Press. doi: https://doi.org/10.1017/9781316106754.010.

Torfstein, A., Goldstein, S. L., Kushnir, Y., Enzel, Y., Haug, G., & Stein, M. (2015). Dead Sea drawdown and monsoonal impacts in the Levant during the last interglacial. *Earth and Planetary Science Letters*, *412*, 235–244. doi:10.1016/j.epsl.2014.12.013

Torfstein, A., Goldstein, S. L., Stein, M., & Enzel, Y. (2013). Rapid communication impacts of abrupt climate changes in the Levant from Last Glacial Dead Sea levels. *Quaternary Science Reviews*, *69*, 1–7.

Tundisi, J. G., & Tundisi, T. M. (2011). *Limnology*. Boca Raton, FL: CRC Press, 864 pp.

U.S. Geological Survey (2018). *Groundwater storage and the water cycle. Water Science School.* https://www.usgs.gov/special-topics/water-science-school/science/groundwater-storage-and-water-cycle

USGS (2022). The Water Cycle. United States Geological Survey. Website: https://www.usgs.gov/special-topics/water-science-school/science/water-cycle-diagrams

Usiglio, J. (1849a) Analyse de l'eau de la Mediterraneé sur le côtes de France. *Annales de Chemie et de Physique*, 3ème sér, *27*, 92–107.

Usiglio, J. (1849b). Etudes sur la composition de l'eau de la Mediterranee et sur l'exploitation des sel quy'elle conteint. *Annales de Chemie et de Physique*, 3ème sér, *27*, 172–191.

Van Waterschoot Van der Gracht, W. A. J. M. (1926). Geology of the salt domes in the Carpathian region of Rumania: Discussion. In R. C. Moore, W. E. Pratt, D. C. Barton, A. Deussen, & J. P. D. Hull (Eds.), *Geology of Salt dome oil fields* (Vol. *1*, pp. 123–128). American Association of Petroleum Geologists, Special Publications, USA.

Vinogradova, N., Lee, T., Boutin, J., Drushka, K., Fournier, S., Sabia, R., et al. (2019). Satellite salinity observing system: Recent discoveries and the way forward. *Frontiers in Marine Sciences*, *6*, *Article 243*, 1–23. https://doi.org/10.3389/fmars.2019.00243

Voitesti, I. P. (1926). Geology of the salt domes in the Carpathian region of Rumania. In R. C. Moore, W. E. Pratt, D. C. Barton, A. Deussen, & J. P. D. Hull (Eds.), *Geology of Salt Dome Oil Fields* (Vol. *1*, pp. 87–122). American Association of Petroleum Geologists, Special Publication.

Waldmann, N., Stein, M., Ariztegui, D., & Starinsky, A. (2009). Stratigraphy, depositional environments and level reconstruction of the last interglacial Lake Samra in the Dead Sea basin. *Quaternary Research*, *72*(1), 1–15. https://doi.org/10.1016/j.yqres.2009.03.005

Warren, J. K. (1991). Sulfate dominated sea-marginal and platform settings. In J. L. Melvin (Ed.), *Evaporites, petroleum and mineral resources. Developments in Sedimentology* (Vol. *50*, Chapter 2, pp. 69–187). Amsterdam: Elsevier.

Warren, J. K. (2006). *Evaporites: Sediments, resources and hydrocarbons*. Berlin: Springer Verlag, 1035p.

Warren, J. K. (2010). Evaporites through time: Tectonic, climatic and eustatic controls in marine and nonmarine deposits. *Earth and Science Review*, *98*, 217–268.

Warren, J. K. (2016). *Evaporites—A geological compendium*. Berlin: Springer, 1807 pp. ISBN 978-3-319-13511-3.

Warren, J. K. (2021). Evaporite deposits. In S. Elias, & D. Alderton (Eds.), *Encyclopedia of Geology* (2nd ed., pp. 945–977). Amsterdam: Elsevier. https://doi.org/10.1016/B978-0-08-102908-4.00165-X.

Webb, P. (2021). *Introduction to Oceanography*. Pressbooks, 383 pp. https://rwu.pressbooks.pub/webboceanography/

Weinberger, R., Begin, Z. B., Waldmann, N., Gardosh, M., Baer, G., Frumkin, A., & Wdowinski, S. (2006), Quaternary rise of the Sedom diapir, Dead Sea basin. In Y. Enzel, A. Agnon, & M. Stein (Eds.), *New frontiers in Dead Sea paleoenvironmental research* (Vol. *401*, pp. 33–51). Geological Society of America Special Paper. doi: 10.1130/2006.2401(03).

Wetzel, R.G. (2001). *Limnology: Lake and river ecosystems* (3rd ed.), 1005 pp. San Diego, CA: Academic Press.

Wisniak, J. (2002). The Dead Sea—A live pool of chemicals. *Indian Journal of Chemical Technology*, *9*, 79–87.

Wong, K., Mason, E., Brune, S., East, M., Edmonds, M., & Zahirovic, S. (2019). Deep Carbon cycling over the past 200 million years: A review of fluxes in different tectonic settings. *Frontiers in Earth Science*, *7*. https://doi.org/10.3389/feart.2019.00263

Yechieli, Y., Gavrieli, I., Berkowitz, B., & Ronen, D. (1998). Will the Dead Sea die? *Geology*, *26*(8), 755–758. https://doi.org/10.1130/0091-7613(1998)026<0755:WTDSD>2.3.CO;2

Zak, I. (1967). *The geology of Mount Sedom* (PhD thesis). The Hebrew University of Jerusalem, Israel.

Zak, I. (1974). Sedimentology and bromine geochemistry of marine and continental evaporites in the Dead Sea basin. In *Fourth International Symposium on Salt*. Northern Ohio Geological Society. USA, pp. 349–361.

Zak, I. (1997). Evolution of the Dead Sea brines. In T. M. Niemi, Z. Ben-Avraham, & J. R. Gat (Eds.), *The Dead Sea: The lake and its setting* (pp. 133–144). Oxford, U.K.: Oxford University Press.

Zak, I., & Freund, R. (1980). Strain measurements in eastern marginal shear zone of Mount Sedom salt diapir, Israel. *American Association of Petroleum Geologists Bulletin*, *64*(4), 568–581.

Zak, I., & Freund, R. (1981). Asymmetry and basin migration in the Dead Sea rift. *Tectonophysics*, *80*, 27–38.

3
Salt Structures and Salt Outcrops

ABSTRACT

Researchers have studied evaporite deposition and salt layer deformation in outcrops worldwide, including in the Arabian Peninsula in the Middle East, especially the Zagros region; north Africa, specifically the Atlas Mountains; and southern Europe, specifically the Alps and the Pyrenees. Salt mines, such as the Hallstatt and Cardona mines, offer three-dimensional views of subsurface salt structures. In North America, the Paradox Basin showcases anticlinal salt structures exposed due to salt dissolution. Red Sea marine ingressions in the Quaternary, which flooded the depression in the Danakil region of the Afar Province, formed a lagoon that desiccated and accumulated a thick sequence of evaporite rocks. Generations of earth scientists have studied the iconic salt structure known as the Sedom Mountain, located at the margins of the Dead Sea, as a paradigm of a salt diapir that is still rising.

Salt structures display an immensely complex variety of geometries in surface outcrops. The genesis of evaporite deposits and the mechanisms of salt deformation have fascinated researchers since the birth of the earth sciences. The majority of researchers believe that salt layers are precipitated as sedimentary rocks and subsequently deformed by gravitational and tectonic forces (see Chapter 2). Salt mobility is responsible for some of the most bewildering geological structures in sedimentary basins worldwide, as observed in outcrops and salt mines. In the subsurface, salt structures are interpreted using geophysical methods and the drilling of exploratory boreholes by the mining and petroleum industries. The relationship between salt structures and hydrocarbon accumulations was first recognized after the drilling of the famous Spindletop well in the Gulf of Mexico in 1901 (see Chapter 1).

The association between salt deposits and hydrocarbon accumulations is evident in the Middle East salt basins, the main target for petroleum exploration worldwide during the first half of the 20th century (Beydoun, 1986). In addition, several landmark publications have highlighted the importance of salt basins for the development of the earth sciences (e.g., Archer et al., 2012; Davison, Alsop, & Blundell, 1996; Warren, 2010, 2016).

This chapter addresses the evaporite deposition episodes in the Arabian Peninsula, northern Africa, Iberia, southern Europe, and North America. Selected examples of salt outcrops in several sedimentary basins are briefly described based on regional maps, satellite images, and detailed photographs. Regional topographic relief maps based on global altimetry (e.g., Amante & Eakins, 2009; GEBCO, 2019; NOAA, 2022) illustrate some major salt features in evaporitic basins. Most salt outcrops are related to evaporite deposits that have been uplifted by orogenic processes, forming salt diapirs that reach the surface. Most structures were controlled by the Cenozoic Alpine Orogeny, which affected large portions of northern Africa, southern central Europe, the Arabian Peninsula, and Asia (Figure 3.1).

The salt layers in rift basins and divergent margin settings are commonly overlain by a few hundred to several thousand meters of water and sediments. However, in some rifts and continental margins that are presently on land, many diapirs are characterized by shallow tops. Due to uplift and erosion, some diapirs have locally extruded salt masses on the depositional surface. This model has been suggested for some salt structures in the Zagros ranges in Iran and the Atlas Mountains in Tunisia (Alavi, 1994; Jacobshagen et al., 1988; Jaillard et al., 2017; Kent, 1979; Masrouhi & Koyi, 2012; Najafi et al., 2018). The Middle East salt basins were the main target for petroleum exploration worldwide during the first half of the 20th century. Salt basins in Europe,

Salt in the Earth Sciences: Evaporite Rocks and Salt Deposition, Geophysical Monograph 286, First Edition. Webster Mohriak.

DOI: 10.1002/9781119405245.ch03

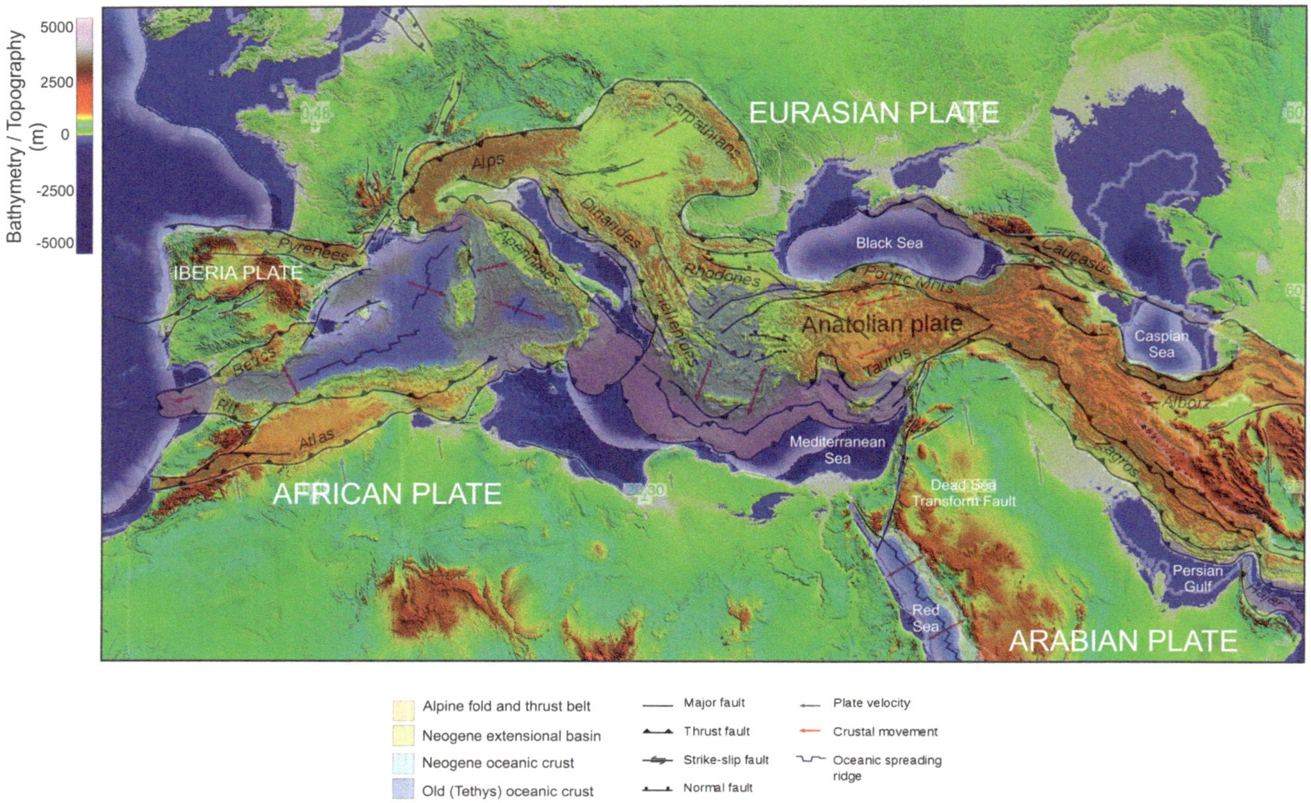

Figure 3.1 The Alpine Orogeny extends from the Iberian and African plates in the west to the Arabian-Asian plates in the east. The orogeny resulted in the formation of several mountain ranges, such as the Betics, Atlas, Pyrenees, Alps, Carpathians, Dinarides, Hellenides, Apennines, and Zagros. The deformation zone is located both south and north of the Mediterranean Sea, across northern Africa, and from Iberia, Europe, Turkey, and Iran. Compressional structures extend across the Mediterranean Sea, forming subduction zones. The Dead Sea transform fault links the Red Sea spreading center with the Palmyride deformation zone in Syria and thrust and strike-slip faults in Anatolia. The Zagros and Elburz fold-thrust belts are located north of the Arabian Plate. Credit: Bally et al. (2020)/Elsevier.

North Africa, and the Middle East have been crucial for the development of the earth sciences, with several landmark publications highlighting their importance (e.g., Soto et al., 2017).

Most of the salt outcrops discussed in this chapter initially developed salt pillow structures sourced from a deeper salt layer, and the diapiric structures resulted from halokinesis (see Chapter 6 in Mohriak, 2025). Subsequently, compressional forces during orogenic events caused the salt diapirs to deform, which locally extruded salt masses on the surface. In several outcrops, salt dissolution processes have removed the halite, leaving a residue of gypsum, except in extremely arid regions, where halite is observed on the present-day surface, as in the Fars Province of Iran (Jahani et al., 2007, 2009). Salt outcrops in several sedimentary basins are associated with evaporite deposits that orogenic processes have uplifted, creating salt diapirs that reach the surface. The Cenozoic Alpine Orogeny, which significantly affected large portions of northern Africa, southern central Europe, the Arabian Peninsula, and Asia, controlled the development of most structures. Due to supra-salt overburden rocks covering them, salt deposits in the majority of sedimentary basins frequently remain hidden from view. Some diapirs in rift basins and divergent margin settings are characterized by shallow tops due to uplift and erosion (see Chapter 8 in Mohriak, 2025).

Two of the most important salt outcrops in southern Europe are related to the Alpine and Pyrenean orogenies, which affected large areas from the Mediterranean to the Eurasian mainland (e.g., Canérot et al., 2005; Dewey et al., 1989; Schmid et al., 2004; Seton et al., 2012). The Hallstatt salt mine in Austria is associated with the Alpine orogeny, which affected the Triassic and Jurassic salt deposits in southern central Europe. The Hallstatt salt mine is considered one of the oldest underground salt production centers in the world, with the first settlements dating around 5000 BCE (before the common era). The mine recorded active production from prehistoric times until the 20th century (see Chapter 1).

The Cardona salt mine in northern Spain constitutes a unique salt structure outcropping in the southern part of the Pyrenees (see Chapter 1). In North America, the Paleozoic orogenies are related to the terminal collision of Gondwana and Laurussia, which closed the Rheic Ocean and developed several Paleozoic basins. Paleozoic evaporite deposits are recorded in the Paradox and Appalachian basins in the United States and the Maritimes Basin in Canada (see Chapter 9 in Mohriak, 2025). Several other evaporite outcrops occur in North America, such as the Jurassic salt structures in the La Popa Basin of northern Mexico (see Chapter 4 in Mohriak, 2025).

The last examples of salt outcrops are recorded in the Red Sea and the Dead Sea, corresponding to the youngest salt deposition events of global importance. These salt basins are related to late Neogene extensional and strike-slip tectonic settings. The Danakil Depression accumulated salt deposits in a depression formed near a volcanic region with active magmatism in the southern Red Sea (see Chapter 9). The Dead Sea salt deposits formed in rift troughs controlled by transform-related fault zones, accumulating marine evaporites during the Pliocene. The Dead Sea basin is still depositing halite in a continental environment, far from marine influence (see Chapter 2).

Evaporite deposits are also found in the arid highlands of mountainous regions, such as the Tibetan Plateau of west China and the Andean Mountains in South America. These Holocene salt deposits are formed by the evaporation of continental, nonmarine waters in lacustrine environments associated with convergent settings (see Chapter 10).

The evaporite deposits connected to the African, Arabian, and Eurasian plates are mostly from times when continents broke apart, creating rift basins. These basins then developed divergent continental margins that subsequently formed convergent margins, subduction zones (e.g., Scarfì et al., 2018), and orogenic belts, as recorded in the Zagros Orogen (e.g., Jafari et al., 2023; Warren, 2010). Wilson (1966) first put forth the Wilson Cycle hypothesis, which many authors (e.g., Burke & Dewey, 1974; Burke, 2011) further developed as a unifying theory in geology. The Wilson Cycle explains the development of salt basins during the various stages of extensional and compressional tectonics (see Chapter 9 in Mohriak, 2025). The salt outcrops discussed in this chapter are related to different stages of the Wilson Cycle (Warren, 2010, 2016).

3.1. SALT DEPOSITION EPISODES IN THE ARABIAN PENINSULA, NORTH AFRICA, AND SOUTHERN EUROPE

Salt outcrops in the European, Middle East, and North African regions are the most studied examples of evaporite rocks on Earth's surface (e.g., Bally et al., 2012; Soto et al., 2017). Early models of salt deposition and deformation were first proposed by earth scientists based on observations in modern and ancient evaporite sequences that outcrop in northern Africa, southern central Europe, and the Middle East. In the desert regions of the Saharan Atlas, the Carpathians, and the Zagros thrust-fold belts, the salt deposits have been preserved by efficient cap rocks or an arid environment that reduced the risk of dissolution by exposing the rocks at the surface (see Chapter 2). The extensional rift basins, formed in the Arabian Peninsula during the Neoproterozoic, accumulated thick deposits of evaporite sediments at the transition to the Paleozoic (Alavi, 1994; Kent, 1979; Najafi et al., 2018).

The rift salt deposits in the onshore and offshore regions of the Arabian plate are related to the Zagros Basin, straddling the region from Saudi Arabia and Oman to the Arabian Gulf (also called the Persian Gulf) and some Middle Eastern countries, such as Iraq and Iran. Most of the salt structures outcropping in the Zagros region correspond to Neoproterozoic Hormuz salt. The Neoproterozoic basins formed two depocenters with evaporite sediments, which are separated by the Qatar-Fars structural high (Figure 3.2). The equivalent Neoproterozoic salt sequences in the eastern Arabian Plate (Ghaba Basin in Oman) are assigned to the Ara salt (Reuning et al., 2009).

The evaporite deposits in the Arabian Gulf and Zagros region were overlain by sedimentary sequences associated with the Tethys Ocean passive continental margins in the Paleozoic and Mesozoic (Berra & Angiolini, 2014). The tectonic setting changed to a convergent phase that started in the late Mesozoic and continues to the present day, forming the Zagros Orogeny (Figure 3.1). The Main Zagros Thrust is a deformation belt that resulted from the Tethys passive margins becoming convergent margins and entering the Zagros subduction zone in the Mesozoic (Figure 3.2). The Arabian plate movement vectors indicate a roughly N-S direction as the continental blocks of the Arabian plate approached the Eurasia plate during the Cenozoic (Ahmadian et al., 2014). The Zagros fold-thrust belt represents the convergent phase of the Wilson cycle, resulting in the subduction of the Tethys oceanic crust, the collision of continental blocks, the development of magmatic arcs, and the formation of suture zones in Iran (Agard et al., 2011; Schettino & Turco, 2011).

According to current earthquake activity and subsidence analysis in Iraq and Iran (such as Lalami et al., 2020), the Persian Gulf today is a shallow-water depression that corresponds to the active foreland basin of the convergent system. In addition to the Neoproterozoic salt, two other phases of evaporite deposition are recognized in the northwestern and southeastern sectors of the Zagros belt. Neogene evaporite sediments are observed in the Dezful embayment, located northwest of the Fars High, corresponding to the Gachsaran Fm. evaporites (Bahroudi & Koy, 1990). Neogene salt sediments are

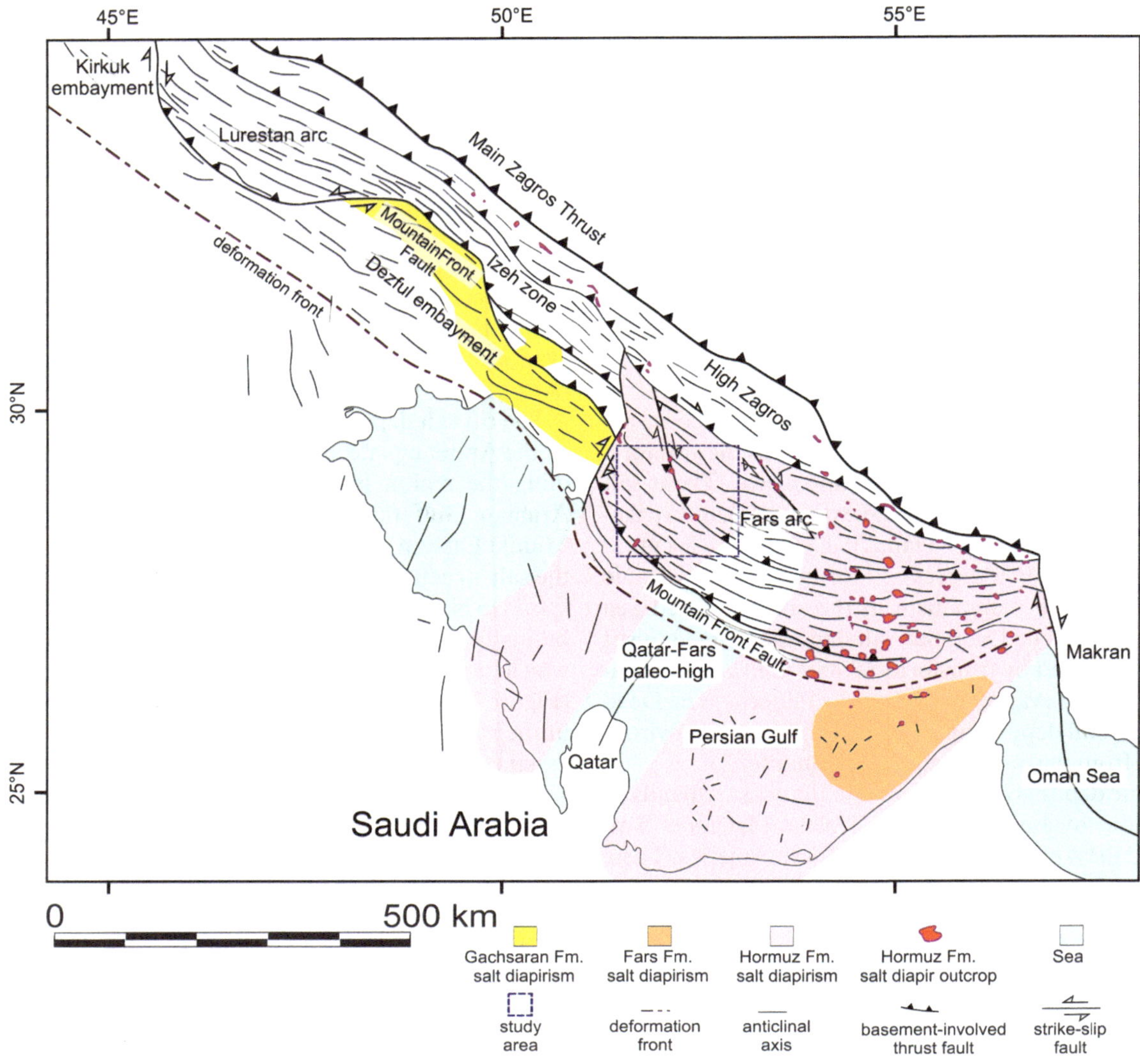

Figure 3.2 The Zagros thrust-and-fold belt, between the Arabian and Eurasian plates, is divided by the Qatar-Fars paleo-high into two main salt provinces associated with Neoproterozoic Hormuz salt. The Gachsaran Fm. (Dezful Embayment, north of the Persian Gulf) and Fars Fm. (southeastern part of the Zagros Basin, near the Oman Peninsula) are two main salt deposition intervals in the Neogene. Many salt structures exposed at the surface are associated with strike-slip fault zones. Credit: Adapted from Najafi et al. (2018).

observed in the southeastern province, near the Hormuz Strait, corresponding to the Fars Fm. (Jahani et al., 2007). The younger evaporite layers locally form salt diapirs that reach shallow depths or are exposed at the surface in the Dezful Embayment and eastern Zagros region (Figure 3.2).

The Zagros Orogeny fold-thrust belt is marked by different evaporite deposits, making anticlinal structures with axes that predominantly exhibit NW-SE and E-W trends (Jahani et al., 2007, 2009). Some salt features can be interpreted on satellite images and geological maps as exposed lithological sequences composed of evaporite rocks (e.g., Chorlton, 2007), but some subsurface structures are only interpreted by geophysical methods. Some transcurrent faults have displaced the axes of many anticlinal and synclinal structures in the Zagros fold-thrust belt during the Neogene, with variable directions and motions (e.g., Hessami et al., 2001; Rahnama-Rad et al., 2009). Some salt structures outcrop at the intersection of NW-trending and NE-trending fault zones (Figure 3.2). Regional geological maps indicate the anticlinal structures involving the post-Neoproterozoic sedimentary sequences exhibit a predominantly NW-SE trend (Figure 3.3).

3.1.1. Salt Deformation Styles in the Zagros Fold-thrust Belt

The salt structures in the Zagros fold belt have been studied for more than a century because they form the most spectacular outcrops in the world. In the late 19th century, several authors visited the region and described the folds and anticlines that are locally marked by salt fountains extruded from the mountaintops (see review by Jackson, 1995). Some of these structures were initially described as melted masses of flowing salt, with their mobility resulting from higher temperatures during their emplacement at depth (e.g., Gussow, 1966, 1968). However, several studies established that salt masses were mobilized and extruded on the surface under normal temperatures (e.g., Talbot, 1981, 1998). The association between the Hormuz Neoproterozoic salt and the Neogene diapirism was accepted only in the early years of the 20th century, when detailed geological maps were published by numerous researchers (such as Alavi, 1994; Ala, 1974; Bosák et al., 1998; de Böckh et al., 1929; Jahani et al., 2007, 2009, 2017; Kent, 1958, 1970, 1979; Lees, 1927; Sherkati et al., 2006).

The geological history of the Zagros region is extremely complex and involves several tectonic phases. Intracratonic rift basins developed in the Arabian plate during the Neoproterozoic, then Paleozoic to Mesozoic passive margins changed into convergent margins and entered subduction zones toward Eurasia in the Cenozoic (Alavi, 1994; Sherkati & Letouzey, 2004; Sherkati et al., 2006; Vergés et al., 2011). The Neoproterozoic Hormuz evaporite sediments were deposited in rift basins that subsequently evolved as divergent and convergent margins, resulting in several sedimentary sequences that reached a total thickness of more than 10–15 km in the Fars region of eastern Iran (Hassanpour et al., 2020; Kent, 1958; 1979). The salt layer deformation involved the basement (thick-skinned tectonics) or developed a thin-skinned tectonic style, with faults detaching at several decollement zones (Molinaro et al., 2005). Most of the anticlinal structures in the Fars region were formed by the folding of Paleozoic, Mesozoic, and Cenozoic strata that covered the Neoproterozoic Hormuz salt, but some structures involved a second detachment level associated with the Neogene salt layers (Hassanpour et al., 2020; Jahani et al., 2007, 2009, 2017; Letouzey & Sherkati, 2004).

The Alpine orogenic events are related to the closure of the Tethys Ocean and the convergence between the African, Arabian, and Eurasian plates. The plate convergence resulted in the subduction of the Tethys oceanic crust and block collision during the late Mesozoic and Cenozoic, resulting in the development of the Zagros fold-thrust belts at the boundary between the Arabian and Eurasian plates (Letouzey, 1990; Letouzey et al., 1995; Vergés et al., 2011). The contractional forces resulted in the squeezing of the previously developed salt structures in the rifts and continental margins. The salt masses were uplifted and locally extruded on the depositional surface, resulting in several outcrops of evaporite sediments, mainly from the Neoproterozoic Hormuz salt layer, but also involving some younger evaporite deposits (Hassanpour et al., 2020; Kent, 1958, 1970, 1979; Sherkati & Letouzey, 2004; Sherkati et al., 2006).

The Hormuz evaporite rocks, which occur in the subsurface at great depths but locally outcrop at the surface, form salt domes and salt glaciers. These salt structures correspond to salt tongues or sheets extruded on ancient and modern depositional surfaces, either subaqueous or subaerial (Talbot, 1979, 1981; Talbot & Rogers, 1980; Talbot & Jarvis, 1984).

The Hormuz salt horizon separates the pre-salt and the overburden regionally, forming a detachment surface. Most folds are formed by the propagation of the detachment faults in the subsurface. The fault propagation fold (also known as detachment fold) model is characterized by compressional forces that deform the overburden, while the autochthonous mother salt layer is affected by flow and locally by thrust faults that are blind or do not reach the surface (Burberry et al., 2007, 2008, 2011; Setchell et al., 2007; Trigui et al., 2016). Another mechanism that is commonly seen in thrust-and-fold belts is associated with faults that detach at stratigraphic levels above the lower evaporites. This mechanism results in faults that cut the overburden stratigraphy and flatten again on a younger horizon, forming a subhorizontal detachment level that controls a fault-bend fold (Burberry et al., 2007, 2008, 2011; Setchell et al., 2007; Trigui et al., 2016).

Some salt diapirs reached the depositional surface by Triassic times and remained close to the seafloor until the Neogene, while others only reached the surface in the Miocene to Pliocene (Hassanpour et al., 2020). The emergence of salt structures and their extrusion on ancient depositional surfaces were locally accelerated by the lateral strike-slip movements associated with the Zagros Orogeny (Ahmadian et al., 2014; Jahani et al., 2017; Rahnama-Rad et al., 2009).

According to earthquake activity, GPS readings, and the InSAR technique, which uses satellite images to measure the uplift of salt mountains, many salt diapirs in the Zagros region are still active (Ahmadian et al., 2014; Aftabi et al., 2020). Some salt mountains are actively rising and extruding salt, as indicated by the salt glaciers in the Fars region of Iran (Talbot & Rogers, 1980). Geological mapping and quantitative analysis of flow rates in the extrusive salt masses indicate that rainfall and climate control the salt movement in many active salt glaciers (Talbot & Aftabi, 2004).

The salt bodies in the Fars region normally spread outwards from their vertical sources, a diapiric root zone that corresponds to a feeder or conduit connecting the Neoproterozoic Hormuz salt at depth with the salt bodies at the surface. The salt mountains that source the flowing salt are characterized by high-relief profiles, which indicate the diapirs are rising faster than the rate of salt dissolution. There are several examples of salt masses presently advancing toward the valleys between salt mountains in the Fars region. For example, the Kuh-e-Gach, Kuh-e-Siah, and Kuh-e-Namak are salt glaciers

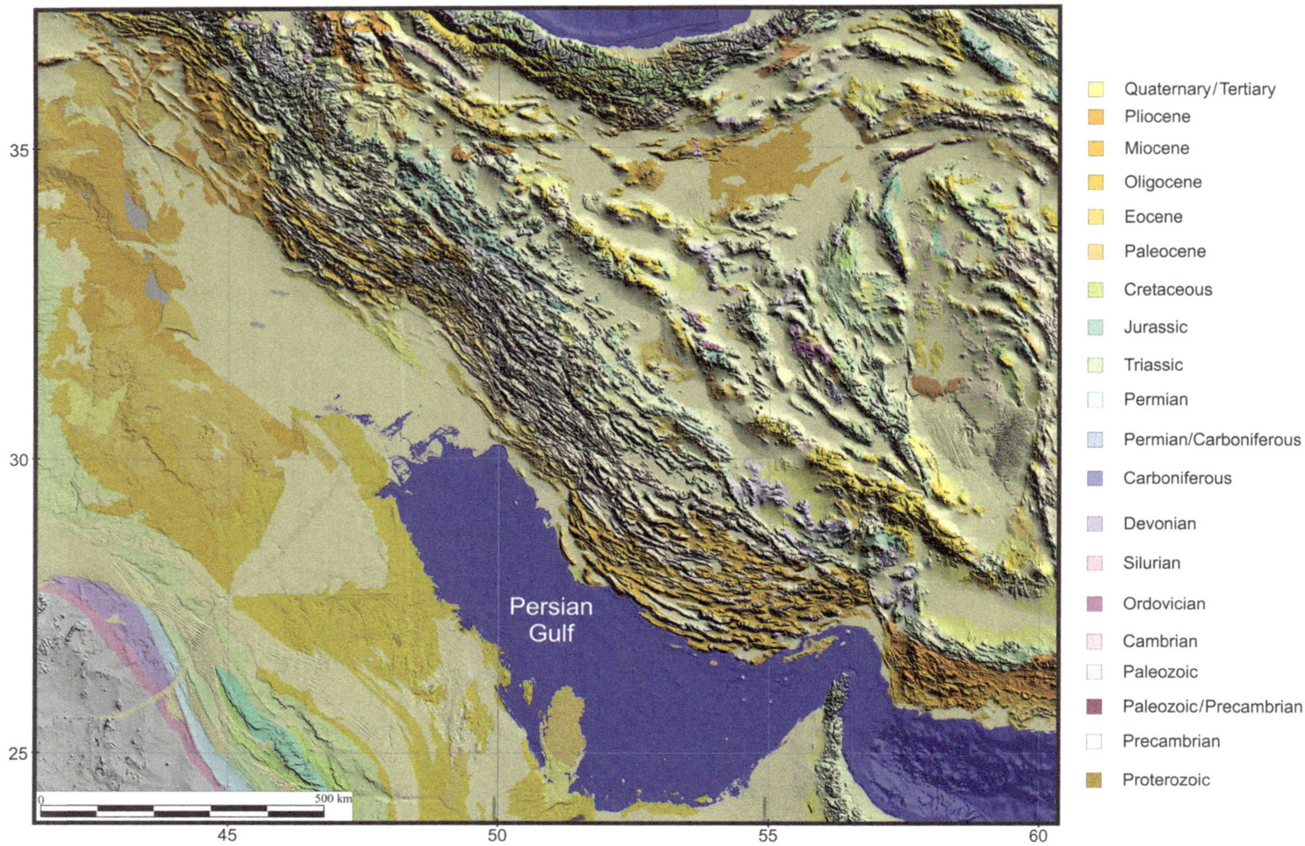

Figure 3.3 The simplified geological map of the Persian Gulf region shows the age of outcropping lithological units draped on topography. The Zagros thrust-and-fold belt has an NW-SE trend, influenced by the convergence of Arabian and Eurasian plates. Quaternary rocks cover the depression northwest of the Persian Gulf, corresponding to the Mesopotamian foredeep. Mesozoic and Cenozoic rocks outcrop at the Dezful Embayment and the northern part of the Zagros deformation zone. Paleogene and Neogene sediments outcrop in the Fars region north of the Hormuz Strait. The Hormuz salt has breached several anticlinal and synclinal structures, forming salt outcrops. The Kavir Desert, located south of the Elburz Mountains, has a large plain covered by Neogene to Quaternary sediments. Some subcircular structures correspond to the tops of salt diapirs exposed at the surface. Credit: Chorlton (2007)/with permission of Natural Resources Canada (NRCAN).

fed by a salt diapir that breached the anticlinal structures (e.g., Maleki & Jahadgar, 2015; Talbot, 1979, 1981, 1998; Zhang et al., 2021).

3.1.2. Salt Outcrops in the Fars Region

Over 200 salt domes outcrop in the central Zagros fold-thrust belt of Iran (Kent, 1979). Some structures are characterized by a central vent feeding salt glaciers or salt fountains that extrude salt on younger sedimentary sequences (Talbot, 1998; Talbot & Pohjola, 2009). Most salt outcrops correspond to topographic protuberances that form hills in the middle of a crater, which can be filled with debris derived from the overlying sedimentary sequences or covered by less soluble material after the salt has been dissolved (see Chapter 8 in Mohriak, 2025).

The original Neoproterozoic Hormuz salt sequences lie at depths of 4–12 km in the Zagros Plateau, but at the eastern segment of the Persian Gulf, near the Strait of Hormuz, the depth to the base of the salt horizon may reach 20 km (Hassanpour et al., 2020). The evaporite layers can only be seen in outcrops when salt is remobilized into diapirs that pierce the cover rocks and reach depths close to the depositional surface. Due to continual uplift and erosion, the overburden is removed and salt is exposed (Figure 3.2).

Locally, the salt stems form the core of some islands near the coastline, such as Qeshm, Hormuz, and Larak islands (Figure 3.4). Qeshm Island, the largest island in the Persian Gulf, is the first Middle East UNESCO Global Geopark due to its geological, biological, cultural, and natural attractions (e.g., Yazdi & Dabiri, 2018). The

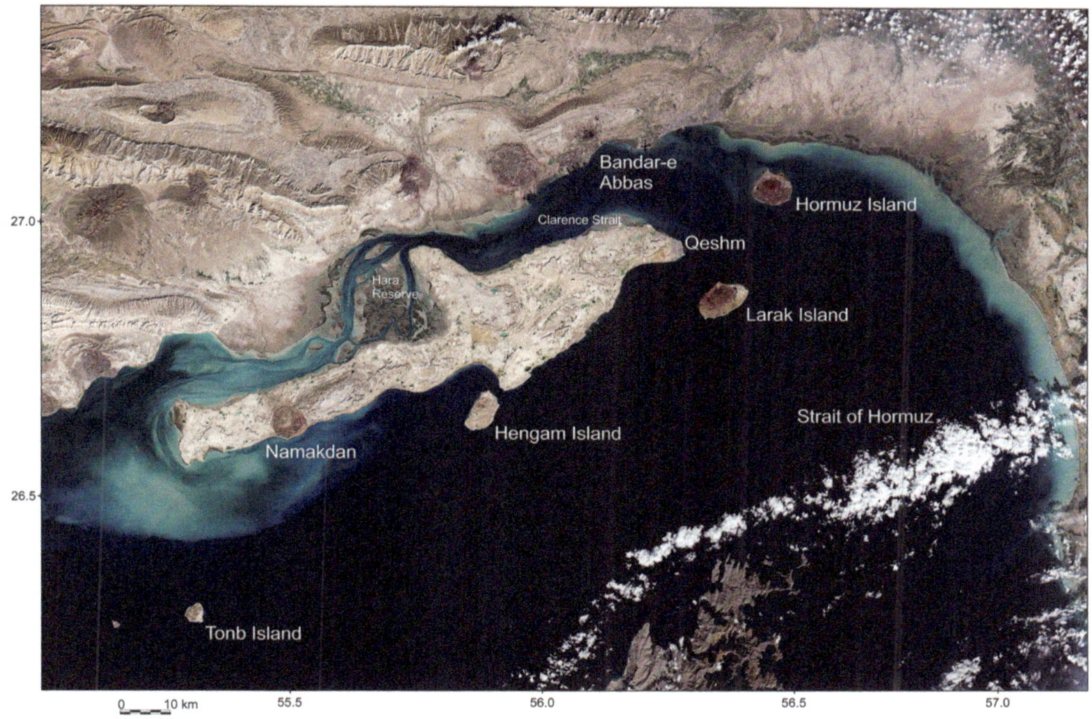

Figure 3.4 The satellite image (natural colors) depicts the geomorphic expression of anticlinal structures in the Fars region of southeastern Iran. The brownish reflectivity on land resembles salt diapirs. A narrow channel separates Qeshm Island from the mainland in the eastern Persian Gulf, near the Strait of Hormuz. A circular structure in the southwestern part of the island corresponds to the Namakdan salt diapir. Credit: Courtesy of NASA.

Namakdan salt diapir, located in the southwestern part of the island, forms a mountain with an altitude of about 400 m above sea level and is still rising at a rate of about 6 mm per year, according to GPS and radar interferometry measurements (Hajiesmaeili & Rashidi, 2019; Hassanpour et al., 2020). Some Hormuz salt outcrops were exploited in salt mines for centuries before being designated as a natural protection area and a World Heritage Site. One of the Namakdan salt caves is considered the longest in the world, with a length of about 6.5 km (IUGS, 2022).

The Zagros fold-thrust belt in southeast Iran is characterized by anticlinal and synclinal structures with NW-SE to E-W axes (Figure 3.5). However, compressional strain has deformed these structures, resulting in contorted anticlinal structures with eroded flanks. These structures are mainly related to fault-bend folds and detachment or propagation folds (Burberry et al., 2008, 2010, 2011; Setchell et al., 2007). Salt structures have breached these folded anticlines, extruding salt masses on the surface and forming salt glaciers (Figure 3.5).

Several anticlinal ridges were breached by Hormuz salt diapirs and have formed prominent regional structures, with topographic heights exceeding 1,000 m. The high topographic elevations indicate that the rate of salt flow from the source is greater than salt extrusion and dissolution. Most salt domes in the Fars region are sourced from Hormuz mother salt, but some diapirs in the Dezful Embayment or in the eastern Fars province involve younger salt deposits (Hassanpour et al., 2020).

Eroding and dissolved salt domes have resulted in karst topography and a crater above the salt stem. Fluvial drainage has transported these evaporitic sediments toward valleys between uplifted mountain ranges, where salt is redeposited with other siliciclastic sediments, mainly silt and clay. The valley floors constitute more fertile land for agricultural use. For example, the valley between the Kuh-e-Gach and Kuh-e-Siah shows a drainage system that allows the cultivation of some agricultural products in a desertic land environment (Figure 3.5).

Extruded salt may form salt tongues when encased between sedimentary layers or salt glaciers when extruded onto the depositional surface (Hassanpour et al., 2020; Talbot, 1981, 1998). The salt glaciers in Iran are sourced from diapirs that rise from great depths in the subsurface. The extruded salt masses flow along the mountain flanks, following sinuous canyons, and finally forming alluvial fans that spread in the valleys or depressions (Figure 3.5). These features serve as distinctive contemporary analogs for salt extrusion phenomena that geophysical techniques (seismic data) have detected in the subsurface of numerous giant salt basins, including the Gulf of Mexico (Talbot, 1998) and the North Sea (Mohr et al., 2007).

Figure 3.5 Landsat satellite image (enhanced colors) showing the geomorphic expression of the anticlinal ridges and salt structures of the Zagros fold-and-thrust belt in the Fars region. The Landsat satellite image (enhanced colors) in the Fars region shows the geomorphic expression of anticlinal ridges and salt structures in the Zagros fold-thrust belt. The image shows salt structures breaching the anticlines, with the main axes trending from NW-SE to E-W. White reflectivity patches correspond to residual salt deposits dissolved in the Hormuz salt, while greenish colors indicate agricultural areas. The rectangles highlighted by thin yellow lines show the location of the zoom images of the salt structures in the northern and southwestern parts of the image. The northern structure shows salt masses extruded from the central feeder, forming a topographic elevation. The ellipsoidal salt mass has been cut by a drainage system that forms alluvial fans in the valleys. South of the salt structure, a shallow lake with dissolved halite exhibits a light blue color. The lower left yellow rectangle shows two salt tongues associated with two E-W trending anticlinal structures. The drainage was diverted by the salt tongue fed by the Gach (Kamostaj) salt diapir. The allochthonous salt glaciers flow toward a valley with cultivated fields that appear green at the center of the image. The salt tongues appear dark brown due to impurities mixed with halite minerals. Credit: Scientific Research Publishing Inc./CC BY 4.0.

One didactic example of an active salt glacier in the eastern Fars Province corresponds to a topographic high reaching an altitude of circa 1,300 m at the summit (Figure 3.6). The salt diapir breached a NW-trending structure, forming a salt glacier that extruded from a central dome. The salt glacier flows toward the valley and displays a pancake shape, suggesting a radial flow (Figure 3.6B). The ellipsoidal salt mass is cut by a drainage system that forms alluvial fans in the valleys (Figure 3.5).

During the Zagros Orogeny, the Hormuz salt diapirs have been deformed by compressional forces. The stems have been squeezed, forming mushroom-shaped structures, resulting in fountains that extruded salt on the modern surface of the Zagros Plateau. Two E-W trending anticlines, located near the Kamostaj village (southwest sector of Figure 3.5), illustrate the formation of salt glaciers flowing down the valley. The northern structure is the Kuh-e-Gach anticline, which is breached by the Gach (or Kamostaj) salt diapir. This diapir sources a salt glacier that flows to the south, toward a valley that extends into an east-west direction (Figure 3.7). The southern anticline, corresponding to the Kuh-e-Siah Mountain, was breached by the Burkh salt diapir, which extruded evaporite masses and formed a salt glacier that flows to the north (Jahani et al., 2007; Razaghian & Arian, 2015; Sarkarinejad et al., 2018; Talbot, 1998; Waltham, 2008). The salt masses are currently flowing toward the valley floor between the surrounding mountain ranges due to gravity forces acting on the mobile evaporites. The valley floor is filled with remnants of dissolved salt and silt deposited from the small streams that flow eastward, surrounding the salt glaciers. The fluvial waters are used for agricultural purposes in farmlands, which appear green in satellite images (Figure 3.5).

NASA's computer-enhanced 3D visualization of two salt glaciers in the Fars region reveals that the salt flows

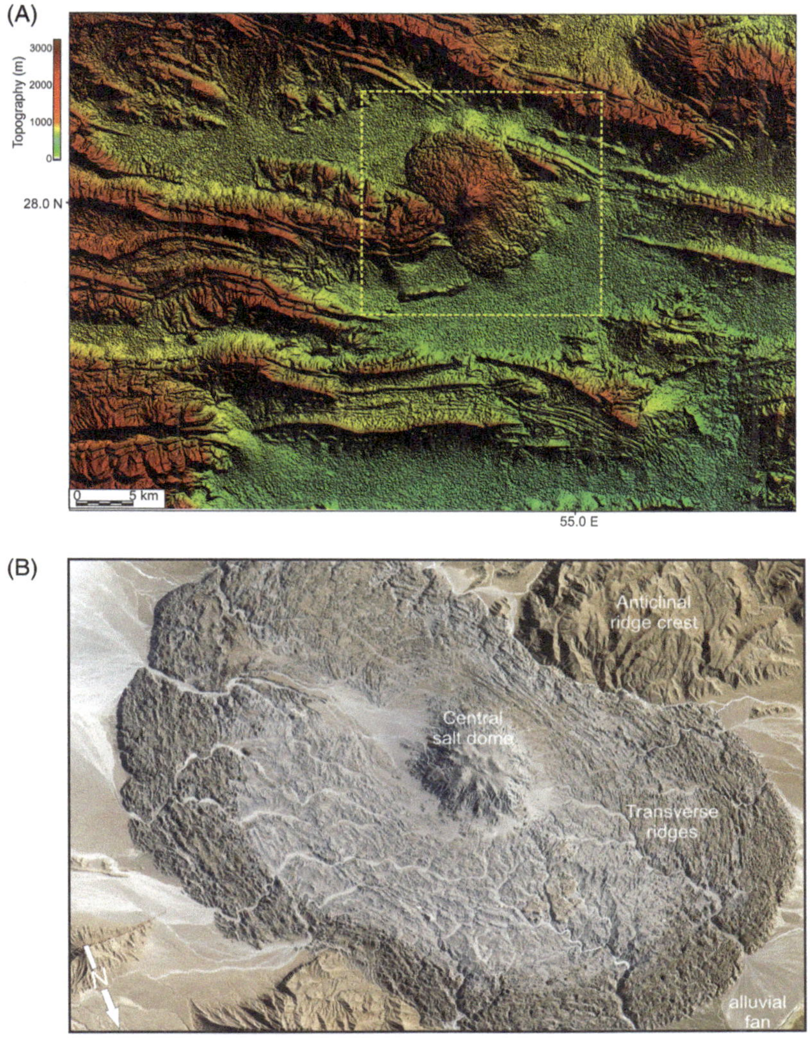

Figure 3.6 (A) The satellite image draped on topography (SRTM) shows an anticlinal structure breached by a salt diapir in the northern part of the Fars region of Iran. The salt mountain, reaching 1,300 m, extrudes salt masses, forming salt glaciers with an ellipsoidal shape. (B) The satellite image (natural color) shows a pancake-shaped structure with a central dome feeding a salt glacier that spreads laterally. The dome is cut by a radial drainage system, forming alluvial fans in valleys. White reflectivity patches correspond to residual salt deposits from the Hormuz evaporite sediments, which were extruded from the salt mountain and redeposited in the valleys. Credit: NASA/Public domain.

from two diapir sources opposite each other, similar to lava flows (Figure 3.8A). The salt glaciers are marked by dark brownish hues due to silts and clays from older sedimentary sequences pierced by the salt diapir. These sediments are mixed with the salt mass, which is predominantly composed of Hormuz halite. Some exotic blocks of various lithologies have been incorporated into the flowing salt mass (Figure 3.8B).

The salt glaciers in the Fars region are characterized by folding inside layered evaporite sequences. They flow slowly downhill toward valleys, developing thrust faults and small-scale folding as they move over terrain obstacles (Talbot & Aftabi, 2004). When their flow velocity is reduced by an obstacle, compression forces result in crenulations in the stratification (Figure 3.8B). Allochthonous sheets advance over underlying sedimentary sequences with different stratigraphic ages, covering Paleogene to Neogene sediments and Quaternary alluvium, as observed in the Kuh-e-Gach and Kuh-e-Siah salt glaciers (Figure 3.8A).

After emerging at the surface, the allochthonous salt masses are affected by gravity, spreading beneath a brittle carapace that thickens downstream. Laterally, the salt spreads, forming tongues with snouts that advance over sediment-infilled valleys. This process is similar to a tank

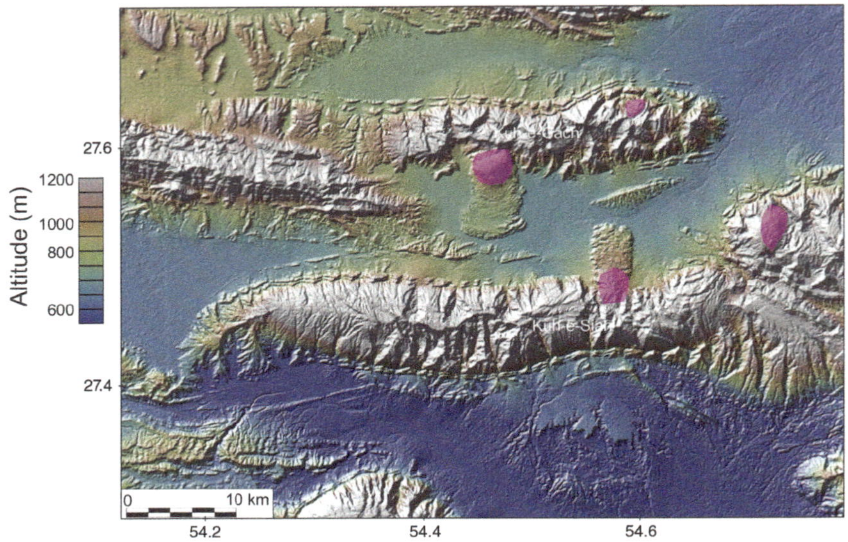

Figure 3.7 The topographic map depicts the E-W trending Kuh-e-Gach and Kuh-e-Siah anticlines, with two salt glaciers flowing toward a valley near Kamostaj village. The salt tongues are shown in yellowish tones, and the salt feeders are in pink. Credit: NASA/Public domain.

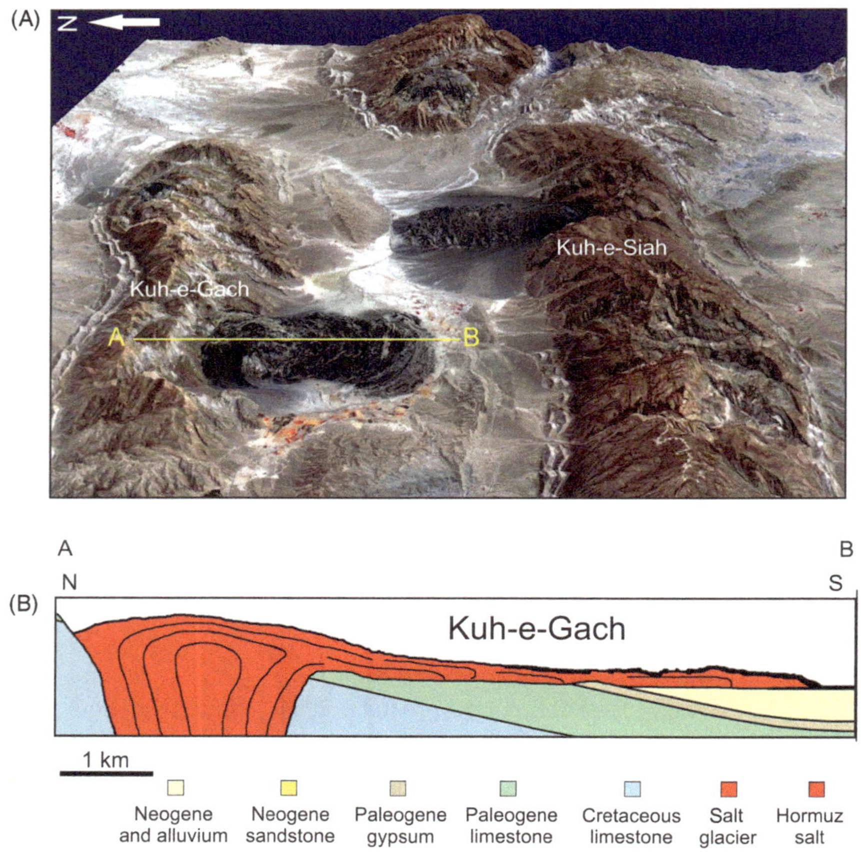

Figure 3.8 (A) False-color 3D view with a perspective toward the east, created by draping a NASA ASTER satellite image on the digital elevation model (DEM). The Kuh-e-Gach and Kuh-e-Siah anticlinal structures were breached by salt diapirs, forming salt glaciers that flow toward the valley. The salt glaciers appear dark brown due to impurities mixed with evaporite sediments, primarily composed of halite. Credit: NASA/Public domain. (B) Schematic geological cross-section showing the interpretation of the Kuh-e-Gach structure. The salt diapir intruded Paleozoic, Mesozoic, and Cenozoic layers and extruded over Cretaceous to Neogene sediments. The distal part of the salt glacier advanced over younger sediments by moving over alluvium and breccia, which are rubble sediments and fragments of evaporitic rocks. When the rate of extrusion exceeds the rate of deflation caused by erosion or dissolution, the salt glacier advances toward the valley like a lava flow. The salt glacier front may retreat during wet periods. Credit: Waltham (2008)/with permission of John Wiley & Sons.

track, with the top salt layer rolling over the snout and advancing over stationary lower limbs. During the process, the evaporite masses are smeared into finer-grained mylonitic fragments. The salt glacier front advances toward topographic lows when the rate of extrusion is greater than the deflation caused by erosion or dissolution. However, during wet periods, the salt front may retreat as the salt is eroded or dissolved (Talbot, 1979, 1998; Talbot & Aftabi, 2004).

3.1.3. The Kuh-e-Namak Anticline and the Dashti Salt Diapir Outcrops

The satellite image of the western segment of the Fars province in central Iran shows Cenozoic to Mesozoic strata affected by the Zagros thrust-and-fold belt (Figure 3.9). Major anticlines near the Arabian Gulf coastline trend NW-SE and are related to fault-bend folds and detachment or propagation folds (Arian & Noroozpour, 2015).

Some salt diapirs sourced from the Hormuz salt locally breached anticlinal structures and developed extrusive salt glaciers, such as the famous Kuh-e-Namak, which means "Mountain of Salt." This classical anticlinal structure in the Zagros fold-thrust belt is marked by a tall mountain (almost 1,500 m high), which is the source of the salt glaciers that flow downslope (Talbot, 1979, 1981; Talbot & Jarvis, 1984; Zhang et al., 2021). The satellite image shows the NW-SE trending anticlinal structure with a salt fountain at the northwest extremity, which is the source of the salt glaciers that flow downslope (Figure 3.9A).

The Kuh-e-Namak anticline and the Dashti salt tongue display a rugged topography deeply dissected by canyons and marked by salt dissolution features (Figure 3.10). The whitish reflectivity in the middle of the eroded fold structure corresponds to the evaporite masses exposed at the surface. The Jashak (or Dashti) salt dome, one of the best-studied Hormuz evaporite outcrops in Iran, is associated with two salt glaciers that flow toward the adjacent valleys in opposite directions. The alluvial fans in the depressions are characterized by a Quaternary to Holocene sedimentary plain composed of outwash gravels mixed with evaporite minerals. The extruded evaporites are mainly composed of halite, derived from Neoproterozoic to Cambrian Hormuz salt, mixed with silt, clay, and evaporite fragments (Figure 3.10). At the center of the crater, the banded salt layers that feed the salt diapir are highly tilted or subvertical, resulting from salt flow from great depths to the surface. Small-scale, conspicuous dissolution features affecting the halite layers can be seen in the foreground of the salt glacier (Figure 3.10).

3.1.4. Salt Exposures in the Great Kavir Desert of Iran

The Great Kavir, also known as Dasht-e-Kavir, is a low-altitude plain north of the Zagros main thrust fault, extending from the Garmsar Basin to the Lut Plain south of the Elburz Mountains (Figure 3.11A). It is associated with Neogene sedimentary sequences covered by desert dunes (Arian, 2012; Jackson et al., 1990). During the Oligocene and Miocene, evaporites in the Garmsar Basin created salt diapirs and allochthonous salt glaciers. These shaped the modern depositional surface in the Great Kavir desert and can be seen as subcircular structures on satellite images (Arian, 2012; Talbot & Aftabi, 2004).

The Great Kavir sedimentary basin is an intercontinental foreland basin with a thickness of about 6–7 km. It comprises two poorly dated evaporite sequences, the Older Salt and the Younger Salt, which are associated with Eocene and Miocene evaporite deposits. These layers have been deformed in the Neogene, resulting in clusters of salt canopies that outcrop at the surface (Figure 3.11B). The dissolution of these evaporite rocks in the Holocene resulted in the redeposition of halite in the modern salt flats that overlie post-Miocene strata (Jackson et al., 1990).

The lower salt unit corresponds to a clean white halite layer at the base of ductile marls and shales, while the upper evaporitic layer (Younger Salt) is associated with rhythmic sequences of multicolored halite and gypsum minerals. These salt layers are intercalated with Oligo-Miocene limestones and marls of the Qum Formation, which do not outcrop on land. Halitic to gypsiferous mudstones and siltstones of the Upper Red Formation, deposited in the Middle and Upper Miocene, cover the upper evaporites. Pliocene conglomeratic siliciclastic rocks overlie the Miocene evaporitic-siliciclastic layers (Jackson et al., 1990).

Geological maps in the Kavir region indicate a sedimentary cover involving mainly Paleogene to Neogene sedimentary sequences capped by the modern salt flats (Figure 3.11A). Dozens of salt diapirs outcrop in the Great Kavir Basin, forming clusters of elongated to subcircular salt bodies encased in sedimentary rocks, which formed mini-basins between the salt stems (Figure 3.12A). Some of the salt exposures are characterized by outstanding geomorphological features, such as subcircular to ellipsoidal salt ridges surrounded by banded siliciclastic rocks (Figure 3.12B). The tall salt diapirs formed extrusive protuberances that look like mushrooms in cross-section, similar to the canopy structures imaged by modern seismic data in the Gulf of Mexico (see Chapter 4 in Mohriak, 2025).

Enhanced-color satellite images in the Kavir desert exhibit clusters of diapiric structures formed by the two Neogene salt sequences (Old and Young salt layers). These emergent salt diapirs commonly exhibit a halite core surrounded by evaporitic-siliciclastic sediments. The salt extrusion process generally inverts the initial depositional stratigraphy, as the older salt layers, composed of white halite, overlie the younger salt deposits, composed of multicolored layered evaporites (Jackson et al.,1990).

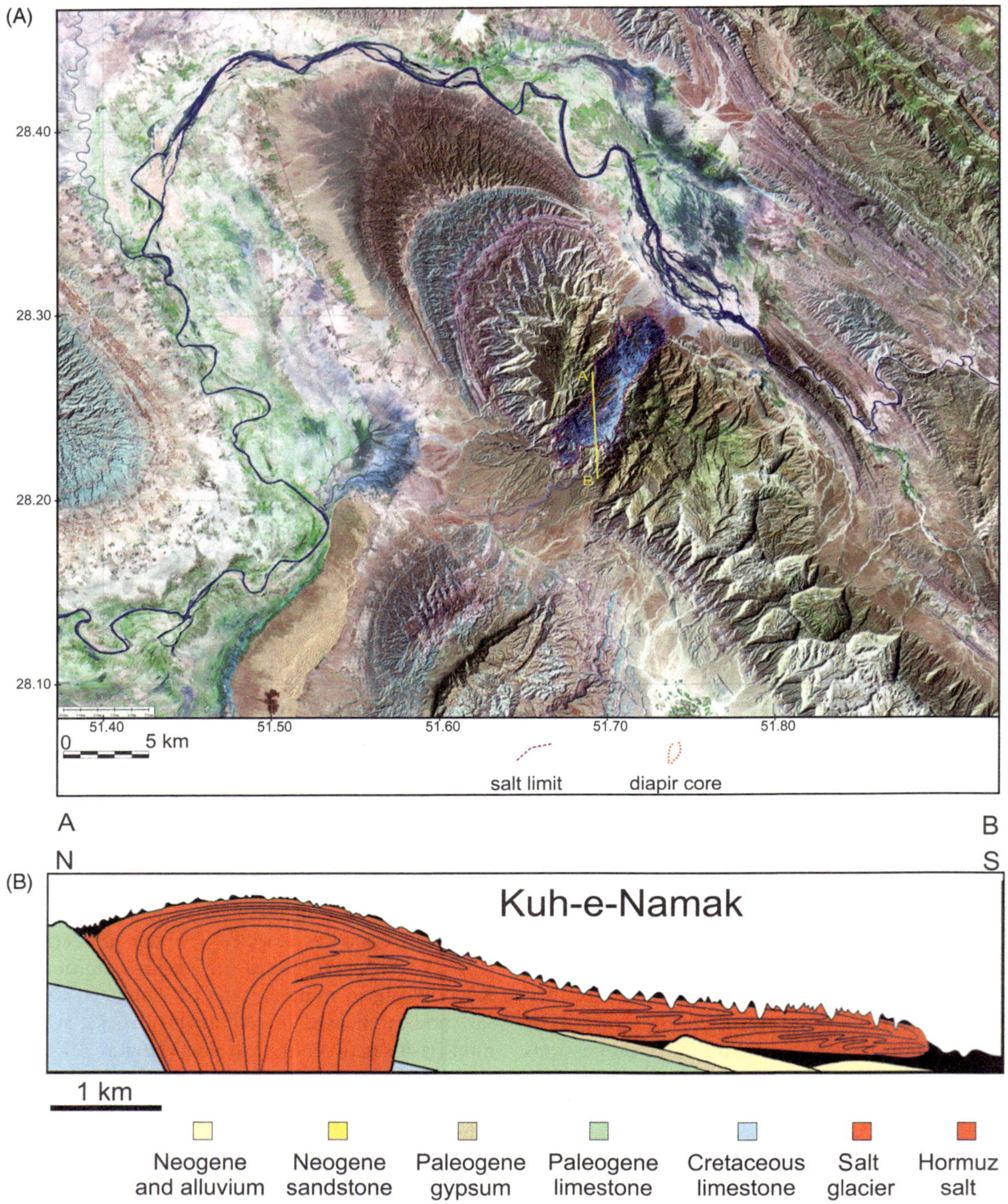

Figure 3.9 (A) Landsat satellite image (enhanced colors) showing the Kuh-e-Namak anticlinal structure, with the main axis trending NW-SE. The salt glacier is characterized by a light blue reflectivity, whereas the river floodplains are light green. The salt glacier shows a general NE-SW trend, following the faults that breached the anticline in the Cenozoic. A thin yellow line highlights the north-south cross-section (profile A-B). Credit: Courtesy of NASA. (B) Schematic diagram (cross-section A-B) illustrating the Kuh-e-Namak salt glacier model. The salt core feeds the glaciers, which flow toward valleys and advance over sedimentary strata. The Dashti salt dome has vertical bedding planes due to salt masses flowing from the mother salt layer, extruding at the surface. Credit: Waltham (2008)/with permission of John Wiley & Sons.

(A)

(B)

Figure 3.10 (A) The Dashti salt diapir in the Kuh-e-Namak anticline, along with extrusive sheets of allochthonous salt that extend over the surrounding country rock. The salt, colored gray to pink, extrudes over stratigraphic sequences uplifted by compressional forces during the Zagros Orogeny. Small-scale dissolution features affecting the halite layers are seen in the foreground. (B) The layered salt sequences in the Dashti salt diapir. Credit: Courtesy of P. Crochet, France.

The Holocene salt flats overlie the Neogene (mainly Miocene) halitic to gypsiferous fine-grained siliciclastic rocks of the Upper Red Formation, which exhibit marked layering in satellite images and geological maps (Figure 3.12B). The dissolution of evaporites by sparse rains results in the development of Holocene mud flats with salt crust sediments. These salt flats appear with a light blue color in enhanced satellite images (Figure 3.13A).

A detailed aerial photograph showing one of these salt bodies shows evaporite sequences characterized by whitish patches encased in Miocene sedimentary sequences (Figure 3.13B). Sedimentary layers are discordant with the salt stem, with the strata abruptly interrupted at the salt body's flanks (Mohriak & Szatmari, 2009). Subparallel concordant layering, expressed as light to dark gray bands, characterizes sedimentary sequences away from

(A)

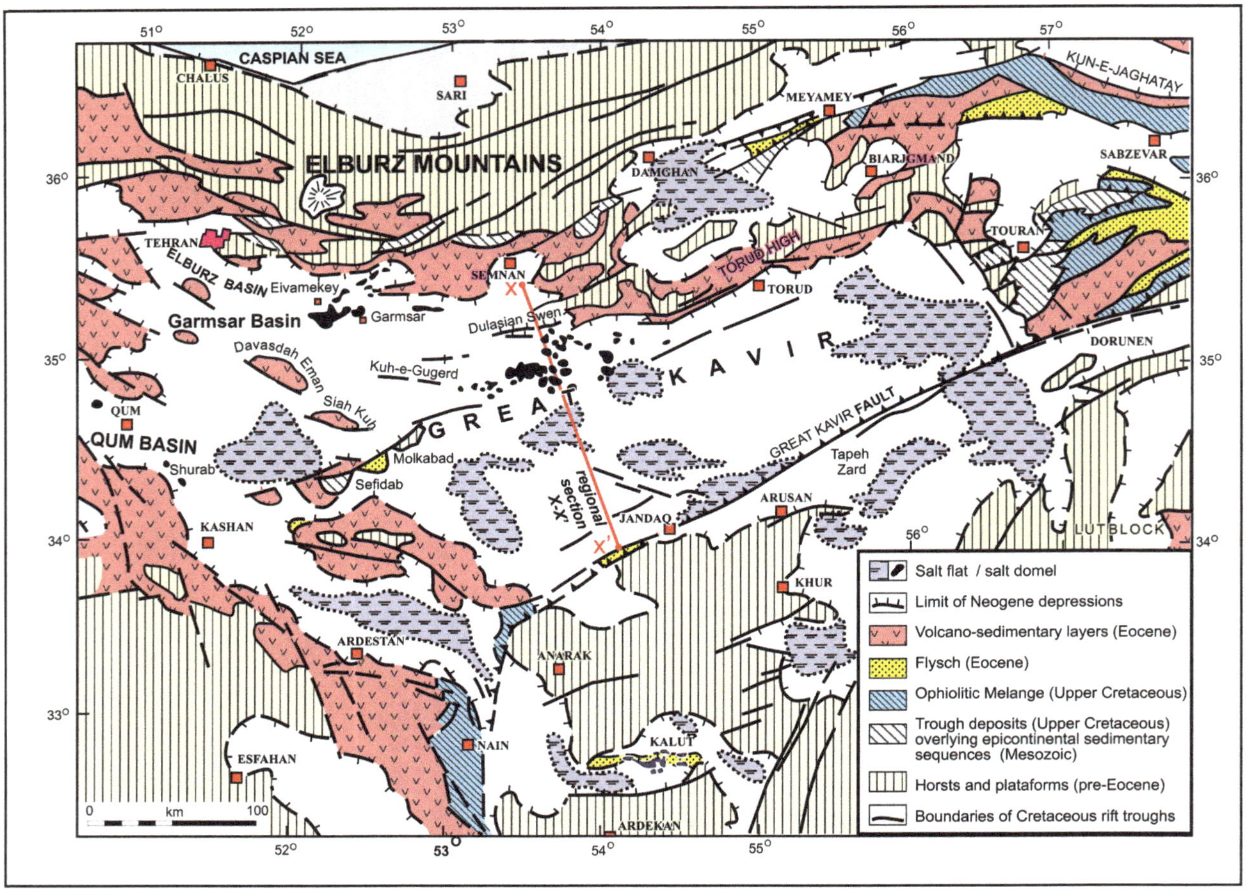

(B)

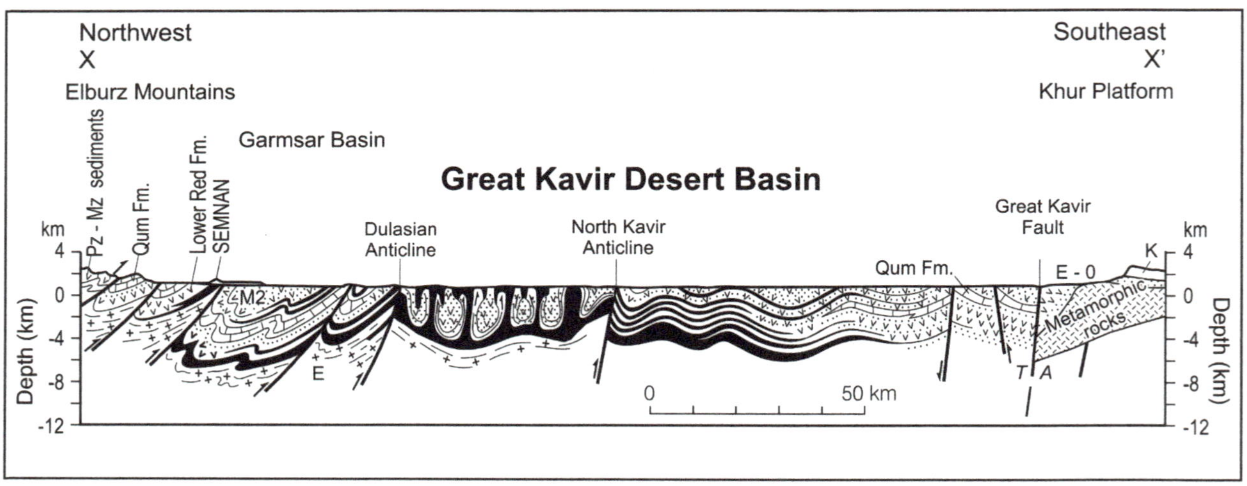

Figure 3.11 Geological map and cross-section in the Great Kavir region of northern Iran. (A) The geological map of the Great Kavir region of northern Iran shows a regional cross-section X-X′ extending from the Elburz Mountains to the central part of the Great Kavir Desert. The map also shows the Great Kavir Fault at the southeastern extremity. The map shows Neogene salt layers, forming a cluster of salt structures in the middle of the profile. (B) Schematic geological cross-section X–X′, showing the Great Kavir Fault at the southeastern extremity. The Neogene salt layers are shown in black, forming a cluster of salt structures in the middle of the profile. The tall diapirs were affected by compressional forces, resulting in extrusive bodies interpreted as mushroom-like structures. Some salt bodies outcrop in the desert, forming clusters of elongated to subcircular structures. Credit: Adapted from Jackson et al. (1990).

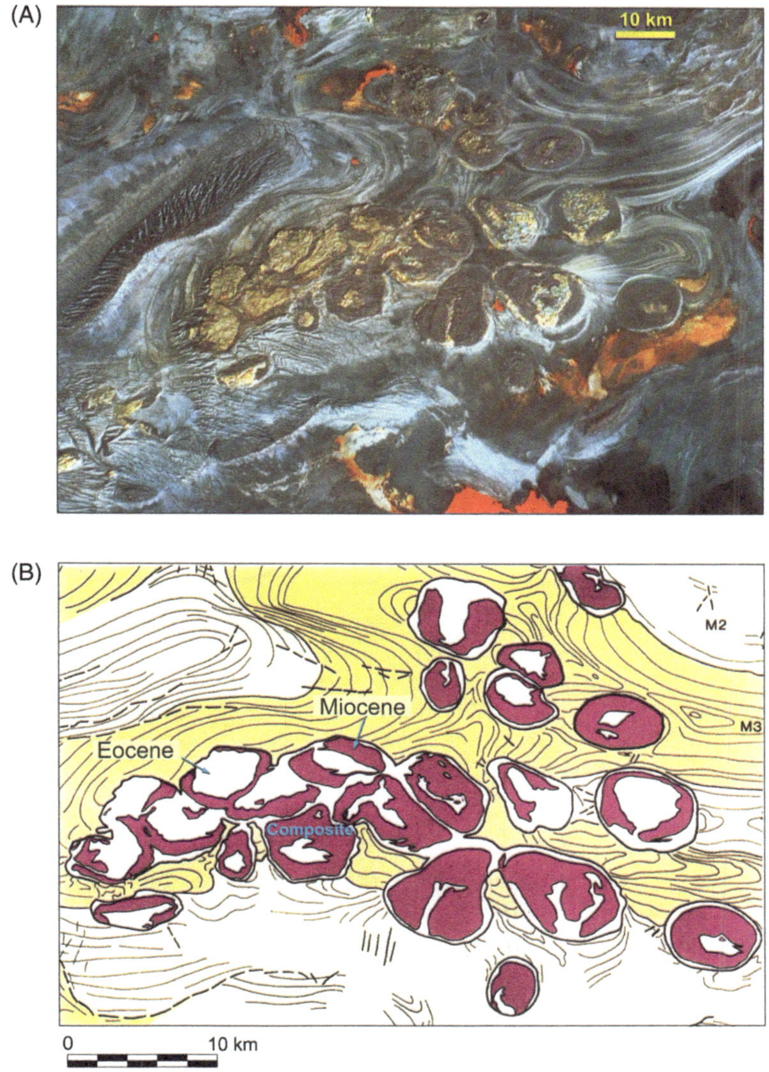

Figure 3.12 Satellite image (enhanced colors) with interpretation. (A) The satellite image (enhanced colors) shows a cluster of salt structures at the northwest corner of the geological map, corresponding to the stems of tall salt diapirs exposed at the surface. The evaporite deposits dissolve, forming Holocene salt flats overlying Neogene sedimentary sequences. The dissolved and redeposited evaporites exhibit light blue reflectivity due to the presence of halite. Credit: Courtesy of M. Jackson. (B) Schematic interpretation of the cluster of salt structures exposed in the Great Kavir Desert. The cores of the circular structures correspond to the Older Salt (Eocene), while the borders are associated with the Younger Salt (Miocene). Credit: Adapted from Jackson et al. (1990).

the diapir stem (Figure 3.13B). Modern erosional surfaces on satellite images resemble time-slices from 3D seismic volumes (see Chapter 7 in Mohriak, 2025).

3.2. SALT OUTCROPS IN THE ALPS AND PYRENEES MOUNTAINS OF SOUTHERN CENTRAL EUROPE

Evaporite deposits from the breakup of Pangea are found on the European mainland in various outcrops, from Iberia to the eastern Alps (Figure 3.14). These salt layers are associated with small extensional rift basins formed in the Triassic and preserved onshore and offshore (e.g., Soto et al., 2017). The stratigraphic sequences include evaporite layers formed during a Late Permian to Early Triassic rifting event in the Northern Calcareous Alps region (Granado et al., 2018). The evaporite deposits are overlain by platform carbonates that grade into deep-water sequences, which currently outcrop in mountain ranges (Figure 3.15). The Alpine Orogeny resulted from the convergence of African, Iberian, and Eurasian plates in the Mesozoic and Cenozoic, resulting in the inversion

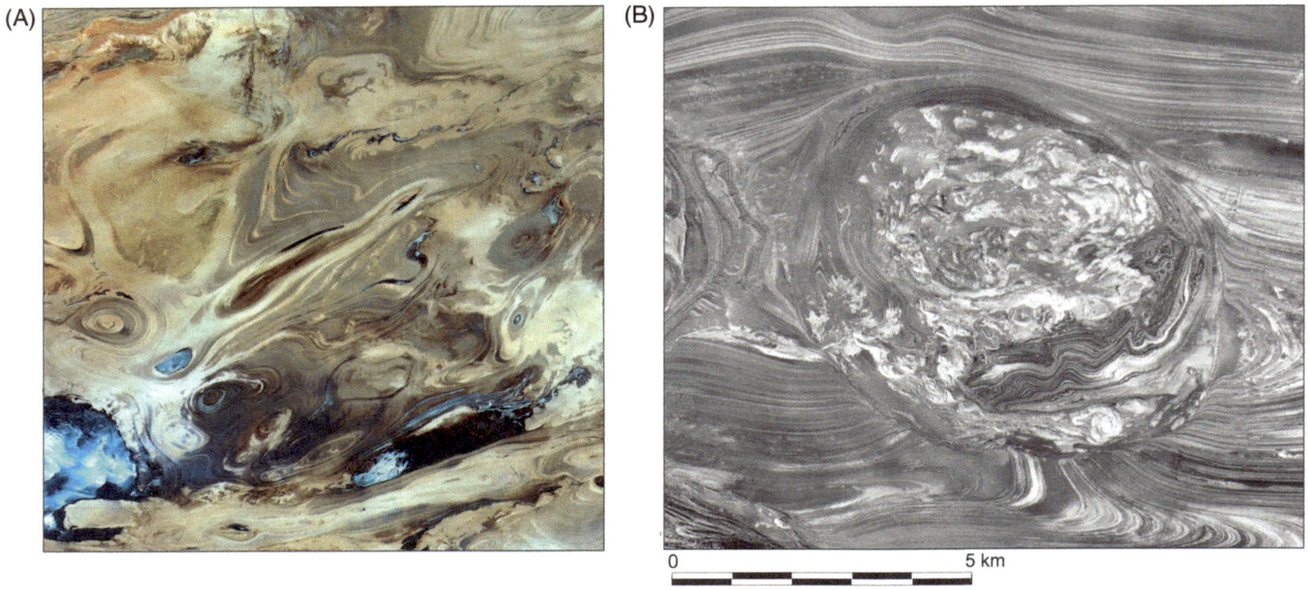

Figure 3.13 Satellite images of an outcropping salt structure in the Great Kavir Desert of northern Iran. (A) Satellite image (enhanced colors) illustrating elongated and subcircular salt structures. Credit: NASA Goddard Space Flight Center/Flickr/CC BY 2.0. (B) The aerial photograph shows an elongated structure corresponding to an outcropping salt body. The salt structure is encased in Miocene sedimentary sequences that are discordant with the vertical diapir stem. The Miocene strata are abruptly interrupted at contact with the salt body, while the Miocene sedimentary sequences north of the salt body (Upper Red Fm.) are characterized by subparallel concordant layering, expressed by light to dark gray bands. Credit: M. Jackson et al. (1990)/with permission of Geological Society of America.

of sedimentary basins by folding and thrusting (e.g., Dewey et al., 1989; Frizon de Lamotte et al., 2015). Several basins developed salt diapirs that were squeezed during these subsequent phases of compressional tectonics.

The deformational structures in Austrian salt mines suggest two collisional events (Leitner & Spötl, 2017). The ocean that deposited Triassic salt layers began to close by Jurassic times, leading to the Astroalpine continental crust forming nappes that stacked Austroalpine sedimentary units on the European foreland (Granado et al., 2018). After the subduction of the oceanic crust slab below the Apulian block, a collisional event involving continental blocks resulted in thrust faults and nappes, mainly involving platform carbonate rocks. These nappes were transported northwards over the European block, covering previously deposited sediments (Granado et al., 2018). The second main collisional episode in the Alpine region occurred in the Cenozoic, with a climax in the Late Eocene to Oligocene, associated with the collision of the Adriatic and European continental crust blocks. This event is associated with the Pyrenean Orogeny in Iberia, which formed a deformation zone between Spain and France (Figure 3.14) and affected the salt and molasse deposits in the northern Alps (Figure 3.15).

The basement-involved compressional structures inverted the Permo-Triassic rift sediments and deformed the Late Triassic salt layers by detachment faults that transported the supra-salt sediments toward the foreland as nappes (López-Mir et al., 2015).

3.2.1. The Hallstatt Salt Diapir in the Austrian Alps

The Austrian Salt District, located in western Austria, is a famous region where salt outcrops have been recognized and explored since prehistoric times (see Chapter 1). The Hallstatt salt mines, located at the foot of the Northern Calcareous Alps (Figure 3.15), are among the oldest active salt mines worldwide, with production dating back to 7,000 years, as in the Hallein salt mine (Harding, 2013). The salt deposits exploit a salt diapir dated as late Permian to early Triassic, and the evaporite layers correspond to the oldest sedimentary rocks in the area, which have been affected by extensional processes in the Mesozoic and compressional tectonics in the Cenozoic (Schorn & Neugebauer, 2014; Soto et al., 2017).

The evaporites mark the transition from mostly terrestrial environments in the Permian to a lagoonal environment in the Late Permian-Early Triassic. The salt layers deposited under subtidal to supratidal marine conditions evolved into shallow and deep marine carbonate-dominated sequences by Middle to Late Triassic times. During the orogenic events that climaxed with the Alpine

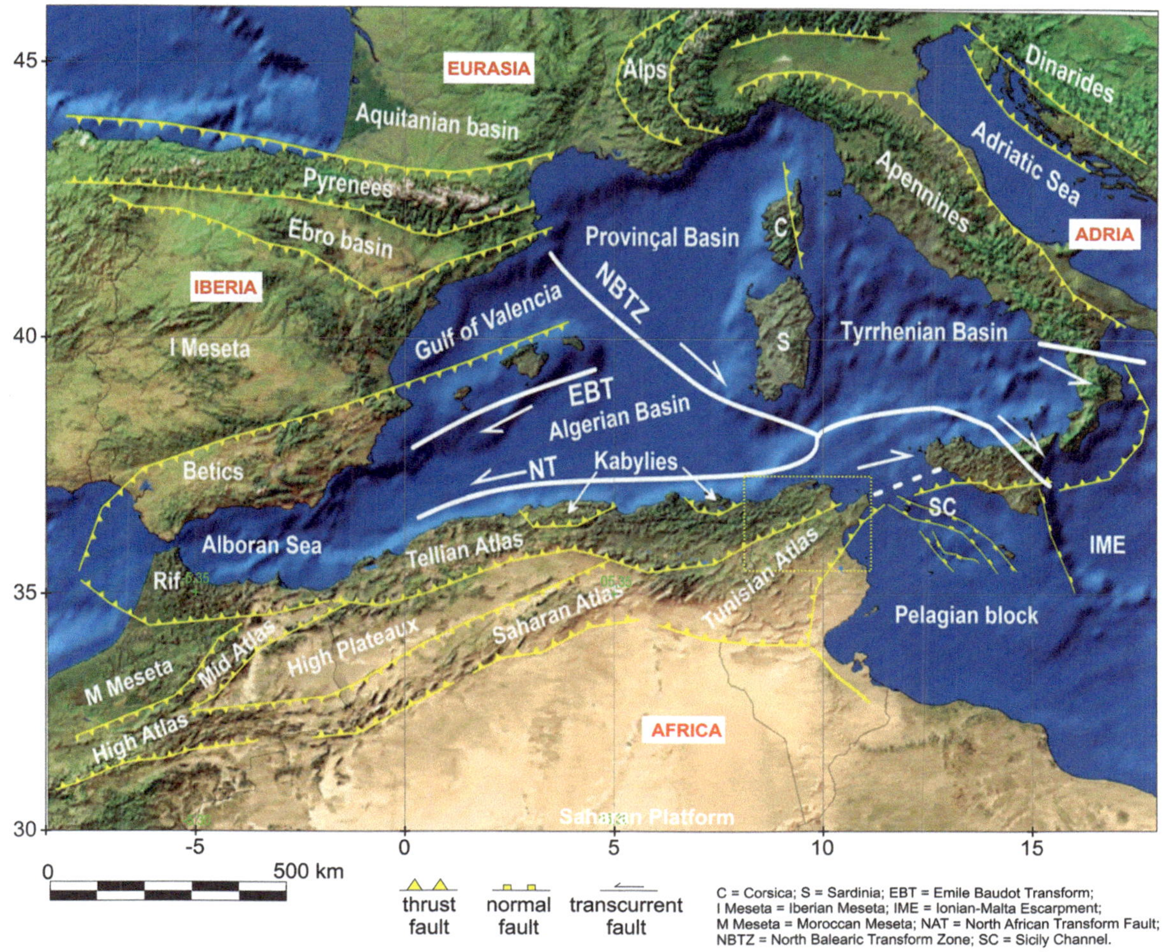

Figure 3.14 The simplified tectonic map of the Mediterranean Sea region, southern Europe, and northern Africa shows the peri-Mediterranean orogenic belts. The Alpine Orogeny involves the African, Iberian, Eurasian, and Adriatic plates. The Alps Mountains separate foreland basins in southern central Europe, while the Adriatic Sea is bordered by fold-thrust belts, the Apennines, and the Dinarides. The Betic Ranges extend from western Iberia to the Gulf of Valencia, while the Pyrenees Mountains separate foreland basins in northern Spain and southern France. The Tellian Atlas and Kabyles in northern Africa are marked by reverse and thrust faults with a south-directed vergence. The Atlas Mountains correspond to an elongated, inverted rift trough from Morocco to Tunisia. The deformation zone extends to Sicily and southern Italy. Evaporite rocks outcrop in many of these fold belts. Credit: Belkhiria et al. (2017)/with permission of John Wiley & Sons.

Orogeny in the Cenozoic, the Hallstatt salt diapir and the overlying Jurassic limestones were pushed northward as thrust sheets or nappes (Granado et al., 2018).

The structural and stratigraphic interpretation of the outcrops with evaporite rocks in the Alpine fold belt is more complicated than in the Zagros region. Most of the evaporite exposures are best seen in underground salt mines, as the climate in Europe is not as arid as in the Arabian Peninsula (see Chapter 2). The salt district in Austria is characterized by a more humid climate than in the Middle East, and the salt minerals are easily dissolved. However, when protected by a carbonate cover, as in the Hallstatt mine, the salt is preserved in relatively small sedimentary pods. These pods have been strongly affected by compressional tectonics, resulting in highly deformed evaporite layers. The evaporite rocks that outcrop in the Hallstatt region are highly sheared and can be more appropriately interpreted as metamorphic rocks or mylonitized salt rocks (Leitner & Spötl, 2017).

These metasedimentary rocks are included in the Haselgebirge Fm., a lithostratigraphic unit characterized by several rock types forming a mélange of siliciclastic, carbonate, and evaporite minerals. The Hallstatt salt diapir exhibits some layers rich in evaporite minerals that

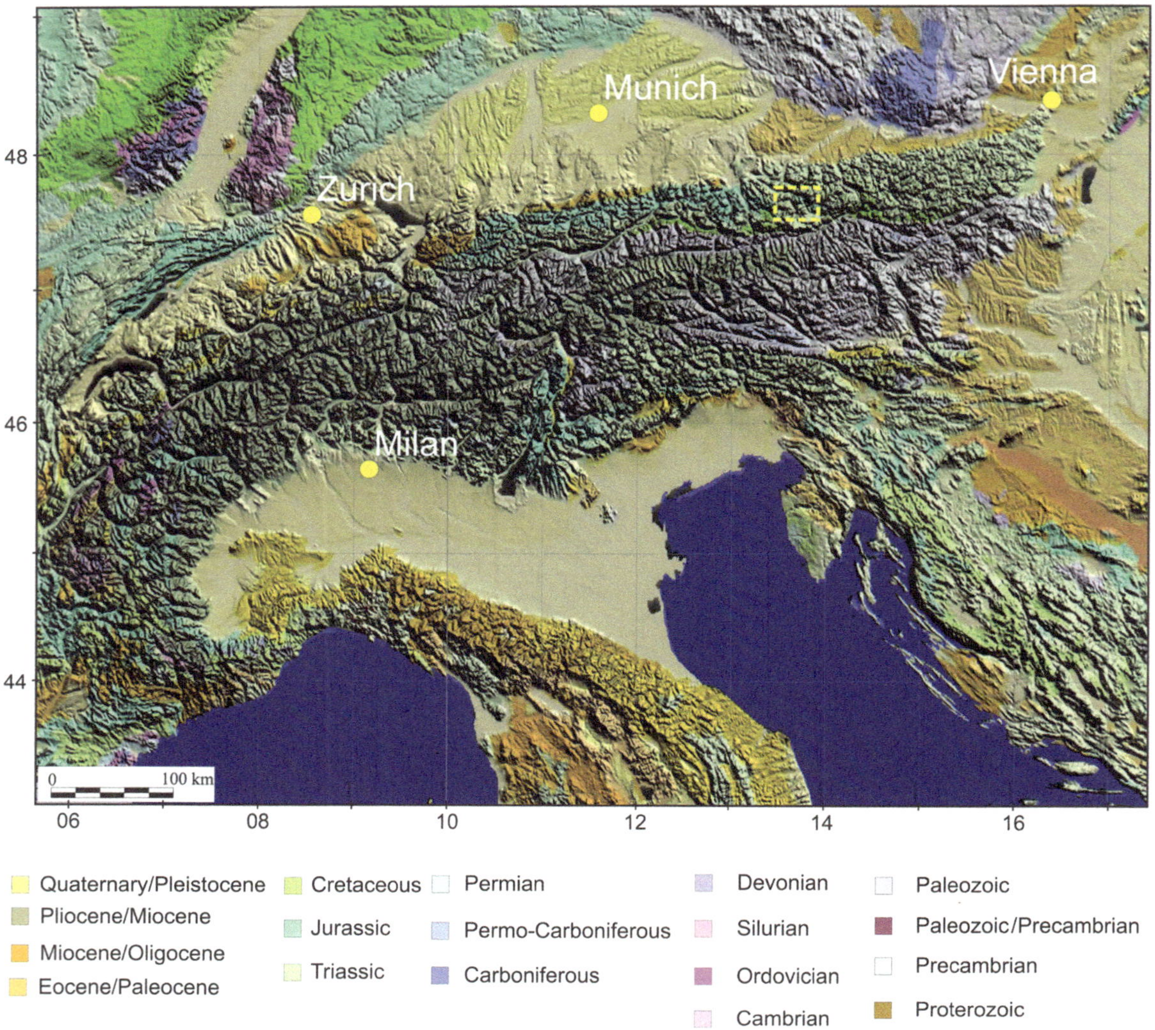

Figure 3.15 The simplified geological map depicts the Alpine Orogeny in southern central Europe. The dashed yellow rectangle shows the location of the Salt District of western Austria, where several deformed salt zones are recognized by geological mapping and are exposed in outcrops. The Hallstatt salt mine in Austria has been exploited since prehistoric times and is considered one of the oldest salt mines in the world. Credit: Image by Jeff Schmaltz, LANCE/EOSDIS Rapid Response/NASA/Public Domain.

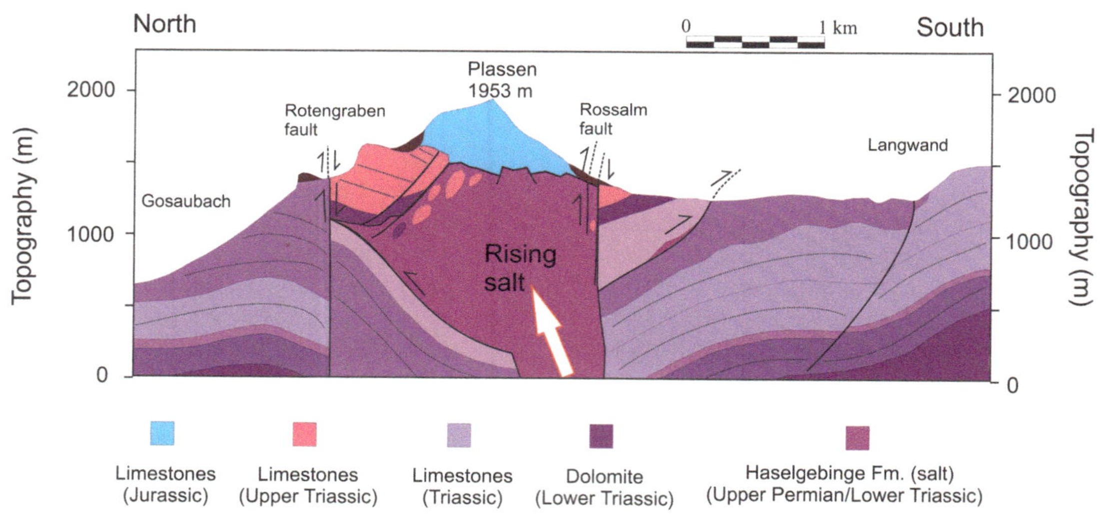

Figure 3.16 The schematic geological cross-section A–A′ illustrates the interpretation of the Hallstatt salt body. The deformed sediments correspond to mylonitized evaporites of the Haselgebinge Fm., forming pods rich in halite minerals. These deposits are associated with Triassic limestones and dolomites. The Plassen Peak, composed of Jurassic limestones, reaches an altitude of almost 2 km. Credit: Adapted from Schorn and Neugebauer (2014).

are commonly seen in salt basins, such as halite, gypsum, anhydrite, sylvite, and several others (Figure 3.16). The salt masses are surrounded by Triassic limestones and capped by Jurassic carbonates that form the highest elevations in the mountains, for example, the Plassen Peak, which rises to an altitude of 1,953 m (Figure 3.16). Examples of the complex structures observed in the salt mines of the Hallstatt region are shown in Figure 3.17.

Both the Cretaceous and Late Eocene-Early Oligocene deformation episodes are registered in the rocks that outcrop in the Austrian salt district. The salt-related structures exhibit nappes developed during the early phase and contractional faults during the Neogene compression. The salt masses are presently considered part of the Hallstatt nappe, with mylonite containing about 55% halite but varying salt content in different parts of the diapiric structure (Schorn & Neugebauer, 2014).

3.2.2. The Cardona Salt Diapir in the Southern Pyrenees, Spain

The convergence of the Iberian and Eurasian plates during the Cenozoic created foreland basins in the south and north of the Pyrenees mountains (Figure 3.18). The

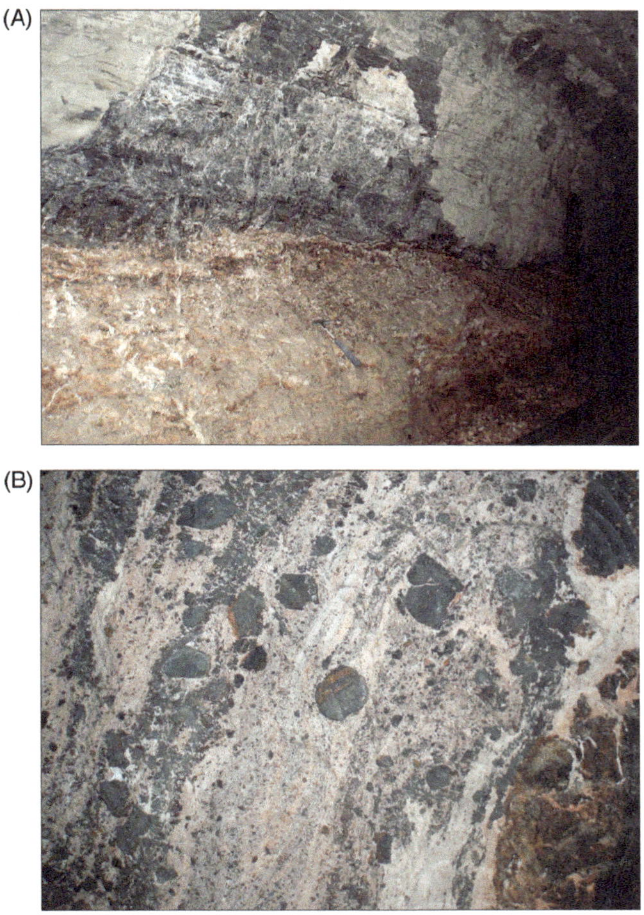

Figure 3.17 The photographs show evaporite rocks in Austria's salt mines. (A) Outcrop of evaporite minerals (anhydrite, black laminated) in the Dürrnberg mine in Austria's Salt District. The hammer in the rocks below the anhydrite layer provides the scale. (B) Outcrop of evaporite minerals in a sheared zone of the Altaussee mine in Austria's Salt District. Halite is characterized by gray to whitish colors and is associated with angular, dark gray mudstone clasts that appear interspersed within the evaporitic matrix. Credit: Courtesy of C. Leitner (2018).

Figure 3.18 (A) Photograph of the Cardona salt mine in northern Spain. The inset map shows the Pyrenees Mountains and the location of the Cardona diapir. The trees at the top of the salt mountain provide an approximate scale. (B) Photograph of the interior of the Cardona salt mine, showing a vertical wall in a ground-floor gallery. The layered evaporites exhibit almost vertical dips due to salt flow within the diapir. The white mass to the left corresponds to the alteration products of the evaporite minerals that have been dissolved. Credit: Mohriak (2022).

Ebro Basin in northern Spain is characterized by continental and marine sedimentary deposits, including evaporite sediments from the Paleogene (Sans, 2003).

The Eocene evaporites correspond to a detachment level for the thrust faults that exhibit a mainly southern vergence. Several anticlinal structures are cored by the Eocene salt, including the salt diapir that outcrops in Cardona (Sans, 2003). The salt diapir has been mined since prehistoric times and is a unique example of an exposed Paleogene salt structure in southern Europe (see Chapter 1).

The Cardona salt diapir, located south of Cardona, is a hill with a height of 120 m above the valley floor. However, geophysical studies indicate that salt diapir reaches a depth of about 2 km below the surface (Sans, 2003). The layered evaporites inside the mine are characterized by almost vertical dips, indicating vertical salt flow. The evaporite rocks have been dissected by weathering and erosion, resulting in karst topography in the valley and alteration of evaporite minerals within the mine galleries (Lucha et al., 2008).

The evaporite sediments are mainly composed of halite minerals, which have been exploited since the Neolithic (see Chapter 1). The mine was deepened in the 19th century and explored in the early 20th century, revealing the presence of sylvinite layers deeper in the subsurface. The new mine galleries exploited these deeper salt lenses, excavating tunnels more than 1 km deep underground. The Cardona salt mine was considered one of the most important production centers of potassium minerals in southern Europe during the second half of the 20th century. The sylvinite deposits were exploited until the 1990s when the mine was converted into a protection area. The remnants of Cardona's ancient salt industry now constitute a tourist attraction in northern Spain.

3.3. SALT DEPOSITION IN THE ATLAS MOUNTAINS OF MOROCCO AND TUNISIA

The Northern Africa region is characterized by Mesozoic sedimentary troughs formed during the Pangea breakup, which were inverted in the Late Cretaceous and Cenozoic, forming the Atlas mountain belt (Beauchamp et al., 1999; Vergés et al., 2017). The Atlas Mountains result from the Alpine Orogeny, which involved the convergence of the Iberian, Eurasian, and North African plates (Figure 3.14). These interactions resulted in transferring compressional stresses through the crust to an intraplate region in Africa, forming deformation belts from the southern Mediterranean coast to regions 200 to 300 km inland (Figure 3.1).

The Atlas Mountains are associated with compressional forces that deformed the previously developed rift basin sedimentary troughs. These Triassic rift depocenters were overlain by Late Triassic evaporite deposits and Jurassic carbonate platforms. The Atlas Orogeny inverted the stratigraphic sequences that outcrop in anticlines and overturned folds, characterized by detachment faults at the base of the evaporite layers.

Most of the Saharan region is covered by Tertiary sediments, including the Quaternary desert dunes that hide a large part of the subsurface geology (Figure 3.19). Some important mountain ranges are observed in northern Africa: the Rif Mountains, the Tellian Atlas, the Kabylies, and the Saharan Atlas (Figure 3.14). The Rif and Tellian Atlas form a northern belt adjacent to the Mediterranean Sea and also exhibit a predominantly southern vergence (Beauchamp et al., 1999; Belkhiria et al., 2017). The fold belts are characterized by the outcrop of Mesozoic rocks, including Triassic carbonates and evaporites, overlain by Jurassic carbonate and evaporite sediments, locally associated with the widespread Cretaceous and Paleogene sediments (Evans & Kendall, 1997; Frizon de Lamotte et al., 2008; Slama et al., 2009).

The topographic relief of the Quaternary cover in the Saharan Desert displays relatively low altitudes, usually less than 1 km. However, the Atlas Mountains in Morocco and the Tellian Atlas to the north exhibit an expressive topographic relief associated with the fold-thrust belt, where altitudes often exceed 2 km (Figure 3.19). The Atlas Mountains follow a general SW-NE trend and constitute the largest Phanerozoic mountain belt in Africa. The mountain ranges extend for about 2,000 km from the Atlantic coast of Morocco, reach Algeria and Tunisia across the Saharan Desert, and terminate near the city of Tunis, located at the southern shores of the Mediterranean Sea (Figure 3.14). The Atlas Mountains correspond to a fold-thrust belt that is similar in size to the Appalachian Mountains in the United States, the Urals in Russia, and the Zagros Mountains in the Persian Gulf region (Frizon de Lamotte et al., 2008).

The extensional episodes started during the transition from the late Permian to the early Triassic (Frizon de Lamotte et al., 2008), and the Pangea breakup resulted in a series of rift basins between northeastern North America, northwestern Africa, and southwestern Europe (Olsen, 1997). The rift troughs in the Atlas region were filled by Mesozoic continental siliciclastic sediments, particularly the Triassic red beds. According to some authors, the original width of the Atlas Rift basin was about 113 km, comparable to the width of the Red Sea Rift (Beauchamp et al., 1999; Belkhiria et al., 2017). The evaporite deposits accumulated during a late syn-rift phase (e.g., Vergés et al., 2017).

A widespread marine invasion in the Late Triassic to Early Jurassic deposited extensive platform carbonates in a seaway that was oriented east-west across southern Morocco and NE-SW across Algeria and Tunisia. During the Jurassic, these troughs were filled by fine-grained carbonate sediments deposited in a deep seaway (Evans & Kendall, 1977). The main structural directions in the

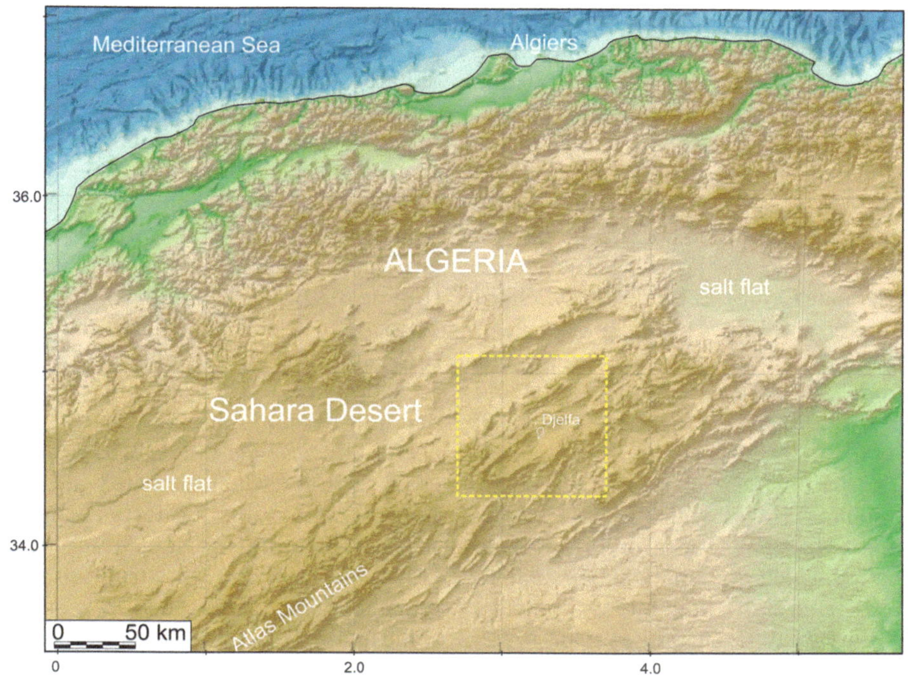

Figure 3.19 The relief image of the topographic map shows the Atlas Mountains in northern Africa, extending across Algeria. The yellow dashed line highlights the location of the zoom image of the fold-thrust belt near the city of Djelfa in the Sahara Desert. The Atlas Mountains in Algeria are characterized by outcrops of Mesozoic sediments following the SW-NE regional trend inherited from the Atlas rift structures. The Triassic evaporites and the overlying Jurassic carbonate sediments occur within the Atlas rift troughs that extend from Morocco to Algeria and Tunisia. Credit: GEBCO (2019)/Public domain.

Saharan Atlas (Figure 3.19) indicate a late Mesozoic to Cenozoic reactivation of the master faults that controlled the Mesozoic rift troughs (Manspeizer, 1981, 1988).

During the Jurassic, oceanic spreading centers formed between Laurentia, North Africa, and Eurasia, resulting in divergent continental margins (Hafid et al., 2008). The extensional stresses focused on the offshore spreading ridges, aborting the onshore troughs in the Atlas region. During the Early Cretaceous, plate motions reversed, resulting in the convergence between Eurasia and Africa, and compressional tectonics started to invert the previously developed rift basins (Frizon de Lamotte et al., 2011).

The Alpine compressional event resulted in the folding of siliciclastic, carbonate, and evaporite rocks, forming large structures such as uplifted anticlinal ridges. Similar to the Zagros thrust-and-fold belt in Iran, the highly elevated regions were deeply eroded, while sedimentary sequences accumulated in synclinal troughs (Ayed-Khaled et al., 2015, 2017; Gharbi et al., 2015). A major phase of uplift in the Tell and Atlas Mountains is recorded between 30 and 20 Ma (Oligocene-Miocene), corresponding to the climax of the Alpine Orogeny in northern Africa (Beauchamp et al., 1999; Schmid et al., 2004). Modern earthquake activity indicates that active compressional tectonics still prevails in this area (e.g., Frizon de LaMotte et al., 2008, 2011).

According to Beauchamp et al. (1999), the high-angle reverse and low-angle thrust faults in the Atlas Mountains resulted from compressional reactivation of the syn-rift normal faults. These faults uplifted the syn-rift and post-rift sedimentary sequences by a few kilometers, locally exposing the Triassic rocks and exhuming the syn-rift extensional structures and the late Triassic evaporite sediments. These evaporite outcrops, predominantly composed of gypsum, are observed in many segments of the Tellian and Atlas Mountains, from Morocco to Algeria and Tunisia, and at the southern margin of the Eastern Mediterranean (Ahmadi et al., 2013; Belkhiria et al., 2017; Michard et al., 2008). In the Atlas region, the Late Triassic evaporite rocks are mainly exposed on top of eroded anticlinal structures (Soto et al., 2017; Troudi et al., 2017; Vergés et al., 2017).

An example of the surface expression of this style of fold-thrust belt with evaporite outcrop (Rocher de Sel) is shown near the city of Djelfa in Algeria (Figure 3.19). The Rocher de Sel is considered by some researchers to be the first scientific description of a salt diapir in the world (Jackson, 1995). The Rocher de Sel structure stands out in the earth sciences as the first salt structure to be described and interpreted using modern geological concepts involving compressional tectonics (see Chapter 6 in Mohriak, 2025).

There is field and seismic evidence for several compressional episodes affecting the Atlas fold-thrust belt in the Cenozoic, particularly in the Late Eocene and the Late Miocene (Soto et al., 2017; Vergés et al., 2017). Many researchers consider these intervals to be important phases for halokinesis in the Alps, Pyrenees, and Atlas ranges (e.g., Frizon de Lamotte et al., 2000; Michard et al., 2008). Late Triassic salt layers played a key role in controlling the folds and detachment zones of Cretaceous to Tertiary thrust faults, forming propagation folds and fault-bend folds (Bracène & Frizon de Lamotte, 2002; Frizon de Lamotte et al., 2008, 2011).

In the Saharan Atlas region, there are two main types of salt structures: (1) anticlinal ridges, which are long structures with elliptical cross-sections that mostly run NE to SW; and (2) salt diapirs, which are small, subcircular to ellipsoidal bodies that are no more than a few kilometers across (Perthuisot & Rouvier, 1992; Perthuisot et al., 1998).

The first structural type (1) is characterized by carbonate and sulfate rocks in outcrops, with salt (halite and other soluble evaporites) forming the cores of anticlines in the subsurface. These compressional structures are composed of Mesozoic to Cenozoic sequences and are affected by thrust faults that detach at deeper stratigraphic levels, particularly Triassic evaporites. These faults result in fault propagation folds, with anticlines eroded at the apex (Ahmadi et al., 2013; Trigui et al., 2016). Locally, the faults involve the basement, resulting in thick-skin tectonics, indicating reactivation of rift-phase faults. In some outcrops, the folds are connected to faults that detach at the supra-salt horizons, corresponding to fault-bend folds that are locally vertical or recumbent, forming asymmetric structures (Trigui et al., 2016).

The second structural type (2) is characterized by salt diapirs that outcrop at the surface with several evaporite minerals derived from the thick Triassic salt-bearing mass. The diapir flanks are covered by sediments that exhibit high rotation, approaching the vertical, due to the uplift of the salt structure. Exhumed evaporite deposits at the diapir's core are usually associated with gypsum, but locally may include halite. One example of such a diapiric structure is the Rocher de Sel near the city of Djelfa in Algeria (Jackson, 1995; Kulke, 1972).

3.4. SALT OUTCROPS IN THE SAHARAN ATLAS, ALGERIA

The Rocher de Sel structure is a famous salt diapir outcrop in Algeria, located near the northeastern fringe of the Saharan Atlas (Figure 3.19). This famous salt diapir outcrop is located approximately 500 km south of Algiers city, forming a subcircular structure encased by Cretaceous sediments that have been exhumed by the Saharan Atlas uplift (Jackson, 1995; Kulke, 1972).

South of the Atlas front fault, the Saharan Atlas fold-thrust belt is marked by subdued topographic elevations, around 1 km, showing little relief in satellite images and topographic maps (Figure 3.19). Most structures trend from NE-SW to E-W, with elongated, deeply eroded anticlinal ridges covered by Quaternary desert dunes (Figure 3.20).

Rocher de Sel is considered the first halokinetic structure analyzed by scientific methods (Kulke, 1972; Jackson, 1995). The salt outcrop was first described in detail in the mid-19th century by a French mining engineer, Ludovic Ville. Ville (1859) published several regional cross-sections, which depicted the structural interpretation of highly dipping Tertiary-age strata that encased the salt core and its subsurface continuation. Gautier (1914) also described the structure and compared its geomorphological appearance to a salt laccolith, bringing some analogies with igneous structures.

This outstanding salt diapir outcrop near the city of Djelfa is a relatively small ellipsoidal feature with a diameter of about 1 km and smooth relief, forming a hill that is about 100 m above the Saharan desert plain (Figure 3.20A). The cross-section illustrating the subsurface geology insightfully assumed vertical beds limiting the salt core, with some recumbent folding (Kulke, 1972; Perthuisot & Rouvier, 1992). The salt diapir is still rising, as indicated by the drainage diversion around the outcropping evaporite rocks (Figure 3.20A).

3.5. SALT STRUCTURES IN THE TUNISIAN ATLAS MOUNTAINS

The Triassic salt outcrops in the Tunisian Atlas Mountains differ from the Rocher de Sel in Algeria due to the absence of halite on the surface. In Tunisia, most evaporite outcrops are made up of sulfate rocks (gypsum) interbedded with siliciclastic and carbonate rocks. A regional map of surface rocks exposed in the Tunisian Atlas (Figure 3.21) shows structural trends indicating a general NE-SW direction for main salt structures, similar to the Atlas Mountains in Algeria (Figure 3.19). The Tunisian Atlas rift basin includes several depositional sequences, including Mesozoic siliciclastic continental sediments, late Triassic evaporites, Jurassic carbonates, and marine Cretaceous sedimentary rocks. During the Alpine Orogeny, these layers were affected by compressional structures that deformed the late Triassic salt diapirs (Slama et al., 2009).

The Atlas Mountains of northern Tunisia have developed several anticlinal structures resulting from compression involving African and Eurasian plates

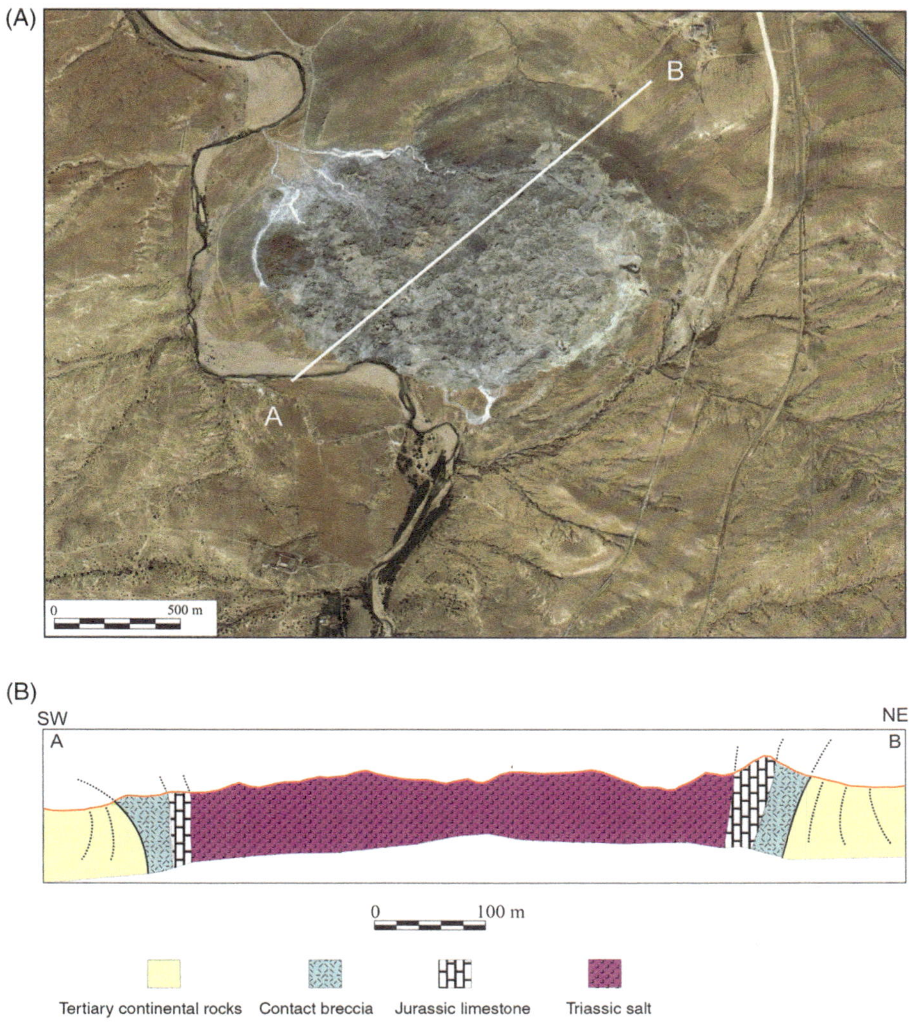

Figure 3.20 Detail of a satellite image and geological cross-section of the Rocher de Sel structure in the Atlas region. (A) Satellite image (natural colors) showing the Rocher de Sel outcrop in the Sahara Desert. The drainage has been strongly diverted by the geomorphology of the exposed evaporitic rocks. The dashed white line A-B shows the location of a schematic geological cross-section with an SW-NE direction. (B) Schematic geological cross-section (profile A–B) showing the interpretation of the Rocher de Sel structure. The outcrop is interpreted as a salt diapir with almost vertical beds flanked by Jurassic carbonate rocks and breccia at the contact with the Tertiary continental rocks. Credit: Adapted from Perthuisot and Rouvier (1992).

(Figure 3.1). These structures are marked by multiple deformation phases involving the basement, Triassic rift sequences, Triassic evaporites, and post-Triassic overburden (Figure 3.21).

Based on field geology, subsurface interpretation of seismic lines, results of borehole drilling, and structural modeling, several authors (e.g., Frizon de Lamotte et al., 2000; Khalfi et al., 2014; Michard et al., 2008) proposed that these structures were formed by a combination of two mechanisms: halokinesis, resulting from the passive rising of the mobile salt, and fault propagation model, with thin-skin deformation related to the movement of overburden layers above a subhorizontal fault zone detaching at the Triassic evaporites. These models assume that the anticlinal structures were formed during multiple compressional events associated with the Alpine Orogeny (Khalfi et al., 2014).

In the Tunisian Atlas (Figure 3.21), the northern thrust fault is marked by a southern vergence, while the southern fault displays a northern vergence. The Medjerda Valley is located between these thrust faults with reverse polarity. The Thibar salt structure is the largest evaporite body outcropping eastward of the valley (Ayed-Khaled et al., 2015; Slama et al., 2009). Several features with similar trends are observed along a wide fold-thrust belt that extends from Algeria's border

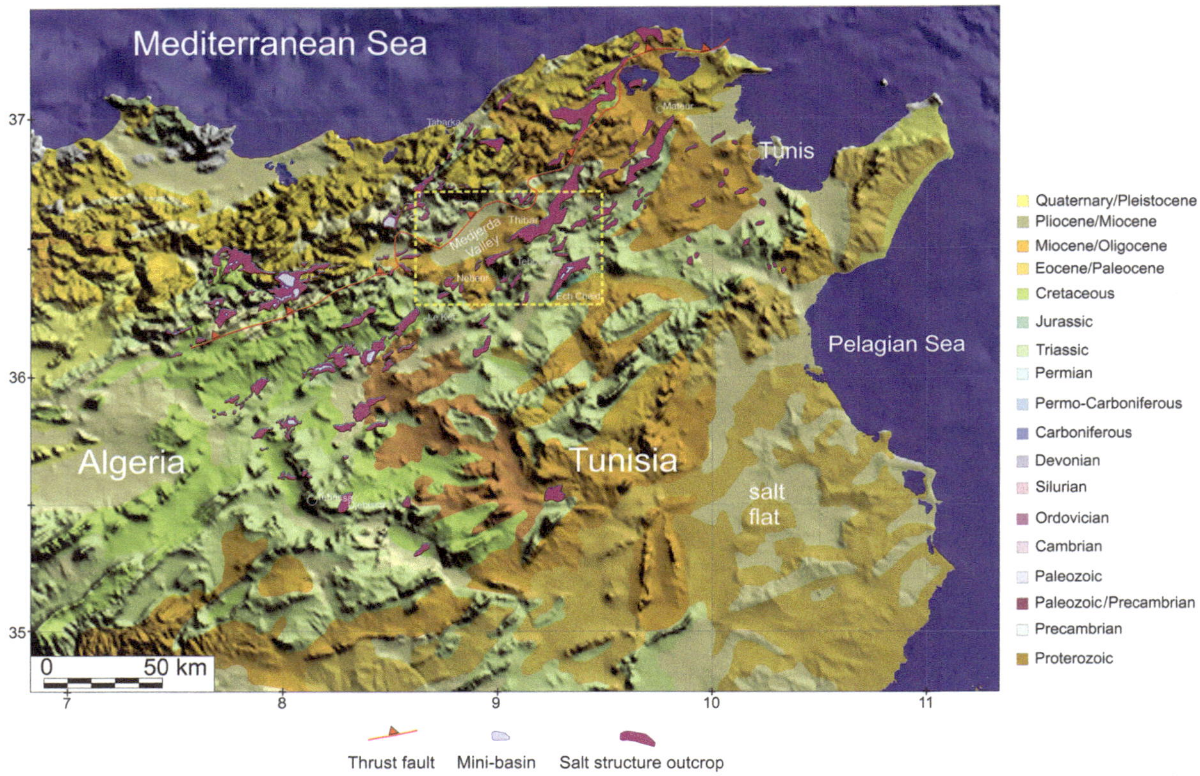

Figure 3.21 The geological map of northeastern Tunisia, at the southern margin of the Mediterranean Sea, shows Cretaceous and Paleogene sedimentary sequences with a regional SW-NE trend. Triassic salt evaporite rocks are highlighted in pink, with sulfate and carbonate minerals outcropping at the surface. A yellow rectangle denotes the Medjerda Valley, which is located south of a thrust fault with a southern vergence. The Thibar salt structure is the largest evaporite body eastward of the valley. Credit: Slama et al. (2009)/with permission of Elsevier.

to northeast Tunisia near the southern margin of the Mediterranean Sea (Figure 3.14).

3.5.1. Salt Structures in the Medjerda Valley, Northeastern Tunisian Atlas

The northeastern Tunisian Atlas features synclinal and anticlinal structures with axes trending NE (Figure 3.21). Evaporite deposits are exposed in hills called jebels, which are associated with anticlinal structures. Topographic elevations are related to thrust and transcurrent faults. Triassic evaporite sediments outcrop in the hills but not in the Medjerda Valley, which is covered by Quaternary alluvial plain sediments. An exploratory well (Med-1) was drilled in the valley and penetrated Triassic evaporites in the subsurface, overlain by a thin sequence of Tertiary sediments (Ayed-Khaled et al., 2015).

The Thibar structure corresponds to a large, elongated anticline trending NE-SW, controlled by a large thrust fault with a SE vergence (Figure 3.22A). The Thibar structure exposes Triassic evaporites on Jebel El-Matria, a mountain range marked by rough topography. The geological cross-section B-B' across the jebel indicates the rugged topography derives from Triassic evaporitic rocks exposed at the surface. The evaporite rocks are enveloped by Cretaceous to Paleogene sedimentary layers, displaying a steep dip at the flank of the salt structure (Figure 3.22B).

The interpretation of the cross-section B-B' suggests a major unconformity marked by the direct contact of Triassic evaporites with Cretaceous (Aptian) strata. The almost vertical contact between the evaporites and the overlying sediments indicates a diapiric salt wall that rotates the overlying sediments. This interpretation is similar to the interpretation of the Rocher de Sel structure in Algeria (Figure 3.20) but has been challenged by many researchers.

3.5.2. Interpretation of the Salt Structures in the Tunisian Atlas

The geological interpretation of Triassic evaporite outcrops in Tunisia is still debated due to differing opinions on salt emplacement mechanisms. Some researchers

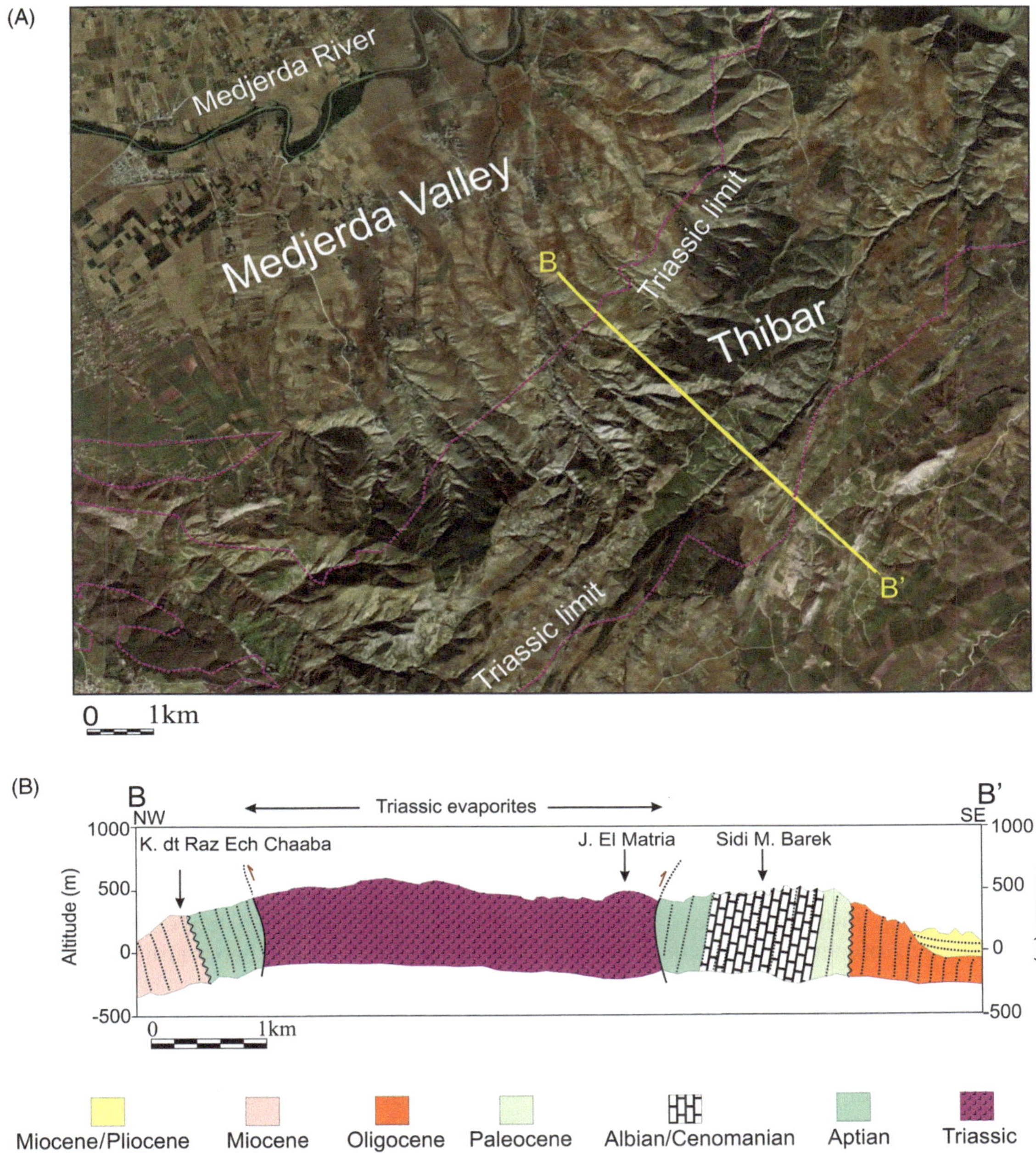

Figure 3.22 Satellite image and geological cross-section of the Thibar salt structure in the Medjerda Valley in Tunisia. (A) The satellite image of the Thibar salt structure shows a rugged topography associated with Triassic rocks. The evaporitic sediments are characterized by gypsum mixed with carbonate sediments. An exploratory well (Med-1) drilled in the valley penetrated evaporite sediments overlain by Tertiary sequences. Credit: Ayed-Khaled et al. (2015)/Annals of Geophysics/CC BY 4.0. (B) The geological cross-section shows Triassic carbonate and evaporite rocks enveloped by Paleogene to Cretaceous sedimentary layers with a steep dip. The contact between Triassic evaporites and overlying sediments is possibly related to a diapiric salt wall. Some researchers postulate that the salt layer corresponds to an allochthonous salt sheet overlying Cretaceous sedimentary sequences. Credit: Ayed-Khaled et al. (2015)/Annals of Geophysics/CC BY 4.0.

suggest that the exposed Triassic evaporite bodies result from diapiric intrusions forming salt plugs, and the vertical uplift mechanism is associated with regional compression (e.g., Amor, 1999; Jallouli et al., 2005). Others believe the Triassic evaporite outcrops are remnants of allochthonous salt sheets that extruded as salt glaciers in submarine environments (e.g., Masrouhi & Bellier, 2013; Slama et al., 2009; Talbot, 2005). This hypothesis explains the mixing of evaporite sediments with deepwater lithologies. Some boreholes have confirmed that salt layers occur interbedded within marine Late Cretaceous strata (Amri et al., 2020; Jaillard et al., 2017; Zouaghi et al., 2013).

Perthuisot et al. (1998) analyzed the Monts du Mellegue Triassic outcrops near the Algeria-Tunisia border and rejected the interpretation of a salt glacier passively extruded during the Albian. According to the interpretation, an extensional salt diapir developed in the Cretaceous, resulting in the superficial spreading of a salt dome. The structure was subsequently deformed by Neogene to Quaternary compression, as indicated by thrust faults, regional folding, and squeezing of the salt diapir stem (Perthuisot et al., 1998).

More recently, several authors have emphasized the passive extrusion model in the interpretation of salt glaciers sourced by a central diapir stem (e.g., Masrouhi & Koyi, 2012; Masrouhi et al., 2013, 2014). According to the model, most Tunisian Atlas structures correspond to extrusive salt bodies formed by compressional events related to the Africa-Eurasia convergence in the late Cretaceous. The squeezing of salt diapirs formed salt canopies and salt glaciers that extruded onto the late Cretaceous submarine depositional surfaces. The model is corroborated by outcrops and boreholes that show the Triassic evaporites are intercalated with marine shales that are much younger than the salt layers (Masrouhi & Koyi, 2012, Masrouhi et al., 2013, 2014).

The exploratory well (Mej-1) drilled in the Medjerda Valley plain, at the western part of the Triassic outcrop area, penetrated Quaternary and Pliocene strata overlying the Triassic evaporite sequences, which are dominated by carbonate and evaporite rocks, including halite (Ayed-Khaled et al., 2015). According to Perthuisot et al. (1998), the results of the borehole and analogies with other structures in the Atlas Mountains suggest the outcrop corresponds to the top of a salt diapir. However, other authors (e.g., Troudi et al., 2017) interpreted that the borehole penetrated the Triassic carbonates and evaporite layers overlying a deeper sequence of Late Cretaceous sediments. This inverted stratigraphic column suggests an allochthonous salt layer spreading as a salt glacier or forming a nappe structure overriding the post-salt layers (Masrouhi & Koyi, 2012; Troudi et al., 2017).

Triassic outcrops in northeastern Tunisia are characterized by diverse extrusion features, such as a chaotic mixture of lithologies composed of gypsum, clay, marl, dolomite, and sandstone, with occasional fossiliferous carbonate rocks within or underlying the sequence. These observations indicate a deepwater setting for the Atlas rift basin during the deposition of the Jurassic carbonates (Michard et al., 2008). The model assumes that the Triassic evaporite rocks were extruded as salt canopies and salt nappes in deepwater settings, covering younger sediments (Masrouhi & Koyi, 2012; Masrouhi et al., 2013, 2014).

Talbot (1993) hypothesized that the salt overhangs in the deepwater region of the Gulf of Mexico are caused by salt spreading on pelagic mud during periods of slow sedimentation. In the case of the Saharan Atlas, the cause of the salt spreading over younger sediments is related to the compressional tectonic forces associated with the plate convergence in the Cenozoic (Troudi et al., 2017). Contraction of the original extensional salt diapirs resulted in squeezing the salt structures, forming salt glaciers similar to the structures observed in the Zagros fold-thrust belt (see Chapter 7 in Mohriak, 2025).

3.6. TECTONIC EVOLUTION OF THE PARADOX BASIN, SOUTHWEST UNITED STATES

The Paradox Basin in the southwestern United States is one of the best examples of exposed salt structures (Trudgill et al., 2004). Other well-known salt outcrops occur on land at the periphery of the Gulf of Mexico, such as the La Popa Basin in northwestern Mexico (Rowan et al., 2003). In Eastern Canada, Carboniferous evaporites form a salt diapir province between Nova Scotia and the mainland. Some salt diapirs outcrop on land, forming evaporite masses mainly composed of gypsum (Alsop et al., 2000). Near-surface Carboniferous evaporite deposits have been exploited in salt mines onshore Nova Scotia since the beginning of the 20th century (Bohener, 1986).

The genesis of the Paradox Basin is related to plate tectonic reorganizations that resulted in the breakup of Gondwana and the amalgamation of several blocks that formed Pangea in the late Paleozoic. The Iapetus Ocean separated three main landmasses: Laurentia, Baltica, and the northern part of Gondwana, a large continental block corresponding to the modern Africa and South America landmasses (Dalziel, 1997). After the Iapetus Ocean closure, two main landmasses dominated the globe during the late Paleozoic: the Laurasia plate in the north and the Gondwana plate in the south, which were separated by a small seaway that connected with the much larger Paleo-Tethys Ocean to the east (Blakey & Ranney, 2018). The large Panthalassa Ocean to the west formed a subduction zone along the western margin of Pangea during the Carboniferous (Domeier & Torsvik, 2014).

Continued subduction of the Paleo-Tethys margins in the Paleozoic resulted in changes in plate configuration, leading to the convergence of continental blocks, forming mountain ranges and suture zones associated with the Variscan Orogeny in Europe and North America (Dalziel, 1997; Domeier & Torsvik, 2014; Scotese, 2001, 2008, 2011; Stampfli & Borel, 2002). The suture zone that resulted from the continent-continent collision is expressed in North America as the Ouachita and Marathon orogenic belts (Kluth & Coney, 1981; Kluth, 1986).

The Ancestral Rocky Mountains consisted of several mountain ranges (basement uplifts and highlands) associated with deep, asymmetrical basins or troughs developed in response to stresses during the Carboniferous plate collision. The Alleghenian Orogeny formed a fold-thrust belt in North America, extending from the United States to Canada. The contraction associated with the Alleghanian Orogeny resulted in the squeezing of the Carboniferous salt structures in the Maritimes Basin, forming salt diapirs that presently outcrop on the northern shores of Nova Scotia (e.g., Gibling et al., 2019). Compressional tectonics resulted in the Uncompahgre Uplift and a major flexural depression corresponding to the Paradox Basin (Baars & Stevenson, 1981; Barbeau, 2003; Blakey & Ranney, 2018; Kluth & Coney, 1981; Kluth, 1986).

3.6.1. Geological Evolution of the Paradox Basin

The Paradox Basin is about 330 km long and 130 km wide (Figure 3.23). The origin of the basin is still controversial, as several tectonic events have been recorded since its development in the Late Carboniferous. Case (1966) suggested that the basin depocenter was mainly controlled by extensional normal faults. Baars (1976) suggested the Paradox Basin could be considered an aulacogen or an aborted rift. Baars and Stevenson (1981) proposed the basin was formed along reactivated northwest-trending faults of pre-Pennsylvanian age. Some authors have postulated that the elevated topography of the Uncompahgre Plateau resulted from the Laramide Orogeny, which spanned the Late Cretaceous-Eocene interval (White & Jacobson, 1983).

Some authors have interpreted the basin as a Paleozoic foreland trough formed in a compressional setting (e.g., Barbeau, 2003; Frahme & Vaughn, 1983; Huffman Jr. & Taylor, 2002). Others have suggested a pull-apart basin formed in a transtensional setting (Stevenson & Baars, 1986). Some authors postulated that the Paradox Basin was formed by Precambrian extensional structures that were inverted as thrust faults in the Carboniferous (e.g., Hudec, 1995). The southwest direction of many thrusts and the presence of a depocenter 5 km thick next to the thrust front show that intraplate shortening developed a foredeep trough that accumulated lacustrine sediments during the Late Carboniferous (Figure 3.23). These sediments were overlain by thick sequences of evaporite sediments across the foreland, which were deformed by compression from the Uncompahgre thrust faults (Hudec, 1995; Ye et al., 1996).

Tectonic reconstructions indicate that a shallow embayment extended across Utah and Colorado in the southwestern United States in the Paleozoic (Baars & Stevenson, 1981; Trudgill et al., 2004; Trudgill, 2011; White & Jacobson, 1983; Whidden et al., 2014). Sedimentation in the Paradox Basin includes a thin layer of Early Paleozoic to Carboniferous rocks. The main depositional phase was related to the accumulation of thick sequences of marine sediments deposited in a seaway or gulf during the Late Carboniferous (Ye et al.,1996).

3.6.2. Evaporite Deposition in the Paradox Basin

During the Late Carboniferous, the central part of North America was located close to the equator (Domeier & Torsvik, 2014; Scotese, 2001, 2008, 2011). Due to climate aridity and intense evaporation of seawater, the ancient seaway dried up, and evaporite minerals precipitated in the hypersaline basin. Repeated marine ingressions from the Panthalassa Ocean refilled the gulf, forming several cycles of carbonate and evaporite deposition. The Paradox Basin accumulated at least 1,500 m of evaporite sediments during the Pennsylvanian Period, about 315–310 million years ago (Baars & Stevenson, 1981; Baars & Doelling, 1987; Carter, 1970; Hite, 1968; Leary et al., 2017). The maximum thicknesses of the salt deposits are located at the northeastern depocenter of the Paradox Basin, close to the Uncompahgre Uplift (Figure 3.24).

In the subsurface, the Paradox Fm. evaporite sequences include mainly halite beds interlayered with gypsum and anhydrite (Lawton & Buck, 2006; Trudgill et al., 2004; Ye et al., 1996). These mobile halite sequences form the core of the elongated structures known as the "Salt Anticlines" of the Paradox Basin (Carter, 1970). Petroleum exploration wells have drilled more than 4 km of evaporite sediments in some salt structures, corresponding to tall, elongated salt walls that are subparallel to the main NW trend of the thrust front (Figure 3.24). Some boreholes drilled a few structural highs in the search for hydrocarbons, although with limited success. One of the largest structures is a major anticline associated with a large salt wall in the Gypsum Valley. The adjacent mini-basin, filled with sediments that might contain reservoir rocks, constituted the prime target for petroleum exploration. Due to exploration failure, it is known as the "Disappointment Mini-Basin" (Escosa et al., 2019).

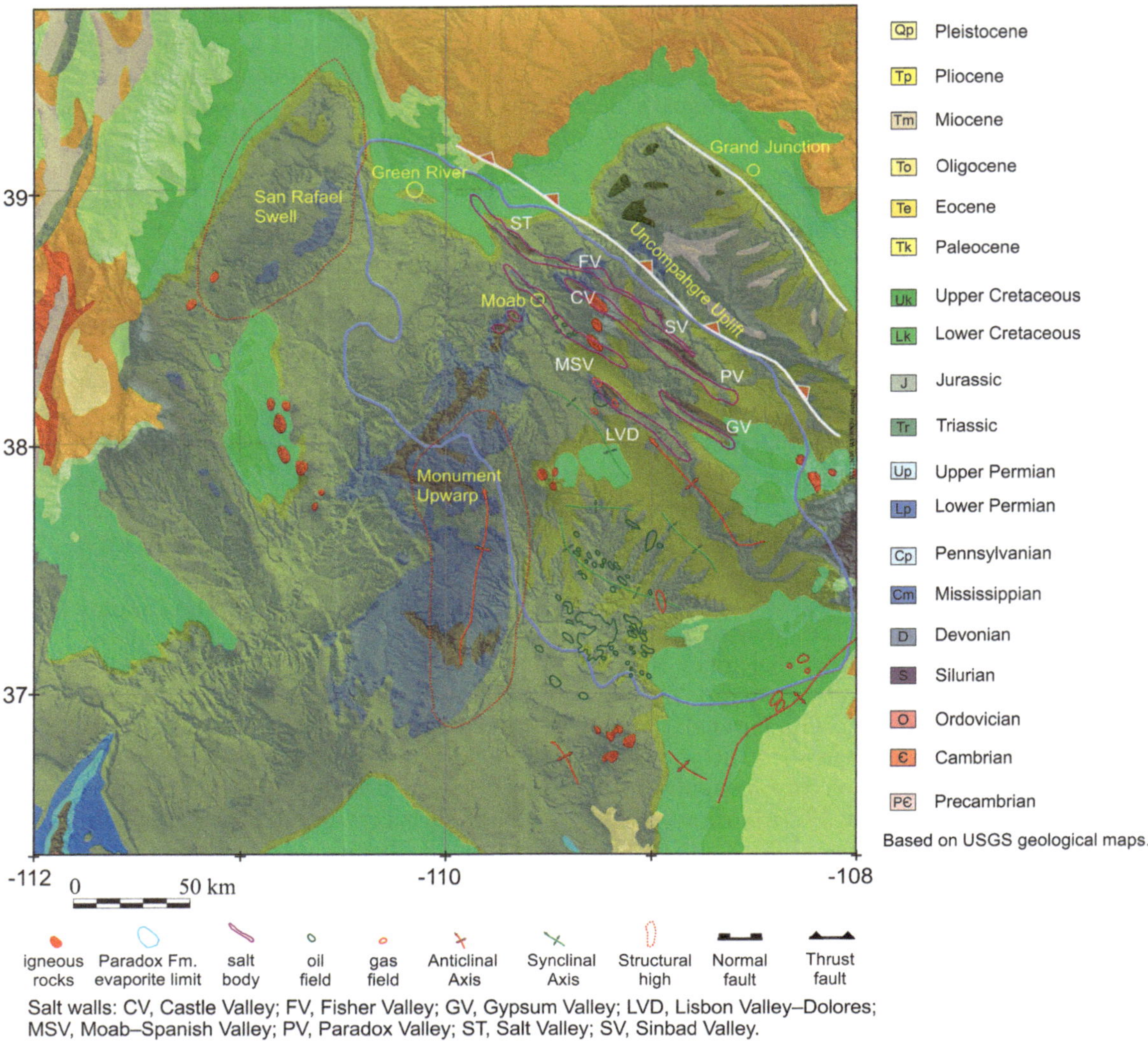

Figure 3.23 The simplified geological map shows the Paradox Basin and adjacent elevated regions in the Uncompahgre Uplift. The basin is surrounded by Cretaceous outcrops and two regional uplifts with Carboniferous rock outcrops. The Uncompahgre Uplift is a major basement-involved thrust fault with a southwest vergence. The map shows the main structures, salt walls, and hydrocarbon fields in the basin, aligned along an NW-SE regional trend. The evaporite lithologies in the basin depocenter correspond to the Upper Carboniferous Paradox Formation, deformed by sedimentary loading and compressional forces, resulting in elongated salt walls. Most oil fields are located in the southeastern corner of the basin, with reservoirs corresponding to Upper Carboniferous carbonate rocks sealed by evaporite rocks. Tertiary igneous structures intruded the salt walls, forming clusters of igneous plugs south of the San Rafael Swell and eastward of the Monument Swell. Credit: Modified from Lawton and Buck (2006).

The Paradox Fm. evaporite sediments grade into shelf carbonates along the western and southern basin margins, forming buildups or reefs that constitute the main petroleum reservoirs in several small oil fields (Whidden et al., 2014). Most of the oil fields are located at the southeastern corner of the Paradox Basin, near the salt limit, and are associated with facies changes from evaporites to carbonates (Peterson, 1989).

The late Carboniferous seaway disappeared in the Permian and the geological setting changed to alluvial, fluvial, lacustrine, and desert environments. The Permian sequences consist of heterogeneous siliciclastic sediments derived from the erosion of the uplifted Uncompahgre thrust front, forming clastic wedges that prograded basinward, covering the salt sequences (Escosa et al., 2019). The Permian carbonate rocks and continental red beds that make up the alluvial, fluvial, and lacustrine deposits are part of the Cutler Group (Trudgill, 2011). The Triassic phase of deposition includes tidal and continental sediments deposited across the basin. The Jurassic phase of

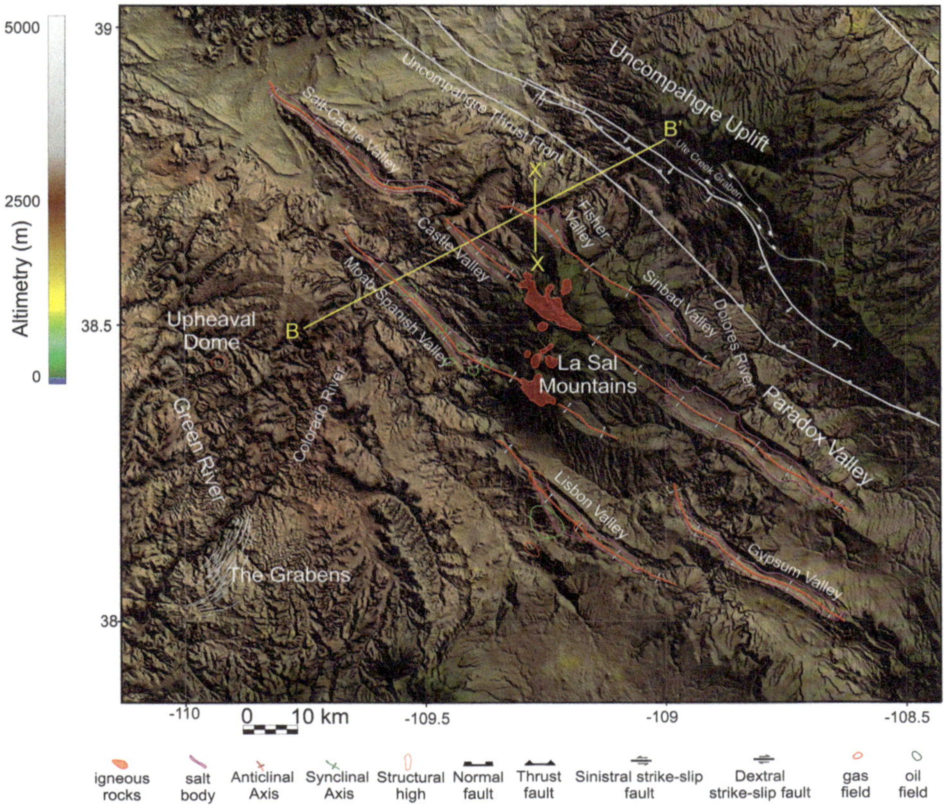

Figure 3.24 The Paradox Basin's salt structures are highlighted using topography draped on the satellite image. The highest elevations are found in the Uncompahgre Uplift and igneous massifs of Tertiary age. The Uncompahgre thrust faults and elongated salt walls follow a regional NW-SE trend. Some Paleogene igneous intrusions, such as the La Sal Mountains, formed plugs along anticlinal axes. The B-B′ profile shows a geological cross-section from the SW part of the basin toward the Uncompahgre Uplift. The X-X′ profile shows a north-south cross-section across the Fisher Valley anticline, associated with a large salt wall. The western part of the map shows the Upheaval Dome, a circular structure with alternative interpretations of an asteroid impact and a collapsed salt diapir. The southwestern part of the map shows "The Grabens," a large area characterized by extensional structures that control present-day geomorphology. Credit: Modified from Gutierrez (2004) and Trudgill (2011).

deposition was characterized by fluvial to desertic environments, where eolian sandstone and siltstone deposits constitute the predominant lithologies. The Cretaceous rocks that cover these fluvial to desertic settings were deposited in continental to marine environments, such as the Mancos shale (Ohlen & McIntyre, 1965).

Currently, the Paradox Basin is encircled by Cretaceous rocks and is limited by two regional upwarps where Carboniferous rocks outcrop at the surface: the San Rafael Swell to the northeast and the Monument Swell, located in the southern part of the basin (Figure 3.23). The region is also marked by igneous intrusive plugs dated as Paleogene, which intruded the salt walls, forming clusters of igneous intrusions south of the San Rafael Swell and eastward of the Monument Upwarp (Figure 3.24). The salt structures and the igneous intrusions are recognized in satellite images by their elevated topography, such as the La Sal Mountains, southeast of the Moab-Spanish Valley (e.g., Doelling et al., 1988).

A circular structure known as the Upheaval Dome marks the western part of the basin (Figure 3.24). Some authors interpret this structure as a collapsed salt diapir or an asteroid impact (Buchner & Kenkmann, 2008; Geesaman et al., 2015; Jackson et al., 1998). The regional cross-section extending from the Uncompahgre thrust to the basin border illustrates the elongated salt structures formed in the basin depocenter (Figure 3.25).

3.6.3. Salt Outcrops in the Paradox Basin

The Paradox Basin has been shaped by salt flow and the formation of elongate, northwest-trending valleys above salt-intruded anticlines (Figure 3.24). This unique geometry has resulted in shallow depressions overlying

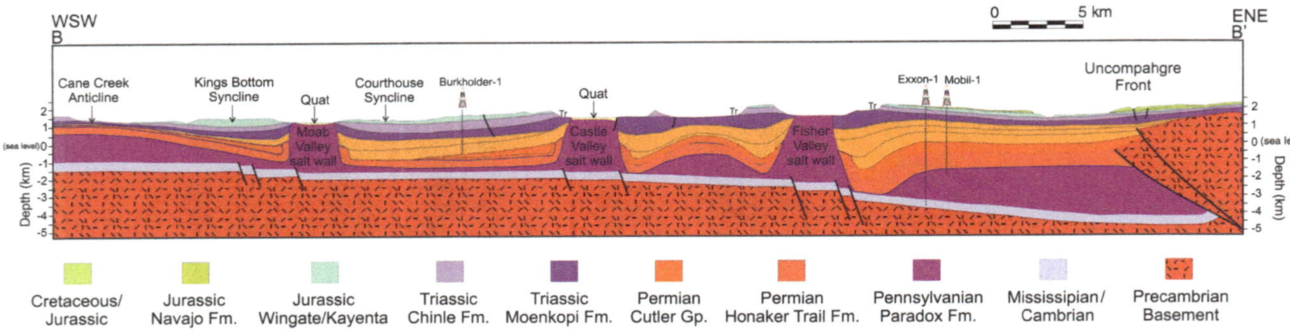

Figure 3.25 The geological cross-section B–B′ extends from the Cane Creek anticline to the Uncompahgre Uplift, characterized by a major reverse fault with a SW vergence. Salt structures in the Paradox Basin form salt walls aligned parallel to the thrust front zone. Three major anticlinal structures are identified: Moab Valley, Castle Valley, and Fisher Valley salt walls. A thin cover of Quaternary sediments occurs in the valleys, overlying the salt structures. The basement becomes shallower to the west, forming a bulge. The thickest salt deposits are located near the Uncompahgre Uplift Front, forming the Paradox Basin foredeep. Credit: Adapted from Trudgill (2011).

the tall salt structures and thick sedimentary depocenters between the salt walls. The Colorado River, a more recent drainage system, runs across the salt anticlines, cutting their collapsed valleys nearly at right angles instead of flowing down along the depressions. This geomorphological phenomenon was considered "paradoxical" by early geologists, leading to the name of the valley. This term has been applied to name the basin and to designate the geological unit that includes the main evaporite sequences (Baars & Doelling, 1987).

The large salt bodies that form the elongated walls in the Paradox Basin are associated with depressions adjacent to the anticlinal structures. These sedimentary-filled depressions correspond to mini-basins formed by loading the salt and creating accommodation space due to salt flow (see Chapter 7 in Mohriak, 2025). The sediments accumulated in the mini-basins record the complex history of basin subsidence and salt flowage toward the foreland (Banham & Mountney, 2013a, 2013b; Escosa et al., 2019; Paz et al., 2009; Trudgill et al., 2004). The mini-basin depocenters have migrated with time due to differential sedimentary loading affecting the evaporite layers (Figure 3.25). Paz et al. (2009) suggested that halokinesis was diachronous, initially forming diapirs and mini-basins first in the east, close to the Uncompahgre Uplift. Salt deformation subsequently migrated to the west, forming the Moab salt wall and associated mini-basins (Figure 3.25).

Most of the salt activity was concentrated in the Late Permian, Triassic, and Jurassic periods, as indicated by the thinning of the sedimentary sequences toward the anticlinal axes of the rising salt structures (Figure 3.25). The Upper Cretaceous Mancos shale was deposited in the Western Interior Cretaceous Seaway and is primarily composed of uniform, dark gray mudstone. During the Tertiary, collapse troughs were formed on the salt walls due to salt dissolution and subsidence. However, the Cretaceous layers do not show a significant increase in thickness when compared to equivalent lateral layers adjacent to the troughs (Trudgill, 2011).

In the Oligocene, granitic rocks formed intrusions that preferentially penetrated the weaker evaporite layers in the salt walls. Some of these igneous intrusions presently outcrop at the La Sal Mountains, located to the west of Paradox Valley (Figure 3.24). Igneous structures can be seen as laccoliths that truncate some of the preexisting salt anticlines, such as in the Spanish-Moab Valley and Castle Valley salt walls (Doelling et al., 1988).

The regional uplift of the Colorado Plateau in the Neogene resulted in the erosion of the sedimentary successions that covered the elongated anticlines with salt cores. These salt walls collapsed when the halite layer was partially dissolved, and the overburdened rocks were downfaulted. Nevertheless, a substantial amount of Pennsylvanian salt remains in the basin, forming tall salt walls (Figure 3.24). Locally, there is evidence for total salt withdrawal, as the Honaker Trail Fm. carbonate rocks directly overlie the pre-salt sediments (Paz et al., 2009).

Several exploratory wells have penetrated the Pennsylvanian evaporites in the basin, and a few, such as the Exxon-1 well, have drilled through the salt (Figure 3.26). The pre-salt sequences deepen toward the Uncompahgre Front, reaching depths of more than 7 km below the surface (Molenaar, 1981; Trudgill, 2011). At the surface, the salt diapirs are often covered by a thin Quaternary layer of siliciclastic rocks, as in the Moab and Castle Valley anticlines (Figure 3.25). Gypsum can be seen in several outcrops, as in the Onion Creek diapir, located to the northeast of the city of Moab, east of the Colorado River (Figure 3.24).

The Canyonlands region, located across the border from Utah and Colorado, is characterized by salt valleys

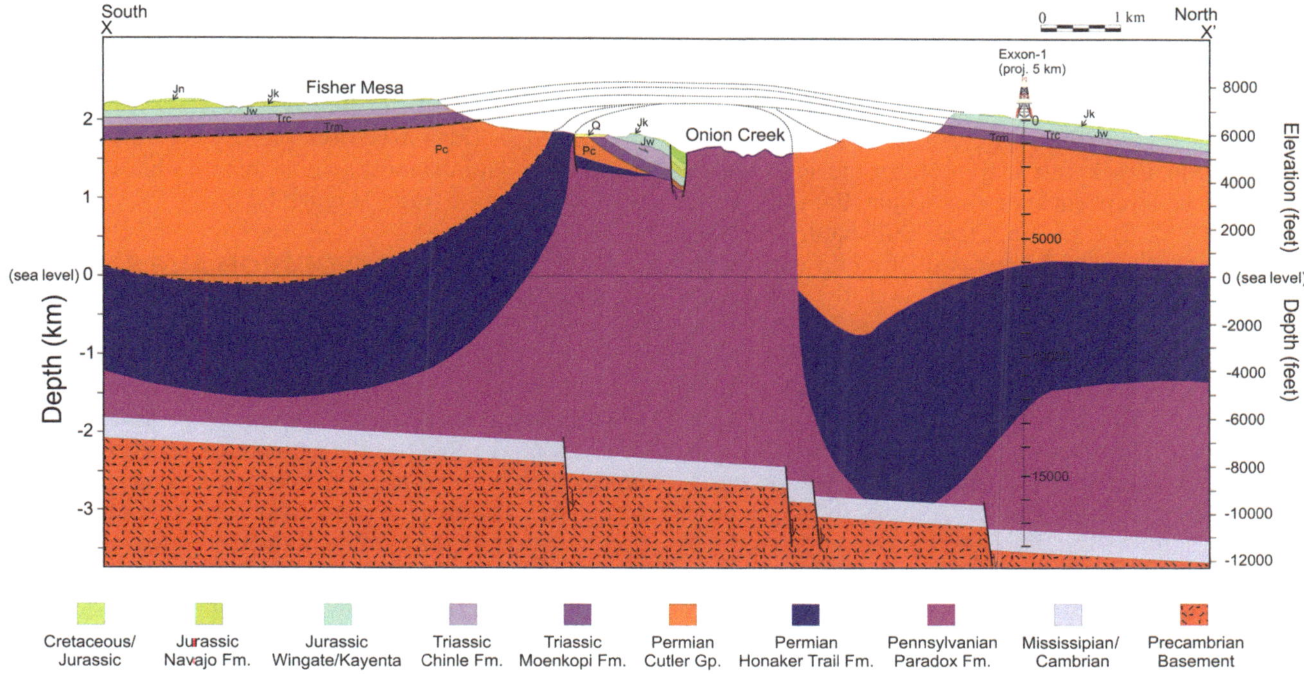

Figure 3.26 Detail of the cross-section X–X′ extending across the Fisher Valley salt wall. The Onion Creek salt diapir, a tall structure with over 4 km in the salt column, is characterized by asymmetric carbonate layers over the Paradox Fm. evaporites. The carbonate overburden was uplifted by salt rise on the southern flank, while the folded carbonate overburden was depressed by sedimentary loading. The Permian Cutler Group sedimentary sequences show a thickening north of the salt wall. The Exxon-1 exploratory borehole was drilled through Honaker Trail Fm., composed of Permian carbonate and siliciclastic rocks. The borehole penetrated the thick Pennsylvanian halite layers below the Permian redbeds. Credit: Adapted from Trudgill (2011).

developed over the axis of the salt-cored anticlines (Gutierrez, 2004). The canyonlands correspond to a deformation belt associated with compressional folds and extensional faults. Some of the salt anticlines have been breached, exposing the evaporite layers at the surface. The region known as "The Grabens" is located in the southwestern part of the Paradox Basin (Figure 3.24) and is characterized by several extensional faults with geomorphological expression at the surface (Trudgill, 2011). The salt diapirs pierced the overlying strata, forming protrusions with tops close to the surface, as in the region where the Colorado River has deeply incised a Neogene canyon (Figure 3.24).

The Paradox Valley is the largest of the breached anticlinal valleys in the northeastern part of the Paradox Basin, extending for about 37 km along an NW direction (Figure 3.24). It has been suggested that this valley is an erosional artifact due to the dissolution of the Paradox salt wall core and subsequent collapse of the anticline (e.g., Gutierrez, 2004). Alternative explanations have been proposed for these collapse structures, including salt dissolution and regional extension of the overburden (Ge & Jackson, 1998).

According to the extensional model, the subsidence is caused by normal faults that rupture the anticline crests and form axial depressions on top of the salt walls (Ge et al., 1996). Other authors have attributed the collapse of the anticlinal crests to the dissolution of the salt walls, resulting in rapid subsidence above the anticline cores and the formation of the valleys (e.g., Gutierrez, 2004; Randles et al., 2012). Naqi et al. (2016) suggested a combination of multiple factors (extensional forces, salt dissolution, and internal salt flow) as the mechanisms responsible for the formation of the salt valleys in the northeastern Paradox Basin. Salt dissolution is an important factor in the structural evolution of salt basins and has been characterized in many basins worldwide (see Chapter 8 in Mohriak, 2025).

An excellent example of a collapsed salt wall can be seen in the Fisher Valley anticline, where the salt structure is flanked by quite asymmetric post-salt layers (Figure 3.26). Recently, structures in the Gypsum Valley have been compared to megaflaps observed in the Gulf of Mexico's offshore salt basin, where highly asymmetric structures are linked to differential salt flow (Escosa et al., 2019; Giles & Rowan, 2012). The elevated rims of the Fisher Valley salt structure consist of Jurassic sandstones that form the Fisher Mesa to the south of the salt wall. The valley floor is composed of remnants of the evaporite rocks that are included in the Paradox Formation (Figure 3.27).

The Onion Creek salt diapir offers the best surface exposures of evaporites in the whole Paradox Basin (Hudec, 1995). The Onion Creek salt diapir reaches a total column of circa 3.5 km, considering the surface elevation of about 1.5 km, and the base of the salt mass is located at a depth of 2 km below sea level. The evaporites of the Onion Creek valley are exposed as a small protuberance on the Fisher-Sinbad Valley salt wall, a much larger salt structure that occurs in the subsurface (Figure 3.26). The salt wall extends across the valley, following the NW-SE trend typical of most structures basinward of the Uncompahgre thrust fault. In the outcrop, the evaporite rocks that have been preserved from dissolution are exposed as a patch of white rocks (gypsum) that correspond to the caprock. A residual "leached gypsum caprock" can be observed in some outcrops along the bottom of the valleys, reaching thicknesses exceeding 100 m (Figure 3.27). Sulfate-reducing bacteria chemically attacked gypsum layers, releasing sulfur and hydrogen sulfide gases as a result. These gases, which have a foul smell, form sulfurous brine that flows through the valley floor. The small fountain is known as Stinking Spring and the valley was named "Onion Creek" by early explorers.

The floor of the Onion Creek salt valley contains large, isolated blocks of supra-salt Triassic to Cretaceous rocks exposed at the anticline crests. These small blocks have foundered into the Paradox Formation either by extensional fault activity or by salt dissolution (Gutierrez, 2004). Early workers in the region attributed the collapse of the salt wall entirely to halite dissolution (Doelling et al., 1988). However, other authors suggest that regional extension might have played an important role in triggering normal faults and salt withdrawal (e.g., Ge et al., 1996, 1997; Ge & Jackson, 1998; Hudec, 1995). Trudgill (2011) proposed that the anticlinal structure involved the folding of overlying Jurassic to Cretaceous sediments that initially covered the Paradox Fm., and during the uplift of these rocks, deep erosion incised the Mesozoic sediments, including the Jurassic sandstone. At present, these blocks are only preserved in fault-controlled troughs that collapse into the salt diapir top (Figure 3.26).

Evaporite dissolution and fluid migration have significantly influenced halokinesis in the Paradox Basin. Salt features in the area are linked to some extensional faults that have allowed the evaporite mass to communicate with shallower aquifers. The salt outcrops are related to thick Carboniferous evaporite masses, which form salt diapirs underground. These deformed salt deposits have risen to the surface, affecting the modern geomorphology and drainage system in the Paradox Basin. At present, the exposed salt structures have been affected by weathering, and the most soluble minerals have been dissolved, forming a crater (Figure 3.27).

3.7. TECTONIC EVOLUTION AND SALT DEPOSITION IN THE DANAKIL DEPRESSION, SOUTHERN RED SEA

The Danakil Depression, located in the southern Red Sea, is home to the youngest evaporite deposits in the Red Sea region. The Red Sea is an elongated rift basin formed during the Oligo-Miocene, with salt deposition occurring from the Middle to Late Miocene, leading to the accumulation of thick halite sequences in the deep basin (Bosworth et al., 2005; Bosworth, 2015). Sedimentation continued from the Pliocene to the Quaternary, accumulating carbonate and siliciclastic sediments in the offshore region (see Chapter 1 in Mohriak, 2025).

The surface geology of the Danakil Depression is characterized by the widespread occurrence of igneous rocks associated with the Afar mantle plume, which is expressed by a cluster of volcanoes in the northern part of the Afar rift (Figure 3.28). The simplified geological map integrating data from Hartmann and Moosdorf (2012) indicates the southern part of the Danakil Alps and large areas of the Afar rift are covered by volcanic rocks (predominantly basalts) that extend from the Ethiopian Plateau to the Yemen highlands in the conjugate margin (Figure 3.28). The Erta Ale volcanic complex, an active magmatic center in the Afar region, borders the Danakil Depression to the south (Nobile et al., 2012). The volcanic buildups at the northern part of the depression blocked communication with the southern Red Sea during the Quaternary and Holocene (Bonatti et al., 1971; López-Garcial et al., 2020).

The evaporite sediments in the Danakil Depression accumulated during the divergent phase of basin development in the Red Sea, after the inception of oceanic spreading centers that separate conjugate margins (Figure 3.28). During the Quaternary, a marine incursion from the southern Red Sea flooded the depression, forming a lagoon with the main entrance south of the Dahlac Islands (Bonatti et al., 1971; Bastow et al., 2018; García-Veigas et al., 2009; Holwerda & Hutchinson, 1968). Holocene saline pans, including evaporitic mud, gypsum, anhydrite, and halite, cover the depression's floor at a depth of 120 m below sea level (López-Garcia et al., 2020; Rime et al., 2020). The salt precipitation is mainly related to periodic marine

(A)

(B)

Figure 3.27 Photographs of outcrops in the Fisher Valley anticline. (A) 3D visualization of the Onion Creek outcrop in the Fisher Valley, located south of the Colorado River. The anticlinal structure along the salt wall is characterized by sedimentary beds dipping in opposite directions from the axial line. The Cutler Group sediments are Permian siliciclastic rocks (red beds) that overlie the Paradox Fm. evaporites. The evaporite deposits in the outcrop are white gypsum minerals that form the caprock of thick halite layers in the subsurface. (B) A detailed photograph of the Onion Creek salt outcrop shows the diapir structure is characterized by a light gray cap rock composed of gypsum minerals. The Cutler Group sedimentary sequences (Permian red beds) are in direct contact with the evaporites. Credit: Modified from Trudgill (2011).

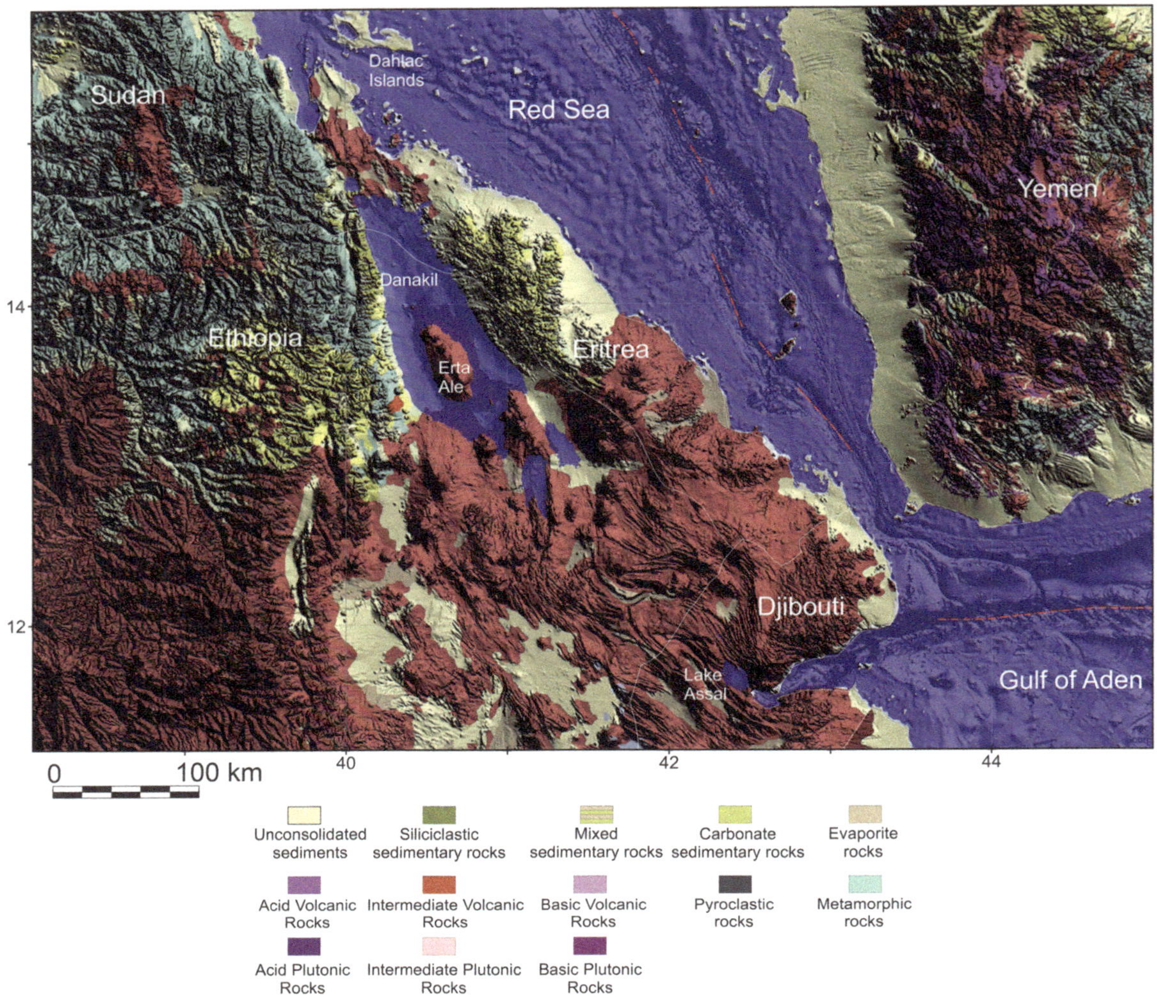

Figure 3.28 Simplified lithological map of the Afar region and the Danakil Depression in the southern Red Sea. The salt flats north of the Erta Ale Volcanic Complex are located between the Ethiopian highlands and the Danakil Alps in Eritrea. The Danakil paleo-lagoon was connected to the southern Red Sea via a channel located south of the Dahlac Islands. The volcanic buildup between the lagoon and the islands formed a barrier that isolated the lagoon from the Red Sea during the Quaternary. The lagoon was desiccated periodically, resulting in the accumulation of thick evaporite sequences. Credit: Modified from Hartmann and Moosdorf (2012).

incursions, basin isolation episodes, and brine concentration by solar evaporation. However, some contributions by hydrothermal springs have been proposed by some researchers (e.g., Hovland et al., 2008; Hovland et al., 2018a, 2018b; Talbot, 2008).

The evaporite deposits form a thick sedimentary layer with over 1 km of halite intercalated with lacustrine deposits (Bastow et al., 2018; Bekele & Schmerold, 2020; Holwerda & Hutchinson, 1968). A study of a drill core on the northern shore of Lake Karum (also known as Lake Assale) revealed a thick sequence of about 625 m of evaporite sediments deposited in the last 130 years, including lenses of potassium-bearing evaporite minerals (Bastow et al., 2018; Rime et al., 2020).

Satellite images and geological maps reveal salt flats between the Danakil Alps and the rift master fault (Figure 3.28). Mount Dallol, located in the middle of the Danakil depression, has been interpreted as a volcanic buildup, but no lava flows are found at its flanks (Warren, 2016). The crater of the Dallol structure is characterized by abundant hydrothermal springs, suggesting a magma chamber in the subsurface (e.g., López-Garcia et al., 2020). Most rocks exposed at the surface are Holocene evaporite deposits, intercalated with lacustrine sedimentary deposits (Figure 3.29). The Danakil Depression is a modern analog for salt deposition in tectonic settings associated with extensional rift basins and active magmatism (see Chapter 9).

Figure 3.29 (A) Photograph of hydrothermal springs in the crater of Mount Dallol, a buildup located at the central part of the Danakil Depression. (B) Photograph of an outcrop of bedded anhydrite intercalated with thin layers of lacustrine sediments. The hammer length (about 30 cm) provides the scale. Credit: Courtesy of López-Garcia (2020).

Figure 3.30 The Dead Sea Transform fault zone and Dead Sea Lake are depicted in a satellite image, with artificial salt ponds at the southern basin and Sedom Mount near the western margin. The tongue-shaped white patch separates the lakes, corresponding to the Lisan Peninsula. The saltworks evaporation pools in the southern basin are visible on the Israeli side and Jordanian side, south of the Lisan Peninsula. Mount Sedom, the most iconic salt structure in the region, reaches an altitude of about 200 m above the lake level. Credit: Jesse Allen/NASA/Public domain.

3.8. TECTONIC EVOLUTION AND SALT DEPOSITION IN THE DEAD SEA BASIN

Evaporite rocks are found on the surface of various locations along the Arabian Peninsula, including the Middle Miocene sequences in the Midyan Basin of the northern Red Sea and onshore Yemen in the southern Red Sea (Davison, Bosence, Alsop, & Al-Aawah, 1996). In the Dead Sea Basin, Pliocene evaporites form caprocks, which are mainly composed of gypsum (Figure 3.30).

The Dead Sea transform (DST) is a left-lateral transform fault along the boundary of the African and Arabian plates, running for over 1,000 km from the Red Sea spreading center offshore of the southern tip of the Sinai Peninsula (see Chapter 5 in Mohriak, 2025). This fault links with the East Anatolian Fault Zone in southern Turkey, forming a depression that extends from the Gulf of Aqaba to the Sea of Galilee. The transpressional deformation zones are marked by thrust and strike-slip faults. These structures can be seen in the Taurus Mountains in Turkey, the Palmyride fold belt in Syria, and the Zagros Mountains in Iran (Beydoun, 1999; Ben-Avraham et al., 2008, 2012; Lazar, 2019).

The region adjacent to the DST fault zone includes pre-rift sedimentary units deposited during the Mesozoic and Cenozoic, indicating shallow seas until the Late Eocene (Ben-Avraham et al., 2008; Garfunkel & Ben-Avraham, 1996). The Dead Sea depression was formed around 18–15 Ma ago between the left-lateral Arava Fault to the southeast and the Jericho Fault on the northwest part of the depression (Garfunkel, 1981; Weinberger et al., 2020).

The Dead Sea Basin is located in a deep depression formed by a transtensional fault system that has been active since the Miocene (see Chapter 2). Miocene to Quaternary sediments overlie the Mesozoic and Paleozoic rocks accumulated in the pre-rift basin. The transcurrent fault develops some tectonically controlled valleys, such as the Arava Valley in the south and the Jordan Valley in the north. The topography of the DST fault zone is quite asymmetric, with the Jordan side showing more elevated highlands to the east (Figure 3.30). The transcurrent fault shifts to the west near the southern Dead Sea, leading to the development of a pull-apart basin in a releasing bend of the strike-slip fault system (see Chapter 2).

The Dead Sea depression corresponds to the lowest place on earth, with the lake level at 436 m below the global sea level in 2021 (see Chapter 2). The lake straddles the international borders of Jordan to the east and Israel to the west. The rift basin fill is relatively thin in the southern and northern extremities, and the maximum sedimentary thickness is located in the region of the Lisan Peninsula (Al-Zoubi & ten Brink, 2001; Al-Zoubi et al., 2002). Evaporite rocks occur at shallow depths on the Lisan Peninsula and outcrop along the western margins of the Dead Sea, forming the Mount Sedom salt diapir, where halite minerals can be seen at the surface (Weinberger, Begin, et al., 2006; Weinberger, Lyakhovsky, Baer & Begin, 2006; Zak & Bentor, 1972). In the northern Dead Sea Lake, some shallow salt diapirs occur in the subsurface, forming protuberances affecting the lake floor (see Chapter 11).

3.8.1. Origin of Evaporite Deposits in the Dead Sea Basin

The Dead Sea Basin's evaporite sediments are attributed to a deep lake that formed in the late Miocene or lower Pliocene (5–3 Ma). The ancient lake evolved into Lake Amora and then Lake Lisan in the Pleistocene, which extended from the northern Gulf of Aqaba toward the Jordan River valley, reaching the Sea of Galilee and the northern Wadi Arabah valley (Zak, 1974). Some authors postulate that the Dead Sea evaporites were deposited coevally with the upper evaporites in the Mediterranean Sea (e.g., Charrach, 2018). However, most researchers assume that salt deposition in the Dead Sea was limited to the Pliocene, post-dating the Late Miocene salt deposition episode that affected the Mediterranean Sea region during the Messinian Salinity Crisis (see Chapter 5 in Mohriak, 2025).

The Pliocene evaporites (Sedom Fm.) were deposited in a marine embayment that extended from the eastern Mediterranean to the rift valley depression (Zak, 1974, 1997). Marine waters from the Mediterranean Sea entered the Sedom lagoon through the Jezreel Valley in Israel, repeatedly inundating the deep lakes (Zak & Freund, 1981). The Sedom Lagoon was desiccated several times in the Pliocene, depositing thick sequences of layered evaporites, reaching thicknesses exceeding 2,000 m (Zak & Freund, 1980). Most researchers admit that during the Pliocene (about 3–2 Ma), the land between the Dead Sea rift and the Mediterranean Sea was uplifted. The connection with the Mediterranean Sea was closed, interrupting the flow of ocean waters into the Sedom Lagoon (Gavrieli & Stein, 2006).

During the Pliocene, the land between the Dead Sea rift and the Mediterranean Sea was uplifted, leading to disconnection with the Mediterranean Sea. The Dead Sea rift valley was a landlocked depression occupied by continental lakes whose sizes varied according to climatic fluctuations and glaciation episodes (Ben-Dor et al., 2019). These lakes were intermittently flooded during global sea level rises or more wet periods and shrank in size during sea-level falls or more arid periods. The lacustrine sediments indicate that annual temperature fluctuations resulted in the deposition of centimeter-thick layers of evaporite minerals. The Dead Sea is considered the only modern, deep hypersaline lake in the world, with varved sediments forming evaporite sequences that are seasonally controlled (see Chapter 2).

The first prehistoric lake to follow the Sedom Lagoon was Amora, which deposited siliciclastic sediments from the Pliocene to the Pleistocene, corresponding to the Amora Fm. (Torfstein, 2017). The sedimentary loading squeezed salt laterally and upward, forming salt diapirs at the Lisan Peninsula and Mount Sedom (Figure 3.30). Lacustrine deposits from the Lisan Fm. can be seen at the margins of the modern Dead Sea Lake, particularly in the Amiaz Plain, where spectacular outcrops exhibit the carbonate and gypsum laminated sequences (Alsop & Marco, 2011; Alsop et al., 2020; Bartov et al., 2002).

As the climate became drier, Lake Lisan continued to shrink and become saltier. Several layers of evaporite minerals were deposited during the Lisan Lake period, indicating periods of high salt concentrations in the lake brine. Halite minerals precipitate at the lake's margins and bottom, but these thin lenses are only a fraction of the evaporite deposits that accumulated on the lake's floor during the Pliocene. The Dead Sea corresponds to a remnant of a larger lake system filled with less concentrated brines in the Pleistocene, with sporadic deposition of evaporite sediments. Even today, halite minerals precipitate at the lake's margins and the lake's bottom (Sirota et al., 2017). Nonetheless, these thin lenses are but a fraction of the evaporite deposits that accumulated on the lake's floor during the Pliocene, resulting in the kilometer-thick Sedom Fm. (Zak & Bentor, 1972).

3.8.2. Salt Outcrops in the Dead Sea Basin

Salt deposits are found on the Dead Sea's shores and in the subsurface, within the rift depocenter. Shallow salt bodies form salt diapirs below the lake surface, covered by a thin cap rock made of aragonite, gypsum, and detritic marls and clays (Closson et al., 2007, 2011; Coianiz et al., 2019, 2020). The most spectacular salt structures are observed near Lisan Peninsula, a strip of land separating the northern lake from the artificial salt ponds in the southern basin (Figure 3.30). The Lisan salt diapir is located below a triangular-shaped caprock exposed at the surface, with an ellipsoidal top measuring about 10 × 13 km and a base deeper than 7 km. The roof of the salt diapir is about 125 m below the present-day water surface on the Lisan Peninsula (Al-Zoubi & ten Brink, 2001).

The Mount Sedom salt diapir at the western margin of the southern basin is one of the most iconic salt structures on Earth (see Chapter 1). Mount Sedom is an elongated ridge with an N-S trend, corresponding to a salt structure with a flat top reaching a height of about 200–240 m above the western shores of the salt ponds (Figure 3.31). The diapiric salt wall measures about 10 km in length and a width of up to 2 km (Alsop et al., 2015; Weinberger, Begin, et al., 2006; Weinberger, Lyakhovsky, Baer & Begin, 2006; Zak & Freund, 1980, 1981). The Sedom salt diapir has been rising since the Pleistocene, with its caprock reaching a maximum topographic altitude of −190 m, still below the Mediterranean Sea level (Alsop et al., 2016).

Mount Sedom is a small geomorphological feature compared with the much larger salt structures that occur in the subsurface of the Dead Sea Basin, which is imaged by several seismic profiles acquired by the petroleum exploration industry in Israel and Jordan (Al-Zoubi & ten Brink, 2001; Al-Zoubi et al., 2002). However, Sedom hill is famous for the evaporite columns that have been attributed in biblical stories to the petrification of Lot's wife (see Chapter 1).

The Sedom salt diapir consists of steeply dipping to vertical layers of evaporite sequences, originally mainly composed of halite, interbedded with layers of anhydrite, marl, sand, and shale of the Sedom Fm. (Figure 3.32). Uplifted Pleistocene carbonates, sulfates, and clastic rocks associated with the Amora and Lisan formations surround and partially cover the salt structure (Weinberger, Lyakhovsky, Baer, & Begin, 2006). The diapir top is covered by Quaternary alluvial sediments and a caprock that can reach thicknesses of a few meters to about 40 m (Figure 3.32). The caprock in Mount Sedom corresponds to remnants of the dissolved evaporites, including undissolved components of anhydrite, gypsum, marl, dolomite, and clastic rocks (Closson et al., 2007).

The Sedom salt structure is a unique example of active salt diapirism, with subvertical to overturned layers resulting from the rise of the salt structure in modern times (Figure 3.32). It is characterized by a subhorizontal dissolution surface below the top of the hill, which corresponds to the water table that dissolved the salt below (Figure 3.33). This water table, also known as the salt

Figure 3.31 A picture of Mount Sedom, viewed toward the southeast, shows the topographic elevation relative to the Dead Sea depression. The mountaintop corresponds to a caprock overlying the dissolved salt. Both the hill and the saltworks are below the Mediterranean sea level. Credit: Courtesy of Poprawski (2018).

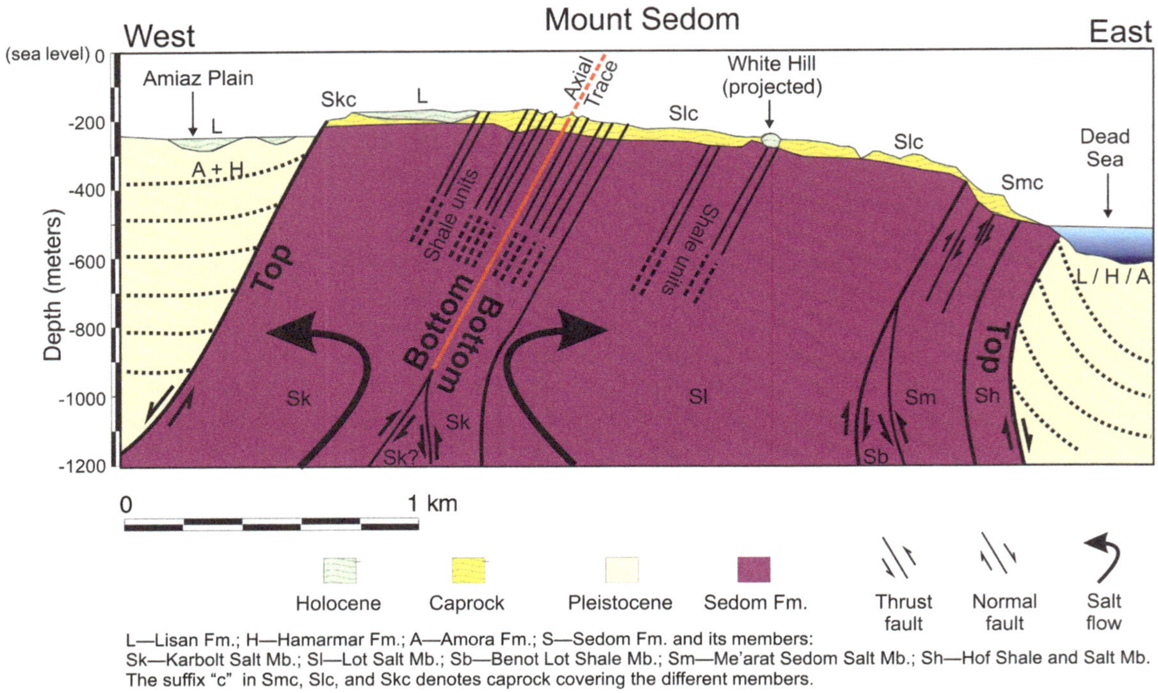

Figure 3.32 Geological cross-section illustrating the Mount Sedom salt diapir at the western border of the Dead Sea. The Amiaz Plain was drilled by the Amiaz borehole, located to the west of Mount Sedom. The Sedom Deep borehole drilled the Dead Sea Basin depocenter. The deep well penetrated Pleistocene sediments, siliciclastic sediments, and underlying Sedom Fm. evaporites, deposited from the Middle to Early Pliocene. The evaporite layers in the salt diapir have an almost vertical axial trace due to upward salt flow, and the eastern flank shows vertical and upturned beds. Credit: Adapted from Weinberger et al. (2006).

Figure 3.33 A detailed photograph shows the leached salt rock column known as "Lot's wife," outcropping at the margin of the saltworks in the southern Dead Sea depression. The ancient water table (salt mirror) was uplifted in Holocene times. The salt mirror separates the residual evaporite caprock from the leached salt rock. The evaporite layers are locally characterized by vertical bedding due to upward salt flow within the diapir. Credit: Mohriak (2018).

mirror, separates the leached salt rock from the overlying residual caprock. It was formed when the diapir was close to the lake level but now corresponds to a fossilized, hanging salt table. At present, the salt mirror is placed at elevations between 30 and 180 m above the Dead Sea lake level, due to the diapir rise in the Holocene (Weinberger, Lyakhovsky, Baer & Begin, 2006; Weinberger, Begin, et al., 2006). This surface is usually subhorizontal or slightly inclined to the east, toward the southern Dead Sea Lake. Locally, the salt mirrors are affected by faults that offset the evaporite beds by a few meters (Zak & Freund, 1980).

The leached salt rocks under the salt table (Figure 3.33) are commonly affected by a karst system forming caves and chimneys, which form vertical shaft-like caverns, reaching dimensions of more than 15 m in diameter and heights up to 80 m, thus resulting in dangerous hazards for people trekking along the top of the hill (Figure 3.31).

The Sedom salt diapir is located close to the western boundary fault of the Dead Sea rift basin (Al-Zoubi et al., 2002). The opposing border fault is located on the eastern side of the rift, across the boundary with Jordan. This indicates a marked asymmetry in the development of the Dead Sea Basin, which forms a deep rift trough filled by sediments that reach a total thickness of about 8–10 km (Zak & Freund, 1980).

The Dead Sea Basin sedimentary depocenter is located near the political boundary between Israel and Jordan. Most of the mobile salt in the southern lake is located in the western depocenter of the basin, in Israel. The thick post-salt depocenter on the Jordan side indicates that much of the salt has migrated upwards, toward the more elevated footwall blocks between the Western Boundary Fault and the Sedom Fault (Al-Zoubi et al., 2002; Lazar et al., 2006; Lazar, 2019; Zak & Freund, 1981). The Sedom salt diapir has been actively rising from the deeper salt layer since the Pleistocene, with an estimated uplift rate of about 3-5 mm/year since 14 ka (Alsop et al., 2016; Weinberger, Lyakhovsky, Baer & Begin, 2006; Weinberger, Begin, et al., 2006; Zak & Freund, 1980).

Figure 3.34A shows an oblique aerial view of Mount Sedom at the western shores of the Dead Sea's artificial salt ponds. The hill corresponds to an elevated plateau covered by a caprock made of residual carbonate and sulfate minerals mixed with siliciclastic fragments, with halite layers visible near the base (Figure 3.34B). The high solubility of halite minerals led to salt dissolution and the development of karst topography and large vadose caves on the hill's flanks (Figure 3.34B). Some of these caves were used by ancient inhabitants to store sacred documents. The famous Dead Sea scrolls were discovered further north in limestone caves in the Qumran region (see Chapter 1).

Mount Sedom salt diapir is flanked by post-Miocene sedimentary layers that are almost vertical, with some caves characterized by alternations of gypsum and halite layers (Figure 3.34C). The interior parts of the Sedom salt diapir are characterized by subvertical beds to recumbent folding, with some small extrusive salt tongues covering upturned post-salt sedimentary sequences (Alsop et al., 2015). Mount Sedom is the most conspicuous outcropping salt diapir in the Dead Sea (see book cover). It provides a unique example of active salt diapirism, with subvertical to overturned layers resulting from the rise of the salt structure in modern times. The Sedom salt diapir constitutes a paradigm for similar salt structures that occur in giant salt basins, such as the South Atlantic and the Gulf of Mexico (Alsop et al., 2016).

3.9. SYNTHESIS OF THE SELECTED SALT OUTCROPS

Figure 3.35 presents a comprehensive analysis of salt deposition intervals and tectonic events that led to the deformation of salt layers and the development of salt diapirs in various basins across the Middle East, Europe, northern Africa, and North America. The Atlas in northern Africa, the Alps in southern central Europe, and the Zagros in Iran form a complex boundary between the African, Eurasian, and Arabian plates (Figure 3.1).

The main stages of diapir contraction, uplift, and eventual exposure on the depositional surface of the selected basins are related to episodes of basin inversion and contraction. The tectonic episodes are mostly linked to the Alpine Orogeny's Cenozoic plate convergence episodes (Najafi et al., 2018; Warren, 2010). Salt deposits were brought to the surface due to halokinetic processes and contractional tectonics. The summary of the salt deposition events is based on the analysis of specific regions and several reference works (e.g., Warren, 2016). Some outcrops are discussed in more detail in other chapters, such as the Permian Zechstein evaporite deposits in northeast England (Chapter 7) and the Messinian evaporites in Sicily (Chapter 5 in Mohriak, 2025).

Evaporite outcrops in Iran, northwest Africa, and southern central Europe indicate that salt layers from several stratigraphic sequences were deformed during the Alpine Orogeny. In the Zagros region of the Middle East, the Neoproterozoic salt occurs deep in the subsurface, while several outcrops are observed in the Zagros thrust-and-fold belt of Iran. The Neoproterozoic to Cambrian rift basins accumulated the Hormuz salt, which developed large diapiric structures during the Paleozoic and Mesozoic.

Salt diapirs were squeezed during the Alpine Orogeny, forming salt diapirs and salt glaciers in many outcrops in

Figure 3.34 Photographs showing an aerial view of Mount Sedom and detailed images of the outcrops located near the Dead Sea's saltworks. (A) Oblique aerial image of Mount Sedom. The opposite margin of the lake shows an uplifted flank, which corresponds to the eastern border fault of the Dead Sea Graben, located in Jordan. Credit: Courtesy of Tsabar (2018). (B) A detailed image of an outcrop, located near the lake's eastern flank and adjacent to the salt diapir, shows some blocks of leached evaporitic sedimentary rocks, locally enriched in halite. The dissolution of evaporite minerals formed small caves at the mountain's flanks. (C) A detailed image inside a cave shows the evaporite rocks displaying vertical bedding, indicating shearing due to upward salt flow. Credit: Mohriak (2018).

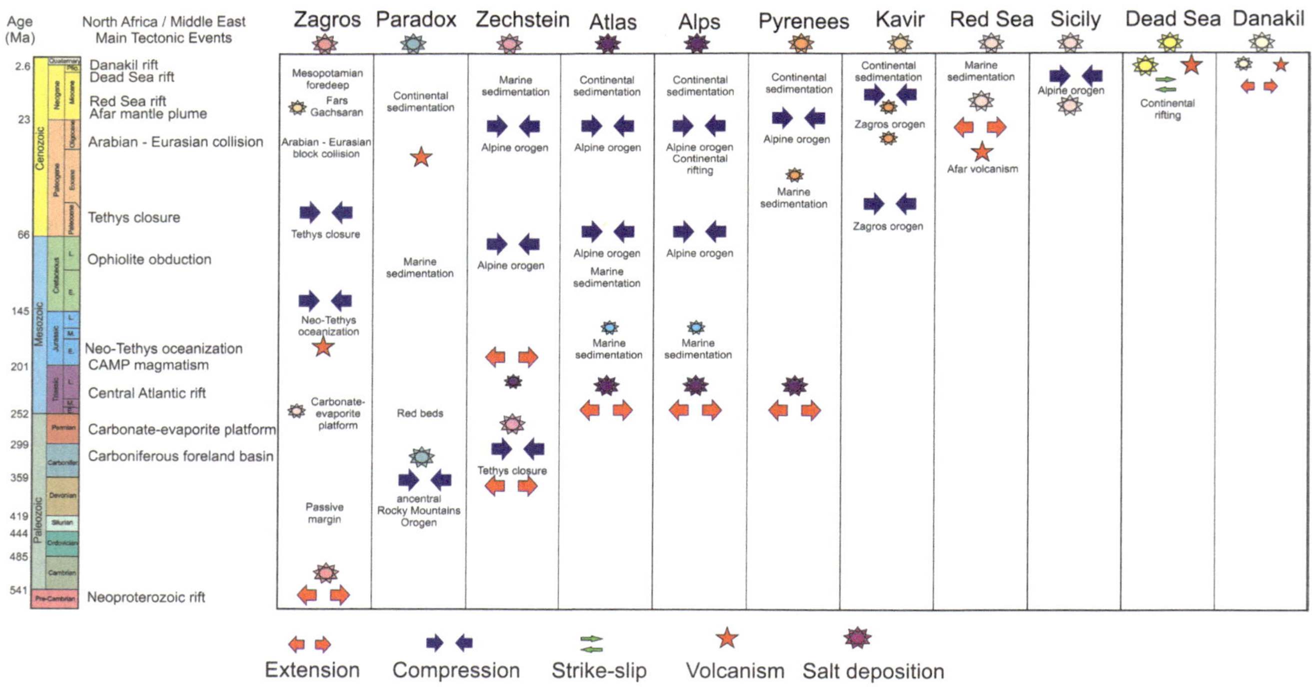

Figure 3.35 The simplified tectonostratigraphic chart of selected basins with salt exposed at the surface shows the main geological events and salt deposition intervals. Credit: (Left) Adapted from Najafi et al. (2018) and (Right) Data from Warren (2010).

southern central Iran. In the offshore region of the Persian Gulf, many islands are characterized by salt masses with tops near the surface, indicating squeezed salt structures sourced from the autochthonous Hormuz salt (Talbot, 1998). In the offshore region of the Persian Gulf, many islands are characterized by salt masses with tops near the surface. This indicates the islands correspond to squeezed salt structures sourced from the autochthonous Hormuz salt. Also, some secondary, younger evaporitic layers were later deformed by compressional and strike-slip tectonics in the Neogene, forming outcrops of younger salt (Hassanpour et al., 2020).

In NW Africa, the Alpine deformation zones are observed in the Atlas thrust-and-fold belts, straddling Morocco, Algeria, and Tunisia. Some exposed salt structures correspond to the tops of salt diapirs, while others correspond to allochthonous salt masses, similar to the salt glaciers in Iran or the extrusive salt tongues in the Gulf of Mexico (e.g., Amri et al., 2020).

In Iberia, the Cantabrian and Pyrenean mountain ranges separate the southern block (African plate) from the northern plate (European plate). The Cardona salt diapir of northern Spain is a unique example of Paleogene salt deposits that accumulated in a foreland basin. The salt layers were uplifted in the Neogene, breaching the Cardona anticlinal structure. The Cardona salt diapir has been dissected by erosion, and the valley was affected by karstification, similar to other regions in Spain (see Chapter 8 in Mohriak, 2025).

The Paradox Basin salt diapirs are linked to Carboniferous salt deposits that formed large, elongated salt wall structures due to foreland basin deformation and salt mobilization. These outcrops are located in valleys above anticlines associated with the salt walls. In the late Paleozoic, compressional tectonics had an impact on the post-salt sedimentary sequences. Some authors use the modern convergent tectonic setting of the Zagros region as an analog for the Alleghanian Orogeny in eastern Canada (Gibling et al., 2019).

The salt outcrops in the Zagros region and the Dead Sea rift zone provide clues for the interpretation of salt masses reaching the surface in areas affected by active strike-slip tectonics. Modern outcrops indicate that transcurrent fault zones might have played an important role in controlling the deformation of salt structures in ancient salt basins. This interpretation applies to the late Paleozoic Maritimes and Paradox basins in North America, as well as many Mesozoic salt basins formed during the evolution of the Tethys Ocean. The African, Arabian, and Eurasian plates' diachronous convergence during the Cenozoic caused the Alpine Orogeny, which had an impact on these basins (see Chapter 10 and Chapter 5 in Mohriak, 2025).

The last examples of outcropping salt structures discussed in the chapter correspond to the Quaternary salt deposits in the Danakil Depression and the Sedom salt diapir in the Dead Sea. These younger salt deposits have not been affected by orogenic forces. The Danakil Depression corresponds to a rift basin that developed a lagoon at the southern margin of the Red Sea during the Quaternary. The basin was partially disconnected from the Red Sea by volcanic barriers. The brine was concentrated and precipitated thick layers of halite and other evaporitic sediments during the Quaternary and Holocene. Active volcanism and hydrothermal springs occur in the northern part of the Afar rift, suggesting an association between magmatism and evaporite deposition (see Chapter 9).

The Dead Sea Basin originated in a transtensional setting that developed rift basins along a major strike-slip fault zone linking the Red Sea to the orogenic belt extending from Turkey to Iran. The northern part of the transform fault zone is associated with a major thrust-and-fold belt that links with the Zagros thrust-and-fold belt. Salt diapirs in the Dead Sea Basin are associated with a transtensional tectonic setting. In the Zagros region of Iran, several salt glaciers are associated with the strike-slip fault zones developed during the Neogene to Recent compressional phases (see Chapter 7 in Mohriak, 2025).

Some of the halokinetic features observed in selected outcrops have counterparts in the subsurface of many salt basins. However, the interpretation of salt tectonic features deep in the subsurface relies mostly on indirect geophysical methods, such as the analysis of seismic reflection datasets (see Chapter 5).

REFERENCES

Aftabi, P., Roustaie, M. G., Alsop, I. G., & Talbot, C. J. (2010). InSAR mapping and modelling of an active Iranian salt extrusion. *Journal of the Geological Society, London*, *167*, 155–170. doi: 10.1144/0016-76492008-165.

Agard, P., Omrani, J., Jolivet, L., Whitechurch, H., Vrielynck, B., Sparkman, W., et al. (2011). Zagros orogeny: A subduction-dominated process. *Geological Magazine, 148(5–6), 692–725.* Cambridge University Press. doi: 10.1017/S001675681100046X

Ahmadi, R., Mercier, E., & Ouali, J. (2013). Growth-strata geometry in fault-propagation folds: A case study from the Gafsa basin, southern Tunisian Atlas. *Swiss Journal of Geosciences*, *106*(1), 91–107.

Ahmadian, J., Murata, M., Nadimi, A., Ozawa, H., & Kozai, T. (2014). Active tectonics of Iran deduced from earthquakes, active faulting and GPS evidences. *Bulletin of Center for Collaboration in Community Naruto University of Education*, *28*, 11–22.

Ala, M. A. (1974). Salt diapirism in southern Iran. *American Association of Petroleum Geologists Bulletin*, *58*(9), 758–770.

Alavi, M. (1994). Tectonics of the Zagros orogenic belt of Iran: New data and interpretations. *Tectonophysics*, *229*, 211–238.

Alsop, G. I., Brown J. P., Davison, I., & Gibling, M. R. (2000). The geometry of drag zones adjacent to salt diapirs. *Journal of the Geological Society, London*, *157*, 1019–1029.

Alsop, G. I., & Marco, S. (2011). Soft-sediment deformation within seismogenic slumps of the Dead Sea basin. *Journal of Structural Geology*, *33*, 433–457.

Alsop, G. I., Weinberger, R., Levi, T., & Marco, S. (2015). Deformation within an exposed salt wall: Recumbent folding and extrusion of evaporites in the Dead Sea basin. *Journal of Structural Geology*, *70*, 95–118.

Alsop, G. I., Weinberger, R., Levi, T., & Marco, S. (2016). Cycles of passive versus active diapirism recorded along an exposed salt wall, *Journal of Structural Geology*, *84*, 47–67.

Alsop, G. I., Weinberger, R., Marco, S., & Levi, T. (2018). Fault and fracture patterns around a strike-slip influenced salt wall. *Journal of Structural Geology*, *106*, 103–124.

Alsop, G. I., Weinberger, R., Marco, S., & Levi, T. (2020). Fold and thrust systems in mass-transport deposits around the Dead Sea basin. In Ogata, K., Festa, & Pini, G. A. (Eds.), *Submarine landslides: Subaqueous mass transport deposits from outcrops to seismic profiles, American Geophysical Union Monograph Series* (Vol. 246, pp. 139–154). Hoboken, *NJ: John Wiley & Sons Inc.*, 362 pp. ISBN: 978-1-119-50058-2.

Al-Zoubi, A., Shulman, H., & Ben-Avraham, Z. (2002). Seismic reflection profiles across the southern Dead Sea basin. *Tectonophysics*, *346*, 61–69.

Al-Zoubi, A., & ten Brink, U. S. (2001). Salt diapirs in the Dead Sea basin and their relationship to Quaternary extensional tectonics. *Marine and Petroleum Geology*, *18*, 779–797.

Amante, C., & Eakins, B. W. (2009). *ETOPO1 1 arc-minute global relief model: Procedures, data sources and analysis. NOAA Technical Memorandum NESDIS NGDC-24. National Geophysical Data Center Marine Geology and Geophysics Division*, Boulder, Colorado, 19 pp.

Amor, H. (1999). Halokinesis and structural evolution of the major features in eastern and southern Tunisian Atlas. *Tectonophysics*, *306*, 79–95.

Amri, Z., Naji, C., Masrouhi, A., & Bellier, O. (2020). Interconnection salt diapir–allochthonous salt sheet in northern Tunisia: The Lansarine–Baoula case study. *Journal of African Earth Sciences*, *170*, 103876. https://doi.org/10.1016/j.jafrearsci.2020.103876

Archer, S. G., Alsop, G. I., Hartley, A. J., Grant, N. T., & Hodgkinson, R. (2012). *Salt tectonics, sediments and hydrocarbon prospectivity*. In G. I. Alsop, S. G. Archer, A. J. Hartley, N. T. Grant, & R. Hodgkinson (Eds.), *Salt tectonics, sediments and prospectivity* (Vol. 363, pp. 1–6). Geological Society, London, Special Publications. http://dx.doi.org/10.1144/SP363.1.

Arian, M. (2012). Clustering of diapiric provinces in the Central Iran Basin. *Carbonates Evaporites*, *27*, 9–18.

Arian, M., & Noroozpour, H. (2015). Tectonic geomorphology of Iran's salt structures. *Open Journal of Geology*, *5*, 61–72. http://dx.doi.org/10.4236/ojg.2015.52006

Ayed-Khaled, A., Ben Salem, M. S., Ezzine, I., & Ghanmi, M. (2017). Halocinese and Halotectonics in the Northern Atlas: Case study of the Triassic outcrop of *J. El Matria (Oued Zarga Area). Journal of Geology and Geophysics*, *6(5), 310*, 6 pp. doi: 10.4172/2381-8719.1000310

Ayed-Khaled, A., Zouaghi, T., Atawa, M., & Ghanmi M. (2015). New evidence on the geologic setting of Medjerda Valley plain (northern Tunisia) from integrated geophysical study of Triassic evaporite bodies. *Annals of Geophysics*, *58*(3), S0326. doi:10.4401/ag-6567.

Baars, D. L. (1976). The Colorado Plateau aulacogen—Key to continental-scale basement rifting. *Proceedings of the Second International Conference on Basement Tectonics*, pp. 157–164.

Baars, D. L., & Doelling, H. H. (1987). Moab salt-intruded anticline, east-central Utah. *Geological Society of America Centennial Field Guide—Rocky Mountain Section*, *61*, 275–280.

Baars, D. L., & Stevenson, G. M. (1981). Tectonic evolution of the Paradox Basin, Utah and Colorado. In D. L. Wiegand (Ed.), *Geology of the Paradox basin* (pp. 23–31). Denver: Rocky Mountain Association of Geologists.

Bahroudi, A., & Koy, H. A. (1990). Tectono-sedimentary framework of the Gachsaran Formation in the Zagros foreland basin. *Marine and Petroleum Geology*, *21*, 1295–1310.

Bally, A. W., Roberts, D. G., Sawyer, D., & Sinkewich, A. (2012). Tectonic and basin maps of the world. In D. G. Roberts & A.W. Bally (Eds.), *Regional geology and tectonics: Phanerozoic passive margins, cratonic basins and global tectonic maps* (Vol. *1C*, pp. 973–1151). Elsevier Science Ltd.

Bally, A. W., Sawyer, D., and Sinkewich, A. (2020). Global Tectonic and Basin Maps. *AAPG Search and Discovery*, Article #30444 (2020).

Banham, S. G., & Mountney, N. P. (2013a). Evolution of fluvial systems in salt-walled mini-basins: A review and new insights. *Sedimentary Geology*, *296*, 142–166.

Banham, S. G., & Mountney, N. P. (2013b). Controls on fluvial sedimentary architecture and sediment-fill state in salt-walled mini-basins: Triassic Moenkopi Formation, Salt Anticline Region, SE Utah, USA. *Basin Research*, *25*(6), 709–737. doi:10.1111/bre.12022

Barbeau, D. L. (2003). A flexural model for the Paradox Basin: Implications for the tectonics of the ancestral Rocky Mountains. *Basin Research*, *15*, 97–115.

Bartov, Y., Stein, M., Enzel, Y., Agnon, A., & Reches, Z. (2002). Lake levels and sequence stratigraphy of lake Lisan, the late Pleistocene precursor of the Dead Sea. *Quaternary Research*, *57*, 9–21.

Bastow, I. D., Booth, A. D., Corti, G., Keir, D., Magee, C., Jackson, C. A.-L., et al. (2018). The development of late-stage continental breakup: Seismic reflection and borehole evidence from the Danakil Depression, Ethiopia. *Tectonics*, *37*, 2848–2862. https://doi.org/10.1029/2017TC004798

Beauchamp, W., Allmendinger, R. W., Barazangi, M., Demnati, A., Alji, M. E., & Dahmani, M. (1999). Inversion tectonics and the evolution of the High Atlas Mountains, Morocco, based on a geological-geophysical transect. *Tectonics*, *18*(2), 163–184.

Bekele, A., & Schmerold, R. (2020). Characterization of brines and evaporite deposits for their lithium contents in the northern part of the Danakil Depression and in some selected areas of the Main Ethiopian Rift lakes. *Journal of African Earth Sciences*, *170*, 103904.

Belkhiria, W., Boussiga, H., & Inoubli, M. H. (2017). Thick-skinned tectonics within the intracontinental easternmost Atlas foreland-and-thrust belt (Tunisia): Meso-Cenozoic kinematics and implications for regional geodynamics. *Tectonics*, *36*, 981–1004. doi:10.1002/2016TC004340.

Ben-Avraham, Z., Garfunkel, Z., & Lazar, M. (2008). Geology and evolution of the southern Dead Sea fault with emphasis on subsurface structure. *Annual Review of Earth and Planetary Sciences*, *36*, 357–387.

Ben-Avraham, Z., Lazar, M., Garfunkel, Z., Reshef, M., Ginzburg, A., Rotstein, Y., et al. (2012). Structural styles along the Dead Sea fault. In D. G. Roberts & A. W. Bally (Eds.), *Regional geology and tectonics: Phanerozoic rift systems and sedimentary basins* (Vol. *1B*, pp. 617–633).

Ben-Dor, Y., Neugebauer, I., Enzel, Y., Schwab, M. J., Tjallingii, R., Erel, Y., & Brauer, A. (2019). Varves of the Dead Sea sedimentary record. *Quaternary Science Reviews*, *215*, 173–184.

Berra, F., & Angiolini, L. (2014). The evolution of the Tethys region throughout the Phanerozoic: A brief tectonic reconstruction. In L. Marlow, C. Kendall, & L. Yose, (Eds.), *Petroleum systems of the Tethyan region, American Association of Petroleum Geologists Memoir* (Vol. *106*, pp. 1–27).

Beydoun, Z. R. (1986). The petroleum resources of the Middle East: A review. *Journal of Petroleum Geology*, *9*(1), 5–27. doi:10.1111/j.1747-5457.1986.tb00370.x

Beydoun, Z. R. (1999). *Evolution and development of the Levant (Dead Sea Rift) Transform System: A historical-chronological review of a structural controversy*. In C. MacNiocaill and P. D. Ryan (Eds.), *Continental Tectonics* (Vol. 164, pp. 239–255). Geological Society, London, Special Publications.

Blakey, R. C., & Ranney, W. D. (2018). *Ancient landscapes of western North America: A geologic history with paleogeographic maps*. Switzerland: Springer International Publishing, 228 pp.

Bohener, R. C. (1986). Salt and Potash resources in Nova Scotia. *Nova Scotia Department of Natural Resources, Mineral Resources Branch, Bulletin ME 5*. https://novascotia.ca/natr/meb/pdf/bull05.asp

Bonatti, E., Emiliani, C., Ostlund, G., & Rydell, H. (1971). Final desiccation of the Afar Rift, *Ethiopia. Science*, *172*(3982), 468–469. doi:10.1126/science.172.3982.468

Bosák, P., Jaros, J., Spudil, J., Sulovsky, P., & Václaverk, V. (1998). Salt plugs in the eastern Zagros, Iran: Results of regional geological reconnaissance. *GeoLines (Praha)*, *7*, 3–174.

Bosworth, W. (2015). Geological evolution of the Red Sea: Historical background, review, and synthesis. In N. M. A. Rasul & I. C. F. Stewart (Eds.), *The Red Sea* (pp. 45–78). Springer Earth System Sciences. Berlin: Springer-Verlag.

Bosworth, W., Huchon, P., & McClay, K. (2005). The Red Sea and Gulf of Aden basins. *Journal of African Earth Sciences*, *43*, 334–378.

Buchner, E., & Kenkmann, T. (2008). Upheaval Dome, Utah, USA: Impact origin confirmed. *Geology*, *36*(3), 227–230. doi:10.1130/G24287A.1

Burberry, C. M., Cosgrove, J. W., & Liu, J. G. (2007). Stream network characteristics used to infer the distribution of fold types in the Zagros Simply Folded Belt, *Iran. Journal of Maps Student Edition*, *2007*, 32–45.

Burberry, C. M., Cosgrove, J. W., & Liu, J. G. (2008). Spatial arrangement of fold types in the Zagros Simply Folded Belt, Iran, indicated by landform morphology and drainage pattern characteristics. *Journal of Maps*, *2008*, 417–430.

Burberry, C. M., Cosgrove, J. W., & Liu, J. G. (2010). A study of fold characteristics and deformation style using the evolution of the land surface: Zagros Simply Folded Belt, Iran. In P. Leturmy and C. Robin (Eds.), *Tectonic and stratigraphic evolution of Zagros and Makran during the Mesozoic-Cenozoic* (Vol. *330*, pp.139–154). The Geological Society of London, Special Publications.

Burberry, C. M., Jackson, C. A. L., & Cosgrove, J. W. (2011). Late cretaceous to recent deformation related to inherited structures and subsequent compression within the Persian Gulf: A 2D seismic case study. *Journal of the Geological Society, London*, *168*, 485–498.

Burke, K. (2011). Plate tectonics, the Wilson cycle, & Mantle Plumes: Geodynamics from the Top. *Annual Review of Earth and Planetary Sciences*, *39*, 1–29. doi: 10.1146/annurev-earth-040809-152521

Burke, K., & Dewey, J. F. (1974). Hot spots and continental breakup: Implications for collisional orogeny, *Geology*, *2*, 57–60.

Canérot, J., Hudec, M. R., & Rockenbauch, K. (2005). Mesozoic diapirism in the Pyrenean orogen: Salt tectonics on a transform plate boundary. *American Association of Petroleum Geologists Bulletin*, *89*(2), 211–229.

Carter, F. W. (1970). Geology of the Salt Anticline Region in southwestern Colorado. *U. S. Geological Survey Professional Paper 637*, 80 pp.

Case, J. E. (1966). Geophysical anomalies over Precambrian rocks, Northwestern Uncompahgre Plateau, Utah and Colorado, *American Association of Petroleum Geologists Bulletin*, *50*(7), 1423–1443.

Charrach, J. (2018). The Sdom evaporite formation in Israel and its relationship with the Messinian salinity crisis. *Carbonates and Evaporites*, *33*, 727–766. https://doi.org/10.1007/s13146-017-0410-1

Chorlton, L. B. (2007). Generalized geology of the world: Bedrock domains and major faults in GIS format: A small-scale world geology map with an extended geological attribute database. *Geological Survey of Canada, Open File 5529*, 2007, 48 pages.

Closson, D., Karaki, N. A., Milisavljević, N., Hallot, F., & Acheroy M. (2011). Salt tectonics of the lisan diapir revealed by synthetic aperture radar images. In D. Closson (Ed.), *Tectonics*, pp. 303–328, Intech.

Closson, D., LaMoreaux, P. E., Karaki, N. A., & al-Fugha, H. (2007). Karst system developed in salt layers of the Lisan Peninsula, Dead Sea, Jordan. *Environmental Geology*, *52*, 155–172. doi 10.1007/s00254-006-0469-9

Coianiz, L., Bialik, O. M., Ben-Avraham, Z., & Lazar, M. (2019). Late Quaternary lacustrine deposits of the Dead Sea basin: High-resolution sequence stratigraphy from downhole logging data. *Quaternary Science Reviews*, *210*, 175–189.

Coianiz, L., Schattner, U., Lang, G., Ben-Avraham, Z., & Lazar, M. (2020). Between plate and salt tectonics—New stratigraphic constraints on the architecture and timing of the Dead Sea basin during the Late Quaternary. *Basin Research*, *32*(4), 636–651.

Dalziel, I. W. D. (1997). Neoproterozoic—Paleozoic geography and tectonics: Review, hypothesis, environmental speculation. *Geological Society of American Bulletin*, *109*(1), 16–42.

Davison, I., Alsop, G. I., & Blundell, D. (1996). Salt tectonics: Some aspects of deformation mechanics. In G. I. Alsop,

D. Blundell, & I. Davison (Eds.), *Salt tectonics* (Vol. *100*, pp. 1–10). Special Publication of the Geological Society of London.

Davison, I., Bosence, D. W. J., Alsop, G. I., & Al-Aawah, M. (1996). Deformation and sedimentation around exhumed Miocene salt diapirs, Al-Salif and Jabal al Milh, NW Yemen. In G. I. Alsop, D. Blundell, & I. Davison (Eds.), *Salt tectonics* (Vol. *100*, pp. 23–39). Geological Society of London, Special Publication.

de Böckh, H., Lees, G. M., & Richardson, F. D. S. (1929). Contribution to the stratigraphy and tectonics of the Iranian ranges. In J. W. Gregory (Ed.), *The structure of Asia* (pp. 58–176). London: Methuen. ISBN: 1138567221, 9781138567221

Dewey, J. F., Helman, M. I., Turco, E., Hutton, D. H. W., & Knott, S. D. (1989). Kinematics of the western Mediterranean. In M. P. Coward, D. Dietrich, & R. G. Park (Eds.), *Alpine tectonics* (Vol. *45*, pp. 265–283). Geological Society of London, Special Publication.

Doelling, H. H., Oviatt, C. G., & Huntoon, P. W. (1988). Salt deformation in the Paradox region. *Utah Geological Survey; Natural Resources*, Bulletin 122, 98pp.

Domeier, M., & Torsvik, T. H. (2014). Plate tectonics in the late Paleozoic. *Geoscience Frontiers*, *5*, 303–350. doi: 10.1016/j.gsf.2014.01.002.

Escosa, F. O., Rowan, M. G., Giles, K. A., Deatrick, K. T., Mast, A. M., Langford, R. P., et al. (2019). Lateral terminations of salt walls and megaflaps: An example from Gypsum Valley Diapir, Paradox Basin, Colorado, USA. *Basin Research*, *31*, 191–212.

Evans, I., & Kendall, C. G. St. C. (1977). An interpretation of the depositional setting of some deep-water Jurassic carbonates of the Central High Atlas Mountains, Morocco. In H. E. Cook & Paul Enos (Eds.), *Deep-water Carbonate environments* (Vol. *25*, 249–261). SEPM Special Publication. doi: 10.2110/pec.77.25.0249

Frahme, C. W., & Vaughn, E. B. (1983). Paleozoic geology and seismic stratigraphy of the northern Uncompahgre Front, Grant County, Utah. In J. D. Lowell & R. Gries (Eds.), *Rocky mountain foreland basins and uplifts* (pp. 201–211). Rocky Mountain Association of Geologists.

Frizon de Lamotte, D., Bezar, B. S., Bracène, R., & Mercier, E. (2000). The two main steps of the Atlas building and geodynamics of the western Mediterranean. *Tectonics*, *19*(4), 740–761. https://doi.org/10.1029/2000TC900003

Frizon de Lamotte, D., Fourdan, B., Leleu, S., Leparmentier, F., & de Clarens P. (2015). Style of rifting and the stages of Pangea breakup. *Tectonics*, *34*, 1009–1029. doi:10.1002/2014TC003760.

Frizon de Lamotte, D., Raulin, C., Mouchot, N., Wrobel-Daveau, J.-C., Blanpied, C., & Ringenbach, J.-C. (2011). The southernmost margin of the Tethys realm during the Mesozoic and Cenozoic: Initial geometry and timing of the inversion process. *Tectonics*, *30*, TC3002. http://dx.doi.org/10.1029/2010TC002691

Frizon de Lamotte, D., Zizi, M., Missenard, Y., Hafid, M., El Azzouzi, M. R., Maury, C., et al. (2008). The Atlas system. In A. Michard., O. Saddiqi, A. Chalouan, & D. Frizon de Lamotte (Eds.), *Continental evolution: The Geology of Morocco* (Vol *116*, pp. 133–202). Lecture Notes in Earth Sciences. Berlin, Heidelberg: Springer.

García-Veigas, J., Rosell, L., Zak, I., Playà, E., Ayora C., & Starinsky, A. (2009). Evidence of potash salt formation in the Pliocene Sedom Lagoon (Dead Sea Rift, Israel). *Chemical Geology*, *265*, 499–511.

Garfunkel, Z. (1981). Internal structure of the Dead Sea leaky transform (rift) in relation to plate kinematics, *Tectonophysics*, *80*, 81–108.

Garfunkel, Z., & Ben-Avraham, Z. (1996). The structure of the Dead Sea basin. *Tectonophysics*, *226*, 155–176.

Gautier, E.-F. (1914). Le rocher de sel de Djelfa. *Annales de Géographie*, *23*(129), 245–260. doi: 10.3406/geo.1914.8099

Gavrieli, I., & Stein, M. (2006). On the origin and fate of the brines in the Dead Sea basin. In Y. Enzel, A. Agnon, & M. Stein (Eds.), *New frontiers in Dead Sea paleoenvironmental research* (Vol. *401*, pp. 183–194). Geological Society of America, Special Paper.

Ge, H., & Jackson, M. P. A. (1998). Physical modeling of structures caused by salt withdrawal: Implications for deformation caused by salt dissolution. *American Association of Petroleum Geologists Bulletin*, *82*, 228–250.

Ge, H., Jackson, M. P. A., & Vendeville, B. C. (1996). Extensional origin of breached Paradox diapirs, Utah and Colorado: Field observations and scaled physical models. In A. C. Huffman, Jr., W. R. Lund, L. H. Godwin (Eds.), *Geology and resources of the Paradox basin* (Vol. *25*, pp. 285–293). Utah Geological Association Guidebook.

Ge, H., Jackson, M. P. A., & Vendeville, B. C. (1997). Kinematics and dynamics of salt tectonics driven by progradation. *American Association of Petroleum Geologists Bulletin*, *81*(3), 393–423.

GEBCO (2019). *General Bathymetric Chart of the Oceans*, British Oceanographic Data Centre, U.K. website www.gebco.net.

Geesaman, P. J., Trudgill, B. D., Hearon, T. E., & Rowan, M. G. (2015). New evidence for long-term, salt-related deformation at upheaval Dome, SE Utah. *AAPG Search and Discovery*, Article #10756 (2015).

Gharbi, M., Espurt, N., Masrouhi, A., Bellier, O., & Amari, E. A. (2015). Style of Atlassic tectonic deformation and geodynamic evolution of the southern Tethyan margin, Tunisia. *Marine and Petroleum Geology*, *66*, 801–816.

Gibling, M. R., Culshaw N., Pascucci, V., Waldron, J. W. F., and Rygel, M. C. (2019). The Maritimes basin of Atlantic Canada: Basin creation and destruction during the Paleozoic assembly of Pangea. In A. D. Miall (Ed.), *The sedimentary basins of the United States and Canada* (pp. 267–314). Elsevier. https://doi.org/10.1016/B978-0-444-63895-3.00006-1

Giles, K. A., & Rowan, M. G. (2012). *Concepts in halokinetic-sequence deformation and stratigraphy*. In G. I. Alsop, S. G. Archer, A. J. Hartley, N. T. Grant, & R. Hodgkinson (Eds.), *Salt tectonics, sediments and prospectivity* (Vol. 363, pp. 7–31). Geological Society, London, Special Publications. doi:10.1144/SP363.8

Granado, P., Roca, E., Strauss, P., Pelz, K., & Muñoz, J. A. (2018). Structural styles in fold-and-thrust belts involving early salt structures: The Northern Calcareous Alps (Austria), *Geology*, *47*(1), 51–54. https://doi.org/10.1130/G45281.1

Gussow, W. C. (1966). Salt temperature: A fundamental factor in salt dome intrusion. *Nature*, *210*, 518–519.

Gussow, W. C. (1968). Salt diapirism: Importance of temperature, and energy source of emplacement. In J. Braunstein & G. D. O'Brien (Eds.), *Diapirism and diapirs, American Association of Petroleum Geologists Memoir* (Vol. *8*, pp. 16–52).

Gutierrez, F. (2004). Origin of the salt valleys in the Canyonlands section of the Colorado Plateau: Evaporite-dissolution collapse versus tectonic subsidence. *Geomorphology*, *57*, 423–435.

Hajiesmaeili, L., & Rashidi, K. (2019). Investigation of active geodynamics caused by performance of salt dome using radar interferometry (Case Study: The Salt Dome of Qeshm Island). *Journal of Radar and Optical Remote Sensing, JRORS*, *3*, 44–54.

Harding, A. (2013). *Salt in prehistoric Europe, 162 pp*. Leiden: Sidestone Press.

Hartmann, J., & Moosdorf, N. (2012). The new global lithological map database GLiM: A representation of rock properties at the Earth surface. *Geochemistry, Geophysics, Geosystems*, *13*, Q12004. doi:10.1029/2012GC004370

Hassanpour, J., Yassaghi, A., Muñoz, J. A., & Jahani, S. (2020). Salt tectonics in a double salt-source layer setting (Eastern Persian Gulf, Iran): Insights from interpretation of seismic profiles and sequential cross-section restoration. *Basin Research*, *33*,159–185. doi:10.1111/bre.12459

Hessami, K., Koyi, H. A., & Talbot, C. J. (2001). The significance of strike-slip faulting in the basement of the Zagros fold and thrust belt. *Journal of Petroleum Geology*, *24*(1), 5–28.

Hite, R. J. (1968). *Salt deposits of the Paradox basin, Southeast Utah and Southwest Colorado* (Vol. *88*, pp. 319–330). Geological Society of America, Special Paper.

Holwerda, J. G., & Hutchinson, R. W. (1968). Potash-bearing evaporites in the Danakil area, Ethiopia. *Economic Geology*, *63*, 124–150.

Hovland, M., Rueslaatten, H., & Johnsen, H. K. (2008). Hydrothermal salt—But how much? Reply to Christopher Talbot on his comments to our articles. *Marine and Petroleum Geology*, *25*(2), 203–204. https://doi.org/10.1016/j.marpetgeo.2007.05.006.

Hovland, M., Rueslåtten, H., & Johnsen, H. K. (2018a). Large salt accumulations as a consequence of hydrothermal processes associated with "Wilson cycles": A review, Part 1: Towards a new understanding. *Marine and Petroleum Geology*, *92*, 987–1009. http://dx.doi.org/10.1016/j. marpetgeo.2017.12.029.

Hovland, M., Rueslåtten, H., & Johnsen, H. K. (2018b). Large salt accumulations as a consequence of hydrothermal processes associated with "Wilson cycles": A review, Part 2: Application of a new salt-forming model on selected cases. *Marine and Petroleum Geology*, *92*, 128–148.

Hudec, M. R. (1995). *The Onion Creek salt diapir: An exposed diapir fall structure in the Paradox basin, Utah*. In Salt, sediment and hydrocarbons. In C. J. Travis, H. Harrison, M. R. Hudec, B. C. Vendeville, F. J. Peel, & B. F. Perkins, *Gulf Coast Section Society of Economic Paleontologists and Mineralogists* (pp. 125–134), 16th Annual Research Conference. Gulf Coast Section SEPM Foundation, Houston, Texas.

Huffman, A. C., Jr., & Taylor, D. J. (2002). The origin and nature of the uncompahgre fault. In "Late Paleozoic tectonics and hydrocarbon systems of western North America—The greater ancestral Rocky Mountains." *AAPG Search and Discovery Article #90012*, 2003.

IUGS (2022). The first 100 IUGS Geological Heritage sites. *International Union of Geological Sciences, International Geosciences Program, IGCP 731*, 188–189.

Jackson, M. P. A. (1995). Retrospective salt tectonics. In M. P. A. Jackson, D. G. Roberts, & S. Snelson (Eds.), *Salt tectonics: A global perspective* (Vol. *65*, pp. 1–28). American Association of Petroleum Geologists Memoir.

Jackson, M. P. A., Cornelius, R. R., Craig, C. H., Gansser, A., Stöcklin, J., & Talbot, C. J. (1990). Salt diapirs of the Great Kavir, Central Iran. *Geological Society of America Memoirs*, *177*, 139 pp.

Jackson, M. P. A., Schultz-Ela, D. D, Hudec, M. R., Watson, I. A., & Porter, M. L. (1998). Structure and evolution of Upheaval Dome: A pinched-off salt diapir. *Geological Society of America Bulletin*, *110*(12), 1547–1573.

Jacobshagen, V., Görler, K., & Giese, P. (1988). Geodynamic evolution of the Atlas system (Morocco) in post-palaeozoic times. In V. H. Jacobshagen (Ed.), *The Atlas system of Morocco, Lecture Notes in Earth Sciences* (Vol. *15*, pp. 481–499). Berlin: Springer.

Jafari, A., Ao, S., Jamei, S., & Ghasemi, H. (2023). Evolution of the Zagros sector of Neo-Tethys: Tectonic and magmatic events that shaped its rifting, seafloor spreading and subduction history. *Earth-Science Reviews*, *241*, 104419. https://doi.org/10.1016/j.earscirev.2023.104419

Jahani, S., Callot, J. P., Lamotte, D. F., Letouzey, J., & Leturmy P. (2007). The Salt Diapirs of the Eastern Fars Province (Zagros, Iran): A brief outline of their past and present. In O. Lacombe, F. Roure, J. Lavé, & J. Vergés (Eds.), *Thrust belts and Foreland basins, Frontiers in Earth Sciences* (pp. 289–308). Berlin: Springer.

Jahani, S., Callot, J. P., Letouzey, J., and Lamotte, D. F. (2009). The eastern termination of the Zagros Fold-and-Thrust Belt, Iran: Structures, evolution, and relationships between salt plugs, folding, and faulting, *Tectonics*, *28, TC6004*, 22 p. doi:10.1029/2008TC002418.

Jahani, S., Hassanpour, J., Mohammadi-Firouz, S., Letouzey, J., Frizon de Lamotte, D., Alavi, S. A., & Soleimany, B. (2017). Salt tectonics and tear faulting in the central part of the Zagros Fold-Thrust Belt, *Iran. Marine and Petroleum Geology*, *86*, 426–446.

Jaillard, E., Bouillin, J.-P., Ouali, J., Dumont, T., Latil, J.-L., & Chihaoui, A. (2017). Albian salt-tectonics in Central Tunisia: Evidences for an Atlantic-type passive margin. *Journal of African Earth Sciences*, *135*, 220–234.

Jallouli, C., Chikhaoui, M., Braham, A., Turki, M. M., Mickus, K., & Benassi, R. (2005). Evidence for Triassic salt domes in the Tunisian Atlas from gravity and geological data. *Tectonophysics*, *396*, 209–225.

Kent, P. E. (1958). Recent studies of south Persian salt plugs. *American Association of Petroleum Geologists Bulletin*, *42*(12), 2951–2972.

Kent, P. E. (1970). The salt of the Persian Gulf region. *Transactions of the Leicester Literary & Philosophical Society*, *64*, 56–88.

Kent, P. E. (1979). The emergent Hormuz salt plugs of southern Iran. *Journal of Petroleum Geology*, *2*(2), 117–144.

Khalfi, C., Ahmadi, R., Rhekiss, F., Trigui, H., & Ouali, J. (2014). Modeling of Mrihla anticline (Central Tunisian Atlas), combination fault propagation fold and halokinetic model. *Arabian Journal of Geosciences*, *7*, 811–818.

Kluth, C. F. (1986). Plate tectonics of the ancestral Rocky Mountains. In J. A. Peterson (Ed.), *Paleotectonics and sedimentation, Rocky Mountain region, USA, American Association of Petroleum Geologists Memoir*, *41*, 353–369.

Kluth, C. F., & Coney, P. J. (1981). Plate tectonics of the Ancestral Rocky Mountains, *Geology*, *9*, 10–15.

Kulke, H. (1972). Le rocher de sel de Djelfa (Atlas saharien algérien), géologie et morphologie. *Bulletin de la Société d'histoire naturelle de l'Afrique du Nord*, *63*(3–4), 81–134. Alger.

Lalami, H. R. K., Hajialibeigi, H., Sherkati, S., & Adabi, M. H. (2020). Tectonic evolution of the Zagros foreland basin since Early Cretaceous, SW Iran: Regional tectonic implications from subsidence analysis. *Journal of Asian Earth Sciences*, *204*, 104550.

Lawton, T. F., & Buck, B. J. (2006). Implications of diapir-derived detritus and Gypsic Paleosols in Lower Triassic strata near the Castle Valley salt wall, Paradox Basin, *Utah. Geology*, *34*(10), 885–888. doi: 10.1130/G22574.1.

Lazar, M. (2019). Tectonic segmentation of the Dead Sea fault system: A review of geophysical evidence. In João C. Duarte (Ed.), *Transform plate boundaries and fracture zones* (pp. 417–436). Amsterdam: Elsevier.

Lazar, M., Ben-Avraham, Z., & Schattner, U. (2006). Formation of sequential basins along a strike–slip fault-geophysical observations from the Dead Sea basin. *Tectonophysics*, *421*, 53–69.

Leary, R. J., Umhoefer, P., Smith, M. E., & Riggs, N. (2017). A three-sided orogen: A new tectonic model for Ancestral Rocky Mountain uplift and basin development, *Geology*, *45*(8), 735–738. https://doi.org/10.1130/G39041.1

Lees, G. M. (1927). Salzgletscher in Persien. *Mitteilungen der deutschen Gesellschaft*, *20*, 29–34.

Leitner, C., & Spötl, C. (2017). The eastern Alps: Multistage development of extremely deformed evaporites. In I. Soto, J. F. Flinch, & G. Tari (Eds.), *Permo-Triassic salt provinces of Europe, North Africa and the Atlantic margins, tectonics and hydrocarbon potential* (pp. 467–482). Amsterdam: Elsevier.

Letouzey, J. (1990). Fault reactivation, inversion and fold-thrust belt. In J. Letouzey (Ed.), *Petroleum and tectonics of mobile belts* (pp. 101–128). Editions Technip, Paris.

Letouzey, J., Colletta, B., Vially, R., & Chermette, J. C. (1995). Evolution of salt-related structures in compressional settings. In M. P. A. Jackson, D. G. Roberts, & S. Snelson (Eds.), *Salt tectonics: A global perspective*, *AAPG Memoir* (Vol. *65*, pp. 41–61).

Letouzey, J., & Sherkati, S. (2004) Salt movement, tectonic events, and structural style in the central Zagros Fold and Thrust Belt (Iran). In P. J. Post, D. L. Olson, K. T. Lyons, S. L. Palmes, P. F Harrison, & N. C. Rosen (Eds.), *Salt–Sediment interactions and hydrocarbon prospectivity: Concepts, applications, and case studies for the 21st century*, Proceedings of the 24th Annual GCSSEPM Foundation, Bob F. Perkins Research Conference, pp. 753–778.

López-García, J. M., Moreira, D., Benzerara, K., Grunewald, O., & López-García, P. (2020). Origin and evolution of the halo-volcanic complex of Dallol: Proto-volcanism in Northern Afar (Ethiopia). *Frontiers in Earth Science*, *7*, 351. https://doi.org/10.3389/feart.2019.00351

López-Mir, B., Muñoz, J. A., & García-Senz, J. (2015). Extensional salt tectonics in the partially inverted Cotiella post-rift basin (south-central Pyrenees): Structure, and evolution. *International Journal of Earth Sciences (Geol Rundsch)*, *104*(2), 419–434. doi 10.1007/s00531-014-1091-9

Lucha, P., Cardona, F., Gutierrez, F., & Guerrero, J. (2008). Natural and human-induced dissolution and subsidence processes in the salt outcrop of the Cardona Diapir (NE Spain). *Environmental Geology*, *53*, 1023–1035. doi 10.1007/s00254-007-0729-3

Maleki, Z., & Jahadgar, F. (2015). Folding Style of the Kuh-e Siah Anticline in the Sarvestan Area, Interior Fars, Zagros, *Iran. Open Journal of Geology*, *5*, 552–561. http://dx.doi.org/10.4236/ojg.2015.58050

Manspeizer, W. (1981). Early Mesozoic basins of the Central Atlantic passive margins. In *Geology of passive continental margins, American Association of Petroleum Geologists Education Course Note Series* (Vol. *19*, 4.1–4.60).

Manspeizer, W. (1988). Triassic—Jurassic rifting and opening of the Atlantic: An overview. In W. Manspeizer (Ed.), *Triassic-Jurassic rifting: Continental breakup and the origin of the Atlantic Ocean and Passive Margins, Developments in Geotectonics* (Vol. *22*, pp. 41–79). Elsevier. ISBN: 978-0-444-42903-2.

Masrouhi, A., & Bellier, O. (2013). Allochthonous Salt sheets associated with cretaceous rifting in the North African Margin, Examples from North Tunisia and Petroleum Implication. *American Association of Petroleum Geologists Search and Discovery*, Article #30270 (2013).

Masrouhi, A., Bellier, O., Koyi, H., Vila, J. M., & Ghanmi, M. (2013). The evolution of the Lansarine–Baouala salt canopy in the North African Cretaceous passive margin in Tunisia, *Geological Magazine*, *150*(5), 835–861. doi:10.1017/S0016756812000763

Masrouhi, A., Bellier, O., Youssef, M. B., & Koyi, H. (2014). Submarine allochthonous salt sheets: Gravity-driven deformation of North African Cretaceous passive margin in Tunisia—Bled Dogra case study and nearby salt structures. *Journal of African Earth Sciences*, *97*, 125–142.

Masrouhi, A., & Koyi, H. A. (2012). *Submarine "salt glacier" of Northern Tunisia, a case of Triassic salt mobility in North African Cretaceous passive margin*. In G. I. Alsop, S. G. Archer, A. J. Hartley, N. T. Grant, & R. Hodgkinson (Eds.), *Salt tectonics, sediments and prospectivity* (Vol. 363, pp. 579–593). Geological Society, London, Special Publications. https://doi.org/10.1144/SP363.29

Michard, A., Frizon de Lamotte, D., Saddiqi, O., & Chalouan A. (2008). An outline of the Geology of Morocco. In A. Michard., O. Saddiqi, A. Chalouan, & D. F. Lamotte (Eds.), *Continental evolution: The geology of Morocco, Lecture Notes in Earth Sciences* (Vol. *116*, pp. 1–31). Berlin, Heidelberg: Springer.

Mohr, M., Warren, J. K., Kukla, P. A., Urai, J. L., & Irmen, A. (2007). Subsurface seismic record of salt glaciers in an extensional intracontinental setting (Late Triassic of northwestern Germany). *Geology*, *35*, 963–966.

Mohriak, W. (2025). *Salt in the Earth sciences: Basin analysis and salt tectonics*. Washington, DC: American Geophysical Union. doi:10.1002//9781394307067

Mohriak, W. U., & Szatmari, P. (2009). Tectônica de sal. In W. Mohriak, P. Szatmari, & S. M. C. Anjos (Eds.), *Sal: Geologia e Tectônica* (2nd ed., pp. 92–165). São Paulo: Editora Beca.

Molenaar, C. M. (1981). Mesozoic stratigraphy of the Paradox basin: An overview. In *D.* L. Wiegand (Ed.), *Geology of the Paradox basin:* Rocky Mountain Association of Geologists Field Conference, pp. 119–127.

Molinaro, M., Leturmy, P., Guezou, J.-C., & Frizon de Lamotte, D. (2005). The structure and kinematics of the southeastern Zagros foldthrust belt, Iran: From thin-skinned to thick-skinned tectonics. *Tectonics*, *24*, TC3007, 1–19. doi: 10.1029/2004TC001633

Najafi, M., Vergès, J., Etemad-Saeed, N., & Karimnejad, H. R. (2018). Folding, thrusting and diapirism: Competing mechanisms for shaping the structure of the north Dezful Embayment, Zagros, *Iran. Basin Research*, *30*, 1200–1229. https://doi.org/10.1111/bre.12300

Naqi, M., Trudgill, B., & Kluth, C. (2016). A new insight on the mechanism of salt wall collapse in northeastern Paradox basin, Utah, *American Association of Petroleum Geologists Search and Discovery Article* #10837 (2016)

NOAA (2022). National Centers for Environmental Information, ETOPO 2022 15 Arc-Second Global Relief Model. *NOAA National Centers for Environmental Information.* doi: 10.25921/fd45-gt74.

Nobile, A., Pagli, C., Keir, D., Wright, T. J., Ayele, A., Ruch, J., & Acocella, V. (2012). Dike-fault interaction during the 2004 Dallol intrusion at the northern edge of the Erta Ale Ridge (Afar, Ethiopia). *Geophysical Research Letters*, *39*(19), 2–7. doi: 10.1029/2012GL053152

Ohlen, H. R., & McIntyre, L. B. (1965). Stratigraphy and tectonic features of paradox basin, Four Corners Area. *American Association of Petroleum Geologists Bulletin*, *49*(11), 2020–2040.

Olsen, P. E. (1997). Stratigraphic record of the Early Mesozoic breakup of Pangea in the Laurasia—Gondwana rift system. *Annual Review of Earth Planetary Sciences*, *25*, 337–401.

Paz, M., Trudgill, B., & Kluth, C. (2009). Salt system evolution of the northern Paradox basin. *American Association of Petroleum Geologists Search and Discovery*, Article #30078 (2009).

Perthuisot, V., Aoudjehane, M., Bouzenoune, A., Hatira, N., Laatar, E., Mansouri, A., Rouvier, H., Smati, A., and Thibieroz, J. (1998). Les corps triasiques des monts du Mellegue (confins algero-tunisiens) sont-ils des diapirs ou des "glaciers de sel"? *Bulletin de la Société géologique de France*, *169*(1), 53–61.

Perthuisot, V., & Rouvier, H. (1992). Les diapirs du Maghreb central et oriental: des appareils variés, résultats d'une evolution structural et pétrogénétique complexe. *Bulletin de la Société géologique de France*, *163*(6), 751–760.

Peterson, J. A. (1989). Geology and petroleum resources, Paradox basin Province, *USGS Open-File Report* 88-450 U, 69 pp.

Rahnama-Rad, J., Farhoudi, G., Ghorbani, H., Mood, S. H., & Derakhshani, R. (2009). Pierced salt domes in the Persian Gulf and in the Zagros mountain ranges in southern Iran and their relationship to hydrocarbon and basement tectonics. *Iranian Journal of Earth Sciences*, *1*, 57–72.

Randles, T., Clarke, S., & Richards, P. (2012). Development of crestal collapse structures above dissolving salt anticlines: Application to Seismic interpretation within salt-controlled basins. *American Association of Petroleum Geologists, Search and Discovery Article* #41033 (2012).

Razaghian, G., & Arian, M. (2015). The emergent salt diapirs in the east Zagros, *Iran. Open Journal of Geology*, *5*, 718–726.

Reuning, L, Schoenherr, J., Heimann, A., Urai, J. L., Littke, R., Kukla, P. A., & Rawahi, Z. (2009). Constraints on the diagenesis, stratigraphy and internal dynamics of the surface-piercing salt domes in the Ghaba Salt Basin (Oman): A comparison to the Ara Group in the South Oman Salt Basin, *GeoArabia*, *14*(3), 83–120.

Rime, V., Foubert, A., Fentimen, R., Negga, H., El Korh, A., Adatte, T., et al. (2020). Evaporites reveal Pleistocene basin dynamics in the Danakil depression (northern Afar, Ethiopia). *EGU General Assembly 2020*, EGU2020-14917. https://doi.org/10.5194/egusphere-egu2020-14917

Rowan, M. G., Lawton, T. F., Giles, K. A., & Ratliff, R. A. (2003). Near-salt deformation in La Popa basin, Mexico, and the northern Gulf of Mexico: A general model for passive diapirism. *American Association of Petroleum Geologists Bulletin*, *87*(5), 733–756.

Sans, M. (2003). From thrust tectonics to diapirism. The role of evaporites in the kinematic evolution of the eastern South Pyrenean front. *Geologica Acta*, *1*(3), 239–259.

Sarkarinejad, K., Sarshar, M. A., & Adineh, S. (2018). Structural, micro-structural and kinematic analyses of channel flow in the Karmostaj Salt Diapir in the Zagros foreland folded belt, Fars province, *Iran. Journal of Structural Geology*, *107*, 109–131.

Scarfì, L., Barberi, G., Barreca, G., Cannavò, F., Koulakov, I., & Patanè, D. (2018). Slab narrowing in the central Mediterranean: The Calabro-Ionian subduction zone as imaged by high resolution seismic tomography. *Nature Science Reports*, *8*, 5178 (2018). https://doi.org/10.1038/s41598-018-23543-8

Schettino, A., & Turco, E. (2011). Tectonic history of the western Tethys since the Late Triassic. *Geological Society of America Bulletin*, *123*(1/2), 89–105. doi: 10.1130/B30064.1.

Schmid, S. M., Fugenschuh, B., Kissling, E., & Schuster, R. (2004). Tectonic map and overall architecture of the Alpine orogeny. *Eclogae Geologicae Helvetiae*, *97*, 93–117.

Schorn, A., & Neubauer, F. (2014). Hallstatt evaporite body of the Upper Permian to Lower Triassic Haselgebirge Fm. within the Northern Calcareous Alps (NCA). *Journal of Structural Geology*, *60*, 70–84.

Scotese, C. R. (2001). *Atlas of Earth history, Vol. 1, Paleogeography*. PALEOMAP Project, Arlington, Texas, 52 pp.

Scotese, C. R. (2008). *The PALEOMAP project PaleoAtlas for ArcGIS, Version 1, Volume 3, Triassic and Jurassic Paleogeographic, Paleoclimatic, and Plate Tectonic Reconstructions*, *PALEOMAP Project*, Arlington, Texas.

Scotese, C. R. (2011). Paleogeographic and paleoclimatic Atlas. *AAPG Search and Discovery, Article #30192*.

Setchell, C. M., Cosgrove, J. W., & Liu, J-G. (2007). *The distribution of fold types in the Zagros Simply Folded Belt, Iran. Journal of Maps*, Student Edition, Imperial College, London, 1p. http://journalofmaps.com/student/07_05_Burberry_M.pdf

Seton, M., Müller, R. D., Zahirovic, S., Gaina C., Torsvik, T. H., Shephard, G., et al. (2012). Global continental and ocean basin reconstructions since 200 Ma. *Earth-Science Reviews*, *113*(3–4), 212–270, ISSN 0012-8252. doi: 10.1016/j.earscirev.2012.03.002

Sherkati, S., & Letouzey, J. (2004). Variation of structural style and basin evolution in the central Zagros (Izeh zone and Dezful Embayment), *Iran. Marine and Petroleum Geology*, *21*(5), 535–554. doi: 10.1016/j.marpetgeo.2004.01.007.

Sherkati, S., Letouzey, J., & Frizon de Lamotte, D. (2006). Central Zagros fold-thrust belt (Iran): New insights from seismic data, field observation, and sandbox modeling. *Tectonics*, *25*, TC4007. doi: 10.1029/2004TC001766.

Sirota, I., Enzel, Y., & Lensky, N. G. (2017). Temperature seasonality control on modern halite layers in the Dead Sea: In situ observations. *Geological Society of America Bulletin*, *129*(9/10), 1181–1194. doi: 10.1130/B31661.1.

Slama, M. M. B., Masrouhi, A., Ghanmi, M., Youssef, M. B., & Zargouni F. (2009). Albian extrusion evidences of the Triassic salt and clues of the beginning of the Eocene Atlasic phase from the example of the Chitana-Ed Djebs structure (N. Tunisia): Implication in the North African Tethyan margin recorded events, comparisons. *Comptes Rendus Geoscience*, *341*, 547–556.

Soto, J. I., Flinch, J. F., & Tari, G. (2017). Permo-Triassic basins and tectonics in Europe, North Africa and the Atlantic margins: A synthesis. In J. I. Soto, J. F. Flinch, & G. Tari (Eds.), *Permo-Triassic salt provinces of Europe, North Africa and the Atlantic Margins tectonics and hydrocarbon potential* (pp. 3–41). Amsterdam: Elsevier.

Stampfli, G. M., & Borel, G. D. (2002). A plate tectonic model for the Paleozoic and Mesozoic constrained by dynamic plate boundaries and restored synthetic oceanic isochrones. *Earth and Planetary Science Letters*, *196*, 17–33.

Stevenson, G. M., & Baars, D. L. (1986). The Paradox: A pull-apart basin of Pennsylvanian age. In Peterson, J. A. (Ed.), *Paleotectonics and sedimentation in the Rocky Mountain Region: American Association of Petroleum Geologists Memoir* (Vol. *41*, pp. 513–539). United States.

Talbot, C. J. (1979). Fold trains in a glacier of salt in southern Iran. *Journal of Structural Geology*, *1*(1), 5–18.

Talbot, C. J. (1981). Sliding and other deformation mechanisms in a glacier of salt, S Iran. In N. J. Price & K. R. McClay (Eds.), *Thrust and Nappe tectonics* (Vol. *9*, pp. 173–183). The Geological Society of London Special Publications.

Talbot, C. J. (1993). Spreading of salt structures in the Gulf of Mexico. *Tectonophysics*, *228*, 151–166.

Talbot, C. J. (1998). *Extrusions of Hormuz salt in Iran*. In D. J. Blundell & A. C. Scott (Eds.), *Lyell: The past is the key to the present* (Vol. 143, pp. 315–334). Geological Society, London, Special Publication.

Talbot, C. J. (2005). Discussion "Evidence for Triassic salt domes in the Tunisian Atlas from gravity and geological data." In C. Jallouli et al. (Eds.), *Tectonophysics*, *406*, 249–254.

Talbot, C. J. (2008). Hydrothermal salt—But how much? Discussion. *Marine and Petroleum Geology*, *25*, 191–202.

Talbot, C. J., & Aftabi, P. (2004). Geology and models of salt extrusion at Qum Kuh, central Iran. *Journal of the Geological Society*, *161*, 321–334.

Talbot, C. J., & Jarvis, R. J. (1984). Age, budget and dynamics of an active salt extrusion in Iran. *Journal of Structural Geology*, *6*, 521–533.

Talbot C. J., & Pohjola, V. (2009). Subaerial salt extrusions in Iran as analogues of ice sheets, streams and glaciers. *Earth-Science Reviews*, *97*, 167–195.

Talbot, C. J., & Rogers, E. A. (1980). Seasonal movements in a Salt Glacier in Iran. *Science*, *208*, 395–397. doi: 10.1126/science.208.4442.395

Torfstein, A. (2017). The Amora formation, Dead Sea basin. In Y. Enzel & O. Bar (Eds.), *Quaternary of the Levant—Environments, climate change, and humans* (pp. 91–98). U.K.: Cambridge University Press. https://doi.org/10.1017/9781316106754.010

Trigui, H., Ahmadi, R., Ouali, J., Khalfi, C., & Mercier, E. (2016). Evidence of fault-propagation folds in foreland basin: the case of Chemsi and Belkhir anticlines of southern Tunisian Atlas. *Arabian Journal of Geosciences*, *9*, 706. https://doi.org/10.1007/s12517-016-2748-4

Troudi, H., Tari, G., Alouani, W., & Cantarella, G. (2017). Styles of Salt tectonics in Central Tunisia: An overview. In I. Soto, J. F. Flinch, G. Tari (Eds.), *Permo-Triassic salt provinces of Europe, North Africa and the Atlantic margins: Tectonics and hydrocarbon potential* (pp. 543–561). Amsterdam: Elsevier.

Trudgill, B. (2011). Evolution of salt structures in the northern Paradox Basin: Controls on evaporite deposition, salt wall growth and supra-salt stratigraphic architecture. *Basin Research*, *23*, 208–238.

Trudgill, B., Banbury, N., & Underhill, J. (2004). Salt evolution as a control on structural and stratigraphic systems: Northern Paradox Foreland Basin, Southeast Utah, USA. In P. J. Post, D. L. Olson, K. T. Lyons, S. L. Palmes, P. F. Harrison, N. C. Rosen, & F. Bob (Eds.), *Salt-sediment interactions and hydrocarbon prospectivity: Concepts, applications, and case studies for the 21st century*. 24th Annual Gulf Coast Section SEPM Foundation, Perkins Research Conference, CD-ROM, pp. 669–700.

Vergés, J., Moragas, M., Martín-Martín, J. D., Saura, E., Casciello, E., Razin, Ph., et al. (2017). Salt tectonics in the Atlas mountains of Morocco. In I. Soto, J. F. Flinch, & G. Tari (Eds.), *Permo-Triassic salt provinces of Europe, North Africa and the Atlantic margins, Tectonics and Hydrocarbon potential* (pp. 563–579). Amsterdam: Elsevier.

Vergés, J., Saura, E., Casciello, E., Fernàndez, M., Villaseñor, A., Jiménez-Munt, I., & Garcia-Castellanos, D. (2011). Crustal-scale cross-sections across the NW Zagros belt: Implications for the Arabian margin reconstruction. *Geological Magazine*, *148*(5–6), 739–761. doi:10.1017/S0016756811000331

Ville, L. (1859). Notice géologique sur les salines des Zahrez et les gîtes de sel gemme du Rang-el-Melah et d'Aïn Hadjera (Algérie). *Annales des Mines, Paris*, *15*, 351–410.

Waltham, T. (2008). Salt terrains of Iran. *Geology Today*, *24*(5), 188–194.

Warren, J. K. (2010). Evaporites through time: Tectonic, climatic and eustatic controls in marine and nonmarine deposits. *Earth and Science Review*, *98*, 217–268.

Warren, J. K. (2016). *Evaporites: A geological compendium* (2nd ed., revised and expanded), 1807 pp. Berlin: Springer. ISBN ISBN 978-3-319-13511-3.

Weinberger, R., Begin, Z. B., Waldmann, N., Gardosh, M., Baer, G., Frumkin, A., & Wdowinski, S. (2006). Quaternary rise of the Sedom diapir, Dead Sea basin. In Y. Enzel, A. Agnon, & M. Stein (Eds.), *New frontiers in Dead Sea paleoenvironmental research* (Vol. *401*, pp. 33–51). Geological Society of America Special Paper. doi: 10.1130/2006.2401(03)

Weinberger, R., Lyakhovsky, V., Baer G., & Begin, Z. B. (2006). Mechanical modeling and InSAR measurements of Mount Sedom uplift, Dead Sea basin: Implications for effective viscosity of rock salt. *Geochemistry, Geophysics, Geosystems*, *7*, Q05014, doi:10.1029/2005GC001185

Weinberger, R., Nuriela, P., Kylander-Clark, A. R. C., & Craddock, J. P. (2020). Temporal and spatial relations between large-scale fault systems: Evidence from the Sinai-Negev shear zone and the Dead Sea Fault. *Earth-Science Reviews*, *211*, 103377. https://doi.org/10.1016/j.earscirev.2020.103377

Whidden, K. J., Lillis, P. G., Anna, L. O., Pearson, K. M., & Dubiel, R. F. (2014). Geology and total petroleum systems of the Paradox Basin, Utah, Colorado, New Mexico, and Arizona. *The Mountain Geologist*, *51*(2), 119–138.

White, M. A., & Jacobson, M. I. (1983). Structures associated with the southwest margin of the ancestral Uncompahgre uplift. In W. R. Averett (Ed.), *Northern Paradox basin—Uncompahgre uplift: Grand Junction Geological Society field trip, October 1–2, 1983*, Grand Junction Geological Society, Grand Junction, Colorado, pp. 33–39.

Wilson, J. T. (1966). Did the Atlantic close and then re-open? *Nature*, *211*, 676–681.

Yazdi, A., & Dabiri, R. (2018). Investigating the Geotourism phenomena in eroded land of Iran, *Qeshm Island. Revista Publicando*, *5*(16), 35–49.

Ye, H., Royden, L., Burchfiel, C., & Schuepbach, M. (1996). Late Paleozoic deformation of interior North America: The greater ancestral Rocky Mountains. *American Association of Petroleum Geologists Bulletin*, *80*(9), 1397–1432.

Zak, I. (1974). Sedimentology and bromine geochemistry of marine and continental evaporites in the Dead Sea basin. In *Fourth International Symposium on Salt*. Northern Ohio Geological Society, pp. 349–*361*.

Zak, I. (1997). Evolution of the Dead Sea brines. In T.M., Niemi, Z. Ben-Avraham, & J. R. Gat (Eds.), *The Dead Sea: The lake and its setting* (pp. 133–144). U.K.: Oxford University Press.

Zak, I., & Bentor, Y. K. (1972). Some new data on the salt deposits of the Dead Sea area, Israel. In G. Richter-Bernburg (Ed.), *Geology of saline deposits*, Proceedings of the Hanover Symposium, 15–21 May 1968, International Union of Geological Sciences, United Nations Educational, Scientific and Cultural Organization. *Earth Sciences*, *7*, 137–*146*.

Zak, I., & Freund, R. (1980). Strain measurements in eastern marginal shear zone of Mount Sedom salt diapir, Israel. *American Association of Petroleum Geologists Bulletin*, *64*(4), 568–581.

Zak, I., & Freund, R. (1981). Asymmetry and basin migration in the Dead Sea Rift. *Tectonophysics*, *80*, 27–38.

Zhang, S., Jiang, Q., Shi, C., Xu, X., Gong, Y., Xi, J., et al. (2021). Application of Sentinel-1 and-2 images in measuring the deformation of Kuh-e-Namak (Dashti) Namakier, *Iran. Remote Sensing*, *13*, 785. https://doi.org/10.3390/rs13040785

Zouaghi, T., Bédir, M., Ayed-Khaled, A., Lazzez, M., Soua, M., Amri, A., & Inoubli, M. H. (2013). Autochthonous versus allochthonous Upper Triassic evaporites in the Sbiba graben, central Tunisia. *Journal of Structural Geology*, *52*, 163–182.

4
Evaporite Minerals in Salt Formations

ABSTRACT

Evaporite minerals are formed by the evaporation of continental, marine, and mixed brines. The evaporation of brines enriched with ions forms several chemical compounds, including carbonates, sulfates, and chlorides, which precipitate salts based on their solubility. Continental evaporite basins also form borates and nitrates, while seawater brine concentration primarily influences salt deposition in giant salt basins. Isotope ratio measurements are crucial in geochemistry for identifying brines from marine and nonmarine evaporite sequences.

Unlike most rocks that occur in sedimentary basins, evaporites are formed by minerals that can dissolve easily and change their original depositional shape in geologically short time intervals. Due to its high mobility, halite layers are deformed into salt domes similar to igneous intrusions, forming narrow and tall structures, with the base several kilometers below the top. Salt minerals can dissolve and disappear from the stratigraphic succession when exposed to the surface. The physical and chemical instability of the salt rocks increases in the presence of volcanic or subvolcanic processes, resulting in extremely high rates of deformation and dissolution.

In some regions, such as Iran, marine evaporites deposited millions of years ago have been exhumed by tectonic forces and are presently exposed at the surface (see Chapter 3). These salt masses outcrop as salt diapirs or salt glaciers, both onshore and offshore, suffer the effects of weathering and dissolution. The erosion and transportation of preexisting evaporites may result in the deposition of new salt layers. These neo-formed salt deposits derive the chemical elements from the ancient evaporites and redeposit salt layers at various ages. Several examples of this phenomenon are provided by the Mesozoic and Cenozoic evaporite deposits derived from the erosion and dissolution of the Neoproterozoic Hormuz salt in Iran. These different layers formed multiple detachment levels that controlled the structural style of deformation of the parent salt rock. Salt diapirs formed in the Early Paleozoic were involved in multiple deformation episodes during the Paleozoic, Mesozoic, and Cenozoic. At present, the Hormuz evaporites are locally seen in outcrops and exposed to weathering and dissolution (see Chapter 3).

What makes evaporites easy to deform, dissolve, or flow over large distances? Understanding the anomalous behavior of evaporites is based on the study of the chemical and physical properties of different salt minerals, including their atomic structure and ionic properties. This chapter presents a summary of the chemical composition and ionic properties of common evaporite minerals in salt formations, including their chemical properties and selected images to illustrate their macroscopic characteristics. The microscopic analysis of the most important evaporite minerals will be addressed in Chapter 7.

4.1. BASIC CHEMISTRY PRINCIPLES APPLIED TO EVAPORITE MINERALS

The essential elements that are present in evaporite minerals, such as alkaline metals, are generally found in nature dissolved in seawater but also occur in igneous rocks that are consolidated from magma (Grotzinger et al., 2007; Press et al., 2003). The main elements that form the common salt (halite) are sodium, whose ion has an electric charge of +1, and chlorine, whose ion has a charge of −1. These two ions are normally dissolved in

Salt in the Earth Sciences: Evaporite Rocks and Salt Deposition, Geophysical Monograph 286, First Edition. Webster Mohriak.

DOI: 10.1002/9781119405245.ch04

seawater and exhibit chemical and physical properties that are complementary to each other, forming a solid chemical compound through a process known as an ionic bond. The salt crystals precipitate when the seawater is concentrated into a brine, forming an evaporite deposit.

Most authors assume that most of the evaporite minerals found in sedimentary basins precipitated from brines after seawater evaporation and concentration (Kendall, 1978, 1984, 1989; Perthuisot, 1989; Richter-Bernburg, 1972a, 1972b; Voigt, 2015; Warren, 1986, 2006, 2016). When dissolved in a saline solution, the Na^{+1} and Cl^{1-} ions (where the superscript numbers correspond to the electric charge) can move around in the brine until the concentration reaches a value higher than about 300,000 parts per million (ppm). At this concentration level, these ions can form a bond that results in the precipitation of the mineral halite, which is expressed by the chemical formula NaCl. Another very important element in seawater is magnesium, which occurs as the Mg^{2+} ion, generally forming evaporites that contain water molecules in their chemical composition. This element generally gives a bitter taste to the compounds in which it is present. In addition to many industrial uses, magnesium has some important medicinal properties.

In the sodium chloride atomic model (Figure 4.1), the sodium atom is linked to the chlorine atom by sharing the outermost electron so that the chlorine ion fills its outermost layer, thereby completing the orbital. The process of transforming the two atoms into ions is called an ionic reaction or ionic bond. There is an electric attraction between these ions. When linked together, they form sodium chloride, a more stable solid compound, corresponding to the evaporite mineral halite (NaCl), known as common salt. Ionic bonds between positively charged cations and negatively charged anions are responsible for the majority of evaporite minerals.

The arrangement of the ions in the mineral is called the crystal lattice. The sodium and chloride ions in halite crystals are arranged alternately in the vertices of the cubes, resulting in a cubic crystalline reticulate in which the faces of the crystals have the same lengths and are at right angles to one another (Figure 4.1). This structure is reflected in the formation of halite crystals, which are present in natural rocks with a cubic shape. The cubic shape of the halite crystal, or its crystalline habit, is determined by the atomic arrangement of the sodium and chloride ions (Figure 4.1).

Evaporite minerals are formed by various elements, with the ionic bond being controlled by the atomic and ionic radius of the elements forming the crystalline lattice. Atoms and ions have no sharp boundaries due to electron distribution in a cloud around the nucleus. They are characterized by radii that correspond to the sum of the ionic

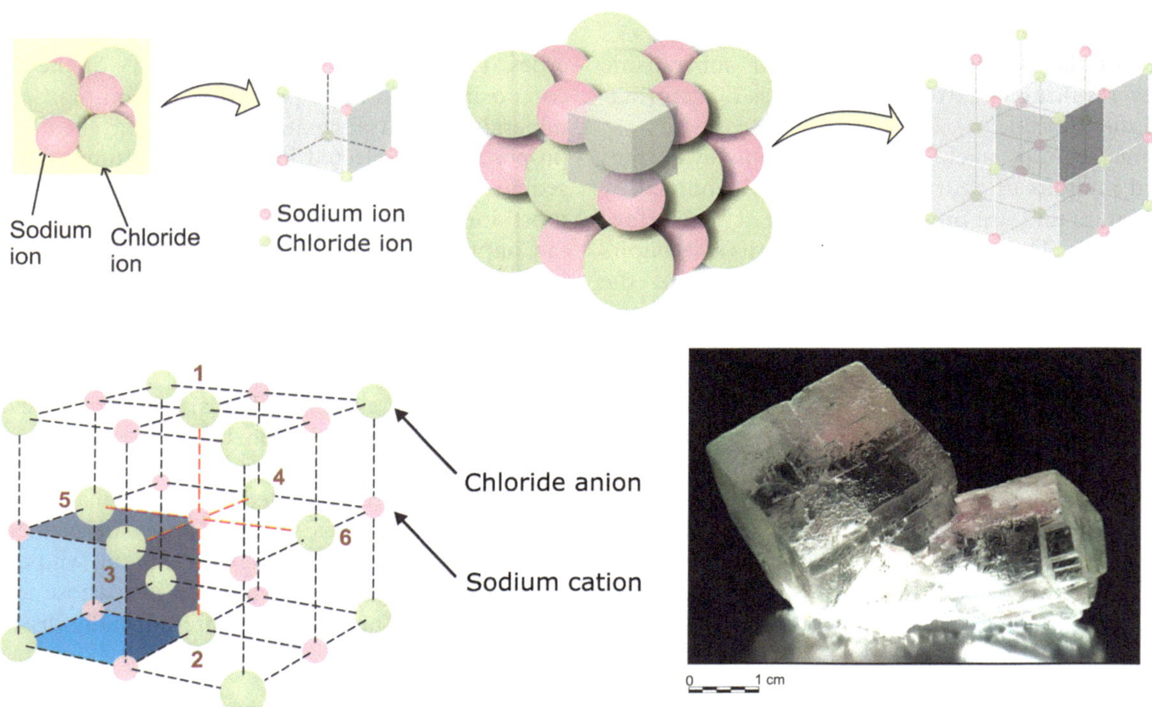

Figure 4.1 Crystalline lattice structure of the mineral halite (sodium chloride) with a cubic geometry. The chloride anion and the sodium cation are placed at the vertices of the cubic structure that forms the mineral. Credit: Adapted from Press et al. (2003).

radii of the cation and anion, or the distance between the ions in a crystal lattice. Ionic radii are typically presented in Ängstroms (Ä), with 1Å = 10^{-10} m.

The size of an atom corresponds to the distance between the nucleus and the outermost shell of electrons. The atom radius is a probabilistic measurement between the nucleus and the cloud of electrons at the outermost shell (Shannon, 1976). The ionic radius depends on the valence of the element. Three main classes of radii can be identified: (1) covalent radii, (2) van der Waals radii, and (3) ionic radii, which measure the distance between the nucleus and the electron cloud in the outermost shell, which is characterized by some probabilistic uncertainty (Clementi et al., 1967; Shannon, 1976).

A positive ion is formed by the removal of part of the electrons in the outermost shell, resulting in a decrease in the ion size. For example, Na has an atomic radius of 1.90 Å, but when it loses the electron, it has an ionic radius of about 0.99 Å. The more electrons are removed from the outer shell, the smaller the ion size becomes (Figure 4.2).

Most evaporite minerals are characterized by cations with large sizes, such as sodium, calcium, and potassium, whereas anions are much larger, as in chlorides (Figure 4.3). For example, the halite mineral is formed by sodium and chlorine atoms that share one electron and become ions with a positive charge (the sodium cation) and a negative charge (the chloride anion). Cations lose the outer electrons, and the radius decreases, while anions receive electrons, and the radius increases. The highest values of atomic radius of some common evaporite minerals (Figure 4.2) occur in the elements displayed in column 1 of the periodic table. The atomic structures of K, Rb, and Cs have a radius greater than 2 Å, but the cations are all less than 1.7 Å (Figure 4.3).

Negative ions are formed by adding one or more electrons to the atom, expanding the electron cloud, and decreasing the force of attraction between the nucleus and the cloud of electrons. The elements that will receive electrons to complete their outermost orbitals are positioned on the right-side column of the periodic table. These chemical elements are in group 7 of the periodic table. They need 1 electron to complete the 8 spaces in the outermost orbital, forming the fluoride, chloride, and bromide anions in chemical compounds.

The most important ions that form evaporite minerals and are present in the oceans (seawater) and in the continents (rivers and lakes) are listed in Figures 4.4 and 4.5. They are as follows:

Positive ions: Na^{1+}, K^{1+}, Mg^{2+}, Ca^{2+}

Negative ions: Cl^{1-}, Br^{1-}, SO_4^{2-}, CO_3^{2-}

Potassium and magnesium are important constituents of evaporite minerals, forming compounds with different anions. Sylvite (potassium chloride) is a mineral that has an atomic arrangement similar to halite. The effective ionic radii of K^{1+} and Cl^{1-} are 1.37 Å and 1.81 Å, respectively, resulting in a crystalline habit with cubic symmetry.

Chemical element	Atomic radius	Crystal radius	Effective ionic radius	Valence
H	0.53	0.24	0.38	1+
Li	1.67	0.73	0.76	1+
Na	1.90	1.13	0.99	1+
K	2.43	1.51	1.37	1+
Rb	2.65	1.66	1.52	1+
Cs	2.98	1.81	1.67	1+
Be	1.12	0.30	0.16	2+
Mg	1.45	0.71	0.57	2+
Ca	1.94	1.14	1.00	2+
Sr	2.19	1.32	1.18	2+
Ba	2.53	1.49	1.35	2+
Fe	1.56	0.77	0.63	2+
Ni	1.49	0.69	0.55	2+
Zn	1.42	0.74	0.60	2+
B	0.87	0.25	0.11	3+
Al	1.18	0.53	0.39	3+
C	0.67	0.29	0.15	4+
Si	1.11	0.40	0.26	4+
N	0.56	1.32	1.46	3–
p	0.98	0.58	0.44	3+
As	1.14	0.72	0.58	3+
O	0.48	1.21	1.35	2–
S	0.88	1.70	1.84	2–
Se	1.03	1.84	1.98	2–
F	0.42	1.145	1.285	1–
Cl	0.79	1.67	1.81	1–
Br	0.94	1.82	1.96	1–
I	1.15	2.06	2.20	1–

Figure 4.2 Atomic and ionic radius (in Ängstrom) of common chemical elements. Typical values for the ionic radius for common elements range from about 0.2 Å to 2 Å. Credit: Data from Clementi et al. (1967).

The most common ions that are present in seawater are Na^{1+}, Mg^{2+}, Ca^{2+}, K^{2+}, Cl^{1-}, and $SO4^{2-}$ and their concentrations differ greatly from the chemical content of the water in the main rivers of the world (Figure 4.4). Therefore, the entrance of fluvial or rainwater into an evaporating basin will result in changing the chemical content of the brine, depending on the chemical composition of the incoming waters (e.g., Baar, 1977). Consequently,

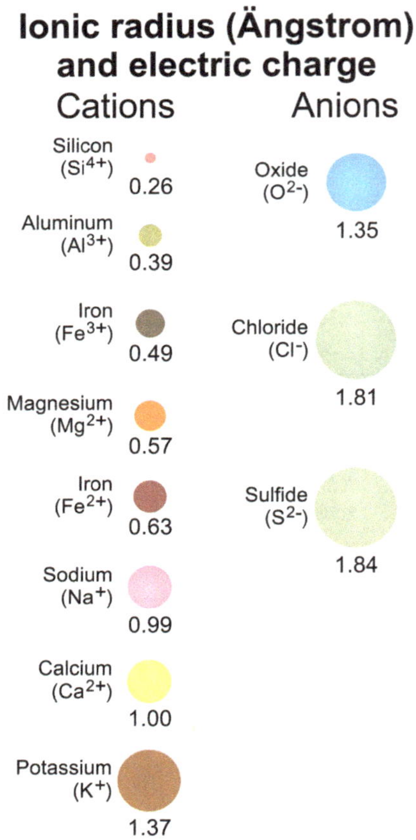

Figure 4.3 Effective ionic radius and electric charge of chemical elements commonly found in rock-forming minerals. Credit: Data from Shannon (1976).

Ion	Seawater	Rivers
Cl^{-}	18,980	7.5
Na^{+}	10,560	6.3
SO_4^{--}	2,650	11
Mg^{++}	1,270	4.1
Ca^{++}	400	15
K^{+}	380	2.3
HCO_3^{-}	140	58
Br^{-}	65	13

Figure 4.4 Average content of the various ions or salts in seawater and rivers (in ppm, parts per million). Credit: Holser (1979), Baar (1977) and Livingstone (1963).

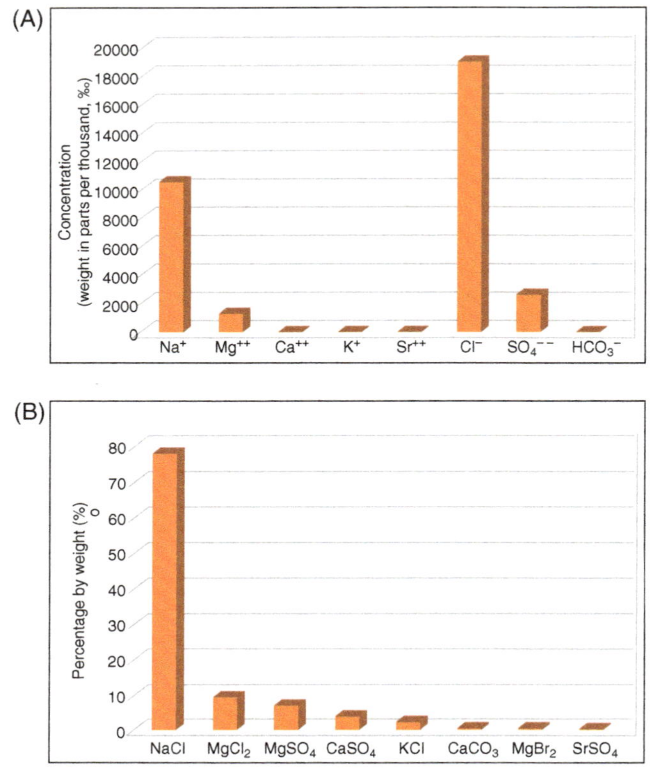

Figure 4.5 Mean contents of ions (A) and salts (B) in seawater with normal salinity (in parts per thousand and percentage by weight). Credit: Holser (1979), Baar (1977) and Livingstone (1963).

the precipitation of chemical compounds will be very different from those that were being deposited when the basin had little water input from rainfall or rivers (e.g., Livingstone, 1963). These changes will be registered in the rocks that make up the lithological sequences of the evaporite succession. Different brine compositions will result in diverse lithological associations, forming distinct evaporite cycles in salt basins (Figure 4.4).

Seawater, with an average salinity of 36,500 mg/l, is a constant composition across different oceans, with predominance of chlorine and sodium ions (Figure 4.5A). In arid regions, evaporation dominates over precipitation, increasing ion concentrations in brines. Conversely, continental seas with a significant influx of continental waters experience brine dilution, resulting in variable ionic concentrations and chemical compositions. The presence of carbonate ions in areas with limestone rocks may increase ion abundance, resulting in the deposition of carbonate minerals.

The main salts in seawater (Figure 4.5B) are sodium chloride (78%), magnesium chloride (9%), and magnesium sulfate (6%). These chemical proportions suggest that common minerals derived from seawater should correspond to halite and magnesium-bearing minerals like kainite and carnallite. However, this is not the case in natural salt deposits, which have a significant deficiency of magnesium minerals. There are chemical compounds that appear with very small contents in seawater, such as bromine. These compounds represent trace elements that are quite important for geochemistry and evaporite studies (Holser, 1979b).

4.2. IONIC POTENTIAL AND THE PHYSICAL PROPERTIES OF MINERALS

The ionic potential is a crucial chemical property that determines the strength of bonds between ions in mineral compounds. The equation is: ionic potential = charge/surface area, where charge is the ion electric charge and the surface area depends on the ionic radius. The equation shows that the higher the charge or valence, the greater the ionic potential of the chemical element. Conversely, the larger the ion size, the lower the ionic potential. That is, as the electric charge is distributed over a larger area, the force of attraction becomes more diffused. The ionic potential gives a sense of how strongly or weakly the ion can be electrostatically attracted to ions of opposite charge and to what extent an ion will repel other ions with the same charge.

In the case of silica (silicon oxide, SiO_2), the silicon has a high charge (+4) and a small radius (0.27 Å). Its ionic potential is extremely high (~14.8), resulting in a highly resistant atomic structure; that is, there is a high resistance to the displacement of the ions in the atomic framework (lattice). This is reflected in the physical properties of quartz, which does not dissolve when placed in water. Silicon oxide and aluminum oxide are also resistant to dissolution in water due to their high ionic potential.

Resistates are minerals that do not dissolve during weathering processes and form the main components of sedimentary rocks like sandstones. Evaporite minerals, on the contrary, have ions (Na^{1+}, K^{1+}, and Cl^{1-}) that are characterized by a very low ionic potential due to their small electric charge and large ionic radius (Figure 4.3). This results in the easy dissolution of these compounds in water and also in the high mobility of the ions within the crystal lattice. In halite crystals, the ions are arranged in a 1:1 ratio in the lattice, allowing a linear displacement of one row of ions to encounter other equivalent ions along the same column (Figure 4.6). This allows for easy movement in the planes that make up the faces of the cubic halite crystal.

The ions (Na^{1+} and Cl^{1-}) have low ionic potential, so they are easy to move in the planes that make up the faces of the cubic halite crystal. Over time, rock salt can flow with a rheology that is comparable to glass, as will be discussed in Chapter 5. Glass panes deform within a few centuries, as observed in the windows of cathedrals built in Europe during the Middle Ages. The easy translation of sodium and chlorine ions in the atomic arrangement of mineral halite makes rock salt easily deformable, flowing easily under low stresses (Figure 4.6). Halite layers can be compared to a solid compound with a rheology intermediate between a liquid and a solid. When submitted to small stresses, the crystals rearrange internally to form the same atomic structure, with each sodium ion binding to a different chlorine ion. Evaporites formed by halite can flow even under low stresses and at surface temperatures. However, when the stresses are large and the confining pressure is small, halite can behave as a brittle material, and faults can be formed to break the structures, as observed in salt mines (Davison, 2009).

4.2.1. The Common Ion Effect on Salt Deposition

Over the past two centuries, chemists have conducted numerous experiments to study the precipitation of evaporite minerals from brine. Van't Hoff's classic works analyzed the ionic properties of chemical compounds found in salt rocks (e.g., Van't Hoff, 1905). The precipitation of evaporite crystals from a liquid follows the laws of

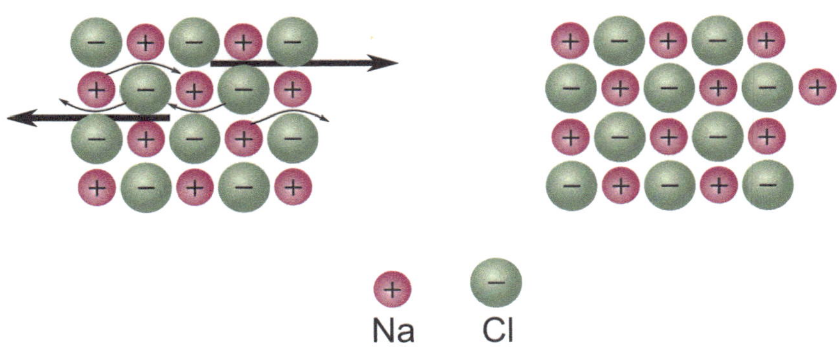

Figure 4.6 Translation of the sodium and chloride ions during salt flow. As one row is dislocated relative to the adjacent row, the ions keep the same distribution in a column. Credit: Adapted from Press et al. (2003).

chemistry, forming sequences of different minerals due to the different solubilities of the chemical compounds.

The transformation of ions from the brine into a solid salt is a dynamic chemical reaction that starts when the brine concentration exceeds the ion solubility. When the reaction is in equilibrium, the number of salt ions that precipitate corresponds to the same number of solid salt crystals that are dissolved. Research has been conducted on how the addition of other ions can affect the solubility of the different chemical compounds present in the brine and precipitate different salts (e.g., Voigt, 2011). The common ion effect is based on the chemical law that salt precipitation is controlled by the solubility of the dissolved evaporites in the brine.

The law of the common ion follows the chemical principles proposed by Le Chatelier in the 19th century (Le Chatelier & Boudouard, 1898). This principle states that when any chemical system is subjected to a change in concentration, temperature, volume, or pressure, the system will change to a new equilibrium. The equilibrium change will partly counteract the effects of the previous change. Applied to aqueous solutions, this principle states that adding a small amount of one of the ions of the dissolved salt in a solution leads to an increase in the precipitation of the salt. The common ion effect results from adding compounds to the solution with the same ion as the original salt that was dissolved in the aqueous system. This effect will reduce the concentration of both ions in the salt by precipitating the solid compound until the solubility equilibrium is again reached.

Laboratory experiments have confirmed the importance of the common ion effect on aqueous solutions with ions commonly found in evaporite minerals. If a vessel is filled with a saturated solution of a single salt, the crystals that precipitate will grow slowly and form large, idiomorphic crystals about 3 to 4 mm in size. If other salts with the same Cl^{1-} ion are added to the brine (e.g., $MgCl_2$), sodium chloride crystals (halite) start to precipitate throughout the volume of water as a cloudy powder raining from above. This fact indicates that the deposition of solid halite crystal was accelerated, as there was an increase in the number of ions of Cl^{1-}. When this happened, there was a cloud of small crystals that precipitated at the same time, since there was great competition for the deposition of the solid material (solute). Consequently, the size of halite crystals will be very small, resulting in amorphous grains with a very fine-grained texture.

The chemical equations for the single salt solution (Equation 1) and the two salts with a common ion (Equation 2) are shown as:

Equation 1:

$$Na^{1+}{}_{(aq)} + Cl^{1-}{}_{(aq)} \leftrightarrow NaCl_{(solid)}$$

Equation 2:

$$Mg^{2+}{}_{(aq)} + 2Cl^{1-}{}_{(aq)} \leftrightarrow MgCl_{2(solid)}$$

The common ion effect in evaporite sediments is a mechanism that causes the precipitation of small crystals of halite. When a compound ($MgCl_2$) is added to the brine, it alters the equilibrium state, leading to the accelerated deposition of halite to counteract the addition of the common ion (Cl^{1-}). Le Chatelier's principle, applied to this simple case, establishes that nature's way to remove the additional ion is to precipitate halite. The magnesium chloride is not precipitated as its solubility is much greater than that of the sodium chloride. Minerals containing magnesium chloride are only deposited during the last stages of salt precipitation. The rate of chemical reactions influences the process, forcing an ionic compound to precipitate as a solid substance from the solution.

When a compound ($MgCl_2$) is added, it changes the balance between the rates at which ions change into solid salt. This makes the first salt (NaCl) less soluble, which leads to the formation of small crystals of halite. This phenomenon is observed in the layers of Aptian evaporites deposited along the eastern Brazilian margin. Petrographic analysis of different salts in the Sergipe Basin indicates that halite layers close to tachyhydrite sequences are typically characterized by small halite crystals, indicating competition for the precipitation of the mineral. Therefore, the larger crystals are deposited first, with enough time to form the cubic faces of the crystals, and there is little competition with the neighboring crystals.

In sedimentary basins with a brine concentration of 300,000 ppm, there is an equilibrium between the transformation of solid salt (halite) and the ions dissolved in the solution (Na^{1+} and Cl^{1-}). The addition of the mineral sylvite (KCl) to the solution dissolves the ions K^{1+} and Cl^{1-} in the brine, resulting in the common ion effect and the immediate precipitation of halite crystals to reach equilibrium again.

4.3. BRINE CHEMISTRY AND DEPOSITION OF EVAPORITE MINERALS

The intense evaporation of seawater in restricted sedimentary basins results in the concentration of several ions, forming brines that fill hypersaline lakes like the Dead Sea (see Chapter 2). Hypersalinity may also occur in shallow seas and gulfs when a barrier limits the connection with an open ocean. Hypersaline lakes may concentrate brines to a point that exceeds the solubility of evaporite minerals, resulting in salt precipitation.

Solubility is defined as the concentration of a chemical compound in a solution that is in equilibrium with a solid phase at the specified temperature and is usually expressed as the mass of the solute per 100 g of brine. When the brine concentration exceeds the solubility value of a given evaporite mineral, the dissolved salt is precipitated, forming mineral crystals that accumulate at the basin floor while more soluble minerals remain dissolved in the brine. Some evaporite minerals are highly soluble in seawater and precipitate only when the brine is concentrated, and the dissolved ions reach a value that exceeds the maximum solubility threshold (Figure 4.7).

The solubility of evaporite minerals in seawater or concentrated brines refers to the ability of a chemical compound to dissolve the ions at a given temperature. For example, halite has a solubility of 26.41 g/100 g of brine at a temperature of 20 °C (Haynes, 2017). Increasing the temperature usually increases the solubility of a given compound, meaning that warm waters can contain a larger quantity of evaporite ions dissolved in solution than cold waters. For example, at a temperature of 100 °C, the same brine will have a NaCl solubility of 28.05 g/100 g of brine. This means that warm waters can contain a larger quantity of evaporite ions dissolved in solution than cold waters. Figure 4.7 (Haynes, 2017) shows the solubility of some common evaporite minerals per 100 g of brine at temperatures of 20, 50, and 100 °C. A few chemical compounds show a decrease in solubility at higher temperatures, a phenomenon known as retrograde solubility. For example, the mineral anhydrite ($CaSO_4$) shows solubility in freshwater that is about 2 mol/kg at 20 °C, but at 60 °C the solubility is only 1 mol/kg (Figure 4.7). Calcium carbonate minerals (e.g., calcite and aragonite, $CaCO_3$) are sparingly soluble in freshwater but are more soluble in acidic water (Haynes, 2017).

Compound solubility (g/100 g of brine) / Temperature (°C)	0	20	50	100
$CaCl_2$	36.70	42.13	56.05	59.94
$CaCO_3$		0.00066		
$CaSO_4$	0.174	0.202	0.207	0.163
KBr	35.0	39.4	44.8	50.8
KCl	21.74	25.39	30.04	36.05
$MgBr_2$	49.3	50.3	52.1	55.7
$MgCl_2$	33.96	35.58	37.34	42.15
$MgSO_4$	18.2	25.1	33.4	33.3
NaCl	26.28	26.41	26.84	28.05
Na_2CO_3	6.44	17.9	32.2	30.9

Figure 4.7 Aqueous solubility of inorganic compounds at various temperatures (weight of minerals in grams per 100 g of brine, temperature in degrees Celsius). Credit: Adapted from Haynes (2017).

In aqueous solutions where evaporite minerals are dissolved, there is an equilibrium system between the solid compounds and the ions. Some ions precipitate as solid salts depending on the concentration, while the chemical compounds may dissolve into their ions. Solubility is an equilibrium phenomenon, as solutions constantly change their characteristics and chemical reactions occur. If the system is in equilibrium, the solution is saturated with respect to some salts, allowing no more solid compounds to be dissolved. This means that the maximum number of free ions is reached, and no more solid compounds can be dissolved. But if the solution is undersaturated, then more solids can dissolve and form ions that will be added to the solution.

Most of the chemical compounds containing chlorine ions (Cl^{1-}) are soluble in water except for AgCl, $PbCl_2$, and Hg_2Cl_2, which do not dissolve in the brines but precipitate, forming these salts when the ions are present. Chemical compounds containing bromine ions (Br^{1-}), fluorine (F^{1-}), and iodine (I^{1-}) have solubilities similar to chlorides, but they show negligible concentrations in normal seawater and do not form salts (see Figure 4.6). Many chemical compounds containing the sulfate anion are common in seawater, and some of them are only slightly soluble, such as $CaSO_4 2H_2O$ (gypsum).

Carbonate compounds are almost insoluble in water; they precipitate, forming the minerals calcite and aragonite. The low-solubility compounds are characterized by a very small amount of ions dissolved in water. In evaporite basins, most of the carbonate minerals precipitate at a low degree of evaporation before the start of the main salt deposition event. For instance, the first sediments deposited in the Mediterranean Sea during the Messinian salinity crisis are carbonate sequences called Calcare di Base, which occur interdigitated with the first gypsum layers (Roveri et al., 2008a, 2008b).

Jacobus Henricus van't Hoff, a 19th-century German chemist, proposed the study of minerals formed by evaporation through laboratory experiments. His groundbreaking work earned him the first Nobel Prize in Chemistry in 1901. In a classical work (Van't Hoff, 1905), he used a solution of two salts mixed with various chemical compounds and analyzed their effects on the minerals that precipitated at the bottom of the container.

According to a summary by Warren (2006), van't Hoff's basic chemical principles include the following:

1. Solubility depends on the relative solubilities of the salts and their concentrations.
2. If two salts have a common ion, the solubility of one salt dissolved in a brine that contains the other salt is less than its solubility in pure water.
3. If two salts do not have a common ion, the solubility of one salt dissolved in a brine that contains the other salt is greater than its solubility in pure water.
4. Solubility is markedly influenced by increases in temperature, generally increasing with higher temperatures.
5. Solubility is much less influenced by increases in pressure.

The solubility of evaporite minerals in freshwater is expressed in moles per kilogram of brine (mol/kg or mol/l) and generally increases with increasing temperature, known as prograde solubility. Some minerals, such as natron, can be dissolved more effectively in higher temperatures. Fresh water at 10 °C can dissolve only 1 mole of natron, but the solubility at 30 °C is almost 3 moles/l.

Evaporite minerals like gypsum and halite are more than three orders of magnitude more soluble than limestone when exposed to meteoric water (Figure 4.8; Haynes, 2017). Outcrops of halite and anhydrite are rare, with gypsum being the only remaining mineral that survives at the surface from the weathering and dissolution of salt diapirs. In northeastern Tunisia, the outcrops are mainly composed of gypsum, whereas thick halite layers occur in the subsurface. Halite is primarily observed at the surface in the most hyperarid environments, such as the Zagros Ranges in Iran or the Dead Sea in Israel, but a few regions worldwide exhibit the mineral in outcrops, as in Romania and Spain. Gypsum is commonly the only remaining mineral that survives at the surface from weathering and dissolution of salt diapirs, as in the case of the Paradox Basin in Utah (see Chapter 3).

Based on chemistry experiments in the 19th century, several researchers suggested that evaporite minerals form when the seawater becomes supersaturated with evaporite mineral ions (e.g., Usiglio, 1849a, 1849b). The concentrated brines form various minerals that are naturally deposited following a sequence that depends on the solubility (Figure 4.8). Thus, each time the concentration of these ions exceeds their solubility in seawater, the brines precipitate diverse chemical compounds, according to the genetic mineral sequence zones, as shown in Figure 4.9 (Strakhov, 1969; Valyashko, 1956, 1972a, 1972b). The evaporite minerals discussed in this chapter follow the nomenclature and chemical composition adopted by most international scientific organizations (e.g., International Mineralogical Association, 2022).

4.3.1. Sequence of Deposition of Evaporite Minerals

Assuming the commonly accepted model of evaporite formation by solar evaporation (e.g., Arrhenius & Lachmann, 2002; Talbot, 2008; Usiglio, 1849a, 1849b), mineral precipitation occurs by reducing the quantity of water in the sedimentary environment. This occurs when the net rate of evaporation is higher than the water input

Mineral	Composition
Calcite	$CaCO_3$
Magnesite	$MgCO_3$
Dolomite	$CaMg(CO_3)_2$
Trona	$Na_3(HCO_3)(CO_3).2H_2O$
Gypsite	$CaSO_4.2H_2O$
Anhydrite	$CaSO_4$
Halite	NaCl
Kainite	$KMg(SO_4)Cl._3H_2O$
Kieserite	$MgSO_4.H_2O$
Langbeinite	$K_2Mg_2(SO_4)_3$
Sylvite	KCl
Carnallite	$KMgCl_3.6H_2O$
Tachyhydrite	$CaMg_2.Cl_6.12H_2O$
Bischofite	$MgCl_2.6H_2O$
Polyhalite	$K_2CaMg_2(SO_4)_4.2H_2O$

Figure 4.8 Chemical composition of the main evaporite minerals. Credit: Adapted from International Mineralogical Association (2022).

into the system by rainfall, surface run-off, rivers, or marine waters. This results in a negative balance that leads to brine concentration and eventually basin desiccation. The necessary conditions for forming evaporite basins include the existence of a desert environment, isolation from global oceans, and a progressive concentration of seawater into a brine. The concentrated brine will precipitate the evaporite minerals in an ordered sequence (Figure 4.9).

In an evaporitic basin, lithological successions involving chemical sediments indicate salinity variations affecting brine concentration throughout the geological time. These phases of high and low salinities alternate periodically due to climatic aridity and variable water supply into the basin. These differences in brine concentration result in dividing the lithological sequences into time intervals or evaporation cycles, which can be correlated across large parts of the depositional area. The Dead Sea provides modern examples of these cycles, with alternations of carbonate muds, gypsum, and halite forming thin laminae that some authors interpret as varves controlled by seasonal variations (e.g., Ben Dor et al., 2019).

Evaporite minerals start to precipitate when their brine concentration reaches such a level that they can no longer exist as solutes. The abundance of evaporites formed by solar evaporation from seawater follows the same pattern as the precipitation sequence in laboratory experiments:

1. Calcite ($CaCO_3$)
2. Gypsum ($CaSO_4.2H_2O$)
3. Halite (i.e., common salt, NaCl)
4. Potassium and magnesium salts (e.g., carnallite, $KMgCl_3.6H_2O$, or sylvite, KCl)

Figure 4.10 shows the lithological facies distribution in evaporite basins, which corresponds to several cycles of the precipitation of minerals according to the brine concentration (Garrett, 1996; Richter-Bernburg, 1972a, 1972b). The calcium carbonate precipitation stage is reached when the concentration exceeds the average seawater salinity and density (>35‰, brine density: 1.0258 g/cm^3 at 12 °C). The carbonate precipitation extends up to a brine concentration characterized by the first precipitation of gypsum crystals (i.e.,] 140–200‰, brine density: 1.11–1.13 g/cm^3 at 12 °C). However, the amount of calcite precipitated during the evaporative concentration is negligible in comparison with the other salts that follow the same sequence (gypsum, halite, and K–Mg salts). For example, the Messinian salinity crisis in the Mediterranean Sea resulted in the accumulation of a few tens to hundreds of meters, while

No. of the zone (in order of precipitation)	Zone (normal)	Primary rock-forming mineral	Secondary minerals and other diagenetic and epigenetic associated minerals
6	Bischofite	Eutonic Borate ($MgO.B_2O_3.xH_2O.yMgCl_2$) Bischofite Magnesium sulfate ($MgSO_4.xH_2O$, where x is 6-4) Carnallite, Halite, Gypsum	Boracite Kieserite Anhydrite
5	Carnallite	Carnallite Magnesium sulfates ($MgSO_4.xH_2O$), where is x 6-4) Kainite (in incipient stages of development of the zone) Halite	Kieserite Langbeinite Kainite (in the lower parts of the zone) Polyhalite (in the lower parts of the zone, together with sylvite) Anhydrite
4	Sylvite	Sylvite Hexahydrite Epsomite Polyhalite Halite	Kainite Langbeinite Kieserite Polyhalite Leonite (?)
3	Magnesium sulfate	Epsomite Hexahydrite polyhalite Halite	Kieserite Polyhalite Blödite
2	Halite	Halite Gypsum	Anhydrite Polyhalite (only in the upper parts of the zone)
1	Gypsum - Anhydrite	Gypsum	Anhydrite

Figure 4.9 Sequence of deposition of genetic components of evaporite rocks, with order of precipitation and primary and secondary minerals associated with the evaporite zones. Credit: Adapted from International Mineralogical Association (2022).

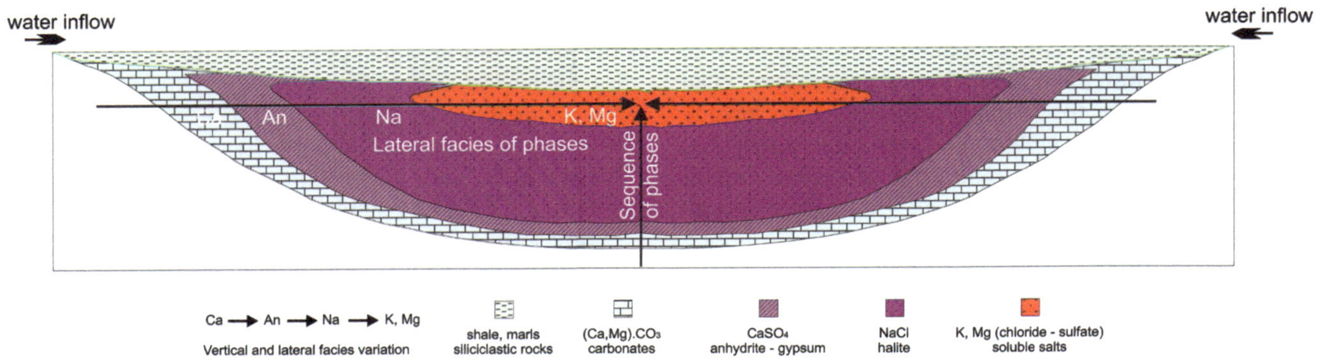

Figure 4.10 Schematic diagram showing the lithological facies distribution in evaporite basins. The chemical phases in saline formations are characterized by the same vertical and lateral sequences. The carbonate rocks grade into sulfates, chlorides, and soluble salts of potassium and magnesium with increasing brine concentration. Credit: Adapted from Richter-Bernburg (1972).

the massive salt deposits reached a thickness of several kilometers (Roveri et al., 2008a, 2008b).

A common phenomenon within the calcium carbonate stage of deposition is the appearance of microbial mats (formed mainly by cyanobacteria) at salinities greater than 110 $\%_0$, with brine density exceeding 1.087 g/cm^3. Carbonate precipitation dominates until ceasing at a salinity of about 150$\%_0$, followed by the onset of the precipitation of the first gypsum crystals (Babel & Schreiber, 2014).

The calcium sulfate precipitation stage continues up to about 300%o when halite crystals start to precipitate. At this evaporitic stage, the seawater brine density is about 1.20–1.26 g/cm^3 as a result of concentrating the brine 11 times the original seawater salinity (see Chapter 2). Marine incursions in basins decrease brine salinity, causing gypsum to be deposited. Evaporite deposits in giant saline basins often show intercalations of anhydrite with halite, forming layered evaporite sequences that reflect brine composition during salt deposition. Anhydrite layers often result from the dehydration of gypsum deposits (Douglas & Goodman, 1957).

Evaporite sequences in giant salt basins are commonly marked by the predominance of halite deposits, reflecting the larger proportion of sodium and chlorine ions in seawater (Figure 4.6). Halite beds exhibit impurities that result in different colors in layered halite sequences (e.g., Figure 4.11). In many evaporite basins, salt layers are often characterized by soft-sediment deformation, especially when the carbonate mud and salt sequences are affected by gravity flow when unconsolidated. In addition, evaporite layers are deformed by the diagenetic recrystallization and growth of halite crystals during burial or flowage (see Chapter 6).

Halite crystals precipitate when the concentration of brine exceeds the solubility of sodium chloride. This happens when the brine volume is only 10% of the initial volume. Halite crystals are formed at the air–brine interface and grow while floating on the brine surface, often exhibiting a peculiar structure resembling an inverted pyramid called a hopper (Dellwig, 1955; Shearman, 1978). Hopper crystals exhibit a depression at the center of the upper face, relative to the crystal borders (Figure 4.12).

Sodium chloride is the most common halide mineral, but potash salts, which are often part of the bittern series that result from the complete desiccation of saline ponds, are also registered in evaporite sequences. Potassium chloride minerals such as sylvite are relatively rare in nature because of their extremely high solubility (Figure 4.8).

Modern deposits of potash salts are found only in extremely restricted continental environments, such as the Qaidam Basin in China or the Danakil Depression in Ethiopia (Garrett, 1996). In these basins, the mineral that is precipitated when the brine is concentrated to more than 95% compared to the initial saltwater salinity is carnallite, not sylvite.

When carnallite, which corresponds to chloride of magnesium and potassium and halite ($MgCl_2$.KCl.6H_2O + NaCl), suffers an incongruent break-down by dissolution with unsaturated brines, the resulting minerals are sylvite and halite. The reaction is expressed as follows:

$$KMgCl_3.6H_2 \rightarrow OKCl + MgCl_2 + 6H_2O$$

$$\text{Carnallite} \rightarrow \text{Sylvite solution}$$

Sylvite is the most important K-bearing salt mineral used in fertilizer products for the agricultural industry. It corresponds to pure potassium chloride, which often occurs mixed with sodium chloride, forming the rock known as sylvinite.

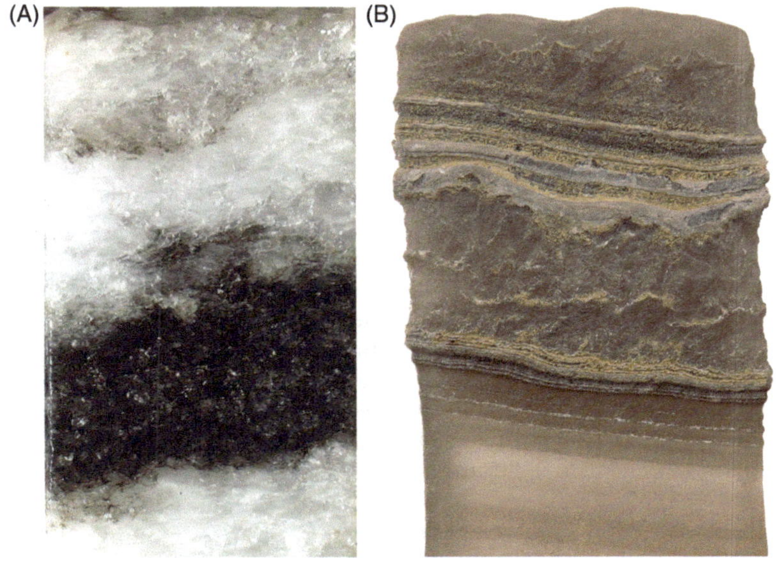

Figure 4.11 Photographs of evaporite cores from exploratory boreholes in the South Atlantic Late Aptian salt basin. (A) Core slab (diameter circa 6 cm) showing stratified evaporites with white halite intercalated with dark halite layers, both precipitated in subaqueous environments. (B) Core slab (diameter circa 6 cm) with stratified evaporites represented by gray halite intercalated with thin beds of whitish, brownish, and grayish anhydrite. Some thin lenses are disrupted by salt flow, forming convex and concave structures in the middle of the image. The lower part of the slab is characterized by more homogeneous laminae of evaporites with a light cream-to-beige color.

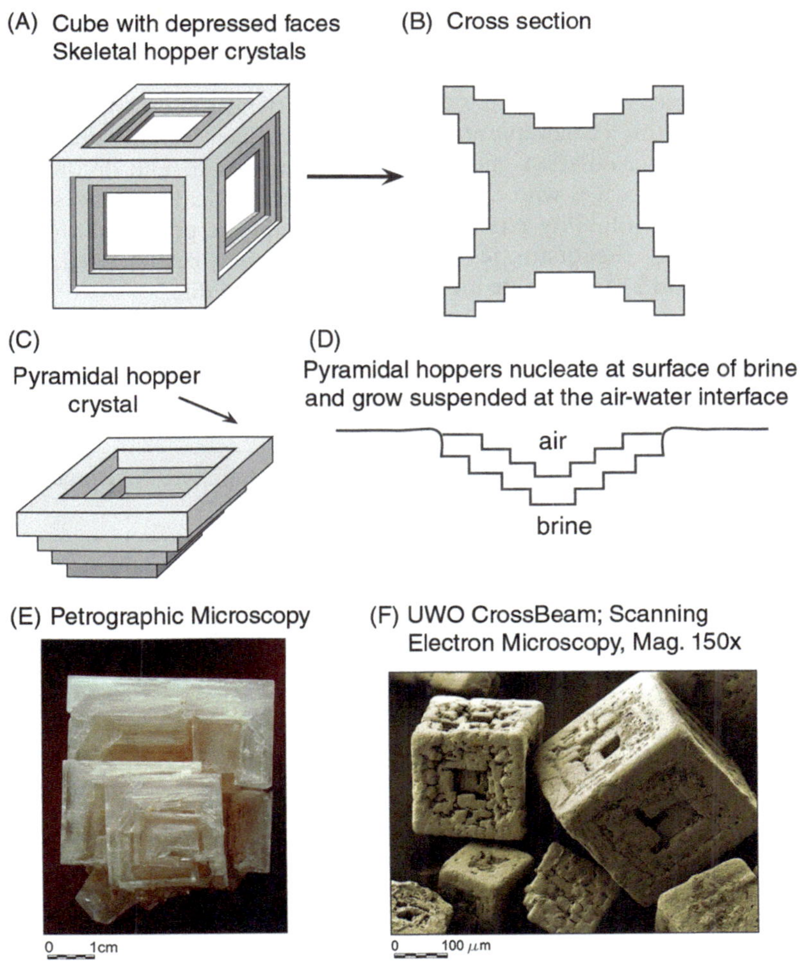

Figure 4.12 Schematic diagrams of cubic halite structure with cross-section. Top: Halite crystal with depressed central part of the cube (A), forming a hopper structure with a rectangular cross-section (B). Middle: Pyramid hopper crystal (C) growing at the air–water interface (D). Bottom: photograph of hopper crystal (E), and microscopic image of the hopper structure (F). Credit: ZEISS Microscopy/flickr/CC BY-NC-ND 2.0.

4.3.2. Experimental Precipitation of Evaporite Minerals

The precipitation of the evaporite minerals in laboratory experiments follows a well-defined sequence that was first described by Usiglio (1849a, 1849b) using Mediterranean seawater. These experiments showed that evaporite minerals in salt basins precipitate according to chemical laws. Different chemical compounds are successively formed according to their different solubilities (Babel & Schreiber, 2014; Braitsch, 1971; Hardie, 1984; Hardie & Eugster, 1971; Holser, 1979a; Richter-Bernburg, 1972a, 1972b; Sonnenfeld, 1979, 1989, 1992).

Braitsch (1971) described the results of classical chemical experiments using seawater as a source for the deposition of salts. The experiments were conducted while keeping the temperature constant at 25 °C, resulting in an increasing loss of water molecules with evaporation. Under these laboratory environmental conditions (which some authors postulate are different from natural conditions in sedimentary basins), the mineral precipitation sequence is established based on the brine chemical composition (Figure 4.13).

The values in Figure 4.13 are shown in molecules of water and remaining dissolved ions. The quantities of salts deposited during the evaporation of the brine are represented through standardized molecular quantities for the water, according to the calculation method adopted by the German chemist Jänecke, who applied chemical diagrams to the study of solubility equilibria (Voigt, 2015). In the calculations, the quantities are expressed as double ions of Na_2 and K_2, and halite molecules as Na_2Cl_2, to be equivalent to Mg and SO_4, which have a valence of 2.

The chemical experiments indicated that an original mass of 1,000 g of seawater (about 1 liter), corresponding to 70,700 moles of water, when submitted to isothermal evaporation (at the 25 °C isotherm), progressively concentrates. The quantities of the salts remain constant in the solution until the number of water molecules decreases by evaporation and the first salts are precipitated (Figure 4.13). When the weight of the brine is reduced to 322 g, the hydrated calcium sulfate (gypsum) will precipitate. The mineral halite (NaCl) will start precipitating when the brine weight is only 121 g, corresponding to about 1/9 of the original water weight (Richter-Bernburg, 1973). There will be no precipitation of any salt containing potassium until the brine concentration reaches very high levels, with less than 1,000 moles of water remaining, corresponding to only 1.5% of the initial water volume.

The order of the deposition of the salts depends on two factors: their solubility and the number of ions available in seawater. This genetic sequence (Figure 4.10) is discussed as follows, with a short description of the main mineral components of the evaporitic rocks. The main evaporite minerals include the following types:

- Calcium carbonates (as aragonite and calcite, $CaCO_3$). These minerals are not exactly evaporites but often precede and accompany the deposition of the chemical precipitates. Several basins include calcite, aragonite, and dolomite deposits, usually forming lithological sequences predating the main phase of evaporite precipitation. The precipitation of carbonates may be enhanced by biological activity, forming microbialite.
- Calcium sulfates (as gypsum, $CaSO_4.2H_2O$, and anhydrite, $CaSO_4$). Gypsum is primarily responsible

Initial precipitation of:

	Seawater	Gypsum	Halite	Blödite	Epsomite	Kainite	Hexahydrite	Kieserite	Carnallite	Bischofite
H_2O (x)	70,700	21,100	6,450	1,170	1,080	965	800	585	576	428
K_2	6.4	6.4	6.4	6.4	6.4	6.4	4.8	2.0	1.0	0.1
Mg	69.0	69.0	69.0	69.0	67.3	63.4	58.1	52.1	49.9	44.9
SO_4	24.6	24.6	24.6	24.6	21.0	17.1	11.6	5.7	3.6	0.4
Na_2Cl_2	303	303	303	26	11.45	9.1	4.7	1.5	1.4	0.25
$CaCO_3$	1.5	–	–	–	–	–	–	–	–	–
$CaSO_4$	11.7	11.7	1.8	–	–	–	–	–	–	–
Weight of solutions (*)	1,000	322.4	121.3	24.46	21.75	19.61	16.30	12.22	11.78	9.27

(*) Weight of solutions in grams per 1,000 g of seawater.

Figure 4.13 Chemical experiment showing sequential precipitation of evaporite minerals during isothermal evaporation at 25 °C of seawater with standard composition. Credit: Data from Braitsch (1971).

for forming the majority of sulfate-rich deposits. Gypsum is usually the first true evaporite mineral to form in evaporitic sequences and the last to remain after the evaporite basin is affected by marine incursions and the dissolution of the more soluble salts. Gypsum is typically later transformed into anhydrite, to the extent that all crystalline water molecules are lost with increasing pressure and temperature. A water-filled porosity volume loss of 38% is commonly observed when gypsum is transformed into anhydrite and the rock is compacted (Warren, 1991). Several authors have questioned the common assumption that the conversion of anhydrite to gypsum by adding two water molecules is accompanied by an overall increase in rock volume of about 60%, but most researchers have estimated increases varying from 18% to 67% (see Klimchouk, 2000).

- Halite (NaCl) is the most abundant evaporite mineral in nature, both in salt deposits and as dissolved ions in seawater. It starts to precipitate when its concentration in the brine reaches 27% or exceeds 300,000 ppm. Halite layers are easily deformed, allowing halokinesis and salt tectonics to occur even under low pressures and temperatures.
- Carnallite ($KMgCl_3.6H_2O$) precipitates only in highly concentrated brines. This mineral is deposited simultaneously with other chlorides, particularly halite. With the dissolution of magnesium chloride ($MgCl_2$), it can be converted into sylvite (KCl).
- Tachyhydrite ($CaMg_2Cl_6.12H_2O$) is a mineral whose name, sometimes also spelled as tachydrite, is derived from the Greek words *tachys* and *hydro*, meaning quick and water, reflecting its rapid hydration and transformation into water when exposed to moist air at room temperature. One crystal placed on a table can absorb the humidity from the air and, overnight, is transformed into a puddle as the mineral is easily dissolved.
- Bischofite ($MgCl_2.6H_2O$) normally occurs only in the last precipitation phase of concentrated seawater. In the evaporitic basins of the Brazilian and West African margins (Sergipe and Gabon basins), its place is taken by calcium and magnesium chlorides (tachyhydrite).

Minerals are deposited in sedimentary basins in cycles influenced by brine concentration during compound precipitation. First-order cycles increase salinity during high aridity, while when seawater replenishes the basin, the brine returns to low salinity (Jenyon, 1986; Warren, 1989). Each cycle is marked by a continuous layer of black, organic shale, which is favorable for hydrocarbon generation and metal sulfide accumulation. The top of an evaporitic cycle is marked by the deposition of the most soluble salts, such as magnesium and potassium salts.

4.3.3. Residual Brines in Deposition of Highly Soluble Minerals

Braitsch (1971) discussed the chemistry of the residual brine that remains with the progressive evaporation of seawater after halite is precipitated (Figure 4.13). New minerals will precipitate with a much higher increase in brine salinity. For example, magnesium sulfate (blödite) is deposited when the brine weight is 24.5 g. This mineral is followed by epsomite (21.7 g), kainite (19.6 g), hexahydrite (16.3 g), and kieserite (12.2 g). The first magnesium chloride to precipitate is carnallite (magnesium and potassium chloride) and it only forms when the brine is reduced to 11.8 g, a little more than 1% of the original weight. Carnallite is followed by sylvite and tachyhydrite. These extremely soluble salts are hydrated compounds of chlorides, with 6 or 12 molecules of water attached to the structure, as shown in the following formulas:

- Carnallite: $KMgCl_3.6H_2O$
- Sylvite: KCl
- Tachyhydrite: $CaMg_2Cl_6.12H_2O$

Sylvite is the main K-bearing mineral exploited by the salt industry worldwide. Sylvinite is not a mineral but a rock predominantly composed of sylvite (KCl) and halite (NaCl) and usually contains various mineral admixtures. Carnallite and sylvite are deposited only when a few hundred free water molecules are present in the evaporite system. The mineral tachyhydrite exhibits a chemical composition similar to that of carnallite, but the potassium chloride is replaced by calcium chloride. There are important carnallite and tachyhydrite deposits in the onshore part of the Sergipe Basin in Brazil, which are associated with large deposits of economic ores such as sylvinite (Szatmari et al., 1979; Wardlaw, 1972).

The deposition of tachyhydrite indicates conditions of extreme dryness and the highest brine concentration in the basin. Besides, the presence of calcium indicates that there was an enrichment of this ion during the final phase of brine evaporation since all calcium is normally consumed during the precipitation of gypsum and anhydrite ($CaSO_4.2H_2O$ and $CaSO_4$). Hardie (1987, 1990) suggested that the enrichment in calcium chloride may be due to the processes of dolomitization of previously deposited carbonate rocks, resulting from the exchange of cations in clays.

The residual brine in Brazilian salt basins is rich in calcium ions, which are used in the deposition of calcium sulfate (gypsum). However, excess calcium is consumed by bacterial activity, leading to the precipitation of calcium chloride and magnesium chloride (tachyhydrite). The mineral bischofite is an extremely concentrated brine with no potassium or calcium ions. Bischofite is only deposited when the initial weight of 1,000 g has been reduced to about 9.3 g, corresponding to only 0.5% of the original volume (Figure 4.14).

Figure 4.14 Evaporite rocks from evaporite sequences in the Sergipe Basin in northeast Brazil, in hand samples and core slabs. The selected examples are typical rocks from depositional environments with growing aridity.

The presence of magnesium chloride in evaporite sequences indicates the nearly total dryness of the depositional environment. In this situation, there are still several chloride ions in the brine, but there is intense competition for the precipitation of soluble minerals. Halite minerals precipitate forming layers constituted by minute crystals due to the common ion effect. In Brazil, the level of evaporation is predominantly limited to the anhydrite and halite phases, but locally highly soluble minerals like tachyhydrite are also observed (Figure 4.14). The mineral bischofite, however, has not been recognized in any of the Brazilian salt basins (Szatmari et al., 1979, 2009, 2021).

4.3.4. Evaporite Sequences in the South Atlantic Salt Basins

Tachyhydrite layers, reaching up to 200 m in the Sergipe Basin in northeast Brazil, are found in the subsurface of South Atlantic salt basins, unlike most evaporite sequences worldwide, where this mineral commonly forms only centimeter-thick layers. The last phases of salt precipitation in the Sergipe Basin salt sequences record tachyhydrite instead of bischofite minerals. This is due to a sulfate deficiency in concentrated brines, which precipitates calcium, resulting in the presence of tachyhydrite instead of bischofite minerals (Szatmari et al., 1979; Wardlaw, 1972). In contrast, other basins, like the Zechstein basin in Europe, have an excess of sulfates in brine, leading to the deposition of highly soluble minerals containing magnesium or potassium, such as kainite and polyhalite, which precipitate with potassium and magnesium chlorides. The Boulby Mine in northeast England has an anomalous proportion of sulfate minerals with potassium, resulting in economic ores (Kemp et al., 2016).

Onshore basins in Brazil, such as the Amazon Basin and Sergipe Basin, have important levels of soluble salts such as carnallite, sylvite, and tachyhydrite (Szatmari et al., 1979). Only a few places in the world have a sedimentary evaporitic basin that has dried up to the point of depositing tachyhydrite, indicating a final stage of the evaporation cycle with extremely high brine concentrations. Sylvite and tachyhydrite minerals also occur in the deepwater region of the Santos Basin on the continental shelf (Mohriak et al., 2012).

There are only minor occurrences of sulfate minerals in the sylvinite salt mine of the Sergipe Basin. A possible reason for this is that the Mg^{2+} and SO_4^{2-} ions might have reacted with the carbonates previously deposited, forming anhydrite and dolomite, as shown by the following equation:

$$CaCO_3 + Mg^{2+} + SO_4^{2-} = CaSO_4 + CaMg(CO_3)_2$$

Another process that could be suggested is the desulfurization (sulfate reduction) of seawater by bacteria that attack hydrocarbons and produce acids, as shown by the following equation:

Bacteria + Petroleum + Sulfate ion =
Carbonic acid + Hydrogen sulfide

$$(C + H) + SO_4^{2-} = H_2CO_3 + H_2S$$

Hydrogen sulfide is a colorless, poisonous, corrosive, and flammable gas.

4.4. TEXTURE AND GRAIN SIZE OF SALT CRYSTALS

Evaporite minerals are formed in various environments and tectonic settings, with salt crystals being formed in three main locations: at the air–water interface, the water–sediment interface, and within sediments (e.g., Handford, 1991; Shearman, 1978; Sonnenfeld, 1989, 1992). The growth of salt crystals is characterized by an arrangement in which one side preferably grows in one direction, using as its base other crystals at the bottom of the lagoon. The competition for space at the bottom of the lagoon will generate new crystal faces, forming a morphology known as the skeletal type (Szatmari et al., 1974).

In general, the skeletal crystals are developed from the initial hopper crystals, which were carried by the wind as rafts or flakes, floating toward the central part of the lagoon. Eventually, these crystals sank and accumulated one above the other, growing with the characteristic cubic form of the halite mineral. The salt flakes can dissolve in the water column if the course to the lake bottom is long or the lake salinity is small. Ghosts of large skeletal crystals are commonly associated with the partial dissolution of halite, which depends on the distance covered by the saturated brines and also on the depth of the lagoon (Mohriak & Szatmari, 2008).

The growth morphology of halite crystals depends, in part, on the brine saturation level (Dellwig, 1955; Handford, 1991; Shearman, 1978; Szatmari et al., 1974, 1978). When generated under conditions of low brine saturation, the growth of NaCl crystals is uniform, forming perfect and transparent cubes without fluid inclusions. Under conditions of high brine saturation, the brines precipitate crystals of the hopper type, characterized by abundant fluid inclusions (Warren, 1989, 2006).

Evaporite minerals may grow at the sediment–water interface, as suggested for the gypsum deposits in the Carpathians foredeep, located between southern Poland, Ukraine, and Romania. During the Middle Miocene, the region was invaded by the Paratethys Ocean, forming platform carbonate-sulfate deposits in the margins and halite-rich deposits in basin depocenters, as in the Transylvania Basin in Romania (Kasprzyk, 2003). The sulfate facies are characterized by large selenite crystals arranged vertically and horizontally in the bedded evaporite sequences. Some clastic deposits, formed by selenite detritus, microcrystalline gypsum, and microbialite buildups, are also recognized in the Badenian evaporite sequences of the Transylvania Basin (Kasprzyk, 2003).

Salt crystals may grow within sediments, as observed in coastal and inland depressions known as sabkhas (also spelled sebkha by some authors), which are associated with extensive salt flats. The growth of salt crystals and evaporite nodules in these pore interstices tend to disrupt the texture and structure of the host sediment (carbonate mud). With time, the matrix becomes a secondary element, totally disrupted and restricted to the remaining spaces around the neo-formed bodies of carbonates, sulfates, and chlorides.

Evaporite minerals often form efflorescence crusts on the muddy surfaces of salty lakes, playas, and coastal sabkhas (also spelled sebkha by some authors), which are associated with extensive salt flats. The salt crystals are formed by chemical precipitation from the brine that occupies the pore spaces within sediments. The growth of salt crystals and evaporite nodules in these pore interstices tends to disrupt the texture and structure of the host sediment (carbonate mud). With time, the matrix becomes a secondary element, totally disrupted and restricted to the remaining spaces that are left around the neo-formed bodies of carbonates, sulfates, and chlorides (Smoot & Lowenstein, 1991). The mineral composition of the crusts reflects the composition of the ions dissolved in the interstitial pores of the host rock, which normally are brines enriched in sulfates and chlorides (Eugster & Hardie, 1978; Shearman, 1978; Smoot, 1991). These crusts are important components of the saline soils in the United Arab Emirates (UAE) coastal plains (e.g., Handford, 1991).

Efflorescence crusts result from the complete evaporation of the interstitial water that fills the pore spaces of superficial rocks or soil. The evaporation of the pore waters in muddy sediments locally generates contraction cracks and tepees (Warren, 1982a, 1982b, 1983a, 1983b). Tepees are sedimentary structures that form in carbonate and evaporite crusts by deformation and overthrust of the sedimentary layer, resulting in an antiformal shape that reminds of an Indian tent (Handford et al., 1984; Lokier & Steuber, 2009). The most common efflorescences are formed by halite and gypsum, but saline crusts and tepee structures may also be formed by carbonates (trona), borates, and nitrates (Warren, 2006, 2010, 2016). As a result of their high solubility, salt crusts are rarely preserved in the geological record and have received little attention in studies of current or ancient saline environments, except for a few places in Australia and along the Arabian Gulf. In South America, there are excellent examples of evaporite environments with the accumulation of salt crusts, particularly Salar de Uyuni in Bolivia and Salar de Atacama in Chile (see Chapter 10).

The size of the crystals of an evaporite mineral depends mainly on the rate of crystallization. Crystals that precipitate on the bottom of a lagoon or inside the pores of previously deposited sediments grow slowly and form large crystals, as in selenite. Crystals that precipitate rapidly when competing with several other salts with the same ions are characterized by very small crystals. When evaporite minerals grow within host sediment, the crystals deform the original rock matrix and change the shape of the mineral crystals that were previously deposited, resembling the porphyroblastic texture in igneous rocks. In contrast, the salt crystals that precipitate simultaneously throughout the brine volume form massive sedimentary layers, which are typically composed of sand-sized crystals, but some facies are characterized by microcrystalline gypsum (Babel, 2004, 2005). Erosion and transport of gypsum deposits may result in clastic sediments known as gypsarenites, as observed in the peripheral basins of the Mediterranean Sea (Barone et al., 2007; Orti et al., 2007; see Chapter 5 in Mohriak, 2025).

In the last phases of precipitation, the mineral tachyhydrite forms a massive rock layer. The syngenetic or epigenetic partial dissolution can transform the carnallite into sylvite, forming the sylvinite rock, which is characterized by an amoeboid intergrowth between sylvite and halite crystals. In the Sergipe Basin, the sylvinite layers are overlain by halite, which is partly fine-grained, with thicker crystals forming a coarser texture (5 mm on average). In other sylvinite deposits, as in the Amazon Basin, syngenetic dissolution causes the potassium salts to disappear completely, locally leaving only layers of residual non-soluble minerals, particularly hematite-rich muds (Szatmari et al., 1979).

4.5. EVAPORITIC CARBONATES AND PRE-SALT MICROBIALITES

Carbonate minerals are present in many evaporite basins, either preceding the main phase of salt deposition or forming lenses within the salt sequences, as in the Oman salt basins (Grotzinger & Al-Rawahi, 2014; Schröder et al., 2003). The Mediterranean peripheral basins often exhibit carbonate layers associated with the lower gypsum deposits. In many South Atlantic basins, carbonate rocks are associated with reefs and microbialite buildups, as in the Campos and Santos basins offshore Brazil, where these rocks constitute important reservoirs for giant hydrocarbon fields (Muniz & Bosence, 2015; Wright & Rodriguez, 2018).

Microbialite reservoirs are found in many ancient evaporite basins, such as the Arabian region, Permian, Paradox, Green River basins, Zechstein salt basin, and North Caspian Sea basins (e.g., Bosence et al., 2015; Seard et al., 2013; Słowakiewicz et al., 2013). Platform carbonates are often deposited on continental margins after the main phase of salt deposition, such as in the South Atlantic salt basins or the Gulf of Mexico (Alves et al., 2020). Carbonate minerals occur in sedimentary basins with various forms and cations, with the most common being Ca, Mg, and Na. The chemical composition and mode of occurrence of these minerals are shown in Table 4.1.

Evaporite deposits with a high proportion of sodium carbonates cannot be linked to the solar evaporation of marine brines due to the small proportion of carbonate and bicarbonate ions in normal seawater. Trona, a sodium carbonate-bicarbonate evaporite mineral, is found in nonmarine continental environments like the East African Rift System. Some lakes, like Magadi and Natron, are characterized by the deposition of sodium carbonate minerals associated with Pleistocene to Holocene sediments (Eugster, 1980; Eugster & Hardie, 1978). The occurrence of primary trona or its pseudomorphs in ancient evaporite sequences provides

Table 4.1 Carbonate Minerals

Mineral	Chemical composition	Occurrence
Calcite	$CaCO_3$	Limestone, soil, igneous rocks
Aragonite	$CaCO_3$	Biological deposits (shells, mollusks)
Dolomite	$CaMg(CO_3)_2$	Limestone, dolostone
Trona	$Na_3(HCO_3)(CO_3).2H_2O$	Nonmarine evaporite deposits

chemical evidence that nonmarine environments resulted in brine with an excess of the bicarbonate ion. Basin desiccation led to brine concentration and precipitation of trona and other carbonate rocks.

Some of the best-known trona occurrences in the world are found in continental lakes far from any marine influence, such as Owens Lake and Searles Lake in California (Pleistocene to Holocene evaporite deposits) and the Green River Formation of Wyoming and Utah (Eocene evaporite deposits). Several continental lacustrine environments record important evaporite deposits in the Neogene (see chapters 8 and 9).

Calcite is the most common carbonate mineral, found in sedimentary sequences, igneous rocks, and hydrothermal deposits, forming carbonate layers associated with a chemical precipitate known as travertine (Figure 4.15A). Carbonate minerals, especially calcite and aragonite, are also commonly found in biological deposits such as coquinas (fragments of mollusk shells) and reefs (e.g., Oliveira et al., 2019). Coquinas are frequently recorded as intercalations within lacustrine deposits and microbialite sequences (Figure 4.15B), whose origin may be hybrids of organic and inorganic precipitation (Riding, 2000).

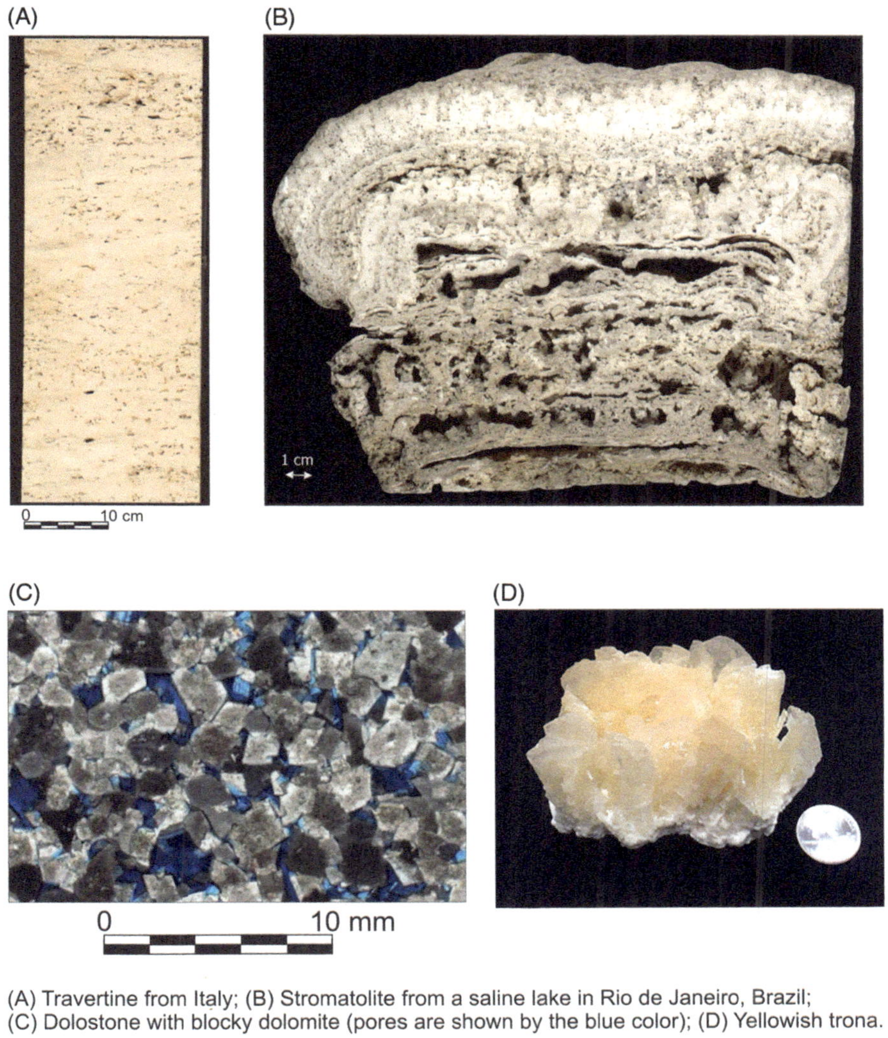

Figure 4.15 Photograph images of carbonate rocks. (A) Rock slab showing travertine from Italy, formed by $CaCO_3$ precipitated from hydrothermal springs. (B) Hand sample showing stromatolite from Lagoa Salgada, Rio de Janeiro, Brazil, formed by $CaCO_3$ derived from Quaternary organisms. Credit: Mohriak (2015). (C) Photomicrograph showing dolostones with blocky dolomite from the Early Cretaceous lacustrine sequences in the Campos Basin, offshore Brazil. (D) Hand sample showing yellowish trona from the Neogene continental lacustrine deposits in the southwestern United States. Credit: E65vern/Wikimedia Commons.

Microbialites are carbonate rocks that form algal mats or buildups in fresh to hypersaline lakes and bays. These rocks are characterized by a layered biochemical accretionary structure with sedimentary grains cemented by biofilms of microorganisms, particularly cyanobacteria. Microbialite deposits are widespread in the pre-salt sequences of several sedimentary basins since the Neoproterozoic. These sediments are genetically associated with carbonate precipitation from CO_2-rich waters sourced by hydrothermal vents and hot springs, forming travertine deposits in active volcanic regions (Bosence et al., 2015).

Microbialite carbonate buildups are found in several saline lakes worldwide, such as the Great Salt Lake in Utah, Shark Bay in West Australia, and Lagoa Salgada in southeast Brazil (Iespa et al., 2012). The study of microbialite carbonates has grown significantly in the 21st century due to the discovery of giant oil fields with highly productive reservoirs (Ahr et al., 2011; Mohriak, 2015).

Mohriak (2015) discussed some analogies between modern carbonate rock analogs and microbialites from oil fields in the South Atlantic and northern Caspian Sea. Pre-salt microbialites in the Campos and Santos basins show features resembling lacustrine stromatolites and travertines derived from calcite precipitation in hydrothermal springs. Two models for pre-salt microbialites in the Santos Basin are suggested: one assumes a deep lake with microbial buildups with high relief relative to the basin floor, and the other suggests a very shallow hyper-alkaline saline lake with predominantly abiotic carbonate and Mg-silicate precipitates (Wright & Barnett, 2015, 2017).

Carboniferous carbonate reservoirs within the Paradox Fm. are associated with an evaporitic stratigraphic sequence (e.g., Chidsey Jr., 2002). During the Pennsylvanian, the basin formed shallow-shelf carbonate buildups and mounds, which are productive reservoirs in the Aneth oil field, the largest hydrocarbon discovery in the Paradox Basin (Peterson, 1989). Several examples of carbonate reservoirs associated with evaporite sequences are registered worldwide, particularly in the region of the Middle East (Alsharhan, 2014; Grotzinger & Al-Rawahi, 2014).

The microbialite carbonate reservoirs associated with the Neoproterozoic-Cambrian evaporites of the Ara Group in Oman are encased in thick salt layers at depths of 3 to 7 km (Schröder et al., 2003). These carbonate rocks form excellent reservoir facies and are economically viable drilling targets for petroleum plays in the South Oman Salt Basin (Grotzinger & Al-Rawahi, 2014).

In the Levantine Basin, offshore the Arabian Peninsula, carbonate buildups, interpreted as reefs and microbialite buildups, are sealed by the Messinian salt and form major reservoirs for large gas accumulations in the eastern Mediterranean (Esestime et al., 2016). Dolomite sequences result from the dolomitization of basinal limestones in evaporite basins, with calcite and aragonite minerals often replaced by microcrystalline dolomite. Microbial dolomites are associated with Quaternary sediments in the Persian Gulf and recognized in the pre-salt sequences of the Campos and Santos basins in the South Atlantic (e.g., Lima & De Ros, 2019).

Trona, a nonmarine evaporite mineral, is the main primary source of sodium carbonate and is exploited in the southwestern United States, particularly in the Owens and Searles lakes in California and the Green River Fm. of Wyoming and Utah (e.g., Dyni, 1996; Smith & Stuiver, 1979). Trona occurs intercalated with other carbonate and evaporite rocks in several basins, forming layered carbonate-evaporite sequences below the surface. Trona deposits are also known in the East African Rift System, particularly at Lake Magadi in the Kenyan Rift Valley (Eugster, 1980).

4.6. SULFATE MINERALS IN EVAPORITE ROCKS

Sulfate minerals (Table 4.2) are found in most evaporite basins worldwide, often preceding the main phase of salt deposition. In some evaporite basins, where brines do not reach the chloride concentration, sulfates correspond to the main mineral phase (Sonnenfeld, 1992; Warren, 2010, 2016). This is the case of evaporite platforms in many giant salt basins worldwide, for example, in the Mesozoic evaporitic sequences of the Arabian Gulf (Khuff and Hith anhydrite intervals). Figure 4.16 shows examples of some common sulfate minerals found in nature. Figure 4.17 shows several sulfate minerals from

Table 4.2 Sulfate Minerals

Mineral	Chemical composition	Occurrence
Gypsum	$CaSO_4 \cdot 2H_2O$	Soils and gypsiferous layers in evaporite sequences
Anhydrite	$CaSO_4$	Evaporite sequences
Bassanite	$CaSO_4 \cdot 0.5H_2O$	Soils in arid regions with evaporite sequences
Kainite	$KMg(SO_4)Cl \cdot 2.75H_2O$	Granular masses in cavities and fissures
Mirabilite	$Na_2SO_4 \cdot 10H_2O$	Preferably nonmarine evaporite deposits
Thénardite	Na_2SO_4	Evaporite sequences
Epsomite	$MgSO_4.7H_2O$	Evaporite sequences
Blödite	$Na_2Mg(SO_4)_2.4H_2O$	Evaporite sequences
Kieserite	$MgSO_4.H_2O$	Evaporite sequences
Langbeinite	$K_2Mg_2(SO_4)_3$	Evaporite sequences
Polyhalite	$K_2Ca_2Mg(SO_4)_4.2H_2O$	Evaporite sequences
Barite	$BaSO_4$	Evaporite sequences

(A) Gypsum nodules; (B) Bassanite;
(C) Gypsum (rose color); (D) Acicular gypsum;
(E) Anhydrite (Keuper); (F) Anhydrite (bluish color);
(G) Gypsum (Zechstein, Poland);
(H) Gypsum after anhydrite (Keuper).

Figure 4.16 Photograph images of rocks with sulfate minerals (gypsum, bassanite, and anhydrite). Photographs (E) and (H) by C. Leitner.

diverse evaporite basins. Examples of common sulfate minerals include potassium sulfate (K_2SO_4) and potassium-magnesium sulfate ($K_2SO_4.MgSO_4$), also known as sulfate of potash and magnesia.

Gypsum, a soft sulfate mineral with a hardness of 2 on the Mohs scale, is a popular choice for sculpture in ancient cultures in Egypt, Mesopotamia, Greece, and the Roman Empire. Some translucid varieties of gypsum, known as selenite and alabaster, are characterized by large crystals (> 2 mm, with lengths reaching more than 1 m) growing directly on the basin floor (e.g., Babel, 2004, 2005; Ortí, 2011).

When gypsum is heated to higher temperatures, it loses water, forming the calcium sulfate hemihydrate mineral known as bassanite. Bassanite is petrographically characterized by microscopic acicular crystals in parallel aggregates. If heated further and compacted by sedimentary layers, it loses water molecules and turns into the anhydrous calcium sulfate mineral anhydrite. Gypsum, bassanite, and anhydrite are commonly found in the subsurface of many sedimentary basins worldwide. Primary precipitation of anhydrite is rare, and most occurrences are related to the dehydration of previously precipitated gypsum (Warren, 2010, 2016).

Kainite is a sulfate-chloride mineral with water molecules, characterized by a dull color and low hardness (2.5–3.0 on the Mohs scale). It occurs in nature as irregular, granular masses filling cavities and fissures. Kainite was the first evaporite mineral discovered in Germany's Stassfurt salt mines. Mirabilite is an evaporite mineral deposited from sodium sulfate-bearing brines in continental settings. It is often associated with other sulfate minerals like gypsum, glauberite, thénardite, epsomite, kieserite, and blödite (Table 4.2).

Mirabilite corresponds to a hydrated sodium sulfate composed of colorless grains and indistinct crystal grains. Mirabilite is unstable at normal temperatures and dehydrates quickly in dry air, turning into a white powder called thénardite (Na_2SO_4).

Blödite corresponds to a hydrated sulfate of sodium and magnesium (Figure 4.17). Alternating blödite-halite couplets in lakes of central Asia (as in the Qaidam Basin in China) represent seasonal intercalations of blödite, which is deposited in winter, well before halite saturation is reached in summer (Sonnenfeld & Perthuisot, 1989).

Epsomite is a hydrous magnesium sulfate mineral found in limestone caverns and often results from weathering processes. It is colorless to white with yellowish-to-greenish tints. Kieserite is the only magnesium sulfate mineral found in sedimentary basins and is commonly found in marine evaporites. Langbeinite is a potassium-magnesium sulfate mineral found in marine evaporite deposits and is often associated with carnallite, halite, and sylvite.

Polyhalite is a hydrated sulfate evaporite mineral with potassium, calcium, and magnesium as cations. Barite is a heavy nonmetallic sulfate mineral with a density of 4.5 g/cm^3 and is formed through various processes, including biogenic and hydrothermal processes. It is often found in lead-zinc veins in limestones and hot spring deposits and is used in the petroleum industry as a weighting agent for mud used in drilling for oil and gas wells.

4.7. HALIDE (CHLORIDE) MINERALS IN EVAPORITE ROCKS

Halides are minerals containing anions Cl^-, Br^-, F^-, or I^-, with chloride being the most important class (Table 4.3). Chloride minerals are highly soluble in water

(A) Polyhalite (Boulby mine, England); (B) Barite;
(C) Mirabillite(cream color); (D) Thernadite (beige color);
(E) Bloedite (Haselgebirge);
(F) Selenite (Sorbas basin);
(G) Kainite (Zechstein); (H) Kieserite.

Figure 4.17 Photograph images of rocks with sulfate minerals. Credits: Photographs (A), (E), (F), and (G) by C. Leitner and (D) Rob Lavinsky/Wikimedia Commons/CC BY-SA 3.0.

(A) Whitish to gray halite; (B) Transparent halite;
(C) Pink halite; (D) Purple halite.

(E) Gray halite (Zechstein); (F) Blue halite (Zechstein);
(G) Gray halite (Hormuz); (H) Red and gray halite (Haselgebirge).

Figure 4.18 Photograph images of rocks with chlorides (mainly halite). Top: Examples of halite samples with variable colors; Bottom: Examples of halite samples from several salt basins. Photographs (E), (F), (G), and (H) by C. Leitner.

Table 4.3 Chloride Minerals

Mineral	Chemical composition	Occurrence
Halite	NaCl	Evaporite sequences
Sylvite	KCl	Evaporite sequences
Carnallite	$KMgCl_3.6H_2O$	Evaporite sequences
Tachyhydrite	$CaMg_2Cl_6.12H_2O$	Evaporite sequences
Bischofite	$MgCl_2.6H_2O$	Evaporite sequences
Kainite	$KMg(SO_4)Cl.3H_2O$	Evaporite sequences

and can be easily leached from soils or near-surface layers. They are derived from marine brines or precipitated from brines in continental settings. Halite (NaCl), the most common chloride mineral, forms isometric crystals (Figure 4.18) and is found in many sedimentary basins worldwide. It is typically colorless or white but may show other colors depending on the quantity of impurities present. Extremely thick halite layers are registered in giant salt basins, such as the Zechstein basin in Europe, the Gulf of Mexico, the Atlantic margins, the Appalachian basin of western New York, the Michigan Basin (e.g., Dellwig, 1955; Dellwig & Evans, 1969), and several basins in the Middle East, Russia, and China (Warren, 2006, 2010, 2016).

According to the chemical principles that control evaporite precipitation, potassium- and magnesium-rich chlorides are deposited after halite. A few examples of these soluble minerals are shown in Figure 4.19. Sylvite, a potassium chloride (KCl), forms crystals similar to

(A) Yellowish carnallite; (B) Red sylvite with white halite;
(C) Pinkish carnallite; (D) Cream to beige sylvite.

Figure 4.19 Photograph images of rocks with chlorides forming soluble salts (mainly carnallite and sylvite). Credits: (A) Thomas Witzke and Abraxas Verlag, (B) maksim_e/Adobe Stock Photos. Photographs (C) and (D) by C. Leitner.

halite but may be colored with shades of yellow and red due to impurities and inclusions. It is one of the last evaporite minerals to precipitate from marine brines and is registered only in desiccated basins. Sylvite layers are exploited in many basins worldwide, including salt provinces in North America, South America, Europe, Russia, and China (Yager, 2016).

Carnallite, a highly soluble evaporite mineral, is a hydrated double chloride of potassium and magnesium. It is deliquescent and dissolves naturally at room temperature. Carnallite usually exhibits a reddish color with variable shades of yellow to white. Carnallite is found in highly concentrated marine brines and is recorded in Permian deposits in the northwest USA, the Perm Basin of Russia, and Europe (Zechstein salt). It also occurs in Devonian deposits of the Williston Basin in Saskatchewan and the Late Aptian salt basins of the South Atlantic.

Tachyhydrite is an unstable, highly hydrated calcium- and magnesium-chloride mineral found in desiccating basins. It hydrates quickly at room temperature and flows rapidly under pressure gradients, making it a drilling risk for petroleum exploration boreholes (Poiate Jr. et al., 2006). Bischofite is a hydrated magnesium chloride mineral that is extremely soluble in water and hygroscopic, dissolving with air moisture at room temperature. It is not found in natural salt sequences, but near-pure bischofite is observed in artificial salinas when seawater evaporates and reaches a density higher than 1.35 g/cm^3 (Sonnenfeld & Perthuisot, 1989). It has also been observed in the Dallol region of northern Ethiopia, precipitating as a greenish material when $MgCl_2$-rich hot spring brines cool below 115 °C. Kainite is a hydrated potassium-magnesium sulfate-chloride that forms irregular granular masses of crystalline coatings in cavities or fissures. Kainite is often accompanied by other sulfate and chloride minerals like anhydrite, carnallite, halite, and kieserite.

4.8. BORATES AND NITRATES

Borate and nitrate minerals are typically found in arid continental environments such as the salt-lake deposits in the deserts of the southwestern United States, mainly in California and Nevada, or in the highlands of the South American Andean ranges (Okrusch & Frimmel, 2020). Some examples of borate and nitrate minerals are shown in Figure 4.20.

The most important deposits worldwide are found in Turkey, the United States (Boron and Searles Lake in California), and the desert regions of Salar de Atacama in Chile. Borax, a hydrated sodium borate, is used in the chemical industry (Figure 4.20A).

Nitrate evaporite minerals are highly soluble and found in extremely arid desertic regions with annual rainfall below 10 mm, i.e., in areas sheltered from the rain by mountains that create a rain shadow, such as in the Andean highlands in South America. These minerals are economically important in Peru and Chile for fertilizer

(A) Borax; (B) Nitratine; (C) Nitramagnesite.

Figure 4.20 Photograph images of rocks with borate and nitrate minerals in hand specimens. Credits: (A) Aram Dulyan/Wikimedia Commons/Public domain, (B) John Sobolewski/Wikimedia Commons/CC BY-SA 3.0 and (C) JGW/mindat.org.

production in the agricultural industry. Niter (also spelled nitre in the UK) corresponds to potassium nitrate, which is also known as saltpeter (spelled saltpetre in the UK). The mineral nitratine (or nitratite) is sodium nitrate, also known as soda niter or Chile saltpeter. The evaporite rock corresponds to a mixture of potassium and sodium nitrates ($NaNO_3$ and KNO_3). It is formed as a residual water-soluble surface deposit, or caliche, mixed with clay in extremely arid environments.

Nitrocalcite is a hydrated calcium nitrate mineral that forms an efflorescence when manure contacts limestone rocks in dry desertic environments. Magnesium nitrate is a hydrated inorganic nitrate compound with magnesium as the cation, highly soluble in water, mainly occurring in mines and caverns. This rather uncommon salt is highly soluble in water, so it mainly occurs in mines and caverns, forming colorless to white transparent low-density (specific gravity 1.46 g/cm^3) crystals with a vitreous luster.

4.9. THE SULFATE PROBLEM IN MARINE EVAPORITES

The chemical analysis of evaporite minerals has been a topic of interest since early experiments involving salt deposition (Braitsch, 1971; Richter-Bernburg, 1972a, 1972b). In 1849, Italian chemist Giulio Usiglio conducted classical evaporation experiments using seawater from the French Riviera along the northern Mediterranean Sea (Fischer & Garrison, 2009; Usiglio, 1849a, 1849b). However, the laboratory results do not agree with the natural evaporite sequences observed in most saline giants, such as the Zechstein salt basin (Sonnenfeld & Hardie, 1985). The 19th-century chemical experiments were conducted by evaporating the Mediterranean Sea brine in a laboratory room that was well-oxygenated and practically sterile. In this environment, both sodium chloride and magnesium sulfate precipitated simultaneously upon halite saturation. This contrasts with the real situation in natural evaporite sequences, where magnesium sulfates and magnesium chlorides are not coprecipitates with halite but form layers in different stages of the evaporite cycle.

The origin of evaporites containing highly soluble salts (such as carnallite, sylvite, tachyhydrite, or bischofite) has been discussed in geological and chemical literature dealing with evaporite sedimentation (e.g., Ayora et al., 2001). Many of these evaporites lack Mg-sulfate-bearing minerals such as polyhalite, kainite, and kieserite, which theoretically should be formed during the precipitation of soluble chlorides. This indicates a marked deficiency in magnesium sulfates and chlorides, which theoretically should be present in proportions according to their abundance in seawater.

Several researchers have recognized the discrepancies between the mineral phases observed in nature and the experimental precipitation sequence determined in the laboratory using modern seawater (e.g., Hardie, 1984, 1990, 1991; Harvie & Weare, 1980; Spencer et al., 1990; Warren, 2016). Some authors have classified many of these deposits as "$MgSO_4$-deficient" and named some playa-lake deposits as "abnormal" because they do not follow the mineral crystallization sequence expected for seawater brines (e.g., Braitsch, 1971; Valyashko, 1972a, 1972b).

In nature, evaporite sequences containing halite are largely formed by pure sodium chloride. These sequences normally do not coexist with magnesium sulfate layers or form mixed deposits (Figure 4.13). This indicates a marked deficiency in magnesium sulfates and chlorides, which theoretically should be present in proportions according to their abundance in seawater (Sonnenfeld, 1979).

The deficiency of potassium and magnesium salts in natural evaporite sequences might be explained by an outflow of the enriched brines after the start of the precipitation process (e.g., Sonnenfeld, 1979). The outflow of denser brines with high-solubility minerals would deplete the remaining brine of potassium and magnesium ions. Chlorides and sulfates of potassium and magnesium would only form when outflow from the basin stopped, resulting in the precipitation of highly soluble minerals during the final stages of basin desiccation (Braitsch, 1971; Sonnenfeld, 1979).

Several authors (e.g., Garrett, 1996; Warren, 2016) suggest that the deficiency in magnesium sulfate in natural evaporite sequences occurs due to the reduction of sulfur by the work of anaerobic bacteria, particularly the *Desulfovibrio* and *Desulfotomaculum* bacteria that occur in anoxic environments. These bacteria deplete the water in oxygen and sulfur by using these elements in their metabolism. In the Dead Sea, for example, sulfates in the form of gypsum are formed at the surface of the lake, but these minerals do not reach the substrate because of the sulfur reduction by bacteria. The gypsum crystals ($CaSO_4.2\ H_2O$) that are formed near the lake's surface are transformed into a $CaCO_3$ rain of fine-grained crystals that precipitate at the bottom of the lake, while the sulfur returns to the surface as H_2S to restart the process (e.g., Warren, 1986, 2016).

This fact is well known by people working in solar saltworks. The workers involved in assessing the quality of the precipitated salts are aware that sulfur consumption by bacterial activity is an important factor for salt production as it results in negative redox potential. Many salt industry companies have adopted, for several centuries, a workflow that results in the production of pure white halite crystals, which are more valuable than dirty brown salt mixed with gypsum. The sulfur reduction

Table 4.4 Borate and Nitrate Minerals

Mineral	Composition	Occurrence
Borax	$Na_2B_4O_7 \cdot 10H_2O$	Continental evaporites
Niter	KNO_3	Continental evaporites
Nitratine	$NaNO_3$	Continental evaporites
Nitrocalcite	$Ca(NO_3)_2.4H_2O$	Continental evaporites
Nitromagnesite	$Mg(NO_3)_2.6H_2O$	Continental evaporites

processes involve plowing up black mud from the bottom of the salinas and mixing the bacteria that live in this anoxic environment with the water column. These bacteria feed on the sulfates and favor the deposition of homogeneous white crystals of halite.

The deficiency in sulfate evaporites from the conjugate margins of Brazil and West Africa has called the attention of many researchers for several decades (e.g., Babel & Schreiber, 2014; Braitsch, 1971; Wardlaw, 1972). Some authors have questioned the commonly held assumption that the highly soluble evaporites observed in the Sergipe Basin and the Gabon Basin are derived from the evaporation of normal marine seawater (e.g., Hardy, 1991). These potash-rich Late Aptian evaporites lack significant amounts of calcium and magnesium sulfate minerals, such as polyhalite or kieserite. There is a predominance of the highly unusual tachyhydrite mineral, a calcium chloride salt that precipitates during advanced stages of evaporation (Table 4.4). This indicates that these evaporites were formed from a parent brine with a composition radically different from present-day seawater or might even correspond to nonmarine evaporites (Brice et al., 1982; Evans, 1978; Hardie, 1990; Wardlaw, 1972). The South Atlantic Mg-rich evaporite deposits might be associated with the upwelling of hydrothermal brines that fed the basin during the inception of the oceanic crust. According to some authors, the volcanic basement was exposed and weathered during the final ingression of normal seawater that flooded the marginal basins in the Aptian and Albian (Hardie, 1991; Jackson et al., 2000).

4.10. RARE MINERALS IN EVAPORITE SEQUENCES

Nonmarine evaporite minerals are found in continental settings like playa lakes, which are derived from various sources such as lakes, rivers, meteoric groundwaters, and hydrothermal brines (see, e.g., Warren, 2006, 2016). These minerals are typically different from those found in marine environments, but the distinction between marine and nonmarine evaporites is complex (e.g., Hardie, 1984, 1991).

Continental evaporite deposits are sediments associated with lacustrine environments. They are mostly made up of sodium and magnesium carbonates and sulfate minerals, with a subordinate proportion of halite and other chloride minerals. Some environments are hybrids, involving a mixture of waters from marine and continental settings (Figure 4.5). Common minerals formed from brines with a continental source include blödite, borax, epsomite, gaylussite, glauberite, mirabilite, thénardite, and trona.

Nonmarine evaporite deposits often contain halite, gypsum, and anhydrite minerals, which can be derived from previously existing marine salt deposits that have been dissolved from outcropping salt diapirs and transported to continental depositional settings. For example, in the Zagros region of Iran, halite layers accumulated in Tertiary and Quaternary salt pans are products of the evaporation of brines that received ions derived from chemical weathering, leaching, dissolution, and local redeposition of Hormuz salt diapirs (see Chapter 3).

A large number of rare minerals can be found locally in nonmarine settings, such as hydrous borates in the desert playas of the southwestern United States. These soft minerals exhibit whitish colors and have been exploited since the late 19th century in California and Nevada. Borates were formed by precipitation in shallow playa lakes in the Holocene. In the Holocene, boron-bearing minerals from volcanic rocks were either leached into ancient lakes through hot springs connected to Late Tertiary to Quaternary volcanism or formed as a result of precipitation in shallow playa lakes formed along the Death Valley (Schaller, 1929). Borates are recorded in the Furnace Creek Fm. of Pleistocene age (6–4 Ma BP), associated with mudstones and limestones deposited in deep perennial lakes (Tanner, 2002). The borate ions entered these ancient lakes through hot springs and by leaching boron-bearing minerals from volcanic rocks, such as altered rhyolites. The Kramer District in California holds a large amount of borate minerals associated with Miocene laminated claystone. These minerals were probably formed by thermal springs in a closed, fault-bounded basin that was active during the Miocene (Christ & Garrels, 1959).

4.11. TRACE ELEMENTS IN EVAPORITE MINERALS

Over 100 mineral species have been recorded in marine and nonmarine evaporite successions, primarily related to marine brines that formed chlorides, sulfates, carbonates, and borates (Figure 4.10). However, several minor chemical constituents of seawater also occur in evaporite brines and minerals, including important trace elements such as strontium, barium, and fluorine (Figure 4.21). These rare elements are associated with evaporite minerals deposited during the early phases of salt precipitation as iron compounds with low solubility.

Element	Weight percent of dissolved constituents in sea water	Common minerals
Sr	0.03	Celestine, strontianite
B	0.01	Many borates
Si	0.008	Clay minerals, opal, quartz
F	0.004	Fluorite
Al	0.0004	Clay minerals, zirklerite
Ba	0.0002	Barite
P	0.0002	Lüneburgite, goyazite
Fe	0.00001	Ankerite, hematite, limonite, magnetite, pyrite, rinneite, siderite
Mn	0.000009	Zirklerite

Figure 4.21 Minor elements in seawater that form chemical compounds in marine evaporite minerals. Credit: Adapted from Stewart (1963).

Some chemical elements, such as bromine and rubidium, are mainly concentrated in later residual brine products and have been the focus of numerous studies (e.g., Stewart, 1963).

Normal seawater has a mean content of bromine of around 60–70ppm. Bromine is considered an important component of marine-derived brines. For example, the basal halite in the Zechstein Z2 evaporite cycle shows a bromine content of about 65 ppm (Kuster et al., 2009). However, some continental hypersaline lakes, such as the Dead Sea, exhibit extremely high concentrations of bromine when compared to seawater (see Chapter 2). The bromine content in the Dead Sea is more than 500 times that of the global seawater, reaching a concentration of 5.35 g/l (Figure 2.7). Only a few lakes on Earth exhibit calcium chloride brines, contrasting with the evaporation of normal seawater (Krumgalz et al., 2002). The Dead Sea is a unique example of a deep lake with an unusual chemical composition that is abnormally high in bromine and calcium chloride.

Seawater shows a bromine concentration of 0.671 g/1,000 g (equivalent to 67 ppm Br), so it is the fourth most common anion after Cl^-, SO_4^{2-}, and HCO_3^{1-} (Figure 4.5). With an ionic radius of 1.96 A, the bromine ion easily substitutes for chlorine (with a radius of 1.81 A) in the halite crystal lattice as well as in the other chloride salts. Bromine-bearing brines are normally associated with marine saline deposits (Hardie, 1984; Warren, 2006, 2016).

When the brine is concentrated to about nine times the original salinity, halite crystals precipitate, and the brine's bromine content is enhanced, reaching about 250–500 ppm, but bromides are so soluble that they do not form saline crystals. After halite minerals are precipitated, the remaining brine, which is called bittern, contains about 2,700 ppm of dissolved bromine. If the evaporation process continues, the bittern is concentrated to a point where magnesium and potassium minerals can precipitate. The remaining brine may contain up to 6,000 ppm of bromine. This concentration is high enough to have commercial value for the chemical industry (Krumgalz et al., 2002; Warren, 2016).

Israel and Jordan are the world's largest bromine producers, artificially concentrating brines in saltworks in the southern Dead Sea. The U.S., responsible for 30% of the world's bromine consumption (U.S. Geological Survey, 2020), produces it from extraction plants located in a few places across the country, particularly in the Smackover Trend in southern Arkansas, where the metal is recovered from formation waters that fill the pores of carbonate rocks. The origin of these enriched brines is probably associated with the dissolution of the Jurassic Louann salt that underlies the Smackover carbonates (Hanor & Mcintosh, 2007).

The element lithium is also concentrated during the evaporation process in many arid regions, principally in connate waters and clays associated with clastic sediments formed during the last phases of basin desiccation (Bradley et al., 2013; Robichaud & Robichaud, 2019). Lithium is mainly produced by recovering the metal from enriched brines. Some salares (also spelled salars) in the Andean mountainous regions are characterized by an extremely high concentration of lithium compounds in subsurface brines (Risacher et al., 2003). These occurrences will result in an important economic asset for the 21st-century salt exploitation industry (see Chapter 9 in Mohriak, 2025).

4.12. ISOTOPIC GEOCHEMISTRY OF EVAPORITE MINERALS

Isotope geochemistry is a field of chemistry that examines the variations in the abundances of isotopes of different chemical elements in evaporite minerals. Stable elements have varying numbers of neutrons in the nucleus, resulting in different atomic masses. For instance, oxygen with an atomic number of 8 may have 8, 9, or 10 neutrons, resulting in atomic masses of 16, 17, and 18.

Chemistry and physics laboratories use a piece of equipment known as a ratio mass spectrometer. Mass spectrometers can calculate the mass ratios of the different oxygen isotopes, which are designated as ^{16}O, ^{17}O, and ^{18}O. Oxygen isotope ratios are measured relative to Vienna Standard Mean Ocean Water (VSMOW) or Vienna Pee Dee Belemnite (VPDB), which provide standard values for the quantity and mass of isotopes that are present in the samples. The variation between the measured ratios in a rock sample and the ratios that are known in the standard samples can provide important clues on the age and origin of the rock sample.

The element sulfur (atomic number 16) has four stable isotopes, but the isotope with the number of neutrons equal to the number of protons is by far more common in nature than the other isotopes. According to Nielsen (1979), sulfur isotopes occur in the following abundances in terrestrial materials: ^{32}S (95%), ^{33}S (0.77%), ^{34}S (4.2%), and ^{36}S (0.017%). Each geologic sample of homogeneous isotopic composition is characterized by a unique ratio value. According to most Earth science researchers, it is convenient to measure the ratio between the two most abundant isotopes, i.e.,

$$R = {}^{34}S/{}^{32}S$$

These abundances are compared to those found in the mineral troilite, an iron sulfide that was obtained from a metallic meteorite that was found in the Canyon Diablo. These ratios, which are very similar to the ratios in native sulfur-bearing minerals on Earth, have been chosen as the international isotope ratio standard for sulfur. Variations in sulfur isotope ratios from a given sample, as compared to the standard values, are used to study the origin of the mineral, including its temperature of formation. Several other chemical elements in the periodic table can also provide isotope ratios that are useful in the study of evaporites, for example, the strontium isotope ratio.

The study of the isotope composition of sulfate minerals in evaporite sequences is commonly employed by researchers aiming to characterize the most relevant geochemical parameters that might be related to the genetic signatures. Sulfate minerals, such as anhydrite, are characterized by the elements sulfur and oxygen, and sulfur is often replaced in the crystal lattice by the element strontium. The difference between the isotope ratio in a sample from the isotope ratio in the standard material is expressed by the Greek letter d (delta). These differences are, in general, very small, so they are reported in parts per thousand ($\%_0$), not in percentage.

The difference in sulfate isotope compositions of sulfur and oxygen ($\delta^{34}S$ and $\delta^{18}O$) and the strontium isotope ratios ($^{87}Sr/^{86}Sr$) are commonly used as geochemical markers to determine the marine or nonmarine origin of the parent brine and to assess the contribution of different water types that sourced an evaporite basin. A comparison of the sulfur and oxygen isotopes in sulfates, particularly the ones derived from seawater, can be compared with the established age curves of these isotopes, which are available in published reports (see, for example, Bernasconi et al., 2017).

The determination of isotope ratios provides a useful geochemical tool to constrain the stratigraphic age of the studied evaporites. This procedure can also be done with the age curve of marine strontium isotope ratios, which may also provide clues on the diagenetic changes in the evaporites since they were originally deposited. When sulfate-bearing minerals precipitate in an evaporitic basin, the $\delta^{34}S_{sulfate}$ and $\delta^{18}O_{sulfate}$ values can be almost coincident with those of the parent brine. Similarly, the $^{87}Sr/^{86}Sr$ ratio of the strontium incorporated into the sulfate minerals (Sr usually replacing Ca and also precipitating as inclusions of the mineral celestine) reflects those of the dissolved Sr^{2+} in the original mother brine.

For example, Orti et al. (2014) calculated the sulfate isotope compositions ($\delta^{34}S$ and $\delta^{18}O$) and strontium isotope ratios ($^{87}Sr/^{86}Sr$) of Triassic evaporites (gypsum and anhydrite samples, with minor amounts of celestine) in the Betic Cordillera in Spain (Figure 4.22). These evaporite rocks occur intercalated with carbonate and siliciclastic rocks that range in age from Middle Triassic (Anisian) to Early Jurassic (Hettangian). These rocks appear as intrusion-like lenses within the compressional structures formed during the Alpine Orogeny. Orti et al. (2014) obtained isotope values for the sulfates ranging from 0.707815 to 0.708114 (Figure 4.22).

Sulfate isotopes	Variation (‰)	Average value (‰)
$\delta^{34}S$	12.5 – 16.6	14.9 ± 1.0
$\delta^{18}O$	8.9 – 16.9	12.6 ± 2.1

Isotope ratios	Variation (‰)
$^{87}Sr/^{86}Sr$	0.707615 – 0.708114

Figure 4.22 Sulfate isotopes determinations for the Triassic evaporites outcropping in the Betic Cordillera in Spain. Credit: Adapted from Orti et al. (2014).

Orti et al. (2014) concluded that the obtained values, as a whole, are in agreement with those of worldwide Triassic marine evaporites, which are characterized by a higher ratio of values ($^{87}Sr/^{86}Sr$ higher than 0.7076, whereas Permian evaporites show the same ratio varying from 0.7070 to 0.7077).

The Offshore Drilling Program (ODP) wells in the Tyrrhenian Sea have provided valuable insights into the evolution of salt basins (e.g., Muller et al., 1990). Samples of Messinian evaporites, deposited in the Upper Miocene, were found to be mainly gypsum layers intercalated with calcareous mudstones. The strontium isotopic ratios of these sulfates were lower than expected, suggesting a greater influence of fresh water on the Sr isotopic composition of the desiccating Tyrrhenian Sea during the Messinian. This could be due to the diagenetic alteration of the evaporites or the freshwater input associated with rivers entering the desiccating Mediterranean Sea during the Messinian salinity crisis (Muller et al., 1990).

Müller and Mueller (1991) proposed that the Sr isotope ratios allowed for a division of the Messinian evaporites into two sequences: one with constant marine inflow during the early stages of salt formation and another with isolated desiccation events with a strong but variable continental influence. Strontium isotopic measurements from oyster fragments and foraminifers from pre-evaporitic and post-evaporitic sedimentary layers were used to constrain the age of the Messinian salinity crisis (see Chapter 5 in Mohriak, 2025).

Palmer and Edmond (1989) measured the strontium concentration and isotopic composition of most of the world's major rivers, which were then compared with the Sr concentration and $^{87}Sr/^{86}Sr$ ratio of oceanic hydrothermal vent waters from diverse tectonic and volcanogenic environments. These determinations indicated the best estimate for the flux of seawater through the high-temperature mid-ocean ridge crests, which vary through time depending on whether landmasses are amalgamated or split by oceanic ridges. These variations influenced the type and quantity of major chemical elements in the seawater, affecting the composition of sediments and chemical precipitates in sedimentary basins. For example, during the Permian, Pangea constituted a single continental mass, but during times of continental breakup, as in the Mesozoic, the oceanic ridges formed hydrothermal systems that extended from the North Atlantic to the South Atlantic (see Chapter 9 in Mohriak, 2025). The breakup event affected not only the sea level but also the major and minor element balances in the oceans. These variations influenced the type and quantity of major chemical elements in the seawater, affecting the composition of sediments and chemical precipitates in the sedimentary basins, as discussed by Edmond et al. (1979) and Warren (2010).

Several researchers have suggested that the concentration of magnesium in seawater has varied through time as the ion is removed from seawater during hydrothermal circulation through basaltic mid-ocean ridges. For example, Babel and Schreiber (2014) postulated that the Mg^{2+}/Ca^{2+} molar ratio has drastically varied in the Phanerozoic, from a minimum of about 1 in the Early Cretaceous (continental breakup and formation of the South Atlantic) to the maximum value of ~5 for the average seawater in today's oceans.

The chemical element bromine often substitutes the chlorine anion in the halite lattice because these ions share a great similarity in size and valence. Several studies have used the bromine isotope ratio aiming to distinguish marine from nonmarine salts (see Hardie, 1984). The basis for the assumption is that continental evaporites have a lower bromine content than halite deposited in marine environments. However, some authors have questioned the validity of the bromine isotope ratio as a reliable indicator of brine origin (Hardie, 1984; Warren, 2016).

The ratio of sulfur and boron isotopes is considered by some authors a reliable indicator of sedimentary environments, particularly to decipher the origin of the brines (Warren, 2006, 2010, 2016). Boron (atomic number 5) naturally occurs with the isotopes ^{10}B and ^{11}B, the latter of which makes up about 80% of natural boron in some minerals. The difference between the isotope ratio in boron-bearing minerals in a sedimentary rock sample and the isotope ratio in the standard material used as a reference is known as $d^{11}B$ (Figure 4.23).

The d^{11} B values in marine evaporites are considerably higher than in continental salts and brines (Figure 4.23). Many researchers have suggested that the difference in isotopic ratios might help to distinguish between marine and nonmarine evaporites in sedimentary basins (e.g., Swihart et al., 1986). Figure 4.23 shows that borates formed by the evaporation of marine brines yielded a mean $\delta^{11}B$ value of about 25 ± 4 $\%_{o,}$ whereas nonmarine evaporite borates yielded values of only 7 ±10 $\%_{o.}$

Boron isotope	Average value (‰) Marine borates	Average value (‰) Non marine borates
$\delta^{11}B$	25 ± 4	7 ± 10

Isotope ratio	Marine borates	Non marine borates
$^{11}B/^{10}B$	4.15 ± 0.02	4.02 ± 0.04

Figure 4.23 Boron isotopes determinations for borates deposited in marine and nonmarine settings. Credit: Adapted from Swihart et al. (1986).

4.13. IONIC PROPERTIES OF EVAPORITE MINERALS AND THE IMPLICATION FOR SALT TECTONICS

The low resistance of evaporite minerals to weathering and erosion is linked to their ionic properties. This is why halite dissolves more easily than anhydrite (Haynes, 2017; Figure 4.7). Compounds with a large ionic radius and small valence are easily removed from evaporite sequences by chemical weathering and transported toward the sea when dissolved in water. These properties also influence the rheological behavior during extension and compression.

The most common ions in evaporites (chlorine, sodium, and potassium) have low ionic potential, depending on their valence and ionic radius (Figure 4.3). Carbonate minerals, on the other hand, are characterized by more complex crystal lattices, with the valence of Ca and Mg being 2+, twice the sodium and potassium ions (1+). This results in higher ionic potentials for carbonate anion bonds with alkaline earth elements, making solid flow phenomena in evaporite rocks more difficult to observe.

Another significant difference in rheological behavior is observed between calcareous and dolomitic sequences. In carbonate rocks, all cations of calcite ($CaCO_3$) are the same, facilitating reticulate reorganization during deformation. In dolomitic sequences, the mineral dolomite contains two different cations, resulting in a crystal lattice structure that is not preserved when subjected to tectonic forces. These differences are particularly evident at the surface of outcrops involving deformed dolomites, which show a brittle disintegration, resulting in many polyhedrons, 1 to 2 centimeters in size, forming a texture that reflects the difficulty of deformation by folding, as evidenced by laboratory experiments (e.g., Kushnir et al., 2015; Mead, 1940) and observations in outcrops (see Chapter 6).

Dolomitization of carbonate rocks in the Zagros thrust-fold belt in Iran results from diagenesis in hypersaline conditions. Fracturing and diagenetic processes resulted in porous reservoirs that are suitable for oil accumulation (Esrafili-Dizaji & Rahimpour-Bonab, 2019; Swart et al., 2005; Warren, 2000). Fine-grained limestones, such as calcilutites, do not show the same fracture intensity; these rocks are more easily deformed into folds. Several methods have been used to assess the influence of tectonics in the petroleum geology of the Zagros oil fields. Deformed and elongated halite crystals can be measured using methods of structural geology, including foliation, lineation, fold axis, and veins (e.g., Leitner & Neubauer, 2011). The microscale analysis of deformed evaporites will be further discussed in Chapter 6, which focuses on the rheology of evaporite minerals.

REFERENCES

Ahr, W. M., Mancini, E. A., & Parcell, W. C. (2011). Pore characteristics in microbial carbonate reservoirs. *AAPG Search and Discovery*, Article #30167 (2011).

Alsharhan, A. S. (2014). Petroleum systems in the Middle East. In H. R. Rollinson, M. P. Searle, I. A. Abbasi, A. Al-Lazki, & M. H. Al Kindi (Eds.), *Tectonic evolution of the Oman mountains* (Vol. 392, pp. 361–408). Geological Society, London, Special Publications. http://dx.doi.org/10.1144/SP392.19

Alves, T., Fetter, M., Busby, C., Gontijo, R., Cunha, T. A., & Mattos, N. H. (2020). A tectono-stratigraphic review of continental breakup on intraplate continental margins and its impact on resultant hydrocarbon systems. *Marine and Petroleum Geology*, *117*, 104341, 1–39. https://doi.org/10.1016/j.marpetgeo.2020.104341

Arrhenius, S., & Lachmann, R. (2002). The physical-chemical conditions relating to the formation of salt deposits and their application to geologic problems. *International Journal of Earth Sciences*, s62–s74; *International Journal of Earth Sciences (Geol Rundsch)*, *91*, S62–S74 English translation from *Geol Rundsch*, *3*, 139–157. doi 10.1007/s00531-002-0270-2

Ayora, C., Cendón, D. I., Taberner, C., & Pueyo, J. J. (2001). Brine-mineral reactions in evaporite basins: Implications for the composition of ancient oceans. *Geology*, *29*(3), 251–254.

Baar, C. A. (1977). *Applied salt-rock mechanics. 1, The in-situ behavior of salt rocks. Developments in Geotechnical Engineering, 16A. Amsterdam: Elsevier*, 294 pp.

Babel, M. (2004). Models for evaporite, selenite and gypsum microbialite deposition in ancient saline basins. *Acta Geologica Polonica*, *54*(2), 219–249.

Babel, M. (2005). Selenite-gypsum microbialite facies and sedimentary evolution of the Badenian evaporite basin of the northern Carpathian Foredeep. *Acta Geologica Polonica*, *55*(2), 187–210.

Babel, M., & Schreiber, B. C. (2014). Geochemistry of evaporites and evolution of seawater. *Treatise on Geochemistry* (2nd ed., Chapter 9.17, pp. 483–560). Amsterdam: Elsevier.

Barone, M., Rocco, D., & Lugli, S. (2007). Interpreting Gypsarenites in the Rossano basin (Calabria, Italy): A contribution to the characterization of the Messinian salinity crisis in the Mediterranean. In J. Arribas, M. J. Johnsson, & S. Critelli (Eds.), *Sedimentary Provenance and Petrogenesis: Perspectives from Petrography and Geochemistry* (Vol. *420*, pp. 135–148). Geological Society of America, Special Paper. doi:https://doi.org/10.1130/2006.2420(09)

Ben Dor, Y., Neugebauer, I., Enzel, Y., Schwab, M. J., Tjallingii, R., Erel, Y., & Brauer, A. (2019). Varves of the Dead Sea sedimentary record. *Quaternary Science Reviews*, *215*, 173–184. doi:10.1016/j.quascirev.2019.04.011

Bernasconi, S. M., Meier, I., Wohlwend, S., Brack, P., Hochuli, P. A., Bläsi, H., et al. (2017). An evaporite-based high-resolution sulfur isotope record of Late Permian and Triassic seawater sulfate. *Geochimica et Cosmochimica Acta*, *204*, 331–349. doi:10.1016/j.gca.2017.01.047

Bosence, D., Gibbons, K., Le Heron, D. P., Morgan, W. A., Pritchard, T., & Vining, B. A. (2015). *Microbial carbonates in space and time: Introduction*. In D. W. J., Bosence, K. A., Gibbons, D. P. Le Heron, W. A., Morgan, T. Pritchard, & B. A. Vining (Eds.), *Microbial carbonates in space and time: Implications for global exploration and production* (Vol. 418, pp. 1–15). Geological Society, London, Special Publications.

Bradley, D., Munk, L., Jochens, H., Hynek, S., & Labay, K. (2013). A preliminary deposit model for lithium brines: *U.S. Geological Survey Open-File Report 2013-1006*, 6 p.

Braitsch, O. (1971). *Salt deposits: Their origin and composition*. New York: Springer-Verlag, 279 pp.

Brice, S. E., Cochran, M. D., Pardos, G., & Edwards, A. D. (1982). Tectonics and sedimentation of the South Atlantic rift sequence: Cabinda, Angola. In J. S. Watkins & C. L. Drake (Eds.), *Studies in continental margin geology, AAPG Memoir* (Vol. *34*, pp. 5–18).

Chidsey, T. C., Jr. (2002). Increased oil production and reserves utilizing secondary/tertiary recovery techniques on small reservoirs in the Paradox Basin, Utah. *Utah Geological Survey, Salt Lake City*, 166 pp.

Christ, C. L., & Garrels, R. M. (1959). Relations among sodium borate hydrates at the Kramer deposit, Boron, California. *American Journal of Science*, *257*, 516–528.

Clementi, E., Raimondi; D. L., & Reinhardt, W. P. (1967). Atomic screening constants from SCF functions. II. Atoms with 37 to 86 Electrons. *The Journal of Chemical Physics*, *47*(4), 1300–1307. doi:10.1063/1.1712084.

Davison, I. (2009). Faulting and fluid flow through salt. *Journal of the Geological Society, London*, *166*, 205–216. doi: 10.1144/0016-76492008-064.

Dellwig, L. F. (1955). Origin of the Salina salt of Michigan. *Journal of Sedimentary Petrology*, *25*, 83–110. doi:10.1306/d4269819-2b26-11d7-8648000102c1865d

Dellwig, L. F., & Evans, R. (1969). Depositional processes in Salina salt of Michigan, Ohio, and New York. *American Association of Petroleum Geologists Bulletin*, *53*(4), 949–956.

Douglas, G. V., & Goodman, N. R. (1957). The deposition of gypsum and anhydrite. *Economic Geology*, *52*(7), 831–837. doi:10.2113/gsecongeo.52.7.831

Dyni, J. R. (1996). Sodium carbonate resources of the Green River Formation. *USGS, Open-File Report* 96–729. https://pubs.usgs.gov/of/1996/ofr-96-0729/ofr-96-0729.pdf

Edmond, J. M., Measures, C., McDuff, R. E., Chan, L. H., Collier, R., Grant, B., et al. (1979). Ridge crest hydrothermal activity and the balances of the major and minor elements in the ocean: The Galapagos data. *Earth and Planetary Science Letters*, *46*(1), 1–18.

Esestime, P., Hewitt, A., & Hodgson, N. (2016). Zohr—A new-born carbonate play in the Levantine Basin, East-Mediterranean. *First Break*, *34*, 87–93.

Esrafili-Dizaji, B., & Rahimpour-Bonab, H. (2019). Carbonate reservoir rocks at giant oil and gas fields in SW Iran and the adjacent offshore: A review of stratigraphic occurrence and poro-perm characteristics. *Journal of Petroleum Geology*, *42*(4), 343–370.

Eugster, H. P. (1980). Lake Magadi, Kenya, and its precursors. In A. Nissenbaum (Ed.), *Hypersaline brines and evaporitic environments, Developments in Sedimentology* (Vol. *28*, pp. 195–232). Amsterdam: Elsevier. doi:10.1016/s0070-4571(08)70239-5

Eugster, H. P., & Hardie, L. A. (1978). Saline lakes. In A. Lerman (Ed.), *Lakes: Chemistry, Geology, Physics* (pp. 237–293). Berlin, Germany: Springer-Verlag.

Evans, R. (1978). Origin and significance of evaporites in basins around Atlantic margins. *American Association of Petroleum Geologists Bulletin*, *62*(2), 223–234.

Fischer, A. G., & Garrison, R. E. (2009). The role of the Mediterranean region in the development of sedimentary geology: A historical overview. *Sedimentology*, *56*, 3–41.

Garret, D. E. (1996). Origin of Potash deposits. In D. E. Garret (Ed.), *Potash deposits, processing, properties and uses* (pp. 1–80). London, UK: Chapman & Hall.

Grotzinger, J., & Al-Rawahi, Z. (2014). Depositional facies and platform architecture of microbialite-dominated carbonate reservoirs, Ediacaran-Cambrian Ara Group, Sultanate of Oman. *American Association of Petroleum Geologists Bulletin*, *98*(8), 1453–1494.

Grotzinger, J., Jordan, T. H., Press, F., & Siever, R. (2007). *Understanding Earth*, 5th ed.). New York, USA: W. H. Freeman and Company, 579 pp + Appendices.

Handford, C. R. (1991). Marginal marine halite: sabkhas and Salinas. In J. L. Melvin (Ed.), *Evaporites, petroleum and mineral resources, Developments in Sedimentology* (Vol. *50, Chapter 1*, pp. 1–66). Amsterdam: Elsevier.

Handford, C. R., Kendall, A. C., Prezbindowski, D. R., Dunham, J. B., & Logan, B. W. (1984). Salina margin tepees, pisoliths and aragonite cements, Lake MacLeod, Western Australia: Their significance in interpreting ancient analogs. *Geology*, *12*, 523–527.

Hanor, J. S., & Mcintosh, J. C. (2007). Diverse origins and timing of formation of basinal brinesin the Gulf of Mexico sedimentary basin. *Geofluids*, *7*, 227–237.

Hardie, L. A. (1984). Evaporites: Marine or non-marine. *American Journal of Science*, *283*, 193–240.

Hardie, L. A. (1987). Dolomitization: acritical view of some current views. *Journal of Sedimentary Petrology*, *57*, 166–183.

Hardie, L. A. (1990). The roles of rifting and hydrothermal $CaCl_2$ brines in the origin of potash evaporites: An hypothesis. *American Journal of Science*, *290*, 43–106.

Hardie, L. A. (1991). On the significance of evaporites. *Annual Review Earth and Planetary Science*, *290*, 43–106.

Hardie, L. A., & Eugster, H. P. (1971). The depositional environment of marine evaporites: a case for shallow, clastic accumulation. *Sedimentology*, *16*, 187–220.

Harvie, C. E., & Weare, J. H. (1980). The prediction of mineral solubilities in natural waters: the Na-K-Mg-Ca-Cl-$S0_4$-H_2O system from zero to high concentration at 25°C. *Geochimica et Cosmochimica Acta*, *44*, 981–997.

Haynes, W. M. (2017). *CRC Handbook of Chemistry and Physics* (97th ed.). Boca Raton, FL: CRC Press, 2643 pp.

Holser, W. T. (1979a), Mineralogy of evaporites. In R. G. Burns (Ed.), *Marine Minerals: Reviews in Mineralogy* (Vol. *6*, pp. 211–294). Chelsea, Michigan: Mineralogical Society of America. doi: 10.2138/rmg.1979.6.8

Holser, W. T. (1979b). Trace elements and isotopes in evaporites. In R. G. Burns (Ed.), *Marine Minerals: Reviews in Mineralogy* (Vol. *6*, pp. 295–346). Chelsea, Michigan: Mineralogical Society of America. doi: 10.2138/rmg.1979.6.9

Iespa, A. A. C., Iespa, C. M. D., & Borghi, L. (2012). Evolução paleoambiental da Lagoa Salgada utilizando microbialitos, com ênfase em microfácies carbonáticas, *UNESP, Geociências*, *31*(3), 371–380.

International Mineralogical Association (2022). IMA Database of Mineral Properties. https://rruff.info/ima/.

Jackson, M. P. A., Cramez, C., & Fonck, J. M. (2000). Role of subaerial volcanic rocks and mantle plumes in creation of South Atlantic margins: Implications for salt tectonics and source rocks. *Marine and Petroleum Geology*, *17*, 477–498.

Jenyon, M. K. (1986). *Salt tectonics*. Amsterdam: Elsevier Applied Science Publishers Ltd, 191 pp.

Kasprzyk, A. (2003). Sedimentological and diagenetic patterns of anhydrite deposits in the Badenian evaporite basin of the Carpathian Foredeep, southern Poland. *Sedimentary Geology*, *158*(3–4), 167–194. doi:10.1016/s0037-0738(02)00265-8

Kemp, S. J., Smith, F. W., Wagner, D., Mounteney, I., Bell, C. P., Milne, C. J., et al. (2016). An improved approach to characterize Potash-bearing evaporite deposits, evidenced in north Yorkshire, *United Kingdom, Economic Geology*, *111*, 719–742.

Kendall, A. C. (1978). Facies Models 12. Subaqueous evaporites. In R. G. Walker (Ed.), *Facies Models, Geoscience Canada*, *5*(3), 124–139.

Kendall, A. C. (1984). *Evaporites*. In R. G. Walker (Ed.), *Facies Models* (Vol. 1, pp. 259–296). Geoscience Canada, Reprint Series.

Kendall, A.C. (1989). Brine mixing in the Middle Devonian of western Canada and its possible significance to regional dolomitization. *Sedimentary Geology*, *64*, 271–285.

Klimchouk, A. B. (2000). Dissolution and conversions of gypsum and anhydrite. In: A. B. Klimchouk, D. C. Ford, A. N. Palmer, & W. Dreybrodt (Eds.), *Speleogenesis: Evolution of Karst Aquifers* (pp. 160–168). National Speleological Society.

Krumgalz, B. S., Miigdal, E., & Starinsky, A. (2002). The evolution of a chloride sedimentary sequence-simulated evaporation of the Dead Sea. *Israel Journal of Earth Sciences*, *51*, 253–267.

Kushnir, A. R. L., Kennedy, L., Misra, S., & Clancy, J. C. (2015). The mechanical and microstructural behaviour of calcite-dolomite composites: An experimental investigation. *Journal of Structural Geology*, *70*, 200–216. doi: 10.1016/j.jsg.2014.12.006

Kuster, Y., Schramm, M., Bornemann, O., & Leiss, B. (2009). Bromide distribution characteristics of different Zechstein 2 rock salt sequences of the Southern Permian Basin: A comparison between bedded and domal salts. *Sedimentology*, *56*, 1368–1391.

Le Chatelier, H., & Boudouard, O. (1898). Limits of flammability of gaseous mixtures. *Bulletin de la Société Chimique de France (Paris)*, *19*, 483–488.

Lima, B. E. M., & De Ros, L. F. (2019). Deposition, diagenetic and hydrothermal processes in the Aptian Pre-Salt lacustrine carbonate reservoirs of the northern Campos Basin, offshore Brazil. *Sedimentary Geology*, *383*, 55–81. doi:10.1016/j.sedgeo.2019.01.006

Livingstone, D. A. (1963). Chemical composition of rivers and lakes. *U. S. Geological Survey, Professional Paper 440-G*, 64 p.

Lokier, S. W., & Steuber, T. (2009). Large-scale intertidal polygonal features of the Abu Dhabi coastline. *Sedimentology*, *56*(3), 609–621. doi:10.1111/j.1365-3091.2008.00988.x

Mead, W. J. (1940). Folding, rock flowage, and foliate structures. *The Journal of Geology*, *48*(8), 1007–1021.

Mohriak, W. U. (2015). Pre-salt carbonate reservoirs in the South Atlantic and World-wide Analogs. *AAPG Search and Discovery, Article #51086 (2015). Posted* April 6, 2015.

Mohriak, W. (2025). *Salt in the Earth sciences: Basin analysis and salt tectonics*. Washington, DC: American Geophysical Union. doi:10.1002/9781394307067

Mohriak, W. U., & Szatmari, P. (2008). Introdução às propriedades químicas e físicas dos evaporitos. In W. Mohriak, P. Szatmari, & S. M. C. Anjos (organizadores). *Sal: Geologia e Tectônica* (pp. 18–41). *São Paulo: Editora Beca.*

Mohriak, W. U., Szatmari, P., & Anjos, S. (2012). *Salt: Geology and tectonics of selected Brazilian basins in their global context*. In G. I. Alsop, S. G. Archer, A. J. Hartley, N. T. Grant, & R. Hodgkinson (Eds.), *Salt tectonics, sediments and prospectivity* (Vol. 363, pp. 131–158). Geological Society, London, Special Publications.

Müller, D. W., & Mueller, P. A. (1991). Origin and age of the Mediterranean Messinian evaporites: Implications from Sr isotopes. *Earth and Planetary Science Letters*, *107*(1), 1–12.

Müller, D. W., Mueller, P. A., & McKenzie, J. A. (1990). 37. Strontium isotopic rations as fluid tracers in Messinian evaporites of the Tyrrhenian Sea (Western Mediterranean Sea). In K. A. Kastens, J. Mascle et al. (Eds.), *Proceedings of the Ocean Drilling Program, Scientific Results* (Vol. 107, pp. 603–614).

Muniz, M. C., & Bosence, D. W. J. (2015). *Pre-salt microbialites from the Campos Basin (offshore Brazil): Image log facies, facies model and cyclicity in lacustrine carbonates*. In D. W. J. Bosence, K. A. Gibbons, D. P. LeHeron, W. A. Morgan, T. Pritchard, & B. A. Vining (Eds.), *Microbial carbonates in space and time: Implications for global exploration and production* (Vol. 418, pp. 221–242). Geological Society, London, Special Publications.

Nielsen, H. (1979). Sulfur Isotopes. In E. Jäger & J. C. Hunziker (Eds.), *Lectures in Isotope Geology* (pp. 283–312). Berlin Heidelberg: Springer-Verlag.

Okrusch, M., & Frimmel, H. E. (2020). Carbonates, Nitrates and Borates. In *Mineralogy. Springer Textbooks in Earth Sciences, Geography and Environment*. Berlin, Heidelberg: Springer. https://doi.org/10.1007/978-3-662-57316-7_8

Oliveira, V. C. B., Silva, C. M. A., Borghi, L. F., & Carvalho, I. S. (2019). Lacustrine coquinas and hybrid deposits from rift phase: Pre-Salt, lower cretaceous, Campos Basin, Brazil. *Journal of South American Earth Sciences*, *95*, 102254. doi:10.1016/j.jsames.2019.102254

Ortí, F. (2011). Selenite facies in marine evaporites: A review. In C. G. St. C. Kendall & A. S. Alsharhan (Eds.), *Quaternary carbonate and evaporite sedimentary facies and their ancient analogues: A tribute to Douglas James Shearman* (Vol. *43*, pp. 431–463), International Association of Sedimentologists Special Publication. doi:10.1002/9781444392326.ch20

Ortí, F., Pérez-López, A., García-Veigas, J., Rosell, L., Cendón, D. I., & Pérez-Valera, F. (2014). Sulfate isotope compositions ($\delta^{34}S$, $\delta^{18}O$) and strontium isotopic ratios

($^{87}Sr/^{86}Sr$) of Triassic evaporites in the Betic Cordillera (SE Spain). *Revista de la Sociedad Geológica de España*, *27*(1), 79–89.

Ortí, F., Rosell, L., Inglès, M., & Playà, E. (2007). Depositional models of lacustrine evaporites in the SE margin of the Ebro Basin (Paleogene, NE Spain). *Geologica Acta*, *5*(1), 19–34.

Palmer, M. R., & Edmond, J. M. (1989). The Strontium isotope budget of the modern ocean. *Earth and Planetary Science Letters*, *92*(1), 11–26.

Perthuisot, J. P. (1989). *Part II.* Recent evaporites. In P. Sonnenfeld & J.-P. Perthuisot (Eds.), *Brines and evaporites, Short Course in Geology* (Vol. 3, pp. 65–118), *28th International Geological Congress*. American Geophysical Union, Washington, DC.

Peterson, J. A. (1989). *Geology and petroleum resources, Paradox basin Province, USGS Open-File Report 88-450 U*, 69 pp.

Poiate, E., Jr., Costa, A. M., & Falcao, J. L. (2006). Well design for drilling through thick evaporite layers in Santos Basin–Brazil. *IADC/SPE Drilling Conference*, Florida, U.S.A, 21–23 February 2006, *Abstracts*, *99161*, 1–16.

Press, F., Siever, R., Grotzinger, J., & Jordan, T. H. (2003). *Understanding Earth* (4th ed.). New York, USA: W. H. Freeman, 568 pp.

Richter-Bernburg, G. (1972a), *Geology of saline deposits*, Proceedings of the Hanover Symposium, 15–21 May 1968, International Union of Geological Sciences, United Nations Educational, Scientific and Cultural Organization, *Earth Sciences*, 7, 316 pp.

Richter-Bernburg, G. (1972b). Sedimentological problems of saline deposits. In G. Richter-Bernburg (Ed.), *Geology of saline deposits*, Proceedings of the Hanover Symposium, 15–21 May 1968, International Union of Geological Sciences, United Nations Educational, Scientific and Cultural Organization, *Earth Sciences*, *7*, 31–39.

Richter-Bernburg, G. (1973). Paleogeographical pre-conditions for the genesis of evaporites. *Anais do XXVII Congresso Brasileiro de Geologia*, *2*, 163–169. Aracaju, SE.

Riding, R. (2000). Microbial carbonates: The geological record of calcified bacterial-algal mats and biofilms. *Sedimentology*, *47*(Suppl. 1), 179–214.

Risacher, F., Alonso, B., & Salazar, C. (2003). The origin of brines and salts in Chilean salars: A hydrochemical review. *Earth-Science Reviews*, *63*, 249–293.

Robichaud, S. R., & Robichaud, N. W. (2019). Mineral Resource–Lithium–Recovery from naturally occurring Permian basin waste water. *AAPG Search and Discovery Article #42475* (2019), 27 pp.

Roveri, M., Lugli, S., Manzi, V., & Schreiber, B. C. (2008a). The Messinian Sicilian stratigraphy revisited: New insights for the Messinian salinity crisis. *Terra Nova*, *20*, 483–488. doi: 10.1111/j.1365-3121.2008.00842.x

Roveri, M., Lugli, S., Manzi, V., & Schreiber, B. C. (2008b). The Messinian salinity crisis: A sequence-stratigraphic approach. *GeoActa*, *7984*, 169–190.

Schaller, W. T. (1929). Borate minerals from the Kramer District, Mohave Desert, California. *Shorter contributions to general geology, USGS Report 158*, 137–170.

Schröder, S., Schreiber, B. C., Amthor, J. E., & Matter, A. (2003). A depositional model for the terminal Neoproterozoic–Early Cambrian Ara Group evaporites in south Oman. *Sedimentology*, *50*(5), 879–898.

Seard, C., Camoin, G., Rouchy, J.-M., & Virgone, A. (2013). Composition, structure and evolution of a lacustrine carbonate margin dominated by microbialites: Case study from the Green River formation (Eocene; Wyoming, USA). *Palaeogeography, Palaeoclimatology, Palaeoecology*, 381–382, 128–144.

Shannon, R. D. (1976). Revised effective ionic radii and systematic studies of interatomic distances in halides and chalcogenides. *Acta Crystallographica*, *A32*, 751–767. https://doi.org/10.1107/S0567739476001551

Shearman, D. J. (1978). Evaporites of coastal sabkhas. In W. E. Dean & B. C. Schreiber (Eds.), *Marine evaporites*, *Lecture Notes for Short Course* (Vol. *4*, pp. 6–42). Tulsa, Okla: Society of Economic Paleontologists and Mineralogists.

Słowakiewicz, M., Tucker, M. E., Pancost, R. D., Perri, E., & Mawson, M. (2013). Upper Permian (Zechstein) microbialites: Supratidal through deep subtidal deposition, source rock, and reservoir potential. *American Association of Petroleum Geologists Bulletin*, *97*(11), 1921–1936.

Smith, G. I., & Stuiver, M. (1979). Subsurface stratigraphy and geochemistry of late Quaternary evaporites, Searles Lake, California, with a section on radiocarbon ages of stratigraphic units. *USGS Professional Paper 1043*, 130 pp.

Smoot, J. P. (1991). Sedimentary facies and depositional environments of early Mesozoic Newark Supergroup basins, eastern North America. *Palaeogeography, Palaeoclimatology, Palaeoecology*, *84*(1–4), 369–423. doi:10.1016/0031-0182(91)90055-v

Smoot, J. P., & Lowenstein, T. K. (1991). Depositional environments of non-marine evaporites. In J. L. Melvin (Ed.), *Evaporites, petroleum and mineral resources, Developments in Sedimentology* (Vol. *50, Chapter 3*, pp. 189–347). Amsterdam: Elsevier.

Sonnenfeld, P. (1979). Brines and evaporites. *Geoscience Canada*, *6*(2), 83–90.

Sonnenfeld, P. (1989). Part I. Genesis of evaporites. In P. Sonnenfeld and J.-P. Perthuisot (Eds.), *Brines and evaporites, Short Course in Geology: Volume 3, 28th International Geological Congress*, American Geophysical Union, Washington, DC, pp. 1–64.

Sonnenfeld, P. (1992). Genesis of marine evaporites—A summation. *Geologica Carpathica*, *43*(5), 259–274.

Sonnenfeld, P., & Hardie, L. A. (1985). Evaporites; marine or non-marine; discussion and reply. *American Journal of Science*, *285*, 661–672.

Sonnenfeld, P., & Perthuisot, J. -P. (1989). Brines and evaporites. *Short Course in Geology, Volume 3, 28th International Geological Congress*. American Geophysical Union, Washington, DC, 126 pp.

Spencer, R. J., Lowenstein, T. K., Casas, E., & Pengxi, Z. (1990). Origin of potash salts and brines in the Qaidam Basin, China. In R. J. Spencer & I-Ming Chou (Eds.), *Fluid-Mineral interactions: A tribute to H. P. Eugster* (Vol. 2, pp. 395–408). The Geochemical Society, Special Publication.

Stewart, F. H. (1963). Marine evaporites. In M. Fleischer (Ed.), *Data of Geochemistry* (6th ed., Chapter Y, pp. Y1–Y53). Geological Survey Professional Paper 440-Y.

Strakhov, N. M. (1969). *Principles of Lithogenesis* (Vol. *3*). Edinburgh, London: Oliver and Boyd, 577 pp.

Szatmari, P., Carvalho, R. S., & Simões, I. A. (1979). A comparison of evaporite facies in the late Paleozoic Amazon and the Middle Cretaceous South Atlantic salt basins. *Economic Geology*, *74*(2), 432–447.

Szatmari, P., Carvalho, R. S., Simões, I. A., Tibana, P., & Leite, D. C. (1974). Evaporitos de Sergipe: Geologia e geoquímica. *PETROBRAS, Internal Report*, *1*, 1–160.

Szatmari, P., Carvalho, R. S., Simões, I. A., Tibana, P., & Leite, D. C. P. (2009). Atlas petrográfico dos evaporitos. In W. Mohriak, P. Szatmari, & S. M. C. Anjos (Eds.), *Sal: Geologia e Tectônica* (2nd ed., pp. 42–65). São Paulo: Editora Beca.

Szatmari, P., Moré de Lima, C., Fontaneta, G., de Melo Lima, N., Zambonato, E., Menezes, M. R., et al. (2021). Petrography, geochemistry and origin of South Atlantic evaporites: The Brazilian side. *Marine and Petroleum Geology*, *127*, 104805, https://doi.org/10.1016/j.marpetgeo.2020.104805

Swart, P. K., Cantrell, D. L., Westphal, H., Handford, C. R., & Kendall, C. G. (2005). Origin of dolomite in the Arab-D reservoir from the Ghawar Field, Saudi Arabia: Evidence from petrographic and geochemical constraints. *Journal of Sedimentary Research*, *75*(3), 476–491.

Swihart, G. H., Moore, P. B., & Callis, E. L. (1986). Boron isotopic composition of marine and nonmarine evaporite borates. *Geochimica et Cosmochimica Acta*, *50*(6), 1297–1301.

Talbot, C. J. (2008). Hydrothermal salt—But how much? Discussion. *Marine and Petroleum Geology*, *25*, 191–202.

Tanner, L. H. (2002). Borate formation in a perennial lacustrine setting: Miocene-Pliocene Furnace Creek Formation, Death Valley, California, USA. *Sedimentary Geology*, *148*(1), 259–273.

Usiglio, J. (1849a). Analyse de l'eau de la Mediterraneé sur le côtes de France. *Annales de Chemie et de Physique*, *3ème sér*, *27*, 92–107.

Usiglio, J. (1849b). Etudes sur la composition de l'eau de la Mediterranee et sur l'exploitation des sel quy'elle conteint. *Annales de Chemie et de Physique*, *3ème sér*, *27*, 172–191.

Valyashko, M. G. (1956). Geochemistry of bromine in the processes of salt deposition and the use of the bromine content as a genetic and prospecting criterion. *Geochemistry*, *6*, 570–589.

Valyashko, M. G. (1972a). Playa lakes-a necessary stage in the development of a salt-bearing basin. In G. Richter-Bernburg (Ed.), *Geology of saline deposits*, Proceedings of the Hanover Symposium, 15–21 May 1968, International Union of Geological Sciences, United Nations Educational, Scientific and Cultural Organization, *Earth Sciences*, *7*, 41–51.

Valyashko, M. G. (1972b). Scientific works in the field of geochemistry and the genesis of salt deposits in the U.S.S.R. In G. Richter-Bernburg (Ed.), *Geology of saline deposits*, Proceedings of the Hanover Symposium, 15–21 May 1968, International Union of Geological Sciences, United Nations Educational, Scientific and Cultural Organization, *Earth Sciences*, *7*, 289–311.

Van't Hoff, J. H. (1905). *Zur Bildung der ozeanischen Salzablagerungen*. Braunschweig: Vieweg, 85 pp.

Voigt, W. (2011). Chemistry of salts in aqueous solutions: Applications, experiments, and theory. *Pure and Applied Chemistry*, *83*(5), 1015–1030. doi:10.1351/PAC-CON-11-01-07

Voigt, W. (2015). What we know and still not know about oceanic salts. *Pure and Applied Chemistry*, *87*(11–12), 1099–1126. doi 10.1515/pac-2015-0606

Wardlaw, N. C. (1972). Unusual marine evaporites with salts of calcium and magnesium chloride in Cretaceous basins of Sergipe, Brazil. *Economic Geology*, *67*, 156–168.

Warren, J. K. (1982a). The hydrological setting, occurrence and significance of gypsum in late Quaternary salt lakes in South Australia. *Sedimentology*, *29*, 609–637.

Warren, J. K. (1982b). The hydrological significance of Holocene tepees, stromatolites, and boxwork limestones in coastal salinas in South Australia. *Journal of Sedimentary Petrology*, *52*, 1171–1201.

Warren, J. K. (1983a). Tepees, modern (Southern Australia) and ancient (Permian-Texas and New Mexico)—A comparison. *Sedimentary Geology*, *34*, 1–19.

Warren, J. K. (1983b). On the significance of evaporite lamination. In B. C. Schreiber (Eds.), *Proceedings of the sixth international symposium on salt* (Vol. 1, pp. 161–170). Toronto, Canada.

Warren, J. K. (1986). Source rock potential of shallow water evaporitic settings. *Journal of Sedimentary Petrology*, *56*, 442–454.

Warren, J. K. (1989). *Evaporite sedimentology: Importance in hydrocarbon accumulation*. Prentice Hall Advanced Reference Series, 285 p.

Warren, J. K. (1991). Sulfate dominated sea-marginal and platform settings. In J. L. Melvin (Ed.), *Evaporites, petroleum and mineral resources, Developments in Sedimentology* (Vol. *50*, *Chapter 2*, pp. 69–187). Amsterdam: Elsevier.

Warren, J. K. (2000). Dolomite: occurrence, evolution and economically important associations. *Earth-Science Reviews*, *52*, 1–81.

Warren, J. K. (2006). *Evaporites: Sediments, resources and hydrocarbons*. Berlin: Springer Verlag, 1035 pp.

Warren, J. K. (2010). Evaporites through time: Tectonic, climatic and eustatic controls in marine and nonmarine deposits. *Earth and Science Review*, *98*, 217–268.

Warren, J. K. (2016). *Evaporites—A geological compendium*. Berlin: Springer, 1807 pp. ISBN 978-3-319-13511-3.

Wright, P., & Rodriguez, K. (2018). Reinterpreting the South Atlantic Pre-Salt "Microbialite" reservoirs: Petrographic, isotopic and seismic evidence for a shallow evaporitic lake depositional model. *First Break*, *36*(5), 71–77.

Wright, V. P., & Barnett, A. J. (2015). An abiotic model for the development of textures in some South Atlantic early Cretaceous lacustrine carbonates. In D. W. J. Bosence, K. A. Gibbons, D. P. Le Heron, W. A. Morgan, T. Pritchard, & B. A. Vining (Eds.), *Microbial carbonates in space and time: Implications for global exploration and production* (Vol. 418, pp. 209–219). Geological Society, London, Special Publications.

Wright, V. P., & Barnett, A. (2017). Critically evaluating the current depositional models for the Pre-Salt Barra Velha Formation, *offshore Brazil. AAPG Search and Discovery, Article #51439 (2017), Posted* November 20, 2017.

Yager, D. B. (2016). Potash—A vital agricultural nutrient sourced from geologic deposits. *U.S. Geological Survey Open-File Report 2016-1167*, 28 p. http://dx.doi.org/10.3133/ofr20161167.

5
Basic Chemical and Physical Properties of Evaporite Rocks

ABSTRACT

Evaporite minerals' chemical compositions and ionic properties influence their physical properties, like crystal shape, color, density, viscosity, thermal conductivity, melting point, seismic velocity, and magnetic susceptibility. Electric logs and geophysical surveys, such as seismic reflection methods, study evaporite minerals at microscopic to metric scales. These geophysical tools help identify salt layers in the subsurface by mapping their distinctive geometries after deformation.

Evaporite minerals are solids with a crystalline habit and specific chemical and physical properties. These properties are crucial for recognizing different depositional environments in sedimentary basins. The chemical composition and atomic arrangement of the crystalline lattice control properties like density and sonic velocity. These properties are obtained through laboratory analysis of rock samples or indirect geophysical methods in basin analysis projects. Petrophysical logs are acquired during the drilling of deep boreholes in sedimentary basins (e.g., Serra, 1984, 1986). The main properties of sedimentary layers, such as rock density, formation resistivity, gamma-ray intensity, and sonic velocity, are obtained by analyzing electric curves from wireline logs (e.g., Schön, 1996, 2015; Tiab & Donaldson, 2004). Several tables that illustrate the diverse physical properties of evaporite minerals are included in subsequent sections.

The study of evaporite deposits in sedimentary basins relies heavily on the identification of mineralogical, petrographical, and petrophysical properties of different minerals. Many textbooks discuss the crystallographic methods applied to mineralogy and petrography (e.g., Okrusch & Frimmel, 2020). This section provides an overview of the fundamental aspects involving the characterization of evaporite minerals by their chemical and physical properties. Chapter 7 discusses the microscopic petrographic characterization of selected minerals, as this method is widely employed in laboratories. In addition, more advanced methods of mineral identification, such as X-ray crystallography, will also be summarized.

Seismic velocity is a fundamental property of sedimentary layers and is widely used in geophysical methods. The reflection seismic method is the most common technique applied to the investigation of sedimentary basins. This chapter briefly introduces the seismic method, presents the typical geometries of reflectors in reflection seismic lines, and outlines some basic principles related to basin analysis studies, including the characterization of different lithological units by integrating seismic methods and petrophysical log interpretation.

5.1. EVAPORITE MINERALS AND CRYSTALLOGRAPHY

Minerals in evaporite rocks are formed by crystals, solids characterized by a periodic arrangement of atoms, ions, or molecules. These crystals form symmetric patterns along a three-dimensional framework that repeats the unit cell of the structure. The crystal lattice results from the repetition of the unit cell along the principal axes, which correspond to three axes with a definite angle to each other. The mineral halite, for example, exhibits a crystal lattice with a regular arrangement of the sodium and chlorine ions in the vertices of a cubic structure (see Chapter 4).

Sir William Hallowes Miller (1801–1880) was an English mineralogist who proposed that minerals could be classified

Salt in the Earth Sciences: Evaporite Rocks and Salt Deposition, Geophysical Monograph 286, First Edition. Webster Mohriak.

DOI: 10.1002/9781119405245.ch05

by measuring the distance between the crystal planes. The Miller indices correspond to three parameters that identify the crystal planes in a mineral (e.g., Miller, 1839). For example, the frontal vertical face of a cube is expressed as (1, 0, 0) using the Miller indices. The right-side face of the cube is denoted as (1, 0, 0), as the right vertical side is located parallel to the a1 (horizontal x-axis) and the vertical side at the a2 axis (horizontal y-axis). Halite crystals commonly display these orthogonal faces, but on rare occasions, the face is characterized by the Miller Index (1, 1, 1). Most textbooks addressing crystallography and mineralogy provide tables with these indexes for common minerals (e.g., Hammond, 2015).

The chemical composition of evaporite minerals is based on the arrangement of the atoms in the crystal lattice or unit cell, giving each mineral a characteristic shape or crystal habit. The crystal habit refers to the typical external shapes of the mineral, while the crystal form is associated with the arrangement of the atoms that make up the crystal structure, known as the crystal lattice.

Evaporite minerals are characterized by diverse crystal shapes, which in crystallography are divided into a small number of different classes. All imaginable crystal shapes of the minerals found in nature belong to one of the seven possible crystallographic classes based on six cell or lattice parameters. These classes are (A) cubic, (B) tetragonal, (C) orthorhombic, (D) trigonal, (E) hexagonal, (F) monoclinic, and (G) triclinic (e.g., Hammond, 2015).

The crystallographic classification of minerals is based on the symmetry of the crystals, which is related to the distribution of atoms in the crystal lattice. The crystal habit of the mineral is associated with the crystal structure of the atoms inside the mineral (e.g., Hammond, 2015). For example, the mineral halite (NaCl) is characterized by the arrangement of the Na and Cl ions in a cell unit characterized by the same length along each axis within a three-dimensional structure. The faces of the halite crystals are characterized by an orthogonal angle (90°), forming a perfect cube (see Chapter 4). Crystal classes are discussed in more detail in mineralogy textbooks (e.g., Dana & Ford, 1922; Hammond, 2015; Nesse, 2000).

Cleavage is the tendency of some mineral crystals to split along the weaker crystallographic planes. This property results from the structural location of atoms and ions in the crystal lattice, resulting in planes of relative weakness when the mineral is subject to stress. For example, the mineral halite has three planes of cleavage when it is broken, all at right angles to each other, forming a cube (see Chapter 4).

Evaporites are rocks with unusual mineralogical and petrophysical properties when compared to other sedimentary rocks of the Earth's crust (Nurmi, 1978; Schön, 1996, 2015). The evaporite minerals are identified by their chemical composition, crystal system, density, hardness, and other chemical and physical properties (Figure 5.1). Most evaporite sequences in sedimentary basins are characterized by a few relatively common minerals, such as gypsum, anhydrite, and halite. The mineral name and formula in the list follow the internationally accepted nomenclature of the International Mineralogical Association (2022). Halite (sodium chloride, NaCl) is the most abundant and volumetrically important compound derived from the evaporation of seawater, corresponding to 78% of all salts precipitated from brines of marine origin (see Chapter 4).

Mineral class	Mineral name	Chemical composition	Crystal system	Density (g/cm^3)	Hardness (mohs scale)
Chlorides	Halite	$NaCl$	cubic	2.17	2.5
	Sylvite	KCl	cubic	1.99	2
	Carnallite	$KMgCl_3.6H_2O$	orthorhombic	1.60	2.5
Sulfates	Anhydrite	$CaSO_4$	orthorhombic	2.98	3.5
	Gypsum	$CaSO_4.2H_2O$	monoclinic	2.32	2
	Kieserite	$MgSO_4.H_2O$	monoclinic	2.57	3.5
	Langbeinite	$K_2Mg_2(SO_4)_3$	cubic	2.83	3.5
	Polyhalite	$K_2Ca_2Mg(SO_4)_4.2H_2O$	triclinic	2.77	3
Carbonates	Dolomite	$CaMg(CO_3)_2$	rhombohedral	2.86	3.5
	Calcite	$CaCO_3$	hexagonal	2.71	3
	Magnesite	$MgCO_3$	hexagonal	3.02	4

Figure 5.1 Chemical composition, crystal system, density, and hardness of the main minerals in evaporite basins. Credit: Data from Mindat (2022).

Property/ units	Mean value	Variation	Comment
Density (g/cm^3)	2.2	2.11–2.22	Halite density is about 2.165 g/cm^3, but varies with the impurities
Viscosity (Pa.s)	10^{+14}	10^{+14}–10^{+17}	No more than 1/100 of the viscosity of other sedimentary rocks
Thermal Conductivity (W/m/°C)	4.5	–	Two to four times greater than that of other sedimentary rocks
Melting point (°C)	804	–	Halite can be melted by intrusive igneous rocks
Seismic velocity (m/s)	4,500	4,200–5,000	Halite seismic velocity is almost constant with depth
Magnetic Susceptibility (cgs)	0	–	25–100 cgs for other rocks

Figure 5.2 Physical properties of halite. Credit: Data from Mindat (2022).

According to the classical works of Dana and Ford (1922) and Braitsch (1971), the mineralogy of a halite crystal has the following characteristics:

- When pure, it is colorless or white with a vitreous luster.
- It presents a very low density compared to other evaporite minerals (2.16 g/cc).
- It presents a normally cubic crystalline habit, rarely {111}.
- It presents a perfect cleavage {100}, in addition to the conchoidal fracture.
- It shows a low refractive index.

The most important physical properties of halite for basin analysis studies are the following (Figure 5.2):

- Great solubility in water (circa 26–36 g/l).
- Low-to-moderate density (2.17 g/cm^3).
- A relatively low hardness (2.5 on the Mohs scale).
- Low viscosity (10^{+14} Pa.s).
- High thermal conductivity (4.5 W/m/°C).
- High seismic velocity (Vp = 4,500 m/s).
- Low melting point (804 °C).
- Lack of magnetism.

Several modern websites are continually updated with physical parameters and data on halite and other evaporite mineral occurrences in the world (e.g., Mindat, 2022). Halite is characterized by great solubility in water (circa 26–36 g/l), forming the ions of sodium and chlorine, the most abundantions in seawater (see Chapter 4). The viscosity of halite is extremely low, resulting in salt flow at normal temperatures (see Chapter 6). The melting temperature of halite is relatively low, so evaporite sequences are strongly affected by igneous intrusions during magmatic episodes (Schofield et al., 2014).

5.2. COMPOSITION AND DENSITY OF EVAPORITE MINERALS

Density is the ratio of mass per volume, and water has a maximum density of 1 g/cm^3 at 4 °C. At this temperature, water is at its minimum volume due to the cohesion of the hydrogen-bonded network, resulting in its maximum density (Chaplin, 2010). At this temperature, water sinks to the bottom of the water mass. This is why the temperature of water at the bottom of the oceans or in deep lakes such as Baikal is around 4 °C. Saltwater, characterized by large amounts of dissolved minerals, has much higher densities than freshwater or seawater. The brine density in hypersaline lakes can be over 20% higher than normal seawater (see Chapter 2). In addition, the density of saline water increases with depth as a consequence of the increase in pressure (Chaplin, 2010).

The densities of evaporite minerals are directly related to their chemical composition and crystal system. Some minerals with the same chemical composition but different crystal systems exhibit different densities, such as calcite ($CaCO_3$) and aragonite ($CaCO_3$). These minerals are called polymorphs and are present in nature with the same chemical composition but in multiple forms. Calcite is characterized by hexagonal crystals and a density of 2.653 g/cm^3, whereas aragonite belongs to the orthorhombic class and shows a density of 2.930 g/cm^3 (Haynes, 2017).

The densities of most solid materials vary as the temperature increases or decreases. Density can be estimated at any temperature based on the density of the material at the melting point (Haynes, 2017). The following equation can be applied to most minerals and chemical compounds:

$$\rho(t) = \rho m - k(t - tm)$$

where ρ(t) corresponds to the density at temperature t, ρm is the density at the melting point, tm is the melting point, and k is the derivative of density relative to temperature.

The three main classes of evaporites (chlorides, sulfates, and carbonates) include minerals with various chemical compositions and crystal systems, which control several physical parameters such as color, density, and hardness. Chlorides are characterized by lower density and hardness compared to sulfates and carbonates, but the most important physical properties that distinguish these mineral classes are viscosity and rheology, which are directly related to the deformation of salt layers or salt tectonics (Hudec & Jackson, 2007; Jackson & Hudec, 2017).

There is an empirical relationship between the melting point of minerals, their hardness, and their density. For example, halite has a melting point of about 804 °C, its hardness is about 2 on the Mohs scale, which varies from 1 (talc) to 10 (diamond), and the density is 2.17 g/cc (Figures 5.1 and 5.2). On the other hand, most sulfate minerals exhibit a hardness higher than 3, except gypsum, which contains water in the molecular structure and exhibits low hardness and density values. Anhydrite has a melting point of about 1,500 °C, its hardness is about 3.5, and a density of about 2.96 g/cc (Figure 5.1).

While most siliciclastic rocks become denser when loaded by sedimentary overburden, salt layers maintain almost the same density both near the surface and at great depths (Jackson & Talbot, 1986). Halite is characterized by minor variations in density with increasing depth or temperature (Jackson et al., 2010). For example, typical rock salt in the Gulf of Mexico exhibits a density of around 2,200 kg/m^3 near the depositional surface but decreases slightly with depth due to the high thermal expansion coefficient of halite. On the other hand, most sulfate minerals exhibit a hardness higher than 3, except gypsum, which contains water in its molecular structure and exhibits low hardness and density values. Anhydrite has a melting point of about 1 500 °C, a hardness of about 3.5, and a density of about 2.96 g/cc (Figure 5.1).

5.3. MAIN PHYSICAL PROPERTIES OF EVAPORITE MINERALS

Evaporite minerals are characterized by physical properties that control their behavior during halokinesis, or salt flow, and salt tectonics, the deformation of salt layers by gravity and tectonic forces. The high density of carbonate and sulfate minerals, such as calcite (2.71 g/cc) and anhydrite (2.96 g/cc), when compared to chloride minerals, favors the buoyancy of the halite layer in an evaporite sequence (see Chapter 6). Other important physical properties of evaporite minerals include viscosity, thermal conductivity, melting point, seismic velocity, and magnetic susceptibility (e.g., Nettleton, 1962, 1976 ; Van der Pluijm and Marshak, 2004).

Sedimentary layers also have important physical properties such as porosity and permeability. The extremely low porosity and permeability of evaporite layers make them excellent seals for hydrocarbons and underground storage. The strategic oil reserves of the USA are stored in salt diapirs in the Gulf Coast salt basin (Seni et al., 1985), while some salt diapirs in the North German Basin were targeted for building salt caverns for storing nuclear waste (see Chapter 2 in Mohriak, 2025).

5.3.1. Viscosity

Viscosity measures the internal resistance to flow, with water flowing rapidly in rivers due to its low viscosity, while honey has a higher resistance to flow, causing it to mobilize more slowly. Viscosity is measured in Pa.s or centipoise (cp). Water has a viscosity of about 0.001 Pa.s or 1 cp. Honey is much more viscous, with a viscosity of 4 Pa.s, or 4,000 cp (Figure 5.3). The low viscosity of salt rocks at low temperatures is one of the main factors controlling the deformation of evaporite layers in orogenic belts (Jackson & Hudec, 2017).

The rheology of evaporite units involved in thrust and fold belts control the deformation zones in many orogenic belts worldwide (see Chapter 6). The structural styles observed in the Zagros region of Iran are variable, depending on several factors, such as the plate velocity, the physical properties of the involved lithologies, and the presence of weak zones forming decollements, such as layers of mobile salt at different stratigraphic levels (see Chapter 3). Deformation of salt layers in the Zagros thrust-fold belt may occur in the subsurface, forming salt domes and salt diapirs, or on the surface, where the exposed salt layers are subject to gravity forces. Some anticlinal structures have been breached by salt diapirs,

Material	Viscosity (Pa.s)
Liquid water	10^{-3}
Petroleum	0.001 to 1
Honey	4.0
Silicon polymer	10^{+4}
Asphalt	10^{+5}
Ice	10^{+12}
Glass	10^{+14}
Halite	10^{+17}
Asthenosphere	10^{+22}

Figure 5.3 Mean viscosity of fluid materials (1 Pa.s = 1000 centipoise or cp). Credit: Adapted from Van der Pluijm and Marshak (2004).

resulting in salt glaciers (also known as namakiers) that are extruded from the tops of salt mountains and flow toward the valleys (see Chapter 3).

The Earth's crust is composed of materials with large variations in viscosity (Figure 5.3). The asthenosphere viscosity is very high, resulting in a very slow transfer of mass, forming convection currents in the mantle. Halite is the main form of common rock salt and presents a plastic behavior, making it very different from other sedimentary rocks (Weijermars et al., 1993).

Halite is a mineral with a viscosity of up to 10^{+17} Pa and exhibits extraordinary mobility and fluidity under stress. Several authors (e.g., Talbot & Rogers, 1980) compare salt flow on the Earth's surface with ice glaciers, whose flow velocity can be measured within a time frame of a few years (see Chapter 6). Ice glaciers in the Alps are reported to have moved downslope at velocities that exceed 100 m/year (Nishimura et al., 2018).

The Hormuz salt in the Kuh-e-Namak structure rises through the salt diapir vent at a rate of 170 mm/year, while the horizontal flow velocity of the extruded salt that advanced into the head of the north salt glacier may reach values averaging up to 2,000 mm/year (Talbot & Jarvis, 1984). During rainy periods, the salt glacier may flow at extremely high velocities of 500 mm/day (Talbot & Jarvis, 1984). The decrease in viscosity results from minor amounts of water incorporated into the salt matrix (see Chapter 6).

The horizontal movement of salt glaciers corresponds to a very fast flow velocity compared to other tectonic velocities. For example, the lithospheric plate velocity depends on the flow of mantle rocks in the asthenosphere (e.g., Doglioni et al., 2011). The asthenosphere corresponds to a mobile zone in the upper mantle, and these rocks display a viscosity of around 10^{22} Pa·s (Figure 5.3). The Arabian lithospheric plate moves northwards relative to Eurasia at a speed of 31 mm/year, as shown by studies of seafloor spreading, fault systems, and earthquake activity (Vernant et al., 2004). This plate motion causes the Zagros and Alborz Mountains to deform in modern times. However, direct measurements based on GPS instruments indicate that the deformation rates between the Persian Gulf and the Caspian Sea are less than 10 mm/year (Vernant et al., 2004). Salt glaciers in Iran move at velocities more than 70 times faster than the movement of the Arabian plate relative to the Eurasian plate (see Chapter 3).

An increase in temperature results in a decrease in the viscosity of halite layers, facilitating salt mobilization and favoring halokinesis or salt diapirism. Evaporite layers at great depths are at higher temperatures and may flow easily, favoring the formation of salt pillows and salt diapirs. The low viscosity of halite at great depths also favors the formation of listric faults, which have a curved shape and become subhorizontal at the detachment horizon. Most listric faults detach at mobile layers, particularly shale and salt, resulting in rollover structures or salt rollers (Bally, 1981; Ellis & McClay, 1988; Nelson, 1991).

Evaporite rocks can behave as viscous during geological time intervals or as brittle rocks when affected by stress during a short interval, such as earthquakes or salt mine excavations (Davison, 2009). Examples include the Messinian salt layers in the Levant Basin of the eastern Mediterranean Sea and the hydraulic fracturing of salt layers underground in several salt mines worldwide (Reiche et al., 2014). The transformation of gypsum to anhydrite results in the release of large quantities of compositional water, increasing the fluid pressure in the salt layers deep in the subsurface. The hydraulic fracturing of salt layers underground has been recognized in several salt mines, such as the potash salt mine in the Sergipe Basin of northeast Brazil and the Boulby Mine in northeast England (Davison, 2009).

5.3.2. Thermal Conductivity

Thermal conductivity is a crucial parameter in the analysis of heat transfer in sedimentary basins and geothermal investigations. The specific heat capacity (cp) is the ability of a given material to store heat (e.g., Van der Pluijm & Marshak, 2004). Thermal conductivity (or heat conductivity) is the capability of a given material to conduct or transfer heat when subjected to changes in temperature, such as water, rocks, and metals (Figure 5.4). When different parts of a body are at

Material	Specific heat (Joules/gram °C)
Liquid water	4.18
Solid water (ice)	2.11
Vapor water	2.00
Dry air	1.01
Salt	0.88
Basalt	0.84
Granite	0.79
Iron	0.45
Copper	0.38
Lead	0.13

Figure 5.4 Specific heat of common materials. Credit: Adapted from Van der Pluijm and Marshak (2004).

Mineral	Formula	Decomposition/melt point (°C)	Thermal conductivity ($W. m^{-1}\ °C^{-1}$)
Epsomite	$MgSO_4·7H_2O$	>40–80 °C, loss of water	0.6
Gypsum	$CaSO_4·2H_2O$	100–150 °C, loss of water	2.5
Bischofite	$MgCl_2·6H_2O$	>120 °C, loss of water	2
Carnallite	$KMgCl_3·6H_2O$	140–170 °C, loss of water	0.8
Kieserite	$MgSO_4·H_2O$	150–200 °C, loss of water	2
Sylvite	KCl	750–790 °C, anhydrous melt	6.7
Halite	NaCl	804 °C, anhydrous melt	6.1
Anhydrite	$CaSO_4$	1460 °C, anhydrous melt	5.7

Figure 5.5 Decomposition temperature (loss of structural water), melting point, and thermal conductivity of common evaporite minerals. Credit: Data from Schofield et al. (2014).

different temperatures, heat flows from the hotter parts to the cooler parts. Heat is a form of energy that can be transformed into work and is commonly measured in units of work, such as joules, kilowatt-hours, or kilocalories per mole (kcal/mol). Thermal conductivity is measured in $W.m^{-1}$. K^{-1} or $m.kg.\ s^{-3}.K^{-1}$ (Schön, 1996, 2015). Some reference tables use °C instead of degrees Kelvin for the temperature (e.g., Figure 5.5). Thermal conductivity is commonly represented by the Greek letter d (delta). Thermal diffusivity (a) refers to a measure of the penetration of temperature changes into a given material, controlling the time-dependent temperature distribution (Schön, 1996, 2015).

The high heat conductivity of halite layers controls the geothermal gradient in sedimentary basins, impacting diagenesis, organic matter maturation, and the prospectivity of geothermal resources. The analysis of thermal conductivity variations in the geological units of evaporitic basins is a complex theme in the salt sciences (e.g., Mello et al., 1995; Schön, 2011). In this section, only the basic aspects of the thermal properties of minerals and rocks are discussed, aiming for a concise and didactic overview. First, the physical units and parameters that are employed in the analysis of temperature variations that affect crustal materials are presented. Second, the effects of temperature changes on the deformation of evaporite sequences are discussed. Finally, the influence of thick salt layers on the thermal evolution of source rocks and petroleum generation is briefly addressed, highlighting the importance of salt for many segments of the energy industry.

Some physical units and parameters are frequently used in the analysis of temperature effects on sedimentary deposits. The measurement of thermal energy uses the joule, which corresponds to the amount of energy required to increase the temperature of a given material by 1°C. The increase in temperature caused by the same amount of heat depends on the specific heat of the substance. The higher the specific heat of a material, the more slowly it will heat up and cool down. Water has a high specific heat, so a body of water retains its temperature for a longer time when compared to other substances, such as rocks. That is why the temperatures near the seaside are warmer during the night, as the ocean waters retain the daytime temperatures longer than mountainous regions, which cool more rapidly.

The smaller the specific heat of a substance, the faster it will heat up and cool down. For example, lead has a very small specific heat (0.13 J/g °C) and requires very small amounts of energy to increase its temperature. Copper has a very small specific heat. Increasing the temperature of 1 g of this metal requires only 0.38 J of energy (Figure 5.4).

More than 4 J are required to increase the temperature of 1 g of water by 1°C. However, less than 1 joule is necessary to increase the temperature of 1 g of salt (Figure 5.4). This implies that for a given amount of heat input, the temperature variation in saline water will be greater than the temperature variation in freshwater.

Density-stratified lakes, characterized by denser water at the bottom, have higher temperatures than surface temperatures for the same degree of insolation. Some density-stratified perennial desert lakes, such as Solar Lake in the Sinai Peninsula, may display temperatures as high as 60 °C at the bottom waters, much higher than the temperature of the surficial waters (Cohen et al., 1977).

This may have a detrimental effect on many thermally intolerant organisms living in these highly saline environments.

Thermal effects can cause significant transformations in evaporite minerals, such as the dewatering of gypsum when it is converted into nodular or laminar anhydrite. This occurs when the temperature rises above 60 °C, corresponding to a burial of approximately 1 to 2 km below the surface. The loss of water molecules in this transformation results in a rock volume reduction of 39%. The conversion also changes the thermal conductivity of the rock layer by four to six, as gypsum has a small thermal conductivity (0.8 W/m ·K), while anhydrite has a thermal conductivity of 5.6 W/m ·K. The geothermal gradient through a bedded gypsum layer is about 63 °C/km, while the geothermal gradient through anhydrite is only 18 °C/km (Jowett et al., 1993). This difference results from the large thermal conductivity of anhydrite, which is more than double the thermal conductivity of gypsum (Figure 5.5).

There is a complex relationship between subsidence and the thermal evolution of sedimentary basins. The temperature increases with depth, following the geothermal gradient, and the pressure increases, following the overburden density. These factors influence the transformations of some sedimentary rocks, usually leading to a reduction in porosity and permeability due to a greater burial and the diagenesis of some sequences, particularly carbonate and evaporite layers (Schön, 2011).

The transformation from gypsum ($CaSO_4.2H_2O$) to anhydrite ($CaSO_4$) also releases heat. At 25 °C, the heat release is about 4 kcal/mol. With complete dewatering, a bedded gypsum layer that was originally 100 m thick is reduced to a 62 m bedded anhydrite layer. This conversion releases about 50 m^3 of water per square meter of cross-sectional area. At an ambient temperature of 30 °C, this volume of $CaSO_4$-saturated water can dissolve more than 8 m^3 of halite, or 54 m^3 of carnallite, a much more soluble evaporite, which is transformed into sylvite following the equation:

$$KMgCl_3 \cdot 6H_2O + 4H_2O \rightarrow KCl + Mg^{2+} + 2Cl^- + 10H_2O$$

Gypsum layers are transformed into anhydrite when buried to great depths, causing water to dissolve carnallite and produce mineral sylvite. These chemical processes occur in the subsurface, where pressure and temperature are increased. These processes are not equivalent to the dissolution of soluble minerals when exposed to the depositional surface, as in the salt glaciers in Iran, where the conditions are normal temperature and 1 atm pressure (see Chapter 3).

Halite, with a thermal conductivity three times greater than other sedimentary rocks (Figure 5.2), can serve as conduits for heat transport and affect the temperature of sedimentary layers above, adjacent to, or below a salt diapir or thick evaporite layer (Jensen, 1990; Selig & Wallick, 1966). Halite transmits heat more efficiently than most sedimentary rocks, and a diapir surrounded by siliciclastic sediments in the subsurface has a higher temperature above the diapir top than sediments at the same depth when salt is absent. The temperature variation around a salt body is positive at the top and negative at the bottom, favoring the generation of hydrocarbons from source rocks deposited above the salt or post-salt sequences (Mello et al., 1995).

The expected temperature reduction at the base of a thick evaporite layer tends to retard maturation in sedimentary successions deposited below the salt or pre-salt source rocks (Mello et al., 1995). In Atlantic-type continental margin basins, oil is mainly generated from stratigraphic sequences below the evaporite layers (see Chapter 3 in Mohriak, 2025). Giant oil fields have been discovered in the deepwater region of the Santos Basin, offshore Brazil, with reservoirs associated with stratigraphic sequences below the autochthonous salt layer, corresponding to pre-salt reservoir rocks (e.g., Carminatti et al., 2008; Mohriak et al., 2012). The presence of liquid hydrocarbons in very deep pre-salt reservoirs is only possible due to the presence of a thick salt layer, which prevents the transformation of hydrocarbons into gas accumulations (Mello et al., 1995).

Igneous intrusions in some basins positively affect immature source rocks, leading to hydrocarbon generation and heating hypersaline formation brines associated with salt deposits (Senger et al., 2017). Magmatism may also result in heating the hypersaline formation brines associated with salt deposits, resulting in the diagenetic increase of porosity in adjacent carbonate rocks. The higher heat conductivity of salt allows for the harvesting of positive temperature anomalies above shallow salt domes. In the Netherlands sector of the Zechstein salt basin, shallow diapirs show temperature differences of up to 25 °C above the salt top, allowing for more efficient geothermal energy extraction (Daniilidis & Herber, 2017).

5.3.3. Melting Point

Melting is a phase change from solid to liquid, requiring heat energy to be added to the system. For example, solar heat can transform solid ice into meltwater, requiring 334,000 J of heat energy for every kilogram of ice that is melted. The melting point of halite (sodium chloride) is relatively low, around 804 °C, and that of more soluble salts, such as sylvite, is even lower when compared to other minerals (Figure 5.5). Consequently, igneous intrusions and volcanic extrusions, which reach the surface at temperatures between 900°C and 1,000°C, result in the melting of the evaporite layers, increasing their fluidity (Gussow, 1968).

Several authors suggest that igneous intrusions in the Amazon Basin in northern Brazil caused partial melting and mobilization of the Carboniferous salt layers (e.g., Heimdal et al., 2019; Mohriak et al., 2012; Szatmari et al., 1979). The Carboniferous evaporite sequences in the basin correspond to bedded salt layers. Some boreholes indicate that at the same stratigraphic level, there is a thickness of tens of meters of halite deposits underlying the carbonate and anhydrite layers. A nearby well shows the same stratigraphic interval with an equivalent thickness of diabase, suggesting the igneous intrusion replaced the original salt layer. However, an alternative hypothesis is valid. During the magmatic episode, the direct contact between the igneous rocks and the formation water, hosted in the subsurface aquifer, might have caused instantaneous heating, forming a vaporwave and leading to the dissolution of the soluble evaporite minerals in the subsurface. Either by melting or dissolution, the borehole evidence indicates that some evaporite rocks disappeared during the magmatic event.

There are several examples worldwide of evaporite layers intruded by igneous rocks, based on analysis of seismic data, borehole data, and outcrops. In the central High Atlas region of Morocco, the Tazoult salt wall is exposed on the surface, showing the evaporite layers intercalated with basalts and diabase (Martín-Martín et al., 2017). The salt deposition in the Morocco region is Late Triassic to Early Jurassic, almost coeval with the magmatism that extended into the Middle Jurassic and intruded the carbonate sequences overlying the salt deposits. The magmatic event in Morocco corresponds to the same Central Atlantic Magmatic Province (CAMP) event observed in the Amazon Basin, with radiometric ages around 200 Ma (Marzoli et al., 1999). The evaporite deposits in the Amazon region are Carboniferous, much older than the CAMP magmatism (Heimdal et al., 2019).

The salt deposition in the Morocco region is Late Triassic to Early Jurassic, almost coeval with the magmatism, which extended into the Middle Jurassic and intruded the carbonate sequences overlying the salt deposits (Martín-Martín et al., 2017; Vergés et al., 2017). The igneous lithologies observed in the High Atlas outcrops vary from fine- to coarse-grained gabbros, dolerites, and syenites to basalts, indicating intrusive and extrusive rocks. However, the effects of the intrusion on the sedimentary rocks are local, affecting only the contact with the evaporite and carbonate sequences (Martín-Martín et al., 2017; Vergés et al., 2017).

Schofield et al. (2014) suggest that the heating and subsequent dehydration reaction of carnallite, a hydrated mineral with very low thermal conductivity, results in transforming carnallite-rich layers into viscous fluidal horizons. Carnallite starts to decompose and lose the water molecules at temperatures as low as 150 °C (Figure 5.5). Carnallite layers are preferential sites for the non-brittle emplacement of magma, forming spectacular textures mingling salt and magma, known as peperitic textures (Schofield et al., 2014).

5.3.4. Seismic Velocity

Seismic waves are particles that travel in the interior of solid materials, with the first type being the primary wave (P-wave), which arrives first in seismometers. The second type is the secondary wave (S wave), which involves particles moving perpendicular to the wave propagation direction, resulting in rotation or shear movements (Sheriff & Geldart, 1995; Telford et al., 1990). The P-wave velocity is higher than the S wave, with a ratio between the P-wave velocity (Vp) and the S wave velocity (Vs) of about 1.8 in halite layers (Figure 5.6). Primary waves travel through solids and liquids, while secondary waves do not. The Vp and Vs velocities are determined by the density and elastic moduli of rock (see Chapter 6), with low-density minerals typically having low elastic moduli. Salt is an anomaly with a low density and high seismic velocity (Figure 5.6).

The seismic velocity of sound waves penetrating an evaporite sequence is very high, with the primary compressional wave velocity (Vp) in halite reaching about 4,500 m/s or 15,000 ft/s (Bourbié et al., 1987). This results in a large impedance contrast in reflection seismic profiles acquired in regions with salt diapirs penetrating Neogene sediments, such as the Gulf of Mexico, the Mediterranean Sea, and the Red Sea. In these basins, the salt diapirs are surrounded by low-velocity rocks, such as shale or marl, with a velocity typically less than 3,000 m/s. The extremely high seismic velocity of halite layers, characterized by a rather low density (Figure 5.6), results in an impedance value that is an important factor in discriminating salt from other evaporitic and sedimentary rocks (Bourbié et al., 1987; Lenoir, 1973; Sheriff & Geldart, 1995; Taylor, 1985).

Normally, the impedance contrast is positive, with salt having a higher velocity than the surrounding sediments. However, some basins are characterized by high-velocity dolomitized carbonate rocks overlying salt sequences, resulting in anomalous negative impedance contrast at the top of the salt horizon (Sheriff & Geldart, 1995).

Using the standard seismic display, which represents a positive variation in impedance from the lower to the upper layer as a peak, the boundary between uncompacted mud sediments (with a seismic velocity of around 1,800 m/s) and the seafloor (water with a seismic velocity of 1,500 m/s) corresponds to a positive anomaly. Similarly, the contact between low-velocity shale and high-velocity salt is

Type of formation	P wave velocity (m/s)	S wave velocity (m/s)	Density (g/cm^3)	Matrix and density of constituent crystal (g/cm^3)
Scree, vegetal soil	300–700	100–300	1.7–2.4	–
Dry sands	400–1200	100–500	1.5–1.7	SiO_2 – 2.65
Oil	1200–1250	–	0.6–0.9	–
Water	1450–1500	–	1.0	–
Wet sands	1500–2000	400–600	1.9–2.1	SiO_2 – 2.65
Saturated shales and clays	1100–2500	200–800	2.0–2.4	–
Coal	2200–2700	1000–1400	1.3–1.8	
Marls	2000–3000	750–1500	2.1–2.6	–
Saturated shale and sand sections	1500–2200	500–750	2.1–2.4	–
Porous and saturated sandstones	2000–3500	800–1800	2.1–2.4	SiO_2 – 2.65
Chalk	2300–2600	1100–1300	1.8–3.1	$CaCO_3$ – 2.71
Ice	3400–3800	1700–1900	0.9	–
Limestones	3500–6000	2000–3300	2.4–2.7	$CaCO_3$ – 2.71
Dolomite	3500–6500	1900–3600	2.5–2.9	$(Ca, Mg)CO_3$ – 2.8–2.9
Salt (Halite)	4500–5500	2500–3100	2.1–2.3	NaCl –2.1
Anhydrite	4000–5500	2200–3100	2.9–3.0	$CaSO_4$ – 2.9
Gneiss	4400–5200	2700–3200	2.5–2.7	–
Granite	4500–6000	2500–3300	2.5–2.7	–
Basalt	5000–6000	2800–3400	2.7–3.1	–

Figure 5.6 Seismic velocities of primary and secondary waves (Vp and Vs) and the mean density of some common materials and lithological types.

represented by a strong positive peak using this convention. The contact between the shale layer and the halite layer (velocity of 4,500 m/s) results in a much stronger seismic signal than the seafloor horizon (Sheriff & Geldart, 1995).

The interval transit time for pure halite layers is about 221.1 µs/m (67 µs/ft), resulting in very high seismic velocities that are only lower than anhydrite layers (Bourbié et al., 1987; Sheriff & Geldart, 1995; Telford et al., 1990). The transit time for transmission of sound waves across sedimentary layers is commonly recorded by petrophysical logs, particularly the sonic log, which is normally acquired during the drilling of petroleum exploration wells (Serra, 1984, 1986).

The seismic time interval for the propagation of sound waves across a salt diapir is much smaller than the seismic time interval for the propagation across siliciclastic sedimentary layers adjacent to the structure. This difference in seismic velocity causes an apparent rise of the reflectors below the base of the diapir structure, which corresponds to a common pitfall in seismic interpretation. Anhydrite layers, characterized by an even higher seismic velocity (5,500 m/s), typically exhibit a stronger impedance contrast in seismic profiles. Anhydrite layers are normally not deformed into salt diapirs and the pre-salt reflectors below bedded salt are affected more homogeneously. However, there are some fold-and-thrust belts with anticlines detaching on gypsum and anhydrite layers, forming compressional structures that generate pull-ups on seismic profiles. For example, the propagation folds in the Tarim Basin, northwest China, are controlled by reverse faults detaching on the Cambrian gypsum layers, stacking Lower Paleozoic high-velocity carbonate rocks that are involved in the fold-thrust belt (Zhang et al., 2021).

5.3.5. Magnetic Susceptibility

Salt has a null magnetic susceptibility due to the absence of magnetic minerals in evaporite sequences (Figure 5.2). Halite is considered a non-magnetic mineral with very low values of magnetization (Nettleton, 1962, 1976; Sheriff & Geldart, 1995). The low magnetic susceptibility of halite (around –0.8 in micro-cgs units) makes it a paramagnetic mineral, unlike other rocks that are diamagnetic, such as sandstone and shale, which display magnetic susceptibility values between +30 and +50 units of micro-cgs (Dobrin & Savit, 1988; Kearey et al., 2002).

Halite forms salt diapirs, resembling igneous plugs in seismic profiles. These diapirs can be distinguished from magmatic intrusions using geophysical methods like gravity and magnetics. Igneous intrusive rocks have a high density, often exceeding 3 g/cc, and high values of magnetic anomaly (Figure 5.6). Mafic rocks are rich in magnetic minerals and exhibit well-defined magnetic anomalies. Geophysical surveys incorporating gravity and magnetic anomalies are crucial for distinguishing igneous structures from salt diapirs.

5.3.6. Integration of Geophysical Data in the Interpretation of Salt Structures

The Atlas Mountains in northern Africa are home to salt structures that demonstrate the application of evaporite sequences to constrain geological interpretation. These structures extend from Morocco, Algeria, and Tunisia and correspond to an inverted rift trough with Triassic to Jurassic sediments. They are overlain by Cretaceous and Cenozoic post-rift deposits in depressions like the Guercif Basin (Rezouki et al., 2020).

Seismic and potential field datasets were used to map subsurface structures in northeastern Tunisia, revealing the formation of salt pillows, salt domes, and salt plugs (Azaiez et al., 2008). The vintage seismic data acquired in the past decades provide poor images of the salt structures under the mountains. The geological interpretation was supported by the integration of gravity and magnetic anomalies, which allowed the distinction between salt pillows, salt domes, and salt plugs (Azaiez et al., 2008). These plugs have been interpreted as Triassic salt diapirs formed during the deposition of the Jurassic, Cretaceous, and Cenozoic sequences (Azaiez et al., 2008; Vergés et al., 2017). Gravity lows were interpreted by a negative density contrast with lateral siliciclastic and carbonate rocks that outcrop on the elevated Atlas Plateau. These negative anomalies were modeled as large, tall structures associated with low-density minerals (halite) forming salt diapirs, some with tops located close to the surface (Azaiez et al., 2008). Geological models that combine seismic, gravity, magnetic, and borehole data (e.g., Djebbi & Gabtni, 2015) indicate that some structures have a squeezed salt diapir stem with a mushroom-shaped canopy. This suggests that the salt was extruded as allochthonous salt glaciers fed by a central vent (see Chapter 3).

The sedimentary troughs in Tunisia host a large salt province characterized by elongated salt walls intruded by Late Triassic to Late Jurassic igneous rocks (Martín-Martín et al., 2017; Saura et al., 2014; Vergés et al., 2017). The peak of magmatic activity, corresponding to the CAMP magmatic event, straddled northwest Africa, eastern North America, and northern South America (Marzoli et al., 1999). Subsequently, the basin was inverted by the Cenozoic Alpine orogeny, which likely resulted in salt extrusion onto the depositional surface during later phases of tectonic compression (Martín-Martín et al., 2017; Vergés et al., 2017). Most of the halite in the outcrops has been dissolved and transported, resulting in the gypsum-rich outcrops typical of the Saharan Atlas in Tunisia (Azaiez et al., 2008). These outcrops contrast with the salt outcrops related to the Zagros orogeny in Iran, where halite has been preserved due to extreme climate aridity (see Chapter 3).

Field geology has contributed to the detailed mapping of the salt structures and associated mini-basins (e.g., Célini et al., 2022). Some salt structures occur underground and can only be interpreted by geophysical methods. The integration of geophysical datasets indicates that the rift troughs in the Tunisian Atlas are related to a system of SW-NE trending basement-involved faults, interpreted by gravity and magnetic anomalies. The rift sequences and the late rift evaporite deposits are characterized by intrusive and extrusive igneous rocks (Saura et al., 2014). The CAMP and subsequent magmatic events formed dikes and sills within the salt walls and affected the overlying carbonate rocks dated as Early, Middle, and Late Jurassic (Martín-Martín et al., 2017; Vergés et al., 2017).

5.4. PETROPHYSICAL EVALUATION OF EVAPORITE SEQUENCES

There are more than 4,000 officially recognized mineral species described in geological literature. However, less than a hundred are classified as evaporite minerals, and fewer than a dozen are normally recorded in sedimentary basins. Their physical and chemical properties are crucial to identify different lithologies (Figure 5.7). The rarity of these minerals is due to their high solubility, which is not preserved in sedimentary successions due to postdeposition dissolution. However, some minerals forming thin layers can be recognized using gamma-ray curves in petrophysical logs. Identifying rare minerals often requires advanced techniques like microscopic petrography and X-ray diffractometry (see Chapter 7).

Evaporite layers contain minerals that depend on the composition of the original brine and the chemical and physical conditions for crystal precipitation and accumulation. Some minerals, like gypsum, are preserved in evaporite sequences due to the pressure and temperature stability fields. Anhydrite is recognized below 600 m in petrophysical logs due to its high-density value (Figure 5.7).

The identification of different evaporite minerals in a sedimentary basin is generally conducted by the description of ditch samples, which are recovered while drilling a well during petroleum exploration, or from cores obtained from specific intervals when searching for sylvite, a valuable potassium chloride (Figure 5.8). When direct

Mineral	Chemical Composition	Density (g/cm^3)	Transit Time (microsec/ft)	Neutron porosity	Gamma Ray (API)
Halite	$NaCl$	2.03	67.0	0	0
Sylvite	KCl	1.86	74.0	0	500
Carnallite	$KMgCl_3 * 6H_2O$	1.57	78.0	65	200
Tachyhydrite	$CaMg_2Cl_6 * 12H_2O$	1.72	92.0	> 50	200
Anhydrite	$CaSO_4$	2.98	50.0	0	0
Gypsum	$CaSO_4 * 2H_2O$	2.35	52.5	49	0
Kieserite	$MgSO_4 * H_2O$	2.57	–	43	0
Langbeinite	$K_2Mg2(SO_4)3$	2.82	52.0	0	275
Polyhalite	$K_2Ca_2Mg(SO_4)4 * 2H_2O$	2.79	57.5	15	180
Dolomite	$CaMg(CO_3)2$	2.88	43.5	4	0
Calcite	$CaCO_3$	2.71	49.7	0	0

Figure 5.7 Petrophysical characteristics of common evaporitic minerals.

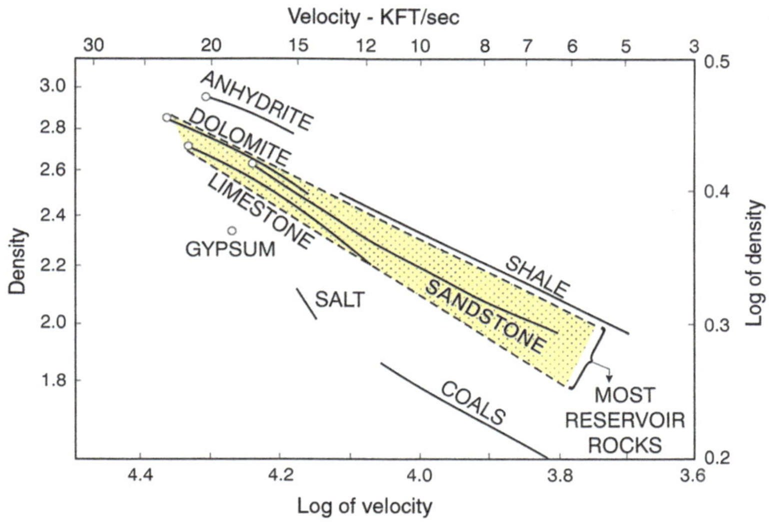

Figure 5.8 Graph showing the density and the logarithmic value of the sonic velocity measured by wireline logs. The plot is used to discriminate between different evaporite minerals in common sedimentary rocks.

rock evidence is unavailable, interpretation is often based on well logs, which can discriminate between different salts based on physical properties like density, transit time, and radioactivity (Lenoir, 1973; Nurmi, 1978; Taylor, 1985). Cross-plots of different properties, such as density and sonic velocity, are often used to interpret different evaporite minerals in sequences (Gardner et al., 1974; Taylor, 1985).

5.5. GEOPHYSICAL METHODS APPLIED TO THE INTERPRETATION OF EVAPORITE SEQUENCES

Geophysical methods, such as gravity and magnetic anomalies, are crucial in basin analysis projects involving the interpretation of evaporite sequences. The gravity method, one of the first techniques, was used to map salt structures in the subsurface due to the low density of halite masses in the subsurface. It was widely employed on the United States Gulf Coast during the early phases of petroleum exploration, leading to the discovery of salt domes in the subsurface. Sophisticated computer algorithms are now used to filter potential field datasets and obtain residual gravity anomalies, upward and downward continuations, and vertical and horizontal derivatives of the anomalies (e.g., Aghajani et al., 2009; Nabighian et al., 2005). These methods help determine the depth and geometry of anomalous bodies in the subsurface (e.g., Filina et al., 2015).

The magnetic method is useful in interpreting basement depth in sedimentary basins, as igneous and metamorphic rocks are more magnetic than the overlying siliciclastic, carbonate, and evaporite layers. In continental margins, it is useful to determine regional structures such as transform fault zones, basement highs, igneous intrusions, tectonic lineaments, and the limit between the continental and oceanic crust (Sheriff & Geldart, 1995).

Potential field methods are less expensive than seismic methods, with less than 5% of oil companies' exploration budgets allocated to gravity and magnetic surveys. Geophysical methods have relevance beyond hydrocarbons, with subsurface geology being useful for various Earth sciences applications, including mineral exploration, geothermal energy, carbon storage, nuclear waste disposal, and underground storage (e.g., English & English, 2022; Verweij, 2012).

5.5.1. Introduction of the Seismic Reflection Method in the Salt Sciences

The seismic reflection method is a widely used geophysical method for basin analysis and petroleum exploration (e.g., Herron, 2011). It involves the generation of a sound wave on the ground surface or underwater, transmitted through sedimentary layers, and reflected at geological horizons in the subsurface (Figure 5.9). The interface between two layers with different physical properties corresponds to a seismic reflector. The reflected energy is detected on the surface using instruments that receive the seismic wave impulse and convert the signal into a digital format, which is processed using mathematical algorithms (Kearey et al., 2002; Sheriff & Geldart, 1995).

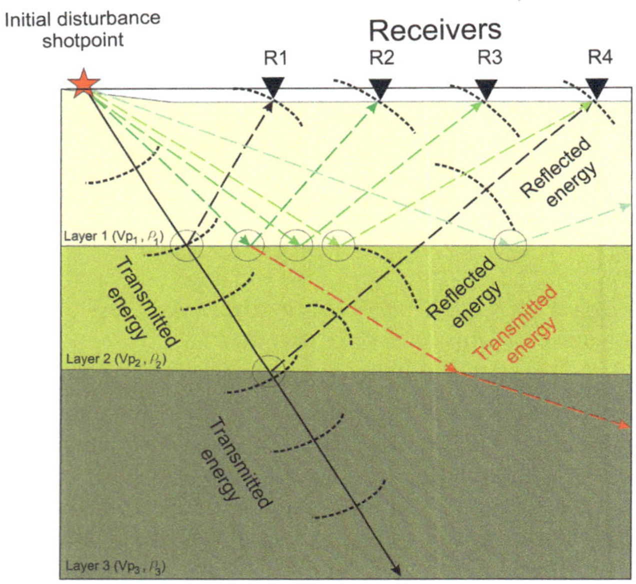

Figure 5.9 The diagram illustrates the process of wave energy transmission in a sedimentary basin, where reflected and refracted waves are transmitted across different layers. The energy pulses are recorded by receivers at different positions. The interface between layer 1 and layer 2 is recorded by multiple receivers, and the distance from the shot point to the receiver position results in a delay in recording the reflected energy. This results in the transmission of the energy ray path in the subsurface, which corresponds to a seismic wave refracted by different layers. Credit: Adapted from Sheriff and Geldart (1995).

Seismic reflection surveys can be acquired along profiles (2D seismic lines) or covering an area (3D seismic cube). The method provides reliable images of the reflectors in the subsurface with an adequate vertical and horizontal resolution, up to a few tens of meters, depending on the sedimentary rock velocities (e.g., Aminzadeh & Dasgupta, 2013; Dobrin & Savit, 1988). The reflection seismic profiles are useful tools for interpreting the stratigraphic and structural relationships of sedimentary strata (e.g., Mitchum Jr. et al., 1977). The interpretation of reflection seismic data profiles is commonly calibrated by results from nearby boreholes using a depth-to-time conversion, in case the seismic data has not been converted to a depth profile. This procedure allows the interpreter to determine the correspondence between geological units and stratigraphic markers with the seismic horizons (e.g., Herron, 2011; Misra & Mukherjee, 2018).

Salt basins are characterized by stratigraphic sequences associated with evaporite deposits. These layers are typically well-imaged using the seismic reflection method due to the physical properties of the evaporite rocks. Evaporite sequences are characterized by chemical compositions and physical properties that allow their distinction from other lithologies by wireline log responses (petrophysical logs) and by their deformation after deposition, forming salt pillows, domes, and diapirs. Highly deformed salt layers are less well-imaged, especially when affected by contraction, and the geological interpretation is based on analogies with outcrops, physical models, and subsidence analysis (Célini et al., 2022; Fernández et al., 2021; Herron, 2011; Jackson & Hudec, 2017; Jones & Davison, 2014; Strauss et al., 2021).

Salt deposits can form in different depositional settings, from continental sabkhas and playa lakes to epeiric seas, gulfs, continental platforms, and deep marine environments (Warren, 2006, 2010, 2016). Evaporitic stratigraphic sequences are characterized by seismic reflection patterns that allow distinction between sedimentary environments, particularly in shallow and deepwater settings, and may provide constraints on the lithological interpretation. For example, evaporite deposits associated with carbonate and sulfate minerals usually display a stratified seismic signature, whereas thick halite deposits may form diapiric structures (see Chapter 7 in Mohriak, 2025).

Direct methods in Earth sciences are based on observation and description of outcrops and mines, allowing sampling of the rocks. Indirect methods, such as geophysical surveys, provide some information on the rock parameters, which may result in nonunique interpretations. The analysis of rocks on the surface or subsurface allows the characterization of depositional environments and the geometry of salt structures. Although basic field geology techniques are extremely useful for the preparation of maps and cross-sections, they lack the depth control afforded by the drilling of exploratory wells, which provides fundamental stratigraphic information for the subsurface units. Indirect geophysical methods are useful to determine the attributes of different environments and tectonic settings of sedimentary strata deeper in the subsurface, particularly the seismic method (e.g., Sheriff, 1976; Sheriff & Geldart, 1995). The use of 3D reflection seismic images allows the visualization of subsurface geological structures in great detail (e.g., Herron, 2011; Nanda, 2016).

Innumerable academic and applied textbooks include seismic examples of salt structures in sedimentary basins, such as Jenyon (1986) and Jackson and Hudec (2017). There are several specialized books published by geological associations, such as the American Association of Petroleum Geologists (AAPG), the Geological Society of London (GSL), the Society of Exploration Geophysicists (SEG), and the American Geophysical Union (AGU).

Trusheim (1960) was one of the first geoscientists to apply the seismic method for visualization of subsurface geology and investigate the evolution of simple and complex salt structures in the North German Basin. His work paved the way for the use of seismic data for the interpretation of salt structures worldwide. Trusheim (1960) proposed a halokinetic model related to gravity forces rather than compressional forces, as previously proposed by European Earth scientists before the 1960s. Some of the landmark publications that discuss the interpretation of salt tectonics based on the analysis of reflection seismic profiles and physical models include Alsop et al. (1996, 2012), Braunstein and O'Brien (1968), Jackson and Hudec (2017), and Jackson et al. (1995).

5.5.2. Basic Principles of the Seismic Method

In the past 50 years, the seismic reflection method has been widely used in the petroleum industry, mineral exploration, and academic research projects for basin analysis, particularly in inferring stratigraphic and structural elements from seismic data (e.g., Bally, 1983; Mitchum et al., 1977; Sheriff, 1976). It has also been used in projects involving carbon dioxide sequestration, characterization of salt domes for nuclear waste and gas storage, engineering, environmental studies for seafloor hazards, and characterization of archaeological sites by high-resolution investigations.

The fundamental principles of the seismic method involve the propagation of sound waves through the Earth's crust, which is characterized by different layers with different physical properties, such as density and seismic velocity (Misra & Mukherjee, 2018; Sheriff & Geldart, 1995). Density is related to the chemical composition of rock layers and porosity, while seismic velocity corresponds to the wave speed across a given medium. When a seismic wave encounters the boundary between two adjacent layers with different physical properties, part of the wave energy reflects off the interface, forming a reflected wave, and continues to propagate across the new medium as a refracted wave (Figure 5.9).

The seismic method is based on generating sound waves at the surface using multiple sources, such as dynamite explosions or vibrators on land and sparkers and shots from an air gun offshore. These sound waves propagate into the subsurface until an interface separates layers with different velocities and densities. When this interface is encountered, the reflected wave returns to the surface, following Snell's law of reflection, which means that the angle of the reflected wave with the normal to the surface is equal to the angle of incidence (Figure 5.9).

In sedimentary basins with horizontal layers, the normal to the surface is a vertical line. If a sound wave reaches the interface with an angle of 45°, then the reflected angle follows Snell's law and is reflected in the same angle (Figure 5.9). The refracted angle with which sound will penetrate the underlying layer will change its path depending on the contrast in seismic velocities of the units above and below the interface (e.g., Sheriff & Geldart, 1995). The transmitted wave will follow a path that depends on the following equation:

$$\sin(o_1)/V_1 = \sin(o_2)/V_2$$

where o_1 is the angle of incidence (measured between the ray path and the normal to the surface), and V_1 is the velocity of the sound wave in the upper medium (1). Consequently, o_2 is the angle of refraction (measured between the ray path and the normal to the surface), and V_2 is the velocity of the sound wave in the lower medium (2).

In a scenario where layer 1 has a P-wave velocity of 2,000 m/s, it propagates through vibrations until the incident ray reaches the interface between layers 1 and 2. If the incident ray has an angle of 30° with the plane, the refraction angle for a layer with a P-wave velocity of 2,500 m/s will be 39°.

According to Snell's law, a medium with a higher velocity results in a higher refraction angle, causing the wave path to approach the horizontal line at higher velocities. When the angle of incidence is too high, the energy is refracted along the interface, resulting in no transmitted wave (Sheriff & Geldart, 1995).

For sound waves approaching the interface with an almost vertical angle, most energy is transmitted or refracted, but a small part is reflected back to the surface (Sheriff & Geldart, 1995). This energy pulse is measured by equipment placed on the Earth's surface. These receivers correspond to geophones in land seismic acquisition and hydrophones in marine seismic acquisition (e.g., Aminzadeh & Dasgupta, 2013).

The reflection coefficient is a fundamental concept in reflection seismology, representing the difference in density and velocity between layers below (layer 2) and above the interface (layer 1). For normal incidence waves, the reflection coefficient (RC) is given by the following equation:

$$RC = \left(r_2 V_2 - r_1 V_1\right) / \left(r_2 V_2 + r_1 V_1\right)$$

where r_2 and V_2 are the density and compressional wave velocity of the medium below a reflecting interface, and r_1 and V_1 are the density and compressional wave velocity of the medium above a reflecting interface.

The reflection coefficient of a shale layer (density 2.2 g/cc, velocity 2,500 m/s) above a carbonate layer (density 2.5 g/cc, velocity 3,500 m/s) is 0.228 (Figure 5.6).

The time it takes for the sound waves to travel across the layers increases as the distance between the shot point and the receivers increases (Figure 5.9). The processing of seismic signals includes a correction for an increasing offset between the shot point and receivers. The near and far offsets are characterized by an increase in the reflection time. The measured energy is recorded after the waves are reflected by a given seismic horizon (Figure 5.10).

The simplified diagram illustrating a distance versus two-way time plot for different reflectors (Figure 5.11) shows energy peaks for different receivers that picked up the energy reflected from the boundary between the layer 1 and layer 2. The seismic profile consists of multiple traces, corrected by processing techniques, representing the interface between layers. The seismic reflector is characterized by a series of peaks and troughs that define a horizon, also known as a seismic horizon. In geological terms, the seismic horizon represents an interface between sedimentary layers with different impedance contrasts, forming a seismic profile (e.g., Sheriff & Geldart, 1995).

The impedance contrast is the difference in density and velocity between sedimentary layers, calculated as the product of density and velocity. For instance, a salt layer above a carbonate layer has twice the impedance of a shale and carbonate layer. The strength of seismic reflectors is related to the impedance contrast and seismic amplitude, which are attributes of the seismic method used to identify different lithological types (Aminzadeh & Dasgupta, 2013). The seismic signal in a processed reflection seismic profile is directly related to the

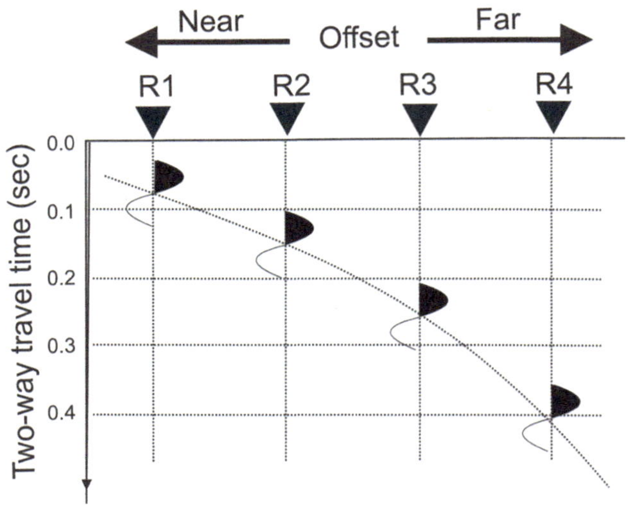

Figure 5.10 The interface between layer 1 and layer 2 is recorded with varying travel times for each receiver due to different wave paths as the distance from the shot point increases. Near receivers record the seismic pulse as an energy peak above the corresponding energy peak, whereas far receivers record the energy farther from the shot point. Arrival times increase with distance, forming a hyperbola. Near receivers have a smaller two-way travel time than far receivers. The seismic processing accounts for offset distance and corrects the computation of two-way travel time for reflectors. The seismic profile has different traces, reinforced by the addition of different receivers, known as the fold of the seismic profile. Credit: Adapted from Sheriff and Geldart (1995).

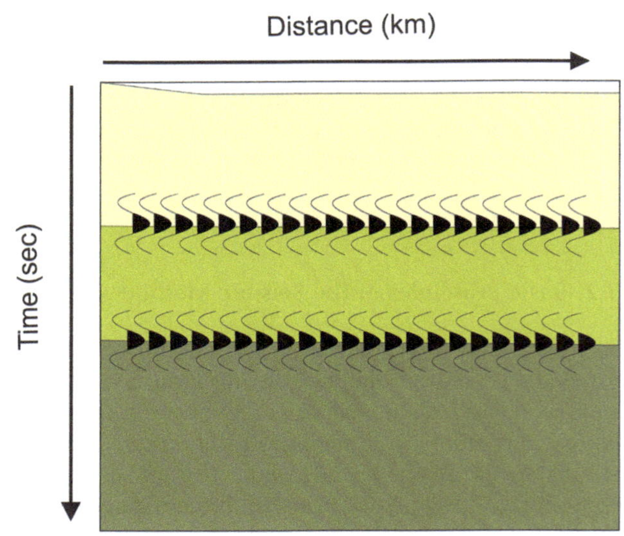

Figure 5.11 The computational algorithms in seismic processing consider the different energy pulse recordings between layer 1 and layer 2, resulting in a seismic profile with accurately positioned wavelets, enabling a more realistic visualization of the seismic horizon, crucial for geological interpretation. Credit: Adapted from Sheriff and Geldart (1995).

impedance contrast between sedimentary layers, as given by the equation:

$$I = V \cdot r \text{ (where V is measured in m/s and r in g/cc or kg/m}^3\text{)}$$

Salt has a density of 2,000 kg/m3 and a velocity of 4,500 m/s (Figure 5.6), making it a strong contrast with a shale sequence with a density of 2,200 kg/m^3 and a velocity of 2,500 m/s, resulting in an impedance contrast of about 3.5×10^6 kg/m^2·s. This results in a strong seismic signal at the boundary between lithologies. If the salt layer overlies a carbonate layer, the impedance contrast is smaller, assuming the carbonate layer has a density of 2,500 kg/m^3 and a velocity of 3,500 m/s (Figure 5.6). A salt–carbonate interface exhibits an impedance contrast that results in a stronger reflector compared to the equivalent shale–carbonate interface. However, the impedance depends on the fluid content in the rock, and this property has been used to discriminate between hydrocarbon-saturated and water-saturated reservoirs (Aminzadeh & Dasgupta, 2013).

Reflected waves in seismic profiles do not show the lithology of the layers, but rather the impedance contrast between the sedimentary layers. Reflection seismic techniques are powerful in determining the layering of sedimentary sequences and other stratigraphic or structural elements, such as unconformities, faults, igneous intrusions, and fluids present in reservoir rocks. Processing seismic data yields a value for seismic velocity for intervals, which can be used to work back elastic properties and densities, which are useful parameters in determining possible lithological types (Aminzadeh & Dasgupta, 2013). Most oil companies use reflection seismic profiles as their main geophysical tool for exploring hydrocarbon resources in sedimentary basins, either on land or offshore, aiming to reduce exploratory risks (Herron, 2011; Nanda, 2016).

Marine seismic acquisition surveys employ a source (air gun) and a long cable (streamer) with hydrophones to record the reflected seismic signal. The receivers (hydrophones) are towed by one or more cables, known as streamers (Figure 5.12). In the subsurface, the seismic horizons are sampled by several ray paths using the same shot. These seismic waves are reflected by the sedimentary layers with impedance contrasts that depend on the physical properties of the sediments, and the signal is recorded at the surface by several receivers. The precise positioning of the shot points and receivers is usually obtained by satellite Global Positioning Systems (GPS).

The processing of seismic data involves removing various artifacts, such as increasing offsets, noises, and multiples that are not related to the reflected wave energy. These spurious signals must be removed from the data, and stacked multiples can appear in the seismic profile as an artifact of the technique (Aminzadeh & Dasgupta, 2013; Herron, 2011).

One of the most commonly used techniques is the common midpoint (CMP) method, which sums the different energies recorded in several receivers symmetrically

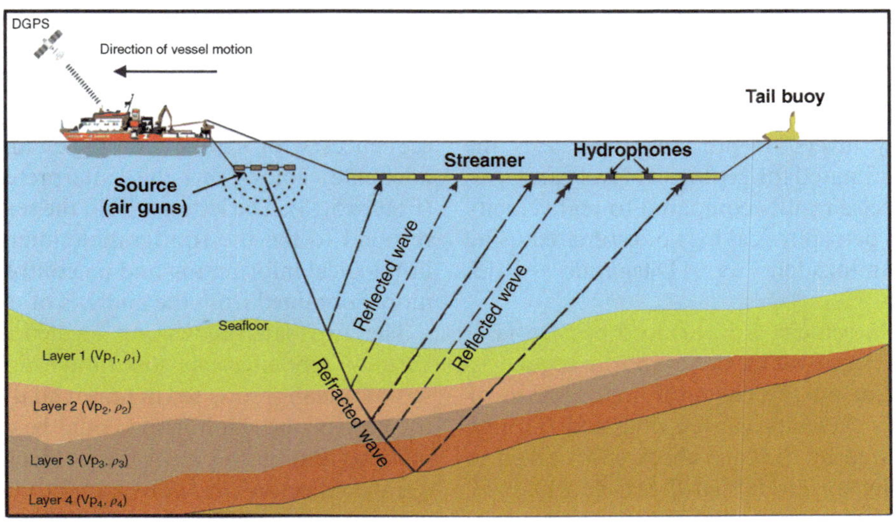

Figure 5.12 Marine seismic surveys use air guns to generate seismic waves, causing explosions or vibrations in the water. Marine seismic surveys utilize special ships with satellite navigation systems and employ air guns as the source of seismic energy. The impedance contrasts at subsurface layer interfaces result in reflected wave signals recorded by hydrophones towed along a streamer cable. Credit: Adapted from Sheriff and Geldart (1995).

from the midpoint receiver that will gather the signals (Sheriff & Geldart, 1995). The data recovered from the shot point gathers are summed or combined, forming a single CMP gather and reinforce the signal by repeating the same subsurface position at the interface. The number of traces stacked is known as the fold number in the seismic data. A well-processed seismic profile is essential for realistic visualization of seismic horizons and contributes to reliable geological interpretation.

Migration is another processing technique used to correctly image subsurface interfaces. Sedimentary layers usually show some dip, which can be extremely high, more than 80° relative to the horizontal plane, and correspond to a subvertical bed. Any interface steeper than 45° will reflect the seismic signal off to the infinite horizontally, resulting in refraction, and the ray path will not be recorded by the receivers. The reflected signals can come from below or laterally if the beds are dipping.

The summing of seismic traces is complicated when the beds are steeply dipping, as there is a shift between the reflected point and the true point of the reflector in the sedimentary layer. The subsurface points are shifted to their correct position by numerical methods known as migration. Most seismic profiles are migrated to allow for a correct structural interpretation. Migration depends on a velocity model of the subsurface, and modern techniques of seismic migration can accurately mimic wave propagation across different media with different geometries using sophisticated algorithms based on the wave equation (Aminzadeh & Dasgupta, 2013; Herron, 2011).

Deep migration results in a seismic profile that shows the depth (in meters) instead of time (seconds). This type of seismic section can integrate information directly obtained from boreholes, which use depth (meters or feet) in the vertical axis. Migration is often carried out in the time domain, using a root mean square velocity for depth intervals as an approximation for the real velocity of the sedimentary layers. Depth migration uses the interval velocities estimated for geological layers and usually yields more precise results compared to real velocity data obtained from petrophysical logs or estimated using check shots (Aminzadeh & Dasgupta, 2013; Herron, 2011).

Sedimentary strata velocity is influenced by factors in carbonate and siliciclastic rocks, but evaporite sequences are mainly dependent on the chemical composition of constituent minerals. Velocity can be obtained through sonic logs or measured by the time check shots taken to reach the surface (Herron, 2011; Sheriff & Geldart, 1995). After processing and quality control checks, seismic lines can be used by interpretation teams. Synthetic seismograms are often used to calibrate seismic profiles with boreholes, involving the correlation between stratigraphic markers and seismic horizons (Herron, 2011).

The geological age is typically based on biostratigraphic analysis of drilled lithologies recovered from boreholes and cores. Correlating well-log information with nearby wells helps identify the ages of reflectors and compare different sedimentary successions. An example of a seismic profile in the North Sea is shown in Figure 5.13, part of a regional seismic survey in the North Sea (Taylor, 1981).

The reflectors or seismic horizons observed in the North Sea (Figure 5.13-top) correspond to the interfaces between sedimentary layers with different impedance contrasts. The interpretation of the seismic line, based on results of exploratory wells drilled by the petroleum industry in the 1970s and 1980s, shows undulating reflectors, forming anticlines and synclines. A major unconformity occurs at the top of the seismic profile, with an erosional truncation affecting the Tertiary strata. The line interpretation uses stratigraphic information to infer the ages of several sequences, for example, the Lower Tertiary, Upper Cretaceous, or Upper Jurassic sequences, which are identified by orange, green, and blue lines. The Zechstein Group (marked by a magenta color line) includes the Permian evaporites and is characterized by thickness variations. The anticlinal structures are characterized by a greater thickness of the salt layer, whereas the synclinal structures exhibit less salt. These variations are not related to the sedimentary environment at the time of deposition but to salt flow or halokinesis after the sedimentary layers were deposited (see Chapter 7 in Mohriak, 2025). The salt layer was deposited as a horizontal bed, but salt flow resulted in the development of a pillow, a salt structure characterized by a small amplitude fold, resulting in a thicker salt layer (Figure 5.13).

The seismic profile is a geological interpretation of a geoseismic profile, which is a geological cross-section based on seismic data displayed with the vertical axis in time (Figure 5.14). The seismic reflectors are interpreted as lines that correspond to the top of stratigraphic sequences with similar lithologies and ages, filled with adequate colors in the interpreted seismic profile (Figure 5.14). The letters within the seismic sequences correspond to the the stratigraphic interpretation based on lithological information and paleontological dating, commonly obtained from the analysis of exploratory wells.

The two salt pillows are located above the pre-salt, Permo-Carboniferous structural highs. However, the seismic data is processed in time, and the apparent uplift of the base of the salt horizon could be caused by a velocity pull-up, resulting in a seismic artifact. Discontinuities in seismic reflectors may correspond to faults offsetting geological units, such as the fault below the southwest salt pillow, which offsets the base of the salt reflector. Other seismic discontinuities may correspond to periods of erosion that removed parts of the stratigraphic sequences (Figure 5.14). The Lower Tertiary geological unit is missing

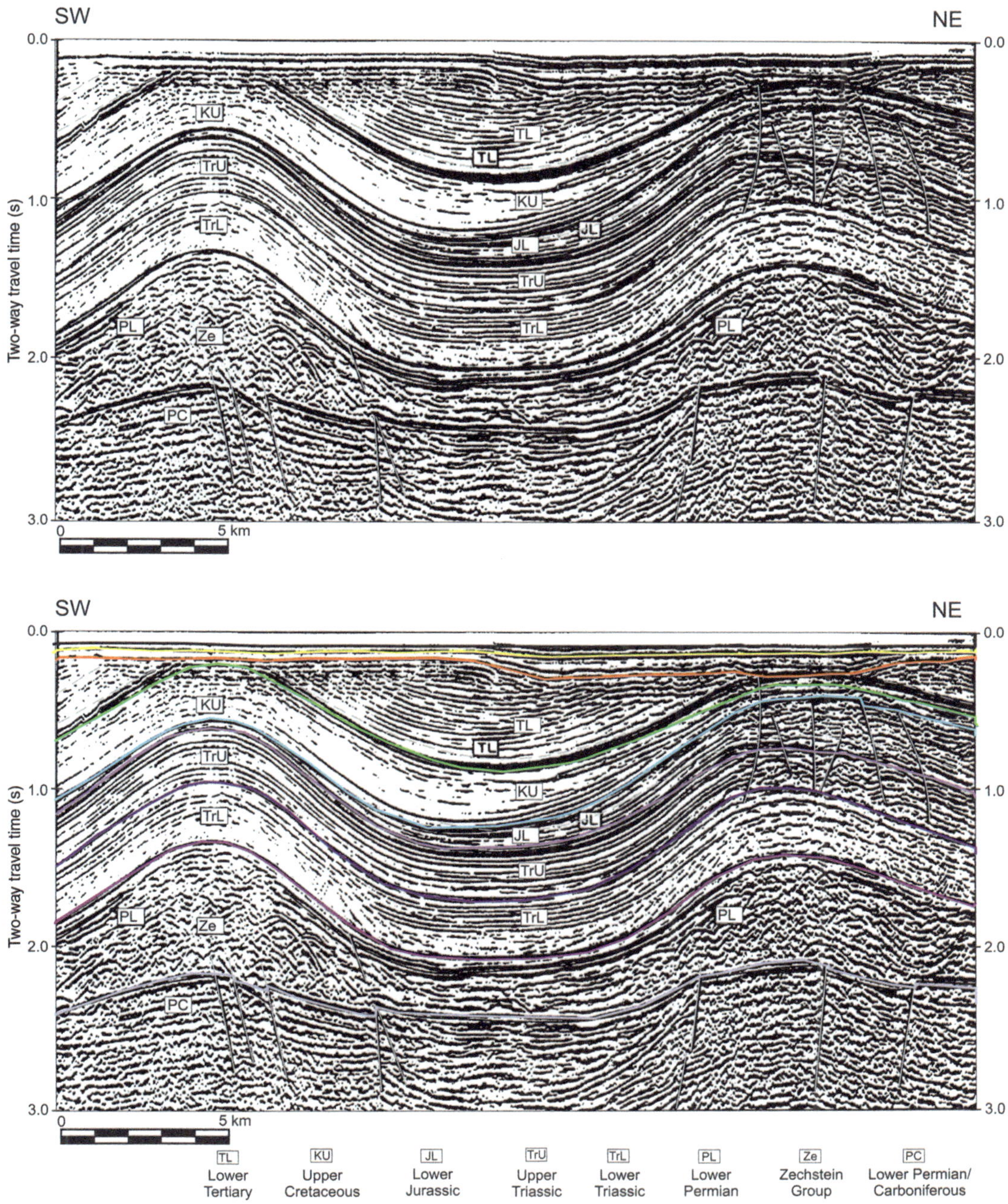

Figure 5.13 A 2D seismic profile from the North Sea displays uninterpreted data (top) and a simplified line interpretation (below). The axis on the left side measures the travel time of seismic waves from an air gun to the seismic horizon, which is recorded in hydrophones. The two-way travel time is determined by the downward and upward ray paths. The datum for the seismic line corresponds to sea level, where hydrophones were placed to receive signals. The bottom image identifies the main reflectors, which correspond to geological units. The pink horizon represents the top of the Permian Zechstein salt, while the light blue reflector represents the top of the Permo-Carboniferous strata.

above the southwest salt pillow, as the Upper Cretaceous sedimentary sequences are directly overlain by Upper Tertiary deposits. This suggests an angular unconformity associated with the uplift of the post-salt strata during the Tertiary (Taylor, 1981).

The Zechstein Group in the seismic profile is interpreted as a mass of evaporites deposited in the Permian and covered by Lower Triassic sediments (Figure 5.14). There is almost no variation in the Lower Triassic layer thickness, indicating uniform sedimentation and

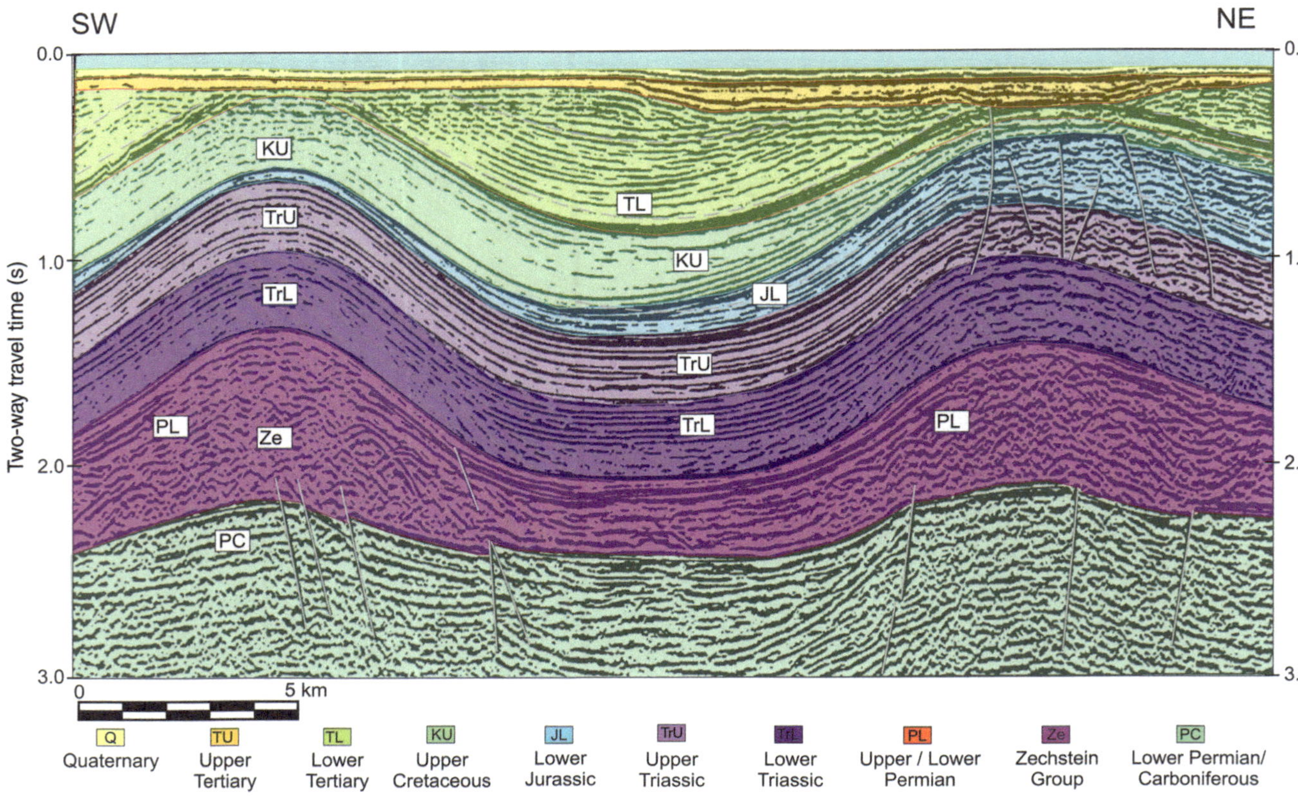

Figure 5.14 The interpretation of a 2D seismic profile in the North Sea using stratigraphic information and paleontological dating from exploratory well analysis shows the seismic sequences marked with different colors. Two salt pillows are located on the pre-salt Permo-Carboniferous highs. The seismic data is processed over time, and the uplift of the salt horizon is likely due to a velocity pull-up. Discontinuities in seismic reflectors may correspond to faults offsetting geological units, such as the fault below the southwestern salt pillow. Other discontinuities may be due to erosion of geological units, as seen in the Upper Cretaceous unit overlain by Upper Tertiary sequences.

negligible salt flow at the time of deposition of these sediments. The Lower Jurassic layers show marked variations in thickness along the profile, with a much thicker sequence at the NE extremity of the seismic line, indicating changes in accommodation space. Some small faults are interpreted below the Zechstein salt reflector, indicating that the Permo-Carboniferous sequence was affected by extensional tectonic events. However, the seismic data was processed in time and the apparent uplifted horst structures below the salt pillows might be caused by velocity artifacts (e.g., Herron, 2011; Sheriff & Geldart, 1995).

5.5.3. Reflection Seismic Methods Applied to Basin Analysis and Salt Tectonics

Seismic profiles and petroleum exploration results are often used to determine the depositional environments of carbonate and evaporitic sequences in sedimentary basins (e.g., Fontaine et al., 1987). The internal platform and intertidal zone of continental margins are examples of depositional environments where evaporites are commonly found (Figure 5.15).

This type of evaporite environment with bedded salt is observed in modern sedimentary basins like the Arabian Gulf and recognized in many ancient sedimentary basins, such as the Permian Delaware Basin in the southwestern United States (see Chapter 8). In ancient divergent margins, like the South Atlantic or Red Sea, salt structures are distributed in distinct zones, corresponding to salt pillows, diapirs, and walls (Figure 5.16). The distal parts of many passive margins are characterized by allochthonous salt sheets that advance toward the oceanic crust (Demercian et al., 1993; Hudec & Jackson, 2007; Jahnert et al., 1998; Mohriak, 2005).

A classical paper by Sheriff (1976) highlights the use of seismic data in the petroleum industry to infer the structural and stratigraphic development of sedimentary basins. In summary, seismic profiles are useful tools for basin analysis projects, determining elements such

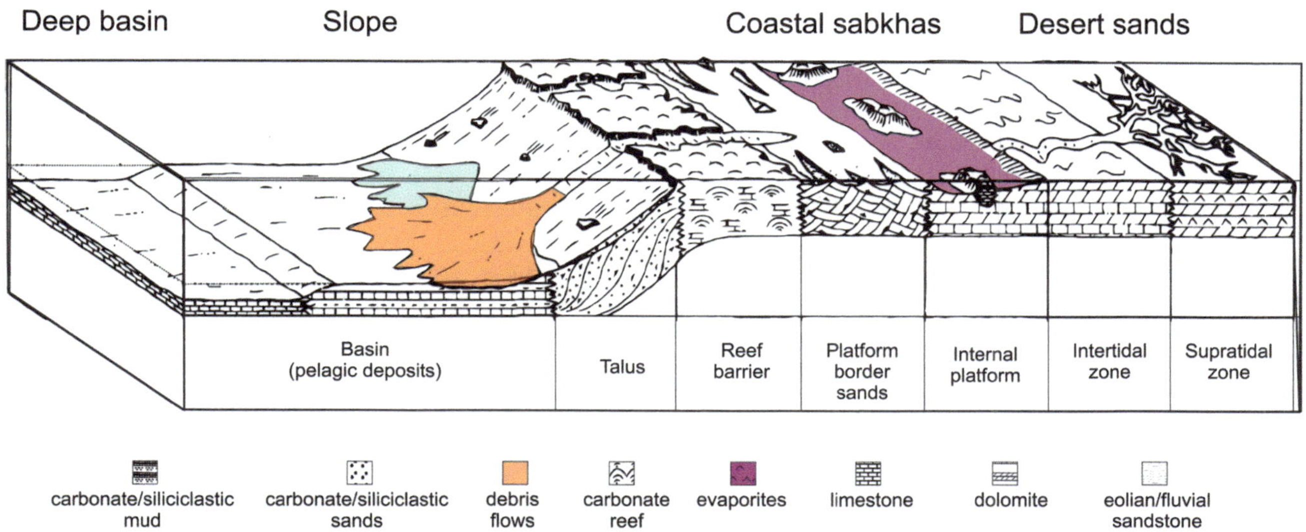

Figure 5.15 The diagram illustrates depositional environments based on seismic interpretation and petroleum exploration boreholes, comparing them with modern examples of evaporite basins and outcropping sequences in ancient salt basins, using paleoenvironmental interpretations. Credit: Adapted from Fontaine et al. (1987).

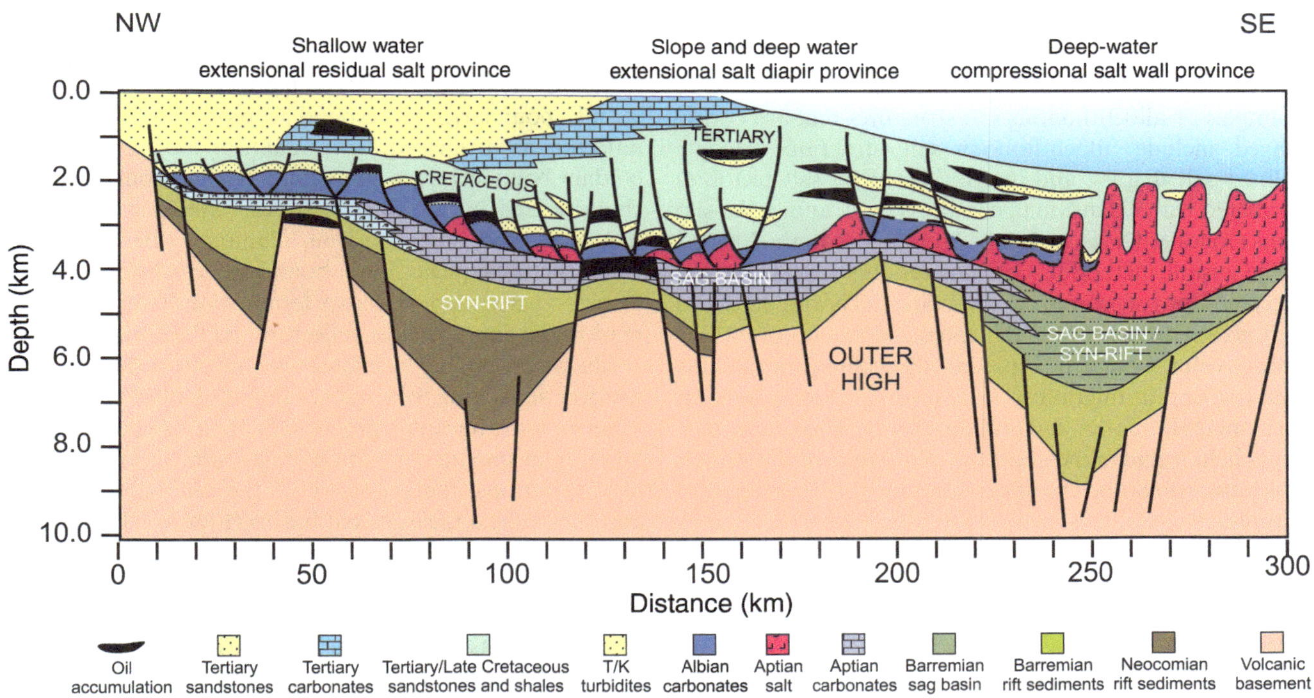

Figure 5.16 The diagram depicts the rift and salt structures in a typical South Atlantic divergent margin, with main source rocks from the Barremian to Neocomian rift sequence. Reservoirs are mainly Tertiary and Late Cretaceous turbidite sandstones. Pre-salt discoveries have been made in the sag basin carbonate rocks. The proximal basin has extensional listric faults, while the slope and deep basin have an extensional salt province with salt diapirs. The distal margin is characterized by salt walls and compressional salt structures. Credit: Adapted from Mohriak (2005) and Jahnert et al. (1998).

as the depth of subsurface horizons, the structural dip of sedimentary layers, the interval velocity, the reflectivity of sedimentary layers, geological history, depositional settings of sedimentary packages, lithological and fluid content of sedimentary sequences, stratigraphic events, and the sequential development of structural features such as faults, folds, salt diapirs, and the emplacement of intrusive and extrusive igneous rocks. These elements are determined through analysis of the arrival time and the seismic velocity of sedimentary layers, the dip of sedimentary layers, interval velocity, reflectivity, geological history, depositional settings, lithological and fluid content, stratigraphic events, and the sequential development of structural features. The use of 2D and 3D seismic data in basin analysis projects helps in understanding the geological history, depositional settings, and stratigraphic events (e.g., Herron, 2011; Sheriff & Geldart, 1995).

The seismic interpretation of salt structures is hindered by the unusual physical properties of evaporite rocks, particularly their high velocities, which create seismic pitfalls and artifacts in imaging complex features. Salt structures can vary rapidly in space and time, and their geometrical shape can be intricate and highly variable, resulting in poor imaging (Taylor, 1985; Trusheim, 1960). Structural models interpreted in areas with excellent 3D visualization cannot be directly extrapolated to areas with poor data or low-quality 2D visualization (Tearpock & Bischke, 2002). Examples of allochthonous salt structures that are poorly imaged include allochthonous salt structures, steeply inclined salt diapirs, and areas affected by salt tectonics associated with extensional, compressional, and strike-slip faults (Jones & Davison, 2014; Trudgill & Rowan, 2004).

The seismic method has been increasingly applied to the interpretation of complex salt bodies. The method has been greatly improved by the enhanced resolution of 3D seismic volumes and new processing techniques, such as prestack depth migration and reverse time migration (Herron, 2011; Jones & Davison, 2014). These improvements help improve the accuracy of seismic interpretation of salt structures and the understanding of their geological features.

5.6. BASIN ANALYSIS METHODS USING SEISMIC AND PETROPHYSICAL LOG INTERPRETATION

Basin analysis projects involve interpreting stratigraphic sequences, structural elements, and lithological facies variations in sedimentary succession to determine different environments of deposition and tectonosedimentary evolution. These studies are crucial in assessing exploration potential and characterizing exploratory plays (Nanda, 2016; Sheriff & Geldart, 1995).

Seismic stratigraphic methodologies are used in seismic interpretation workflows to subdivide seismic sections into constituent sequences based on the termination of reflection events, interpreted as geological boundaries (Mitchum & Vail, 1977; Mitchum et al., 1977). Upper seismic boundaries correspond to erosional events when reflectors are truncated, concordant, or toplap, while concordant boundaries are subparallel. Toplap boundaries are characterized by an angle between reflectors in a lower sequence and the horizon, limiting the sequence. Depositional sequence lower boundaries can be concordant or exhibit downlap or onlap terminations, with downlap marking an angle between reflectors and the horizon. Onlap is characterized by terminations that advance over the horizon, separating the sequence from its base (Figure 5.17).

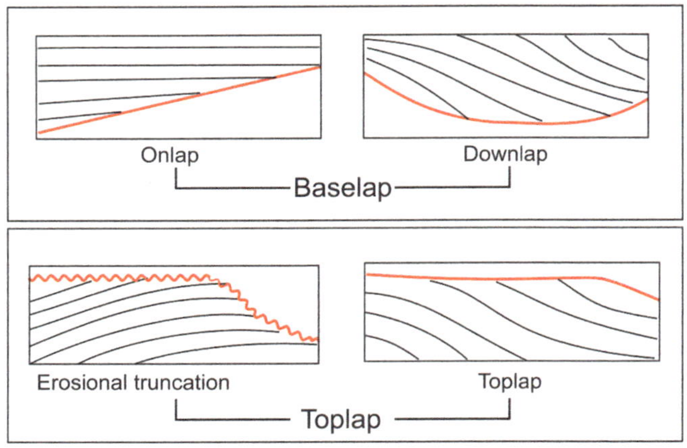

Figure 5.17 The diagram depicts the stratigraphic relations of sedimentary strata in seismic data. It includes baselap geometry, downlap terminations, and toplap geometry. Baselap geometry involves the onlap of reflectors on a climbing reflector, while downlap terminations indicate sedimentary progradational units filling the accommodation space. Toplap geometry refers to the termination of underlying reflectors on a depositional surface. Credit: Nanda (2016)/with permission of Springer Nature.

Seismic stratigraphic procedures are a valuable tool for defining evaporite deposition stages, stratigraphic thickness variations, the geometry of the sedimentary units, and lithological facies in salt basins (Figure 5.18). For instance, the Mediterranean salt deposits associated with the Messinian Salinity Crisis in Sicily have been divided into 12 sequences based on the terminations of different geological units (see Chapter 5 in Mohriak, 2025).

These Messinian evaporite sequences are affected by an unconformity that separates them from the overlying transgressive gypsarenites and gypsum layers intercalated with marls, which are concordant with the Pliocene Arenazollo siliciclastic rocks and Trubi Fm. carbonates (Schreiber et al., 1976). The correlation of these onshore geological units as described in outcrops can be correlated with the much thicker evaporite sequences in the several basins across the Mediterranean Sea (see Chapter 5 in Mohriak, 2025).

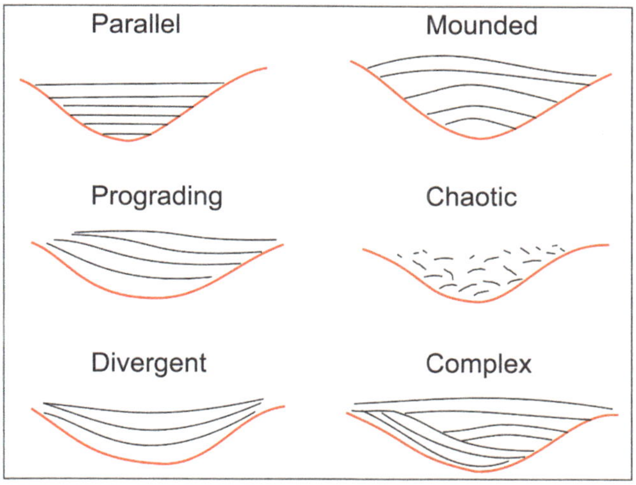

Figure 5.18 The diagram depicts the stratigraphic relations of sedimentary strata in seismic data. It shows parallel geometry, characterized by subhorizontal sequences filling the accommodation space, a mounded configuration, prograding geometry with sigmoidal reflectors, chaotic fill with packages and depositional boundaries lacking reflectors, a divergent pattern with thicker sequences in the basin center, and a complex pattern with irregular sedimentary units infilling the depression. Credit: Adapted from Nanda (2016).

The integration of outcrops with borehole data provides an efficient method for mapping lithological facies variations in onshore basins (Miall, 2016). Different belts from evaporite deposits in the platform interior, restricted lagoons, carbonate banks at the shelf edge, and deep marine environments basinward of the slope can be identified and mapped regionally (e.g., Alsharhan & Kendall, 2003).

Seismic datasets are mapped using 2D and 3D datasets that commonly constitute regional and local surveys. The analysis of the top of seismic horizons calibrated by petrophysical logs from deep boreholes allows for structural interpretation, identifying the main extensional and contractional faults. The interpretation of different stages of halokinesis is conducted by interpreting the salt structure, geomorphology, and configuration of the overlying strata (Figure 5.19).

The integration of geological and geophysical data leads to the determination of the main tectonic events that affected the basin, such as basement-involved rift structures, basement-detached salt structures, and the inception of oceanic crust on divergent continental margins (see Chapter 7 in Mohriak, 2025).

5.6.1. Structural Interpretation Integrating Seismic and Petrophysical Logs

Salt deformation produces various structures that can be characterized by well logs and seismic profiles. 2D and 3D seismic datasets are commonly used to interpret structural surfaces affected by extensional or contractional faults. Offset horizons are correlated using boreholes that drill these horizons. Many of these seismic horizons or stratigraphic surfaces are involved in folds, which can be identified using geological and geophysical information, particularly reflection seismic profiles. Well data (ditch samples, cores, and petrophysical logs) allow the interpretation of main sedimentary sequences, environments of deposition, and stratigraphic evolution (Herron, 2011; Nanda, 2016).

Stratigraphic and structural mapping of sedimentary horizons using contour maps can be associated with fault maps and integrated with lithological and sedimentary facies maps, leading to a comprehensive analysis of the tectono-sedimentary evolution of the basin and a better understanding of possible exploratory plays (Jackson & Vendeville, 1994; Jackson et al., 1994; Stewart & Clark, 1999; Stewart & Coward, 1996; Tearpock & Bischke, 2002; Thompson et al., 1989).

A selected example of a 3D seismic line in the eastern Brazilian margin shows different styles of basement-involved rift-phase faults and basement-detached faults related to salt tectonics (Figure 5.20). The seismic line shows extensional salt structures and a salt diapir basinward of the exploratory well. The salt diapir corresponds to an intrusive feature that pierced the post-salt Albian to Late Cretaceous sedimentary sequences (compare with Figure 5.24). The stratigraphic relationships in the seismic data indicate thinning of early post-salt sediments toward the salt diapir, a marked thickening of late post-salt sequences near the diapir top, and a late phase of contraction affecting the salt structure. The boundary between the Late Cretaceous and the Paleogene is characterized by a regional erosional unconformity related to tectono-magmatic instability on land. This post-rift magmatic event is indicated by some important volcanic episodes affecting the continental shelf and slope of several basins in the eastern Brazilian margin (e.g., Mohriak & Fainstein, 2012).

The Campos Basin, a prolific hydrocarbon-producing basin in the South Atlantic Ocean, was responsible for most oil production offshore Brazil until the late 20th century (Carminatti et al., 2008; Mohriak & Szatmari, 2009; Waisman, 2009). After discovering super-giant oil fields in pre-salt carbonate reservoirs, the Santos Basin became the most important oil province (Carminatti et al., 2008; Mohriak et al., 2012). Currently, the pre-salt carbonate reservoirs in the Santos Basin are the most important target for petroleum exploration in the South Atlantic (see Chapter 3 in Mohriak, 2025).

The Campos Basin's 3D seismic profile (Figure 5.21) reveals a major rollover structure linked to a listric fault near the northwest extremity. This fault detaches at the Aptian salt and controls the growth of the Upper

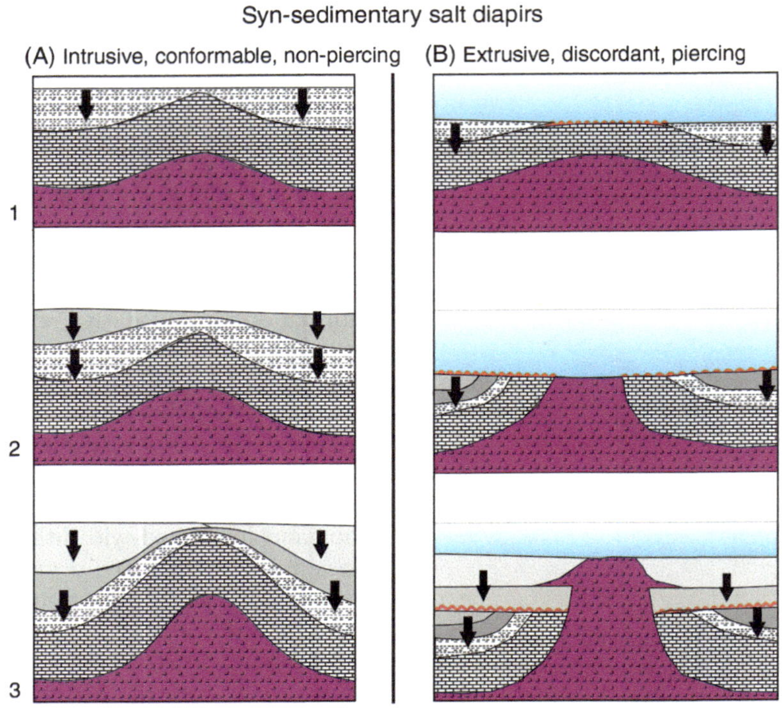

Figure 5.19 The diagram shows the sequential evolution of diapiric salt structures, both concordant and discordant with the overburden. The diagram to the left (A) represents an intrusive, conformable salt structure formed by sedimentary loading on the salt layer. The diagram to the right (B) illustrates the sequential development of an extrusive, discordant salt structure that pierced the overburden and extruded salt on the depositional surface. The sedimentary overburden was affected by angular unconformity when sedimentary layers were located above the regional level. The erosion may occur subaerially or subaqueously. Credit: Adapted from Nanda (2016).

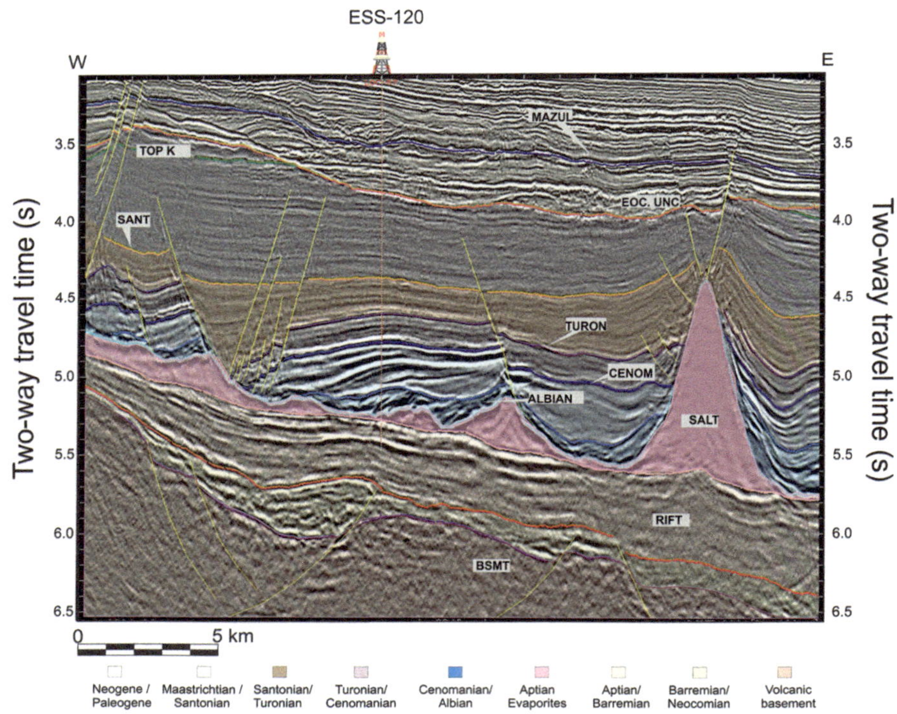

Figure 5.20 The 3D seismic line from the exploratory borehole ESS-120, located in the eastern Brazilian margin, reveals the transition from extensional salt tectonics to the salt diapir province. As indicated by the vertical scale with two-way travel time in seconds, the seismic line was processed in time. A depth conversion of the seismic line indicates a vertical exaggeration of about 3:1. The seismic line shows the transition from the extensional salt tectonics domain to the salt diapir province. The Eocene unconformity has deeply incised the Late Cretaceous sedimentary sequences, resulting in the erosional truncation of the reflectors. The Mazul is an Oligocene stratigraphic marker of regional occurrence. Credit: Adapted from Mohriak (2009)/Servico Geologico do Brasil.

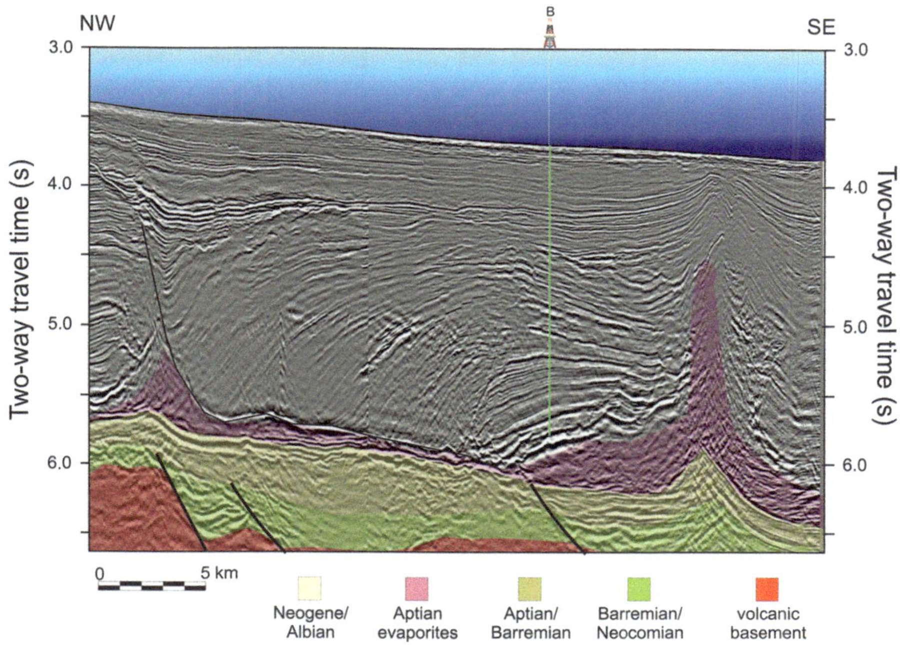

Figure 5.21 A 3D seismic line in Brazil's Campos Basin reveals salt structures in the deepwater region. Post-salt strata have highly inclined seismic reflectors associated with a residual salt layer. Post-salt horizons have steep dips, while pre-salt seismic horizons have subhorizontal reflectors. A rollover structure is associated with a listric fault. The structural discontinuity between pre-salt and post-salt strata is often characterized by drilling logs. Erosional unconformities near the salt structures are observed between 4.0 and 4.5 seconds two-way travel time (TWTT), with erosional truncation indicating canyon excavation by submarine currents. The exploratory borehole B drilled the Albian carbonate rocks above the salt structure. Credit: Waisman (2009)/with permission of The State University of Campinas.

Cretaceous sedimentary sequences. Rollover structures in the basin are typically associated with salt flow and are typically related to extensional listric faults that dip oceanward (e.g., Demercian et al., 1993; Mohriak & Fainstein, 2012; Mohriak & Szatmari, 2008).

The seismic profile shows a progressive increase in the structural dip as the post-salt horizons approach the detachment zone, which corresponds to a residual salt layer (Aptian evaporites). The salt layer is a detachment zone between post-salt strata (Albian and Late Cretaceous) and upper pre-salt strata (Aptian sag basin). The Albian strata formed a block that was rotated and moved basinward during the salt flow, as indicated by the landward thickening of the Albian to Cenomanian layers. The geometry of the post-salt strata suggests a thicker salt layer was initially present, and a salt pillow controlled the oceanward thinning of sedimentary sequences before the formation of the salt diapir (Figure 5.21).

The dipmeter is a petrophysical log that measures the dip of sedimentary layers (Serra, 1984, 1986). This tool has been widely used for structural interpretation of evaporitic basins until the last decades of the past century. The dipmeter tool aims to measure the dips of sedimentary strata with extremely high resolution, up to a few centimeters. However, for some complex structures, the results were not accurate. More recently, acoustic image logs have been increasingly used in the interpretation of structural dips (e.g., Garcia-Carballido et al., 2009).

The interface between the salt mass and the overlying and underlying stratigraphic sequences is generally characterized by marked angular unconformities that are readily identified on seismic displays and petrophysical logs (Figure 5.22).

In some intracratonic basins, there is a conformable contact between the post-salt and the pre-salt sedimentary layers, as observed in the North Sea seismic example (Figure 5.14). In most basins, the petrophysical logs indicate a lithological contrast between the salt layer and the overlying and underlying strata. The seismic profile in the Campos Basin (Figure 5.21) shows a change in the dip angle of the overlying strata near the contact with the evaporite deposits. The salt layer represents a detachment plane, separating the basement-involved rift faults from the listric faults affecting the overburden. In most basins, the original bedding planes of the evaporite

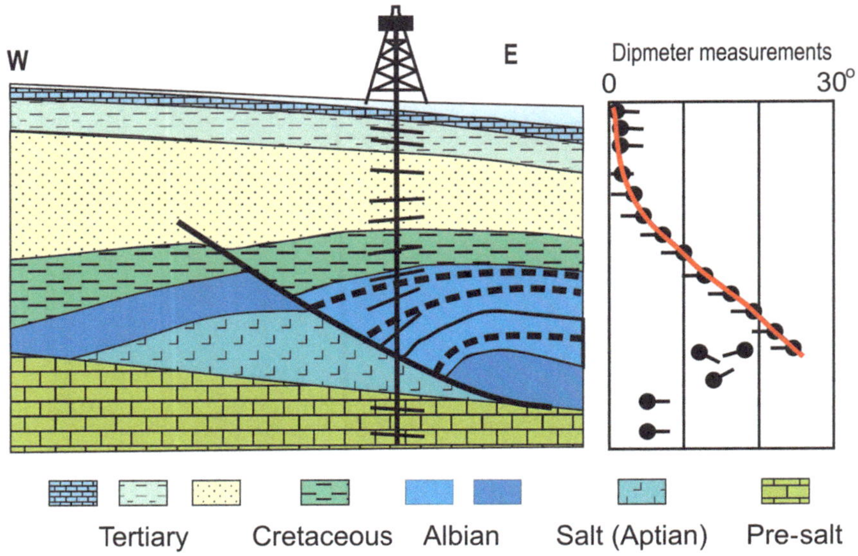

Figure 5.22 The diagram illustrates a rollover structure using dipmeter log measurements. The pre-salt strata have small eastward dips, while the Cretaceous to Albian strata have westward dips that increase with depth, reaching 25° at the base. The maximum dip values occur at the interface between the Albian strata and the Aptian salt layer. A major listric fault with eastward vergence controls the growth of the Cretaceous–Albian stratigraphic sequences. The bedding of the pre-salt layers is characterized by small (about 5°) eastward dips.

layers have been deformed by the salt flow (e.g., Hudec & Jackson, 2007).

The comparison of the Campos Basin seismic line (Figure 5.21) with a geological cross-section illustrating similar features (Figure 5.22) highlights the stratigraphic expansion of the Upper Cretaceous to Albian strata, which was controlled by the listric fault activity during sedimentation (Kukla et al., 2018; Mohriak & Szatmari, 2008; Waisman, 2009). The graph showing the dipmeter measurements indicates a marked discontinuity in the bedding plane inclination at the salt detachment zone (Figure 5.22). These variations in dips are related to salt flow and the concomitant formation of a listric fault that detached on the Aptian salt and controlled the rollover structure. The petrophysical logs acquired by the borehole that drilled the sedimentary wedges indicate a rapid increase in the structural dip for the sedimentary sequences immediately above the evaporite layer.

The seismic line and the dipmeter log indicate a rotation of the post-salt blocks by a listric fault that formed adjacent to a salt diapir. However, the strata below the salt deposits exhibit small dip angles, indicating the rift-phase blocks were affected by relatively small rotations. The subsequent sag basin subsidence resulted in the accumulation of lacustrine carbonate and siliciclastic sediments. During the continental breakup, the basin was flooded by marine incursions and precipitated the Late Aptian evaporite sediments (see Chapter 3 in Mohriak, 2025).

Petrophysical logs are routinely used for lithological and stratigraphic interpretations in basin analysis projects, bearing important implications for sedimentological studies involving facies analysis and reservoir prediction (e.g., Nanda, 2016; Tearpock & Bischke, 2002). The calibration of seismic horizons with the petrophysical logs allows the interpreter to solve complex problems in geology, such as the lack of a regional horizon or the repetition of stratigraphic sequences in a borehole. Petrophysical logs are fundamental in the assessment of the sealing capacity of faults in trapping hydrocarbons and are used in the characterization of pre-salt and subsalt reservoirs associated with autochthonous and allochthonous salt layers in evaporite basins (Tearpock & Bischke, 2002; Tiab & Donaldson, 2004).

5.6.2. Stratigraphic Interpretation Integrating Seismic and Petrophysical Logs

Stratigraphic markers are crucial in understanding the short-duration and regional events in sedimentary basins. Some markers provide important constraints for structural and paleogeographic interpretations. For example, some basins are characterized by stratigraphic markers that provide geochronological timelines. The repetition or absence of a marker in a borehole is commonly associated with reverse and extensional faults (Tearpock & Bischke, 2002; Tiab & Donaldson, 2004).

Stratigraphic markers provide constraints to structural and paleogeographic interpretations. For example, the central to northern parts of the Campos Basin in Brazil are characterized by a layer of volcanic ash deposited in a very short time. This stratigraphic event is known as the "Three Fingers" due to its characteristic triad formed by gamma-ray peaks in electric logs (Alves, 2006). This marker is associated with a magmatic event that deposited volcaniclastic rocks throughout the northern part of the basin during the Late Cretaceous (Santonian), separating Paleogene and Upper Cretaceous reservoir rocks. The marker is characterized by a regional distribution in the basin, as suggested by its presence in the petrophysical logs of several wells (Alves, 2006).

The salt layer can also be interpreted as a short-term depositional event with a duration estimated by the age of the underlying and overlying sediments, as the salt rocks are devoid of fossils. Evaporite deposition is normally associated with a phase of climatic aridity throughout the basin, making salt rocks an important factor in reconstructing the paleogeographic and stratigraphic framework of sedimentary basins (Warren, 2010, 2016).

Stratigraphic markers obtained from the interpretation of potentially equivalent peaks in petrophysical logs may be correlated from the system scale (e.g., fluvial systems with hundreds of kilometers) to the laminar scale, as identified in cores with varve sedimentation (centimeter-scale cycles). In evaporite sequences, such as the Permian Basin in the southwestern United States, the Late Aptian salt basin in the South Atlantic, or the Late Permian Zechstein salt basin in Europe, correlations can be made from the basin scale to the outcrop scale, or even to the hand sample and microscopic scale, using petrographic analysis (Miall, 1990, 2016). The changes in the petrophysical log responses can be associated with lithological facies variations within a depositional sequence, indicating changes in tectonic settings or major climatic or depositional changes in the environment. Gamma-ray profiles, for example, can be used for the correlation of chronostratigraphic markers, the determination of maximum flood surfaces, and lithological facies interpretation (e.g., Schön, 1996, 2015).

Gamma-ray wireline logs are useful for the identification of different evaporite minerals, as many are characterized by a low potassium content (e.g., halite), whereas others exhibit anomalously high peaks (e.g., sylvite). Potassium-bearing minerals, particularly sylvite and carnallite beds, are easily identified by a characteristic peak in the gamma-ray curve and the density values (Figure 5.7). However, several pitfalls occur in the interpretation of petrophysical logs. For example, Warren (2006) cites the example of the dissolution of halite layers in an evaporite sequence with minor amounts of intercalated shales. When halite is dissolved in an evaporite cycle, the only residual sediments that may remain in cores are clay-rich layers. These thin layers are often radiogenic and show up on gamma-ray log readings as conspicuous spikes. This may be equivocally associated with thin shale or mud, which might suggest a maximum flooding surface within a carbonate platform. However, these residual deposits may indicate the desiccation of a shallow-water platform followed by the dissolution of soluble minerals, which may have happened millions of years after deposition.

Figure 5.23A shows the application of electric profiles in the regional interpretation of stratigraphic sequences. The interpretation indicates that in the time interval between markers M2 and M3, represented by sedimentary sequences penetrated by wells P1 and P2, the sedimentary environment changed from a continental shelf to a deepwater setting. The figure also suggests a change in depositional facies, varying from pelitic rocks (shales), carbonates (limestone and dolomite), and evaporites. The paleogeographic setting evolved from a deep lacustrine or marine environment to a transitional marine environment, where sulfate (anhydrite) and chloride (halite) deposits accumulated in a desiccating basin.

In this schematic diagram, the thick sequences of carbonate rocks and marine shale are observed in well P2, just below the halite layer. However, well P1 displays only one thin sequence of residual halite and anhydrite above the sandstone layer. The anhydrite may represent a sulfate facies in a lagoon originally located toward the interior of the continent, relative to the more distal shale and carbonate deposits. The variation of facies from a deep marine setting to a transitional, evaporitic setting is marked by a carbonate platform formed in shallow waters, followed by evaporite sediments at the top (Figure 5.23B). This indicates a depositional sequence related to a drying-up event, forming an evaporite cycle during a regression episode (Taylor, 1985; Warren, 1989). After deposition of the evaporite sediments the basin was desiccated. The marks M2 and M3 suggest another episode of marine transgression, with a carbonate platform subsequently covered by siliciclastic sediments, forming a shelf-slope-basin system. Petrophysical logs are also useful in analyzing subsidence rates (Figure 5.23C) and the characterization of stratigraphic sequences limited by timelines defined by markers (Figure 5.23D).

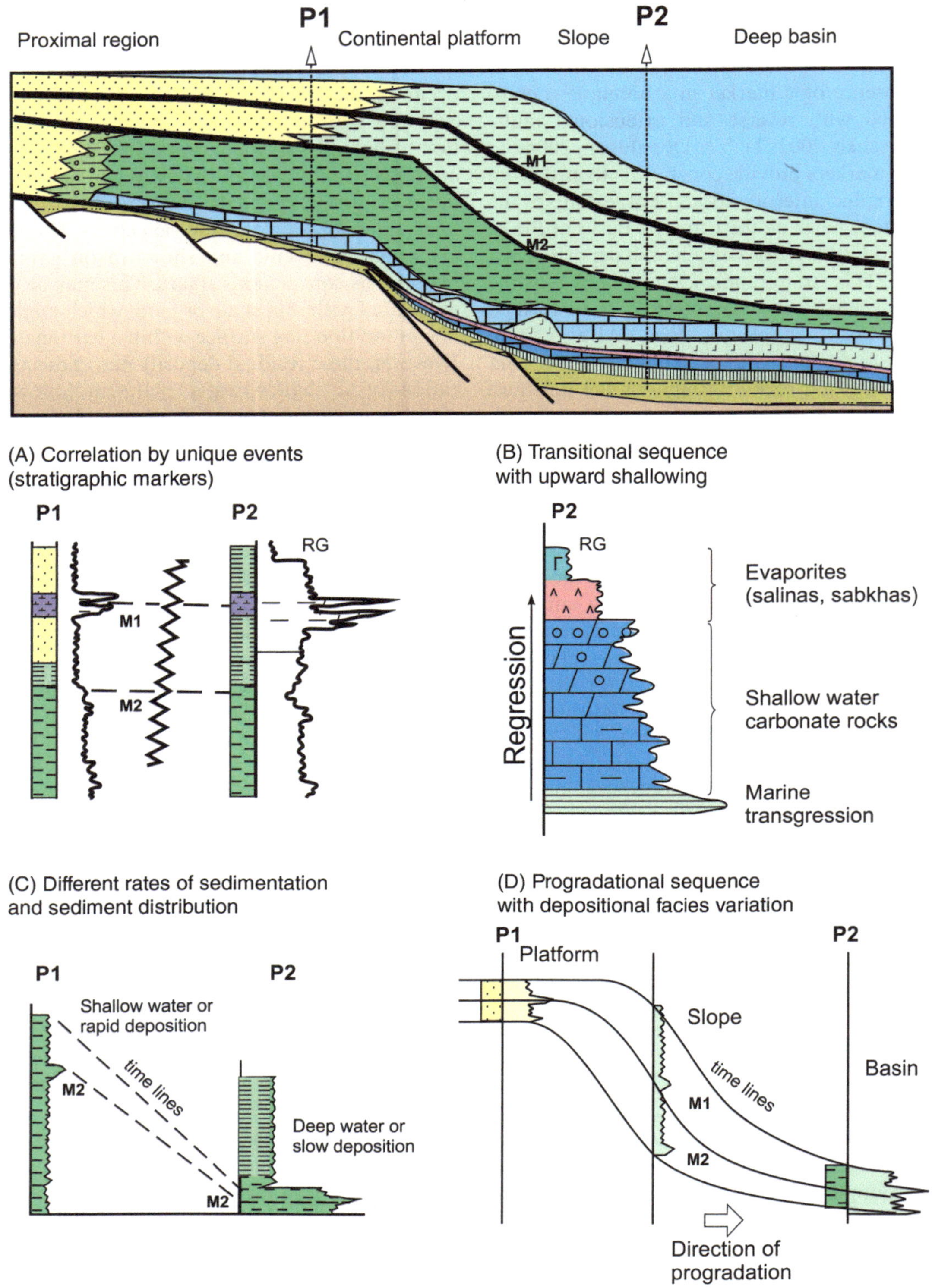

Figure 5.23 The diagram illustrates the structural and stratigraphic interpretation of sedimentary sequences in a passive margin using well logs and stratigraphic markers in boreholes. Top: wireline log correlation of boreholes on the continental shelf and in the deepwater region. (A) shows the correlation of gamma-ray peaks to define stratigraphic markers. (B) shows the interpretation of the paleogeography and sedimentary environments. (C) shows the interpretation of the subsidence and sedimentation rates using stratigraphic markers. (D) shows the interpretation of a shelf-slope-basin profile using stratigraphic markers, indicating progradational sequences with variable thickness and depositional facies.

5.6.3. Practical Applications of Seismic and Petrophysical Logs in Basin Analysis

Basin analysis projects aim to identify stratigraphic and structural elements in the subsurface based on indirect geophysical methods, such as potential field and seismic methods. However, there are several direct methods based on classical outcrop geology that have contributed to the geological knowledge of salt structures (e.g., Davison et al., 2017). Major salt structures are not only encountered in offshore regions but also in many locations on land. Standard field geology methods and modern techniques have been applied to investigate salt structures, such as drone imagery, 3D visualization of outcrops, hyperspectral mapping, and galleries in salt mines (e.g., Pavlis & Mason, 2017; Whitmeyer & Nicoletti, 2010). The integration of different methods is key to a better understanding of the salt systems from salt deposition, deformation, and exposure at the surface (see Chapter 3).

The petrophysical log responses in giant salt basins, such as the Zechstein Basin in Germany, the South Atlantic, the Gulf of Mexico, and the Mediterranean Sea, are associated with different lithologies and stratigraphic sequences (e.g., Taylor, 1981, 1985, 1998). These variations allow for the identification of evaporitic cycles with carbonate, sulfate, and chloride minerals. Several evaporite cycles are recognized in giant salt basins, such as the Zechstein Basin in Germany (see Chapter 2 in Mohriak, 2025), the South Atlantic (Chapter 3 in Mohriak, 2025), the Gulf of Mexico (Chapter 4 in Mohriak, 2025), and the Mediterranean Sea (Chapter 5 in Mohriak, 2025).

Important hydrocarbon accumulations are controlled by these evaporite sequences, which can result in structural traps formed by the deformation of salt layers. Stratigraphic trapping can be created under depositional conditions that result in lithological facies variations, particularly when limestone or porous dolomite rocks occur below a salt layer, forming reefs. The structural style is influenced by the evaporite zones. For example, the composition of the Zechstein salt controls the structural style of extensional fault systems in the North Sea, with direct implications for hydrocarbon exploration (Jackson & Stewart, 2017). In the southwestern Iberian margin, the Algarve Basin displays several salt-related structures with geometries controlled by the distinct evaporite facies in the basin (Ramos et al., 2017).

The structural style is influenced by the composition of the evaporite zones, such as the composition of the Zechstein salt, which controls the extensional fault systems in the North Sea. In the southwestern Iberian margin, the Algarve Basin displays several salt-related structures with geometries controlled by the distinct evaporite facies in the basin. The salt thickness increases in the basin depocenters located in the offshore region, where the boreholes indicate halite-rich intervals (Ramos et al., 2017). The structural style is marked by listric normal faults that detach at the base of the autochthonous salt, developing a domain of salt pillows, salt rollers, and isolated diapirs. The salt structures were subsequently affected by the Cenozoic Alpine orogeny, resulting in contractional structures and allochthonous salt wings and nappes (Ramos et al., 2017). Based on the integration of seismic data and petrophysical logs, deformed salt layers can be restored to a pre-deformation stage (e.g., Rowan & Ratliff, 2012).

Evaporitic cycles involving salts of increasing solubility are identified in various South Atlantic basins, such as the Santos Basin offshore Brazil. One of the first boreholes drilled in the basin (SPS-4) encountered bedded salt deposits at the proximal part of the basin, near the pre-Aptian hinge line, which corresponds to the rift limit (Gordon et al., 2023; Mohriak & Fainstein, 2012). The proximal basin is characterized by some small blocks of Barremian to Neocomian sediments deposited in troughs above the volcanic basement (Figure 5.24). The exploratory well SPS-4 was located above a structural high near the rift border fault. The borehole recovered basalt conglomerates overlain by a thick salt layer, as indicated by the seismic data and interpretation of the well logs (Figure 5.25).

The interpretation of petrophysical data allowed the identification of different minerals within the stratified evaporites of the Santos Basin, which are key to determine the evaporitic cycles as a function of the salinity variations in the brine that filled the basin, precipitating sulfates and chlorides with increasing salt concentrations (see Chapter 4).

The computer-aided interpretation of petrophysical logs in exploratory wells results in important geological applications for basin analysis projects. Some workflows include the correlation of chronostratigraphic markers, estimation of depositional rates, the spatial distribution of lithological units, characterization of transgressive and regressive sequences, paleoenvironmental analysis from petrophysical log variations, and mapping of erosional channels in hemipelagic depositional settings. For example, a deepening of the depositional environment is commonly associated with an increase in shale content, which is indicated by an increase in the gamma-ray value (e.g., Schön, 1996, 2015).

Recent advances in 3D seismic acquisition and processing, combined with the calibration of intra-salt seismic horizons with the mineralogy of the salt layers, have resulted in the quantitative evaluation of the evaporitic sequences in several basins. Several methods and techniques, such as rock physics, seismic attributes,

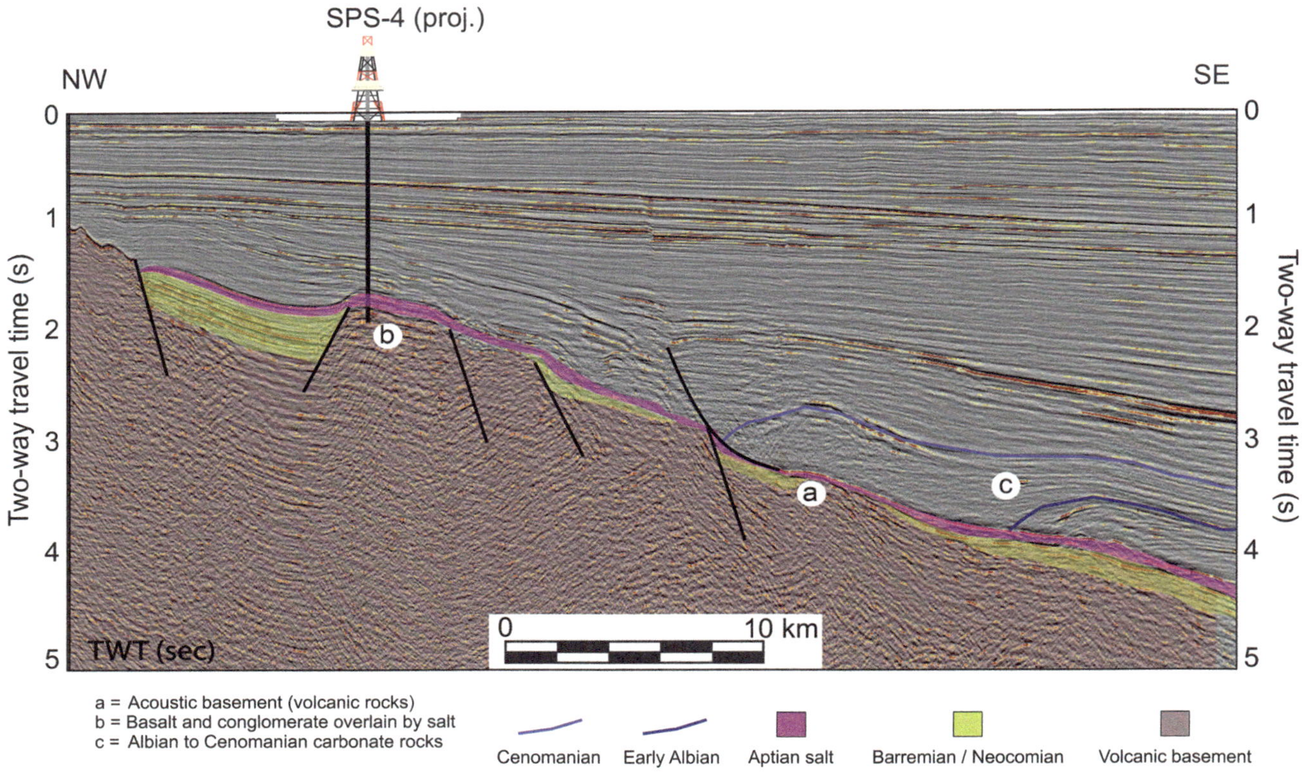

Figure 5.24 The 2D regional seismic line in the northern Santos Basin shows the location of the proximal borehole SPS-4, which drilled a thick sequence of bedded evaporites over basalt conglomerate rocks. The borehole penetrated basalts at total depth, showing basement-involved, planar, normal faults affecting the presalt sequences, and listric normal faults detaching on the salt horizon. The upper parts of the borehole correspond to Neogene, Paleogene, and Late Cretaceous siliciclastic sequences. The borehole drilled some Cenomanian to Albian siliciclastic and carbonate rocks above the bedded salt layers. Credit: Adapted from Gordon et al. (2023).

seismic inversion, and statistical methods, have been recently applied to the interpretation of different salt facies in the Zechstein Group of the North Sea (Jackson & Lewis, 2014; Jackson & Stewart, 2017; Jackson et al., 2019) or the Santos Basin, offshore Brazil (Teixeira et al., 2020). The seismic-derived 3D facies volumes display the geometry and composition of the salt layers. The final result from the workflow is a more reliable estimate of the percentages of different minerals (halite, anhydrite, and soluble salts) in the different cycles of the stratified evaporites, with direct implications for salt flow and geomechanics (see Chapter 6).

There is a growing tendency to use computer-aided techniques and artificial intelligence (AI) algorithms in the predictive analysis of lithological facies distribution in sedimentary basins and reservoir characterization. These AI techniques have been used to improve the visualization and interpretation of geological data in 3D, map the occurrence of possible reservoir rocks, such as sandstone turbidites and other conventional and unconventional reservoirs, and optimize hydrocarbon production (e.g., Ertekin & Sun, 2019; Gharavi et al., 2022; Kinsland & Borst, 2015; Ogiesoba & Zeng, 2022; Pavlis & Mason, 2017).

Salt mineralogy is crucial for well design and borehole drilling, as some salts, like tachyhydrite, pose a geological hazard for drilling operations that target the pre-salt reservoirs. Risks include borehole closure due to mobile salt, dissolution of soluble salts in evaporite cycles, and the unexpected presence of carbonate and sulfate stringers in halite sequences. These events can lead to stuck drilling tools, expensive engineering solutions, borehole abandonment, and significant economic losses. The precise determination of mineralogical components and evaporite sequences is also important for engineering investigations, including salt mine construction and storage cavern construction (e.g., Giambastiani, 2020; Habibi, 2019).

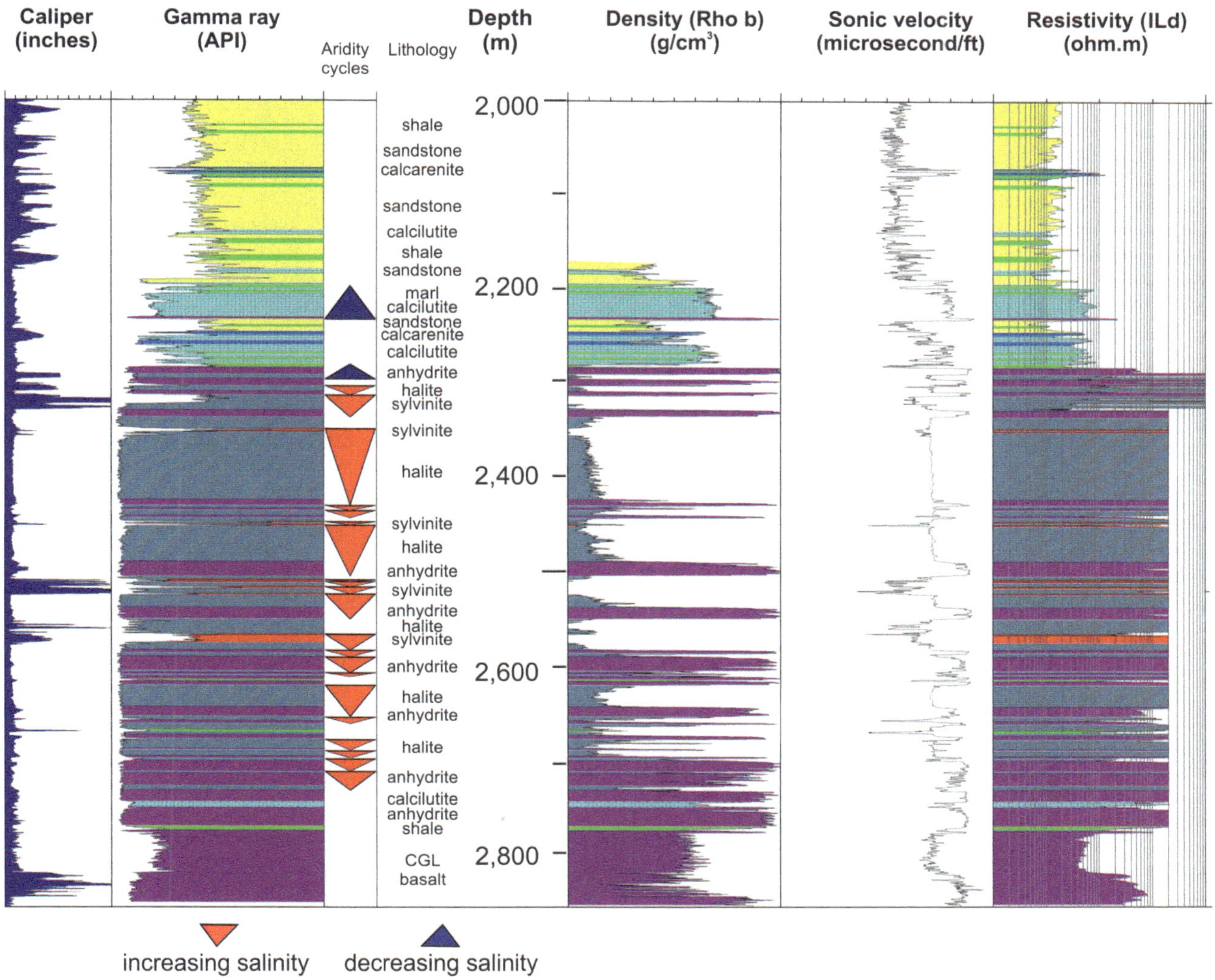

Figure 5.25 The composite log of the exploratory borehole SPS-4 displays the Late Aptian evaporite sequences in the Santos Basin. The petrophysical logs show the main physical parameters (density, sonic velocity, and resistivity). The lithologies are interpreted using computer algorithms. The evaporite minerals were deposited following several aridity cycles, which are associated with the brine concentration. Sylvinite layers correspond to gamma-ray curve peaks, while sea level rise indicates marine incursions and brine dilution. Red triangles indicate increasing aridity, while blue triangles indicate sea level rise from the horizontal base to the triangle top, suggesting dilution of the brine. Credit: Courtesy of C. Sombra (2005).

REFERENCES

Aghajani, H., Moradzadeh, A., & Zeng, H. (2009). Estimation of depth to salt domes from normalized full gradient of gravity anomaly and examples from the USA and Denmark. *Journal of Earth Sciences*, *20*, 1012. https://doi.org/10.1007/s12583-009-0088-y

Alsharhan, A. S., & Kendall, C. G. St. C. (2003). Holocene coastal carbonates and evaporites of the southern Arabian Gulf and their ancient analogues. *Earth Science Reviews*, *61*, 191–243.

Alsop, G. I., Archer, S. G., Hartley, A. J., Grant, N. T., & Hodgkinson, R. (Eds.). (2012). *Salt tectonics, sediments and prospectivity (Vol. 363)*. Geological Society, London, Special Publications, 624 pp. http://dx.doi.org/10.1144/SP363.1

Alsop, G. I., Blundell, D. J., & Davison, I. (Eds.). (1996). *Salt tectonics* (Vol. *100*). Geological Society Special Publication, 310 pp.

Alves, D. B. (2006). Sedimentação vulcanoclástica do Cretáceo Superior da Bacia de Campos, Sudeste do Brasil. *Boletim de Geociências da Petrobras*, *14*(1), 149–154.

Aminzadeh, F., & Dasgupta, S. N. (2013). *Geophysics for Petroleum Engineers, Developments in Petroleum Science* (Vol. *60*), 282 pp.

Azaiez, H., Amri, D. T., Gabtni, H., Bedir, M., & Soussi, M. (2008). Integrated geophysical study of the Triassic salt bodies' geometry and evolution in central Tunisia. *Comptes Rendus Geoscience*, *340*(1), 10–19.

Bally, A. (1981). Thoughts on the tectonics of folded belts. In K. McClay & N. Price (Eds.), *Thrust and Nappe tectonics* (Vol. *9*, pp. 13–32). Geological Society, Special Publications, London.

Bally, A. W. (Ed.). (1983). Seismic expressions of structural styles: A picture and work Atlas. American Association of Petroleum Geologists, *Studies in Geology*, *15*, 3 Volume Set.

Bourbié, T., Coussy, O., & Zinszner, B. (1987). *Acoustics of Porous Media*. Editions Téchnip, Institut Français du Pétrole Publications, 333 pp.

Braitsch, O. (1971). *Salt deposits: Their origin and composition*. New York: Springer-Verlag, 279 pp.

Braunstein, J. & O'Brien, G. D. (1968). *Diapirism and Diapirs, American Association of Petroleum Geologists, Memoir (Vol. 8)*. Tulsa, Oklahoma, 357 pp.

Carminatti, M., Wolff, B., & Gamboa, L. A. P. (2008). New exploratory frontiers in Brazil. *19th World Petroleum Congress*, Madri, Spain, WPC Proceedings, Abstracts CD, 11pp.

Célini, N., Callot, J.-P., Pichat, A., Legeay, E., Graham, R., & Ringenbach, J.-C. (2022). Secondary minibasins in orogens: Examples from the Sivas Basin (Turkey) and the sub-Alpine fold-and-thrust belt (France). *Journal of Structural Geology*, *156*, 104555. https://doi.org/10.1016/j.jsg.2022.104555

Chaplin, M. F. (2010). Water's hydrogen bond strength. In Lynden-Bell, R. M., Morris, S. C., Barrow, J. D., Finney, J. L., & Harper, C. (Eds.), *Water and Life: The unique properties of H2O* (pp. 69–86). Boca Raton, FL: CRC Press.

Cohen, Y., Krumbein, W. E., Goldberg, M., & Shilo, M. (1977). Solar Lake (Sinai).1. Physical and chemical limnology. *Limnology and Oceanography*, *22*(4), 597–608. doi:10.4319/lo.1977.22.4.0597

Dana, E. S., & Ford, W. E. (1922). *A Text-Book of Mineralogy with an extended treatise on Crystallography and Physical Mineralogy* (3rd ed.). New York: John Wiley & Sons, Inc., 720 pp.

Daniilidis, A., & Herber, R. (2017). Salt intrusions providing a new geothermal exploration target for higher energy recovery at shallower depths. *Energy*, *118*, 658–670. http://dx.doi.org/10.1016/j.energy.2016.10.094

Davison, I. (2009). Faulting and fluid flow through salt. *Journal of the Geological Society,* London, *166*, 205–216. doi: 10.1144/0016-76492008-064.

Davison, I., Barreto, P., & Andrade, A. J. M. (2017). Loulé: The anatomy of a squeezed diapir, Algarve Basin, southern Portugal. *Journal of the Geological Society*, *174*, 41–55. https://doi.org/10.1144/jgs2016-035

Demercian, L. S., Szatmari, P., & Cobbold, P. R. (1993). Style and pattern of salt diapirs due to thin-skinned gravitational gliding, Campos and Santos basins, offshore Brazil. *Tectonophysics*, *228*, 393–433.

Djebbi, M., & Gabtni, H. (2015). 3D gravity modeling of a salt structure associated to the Trozza–Labaied lineament (Central Tunisia) constrained by seismic and borehole data. *Journal of African Earth Sciences*, *103*, 71–80. http://dx.doi.org/10.1016/j.jafrearsci.2014.11.012

Dobrin, M. B., & Savit, C. H. (1988). *Introduction to geophysical prospecting* (4th ed.). New York, USA: McGraw-Hill, 867 pp.

Doglioni, C., Ismail-Zadeh, A., Panza, G., & Riguzzi, F. (2011). Lithosphere–asthenosphere viscosity contrast and decoupling. *Physics of the Earth and Planetary Interiors*, *189*, 1–8. doi:10.1016/j.pepi.2011.09.006

Ellis, P. G., & McClay, K. R. (1988). Listric extensional fault systems; Results of analogue model experiments. *Basin Research*, *1*, 55–70.

English, J. M., & English, K. L. (2022). An overview of Carbon capture and storage and its potential role in the energy transition. *First Break*, *40*(4), 35–40. https://doi.org/10.3997/1365-2397.fb2022028

Ertekin, T., & Sun, Q. (2019). Artificial intelligence applications in reservoir engineering: A status check. *Energies*, *12*, 2897. doi:10.3390/en12152897

Fernández, O., Habermüller, M., & Grasemann, B. (2021). Hooked on salt: Rethinking Alpine tectonics in Hallstatt (Eastern Alps, Austria). *Geology*, *49*, 325–329. https://doi.org/10.1130/G47981.1

Filina, I., Delebo, N., Mohapatra, G., Coble, C., Harris, G., Layman, J., et al. (2015). Integration of seismic and gravity data—A case study from the western Gulf of Mexico. *Interpretation*, *3*(4), SAC99–SAC106. doi: 10.1190/int-2015-0050.1.

Fontaine, J. M., Cussey, R., Lacaze, J., Lanaud, R., and Yapaudjian, L. (1987). Seismic interpretation of carbonate depositional environments. *American Association of Petroleum Geologists Bulletin*, 71(3), 281–297.

García-Carballido, C., Styles, A., & Pöppelreiter, M. (2009). Reduction of gross rock volume uncertainty in a salt flank structure using dipmeter and image-log data. In M. Pöppelreiter, C. García-Carballido, & M. Kraaijveld (Eds.), *Dipmeter and borehole image log technology, AAPG Memoir* (Vol. *92*, pp. 113–127).

Gardner, G. H. F., Gardner, L. W., & Gregory, A. R. (1974). Formation velocity and density—The diagnostic basics for stratigraphic traps. *Geophysics*, *39*(6), 770–780.

Gharavi, A., Hassan, M., Gholinezhad, J., Ghoochaninejad, H., Barati, H., Buick, J., & Abbas, K. A. (2022). Application of machine learning techniques for identifying productive zones in unconventional reservoir. *International Journal of Intelligent Networks*, *3*, 87–101. https://doi.org/10.1016/j.ijin.2022.08.001

Giambastiani, M. (2020). Geomechanical characterization of evaporitic rocks. In Kanji, M., He, M., & Ribeiro e Souza, L. (Eds.), *Soft rock mechanics and engineering* (Chapter 6, pp. 129–161). Switzerland: Springer Nature.

Gordon, A. C., Mohriak, W. U., Stanton, N., & Santos, A. C. (2023). Magmatic cycles in Santos Basin (S.E. Brazil): Tectonic control in the temporal-spatial distribution and geophysical signature. *Journal of South American Earth Sciences*, *121*, 104111. https://doi.org/10.1016/j.jsames.2022.104111

Gussow, W. C. (1968). Salt diapirism: Importance of temperature, and energy source of emplacement. In J. Braunstein & G. D. O'Brien (Eds.), *Diapirism and diapirs, American Association of Petroleum Geologists Memoir* (Vol. *8*, pp. 16–52).

Habibi, R. (2019). An investigation into design concepts, design methods and stability criteria of salt caverns. *Oil & Gas Science and Technology—Revue d'IFP Energies nouvelles, 74, 14*, 1–17. https://doi.org/10.2516/ogst/2018066

Hammond, C. (2015). *The basics of crystallography and diffraction* (4th ed.). UK: Oxford University Press, 519 pp.

Haynes, W. M. (2017). *CRC Handbook of Chemistry and Physics* (97th ed.). Boca Raton, FL: CRC Press, 2643 pp.

Heimdal, T. H., Callegaro, S., Svensen, H. H., Jones, M. T., Pereira, E., & Planke, S. (2019). Evidence for magma–evaporite interactions during the emplacement of the Central Atlantic Magmatic Province (CAMP) in Brazil. *Earth and Planetary Science Letters, 506*, 476–492.

Herron, D. A. (2011). *First steps in seismic interpretation, Society of Exploration Geophysicists, Geophysical Monograph.* (Vol *16*), 203 pp. Tulsa, OK.

Hudec, M. R., & Jackson, M. P. A. (2007). Terra infirma: Understanding salt tectonics. *Earth-Science Reviews, 82*(1–2), 1–28. doi: 10.1016/j.earscirev.2007.01.001

International Mineralogical Association (2022). The new IMA list of minerals—A work in progress. *International Mineralogical Association.* http://cnmnc.main.jp/, updated May 2022.

Jackson, C. A. -L., Elliott, G. M., Royce-Rogers, E., Gawthorpe, R. L., & Aas, T. E. (2019). Salt thickness and composition influence rift structural style, northern North Sea, offshore Norway. *Basin Research, 31*(3), 514–538. doi: 10.1111/bre.12332

Jackson, C. A.-L., & Lewis, M. M. (2014). Structural style and evolution of a salt-influenced rift basin margin; the impact of variations in salt composition and the role of polyphase extension. *Basin Research, 26*, 1–22. doi: 10.1111/bre.12099.

Jackson C. A. -L., & Stewart, S. A. (2017). Composition, tectonics, and hydrocarbon significance of Zechstein Supergroup Salt on the United Kingdom and Norwegian Continental Shelves: A review. In I. Soto, J. F. Flinch, G. Tari (Eds.), *Permo-Triassic salt provinces of Europe, North Africa and the Atlantic margins, tectonics and hydrocarbon potential* (pp. 175–201). Amsterdam: Elsevier.

Jackson, M. P. A., & Hudec, M. R. (2017). *Salt tectonics: Principles and practice.* Cambridge, UK: Cambridge University Press, 510 pp.,

Jackson, M. P. A., Roberts, D. G., & Snelson, S. (Eds.). (1995). *Salt tectonics: A global perspective, American Association of Petroleum Geologists, Memoir* (Vol. *65*), 454 pp.

Jackson, M. P. A., Hudec, M. R., & Dooley, T. P. (2010). Some emerging concepts in salt tectonics in the deepwater Gulf of Mexico: intrusive plumes, canopy-margin thrusts, minibasin triggers and allochthonous fragments. In B. A. Vining & S. C. Pickering (Eds.), *Petroleum Geology: From Mature Basins to New Frontiers—Proceedings of the 7th Petroleum Geology Conference* (Vol. 7, pp. 899–912). Geological Society, London. ISBN 9781862392984.

Jackson, M. P. A., & Talbot, C. J. (1986). External shapes, strain rates, and dynamics of salt structures. *Geological Society of America Bulletin, 97*, 305–323.

Jackson, M. P. A., & Vendeville, B. C. (1994). Regional extension as a geologic trigger for diapirism. *Geological Society of America Bulletin, 106*, 57–73.

Jackson, M.P.A., Vendeville, B.C., & Schultz-Ela, D.D. (1994). Salt-related structures in the Gulf of Mexico: A field guide for geophysicists. *The Leading Edge, August 1994*, pp. 837–842.

Jahnert, R., França, A., Trindade, L., Quintaes, C., Santos, P., Pessoa, J., and Bedregal, R. (1998). The petroleum system of Campos Basin. In M. R. Mello and P.O. Yilmaz (Eds.), *1998 AAPG International Conference and Exhibition*, Rio Janeiro, Brazil, Extended Abstract Volume, 600–601.

Jensen, P. K. (1990). Analysis of the temperature field around salt diapirs. *Geothermics, 19*(3), 273–283. doi:10.1016/0375-6505(90)90047-f

Jenyon, M. K. (1986). *Salt tectonics.* Amsterdam: Elsevier Applied Science Publishers Ltd., 191 pp.

Jones, I. F., & Davison, I. (2014). Seismic imaging in and around salt bodies. *Interpretation, 2*(4), SL1–SL20. http://dx.doi.org/10.1190/INT-2014-0033.1

Jowett, E. C., Cathles, L. M., III, & Davis, B. W. (1993). Predicting depths of gypsum dehydration in evaporitic sedimentary basins. *American Association of Petroleum Geologists Bulletin, 77*(3), 402–413.

Kearey, P., Brooks, M., & Hill, I. (2002). *An introduction to geophysical exploration* (3rd ed.). Oxford, UK: Blackwell Science, 268 pp.

Kukla, P. A., Strozyk, F., & Mohriak, W. U. (2018). South Atlantic salt basins—Witnesses of complex passive margin evolution. *Gondwana Research, 53*, 41–57. https://doi.org/10.1016/j.gr.2017.03.012

Lenoir, S. (1973). Location and identification of evaporite deposits through well logging, *Congresso Brasileiro de Geologia, 27, 291–315.* Aracajú. Anais, Sociedade Brasileira de Geologia.

Martín-Martín, J. D., Vergés, J., Saura, E., Moragas, M., Messager, G., Baqués, V., et al. (2017). Diapiric growth within an Early Jurassic rift basin: The Tazoult salt wall (central High Atlas, Morocco). *Tectonics, 36*, 2–32. doi:10.1002/2016TC004300

Marzoli, A., Renne, P. R., Piccirillo, E. M., Ernesto, M., Bellieni, G., & De Min, A. (1999). Extensive 200 Ma-year-old continental flood basalts of the Central Atlantic Magmatic Province. *Science, 284*, 616–618.

Mello, U. T., Karner, G. D., & Anderson, R. A. (1995). Role of salt in restraining the maturation of subsalt source rocks. *Marine and Petroleum Geology, 12*, 697–716.

Miall, A. D. (1990). *Principles of sedimentary basin analysis.* New York, USA: Springer-Verlag, 490 pp.

Miall, A. D. (2016). *Stratigraphy: A modern synthesis.* New York, USA: Springer, 454 pp.

Miller, W. H. (1839). *A treatise on crystallography.* Cambridge, UK: Cambridge, *139* pp. + plates.

Mindat (2022). Halite. https://www.mindat.org/min-1804.html

Misra, A. A., & Mukherjee, S (2018). *Atlas of structural geological interpretation from seismic images*. Hoboken, NJ: Wiley Blackwell, 288 pp.

Mitchum, R. M., Jr., & Vail, P.R. (1977). Seismic stratigraphy and global changes of sea level, Part 7. Seismic stratigraphic interpretation procedure. In: C. E. Payton (Ed.), *Seismic stratigraphy—Applications to hydrocarbon exploration, American Association of Petroleum Geologists, Memoir* (Vol. *26*, pp. 135–143).

Mitchum, R. M., Jr., Vail, P. R., & Sangree J. B. (1977). Seismic stratigraphy and global changes of sea level: Part 6. Stratigraphic interpretation of seismic reflection patterns in depositional sequences. In C. E. Payton (Ed.), *Seismic stratigraphy—Applications to hydrocarbon exploration, American Association of Petroleum Geologists, Memoir* (Vol. *26*, pp. 117–133).

Mohriak, W. (2025). *Salt in the Earth sciences: Basin analysis and salt tectonics. Washington,* DC: American Geophysical Union. doi:10.1002/9781394307067

Mohriak, W. U., (2005). Salt tectonics in Atlantic-type sedimentary basins: Brazilian and West African perspectives applied to the North Atlantic Margin. In Post, P. J., Rosen, N. C., Olson, D.L., Palmes, S. L., Lyons, K. T. & Newton, G. B. (Eds.), GCSSEPM 25th Annual Bob F. *Perkins Research Conference, Petroleum Systems of Divergent Continental Margin Basins*, p. 375–413.

Mohriak, W.U. (2009). Tectônica de sal autóctone e alóctone na margem sudeste brasileira. In W. Mohriak, P. Szatmari & S. M. C. Anjos (organizadores), *Sal: Geologia e Tectônica.* 2ed., pp. 302–315. São Paulo: Editora Beca.

Mohriak, W. U., & Fainstein, R. (2012). Phanerozoic regional geology of the Eastern Brazilian margin. In D. G. Roberts & A. W. Bally (Eds.), *Phanerozoic passive margins, cratonic basins and global tectonic maps* (Vol. *1C*, Chapter 7, pp. 223–282). Amsterdam: Elsevier B.V. doi: 10.1016/B978-0-444-56357-6.00006-8

Mohriak, W. U., & Szatmari, P. (2008). Introdução às propriedades químicas e físicas dos evaporitos. In W. Mohriak, P. Szatmari, & S. M. C. Anjos (organizadores), *Sal: Geologia e Tectônica* (pp. 18–41). *São Paulo: Editora Beca.*

Mohriak, W. U., Szatmari P., & Anjos, S. (2012). *Salt: Geology and tectonics of selected Brazilian basins in their global context*. In G. I. Alsop, S. G. Archer, A. J. Hartley, N. T. Grant, & R. Hodgkinson (Eds.), *Salt tectonics, sediments and prospectivity* (Vol. 363, pp. 131–158). Geological Society, London, Special Publications.

Nabighian, M. N., Ander, M. E., Grauch, V. J. S., Hansen, R. O., LaFehr, T. R., Li, Y., et al. (2005). Historical development of the gravity method in exploration. *Geophysics*, *70*(6), 63ND–89ND. doi:10.1190/1.2133785

Nanda, N. C. (2016). *Seismic data interpretation and evaluation for hydrocarbon exploration and production: A practitioner's guide*. Switzerland: Springer., 224 pp.

Nelson, T. H. (1991). Salt tectonics and listric-normal faulting. In A. Salvador (Ed.), *The geology of North America, Vol. J, The Gulf of Mexico Basin* (pp. 73–89). Boulder, CO: Geological Society of America.

Nesse, W. D. (2000). *Introduction to mineralogy*. UK: Oxford University Press, 442 pp.

Nettleton, L. L. (1962). Gravity and magnetics for geologists and seismologists. *American Association of Petroleum Geologists Bulletin*, *46*(10), 1815–1838.

Nettleton, L. L. (1976). *Gravity and magnetics in oil prospecting*. New York: McGraw-Hill, 464 p.

Nishimura, D., Sugiyama, S., Bauder, A., & Funk, M. (2018). Changes in ice-flow velocity and surface elevation from 1874 to 2006 in Rhonegletscher, Switzerland. *Arctic, Antarctic, and Alpine Research*, *45*(4), 552–562. https://doi.org/10.1657/1938-4246-45.4.552

Nurmi, R. D. (1978). Use of well logs in evaporite sequences. In W. E. Dean & B. C. Schreiber (Eds.), *Marine evaporites, Society of Economic Paleontologists and Mineralogists. Lecture Notes for Short Course* (Vol. *4*, pp. 144–176). Tulsa, Okla: Society of Economic Paleontologists and Mineralogists.

Ogiesoba, O. C., & Zeng, H. (2022). Identification of sandstone-rich zones in upper bathyal, deep water environment on South Texas Gulf Coast. *SEG Interpretation*, *10*(2), T265–T278. https://doi.org/10.1190/INT-2021-0139.1

Okrusch, M., & Frimmel, H.E. (2020) *Mineralogy. Springer Textbooks in Earth Sciences, Geography and Environment*. Berlin, Heidelberg: Springer. https://doi.org/10.1007/978-3-662-57316-7_8

Pavlis, T. I., & Mason, K. A. (2017). The New World of 3D Geologic Mapping. *Geological Society of America. GSA Today*, *27*. doi: 10.1130/GSATG313A.1

Ramos, A., Fernández, O., Muñoz, J. A., & Terrinha, P. (2017). Impact of basin structure and evaporite distribution on salt tectonics in the Algarve Basin, Southwest Iberian margin. *Marine and Petroleum Geology*, *88*, 961–984. https://doi.org/10.1016/j.marpetgeo.2017.09.028

Reiche, S., Hübscher, C., & Beitz, M. (2014). Fault-controlled evaporite deformation in the Levant Basin, Eastern Mediterranean. *Marine Geology*, *354*, 53–68. http://dx.doi.org/10.1016/j.margeo.2014.05.002

Rezouki, I., Boujamaoui, M., Hafid, M., Bbac, A. N., Amiri, A., Inoubli, M. H., et al. (2020). Contribution of gravity and aeromagnetic data to the structural modeling of the hidden faults in Guercif Basin, northeastern Morocco. *Journal of African Earth Sciences*, *164*, 103797. https://doi.org/10.1016/j.jafrearsci.2020.103797

Rowan, M. G., & Ratliff, R. A. (2012). Cross-section restoration of salt-related deformation: Best practices and potential pitfalls. *Journal of Structural Geology*, *41*, 24–37. doi:10.1016/j.jsg.2011.12.012.

Saura, E., Vergés, J., Martín-Martín, J. D., Messager, G., Moragas, M., Razin, P., et al. (2014). Syn- to post-rift diapirism and minibasins of the Central High Atlas (Morocco): The changing face of a mountain belt. *Journal of the Geological Society*, *171*, 97–105. doi: 10.1144/jgs2013-079.

Schofield, N., Alsop, I., Warren, J., Underhill, J. R., Lehné, R., Beer, W. & Lukas, V. (2014). Mobilizing salt: Magma-salt interactions. *Geology*, *42*(7), 599–602.

Schön, J. H. (1996). *Physical properties of rocks: Fundamentals and principles of petrophysics* (*Handbook of geophysical exploration: Seismic exploration*, Vol. *18)*. Amsterdam: Elsevier, 583 pp.

Schön, J. H. (2011). *Physical Properties of rocks (Handbook of petroleum exploration and production*, Vol. *8*). Amsterdam: Elsevier, 481 pp.

Schön, J. H. (2015). *Physical properties of rocks: Fundamentals and principles of petrophysics*. (2nd ed.), *Developments in Petroleum Science* (Vol. *65*). Amsterdam: Elsevier, 497pp.

Schreiber, B. C., Friedman, G. M., Decima, A., & Schreiber, E. (1976). Depositional environments of Upper Miocene (Messinian) evaporite deposits of the Sicilian basin. *Sedimentology*, *23*(6), 729–760.

Selig, F., & Wallick, G. C. (1966). Temperature distribution in salt domes and surrounding sediments. *Geophysics*, *31*(2), 346–361. doi:10.1190/1.1439777

Senger, K., Millett, J., Planke, S., Ogata, K., Haug Eide, C., Festøy, M., et al. (2017). Effects of igneous intrusions on the petroleum system: A review. *First Break*, *35*(6), 47–56.

Seni, S. J., Mullican, W. F., III, & Hamlin, H. S. (1985). Texas salt domes: Natural resources, storage caverns, and extraction technology. *Report for Texas Department of Water Resource*. The University of Texas at Austin Bureau of Economic Geology, 161 pp.

Serra, O. (1984). *Fundamentals of well-log interpretation 1. The acquisition of logging data. Developments in Petroleum Science* (Vol. 15, Part A). Amsterdam: Elsevier, 423 pp.

Serra, O. (1986). *Fundamentals of well-log interpretation. 2. The interpretation of logging data. Developments in Petroleum Science* (Vol. 15B). Amsterdam: Elsevier, 679 pp.

Sheriff, R. E. (1976). Inferring stratigraphy from seismic data. *American Association of Petroleum Geologists Bulletin*, *60*(4), 528–542.

Sheriff, R. E., & Geldart, L. P. (1995). *Exploration seismology*. Cambridge: Cambridge University Press, 592 pp.

Stewart, S. A., & Clark, J. A. (1999). Impact of salt on the structure of the Central North Sea hydrocarbon fairways. In A. J. Fleet & S. A. R. Bouldy (Eds.), *Petroleum geology of northwest Europe: Proceedings of the 5th Conference* (pp. 179–200). doi:10.1144/0050179

Stewart, S. A., & Coward, M. P. (1996). Genetic interpretation and mapping of salt structures. *First Break*, *14*, 135–141.

Strauss, P., Granado, P., & Muñoz, J. A. (2021). Subsidence analysis of salt tectonics-driven carbonate minibasins (Northern Calcareous Alps, Austria). *Basin Research*, *33*, 968–990. https://doi.org/10.1111/bre.12500

Szatmari, P., Carvalho, R. S., & Simoes, I. A. (1979). A comparison of evaporite facies in the late Paleozoic Amazon and the Middle Cretaceous South Atlantic salt basins. *Economic Geology*, *74*(2), 432–447.

Talbot, C. J., & Jarvis R. J. (1984). Age, budget and dynamics of an active salt extrusion in Iran. *Journal of Structural Geology*, *6*, 521–533.

Talbot, C. J., & Rogers, E. A. (1980). Seasonal movements in a Salt Glacier in Iran. *Science*, *208*, 395–397. doi: 10.1126/science.208.4442.395

Taylor, J. C. M. (1981). Late Permian—Zechstein. In *Introduction to the petroleum geology of the North Sea, London*, Joint Association for Petroleum Exploration Courses, Japec Course Notes No. 01.

Taylor, J. C. M. (1985). The role of evaporites in hydrocarbon exploration. *Joint Association of Petroleum Exploration Courses (UK)*, Japec Course No. 39.

Taylor, J. C. M. (1998). Upper Permian—Zechstein. In K. W. Glennie (Ed.), *Petroleum geology of the North Sea: Basic concepts and recent advances* (4th ed., pp. 174–211). Oxford, UK: Blackwell Publishing Limited.

Tearpock, D. J., & Bischke, R. E. (2002). *Applied subsurface geological mapping with structural methods* (2nd ed.). New Jersey, USA: Prentice Hall, 822 pp.

Teixeira, L., Lupinacci, W. M., & Maul, A. (2020). Quantitative seismic-stratigraphic interpretation of the evaporite sequence in the Santos Basin. *Marine and Petroleum Geology*, *122*, 104690.

Telford, W. M., Geldart, L. P., & Sheriff, R. E. (1990). *Applied geophysics* (2nd ed.). UK: Cambridge University Press, 744 pp.

Thompson, D. E., Lerche, I., & O 'Brien, J. J. (1989). Salt basins of Western Europe and Gabon, West Africa: Dynamical aspects and hydrocarbon production. In *GCSSEPM Foundation Tenth Annual Research Conference Program and Abstracts*, Houston, Texas, pp. 122–130.

Tiab, D., & Donaldson, E. C. (2004). *Petrophysics: Theory and practice of measuring reservoir rock and fluid transport properties* (2nd ed.). Gulf Professional Publishing, Elsevier, 889 pp.

Trudgill, B. D., & Rowan, M. G. (2004). Integrating 3D seismic data with structural restorations to elucidate the evolution of a stepped counter-regional salt system, eastern Louisiana Shelf, Northern Gulf of Mexico. In R. J. Davies, J. A. Cartwright, S. A. Stewart, M. Lappin, & J. R. Underhill (Eds.), *3D seismic technology: Applications to the exploration of sedimentary basins. Geological Society, London, Memoirs* (Vol. 29, pp. 165–176).

Trusheim, F. (1960). Mechanism of salt migration in northern Germany. *American Association of Petroleum Geologists Bulletin*, *44*(9), 1519–1540.

Van der Pluijm, B. A., & Marshak, S. (2004). *Earth structure: An introduction to structural geology and tectonics (2nd ed.)*. New York: W. W. Norton, 656 pp.

Vergés, J., Moragas, M., Martín-Martín, J. D., Saura, E., Casciello, E., Razin, Ph., et al. (2017). Salt tectonics in the Atlas Mountains of Morocco. In J. I. Soto, J. F. Flinch, & G. Tari (Eds.), *Permo-Triassic salt provinces of Europe, North Africa and the Atlantic margins, tectonics and hydrocarbon potential* (Chapter 26, pp. 563–579). http://dx.doi.org/10.1016/B978-0-12-809417-4.00027-6

Vernant, Ph., Nilforoushan, F., Hatzfeld, D., Abbassi, M. R., Vigny, C., Masson, F., et al. (2004). Present-day crustal deformation and plate kinematics in the Middle East constrained by GPS measurements in Iran and northern Oman. *Geophysical Journal International*, *157*(1), 381–398. https://doi.org/10.1111/j.1365-246X.2004.02222.x

Verweij, J. M. (2012). Using basin modeling for geothermal energy exploration in the Netherlands. Conference Proceedings, *74th EAGE Conference and Exhibition incorporating EUROPEC 2012*, European Association of Geoscientists & Engineers, cp-293-00051. doi: https://doi.org/10.3997/2214-4609.20148148

Waisman, G. (2009). Tectônica de sal da Bacia de Campos. In W. Mohriak, P. Szatmari. S. M. & C. Anjos (Eds.), *Sal: Geologia e Tectônica* (pp. 314–339). São Paulo: Editora Beca.

Warren, J. K. (1989). *Evaporite sedimentology: Importance in hydrocarbon accumulation*. New Jersey, USA: Prentice-Hall Advanced Reference Series, 285 pp.

Warren, J. K. (2006). *Evaporites: Sediments, resources and hydrocarbons*. Berlin: Springer Verlag, 1035p.

Warren, J. K. (2010). Evaporites through time: Tectonic, climatic and eustatic controls in marine and nonmarine deposits. *Earth and Science Review*, *98*, 217–268.

Warren, J. K. (2016). *Evaporites—A geological compendium*. Berlin: Springer, 1807 pp. ISBN 978-3-319-13511-3.

Weijermars, R., Jackson, M. P. A., & Vendeville, B. C. (1993). Rheological and tectonic modeling of salt provinces. *Tectonophysics*, *217*, 143–174.

Whitmeyer, S. J., & Nicoletti, J. (2010). The digital revolution in geologic mapping. *GSA Today*, *20*(4/5), 4–10. doi: 10.1130/GSATG70A.1

Zhang, Y., Dengfa, H., Wu, B., & Gao, H. (2021). Kinematic evolution of fold-and-thrust belts in the Yubei-Tangbei area: Implications for tectonic events in the southern Tarim Basin. *Geoscience Frontiers*, *12*, 101233. https://doi.org/10.1016/j.gsf.2021.101233

6

Rheology of Evaporite Rocks, Salt Deformation, and Fluid Flow

ABSTRACT

Minerals like halite, characterized by weak rheology, influence the deformation of evaporite layers. Most evaporite rocks show varying resistance to deformation under extensional or compressional stress. Studying mineral deformation in evaporite sequences involves petrographic microscope analysis and rock mechanics laboratory experiments. Flow mechanisms are based on fluid dynamics and rheological principles. Fluid hydrodynamics affects mineralization in sedimentary sequences, which leads to economic resources like copper, zinc, and emeralds.

Evaporite rocks are solid substances at room temperature and pressure, but compared with other rocks, they are extremely ductile, as they deform easily when relatively low stresses are applied (Baar, 1977; Carter & Hansen, 1983; Schön, 2015; Weijermars et al., 1993; 2014). At first sight, a hand sample of evaporite rock looks like any other rock: it is hard, relatively dense, and breaks when hit by a hammer. But if salt is subjected to stress for thousands of years, even at surface temperatures, it will slowly flow, like ice in glaciers (Figure 6.1). The strata in the glacial ice are formed by polycrystalline ice, which is deformed in a relatively short geological time, as some glaciers flow at a rate of a few meters to a few kilometers per year (Hambrey & Glasser, 2005). Glaciers flow under the influence of gravity and the motion is achieved by internal deformation (spreading), basal sliding, and subglacial deformation of the sedimentary beds. During ice flow, the basal gliding results in primary creep and dynamic recrystallization of the ice crystals. The process is facilitated by ice melting at the bottom of the glacier, where temperatures approach 0 °C. The temperature increase is related to the geothermal heat from below the base of the glacier and frictional heat, which results from basal sliding and internal deformation (Hambrey & Glasser, 2005).

The low ionic potential of the elements that make up the evaporite rocks also strongly affects their deformation after they are buried. Evaporite minerals are easily deformed by solid flow because the attraction between their ions is relatively weak (see Chapter 4). Even without a significant elevation in temperature, evaporite rocks can display the same structures that are observed in gneisses and other metamorphic rocks formed at high temperatures and pressures (Mohriak & Szatmari, 2009).

Rock salt deformation is a phenomenon observed in various geological formations, including the folded layers of Triassic evaporites in the Alps (Figure 6.2) and the Zagros Mountains in Iran (see Chapter 3). Huge salt diapirs, often exceeding several kilometers in height, can be seen in the subsurface of several sedimentary basins, even in situations of small tectonic stresses. Some evaporite minerals (such as halite, carnallite, and sylvite) can flow easily because they are less resistant to stress (see Chapter 5). These deformations are influenced by the weak rheology of the evaporite layers at various stratigraphic levels, leading to salt extrusion at the surface. Evaporite minerals, such as halite, carnallite, and sylvite, can flow easily due to their lower viscosity and density, making them distinct from other minerals in sedimentary layers (e.g., Schön, 2015; Weijermars et al., 1993). Quartz-rich rocks, on the other hand, do not deform under normal pressure and temperature, but when subjected to stresses, they deform through brittle processes, forming fractures and faults. Rocks with intermediate rheological behavior, such as limestone, usually deform by folding and faulting, as seen in mountain ranges formed by orogenic processes, such as the Zagros, Andes, Atlas, Rocky Mountains, and the Alps (Figure 6.2). The weak rheology of the evaporite layers at diverse stratigraphic levels

Salt in the Earth Sciences: Evaporite Rocks and Salt Deposition, Geophysical Monograph 286, First Edition. Webster Mohriak.

DOI: 10.1002/9781119405245.ch06

Figure 6.1 An ice glacier is rigid enough to walk over and appears stationary. However, it flows under the action of gravity due to its weight, advancing from mountain tops toward lower elevations. The image on top (A) is an aerial photo of the Crusoe glacier on the island of Axel Heiberg in northern Canada. It corresponds to a valley glacier with a total surface area of about 70 km^2 and is presently advancing toward the valleys. The bottom image (B) shows the glacier front forming a cliff where the ice strata's internal deformation is characterized by recumbent folding. Salt glaciers (or namakiers) also deform by flowing, as the halite minerals are characterized by low viscosity compared to other Earth materials. Credit: Courtesy of Jürg Alean and Michael Hambrey.

influences the deformation of the sedimentary strata under compressional stresses, resulting in salt extrusion at the surface (see Chapter 3).

6.1. BASIC PRINCIPLES OF RHEOLOGY APPLIED TO EVAPORITE ROCKS

Rock salt sequences flow toward less elevated regions like a highly viscous fluid due to the low ionic strength of halite (see Chapter 5, Figure 5.3). Researchers have compared the mechanical behavior of halite to that of a Newtonian fluid, which is influenced by variations in temperature and pressure (e.g., Carter & Hansen, 1983; Schultz-Ela et al., 1993).

Viscosity is a physical property of fluids that measures their resistance to flow. High-viscosity materials flow slowly, while low-viscosity materials, like water, flow quickly. An ideal Newtonian liquid has no variation in viscosity with deformation rate or time but is influenced by temperature and pressure. For example, bitumen on a hot summer day becomes less viscous and soft.

Some rocks, such as mobile shale and halite layers, can easily deform due to the weight of the overburden rocks. Salt deforms and rises to the surface when buried below other sedimentary rocks and subjected to compressional forces, forming salt diapirs that are observed in outcrops in many regions of the globe, such as the Zagros Mountains in Iran or the Atlas Mountains in Algeria and Tunisia (see Chapter 3). In submarine regions of many continental margins, such as the Gulf of Mexico (GOM) and the South Atlantic salt basins, the evaporite layers developed diapiric structures that are imaged by seismic profiles (see Chapter 5).

Rheology is the study of the mechanical behavior of rocks under stress, which may be associated with extensional, compressional, or shearing forces (Baar, 1977; Fossen, 2016). These forces deform the evaporites by a solid-state flow. Plastic flow results in permanent deformation in rocks that exhibit ductile behavior. Fracturing and faulting also result in permanent deformation in rocks that typically exhibit rigid or brittle behavior (Figure 6.3).

Brittle rocks rupture and lose cohesion between the deformed parts when the forces exceed the rock's

Figure 6.2 Evaporite layers (mainly gypsum, probably of Triassic age) folded during the Alpine Orogeny in an outcrop in the Alps of northern Italy. The evaporite sequences are deformed by gravity and compressional forces, resulting in plastic deformation and ductile flow. The trees in the foreground and on the top of the hill provide an approximate scale. Credit: Photo by H. Goldhammer, Courtesy of M. Jackson (University of Texas). Adapted from Mohriak and Szatmari (2009).

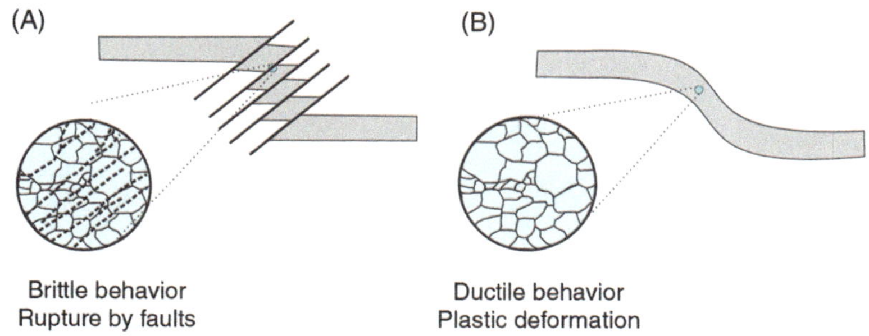

Figure 6.3 Rheological behavior of rocks under stress as seen at macro- and microscales: (A) Brittle behavior is associated with ruptures caused by fractures and faults. (B) Ductile behavior is associated with plastic deformation. Credit: Data from Van der Pluijm and Marshak (2004).

strength (Figure 6.3). Rock failure by faulting is characterized by extensional, compressional, or strike-slip faults (Figure 6.3A), whereas ductile rocks deform by flowing and folding (Figure 6.3B).

Rheology is derived from the Greek term *rheos*, which means stream, analogous to water, which flows due to the action of gravity. Salt rocks may deform due to several reasons, including the action of gravity on the salt mass, the weight of the overburden rocks, or tectonic forces related to plate motion.

Geomechanics laboratories use rock sample plugs in physical experiments to characterize the rheological properties of rocks. These plugs are cylinders cut from rock and subjected to stress using special equipment. The gradual deformations of these plugs are measured and plotted in graphs, which are then applied to engineering studies, mining operations, and petroleum well drilling (e.g., Aadnøy & Looyeh, 2019; Baar, 1977). The plug deformation tests provide stress-strain relationships that can be graphically interpreted, determining rheological parameters for different lithologies. Researchers also use numerical modeling methods to simulate the behavior of salt layers subjected to diverse tectonic forces (e.g., Li & Urai, 2016; Li et al., 2012). These numerical experiments use algorithms based on physical and mechanical laws, involving complex equations, to calculate the behavior of evaporites under different strain rates, temperatures, and pressures. The results are often compared to mesoscale structures observed in outcrops, aiming to constraining theoretical results with real-world measurements (e.g., Kukla et al., 2011; Li & Urai, 2016).

Evaporite sequences are influenced by the brittle and ductile behavior of the materials in these rocks. Numerous rheological studies investigate these deformation mechanisms at both macroscale (folds and faults) and microscale (Figure 6.3), using optical microscopes for microstructure analysis. Petrographic studies involve routine descriptions of sedimentary rocks and techniques used in metamorphic rock geology. Evaporite rocks deformed in orogenic belts or salt diapirs form metamorphic rocks that can be better analyzed using microtectonic techniques. A useful reference for this advanced topic is the textbook by Passchier and Trouw (2005).

Rocks and other materials under stress are analyzed in laboratories using three end-members of solid deformation: elastic, plastic, and viscous (Figure 6.4A). The different rheological behavior of rocks leads to three deformation models: rupture by brittle deformation, ductile deformation in the solid state, and fluid flow (Figure 6.4B).

Salt layers exhibit varying deformation behavior under stress, ranging from brittle to ductile regimes (Figure 6.4). Salt is extremely weak and creeps under deviatoric stress even at surface temperatures, as observed in the salt glacier of the Zagros Mountains (see Chapter 3). However, under high-stress rates during instantaneous events, such as earthquakes, salt rocks behave as brittle material, developing fractures and faults (Davison, 2009).

Salt behavior under stress is commonly analyzed in laboratories using special equipment and techniques. The uniaxial or unconfined compressive strength (UCS) test is the most common laboratory test undertaken for rock mechanics studies. However, sometimes unconfined tensile strength (UTS) testing is also performed to assess the tensile strength (Pells, 1995).

The apparatus for uniaxial tests consists of a cylinder where the sample is placed and subjected to increasing stress by a movable piston. Under uniaxial load tests, a cylindrical sample of an elastic rock will shorten under compression while expanding in an orthogonal direction (e.g., Hunsche, 1994; Pells, 1995).

In many laboratory experiments, stress is measured in megapascals (MPa). One megapascal (1 MPa) corresponds to about 145 psi, or 10 atm. In some uniaxial experiments, the stress may be increased by 1 MPa every minute, resulting in a stress rate of 1 MPa/min. The deformation of the salt specimen (strain) is measured for different stresses. In nature, salt rocks in the subsurface are subject to overburden. The weight of the overburden on the subsurface rocks results in lithostatic pressure. Porous

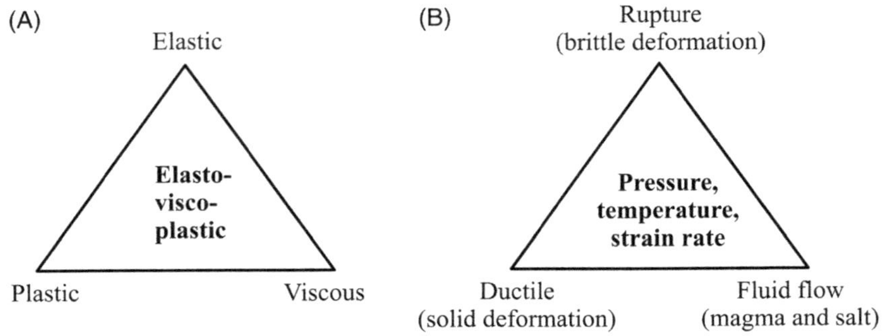

Figure 6.4 The behavior of different materials under stresses can be analyzed in laboratories by considering three rheological end-members: elastic, viscous, and plastic (A). The behavior of rocks affected by extensional, compressional, or strike-slip tangential forces depends on several factors, particularly pressure, temperature, and strain rate (B). Rocks can deform by rupture (brittle deformation associated with fractures and faults), ductile process (solid deformation with material strained but not ruptured), or fluid flow, as in high-temperature magma or low-temperature salt.

rocks are characterized by fluids filling the interstitial pores. In the subsurface, these fluids exhibit different pressures depending on several geological factors, resulting in hydrostatic pressure (Van der Pluijm & Marshak, 2004).

Triaxial experiments simulate confining pressure in laboratories using more sophisticated apparatus than the uniaxial tests, such as pressurized fluid vessels and hydraulic cylinders. When placed within the apparatus, the specimen is surrounded by the fluid, and the confining pressure is uniform, affecting the specimen in all directions. Similar to the uniaxial test, the axial force is obtained by a hydraulic cylinder with a piston, usually representing a vertical force (e.g., Hunsche, 1994).

Salt under low stress and at room temperature exhibits elastic behavior up to 7 MPa. For stresses larger than this, salt flows plastically, characterized by partial ductility and halite crystal sliding. This elastoplastic behavior is predominant for stresses up to 14 MPa. However, for stresses larger than 14 MPa, salt becomes virtually plastic and exhibits ductile behavior, deforming easily as a viscous fluid, similar to molten magma at high temperatures. This ductile deformation is irreversible, contrasting with the elastic deformation, which recovers strain when stress is removed. For larger stresses, salt's ductile behavior results in permanent deformation, which cannot be reversed when the stresses are removed. A salt diapir forced to rise due to a stress increase at depth does not recover its previous shape when extruded at the surface, where pressure is 1 atm or 1/10 of an MPa.

Solid rock deformation processes are influenced by the physical properties of materials. These parameters control how the materials react to the stresses, either by rupturing (brittle behavior), by ductile plastic deformation (solid-state), or by fluid flow deformation (such as in magma, ice, or salt). Materials that display brittle behavior under stress may deform or strain by a few percent until they break up and rupture when the stress reaches a critical value. This maximum stress is called the maximum strength of the rock under the forces. Rock failure in nature and in the laboratory reflects the occurrence of fracture mechanisms that lead to rupture or breakup. The material loses its physical continuity (cohesion) after the body has already undergone some deformation by elastic and/or plastic behavior. Brittle deformation is often observed in rigid materials, such as concrete, sandstones, and limestones. When subjected to stresses, these rocks are deformed and commonly affected by folds, fractures, and faults (Van der Pluijm & Marshak, 2004).

Solid materials that exhibit ductile behavior start to deform when the stress reaches a critical value, known as yield stress. The applied stress results in plastic deformation or solid-state flow. Mesoscopically, the process is similar to a viscous fluid flow. For example, a pile of sand can accumulate grains until a critical angle on the flanks is reached. Adding more sand to the pile will result in the sand grains falling along the flanks, similar to an avalanche. A mountain glacier will accumulate ice until a certain thickness is reached. The weight of the glacier will result in pressure that will deform the ice layers by ductile flow after a critical load value is reached (Figure 6.1). The salt glaciers in Iran deform by flowing downslope along the flanks of the elevated anticlinal structures. After the salt diapirs rise to a certain height, the gravity forces result in the collapsing of the salt masses by gravity spreading (Talbot, 1981, 1998).

After a rock starts to deform by ductile behavior, the strain is distributed throughout the body by internal deformation of the particles, i.e., strain develops without the formation of mesoscopic discontinuities such as fractures or faults. Ductile behavior can involve brittle (cataclastic flow) or plastic deformation mechanisms. Ductile deformation is a major process in rocks like mud, undercompacted shale, and mobile salt, where strain is distributed throughout the body through the internal deformation of particles. The plasticity of a homogeneous material depends on factors like pressure, temperature, and strain rate (Van der Pluijm & Marshak, 2004).

The deformation of the Triassic evaporitic rocks in the Apennines or the Alps (Figure 6.2) displays some ductile zones where the rocks are folded due to the compressional forces associated with the Alpine Orogeny (see Chapter 3). However, there are some deformed zones where the evaporites have been sheared and ruptured by thrust and strike-slip faults (e.g., Schorn & Neubauer, 2014; Sommaruga et al., 2017). Damaged zones and brecciated material (fault breccia, gouge, and cataclasites of different grain sizes) occur along the discontinuity planes. These zones can be studied at several scales, from macroscopic (aerial photographs, outcrops, and hand samples) to microscale (petrographic thin sections of the hand samples and cores), which helps in elucidating the mechanisms of deformation (De Paola et al., 2008).

6.1.1. Basic Equations of Solid Material Deformation

In structural geology, deformation results from forces acting on a solid material. Deformation refers to changes in the shape, position, or orientation of a body. More specifically, deformation involves three components: rotation, corresponding to the pivoting of a body around a fixed axis; translation, corresponding to a change in position; and strain, corresponding to the distortion or change in the shape of a body (Van der Pluijm & Marshak, 2004).

Stress is defined as the force acting on a plane per unit area. Stress is commonly represented by the Greek letter sigma (s). The equation is shown as

$$\text{Stress} = s = F / A$$

where F = force (measured in $kg/m^2{\cdot}s^2$ or N/m^2) and A = area (measured in square meters). Stress is measured as a force per unit area (N/m^2 or $kg/m{\cdot}s^2$). The stress of 1 $kg/m{\cdot}s^2$ (1 N/m^2) is called a pascal (Pa). One million pascals correspond to 1 MPa. Some researchers continue to use the term bar, which corresponds to a pressure of approximately 1 atm, which is equivalent to 10,000 Pa (Van der Pluijm & Marshak, 2004).

$$1 \text{ bar} = 10^5 \text{ Pa} \sim 1 \text{ atm}.$$

Pressure corresponds to stress acting on a body. For example, at the Earth's surface, the pressure is 1 atm, corresponding to one-tenth of a megapascal (0.1 MPa). Pressures in the subsurface are much larger and are usually expressed by 1,000 bar or kilobar (kbar), where

$$1 \text{ kbar} = 1000 \text{ bar} = 10^8 \text{ Pa} = 100 \text{ MPa}.$$

The weight of a sedimentary column exerts pressure on its base, resulting in lithostatic (or geostatic) stress (Tiab & Donaldson, 2016). The lithostatic stress, also called lithostatic (or geostatic) pressure, corresponds to the weight of a column of rock in the subsurface. The lithostatic pressure (P_L) of a rock column with a length of h is given by the rock density multiplied by gravity multiplied by the column length:

$$P_L = r \cdot g \cdot h$$

where r corresponds to density (kg/m^3), g to gravity (N/m^2), and h to length (m).

For example, the lithostatic pressure of a 1-km column of rock with a density of 2000 kg/m^3 is given by

$$P_L = r \cdot g \cdot h = 2000.9.8.1000(kg \cdot m^3 \cdot N \cdot m^{-2} \cdot m) \\ = 1.96 \times 10^7 = 19.6\,\text{MPa}.$$

Strain, usually represented by the Greek letter epsilon (e), is the amount of deformation that occurs when a solid material is subjected to a force. For example, when a rubber band is gradually stretched by pulling at the ends, the strain corresponds to the ratio between the increase in length of the stretched rubber band after a time t and divided by the original (unstretched) length of the rubber band. The following equation expresses this relationship:

$$\text{Strain} = e = (DL)/L = (L(t) - L(0))/(L(0) \\ = L(t)/L(0) - 1.$$

The strain rate of a material is a measure of its elongation observed during a given time, as shown in the following equation:

$$e^{\cdot} = e/t = DL/(Lot)$$

where $e^{\cdot}$ = strain rate = elongation/time, DL = (L – Lo), and e = relative elongation, being dimensionless.

For example, the elongation of a 1-m metal rod by 30 cm is given by the strain:

$$e = (L - Lo)/Lo, \text{ or } (1.30 - 1.0)/1.0$$

resulting in a dimensionless value of 0.30 (30%). If this deformation occurred in 1 h (3,600 s), a strain rate of 0.3/3600 s, or $8.3 \times 10^{-5}\ s^{-1}$, is obtained. If this elongation was produced during one year (3.15×10^{7} s), the deformation rate would be $9.5 \times 10^{-9}\ s^{-1}$.

Assuming that a rubber band with a length of 1 m is stretched by 10 cm in 1 h (3600 s), the strain rate is $2.7 \times 10^{-5}\ s^{-1}$

$$(1.10-1.0)/1.0 = 0.1/3600 = 10\times10^{-1}/3.6\times10^{3} = 2.7\times10^{-5}\ s^{-1}.$$

A strain rate of $10^{-5}\ s^{-1}$ means that 1 m of some material subjected to stress deforms by 1/100,000 in 1 s.

If constant stress is applied for 12 months and the 1-m deformation occurs after one year (or $24 \times 3{,}600 \times 365 = 3.15 \times 10^{7}$ s), then the strain rate is 3×10^{-8}.

If a strain rate of 10^{-5} /s is kept constant for one year, the 1-m rubber band would be stretched to

$$L = 1.00001\,m,\ \text{or}\ 1.00001\times3.15x10^{7}\,s = 315\,m.$$

This is a very large deformation, and probably the rubber band will rupture before it reaches the length of a few meters. The maximum strain before the rubber band breaks is called the ultimate strength. If the strain rate is much smaller, such as 10^{-10}, then the deformation will be only 0.003 m, or 3 mm.

Rocks in nature present extremely small deformation rates, as the deformations are measured in millions of years. On the other hand, the pressures that affect the rocks in the crust are huge and are measured in megapascals. Pascal is a rather small unit; for example, the atmosphere exerts a pressure of about 1 bar at the surface, which is equivalent to 101,325 Pa, or about 10^{5} Pa. Thus, 1 atm is equivalent to only 0.1 MPa, which corresponds to 14.5 psi. As a comparison, the pressure in car tires is circa 30 psi, corresponding to 2 atm. For differential stresses of only 0.5 MPa (about 75 psi or just over 5 atm), the halite deformation rate is $10^{-13}\ s^{-1}$. This value indicates a relatively rapid deformation. To give a quantitative idea, 10^{13} s corresponds to approximately 300,000 years. This means that in less than one-third of a million years, a cube of salt with a length of one unit will double this dimension by plastic deformation (Odé, 1968).

Different materials exhibit different mechanical behaviors when subjected to stresses, resulting in very different stress-strain curves (Jeremic, 1994). The changes in these materials are usually measured in the laboratory as a percentage of deformation compared to the original dimensions of the sample that was tested. This deformation or strain is often expressed as a percentage (for example, 5%) or as a decimal digit, such as 0.1 or 0.15 (Van der Pluijm & Marshak, 2004).

The main forces that induce salt masses to flow are gravitational forces due to sedimentary loading, with the main stress (s_1) acting in the vertical direction. Other important forces related to salt movement are tectonic and magmatic loading. Tectonic loading involves plate tectonics in extensional and compressional settings, for example, loading a salt basin with nappes. Magmatic loading is related to volcanic rocks piling above evaporite sequences. Some authors (e.g., Talbot, 1978) also suggest that thermal effects may result in salt movement, particularly when heat sources occur below mobile evaporites. Some early models for the salt glaciers in Iran postulated that the salt flow resulted from convection currents within the salt diapir, which extruded on the surface due to its reduced viscosity (see Chapter 3).

Gravitational forces are associated with the weight of the overburden above the salt layer. The sedimentary load creates a pressure that forces the salt to flow laterally toward regions of lesser pressure. This basic physical principle was adopted by many researchers who proposed differential loading as a halokinetic mechanism in the past century (see Chapter 6 in Mohriak, 2025). According to these early models, most salt structures are related to the density difference between salt and sediment. As salt behaves as a highly viscous fluid, the loading of a salt layer with the weight of the overburden rocks results in the formation of salt domes (e.g., Barton, 1933; Nettleton, 1934).

The evaporite layers in the Zagros thrust and fold belt were deformed in the Paleozoic and Mesozoic mainly by the weight of the superimposed sedimentary column, which resulted in the development of salt diapirs by differential loading (Callot et al., 2012). Only in the Cenozoic, when continental blocks of the Arabian plate collided with the Eurasian plate during the Alpine Orogeny (see Chapter 3), the diapirs were deformed by the compressional forces associated with the plate convergence and block collision (Hassanpour et al., 2020; Sarkarinejad et al., 2018).

6.1.2. Basic Types of Solid Material Deformation

According to several rock mechanics and structural geology textbooks (e.g., Fossen, 2016; Jeremic, 1994; Van der Pluijm & Marshak, 2004), there are three basic types of deformation in natural materials: elastic, plastic, and viscous (Figure 6.4). These end-member rheological behaviors can be compared with simple physical apparatus that simulates the deformation when forces are applied. These correspond to a spring that is pulled by a force F, a block that is pulled by a force parallel to the floor, and a dashpot that is filled with a liquid with a piston that is pulled by a force F (Figure 6.5).

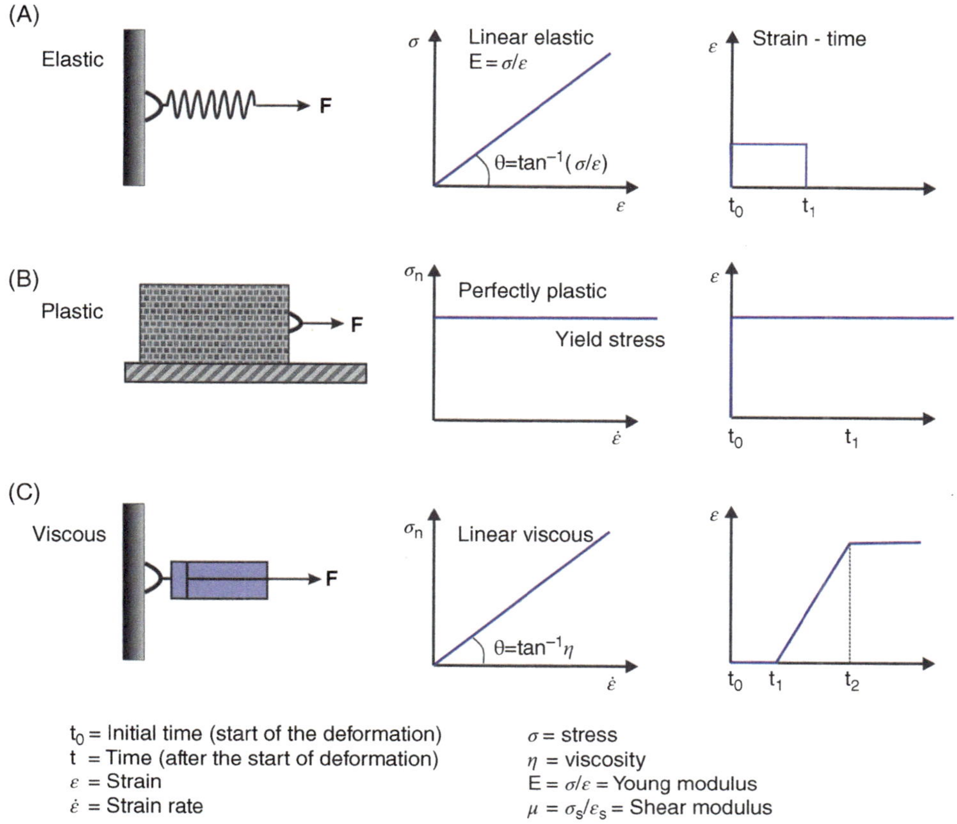

Figure 6.5 Schematic diagrams illustrating the rheological behavior of materials. (A) Elastic behavior is characterized by a strain (e) that is linearly proportional to the applied stress (s). This parameter is a constant that is equivalent to the tangent of the straight line that relates the stress with the strain, and in linear elastic bodies, this constant (s/e) is known as the Young modulus (E). The deformation occurs only during the time that the stress is applied, and the spring returns to its undeformed length (strain = 0) when the force is removed. (B) Plastic behavior is compared to that of a block that is pulled by a force. When the force reaches the yield stress of the block movement (exceeding the friction force that resists the movement), the block starts to move with a constant velocity. (C) The viscous behavior is compared with a dashpot filled with liquid. The force that is applied to a piston will result in a strain rate that is proportional to the applied force in Newtonian fluids. Linear viscous materials display a straight line relating the strain rate and the applied force. After the force is removed, the deformation remains permanent. Credit: Data from Van der Pluijm and Marshak (2004).

Several parameters in rheology are used in equations applied to rock deformation. For example, Young's modulus E is defined as the ratio of stress to strain in a uniaxial stress state. The shear modulus μ is defined as the ratio of shear stress to shear strain (Van der Pluijm & Marshak, 2004):

$$E = s/e$$

$$\mu = s_s/e_s.$$

The elastic behavior (Figure 6.5) is characterized by a strain (e) that is proportional to the stress (s). In rock mechanics, the elastic behavior is often compared to the mechanics of a spring that is pulled by a force F, resulting in the elongation of the spring by an amount that is proportional to the applied force (e.g., Jeremic, 1994; Van der Pluijm & Marshak, 2004).

Elastic deformation lasts only during the application of force. When the stress is removed, the material returns to its previous undeformed state, similar to a spring that is pulled and extends some distance but returns to its previous length when released (Figure 6.5A). Elastic deformation is expressed in the stress-strain graph as a linear function, with the force applied to a material (stress) proportional to the measured deformation (strain).

Figure 6.5A shows, in the center graph, the representation of the strain on the horizontal axis and the stress on the vertical axis. The elastic relationship is represented by a straight line where the inclination corresponds to an angle that is expressed by the Greek letter theta (Θ). This angle depends on the stress and strain values but is constant for a given material. The linear elastic equation can be expressed as Hooke's Law, where the constant E in the equation is Young's modulus.

$$s = E.e$$

where s in the y-axis corresponds to the stress and e in the x-axis corresponds to the strain (Figure 6.5A).

Plastic deformation is expressed in the graph (Figure 6.5B) with the x-axis representing the strain rate (variation of deformation with time) and the y-axis representing the stress values. Plastic materials show a constant value after a critical force is applied (known as the yield stress) for a given deformation rate. Plastic deformation is permanent and deforms the material after reaching a yield stress value that depends on the material's physical properties. Some materials deform by brittle deformation under low confining pressures but show ductile strain under high confining pressures. In physics, the plastic behavior is compared to a block pulled by a force (Figure 6.5B). When the amount of force reaches the yield stress required to move the block, thus exceeding the frictional force that resists the movement, the block starts to move with a constant velocity.

Viscous deformation is expressed in the stress-strain rate graph (Figure 6.5C) by an inclined line with the Greek symbol êta (η), whose tangent corresponds to the inclination of the strain rate line when the rheology behavior is linear. In physics, the viscous behavior is compared to a dashpot filled with a liquid (Figure 6.5). A dashpot is a simple mechanical device consisting of a cylinder with enclosed fluid. A piston operates to create pressure to move the fluid out of the chamber, usually through narrow openings in the cylinder. The force that is applied to the piston will result in a strain rate that is proportional to the applied force (Figure 6.5C).

Viscous materials that show a curve with different angles rather than a straight line display nonlinear rheology. Viscous strain is a permanent deformation involving flow, and the strain rate is proportional to the applied stress. This means that the deformation depends on the duration of the stress applied to the material. The velocity of the flow depends on the viscosity of the material. Newtonian fluids, such as water, are characterized by viscous behavior with linear rheology. Some materials, such as halite minerals, deform with a brittle behavior under high strain rates but show a ductile or viscous behavior under low strain rates (Baar, 1977; Van der Pluijm & Marshak, 2004).

The deformation of natural salt deposits can be expected after critical pressure from the overburden is applied to the strata. The evaporite rocks will flow when the critical lithostatic pressure is reached. This threshold depends on the nature of the salt, with mean values ranging from about 100 to 150 bar (from 10 to 15 MPa). The pressure of 10 MPa equals the weight of a water column of 1000 m. When a layer of salt (with twice the density of water) reaches a thickness of 500 m, equivalent to a lithostatic pressure of about 11 MPa, the salt begins to flow. In most basins, the salt flow is triggered by some mechanism, such as differential loading, overburden extension, or regional contraction (see Chapter 7 in Mohriak, 2025). If there is an accumulation of 1000 m of salt, as is commonly registered in giant salt basins (e.g., South Atlantic, GOM, Mediterranean Sea), some of that salt is already flowing at the base during deposition at the surface. Consequently, the top and base of the salt layer will exhibit different rheological behaviors at the same time, with different salt flow velocities at different depths (see Chapter 7 in Mohriak, 2025).

Some plastic deformation may occur in evaporite layers that have less than 500 m in thickness if gravity or tectonic forces act on the salt mass. Costa et al. (2004) compared seismic profiles with deformed evaporite sequences in the Mediterranean Sea with results from physical models. Although the suprasalt sediments were only a few hundred meters thick, these layers were locally folded and faulted by thrust and normal faults. According to Costa et al. (2004), these observations and the physical experiments indicated: (1) the halite sequences were deformed mainly by viscous behavior at the base of a salt layer; and (2) the overburden was ruptured by extensional faults, resulting in diapirism (see Chapter 7 in Mohriak, 2025).

Mechanical rock properties are evaluated using uniaxial and triaxial strength tests conducted in rheological laboratories (e.g., Dusseault & Fordham, 1995; Elliott, 1995). The effective stress in rheological studies is the sum of the geostatic stress, the pore pressure, and the tectonic stress. In laboratory tests, the difference between the maximum stress and the neutral, confining pressure represents the effective stress. Differential stress corresponds to the difference between two principal stresses (e.g., the maximum and minimum, $\sigma 1 - \sigma 3$), which, by definition, is ≥ 0. Several factors control the rheological behavior of natural rock in laboratory tests, particularly confining pressure, temperature, strain rate, and boundary conditions (Van der Pluijm & Marshak, 2004). An example of a rock mechanical test of rock salt from a borehole in the U.K. (Figure 6.6) shows the sample deformed at room temperature (Dûsterloh & Lux, 2007). The triaxial compressional test was performed with a confining pressure of 3 MPa and compressional stress increments of 0.25%. The test reached a pressure of 50.87 MPa for the uniaxial test and 45.91 MPa for the triaxial test, resulting in a strain of 9%, corresponding to a deformed length of about 13.8 cm (Figure 6.6).

Core from Holford H401
Depth: 604.68 - 605.67 m
Diameter (mm): 80.00
Length (mm): 160.00
Temperature (°C): 30
TC (Triaxial Compression)
UC (Uniaxial Compression)
Testing technique: TC - 0,25 %/min
σ3 (MPa): 3
σ1u (MPa): 50.7
σ1uc (MPa): 45.91

Figure 6.6 Photographs of rock mechanics tests show the deformation of a rock salt sample obtained from a borehole. The test plug was subjected to uniaxial and triaxial tests in the laboratory, resulting in a deformed rock sample. The test resulted in a length shortening from circa 16 cm to about 13.8 cm, with a contractional strain of 9%. Some of the parameters used in the rock mechanics tests are shown below the images.

Rheological laboratories studying rock mechanics can characterize several types of rocks or materials by their stress-strain paths, measuring the deformation that results from applying unidirectional forces to unconfined samples, which are usually cores (elongated cylinders) or cubes. When the tested material breaks up, the test is interrupted, and the maximum stress supported by the sample before rupturing can be determined. Figure 6.7 illustrates the behavior of four hypothetical materials showing different stress-strain paths.

Materials that deform with ductile behavior are characterized by solid-state changes in the shape or crystal orientation without a loss of physical continuity, as opposed to brittle behavior. Ductile behavior is often associated with the flow of igneous rocks near their melting point, such as basaltic lava that flows from volcanoes, granitic magma that forms large batholiths in the lower crust, or slightly consolidated shales that are mobilized on continental shelves and slopes, forming mud diapirs. This plastic behavior can be observed in glaciers (Figure 6.1), where ice is deformed by dislocation glide and dynamic recrystallization in temperatures that normally range between –50 and –5 °C (Hambrey & Glasser, 2005).

Salt glaciers (also known as namakiers, derived from the Persian word *namak*, meaning salt) are formed by salt masses that are associated with a salt feeder in a diapir that extrude on the surface or the subsurface, forming allochthonous salt sheets (see Chapter 3). Salt glaciers are characterized by low-temperature deformation of the halite crystals, which change their shape and orientation as they flow under low pressure and low temperatures (but normally above the freezing point of water). Similar to ice glaciers (Figure 6.1), the deformation observed in the macroscale analysis of salt glaciers reflects a solid-state flow resulting from the deformation of the halite crystals in the microscale (Urai et al., 1987).

6.2. STRENGTH OF EVAPORITE ROCKS UNDER EXTENSIONAL AND COMPRESSIONAL STRESSES

Several physical properties of halite (see Chapter 5, Figure 5.2) are of great importance for rock mechanics and salt tectonics. The salt density is almost invariable over a wide range of temperatures and pressures compared to other rocks (Gussow, 1968). Salt is characterized by its plastic behavior, and some authors suggest that in geological time it may be considered almost a fluid. Halite has an average viscosity of around 10^{+16} to 10^{+17} Pa·s. (Weijermars et al., 1993, 2014), which in the viscosity scale makes it fall in the range between the glass in the window's panes and the mantle rocks in the asthenosphere (see Chapter 5, Figure 5.3). Some glass windows in medieval-age cathedrals in Europe show a protuberance at the base of the glass pane, and this deformation results from the glass slowly flowing due to the action of gravity over a few centuries. The asthenosphere flows during plate tectonic processes, as convection currents in the mantle result in drifting the continents at rates that are measured in a few centimeters per year.

Observations of natural salt flowing in several outcrops in the Zagros region, as well as experimental data, suggest that salt with very small percentages of water is even less viscous than dry salt and is mobilized more easily, increasing the strain rate (Talbot & Rogers, 1980). The addition of small amounts of water in an evaporite layer results in a viscosity three to four times smaller than the dry salt, greatly increasing the creep. Creep is defined as the gradual flow of rocks or crystalline material when subjected to gravity forces. In rock mechanics, creep is time-dependent continued deformation without a stress change (Dusseault & Fordham, 1995).

Creep is a process that leads to permanent deformation of the rocks, forming folds and faults. This phenomenon

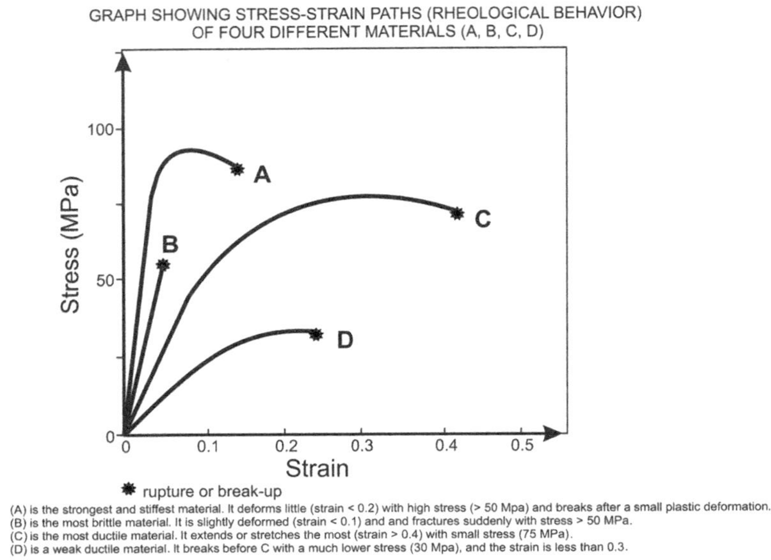

Figure 6.7 Stress-strain paths of four different materials (A, B, C, and D). The materials display elastic behavior when the stress-strain curve is linear. When the curve departs from a linear path, the materials are characterized by plastic or viscous behavior. The yield strength is the stress at which the curve changes slope. When the strain reaches a maximum, the material is ruptured. The rupture strength is defined as the maximum stress difference before the rocks break up by fracturing. Credit: Adapted from Jeremic (1994).

can be observed in ice glaciers that slowly move downhill in the Alps, Alaska, or Canada (Figure 6.1), or in salt glaciers that emerge in the Zagros Mountains of Iran (see Chapter 3). In evaporite rocks, creep can occur at surface temperatures, particularly if the salt contains traces of water at the boundary between grains (Urai et al., 1986). A few ppm (5–20 ppm) of water is enough to facilitate the flow in the salt glaciers in Iran, which extrude faster after rainy seasons (Talbot & Jarvis, 1984).

The strain rate is related to the deformation of salt layers and diapiric structures. The strain rate can be determined by analyzing the uplift of sedimentary layers that overlie the salt. The uplift of salt diapirs may be estimated by mapping and dating marine terraces that have accompanied the salt rise. For example, it is estimated that the Al Salif salt diapir in Yemen has uplifted about 4.6 mm per year in the past 3,700 years (Davison et al., 1995, 1996). Assuming that the diapir has a vertical height of 4500 m, corresponding to the initial length (Lo), this gives a strain rate of

$$e^{\bullet} = e/t = DL/(Lot)$$

$$DL/(Lo) = 3700 \times 4.6 \times 10^{-3} / 4500 = 3.782 \times 10^{-3}$$

$$t = 3700 \times 3.15 \times 10^{7}\,s = 1.2 \times 10^{11}\,s$$

$$\text{so } e/t = (3.782 \times 10^{-3}) / (1.2 \times 10^{11}) = \sim 3 \times 10^{-14}\,s^{-1}.$$

It is interesting to observe that the uplift of Mt. Sedom in the Dead Sea, one of the most iconic salt diapirs in the world (see book cover), is characterized by similar strain rates (e.g., Weinberger et al., 2006; Weinberger, Lyakhovsky, Baer, & Begin, 2006). The Sedom salt diapir rise can be estimated by the offset of the phreatic table dissolution surfaces, which are known as salt mirrors and have been uplifted in the Holocene (see Chapter 3, Figure 3.33).

Frumkin (1996) estimated that the Sedom salt mass has been rising at a rate of 6–7 mm per year for the past 8,000 years. This uplift rate was based on the measurement of terraces and caves excavated in the salt mass. The caves and terraces were initially at lake level, but at present, these features are located high and dry. The salt uplift in the Dead Sea can also be estimated from the uplift from interferometric synthetic aperture radar (InSAR) surveys, which measure the absolute uplift motion of the diapir top. The estimates are based on InSAR, a technique that has been recently used to measure subtle displacements of the ground surface (Weinberger, Lyakhovsky, Baer, & Begin, 2006). The uplift measurements based on geological markers indicate a strain rate ranging from 2 to 6×10^{-14} (Frumkin, 1996; Zak & Freund, 1980), similar to the strain rate for the Al Salif salt diapir in Yemen. According to the measurements and modeling by Weinberger et al. (Weinberger et al., 2006; Weinberger, Lyakhovsky, Baer, & Begin, 2006), the average strain rate since the salt diapir was formed in the Pleistocene (about 2 Ma) is $3 \times 10^{-14}\,s^{-1}$, equivalent to the strain rate values calculated at the Al

Salif salt diapir in Yemen (Davison et al., 1995, 1996). The strain rate obtained by measuring the displacement of the salt mirrors since the time of diapir emergence (14 ka), a much shorter interval, corresponds to $9.0 \times 10^{-13}\ s^{-1}$. The strain rate determined by InSAR data measurements resulted in values of $5–6 \times 10^{-13}\ s^{-1}$, as calculated by measuring the present-day surface uplift rates. Based on modern topography and modeling, the effective viscosity of the salt diapir was estimated at $2–3 \times 10^{18}$ Pa·s (Weinberger, Lyakhovsky, Baer, & Begin, 2006).

6.2.1. Resistance to Compressional and Extensional Stress

Salt is an extraordinarily ductile material when compared to other sedimentary rocks, so it is reasonable to analyze its deformation in terms of fluid mechanics (Carter & Hansen, 1983). Rock salt minerals exhibit rigid or fragile behavior at the surface, as evidenced by fractures and faults affecting salt layers in many salt mines worldwide (e.g., the Sergipe Basin onshore Brazil and the Boulby mine in England; see Davison, 2009). However, when the isostatic (confining) stress exceeds 10 MPa, salt displays a plastic or ductile behavior, as indicated by seismic interpretation and laboratory experiments. Salt already becomes mobile when the elastic threshold is surpassed under low differential stresses of ~1 MPa, and the deformation rate is slow enough to accommodate further strain without rupturing the layers. Laboratory experiments with low strain rates ($10^{-16}–10^{-14}.s^{-1}$) corroborate the differential stress of 1–2.5 MPa for dry rock salt deformation (creep). The brittle/ductile transition occurs at depths between 10 and 130 m (Weijermars et al., 1993). The plastic behavior of evaporite sequences results in salt flow forming pillows, domes, and diapirs (Jackson & Hudec, 2017; Van der Pluijm & Marshak, 2004).

The elastic or plastic behavior of salt layers also depends on the rate of deformation. Under the pressure of 1 atm and ambient temperature, salt exhibits a behavior that is as rigid as concrete for short-time stresses, for example, during an earthquake. However, during geological time, measured in millions of years, salt typically exhibits ductile behavior and can flow in the subsurface or at the depositional surface, forming salt glaciers (see Chapter 3).

Salt is more resistant to deformation than the surrounding rocks when they are poorly lithified. For example, Neogene clays on the seabed, when accumulated in the depressions adjacent to a salt diapir, are much weaker than halite and deform easily. After the clay layers are buried beneath the overburden, they become indurated shales, with a resistance to stress much greater than salt (see Chapter 8 in Mohriak, 2025).

Rocks undergo different behaviors when subjected to compressional, extensional, or shear stresses, deforming elastically and breaking when rigidity is exceeded (Pluijm & Marshak, 2004). The ultimate fracture strength is the maximum stress a rock can withstand without breaking. Laboratory experiments show that rocks are weaker during extension, as the tensile strength is about 10% of the compressional strength. These strengths vary greatly for different rock types (Figure 6.8). For example, laboratory uniaxial tests from cores obtained from a salt mine in the Sergipe Basin, northeast Brazil, indicate that carnallite has a compressional strength of 19.7 MPa, while halite samples have a strength of 37.3 MPa under compression (Fairhurst et al., 1979). Similar laboratory mechanical tests, carried out with different evaporite lithologies from the Khorat Basin in Thailand, revealed that the UCS increased from 27 MPa (pure halite samples) to 40 MPa in anhydrite samples (Fuenkajorn et al., 2011).

Basalt is a rock with a strength that is over 10 times greater than its strength under tension, making it difficult to break under compression (Figure 6.8). Rocks can also fracture due to simple shear or flexural slip, when the forces are opposite in direction, resulting in a movement along the shearing plane. The maximum stress a rock can withstand under simple shear without breaking is known as shear strength, which is large for basalt (38 MPa) and very small for ice and salt crystals (Figure 6.9). Gypsum has a shear strength that is one-tenth of that of marble, sandstone, and limestone (Figure 6.9).

Rock	Strength under tension (MPa)	Strength under compression (MPa)
Quartzite	25	252
Gneiss	14	159
Basalt	13	150
Granite	12	166
Limestone	12	102
Sandstone	5	96
Shale	3	95
Salt	1.5	26
Coal	2	22

Figure 6.8 Mean value of ultimate strength of common materials under uniaxial tension and compression (in MPa, 1 MPa ~10 atm, 1 MPa = 145 psi; 1 MPa = 10.2 kg/cm^2). Salt exhibits an elastic behavior for an extensional stress field up to 7 MPa and deforms with an elastoplastic behavior for stresses from 7 up to 14 MPa. For larger stresses (>14 MPa), salt becomes virtually plastic and behaves as a ductile material. Credit: Data from Van der Pluijm and Marshak (2004).

Rock	Shear strength (MPa)
Basalt	38
Granite	31
Limestone	28
Sandstone	23
Marble	22
Gypsum	1.2–3.4
Shale (mobile)	0.7–2.8
Chalk	0.5
Ice	0.1

Figure 6.9 Mean value of shear strength of common rocks and materials under shear or flexural stresses (in MPa, 1 MPa ~10 atm, 1 MPa = 145 psi; 1 MPa = 10.2 kg/cm^2). Credit: Data from Van der Pluijm and Marshak (2004).

One example of the changes in the physical properties of salt rock is presented by Li et al. (2022), based on laboratory tests of rock salt from the Sanshui Basin in Guangdong Province, China. The basin is characterized by thick, high-purity salt deposits, predominantly composed of halite. The salt layers are considered potential sites for the construction of salt caverns for gas storage underground. The results showed that the average tensile strength and UCS of salt rock are 1.51 and 26.04 MPa, respectively. However, under triaxial compression, the compressive strength increases without failure. Under a confining pressure of 30 MPa and a temperature of 60 °C, the strength of salt rock decreases by 8.3% compared with that at room temperature. This indicates that, for the salt rock samples from the Sanshui Basin, the slight increase in temperature resulted in a modest effect on the mechanical properties (Li et al., 2022).

The influence of a much larger range of temperatures on the rheological properties of rock salt was investigated by Sriapai et al. (2012). The salt cores were obtained from the evaporite deposits in the Khorat Basin, Thailand, and correspond to almost pure halite. The uniaxial and triaxial compressive strength and Brazilian tensile strength tests were applied to several rock salt specimens that were subject to temperatures varying from 273 to 467 K (0–194 °C). The UCS decreased from 37.9 MPa to about 25 MPa when the temperature increased from 277 to 455 K (Sriapai et al., 2012). The triaxial compressive strength (s_1) was measured at different temperatures and confining pressures. At the confining pressure (s_3) of 10 MPa, the compressive strength was 96.6 MPa at 274 K, and decreased to 50.0 MPa at 467 K (194 °C). Under the confining pressure of 30 MPa, the compressive strength increased to 135.0 (274 K) and 97.1 (467 K). The Brazilian tensile strengths were much lower, varying from 7.3 to 4.8 MPa, as the temperatures increased from 274 to 467 K (Sriapai et al., 2012).

Geomechanics laboratories analyze the rheological behavior of materials and rocks under various stresses, including extensional, compressional, and shear stresses. Evaporites exhibit rheological behavior that follows exponential laws, with greater rigidity in compression than extension. These stress rigidity parameters are crucial for the stability of salt mines and exploratory well drilling. The rigidity of a material is measured by bulk moduli (K) and shear moduli (G), which are used to measure the material's rigidity.

The rigidity moduli values at 1-atm pressure and room temperature are given in 10^{+10} Pa for various materials in the Earth's crust (Figure 6.10). The mineral olivine, a major component of mantle rocks, is characterized by high-rigidity parameters, whereas quartz, a common mineral in the crust, is characterized by much lower rigidity parameters. Sedimentary basins are formed through extensional tectonics, with crust deformation being easier than upper mantle deformation due to different rigidities. Because of its very low rigidity moduli, ice is a material that is less rigid than salt, and the movement of glaciers provides an interesting analogy for the flow of salt layers (Figure 6.1).

Anhydrite is a stronger mineral than halite but weakens easily with temperature increases, particularly above 100 °C.

Material	Bulk modulus (K, 10^{+10} Pa)	Shear modulus (G, 10^{+10} Pa)
Olivine (Mg_2SiO_4)	12.9	8.1
Calcite ($CaCO_3$)	6.9	3.7
Quartz (SiO_2)	3.0	4.7
Halite (NaCl)	1.4	2.6
Ice (H_2O)	0.73	0.25

Figure 6.10 Bulk and shear moduli of common minerals and materials. The bulk modulus of mantle rocks (mainly composed of the mineral olivine) is more than four times higher than that of the continental crust (mainly composed of silicate minerals). Ice flows easily in glaciers because of the small bulk and shear moduli. Credit: Data from Van der Pluijm and Marshak (2004).

Gypsum, a ductile rock common in sedimentary basins worldwide (Mohammed & Mahmood, 2018), is a key component of evaporite layers in fold-thrust belts in Europe and Northwest Africa, such as the Jura foreland fold-thrust belt and the Atlas Mountains in Tunisia (Sommaruga, 1999).

Carnallite, bischofite, and tachyhydrite are weaker than halite and flow easily during drilling operations in salt basins, making them a hazard for salt mines. The Sergipe Basin, located in northeast Brazil, has several evaporite minerals deposited during the Late Aptian, with different rheological properties and strengths. Geomechanical tests of evaporite rock plugs from the Sergipe Basin Taquari-Vassouras salt mine showed that tachyhydrite plugs have a smaller strength compared to other lithologies, including halite (Fairhurst et al.,1979).

Sylvinite, a rock composed of sylvite and halite, is the most important ore in the Taquari-Vassouras salt mine in the Sergipe Basin (see Chapter 7). It is commonly associated with carnallite, a mineral that has half the strength of halite. The presence of carnallite deteriorates the strength of salt mine walls and pillars, and engineering studies are required to prevent mine collapses where the proportion of carnallite is significant. This is also observed in salt mines in Canada (Saskatchewan), where 5–9% carnallite decreases about half of the UCS of monomineral layers composed of sylvinite.

The mechanical properties and behavior of evaporite rocks under stress are complex, making numerical modeling difficult due to the plastic nature of salt. Salt exhibits elastic and inelastic behaviors depending on stress magnitude, time scale, and loading conditions. Incremental loading or strain increases can cause salt to deform plastically, while instantaneous strain can cause it to behave like brittle rock and fracture. Natural rock salt exhibits various mechanical behaviors, including brittle structures and ductile flow. This is suggested by experiments using the "Kristallbrocken" salt layer of the Zechstein Group in Germany (Küster et al., 2010). If the strain is instantaneous, salt behaves like brittle rock and fractures, as observed in the Boulby salt mine in England (Davison, 2009). The observations of a wide range of mechanical behavior in natural rock salt are compared with laboratory experiments that use complex power-law equations to simulate the deformation of salt plugs (e.g., Li & Urai, 2016).

6.3. MACROSTRUCTURAL ANALYSIS OF SALT DEFORMATION

Salt deformation can be analyzed in outcrops using hand specimens that provide a macro-scale that can be useful as a first approach. However, the proper identification of evaporite textures indicative of deformation mechanisms is based on observation and study of the microstructures under the microscope using petrographic thin sections (see Chapter 7). The literature on this topic is vast, and most researchers who investigate these problems are specialized in complex engineering equations, physics, and mathematics.

These rheological processes are simulated by laboratory experiments and algorithms using numerical solutions that involve equations beyond the scope of this book. Numerical methods, based on finite-element models and the rheological properties of salt and overburden sediments, aim at reproducing salt deformation in different tectonic settings. These methods have been applied to the analysis of salt structures in rift basins, divergent margins, and foreland basins. Key references for numerical modeling of salt tectonics on a basinwide scale include several researchers, such as Albertz and Beaumont (2010), Albertz and Ings (2012), Albertz et al. (2010), Allen and Beaumont (2016), Allen et al. (2020, Ings et al. (2004) and Liang et al. 2012). These works provide a discussion of geological concepts and mathematical methods, helping readers find more advanced techniques.

The degree of salt mobility and associated deformations involve several parameters: (1) salt thickness and purity, as thin layers and evaporites stratified with siliciclastic and carbonate rocks are less mobile; (2) the temperature of the salt layers, which is related to the burial depth; (3) the presence of water in the evaporite minerals, which may be due to dehydration of associated gypsum layers; (4) the presence of free water (brine) in the interstices of halite crystals, which may greatly increase the rate of creep even in small amounts, usually 0.1–1.0 wt% in natural rock salt (Urai et al., 1986); and (5) the magnitude of the extensional and compressional forces involved in the deformation.

Laboratory experiments on halite samples from many salt basins worldwide show a wide range of stresses these rocks can support before rupturing. For example, the compilation of laboratory data by Pells (1995) indicates that the mean UCS of halite is 46 MPa, with values ranging from 21 to 157 MPa depending on the specimen. The laboratory tests indicate that the tensile strength is much lower than the compressive strength of most rocks (Figure 6.8). For example, the laboratory results of UCS tests of halite cores from the Sanshui Basin in China indicate a compressive strength of 26 MPa, contrasting with an average tensile strength of about 1.5 MPa (Li et al., 2022).

The UCS (or maximum compressive stress) for halite in the GOM salt domes shows a typical value of about 23 MPa, similar to the compressive strength of coal beds in many basins (Figure 6.6). However, the tensile strength of GOM salt samples ranges from 0.2 MPa to about 3 MPa, with a mean value typically around 1.0 MPa. Laboratory experiments indicate that the ratio of compressive strength to tensile strength is typically more than 20 for salt rock, as compared to a value of around 10 for most brittle rocks (Figure 6.8).

Halite, the most common evaporite rock, has a lower tensile strength compared to most other evaporite rocks. The Brazilian Tensile Strength Test found a mean value of 2 MPa for halite (Bell, 1981), similar to the Sanshui Basin salt plugs (Li et al., 2022). Sulfate rocks have significantly higher tensile strengths, with anhydrite and gypsum having values of 8.2 and 3.4 MPa, respectively (Bell, 1981). This weak mechanical behavior is crucial for sedimentary basins affected by extensional tectonics (Jackson & Hudec, 2017).

The stress-strain paths of different materials (Figure 6.6) exhibit elastic behavior for small strains, but when stress increases, the curve changes, and the material exhibits ductile behavior (e.g., Fjaer et al., 2021; Pells, 1995). This stress limit is called the yield strength of the material. The yield strength of a solid is defined as the maximum stress difference at the beginning of the deformation by plastic flow. The yield strength of halite is extremely low. Laboratory experiments show that even at stress differences below 1 MPa, halite may deform by creep (Fossum & Frederich, 2002; Weijermars et al., 1993, 2014). The depth of the brittle-ductile transition for dry rock salt lies a few meters below the surface for extension and compression (Weijermars et al., 1993).

The rheology of evaporite rocks is also highly dependent on temperature. Robertson et al. (1958) reported that halite at 20 °C ruptures with a strain of about 1%, corresponding to a stress of about 2.5 MPa (Figure 6.11). This value corresponds to the ultimate tensile strength of a solid material between coal and shale (see Figure 6.8). The GOM salt, when heated to 200 °C, can withstand deformation of up to 5% before breaking up, indicating increased rupture strength (stress greater than 5 MPa).

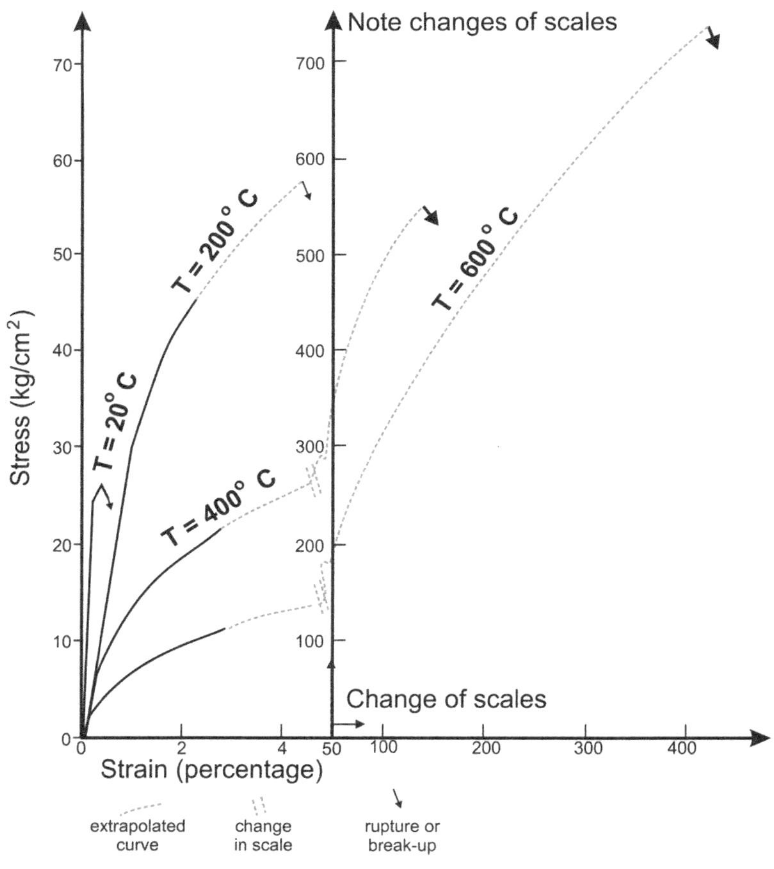

Figure 6.11 Graph with stress-strain paths illustrating the effect of temperature on the deformation of rock salt with increasing stress (10 kg/cm^2 ~ 1 MPa). At surface temperature (20 °C), halite deforms elastically under compression to a maximum stress of about 25 kg/cm^2. The ultimate strength before the rock is ruptured corresponds to a strain of 1%. At 200 °C, the ultimate strength is about 58 kg/cm^2 and the strain is almost 5%. At 600 °C, the material can be extended by more than 400% before rupturing. The ultimate strength at this high temperature corresponds to a stress of 70 MPa (700 kg/cm^2). Credit: Robertson et al. (1958)/USGS/Public domain.

However, experiments for the GOM salt show that deformation is not elastic at these temperatures, and the rock sample cannot recover to predeformation dimensions (Robertson et al., 1958).

Experiments for the effects of temperature on rock salt deformation (Figure 6.11) reveals that increasing temperature leads to a sharp decrease in yield strength and plastic behavior in halite rock, resulting in increased ductility by the increase in temperature (Robertson et al., 1958). The strain curve had to be extrapolated to account for larger temperatures, which resulted in a strain of over 100% before rupturing at stress values above 500 kg/cm^2. Most of the ductile deformation is permanent (Robertson et al., 1958).

The deformation of natural rock salt layers is linked to a series of depositional, diagenetic, and structural processes that evolve over geological time. Depending on the importance of these factors, a wide range of structures, from macroscale to microscales, are formed in evaporite rocks. Some are typically sedimentary or primary structures that can be observed in recently deposited and shallow salt layers, such as desiccation cracks in shallow environments. Some are related to the strongly deformed and completely recrystallized tectonites in salt diapirs, as observed in salt basins in Germany, Austria, Iran, and Oman (Kukla et al., 2008, 2011).

6.4. MICROSTRUCTURAL ANALYSIS OF SALT DEFORMATION USING OPTICAL MICROSCOPY

Evaporite structures in salt basins can be interpreted from regional scales (satellite images and geological maps) to metric scales in outcrops (see Chapter 3). Studies have been conducted using this methodology, including the analysis of slightly deformed bedded salt in the Permian Zechstein Basin in Europe and highly deformed Early Cambrian Hormuz salt in Iran. Field descriptions of bedded salt layers and deformed evaporites in salt diapirs in Iran indicate that they have been strongly deformed by salt flow processes (Talbot, 1981).

Field descriptions of the bedded salt layers and deformed evaporites in salt diapirs in Iran indicate that they have been strongly deformed by salt flow processes that occur in the subsurface or at the depositional surface in the case of extrusive salt (for example, Sarkarinejad et al., 2018). Salt science researchers have used microstructural analysis techniques to complement interpretations based on large-scale deformation observed in outcrops or seismic profiles. These studies have particularly investigated the Zechstein salt basin, where rheological studies involve macroscale outcrop and core analysis of boreholes drilled in salt mines. These studies are complemented by detailed bulk chemical methods, X-ray diffraction, and optical microscopy to determine the mineralogy and microstructure of the rock samples (e.g., Raith et al., 2017). Examples of petrographic analysis of evaporite rocks are discussed in Chapter 7.

In recent decades, major advances have been made in methods and technologies that allow the microstructural analysis of evaporite rocks. Halite minerals, often barely distinguishable with a microscope due to their transparency and low refractive index, can be made visible when thin sections from halite samples are irradiated with gamma rays and become decorated with blue tints. The irradiation of the samples with gamma rays under temperatures of about 120 °C is a technique that has been commonly employed to highlight otherwise invisible microstructures in rock salt (Schléder & Urai, 2005; Urai et al., 1986, 1987). Using this technique, halite samples acquire blue tints due to irradiation-induced lattice defects in the mineral structure, providing important microstructural information on the grain boundaries (Kukla et al., 2011; Urai et al., 2008).

6.4.1. Mechanisms of Salt Deformation

Salt deformation mechanisms in sedimentary rocks are difficult to replicate in laboratories due to the small strain rates in nature, whereas laboratory rheological experiments are based on the quantification of short-term, high-stress tests. However, laboratory experiments have provided valuable information on the microstructures involved in the deformation of sedimentary rocks (Figure 6.12).

The deformation mechanisms in evaporite rocks are dependent on the overburden load, which increases hydrostatic pressure and stress (Figure 6.12). Most deformation mechanisms are related to natural stresses, mainly gravity, sedimentary loading, and tectonic forces. In the twentieth century, halokinetic models associated salt dome formation with the loading of evaporite layers by sediments deposited on top of salt sequences, which behave as a high-viscosity fluid (e.g., Alves et al., 2017; Hudec & Jackson, 2007; Nettleton, 1934; Trusheim, 1960). During the last decades of the 1990s, the extensional model has been widely adopted by most researchers, assuming brittle behavior for the overburden (e.g., Jackson & Hudec, 2017).

Laboratory experiments on evaporite deformation have been conducted in geological and engineering studies, particularly in salt caverns for gas storage (e.g., Hovorka & Nava, 2000; Liu et al., 2020; Zong et al., 2017). The gradual deformation of evaporite minerals is interpreted through stress-strain paths in rheological experiments until rock salt samples transform into quasi-metamorphic rock. However, interpreting the mechanisms and quantifying the factors controlling salt deformation in nature is a challenging task in

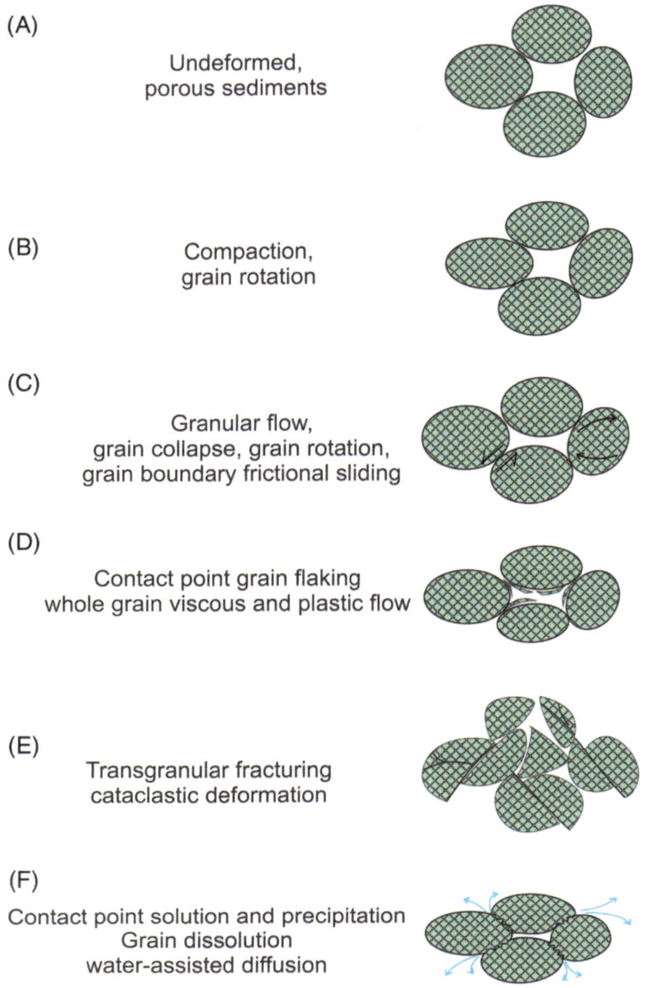

Figure 6.12 A schematic diagram shows compaction creep mechanisms in granular materials. The deformation mechanisms of evaporitic sedimentary rocks are characterized by initially undeformed, porous sediments that become consolidated and increasingly deformed with the sedimentary load and tectonic forces. Credit: Adapted from Dusseault and Fordham (1995).

the Earth sciences. The microstructural deformation is highly dependent on pressure and temperature, which are important parameters in the resultant rock fabric (e.g., Vernon, 2018).

Evaporite microtextures resulting from salt flow and deformation have been analyzed by several researchers (e.g., Kukla et al., 2011; Urai et al., 2008), providing a didactic visualization of microscopic features in deformed rock salt from classical evaporite basins affected by the Alpine Orogeny, such as the Zagros and Alps (see Chapter 3). In nature, the deformation mechanism can be associated with granular flow (as in ice glaciers), grain boundary flaking, cataclastic deformation, and grain dissolution by water-assisted diffusion. Some researchers postulate similar mechanisms for the ice flow in glaciers and the salt flow in evaporite basins (e.g., Talbot, 1979, 1981, 1993; Talbot & Pohjola, 2009). The deformation of ice glaciers is characterized by folds on a macroscopic scale (Figure 6.1). On a microscopic scale, the deformation of the ice crystals is accommodated by a combination of dislocation creep and grain boundary sliding (Duval et al., 2010; Jennings & Hambrey, 2021). In evaporite sequences, several other mechanisms are involved in the complex deformation of layered evaporites, which intercalate mobile salt units with less mobile siliciclastic and carbonate units (Figure 6.2).

Evaporitic sedimentary rocks are initially undeformed, porous sediments that become consolidated and deformed with sedimentary load and tectonic forces. Deformation mechanisms vary depending on burial depth, overburden loading, and tectonic stresses (e.g., Dusseault & Fordham, 1995; Jeremic, 1994). A schematic diagram illustrating the grain deformation under increasing stresses (Figure 6.12) shows some possible mechanisms: A) consolidation and lithification; B) compaction and grain rotation; C) granular flow and grain collapse; D) flaking at the contact points between grains and viscoplastic grain deformation; E) transgranular fracturing and cataclastic deformation; and F) grain dissolution, solution-precipitation mass transfer, and water-assisted diffusion.

Rock mechanics laboratories have studied the deformation mechanisms in rock salt by analyzing macrostructural and microstructural observations on naturally or experimentally deformed samples (e.g., Dûsterloh & Lux, 2007; Mills et al., 2018; Vernon, 2018). Urai et al. (2008) proposed that rock salt deforms mainly by internal deformation of halite grains, with four important processes related to halite deformation. According to the microstructural analyses, the four most important processes related to halite deformation are:

1. brittle deformation involving the microcracking and cataclastic flow of salt grains at low temperatures and confining pressures;
2. dislocation creep involving crystal plastic processes, such as kinking of the halite lattice and grain rotation and gliding;
3. solution-precipitation creep, also known as pressure solution, occurs when mineral grains dissolve at the contact points with other grains, resulting in the formation of a thin film of interstitial fluid due to stress-induced recrystallization;
4. recrystallization, involving the reorganization of the salt mass with changes in grain size, shape, and orientation of the grains.

Cataclastic flow (process 1) is a process that involves the microfracturing and sliding of grains and grain fragments (Baar, 1977). Dislocation creep (process 2) is another process that involves plastic flow at low temperatures and pressures, kinking of the halite crystal lattice during creeping, and various combinations of slip rotation, gliding, and climbing of dislocations through the grains (Carter & Hansen, 1983). The solution-transfer creep (process 3, pressure solution) involves the dissolution and precipitation of salt at grain contacts (Rutter, 1976). Recrystallization (process 4) involves changes within the mineral grain and creep (Poirier & Guillopé, 1979). These processes work together during salt deformation due to halite's high water-soluble nature (Baar, 1977; Carter & Hansen, 1983).

Brittle deformation in sedimentary rocks involves fracturing rocks and frictional sliding of grains, which deform along fault zones (Figure 6.3). Evaporite rocks are rarely affected by brittle flow, resulting in cataclastic or granular flow, as observed in salt glaciers and some salt outcrops (see Chapter 3). Cataclastic flow results in the development of gouge zones, characterized by cataclasites and brecciated rocks. Cataclasites are often observed at the boundary between an evaporite layer and the detachment plane associated with listric faults (Jackson & Hudec, 2017).

Dislocation creep (process 2) is a fundamental mechanism for plastic deformation, dominated by dislocation glide. This process occurs when a crystal defect propagates through a plane in the mineral lattice structure, causing the material to dislocate (Pluijm & Marshak, 2004). These defects, often associated with impurities or atom vacancies, rupture the crystalline structure along a plane that can be used as a gliding surface (Figure 6.13). Dislocation creep is most commonly observed in materials under high temperatures that flow like a viscous fluid, such as migmatites in the lower crust or lavas erupting from volcanoes. However, in salt rocks, dislocation creep occurs under low pressure and temperature, resulting in solid-state flow, like salt glaciers (Figure 6.1). Deformation observed in macroscale analysis, such as sedimentary stratification and brittle and ductile deformation (faults, foliation, folding, and boudinage structures), reflects the microscale changes in crystals (Hambrey & Glasser, 2005; Jennings & Hambrey, 2021; Urai et al., 1987).

Dislocation creep is a process where natural coarse-grained halite crystals in sedimentary layers deform when the differential stress is large (>10 MPa). The strain rates for these high-pressure experiments vary from 10^{-8} to 10^{-4} s^{-1} (e.g., Jackson & Hudec, 2017; Urai et al., 2008). The process is characterized by the presence of subgrains in the halite mass. Petrographic thin sections observed in optical microscopes show polygonal lines that mark the fine cracks along the subgrain boundaries (e.g., Linckens et al., 2016; Urai et al., 2008).

The dislocation creep mechanism becomes dominant when temperature and confining pressure are increased, typically occurring when evaporites are deeper than about 5 km or salt layers are affected by tectonic forces. The crystals deform plastically during the dislocation creep mechanism, and the dislocation recovery mechanism involves crystal reorganization and dynamic recrystallization, resulting in the development of subgrains inside the original grain (Urai et al., 2008). Solution-precipitation creep (process 3) is associated with the dissolution of grain boundaries with increasing pressure at contact with other grains, leading to increased solubility of mineral grains at grain boundaries and the diffusion of ions through a fluid film enveloping the grain (Figure 6.12F).

During pressure solution, the highly stressed part of the halite grain goes into the solution and new material

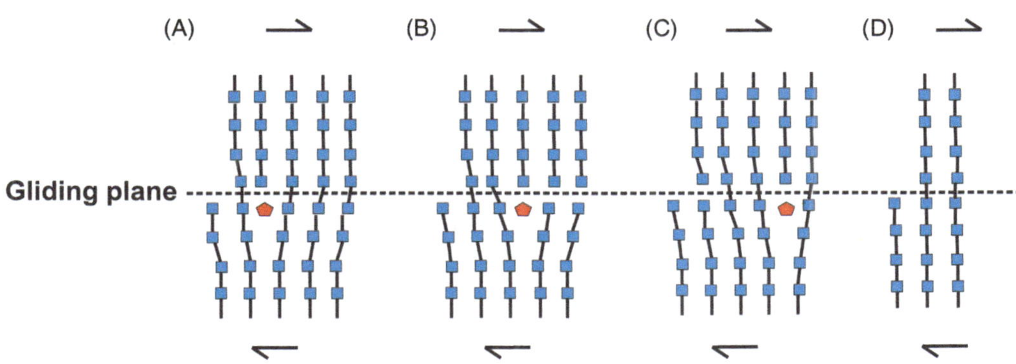

Figure 6.13 The dislocation glide process is related to the dislocation of defects in the crystalline structure. The defect in the structure (red polygon) is characterized by a discontinuity of the aligned crystals above and below the gliding plane (A). The stress is absorbed by the material with the dislocation of the defect from one point to its adjacent neighbor (B and C) until the alignment is restored with the continuity of the crystals across the gliding plane. The movement of the line defect occurs by breaking and reforming bonds with adjacent lines. Note that as the discontinuity sweeps through the crystal on a microscopic scale, the material shift is only one lattice unit at each step. Credit: Data from Van der Pluijm and Marshak (2004).

precipitates on the less stressed part, resulting in a microstructure characterized by grain boundary migration and the formation of new grains (Figure 6.14). This mechanism can be observed in salt glaciers in Iran (Schléder & Urai, 2007). In laboratory experiments, solution creep is observed when the differential stress is very small (less than 1 MPa) and the strain rates are around 10^{-10} s^{-1}. In some experiments, the wet halite at 20 °C displays a strain rate of about 10^{-7} s^{-1}, and the crystals are very small, about 0.1–0.5 mm in diameter (Urai et al., 2008).

Solution-precipitation creep has been observed in laboratory experiments on wet, fine-grained (< 500 mm, or < 0.5 mm) halite at temperatures in the range of 20–200 °C (Urai et al., 2008). The solution-precipitation creep process leads to a quite rapid linear viscous deformation of halite. Salt is compacted within the upper 50 m of burial. Halite deformation always consists of glide, climb, grain rotation, and recrystallization, plus solution and precipitation, depending on the deformation rate and the presence of water (Urai et al., 1987, 2008).

Microscopic analysis of rock salt samples in the Zagros fold-thrust belt suggests that halite undergoes extensive recrystallization with the formation of new grains and the solution-precipitation process. The neoformed crystals are significantly smaller than the original Hormuz halite crystals, which are characterized by large crystals, often exceeding 10–20 mm in diameter. Urai et al. (2008) report that in the salt glaciers of central Iran (for example, the Tertiary salt basin in the Garmsar region), the evaporitic material in the highly deformed mylonitic zones is extremely fine-grained (~0.6 mm).

Salt flow deformation mechanisms have been mainly studied using natural examples of outcropping salt structures, such as the salt glaciers in Iran (see Chapter 3). Nonetheless, some researchers have investigated salt flow mechanisms based on observations from subsurface salt structures, such as in Germany, Iran, and the GOM (Kukla et al., 2011).

Exploratory boreholes drilled in the Hengelo area near the Netherlands-German border drilled Triassic Röt salt deposits at depths of around 400 m (Schléder & Urai, 2005). This bedded-salt sequence is younger, shallower, and less deformed than the Zechstein salt (see Chapter 2 in Mohriak, 2025). The microstructures were analyzed using petrographic thin sections from borehole cores decorated with blue colors by the gamma-irradiation technique. The micrographs showed solution-precipitation and the evolution of subgrains formed from precipitated crystals. The calculated differential stress values indicated deformation at stresses around 0.5–1 MPa, indicating the shallow occurrence of salt layers and the absence of major deformation zones affecting evaporite layers (Schléder & Urai, 2005).

Dynamic recrystallization processes in evaporite rocks can be recognized by the development of smaller mineral grains when exposed to increased pressures (Heege et al., 2005). The relationship between crystal size, temperature, and pressure is not intuitive, as these parameters depend on many factors. Researchers have investigated factors influencing the size of halite grains during deformation based on the analysis of undeformed crystals (e.g., Heege et al., 2005). The starting material used in the laboratory experiments was synthetic polycrystalline halite, containing >10 ppm water by weight and a mean grain size of 0.145 mm. After applying variable flow stresses, strains, and temperatures, the resultant recrystallized grains had a mean size significantly larger, from 2 to 4–5 times the original grain size, with some reaching a mean grain size exceeding 0.600 mm (Heege et al., 2005).

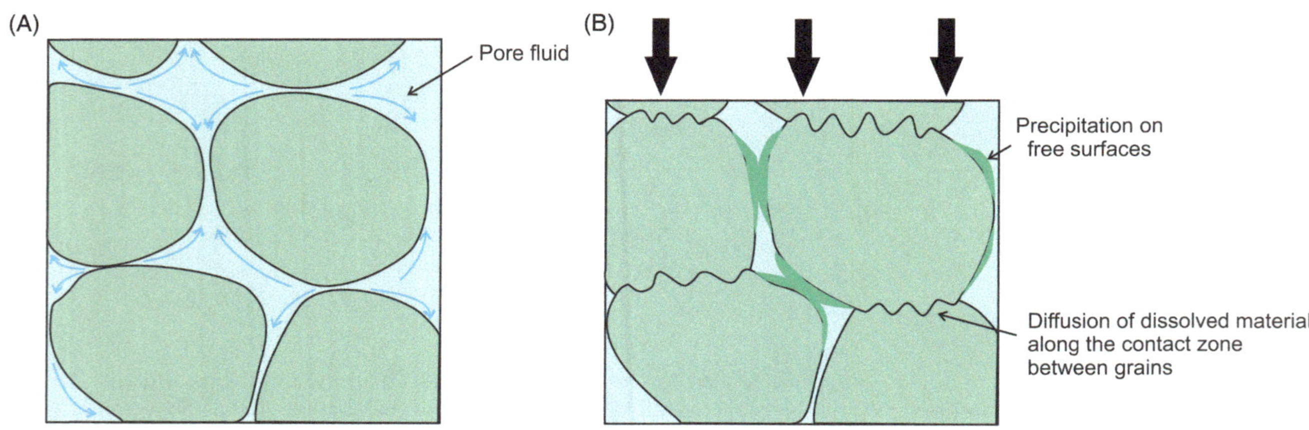

Figure 6.14 Pressure solution mechanism involving the reduction of pore space in uncompacted salt rock (A) by overburden pressure. The load of the sedimentary column results in compaction and reduction of pore space, resulting in sutures at the contacts with other halite grains (B). The precipitation of halite in parts of the grain with smaller pressure (free surface) results in migration of the boundary.

The samples were deformed in axial compression at 50-MPa confining pressure and a temperature of 125 °C (Heege et al., 2005). The laboratory experiments indicate that, statistically, the grain size distribution is characterized by a mean grain size that is inversely related to the flow stress. At the same temperature of 125 °C and strain rate of 5×10^{-7}, the mean grain size increased from 0.188 to 0.774 mm when the flow stress was reduced from 19.7 to 10.9 MPa (Heege et al., 2005, table 1). Higher differential flow stresses in laboratory experiments lead to smaller grain sizes in wet polycrystalline halite deformed at a near-mechanical steady state. The median and mean recrystallized grain sizes decrease with increasing stress and temperature. Other factors like temperature and strain rates may also influence grain size distribution (Figure 6.15). The experiment resulted in neoformed crystals with larger sizes under lower flow stresses (Heege et al., 2005).

Several authors studied grain size and deformation mechanisms using Zechstein Group evaporite sequences in Germany and the Netherlands (e.g., Barabasch et al., 2023; Küster et al., 2009, 2010). The Zechstein Z2 cycle corresponds to the Stassfurt Formation, which includes thick halite deposits in the Permian Basin (see Chapter 2 in Mohriak, 2025). Halite units include coarse-grained beds (Kristallbrockensalz) with grains a few centimeters in diameter and a fine-grained halite matrix formed by submillimetric grains (Küster et al., 2009, 2010). The Kristallbrockensalz layers resemble porphyroblasts in gneissic and mylonitic rocks. These porphyroblasts are formed by the plastic deformation of halite, accompanied by subgrain formation, and by rupturing in extension. This process occurs under low differential stresses, estimated around 1–2 MPa, and a slow strain rate (10–14 s^{-1}) associated with dislocation creep (Barabasch et al., 2023). This contrasts with the fine-grained halite in the rock matrix, which exhibits a faster strain rate during deformation (10^{-10}–10^{-8} s^{-1}, according to the rheological estimates). These observations from the Permian Basin natural salt rocks indicate the matrix salt in the Stassfurt Formation was weaker and more strongly deformed by dislocation creep and dynamic recrystallization than the coarse-grained halite facies (Barabasch et al., 2023).

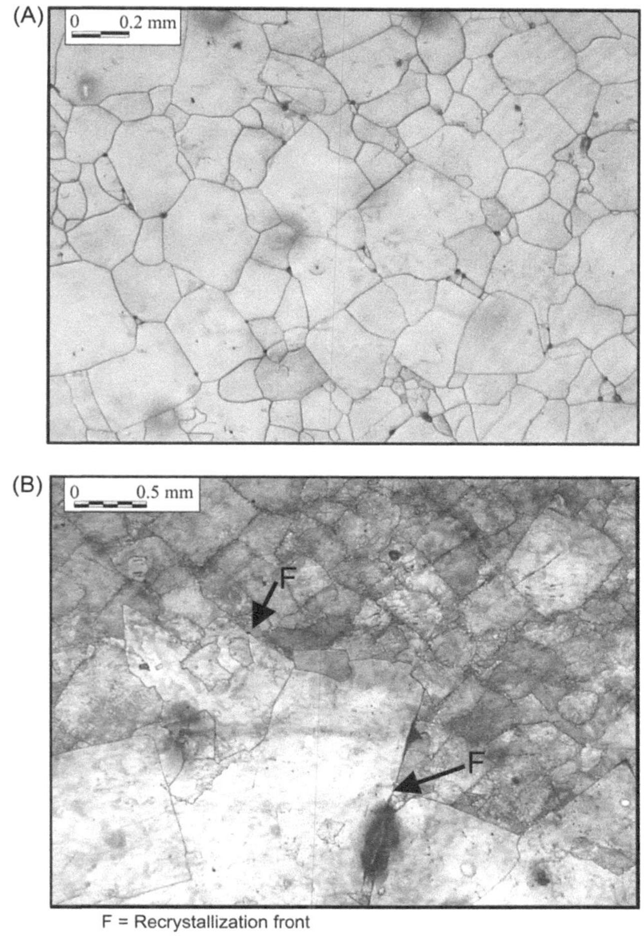

Figure 6.15 Micrograph images of salt rock (synthetic halite crystals) before (A) and after (B) deformation in the laboratory. Observe the change in scale from (A) to (B) and the change in the size of the recrystallized halite minerals. The original grain size was 0.145 mm, and the size of the crystals after the experiments exceeded 0.600 mm, and some crystals reached more than 1 mm in diameter. The recrystallization front is marked by a clear linear boundary between the small original crystals and the neoformed, much larger crystals. Credit: Heege et al. (2005)/with permission of Elsevier.

6.5. MICROSTRUCTURAL PROCESSES AND HALITE DEFORMATION IN THE ZECHSTEIN SALT BASIN

The Zechstein salt basin in northern central Europe is a natural laboratory for tectonic studies, focusing on basin-forming processes, sedimentation, and deformation of evaporite sequences (Kukla et al., 2008, 2011; Maystrenko et al., 2008, 2012; Urai et al., 2008). The Zechstein evaporites belong to a complex and large basin included in the Central European Basin System (CEBS), extending from the North Sea to the Netherlands, Germany, and Poland (Figure 6.16). Researchers have extensively analyzed the basin's geodynamic and tectonic evolution, which has gone through phases of subsidence and uplift, resulting in a complex geological history from the Late Paleozoic to the Neogene (Kukla et al., 2011; Littke et al., 2008).

The Zechstein salt deposits have been deformed by several tectonic episodes that affected the basin in the Paleozoic, Mesozoic, and Cenozoic. These events include the Permian extension that formed rift depocenters filled with salt (Kley et al., 2008; Maystrenko et al., 2008).

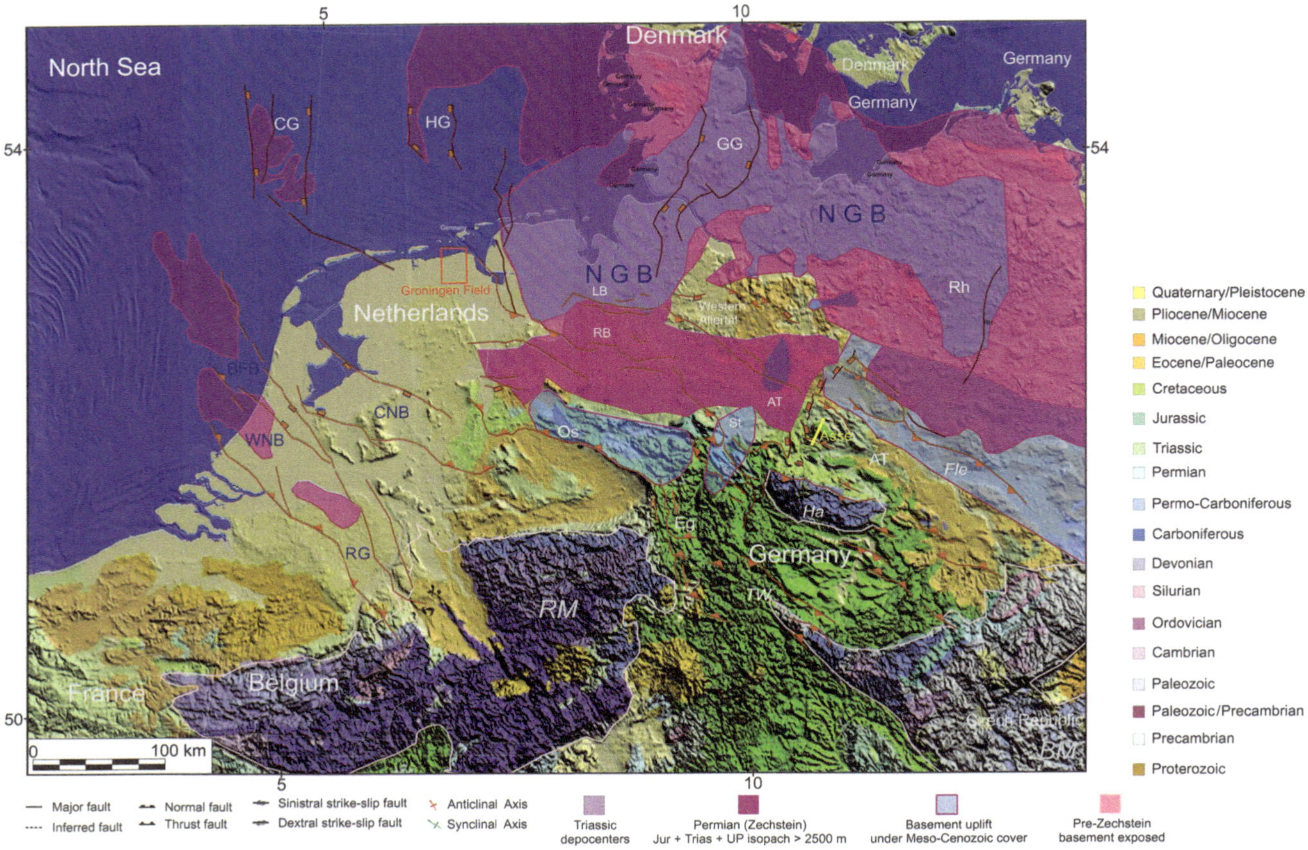

Figure 6.16 A regional geological map of the CEBS, draped on topography, shows the main structural elements related to the salt basins. The Late Permian Zechstein basin (highlighted by transparent magenta color), extended across the North Sea, the Netherlands, Denmark, Germany, and Poland. The Triassic depocenters are highlighted by a transparent pale blue color. The outlines of major basement uplifts, such as the Rhenish Massif and the Bohemian Massif, are highlighted by a pink color. The Harz Mountains basement uplift is located in the central part of the map, south of the main salt basin. The Asse salt mine is located northward of the Harz Mountains, and a thin yellow line shows the position of a geological cross-section illustrating the Asse anticlinal structure, which is associated with an NW-SE thrust fault with a southeast vergence.

Regional extension during the Late Triassic to Jurassic (e.g., Coward, 1995), compressional episodes in the Early Cretaceous to Paleogene (Kley, 2018; Kley et al., 2008; Scheck-Wenderoth, Krzywiec, et al., 2008; Scheck-Wenderoth, Maystrenko, et al., 2008; Voigt et al., 2021), and the reactivation of diapiric structures, originally formed during the Mesozoic by halokinesis, resulted in the contraction of salt diapirs, forming structures that presently exhibit the tops near the depositional surface (e.g., Maystrenko et al., 2012; Stewart & Coward, 1995; Strozyk et al., 2017).

6.5.1. Asse Salt Diapir in Germany

Kley et al. (2008) examined deformation structures and stress-strain relationships in Zechstein evaporite rocks in the Asse salt diapir, located to the north of the Harz Mountains in Germany (Figure 6.16). The rheological-studies focused on the test plugs obtained from halite deposits from the Asse salt mine (Figure 6.17).

The study analyzed salt deformation using a macro-scale and microstructural analysis of salt plugs and petrographic thin sections to determine the kinematic and

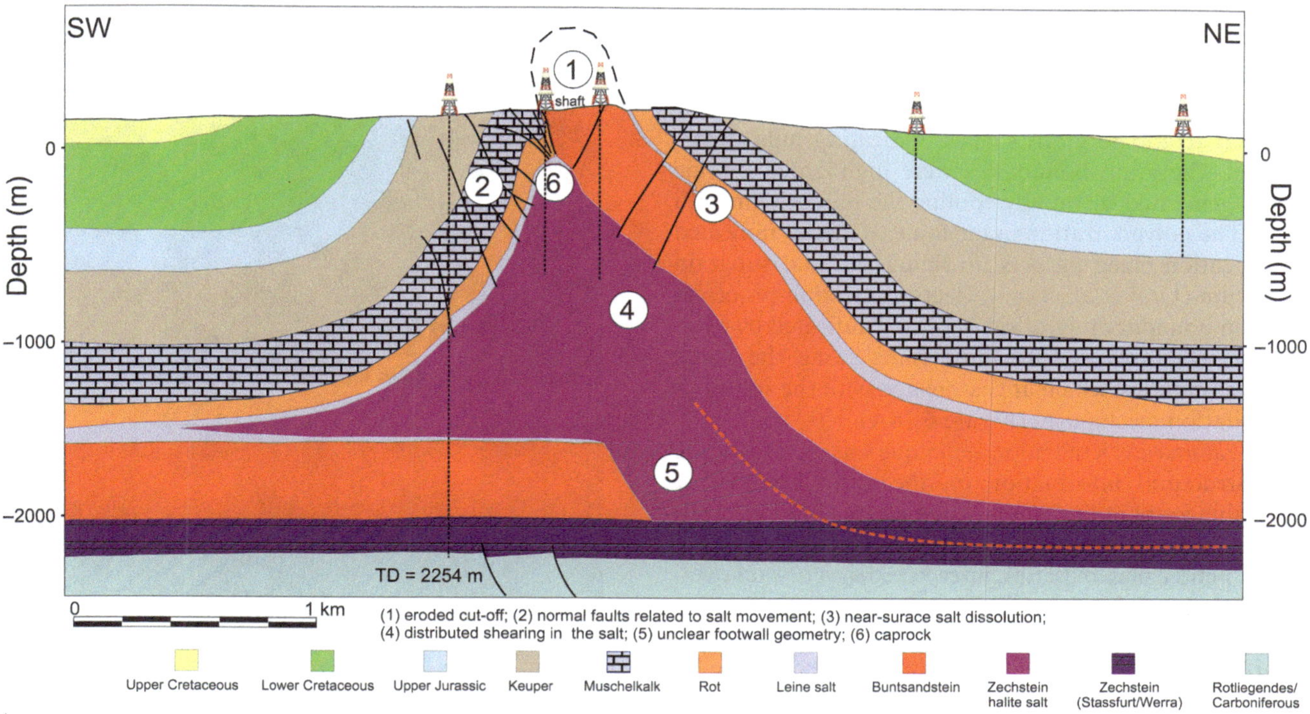

Figure 6.17 Geological cross-section showing the Asse salt structure, a large salt diapir mined for potash and halite minerals in the last century. Subsequently, the salt diapir was investigated as a potential repository for radioactive material. The salt mine was mapped by boreholes that penetrated the salt structure to a depth of about 1 km below the ground. The numbers in the geological cross-section indicate specific features in the salt diapir. Credit: Adapted from Kley et al. (2008).

dynamic factors that controlled the salt flow. The geological-section based on seismic interpretation and borehole data revealed an asymmetric structure due to the uplift of post-salt layers, including the Buntsandstein, Muschelkalk, and Keuper sequences (Figure 6.17). Compressional tectonic forces deformed the salt structure, forming an anticline primarily composed of mobile evaporites. The Zechstein evaporite sediments were deposited above the Rotliegendes sandstones. Salt flow transported the halite mass from below the Buntsandstein to a taper-like structure that overlaps the Buntsandstein layer. During halokinesis, salt flow led to salt pillows, domes, and diapirs, and post-salt sediment accumulation in mini-basins. Overburden loading on the evaporite layer caused salt expulsion under the mini-basins, resulting in direct contact between the overlying Buntsandstein and nonmobile salt facies of Zechstein evaporites, mainly composed of anhydrite (Kley et al., 2008).

The Asse salt diapir in Saxony was excavated as a salt mine between 1906 and 1909, initially intended to exploit potash minerals (BGE, 2022). However, potash ore production ceased after 20 years. The mine then began producing rock salt, mainly halite, which was commercially exploited from 1916 to 1964, when production was discontinued for economic reasons (BGE, 2022). In the late twentieth century, the mine was investigated as a potential storage site for radioactive waste from German nuclear plants (Gärtner et al., 2008). However, plans were abandoned due to the contamination of aquifer brines with radioactive elements like cesium, plutonium, and strontium. The mine and the exploratory wells that penetrated the salt diapir (Figure 6.17) have been used by several researchers focusing on rheology studies (e.g., Kley et al., 2008; Pollok et al., 2018, 2022; Urai et al., 1987; Weijermars et al., 1993).

Urai et al. (1987) conducted a microstructural analysis of petrographic thin sections of naturally deformed rock samples from the Asse salt diapir. They found that the most important processes during halite deformation were dislocation creep and grain-boundary migration, combined with solution-precipitation, which resulted in the recrystallization of the halite masses (Figure 16.18).

The halite minerals were analyzed from cores obtained by drilling the Asse salt diapir. The samples from the mine correspond to the Zechstein Speisesalz, which is almost pure halite (~ 99%) with less than 1% impurities of polyhalite ($K_2MgCa_2(S0_4)_4.2H_2O$), which appear as black grains contrasting with the colorless matrix material (Urai et al., 1987). The microstructural deformation was analyzed using thin sections in dark-field transmission light optical micrographs (Figure 6.18A) and gamma-irradiated thin sections (Figure 6.18B).

The halite mineral grains were characterized by a grain size of 3–10 mm and a moderately developed fabric of irregular to lobate grain shapes, indicating mild deformation. The grain boundaries were highlighted by line cracks generated by rapidly cooling the thin section with ether. The polygonization in the mineral matrix indicated a dislocation creep process involving slip and subgrain formation (Urai et al., 1987). Recrystallization processes by grain boundary migration were also indicated by overgrowth structures as thin brine films along the halite grain boundaries, which are associated with the solution-precipitation mechanism (Figure 6.18A).

The gamma-ray irradiation technique was used to obtain microstructural information on the grain boundaries. The halite grain in the center of the image (Figure 6.18B) is characterized by a deep blue nucleus surrounded by a rim of lighter blue material, interpreted as a crystal overgrowth structure. There are no primary microstructures inside the grain, suggesting that halite was strongly deformed and dynamically recrystallized. The evidence for fluid-grain boundary migration is shown by the core and mantle structures, which are characterized by pale blue colors that smoothly surround the deep blue halite nucleus (Figure 6.18B).

(A)

(B)

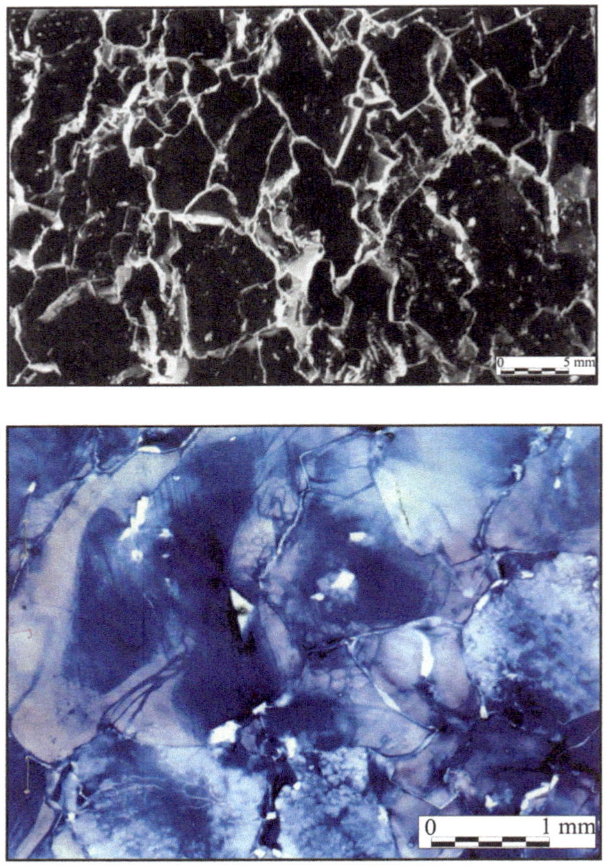

Figure 6.18 Micrographs of halite minerals in the Asse salt structure. (A) Microphotography of the Permian Zechstein salt cores obtained from wells drilled in the Asse salt mine. The microstructural analysis of the rock salt revealed the halite minerals are slightly deformed, and the crystalline structure of the rock framework is characterized by fine cracks along the grain boundaries, indicated by thin white films surrounding the grains. (B) Thin section from an Asse salt mine sample, irradiated with gamma rays at temperatures of about 120 °C. This technique highlights the crystals with black and white lines. This technique provides important additional microstructural information on the internal microstructures, which would be almost invisible under transmitted light. The halite grain in the center of the image shows a deep blue nucleus surrounded by a rim of lighter blue material that is interpreted as a crystal overgrowth structure. The lack of primary microstructures in the neoformed halite grain suggests strong deformation and dynamic recrystallization.

6.5.2. Neuhof Salt Mine in Southern Germany

The Neuhof mine in the Fulda Basin, situated at the southern margin of the Zechstein salt basin, exploits the Hessen potash bed, linked to the Zechstein Z1 sequence (Schléder et al., 2008). The evaporites in the mine are highly folded and cut by halite veins. A core from boreholes was gamma-irradiated and prepared for a detailed petrographic analysis. The core shows halite-filled veins about 1 cm thick and fractures filled with coarse halite crystals (Schléder et al., 2008). The deformation in the core reflects the rheological properties of the halite and anhydrite layers, with strain concentration occurring in the halite layers due to their higher mechanical strength.

6.5.3. Hengelo Salt Structure in the Netherlands

Another example of evaporite microstructures comes from cores from the Hengelo salt structure in The Netherlands, near the border with Germany (Schléder & Urai, 2005; Urai et al., 2008). The gamma-irradiated thin sections of the evaporite rocks from the Middle Triassic Röt Formation were prepared from samples obtained from a bedded halite layer at a depth of 457.5 m. The internal microstructures of the halite are made visible in the thin sections as tints of blue (Figure 6.19).

The microstructural study of halite samples in the Hengelo salt structure revealed that grain boundary migration is the main recrystallization mechanism, resulting in clear, strain-free new halite grains. The newly formed grains have dark bands in the gamma-irradiated thin slide, while subgrains and microfractures are marked by thin white lines (Figure 6.19). Based on piezometer

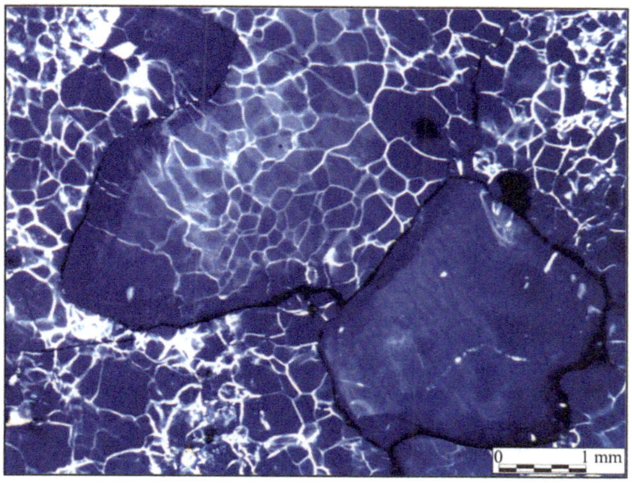

Figure 6.19 Gamma-irradiated thin section from evaporite rocks at the Hengelo salt structure in the Netherlands. The micrograph was prepared from a halite sample obtained from the salt layer at a depth of 457.5 m. The halite internal microstructures are made visible when the thin sections are irradiated with gamma rays and become decorated with tints of blue. The halite grain boundaries are marked by dark bands, whereas the subgrains and microfractures are characterized by thin white lines forming a cellular pattern. The mean grain size in Hengelo halite samples varies from 5 to 25 mm, but the sizes of the internal subgrains are much less than 1 mm. Credit: Urai et al. (2008)/with permission of Springer Nature.

measurements, the study also determined differential stresses and subgrain sizes, with low values for differential stress (0.45–0.97 MPa). This was expected, as the Triassic salt layers occur at shallow depths near the surface, in the interval between 420 and 460 m, whereas the deeper Zechstein salt layers occur at depths of about 1500 m (Schléder & Urai, 2005; Urai et al., 2008).

However, although the estimated differential stress was less than 1 MPa, the interpretation of geological cross-sections in the region indicates that the Triassic (Muschelkalk) and Jurassic layers have been slightly folded, forming synclines and anticlines. The deformation that affected the Triassic salt layers in this part of the basin was likely related to a mild Cretaceous inversion (e.g., De Jager, 2003). The microstructural analysis of rock samples from Triassic salt cores indicated that the grains are mainly characterized by syndepositional structures associated with salt deposition in salt pans. The Triassic salt was not strongly affected by recrystallization processes, unlike samples from large Zechstein salt diapirs in the North Central Europe Basin, such as the Asse structure north of the Harz Mountains (Figure 6.19).

The deposition of natural halite crystals forming shallow bedded sedimentary layers (as in the salt pans in the Hengelo area) is characterized by low differential stress (0.5–1.0 MPa) and low temperatures (20–50 °C). Under these low-stress conditions, the halite crystals mainly deform by dislocation creep. However, Schléder and Urai (2005) point out that solution-precipitation creep also occurs and hypersaline fluids may be released during recrystallization processes.

On the other hand, halite flow in large salt domes and diapirs such as the ones that occur in the depocenter of the Zechstein salt basin in North Central Europe (e.g., Asse and Gorloben salt diapirs in Germany and Klodawa diapir in Poland) is characterized by higher pressures (1–2 MPa) and higher temperatures (40–150 °C). The deformation mechanisms involve mainly dislocation creep, and, secondarily, dynamic recrystallization (mainly fluid-assisted) and solution-precipitation creep (Urai et al., 2008). The microstructure analyses of natural halite samples from salt diapirs in the Zechstein salt basin reveal dislocation creep and fluid-assisted grain boundary migration, as indicated by growth banding in halite grains (Schléder & Urai, 2005; Urai et al., 2008).

6.6. GEOLOGICAL INTERPRETATION OF SALT GLACIERS IN THE ZAGROS REGION

The Neoproterozoic Hormuz evaporite deposits occur at depths of 10–15 km in the Fars region of eastern Iran (Hassanpour et al., 2020; Kent, 1958, 1979). These autochthonous salt deposits are overlain by Paleozoic, Mesozoic, and Cenozoic sedimentary successions locally exceeding 14 km in thickness (Motamedi et al., 2011). The Zagros Basin experienced extension and rifting during the Permo-Triassic period, leading to the Neo-Tethys Ocean, which developed a passive continental margin from the Jurassic to the Cretaceous (e.g., Hassanpour et al., 2020). This phase was followed by the convergence of Northern Africa, the Arabian Peninsula, and the Eurasia plates as part of the Alpine-Zagros Orogeny. The Late Cretaceous and Paleogene periods were marked by the subduction of the Arabian Plate and Neo-Tethys margins under the Eurasian Plate. In the Neogene, continental blocks collided, resulting in crustal shortening (Frizon de Lamotte et al., 2011). During the Paleozoic and Mesozoic, Hormuz evaporite rocks formed tall salt diapirs by halokinesis (Hassanpour et al., 2020; Motamedi et al., 2011). Compressional tectonics associated with the Alpine Orogeny resulted in the contraction of salt structures, with some diapirs breaching anticlines in the Zagros fold-thrust belt and extruding on the depositional surface, forming salt glaciers or namakiers (Husseini & Husseini, 1990; Talbot, 1981, 1993, 1998).

The salt outcrops in the Zagros region have been investigated for several decades because halite can be observed forming extrusive sheets flowing from the elevated salt mountains (see Chapter 3). The salt glaciers flow at fast strain rates, allowing the measurement of deformation in a short time interval (Talbot, 1981).

The study of the salt flow in the Zagros region combines macrostructural analysis of seismic data and outcrops with microstructural analysis of halite samples from bedded salt, salt diapirs, and extrusive salt sheets. These microstructures have been compared with laboratory petrographic techniques and results from other methods such as structural geology, rock mechanics, rheology, and fluid dynamics (e.g., Sarkarinejad et al., 2018; Schléder & Urai, 2007; Sepehr et al., 2006).

Some features observed in the Zagros region share similarities with extrusive, spreading salt sheets in the GOM (e.g., Talbot, 1993). In the Gulf Coast, salt flow led to the development of allochthonous salt layers and salt diapirs, forming a distal salt province extending from the shelf, slope, and deep basin. These structures also developed intrusive and extrusive salt sheets emplaced near the seafloor, resulting in large submarine salt features known as salt tongues or salt nappes, which are associated with the bathymetric escarpments at the transition to the abyssal plain (see Chapter 4 in Mohriak, 2025).

Early theories about the origin of Iranian salt glaciers included controversial hypotheses. Gussow (1968) suggested that salt glaciers formed from diapirs in the Zagros region as molten salt heated at depth. He believed that when a temperature of 205 °C was reached, the halite layers became soft and plastic, flowing indefinitely under pressures of 500–1400 psi, which are equivalent to circa 3.5–10 MPa (Gussow, 1968). This interpretation assumes that halite was plastic during the entire process of diapiric intrusion, and even during extrusion at the surface, it formed "salt glaciers" as long as the salt remained hot. Gussow (1968) discussed these salt structures and concluded that salt glaciers were presently rigid and stationary, concluding that they represented "wasting remnants of catastrophically fast (and hot) flows in the past."

On the other hand, Talbot (1979) described active salt glaciers in Iran and interpreted them as flowing downward from salt domes, creating fold trains and other deformation structures. Based on observations during field campaigns in 1977, Talbot and Rogers (1980) estimated that Iranian salt glaciers might flow at rates approaching 0.5 m/day when the halite was sufficiently wet, but under normal conditions, they flow at very slow rates, estimated to be around 4 cm/year, based on the average of 17 years of observations (Talbot, 1998).

Based on several campaigns of field observations, several researchers have studied the salt deformation in salt diapirs and salt glaciers in Iran since the 1970s (see Chapter 6 in Mohriak, 2025). They found that these structures were not static at low temperatures and pressure but were flowing steadily during the Holocene (e.g., Talbot, 1979, 1981). First, there are some historical records of periodic avalanches of salt debris at the glacier's snout. Second, most salt domes in Iran are topographic highs with elevations varying from 50 to 100 m above the ground to more than a thousand meters above the surrounding plains, as in the Kuh-e-Namak salt structure, where the altitude reaches 1.3 km above sea level and the adjacent plains are at altitudes less than 200 m above sea level (see Chapter 3). The presence of a dynamic topographic elevation despite the salt dissolution that must have degraded the structures over the past few tens of thousands of years indicates a continuous replenishment of salt to the glacier. Third, field measurements of deformation in time frames extending for several years indicate the gradual movement of salt in the glacier, which is greatly accelerated in rainy seasons (Talbot & Rogers, 1980). Fourth, the rather recent methods that involve remote sensing techniques, particularly InSAR, indicate significant salt flow and ground uplift at present. This method is based on the analysis of satellite images acquired in different time frames, providing detailed altimetric information on subtle changes in the Earth's surface, particularly the deformation of salt diapirs and salt glaciers (Aftabi et al., 2010).

Iranian salt glaciers have been compared to allochthonous salt sheets in the GOM (e.g., Talbot, 1998) and the Central European Salt Basin (e.g., Mohr et al., 2007). According to Talbot (1998), the Zagros salt glaciers are considered "a unique laboratory where salt extrusions not only mimic past and present salt extrusions elsewhere, but also represent smaller and faster natural analog models of the gravity spreading of other tectonites extruded as piles of nappes in orogens." The intriguing salt glaciers and the associated foliations and tectonic slides might be analogous to some of the structures observed in orogenic belts worldwide (Talbot, 1998). According to these early interpretations, the salt glaciers formed nappes, thrust sheets, folds, reverse faults, landslides, and some intriguing features that resembled submarine slumps, ice glaciers, and possibly lava flows (Talbot & Pohjola, 2009).

Talbot (1979) proposed a model to explain the fold trains in the Kuh-e-Namak salt glacier. Salt domes on topographic highs consist mainly of coarse crystalline halite, which is characterized by a more viscous behavior and less susceptibility to strain. Salt masses flow downslope, feeding extrusive sheets of fine-grained to microcrystalline halite groundmass, which is characterized by a less viscous behavior and higher susceptibility to mylonitization. Unlike ice glaciers, which are fed by accumulations of snow falling from above, salt glaciers are fed by salt diapirs that were squeezed by tectonic forces. These structures originate from a mother salt layer several kilometers below the ground and are extruded at the surface by differential pressure (Talbot, 1979; Talbot & Pohjola, 2009).

The process is similar to lava flows that form at the surface but are fed from highly pressurized magma chambers at depth. However, the mechanisms are different. Melted lavas flow due to an overpressured magma chamber, with the material in a molten state due to a high temperature. Evaporites flow due to tectonic forces affecting material in a solid state but behaving like a viscous fluid (e.g., Nettleton, 1934). While the velocity of lava flows depends on the silica content and temperature, the velocity of the salt glaciers, which are mainly associated with halite, depends on several factors, including grain size and viscosity, but the water content seems to be a crucial factor in weakening the material (Talbot, 1981, 1998; Urai et al., 1986, 1987).

6.6.1. Salt Glaciers in Iran and Flow Mechanisms

Salt glaciers commonly occur in arid climate regions affected by orogenies, such as in the Zagros Mountains, and they flow downhill under very low stresses (Talbot, 1981, 1998). These structures are fed by a rising salt dome, sourced from the Neoproterozoic Hormuz salt, which rises from depths of over 5 km (probably exceeding 10 km in some places) and locally reaches the surface, forming hills or topographic highs that may exceed 2 km in altitude (Talbot, 1998). One classical example of an extrusive salt sheet is the Kuh-e-Namak structure in western Iran, which is associated with the Dashtir salt diapir that is sourced from the Hormuz salt (see Chapter 3).

Researchers have analyzed the mechanical behavior of evaporite rocks that breach the Earth's surface and flow down toward adjacent valleys like an Alpine ice glacier, moving only a few centimeters to a few meters per year (e.g., Talbot, 1998; Talbot & Aftabi, 2004; Talbot & Jarvis, 1984; Talbot & Pohjola, 2009; Talbot & Rogers, 1980; Urai et al., 1986; Wenkert, 1979). The salt glacier model can also be applied to subsurface structures in the Central European Zechstein salt basin (Mohr et al., 2007), the GOM (Fletcher et al., 1995), and the South Atlantic Ocean (Fainstein & Krueger, 2005), where allochthonous salt tongues have been imaged with seismic profiles (Chapter 7 in Mohriak, 2025).

The Zagros salt glaciers in southern central Iran lack subgrains within small, fine-grained halite crystals, suggesting that the main deformation mechanism is pressure solution acting under low pressures (< 1 MPa). However, the salt diapir feeds the glaciers and exhibits a stem with highly stressed and deformed halite grains. These large grains were deformed by a dislocation climb process under high deviatoric stress, possibly exceeding 4 MPa. These large grains were transported upward from the mother salt layer (Neoproterozoic Hormuz salt) and extruded subaerially using extrusion channels or ducts, which require water. Water comes from sporadic seasonal rains, which are rare in the desertic region of the Middle East (Talbot & Rogers, 1980).

Wenkert (1979) suggests that the stress and strain rates calculated for the Iranian salt glaciers can be better explained by a creep mechanism associated with the diffusion of ions through an interstitial salt-water solution. Thus, the main deformation mechanism for the glacier's salt flow is related to pressure solution (Figure 6.13). The microstructures observed in petrographic thin sections of salt samples associated with flowing salt are strikingly different from those in bedded salt that has not flown significantly, which are characterized by grains only slightly deformed by compaction and creep.

Salt glaciers in Iran are primarily found in the Zagros thrust and fold belt, linked to the Late Precambrian Hormuz salt (see Chapter 3). However, there are other examples of salt glaciers with Tertiary-age salt, such as salt outcrops in the Great Kavir and Qum and Garmsar basins in northern Iran (Figure 6.20). The Garmsar Hills in the Garmsar Basin (Figure 6.21) exhibit extensive rock salt outcrops, characterized by folded evaporite rocks intercalated with Eocene to Oligocene sediments (Jackson et al., 1990; Schléder & Urai, 2007). These Tertiary-age evaporite layers are associated with siliciclastic (shales and marls), carbonate, and volcanic rocks, which can also be seen in the Great Kavir Basin to the south (Figure 6.22).

The Garmsar Basin in northern Iran is a foreland basin influenced by the Alborz Mountains fold-thrust belt with a predominantly E-W structural trend (Figure 6.20). This deformation zone, located south of the Caspian Sea, has been active since the Neogene (Jaberi et al., 2018). The highlands are part of the Alpine-Himalayan orogenic belt, resulting from block collisions during plate convergence (Alavi, 1994, 1996). The Arabian-Eurasian plates are actively deforming, with a convergence rate of about 22 mm/year (Guest, Axen, et al., 2006; Guest, Stockli, et al., 2006). Compressional tectonic styles are associated with NNE-SSW-oriented shortening of the lithosphere by folding or buckling, which involves undulating uplift and subsidence of crustal blocks in the region of the Alborz Mountains (Guest et al., 2007).

The Neogene deformation of sediments in the Garmsar Basin involves the folding of evaporite, carbonate, and siliciclastic sequences deposited in the Paleogene (Baikpour & Talbot, 2012; Baikpour et al., 2010). The extrusive Eocene-Oligocene evaporites in the Eyvanekey Plateau and Garmsar Hills (Figure 6.21) form an extrusive salt sheet system that is considered the largest in Iran (Arian & Noroozpour, 2015). The anticlinal structures in the highlands of the Garmsar Basin can be seen in satellite images (Figure 6.21A) and geological maps (Figure 6.21B). The evaporites exposed on the surface

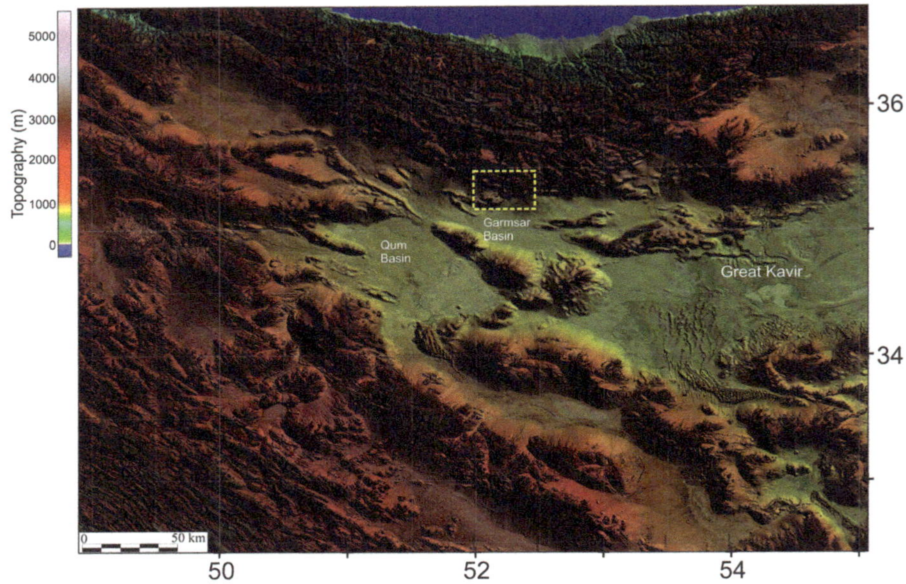

Figure 6.20 Regional satellite image draped on a topographic map of the Garmsar Basin in northern Iran, located south of the thrust and fold belt associated with the Alborz Mountains, near the southern margin of the Caspian Sea (blue color). The elevated topography is characterized by anticlinal structures and E-W thrust faults with predominant vergence to the south. The dashed, thin yellow rectangle shows the zoom area of the Garmsar Basin, where Tertiary evaporites outcrop north of the valley, near the city of Garmsar.

correspond to extrusive salt tongues, extruded by a Neogene compressional event associated with the NW-SE trending thrust fault system. The reverse faults are characterized by a marked vergence to the southwest (Arian & Noroozpour, 2015; Baikpour & Talbot, 2012).

A regional cross-section from the highlands toward the valley where Pliocene and Quaternary sediments outcrop (Figure 6.22) illustrates the interpretation of the thrust faults and the extrusive salt glaciers in the Garmsar Basin. The evaporite deposits outcrop on the Eyvanekey Plateau and Garmsar Hills. Although rock salt is rarely visible at the surface, halite minerals are exploited in open pits and underground mines (e.g., Baikpour & Talbot, 2012).

These salt layers (Figure 6.23) provide exceptional natural examples of deformation structures associated with mylonitic shear zones and recrystallization mechanisms involving halite. The salt mass often contains thin beds of crystalline halite with deformed grains exhibiting a preferred orientation, similar to metamorphic rocks or mylonites (Schléder & Urai, 2007).

The cold diapir stem of salt glaciers in the Garmsar Basin undergoes deformation through dislocation glide at high differential stresses (3–5 MPa), implying strong shearing of the evaporite layers during deformation (Schléder & Urai, 2007). Extrusive salt tongues are characterized by mylonitic halite layers sheared by salt flow, resulting in subhorizontal and bedding-parallel recumbent fold structures (Figure 6.23).

The microstructural analyses of samples from thin halite lenses in banded evaporites reveal a large number of small grains or subgrains dispersed in the flowing mass (Figure 6.24A). These folded evaporite rocks resemble mylonitic rocks, with only a few large grains developing in the fine-grained sedimentary matrix, resembling the porphyroblastic texture observed in metamorphic rocks. Based on the microscopic analysis, Schléder and Urai (2007) concluded that even large grains are full of small subgrains, but no primary microstructures, such as fractures or microfaults, are found inside the halite grains. This suggests that strong deformation during salt flow led to the dynamic recrystallization of the halite minerals (Schléder & Urai, 2007).

The micrograph image of mylonitic evaporites (Figure 6.24B) reveals oriented fibrous microstructures and growth banding on grain boundaries, as seen in gamma-irradiated sections. The crystal fabrics in the thin slide display a weak preferred orientation of the crystal axes, suggesting solution-precipitation mechanisms accommodating grain boundary sliding. Schléder et al. (2008) suggest that the principal deformation mechanism for salt glaciers in the Garmsar Basin was solution-precipitation creep (Figure 6.14). This process involves nonconservative grain boundary migration, implying that

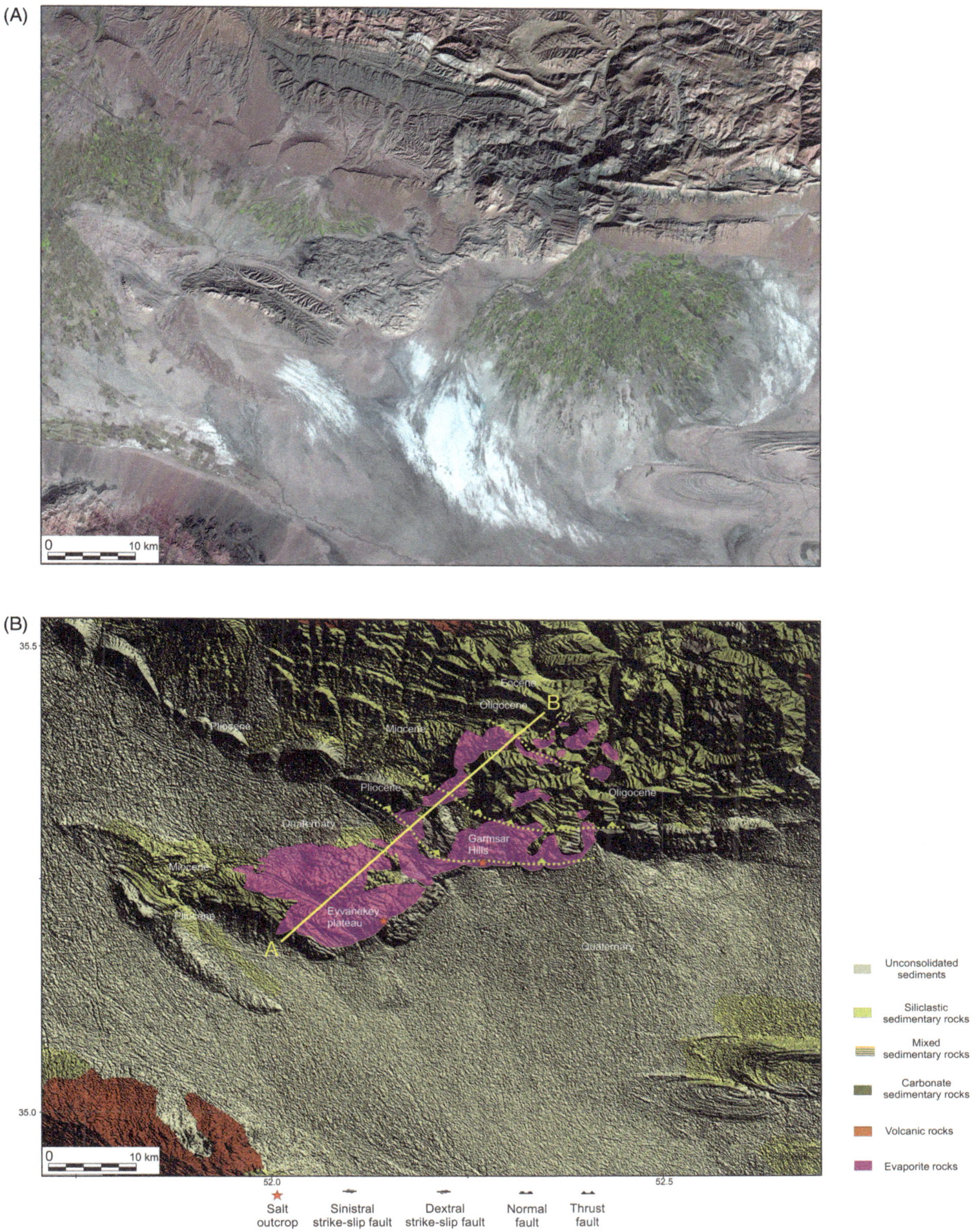

Figure 6.21 Satellite images of the Garmsar Basin in Iran. (A) Landsat satellite image (false colors) of the Garmsar Basin shows the cultivated agricultural fields in green colors. Observe the regional NW-SE trend of the anticlinal structures north of the valley, where the topography reaches more than 3,000 m. (B) Simplified geological map of the area above, with the outcropping evaporites marked by magenta color. The thin yellow line (profile A-B) corresponds to a regional geological cross-section that illustrates the structural style of the fold belt and the extrusive salt glaciers. Observe the characteristic low-relief aspect of the evaporite sediments in the satellite image.

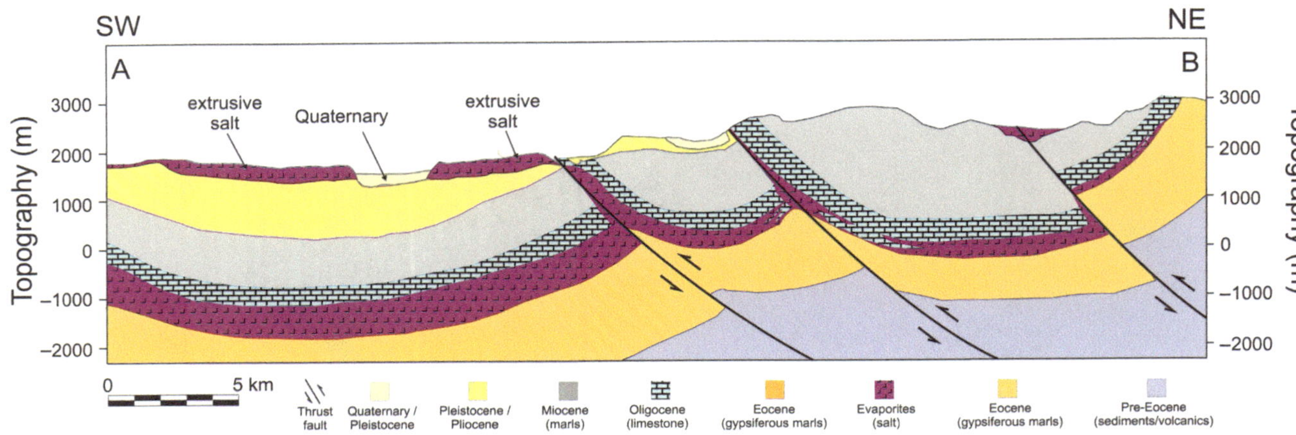

Figure 6.22 Regional geological cross-section (line A-B, see the previous map) from the southwestern portion of the Garmsar Basin to the highlands in the northeast, which are affected by the Neogene thrust and fold belt. Observe that extrusive salt tongues advanced toward the synclinal structure, where Pliocene sediments show an increased thickness. Credit: Adapted from Schléder and Urai (2007).

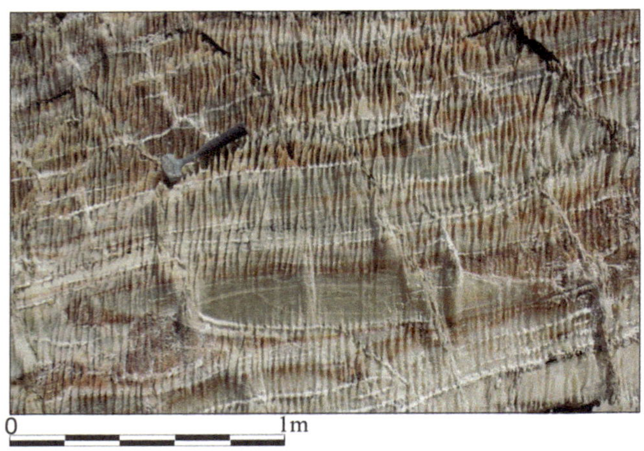

Figure 6.23 An example of mylonitic texture is seen in an outcrop of Neogene-age evaporites in the Garmsar Hills. The bedded evaporites have been involved in folds and shear zones during the extrusive processes. The salt flow resulted in subhorizontal and bedding-parallel recumbent fold structures involving the evaporite layers.

grain boundary sliding was accommodated by solution-precipitation of halite minerals. Grain boundary migration resulted in the development of mineral orientation according to the stress field (Schléder et al., 2008).

This interpretation agrees with the solution-precipitation creep model suggested by Wenkert (1979) to account for the unusual weakening of salt glaciers that can move downhill under very low shear stresses (< 0.25 MPa) and extremely high strain rates (1.1×10^{-11} to 1.9×10^{-9} s^{-1}). The rheology of the salt glaciers has been interpreted as creep due to the diffusion of sodium and chloride ions through an interstitial salt-brine solution (Schléder & Urai, 2007; Wenkert, 1979). This weak behavior has been confirmed by fieldwork observations (Talbot, 1981, 1998; Talbot & Rodgers, 1980) and laboratory experiments that employ fine-grained halite grains that are wet with small quantities of water, usually with a minor content of about 0.05 wt% (e.g., Urai et al., 1986).

6.6.2. Salt Deformation in the Zagros Fold-Thrust Belt

Researchers have compared the deformation observed in the Zagros outcrops with physical models and numerical simulations to reproduce the observations at different scales. Koyi (1988, 1998) presented a review of the rheological controls on the development of salt diapirs in the Zagros fold-thrust belt, focusing on the role of gravity in the deformation. Based on physical experiments, several parameters were considered to control the geometry of active diapirs, including the rate of salt supply, salt dissolution, sediment accumulation, erosion, extension, and shortening (Koyi, 1988, 1998).

More recently, numerical models have been applied to the analysis of salt deformation (e.g., Peric & Crook, 2004). Computer algorithms, based on Lagrangian methods, simulate the active elastoplastic, viscoplastic, and viscoelastic behavior of salt. Based on numerical experiments, Chemia and Koyi (2008) postulated that the dynamic evolution of salt structures is dependent on the rheology of anhydrite layers within the more mobile halite mass.

The crystal sizes of evaporite minerals subjected to stresses vary substantially depending on the deformation mechanisms involved in the process, as illustrated by the salt structures exposed in the Zagros Mountains of Iran.

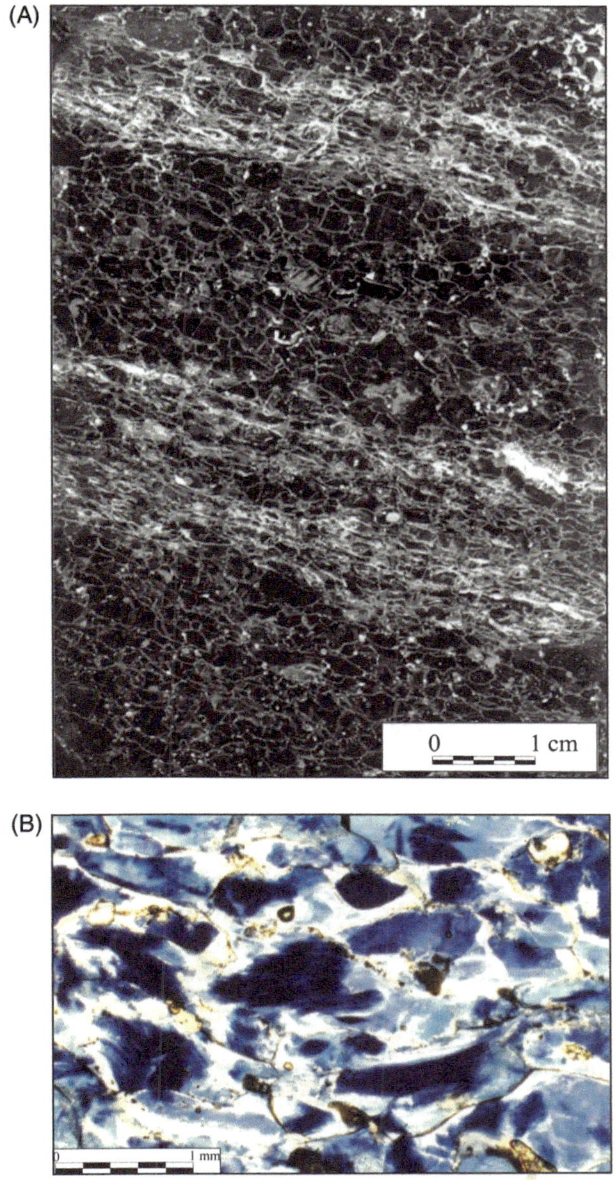

Figure 6.24 Microphotograph of mylonitic evaporite and an irradiated thin section from rock salt samples from the Garmsar Basin in Iran. (A) Micrograph from banded mylonitic evaporites from the Garmsar Basin, central Iran. (B) Gamma-irradiated thin section from a rock salt sample from the mylonitic evaporites from the Garmsar Basin, showing typical microstructures associated with solution-precipitation deformation in extrusive salt glaciers. The material in the highly deformed mylonitic zones displays a very small grain size, with a mean value of about 0.6 mm. There are no primary microstructures inside the halite grains, indicating that the strong deformation during salt flow resulted in a dynamic recrystallization of the halite minerals. The core and mantle structures in the halite grains, characterized by a smooth variation of the deep to pale blue color, indicate fluid-assisted grain boundary migration. Credit: Urai et al. (2007)/with permission of Elsevier.

Microstructural analysis indicates that halite decreases in grain size down the length of the glacier, resembling mylonitization in metamorphic rocks (e.g., Figure 6.24). Laboratory experiments using synthetic halite crystals and microstructural analysis confirm the changes in grain size by changing flow stress, temperature, and strain rate during the deformation (e.g., Desbois et al., 2010; Heege et al., 2005).

The rock salt in the cores of the Zagros fold-thrust belt salt domes is composed of large, transparent halite grains. The diapiric salt structures are typically associated with coarse halite with a uniform grain size in the stem, making it too strong to flow in subaerial surface conditions. This coarse-grained halite forms the cores of many mountains that rise above the desert plains in Iran (see Chapter 3). As the salt is extruded from the feeder, the halite grains decrease in size, forming subgrains around the margins (Desbois et al., 2010). Smaller halite subgrains recrystallize, forming an increasing volume of fine-grained translucent groundmass, while large clear grains are reduced in number. Talbot (1979) suggested that mylonitization processes associated with halite flow in salt glaciers involve a diffusion process. Although diffusion is typically a high-temperature mechanism associated with magma flow, the extrusive salt in Iran namakiers occurred at the surface, under low pressure and temperature, contrary to early hypotheses that assumed that these salt masses were extruded at high temperatures (e.g., Gussow, 1968).

Salt flow processes in nature occur as dominant or combined mechanisms. Deep in the subsurface, as is the case of the autochthonous Hormuz salt, when buried by more than 10 km of sediments, the rocks are subjected to higher confining pressures and higher temperatures, and the halite sequences are mainly affected by plastic flow. Under these conditions, the minerals are deformed by dislocation creep, pressure solution, and grain-boundary diffusion (Kukla et al., 2011; Schléder & Urai, 2005, 2007; Urai et al., 2008). In laboratory experiments, dislocation creep dominates the deformation processes at high-strain rates and differential stresses when temperatures exceed 100 °C.

Rapid salt glacier flow during rainy periods has been observed in field observations and corroborated by laboratory experiments (Talbot & Rogers, 1980). The addition of freshwater significantly accelerates the diffusion process in halite minerals, leading to changes in long-held concepts in microstructural analysis (see, for example, Urai et al., 1986, 1987, 2008). Rainwater penetration into rock salt that flows downdip from the diapir stem in Iran's salt mountains results in enhancing the solution-precipitation creep processes, and thus accelerating salt flow (Urai & Spiers, 2007; Urai et al., 1986, 1987; Urai et al., 2008). The grain size of evaporite rocks also affects viscosity, with larger grain sizes causing more viscous behavior. Based on numerical models of salt diapirs using dislocation and diffusion creep laws, Van Keken et al. (1993) proposed that a change in grain size from 5 to 30 mm increases the effective viscosity by more than two orders of magnitude. For larger grain sizes (diameter > 10 mm), dislocation creep is an important mechanism during the diapir growth stage, effectively reducing the viscosity (Linckens et al., 2016; Van Keken et al., 1993).

6.6.3. Fluid Dynamics Applied to Salt Flow Within Hormuz Salt Diapirs

The Poiseuille flow is related to experiments in fluid dynamics that investigate flow through a cylindrical pipe with length L and a constant cross-section with radius R (Figure 6.25A). The fluid flow in the Poiseuille model occurs between stationary plates, which correspond to the fixed cylinder walls of the tube (e.g., Sutera & Skalak, 1993). A pressure drop in an incompressible Newtonian fluid occurs when a laminar flow is established by applying pressure P1 to one side of the tube. The fluid flow Q corresponds to the velocity multiplied by the time interval and depends on the fluid viscosity. The Poiseuille flow in a viscous salt layer is pressure-induced. A pure Poiseuille flow requires a no-slip surface at the base and top of the channel and a pressure gradient resulting in a maximum velocity at the center of the channel (Evans & Jackson, 2021). As velocity changes to a minimum at the cylinder boundaries, there are opposite shear senses for the fluid flow near the top and bottom of the channel, resulting from the induced pressure gradient (e.g., Godin et al., 2006).

The Couette flow is defined as the flow of a viscous fluid in the space between two parallel plates separated by a distance (h), one fixed (stationary) and the other that can move by applying a tangential force to its top surface. The Poiseuille flow in a viscous salt layer is pressure-induced. A pure Poiseuille flow requires a no-slip surface at the base and top of the channel and a pressure gradient resulting in a maximum velocity at the center of the channel (Evans & Jackson, 2021; Sutera & Skalak, 1993).

The fluid flow velocity (u) depends on the fluid viscosity and distance to the moving plate (Figure 6.25C). The viscous drag results in a velocity decrease toward the boundary with the stationary plate. The largest shear stresses in salt channels occur at the top of the moving plate (Figure 6.25D).

Researchers have explored the application of fluid flow concepts to salt basins (e.g., Brun & Manduit, 2009; Cartwright et al., 2012; Evans & Jackson, 2021; Rowan et al., 2019). Many authors have discussed the salt flow in the Zagros thrust and fold belt based on field geology, microstructural analysis, and physical and numerical models. Sarkarinejad et al. (2018) discussed the salt flow models for Karmostaj salt diapir in the region of the Kuh-e-Gach anticline, including the results of the microstructural analysis of the deformed halite in the salt glaciers.

The Kuh-e-Gach anticline corresponds to a fold with a curved axial plane trending E-W (Figure 6.26). The axis of the anticline plunges to the east and west, forming a structural high in the center, where a salt diapir breached the overburden and extruded salt on the mountain flanks, forming salt tongues that flow toward the valley, near the Karmostaj village (see Chapter 3).

Similar processes are observed in the anticlinal structure located southward of the valley. The Burkh anticline exhibits an axis trending subparallel to the Kuh-e-Gach structure. The apex of the Burkh anticline was breached by a salt diapir in the Neogene (Sarkarinejad et al., 2018). The salt diapir stem formed a salt fountain that flows toward the valley, with salt masses exposed eastward of the Karmostaj village (Figure 6.26).

Salt extrusions in Iran are caused by the flow of metamorphic rocks from a subhorizontal crustal channel corresponding to the mother salt layer (Hormuz salt). The subsurface salt flows toward the stem of the diapiric structure, which is discordant with the adjacent Mesozoic and Cenozoic overburden rocks (Figure 6.27). During the contraction of these sediments, the salt diapirs were nucleated at the cores of the anticlines by a channel flow mechanism. When the overburden rocks were breached by active salt intrusions, the extrusive salt formed salt tongues on the surface, flowing toward the adjacent depressions (Sarkarinejad et al., 2018).

The loading of the Hormuz salt layer by overburden sediments resulted in a salt flow toward the incipient salt dome, following the Poiseuille flow model, with higher velocities at the center of the deforming mother salt layer. As the diapir structure evolved, the salt flow toward the stem formed a bulb- or mushroom-like salt structure (Sarkarinejad et al., 2018). The lower parts of the diapir stem were squeezed by the contraction of the Paleozoic and Mesozoic blocks, which were deformed during the Alpine Orogeny. The Poiseuille flow, marked by a parabolic velocity profile, dominated the salt system when the overburden blocks were stable (Figure 6.27). The model assumes that the salt flow was restricted by friction with the walls, resulting in viscous shear forces. The flow velocity in the central part of the mother salt layer was higher than at the boundary with the pre-salt or post-salt sequences (Sarkarinejad et al., 2018).

The proposed model for the Hormuz salt structure assumes channel flow involving the Poiseuille and Couette mechanisms (Sarkarinejad et al., 2018). When the overburden blocks are unstable, sliding above the

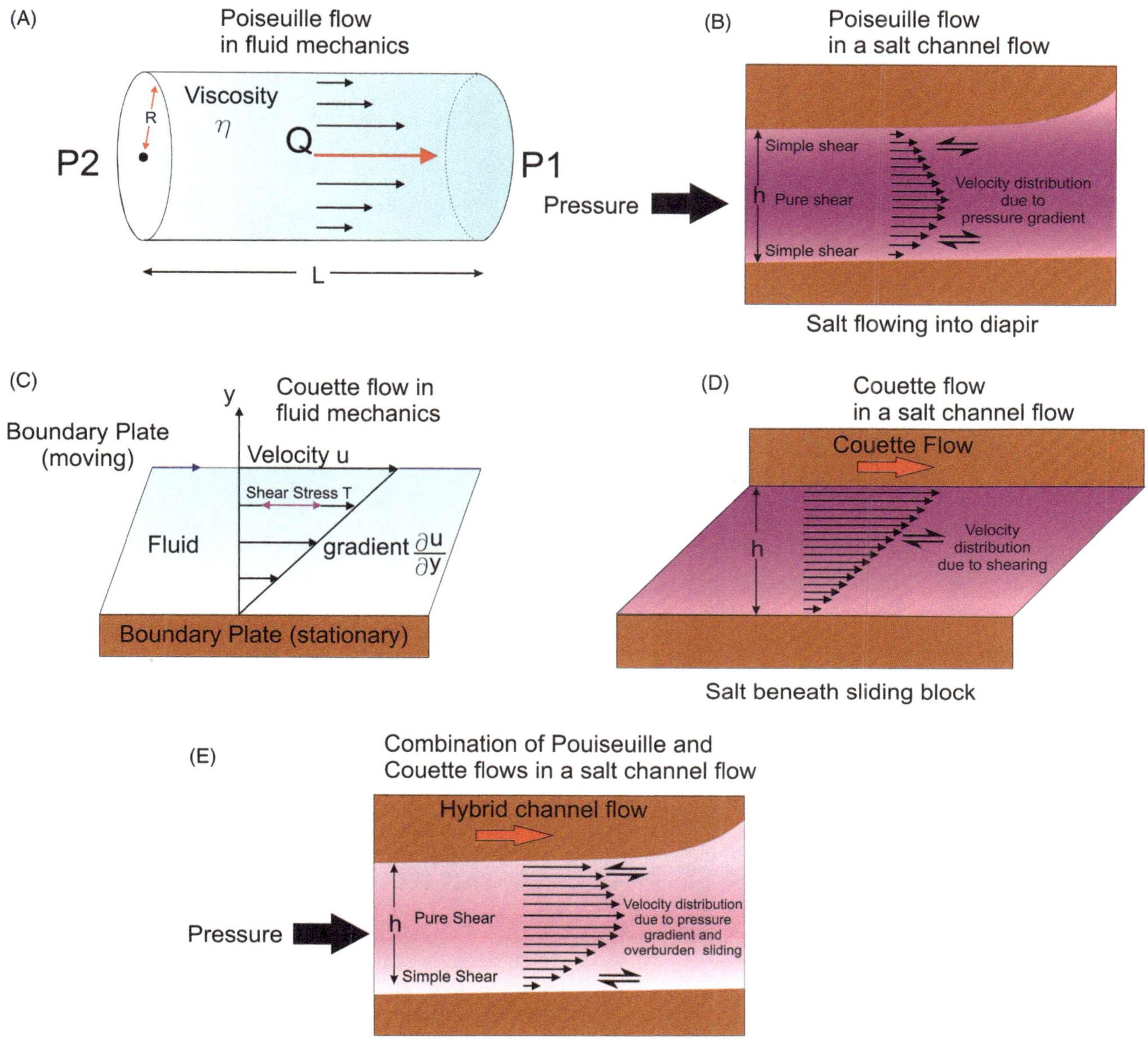

Figure 6.25 The schematic diagrams show the conceptual end-member types of salt flow based on fluid dynamics. The figures illustrate the Poiseuille and Couette fluid flow mechanisms in fluid dynamics experiments (left side) and in natural salt flow (right side). The diagrams to the left (A and C) illustrate the fluid mechanics models based on fluid dynamics theory and experiments. The diagrams to the right (B and D) illustrate the models applied to salt channels. Credit: Sarkarinejad et al. (2018)/with permission of Elsevier.

In the Poiseuille flow experiment (A), the experimental apparatus is based on a cylinder with length L and radius R, filled with a fluid with a given viscosity. Applying a pressure of P2 to the left side of the cylinder results in a fluid flow with a decrease in pressure to P1, in the direction of the fluid flow. The maximum fluid flow Q and velocity values are observed at the center of the pipe, with a velocity decrease toward the edges of the cylinder, resulting in shear stresses with opposite senses due to the differential velocities.

In the Couette flow model (C), the experimental apparatus is based on two infinite flat plates, with the upper plate consisting of a fluid that can be moved by tangential forces applied to its top surface. The plate is translated with a velocity u that decreases toward the lower fixed plate. Diagram (E) illustrates the concept of a hybrid salt flow mechanism based on the Poiseuille and Couette models.

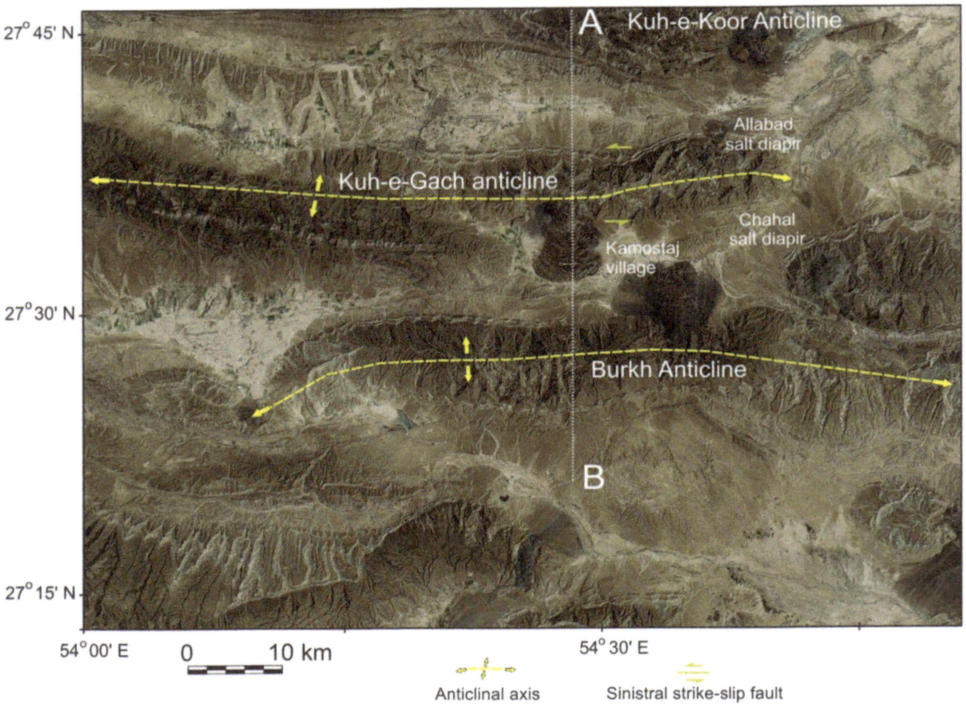

Figure 6.26 Satellite image of the Kuh-e-Gach and Burkh anticlines. These folds are characterized by double-plunging axes trending E-W. The folded strata were breached by salt diapirs that formed allochthonous salt sheets that flow downslope toward the Kamostaj village valley. The dashed white line (A-B) shows the approximate location of the regional cross-section illustrating the deep structure of the fold belt anticlines.

salt layer with different velocities, the Couette flow is probably induced, resulting in a shear-type salt channel flow (Sarkarinejad et al., 2018). This mechanism assumes salt shearing with a linear velocity profile, increasing from a minimum at the contact with the fixed plate (Neoproterozoic pre-salt strata) to a maximum at the boundary with the supra-salt layers (Figure 6.25D).

The hybrid flow model (Figure 6.25E) results in different proportions of salt flow associated with the two mechanisms. According to the proposed fluid dynamics model, the Poiseuille flow is characterized by a high pure shear component in the center of the channel and a high simple shear component along the margins (Figure 6.25B). The Couette flow includes simple shear within the Hormuz salt mass, as the halite layers act as decollements and gravity glide planes (Figure 6.25D). Within the diapir stem, the channel flow is a pure Poiseuille flow, as the salt is rising from the mother salt layer to the surface by a difference in pressure, from the lithostatic pressure at depth to the atmospheric pressure at the surface. The salt glaciers correspond to the extrusion of metamorphic rocks from a subhorizontal crustal channel flow in the deep source layer (Sarkarinejad et al., 2018).

6.7. GEOLOGICAL INTERPRETATION OF SALT TECTONITES IN THE NORTHERN CALCAREOUS ALPS

Salt deposits in the central to eastern Alps of southern Germany and Austria are found in the subsurface and underground mines excavated by the salt industry. Some salt occurrences have been exploited since Celtic times, such as the Hallstatt salt mines, where archaeological evidence indicates mining activities dating back over 7,000 years (see Chapter 3). The Haselgebirge Formation is the lowest unit of the Northern Calcareous Alps, a fold-thrust belt resulting from the Alpine Orogeny, with a climax in the Late Cretaceous to Paleogene (Leitner & Spötl, 2017; Spötl, 1989). The Late Permian salt layers (Haselgebirge Formation) are detached from the basement, which corresponds to Paleozoic metasedimentary rocks deformed during the Variscan Orogeny. The integration of geological maps (e.g., Chorlton, 2007) and topobathymetric datasets (e.g., GEBCO, 2019) indicates that the Northern Calcareous Alps extend in an east-west direction for about 1,000 km, reaching 50–100 km in a north-south direction (Figure 6.28).

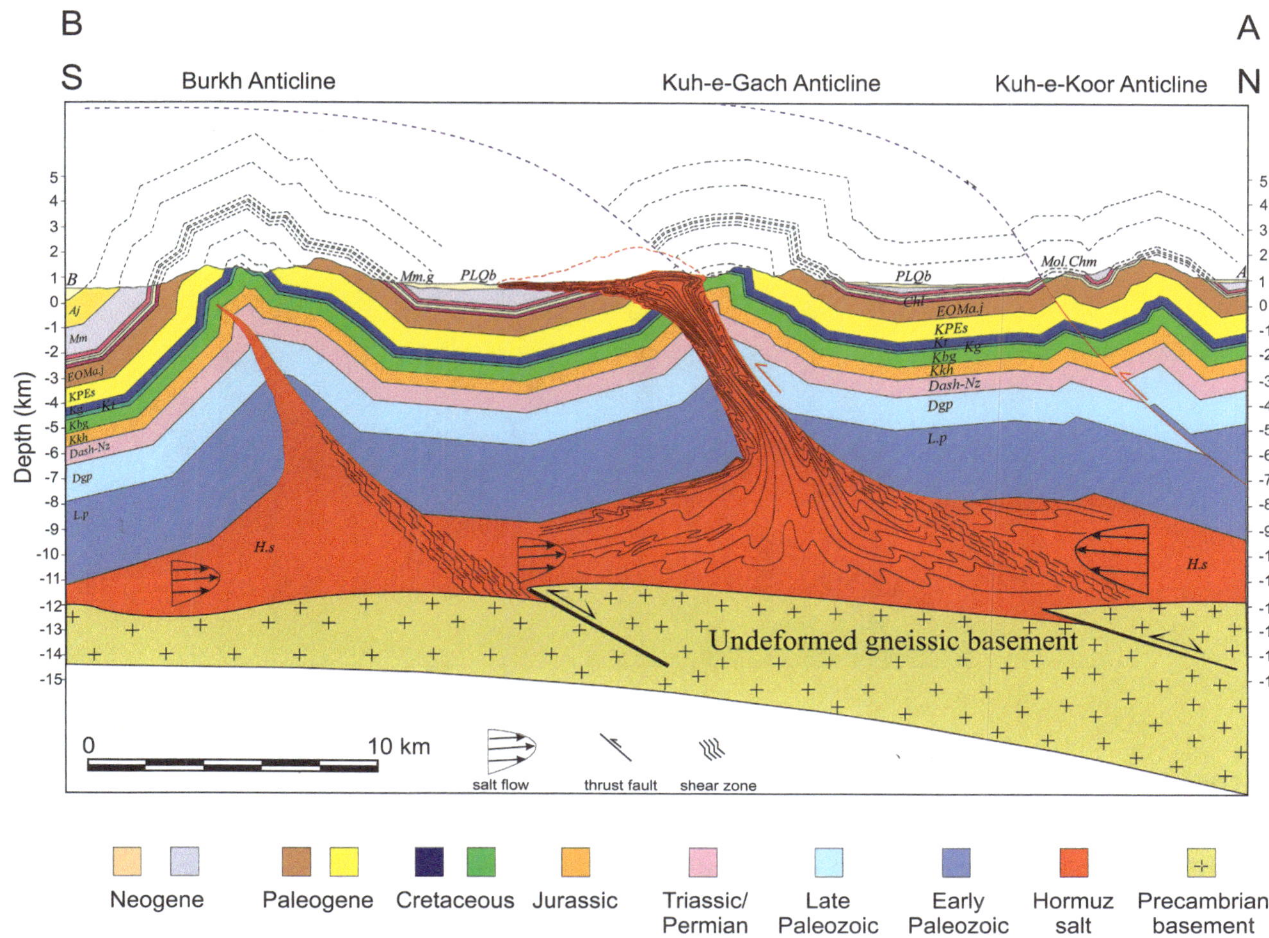

Figure 6.27 A balanced geological cross-section shows the deep structure of the Burk and Kuh-e-Gach anticlines in the Fars region of central Iran. The salt flow interpretation at the base of the Hormuz salt (marked by the red color) is based on detailed geological mapping of the exposed salt tongues. The profile shows a blind thrust in the Burkh anticline, which resulted in squeezing the salt diapir. The salt intrusion was limited to the Cretaceous sequences, but toward the east, the Burkh anticline was breached and the Hormuz salt was extruded on the surface, forming a salt tongue that flows toward the valley. Only the large flow diagram in the mother salt layer is constrained by surface mapping of the Hormuz salt layers, as indicated by the shear zones in the salt diapir that breached the Kuh-e-Gach anticline. The smaller flow diagrams to the south of the diapiric structures are inferred by the thickness variations in the salt layer. Credit: Sarkarinejad et al. (2018)/with permission of Elsevier.

The Late Permian salt layers (Haselgebirge Formation) are detached from the basement, which corresponds to the Paleozoic metasedimentary rocks, deformed during the Variscan Orogeny, and the Precambrian crystalline crust (Mandl, 1999). The Haselgebirge evaporites correspond to the weakest lithological units in the overlying sedimentary succession, with Triassic and Jurassic carbonate units separated by detachments, forming nappes with a northern vergence (Fernández et al., 2021; Mandl, 1999; Strauss et al., 2021). These allochthonous blocks correspond to the Bajuvanic, Tirolic, and Juvavic units composed of Triassic carbonate rocks (Fernández et al., 2021). The Hallstatt salt structure is surrounded by carbonate rocks of the Juvavic unit, which is separated from the Tirolic unit by a detachment zone with evaporite rocks (Figure 6.29).

Fernández et al. (2021) proposed that the salt structure in Hallstatt was formed through downbuilding and upbuilding processes since the salt layers were deposited in the Late Permian. Previous studies suggested that salt layers were formed as allochthonous nappes associated with the Alpine Orogeny, similar to salt glaciers in the Zagros and Atlas fold-thrust belts (see Chapter 3). Recently, some researchers (e.g., Fernández et al., 2021; Granado et al., 2018; Strauss et al., 2021) have suggested

Figure 6.28 Topobathymetric relief map showing the North Calcareous Alps in northern Austria, extending along a regional E-W trend for more than 1,000 km. The basement rocks outcrop south of the Mesozoic carbonate outcrops in the mountain ranges. Cenozoic sediments accumulated in the foreland basins north of the highlands. The thin yellow line A-B corresponds to a regional schematic geological cross-section across the Alps.

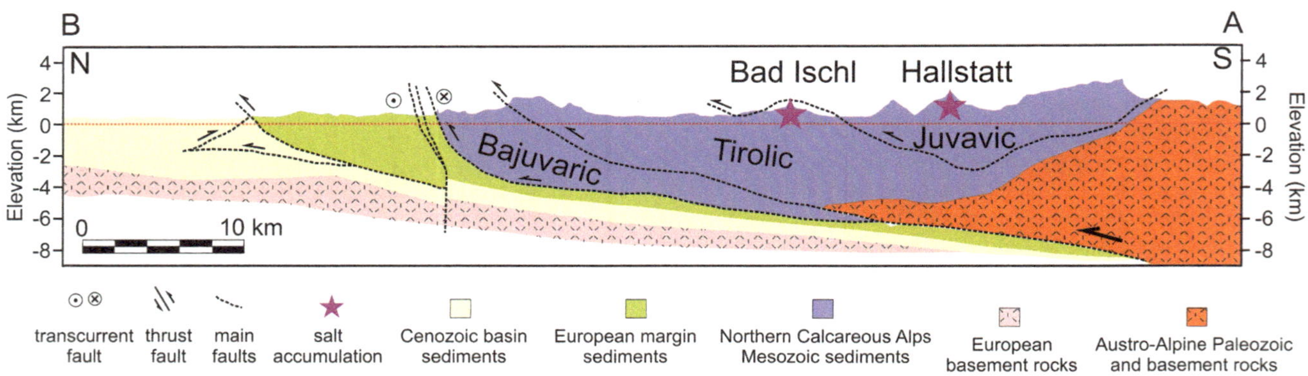

Figure 6.29 A regional schematic geological cross-section shows the Austro-Alpine Paleozoic and basement rocks southward of the North Calcareous Alps and the European margin sediments and molasse basin across southern Germany and northern Austria. North of the highlands, the European margin and the Molasse basins are characterized by Cenozoic sediments accumulated in foreland basins. The cross-section shows some nappes detaching on several levels within the Mesozoic stratigraphic sequences, forming three main units: the Juvavic, which hosts the Hallstatt salt diapir; the Tirolic, with the Bad Ischl salt occurrence; and the Bajuvanic, at the boundary with the European margin sediments. The highest elevations in the Eastern Alps correspond to the carbonate rocks of the Juvavic unit, south of the Hallstatt structure. Basement rocks crop out in the Eastern Alps, associated with basement-involved thrust fault zones. Credit: Adapted from Fernández et al. (2021).

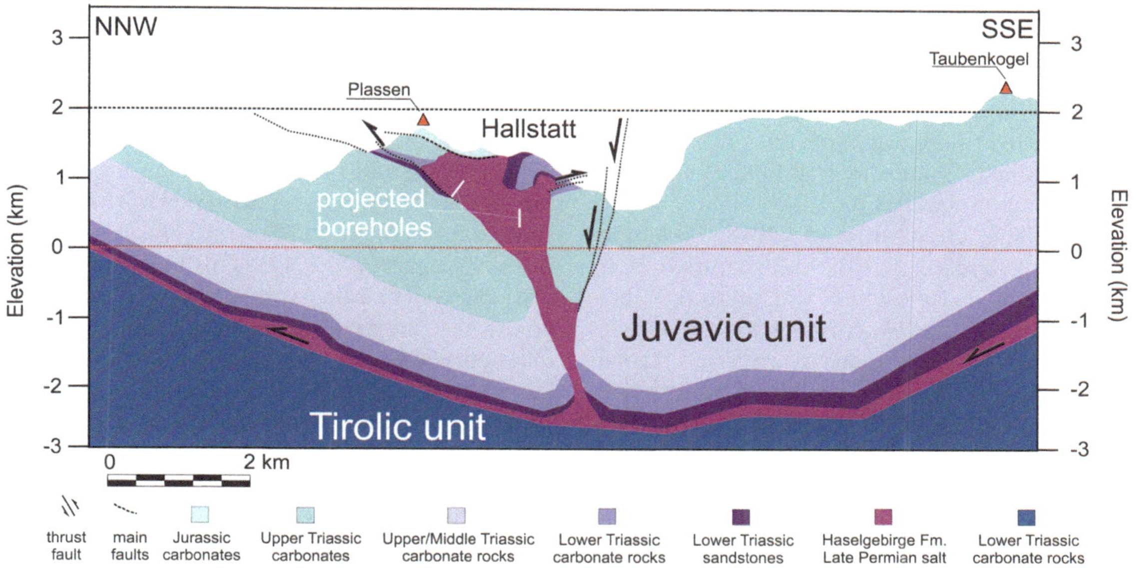

Figure 6.30 Schematic geological cross-section across the Hallstatt salt structure, interpreted as a squeezed salt structure surrounded by Triassic carbonate rocks. The thrust faults are marked by a northward vergence, resulting in high elevations to the south. The exploratory boreholes have penetrated only the top of the diapiric structure. Credit: Adapted from Fernández et al. (2021).

that the contraction of the previously developed salt diapirs resulted from the Alpine Orogeny during the Late Mesozoic and Cenozoic (Figure 6.30).

The Haselgebirge Formation represents a tectonic mélange formed during the Alpine Orogeny and consists of several proportions of mud and salt, varying from less than 10% of halite, resulting in mudrock, to more than 90% of halite, resulting in rock salt. Rocks with 10–70% halite are classified as Haselgebirge, and rocks with 70–90% halite are classified as Kerngebirge. Subordinate rocks include other evaporites (anhydrite and polyhalite), clastic, and volcanic rocks. The evaporitic lithologies exposed in the salt mines form a two-component tectonite, ranging from mudstone to salt rock. These rocks display a wide range of fabrics, mostly corresponding to crushed breccia, protocataclasite, mylonite, and ultramylonite (Leitner & Neubuer, 2011). The Haselgebirge originated from the interaction of halite and mudrock under tectonic stress, with the recrystallization of halite grains and the hydro-fracturing of the mudrock (Leitner et al., 2011).

The Haselgebirge Formation provides a unique natural laboratory for studying the rheology of halite and the interaction between mudstone and salt rock. Petrographic studies show that halite displacive cubes crystallized in soft mud during evaporation after salt precipitation. These cubes were also deformed during the initial stage of compaction of the surrounding mud. This is indicated by the stress direction, which is perpendicular (vertical) to the sedimentary layering. As muds compact within the upper tens to hundreds of meters of sedimentary loading, the halite cubes are also deformed during this early stage of compaction (Leitner et al., 2021).

Some early laboratory studies indicated that, under very low stress and the presence of water or saturated brine, the halite crystals easily deformed without breaking, maintaining a cuboid form (Leitner et al., 2021). These deformed, displacive cubes were later protected by lithified mudstone during halokinetic and orogenic stresses, which occurred during various stages of halite movement and compressional deformation associated with the Alpine Orogeny (Leitner & Neubauer, 2011; Leitner & Spötl, 2017; Leitner et al., 2014, 2017).

Granado et al. (2018) and Fernández et al. (2021) suggest that during the Triassic, the first submarine diapirs formed salt structures with shallow tops due to halokinesis and downbuilding processes. Limestone layers accumulated on top of these diapirs and suffered redissolution. Thicker sequences of siliciclastic and carbonate rocks were deposited in mini-basins between the salt. The richness and diversity of ammonite fossils accumulated on top of salt structures and within mini-basins in the Hallstatt region based on Jurassic stratigraphy divisions. The sedimentary sequences in the Hallstatt region are associated with evaporite sediments, with shallow and deep-water sedimentary facies intercalated with salt layers.

The Alps were subjected to extension during the Early Triassic, leading to a rift system and subsiding continental margins that were associated with the opening of the Tethys Ocean (see Chapter 3). A subsiding continental margin with a thick carbonate platform developed from the Middle Triassic until the Middle Jurassic (Leitner & Spötl, 2017). The Meliata Ocean, corresponding to the western part of the Tethys Ocean, started to close by the Late Jurassic. During the Late Jurassic and Cretaceous, the basin was affected by contractional tectonics, inverting previously developed extensional faults (Leitner & Neubauer, 2011; Leitner & Spötl, 2017).

Some authors (e.g., Granado et al., 2018; Fernández et al., 2021) have proposed that the salt diapirs initially grew by downbuilding during the Triassic and, subsequently, during the phase of compressional tectonics in the Jurassic, formed active piercements when the salt diapirs rose actively by isostasy or lateral contraction. During the upbuilding phase, the diapir stems were squeezed and the salt was extruded on the deep-water Jurassic facies, forming salt glaciers similar to the models proposed for the outcrops in Iran and Tunisia (see Chapter 3). The largest salt bodies in the North Calcareous Alps are surrounded by Triassic and Upper Jurassic carbonate rocks, indicating the evaporite rocks are displaced from the mother salt layer and extruded onto the deep marine carbonate facies (Leitner & Spötl, 2017).

Finally, during the Late Cretaceous and Paleogene, the Alpine orogeny culminated in the thrusting of huge limestone blocks that had previously separated the salt diapirs, forming nappes. Fernández et al. (2021) suggested that some thrust faults affecting the Triassic carbonate units detach on evaporitic layers intercalated within the carbonate sequences (Figure 6.30). In the outcrops, the shear zones separating the carbonate layers exhibit only residual salt, forming welds (Fernández et al., 2021).

Similar to other halite deposits deformed by ductile mechanisms worldwide, microscopic analysis of the deformed Alpine salt in the Hallstatt salt diapir indicates subgrain rotation and recrystallization, implying grain size reduction, and solution/precipitation, suggesting a simultaneous grain size increase and grain boundary migration. Some grains show evidence of one-sided growth, while others are decomposed into subgrains on the other end. Subsequently, the subgrains started to grow and increase in size again (Leitner et al., 2011).

The maximum differential stress ever measured worldwide in halite-rich sediments is about 5 MPa, measured by grain size piezometric techniques on subgrains from the Austrian salt deposits (Leitner et al., 2011, 2021). Based on the paleothermometric techniques applied to the Alpine salt, which assume a relationship between the average grain size, differential stress, and temperature, Ter Heege et al. (2005a, 2005b) deduced a temperature in the range of 100–150 °C during this deformational event (Leitner et al., 2011).

The salt layers recorded many deformation episodes, but some authors postulate that the macroscale structures seen today are mainly associated with the Late Cretaceous and younger events (Leitner & Spötl, 2017). Halite was deformed in a ductile style, similar to high-grade metamorphic rocks in mylonite zones. On the other hand, mudstone is deformed in a brittle style and is characterized by fracturing and faulting, breaking up into pieces that locally reach a few meters in diameter. However, the shear zone rocks were ground down to gravel or sand-size particles (Leitner et al., 2011). This mechanism developed a two-component cataclasite, consisting of the halite and mudstone end-members. The ductile structures include foliation planes and shear sense indicators. The tectonic fabrics can be quantified utilizing classical structural geology methods: foliation, halite mineral lineation, shear sense, fold axes, and fibers in strain shadows. Additionally, the mudstone exposes structures of brittle deformation, such as extension veins, fibers in veins, and several generations of fibers in veins, indicating several opening stages of the veins (Leitner et al., 2011, 2017).

Finally, the deformation of the Haselgebirge salt has also been induced by modern mining. Galleries excavated in halite layers are getting smaller and can disappear totally within a few decades. Caverns of some meters in height may disappear within about 200 years by collapsing. The structures associated with the current halite creep are difficult to see, except for the borehole closure. However, mudstone is still reacting to stress through brittle failure, as evidenced by pieces of mudstone that break off from walls according to the stress field around the open caverns. Chips of mudstone are peeled off in parallel to the excavation walls. In the gaps, fresh efflorescent halite is growing due to the moisture, further destroying the mudstone and the gallery walls (Leitner et al., 2014).

6.8. FLOW REGIMES AND FLUID MECHANICS OF HIGH-SALINITY BRINES

Fluid flow is a crucial aspect of hydrogeology, particularly in petroleum geology and basin analysis. It significantly impacts the thermal, structural, and diagenetic evolution of sedimentary basins (e.g., Bjørlykke, 1993). The paragenesis of valuable minerals can be linked to saline fluids percolating through sedimentary rocks, especially when tectonic events result in overpressuring the fluid system (Eugster, 1985). Many ore deposits, including valuable metals like lead, copper, and zinc, are associated with hydrothermal fluids, as evidenced by the accumulation of hot brines and metal sulfides on the bottom of the Red Sea axial trough (Laurila et al., 2014; Warren, 2000, 2006).

Fluid overpressure occurs when the pressure relative to the hydrostatic pressure is increased, causing pore fluids to support the weight of the rock column above. Fluid overpressure can be associated either with a reduction in pore space (compaction) or with an increase in the fluid volume due to some chemical reactions (e.g., Bjørlykke, 1993).

Fluid volume increases in sedimentary basins can be caused by thermal expansion, hydrocarbon generation, or mineral dewatering, such as the transformation of gypsum into anhydrite (Davison, 2009). When the rate of change in pore space or fluid volume exceeds the reservoir rock's capacity to expel fluids, the process results in overpressure (Bjørlykke, 1993).

Fluid overpressure in sedimentary basins may be generated by compressional deformation associated with orogenic belts (e.g., Chi & Xue, 2011). In compressional settings, the fluid flow system is driven by topographic relief, with recharge areas in more elevated areas and discharge areas in depressions at the distal parts of the basin. Fluid overpressure may also result from magmatic intrusions and fluid exsolution from magma chambers that rise to the upper crust, where the pressure is decreased. Pervasive fluid flow may occur in the deeper, overpressured portions of sedimentary basins. These fluids tend to rise along fractured rocks and fault zones that behave as pathways for fluid migration, which may connect with low-pressure zones that act as suction channels for focused flow (Chi & Xue, 2011).

The hydraulic potential is related to the height difference between the recharge and discharge points, so topographic relief is a major driving force for fluid flow in sedimentary basins (Bjørlykke, 1993; Hubbert, 1940). Water flows downward from high-altitude aquifers in recharge areas to depressions in the basin, while fluids also flow upward in discharge areas, for example, from the basin depocenter to the basin border (Chi & Xue, 2011).

The hydraulic potential gradient can be induced by changes in fluid density, such as increasing salt concentration or generating hydrocarbons. This can result in buoyancy, another force causing fluid flow. Two major parameters control the density of fluids in sedimentary basin reservoirs: the temperature and the salinity. Fluid density decreases with increasing temperature, particularly if the fluid is oil, and increases with increasing brine salinity (Bjørlykke, 1993; Chi & Xue, 2011). Brine density depends on the chemical composition and concentration of common evaporite ions. Salt dissolution in subaqueous settings can result in brine pools at the bottom of basins, as locally observed in the GOM (see Chapter 8 in Mohriak, 2025).

Temperature-related density changes in sedimentary basins can be attributed to normal temperature increases or local thermal anomalies, often linked to magmatic activity like volcanism or igneous intrusions at depth (e.g., Chi & Xue, 2011). Fluid flow can form convection cells, moving upward or downward, similar to the mantle processes, where convection currents move upward in spreading centers and downward in subduction zones.

In sedimentary basins where a magmatic intrusion is present, the thermal anomaly results in increasing the pressure and hydraulic potential above the intrusion, whereas the base of the intrusion is characterized by a lower hydraulic potential. The fluids tend to move horizontally toward the base of the intrusion and upward, away from the top of the intrusion, forming two convection cells. Salt diapirs display the opposite behavior. Due to salt dissolution, fluids become denser above and adjacent to the salt diapir (Ranganathan & Hanor, 1988). The convection cells are characterized by downward movement over the top and along the flanks of salt structures, moving horizontally outward at their base (e.g., Chi & Xue, 2011).

6.9. FLUID HYDRODYNAMICS AND MINERALIZATIONS IN EVAPORITE BASINS

Hydrodynamic indicators like hydraulic fractures, hydrothermal veins, clastic injection structures, and breccias are crucial for analyzing fluid flow in evaporitic sedimentary basins (e.g., Babel & Schreiber, 2014; Batchelor, 2000; Chi & Xue, 2011; Cosgrove, 2015; Sonnenfeld & Perthuisot, 1989). Hydraulic fractures are linked to increasing pore pressure in reservoirs, similar to the artificial fracturing method that is used by the petroleum industry to increase permeability and produce hydrocarbons trapped in shales. This process involves pumping high-pressure fluids (especially water) into reservoirs or low-permeability source rocks, such as shales and silts. When fluid pressure exceeds the minimum principal stress, tensile fractures form, increasing the rock's capacity to transmit fluids. This fracturing can lead to fluid migration, ore, and hydrocarbon accumulations (Cosgrove, 1998, 2015).

Fluid flow is also important for fracture deformation and porosity changes in reservoir rocks that may host hydrothermal mineral deposits (e.g., Cosgrove, 1998, 2015). Carbonate rocks, in particular, are highly susceptible to develop enhanced porosity due to hypersaline or hydrothermal fluids. The hydrothermal overpressured fluid model has been applied to important ore deposits with sulfide minerals, such as copper, lead, and zinc, in the Mississippi Valley and other basins worldwide (Leach et al., 2005, 2010).

The Mississippi Valley-type ore deposits are formed by large-scale fluid flow caused by tectonic squeezing of hypersaline formation waters from sedimentary succession. These fluids migrate updip, carrying dissolved ions that precipitate as ore minerals in carbonate host rocks. This mineralization is often associated with compressional thrusting events that expulse saline brines from evaporite sequences (Ridley, 2013). The brines migrate along fault zones and precipitate the ore minerals in porous host sediments, particularly platform carbonates and cap rocks on salt diapirs.

Leach and Rowan (1986) suggested that hot saline fluids expelled from sedimentary basins affected by orogenetic events, such as the compressional tectonics related to the Ouachita thrust and fold belt in the southern U.S., may provide a genetic link with the Mississippi Valley-type lead-zinc deposits of the uplifted Ozark Mountains, north of the Permian Basin. Some authors postulate that hydrothermal fluids flowing through carbonate rocks control the development of hydrothermal dolomites. This hypothesis has been proposed for several major productive Paleozoic and Mesozoic petroleum reservoirs in the Middle East. For example, Cantrell et al. (2001) speculated that the super-giant Ghawar oil field in Saudi Arabia displays enhanced porosity due to the circulation of hypersaline or hydrothermal fluids through the carbonate rocks.

6.9.1. Hydrothermal Fluids and the Genesis of Emeralds in Evaporite Sediments

There are several examples of mineralization-related ores to the subsurface flow of hypersaline fluids through sedimentary and metamorphic rocks in the Andean region of South America. According to some authors (e.g., Giuliani et al., 1995, 2018, 2019), high-salinity fluids, resulting from the dissolution of evaporites, are believed to be associated with the formation of emerald deposits in Colombia. These semiprecious gemstones were previously thought to be formed by magmatic fluids from Andean igneous plutons (Banks et al., 2000; Branquet, Cheilletz, Giuliani, Laumonier, & Blanco, 1999; Branquet, Laumonier, Cheilletz, & Giuliani, 1999; Laumonier, et al., 1999). Until the mid-1990s, no work had been published to explain the emeralds' origin as associated with evaporite sediments. However, recent research suggests that hydrothermal fluids that affect evaporitic sedimentary rocks are responsible for the genesis of these valuable gemstones, postulating that they are related to hydrothermal fluids interacting with evaporitic sedimentary rocks (Branquet, Cheilletz, Giuliani, Laumonier, & Blanco, 1999; Branquet, Laumonier, Cheilletz, & Giuliani, 1999; Cheilletz & Giuliani, 1996).

Emerald gems are silicates of aluminum and beryl, colored with a beautiful green tint by trace amounts of chromium and sometimes vanadium. In Brazil, these gems are possibly linked with Precambrian orogenies that formed imbricated structures affecting biotite schists and ultrabasic rocks, which reacted with fluids derived from pegmatitic veins (Giuliani et al., 1990). However, several geochemical studies based on fluid inclusions in quartz, calcite, dolomite, fluorite, and even emerald crystals have highlighted the importance of hypersaline and hydrothermal fluids in the genesis of Colombian gemstones. These fluids formed mineralized veins in the Eastern Cordillera emerald belts, which have been exploited for more than a century in two distinct zones located about 80 km apart, north of the country's capital, Bogotá (Banks et al., 2000).

Colombia is the world's largest producer of emeralds, accounting for over half of the global production. Major emerald deposits are found in the Central Cordillera and Eastern Cordillera provinces, both linked to thrust fault zones (Figures 6.31 and 6.32). The gems are hosted in breccias and black shales deposited during an Early Cretaceous rift event that affected the Magdalena Basin, a major back-arc basin associated with the Nazca plate subduction zone under South America (Branquet, Cheilletz, Giuliani, Laumonier, & Blanco, 1999; Branquet, Laumonier, Cheilletz, & Giuliani, 1999; Cooper et al., 1995).

Emerald mining districts in Colombia are situated west and east of the Eastern Cordillera (Figure 6.31). The finest Colombian emeralds are found in Muzo and Coscuez (Western Zone) and Chivor and Gachalá (Eastern Zone). These emerald ores occur within a breccia level at the lower part of the Early Cretaceous black shales, which have been uplifted by compressional tectonics. The Early Cretaceous rift sequences have been uplifted by thrust faults, resulting in inverted rift troughs adjacent to the Magdalena Basin (western zone) and Llanos Basin (eastern zone), near the outer edges of the Eastern Cordillera (Figure 6.31). These inverted rift troughs are overlain by the Late Cretaceous and Tertiary sedimentary successions (Cooper et al., 1995).

After the Early Cretaceous rift phase, the Eastern Cordillera region experienced a significant magmatic event at the Late Cretaceous/Paleogene transition (circa 66 Ma), resulting in several plutonic intrusions in the Central Cordillera (Bayona et al., 2012). This magmatism resulted in hydrothermal fluids that affected the Early Cretaceous evaporitic breccias, which corresponded to remnants of several evaporite layers. The emerald deposits were formed from salt ablation and the dissolution of halite layers during a later hydrothermal event related to Paleogene magmatism. Heated fluids migrated in the subsurface using the faults as conduits and probably dissolved the halite layers (Branquet, Cheilletz, Giuliani, Laumonier, & Blanco, 1999; Branquet, Laumonier, Cheilletz, & Giuliani, 1999; Cheilletz & Giuliani, 1996).

The Lower Cretaceous sediments in the basin depocenters were uplifted by folds and thrust faults due to tectonic inversion, resulting in anticlinal and synclinal structures, with the axes trending predominantly along the NE direction (Figure 6.32). However, the total shortening in the Eastern Cordillera is modest, around 25–30%, and was

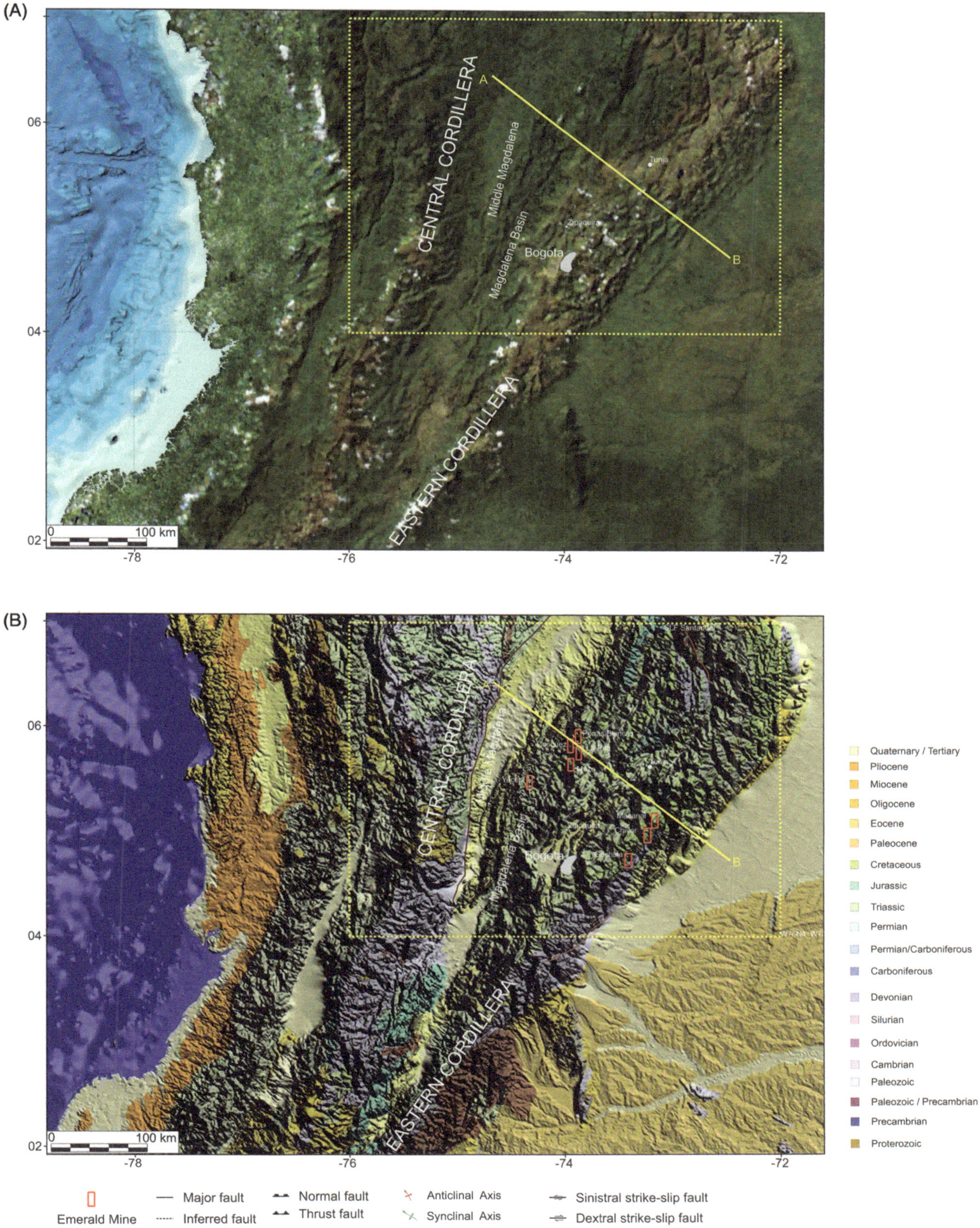

Figure 6.31 Satellite image of the Andean Cordillera and a simplified geological map. (A) Regional satellite image of the NW Colombia Andean region, with the Eastern Cordillera separating the Magdalena and Llanos basins. (B) Regional geological map with ages of the main lithological unities that outcrop in this segment of the Andes. The location of the cross-section A-B is shown with a thick yellow line. The zoom area of the Eastern Cordillera, where emerald mines have been found (red rectangles), is highlighted with a thin dashed yellow line.

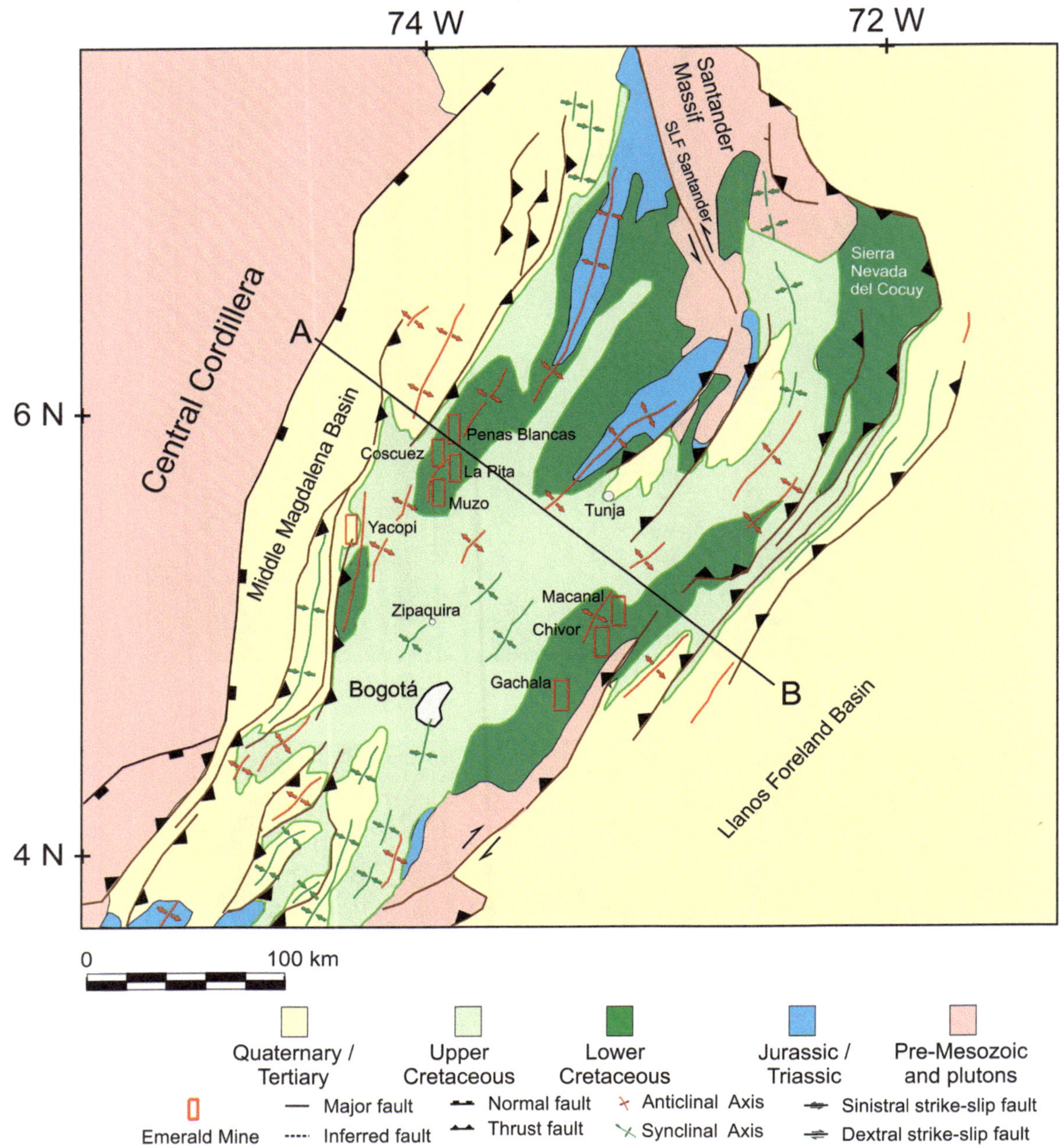

Figure 6.32 Simplified geological map with the main lithological units that outcrop in the Eastern Cordillera. The cross-section A-B extends from the Middle Magdalena Basin, eastward of the Central Cordillera massif, to the Llanos foreland basin. Observe the anticlinorium system affecting the Cretaceous sedimentary sequences, with the fold axes often trending in a NE direction. Compressional tectonics in the Tertiary inverted the rift basin, resulting in thrust faults with opposite polarities affecting the Cretaceous sediments. The emerald mines are located on the Lower Cretaceous outcrops north of Bogotá (the Western Zone of mineralization) and to the northeast of Bogotá (the Eastern Zone of mineralization). The Zipaquirá salt mine is located to the north of Bogotá. Halite minerals from a large salt diapir were exploited for centuries until the mine was converted into a tourist attraction. Credit: Adapted from Giuliani et al. (2019).

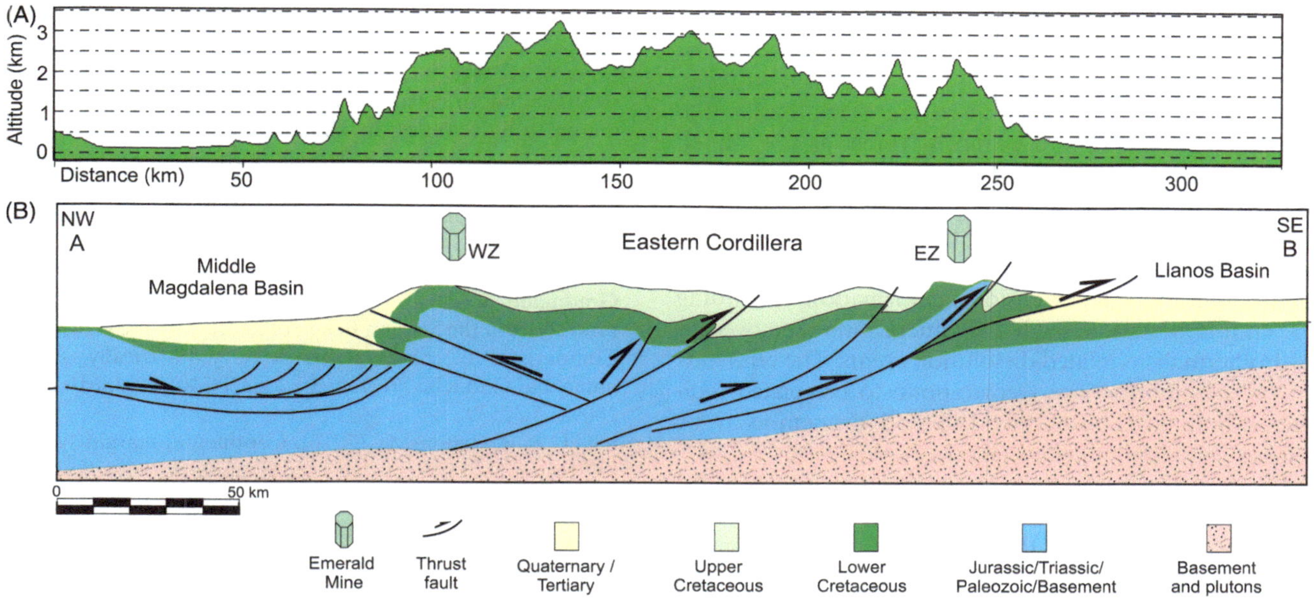

Figure 6.33 Regional profiles across the Eastern Cordillera. (A) Regional topographic profile along the geological cross-section A-B from the Middle Magdalena Basin to the Llanos Basin, across the Eastern Cordillera highlands, which reach more than 3,000 m at the highest peaks. (B) Schematic geological cross-section across the Eastern Cordillera, from the Magdalena to the Llanos basins. The section shows the system of reverse faults with opposite polarities and the emerald mineralization zones (West Zone and East Zone). The emerald ores are hosted in the Lower Cretaceous breccias, which are associated with evaporite sediments affected by hydrothermal fluids. Credit: Adapted from Branquet et al. (1999).

accommodated by reactivation and inversion of previously normal faults, with basement-involved buckling, folding, and footwall shortcuts (Tesón et al., 2013).

The Lower Cretaceous syn-rift depocenters were uplifted and inverted during a major compressional event in the Late Eocene–Early Oligocene, which lasted until the Middle Miocene (Tesón et al., 2013). This event resulted in contractional tectonics and large anticlinoria along the Eastern Cordillera flanks, which are characterized by thrust faults with opposite vergences (Figure 6.33). The fold-thrust belt overthrusts two foreland basins: the middle Magdalena basin to the west, characterized by fault vergence toward the Central Cordillera, and the eastern belt of the Eastern Cordillera, marked by fault vergence toward the Llanos Basin (Babault et al., 2013; Branquet, Cheilletz, Giuliani, Laumonier, & Blanco, 1999; Branquet, Laumonier, Cheilletz, & Giuliani, 1999; Cooper et al., 1995).

The Lower Cretaceous stratigraphic level in the Andes contains evidence of sabkha evaporitic sediments that were destabilized and reworked on an active slope, forming stratiform breccias that host emeralds. Subvertical veins that are mineralized by the hydrothermal fluids probably represent listric faults that detached on the breccia horizon. The main breccia level likely constituted a subhorizontal detachment plane associated with halite that was dissolved by hydrothermal fluids (Branquet, Cheilletz, Giuliani, Laumonier, & Blanco, 1999; Branquet, Laumonier, Cheilletz, & Giuliani, 1999). Some dissolved evaporites developed bedding-parallel fibrous calcite and pyrite veins, known as "beef" and indicative of fluid overpressure (Cobbold et al., 2013).

These deep, hot, and overpressurized fluids are responsible for emerald mineralization. These fluids mixed with basinal brines derived from the dissolution of evaporite layers, which were probably covered by a thick overburden in the Late Cretaceous, estimated to comprise at least 5–6 km of sediments (Branquet et al., 2015). The hydrothermal event predated the main Andean phase of compressional tectonics and magmatism (Middle Miocene). During the Miocene orogenic climax, the Middle Magdalena Basin was uplifted, folded, and thrust over the Llanos foreland to the east (Cooper et al., 1995). Several authors have suggested that compressional tectonics in the Middle Magdalen Basin utilized previously extensional faults of the Early Cretaceous rift event, which were reactivated as thrust structures during the basin inversion episodes in both the Late Cretaceous and the Middle Miocene (e.g., Caballero et al., 2013).

This emerald deposit type is unique due to its petrogenesis associated with hypersaline hydrothermal fluids. Its mesothermal mineralization (brines around 300 °C) occurs in evaporitic sedimentary sequences (Giuliani et al., 2018, 2019). The chemical signatures of the emeralds indicate that they result from fluid dynamics and precipitation mechanisms involving the thermochemical reduction of sulfate-rich brines to hydrogen sulfide. The chemical compounds involve the interaction with organic-rich strata, which correspond to the black shales that underlie the mineralized evaporitic breccias (Branquet, Cheilletz, Giuliani, Laumonier, & Blanco, 1999; Branquet, Laumonier, Cheilletz, & Giuliani, 1999).

Banks et al. (2000) suggested that the mineralization in the eastern belt occurred around 65 Ma, during a phase of salt migration and extensional deformation. According to this interpretation, the halite minerals were removed by hydrothermalism-related dissolution events. The remnants of these evaporitic layers correspond to brecciated sediments that have been reworked and dismembered by subsequent tectonic events (Banks et al., 2000).

Emerald deposits in the western belt, specifically in the Muzo and Coscuez mines, are found in Lower Cretaceous brecciated black shales that deformed under low-grade metamorphic conditions. However, the mineralization in this location is considered a much later event, associated with a Neogene compressive tectonic phase at circa 35 Ma (Banks et al., 2000). The formation of emeralds is linked to thrust and reverse faults that carried hypersaline brines through halite-bearing sedimentary layers.

Emerald mineralizations in Colombia are located at the front of thrust faults, which facilitated the flow of hydrothermal fluids that dissolved halite from the Lower Cretaceous breccias. In some areas of the Eastern Cordillera fold-thrust belt, halite layers have been preserved from dissolution, forming large diapiric salt structures like the Zipaquirá salt mine, located north of Bogotá. Similar to its European counterparts, the old salt mine in Colombia has been converted into a famous tourist attraction by carving a cathedral inside the salt diapir.

REFERENCES

Aadnøy, B. S., & Looyeh, R. (2019). Rock strength and rock failure. In B. Aadnøy & R. Looyeh, *Petroleum rock mechanics: Drilling operations and well design* (2nd ed., pp. 145–163). Cambridge, MA: Gulf Professional Publishing. doi:10.1016/b978-0-12-815903-3.00009-1

Aftabi, P., Roustaie, M. G., Alsop, I. G., & Talbot, C. J. (2010). InSAR mapping and modelling of an active Iranian salt extrusion. *Journal of the Geological Society, London*, *167*, 155–170. doi:10.1144/0016-76492008-165.

Alavi, M. (1994). Tectonics of the Zagros erogenic belt of Iran: New data and interpretations. *Tectonophysics*, *229*, 211–238.

Alavi, M. (1996). Tectonostratigraphic synthesis and structural style of the Alborz mountain system in northern Iran. *Journal of Geodynamics*, *21*, 1–33.

Albertz, M., & Beaumont, C. (2010). An investigation of salt tectonic structural styles in the Scotian Basin, offshore Atlantic Canada: 2. Comparison of observations with geometrically complex numerical models. *Tectonics*, *29*, TC4018. doi:10.1029/2009TC002540.

Albertz, M., & Ings, S. (2012). Some consequences of mechanical stratification in basin-scale numerical models of passive margin salt tectonics. In G. I. Alsop, S. G. Archer, A. J. Hartley, N. T. Grant, & R. Hodgkinson (Eds.), *Salt tectonics, sediments and prospectivity* (Vol. *363*, pp. 303–330). Geological Society, London, Special Publications. http://dx.doi.org/10.1144/SP363.14

Albertz, M., Beaumont, C., Shimeld, J. W., Ings, S. J., & Gradmann, S. (2010). An investigation of salt tectonic structural styles in the Scotian Basin, offshore Atlantic Canada: 1. Comparison of observations with geometrically simple numerical models. *Tectonics*, *29*, TC4017. doi:10.1029/2009TC002539.

Allen, J., & Beaumont, C. (2016). Continental margin syn-rift salt tectonics at intermediate width margins. *Basin Research*, *28*, 598–633. doi: 10.1111/bre.12123

Allen, J., Beaumont, C., & Deptuck, M. E. (2020). Feedback between synrift lithospheric extension, sedimentation and salt tectonics on wide, weak continental margins. *Petroleum Geoscience*, *26*(1), 16–35. https://doi.org/10.1144/petgeo2018-089

Alves, T. M., Fetter, M., Lima, C., Cartwright, J. A., Cosgrove, J., Gangá, A., et al. (2017). An incomplete correlation between pre-salt topography, top reservoir erosion, and salt deformation in deep-water Santos Basin (SE Brazil). *Marine and Petroleum Geology*, *79*, 300–320.

Arian, M., & Noroozpour, H. (2015). Tectonic geomorphology of Iran's salt structures. *Open Journal of Geology*, *5*, 61–72. http://dx.doi.org/10.4236/ojg.2015.52006

Baar, C. A. (1977). *Applied salt-rock mechanics: The in-situ behavior of salt rocks*. Amsterdam: Elsevier, 294 pp.

Babault, J., Teixell, A., Struth, L., Driessche, J. V. D., Arboleya, M. L., & Tesón, E. (2013). Shortening, structural relief and drainage evolution in inverted rifts: Insights from the Atlas Mountains, the Eastern Cordillera of Colombia and the Pyrenees. In M. Nemcok, A. Mora, & J. W. Cosgrove (Eds.), *Thick-skin-dominated orogens: From initial inversion to full accretion* (Vol. *377*, pp. 141–157). Geological Society, London, Special Publications.

Babel, M., & Schreiber, B. C. (2014). Geochemistry of evaporites and evolution of seawater. *Treatise on Geochemistry* (2nd ed., Chapter 9.17, pp. 483–560). Amsterdam: Elsevier.

Baikpour, S., & Talbot, C. J. (2012). The Garmsar salt nappe and seasonal inversions of surrounding faults imaged by SAR interferometry, Northern Iran. In G. I. Alsop, S. G. Archer, A. J. Hartley, N. T. Grant, & R. Hodgkinson (Eds.), *Salt tectonics, sediments and prospectivity* (Vol. *363*, pp. 563–578). Geological Society, London, Special Publications. doi: 10.1144/SP363.28

Baikpour, S., Zulauf, G., Sebti, A., Kheirolahi, H., & Dietl, C. (2010). Analogue and geophysical modelling of the Garmsar Salt Nappe, Iran: Constraints on the evolution of the Alborz Mountains. *Geophysical Journal International*, *182*, 599–612.

Banks, D. A., Giuliani, G., Yardley, B. W. D., & Cheilletz, A. (2000). Emerald mineralization in Colombia: Fluid chemistry and the role of brine mixing. *Mineralium Deposita*, *35*, 699–713.

Barabasch, J., Schmatz, J., Klaver, J., Schwedt, A., & Urai, J. L. (2023). Large grain-size-dependent rheology contrasts of halite at low differential stress: Evidence from microstructural study of naturally deformed gneissic Zechstein 2 rock salt (Kristallbrockensalz) from the northern Netherlands. *Solid Earth*, *14*, 271–291. https://doi.org/10.5194/se-14-271-2023

Barton, D. C. (1933). Mechanics of formation of salt domes with special reference to Gulf Coast salt domes of Texas and Louisiana. *American Association of Petroleum Geologists Bulletin*, *17*(9), 1025–1083.

Batchelor, G. K. (2000). *An introduction to Fluid Dynamics*. Cambridge, UK: Cambridge University Press, 631 pp.

Bayona, G., Cardona, A., Jaramillo, C., Mora, A., Montes, C., Valencia, V., et al. (2012). Early Paleogene magmatism in the northern Andes: Insights on the effects of Oceanic Plateau–Continent convergence. *Earth and Planetary Science Letters*, *331–332*, 97–111.

Bell, F. G. (1981). Geotechnical properties of some evaporitic rocks. *Bulletin of the International Association of Engineering Geology*, *24*(1), 137–144. doi:10.1007/bf02595264

BGE (2022). History of the Asse II mine. https://www.bge.de/en/asse/short-information/history-of-the-asse-ii-mine/

Bjørlykke, K. (1993). Fluid flow in sedimentary basins. In S. Cloetingh, W. Sassi, F. Horvath, & C. Puigdefabregas (Eds.), Basin analysis and dynamics of sedimentary basin evolution. *Sedimentary Geology*, *86*, 137–158.

Branquet, Y., Cheilletz, A., Giuliani, G., Laumonier, B., & Blanco, O. (1999). Fluidized hydrothermal breccia in dilatant faults during thrusting: The Colombian emerald deposits. In K. McCaffrey, L. Lonergan, & J. Wilkinson (Eds.), *Fractures, fluid and mineralization* (Vol. *155*, pp. 183–195). Geological Society, London, Special Publication.

Branquet, Y., Giuliani, G., Cheilletz, A., & Laumonier, B. (2015). Colombian emeralds and evaporites: Tectono-stratigraphic significance of a regional emerald-bearing evaporitic breccia level. *13th SGA Biennial Meeting, Nancy, France, Proceedings* (Vol. *4*, pp. 1291–1294).

Branquet, Y., Laumonier, B., Cheilletz, A., & Giuliani, G. (1999). Emeralds in the Eastern Cordillera of Colombia: Two tectonic settings for one mineralization. *Geology*, *27*(7), 597–600.

Brun, J. P., & Mauduit, T. P.-O. (2009). Salt rollers: Structure and kinematics from analogue modelling. *Marine and Petroleum Geology*, *26*, 249–258. doi:10.1016/j.marpetgeo.2008.02.002.s.

Caballero, V., Parra, M., Mora, A. S., López, C., Rojas, L. E., & Quintero, I. (2013). Factors controlling selective abandonment and reactivation in thick-skin orogens: A case study in the Magdalena Valley, Colombia. In M. Nemcok, A. Mora, & J. W. Cosgrove (Eds.), *Thick-skin-dominated orogens: From initial inversion to full accretion* (Vol. *377*, pp. 343–367). Geological Society, London, Special Publications.

Callot, J. P., Trocme, V., Letouzey, J., Albouy, E., Jahani, S., & Sherkati, S. (2012). Pre-existing salt structures and the folding of the Zagros Mountains. In G. I. Alsop, S. G. Archer, A. J. Hartley, N. T. Grant, & R. Hodgkinson (Eds.), *Salt tectonics, sediments and prospectivity* (Vol. *363*, pp. 545–561). Geological Society, London, Special Publications. http://dx.doi.org/10.1144/SP363.27

Cantrell, D. L., Swart, P. K., Handford, R. C., Kendall, C. G., & Westphal, H. (2001). Geology and production significance of dolomite, Arab-D Reservoir, Ghawar Field, *Saudi Arabia. GeoArabia*, *6*(1), 45–60.

Carter, N. L., & Hansen, F. D. (1983). Creep of rocksalt. *Tectonophysics*, *92*, 275–333.

Cartwright, J., Jackson, M., Dooley, T., & Higgins, S. (2012). Strain partitioning in gravity-driven shortening of a thick, multilayered evaporite sequence. In G. I. Alsop, S. G. Archer, A. J. Hartley, N. T. Grant, & R. Hodgkinson (Eds.), *Salt tectonics, sediments and prospectivity* (Vol. *363*, pp. 449–470). Geological Society of London, Special Publication. doi: 10.1144/SP363.21

Cheilletz, A., & Giuliani, G. (1996). The genesis of Colombian emeralds: A restatement. *Mineralium Deposita*, *31*, 359–364.

Chemia, Z., & Koyi, H. (2008). The control of salt supply on entrainment of an anhydrite layer within a salt diapir. *Journal of Structural Geology*, *30*, 1192–1200. https://doi.org/10.1016/j.jsg.2008.06.004.

Chi, G., & Xue, C. (2011). An overview of hydrodynamic studies of mineralization. *Geoscience Frontiers*, *2*(3), 423–438.

Chorlton, L. B. (2007). Generalized geology of the world: bedrock domains and major faults in GIS format: A small-scale world geology map with an extended geological attribute database. Geological Survey of Canada, Open File *5529*, 2007, 48 pages.

Cobbold, P. R., Zanella, A., Rodrigues, N., & Løseth, H. (2013). Bedding-parallel fibrous veins (beef and cone-in-cone): Worldwide occurrence and possible significance in terms of fluid overpressure, hydrocarbon generation and mineralization. *Marine and Petroleum Geology*, *43*, 1–20.

Cooper, M. A., Addison, F. T., Alvarez, R., Coral, M., Graham, R. H., Hayward, A. B., et al. (1995). Basin development and tectonic history of the Llanos basin, Eastern Cordillera, and Middle Magdalena Valley, Colombia. *American Association of Petroleum Geologists Bulletin*, *79*(10), 1421–1443.

Cosgrove, J. W. (1998). The role of structural geology in reservoir characterization. In M. P. Coward, T. S. Daltaban, & H. Johnson (Eds.), *Structural geology in reservoir characterization* (Vol. *127*, pp. 1–13). Geological Society, London, Special Publications.

Cosgrove, J. W. (2015). The association of folds and fractures and the link between folding, fracturing and fluid flow during the evolution of a fold-thrust belt: A brief review. In F. L. Richards, N. J., Richardson, S. J., Rippington, R. W. Wilson, & C. E. Bond (Eds.), *Industrial structural geology: Principles, techniques and integration* (Vol. *421*, pp. 41–68). Geological Society, London, Special Publications.

Costa, E., Camerlenghi, A., Polonia, A., Cooper, C., Fabretti, P., Mosconi, A., et al. (2004). Modeling deformation and salt tectonics in the eastern Mediterranean Ridge accretionary wedge. *Geological Society of America Bulletin*, *116*(7), 880–894.

Coward, M. P. (1995). Structural and tectonic setting of the Permo-Triassic basins of northwest Europe. In S. A. R. Boldy (Ed.), *Permian and Triassic Rifting in Northwest Europe*

(Vol. *91*, pp. 7–39). Geological Society, London, Special Publications. https://doi.org/10.1144/GSL.SP.1995.091.01.02

Davison, I. (2009). Faulting and fluid flow through salt. *Journal of the Geological Society,* London, *166*, 205–216. doi: 10.1144/0016-76492008-064.

Davison, I., Bosence, D., Alsop, I., & Al-Aawah, M. H. (1995). Deformation and sedimentation above and below Miocene salt diapirs and their overhangs, NW Yemen. In C. J. Travis, H. Harrison, M. R. Hudec, B. C. Vendeville, F. J. Peel, & R. E. Perkins (Eds.), *Salt, sediment and hydrocarbons* (pp. 33–40), GCSSEPM Foundation, 16th Annual Research Conference, Abstracts.

Davison, I., Bosence, D., Alsop, I., & Al-Aawah, M. H. (1996). Deformation and sedimentation around active Miocene salt diapirs on the Tihama Plain, northwest Yemen. In I. Alsop, D. J. Blundell, & I. Davison (Eds.), *Salt tectonics* (Vol. *100*, pp. 23–39). Geological Society of London, Special Publication.

De Jager, J. (2003). Inverted basins in the Netherlands, similarities and differences. *Netherlands Journal of Geosciences (Geologie en Mijnbouw)*, *82*(04), 339–349. doi:10.1017/s0016774600020175

De Paola, N., Collettini, C., Faulkner, D. R., & Trippetta, F. (2008). Fault zone architecture and deformation processes within evaporitic rocks in the upper crust. *Tectonics*, *27*(4), TC4017, 21 pp. https://doi.org/10.1029/2007TC002230

Desbois, G., Závada, P., Schléder, Z., & Urai, J. L. (2010). Deformation and recrystallization mechanisms in actively extruding salt fountain: Microstructural evidence for a switch in deformation mechanisms with increased availability of meteoric water and decreased grain size (Qum Kuh, central Iran). *Journal of Structural Geology*, *32*, 580–594.

Dusseault, M. B., & Fordham, C. J. (1995). Time-dependent behavior of rocks. In Hudson, J. A. (Ed.), *Comprehensive rock engineering. Volume 3, Rock testing and site characterization* (pp. 119–149). Oxford, UK: Pergamon Press.

Dûsterloh, U., & Lux, K.-H. (2007). *Rock mechanical investigations on the strength and creep behaviour of salt and marl core samples from the Holford H401 location.* Clausthal University of Technology—Professorship for Waste Disposal Technologies and Geomechanics Laboratory Report Holford H401, 186 pp.

Duval, P., Montagnat, M., Grennerat, F., Weiss, J., Meyssonnier, J., & Philip, A. (2010). Creep and plasticity of glacier ice: A material science perspective. *Journal of Glaciology*, *56*(200), 1059–1068.doi:https://doi.org/10.3189/002214311796406185

Elliott, G. M. (1995). Triaxial testing for rock strength. In J. A. Hudson (Ed.), *Comprehensive rock engineering. Volume 3, Rock testing and site characterization* (pp. 87–104). Oxford, UK: Pergamon Press.

Eugster, H. P. (1985). Oil shales, evaporites and ore deposits. *Geochimica et Cosmochimica Acta*, *49*(3), 619–635. https://doi.org/10.1016/0016-7037(85)90158-9

Evans, S. L., & Jackson C.A.-L. (2021). Intrasalt structure and strain partitioning in layered evaporites: Implications for drilling through Messinian Salt in the eastern Mediterranean. *Petroleum Geoscience*, *27*(4), petgeo2020-072. https://doi.org/10.1144/petgeo2020-072

Fainstein, R., & Krueger, A. (2005). Salt tectonics comparisons near three continent-ocean boundary escarpments. In P. J. Post, N. C. Rosen, D. L. Olson, S. L. Palmes, K.T. Lyons, & G. B. Newton (Eds.), *GCSSEPM 25th Annual Bob F. Perkins Research Conference: Petroleum Systems of Divergent Continental Margin Basins*, Abstracts CD, pp. 510–540.

Fairhurst, C., John, C. M. St., & Midea, N. F. (1979). *Rock mechanics studies of proposed underground mining of Potash in Sergipe, Brazil.* Proceedings 4th Congress International Society for Rock Mechanics, Montreux, 2–8 Sept 1979, Vol. *1*, P131–137. Publ Rotterdam, A A Balkema.

Fernández, O., Habermüller, M., & Grasemann, B. (2021). Hooked on salt: Rethinking Alpine tectonics in Hallstatt (Eastern Alps, Austria). *Geology*, *49*, 325–329. https://doi.org/10.1130/G47981.1

Fjær, E., Holt, R. M., Horsrud, P., Raaen, A. M., & Risnes, R. (2021). Petroleum related rock mechanics. *Developments in Petroleum Science*, *72*, 1–662.

Fletcher, R. C., Hudec, M. R., & Watson, I. A. (1995). Salt glacier and composite sediment-salt glacier models for the emplacement and early burial of allochthonous salt sheets. In M. P. A. Jackson, D. G. Roberts, & S. Snelson (Eds.), *Salt tectonics: A global perspective, American Association of Petroleum Geologists Memoir* (Vol. *65*, pp. 77–108).

Fossen, H. (2016). *Structural geology* (2nd ed.). Cambridge, UK: Cambridge University Press, 524 pp.

Fossum, A. F., & Fredrich, J. T. (2002). *Salt mechanics primer for near-salt and sub-salt deepwater Gulf of Mexico field developments.* Sandia National Laboratories, Albuquerque, New Mexico, Sandia Report SAND2002-2063, 67 pp.

Frizon de Lamotte, D., Raulin, C., Mouchot, N., Wrobel-Daveau, J.-C., Blanpied, C., & Ringenbach, J.-C. (2011). The southernmost margin of the Tethys realm during the Mesozoic and Cenozoic: Initial geometry and timing of the inversion process. *Tectonics*, *30*, TC3002. http://dx.doi.org/10.1029/2010TC002691

Frumkin, A. (1996). Uplift rate relative to base-levels of a salt diapir (Dead Sea Basin, Israel) as indicated by cave levels. In G. I. Alsop, D. J. Blundell, & I. Davison (Eds.), *Salt tectonics* (Vol. *100*, pp. 41–47). Geological Society, Special Publication.

Fuenkajorn, K., Walsri, C., & Phueakphum, D. (2011). Intrinsic variability of the mechanical properties of Maha Sarakham salt. *Quarterly Journal of Engineering Geology and Hydrogeology*, *44*, 445–456. https://doi.org/10.1144/1470-9236/10-018

Gärtner, G., Stockmann, N., Fricke, H., & Wallmüller, R. (2008). Geological characterization of the Asse salt mine for mining, disposal of radioactive waste, and proof of long-term safety. In Rempe, N.T. (Ed.), *Deep geologic repositories, Geological Society of America, Reviews in Engineering Geology* (Vol. *XIX*, pp. 41–52). doi: 10.1130/2008.4119(05)

GEBCO (2019). General Bathymetric Chart of the Oceans, British Oceanographic Data Centre, U.K., website www.gebco.net.

Giuliani, G., Cheilletz, A., Arboleda, C., Carrillo, V., Rueda, F., & Baker, J. H. (1995). An evaporitic origin of the parent brines of Colombian emeralds: Fluid inclusion and sulphur isotope evidence. *European Journal of Mineralogy*, *7*(1), 151–166.

Giuliani, G., Dubessy, J., Ohnenstetter, D., Banks, D., Branquet, Y., Feneyro, J., et al. (2018). The role of evaporites in the formation of gems during metamorphism of carbonate platforms: A review. *Miner Deposita*, *53*, 1–20. doi 10.1007/s00126-017-0738-4

Giuliani, G., Groat, L. A., Marshall, D., Fallick, A. E., & Branquet, Y. (2019). Emerald deposits: A review and enhanced classification. *Minerals*, *9*, 105, 1–63. doi:10.3390/min9020105

Giuliani, G., Silva, L. J. H. D., & Couto, P. (1990). Origin of emerald deposits of Brazil. *Mineralium Deposita*, *25*, 57–64.

Godin, L., Grujic, D., Law, R. D., & Searle, M. P. (2006). Channel flow, ductile extrusion and exhumation in continental collision zones: An introduction. In R. D. Law, M. P. Searle, & L. Godin (Eds.), *Channel flow, ductile extrusion and exhumation in continental collision zones* (Vol. *268*, pp. 1–23), Geological Society, London, Special Publications,.

Granado, P., Roca, E., Strauss, P., Pelz, K., & Muñoz, J. A. (2018). Structural styles in fold-and-thrust belts involving early salt structures: The Northern Calcareous Alps (Austria). *Geology*, *47*(1), 51–54. https://doi.org/10.1130/G45281.1

Guest, B., Axen, G. J., Lam, P. S., & Hassanzadeh, J. (2006). Late Cenozoic shortening in the west-central Alborz Mountains, northern Iran, by combined conjugate strike-slip and thin-skinned deformation. *Geosphere*, *2*(1), 35–52.

Guest, B., Guest, A., & Grey, A. (2007). Late Tertiary tectonic evolution of northern Iran: A case for simple crustal folding. *Global Planetary Change*, *58*, 435–453.

Guest, B., Stockli, D. F., Grove, M., Axen, G. J., Lam, P. S., & Hassanzadeh, J. (2006). Thermal histories from the central Alborz Mountains, northern Iran: Implications for the spatial and temporal distribution of deformation in northern Iran. *Geological Society of America Bulletin*, *118*, 1507–1521.

Gussow, W. C. (1968). Salt diapirism: Importance of temperature, and energy source of emplacement. In J. Braunstein & G. D. O'Brien (Eds.), *Diapirism and diapirs, American Association of Petroleum Geologists Memoir* (Vol. *8*, pp. 16–52).

Hambrey, M. J., & Glasser, N. F. (2005). Glaciers. In R. C. Selley, L. R. Cocks, & I. R. Plimer (Eds.), *Encyclopedia of geology* (pp. 663–678). Amsterdam: Elsevier.

Hartmann, J., & Moosdorf, N. (2012). The new global lithological map database GLiM: A representation of rock properties at the Earth surface. *Geochemistry Geophysics Geosystems*, *13*, Q12004. doi:10.1029/2012GC004370

Hassanpour, J., Yassaghi, A., Muñoz, J. A., & Jahani, S. (2020). Salt tectonics in a double salt-source layer setting (Eastern Persian Gulf, Iran): Insights from interpretation of seismic profiles and sequential cross-section restoration. *Basin Research*. doi:10.1111/bre.12459

Heege, J. H. T., De Bresser, J. H. P., & Spier, C. J. (2005). Dynamic recrystallization of wet synthetic polycrystalline halite: Dependence of grain size distribution on flow stress, temperature and strain. *Tectonophysics*, *396*, 35–57.

Hovorka, S. D., & Nava, R. (2000). *Characterization of bedded salt for storage salt caverns—A case study from the Midland Basin, Texas*. National Petroleum Technology Office, U.S. Department of Energy, DOE/BC/15030-1, 115 pp.

Hubbert, M. K. (1940). The theory of ground-water motion. *Journal of Geology*, *48*, 785–944.

Hudec, M. R., & Jackson, M. P. A. (2007). Terra infirma: Understanding salt tectonics, *Earth-Science Reviews*, *82*(1–2), 1–28. doi: 10.1016/j.earscirev.2007.01.001

Hunsche, U. (1994). Uniaxial and triaxial creep and failure tests on rock: Experimental technique and interpretation. In N. D. Cristescu & G. Gioda (Eds.), *Visco-plastic behaviour of geomaterials* (Vol. *350*). International Centre for Mechanical Sciences. Vienna: Springer. https://doi.org/10.1007/978-3-7091-2710-0_1

Husseini, M. I., & Husseini, S. I. (1990). Origin of the Infra-Cambrian salt basin of the Middle East. In J. Brooks (Ed.), *Classic petroleum provinces* (Vol. *50*, pp. 279–292). Geological Society, London, Special Publication.

Ings, S., Beaumont, C., & Gemmer, L. (2004). Numerical modeling of salt tectonics on passive continental margins: preliminary assessment of the effects of sediment loading, buoyancy, margin tilt, and isostasy. In P. J. Post & N. C. Rosen (Eds.), *Salt–sediment interactions and hydrocarbon prospectivity: Concepts, applications, and case studies for the 21st century* (pp. 37–68). Society of Economic Paleontologists and Mineralogists Gulf Coast Section, 24th Annual Research Conference.

Jaberi, M., Ghassemi, M. R., Shayan, S., Yamani, M., & Zamanzadeh, S. M. (2018). Interaction between active tectonics, erosion and diapirism, a case study from Habble-Rud in Southern Central Alborz (Northern Iran). *Geomorphology*, *300*, 77–94.

Jackson, M. P. A., Cornelius, R. R., Craig, C. H., Gansser, A., Stöcklin, J., & Talbot, C. J. (1990). *Salt Diapirs of the Great Kavir, Central Iran. Geological Society of America Memoirs, Memoir 177*, 139 pp. doi: 10.1130/MEM177-p1

Jackson, M. P. A., & Hudec, M. R. (2017). *Salt tectonics: Principles and practice*. Cambridge, UK: Cambridge University Press, 510 p.

Jennings, S. J. A., & Hambrey, M. J. (2021). Structures and deformation in Glaciers and Ice Sheets. *Reviews of Geophysics*, *59*(3), e2021RG000743. https://doi.org/10.1029/2021RG000743

Jeremic, M. L. (1994). *Rock mechanics in salt mining*. Brookfield VT., Rotterdam, 532 pp.

Kent, P. E. (1958). Recent studies of south Persian salt plugs. *American Association of Petroleum Geologists Bulletin*, *42*(12), 2951–2972.

Kent, P. E. (1979). The emergent Hormuz salt plugs of southern Iran. *Journal of Petroleum Geology*, *2*(2), 117–144.

Kley, J. (2018). Timing and spatial patterns of Cretaceous and Cenozoic inversion in the Southern Permian Basin. In B. Kilhams, P. A. Kukla, S. Mazur, T. McKie, H. F. Mijnlieff, K. Van Ojik, & E. Rosendaal (Eds.), *Mesozoic resource potential in the Southern Permian Basin area: The geological key to exploiting remaining hydrocarbons whilst unlocking geothermal potential* (Vol. *469*, pp. 19–31). Geological Society, London, Special Publication.

Kley, J., Franzke, H.-J., Jähne, F., Krawczyk, C., Lohr, T., Reicherter, K., et al. (2008). Strain and stress. In R. Littke, U. Bayer, D. Gajewski, & S. Nelskam (Eds.), *Dynamics of complex intracontinental basins—The central European basin system* (Chapter 3.3, pp. 97–124). Berlin, Heidelberg: Springer-Verlag.

Koyi, H. (1988). Experimental modeling of role of gravity and lateral shortening in Zagros mountain belt. *American Association of Petroleum Geologists Bulletin*, *72*(11), 1381–1394.

Koyi, H. (1998). The shaping of salt diapirs. *Journal of Structural Geology*, *20*(4), 321–338.

Kukla, P. A., Urai, J. L., & Mohr, M. (2008). Dynamics of salt structures. In R. Littke, U. Bayer, D. Gajewski, & S. Nelskamp (Eds.), *Dynamics of complex intracontinental basins—The central European basin* (Chapter 5.3, pp. 291–306). Berlin, Heidelberg: Springer-Verlag.

Kukla, P., Urai, J., Warren, J., Reuning, L., Becker, S., Schoenherr, J., et al. (2011). An integrated, multi-scale approach to salt dynamics and internal dynamics of salt structures. *AAPG Search and Discovery,* Article #40703 (2011).

Küster, Y., Leiss, B., & Schramm, M. (2010). Structural characteristics of the halite fabric type "Kristallbrocken" from the Zechstein Basin with regard to its development. *International Journal of Earth Sciences*, *99*(3), 505–526. doi:10.1007/s00531-008-0399-8

Küster, Y., Schramm, M., Bornemann, O., & Leiss, B. (2009). Bromide distribution characteristics of different Zechstein 2 rock salt sequences of the Southern Permian Basin: a comparison between bedded and domal salts. *Sedimentology*, *56*, 1368–1391. doi: 10.1111/j.1365-3091.2008.01038.x

Laurila, T. E., Hannington, M. D., Petersen, S., & Garbe-Schönberg, D. (2014). Trace metal distribution in the Atlantis II Deep (Red Sea) sediments. *Chemical Geology*, *386*, 80–100.

Leach, D. L., & Rowan, E. L. (1986). Genetic link between Ouachita foldbelt tectonism and the Mississippi Valley-type lead-zinc deposits of the Ozarks. *Geology*, *14*, 931–935.

Leach, D. L., Sangster, D. F., Kelley, K. D., Large, R. R., Grant, G., Allen, C. R., et al. (2005). Sediment-hosted lead-zinc deposits: A global perspective. *Society of Economic Geologists, Economic Geology One Hundredth Anniversary Volume*, 1905–2005, 561–607.

Leach, D. L., Taylor, R. D., Fey, D. L., Diehl, S. F., & Saltu, R. W. (2010). A deposit model for Mississippi Valley-type lead-zinc ores. *U.S. Geological Survey*, Mineral Deposit Models for Resource Assessment, Scientific Investigations Report 2010–5070–A, 52 pp.

Leitner, C., Marschallinger, R., & Huber, K. (2021). Deformed displacive halite crystals: Diagenetic or tectonic origin? *Journal of Sedimentary Research*, *91*(1), 21–33. doi:10.2110/jsr.2020.106

Leitner, C., & Neubauer, F. (2011). Tectonic significance of structures within the salt deposits Altaussee and Berchtesgaden–Bad Dürrnberg, Northern Calcareous Alps. *Austrian Journal of Earth Sciences*, *104*, 2–21.

Leitner, C., Neubauer, F., Genser, J., Borojević Šoštarić, S., & Rantitsch, G. (2014). $40^{Ar}/39^{Ar}$ ages of crystallization and recrystallization of rock-forming polyhalite in Alpine rocksalt deposits. In F. Jourdan, D. F. Mark, & C. Verati (Eds.), *Advances in 40Ar/39Ar dating: From archaeology and planetary sciences* (Vol. *378*, pp. 207–224). Geological Society, London, Special Publication. doi: 10.1144/SP378.5

Leitner, C., Neubauer, F., Urai, J. L., & Schoenherr, J. (2011). Structure and evolution of a Rocksalt–Mudrock–Tectonite: The Haselgebirge in the northern Calcareous Alps. *Journal of Structural Geology*, *33*, 970–984. doi:10.1016/j.jsg.2011.02.008

Leitner, C., & Spötl, C. (2017). The Eastern Alps: Multistage development of extremely deformed evaporites. In I. Soto, J. F. Flinch, & G. Tari (Eds.), *Permo-Triassic salt provinces of Europe, North Africa and the Atlantic margins, tectonics and hydrocarbon potential* (pp. 467–482). Amsterdam: Elsevier.

Leitner, C., Wiesmaier, S., Köster, M. H., Gilg, H. A., Finger, F., & Neubauer, F. (2017). Alpine halite-mudstone-polyhalite tectonite: Sedimentology and early diagenesis of evaporites in an ancient rift setting (Haselgebirge Formation, eastern Alps). *Geological Society of America Bulletin*, *129*(11/12), 1537–1553. https://doi.org/10.1130/B31747.1

Li, H., Wanyan, Q., Ding, G., Li, K., Kou, Y., Bai, S., et al. (2022). Geomechanical feasibility analysis of salt cavern gas storage construction in Sanshui Basin, Guangdong Province. *Eng*, *3*, 709–731. https://doi.org/10.3390/eng3040048

Li, S., Abe, S., Reuning, L., Becker, S., Urai, J. L., & Kukla, P. A. (2012). Numerical modelling of the displacement and deformation of embedded rock bodies during salt tectonics: A case study from the South Oman Salt Basin. In G. I. Alsop, S. G. Archer, A. J. Hartley, N. T. Grant, & R. Hodgkinson (Eds.), *Salt tectonics, sediments and prospectivity* (Vol. *363*, pp. 503–520). Geological Society, London, Special Publications. http://dx.doi.org/10.1144/SP363.24

Li, S.-Y., & Urai, J. L. (2016). Rheology of rock salt for salt tectonics modeling. *Petroleum Science*, *13*, 712–724. doi 10.1007/s12182-016-0121-6

Liang, W., Zhang, C., Gao, H., Yang, X., Xu, S., & Zhao, Y. (2012). Experiments on mechanical properties of salt rocks under cyclic loading. *Journal of Rock Mechanics and Geotechnical Engineering*, *4*(1), 54–61.

Linckens, J., Zulauf, G., & Hammer, J. (2016). Experimental deformation of coarse-grained rock salt to high strain. *Journal of Geophysical Research, Solid Earth*, *121*(8), 6150–6171. https://doi.org/10.1002/2016JB012890

Littke, R., Bayer, U., Gajewski, D., & Nelskamp, S. (Eds.). (2008). *Dynamics of complex intracontinental basins: The central European basin system*. Berlin: Springer, 519 pp.

Liu, W., Zhang, X., Li, H., & Chen, J. (2020). Investigation on the deformation and strength characteristics of rock salt under different confining pressures. *Geotechnical and Geological Engineering*, *38*, 5703–5717. https://doi.org/10.1007/s10706-020-01388-1

Mandl, G. W. (1999). The Alpine sector of the Tethyan shelf—Examples of Triassic to Jurassic sedimentation and deformation from the Northern Calcareous Alps. *Austrian Journal of Earth Sciences*, *92*, 61–77.

Maystrenko, Y., Bayer, U., Brink, H.-J., & Littke, R. (2008). The Central European Basin System—An overview. In R. Littke, U. Bayer, D. Gajewski, & S. Nelskamp (Eds.), *Dynamics of complex intracontinental basins. The central European basin system* (Chapter 2, pp. 17–34). Berlin: Springer.

Maystrenko, Y. P., Bayer, U., & Scheck-Wenderoth, M. (2012). Regional-scale structural role of Permian salt within the Central European Basin System. In G. I. Alsop, S. G. Archer, A. J. Hartley, N. T. Grant, & R. R. Hodgkinson (Eds.), *Salt tectonics, sediments and prospectivity* (Vol. *363*, pp. 409–430).

Geological Society, London, Special Publications. http://dx.doi.org/10.1144/SP363.19

Mills, M. M., Stormont, J. C., and Bauer, S. J. (2018). Micromechanical Processes in Consolidated Granular Salt, Sandia National Laboratories, Report SAND2018-3372J https://www.osti.gov/servlets/purl/1444081

Mohammed, A., & Mahmood, W. (2018). Statistical variations and new correlation models to predict the mechanical behavior and ultimate shear strength of gypsum rock. *Open Engineering*, *8*(1), 213–226. doi: https://doi.org/10.1515/eng-2018-0026

Mohr, M., Warren, J. K., Kukla, P. A., Urai, J. L., & Irmen, A. (2007). Subsurface seismic record of salt glaciers in an extensional intracontinental setting (Late Triassic of northwestern Germany). *Geology*, *35*, 963–966.

Mohriak, W. (2025). *Salt in the Earth sciences: Basin analysis and salt tectonics. Washington*, DC: American Geophysical Union. doi:10.1002/9781394307067

Mohriak, W. U., & Szatmari, P. (2009). Introdução às propriedades químicas e físicas dos evaporitos. In: W. Mohriak, P. Szatmari, & S. M. C. Anjos (organizadores), *Sal: Geologia e Tectônica, 2ed.*, pp. 18–41. São Paulo: Editora Beca.

Motamedi, H., Sepehr, M., Sherkati, S., & Pourkermani, M. (2011). Multi-phase Hormuz salt diapirism in the Southern Zagros, SW Iran. *Journal of Petroleum Geology*, *34*(1), 29–44.

Nettleton, L. L. (1934). Fluid mechanics of salt domes. *American Association of Petroleum Geologists Bulletin*, *18*(9), 1175–1204.

Odé, H. (1968). Review of mechanical properties of salt relating to salt-dome genesis. In J. Braunstein & G. D. O'Brien (Eds.), *Diapirism and diapirs: A symposium, American Association of Petroleum Geologists, Memoir* (Vol. *8*, pp. 53–78).

Passchier, C. W., & Trouw, R. A. J. (2005). *Microtectonics* (2nd ed.). Berlin: Springer, 366 pp.

Pells, P. J. N. (1995). Uniaxial strength testing. In J. A. Hudson (Ed.), *Comprehensive rock engineering. Volume 3, Rock testing and site characterization* (pp. 67–85). Oxford, UK: Pergamon Press.

Ranganathan, V., & Hanor, J. S. (1988). Density-driven groundwater flow near salt domes. *Chemical Geology*, *74*, 173–188.

Robertson, E. G., Robie, R. A., & Books, K. G. (1958). *Physical properties of salt, anhydrite, and gypsum—Preliminary report*. United States Department of the Interior, U. S. Geological Survey, Trace Elements Memorandum Report 1048, 38 p.

Peric, D., & Crook, A. J. L. (2004). Computational strategies for predictive geology with reference to salt tectonics. *Computer methods in Applied Mechanics and Engineering*, *193*, 5195–5222.

Pollok, L., Sassnowski, A., Henneberg, M., Pusch, M., Gundelach, V., Thiemeyer, T., et al. (2022). *Progress in the geoscientific exploration of the Asse salt structure as a base for the construction of a new mine for the retrieval of radioactive waste from the Asse II mine*. CRC Press, London: Taylor & Francis, pp. 467–475. In J. H. P. de Bresser, M. R. Drury, P. A. Fokker, M. Gazzani, S. J. T. Hangx, A. R. Niemeijer, & C. J. Spiers (Eds.), *The mechanical behavior of Salt X*, Proceedings of the 10th Conference on the mechanical behavior of salt (SATMECH X), Utrecht, The Netherlands, 06–08 July 2022, 690 pp.

Pollok, L., Sassnowski, M., Kûhnlenz, T., Gundelach, V., Hammer, J., & Pritzkow, C. (2018). Geological exploration and 3D model of the Asse salt structure for SE expansion of the Asse II mine. *Mechanical Behavior of Salt IX, 2018*. Proceedings (ISBN 978-3-9814 108-6-0), pp. 753–763.

Raith, A. F., Urai, J. L., & Visser, J. (2017). Structural and microstructural analysis of K–Mg salt layers in the Zechstein 3 of the Veendam Pillow, NE Netherlands: Development of a tectonic mélange during salt flow. *Netherlands Journal of Geosciences*, *96*(04), 331–351.

Ridley, J. (2013). *Ore deposit geology*. Cambridge, UK: Cambridge University Press, 398 pp.

Rowan, M. G., Urai, J. L., Fiduk, J. C., & Kukla, P. A. (2019). Deformation of intrasalt competent layers in different modes of salt tectonics. *Solid Earth*, *10*, 987–1013. https://doi.org/10.5194/se-10-987-2019

Rutter, E. H. (1976). The kinetics of rock deformation by pressure solution. *Philosophical Transactions of the Royal Society of London. Series A, Mathematical and Physical Sciences*, *283*(1312), 203–219.

Sarkarinejad, K., Sarshar, M. A., & Adineh, S. (2018). Structural, micro-structural and kinematic analyses of channel flow in the Karmostaj salt diapir in the Zagros foreland folded belt, Fars province, *Iran. Journal of Structural Geology*, *107*, 109–131.

Scheck-Wenderoth, M., Krzywiec, P., Zühlke, R., Maystrenko, Y., & Froitzheim, N. (2008). Permian to Cretaceous tectonics of Central Europe. In T. McCann (Ed.), *The geology of Central Europe: Vol. 2, Mesozoic and Cenozoic* (pp. 999–1030). London, UK: Geological Society Publishing House.

Scheck-Wenderoth, M., Maystrenko, Y., Hübscher, C., Hansen, M., & Mazur, S. (2008). Dynamics of salt structures. In R. Littke, U. Bayer, D. Gajewski, & S. Nelskamp (Eds.), *Dynamics of complex intracontinental basins: The Central European basin* (Chapter 5.4, pp. 307–322). Berlin: Springer.

Schléder, Z., & Urai, J. L. (2005). Microstructural evolution of deformation-modified primary halite from the Middle Triassic Röt Formation at Hengelo, the Netherlands. *International Journal of Earth Sciences (Geol Rundsch)*, *94*(5–6), 941–955.

Schléder, Z., & Urai, J. L. (2007). Deformation and recrystallization mechanisms in mylonitic shear zones in naturally deformed extrusive Eocene-Oligocene rocksalt from Eyvanekey plateau and Garmsar hills (central Iran). *Journal of Structural Geology*, *29*(2), 241–255. doi:10.1016/j.jsg.2006.08.014

Schléder, Z., Urai, J. L., Nollet, S., & Hidgers, C. (2008). Solution-precipitation creep and fluid flow in halite: A case study of Zechstein (Z1) rocksalt from Neuhof salt mine (Germany). *International Journal of Earth Sciences (Geol Rundsch)*, *97*, 1045–1056.

Schön, J. H. (2015). *Physical properties of rocks: Fundamentals and principles of petrophysics* (2nd ed.). Amsterdam: Elsevier, 497pp.

Schorn, A., & Neubauer, F. (2014). Hallstatt evaporite body of the Upper Permian to Lower Triassic Haselgebirge Fm. within the Northern Calcareous Alps (NCA). *Journal of Structural Geology*, *60*, 70–84.

Schultz-Ela, D. D., Jackson, M. P. A., & Vendeville, B. C. (1993). Mechanics of active salt diapirism. *Tectonophysics*, *228*, 275–312.

Sepehr, M., Cosgrove, J., & Moieni, M. (2006). The impact of cover rock rheology on the style of folding in the Zagros fold-thrust belt. *Tectonophysics*, *427*, 265–281. doi:10.1016/j.tecto. 2006.05.021

Sommaruga, A. (1999). Décollement tectonics in the Jura foreland fold-and-thrust belt. *Marine and Petroleum Geology*, *15*, 111–134.

Sommaruga, A., Mosar, J., Schori, M., & Gruber, M. (2017). The role of the Triassic evaporites underneath the North Alpine foreland. In J. I. Soto, J. Flinch, & G. Tari (Eds.), *Permo-Triassic salt provinces of Europe, North Africa and the Atlantic margins tectonics and hydrocarbon potential* (pp. 447–466). https://doi.org/10.1016/B978-0-12-809417-4.00021-5

Sonnenfeld, P., & Perthuisot, J.-P. (1989). *Brines and evaporites, Short Course in Geology, Volume 3, 28th International Geological Congress*, American Geophysical Union, Washington, DC, 126 pp.

Spötl, C. (1989). The Alpine Haselgebirge formation, Northern Calcareous Alps (Austria): Permo-Scythian evaporites in an alpine thrust system. *Sedimentary Geology*, *65*, 113–125. https://doi.org/10.1016/0037-0738(89)90009-2.

Sriapai, T., Walsri, C., & Fuenkajorn, K. (2012). Effect of temperature on compressive and tensile strengths of salt. *Science Asia*, *38*, 166–174. doi: 10.2306/scienceasia1513-1874.2012.38.166

Stewart, S. A., & Coward, M. P. (1995). Synthesis of salt tectonics in the southern North Sea, *UK. Marine and Petroleum Geology*, *12*(5), 457–475.

Strauss, P., Granado, P., & Muñoz, J. A. (2021). Subsidence analysis of salt tectonics-driven carbonate minibasins (Northern Calcareous Alps, Austria). *Basin Research*, *33*, 968–990. https://doi.org/10.1111/bre.12500

Strozyk, F., Reuning, L., Scheck-Wenderoth, M., & Tanner, D. C. (2017). The tectonic history of the Zechstein basin in the Netherlands and Germany. In J. I. Soto, J. F. Flinch, & G. Tari (Eds.), *Permo-Triassic salt provinces of Europe, North Africa and the Atlantic margins, tectonics and hydrocarbon potential* (pp. 221–241). doi:10.1016/b978-0-12-809417-4.00011-2

Sutera, S. P., & Skalak, R. (1993). The history of Poiseuille's Law. *Annual Review of Fluid Mechanics*, *25*(1), 1–20. doi:10.1146/annurev.fl.25.010193.000245

Talbot, C. J. (1978). Halokinesis and thermal convection, *Nature*, *273*, 739–741.

Talbot, C. J. (1979). Fold trains in a glacier of salt in southern Iran. *Journal of Structural Geology*, *1*(1), 5–18.

Talbot, C. J. (1981). Sliding and other deformation mechanisms in a glacier of salt, S Iran. In N. J. Price & K. R. McClay (Eds.), *Thrust and Nappe tectonics* (Vol. *9*, pp. 173–183). The Geological Society of London Special Publications.

Talbot, C. J. (1993). Spreading of salt structures in the Gulf of Mexico. *Tectonophysics*, *228*, 151–166.

Talbot, C. J. (1998). Extrusions of Hormuz salt in Iran. In D. J. Blundell & A. C. Scott (Eds.), *Lyell: The past is the key to the present* (Vol. *143*, pp. 315–334). Geological Society, London, Special Publication.

Talbot, C. J., & Aftabi, P. (2004). Geology and models of salt extrusion at Qum Kuh, central Iran. *Journal of the Geological Society*, *161*, 321–334.

Talbot, C. J., & Jackson, M. P. A. (1987). Internal kinematics of salt diapirs. *American Association of Petroleum Geologists Bulletin*, *71*(9), 1068–1093.

Talbot, C. J., & Jarvis, R. J., (1984). Age, budget and dynamics of an active salt extrusion in Iran. *Journal of Structural Geology*, *6*, 521–533.

Talbot, C. J., & Pohjola, V. (2009). Subaerial salt extrusions in Iran as analogues of ice sheets, streams and glaciers. *Earth-Science Reviews*, *97*, 167–195.

Talbot, C. J., & Rogers, E. A. (1980). Seasonal movements in a Salt Glacier in Iran. *Science*, *208*, 395–397. doi: 10.1126/science.208.4442.395

Ter Heege, J. H., De Bresser, J. H. P., & Spiers, C. J. (2005a). Dynamic recrystallisation of wet synthetic polycystalline halite: Dependence of grain size distribution on flow stress, temperature and strain. *Tectonophysics*, *396*, 35–57.

Ter Heege, J. H., De Bresser, J. H. P., & Spiers, C. J. (2005b). Rheological behaviour of synthetic rocksalt: The interplay between water, dynamic recrystallization and deformation mechanisms. *Journal of Structural Geology*, *27*, 948–963. doi:10.1016/j.jsg.2005.04.008

Tesón, E., Mora, A., Silva, A., Namson, J., Teixell, A., Castellanos, J., et al. (2013). Relationship of Mesozoic graben development, stress, shortening magnitude, and structural style in the Eastern Cordillera of the Colombian Andes. In M. Nemcok, A. Mora, & J. W. Cosgrove (Eds.), *Thick-skin-dominated orogens: From initial inversion to full accretion* (Vol. *377*, pp. 257–283). Geological Society, London, Special Publications.

Tiab, D., & Donaldson, E. C. (2016). *Petrophysics: Theory and practice of measuring reservoir rock and fluid transport properties* (4th ed.). Amsterdam: Gulf Professional Publishing, Elsevier, 894 pp.

Trusheim, F. (1960). Mechanism of salt migration in northern Germany. *American Association of Petroleum Geologists Bulletin*, *44*(9), 1519–1540.

Urai, J. L., Schléder, Z., Spiers, C. J., & Kukla, P. A. (2008). Flow and transport properties of salt rocks. In R. Littke, U. Bayer, D. Gajewski, & S. Nelskamp (Eds.), *Dynamics of complex intracontinental basins: The central European basin* (Chapter 5.2, pp. 277–290). Berlin: Springer.

Urai, J. L., & Spiers, C. J. (2007). The effect of grain boundary water on deformation mechanisms and rheology of rocksalt during long-term deformation. In M. Wallner, K.-H. Lux, W. Minkley, & H. Hardy Jr. (Eds.), *The mechanical behaviour of salt—Understanding of THMC processes in salt* (pp. 149–158). London: Taylor & Francis.

Urai, J. L., Spiers, C. J., Franssen, R. C. M. W., & Liezenberg, J. L. (1987). Deformation mechanisms operating in naturally deformed halite rocks as deduced from microstructural investigations. *Geologie en Mijnbouw*, *66*, 165–176.

Urai, J. L., Spiers, C. J., Zwart, H. J., & Lister, G. S. (1986). Weakening of rock salt by water during long-term creep. *Nature*, *324*, 8097, 4pp, 11–17 December 1986.

Van der Pluijm, B. A., & Marshak, S. (2004). *Earth structure: An introduction to structural geology and tectonics* (2nd ed.). New York: W. W. Norton, 656 pp.

Van Keken, P. E., Spiers, C. J., van den Berg, A. P., & Muyzert, E. J. (1993). The effective viscosity of rocksalt: Implementation of steady-state creep laws in numerical models of salt diapirism. *Tectonophysics*, *225*, 457–476.

Vernon, R. H. (2018). *A practical guide to rock microstructure* (2nd ed.). Cambridge, UK: Cambridge University Press, 431 pp. https://doi.org/10.1017/9781108654609

Voigt, T., Kley, J., & Voigt, S. (2021). Dawn and dusk of Late Cretaceous basin inversion in central Europe. *Solid Earth*, *12*(6), 1443–1471. https://doi.org/10.5194/se-12-1443-2021

Warren, J. K. (2000). Evaporites, brines and base metals: Low-temperature ore emplacement controlled by evaporite diagenesis. *Australian Journal of Earth Sciences*, *47*, 179–208.

Warren, J. K. (2006). *Evaporites: Sediments, resources and hydrocarbons*. Berlin: Springer Verlag, 1035 pp.

Weijermars, R., Jackson, M. P. A., & van Harmelen, A. (2014). Closure of open wellbores in creeping salt sheets. *Geophysical Journal International*, *196*, 279–290. doi: 10.1093/gji/ggt346

Weijermars, R., Jackson, M. P. A., & Vendeville, B. C. (1993). Rheological and tectonic modeling of salt provinces. *Tectonophysics*, *217*, 143–174.

Weinberger, R., Begin, Z. B., Waldmann, N., Gardosh, M., Baer, G., Frumkin, A., & Wdowinski, S. (2006). Quaternary rise of the Sedom diapir, Dead Sea basin. In Y. Enzel, A. Agnon, & M. Stein (Eds.), *New frontiers in Dead Sea paleoenvironmental research* (Vol. *401*, 33–51). Geological Society of America Special Paper. doi: 10.1130/2006.2401(03).

Weinberger, R., Lyakhovsky, V., Baer, G., & Begin, Z. B. (2006). Mechanical modeling and InSAR measurements of Mount Sedom uplift, Dead Sea basin: Implications for effective viscosity of rock salt. *Geochemistry, Geophysics, Geosystems*, *7*, Q05014.

Wenkert, D. D. (1979). The flow of salt glaciers. *Geophysical Research Letters*, *6*(6), 523–526. doi: 10.1029/GL006i006p00523

Zak, I., & Freund, R. (1980). Strain measurements in eastern marginal shear zone of Mount Sedom salt diapir, Israel. *American Association of Petroleum Geologists Bulletin*, *64*(4), 568–581.

Zong, J., Stewart, R. R., Dyaur, N., & Myers, M. T. (2017). Elastic properties of rock salt: Laboratory measurements and Gulf of Mexico well-log analysis. *Geophysics*, *82*(5), D303–D317. doi:10.1190/geo2016-0527.1

7

Mineral Analysis of Evaporite Rocks in Sedimentary Basins

ABSTRACT

Laboratories analyze evaporite minerals using various techniques, such as chemical analysis, X-ray diffraction, and petrographic microscopy. Earth scientists collect samples from outcrops, underground mines, or deep exploratory boreholes that drill the subsurface layers. Researchers use advanced geochemical techniques like bromine content and strontium isotope ratios to determine the fluid origin in evaporite basins.

Previous chapters have discussed evaporite minerals by analyzing regions where evaporite rocks are exposed on the surface. This chapter presents some basic principles of mineral identification using modern analytical methods, with an emphasis on microscopic petrography of evaporite rocks in the subsurface. Several examples of common mineral types from different salt basins are discussed based on laboratory analysis. An overview of the geological evolution of the selected salt basins includes the geological setting of evaporite sequences and some images of thin sections obtained from outcrop samples, borehole cores, and salt mines. The integration of datasets from different scales, from seismic profiles to microscopic analysis of evaporite minerals obtained by drilling, constitutes the backbone of exploratory basin analysis projects (e.g., Warren, 2006, 2010, 2016).

The petrographic analysis of the evaporite sequences is complemented by more advanced methods, such as X-ray analysis, geochemistry, and the determination of isotope ratios. The measurement of strontium isotope ratios can be extremely useful to constrain the age and environmental conditions of salt deposits. The chapter presents examples from the Dead Sea and the Mediterranean Sea to illustrate the application of isotope ratio determination in the analysis of salt basins.

The text discusses the genesis of ancient evaporite basins and provides a summary of the sedimentary environments of deposition, including marine to continental salt basins formed during the Paleozoic, Mesozoic, and Cenozoic. The Permian evaporites found in the intracratonic Parnaiba Basin (northern Brazil) are mostly linked to anhydrite layers that formed in dry places on a continent far from the world's oceans. Petrographic thin sections of different evaporite minerals from the Zechstein salt basin in northeast (NE) England are presented as an introduction to this giant salt province that extends from the North Sea toward mainland Europe (see Chapter 2 in Mohriak, 2025).

The Triassic evaporite rocks that occur in the Alpine foreland basins in the Central European Basin are discussed based on the detailed petrographic analysis of cores from several boreholes that penetrated the Triassic salt layers in the Molasse Basin in Switzerland. Multiple cycles of salt deposition during the Cretaceous are recorded in the Brazilian margin, as shown by cores from exploratory boreholes that drilled the syn-rift Aptian rock salt in the onshore part of the Sergipe Basin. This region is characterized by the most important salt deposits in Brazil, which are exploited in a salt mine hundreds of meters below the surface. The salt deposits in the basin have been thoroughly studied by several researchers based on the analysis of borehole cores obtained during petroleum exploration drilling activities in the 1970s (Szatmari et al., 1974, 2009).

The most recent giant salt basin in the world is associated with the Messinian Salinity Crisis (MSC) at the transition from the Miocene to the Pliocene, around 6 Ma (Hsü, 1972). This event resulted in the deposition of thick salt layers in the deep Mediterranean Sea and thinner sequences with gypsum and halite in the peripheral basins, as illustrated by the Late

Salt in the Earth Sciences: Evaporite Rocks and Salt Deposition, Geophysical Monograph 286, First Edition. Webster Mohriak.

DOI: 10.1002/9781119405245.ch07

Miocene deposits onshore Spain. The central part of the Mediterranean Sea is characterized by halite deposits forming thick salt sequences in the deepwater region, which have been drilled by the Deep Sea Drilling Program (DSDP) boreholes (Hsü et al., 1973, 1978). The salt deposits around the Mediterranean Sea are briefly discussed based on recent drilling of salt industry boreholes that penetrated the Miocene evaporites in the subsurface of peripheral basins in Spain and Italy (see Chapter 5 in Mohriak, 2025). Roveri et al. (2008, 2014) correlated the salt deposits recorded on land to the Messinian evaporites in several compartments of the Mediterranean Sea, which were penetrated by several boreholes.

This chapter presents a summary of some geological applications of geochemistry and quantitative determination of isotope ratios, which are widely used in the study of evaporite deposits and fluid distribution in sedimentary basins. The Dead Sea corresponds to a unique example of halite deposition in a modern deepwater lake (see Chapters 2 and 3). Geochemical methods and strontium isotope ratios have been used to determine the origin of the thick salt layers deposited during the Neogene (e.g., Charrach, 2018). The strontium isotope ratio determination has been applied to the paleoenvironmental characterization of the resedimented gypsum deposits in the Calabria region of southern Italy, adjacent to the Ionian Sea (Cipriani et al., 2021).

7.1. LABORATORY ANALYSIS OF EVAPORITE MINERALS

Evaporite sequence analysis involves field geology methods and laboratory analysis of rock samples, including sedimentological, petrographic, and chemical studies. Most evaporite basins contain carbonate, sulfate, and chloride minerals, with the most common deposits consisting of calcite, aragonite, gypsum, and halite (see Chapter 4). Identifying mineral compounds in evaporite sequences involves techniques such as thin-section microscopy, petrographical studies, and bulk mineral composition based on X-ray diffraction (XRD). Petrographic microscopy helps to determine the mineralogical classes of minerals by analyzing crystals in thin sections. These methods also provide information on the genetic mechanisms involved in the mineral formation and subsequent modifications by diagenetic processes (Okrusch & Frimmel, 2020).

Salt basins contain dozens of minerals in evaporite sequences. Even pure rock salt successions, predominantly composed of halite, exhibit a large number of minor components. For example, the Dead Sea brine includes several ions in addition to chlorine, magnesium, sodium, calcium, and potassium (see Figure 2.7). The chemical complexity of mineral compounds and minor trace elements interspersed within rock salt is exemplified by the Himalayan pink salt, obtained from evaporite deposits in Pakistan (e.g., Fayet-Moore et al., 2020). This famous salt is obtained from evaporite deposits from the Punjab region of Pakistan, located more than 300 km from the Himalayan Mountains thrust front. The salt mines in the Khewra region produce salt from the Neoproterozoic evaporite deposits that occur at the foothills of the Salt Range system, near the Indo-Gangetic Plain in Pakistan. The salt blocks are cut from the mine walls using saw machines, resulting in smaller blocks that are exported to many countries. Pink salt is used as a food condiment and in medical treatments, as some people believe that the color of the Himalayan salt results from chemical elements with therapeutic properties.

The halite crystals of Himalayan pink salt have been analyzed in chemical laboratories, revealing that over 97% of the components are sodium and chlorine. However, there are a few chemical compounds that form many other minerals in minute proportions. Most occur as trace elements measured in g/ton (ppm), with many values much less than 1 ppm. According to geochemical analysis (e.g., Fayet-Moore et al., 2020), the only elements measured in g/kg with concentrations above 1 g/kg are oxygen (1.2 g/kg), potassium (3.5 g/ton), calcium (4 g/kg), and sulfur (12.4 g/kg). These determinations are conducted in laboratories that use several methods in the analyses. These include standard and advanced chemical analysis, X-ray and mass spectral analysis, and reflectance spectroscopy.

Various methods are used in chemical laboratories to determine the concentration of chemical elements in salt. Traditional analytical chemistry methods, such as titration, are employed to determine calculate the calcium concentration, while potentiometric methods are used to determine minor constituents like iodine and fluorine. More sophisticated techniques are used to identify minute concentrations of chemical elements in the pink Himalayan salt. Mass spectrometry determination of chemical compounds determines the atomic mass of the elements that compose the salt crystals. Reflectance spectral methods record the reflected energy in different spectrum regions, while atomic absorption spectroscopy (AAS) determines the concentrations of minor elements like lithium, beryllium, magnesium, aluminum, silicon, manganese, iron, cobalt, and other metals. Tantalum, iridium, platinum, and gold are usually determined as trace elements using total reflection X-ray fluorescence (TRXF) methods. A synthesis of instrumental methods for the quantitative characterization of minor minerals and trace elements is presented by Klein and Philpotts (2017).

Diffractometry is a technique used to determine the mineralogical composition of rock samples and the atomic structure of mineral components. X-ray data for most common evaporite minerals have been compiled by Braitsch (1971) and are also available in several books and websites. Advanced methods of mineral analysis, such as scanning electric microscope, transmission electric microscope, and mass spectrometry, are often used to determine the chemical composition and understand the sedimentary settings of evaporite deposits. These methods are beyond the scope of the chapter, which presents only some fundamental principles of mineral identification. The numerous references cited in this section may help readers interested in more advanced techniques.

In the salt and petroleum industry, identifying diverse carbonate evaporite minerals in sedimentary basins is usually done through the petrographic description of small chips recovered from drilling operations, particularly ditch samples obtained by rotary drilling (Scholle & Ulmer-Scholle, 2003). Different mineral assemblages may also be defined based on the systematic description of cores, plugs, or interpretation of wire-line logs.

There is no consistent classification for mixed evaporite sequences. Most researchers describe the lithological units as evaporite cycles with dominant minerals occurring at specific intervals. For example, the upper gypsum layers in the Mediterranean Sea salt basin are predominantly composed of gypsum, whereas other intervals exhibit halite as the main mineral component (Roveri et al., 2014).

The lithological interpretation of the sedimentary successions drilled by an exploratory borehole in the Santos Basin, offshore Brazil, was discussed in Chapter 5 (Figure 5.24). The preliminary interpretation of the lithological units penetrated by the borehole was based on the analysis of ditch samples and electric logs. This methodology is commonly complemented by calibrating the electric log interpretation with systematic chemical and petrographic analysis of drilling cores from key evaporite intervals, particularly the ones associated with potassium-bearing salts (Szatmari et al., 1974, 2009).

7.2. ANALYTICAL METHODS USED IN EVAPORITE MINERALOGY

Chapter 4 discusses the most common evaporite minerals found in various sources, including field outcrops, salt mines, drilling samples, and borehole cores. At the macroscale, rock salt can be described with the naked eye, but proper mineral identification is usually done using petrographic microscope techniques (Adams et al., 1984; Bloss, 1999; Nesse, 2012; Raith et al., 2011). This inexpensive method enables the identification of different minerals in evaporite rocks, their crystallization sequences, and recrystallization processes. A brief review of the basic principles in mineralogy is presented in this chapter so that the table of the main optical properties of evaporite minerals can be understood by nonspecialists in this branch of the salt sciences.

The analytical methods used in mineralogy laboratories cover a wide range of scales, from borehole core samples and plugs to the atomic structure of minerals (Figure 7.1). The workflow in some projects involves the determination of chemical compounds using advanced techniques like XRD and scanning electron microscopy (SEM).

Optical microscopy is a powerful tool for mineral analysis, but it can only provide images of the crystal surface, which is usually hidden from our eyesight. However, in the case of very limpid minerals, visualization can reach up to a few millimeters inside the crystal, allowing for

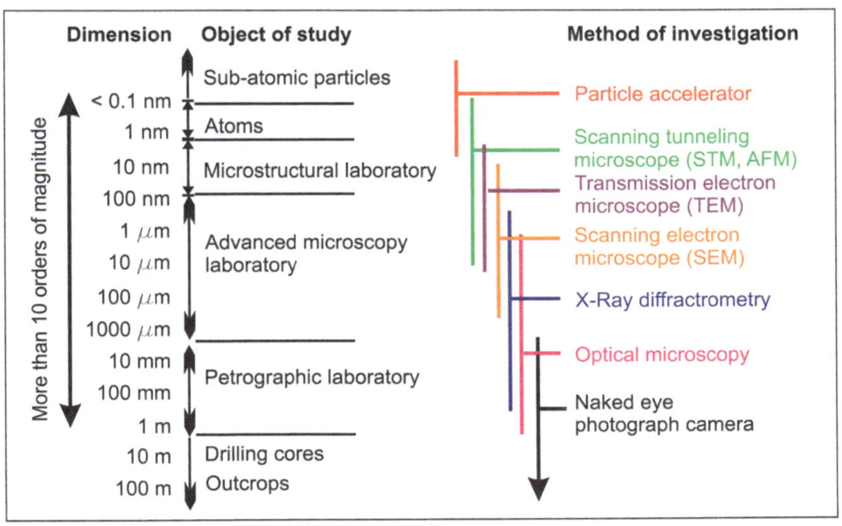

Figure 7.1 Laboratory methods of investigation of rocks, minerals, and chemical elements. Most evaporite basins are sampled by rock outcrops or by exploratory wells drilled for rock salt or petroleum. Specialized laboratories conduct microscopic analyses of sedimentary rocks, aiming at characterizing reservoir properties and their petrophysical parameters.

the characterization of fluid and mineral inclusions. Advanced microscopy and microstructural laboratories are commonly necessary for investigating some properties of evaporite minerals (Figure 7.1). Some microstructures are important for the understanding of the rheological properties of evaporite sequences deformed by folding, thrusting, and salt flow (see Chapter 6).

Minerals are characterized by geometric crystals with complex symmetry properties, which can be studied using petrographic microscopes, which are used up to a scale of about 100 mm (10^{-4} m, or 0.1 mm). Advanced methods like XRD provide a resolution of about 100 nm (10^{-7} m). Scanning electron microscopes can image objects in the range of a few nanometers to less than 1 Å (1 Å corresponds to 10^{-10} m) and are employed to determine the microstructure of deformed minerals (Figure 7.1). Atomic cell properties are mainly obtained through advanced methods like X-ray crystallography, but basic petrographic methods and mineralogy tables are useful for identifying common minerals in evaporite sequences.

7.2.1. Basic Principles of X-Ray Diffraction Methods

The principles of XRD were established more than a century ago, and it has become a key method to determine the atomic arrangement of chemical elements in crystalline compounds. This discovery has resulted in extraordinary advances in several branches of science. The XRD technique has proved to be very reliable and useful to determine the bulk mineralogy of carbonate and evaporite rocks, particularly the ones that contain sulfate and chloride mineral components. By examining a pure mineral crystal in submicroscopic detail, scientists can decipher the internal three-dimensional atomic structure of the chemical elements.

X-rays were discovered by Wilhelm Conrad Roentgen, a German professor, who in 1895 developed a piece of equipment that produced electromagnetic radiation in a wavelength range that could penetrate solid materials. X-rays correspond to a type of electromagnetic radiation that is transmitted as waves with a wavelength of about 1 Å (10^{-10} m). This is a much shorter wavelength than the visible light, which has a wavelength of ~5.5×10^{-7} m at the center of the spectrum, which corresponds to the yellow color. Shorter wavelengths (400 nm or 4.0×10^{-7} m) appear as bluish colors, whereas longer wavelengths (greater than 700 nm) appear with red colors (Halliday et al., 2014).

X-rays are emitted by certain elements (such as copper) when they are bombarded with fast electrons accelerated by a high potential difference. These rays can penetrate solid matter as waves but are partially absorbed by the material they penetrate. Denser materials are composed of chemical elements that have more electrons surrounding their nucleus and thus absorb X-rays more efficiently. The technique developed by Roentgen, which used a photographic plate to register the shadows of denser materials beneath softer ones, provided the first images of hidden parts of solid matter, making the invisible visible to the human eye. This remarkable achievement was a breakthrough in science, and Roentgen was honored with the first Nobel Prize in Physics, awarded in 1901. Since then, the method has been extensively used in medicine because it provides a means of seeing through the fleshy tissues and illuminating the internal parts of the human body.

In 1912, the German physicist Max von Laue (1879–1960) showed that X-rays could be used to decipher the geometry and atomic structure of crystals. When a beam of X-rays is fired at crystals, the waves are diffracted by the internal distribution of atoms inside the crystal. In simple terms, Max von Laue proposed that a crystalline solid, such as the mineral halite, might be interpreted as composed of atoms disposed of in a regular array, forming a natural three-dimensional "diffraction grating" for X-rays. The model envisaged that halite, which corresponds to sodium chloride (NaCl) crystal, is formed by a basic unit of atoms arranged in a cubic structure called the unit cell or lattice (see Chapter 4), which repeats itself throughout the array.

In 1912–1913, William Henry Bragg was the first British scientist to make an X-ray spectrometer. This equipment could throw X-rays at crystalline materials and determine their structural arrangement from the diffraction pattern. Henry Bragg's son, named William Lawrence Bragg, suggested an insightful explanation for the X-ray patterns discovered by Laue. Based on geometric principles, Bragg proposed what is now known as the famous Bragg's Law. He observed that when moving the X-ray source to different angles of incidence, the detector received intense signals at some angles. Bragg interpreted these waves as being in phase, indicating that they experienced constructive interference when summed, resulting in an increased amplitude. Using other angles of incidence, no signals were observed, depending on the distance d of the atomic planes in the structure. According to Bragg, the complete destructive interference of the waves occurs when they are out of phase by exactly one wavelength (Halliday et al., 2014; Klein & Philpotts, 2017).

The XRD method detects the intensity of the waves when the angle of diffraction varies as the detector is rotated around the crystal (Figure 7.2). After determining the angles and intensities of the diffraction beams, the data obtained by single-phase X-ray powder diffraction is plotted in graphs. These graphs display the most intense values measured in the detector, depending on the mineral framework, interplanar spacing D, and the relative intensity of the diffraction (I/Io). The crystallographic information of the minerals in the sample can also be expressed in terms of the lattice structure, lattice parameter, and atomic positions in the unit cell. This information is compared with standard values in

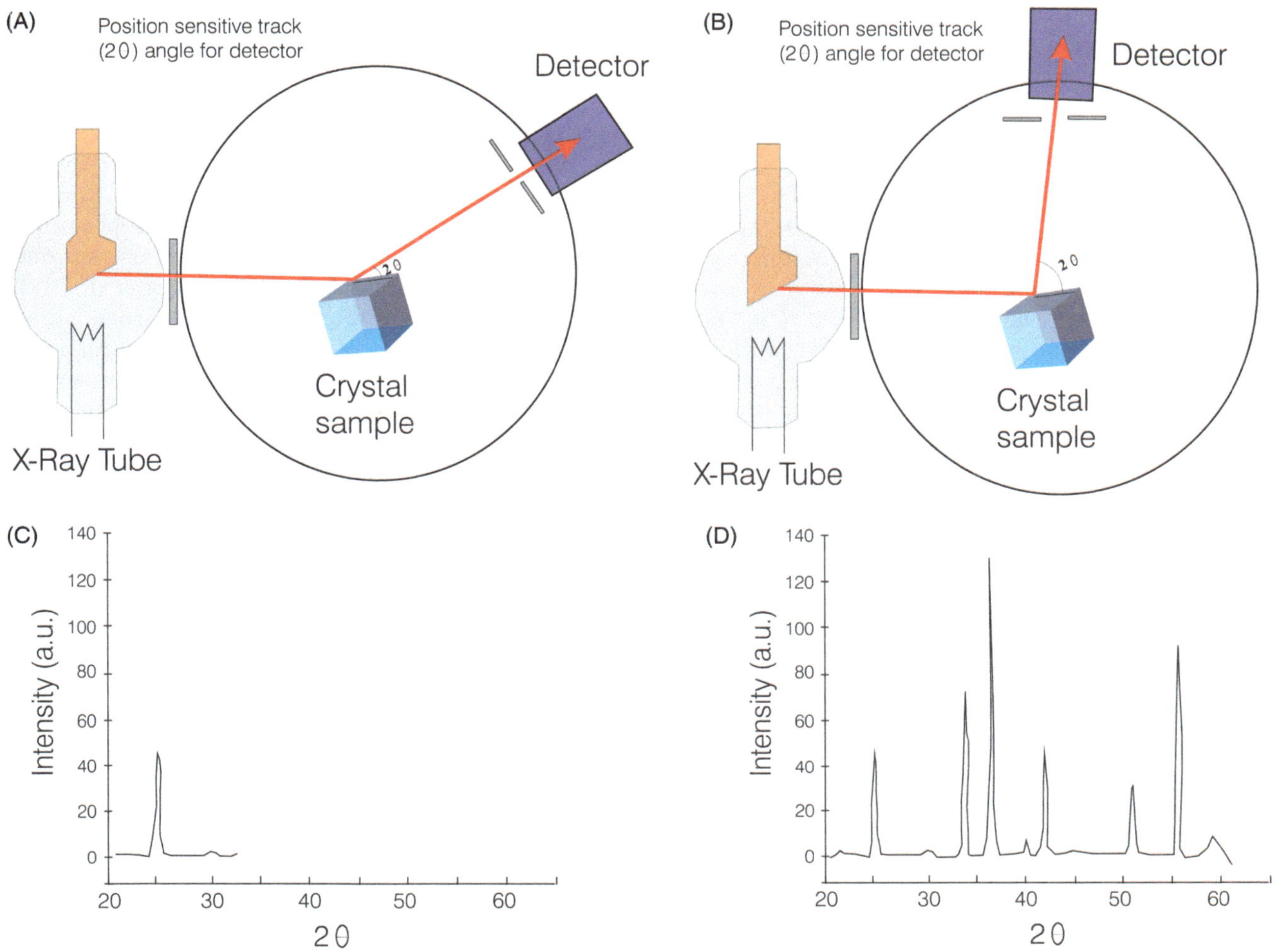

Figure 7.2 The X-ray method detects the intensity of the waves when the angle of diffraction varies as the detector is rotated around the crystal. Diagrams (A) and (B) show the X-ray tube and the detector that records the X-ray intensity. Diagrams (C) and (D) show the plot of intensity versus the angle 2Θ (theta). The peaks in the intensity curve are associated with the crystalline structure of the chemical compounds. Credit: Halliday et al. (2014)/with permission of John Wiley & Sons.

reference books (e.g., Haynes, 2017) or internet sources (e.g., webminerals.com).

Crystalline solids, such as pure minerals, when exposed to monochromatic X-rays will diffract the rays according to the principles of Bragg's law. The technique is based on the physical principles applied to the crystallographic analysis of a single crystal. However, the XRD technique generally uses the finely ground powder of a rock sample, with several minerals randomly oriented in the sample holder. This technique is often called whole rock X-ray powder diffraction analysis (powder XRD), which is applied for multiphase identification and quantitation of several minerals that might be present in the rock sample. The plots display the specific intensity peaks as the detector is rotated by an angle Θ (theta). The intensity versus 2Θ (theta) angle graph is appropriately called a fingerprint technique because the unknown mineral components of rocks are individually compared to standards or XRD references for specific compounds (Lavina et al., 2014; Reyes et al., 2017).

The XRD of the mineral halite, which is mainly composed of the Na and Cl ions (see Chapter 4), shows a maximum number of counts when the 2Θ (theta) angle is around 32. This value is then confirmed in standard tables derived from the analysis of thousands of minerals from salt basins. For example, XRD analyses of the mineral constituents in the polyhalite seam in the Yorkshire region of England have proved an invaluable tool to determine the chemical content of the evaporite layers in the subsurface, which include halite, sylvinite, polyhalite, and several other minerals. The process of quantitatively grading the salt ores established this area as the world's largest known polyhalite resource (Kemp et al., 2016).

Several sophisticated techniques are presently available for the powder XRD method (e.g., Lavina et al., 2014). Evaporite minerals from several regions of the world have been analyzed using this technique, such as in the onshore Dinarides in Croatia (Dedić et al., 2018) and the deepwater region of the Levant Basin in the Mediterranean (Meilijson et al., 2018).

XRD methods are widely used to identify unknown evaporite minerals in sedimentary basins worldwide. In the Parnaíba Basin in northern Brazil, Permian sequences with intercalations of evaporite deposits, mainly corresponding to sulfate minerals (gypsum and anhydrite), are found in exploratory wells and outcrops (Góes & Feijó, 1994; Vaz et al., 2007). These evaporite sequences are intercalated with siliciclastic deposits, which locally exhibit fragments or nodules derived from the erosion and transport of pre-existing sedimentary layers, including mudstone, carbonate, and evaporite rocks deposited in lacustrine environments (Araújo et al., 2016; Barbosa et al., 2016).

XRD of evaporite nodules in sandstones from the Permian sequences in the Parnaíba Basin identified several other minerals, including mica, kaolinite, hematite, and halite (Figure 7.3). Although there are no halite layers outcropping in the basin or described in cores from wells that penetrated the Permian strata in the subsurface, the presence of this mineral is indicated by its presence in the nodules. The XRD indicated that although halite is a mineral that is much more soluble than gypsum and anhydrite, it has been detected in the nodules in the sandstone layers intercalated with the Permian evaporitic sequences (Araújo et al., 2016).

In the future, XRD techniques will be used to identify unknown minerals on remote planets. NASA's Curiosity Rover, a remotely operated vehicle, has sent a sophisticated spectrometer to Mars, which can drill the Martian surface and collect soil and rock samples (Grotzinger et al., 2012). The car-sized NASA rover can drill the Martian surface and pick up samples of soils and rocks, which are analyzed on the spot using the XRD instrument to determine the mineral composition of the red planet's crust (Bish et al., 2014). The methods employed in the analysis of extreme environments on Earth have been recently applied to the study of some regions on Mars which include evaporite deposits (see Chapter 9 in Mohriak, 2025).

7.3. BASIC PRINCIPLES OF MICROSCOPIC PETROGRAPHY

Microscopic petrography is a valuable technique for identifying minerals in rocks, using thin sections from rocks obtained from outcrops or drilled cores to determine the basic mineralogy. The method is based on the fundamental aspects of the method and the optical characteristics of common evaporite minerals. Dozens

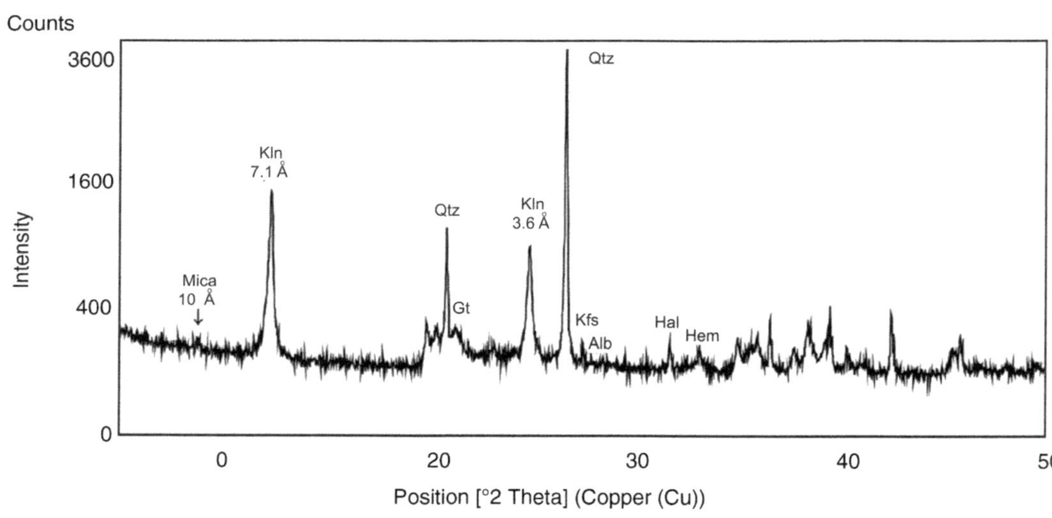

Figure 7.3 XRD plot of diverse minerals in the sandstone nodules collected at the Parnaíba Basin in northern Brazil. The presence of the halite mineral in the sample is indicated by a peak in the intensity curve at 2Θ values of 32°. Halite has not been preserved in sedimentary rock outcrops in this basin, but there is some evidence it might occur in the subsurface, as evidenced by its presence in nodules within the Permian sequences. Credit: Araújo et al. (2016)/with permission of Elsevier.

of introductory course notes and elementary to advanced textbooks deal with this subject (for example, Bloss, 1999; Gribble & Hall, 1985; Klein & Philpotts, 2017; MacKenzie et al., 2017; Nesse, 2012; Selvestone, 2015). This section will only highlight the basic fundamental aspects of the method and subsequently present a summary of the optical characteristics of the most common evaporite minerals.

Light behaves as a wave, characterized by amplitude and wavelength. When light reaches our retina, it transforms into electric signals that transmit to the brain as pulses with different colors depending on their wavelength. Blue colors have smaller wavelengths than red colors. The velocity at which these waves travel in space is the velocity of light, which is 300,000 km/s in a vacuum. However, the velocity of light varies according to the medium. For example, light traveling through a water body has its velocity reduced to 225,000 km/s. This reduction in velocity is characterized by an index number, which is the ratio of the velocity in a vacuum to the velocity in a different medium.

The refractive index (RI) of a material is defined as: $n = c/v$, where n is the refraction index, c is the speed of light in a vacuum, and v is the velocity of light in the material. The refraction index for light waves traveling in water corresponds to 1.33. When light rays cross a glass window, they experience a reduction in velocity due to the refraction index of 1.52, resulting in a velocity of about 200,000 km/s within the window glass. Petrographic microscopes analyze thin sections of minerals, which allow light waves to pass through. These thin sections are mounted on glass (usually with a thickness of about one-third of a millimeter) and glued with a resin called epoxy. The epoxy resin has a refraction index of 1.54, slightly higher than that of glass (Bloss, 1999; Nesse, 2012; Selvestone, 2015).

Opaque minerals do not allow light to travel through them, appearing black in a thin section examined under a microscope. Most minerals are nonopaque or semitransparent, meaning that some light goes through the crystals. Semitransparent minerals can be analyzed and identified in petrographic laboratory microscopes when cut in very thin slices (usually 0.30 mm). Most rock minerals are easily identified by their physical properties and optical characteristics in the microscope, such as calcite, gypsum, anhydrite, halite, or sylvite.

7.3.1. Light Transmission and Optical Microscopy

Because light travels as waves, the light rays vibrate as they travel through a substance. In the petrographic microscope, the source of light is a lamp that produces white light, which is composed of several wavelengths. The smaller wavelengths are perceived by our brain with colors that range from blue to violet, with wavelengths around 400 nm (400×10^{-9} m, equivalent to 4,000 Å). Shorter wavelengths are invisible to human eyes, corresponding to ultraviolet. Yellow is associated with a wavelength of about 600 nm. The red color is perceived by our brain with a wavelength of about 700 nm. Beyond the red end of the spectrum, there are longer waves associated with the infrared radiation rays, which are not perceived by the human eye but can be felt as heat. These waves are used in medical therapies to reduce muscle tension and inflammation, promoting pain relief.

Light vibrates in all planes that contain the light ray, that is, all planes that are perpendicular to the direction of light propagation. This is called unpolarized light, which vibrates in all directions (Bloss, 1999). Polarizers, often referred to as polars, are used by mineralogists in petrographic microscopes. This results in plane-polarized light, and the rays vibrate exclusively in the privileged direction of the polarizer. This results in a light that has its intensity decreased, but it is still white light. If another polarizer is inserted but only allows wave vibrations in the opposite direction (for example, N-S, Figure 7.4A), then no light will reach the ocular, and the image will be dark or extinct. Petrographic microscopes also use a special lens known as the Bertrand lens, which changes the plane of focus, allowing the observation of interference figures, which are helpful to identify minerals (Bloss, 1999; Nesse, 2012; Selvestone, 2015).

In petrographic microscopes, we can observe the color of the minerals in thin sections using plane-polarized light. When a polarizer is inserted into the microscopic optical train (between the ocular and the light source), it allows the light ray to vibrate in one direction only, for example, from left to right (or west to east). However, when this light crosses the minerals in the thin section, some wavelengths are preferentially absorbed according to the direction of the light that is crossing through the mineral. When observed under crossed polars in the microscope, isotropic minerals display the same light absorption in all directions and appear dark whatever optical orientation is applied by rotating the stage that holds the rock thin section (Figure 7.4B).

Other minerals, however, change their color under polarized light when the microscope stage is rotated. These minerals are called anisotropic, as they change color due to the selective absorption of certain wavelengths by different crystallographic directions. The transmitted light is split into ordinary and extraordinary rays that have different refraction indexes and, thus, different speeds and colors. The change in colors with rotation is known as pleochroism, meaning multiple changes in colors. Pleochroism is explained by the unequal absorption of light when the thin section is rotated and the mineral is observed in different orientations (Gribble & Hall, 1985; Klein & Philpotts, 2017).

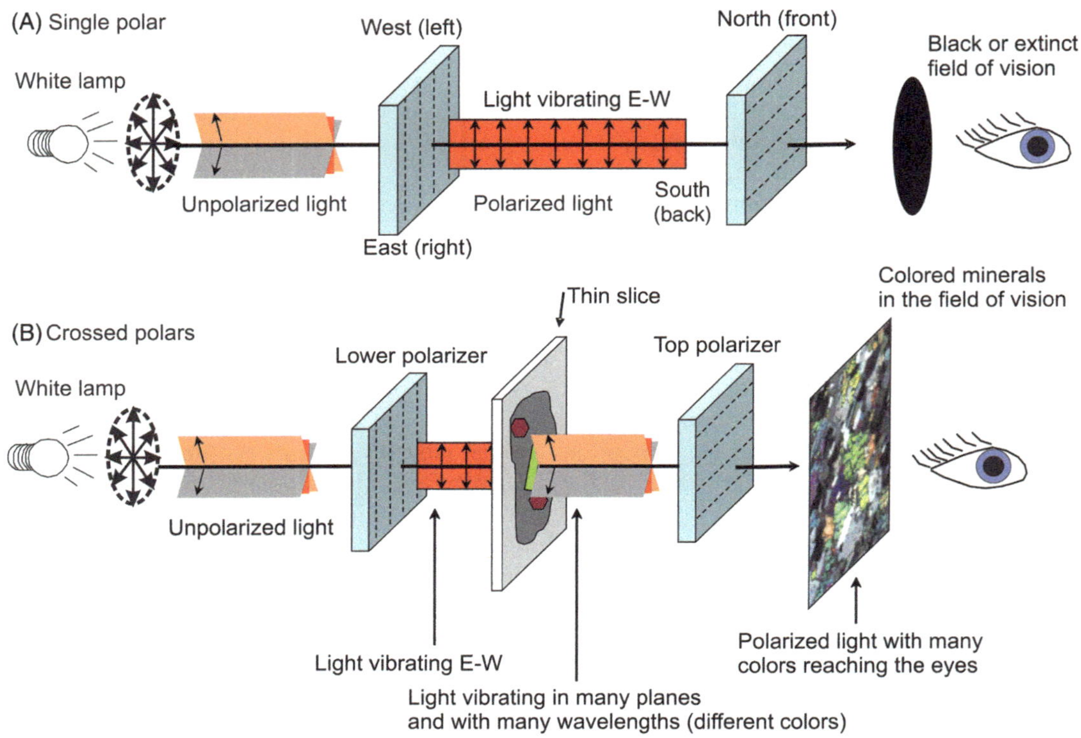

Figure 7.4 Images obtained by petrographic microscopy of mineral thin slices: (A) using single polar; (B) using crossed polars. Credit: With permission of Jane Selvestone (2015).

For example, biotite exhibits colors that change from tan to brown when the stage is rotated 90° (e.g., MacKenzie et al., 2017).

Anisotropic minerals change the velocity of light within the crystal, resulting in a RI that depends on the crystallographic direction. Isotropic minerals have one RI, while uniaxial minerals have two RIs, and biaxial minerals have three RIs, one for each crystallographic direction (expressed as the X, Y, and Z axes). The refraction index, usually called n, varies for different crystalline orientations observed under a microscope (Bloss, 1999; Gribble & Hall, 1985; Nesse, 2012).

When a lower-plane polarizer is inserted into a microscope, light travels in one direction (west to east), while the upper-plane polarizer allows light waves to vibrate only from top to bottom (north to south). When both polarizers are inserted, the light source (white) appears black or becomes extinct. When a thin section is inserted between the lower and upper polarizers, the preferential absorption of light along different crystallographic directions changes the wavelength, resulting in distinct colors under the microscope (Figure 7.4B).

Using the two polarizers in the microscope optical train, the minerals in the thin section reorient the planes in which light is vibrating, and some light passes through the upper polarizer and is seen by our vision with clear, different colors. As most minerals are characterized by particular crystal forms, habits, and cleavages, some internal structures are recognized when examined under the microscope, such as the cleavage planes (e.g., Figure 7.4).

Minerals are typically anisotropic, reorienting light in one or two preferential directions. When the stage is rotated, some minerals become black four times during a 360° rotation. This means that the crystals of these minerals do not allow light to be seen every 90°. Some minerals, like garnet, are always black when the stage is rotated, indicating that they have the same optical characteristics in all directions. These minerals are called isotropic and belong to the cubic class in crystallography (e.g., Bloss, 1999). The sphere diameters represent the refraction index (n), which is constant for all directions.

One of the best examples is garnet, a common mineral in igneous, metamorphic, and sedimentary rocks. However, there are several evaporite minerals that are isotropic, such as, for example, halite and sylvite, both appearing black under the crossed nicols in petrographic microscopes (e.g., MacKenzie et al., 2017). Isotropic minerals are isometric, meaning that they possess the same crystalline properties in all directions and cannot reorient polarized light at perpendicular angles. Therefore, light does not rotate or split; it propagates with the same

Mineral optics	General characteristics
Isotropic	The crystals are isometric and all crystallographic axes (a, b, and c) are equal and the angles between them are all 90°. Cubic minerals belong to this class.
Uniaxial	The crystals have all axes with the same length and are perpendicular to the c axis. However, the c axis is distinct, with a different length. The mineral classes that are associated with these minerals are the hexagonal and tetragonal classes.
Biaxial	The crystals have all the axes with different lengths. The mineral classes that are associated with these minerals are the orthorhombic, monoclinic and triclinic.

Figure 7.5 Mineral optics with unit cell characteristics of isotropic, uniaxial, and biaxial crystals. The general characteristics are identified by examining the thin sections in a petrographic microscope.

velocity in all directions across the crystal. When these minerals are seen with the two polarizers (crossed polars), they always appear black (extinct), even when rotated by any angle (e.g., Bloss, 1999).

Minerals with one special direction are uniaxial, and those with two special directions are biaxial (Figure 7.5). This chapter provides a simplified table of the optical characteristics of some common evaporite minerals and their chemical composition, but most mineralogy textbooks offer more extensive datasets for mineral identification.

7.3.2. Basic Properties of Transmitted Light in Minerals

In isotropic minerals, such as garnet, a source of light will travel through the crystal at the same speed in all directions. Imagine a garnet crystal with a lamp inside. When the light is turned on, the light beams travel the same distance for a fixed amount of time. The indicatrix is the shape of the light rays traveling through a mineral for a fixed amount of time. Thus, for isotropic minerals, light travels the same distance in all directions. Mapping the geometric shape defined by the light rays will result in an indicatrix corresponding to a sphere. This means that the difference between the two refraction indexes (birefringence) is null, as n is the same everywhere.

$$d = n_{high} - n_{low} = 0(\text{image is black}).$$

Anisotropic minerals are also characterized by changing the light rays that enter the crystal. A single ray of the incoming light is split into two rays, which are refracted in different amounts. The changes in wave characteristics are restricted to only two (mutually perpendicular) vibration directions once the light enters an anisotropic crystal. These directions are called privileged directions, which correspond to the ordinary (O or w) and extraordinary (E or e) rays. The refraction index (n) varies for each light ray along with these directions, which are expressed as:

$$\mathbf{w} = n_w$$

$$\mathbf{e} = n_e.$$

These refracted rays have different velocities within the crystal, hence different refraction indexes. One of the rays that are formed is the ordinary ray, which is not deflected when entering the crystal. The other ray, whose plane of vibration is rotated by 90°, is called the extraordinary ray. When the light ray enters an anisotropic crystal, it splits into two rays, causing a change in velocity. However, the frequency of the light remains unchanged regardless of the mineral composition. Since frequency is the ratio between velocity and wavelength (f = v/l), the wavelength will have to change to keep the ratio constant.

Uniaxial minerals are characterized by an optical indicatrix that corresponds to an ellipsoid with one long axis (the Z-axis) and two equal axes (the X- and Y-axes). The horizontal sections across a vertical Z-axis thus correspond to circles, with the radius representing the refraction index n. If the refraction index for the ordinary ray (along the Z-axis) is greater than the extraordinary rays, the optical indicatrix is assigned a positive (+) index, corresponding to positive uniaxial minerals (Bloss, 1999; Nesse, 2012).

In sections cut parallel to the c-axis, we observe the maximum and minimum refraction indexes of the uniaxial mineral. Rotating the thin slice on the microscope stage causes the grain to change color (birefringence). But there are two positions where the grain will go black when the indicatrix axis is E-W or N-S, indicating that the refraction indexes are equal (Bloss, 1999; Nesse, 2012; Selvestone, 2015).

Uniaxial minerals split light that enters from all but one special direction into two plane-polarized components, which vibrate perpendicular to one another and travel with different velocities. But in one particular direction, the light does not get rotated; it propagates with the same velocity in all directions, and along that direction, the mineral behaves as isotropic, so it becomes extinct under the crossed polars. The ordinary rays are characterized by a refraction index n w, which has the same value in all directions, and in the optical indicatrix, it corresponds to a sphere (Nesse, 2012; Selvestone, 2015).

The extraordinary ray vibrates in a direction perpendicular to the plane of the ordinary ray, and the refraction index for this ray is known as n e. The difference in the refraction indexes (maximum and minimum velocity of light within the crystal) results in interference colors, or birefringence, which can be observed when polars are crossed. The difference is known as delta (d) and corresponds to the difference of the n values:

$$d = n_{high} - n_{low}.$$

When the rays with different velocities (and thus different wavelengths) leave the crystal, they are recombined, but the wavelengths will be different, and thus, the combined ray will have a change in wavelength, equivalent to a change in color. The apparent birefringence – (d) is the difference in RI (speed) between the two rays traveling through the crystal. This results in the retardation of the waves that are combined after leaving the crystal. The retardation, or the distance separating the two rays, thus results in variations of the wave amplitude, which can be increased (when in-phase) or decreased (out-of-phase).

The differences in refraction indexes (birefringence) usually range from 0.001 to 0.036 and can be seen in minerals with different colors corresponding to the first-order to the third-order, as shown on the standard Michel Levy's color charts. For example, the mineral quartz is characterized by a refraction index along the normal or ordinary ray (w) of 1.544 and a refraction index along the extraordinary ray (e) of 1.553. This gives a difference (e – w) of 0.009, which corresponds to the birefringence of the mineral in the plane cut in the thin slide. As e is greater than w, the sign is positive (Bloss, 1999; Selvestone, 2015).

Interference figures are obtained in cross-polarized light by the insertion of the Bertrand lens. For uniaxial minerals, these figures resemble a cross separating four quadrants in the optical indicatrix. The insertion of a gypsum accessory plate with a fast direction along the NW-SE direction results in changes in the interference color bands. Positive minerals are characterized by a decrease in the interference colors along the fast direction of the gypsum plate and an increase in the interference colors along the opposite NE-SW direction. Negative minerals show an increase in the interference colors along the fast direction of the gypsum plate and a decrease in the opposite direction (Bloss, 1999; Selvestone, 2015).

In biaxial minerals, the light that enters the crystal is resolved into two plane-polarized components except in two special directions, which become extinct when the stage is rotated under crossed polars. In these special directions, which are called the optic axes, the mineral behaves as isotropic, which means that no splitting of light occurs. However, in most planes that are obtained from mineral cross-sections in thin slides, the refraction indexes are unequal, thus resulting in several possibilities for the cross-sections of the biaxial indicatrix, which is represented by a triaxial ellipsoid (Bloss, 1999; Nesse, 2012).

As mentioned, birefringence is the optical property of some minerals that have more than one index of refraction. This anisotropy depends on the crystallographic projection of the mineral in the thin section, and it can be estimated from the pattern of color interference in a petrographic microscopic thin section. The thin section thickness is normally around 30 μm, and so the bands of different colors depend on the birefringence, which changes with the crystallographic orientation of the mineral observed under the microscope. The greater the birefringence, the higher the order of the interference colors, which can be compared with standard color tables. There are several charts published in the crystallography literature, such as the ones published by Bloss (1999) and Raith et al. (2012), or the revised Michel-Lévy interference color chart published by Sorensen (2013).

Uniaxial and biaxial minerals can be further subdivided into optically positive and optically negative, depending on the orientation of fast and slow rays relative to the crystal axes. This property can be determined using interference figures known as isogyres, which look like small black crosses splitting the different colors resulting from birefringence (Bloss, 1999; Nesse, 2012; Raith et al., 2012). These curved, broad black bands mark areas of extinction in the interference figures relating to the optic axes, and for uniaxial minerals, a simple cross is produced, whereas for biaxial minerals, a more complex cross can be seen (Barker, 2014).

The biaxial minerals have three perpendicular principal axes of different lengths (Figure 7.5). Thus, the indicatrix will have three different RIs to describe a biaxial mineral. The ellipsoid can be determined by one circle with a radius equivalent to the intermediate RI (n B), and three ellipses formed with the minimum, intermediate, and maximum RI (Bloss, 1999; Raith et al., 2012). The acute angle between the indicatrix planes is called 2V, and using interference figures obtained by examining the mineral under crossed polars, the crystals can be classified as positive (when the 2V angle is acute around the Z-axis) or negative (when the 2V angle is acute around the X-axis). These optical properties are included in the standard petrographic tables (e.g., Figure 7.6).

Mineral	Formula	Crystal system	Morphology	Optical data reflection index values	2V Angle	Optical properties	Extinction	Birefringence
Aragonite	$Ca(CO_3)$	Orthorhombic	columnar, fibrous pseudo-hexagonal	$n\alpha$ = 1.529 – 1.530 $n\beta$ = 1.680 – 1.682 $n\gamma$ = 1.685 – 1.686	Measured 18° to 19° Calculated: 16° to 18°	Biaxial (–)		extremely high δ = 0.156
Calcite	$Ca(CO_3)$	Trigonal	Prominent rhombohedral cleavage. Lamellar glide twinning is common.	$n\omega$ = 1.640 – 1.660 $n\varepsilon$ = 1.486		Uniaxial (–)	Symmetrical to cleavage traces	extremely high δ = 0.154 – 0.174
Dolomite	$CaMg(CO_3)_2$	Trigonal	Usually form clusters of small rhombohedral crystals, simple twinning common	$n\omega$ = 1.679 – 1.681 $n\varepsilon$ = 1.500 – 1.503		Uniaxial (–)		extremely high δ = 0.179
Baryte	$Ba(SO_4)$	Orthorhombic	Usually thin to thick tabular	n_α = 1.636 n_β = 1.637 n_γ = 1.648	Measured: 36° to 42° Calculated: 36° to 40°	Biaxial (+)	// best cleavage	weak (cf. quartz) δ = 0.012
Celestine	$Sr(SO_4)$	Orthorhombic	Crystals commonly thin to thick tabular	$n\alpha$ = 1.619 – 1.622 $n\beta$ = 1.622 – 1.624 $n\gamma$ = 1.630 – 1.632	Measured: 50° to 51° Calculated: 54° to 58°	Biaxial (+)	// cleavages and faces	weak (cf. quartz) δ = 0.011
Anhydrite	$Ca(SO_4)$	Orthorhombic	Crystals nearly equant, with large pinacoidal faces	$n\alpha$ = 1.567 – 1.574 $n\beta$ = 1.574 – 1.579 $n\gamma$ = 1.609 – 1.618	Measured: 36° to 45° Calculated: 44°	Biaxial (+)	// cleavages	strong (colours 3rd order) δ = 0.042 – 0.044
Gypsum	$Ca(SO_4){\cdot}2H_2O$	Monoclinic	Thin to thick tabular crystals, "swallow-tail" twinning very common	$n\alpha$ = 1.519 – 1.521 $n\beta$ = 1.522 – 1.523 $n\gamma$ = 1.529 – 1.530	2 V ~58 Measured: 58° Calculated: 58° to 68°	Biaxial (+)	// best cleavage	weak (cf. quartz) δ = 0.010
Halite	NaCl	Isometric	Normally cubic, rarely octahedral. Crystal faces often cavernous and stepped (hopper crystals). Massive. Coarsely granular to compact. Columnar, stalactitic or capillary forms are rare.	n = 1.5443		Isometric	Isotropic/no extinction	Isotropic/no birefringence
Sylvite	KCl	Isometric	Normaly cubic; less commonly octahedral habit.	n = 1.4903		Isometric	Isotropic/no extinction	Isotropic/no birefringence
Carnallite	$KMgCl_3 \cdot 6H_2O$	Orthorhombic	Pseudo-hexagonal crystals with pyramidal habit.	$n\alpha$ = 1.465 – 1.466 $n\beta$ = 1.474 – 1.475 $n\gamma$ = 1.494 – 1.496	Measured: 70° Calculated: 66°	Biaxial (+)		δ = 0.029 – 0.030
Trona	$Na_3(HCO_3)(CO_3){\cdot}2H_2O$	Monoclinic	Crystals elongated and also flattened	n_α = 1.412 n_β = 1.492 n_γ = 1.540	Measured: 72° Calculated: 70°	Biaxial (–)		δ = 0.128
Mirabilite	$Na_2(SO_4)$ $10H_2O$	Monoclinic	Crystals short prismatic to acicular.	$n\alpha$ = 1.396 $n\beta$ = 1.410 $n\gamma$ = 1.419	Measured: 75° Calculated: 74°	Biaxial (–)		δ = 0.023
Bischofite	$MgCl_2 \cdot 6H_2O$	Monoclinic	Short prismatic crystals, commonly foliated, fibrous, crystalline granular, also massive	n_α = 1.495 n_β = 1.507 n_γ = 1.528	Measured: 79° Calculated: 76°	Biaxial (+)		δ = 0.033
Kainite	$KMg(SO_4)Cl \cdot 2.75H_2O$	Monoclinic	Crystals thick tabular to equidimensional	n_α = 1.494 n_β = 1.505 n_γ = 1.516	Measured: 90° Calculated: 88°	Biaxial (–)		δ = 0.022
Kieserite	$Mg(SO_4) \cdot H_2O$	Monoclinic	Crystals are rare, mostly bipyramidal.	n_α = 1.520 n_β = 1.533 n_γ = 1.584	Measured: 55° Calculated: 56°	Biaxial (+)		δ = 0.064
Polyhalite	$K_2Ca_2Mg(SO_4)_4 \cdot 2H_2O$	Triclinic	Crystals small and rare; tabular to elongated.	n_α = 1.546 – 1.548 n_β = 1.558 – 1.562 n_γ = 1.567	Measured: 60° to 62 Calculated: 60° to 80°	Biaxial (–)		δ = 0.021
Tachyhydrite	$CaMg_2Cl_6 \cdot 12H_2O$	Trigonal	Massive; in rounded masses, can show twin lamellae at times.	n_ω = 1.520 n_ε = 1.512		Uniaxial (–)		δ = 0.008

Figure 7.6 Petrographic characteristics of common evaporite minerals. The summary includes the crystal system, common crystal habits, and basic optical properties.

Interference figures are useful techniques used to help identify minerals using a polarizing light microscope. The interference figures are displayed for all minerals except cubic minerals. Two main types of these figures are associated with anisotropic minerals, corresponding to uniaxial or biaxial figures. Uniaxial figures are formed from tetragonal, trigonal, and hexagonal minerals. They are produced by orientating the thin section in a suitable direction. The color of the NE quadrant changes when a Bertrand lens is inserted in the microscope and a first-order red accessory plate is added to the optical train. If the color of this quadrant is blue, the mineral interference figure is noted as positive (denoted +ve), and if it becomes yellow, then the mineral is negative (−ve). Depending on the direction of the length-slow or length-fast direction in the gypsum accessory plate, these parameters are reversed (Klein & Philpotts, 2017; Nesse, 2012). Petrographic charts include the positive or negative sign of uniaxial and biaxial minerals (Figure 7.6).

7.4. PETROGRAPHIC CHARACTERISTICS OF EVAPORITE MINERALS

Some optical microscopy techniques are valuable in determining the mineral species when a thin section slide is observed in the microscope. Figure 7.6 presents a succinct summary of the petrographic characteristics of the most common minerals present in evaporite successions. The general characteristics of the evaporite minerals in salt basins are presented in Figure 7.7. The evaporite mineral names and formulas comply with the internationally accepted publications by the International Mineralogical Association (2022).

Considering that the maximum and minimum RIs of any mineral are oriented along with precise crystallographic directions according to their crystal class, the highest interference colors will be shown by a mineral section that contains the maximum and minimum refraction indexes in the plane of the thin section (Gribble & Hall, 1985). Figure 7.6 shows the refraction indexes of the common evaporite minerals along their crystal axes.

All anisotropic minerals can resolve light into two plane-polarized components that travel at different velocities and vibrate in planes that are perpendicular to one another. When rotated in the microscope under polarized light, these anisotropic minerals will go into extinction four times during a complete rotation (360°) of the thin section. Removing the analyzer from the microscope optical train while the mineral is in extinction allows the determination of some physical properties, such as the cleavage or trace of a crystal face edge (Gribble & Hall, 1985). The extinction angles are usually expressed in the mineral description tables as the angle between the slow (y) or fast (a) ray and the cleavage or face edge of the mineral (Figure 7.6).

7.5. PETROGRAPHIC ANALYSIS OF EVAPORITE MINERALS AND INTEGRATION WITH ADVANCED METHODS

Microscopic petrographic analysis is a valuable tool for determining rock textures, structures, mineral suites, and the characterization of petrological microfacies in carbonate and evaporite sequences. It can analyze parameters such as mineral types, pore distribution, and diagenetic factors such as cementation processes and the replacement of authigenic mineral phases (Adams et al., 1984; Gribble & Hall, 1985; Raith et al., 2012).

Slabs obtained from evaporite sequences exposed on land (outcrops), underground mines, and borehole cores commonly provide the rock material used for petrographic analysis in salt basins. Exploratory boreholes usually sample evaporite sequences at significantly thick (at least a few meters) intervals during drilling operations. Thin sections prepared from outcrop samples or core slabs are usually studied by normal transmitted light. Researchers usually study thin sections using normal transmitted light or crossed polars in a polarizing microscope to enhance visualization and determine optical properties. Some microtectonics research groups irradiate thin sections with gamma rays, highlighting halite grain boundaries within a blue-colored background (see Chapter 6).

Electron microprobe analysis (EMPA) is an advanced technique used for chemical analysis of selected areas of rock samples, focusing an electron beam that excites X-rays in compounds (Reed, 2005). The scanning electron microscope produces detailed images of the crystalline structure of evaporite minerals on a microscale to nanoscale (Reed, 2005).

In addition to petrographic microscopy, cathode luminescence (CL) is commonly conducted in mineralogy laboratories using polished rock chips or thin sections examined with a CL microscope (Adams & Mackenzie, 1998). Luminescent light is emitted when some minerals are excited by electron fluxes, which are associated with electromagnetic radiation with wavelengths lower than ~390 nm, corresponding to the ultraviolet range that is invisible to the human eye. Minerals that are fluorescent when bombarded with electrons (for example, some chlorides and carbonates) emit light in the visible range. Similar processes are associated with light emission when some substances are heated, resulting in thermoluminescence (Lustrino, 2021).

The CL methodology can help reveal further details about different phases of cement generation and secondary diagenetic replacements affecting evaporite minerals, particularly in sedimentary rocks containing carbonate minerals. Cathodoluminescence microscopy is especially useful to provide information on the distribution of trace elements in sedimentary rocks containing carbonate minerals (for example, calcite and dolomite).

Mineral/Formula	Characteristics
Aragonite $Ca(CO_3)$	A common component in recent marine carbonate sediments, precipitated directly from seawater. Many organisms with calcareous skeletons build their hard parts with aragonite, similar to calcite. But as aragonite is less stable than calcite after burial, most fossil shells are transformed into calcite. Easily soluble with effervescence in cold dilute HCl. Much denser than calcite or dolomite, aragonite sinks in bromoform whereas the others float. Using Feigl's solution, aragonite is stained black whereas calcite and dolomite remain with the color unchanged.
Calcite $Ca(CO_3)$	One of most abundant minerals in sedimentary sequences associated with evaporite rocks. It is the principal constituent of limestone, chalk, travertine, calcite tufa, oolitic carbonates and coquinas (shell fragments). It is the major chemical constituent of many organisms, such as foraminifers, sponges, corals, echinoids, crinoids, bryozoans, grastropods, etc. The mineral is easily soluble with effervescence in cold dilute HCl. Calcite displays fluorescent, thermoluminescencent and cathodolominescent characteristics. On heating, calcites show a natural thermoluminescence. Under ultraviolet radiation calcite displays fluorescence colors at various wavelengths. Under electron beam excitation, calcite produces cathodoluminescene. In acid solutions, some organic dyes (such as alizarin red S) stain calcite minerals, whereas other carbonate minerals such as dolomite and magnesite are stained in basic solutions.
Dolomite $CaMg(CO_3)_2$	Dolomite is uniaxial and optically negative. The maximum birefringence of dolomite ($\omega - \varepsilon \sim 0.179$) is very high. The crystals are characterized by a perfect rhombohedral cleavage. Dolomite is soluble with difficulty in cold dilute HCl solution.
Baryte $Ba(SO_4)$	Baryte (also spelled barite) is the least soluble sulfate and the most abundant barium mineral in the earth's crust. It may occur as a residual deposit left after the host rock (evaporite or carbonate rocks) have dissolved. It is the principal source of barium compounds and is used in the petroleum industry as an additive to the drilling mud. Baryte is not easily distinguishable from celestine, except for its greater density, which is a criteria to distinguish from other minerals such as gypsum or calcite. Baryte is insoluble in diluted HCl whereas carbonate rocks such as calcite or aragonite effervesce with the acid solutions. Concretions of baryte in sandstone sometimes form structures known as rosettes, also called sand baryte or desert roses.
Celestine $Sr(SO_4)$	Celestine is member of the baryte group and is by far the principal source of strontium. It produces a crimson flame color and it widely used to manufacture fireworks and flares. It often occurs in sedimentary layers that are rich in calcium sulfate (anhydrite or gypsum deposits).
Anhydrite $Ca(SO_4)$	One of the most common sulfate minerals, lacking water in its composition, in contrast to gypsum, which contains two water molecules in the chemical structure. Gypsum layers are transformed into anhydrite with the increase in temperature and pressure due to burial.
Gypsum $Ca(SO_4).2H_2O$	The most common sulfate mineral. Concretions of gypsum in sandstone sometimes form structures known as rosettes, also called desert roses (similarly to baryte).
Halite NaCl	Easily soluble in water, exhibits a salty taste, and colours a flame deep yellow.
Sylvite KCl	The most important source for potassium. Normally associated with halite and carnallite.
Carnallite $KMgCl_3 \cdot 6H_2O$	Deliquescent and soluble in water. The evaporated solution yields sylvite and bischofite.
Trona $Na_3\,(HCO_3)\,(CO_3) \cdot 2H_2O$	Soluble in water and effervescent in acids. The mineral is associated with continental environments.
Mirabilite $Na_2(SO_4) \cdot 10H_2O$	Quickly dehydrates to thenardite in dry air.
Bischofite $MgCl_2 \cdot 6H_2O$	Deliquescent and readily soluble in alcohol and in water. A relatively rare constituent of saline deposits. Secondary in origin in part. Formed by alteration of carnallite by water.
Kainite $KMg(SO_4)Cl \cdot 2.75H_2O$	Usually formed as a secondary mineral in sulfate deposits.
Kieserite $Mg(SO_4) \cdot H_2O$	Sulfate mineral with an origin associated with marine environments.
Polyhalite $K_2Ca_2Mg(SO_4)_4 \cdot 2H_2O$	Sulfate mineral associated with sulfate deposits. It is the most common source for potassium in New Mexico deposits.
Tachyhydrite $CaMg_2Cl_6 \cdot 12H_2O$	Extremely soluble mineral associated with marine brine. Readily deliquescent in the air and flows rapidly within evaporite sequences.

Figure 7.7 General characteristics of common evaporite minerals.

The CL responses are often described as luminescent or nonluminescent (dull minerals). Some types of equipment can provide quantitative information by measuring the intensity and the spectral analysis of the minerals in the sample (Scholle & Ulmer-Scholle, 2003).

More recently, the microtomography technique has been extensively used in the study of submicroscopic structures, as the method allows 3-D visualization of microscopic features in rock samples. Some interesting applications for the microstructural study of reservoir rocks have been discussed in the geological literature, including the characterization of presalt microbialites in the South Atlantic (Mohriak, 2015).

Microtomography is a technique that has gained significant use in studying submicroscopic structures, providing high-resolution images of pore inclusions within evaporite and non-evaporite minerals (e.g., Thiemeyer et al., 2015). Microtomographic analyses provide high-resolution images of pore inclusions within evaporite and non-evaporite minerals (e.g., Thiemeyer et al., 2015). The method has also been applied to production engineering projects, interpreting evaporite caprock integrity and sealing mechanisms (for example, Smith et al., 2012), and 3-D visualization of micropore distribution in reservoir rocks (e.g., Reyes et al., 2017). Meng et al. (2018) discussed the use of microtomography and SEM to characterize the morphology of halite and gypsum sequences subjected to dissolution by soaking in water at different temperatures. Pace et al. (2016) discussed an example of a modern approach to analyzing the diagenesis of the microbialites in the Great Salt Lake region of Utah. The study involved the integration of several of the techniques mentioned above, including mineralogical analysis, geochemistry, optical microscopy, and SEM.

Modern analytical techniques have been used to interpret the chemical composition of seawater in ancient sedimentary basins and study the paleoclimate evolution in sedimentary basins, as determined from analysis of primary fluid inclusions in halite minerals (e.g., Benison & Goldstein, 1999; Holt et al., 2014). Despite the increasing use of advanced methods, microscopic petrography remains an essential tool for the characterization of evaporite sequences in basin analysis projects. Examples of microscopic petrography are presented below, illustrating the geological analysis of the Permian evaporites in the Parnaiba Basin (northern Brazil) and the North Sea (NE England).

7.6. PERMIAN EVAPORITE DEPOSITS IN SOUTH AMERICA AND CENTRAL EUROPE

The Late Permian to Early Triassic geological period was marked by extreme climate aridity in the northern South America and central Europe (Scotese & Langford, 1995). This event resulted in the development of large evaporite basins in several regions, such as the Parnaíba Basin in northern Brazil and the Southern Permian Basin in North-Central Europe (Figure 7.8). The Permian evaporites in the Parnaíba Basin are associated with the Pedra de Fogo Fm. (Góes & Feijó, 1994; Vaz et al., 2007). The Permian evaporites in North-Central Europe were deposited from the North Sea (U.K. and Norway) to the Netherlands, Germany, and Poland, corresponding to the Zechstein Supergroup (Chorlton, 2007; Glennie, 1998; Słowakiewicz et al., 2013; Strozyk et al., 2017; Taylor, 1984, 1998).

Paleogeographic reconstructions for the Permian suggest that the Pangea supercontinent was formed by two large landmasses, the Laurasia continent in the north and the Gondwana continent in the south, surrounded by active subduction zones (Figure 7.8). The Panthalassa Ocean subducted under Laurasia and Gondwana, while smaller subduction zones associated with the Paleotethys Ocean developed west and east of Laurasia. The Parnaíba Basin was located inland, far from the paleo-oceans, while the Southern Permian Basin was connected to the Paleotethys (Figure 7.8B). Paleogeographic reconstructions indicate that the Parnaiba and North Sea-Central European basins were located close to the paleo-equator (Blakey, 2008; Scotese, 2008; Scotese & Schettino, 2017).

Climate aridity in the North-Central European basins led to the formation of the southern Permian salt basin, which accumulated thick evaporite deposits, including the Zechstein Supergroup (Słowakiewicz et al., 2013; Strozyk et al., 2017; Taylor, 1984, 1998; Ziegler, 1975). These deposits precipitated in basins within a depression extending from the North Sea region to the Netherlands, Germany, and Poland (Figure 7.9).

Desert environments occurred onshore in England, and the evaporite deposits included sedimentary facies such as the carbonate platform and halite layers (Figure 7.9). The North-Central European basins, particularly the North Sea, host significant hydrocarbon accumulations. In the Netherlands, the Groningen High straddles the slope, platform, and desert sedimentary facies of the Zechstein salt basin. The presalt sandstone reservoirs in the Groningen structure correspond to the largest gas discovery in Europe (Glennie, 1998).

In the Western Pangea landmass, the Parnaíba Basin onshore Brazil accumulated evaporite layers during the Permian (Góes & Feijó, 1994; Vaz et al., 2007). However, the salinity did not reach concentration levels similar to the Zechstein basin, where halite and soluble potassium salts were deposited at the end of the evaporite cycles (Taylor, 1984, 1985, 1998). The stratigraphic succession of the intracratonic Parnaiba Basin mainly includes sulfate minerals (gypsum and anhydrite) deposited in a continental desertic environment, forming a relatively thin evaporite sequence in the basin depocenter (Abrantes et al., 2016; Araújo et al., 2016; Barbosa et al., 2016).

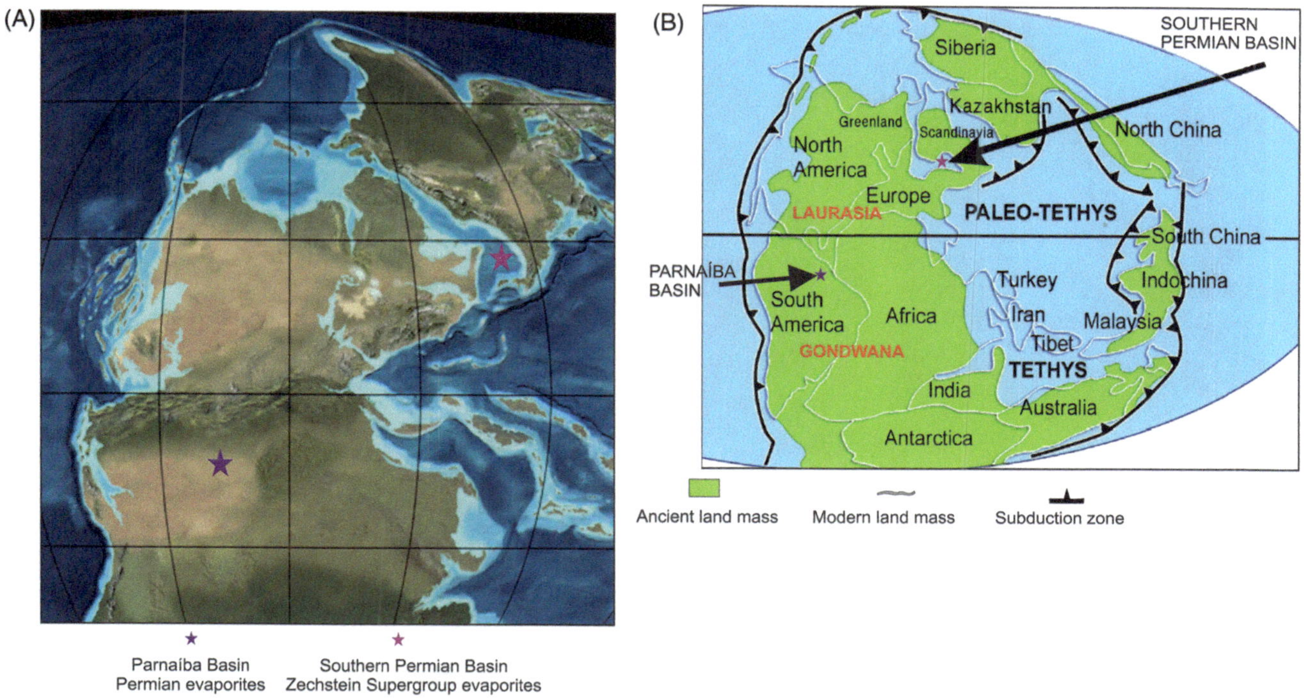

Figure 7.8 Paleoreconstruction map of the Gondwana and Laurasia landmasses during the Permian (A). The pink star shows the Southern Permian Basin (Zechstein evaporites) in northern central Europe. The purple star shows the Parnaiba Basin (Pedra de Fogo Fm.) evaporites in South America. The right-side diagram (B) shows a schematic tectonic map of the continents in the Permian. The landmasses were surrounded by the subduction zones affecting the Panthalassa Ocean (to the west) and the Paleo-Tethys Ocean (to the east). Credit: Adapted from Scotese (2008).

7.6.1. Parnaíba Basin, Onshore Northern Brazil

The Parnaíba Basin, a Paleozoic-Mesozoic basin in the northern Brazilian region (Figure 7.10), spans 600,000 km^2 and is home to an intracratonic sag basin with a maximum total sedimentary thickness of 3,500 m (Góes & Feijó, 1994; Vaz et al., 2007). The Parnaíba Basin has been extensively studied by academic researchers and oil companies since the last century. Geophysical investigations have been conducted, including the acquisition of deep seismic profiles and the drilling of exploratory and stratigraphic wells by oil companies (Manenti et al., 2018; Miranda et al., 2018).

The Brazilian Geological Survey agency, CPRM, has integrated the country's geological maps and conducted campaigns to identify mineral resources in the basin. The Brazilian National Petroleum Agency (ANP) has also drilled stratigraphic wells and acquired regional seismic profiles in underexplored areas. Petrobras and other oil companies have also drilled wells in the basin, and gas accumulations have been discovered in recent years (Abelha et al., 2018; Miranda et al., 2018).

The Parnaíba Basin is an intracratonic sag basin formed above the Neoproterozoic to Early Paleozoic rifts. During the Silurian to Devonian, siliciclastic stratigraphic sequences were deposited within a shallow-water platform and deep marine environments, including organic-rich shales deposited in the Devonian (Pimenteiras Fm.), which display some potential for hydrocarbon generation (Milani & Thomaz Filho, 2000). There are several depositional cycles in the basin, which are limited by regional faults. Some of these cycles are linked to orogenies or tectono-magmatic events (Góes & Feijó, 1994; Vaz et al., 2007).

After the Hercynian Orogeny, which caused a regional unconformity impacting the Silurian and Devonian marine stratigraphic sequences, the Carboniferous and Permo-Triassic (P-T) sequences were deposited in a depression characterized by extreme aridity. The Pedra de Fogo Fm. (Early to Late Permian) is characterized by a large variety of lithologies in several evaporite layers, including silex, limestone, dolomite, sandstone, siltstone, shale, and sulfate minerals (gypsum), which in the electric logs are characterized by the low gamma ray readings and high resistivity (Figure 7.11). These rocks were deposited in shallow-water marine environments and coastal and inland desertic environments, including mudflats and saline lakes that desiccated forming salt pans. The main mineral was gypsum, but probably some halite was also

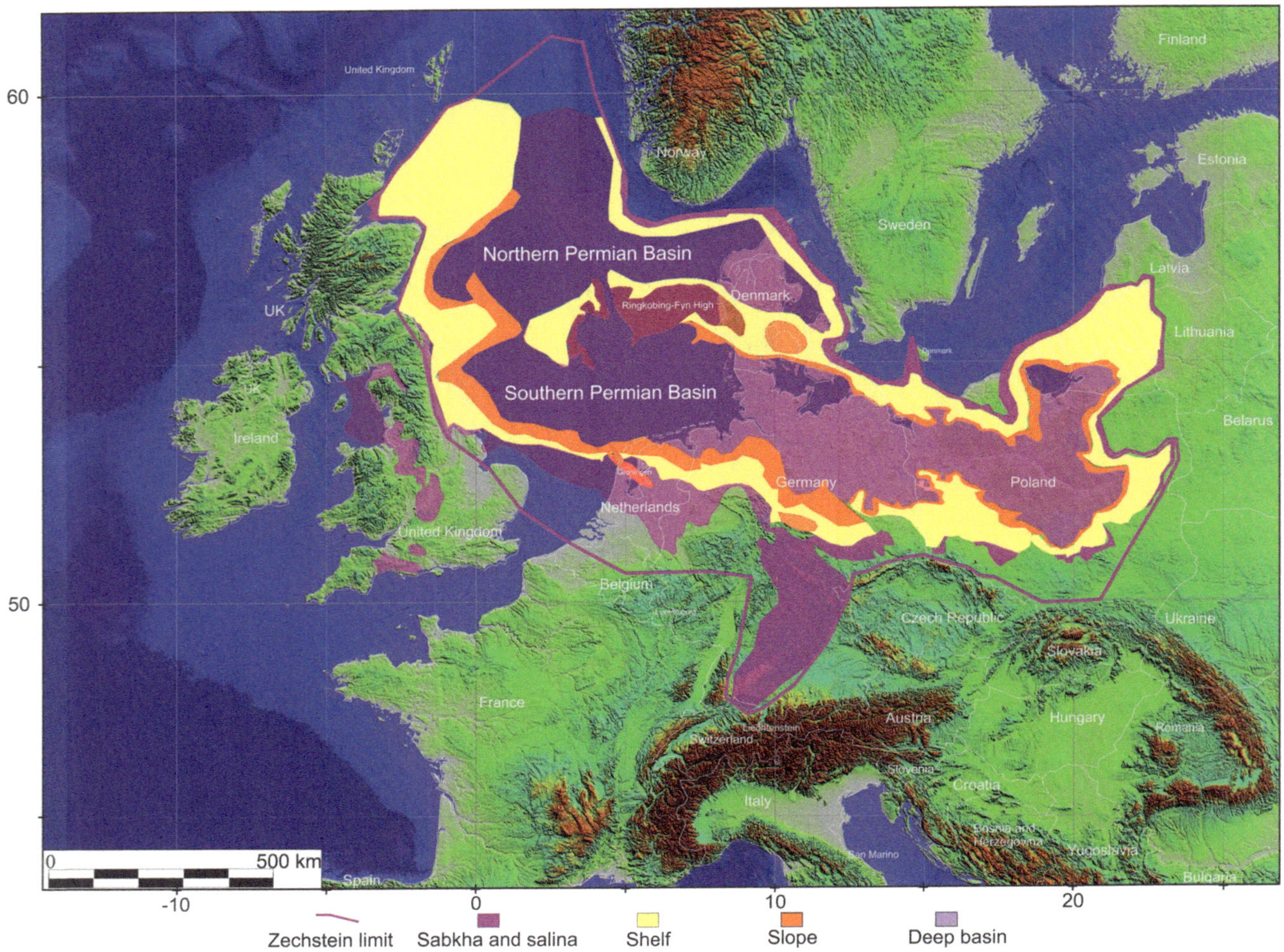

Figure 7.9 Simplified outline of the giant Zechstein salt basin in North-Central Europe. The Northern and Southern Permian basins were separated by the Ringkobing-Fyn High in the Central North Sea. Desert environments (sabkha and salinas) occurred on land in Northeastern England. The halite layers were mainly concentrated in the Zechstein deep basin offshore, which extended from the North Sea toward Germany and Poland. Carbonate rocks at the edge of the Permian platform constitute one of the main exploratory targets in the Zechstein Basin.

deposited, as indicated by the interpretation of the electrical logs from exploratory wells and by the XRD analysis of evaporitic nodules in Permian sandstones (Araújo et al., 2016; Barbosa et al., 2016).

In the Late Triassic/Early Jurassic, the basin was affected by an important magmatic episode that formed intrusive and extrusive rocks included in the Mosquito Fm. (Figure 7.11). These volcanic rocks are associated with the Central Atlantic Magmatic Province (CAMP), a major event that also affected North America and northwest Africa (Marzoli et al., 1999). In the Early Cretaceous, the basin was affected by another magmatic episode. The volcanic rocks of the Sardinha Fm. are related to the breakup of the Western Gondwana and the inception of the South Atlantic Ocean (Heilbron et al., 2018).

The stratigraphic successions of the Parnaiba Basin, as drilled by the deep stratigraphic borehole 2-CP-1-MA (Figure 7.11A) are predominantly composed of siliciclastic sediments. The Permian evaporite sequences in the Pedro de Fogo Fm. record two layers of anhydrite with a thickness of about 10 m each, at depths of approximately 700 and 900 m (Figure 7.11B).

The Permian evaporite rocks outcrop at the central to southern portion of the basin, forming a curved belt that trends NE south of the depocenter and then inflects toward the north direction, approaching the coastline (Figure 7.10). The evaporite sediments at the surface correspond to mudstones and gypsum layers deposited in mudflats and saline pans (Abrantes et al., 2016; Araújo et al., 2016).

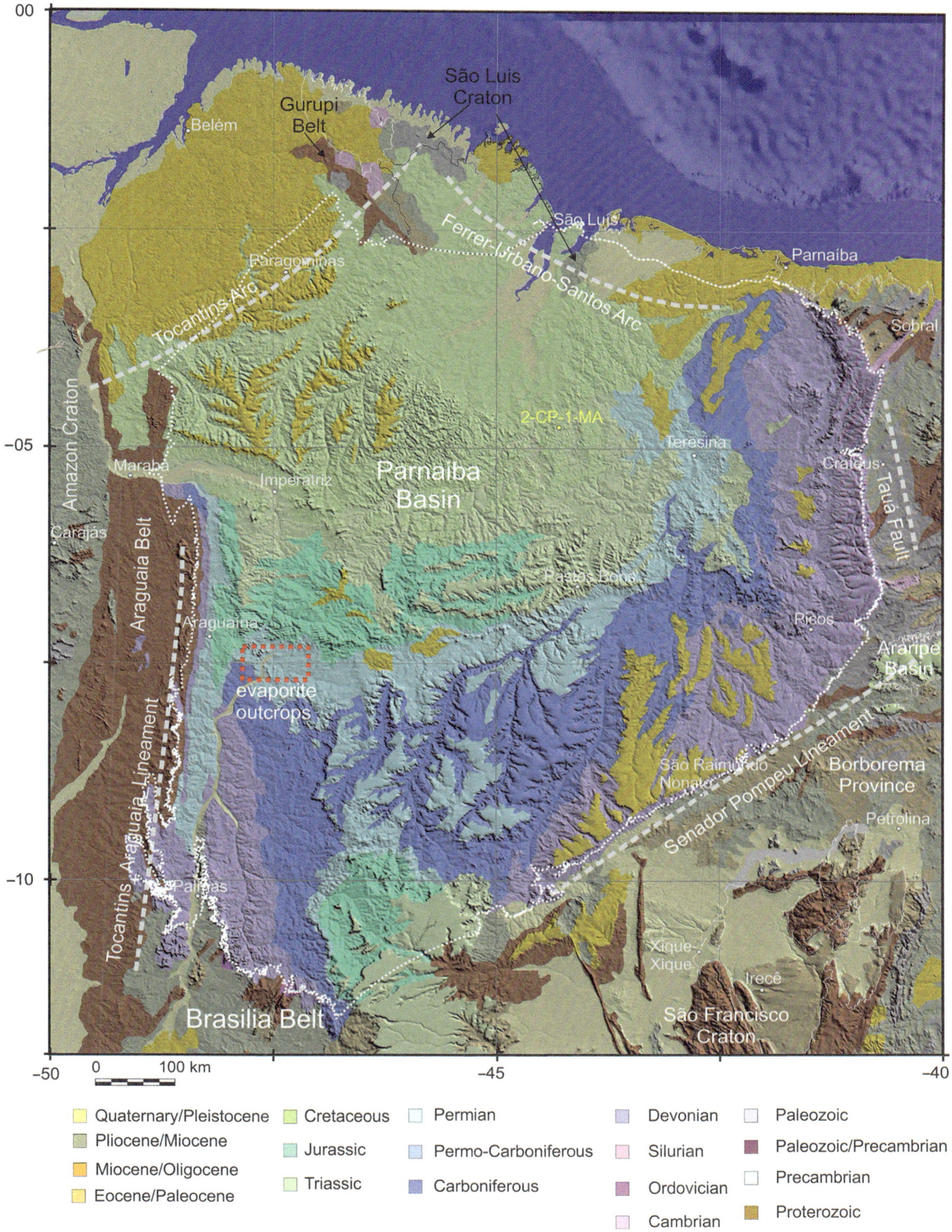

Figure 7.10 A simplified geological map of the Parnaiba Basin, located on land in northern Brazil, is draped on the topobathymetric map. The Paleozoic basin limit is outlined by a thin, white dashed line. The stratigraphic well 2-CP-1-MA, located in the central northern part of the basin, is marked by a yellow dot. The Permian rock outcrops, indicated by light blue color, form an E-W belt south of the basin center but inflect in an N-S direction toward the coastline. The region of Permian evaporite outcrops at the western part of the basin, south of the basin depocenter, is marked by the red dash rectangle.

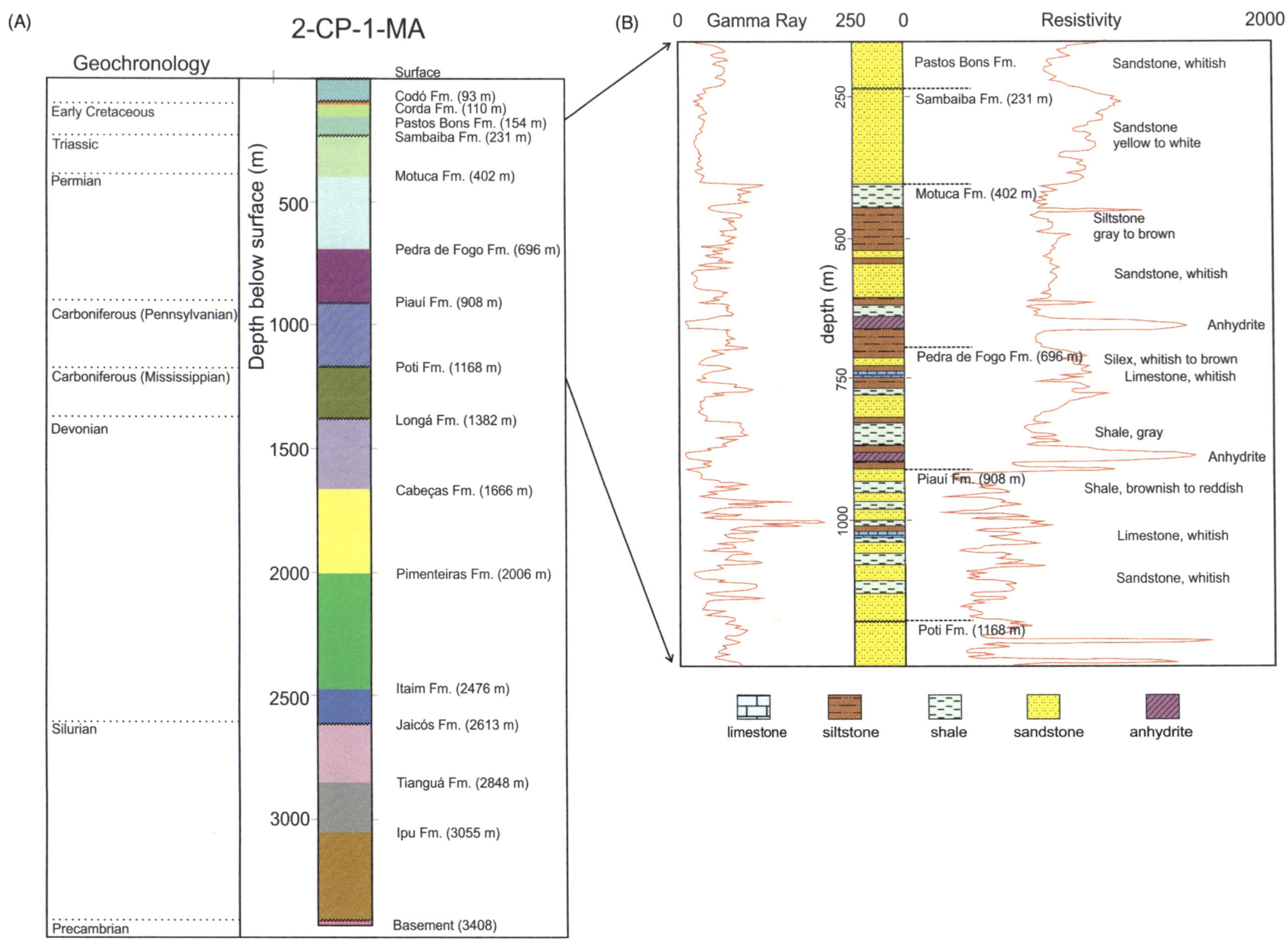

Figure 7.11 Lithostratigraphic column of the stratigraphic well 2-CP-1-MA in the central northern part of the Parnaiba Basin. The borehole penetrated the Precambrian basement at a total depth of 3,408 m below the surface. The left column (A) shows the main depositional sequences in the Mesozoic and Paleozoic. The right column (B) shows the electric logs and main lithostratigraphic units associated with the Permian sequences. The interval with evaporite sediments (magenta color) corresponds to the Early to Late Permian Pedra de Fogo Fm. and the Motuca Fm. (Late Permian to Early Triassic). The anhydrite layers are characterized by low gamma ray (GR) values and high resistivity readings in the electric logs. Credit: Adapted from Góes and Feijó (1994).

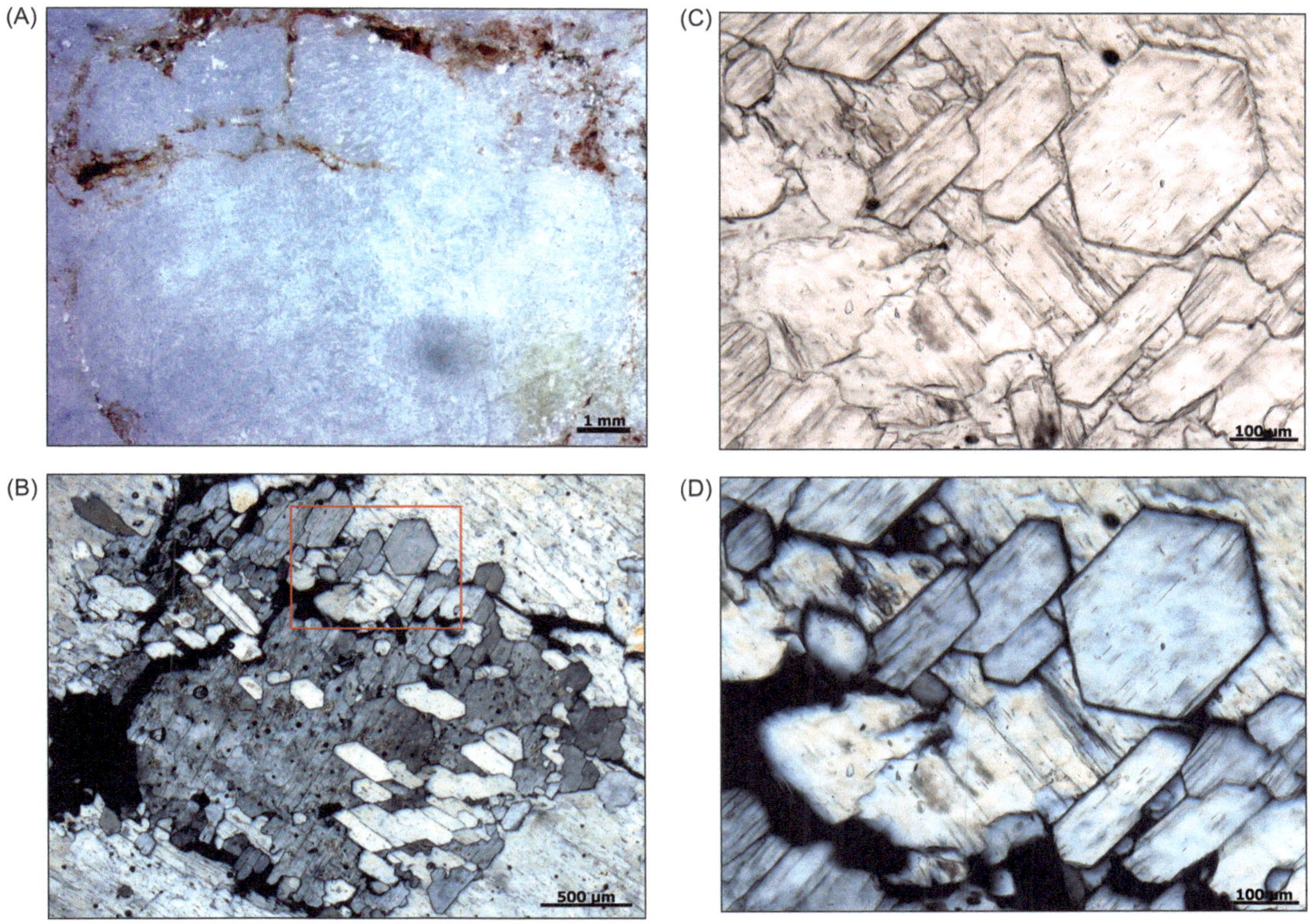

Figure 7.12 Photomicrographs of Permian evaporites from outcrops in the western southern part of the Parnaíba Basin show the petrographic textures of the gypsum-bearing deposits. (A) Nodular or micronodular gypsum is bounded by thin films of brown mudstone in the upper part of the image. The scale bar is 1 mm. (B) cross-polar image showing intergrowth between adjacent gypsum nodules, characterized by euhedral forms and poikilotopic texture. The scale bar is 500 μm (0.5 mm). The rectangle outlined by a thin red line on the central upper part of (B) marks the detail area with gypsum crystals, illustrated in images (C) and (D). These higher magnification microscopic images highlight the relief, birefringence, and the main cleavage planes of the sulfate mineral. The zoom image on the top right (C) shows the hexagonal prismatic gypsum crystals under plane-polarized light. The image on the bottom right (D) shows the same gypsum crystals under crossed polars. The scale bar for the right-column images is 100 μm (0.1 mm). Credit: Abrantes et al. (2016)/with permission of Elsevier.

Petrographic images of the exposed evaporite sediments (Figure 7.12) show that the gypsum minerals occur intercalated with thin films of mudstone, indicating a relatively low brine concentration in a saline lake. Some of the gypsum crystals are well developed, forming hexagonal prisms according to the monoclinic crystalline habit of the mineral (see Figure 7.6). The petrographic thin sections show nodular gypsum bounded by thin films of reddish-colored mudstone (Figure 7.12A) and large intergrowth of gypsum nodules with poikilotopic texture (Figure 7.12B). The detailed microscopic image of the gypsum nodules indicates that gypsum crystals can be recognized by their hexagonal to prismatic habit in cross-section. At higher magnification microscope images, the gypsum crystal can be seen in the microscope both in plane-polarized light (Figure 7.12C) and under crossed polars (Figure 7.12D).

7.6.2. Zechstein Salt Basin, Onshore and Offshore NE England

The North-Central European salt basin is a significant global salt basin, with Late Permian Zechstein evaporite deposits extending from NE England to Southwest Norway in the North Sea (Figure 7.9). The basin is divided

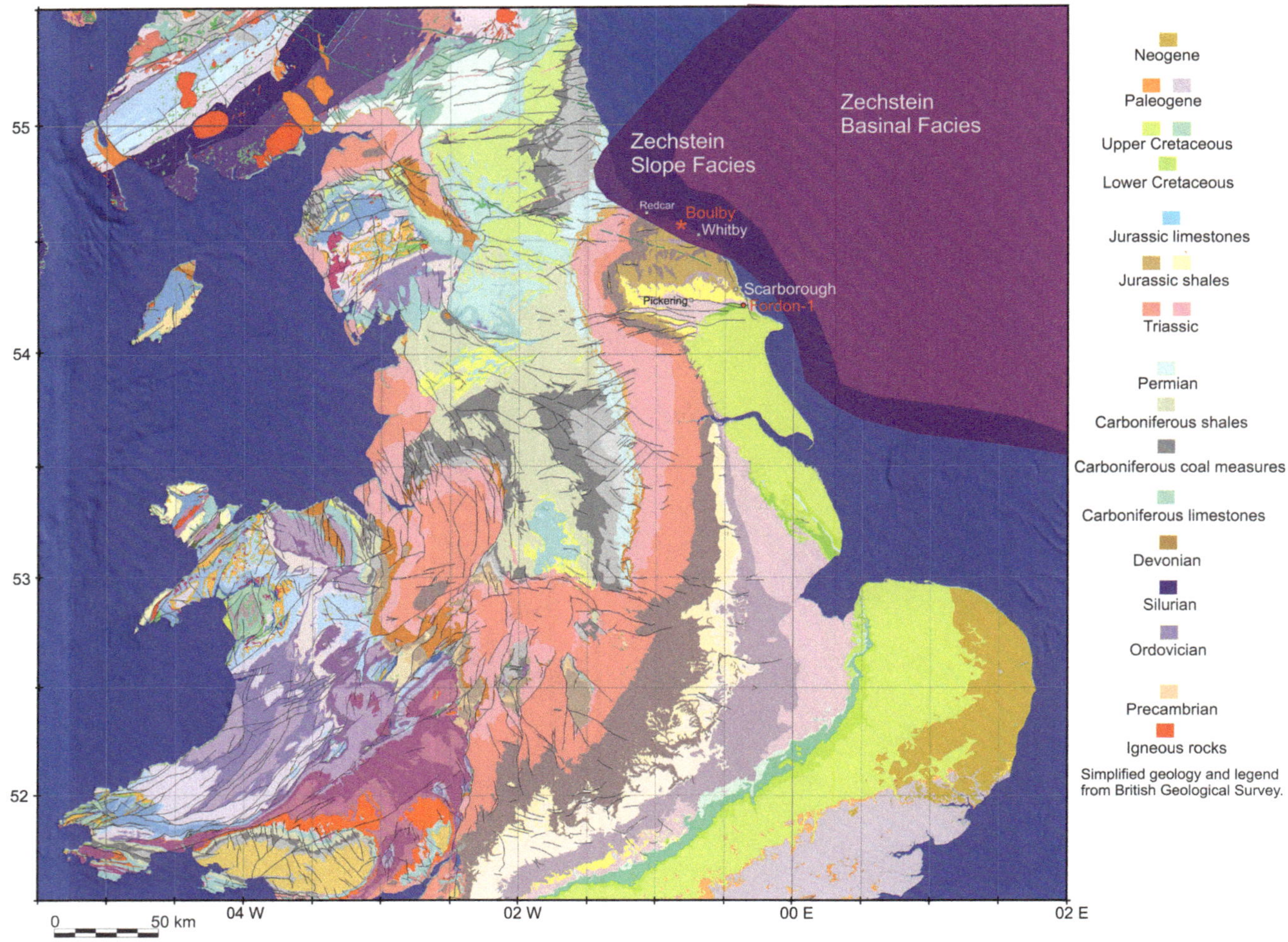

Figure 7.13 Geological map of England with an overlay of the regional distribution of the Zechstein evaporites in the North Sea. Rock salt has been drilled by exploratory wells in the onshore Cleveland Basin, located in the Yorkshire region of NE England. The Fordon well near Scarborough penetrated the Zechstein evaporite layers in the subsurface. The Boulby mine near Whitby produces potassium minerals from the Zechstein salt deposits, exploiting thick seams of sylvite and polyhalite. Credit: Modified from Taylor (1985) and Slowakiewicz et al. (2013).

into two sub-basins: the Northern Permian (or Rotliegend) and the Southern Permian (or Zechstein) sub-basins. The Ringkobing-Fyn High, a regional high that trends E-W offshore northeastern England and southwestern Norway, limits these sub-basins (Glennie, 1998; Taylor, 1984, 1998). The North Sea's Central Graben is split by this regional high, which is capped by a thinner cover of shallow-water carbonates and a thinner layer of evaporites (see Chapter 2 in Mohriak, 2025).

In the northern Permian basin, the evaporite deposits extend across the North Sea and reach the continent in northeastern England (Figure 7.13). These sediments outcrop near the coastline and occur in the subsurface near Yorkshire (BGS, 2014; Stewart, 1963; Tucker, 1991). The proximal and slope evaporite facies of the Zechstein Supergroup, mainly composed of carbonate and sulfate sequences, reach the onshore region near Whitby in the Yorkshire Province. In the offshore region, the deep basin evaporite facies in the North Sea correspond to thicker sequences mainly composed of halite (Figure 7.13).

Rock salt layers are not visible on surface exposures along the NE coast of England but were discovered in the subsurface of the Yorkshire region by petroleum exploration drilling in the 1930s (Stewart, 1963; Tucker, 1991). The Permian carbonate-evaporite sequences in the Cleveland Basin occur at depths between 1,000 and about 1,600 m below ground level (Stewart, 1963). Figure 7.14 shows one example of the Late Permian stratigraphic succession drilled by the exploratory boreholes in the Yorkshire region (Stewart, 1963; Tucker, 1991).

The sediments associated with the Zechstein Supergroup occur in the proximal and slope basins of the Yorkshire

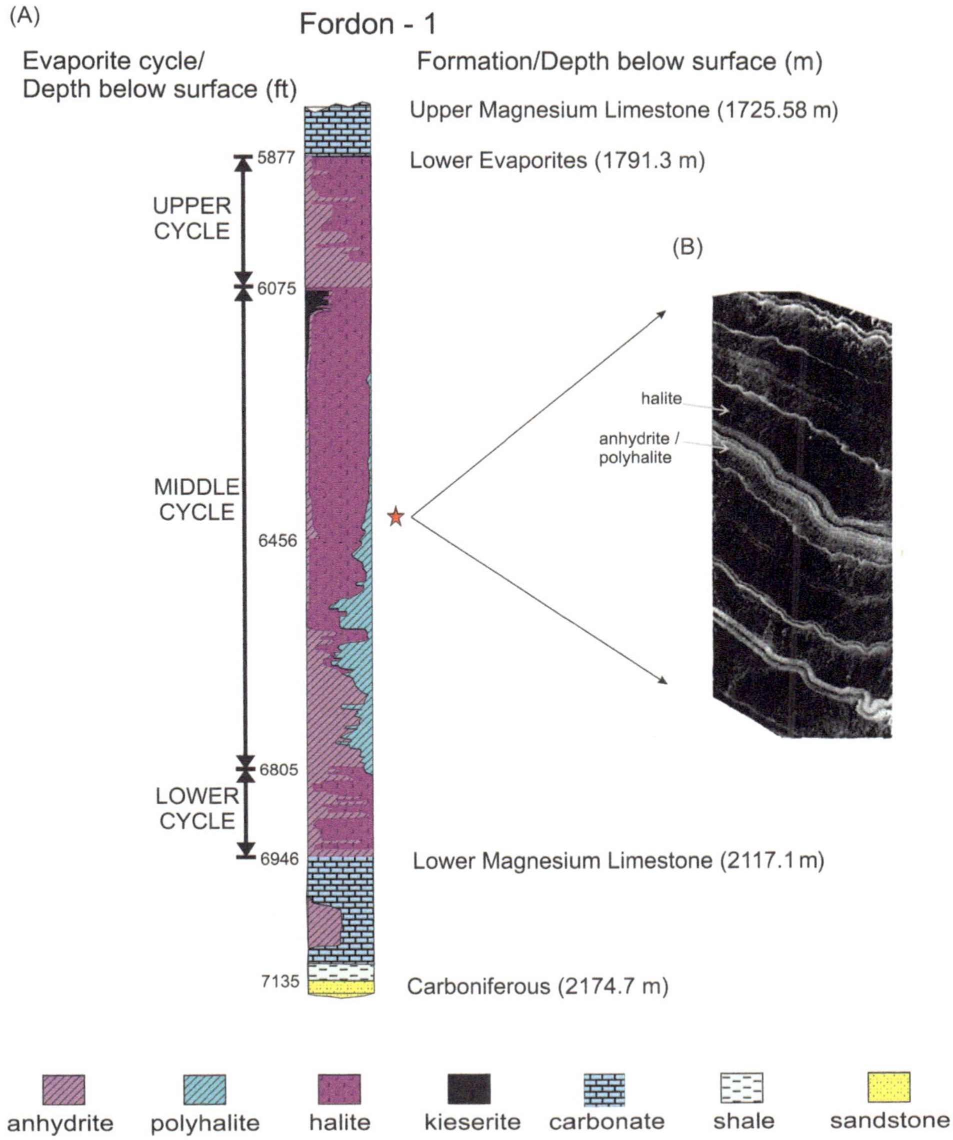

Figure 7.14 The Fordon-1 well was drilled in 1956 to a depth of 7,559 ft (2,304 m) below the surface. The borehole penetrated Chalk carbonates and Millstone Grit sediments from the surface to a depth of 4,700 ft (1,432.6 m) below the surface when saliferous marls were penetrated. The Permian stratigraphic succession (below the top of the Saliferous Marls) indicates several lithologies, particularly dolomite (Upper and Lower Magnesian Limestone), anhydrite, polyhalite, kieserite, and halite. Carboniferous sediments were encountered at a depth of 7,135 ft (2,174.7 m) below the surface. The Permian sequence shows a thickness of about 2,435 ft (742 m). The evaporite lithologies are shown with a magenta color in the lithologic column (A). The evaporite deposits show a total thickness of 1,387 ft (422.7 m), most of which correspond to the Lower Evaporites, drilled from 1,791.3 to 2,117.1 m. The image to the right (B) shows a core slab from the Fordon-1 Lower Evaporites sequences, illustrating the middle cycle lithologies. The length of the slab is about 10 cm. Credit: Adapted from Stewart (1963).

province and adjoining the North Sea (BGS, 2014). The lithological units have been classically divided into five main carbonate-evaporite cycles, resulting in seven stratigraphic sequences deposited during the Late Permian, from about 257 to 254 Ma (Tucker, 1991). The Zechstein evaporite cycles correspond to cycles Z1 to Z5 (Figure 7.15), including diverse carbonate-evaporite units. Different countries in Europe assign several lithostratigraphic names for these cycles. The most common stratigraphic units in the U.K. and Germany are associated with the following cycles: Z1 (Ford Fm. carbonate rocks and coquinas, Werra anhydrite in Germany),

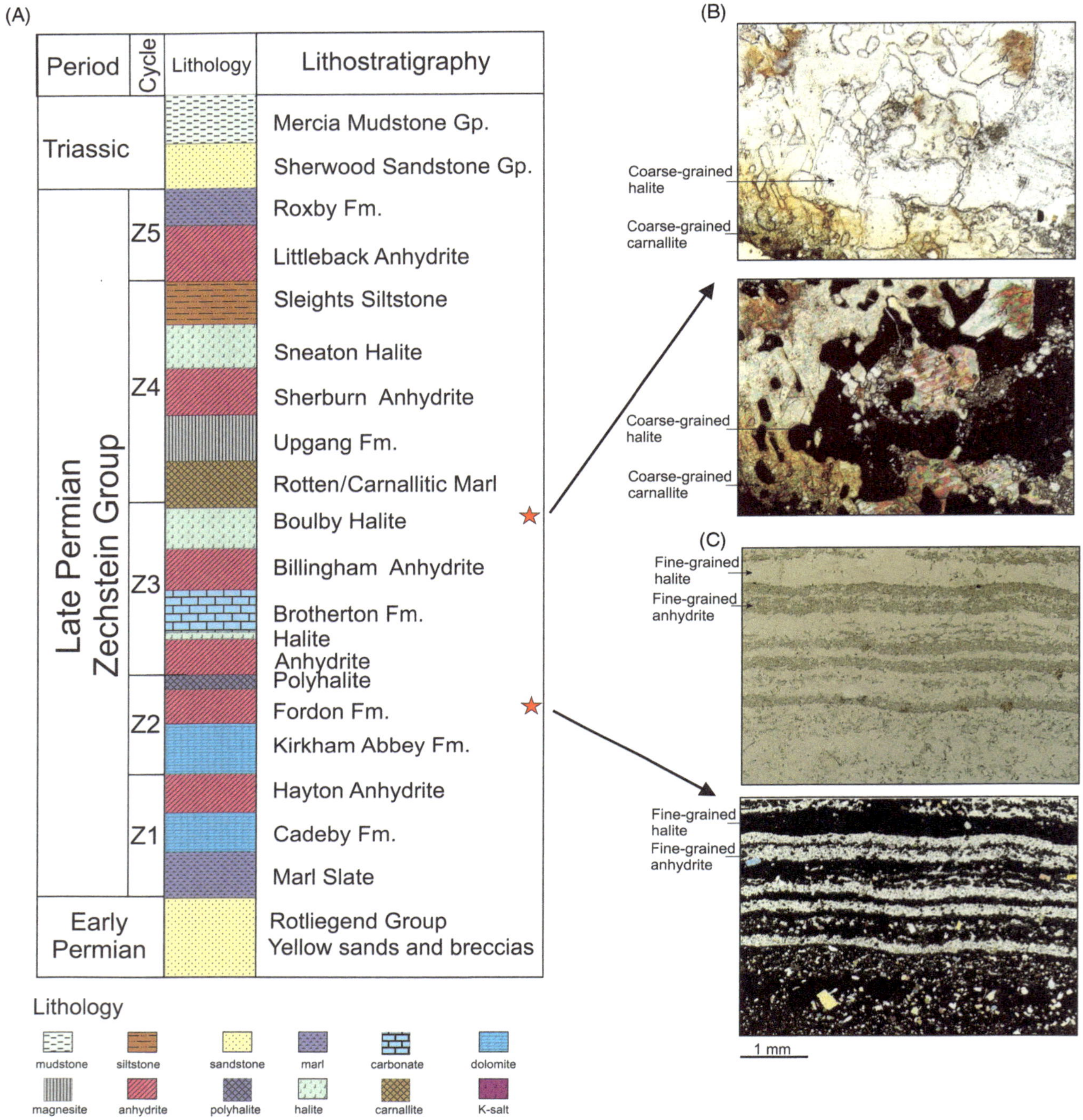

Figure 7.15 The left column (A) shows the lithologic column of the Zechstein Group in the Yorkshire region of Northeast England, characterized by evaporite cycles in the Late Permian. The Fordon evaporites are associated with the Z2 cycle, whereas the Boulby Halite (which occurs in an underground mine near the city of Whitby) is associated with the Z3 cycle. Credit: Adapted from Tucker (1991). The right-side column (B and C) shows the petrographic images of the evaporite facies from the Z2 and Z3 cycles. The top micrographs in (B) show halite and carnallite imaged by plane-polarized light (above) and crossed polars (below). The bottom micrographs in (C) show the thin sections of anhydrite and halite intercalations. The petrographic thin sections are shown as plane-polarized light (above) and with crossed polars (below). The crossed-polar images in (B) and (C) indicate the isotropic halite crystals exhibit a black color under the petrographic microscope.

Z2 (Fordon and Stassfurt salt), Z3 (Leine, Boulby Halite), Z4 (Aller, Pegmatite Anhydrite), and Z5 (Ohre, Salt-Clay). Several carbonate, sulfate, and chloride facies are associated with each cycle, and the higher evaporite cycles indicate a decreasing marine influence in the basin (see Chapter 2 in Mohriak, 2025).

Salt sequences in the basin are mainly found in mid- and lower cycles (Tucker, 1991). Sedimentological models use sequence stratigraphy interpretation and petrographic descriptions of evaporite minerals that accumulated in evaporitic cycles during brine concentration episodes. Incomplete drawdown leads to carbonates and sulfates, implying a relatively reduced brine concentration (see Chapter 4). The wholesale basin desiccation and extreme brine concentration resulted in the deposition of halite and more soluble salts in diverse evaporite cycles in the basin depocenter (Geluk, 2007; Tucker, 1991).

The halite sequences in the Cleveland Basin are associated with the Fordon Fm. evaporites, interpreted at the top of the Z2 cycle (Tucker, 1991). The more soluble salts mainly occur in the subsurface of the North Sea, in the Zechstein basinal facies, forming diapiric salt structures (Figure 7.13). According to Kuster et al. (2009), the Z2 cycle evaporites in the offshore region correspond to the Stassfurt Fm., which includes mainly carbonates, sulfates (anhydrite and polyhalite), and halides (halite and carnallite).

The Fordon No. 1 borehole (Figure 7.14) was drilled in 1956, preceding the exploratory drilling in the offshore part of the North Sea. The borehole was located about 13 km south of the city of Scarborough and reached a total depth of 7,559 ft (2,304 m). The borehole penetrated the presalt Carboniferous sediments after drilling 1,387 ft (423 m) of the Permian evaporite successions. The Zechstein Group evaporites are recorded between 1,791 and 2,117 m (Stewart, 1963).

The lithologic sequences penetrated by the Fordon No. 1 borehole mainly correspond to anhydrite and halite deposits. The main polyhalite seam is recorded at the top of the Z2 cycle (Figure 7.14). The Z2 and Z3 cycles intercalate halite and other soluble minerals, including carnallite and sylvite (Adams et al., 1984). The Boulby Halite is associated with the overlying evaporites at the top of the Z3 cycle, representing the basinal evaporite facies of the Zechstein Group (Figure 7.15).

According to Tucker (1991), the evaporite sequences in NE England correspond to the Z1–Z5 Zechstein cycles from bottom to top (Figure 7.15). The lower part of the Zechstein Group (cycles Z2 and Z3) includes several lithologic types with a predominance of dolomite, anhydrite, polyhalite, kieserite, and halite (Figure 7.15A). The petrographic thin sections of the evaporite layers (Figures 7.15B and 7.15C) illustrate the petrographic characteristics of selected intervals of the Z2 cycle, including halite, carnallite, and anhydrite. The plane-polarized light images (top micrographs of Figures 7.15B and 7.15C) show little contrast between the evaporite minerals. The same image with crossed polars enhances the petrographic differences between the minerals (bottom micrographs of Figures 7.15B and 7.15C). Halite is an isotropic mineral and appears black under crossed nicols, as previously mentioned.

Two main types of minerals are mined from subsurface salt layers in the NE England province: sylvite, or potassium chloride (KCl), and polyhalite, a complex hydrated sulfate ($K_2SO_4.MgSO_4.2CaSO_4.2H_2O$). These minerals are abundant in the Boulby salt mine near Whitby, in the Yorkshire region of NE England (Figure 7.13). This salt mine is famous for exploiting potash salt from galleries below the North Sea. The Boulby salt mine is the only mine in England producing potassium from polyhalite layers, which is mainly used in the agriculture industry as a fertilizer. The polyhalite deposits are associated with the transition from the Z2 to the Z3 cycle of the Zechstein Group, above the Fordon Fm. evaporites (Kemp et al., 2016; Tucker, 1991).

Sylvite and polyhalite minerals in the Fordon Fm. are interpreted as a product of evaporation from highly concentrated brines or alteration of evaporite deposits. Carnallite is decomposed into sylvite, while anhydrite or gypsum minerals are altered to polyhalite. Stewart (1963) interpreted the diagenetic alteration of anhydrite layers as a large-scale diagenetic process, releasing calcium for the formation of large amounts of primary polyhalite, resulting in thick ore sequences at the bottom of the middle cycle (Figure 7.14).

Peryt et al. (1998) postulated polyhalite deposits in the Zechstein of northern Poland, associated with the Werra Anhydrite, formed near the shelf edge of a Late Permian carbonate platform. The process involved chemical reactions between highly concentrated brine in salt pans and sulfate compounds that precipitated from less concentrated marine brine (Peryt et al., 1998).

The most soluble salt minerals in the Boulby mine are associated with the Boulby Potash Member of the overlying EZ3 cycle (Kemp et al., 2016). This unit yields a much larger content (63%) of potassium oxide (K_2O), as compared to polyhalite (15%). Although discovered in the late 1930s by petroleum exploration wells, which targeted at carbonate reservoirs in the Cleveland Basin, the polyhalite layers were overlooked for decades due to their reduced ore content. During the initial phase of salt mining, the exploitation focused on the shallower Boulby Potash seam (sylvite, KCl). However, the polyhalite seam in the Boulby mine exhibits a percentage of over 90% as the primary ore mineral, making it one of the richest polyhalite occurrences to be mined globally (Kemp et al., 2016).

7.7. TRIASSIC SALT BASINS IN CENTRAL EUROPE

The Molasse Basin in Switzerland hosts salt layers that have been studied by several researchers as potential sites for radioactive waste fuel storage (e.g., Dronkert et al., 1990). The cores from several boreholes provide a unique dataset for the interpretation of the evaporite deposits involved in the Alpine orogeny (e.g., Frizon de Lamotte et al., 2011). The northern part of Switzerland is divided into various geotectonic units within the Alpine domain of Central European basins (Figure 7.16). The Precambrian to Early Paleozoic crystalline basement rocks in the Schwarzwald (Black Forest) Massif form the basement for Late Paleozoic to Mesozoic sedimentary basins in Central Europe (Matte, 1986).

After the Variscan Orogeny, the Central European region experienced extensional tectonics, magmatism, and the development of large foreland depressions filled with Permo-Carboniferous (250–300 Ma) sediments and volcanic rocks, from the U.K. and Norway in the north toward the central European mainland in the south. This phase is associated with the Zechstein salt basin in the Late Permian (Glennie, 1998; Ziegler, 1975, 1990).

During the Triassic and Jurassic, a renewed phase of extension resulted in the formation of the North Sea Basin rift and Neotethyan continental margins. The rift sedimentary successions were overlain by evaporite rocks and

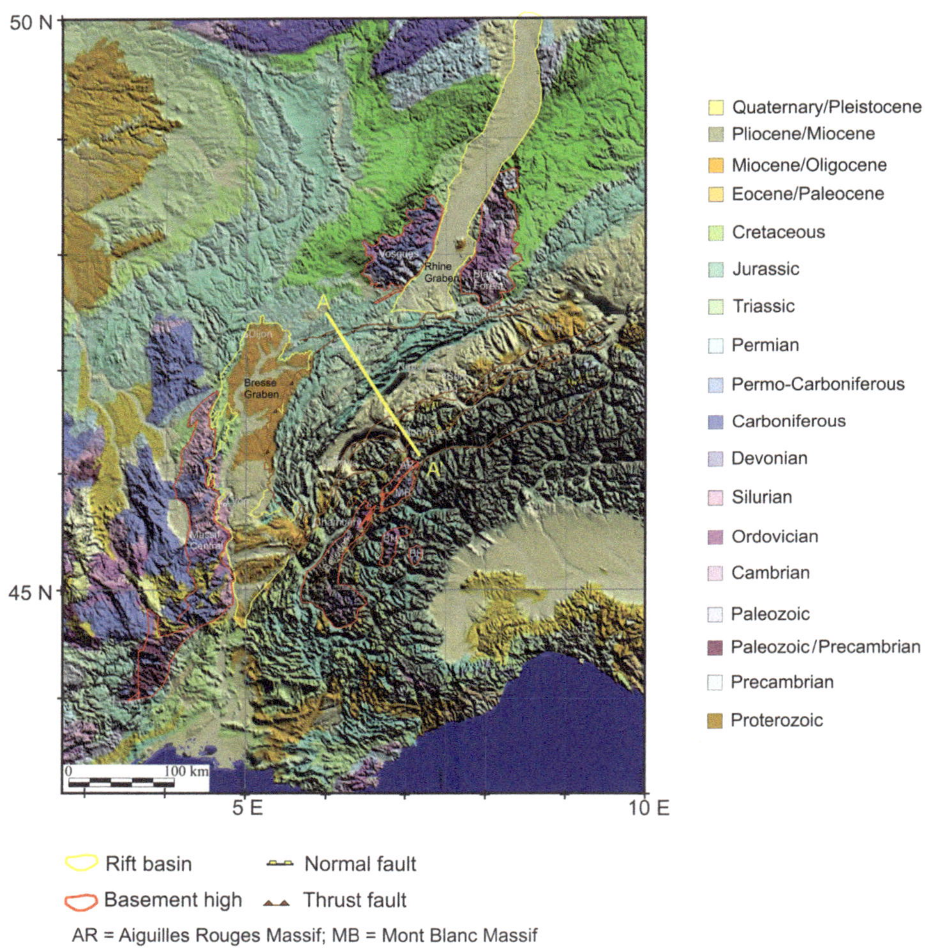

Figure 7.16 Geological map of the Alpine fold-thrust belt draped on topography. The Alpine geological units north of the Pó Basin in Italy show a predominant NE trend. Tertiary rocks are shown in light yellow and light orange colors. These sediments cover the Po Basin, the Molasse Basin, and the Bresse and Rhine grabens. South of the Rhine Graben, the Molasse Basin is characterized by Tertiary sedimentation in a foreland basin trending SW-NE, forming a trough that extends from Switzerland to Germany. The thick yellow line shows the location of profile A-A′ across the Molasse Basin. The Jurassic platform carbonate rocks outcrop at the Jura fold-thrust belt south of the Rhine graben (cyan color). The underlying Triassic succession includes carbonate and evaporite layers in the subsurface, which are mainly associated with sulfate (anhydrite) and chloride (halite) minerals.

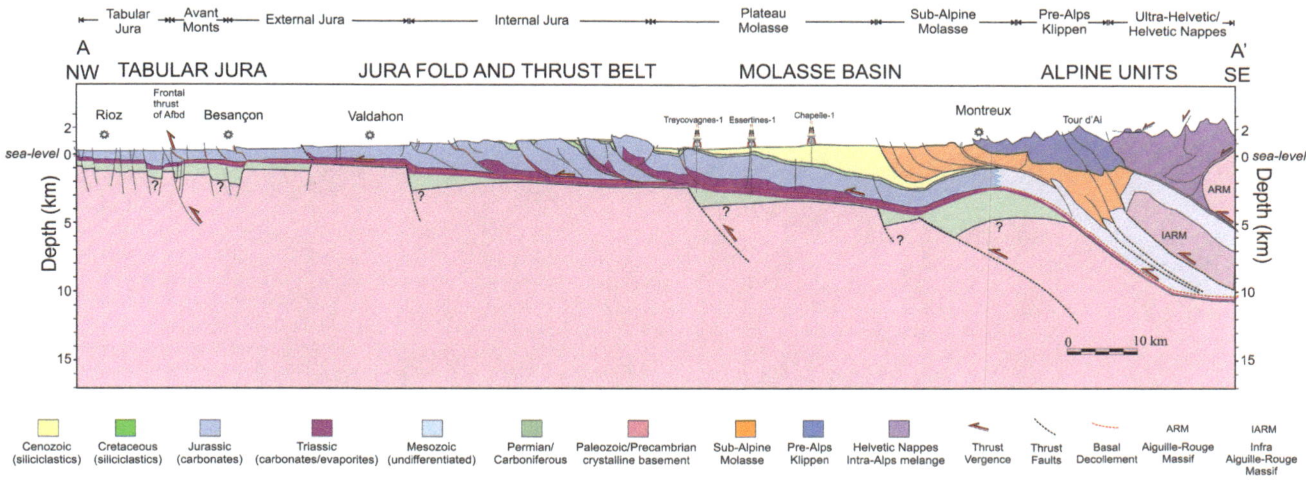

Figure 7.17 Regional geological cross-section A-A′ across the Alpine fold-thrust belt. The internal deformation zones of the Alpine Units are located to the southeast of the profile. The Molasse Basin is the foreland depression associated with the flexural loading of the Alpine nappes. The Jura fold-thrust belt extends toward the tabular Jurassic carbonate platform in the northwest part of the profile. The Triassic evaporite layers in the subsurface of the Molasse Basin were affected by the Jura fold-thrust belt, resulting in reverse faults with an NW vergence. Several wells have drilled the Tertiary sediments of the Molasse Basin in Switzerland, north of the Alpine units. Credit: Sommaruga et al. (2017)/with permission of Elsevier.

platform carbonates that graded into deep water marls and shales (Frizon de Lamotte et al., 2011). The Cretaceous period was marked by the inversion of structural depressions on the continental margins, with the Alpine orogeny marking the closure of the Neo-Tethys Ocean (e.g., Berra & Angiolini, 2014; Frizon de Lamotte et al., 2011; Sommaruga et al., 2017). The Alpine orogeny affected both European and African plates, resulting in the Alps and Atlas mountains, with compressional events recorded from the Middle Cretaceous (105 Ma) and reaching its climax during the Paleogene (~ 40 Ma). In the aftermath of this phase, the Jura Mountains started to acquire their modern geomorphology in the Miocene (11 Ma). The plates are still in convergence, as indicated by modern earthquake activity that is widespread in the Central European region (Camelbeeck et al., 2007).

The Molasse Basin, a foreland trough north of the Alpine chain (Figure 7.16), formed during the Neogene. The Rhine Graben originated in Oligocene times from extensional faults controlling a sedimentary fill that exceeds 3,000 m in thickness (Berger et al., 2005; Hinsken et al., 2007). Magmatic activity with local volcanism accompanied several phases of the rifting process in the Neogene, which seems seismogenically active up to the present day (Fracassi et al., 2005).

The regional transect across the Alps in Switzerland (Figure 7.17) reveals the main structural features of the transition from the Alpine units to the Molasse Basin and Jura fold-thrust belt. The unfolded or tabular Jura comprises Upper Carboniferous, Permian, Triassic, Jurassic, and Tertiary sediments north of the Alps, across Germany and France. The tabular Jura corresponds to a relatively thin platform sequence of carbonate and siliciclastic deposits. The folded Jura is the folded and partly overthrust equivalent of the tabular or unfolded Jura (Figure 7.17).

This section presents a summary of the microscopic petrography of evaporite minerals from boreholes drilled in Switzerland by Nagra, the Swiss National Cooperative for the Storage of Radioactive Waste. The wells penetrated the Jurassic carbonate rocks and the Triassic carbonate and evaporite rocks in the subsurface. The pre-evaporite, Permo-Carboniferous sequences were deposited in basins locally controlled by extensional syn-rift faults, which have been inverted during the convergence between Africa and Europe (Figure 7.18). The Triassic evaporites were deposited in the continental margin of the Neo-Tethys Ocean and deformed by the Alpine Orogeny in the Cenozoic. A major decollément is interpreted along the Triassic sequence, between the Jurassic carbonates and the Permo-Carboniferous sequences (Figure 7.17). The cores and petrographic thin sections were analyzed in seminal projects conducted by the Geologicals Institut der Universitat Bern in the 1980s. The microfacies of the evaporite lithologies in the Nagra wells provide a unique dataset for deformed evaporite sequences in the Alpine fold-thrust belt (Dronkert et al., 1990).

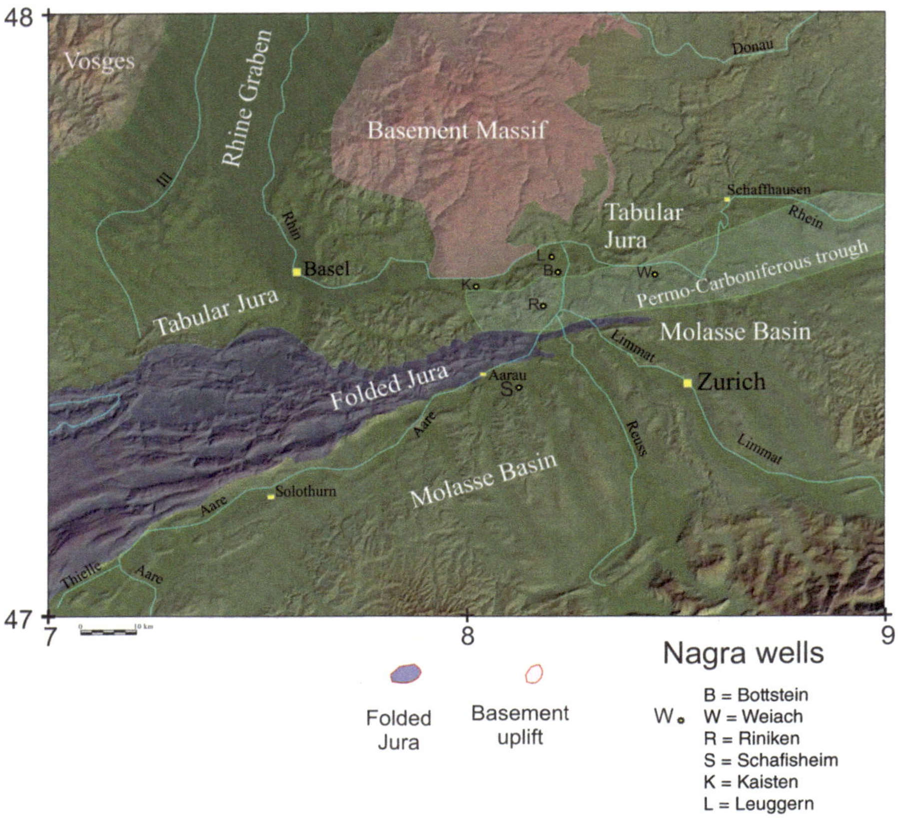

Figure 7.18 Simplified geological map draped on the topographic map of the Alpine region near Zurich in Switzerland. The Rhine Graben is located north of the Folded Jura Mountains and west of the Basement Massif. The Molasse Basin is located to the south of the Folded and Tabular Jura. The Permo-Carboniferous trough, located south of the uplifted basement block, is characterized by an NE trend. The Neogene sediments of the Molasse foreland basin overlie the Jurassic and Triassic carbonate-evaporite deposits recorded in the subsurface. These Mesozoic sequences were drilled by the Nagra wells (marked by yellow dots and letters).

7.7.1. Triassic Evaporites in the Molasse Basin

The Triassic evaporites in the Nagra wells were only indirectly influenced by the Alpine orogenic phase, as they are located far from the internal zones of deformation (Figure 7.18). However, the fold-thrust belts in the Molasse Basin and Jura Mountains are associated with Triassic salt deposits, forming a décollement for low-angle compressional structures (Figure 7.17). Salt tectonics in the Alpine fold-thrust belt reached a peak in the Jurassic and continued into the Cenozoic (Célini et al., 2020). Cenozoic deformation structures include reverse faults with an NW vergence and major detachment faults that use the Triassic evaporite layer as a lubricant (e.g., Sommaruga et al., 2017).

In Central European basins, Triassic sedimentary sequences are associated with two different lithofacies: the Germanic facies in the north, primarily corresponding to continental deposits, and the Alpine facies in the south, mainly corresponding to marine deposits, marking the transition to the Neo-Tethys continental margins (Dronkert et al., 1990; Sommaruga et al., 2017). In the Molasse Basin, the Middle Triassic corresponds to the most important depositional phase, with thick carbonate and evaporite sequences preserved under the Jurassic carbonate rocks. These stratigraphic sequences comprise the passive continental margins associated with the development of the Neo-Tethys Ocean (Figure 7.19).

The Triassic sedimentary successions comprise three groups, following the Germanic nomenclature: the Buntsandstein, Muschelkalk, and Keuper, known as the Germanic Trilogy. Each of these sequences is subdivided into several geological units or formations with local names, resulting in very complicated stratigraphic correlations. Nonetheless, the tripartite division is recognized in most of the Central European basins of the Alpine domain (Pérez-López & Pérez-Valera, 2007).

The Germanic Trilogy comprises geological units from continental and epicontinental deposits, including desert regions, sabkhas, and coastal environments. The

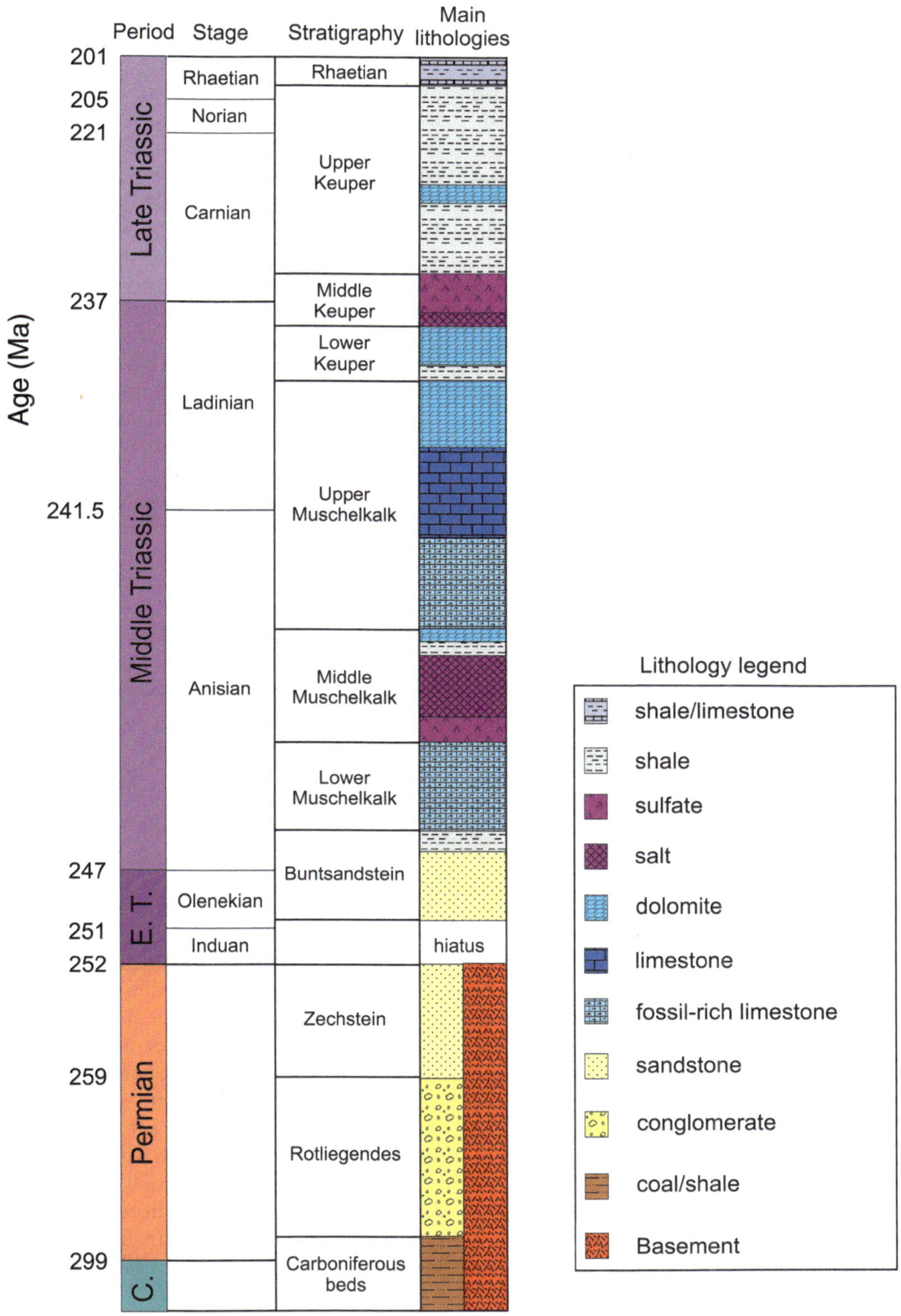

Figure 7.19 Stratigraphic succession of the Triassic carbonate-evaporite rocks in the Molasse Basin region in Switzerland, north of the Alpine units. Evaporite rocks with halite layers interbedded with sulfate and carbonate layers are observed in the Middle Keuper (around 237 Ma) and the Middle Muschelkalk (around 244 Ma). The evaporite-carbonate sequences are underlain by Early Triassic Bundsandstein and Permian Zechstein rocks, which mainly correspond to siliciclastic sequences. Credit: Sommaruga et al. (2017)/with permission of Elsevier.

Buntsandstein, a siliciclastic rock deposited in continental environments, spans the Early to Middle Triassic. Muschelkalk, a marine carbonate rich in shell fragments, is dated as Middle Triassic (Anisian to Ladinian) and divided into Upper and Lower Muschelkalk units (Figure 7.19). The Ladinian carbonate rocks are overlain by the Keuper sequences, consisting of shales, limestones, reddish-brown clays, and evaporites (Sommaruga et al., 2017).

The Triassic evaporites sampled by the six deep wells drilled by Nagra recovered cores that illustrate four lithofacies associations, establishing a pattern that can be recognized in many salt basins worldwide. According to Dronkert et al. (1990), these evaporite facies comprise:

a. Clay lithofacies association, including black sapropelic laminites and greenish to brownish-red calcareous claystone, indicative of a subaqueous origin,
b. Carbonate lithofacies association, consisting of brackish to hypersaline facies of micrites, dolomicrites, and intertidal stromatolitic laminites intercalated with calcarenites;
c. Sulfate lithofacies association, consisting of selenitic sulfate rocks, indicative of subaqueous growth, and clastic evaporites (gypsarenites), with ripples and other sedimentary structures;
d. Chloride lithofacies association, consisting of massive or laminated halite and halite collapse breccias.

All wells drilled by Nagra are located in the unfolded (or tabular) Jura, and the overburden corresponds to tectonically undisturbed layers, except the Schafisheim well, which is located in the Molasse Basin, south of the exposed folded Jura and closest to the Alpine front of deformation (Figure 7.18). The Jurassic rocks have been covered by a thin veneer of Late Mesozoic and Cenozoic sediments in Switzerland. Only one well drilled by Nagra (the Schafisheim well) shows a sedimentary cover greater than 1,000 m in thickness (Dronkert et al., 1990). The other wells had the overburden layer eroded to such an extent that anhydrite layers could not be protected from the water inflow through fractures. As a result of the water infiltration, the subsurface evaporite layers were hydrated, and most of the sulfate rocks transformed into gypsum. Most of the evaporite units drilled by the Nagra wells consist of carbonates and sulfates (Dronkert et al., 1990).

The borehole cores indicate some intervals with clastic evaporitic deposits, formed by the accumulation of sand-size grains of sulfate minerals in submarine settings. The arenites, composed of clasts of calcium sulfate, are known as gypsarenites or sulfarenites. Other intervals indicate shallow water deposition, with mudcracks indicating subaerial exposure. Nodular anhydrite and gypsum are common in several boreholes. The sulfate zones include chicken-wire mosaics and red sulfate nodule stringers (Dronkert et al., 1990).

7.8. APTIAN SALT BASIN IN NORTHEASTERN BRAZIL

The Sergipe-Alagoas Basin, located in northeastern Brazil (Figure 7.20), is part of the South Atlantic rift system and has developed thick salt deposits in the Late Aptian (Kukla et al., 2018; Mohriak & Fainstein, 2012). It is the northernmost continental margin salt basin with active petroleum systems and hydrocarbon discoveries. The basin is divided into two sub-basins: the southern Sergipe sub-basin and the northern Alagoas sub-basin. These sub-basins are characterized by Early Cretaceous rift-basin depocenters both onshore and offshore, with an elongated shape indicating an oblique rift system (Martins et al., 2019; Souza-Lima, 2009). The distal parts of the rift troughs extend offshore to deep waters, with a narrow shelf in the southern part of the Alagoas sub-basin (Figure 7.20).

The basin presents a diverse and well-preserved stratigraphic record that includes pre-rift, syn-rift, and post-rift successions ranging in age from the Paleozoic to the Meso-Cenozoic (Ojeda, 1982; Souza-Lima, 2009). The evaporite layers are dated as Late Aptian and are included within the early syn-rift sedimentary successions, which range in age from Barremian to Aptian. The evaporite layers are dated as Late Aptian and are included within the Early syn-rift sedimentary successions, which range in age from Barremian to Aptian. The post-salt sediments include Albian carbonates and Late Cretaceous to Neogene siliciclastic sediments (Martins et al., 2019; Ojeda, 1982; Souza-Lima, 2009).

The Sergipe-Alagoas Basin was studied during an initial petroleum exploration phase in the 1960s and 1970s, when several wells werer drilled on land and in the continental shelf. Since the late 1970s, the onshore part of the basin has been investigated by drilling deep wells to characterize the exploitation potential of subsurface evaporite deposits (Szatmari et al., 1974, 1979, 2009, 2021). Although the onshore parts of the basin are mature petroleum provinces, the southernmost parts of the rifted crust (Sergipe sub-basin) are characterized by active petroleum systems, with recent discoveries in the deep-water region (Feijó, 2014). Sylvite minerals are produced by subsurface salt mining in the onshore part of the Sergipe Basin, which constitutes the main source of potassium used by the Brazilian agricultural industry (Bobco, 2017; Szatmari et al., 2021).

The basin is characterized by several evaporite layers deposited in the Aptian, assigned local names (Horizonte, Paripueira, and Ibura salt). The Horizonte Salt is interpreted as a halite occurrence below the Ponta Verde Fm., while the Paripueira Salt occurs above the Ponta Verde Fm., forming thick salt deposits within syn-rift troughs (Martins et al., 2019; Souza-Lima, 2009). These sequences

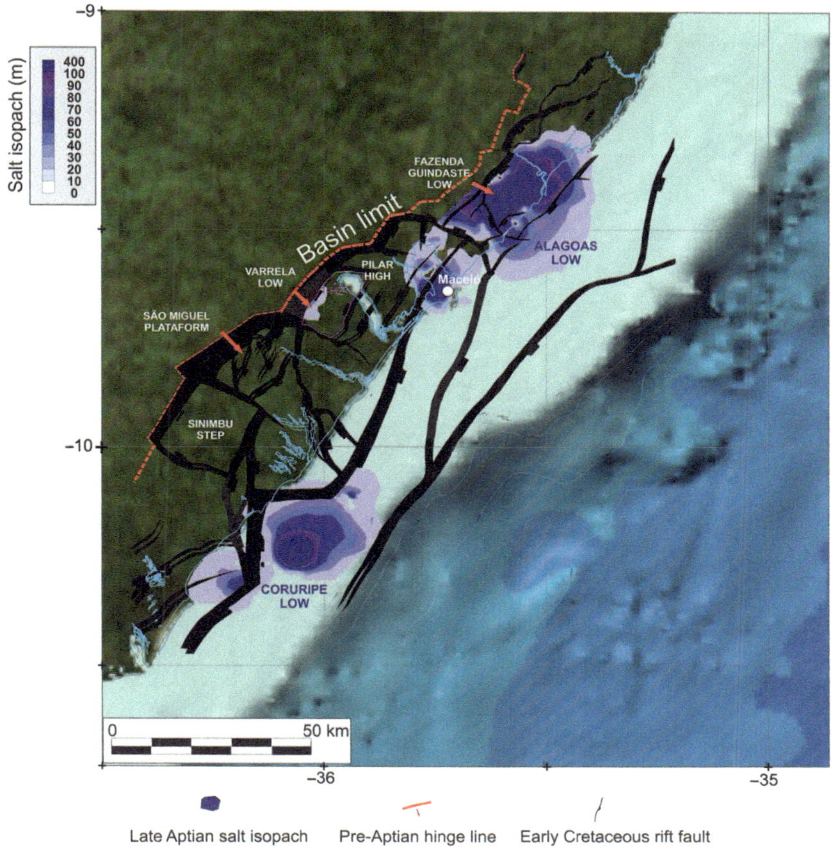

Figure 7.20 A regional map of the Alagoas sub-basin in NE Brazil shows the Late Aptian salt isopachs and the Early Cretaceous rift architecture. The basement-involved rift faults control the structural lows where evaporites have been drilled by exploratory wells. The Varrela Low, west of Maceió, is a smaller depocenter near the basin border fault, where the exploratory well 1-HZ-1-AL was drilled and penetrated halite layers in the syn-rift sequence. These layers are the oldest evaporite deposits in the South Atlantic Aptian salt basin.

are overlain by the Ibura Salt, which is mainly characterized in the Eastern Brazilian margin as the widespread evaporite layer forming salt diapirs in the deepwater region (Mohriak et al., 2012; Ojeda, 1982).

The lowermost evaporite deposits correspond to the Horizonte Salt, a sedimentary sequence characterized by massive halite with large crystals (Souza-Lima, 2009). The Horizonte Evaporites are located above the regional unconformity that separates tilted syn-rift blocks from sag basin deposits (Martins et al., 2019). The log interpretation of the evaporites in the 1-HZ-1-AL well, drilled in the Varrela Low (Figure 7.21), shows rock salt in two main depositional sequences, separated by a thick shale layer with minor sandstone intercalations. The lower evaporites (Horizonte salt), with the top at 1504 m, have been cored by exploratory boreholes. The salt layer is characterized by massive, almost pure halite that is locally banded (Figure 7.21B). The detailed images of the top core (Figures 7.21C and 7.21D) indicate that the predominant lithology is halite with large crystals. Anhydrite layers are absent in the sequence, and only minor shale intercalations occur within the rock salt, raising some questions about the depositional model for this unusually homogeneous halite unit (Martins et al., 2019). The sedimentological interpretation of such coarse-grained halite facies, which appear intercalated with thin shale layers, suggests a hypersaline lake or salt pan in a continental rift system affected by drastic hydrologic and climate fluctuations (Bobco, 2017; Martins et al., 2019; Souza-Lima, 2009).

The subsurface evaporite lithologies in the Taquari-Vassouras salt mine are located near the Carmópolis oil field in the Sergipe sub-basin (Mohriak & Fainstein, 2012). The onshore evaporite deposits have been investigated by drilling several exploratory boreholes aiming to characterize the facies variations. These wells are designated as PKC (Potassium Project of Carmópolis). A detailed report on the geological and petrographical characterization of these evaporites was prepared in the 1970s

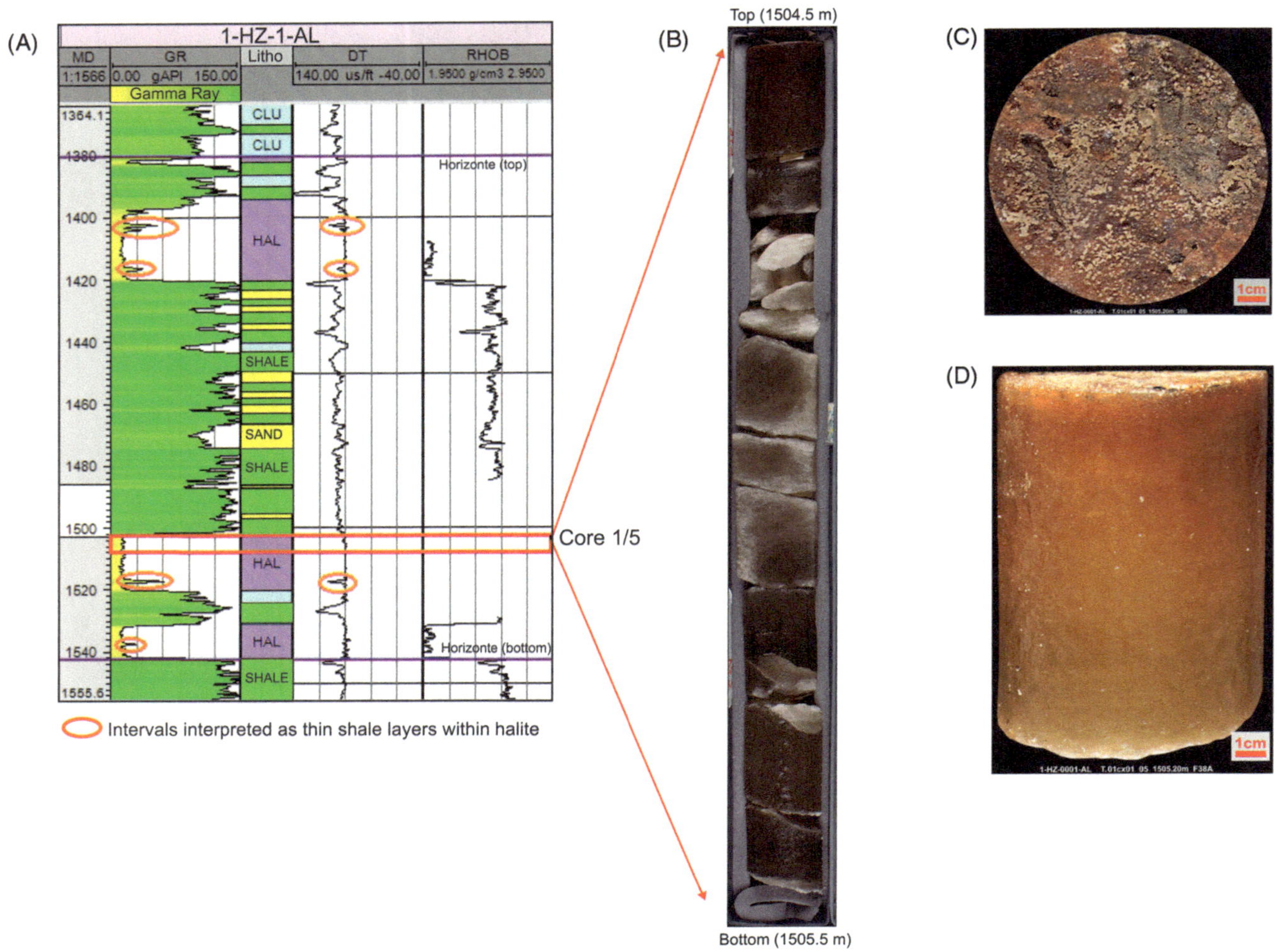

Figure 7.21 Log interpretation of the Horizonte evaporites drilled by the well 1-HZ-1-AL (A); the central column in (B) shows the core image from the lower evaporite interval, mainly composed of halite. The right-column images illustrate photographs of the halite minerals in the core as detailed images seen at the depth of 1,505.2 m. The image (C) corresponds to a view from the top, and (D) corresponds to a view along the core length. The bar scale in the core slab is 1 cm. Credit: Adapted from Martins et al. (2019).

(Szatmari et al., 1974), and selected images from the cores were included in more recent publications (Szatmari et al., 2009, 2021).

These borehole cores illustrate intercalations of halite with shale and carbonate layers, forming cycles that include sulfate layers with anhydrite nodules and are topped by soluble potassium salt minerals (Figures 7.22A–7.22C). Micro-folding associated with the salt flow can be seen in some cores, which are characterized by centimetric alternations of dark grey halite and whitish anhydrite, resulting in stratified, folded evaporite sequences (Figure 7.22E).

The petrographic analysis of the mosaic/massive halite petrofacies suggests the replacement of organic-rich muddy matrix and recrystallization, obliterating features originally related to primary precipitation (Figures 7.23A and 7.23B). The accumulation of sylvinite in salt pans suggests extreme brine concentration and precipitation of highly soluble minerals, such as sylvite and tachyhydrite. The presence of bitter salts suggests the salt basin was almost completely desiccated, forming concentrated brine enriched in potassium (Bobco, 2017; Szatmari et al., 1979; Szatmari et al., 2021).

7.9. NEOGENE SALT BASINS IN THE MEDITERRANEAN SEA

The simplified geological map of the southern Iberian margin in Spain (based on Chorlton, 2007) indicates that several lithological units with evaporite rocks outcrop on the surface, intercalated with Mesozoic and Cenozoic sediments (Figure 7.24). The narrow oceanic gulf that

Figure 7.22 Lithological diversity of the Aptian evaporites in the onshore Sergipe sub-basin. The plates A to F illustrate several evaporite facies observed in borehole cores from exploratory wells: (A) nodular anhydrite immersed in a micritic limestone matrix; (B) crystalloblastic crystal of halite in a shale matrix; (C) intercalations of coarse-grained sylvinite (red color) with fine-grained halite (gray color); (D) small recumbent fold in finely laminated layers of halite intercalated with dark grey shale, immersed in a matrix of carnallite (red color); (E) fine-grained halite crystals in millimetric layers characterized by micro folds related to early salt flow; and (F) nodular carnallite (red color) intercalated with fine halite (gray color). Scale bars are equal to 1 cm.

separated the foreland basin in the Iberian Plate and the branch of the western Tethys Ocean during the Mesozoic was closed in the Cenozoic as a result of the African and Eurasian plate convergence (Frizon de Lamotte et al., 2011; Playà et al., 2000). The compressional tectonics formed several mountain ranges, such as the Alps and the Betics in Europe and the Atlas Mountains in Northwest Africa (Figure 7.24).

The Betic orogenic belt is floored by the Variscan basement rocks, which are separated from the less deformed Mesozoic rocks by a detachment fault zone involving the Triassic evaporites (Sommaruga et al., 2017). The deformed zone is characterized by NE-trending exposures of the Mesozoic sediments that show intercalations of deformed Triassic evaporites intruded into fault planes (Figure 7.25). The fold-thrust belt comprises three main geological units: the External Zone, formed landwards toward the cratonic region; the Internal Zone, where uplifted blocks expose Mesozoic and Paleozoic units; and the flysch units (Campo de Gibraltar Complex) in the Alboran Sea (Figure 7.24).

The Internal Zone includes Paleozoic metamorphic rocks and a cover of Mesozoic rocks involved in nappe structures, particularly carbonate and siliciclastic sequences intercalated with the evaporite layers (Figure7.25). The External Zone in the foreland basin includes Triassic to Jurassic carbonate-evaporitic sequences involved in the compressional tectonics during the Neogene (Pérez-López & Pérez-Valera, 2007; Vera, 2001). The major deformation in the Betic thrust and fold belt is marked by reverse faults with an NW vergence (Figure 7.25).

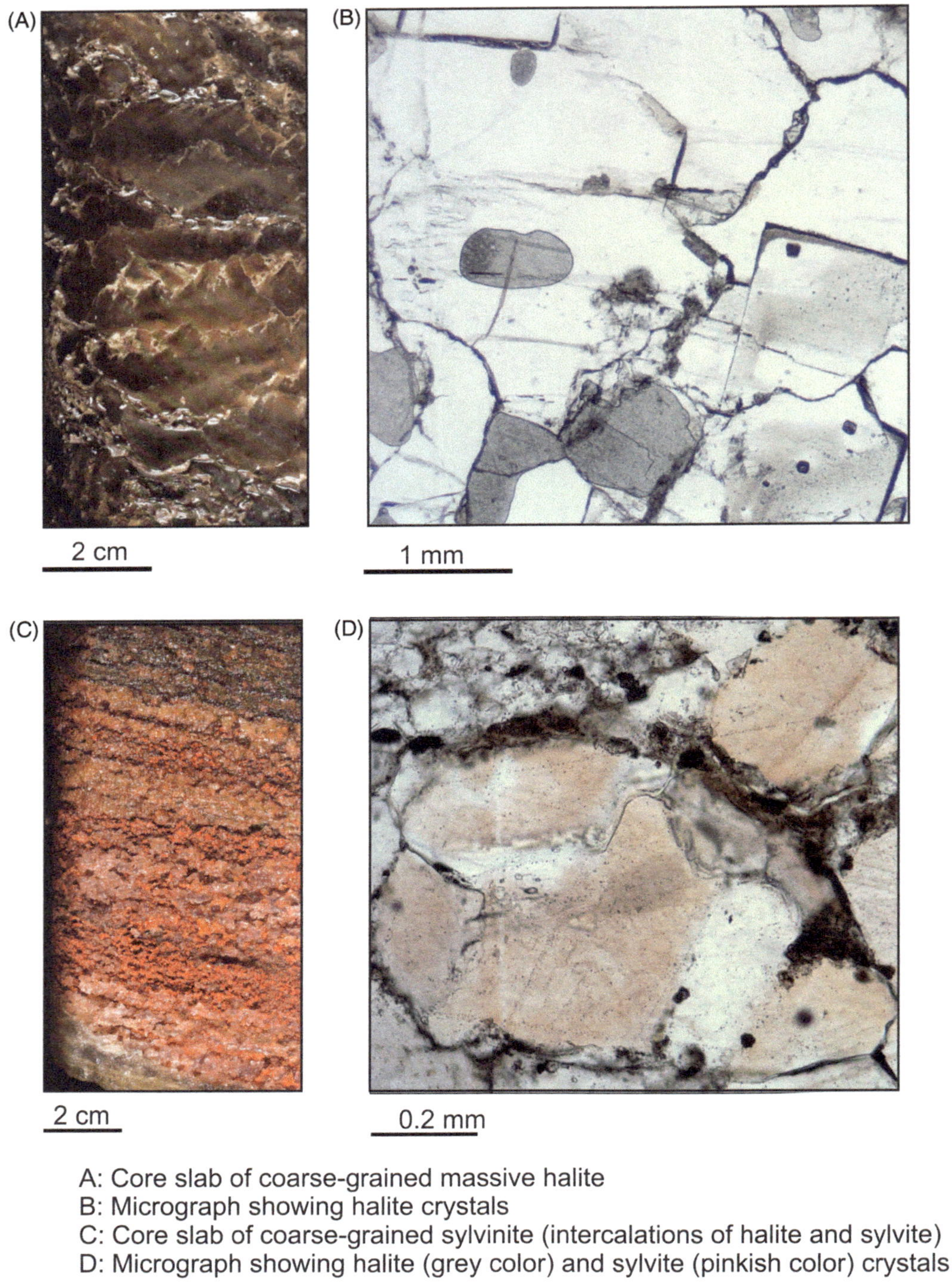

Figure 7.23 Evaporite rocks in the onshore Sergipe sub-basin. Top: a core slab of coarse-grained massive halite (A) and petrographic thin section of halite crystals (B). Bottom: a core slab of laminated sylvinite composed of pink sylvite and grey halite (C) and petrographic thin section showing sylvite (pink) and halite (gray) macrocrystals (D). The dark brown film surrounding the salt crystals corresponds to a laminated, muddy siliciclastic matrix. Credit: Bobco (2017)/with permission of Universidade Federal Do Bo Grande Do Su.

During the Middle to Late Miocene, several small pull-apart and foreland basins were formed in the southeast Iberian margin due to plate rearranging during the Alpine-Betic Orogeny. Some of these basins, like the Guadalquivir Foreland Basin, opened to the Atlantic, while others, like the Granada, Lorca, and Fortuna basins, opened to the Mediterranean Sea (García-Veigas et al., 2020; Playà et al., 2000; Vera, 2001).

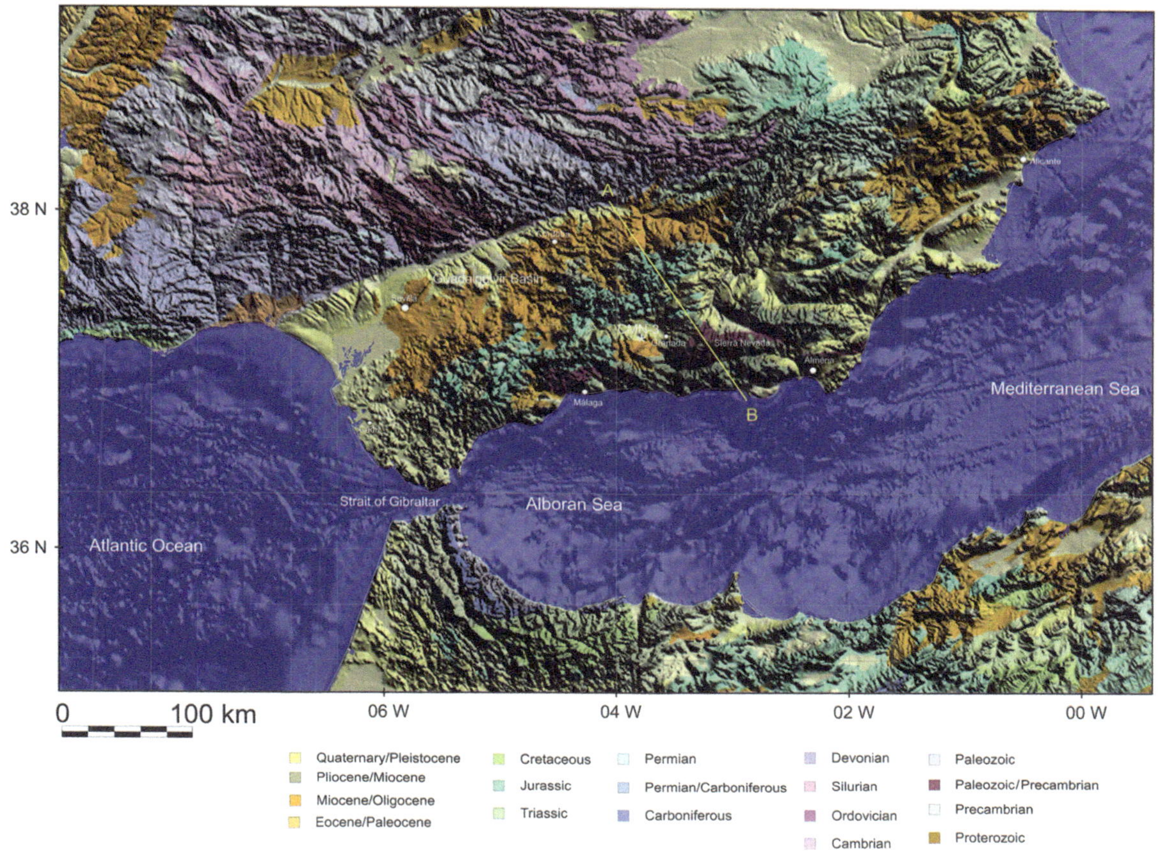

Figure 7.24 A simplified geological map of the southern Iberian margin shows the main lithological units on land. Offshore, the map shows the bathymetry in the Atlantic Ocean, Alboran Sea, and Mediterranean Sea. The peripheral Neogene salt basins in the onshore region are controlled by the NE-trending tectonic lineaments, as observed in the Guadalquivir Basin. The thin yellow line A-B shows the location of a regional cross-section extending from the Guadalquivir Foreland Basin in the north to the coastal region south of the Sierra Nevada. The geological profile crosses the External Zone south of the Guadalquivir Basin, where Cretaceous and Jurassic sediments are exposed and intercalated with Triassic salt layers. The geological section crosses the Internal Zone near the coastline, where the Sierra Nevada uplifted blocks expose Mesozoic, Paleozoic, and Precambrian rocks. The exploratory borehole CMN-3 was drilled in the Neogene Granada Basin, west of the cross-section A-B. The borehole penetrated Miocene sedimentary successions with Messinian and Tortonian evaporite deposits. Credit: Chorlton (2007)/with permission of Natural Resources Canada (NRCAN).

The Granada Basin is a small intramontane basin located at the boundary between the Betic External and Internal zones, close to the Cádiz-Alicante fault zone, a major SW-NE lineament extending from the Atlantic to the Mediterranean margin, subparallel to the boundary fault that limits the Guadalquivir Basin to the north (Figure 7.24). The basin depocenter is filled with Neogene sediments including thick Miocene evaporite sequences (García-Veigas et al., 2013).

The Granada, Lorca, and Fortuna basins are small peripheric rifts affected by transcurrent movements in the Neogene, resulting in depressions infilled by Mediterranean Sea marine waters. The closure of the seaway connecting these basins to the ocean during the Neogene led to hydrologic isolation, resulting in brine concentration and deposition of evaporites during the Tortonian and Messinian (García-Veigas et al., 2013, 2020; Pérez-Lópes & Pérez-Valera, 2007). The dissolution of ancient evaporite sequences likely contributed to the enrichment of salt ions, which were concentrated and redeposited as halite in the basin depocenter and gypsum in the basin margin (Utrilla et al., 1992).

7.9.1. Salt Deposits in the Granada Basin in Spain

The Granada Basin is a small basin located in the boundary between the Betic Internal Zone (Sierra Nevada) and the Betic External Zone (Figure 7.25). A schematic cross-section from the Sierra Nevada to the center of the Granada Basin shows the facies variations

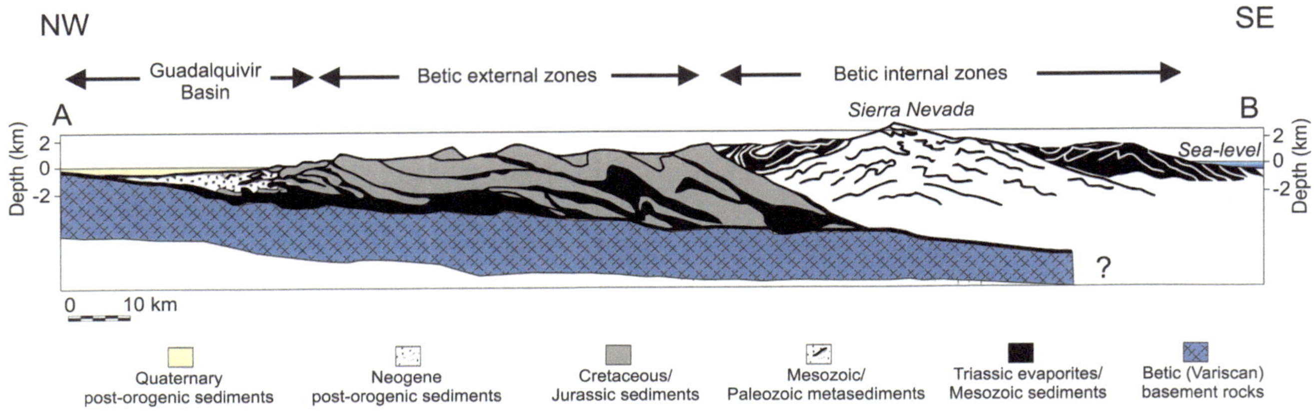

Figure 7.25 Geological cross-section extending from the Guadalquivir foreland basin to the Iberian margin. The profile crosses the Betic External Zones, where Mesozoic sequences (Triassic salt, Jurassic carbonate, and Cretaceous sediments) are affected by compressional faults. The fold-thrust belt extends to the Betic Internal Zones and the offshore region. The Sierra Nevada reaches altitudes of more than 2 km in the Betic Internal Zone, where uplifted Mesozoic and Paleozoic metasediments are involved in the deformation belt. The Betic orogenic belt is floored by the Variscan basement rocks, which are separated from the less deformed Mesozoic rocks by a detachment fault zone involving Triassic evaporites. The Granada Basin is located to the west of the geological cross-section, at the boundary between the Betic External and Internal zones. Credit: Pérez-López and Pérez-Valera (2007)/with permission of Elsevier.

and the stratigraphic sequences penetrated by the exploratory well CMN-3, drilled at the distal margin of the basin (Figure 7.26).

The Granada Basin shows two main stratigraphic sequences above the Variscan basement rocks (Figure 7.26). The lower sequence corresponds to the Miocene rift deposits, which accumulated in extensional rift troughs that were deformed by compression during the Neogene, resulting in the inversion of the rift layers as observed in the Sierra Nevada region (García-Veigas et al., 2013). The basin accumulated a thick sequence of Tortonian evaporite sediments at the central portions of the basin, including thick halite layers that are subdivided into the upper and lower evaporite units (García-Veigas et al., 2013). The upper sequence corresponds to the Tortonian and Messinian evaporites (Figure 7.27).

During the Early to Middle Miocene, the Mediterranean Sea experienced deep marine sedimentation before the MSC, which led to the accumulation of thick evaporite deposits, primarily halite, in the deepwater region (Cita, 1982; Hsü et al., 1973). These deposits, which corresponded to sulfates and chlorides, predate the main phase of evaporite deposition associated with the crisis, estimated between 5.9 and 5.3 Ma by various authors (e.g., CIESM, 2008; Roveri et al., 2008, 2014).

The Granada Basin's thick halite deposits are believed to be a result of a Tortonian interval, based on biostratigraphic analysis and isotope determinations (García-Veigas et al., 2018). Some authors postulate that the Tortonian evaporites were deposited between 7.4 and 7.2 Ma, predating by almost 2 million years the widespread Messinian evaporites in the Mediterranean Sea (García-Veigas et al., 2013). However, there are doubts about the correlation between the Granada and Lorca basins, with some authors suggesting the Granada evaporites are younger than the Lorca evaporites (e.g., Corbí et al., 2012). Some upper evaporite sequences in these basins may also correlate with the Messinian event in the deep Mediterranean Sea (see Chapter 5 in Mohriak, 2025).

The Tortonian evaporite deposits in the Granada Basin are associated with different lithological facies, with lower sequences corresponding to gypsum and anhydrite layers formed in evaporitic lagoons connected to the Mediterranean Sea gulf (García-Veigas et al., 2018). Progressive isolation, while still maintaining a connection with the Mediterranean waters, resulted in the precipitation of marine halite in the basin depocenter. The halite deposits grade laterally into primary gypsum (selenite) and carbonate rocks (stromatolites), which host one of the largest celestine (strontium sulfate) deposits in the world (García-Veigas et al., 2013).

During the later stages of basin evolution, the salt layers were formed by recycling previous evaporites. The Upper Evaporite Unit is characterized by the reworking, transportation, and redeposition of clastic evaporites (gypsarenites) and non-marine halite beds. The Tortonian Upper Evaporites are overlain by the Messinian Evaporites, represented by marls intercalated with non-marine gypsum deposits, indicating a progressive continentalization of the depositional environment during the

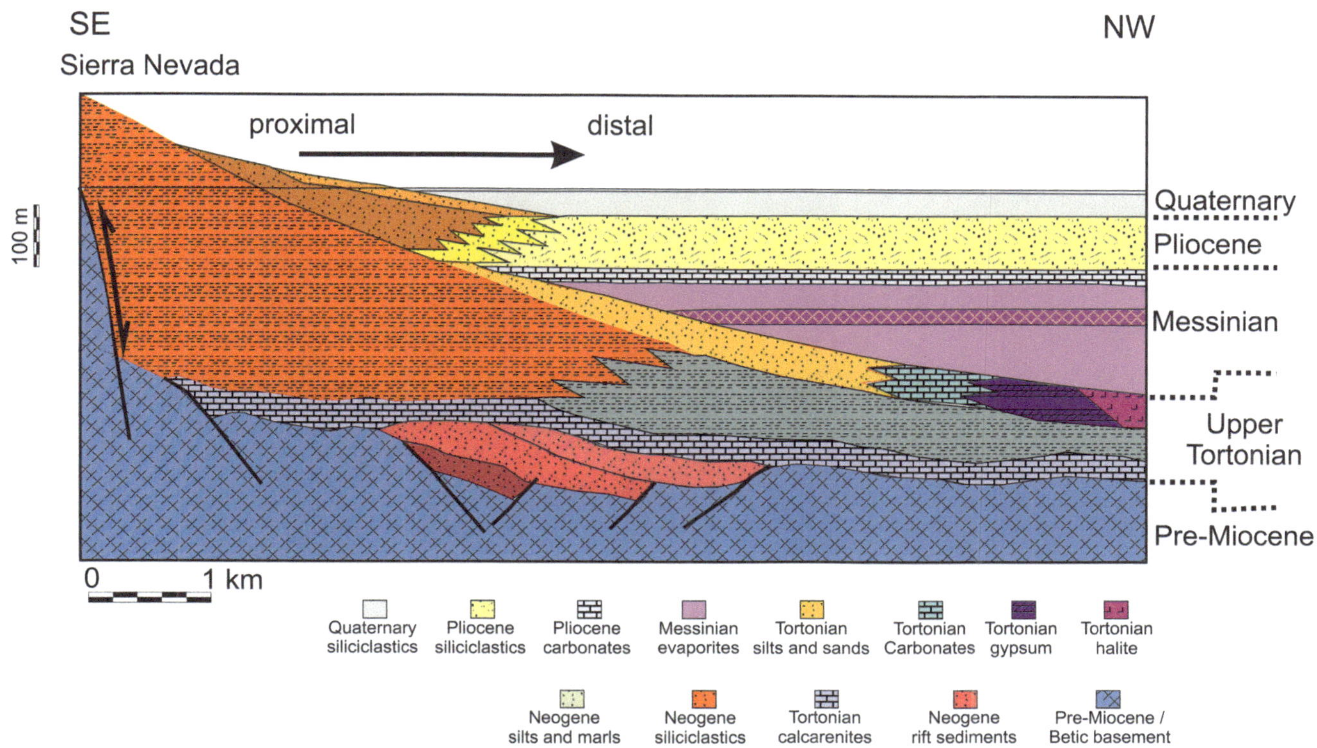

Figure 7.26 A schematic geological cross-section in the Granada Basin, from the Sierra Nevada toward the basin depocenter, shows the simplified stratigraphy of the basin and the lithological facies variation of the Neogene evaporite successions. The distal part of the evaporite basin is characterized by Upper Tortonian evaporite deposits that include thick halite layers in the basin depocenter. These sequences are overlain by Messinian marls intercalated with gypsum layers and by Pliocene to Quaternary siliciclastic deposits. The borehole CMN-3 was drilled at the distal margin. Credit: García-Veigas et al. (2013)/with permission of Elsevier.

Messinian to Pliocene transition (García-Veigas et al., 2013). These successions are capped by Pliocene to Quaternary siliciclastic deposits that accumulated on a shallow lake dominated by non-marine waters, with rare precipitation of continental gypsum beds intercalated with the siliciclastic sediments (Figures 7.27 and 7.28).

Mining companies have drilled exploratory boreholes in the center of the Granada Basin, near the village of Chimeneas, west of Granada (Figure 7.24). The deepest well (CMN-3) reached a total depth of 733 m and penetrated the Lower Evaporite sequences (basal anhydrite) at a depth of 730 m (García-Veigas et al., 2013). The Tortonian Lower Evaporite sequences are characterized by halite deposits overlying the basal anhydrite layer. The Intermediate Sandstone Unit separates the Lower Evaporite from the much thicker Upper Evaporite Unit (García-Veigas et al., 2013). Messinian-age gypsum layers intercalated with marls cap the Upper Evaporites, forming a continental sequence between 100 and 200 m in thickness (Figure 7.27).

The petrographic interpretation and the mineralogical identification of the evaporite deposits in the Granada Basin were carried out by the analysis of thin sections under petrographic microscopes, XRD techniques, and the analysis of polished core samples using electron scanning microscopy (García-Veigas et al., 2013). Evaporite minerals include chlorides (halite) and sulfates (anhydrite and gypsum), but polyhalite spherules are also recognized in the Lower Evaporites sequence (Figures 7.28 and 7.29). Calcite is the dominant carbonate mineral in the Lower Evaporites sequence, although some magnesite is also recognized. Celestine (strontium sulfate) is present in the Intermediate Sandstone Unit and the Upper Halite Unit (García-Veigas et al., 2013).

Some of the sulfate minerals (polyhalite and celestine) are interpreted as having formed during the early diagenesis of the evaporite layers. The polyhalite spherulites in the CMN-3 borehole (Figure 7.29) probably correspond to a diagenetic product replacing Ca sulfates (García-Veigas et al., 2013). The much thicker and more complex evaporite deposits associated with the MSC are mainly developed in the deepwater region of the Mediterranean Sea (see Chapter 5 in Mohriak, 2025).

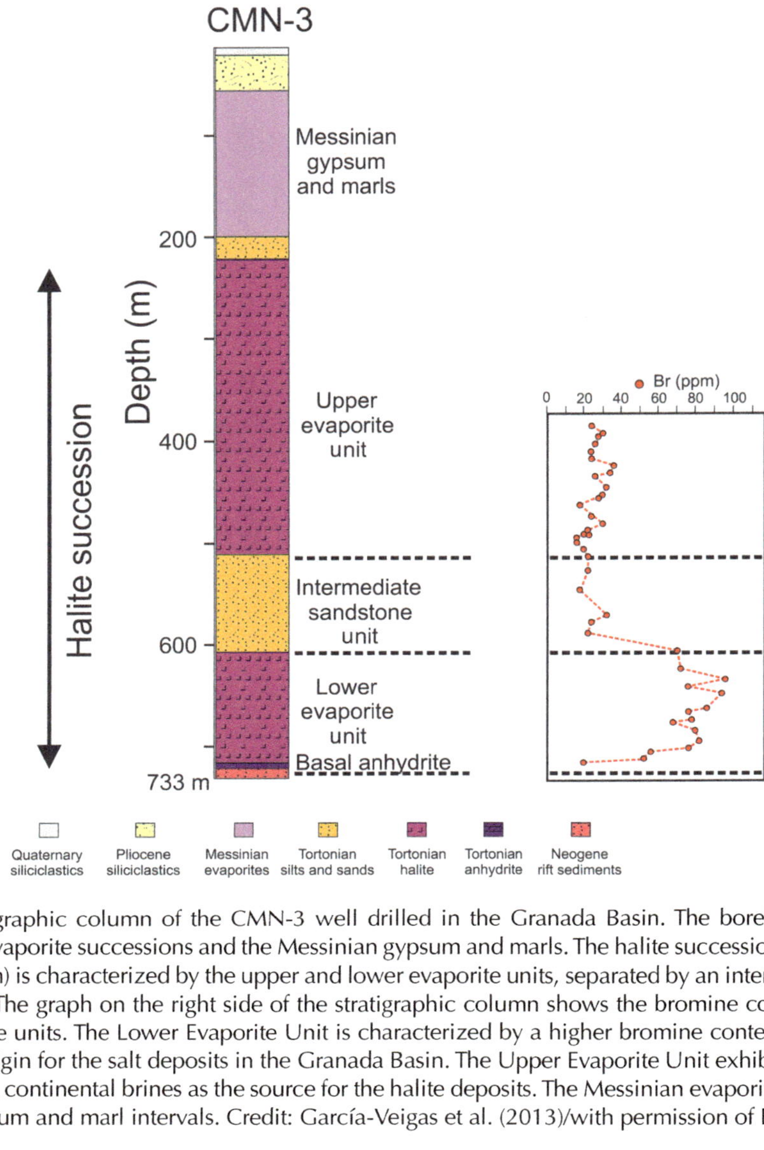

Figure 7.27 Stratigraphic column of the CMN-3 well drilled in the Granada Basin. The borehole penetrated the Upper Tortonian evaporite successions and the Messinian gypsum and marls. The halite succession (with a total thickness of circa 500 m) is characterized by the upper and lower evaporite units, separated by an intermediate siliciclastic unit (sandstones). The graph on the right side of the stratigraphic column shows the bromine content for the Upper Tortonian evaporite units. The Lower Evaporite Unit is characterized by a higher bromine content (> 60 ppm), indicating a marine origin for the salt deposits in the Granada Basin. The Upper Evaporite Unit exhibits a lower bromine content, indicating continental brines as the source for the halite deposits. The Messinian evaporites correspond to an alternation of gypsum and marl intervals. Credit: García-Veigas et al. (2013)/with permission of Elsevier.

7.10. GEOCHEMISTRY AND ADVANCED CHEMICAL METHODS

Stable isotope geochemistry methods in sedimentary rocks involve measuring the ratio of isotopes using mass spectrometers. These methods compare the relative amounts of "heavier" and "lighter" isotopes, which are stable and do not decay into other elements. Stable isotopes are relevant for studying evaporite minerals, such as carbon, sulfur, strontium, and boron. Newton and Bottrell (2007) discuss the determination of the stable isotopes of carbon and sulfur in the analysis of environmental changes in modern and ancient sedimentary basins. Claypool et al. (1980) discussed the application of the oxygen and sulfur isotopes as proxies to determine the paleo-seawater salinity. Isotope ratios are crucial for reconstructing ancient evaporites (see Garcia-Veigas et al., 2018). The source of sulfur is very important to reconstruct the recycling of older evaporites. The $\delta^{34}S$ signature is an important indicator of reducing conditions during the precipitation of evaporite minerals (Newton & Bottrell, 2007).

Scientists apply the mass spectrometry method to study the relative abundance of isotopes in various chemical compounds. For example, the carbon element has two stable isotopes: the heavy one with a mass of 13 (^{13} C) and the lighter one with a mass of 12 (^{12}C). The isotopic determination methodologies are similar, based on the comparison of the measured values with the accepted values obtained from standard samples. For example, carbon, oxygen, sulfur, and boron isotope ratios can be compared with standard measurements in compounds with known ratios determined in geochemistry laboratories.

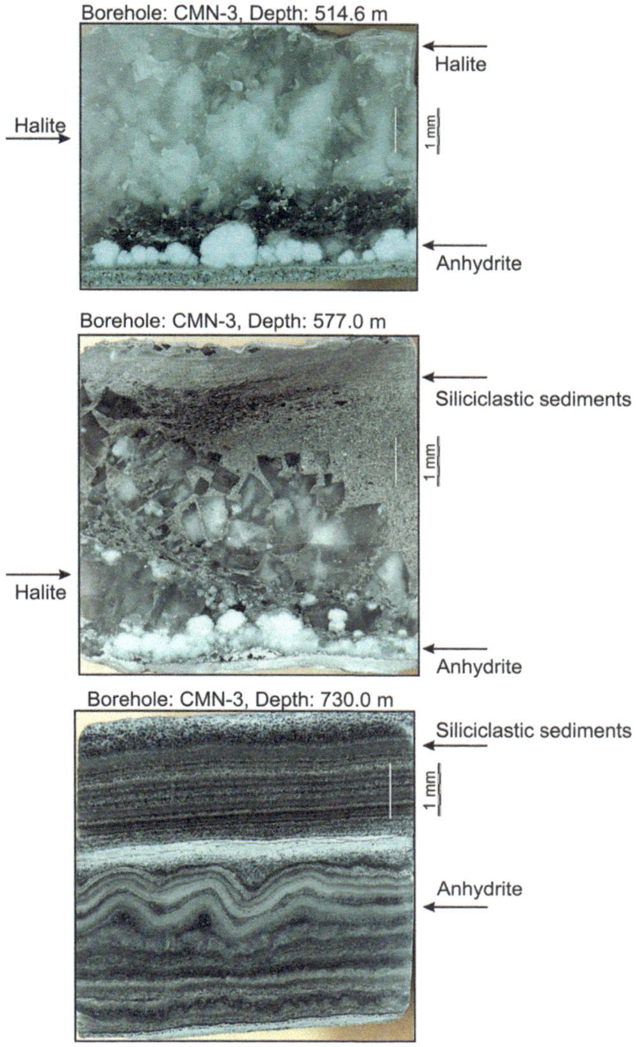

Figure 7.28 Petrographic characterization of the Miocene (Tortonian) evaporite lithologies drilled by the CMN-3 well in the Granada Basin. The Upper Evaporite Unit (depth 514.6 m) is characterized by halite and anhydrite (top image). The Intermediate Sandstone Unit (central image) is marked by siliciclastic sediments (depth 577.0 m). The Lower Evaporite Unit (bottom image) is marked by halite and anhydrite layers with local intercalation of siliciclastic sediments. The basal anhydrite sequence is marked by layered evaporites, locally exhibiting microfolding of the anhydrite lamina (depth 730.0 m). Credit: Courtesy of García-Veigas (2020).

In most carbonate rocks, such as marine limestones deposited in shallow-water environments, the $^{13}C/^{12}C$ can be measured and compared with the equivalent ratio in Pee Dee Belemnite (known as PDB). The PDB corresponds to a marine mollusk from the *Cephalopoda* class whose shell is composed of calcium carbonate. This organism lived in Cretaceous oceans and became extinct before the Cenozoic, but its shells are preserved in several outcrops worldwide (e.g., Craig, 1957).

A fossilized shell of a belemnite was collected several decades ago at the banks of the Pee Dee River in South Carolina, in the southeast United States. The measurements of the carbon ratio in the Pee Dee Fm. sample have been used to compare with other measurements, and this standard value is known as the PDB value (e.g., Craig, 1957). As the original material has been used in laboratory procedures, other belemnites have been used in recent decades, but the results are normalized to the original values. The international reference standard for carbon isotopes is currently one that was measured in the Vienna laboratories and is known as Vienna Pee Dee River Belemnite (VPDB). The original material displays an anomalously high ^{13}C /^{12}C ratio (0.01123720). This value is considered by the International Atomic Energy Agency (IAEA) in Vienna to define a $\delta^{13}C$ value of zero.

The $\delta^{13}C$ parameter is a measure of the ratio of the stable isotopes ^{13}C and ^{12}C and is widely used in several branches of geochemistry, paleoclimatology, and paleoceanography. The carbon isotope ratios are commonly expressed in parts per thousand (or part per mil, ‰), and the variation from the standard carbon ratio established for the PDB is calculated by the following formula (Craig, 1957):

$$\delta^{13}C = \left[\left(\delta^{13}C/\delta^{12}C\right)_{sample} \Big/ \left(\delta^{13}C/\delta^{12}C\right)_{standard} - 1 \right] \times 1000‰.$$

This means that the carbon isotope ratios measured in some materials are compared with the standard ratio. If the measured values are smaller than the standard values, the ratio is a negative number; if the value is higher than the standard, it implies a positive variation. Using different materials in different laboratories, geochemists observed that the resultant $\delta^{13}C$ varies in geological time as a function of organic productivity and the source of organic carbon. These concentrations depend on different vegetation types, organic carbon burial, and also on the inorganic source's signature. For example, mass extinction episodes are marked by a negative $\delta^{13}C$ anomaly, as there is a global decrease in the primary productivity of organic carbon (see Chapter 9 in Mohriak, 2025).

Researchers have well documented this anomaly in the region of Williston Lake in British Columbia, Canada. Wang et al. (1994) based the carbon isotope ratio determination on the geochemical analysis of a marine sequence of P-T sediments deposited in basinal settings, which presently outcrop in the lake margins. The δ ^{13}C values of kerogens in organic-rich carbonate rocks are characterized by a sudden shift at the P-T boundary. The uppermost Permian values correspond to values of −29‰

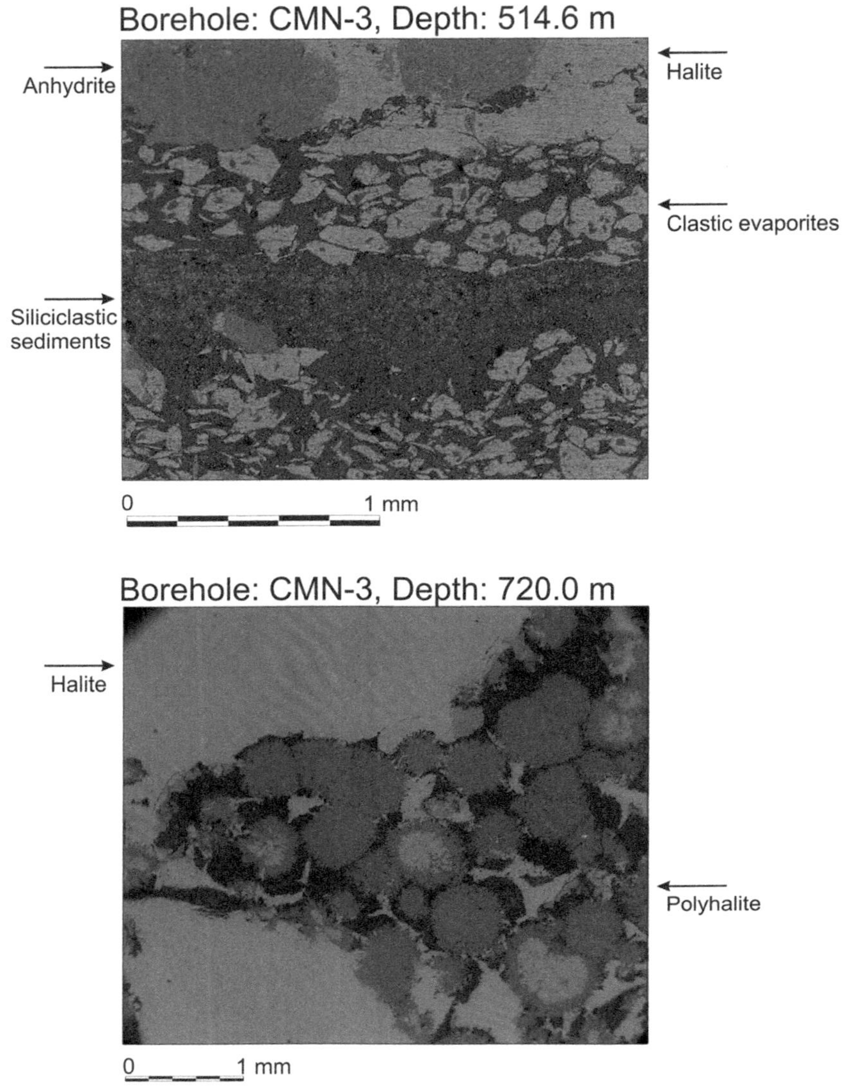

Figure 7.29 Thin section images (plane-polarized transmitted light) of the Miocene (Upper Tortonian) evaporite lithologies in the CMN-3 well in the Granada Basin. The top image shows the evaporite sediments (halite and anhydrite) near the boundary between the Upper Evaporite Unit and the Intermediate Sandstone Unit. The image at the bottom shows the Lower Evaporite Unit with halite (large crystals) and polyhalite (spherulites) minerals. Credit: Courtesy of García-Veigas (2020).

± 1‰ (PDB), reaching a minimum of –32.6‰ measured about 2 m above the P-T boundary. A few meters above the anomaly, the values returned to the mean values for the Permian. Wang et al. (1994) concluded that the abrupt reduction in surface-water primary productivity resulted from the global mass extinction episode at the P-T boundary (see Chapter 9 in Mohriak, 2025).

For brines that occur as fluid inclusions in evaporite minerals, the ^{18}O/^{16}O can be determined and compared with the standard mean ocean water (SMOW) values. The isotopic values are expressed as delta oxygen (δ O) values, which represent the difference between the isotopic ratio obtained from the analyzed sample and the ratio that is considered a constant for the standard sample.

7.10.1. Salt Age and Fluid Origin Based on Strontium Isotopes Measurements

In the case of the strontium isotopes, it is assumed that the ^{87}Sr/^{86}Sr ratio is constant for a given geological age because the global ocean waters are well mixed. Consequently, the Sr isotope ratio can be used to determine the time when the seawater was formed by using fluid inclusions found in evaporite minerals or by measuring the ratio in minerals that contain minor amounts of strontium, which usually substitutes for calcium, as they have similar ionic radius and valence (e.g., McArthur et al., 2001).

This technique has been recently used in the investigation of the age of the Gulf of Mexico evaporites (Snedden et al., 2019). As the Louann salt is devoid of fossils, the age of these evaporites is rather uncertain. However, the sulfate and carbonate minerals in the Louann evaporite cycles present an $^{87}Sr/^{86}Sr$ isotope ratio with an average value of 0.70714–0.70716 (Snedden et al., 2019). The comparison with the marine Sr isotope ratios for the past 509 million years (McArthur et al., 2001) suggests that the age of the Louann salt can be estimated at 170 Ma. This Bajocian age is slightly older than the Callovian age estimated for the salt deposition episode in the Gulf of Mexico by several researchers (e.g., Hudec et al., 2013; Salvador, 1987).

Strontium isotope ratios ($^{87}Sr/^{86}Sr$) are often employed to identify the source of strontium in barite mineralizations, which are the most important sources of barium in the earth's crust, and thus determine the origin of an ore deposit. For example, Barbieri et al. (1987) determined the Sr ratios for barite and associated fluorite from Mesozoic sedimentary rocks in Italy. The carbonate sediments host barite and fluorite as vein and cavity fillings in the brecciated limestones of Jurassic age in northwestern Sicily. These minerals are characterized by rather different values for their Sr isotopic composition (Figure 7.30), with the fluorite ratio slightly less than the barite ratio. This was interpreted as indicating that the fluorite and barite mineralizations were derived from two isotopically different fluids. Based on the similarity of the Sr isotope ratios, the fluorite fluids originated from the Upper Jurassic limestones, whereas the barite mineralization fluids probably originated from the Lower Jurassic cherts and Tertiary flysch (Barbieri et al., 1987).

Celestine minerals are found in various salt basins, including the Betic Cordillera in Spain (García-Veigas et al., 2018) and the Wieliczka Basin in Poland (Nowińska, 2017). One of the principal occurrences of celestine minerals in England is located in the Yate village near Bristol, where strontium sulfate deposits are associated with limestone layers that occur in the Triassic-age Keuper marls (Wood & Shaw, 1976).

Some authors suggest that the celestine in the Bristol area was not directly precipitated from seawater but was rather derived from the diagenesis of anhydrite and gypsum (e.g., Wood & Shaw, 1976). Carbonate minerals precipitated from groundwater evaporation that percolated through the Keuper Marl in arid environments, forming primary minerals like aragonite, which converts to calcite and expels some strontium, which may form compounds with sulfates in the brines (Wood & Shaw, 1976). Four main reasons are suggested for this interpretation. First, if the celestine minerals were directly deposited from seawater, they would occur in evaporite sequences containing halite and anhydrite, but this is not observed in the Bristol area. The celestine deposits are associated with the Triassic limestones and are absent in sedimentary beds formed far from the carbonate layers. Second, in modern sedimentary environments, celestine precipitation as a primary mineral from concentrated seawater (brines) is very rare. Third, the enrichment of the strontium element as compared to the calcium element is more than a thousand times higher than the normal chemical ratio observed in common rocks and normal seawater. Fourth, if the celestine minerals were precipitated from the Triassic seawater, they would have a similar $^{87}Sr/^{86}Sr$ ratio. However, the strontium isotope ratio determined in laboratory analysis (0.7095–0.7110) was significantly higher than the values estimated for the Triassic seas, which range from 0.7070 to 0.7080 (Wood & Shaw, 1976).

The southern margin of Spain is characterized by the Betic External Zone, which forms mountain ranges due

Mineral	$^{87}Sr/^{86}Sr$ ratio	Possible origin of the Sr-rich fluids
Fluorite	0.70785 to 0.70790	Upper Jurassic Limestones (Sr ratio: 0.70752 – 0.70789)
Barite	0.70861 to 0.70879	Lower Jurassic cherts and Tertiary flysch (Sr ratio: 0.71646 and 0.72040)

Figure 7.30 Strontium isotope ratios for the fluorite and barite deposits in Sicily (southern Italy). The strontium isotope ratios for the fluorite are similar to those of the Upper Jurassic limestones. The strontium isotope ratios for the barite are considerably higher, indicating fluid flow through Lower Jurassic cherts and Tertiary flysch, which are characterized by higher Sr ratios. Credit: Barbieri et al. (1987)/with permission of Elsevier.

to compressional tectonics (Figure 7.24). Orti et al. (2014) analyzed the $^{87}Sr/^{86}Sr$ isotopic ratios in the Triassic evaporites in the Betic Cordillera, where the stratigraphic relationships are obscured by the intense tectonic and halokinetic deformation resulting from the Alpine Orogeny (Figure 7.25).

The values in the gypsum and anhydrite samples from the Buntsandstein, Muschelkalk, and Keuper formations, ranging between 0.707615 and 0.708114, agree with worldwide Triassic evaporites sourced from marine waters. This supports the interpretation that the evaporite layers in the Betic Ranges were formed during a marine incursion in the Triassic (Pérez-López & Pérez-Valera, 2007).

According to García-Veigas et al. (2020), the fluid inclusion compositions and bromine concentrations in halite from the Triassic evaporite sequences in the Iberian margin peripheral basins (Granada, Lorca, and Fortuna basins) match the isotope ratios measured in Tortonian salt. The sulfate isotope compositions ($\delta^{34}S$ and $\delta^{18}O$) and $^{87}Sr/^{86}Sr$ ratios indicate a marine origin for the Tortonian evaporites. Based on the interpretation of the petrological and geochemical data, the sedimentary environment of the Lower Halite Unit was associated with Mediterranean seawater in lagoons enclosing perennial brines in shallow marine environments that were periodically isolated and evaporated. The overlying evaporite sequences were associated with continental environments marked by the reworking of the previously precipitated Triassic evaporites. These ancient evaporite successions were exposed to erosion and dissolution, involving mainly non-marine waters, and redeposited in isolated lakes. These dilution events during salt deposition reflect the increasing continentalization of the basin as it was isolated from the Mediterranean Sea (García-Veigas et al., 2020).

7.10.2. Bromine Content in Evaporite Minerals and Origin of Brines

Bromine contents in halite minerals are widely used as a geochemical tracer of marine or continental origin of evaporite deposits (García-Veigas et al., 2013, 2020). Bromide ion concentration in seawater is around 65–70 ppm, with a residence time of ~ 100 Ma, whereas bromide concentrations in fluvial and lacustrine freshwaters are much lower (<1 ppm). As bromide concentrations in seawater have not changed significantly during the Neogene, bromine concentrations in rock salt with values above 70 ppm indicate a marine brine. Lower values indicate non-marine or mixed brines, implying recycling of previously deposited salt either by marine or non-marine waters.

García-Veigas et al. (2013) analyzed the bromine content of the Granada Basin evaporites and observed high values for the Lower Evaporite Unit (60–100 ppm Br) and very low values for the Upper Evaporite Unit (20–40 ppm Br). The analyses of geochemical markers (bromine content and strontium isotope ratios) indicated that the lowermost part of the saline successions in the Granada Basin was highly influenced by marine waters, with the intervals with Br ~ 70 ppm associated with lagoons sourced by the adjacent ocean. The uppermost successions (Upper Evaporites), with a mean Br content of around 20-30 ppm, recorded an increasing continentalization. The isotope ratios suggested a progressive isolation from the oceanic connection and non-marine contributions to the lake system (Figure 7.27).

The bromine isotope ratios suggest that the early phases of salt deposition were coeval with a marine gulf that connected the Granada Basin with the Mediterranean Sea, forming the Betic Seaway. This interpretation is corroborated by the determination of the strontium isotope ratios. The $^{87}Sr/^{86}Sr$ ratios are higher than 0.708800, close to the values estimated for the Tortonian seawater (up to 0.708900). At depths shallower than 720 m, the Lower Halite Unit brines become sulfate-depleted, and the sulfate isotope compositions trend to lower values. The same trend is observed for the $^{87}Sr/^{86}Sr$ ratio, which approach 0.708700. The brine depletion in sulfate minerals is interpreted as the result of the contribution to the basin of $CaCl_2$-rich hydrothermal waters. The brine probably entered the lake via active local fault zones that intersected the Cádiz-Alicante crustal fracture (García-Veigas et al., 2013).

Subsequently, the strontium isotope ratios remained constant and close to 0.708700, suggesting the contribution of non-marine waters to the halite deposits shallower than 500 m in the CMN-3 borehole. This indicates that the salt pan was isolated and desiccated when the water influx from the Mediterranean Sea was interrupted. The upper evaporites indicate water inflow from continental sources, mainly pluvial and fluvial systems (García-Veigas et al., 2013).

7.10.3. The Origin of the Dead Sea Salt Deposits Based on Strontium Isotope Ratio

The Dead Sea provides another example of strontium isotope ratio usage in the interpretation of the genesis of evaporite sequences. Charrach (2018) used the geochemical method as a basis for the analysis of the Dead Sea salt deposits, whose origin has been disputed by several authors in the past decades, proposing an origin derived from marine or continental brine (e.g., Stein et al., 2000; Zak & Bentor, 1972; Zaq, 1974).

Previous models for the Sedom Lagoon admitted that the halite layers in the Dead Sea were deposited during the Pliocene after the interruption of a seaway that connected the lake with the normal seawater from the Mediterranean Sea (see Chapter 3). This hypothesis assumes that the Sedom Fm. salt layers were precipitated as the result of marine water inflow and brine concentration (Zak, 1974). However, the $^{87}Sr/^{86}Sr$ ratio variation for the global seawater (Figure 7.31) has steadily increased from 0.7090 to 0.7092 from 5 Ma to the present day (McArthur et al., 2001). The $^{87}Sr/^{86}Sr$ measured values in halite from the Sedom salt diapir in the Dead Sea resulted in ratios of 0.7083 to 0.7087, much lower than the estimated strontium isotope ratios for the global seawater (Charrach, 2018; Stein et al., 2000).

One possible explanation for the reduced Sr isotope ratio in the Dead Sea is that the seawater brines that filled the lagoon in the Pliocene were modified by mixing with waters derived from the alteration of Cretaceous carbonate rocks that surround the Dead Sea rift and were exposed during the rift flank uplift (Stein et al., 2000). Dolomitization of the Cretaceous carbonates, followed by the reflux of the resulting Ca-chloridic brine into the lagoon, resulted in reducing the Sr isotope ratio. Another explanation is the inflow of Ca-chloride brines from hydrothermal sources mixed with freshwater (Stein et al., 2000).

On the other hand, Charrach (2018) highlighted that the results of Sr isotope ratios suggest that the Sedom Fm. evaporites have a non-marine origin. According to this interpretation, the original Dead Sea brines were sourced from continental waters. This possibility is supported by the sedimentological analysis of the upper 5 km of sediments that overlie the salt layers in the Dead Sea Basin. Based on the biostratigraphic analysis, these sediments are mainly associated with continental fluvial and lacustrine deposits that formed isolated depocenters with no inflow of marine seawater (see Chapter 11).

Charrach (2018) pointed out that the $^{87}Sr/^{86}Sr$ ratios measured by Stein et al. (2000) in the Sr-bearing minerals of the Sedom Fm. indicate that they might fall within the range for the upper Mediterranean continental evaporitic cycle that preceded the Zanclean flooding (Roveri et al., 2008; Ryan, 2008, 2009). The MSC is dated as 5.97 to 5.33 Ma, ending with the opening of the Gibraltar Strait that resulted in the Zanclean flooding event, which infilled the basin with seawater from the Atlantic Ocean (e.g., CIESM, 2008). The upper evaporite cycle, interpreted as formed between 5.6 and 5.5 Ma, is closer in time to the demise of the MSC. The Zanclean flooding is marked by continental water inflow into the Mediterranean Basin lacustrine systems, resulting in the accumulation of gypsum layers during the Lago Mare sedimentation (see Chapter 5 in Mohriak, 2025). The upper evaporite cycle rocks of the Eastern Mediterranean Sea exhibit a strontium ratio that is similar to the ratio measured in halite of the Sedom salt diapir (Figure 7.32).

The similarity of the isotope ratios in both basins, lower than the global sea-level isotope ratio, indicates a major hydrologic change associated with salt deposition and mixing with hydrothermal or continental waters. The Mediterranean Sea changed from an oceanic basin with an abundant supply of seawater at the beginning of the brine concentration in the MSC (Roveri et al., 2014). Due to a very arid climate and excessive evaporation, the Mediterranean Sea brine concentrated to the point of depositing salt layers in a non-marine setting at the transition to the Pliocene. In the Dead Sea, the precipitation of the Sedom salt in the Pliocene occurred on a much smaller scale within a restricted continental basin. However, the geochemical signatures based on Sr isotope ratios show that the brine followed similar hydrological paths (Charrach, 2018; Roveri et al., 2014).

According to Ryan (2008), the wide-ranging values of the Sr isotope ratios for the Mediterranean evaporites, much lower than the 0.7090 to 0.7089 values estimated for the interval 10.0-5.0 My (McArthur et al., 2001), might also indicate a complete closure of the Atlantic spillway around 5.6–5.5 Ma. Following this assumption,

Global ocean seawater	$^{87}Sr/^{86}Sr$ ratio
Pliocene – Recent 5.3 – 0.0 Ma	0.7090 to 0.7092
Tortonian – Messinian 10 – 5 Ma	0.7089 to 0.7090

Figure 7.31 Strontium isotope ratios for the Upper Miocene and Pliocene to Recent intervals based on fitting global values with a polynomial curve with minimum statistical deviations.

Basin	$^{87}Sr/^{86}Sr$ ratio
Eastern Mediterranean Sea Messinian Event Upper Evaporites: 5.6 – 5.5 Ma	0.7085 to 0.7089
Dead Sea Pliocene halite deposits (Sedom Fm.)	0.7083 to 0.7087

Figure 7.32 Strontium isotope ratios for the East Mediterranean Sea (Upper Evaporite Cycle of the Messinian Event, mainly gypsum) and the Dead Sea (Pliocene rock salt). These values are lower than the marine seawater in the global oceans during the Pliocene to Holocene, estimated to vary steadily from 0.7090 (Pliocene, 5 Ma) to almost 0.7092 (Recent, 0 Ma).

only non-marine waters entered the basin during the final stages of the MSC, resulting in the event known as Lago Mare (e.g., Caruso et al., 2020). This hypothesis assumes that the Mediterranean Sea paleoenvironments were associated with a series of isolated but deep hypersaline lakes, similar to the non-marine depositional setting for the Dead Sea Lake in the Pliocene, after the deposition of the Sedom Fm. main salt layers (Charrach, 2018). On the other hand, recent reviews by Carnevale et al. (2019) and Andreetto et al. (2021) considered that marine conditions were still present in the Mediterranean Sea during the Lago Mare phase, due to the intermittent connection with the Atlantic Ocean via the Gibraltar Corridor (see Chapter 5 in Mohriak, 2025).

7.10.4. The Messinian Salinity Crisis in the Mediterranean Sea

The MSC is a topic that has been the focus of many research projects in the past century (see review by Meilijson et al., 2018). Several hypotheses have been proposed to explain the genesis of the thick evaporite deposits in the distal margins of the Mediterranean Sea and peripheral basins (see Chapter 5 in Mohriak, 2025).

Roveri et al. (2008, 2014) proposed distinct stages for the evaporite deposition episodes associated with the MSC, from the onset of the MSC at 5.971 Ma to its demise at the base of the Pliocene at 5.330 Ma (Figure 7.33). The Primary Lower Gypsum (PLG) corresponds to Stage 1, which is deposited above the Tortonian/Messinian carbonates associated with the Calcare di Base. Stage 2 is characterized by the Resedimented Lower Gypsum with the Halite layers. Stage 3.1 represents the Upper Gypsum, and Stage 3.2 is known as the Lago Mare. These evaporite sequences are overlain by deepwater sediments accumulated in the basin after the Zanclean flooding event, which resulted from the reconnection of the Atlantic Ocean with the Mediterranean Sea via the Gibraltar Corridor (Andreetto et al., 2021; CIESM, 2008; Cita, 1982; Meilijson et al., 2018).

The three MSC stages are known as the "Messinian Trilogy" (CIESM, 2008; Roveri et al., 2008, 2014), which seismically correspond to the Lower Evaporites (LE), Messinian Salt (H) or Mobile Unit (MU), and Upper Evaporites (UE) in the distal margin. Several authors have suggested a link between the Messinian Trilogy found on land (for example, in the Caltanissetta Basin in Sicily) and the different sub-basins in the deepwater region (e.g., Roveri et al., 2008, 2014). However, there is no direct link between the evaporite layers in the peripheral basins and the thick salt layers in the basin depocenters. Seismic interpretation, sequence stratigraphy, sedimentological and paleontological analysis, and correlation of isotope measurements were some of the methods used in these studies (Carnevale et al., 2019; CIESM, 2008; Raad et al., 2021).

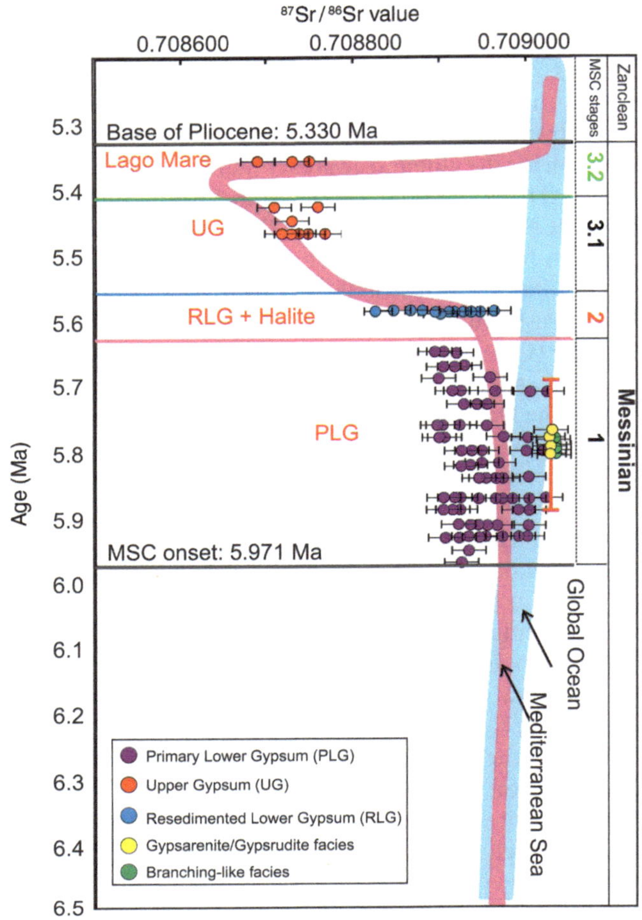

Figure 7.33 Strontium isotope ratios for the Mediterranean Sea and the mean curve for the Global Ocean during the MSC. The pink curve shows the progressive change in the isotopic composition of the Mediterranean waters during the MSC, with a minimum value between 5.3 and 5.4 Ma (Lago Mare interval). The vertical red line defines the range of ages and isotopic ratios for the samples from the Benestari quarry in Calabria (yellow circles), which are compared with the values obtained from the PLG interval (MSC Stage 1) in other outcrops from the Mediterranean Sea peripheral basins (purple circles). Credit: Roveri et al. (2014)/with permission of Elsevier.

Each of the three MSC stages exhibits a characteristic $^{87}Sr/^{86}Sr$ signature (Figure 7.33). The isotopic values are characterized by well-defined ranges that do not overlap between stages 2 and 3 and thus provide a quantitative tool to differentiate the evaporite deposits. The isotopic values record marked changes in the hydrology balance in the basin, with a strong influence on the climate and biology of the Mediterranean Sea. The diminished continental water input during the MSC (derived from pluvial and fluvial run-off in lacustrine environments) resulted in brine concentration. This interval contrasts markedly with the isotopic measurements of strontium ratios in the sedimentary successions deposited above the Miocene/Pliocene boundary. The Zanclean deposits immediately above this boundary reflect the normal marine environments in the global oceans, with isotopic values falling in the range of values above 0.709000 (Cipriani et al., 2021; Roveri et al., 2014).

Cipriani et al. (2021) studied the gypsum samples from the Benestare quarry in the Calabrian fore-arc basin, located in Sicily (southern Italy). The evaporitic deposits associated with the MSC are characterized by selenite crystals in a muddy, organic-rich matrix, forming the gypsrudite and branch-like facies of the PLG. These PLG sediments are overlain by the Resedimented Lower Gypsum (RLG), the Upper Gypsum, and the Lago Mare evaporitic to siliciclastic facies at the transition from the Miocene to the Pliocene (Figure 7.33). The MSC deposits in the studied quarry are covered by the Pliocene Trubi Fm. carbonates and marls, deposited after the Zanclean flooding of the Mediterranean Sea (Cipriani et al., 2021).

The samples of the evaporitic sediments were analyzed by mass spectrometers for strontium isotopes. The gypsrudite and the branching-like facies yielded $^{87}Sr/^{86}Sr$ average values of 0.709045 and 0.709082, respectively (Figure 7.33). These values suggest a depositional interval between 5.9 and 5.6 Ma, compatible with the PLG interval defined by Roveri et al. (2008, 2014). The isotopic ratios indicate an evaporite basin with a reduced freshwater input and a strong connection with the waters from the global ocean, resulting in periodic seawater ingressions (Cipriani et al., 2021).

Roveri (2014) and Cipriani et al. (2021) proposed the following strontium isotope values for the MSC stages (Figure 7.33):

- Primary Lower Gypsum: isotopic values between 0.709024 and 0.708890. These values are in the range of normal seawater or slightly isolated from the global ocean.
- Resedimented Lower Gypsum and Halite: isotopic values between 0.708960 and 0.708820, and between 0.709034 and 0.708801, respectively. These lower values suggest the presence of a more diluted water mass when compared to the previous stage, but the basin was still connected to the global ocean.
- Upper Gypsum and Halite: this interval is characterized by the lowest and most depleted isotopic values, mainly below 0.708800 and as low as 0.708600. Compared with the coeval global ocean, these values suggest basin isolation with extremely reduced seawater inflow.
- Lago Mare: this interval shows a rapid increase in the isotopic values, indicating partial communication with seawater and an influx of continental waters. At the end of the interval, around 5.33 Ma, an almost instantaneous return to marine conditions is indicated by isotopic values reaching 0.709000. This phase of normal marine sedimentation corresponds to the Zanclean flood (see Chapter 5 in Mohriak, 2025).

REFERENCES

Abelha, M., Petersohn, E., Bastos, G., & Araújo, D. (2018). New insights into the Parnaíba Basin: Results of investments by the Brazilian National Petroleum Agency. In M. C. Daly, R. A. Fuck, J. Julià, D. I. M. Macdonald, & A. B. Watts (Eds.), *Cratonic basin formation: A case study of the Parnaíba Basin of Brazil* (Vol. 472, pp. 341–360). Geological Society, London, Special Publications. https://doi.org/10.1144/SP472.13

Abrantes, F. R., Jr., Nogueira, A. C. R., & Soares, J. L. (2016). Permian paleogeography of west-central Pangea: Reconstruction using sabkha-type gypsum-bearing deposits of Parnaíba Basin, Northern Brazil. *Sedimentary Geology*, *341*, 175–188.

Adams, A. E., & Mackenzie, W. S. (1998). *A colour atlas of carbonate sediments and rocks under the microscope*. London: UK Manson Publishing, 180 pp.

Adams, A. E., Mackenzie, W. S., & Guilford, C. (1984). *Atlas of sedimentary rocks under the microscope*. Essex, UK: Longman Group UK Ltd., 104 pp.

Andreetto, F., Alois, G., Raad, F., Heida, H., Flecker, R., Agiadi, K., et al. (2021). Freshening of the Mediterranean Salt Giant: Controversies and certainties around the terminal (Upper Gypsum and Lago-Mare) phases of the Messinian Salinity Crisis. *Earth-Science Reviews*, *216*, 103577. https://doi.org/10.1016/j.earscirev.2021.103577

Araújo, R. N., Nogueira, A. C. R., Bandeira, J., & Angélica, R. S. (2016). Shallow lacustrine system of the Permian Pedra de Fogo Formation, Western Gondwana, Parnaíba Basin, Brazil. *Journal of South American Earth Sciences*, *67*, 57–70. doi: 10.1016/j.jsames.2016.01.009

Barbieri, M., Bellanca, A., Neri. R., & Tolomeo, L. (1987). Use of strontium isotopes to determine the source of hydrothermal fluorite and barite from northwestern Sicily (Italy). *Chemical Geology Isotope Geoscience section*, *66*(3–4), 273–278. doi: 10.1016/0168-9622(87)90046-7

Barbosa, E. N., Córdoba, V. C., & Sousa, D. C. (2016). Evolução estratigráfica da Sequência Neocarbonífera-Eotriássica da Bacia do Parnaíba, *Brasil. Brazilian Journal of Geology*, *46*(2), 181–198. doi: 10.1590/2317-4889201620150021

Benison, K., & Goldstein, R. H. (1999). Permian paleoclimate data from fluid inclusions in halite. *Chemical Geology*, *154*(1), 113–132. doi: 10.1016/S0009-2541(98)00127-2

Berger, J.-P., Reichenbacher, B., Becker, D., Grimm, M., Grimm, K., Picot, L., et al. (2005). Paleogeography of the Upper Rhine Graben (URG) and the Swiss Molasse Basin

(SMB) from Eocene to Pliocene. *International Journal of Earth Sciences (Geol Rundsch)*, *94*, 697–710.

Berra, F., & Angiolini, L. (2014). The evolution of the Tethys region throughout the Phanerozoic: A brief tectonic reconstruction. In L. Marlow, C. Kendall, & L. Yose (Eds.), *Petroleum systems of the Tethyan region, American Association of Petroleum Geologists, Memoir* (Vol. *106*, pp. 1–27).

BGS (2014). Bedrock summary lithologies across the UK Continental Shelf (2014 Version). British Geological Survey. https://webapps.bgs.ac.uk/services/ngdc/accessions/index.html?simpleText=Geological%20map%20UK#item172370

Bish, D., Blake, D., Vaniman, D., Sarrazin, P., Bristow, T., Achilles, C., et al. (2014). The first X-ray diffraction measurements on Mars. *IUCrJ*, *1*, 514–522. doi:10.1107/S2052252514021150

Blakey, R. C. (2008). Gondwana paleogeography from assembly to breakup—A 500 m.y. odyssey. In C. R. Fielding, T. D. Frank, & J. L. Isbell (Eds.), *Resolving the late Paleozoic ice age in time and space* (Vol. *441*, pp. 1–28). The Geological Society of America Special Paper.

Bloss, F. D. (1999). *An introduction to the methods of optical crystallography*. New York: Saunder College Publishing, Holt, Rinehart and Winston, 294 pp.

Bobco, F. E. R. (2017). *Aspectos deposicionais e diagenéticos das rochas evaporíticas aptianas da porção emersa da Bacia de Sergipe-Alagoas*. Universidade Federal do Rio Grande do Sul, Instituto de Geociências. Programa de Pós-Graduação em Geociências, Dissertação de Mestrado, 120 pp.

Braitsch, O. (1971). *Salt deposits: Their origin and composition*. New York: Springer-Verlag, 279 pp.

Camelbeeck, T., Vanneste, K., Alexandre, P., Verbeeck, K., Petermans, T., Rosset, P., et al. (2007). Relevance of active faulting and seismicity studies to assessments of long-term earthquake activity and maximum magnitude in intraplate northwest Europe, between the Lower Rhine Embayment and the North Sea. In S. Stein & S. Mazzotti (Eds.), *Continental intraplate earthquakes: Science, hazard, and policy issues* (Vol. *425*, pp. 193–224). Geological Society of America Special Paper. doi: 10.1130/2007.2425(14).

Carnevale, G., Gennari, R., Lozar, F., Natalicchio, M., Pellegrino, L., & Dela Pierre, F. (2019). Living in a deep desiccated Mediterranean Sea: An overview of the Italian fossil record of the Messinian salinity crisis. *Bollettino della Società Paleontologica Italiana*, *58*(1), 109–140. doi:10.4435/BSPI.2019.04

Caruso, A., Blanc-Valleron, M. -M., Da Prato, S., Pierre, C., & Rouchy, J. M. (2020). The late Messinian "Lago-Mare" event and the Zanclean Reflooding in the Mediterranean Sea: New insights from the Cuevas del Almanzora section (Vera Basin, South-Eastern Spain). *Earth-Science Reviews*, *200*, 102993, 1–17.

Célini, N., Callot, J. -P., Ringenbach, J.-C., & Graham, R. (2020). Jurassic salt tectonics in the SW sub-Alpine fold-and-thrust belt. *Tectonics*, *39*, e2020TC006107. https://doi.org/10.1029/2020TC006107

Charrach, J. (2018). The Sdom evaporite formation in Israel and its relationship with the Messinian salinity crisis. *Carbonates and Evaporites*, *33*, 727–766. https://doi.org/10.1007/s13146-017-0410-1

Chorlton, L. B. (2007). Generalized geology of the world: Bedrock domains and major faults in GIS format: A small-scale world geology map with an extended geological attribute database. *Geological Survey of Canada, Open File 5529*, 2007, 48 pages.

CIESM (2008). The Messinian salinity crisis from mega-deposits to microbiology—A consensus report. In F. Briand (Ed.), *CIESM Workshop Monographs* (Vol. *33*), Monaco, 168 pp.

Cipriani, M., Dominici, R., Costanzo, A., D'Antonio, M., & Guido, A. (2021). A Messinian gypsum deposit in the Ionian forearc basin (Benestare, Calabria, Southern Italy): Origin and paleoenvironmental indications. *Minerals*, *11*(12), 1305. https://doi.org/10.3390/min11121305

Cita, M. B. (1982). The Messinian salinity crisis in the Mediterranean: A review. In H. Berckhemer & K. J. Hsü (Eds.), *Alpine Mediterranean Geodynamics, Geodynamics Series* (Vol. *7*, pp. 113–140). Washington, DC: American Geophysical Union, .

Claypool, G.E., Holser, W. T., Kaplan, I. R., Sakai, H., & Zak, I. (1980). The age curves of sulfur and oxygen isotopes in marine sulfate and their mutual interpretation. *Chemical Geology*, *28*, 199–260.

Corbí, H., Lancis, C., García-García, F., Pina, J.-A., Soria, J. M., Tent-Manclús, J. E., & Viseras, C. (2012). Updating the marine biostratigraphy of the Granada Basin (central Betic Cordillera). Insight for the Late Miocene palaeogeographic evolution of the Atlantic-Mediterranean seaway. *Geobios*, *45*(3), 249–263. doi:10.1016/j.geobios.2011.10.006

Craig, H. (1957). Isotopic standards for carbon and oxygen and correction factors for mass-spectrometric analysis of carbon dioxide. *Geochimica et Cosmochimica Acta*, *12*(1–2), 133–149. doi:10.1016/0016-7037(57)90024-8

Dedić, Z., Ilijanić, N., & Miko, S. (2018). A mineralogical and petrographic study of evaporites from the Mali Kukor, Vranjkovići, and Slane Stine deposits (Upper Permian, Dalmatia, Croatia). *Geologia Croatica*, *71*/1,19–28.

Dronkert, H., Blasi, H. R., & Matter, A. (1990). *Facies and origin of Triassic evaporites from the Nagra boreholes, Northern Switzerland*, Technical Report 87-02, Geologisches Institut der UniversiHit Bern, 120pp.

Fayet-Moore, F., Cinthya Wibison, C., Carr, P., Duve, E., Petocz, P., Lancaster, G., et al. (2020). An analysis of the mineral composition of pink salt available in Australia. *Foods*, *9*, 1490, 1–15. doi:10.3390/foods9101490

Feijó, F. J. (2014). Sergipe basin, old oil province reborn. *AAPG Search and Discovery, Article #110174*.

Fracassi, U., Niviére, B., & Winter, T. (2005). First appraisal to define prospective seismogenic sources from historical earthquake damages in southern Upper Rhine Graben. *Quaternary Science Reviews*, *24*, 403–425.

Frizon de Lamotte, D., Raulin, C., Mouchot, N., Wrobel-Daveau, J.-C., Blanpied, C., & Ringenbach, J.-C. (2011). The southernmost margin of the Tethys realm during the Mesozoic and Cenozoic: Initial geometry and timing of the inversion process. *Tectonics*, *30*, TC3002. http://dx.doi.org/10.1029/2010TC002691

García-Veigas, J., Cendón, D. I., Gibert, L., Lowenstein, T. K., & Artiaga, D. (2018). Geochemical indicators in Western Mediterranean Messinian evaporites: Implications for the salinity crisis. *Marine Geology*, *403*, 197–214.

García-Veigas, J., Cendón, D. I., Rosell, L., Ortí, F., Ruiz, J. T., Martín, J. M., & Sanz, E. (2013). Salt deposition and brine evolution in the Granada Basin (Late Tortonian, SE Spain). *Palaeogeography, Palaeoclimatology, Palaeoecology*, *369*, 452–465.

García-Veigas, J., Gibert, L., Cendón, D. I., Artiaga, D., Corbi, H., Soria, J. M., et al. (2020). Late Miocene evaporite geochemistry of Lorca and Fortuna basins (Eastern Betics, SE Spain): Evidence of restriction and continentalization. *Basin Research*, *32*(5), 926–958. doi: https://doi.org/10.1111/bre.12408

Geluk, M. C. (2007). Permian. In Th.E. Wong, D. A. J. Batjes, & J. de Jager (Eds.), *Geology of the Netherlands* (pp. 63–83). Royal Netherlands Academy of Arts and Sciences.

Glennie, K. W. (1998). *Petroleum geology of the North Sea: Basic concepts and recent advances* (4th ed.). Oxford, UK: Blackwell Publishing Limited, 656 pages.

Góes, A. M. O., & Feijó, F. J. (1994). Bacia do Parnaíba. *Boletim de Geociências da Petrobras*, *8*(1), 57–68.

Gribble, C. D., & Hall, A. J. (1985). *Optical mineralogy principles and practice*. London: George Allen and Unwin (Publishers) Ltd., 249 pp.

Grotzinger, J. P., Crisp, J., Vasavada, A. R., Anderson, R.C., Baker, C. J., Barry, R., et al. (2012). Mars Science Laboratory Mission and Science Investigation. *Space Science Reviews*, *170*, 5–56. doi 10.1007/s11214-012-9892-2

Halliday, D., Resnick, R., & Walker, J. (2014). *Fundamentals of Physics Extended* (10th ed.). Hoboken, NJ: Wiley, 1456 pp.

Haynes, W. M. (2017). *CRC Handbook of Chemistry and Physics* (97th ed.). Boca Raton, FL: CRC Press, 2643 pp .

Heilbron, M., Guedes, E., Mane, M., Valeriano, C. M., Tupinambá, M., Almeida, J., et al. (2018). Geochemical and temporal provinciality of the magmatism of the eastern Parnaíba Basin, NE Brazil. In M. C. Daly, R. A. Fuck, J. Julià, D. I. M. Macdonald, & A. B. Watts (Eds.), *Cratonic Basin formation: A case study of the Parnaíba Basin of Brazil* (Vol. 472, pp. 251–278). Geological Society, London, Special Publications. https://doi.org/10.1144/SP472.11

Hinsken, S., Ustaszewski, K., & Wetzel, A. (2007). Graben width controlling syn-rift sedimentation: the Palaeogene southern Upper Rhine Graben as an example. *International Journal of Earth Sciences*, *96*(6), 979–1002. doi:10.1007/s00531-006-0162-y

Holt, N. M., García-Veigas, J., Lowenstein, T. K., Giles, P. S., & Williams-Stroud, S. (2014). The major-ion composition of Carboniferous seawater. *Geochimica et Cosmochimica Acta*, *134*, 317–334.

Hsü, K. J. (1972). Origin of saline giants: A critical review after the discovery of the Mediterranean evaporite. *Earth-Science Reviews*, *8*, 371–396.

Hsü, K. J., Cita, M. B., & Ryan, W. B. F. (1973). The origin of the Mediterranean evaporites. In W. B. F. Ryan, K. J. Hsü, et al. (Eds.), *Initial reports of the Dead Sea Drilling Project* (Vol. *13*, pp. 1203–1231). Washington, United States Government Printing Office.

Hsü, K. J., Montadert, L., Bernoulli, D., Cita, M. B., Erickson, A., Garrison, R. E., et al. (1978). History of the Mediterranean salinity crisis. In K. J. Hsü, L. Montadert, et al. (eds.), *Initial reports of the Deep Sea Drilling Project* (Vol. *42*, Part 1, pp. 1053–1078). Washington (U.S. Government Printing Office). doi:10.2973/dsdp.proc.42-1.149.1978

Hudec, M. R., Norton, I. O., Jackson, M. P. A., & Peel, F. J. (2013). Jurassic evolution of the Gulf of Mexico salt basin. *American Association of Petroleum Geologists Bulletin*, *97*(10), 1683–1710.

International Mineralogical Association (2022). The new IMA list of minerals—A work in progress. *International Mineralogical Association*. http://cnmnc.main.jp/ , updated May 2022.

Kemp, S. J., Smith, F. W., Wagner, D., Mounteney, I., Bell, C. P., Milne, C. J., et al. (2016). An improved approach to characterize potash-bearing evaporite deposits, evidenced in North Yorkshire, United Kingdom. *Economic Geology*, *111*, 719–742.

Klein, C., & Philpotts, A. (2017). *Earth materials: Introduction to mineralogy and petrology* (2nd ed.). Cambridge, UK: Cambridge University Press, 616 pp.

Kukla, P. A., Strozyk, F., & Mohriak, W. U. (2018). South Atlantic salt basins—Witnesses of complex passive margin evolution. *Gondwana Research*, *53*, 41–57.

Kuster, Y., Schramm, M., Bornemann, O., & Leiss, B. (2009). Bromide distribution characteristics of different Zechstein 2 rock salt sequences of the Southern Permian Basin: A comparison between bedded and domal salts. *Sedimentology*, *56*, 1368–1391.

Lavina, B., Dera, P., & Downs, R. T. (2014). Modern X-ray diffraction methods in mineralogy and geosciences. *Reviews in Mineralogy and Geochemistry*, *78*, 1–31.

Lustrino, M. (2021). Luminescence in the mineral realm to teach basic Physics concepts. *Journal of Education and Learning*, *10*(3), 83–95. doi:10.5539/jel.v10n3p83

MacKenzie, W. S., Adams, A. E., & Brodie, K. H. (2017). *Rocks and minerals in thin section* (2nd ed.). London, UK: CRC Press, Taylor & Francis Group, 232 pp.

Manenti, R. R., Souza, W. E., & Porsani, M. J. (2018). Reprocessing and interpretation of deep structures in a regional transect of the Parnaíba Basin, Brazil. In M. C. Daly, R. A. Fuck, J. Julià, D. I. M. Macdonald, & A. B. Watts (Eds.), *Cratonic Basin formation: A case study of the Parnaíba Basin of Brazil* (Vol. 472, pp. 101–107). Geological Society, London, Special Publications. https://doi.org/10.1144/SP472.17

Martins, G. S., Mohriak, W. U., & Destro, N. (2019). Synrift evaporite deposition and structural characterization of the onshore Alagoas subbasin. *Interpretation*, *7*(4), SH19–SH31.

Marzoli, A., Renne, P. R., Piccirillo, E. M., Ernesto, M., Bellieni, G., & De Min, A. (1999). Extensive 200 Ma-year-old continental flood basalts of the Central Atlantic Magmatic Province. *Science*, *284*, 616–618.

Matte, P. (1986). Tectonics and plate tectonics model for the Variscan belt of Europe. *Tectonophysics*, *126*, 329–374.

McArthur, J. M., Howarth, R. J., & Bailey, T. R. (2001). Strontium isotope stratigraphy: LOWESS version 3: Best fit to the Marine Sr-Isotope Curve for 0–509 Ma and accompanying look-up table for deriving numerical age. *The Journal of Geology*, *109*(2), 155–170. doi:10.1086/319243

Meilijson, A., Steinberg, J., Hilgen, F., Bialik, O. M., Waldmann, N. D., & Makovsky, Y. (2018). Deep-basin evidence resolves a 50-year-old debate and demonstrates synchronous onset of Messinian evaporite deposition in a non-desiccated Mediterranean. *Geology*, *46*(3), 243–246.

Meng, T., Xiangxi, M., Donghu, Z., & Hu, Y. (2018). Using micro-computed tomography and scanning electron microscopy to assess the morphological evolution and fractal dimension of a salt-gypsum rock subjected to a coupled thermal-hydrological-chemical environment. *Marine and Petroleum Geology*, *98*, 316–334.

Milani, E. J., & Thomaz Filho, A. (2000). Sedimentary basins of South America. In U. G. Cordani, E. J. Milani, A. Thomaz Filho, & D. A. Campos (Eds.), *Tectonic evolution of South America*. 31 International Geological Congress, August 2000, Rio de Janeiro, pp. 389–449.

Miranda, F. S., Vettorazzi, A. L., Cunha, P. R. C., Aragão, F. B., Michelon, D., Caldeira, J. L., et al. (2018). Atypical igneous-sedimentary petroleum systems of the Parnaíba Basin, Brazil: seismic, well logs and cores. In M. C. Daly, R. A. Fuck, J. Julià, D. I. M. Macdonald, & A. B. Watts (Eds.), *Cratonic Basin formation: A case study of the Parnaíba Basin of Brazil* (Vol. 472, pp. 341–360). Geological Society, London, Special Publications. https://doi.org/10.1144/SP472.15

Mohriak, W. U. (2015). Pre-salt carbonate reservoirs in the South Atlantic and World-wide analogs. AAPG Search and Discovery, Article #51086 (2015). Posted April 6, 2015.

Mohriak, W. (2025). *Salt in the Earth sciences: Basin analysis and salt tectonics*. Washington, DC: American Geophysical Union. doi:10.1002/9781394307067

Mohriak, W. U., & Fainstein, R. (2012). Phanerozoic regional geology of the eastern Brazilian margin. In D. G. Roberts & A. W. Bally (Eds.), *Phanerozoic passive margins, cratonic basins and global tectonic maps* (Vol. 1C, Chapter 7, pp. 223–-282). Elsevier B.V. doi: 10.1016/B978-0-444-56357-6.00006-8

Mohriak, W. U., Szatmari, P., & Anjos, S. (2012). Salt: Geology and tectonics of selected Brazilian basins in their global context. In G. I. Alsop, S. G. Archer, A. J. Hartley, N. T. Grant, & R. Hodgkinson (Eds.), *Salt tectonics, sediments and prospectivity* (Vol. 363, pp. 131–158). Geological Society, London, Special Publications.

Nesse, W. D. (2012). *Introduction to optical mineralogy* (3rd ed.). Oxford, UK: Oxford University Press, 384 pp.

Newton, R., & Bottrell, S. (2007). Stable isotopes of carbon and sulphur as indicators of environmental change: Past and present. *Journal of the Geological Society*, London, *164*, 691–708.

Nowińska, J. (2017). New data on celestine-rich salts in the Wieliczka salt deposits. *Geology, Geophysics and Environment*, *43*(1), 57–68.

Ojeda, H. A. O. (1982). Structural framework, stratigraphy, and evolution of Brazilian marginal basins. *American Association of Petroleum Geologists Bulletin*, *66*(6), 732–749.

Okrusch, M., & Frimmel, H. E. (2020). *Mineralogy—An introduction to minerals, rocks, and mineral deposits*. Springer Textbooks in Earth Sciences, Geography and Environment. Berlin, Heidelberg: Springer, 719 pp. doi.org/10.1007/978-3-662-57316-7

Ortí, F., Pérez-López, A., García-Veigas, J., Rosell, L., Cendón, D. I., & Pérez-Valera, F. (2014). Sulfate isotope compositions ($\delta^{34}S$, $\delta^{18}O$) and strontium isotopic ratios ($^{87}Sr/^{86}Sr$) of Triassic evaporites in the Betic Cordillera (SE Spain). *Revista de la Sociedad Geológica de España*, *27*(1), 79–89.

Pace, A., Bourillot, R., Bouton, A., Vennin, E., Galaup, S., Bundeleva, I. P., et al. (2016). Microbial and diagenetic steps leading to the mineralisation of Great Salt Lake microbialites. *Scientific Reports*, *6*, 31495. doi: 10.1038/srep31495 (2016).

Pérez-López, A., & Pérez-Valera, F. (2007). Palaeogeography, facies and nomenclature of the Triassic units in the different domains of the Betic Cordillera (S Spain). *Palaeogeography, Palaeoclimatology and Palaeoecology*, *254*, 606–626.

Peryt, T. M., Pierre, C., & Gryniv, S. P. (1998). Origin of polyhalite deposits in the Zechstein (Upper Permian) Zdrada Platform (Northern Poland). *Sedimentology*, *45*, 565–578.

Playà, E., Orti, F., & Rosell, L. (2000). Marine to non-marine sedimentation in the upper Miocene evaporites of the Eastern Betics, SE Spain: Sedimentological and geochemical evidence. *Sedimentary Geology*, *133*, 135–166.

Raad, F., Lofi, J., Maillard, A., Tzevahirtzian, A., & Caruso, A. (2021). The Messinian salinity crisis deposits in the Balearic Promontory: An undeformed analog of the MSC Sicilian basins?? *Marine and Petroleum Geology*, *124*, 104777. https://doi.org/10.1016/j.marpetgeo.2020.104777.

Raith, M. M., Raase, P., & Reinhardt, J. (2012). *Guide to thin section microscopy* (2nd ed.) . Mineralogical Society of America, ebook, 127 pp. http://www.minsocam.org/msa/OpenAccess_publications/Guide_Thin_Sctn_Mcrscpy/Thin_Sctn_Mcrscpy_2_prnt_eng.pdf

Reed, S. J. B. (2005). *Electron microprobe analysis and scanning electron microscopy in geology*. Cambridge, UK: Cambridge University Press, 192 pp.

Reyes, F., Lin, Q., Udoudo, O., Dodds, C., Lee, P. D., & Neethling, S. J. (2017). Calibrated X-ray micro-tomography for mineral ore quantification. *Minerals Engineering*, *110*, 122–130.

Roveri, M., Lugli, S., Manzi, V., Gennari, R., & Schreiber, B.C. (2014). High-resolution strontium isotope stratigraphy of the Messinian deep Mediterranean basins: Implications for marginal to central basins correlation. *Marine Geology*, *349*, 113–125. https://doi.org/10.1016/j.margeo.2014.01.002

Roveri, M., Lugli, S., Manzi, V., & Schreiber, B. C. (2008). The Messinian salinity crisis: A sequence-stratigraphic approach. *GeoActa*, *7984*, 169–190.

Ryan, W. B. F. (2008). Modeling the magnitude and timing of evaporite drawdown during the Messinian salinity crisis. *Stratigraphy*, *5*(3–4), 227–243.

Ryan, W. B. F. (2009). Decoding the Mediterranean salinity crisis. *Sedimentology*, *56*, 95–136.

Salvador, A. (1987). Late Triassic-Jurassic paleogeography and origin of Gulf of Mexico Basin. *American Association of Petroleum Geologists Bulletin*, *71*(4), 419–451.

Scholle, P. A., & Ulmer-Scholle, D. S. (2003). A Color Guide to the Petrography of Carbonate Rocks: Grains, textures, porosity, diagenesis, *American Association of Petroleum Geologists, Memoir* (Vol. 77), 459 pp.

Scotese, C. R. (2008). The PALEOMAP Project PaleoAtlas for ArcGIS, Version 1, Volume 3, *Triassic and Jurassic Paleogeographic, Paleoclimatic, and Plate Tectonic Reconstructions*. PALEOMAP Project, Arlington, Texas.

Scotese, C. R., & Langford, R. P. (1995). Pangea and the Paleogeography of the Permian. In P. A. Scholle, T. M. Peryt, & D. S. Ulmer-Scholle (Eds.), *The Permian of Northern Pangea, Volume 1: Paleogeography, Paleoclimates, Stratigraphy* (pp. 3–19). Berlin, Germany: Springer-Verlag. https://doi.org/10.1007/978-3-642-78593-1_1

Scotese, C. R., & Schettino, A. (2017). Late Permian-Early Jurassic Paleogeography of Western Tethys and the World. In I. Soto, J. F. Flinch, & G. Tari (Eds.), *Permo-Triassic salt provinces of Europe, North Africa and the Atlantic margins, Tectonics and Hydrocarbon potential* (pp. 57–95). Amsterdam: Elsevier.

Selvestone, J. (2015). *Optical mineralogy in a nutshell*. University of New Mexico shortcourse, 3 parts.

Słowakiewicz, M., Tucker, M. E., Pancost, R. D., Perri, E., and Mawson, M. (2013). Upper Permian (Zechstein) microbialites: Supratidal through deep subtidal deposition, source rock, and reservoir potential. *American Association of Petroleum Geologists Bulletin*, *97*(11), 1921–1936. DOI:10.1306/06181312179

Smith, M. M., Sholokhova, Y. Hao, Y., & Carroll, S. A. (2012). Evaporite Caprock integrity: An experimental study of reactive mineralogy and pore-scale heterogeneity during brine-CO_2 exposure. *Lawrence Livermore National Security*, LLNL-JRNL-544136, 27 pp.

Snedden, J. W., Norton, I. O., Hudec, M. R., & Peel, F. (2019). Paleogeographic Reconstruction of the Louann Salt Basin, *37th Annual GCSSEPM Foundation Perkins-Rosen Research Conference, Salt Tectonics, Associated Processes, and Exploration Potential: Revisited: 1989–2019,* Abstracts, pp. 24–27.

Sommaruga, A., Mosar, J., Schori, M., & Gruber, M. (2017). The role of the Triassic evaporites underneath the North Alpine Foreland. In J. I. Soto, J. Flinch, & G. Tari (Eds.), *Permo-Triassic salt provinces of Europe, North Africa and the Atlantic margins, Tectonics and Hydrocarbon potential* (pp. 447–466). https://doi.org/10.1016/B978-0-12-809417-4.00021-5

Sorensen, B. E. (2013). A revised Michel-Lévy interference colour chart based on first-principles calculations. *European Journal of Mineralogy*, *25*, 5–10.

Souza-Lima, W. (2009). Sequências Evaporíticas da Bacia de Sergipe-Alagoas. In W. U. Mohriak, P. Szatmari, & S. C. Anjos (Eds.), *Sal: Geologia e Tectônica, Exemplos nas Bacias Brasileiras* (2nd ed., pp. 232–251). São Paulo: Editora Beca.

Stein, M., Starinsky, A., Agnon, A., Katz, A., Raab, M., Spiro, B., & Zak, I. (2000). The impact of brine-rock interaction during marine evaporite formation on the isotopic Sr record in the oceans: Evidence from Mt. Sedom, Israel. *Geochimica et Cosmochimica Acta*, *64*(12), 2039–2053.

Stewart, F. H. (1963). The Permian lower evaporites of Fordon in Yorkshire. *Proceedings of the Yorkshire Geological Society*, *34*, 1–44.

Strozyk, F., Reuning, L., Scheck-Wenderoth, M., & Tanner, D. C. (2017). The Tectonic history of the Zechstein Basin in the Netherlands and Germany. In J. I. Soto, J. F. Flinch, & G. Tari (Eds.), *Permo-Triassic salt provinces of Europe, North Africa and the Atlantic margins, Tectonics and Hydrocarbon potential* (pp. 221–241). doi:10.1016/b978-0-12-809417-4.00011-2

Szatmari, P., Carvalho, R. S., & Simões, I. A. (1979). A comparison of evaporite facies in the late Paleozoic Amazon and the Middle Cretaceous South Atlantic salt basins. *Economic Geology*, *74*(2), 432–447.

Szatmari, P., Carvalho, R. S., Simões, I. A., Tibana, P., & Leite, D. C. (1974). Evaporitos de Sergipe: Geologia e geoquímica. *PETROBRAS, Internal Report*, Vol. *1*, pp. 1–160.

Szatmari, P., Carvalho, R. S., Simões, I. A., Tibana, P., & Leite, D. C. (2009). Atlas petrográfico dos evaporitos. In W. Mohriak, P. Szatmari, & S. M. C. Anjos (Eds.), *Sal: Geologia e Tectônica* (pp. 42–65). São Paulo: Editora Beca.

Szatmari, P., Moré de Lima, C., Fontaneta, G., de Melo Lima, N., Zambonato, E., Menezes, M. R., et al. (2021). Petrography, geochemistry and origin of South Atlantic evaporites: The Brazilian side. *Marine and Petroleum Geology*, *127*, 104805. https://doi.org/10.1016/j.marpetgeo.2020.104805

Taylor, J. C. M. (1984). Late Permian Zechstein. In K. W. Glennie (Ed.), *Introduction to the Petroleum Geology of the North Sea* (Chapter 4, pp. 61–101). London: Blackwell Scientific Publishers.

Taylor, J. C. M. (1985). The role of evaporites in hydrocarbon exploration. *Joint Association of Petroleum Exploration Courses (UK)*, Japec Course No. 39.

Taylor, J. C. M. (1998). Upper Permian—Zechstein. In K. W. Glennie (Ed.), *Petroleum geology of the North Sea: Basic concepts and recent advances* (4th ed., pp. 174–211). Oxford, UK: Blackwell Publishing Limited.

Thiemeyer, N., Habersetzer, J., Peinl, M., Zulauf, G., & Hammer, J. (2015). The application of high-resolution X-ray computed tomography on naturally deformed rock salt: multi-scale investigations of the structural inventory. *Journal of Structural Geology*, *77*, 92–106.

Tucker, M. E. (1991). Sequence stratigraphy of carbonate-evaporite basins: models and application to the Upper Permian (Zechstein) of northeast England and adjoining North Sea. *Journal of the Geological Society of London*, *148*, 1019–1036.

Utrilla, R., Pierre, C., Orti, F., & Pueyo, J.J. (1992). Oxygen and Sulphur isotope compositions as indicators of the origin of Mesozoic and Cenozoic evaporites from Spain. *Chemical Geology (Isotope Geoscience Section)*, *102*, 229–244.

Vaz, P. T., Rezende, N. G. A. M., Wanderley Filho, J. R., & Travassos, W. A. S. (2007). Bacia do Parnaíba. *Boletim de Geociências da Petrobras*, *15*(2), 253–263.

Vera, J. A. (2001). Evolution of the south Iberian continental margin. In Ziegler, P. A., Cavazza, W., Robertson, A. H. F., & Crasquin-Soleau, S. (Eds.), *Peri-Tethys Memoir 6: Peri-Tethys Rift/Wrench Basins and Passive Margins* (Vol. *186*, pp. 109–143). Mém. Mus. Natn. Hist. Nat. Paris.

Wang, K., Geldsetzer, H. H. J., & Krouse, H. R. (1994). Permian-Triassic extinction: Organic $\delta^{13}C$ evidence from British Columbia, *Canada. Geology*, *22*(7), 580. https://doi.org/10.1130/0091-7613(1994)022<0580:PTEOCE>2.3.CO;2

Warren, J. K. (2006). *Evaporites: Sediments, resources and hydrocarbons*. Berlin: Springer Verlag, 1035 pp.

Warren, J. K. (2010). Evaporites through time: Tectonic, climatic and eustatic controls in marine and nonmarine deposits. *Earth and Science Review*, *98*, 217–268.

Warren, J. K. (2016). *Evaporites—A geological compendium*. Berlin: Springer, 1807 pp. ISBN 978-3-319-13511-3.

Wood, M. W., & Shaw, H. F. (1976). The geochemistry of celestites from the Yate area near Bristol (U.K.). *Chemical Geology*, *17*, 179–193.

Zak, I. (1974). Sedimentology and Bromine geochemistry of marine and continental evaporites in the Dead Sea basin. In *Fourth International Symposium on Salt*, Northern Ohio Geological Society, pp. 349–361.

Zak, I., & Bentor, Y. K. (1972). Some new data on the salt deposits of the Dead Sea area, Israel. In G. Richter-Bernburg (Ed.), *Geology of saline deposits*, Proceedings of the Hanover Symposium, 15–21 May 1968, International Union of Geological Sciences, United Nations Educational, Scientific and Cultural Organization. *Earth Sciences*, *7*, 137–146.

Ziegler, P. A. (1975). North Sea Basin history in the tectonic framework of North-Western Europe. In A. W. Woodland (Ed.), *Petroleum and the continental shelf of North-West Europe* (pp. 131–148). London: Applied Science.

Ziegler, P. A. (1990). Tectonic and palaeogeographic development of the North Sea rift system. In D. J. Blundell & A. D. Gibbs (Eds.), *Tectonic evolution of the North Sea rifts* (pp. 1–36). Oxford: Oxford Science Publications.

Part II

Salt Deposition in Different Sedimentary Environments

8
Salt Deposition Models and Sedimentary Environments

ABSTRACT

Earth scientists have long debated the origin of giant salt basins, with some early models suggesting that they formed from the evaporation of desert-flooded regions. Current theories suggest that climate and plate tectonic processes control the formation of large salt basins, allowing for the opening of rifts and gulfs during plate divergence and the closure of inland seas during plate convergence. Evaporite minerals precipitate and accumulate in various environments, including deserts, arid shorelines, sabkhas, gulfs, playa lakes, inland seas, and remnant oceans.

Evaporite deposits are recorded in sedimentary basins from the Proterozoic to the Cenozoic, primarily in divergent, convergent, and transform plate tectonic settings (e.g., Kendall, 1978, 1984, 1992, 2010; Warren, 2010). Most basins precipitated some evaporite deposits at some geological intervals. However, the "Salt Giants"—basins with evaporite sequences of several kilometers and lateral extents reaching over half a million square kilometers—are limited to a few basins and geological time intervals and have no equivalent in the modern world (Friedman, 1972; Kendall et al., 2009; Schreiber, 1978, 1986; Schreiber & Hsü, 1980; Schreiber et al., 2007; Warren, 2006, 2016, 2021).

Evaporite sediments have been found in various environments throughout geological history, including lacustrine settings, playa lakes, marginal marine settings, and shallow-to-deep marine environments (Figure 8.1). The classical "Salt Giants" are associated with marine environments such as the Red Sea, north-central European basins, the South Atlantic, the Gulf of Mexico, and the Mediterranean Sea. The Dead Sea is the only deepwater setting where halite deposits are observed at present. The Dead Sea's surface geology and tectono-sedimentary evolution are described in Chapters 3 and 11. The modern Dead Sea evaporative systems and lake hydrology were discussed in Chapter 2. Isotope ratios measured in halite minerals from the Dead Sea Lake indicate multiple sources for the brine through geological time (see Chapter 7).

Platform and shallow-water marine evaporites are rare in modern sedimentary environments, except in some regions adjacent to marine gulfs and seaways, such as the supratidal evaporite deposits accumulating at the southern margin of the Persian Gulf (Alsharhan & Kendall, 2003, 2011; Kendall, 1984, 1992, 2010; Kendall et al., 2014). The Kara Bogaz Gol is an important example of a shallow, barred salt basin linked to the Caspian Sea by a channel through a siliciclastic sill, providing a modern analog for an almost complete basin desiccation event affecting a lake fed by brackish water (see Chapter 10).

In the second half of the twentieth century, researchers emphasized supratidal evaporite sedimentation in arid environments as possible modern analogs for salt deposition in the Mediterranean Sea (e.g., Cita, 1983; Hsü, 1972, 1977) and the Red Sea (e.g., Whitmarsh et al., 1974). Early conceptual models for salt deposition episodes in these deep basins were proposed based on drilling the tops of some salt diapirs in the deepwater region of the Mediterranean Sea and the Red Sea. However, subsequent works, based on the interpretation of modern seismic data and the drilling of deep boreholes that penetrated the whole of the salt layers in the distal margins, have questioned the original interpretations (see Chapter 12). In the present century, the depositional models for the thick halite layers accumulated during the Messinian salinity crisis (MSC) in the Mediterranean Sea have been greatly improved by

Salt in the Earth Sciences: Evaporite Rocks and Salt Deposition, Geophysical Monograph 286, First Edition. Webster Mohriak.

DOI: 10.1002/9781119405245.ch08

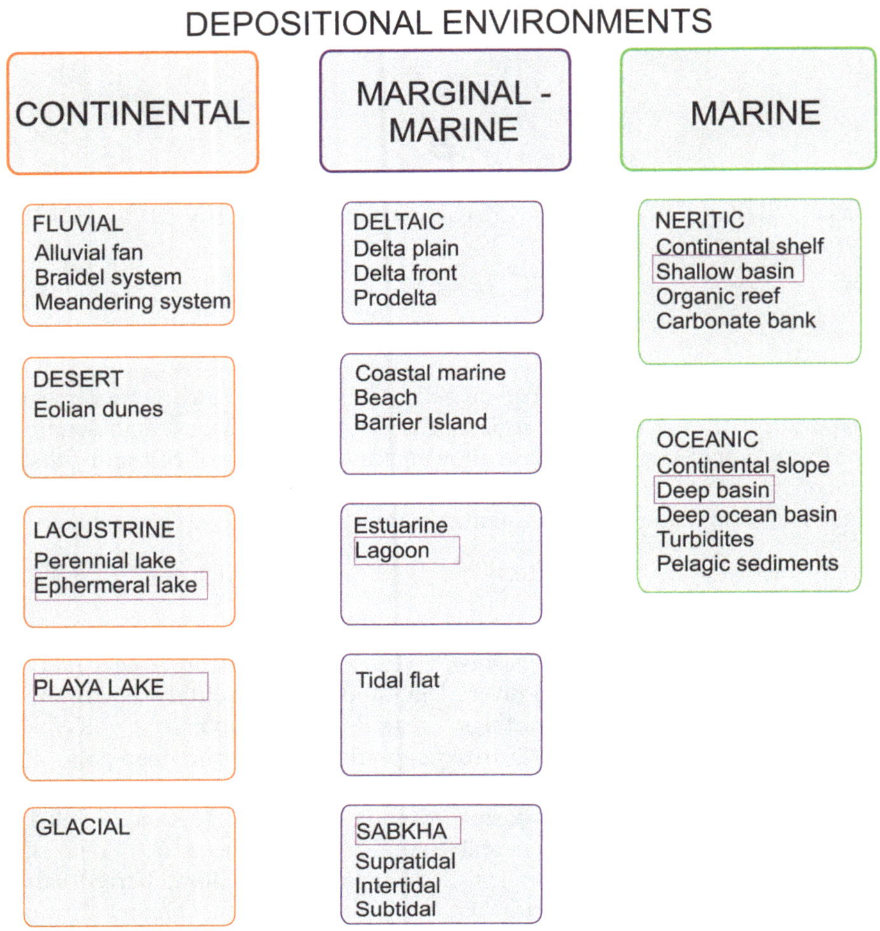

Figure 8.1 Depositional environments in sedimentary basins. Evaporite deposition is observed in continental, marginal-marine, and marine environments. The pink rectangle highlights the main environments associated with significant evaporite deposits. Credit: Adapted from Boggs Jr. (2014).

integrating the sedimentary facies associations of several evaporite basins on land with their deepwater counterparts (e.g., CIESM, 2008). These studies include the precise paleontological dating of the depositional events, high-resolution sequence stratigraphy and geochemistry, and strontium isotope ratios of ostracod shells from outcrops and borehole cores from drilling sites across the Mediterranean Sea (see Chapter 5 in Mohriak, 2025).

Sedimentary facies are defined as the sum of the characteristics of a sedimentary unit resulting from some particular set of physical, chemical, and biological parameters that reflect the environment of deposition. According to several authors (e.g., Blatt et al., 1980; Boggs, 2009, 2014; Serra, 1984, 1986), the sedimentary facies are characterized by lithological units with specific compositional, textural, and structural properties, including the chemical composition, physical properties, grain size distribution, color, biota, trace fossils, and sedimentary structures. The term "facies" may be applied to different scales, from thin petrographic sections to rock samples, outcrops, and basins. The interpretation of the sedimentary environments associated with modern and ancient evaporite deposits is based on the sedimentary structures and facies associations (Figure 8.1).

This chapter addresses the sedimentary models for salt deposition in present-day environments, focusing on deserts, ephemeral lakes in arid coastal regions, and supratidal evaporite deposits (sabkhas). These sedimentological models are based on facies associations and have been applied to the interpretation of platform and basinwide salt deposits in many ancient basins, such as the Permian Basin in the United States or the Persian Gulf in the Middle East (see Chapter 12).

8.1. INTRODUCTION TO SALT DEPOSITION MODELS AND SEDIMENTARY ENVIRONMENTS

The study of the geological and hydrological evolution of sedimentary basins that accumulated significant thick salt sequences in the Holocene has important applications in salt mining and petroleum exploration (Schreiber & Marshak, 1981; Schreiber, 1988; Warren, 1989, 2012, 2014). Sedimentological studies include the interpretation of chemical processes that led to the concentration of brines, their composition, the evolution through geological time, and the mechanisms that resulted in the precipitation of different salt minerals (see Chapter 4).

The deposition of evaporite sequences results from climatic factors and implies a decrease in the water supply in the hydrological cycle of sedimentary basins. Factors controlling the reduction of water in the system include reduction of meteoric precipitation, lack of drainage (rivers), lack of subterranean sources (springs), or isolation of oceanic or continental water masses that were previously connected with larger water bodies. The negative hydrological balance resulted in the almost complete desiccation of water masses in most giant salt basins of the geological past and is observed in a few basins at present. These basins are mainly located in arid regions, such as the Sahara Desert in northern Africa, or the Danakil Depression in the Afar region (Figure 8.2), but also occur in cold, elevated regions, such as the Tibetan Plateau in China and the Andean highlands in South America (e.g., Barbieri et al., 2006; Duan & Hu, 2001; Warren, 2006, 2016; Yechieli & Wood, 2002).

Researchers have been intrigued by the fact that even in the aridest regions of the planet, it is difficult to find a halite layer of a regional extent that is thicker than a few centimeters. This observation contrasts with some ancient basins where thousands of meters of evaporite sediments were accumulated within a short time frame. The salt deposition interval in the South Atlantic basins is extremely short compared to the total sedimentation time

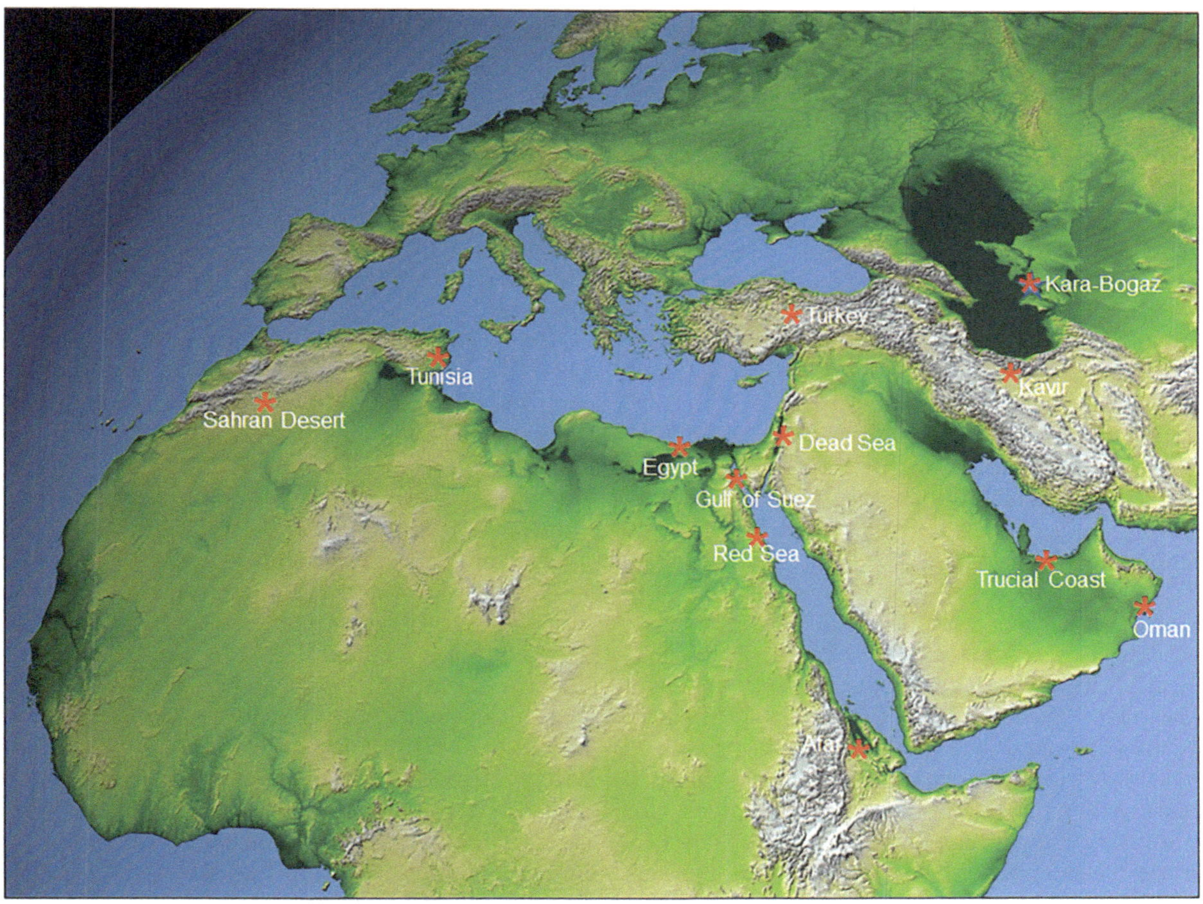

Figure 8.2 Earth's globe satellite image (courtesy of NASA), shows the main localities where evaporite sediments are depositing at present. Significant salt accumulations are observed mainly in desert areas of Africa, the southern Mediterranean Sea, Turkey, Arabia, and regions adjacent to the Red Sea and Caspian Sea. Credit: Courtesy of NASA.

on the continental margins. The answer to this enigma has been investigated by many researchers in the past two centuries, analyzing Holocene settings to understand modern processes and elucidating the past (Hutton, 1788, 1795; Hsü, 1972; Ochsenius, 1877, 1888; Schreiber & Hsü, 1980).

Ochsenius (1877,1888) and Hsü (1972) suggested that the floor of the Zechstein Basin in the Upper Permian might have been about 720 m below sea level. Simple calculations based on mineral solubility in seawater indicate that only 12 m of evaporite sediments can accumulate on the basin floor each time it is filled to the top and the water evaporates completely. As the Zechstein salt layers reached a thickness of hundreds of meters, the depositional mechanism would require a floodgate to be opened and closed more than 40 times to precipitate the amount of salt known in some parts of the basin (Hsü, 1972). The periodic alternation of water inflow and outflow in an almost isolated basin defies belief. That is why Ochsenius (1877) postulated a continuous supply of seawater through a narrow opening or barrier, which became known as the bar theory. A barred basin with a seawater column of 1000 m, if allowed to evaporate completely, would precipitate only 14 m of evaporite minerals, predominantly halite (Lugli, 2009).

Adding to this paradox, most sedimentologists interpret that evaporite sediments are formed very rapidly in geological time, almost corresponding to a volcanic event. This fact led geoscientists to propose different geodynamic models for the formation of large ancient salt basins, called "giant salinas" or "Salt Giants." These models are based on controversial assumptions and interpretations, and some basins, such as the Mediterranean Sea, have been discussed for more than a century without reaching a consensus (CIESM, 2008; Meilijson et al., 2018).

One reason why the formation of giant saline basins is not observed anywhere in the world today is the fact that we are living in a geological time that follows the last glaciation that affected the planet and demised about 26,000 years ago (Kendall, 1978, 1984, 1992; Warren, 2010, 2016).

During the last glaciation episode, ice sheets covered much of the northern and southern hemispheres, profoundly affecting the Earth's climate by removing water from the hydrological cycle (Philip et al., 2013). This glacial period resulted in droughts and a large drop in sea level, exposing large expanses of the seas as land and causing desertification in many regions of the globe, affecting the development of the human race and the establishment of new civilizations (e.g., Degroot et al., 2022).

The demise of the last glaciation episode in the Northern Hemisphere resulted from rising temperatures at approximately 20–15 ka (Degroot et al., 2022; Philip et al., 2013). In Antarctica, located in the Southern Hemisphere, deglaciation can be fingerprinted by the abrupt rise of sea level at about 14.5 ka (Clark et al., 2009). Thus, in the past few millennia (the last 10 ka), the rising sea level in the oceans has not provided favorable conditions for salt deposition and accumulation in large areas across the globe. This hinders the application of present-day observations to explain how thick layers of salt could have been extensively deposited in many evaporite basins throughout geological time (Warren, 2006, 2010, 2016).

Some of the first geological concepts, such as "The Principle of Uniformitarianism," were developed by Scottish naturalist James Hutton in the second half of the eighteenth century (Hutton, 1788, 1795). These concepts were further developed by Lyell (1830) in his book "Principles of Geology," which is considered one of the first textbooks to systematically present a thorough study of the earth by establishing the foundations of the geological sciences.

Several researchers and textbooks refer to the phrase "The present is the key to the past" as a guiding maxim for the earth sciences. Although textbooks link this adage to Hutton's and Lyell's works from the eighteenth and nineteenth centuries, it was probably empirically used as a scientific methodology in earlier centuries. Leonardo da Vinci, the famous painter and inventor who lived in medieval Europe, examined fossil-bearing rocks high in the Apennines Mountains and concluded that they could be accepted as evidence for a former seafloor that existed in the region before it was raised and formed a mountain (Baucon, 2010).

Lyell (1830) postulated that nature follows some basic physical and chemical laws that have not varied in space and time. Nonetheless, the uniformitarianism tenet has been rebuked by some authors who have correctly pointed out that, at present, the planet is characterized by conditions that may be different from the past (Gould, 1965; Romano, 2015; Shea, 1982). For example, the high content of oxygen in the atmosphere has no counterpart in the Archean, and several major catastrophic geological events have affected life on Earth, resulting in massive extinctions (see Chapter 9 in Mohriak, 2025).

Earth scientists have identified small-scale examples of salt deposition in several basins worldwide, such as the salinas along the Dead Sea, the salt pans across the Sahara Desert, and the arid deserts and playa lakes in the southwestern region of the United States. Evaporite rocks can be deposited when continental lakes or shallow seas become disconnected from the main body of water and dry up by solar evaporation (Hsü, 1972; Ochsenius, 1877). However, there is a much larger quantity of salt in the saline giants of the past than what could be accounted for by a single desiccation event of isolated bodies of seawater. Most authors consider that the giant saline basins have been repeatedly desiccated and refilled by marine incursions, either continuously or periodically. This results in a cycle of brine concentration, salt deposition, and basin desiccation, resulting in the piling up of different evaporite minerals on top of older evaporite sequences at the bottom of the basins (see Chapter 4).

Researchers admit that evaporites are deposited following chemical principles applied to brine concentration by solar evaporation. However, there are other more complex and speculative models based on interpretations of catastrophic changes in brine chemistry. Some authors advocate drastic changes in the sedimentary environment (from deep marine to shallow marine and subaerial) to explain the salinity crises that resulted in brine concentration and the rapid and simultaneous accumulation of salt layers that covered the oceanic areas of the Mediterranean Sea (Hsü, 1972; Hsü et al., 1973, 1977, 1978a, 1978b). Several authors have postulated that magmatic and volcanic events are responsible for the precipitation of salt when the seawater reaches a supercritical temperature, losing the ability to dissolve salts (Hovland et al., 2006, 2018a, 2018b). Other authors have suggested mantle exhumation and serpentinization processes in abyssal environments as related to the formation of giant evaporite basins (Scribano et al., 2017, 2018).

The geographic distribution of the main localities where evaporite sediments are accumulating indicates that climate aridity is one of the main controls for salt accumulation and preservation. Modern salt deposition is mainly observed in desert areas of Africa and Asia, along the coastlines of the Mediterranean Sea, Red Sea, and Arabian Peninsula (Figure 8.2). Consequently, evaporites are formed in areas where more water leaves the basin by evaporation than the water that enters the system by a combination of rainfall, surface runoff (rivers), and subsurface inflow. Evaporite sedimentation is mainly restricted to the arid and semiarid deserts of the globe, and these areas are located in two belts between 10° and 45° north and south of the equator (see Chapter 2).

This chapter explores modern settings for evaporite deposition in deserts and coastal regions. The Persian Gulf and North African lowlands display sedimentary environments along the coastal plain, forming continental and coastal sabkhas. Salt accumulations may occur in salt flats at high elevations, as observed in Salar de Uyuni in Bolivia and the Qarhan salt flats in China (see Chapter 10). The extreme aridity that formed bitter salts in ancient evaporite basins is difficult to imagine. The only modern example of significant salt accumulation is in the Qarhan Salt Lake in western China, where the origin of the potash minerals is associated with hydrothermal waters entering the basin through faults or from the dissolution of previously deposited potash evaporites (Hardie, 1984, 1987, 1990; Hardie & Eugster, 1971). Salt layers were formed by the desiccation of small Plio-Miocene lakes formed in a continental setting related to plate convergence (see Chapter 10).

Dissolution of previous evaporite deposits is observed in many basins where the brines receive fresher water input, resulting in the precipitation of less soluble minerals, forming evaporite cycles. This is the case of the Chotts in Tunisia, where the ephemeral lakes are periodically flooded and desiccated, depositing several salt minerals according to brine saturation (Barbieri et al., 2006; Warren, 2006, 2016, 2021).

Salt precipitation on the bottom of a sedimentary basin involves compaction, diagenesis, and lithification. Thick layers of rock salt can deform by geological processes, resulting in structures exposed at the surface. In the Zagros Mountains in Iran, the salt structures were affected by remarkable deformation related to compressional tectonics, forming salt glaciers (see Chapter 3).

Several geological models have been put forward to explain the development of salt structures in the subsurface, interpreting them through indirect geophysical methods such as seismic profiles (see Chapter 5). The early interpretations for salt deformation assumed orogenic compressional forces, forming salt diapirs outcropping on land (as in the Zagros region) or in submarine regions (as in the Mediterranean Sea). The more recent interpretations are based on seismic interpretation and physical and numerical models that aim to explain the subsurface salt structures as a complex interaction of gravitational instability, sedimentary loading, and tectonic forces (see Chapter 6 in Mohriak, 2025).

8.2. SALT DEPOSITION AND SALT DEFORMATION MODELS BEFORE PLATE TECTONICS

The early geological concepts discussing the genesis of salt deposits were divided between two main currents: plutonists and neptunists. In the eighteenth century, plutonists believed that salt rocks were formed by magmatic processes, while neptunists believed that salt minerals were precipitated from seawater. Italian chemist Usiglio conducted experiments on salt precipitation from the evaporation of normal seawater from the Mediterranean Sea in the mid-nineteenth century. These experiments were the first scientific studies to propose a model for salt deposition based on chemical laws (Usiglio, 1849a, 1849b).

Hsü (1972) discussed several early models for salt deposition, including the bar theory (Ochsenius, 1877). The alternative model of evaporite genesis, assuming that salt was as predominantly formed by the evaporation of continental waters in desert basins, was also advocated by several authors (e.g., Grabau, 1924). This theory fell into oblivion after the discovery of thick evaporite layers in oceanic basins, where the salt was predominantly constituted by halite on top of deep marine sediments, as observed in Sicily and other peripheral basins around the Mediterranean Sea (see Chapter 5 in Mohriak, 2025). A classical paper by Branson (1915) discussed the barred basin model and several other alternative hypotheses for the formation of thick deposits of gypsum and halite in sedimentary basins. The bar theory might explain the accumulation of thick salt layers without the need for a huge water column, but the model required that a new supply of seawater be brought in over the barrier as fast as evaporation created space for new water inflows (Branson, 1915). These marine incursions constantly replenished the necessary ions to precipitate the salts. The total lack of fossils in salt sequences was explained by the extremely high concentration of the brines, which would cause most forms of life to become extinct before gypsum crystals started to precipitate.

The barred basin model held sway into the 1960s, when studies of modern sabkha (salt and mudflats) in the Persian Gulf and Baja California began appearing in the geological literature (e.g., Evans et al., 1969; Kinsman, 1969; Shearman, 1966). Sabkhas have been described in various regions worldwide, both in modern and ancient salt basins. Tidal flats adjacent to marine gulfs constitute a key element in the formation of shallow-water evaporite sequences in many ancient salt basins (Kendall, 1978, 1984, 1992, 2010; Shearman, 1966, 1978; Warren, 1991, 1992, 2006, 2016; Warren & Kendall, 1985).

In the 1960s, the discovery of massive salt layers in oceanic basins led to the proposition of deepwater settings for evaporite deposition, based on analogies with the Dead Sea and the Mediterranean Sea (e.g., Schmalz, 1969). Two end-member models were established for salt deposition: the shallow water-deep basin and the deep water-deep basin (see Chapter 12). The Deep Sea Drilling Program (DSDP) drilled the Late Miocene salt diapirs in the Mediterranean Sea to investigate the origin of these structures. The desiccation model was proposed by the DSDP researchers for the deep Mediterranean basin and was widely accepted by the scientific community, although some works disputed some of its tenets (CIESM, 2008; Cita, 1983; Hsü, 1972, 1973; Hsü et al., 1977, Hsü, Montadert, et al., 1978; Hsü, Stoffers, & Ross, 1978; Schreiber & Friedman, 1976).

The desiccation model has been challenged by several researchers (e.g., Meilijson et al., 2018; Schmalz, 1991). The re-examination and analyses of the DSDP salt cores by several researchers (e.g., Hardie & Lowenstein, 2004) indicated a subtidal rather than a supratidal sabkha environment. Lugli et al. (2015), based on detailed sedimentological studies, argued that "no unequivocal evidence of shallow water or even supratidal (sabkha) deposition is in evidence, suggesting that at the very last phase of the salinity crisis the Mediterranean Sea did not experience desiccation, but that deposition took place under permanent subaqueous conditions."

8.3. EARLY MODELS OF SALT DEPOSITION, DEFORMATION, AND DIAPIRISM

Before plate tectonics, scientists proposed various theories to explain the origin of salt deposits in the Gulf of Mexico region (see, for example, Jackson, 1995; Murray, 1966). Some assumed that a primordial ocean covered Earth and deposited salt on the continents, while others suggested that some salt structures near the present-day surface might correspond to old islands emerging in the Cretaceous or Tertiary seas. The first decades of the twentieth century saw a recurring volcanic origin for the salt domes, with the Petite Anse region of the Gulf of Mexico (see Chapter 1) often interpreted as the product of a volcano composed of mud, water, and gas (DeGolyer, 1925).

According to the magmatic hypothesis, the rock salt was derived from the evaporation of seawater by volcanic heat, and even the ponds that were found on top of some domes were related to extinct volcanic craters. DeGolyer (1925) wittingly expressed this interpretation in his famous comment, "The fact that structurally an igneous plug is the only geological feature known to us which even remotely resembles a salt dome, and the similarity of the two is indeed striking." The magmatic theory was probably influenced by the proximity of the U.S. sector of the Gulf of Mexico to the volcanic regions in southern Mexico, where salt and volcanic rocks co-exist in some places. This coincidence is also observed in the Danakil Depression in the Afar region of northern Africa (Hovland et al., 2006, 2018a, 2018b).

During the first half of the twentieth century, the prevailing theories for salt deposition assumed an association with volcanism or that it resulted from the saturation of aqueous solutions with evaporite ions derived from the leaching of igneous rocks. Many researchers proposed salt-flowage models associated with compressional forces to explain diapiric structures in oceanic basins. In these models, salt layers deform from their original horizontal layering to their current position, forming a dome or diapir. Due to contraction, the apex of the dome or diapir may rise above the surface (see Chapter 6 in Mohriak, 2025).

Most of the researchers who advocated the salt dome model to be created by tectonic forces assumed an analogy with the salt domes that are exposed in many orogenic belts from Iberia to the Middle East (e.g., Jackson, 1995). However, DeGolyer (1925) highlighted that the diapiric structures in the Texas and Louisiana salt provinces corresponded to "plugs of salt intruded into previously almost undisturbed Tertiary and Cretaceous formations from deeply buried salt structures." Drilling in the Gulf of Mexico, the Mediterranean Sea, and the Red Sea proved that evaporite structures are common in the deepwater regions of many continental margins. Deep topographic depressions were necessary to form extremely thick salt layers in the distal parts of these margins. Early hypotheses for deeply buried salt structures in Europe and Asia assumed that evaporite sediments were deposited in horizontal layers but were deformed by forces associated with the Alpine Orogeny (see Chapter 3). The hypothesis assumed salt flow in a plastic or semi-plastic state, induced by compressional movements resulting from folding. The orogenic theory would only change when Trusheim (1960) described the salt structures in Germany as related to salt movement caused by sedimentary loading and differential pressures, behaving as a fluid, with salt diapirism often unrelated to tectonic forces (see Chapter 6 in Mohriak, 2025). The salt flow mechanisms in the Gulf of Mexico have been discussed by several researchers (e.g., Kostick, 1993; Murray, 1966; Nettleton, 1934, 1943).

Following the tenets of the plate tectonic concepts developed in the second half of the twentieth century, many authors have proposed that the largest salt accumulations on Earth are related to specific stages of the Wilson cycle, which describes the opening and closing of oceanic basins during plate movements. Kinsman (1975) interpreted that most continental margins (or geosynclines, in pre-Plate Tectonics terminology) might be floored by extensive salt layers that advanced basinward and covered the proto-oceanic crust. It is interesting to note that Kinsman (1975) considered the Gulf of California a young oceanic basin without evaporites. However, this region is characterized by significant Neogene salt deposits in the Guaymas Basin, located at the transition from continental to oceanic crust (see Chapter 12).

Incipient exploratory drilling along several divergent margins indicated that salt deposition extended to the continent-ocean boundary. These basins correspond to divergent margins where orogenic compression can be discarded as the main factor for diapirism, but the original salt layers have been deformed by extension, contraction, and strike-slip movements (see Chapter 6 in Mohriak, 2025). The Red Sea salt basin constitutes one of the most extraordinary examples of Neogene salt deposition in the world (see Chapter 1 in Mohriak, 2025). Several authors consider the Red Sea is a modern analog for salt deposition in ancient basins, such as the Gulf of Mexico and South Atlantic divergent margins (Almalki et al., 2015; Hall, 1995; Mohriak, 2014). According to Hutchinson and Engels (1972), salt deposits in the Red Sea might reach a thickness of about 3 km, immediately west of the present-day axial trough in the oceanic crust. Modern geophysical data indicate that the evaporite layers locally form salt diapirs exceeding 5 km in height (see Chapter 1 in Mohriak, 2025).

Many researchers have proposed that the largest salt accumulations on Earth are related to specific stages of the Wilson cycle, which describes the opening and closing of oceanic basins during plate movements (e.g., Burke, 1975; Hovland et al., 2018a, 2018b; Kendall et al., 2009). During the rifting and breakup of the Pangea supercontinent, several basins were formed in the Atlantic Ocean, such as the Late Triassic-Early Jurassic salt basins in northeastern North America and northwestern Africa margins (e.g., Soto et al., 2017). Several giant salt basins formed during the closure of oceanic basins, such as the northern central European basins and the Pricaspian basin (e.g., Kendall et al., 2009; Wu et al., 2015). The isolation of oceanic basins during subduction and continental block collisions resulted in the development of important salt basins, such as the Mediterranean Sea (e.g., Jolivet et al., 2006). The geographic and temporal distribution of the "Salt Giants" and the relationship between the Wilson cycle, salt deposition, and basin deformation episodes are addressed in Chapter 9 in Mohriak (2025).

8.4. GEOLOGICAL OBSERVATIONS AND MODELS FOR UNDERSTANDING SALT DEPOSITION

Sedimentologists began searching for modern analogs of ancient salt basins in the late twentieth century. Offshore basins, such as the South Atlantic divergent margins, can only be studied using seismic methods integrated with deep boreholes in regional interpretation projects. Salt deposits in offshore basins are rarely cored by exploratory boreholes, but the integration of borehole data with electric logs and seismic profiles provides constraints on the interpretation of the mineralogical composition (see Chapter 7).

At present, salt deposition is observed in a few basins where brine has been concentrated by solar evaporation, exceeding the solubility of several evaporite minerals (see Chapter 4). Ancient salt deposits are recognized in large depressions that formed isolated oceans and in nonmarine settings, within continental landmasses (Hardie, 1984, 1987; Schreiber & Marshak, 1981).

Most sedimentary environments in shallow subtropical regions with marine influences accumulate carbonate or siliciclastic rocks, not evaporites. This is the case of the platforms surrounding islands in the Caribbean Sea or the Mesopotamian depression in the Persian Gulf depressions in Mesopotamia (Warren, 2006, 2016). Many salt deposits are formed by subsurface evaporation from brine-bearing aquifers, as observed in the coastal salt flats along the Persian Gulf (Handford, 1991; Hardie, 1991). Evaporite deposition may also occur in cold, high-altitude mountainous regions, as in the salt flats of the Tibetan Plateau or the Andean Altiplano in South America, or even in polar regions where rainfall is scarce (see Chapter 2).

Salt deposits in continental environments are typically found in isolated lakes where the waters are concentrated into hypersaline brines. Examples of modern nonmarine depositional environments include the Great Salt Lake in Utah, some shallow, hypersaline salt lakes in California, several ephemeral lakes in the Sahara Desert, and coastal regions of the Arabian Peninsula adjacent to the Persian Gulf. The Dead Sea is a perennial lake with deep-water column, whereas the extremely shallow playa lakes in California dry up during some periods of the year, and the Persian Gulf sabkhas are subaerial except when flooded by high sea level events (Kendall, 1978, 1992, 2010; Warren, 2006, 2016).

8.4.1. Geochemical and Physical Properties of Brines

Although the mechanisms for brine concentration have been heatedly debated in the geological literature (e.g., Hovland et al., 2006, 2018a, 2018b; Scribano et al., 2017, 2018), most researchers admit that the primary mechanism is evaporation as a consequence of solar heating (Schreiber, 1988; Warren, 2006, 2016). According to several authors (e.g., Babel & Schreiber, 2014; Sonnenfeld & Perthuisot, 1989), the primary sources of the chemical compounds in brines are:

- influx of seawater;
- minerals carried by rivers as dissolved ions derived from the chemical weathering of rocks in the catchment area;
- hydrothermal fluids related to igneous activity at depth.

In addition, brine concentration in evaporite basins is genetically related to the following secondary processes:

- dissolution of evaporite minerals, particularly halite and more soluble salts;
- expulsion of interstitial fluids from porous and evaporite layers in the subsurface due to sedimentary loading and tectonic forces;
- alteration or diagenesis of hydrous evaporite minerals into anhydrous forms, such as gypsum into anhydrite. Carnallite transforms into sylvite and releases water molecules as salty fluids.

Many basins exhibit several levels of evaporites in the stratigraphic column, such as the salt basins in Iran, which are characterized by the Hormuz salt (Neoproterozoic) but also register other levels of salt deposition during the Mesozoic and Cenozoic (Talbot, 1998). In Iberia, the Neogene evaporites in small intramontane basins are located near Triassic salt outcrops (see Chapter 7). Some of these younger evaporite layers are associated with the dissolution of previously deposited salt layers, transportation, and reprecipitation. Most salt ions in the Great Salt Lake in Utah derive from the dissolution of halite beds from Mesozoic salt sequences that occur in the Wasatch Range. In addition, the various Late Paleozoic (Permian) to Mesozoic evaporite sequences of North Dakota are related to the recycling and redeposition of Devonian Elk Point evaporites. The northern Great Plains of western Canada constitute another example of modern sedimentary brines sourced by the dissolution of ancient saline subsurface formations (Last, 1989).

The origin and migration of brine in an evaporite systems are complex topics in the salt sciences. Sr isotope ratios have been used to analyze the origin of evaporite sequences in many basins as fingerprints of the original environment of deposition. Several sedimentary environments with modern salt deposition are used as models or analogs for the formation of the giant salt basins in the past. The Sr isotope ratios have been used in the analysis of the origin of evaporite sequences in many basins as fingerprints of the original environment of deposition (see Chapter 7). Similar salt layers composed of halite and gypsum may be sourced from different brines, either in marine or continental settings. The next section will briefly discuss the interpretations of these diverse modern, continental environments as a key to understanding the paleoenvironments that accumulated large thicknesses of evaporite minerals.

8.4.2. Sedimentary Environments with Precipitation of Salt Sequences

Evaporite basin studies have evolved over the past 50 years, building upon early concepts of salt deposition in desert environments, sabkhas, and continental lakes. Most researchers believe that ancient evaporite rocks were deposited in various depositional settings, including inland lakes (playas and salinas), tidal flats, supratidal zones (sabkhas), gulfs, and deepwater environments. Evaporite deposition is accepted in regions below sea level, coastal regions close to sea level, and elevated regions like the Andean highlands or the Tibetan Plateau (see Chapter 10).

Based on global observations, several authors (e.g., Kendall, 1992, 2010; Warren, 1992, 2010, 2016; Warren & Kendall, 1985) divided the evaporite facies into three main types:

1. Mudflat evaporites, which are characterized by the displacive growth of minerals in a pre-existing carbonate or siliciclastic matrix, as observed in the continental and coastal sabkhas.
2. Shallow-water evaporites, which are formed by the precipitation of minerals as the brines are concentrated in shallow lakes, epeiric seas, and gulfs, are usually less than tens of meters deep and constitute either permanent or ephemeral water bodies. This can occur in (a) salt flats, where the evaporites are precipitated and accumulate, forming thin laminated sequences, as in the southern margin of the Persian Gulf; (b) salt pans or salinas, which are ephemeral lakes that are periodically desiccated, such as the salt lakes in California or the MacLeod Lake in Western Australia; and (c) shallow perennial salt lakes, where the evaporites are precipitated from permanent shallow brines, as in the Great Salt Lake in Utah. These shallow-water evaporites often show dissolution features indicating subaerial exposure, such as mudcracks or tepees (Warren, 1982a, 1982b, 1983).
3. Deepwater evaporites, which are formed by the precipitation of minerals in lakes, gulfs, and inland seas that are more than 50 m deep. The brines are concentrated and deposit salt minerals when the lakes or inland seas are a few meters deep, according to some models (see Chapter 12). Deepwater evaporite sequences can be divided into (a) basin margin facies, where the evaporite rocks are associated with clastic sedimentation processes involving shallow shelves, slopes, and basin floors, as indicated by turbidites, slumps, and debris flows. For example, clastic evaporites are observed in the marginal areas of the Mediterranean Sea in Sicily; (b) a deep basin floor is characterized by finely laminated evaporite sequences with a basinwide distribution.

Schreiber (1978, 1986, 1988) analyzed several salt basins worldwide, associated with continental and marine environments, and proposed a classification into several global types. Many basin types accumulate salt layers, including continental playa lakes, territorial marine margins, shallow epeiric seas, carbonate marine margins, deep basin epeiric seas, oceanic basins, embryonic divergent margins, and divergent margins. Figure 8.3 shows schematically the main basin types that accumulate salt layers.

A. Continental playa lakes
B. Terrigenous marine margins
C. Shallow epeiric seas
D. Carbonate marine margins
E. Deep basin epeiric seas
F. Oceanic basins
G. Embryonic divergent margins
H. Divergent margins

Continental playa lakes (A) may occur in intracontinental settings, such as desert areas where lakes might concentrate the brines and deposit evaporites. Playa lakes are observed in the Sahara Desert in North Africa, such as the Chotts in Tunisia or the Mojave Desert in California (see Chapter 9). Some high-altitude playas with lacustrine salt deposition occur in Tibet and the Andean Altiplano (see Chapter 10).

Terrigenous marine margins (B) are mainly characterized by siliciclastic sediments and are often associated with a delta. The Gulf of California is a modern example, with salt deposition only in some rift depocenters (see Chapter 12).

Shallow epeiric seas (C) are located near an ocean or form a restricted gulf, such as the Trucial Coast along with the Persian (or Arabian) Gulf, where water depths reach a maximum of only a few tens of meters. This setting is the focus of this chapter.

Carbonate marine margins (D) are characterized by carbonate platform deposits that may form reefs or carbonate islands between the lagoon and the gulf or open sea. The classical example is the margin along the coast of the United Arab Emirates (UAE), where the lagoon is separated from the open ocean by carbonate banks (see Chapter 12).

Deep epeiric seas (E) are sedimentary basins located between continental margins, and their water depth may reach hundreds to a few thousand meters. One possible example of this setting is the Persian Gulf during the Jurassic (see Chapter 12).

Oceanic basins (F) are associated with deep marine settings, where salt deposits may occur intercalated with deep water sediments, such as the Mediterranean Sea basins in the Upper Miocene (Messinian), or overlie redbeds, continental fluvial-lacustrine sediments, and volcanic rocks, such as the Jurassic Gulf of Mexico (see Chapter 4 in Mohriak, 2025).

Embryonic divergent margins (G) precipitate salt during the last stages of rifting, when the oceanic spreading centers are emplaced in the axial trough. One possible example of this setting for salt deposition is the Red Sea (see Chapter 1 in Mohriak, 2025).

Divergent margins (H) are characterized by salt deposits on the volcanic basement or oceanic crust. The salt deposits are either autochthonous when precipitated on the volcanic substratum or allochthonous when clastic evaporite sediments formed aprons overlying a volcanic or pre-salt sedimentary substrate. Possible examples of this setting for salt deposition are the Guaymas Basin in the Gulf of California (see Chapter 12) and the distal margin of the South Atlantic basins (see Chapter 3 in Mohriak, 2025).

This chapter explores the geological characteristics of salt deposition in various regions, including arid regions, continental playa lakes, sabkhas, shallow epeiric seas, and marine gulfs. Classical sabkha models have been developed by research groups in the Persian Gulf since the

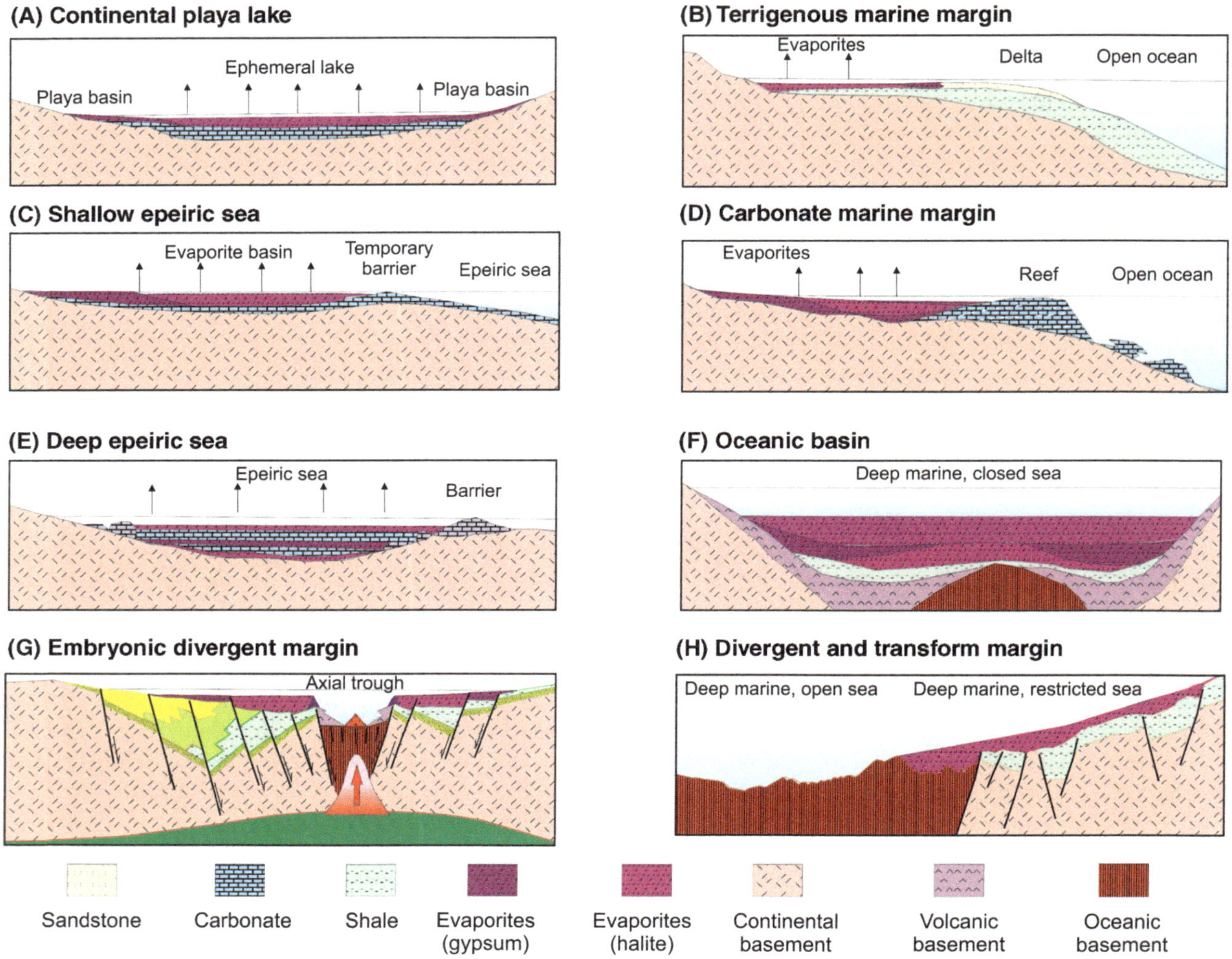

Figure 8.3 Schematic drawing (no scale) showing the classification of evaporite basins according to the main depositional settings.

1960s (Alsharhan & Kendall, 2003; Evans, 2011; Kendall et al., 2014; Schreiber & Hsü, 1980). Shearman (1978) and Warren (1986, 1989) define sabkhas as related to marine and continental mudflats composed of siliciclastic or carbonate rocks, particularly silt, clay, and marl. Evaporite minerals grow through displacive and replacive processes when saline fluids in sedimentary rocks are concentrated in the capillary zone. The sabkha environment is limited to a supratidal zone straddling subaerial desert dunes and subtidal regions of lakes, gulfs, or inland seas (Warren, 2006, 2016). The pore fluids become saturated brine when water is evaporated by solar heating, precipitating evaporite minerals according to their solubility (see Chapter 4).

Handford (1981a) and Warren (1982a, 1982b, 1989) proposed a tripartite classification of sabkhas based on typical geological settings of evaporite deposition. The end-members of continental playa lakes are associated with eolian, lacustrine, and coastal marine settings (Figure 8.4). Modern continental sabkhas include the Sahara Desert of northern Africa, dry lakes in California, and the Trucial Coast south of the Persian Gulf. Some characteristics of these desert and sabkha environments are recognized in ancient salt basins, such as the Late Permian Rotliegendes and Zechstein salt basins of Central Europe, the Late Permian salt basins in the southwestern U.S., the Triassic evaporite deposits overlying the rift basins along the Central Atlantic and Northwest Africa margins, and the proximal margins of the Late Aptian salt basins in the South Atlantic (Chaboureau et al., 2012; Hardie, 1991; Warren, 2006, 2010, 2016).

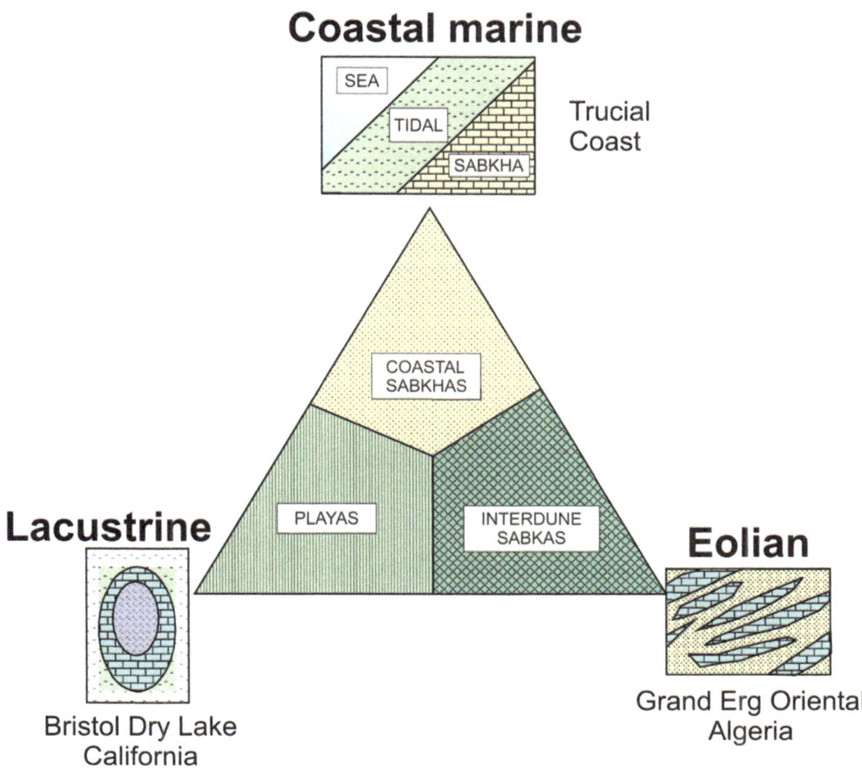

Figure 8.4 The only modern geological settings for the widespread accumulation of evaporite sediments correspond to a tripartite division of arid regions: coastal marine environments (sabkhas), eolian environments in desert regions (playa lakes), and lacustrine environments (hyper saline lakes). Credit: Adapted from Warren (1989).

8.5. DESERT ENVIRONMENTS AND ARID SHORELINES

The continental environments are located far from marine influence and accumulate evaporites in desert regions characterized by dunes, interdune lakes, and depressions (Figure 8.3A). Eolian environments, on the other hand, are characterized by sediment deposition controlled by wind velocity, forming sandy dunes. Deserts are mainly associated with arid environments with less than 250 mm/year of rainfall (see Chapter 2). The lack of water leads to scarce vegetation, facilitating wind-driven erosion and surface deflation, forming depressions in the interdune flats. Desert environments include some fluvial drainage and ephemeral lakes that occupy the depressions resulting from wind deflation (Kendall, 1984, 1992, 2010; Warren, 2006, 2016). Arid shorelines accumulate evaporites in the sabkha environments adjacent to shallow epeiric seas, as observed in the southern margin of the Persian Gulf (e.g., Figures 8.3C and 8.3D).

These modern depositional settings are recognized in ancient basins like the Permian North Sea and North German basins in Europe, the Parnaíba Basin in South America, and the Permian Basin in the southwestern United States (e.g., Silver & Todd, 1969). Some salt basins have desert sandstones forming important pre-salt reservoir rocks sealed by evaporite layers, such as the North Sea (see Chapter 2 in Mohriak, 2025).

8.5.1. Eolian Dune-Interdune Environment

The eolian dune-interdune environment (Figure 8.5) includes desert sands with evaporite minerals precipitated in the interstices of the sediments in the subsurface and evaporitic mudflats. Some shallow lakes and salt flats are observed between dunes, forming evaporite crusts that cover the desert dunes in the subsurface. The halite crust represents an erosional surface that truncates the eolian deposits and indicates a flooding episode (e.g., Renaut & Gierlowski-Kordesch, 2010).

The large eolian dune fields in the Sahara Desert are known as "ergs," an Arabic term meaning an extensive desert region covered by dunes. The interdune sabkhas commonly occupy areas of deflation and are associated with lacustrine systems developed in the depressions. During wet seasons, some ephemeral lakes are filled with water, as in the Chotts in Tunisia (Bryant et al., 1994; Warren, 2006, 2016). During dry seasons, the shallow

Figure 8.5 Satellite images (courtesy of NASA) showing the Sahara Desert between Algeria and Libya, with continental sabkha plains, dried lakes, and mudflats adjacent to the Grand Erg Oriental Desert with hundreds of dunes (A). The eolian dune-interdunal sabkha dry lakes are characterized by whitish to bluish colors. These interdune flats indicate that water was present in the environment but was evaporated, leaving residual evaporitic sediments (halite crusts and halite-rich muds) covering the dried lakes (B). The eastern part of the Grand Erg Oriental is characterized by a transition between the yellow-orange dunes and the dark brown consolidated rocks of the Tinrhert Plateau to the south and east, formed by older sediments. Locally, the remnants of a fluvial system are indicated by incised river channels (C). Credit: Courtesy of NASA.

lakes dry out, and the brine precipitates evaporite minerals, including gypsum and halite. Sabkha El Melah in southeastern Tunisia records the precipitation of several evaporite minerals, which are identified by the X-ray diffraction patterns (Essefi & Tagorti, 2021).

The Grand Erg dune field between Algeria and Libya is formed by hundreds of dune mounds that display several shapes. These include (1) linear or longitudinal dunes, which are oriented parallel to the prevailing wind; (2) transverse dunes, which are oriented more or less perpendicular to the wind; (3) barchans, which are seen with a bow-shape; and (4) star dunes, which display a starfish-like pattern, as they are formed by winds blowing with roughly equal strength from several directions (Figure 8.5B). The Eastern Erg (known as the Grand Erg Oriental) of Algeria includes hundreds of dune mounds in a desert region adjacent to a salt flat where dried lakes are seen in satellite images (Figure 8.5A). Evaporites are deposited in ephemeral dried-up lakes that occupy depressions, forming gypsum- and halite-rich sediments that appear as white patches in satellite imagery (Figure 8.5C).

The present-day climate in the Sahara Desert is extremely arid, with scarce rainfall. However, the sedimentary environment was less dry and much cooler during the episodic glaciations that affected the planet in the Holocene and resulted in a wetter climate, with some lakes and rivers across the plains. These now dried-up lakes and rivers are seen in satellite images as white blobs of evaporite sediments and meandering, incised river channels. These features can be seen in the more elevated regions of the Tinrhert Plateau in the eastern part of the lower right image (Figure 8.5C). This region is characterized by more consolidated rocks and displays a dark brown color that contrasts with the yellow-to-orange color of the unconsolidated eolian dunes in the Grand Erg Oriental (Figures 8.5A and 8.5C).

8.5.2. Continental Playa Lakes and Sabkhas

Sabkha, an Arabic term meaning "salt flat," refers to a subaerial mudflat coastal plain that is occasionally invaded by marine waters from a nearby gulf or inland sea (Briere, 2000; Schreiber, 1986, 1988; Warren & Kendall, 1985). The word "sabkha" commonly refers to a salt flat along a coastal embayment adjacent to a marine or gulf environment. A marine sabkha occurs near an inland sea or marine gulf and develops a coastal salt flat dominated by marine-derived brines. The evaporites that accumulate in the depressions are affected by marine sedimentary processes.

When the salt flat is located inland, far from the sea, it is designated as a continental sabkha or continental playa lake (Kendall, 1978a, 1978b, 1984). These two main types of sabkhas are associated with the hydrological cycle of the environment. The continental sabkhas include permanent or ephemeral lakes that are fed by nonmarine waters (Kendall, 1984, 1992, 2010; Renaut & Gierlowski-Kordesch, 2010).

Playa lakes may precipitate several types of evaporite minerals, including carbonates, sulfates, and halides. Modern examples of continental sabkhas or playa lakes are described in the Australian desert region, located on the central part of the continent, and the California deserts in the southwestern part of the United States (Kendall, 1978a, 1978b, 1984; Warren & Kendall, 1985).

Lake Eyre in central-south Australia is an ephemeral playa-salt lake within one of the largest internal drainage basins in the world, located in an extremely arid setting (Figure 8.6A). The drainage system covers an area of over 1 million km^2, corresponding to more than 15% of the Australian continent. The Simpson Desert is characterized by large linear dunes and a predominant N-NW trend (Figure 8.6B). Lake Eyre, located to the south of this desert, is the largest salt pan in the world, with a surface area of about 9,300 km^2 (Warren, 2006, 2016). At present, the lake's surface lies around 15 m below global sea level, as most of the water in the depression has evaporated.

There is no outlet for the fluvial system in the Simpson Desert and the waters discharge into the terminal lake system, resulting in a large endorheic basin. Most of the fluvial water that reaches the basin quickly evaporates due to the intense solar radiation, leaving behind a thin veneer of evaporitic sediments that form a playa mudflat with a crust of evaporite minerals. The land surface is usually a dry salt pan that corresponds to the lowest point in the arid lowlands of central South Australia. However, during occasional wet periods, the river systems bring fresh water into the basin, and on a few occasions, the lake is filled with water. In the past decades, this has happened only a couple of times, particularly in 1989 and 2009, when heavy rains fell over northern Australia and the rivers filled the lake system with fresh water (Figure 8.7).

The lake is extremely shallow, usually with a water depth of less than 1 m, appearing in satellite images with a blue color (Figure 8.7). A set of satellite images from 2000 to 2009 shows the variable colors and geomorphology of Lake Eyre during the first decade of the 21st century (Figure 8.7). The lake is remarkably colorful, with greenish to bluish tints, which are associated with different water depths and derive from lake sediments, water chemistry, and organisms that proliferate in the saline waters. However, in dry periods, the lake is usually dry, its surface is covered by mud and silt, and the lake surface appears brownish in the satellite images (Figure 8.7). When the lake is desiccated, the surface is covered by evaporite crusts. In satellite images, the salt flat appears white to bluish (Figure 8.7).

Many lake complexes in southern central Australia were freshwater to brackish water lakes surrounded by grassy plains around 35 ka ago, during the last glaciation. About 16–10 ka, climate change resulted in increasing temperatures and aridity, leading to the shrinking of the saline lakes, which covered only a fraction of their previous extent (Warren, 2006, 2016). Today, the dry lake layers are remnants of these saline lakes, with only short-term episodes of water filling the depressions, as in the 2009 rainy season, which resulted in an ephemeral lake level rise (Figure 8.7).

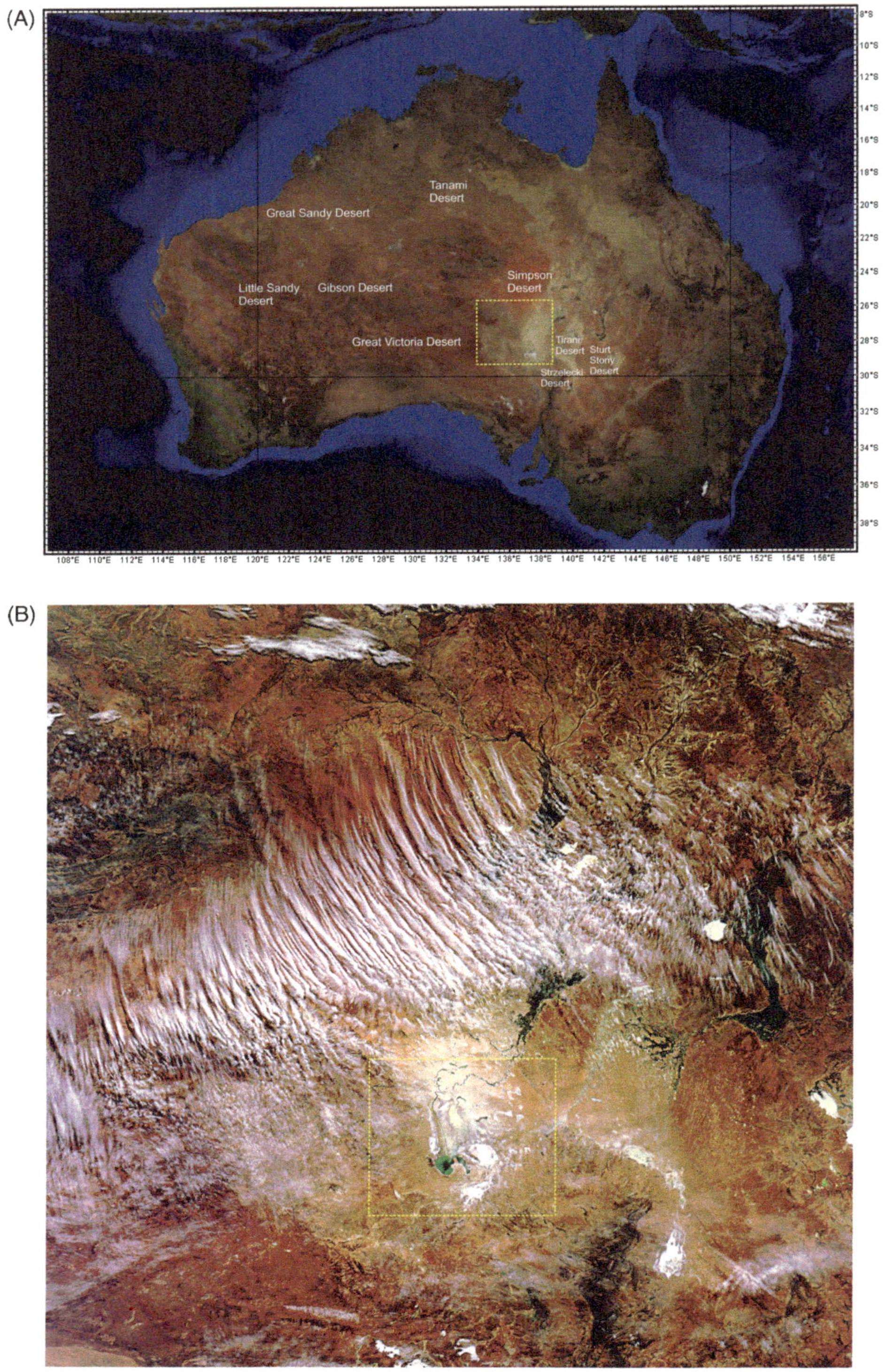

Figure 8.6 Satellite images (courtesy of NASA) showing Lake Eyre in central Australia. (A) Lake Eyre is an endorheic lake located south of the Simpson Desert, highlighted by a thin white rectangle. (B) Zoom of the satellite image showing the Simpson Desert with the longitudinal eolian dunes and the Lake Eyre to the south, at the center of the thin yellow rectangle. The Eyre Lake displays a residual cover of greenish water. Credit: Courtesy of NASA.

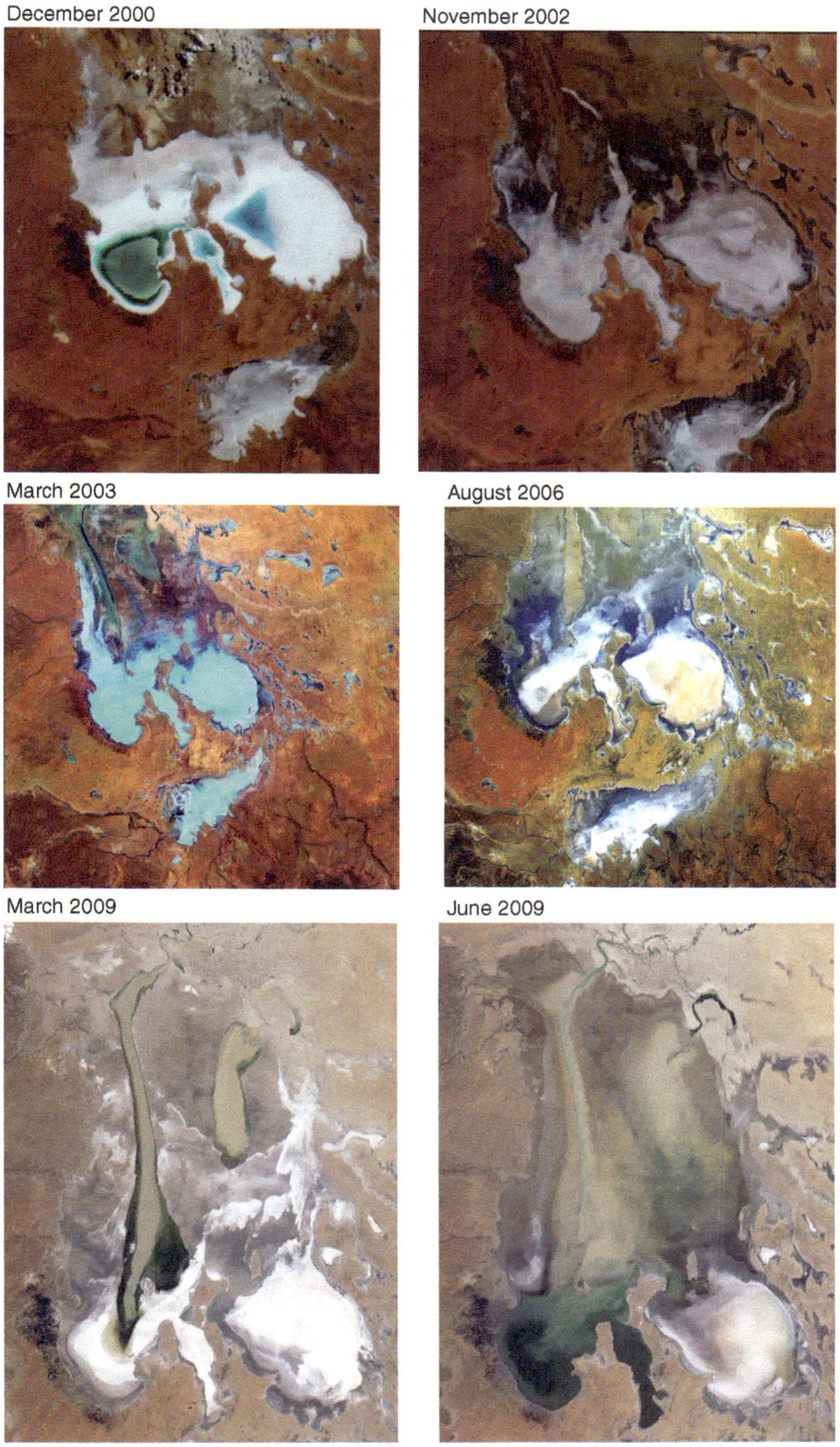

Figure 8.7 Satellite images (courtesy of NASA) showing the Lake Eyre geomorphology from 2000 to 2009, with variations in reflectivity due to different water levels in the depression. In 2002, most of the lake dried up and formed a salt flat. Water inflow in 2009 resulted in filling the western lake with some water, resulting in a greenish color in the satellite image. Credit: Courtesy of NASA.

8.5.3. Arid Shorelines and Coastal Marine Evaporite Lakes

Arid shorelines are associated with coastal marine evaporite environments, including lakes that are periodically flooded by marine waters from an adjacent sea or gulf. These shorelines correspond to coastal sabkhas, which involve the transition from desert dunes to littoral environments along an epeiric sea or marine gulf. Sabkha El Melah, located on Tunisia's southeastern coast near the eastern part of the Mediterranean Sea, is an example of an evaporitic coastal lake in Tunisia (Figure 8.8A).

Sabkha El Melah is a modern example of a coastal depression below sea level that is periodically infiltrated by marine waters. The coastal sabkha extends for an area of about 150 km^2 in a depression that is separated from the sea by dunes composed of oolitic carbonate sands (Perthuisot, 1971, 1981). The carbonate banks bordering the lagoon were formed during the Flandrian period (~ 6 ka BCE, Before Common Era) and constituted a barrier that isolated the lake from the open sea during salt deposition events. These marine incursions carry dissolved salt ions that are later precipitated as evaporites in periods of extreme aridity when the brine is concentrated and saturated with salt minerals. The lake bottom is floored by a thick sequence of evaporite sediments, including carbonates, sulfates (gypsum), and chlorides (halite), that formed when the lake dried up (Essefi & Tagorti, 2021; Perthuisot, 1971, 1981).

Several companies have drilled numerous exploratory boreholes into the Sebkhat El Melah to investigate the possibility of economically producing potash minerals from the subsurface deposits. These exploratory wells indicated that the sabkha has a long-lived geological history in this arid environment at the fringes of the Sahara Desert (Essefi & Tagorti, 2021). However, during the last glacial period (Late Pleistocene to Holocene), the basin held a perennial lake that deposited lacustrine sediments with organic-rich muds. In the Holocene, the lake was affected by periodic desiccation episodes, depositing several layers of carbonate, sulfate, and chloride minerals in several evaporite cycles (Perthuisot, 1971, 1981).

Figure 8.8B is a Landsat image of the Sabkha El Melah from January 1987, which shows the lake filled with dark water that entered the basin as it lies at or slightly below the Mediterranean Sea water level. In December 1999, almost 13 years later, the basin is dry, and only the basin's edges, which appear bluish in the satellite image (Figure 8.8C), are characterized by remnant moisture. The evaporite deposition raised the salt industry's interest in producing salt from artificial salt ponds. The satellite images (Figure 8.8C) show signs of human activity along the shores of the dried lake. The saltworks at the eastern edge of the basin produce potassium-rich salt minerals for the fertilizer industry.

The Mediterranean seawater that occasionally infiltrates the coastal basin has repeatedly filled the depression with marine waters that filled the lake and dissolved the minerals accumulated at the lake's bottom. Due to climate aridity, the brine was subsequently concentrated, starting a new cycle of salt deposition, resulting in the precipitation of diverse evaporite minerals that accumulated on the basin's floor (Figure 8.9). Highly soluble potash-bearing minerals are also observed in the Holocene salt layers. For example, when the brine concentration is increased to a maximum during the summertime, carnallite and polyhalite precipitate on the lake's floor (Perthuisot, 1971).

The coastal sabkhas in northern Africa, around the periphery of the Mediterranean Sea, are highly influenced by sea-level fluctuations in the Holocene. The sediments reflect sea-level variations due to the demise of the last glacial episode on Earth, which waned from circa 20 ka to about 12 ka (Philip et al., 2013). With increasing temperatures, glaciers started to melt, particularly in regions near northern Europe and northern North America. This resulted in a sea-level increase, reaching a maximum at around 6 ka, with a sea-level rise of around 125–130 m, reflecting the retreat of most of the Earth's ice sheets in higher latitudes (Carlson, 2011). This led to a general marine transgression (known as the Flandrian event) that flooded lowland regions across the globe. The peak of the Flandrian transgression occurred circa 5.5 ka, when the sea level was a few meters higher than the present-day sea level. The barrier that separated the lagoon from the open sea corresponds to the oolitic and bioclastic carbonate rocks located to the northeast of the present-day lake (Figure 8.9). The sedimentary barrier was up to 6 m high but allowed periodic marine incursions into the basin. During the Flandrian transgression, the El Melah depression formed a wide, restricted lagoon with a maximum depth of about 30 m (Figure 8.10). As the carbonate barrier restricted communication with the ocean, the lagoon became hypersaline when the sea level started to drop after 5.5 ka, resulting in basin isolation and hypersaline brines.

Evaporite minerals, including the rare carbonate known as huntite ($CaMg_3(CO_3)_4$), which overlies the lagoonal organic-rich mud deposits, precipitated at the lake's bottom around 5 ka (Davaud et al., 1996; Perthuisot, 1981). Due to the hyperaridity in this Saharan region, the lagoon was rapidly filled with evaporites. The salt deposits that accumulated over the past 5,000 years include a thick layer of halite more than 20 m thick (Figure 8.10).

Following sea-level fall in the past millennia, the lagoon evolved into the present-day coastal sabkha plain,

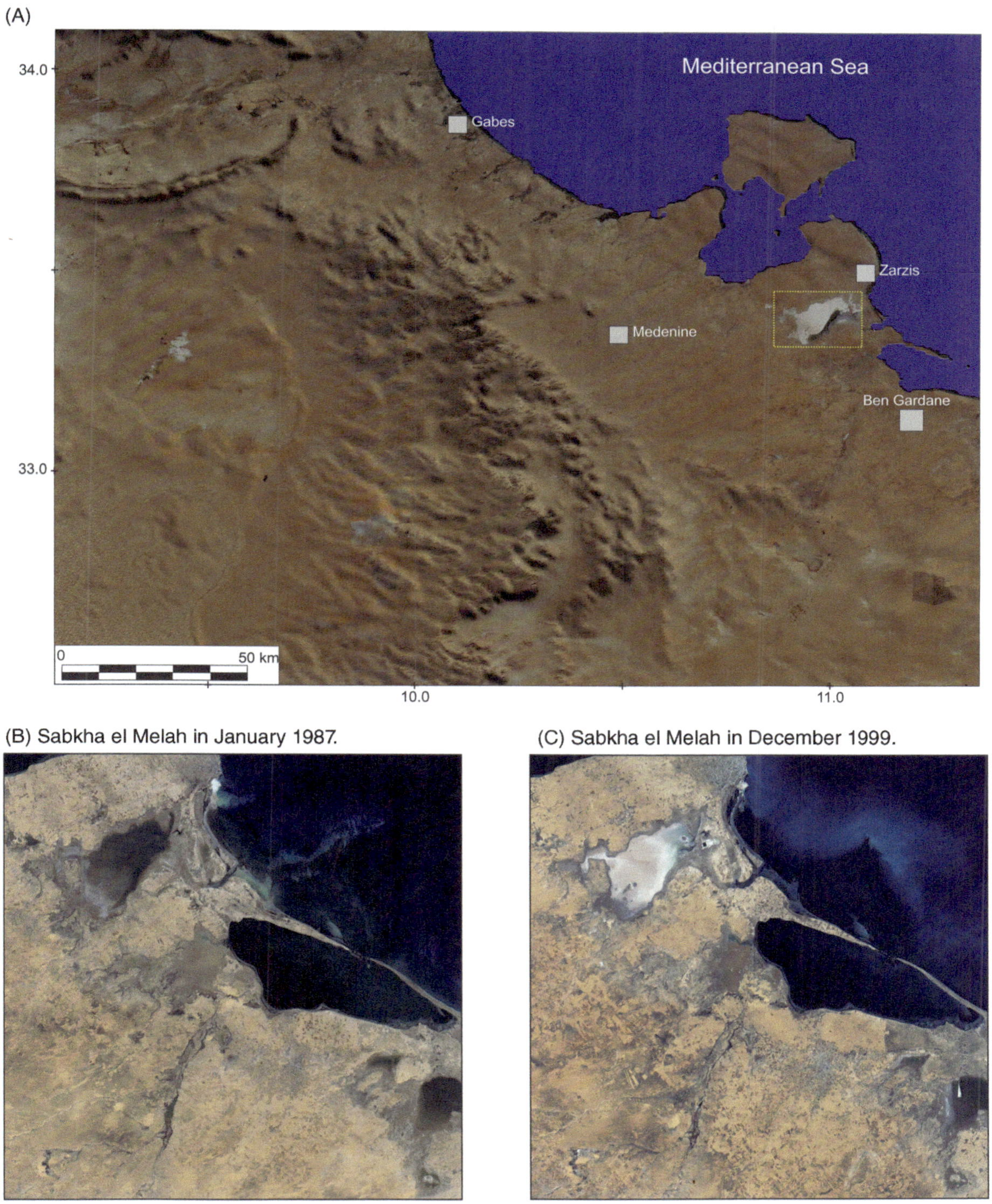

Figure 8.8 Satellite images (courtesy of NASA) showing the coastal sabkha El Melah eastward of the Sahara Desert. The sabkha is located near the northeastern coastal plain of southeast Tunisia, close to the shores of the Mediterranean Sea (A). Dark waters filled the depression in 1987 (B). The climate aridity resulted in the desiccation of the basin. In 1999, the satellite image of the same regions (C) showed whitish colors covering the dry lake surface as a result of evaporite precipitation. There are signs of human industrial activity in the upper right portion of the 1999 image, as seen in the rectangular evaporite pans. These saltworks are used for sea-salt production, including the extraction of potash-bearing evaporite minerals. Credit: Courtesy of NASA.

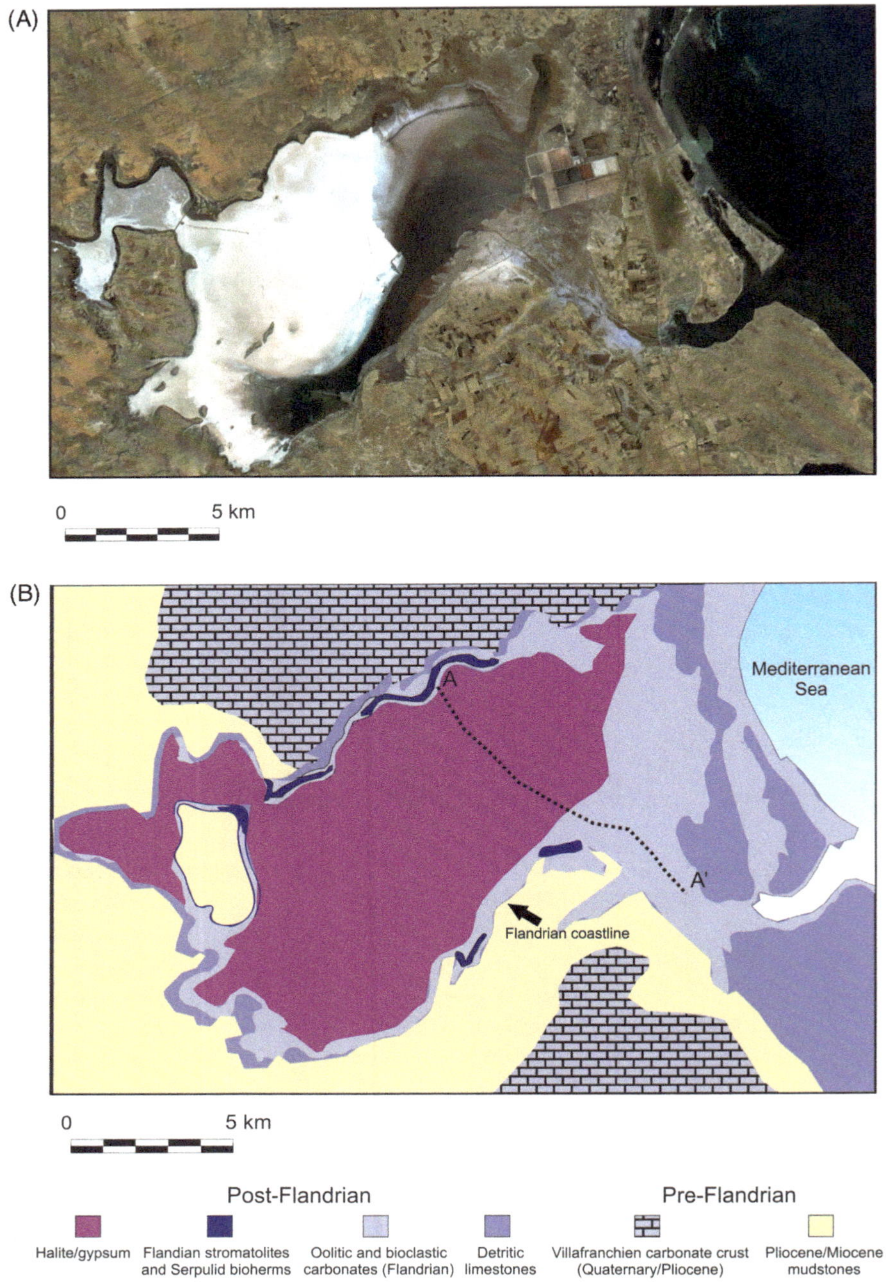

Figure 8.9 Detailed satellite image (courtesy of NASA) and a simplified geological map of the Sabkha El Melah in southeast Tunisia. (A) Satellite image showing the residual lake and the salt flat, with the saltworks seen at the northeastern margin of the lake. (B) Geological map showing the evaporite and carbonate sediment distribution. The geological cross-section A–A′ extends across the salt flat toward the bioclastic sands that form carbonate banks above sea level. The oolitic and bioclastic carbonate rocks probably constituted a barrier that separated the lagoon from the open sea during the Flandrian (~ 6 ka BCE).

exposing the salt deposits and underlying carbonates (Figure 8.9). The surface of the sabkha is presently covered by a thick (up to 1 or 2 m) layer of evaporite-rich mud with halite crystals and a little gypsum, which contains black organic matter (Perthuisot, 1971, 1981).

Given the present-day hydrological parameters, with main water losses due to evapotranspiration (1,400 mm per year) and infiltration, greatly exceeding the water input, the central lagoon water body will eventually dry up (e.g., Ruiz et al., 2006). The fate of the lagoon will

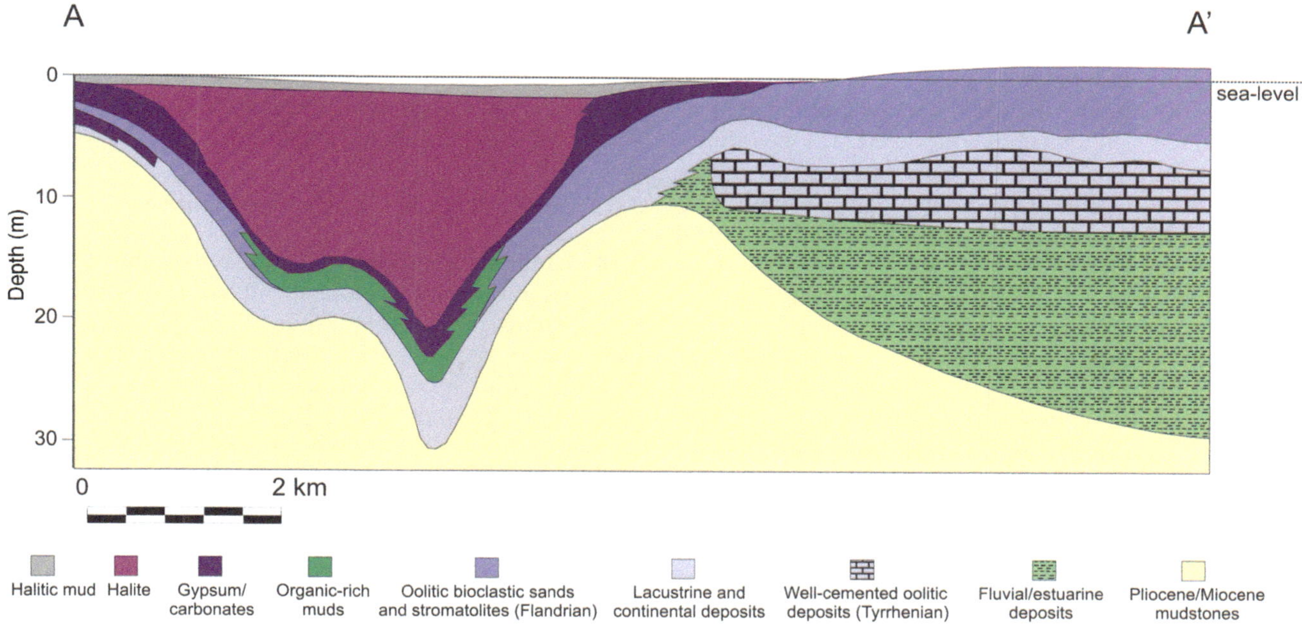

Figure 8.10 Geological cross-section across the El Melah salt flat showing the thick halite layer that was formed in the past 5,000 years after the sea-level fall following the Flandrian transgression (6 ka). The oolitic bioclastic sands and stromatolites formed a carbonate barrier that isolated the lagoon and resulted in basin desiccation. The lagoon was about 30 m deep before the evaporite sediments started to accumulate, and the halite deposits in the saltern were thicker than 20 m.

probably be similar to other sabkha environments in the Sahara Desert, resulting in the dry salt flats or dried-up playa lakes surrounded by almost sterile marshes, sandy ridges, spits, and eolian dunes (Figure 8.5).

8.6. SALT DEPOSITION IN PLAYA LAKES AND COASTAL SABKHAS

Lacustrine environments with salt deposits are found in areas near marine environments and continental regions without marine influence, such as alluvial fans adjacent to mountain ranges, depressions on the continental crust, and elevated plateaus (Kendall et al., 1978, 1992, 2010; Renaut & Gierlowski-Kordesch, 2010). Evaporite sediments form in coastal environments near gulfs and plate margins, forming sedimentary basins that extend from onshore to deep offshore regions in diverse evaporite facies. The sabkha and mudflat facies are characterized by supratidal evaporite deposition, accumulating salt within older sediments. The sabkha environment straddles the coastal region from the regions on land, adjacent to desert dunes, to the mudflats, and commonly extends to the lagoon and deep basin environments, stacking different lithological sequences during progradation and retrogradation episodes (Kendall, 1978, 1992, 2010).

Mudflats are subaerial environments flooded by marine incursions and often covered by efflorescent surface crusts. The shallow-water facies are characterized by evaporite minerals precipitating from thin brine sheets. Mudflats are desiccated due to isolation from the open sea by a siliciclastic or carbonate barrier. Both the sabkha and shallow-water facies contain dissolution surfaces and abundant bottom-grown crusts that accumulate on the bottom of ephemeral lakes and salt pans. The deepwater facies are characterized by laminated and gravity-displaced sediments, forming detrital evaporitic beds. Clastic evaporites often display features associated with turbidity currents in deep lakes or gulfs, such as graded bedding or Bouma sequences (Kendall, 2010).

In the latter half of the twentieth century, sedimentologists studied modern deposition environments in the Middle East and Northern Africa to understand the origin of evaporite sequences in ancient sedimentary basins (see review by Alsharhan & Kendall, 2011). They identified only a few present-day environments where salt minerals were formed, such as inland and coastal sabkhas, small lakes in arid regions, and sea-marginal subaqueous lakes (Figure 8.4). However, salt layer thicknesses in most Holocene basins varied from a few centimeters to a few meters, contrasting with ancient salt basins where evaporites accumulated in depositional settings that extended from the platform to the shelf break, slope, and deep basin. In these basins, the salt layers commonly exceeded 1 or 2 km in thickness and covered areas that extended up to a few million square kilometers (Warren, 2006, 2010, 2016).

The Persian Gulf became a focus for studies on evaporite deposits in the 1960s, particularly the southern margin of the Gulf in the UAE, which is considered the type locality for present-day sabkha environments and a paradigm for ancient coastal sabkha environments (e.g., Alsharhan & Kendall, 2003, 2011; Evans et al., 1964; Kendall et al., 2014).

Evaporite sediments in several salt basins were initially believed to be formed in coastal or shallow-water settings (subtidal to supratidal environments) by the evaporation of interstitial water from sediments. Sedimentological studies showed that evaporite minerals were formed through a diagenetic process, generating evaporitic layers in sabkhas, primarily composed of carbonate, sulfate, and chloride minerals (Evans et al., 1969; Kendall, 1978a, 1978b, 1984, 1992; Kinsman, 1969).

The sabkha host matrix, where evaporite minerals grow by capillary evaporation, is the basis for classifying most sabkha occurrences worldwide as a combination of terrigenous clastic sediments, carbonate-sulfate sediments, or soluble salts with a halite-rich matrix (Evans et al., 1969; Kendall, 1978a, 1978b, 1984, 1992; Kinsman, 1969).

Academic research in the Arabian Gulf margins led to several seminal works focusing on the sedimentological aspects of evaporite deposition. Several authors advocated that most ancient evaporite deposits were not formed in barred basins where water evaporated, as proposed by Ochsenius (1877, 1888), but formed by replacing the previous minerals with the brine expelled from carbonate-sulfate sabkha sediments, as interpreted in the Arabian Gulf sabkhas (e.g., Shearman, 1966).

This proposal faced harsh criticism from sedimentologists involved in the study of evaporites. The hypothesis was generally disbelieved or only partially accepted by researchers and explorationists working in the Gulf of Mexico and South Atlantic salt basins, where most of the salt deposits are in deep waters. The sabkha model became well-established in the Arabian Gulf's southern margin during the 1960s and was widely disseminated among sedimentologists from the 1970s to the 1980s. Since then, it has been discussed in textbooks, academic works, and petroleum exploration projects. The sabkha model has been extensively applied to many other evaporite successions, both modern and ancient, involving projects conducted by academic and petroleum industry researchers (Alsharhan & Kendall, 2003, 2011; Kendall, 1978, 1992, 2010; Shearman, 1978).

The sabkha model has been applied to explain evaporite deposits in ancient evaporite basins worldwide, and evaporitic sequences formed in these environments are crucial components of petroleum systems in several petroliferous salt basins. For instance, sabkha evaporites form the seal for carbonate reservoir rocks in Saudi Arabia and the Persian Gulf (Puls et al., 2009; Taylor, 1985; Warren, 2006, 2016). Eolian sandstones in desert dunes constitute the main reservoir rocks in the Groningen field, the largest gas field in Europe (Jager & Visser, 2017; Stauble & Milius, 1970).

8.6.1. Geological Characteristics of Sabkhas

Most sabkhas are related to coastal environments adjacent to a gulf in arid regions or inland lakes sourced by fluvial-lacustrine systems (Figure 8.11). The lithological facies are associated with continental sabkhas (playa lakes far from marine influence) and coastal sabkhas (adjacent to an inland sea or gulf).

Coastal sabkhas typically occur between a desert environment and an ocean or gulf, with the top surface consisting of a crust with salt minerals. The lithologies associated with coastal sabkhas include several evaporite minerals (calcite, gypsum, and halite). These deposits are classified according to the most important physical processes related to their depositional environment. The main lithological facies include the following:

1. terrigenous clastic sediments, typically associated with eolian environments with dune fields;
2. carbonate-sulfate chemical precipitates (dolomitic mud, gypsum, and anhydrite) formed in tidal flats;
3. soluble salts (halite, sylvite, polyhalite, etc.) formed at the basin depocenter.

Continent sabkhas, also known as continental playa lakes, occur inland, far from the nearest ocean or gulf. They are characterized by a depression filled with fluvial-lacustrine sediments and surrounded by topographic elevations (Figures 8.11 and 8.12). During periods of climate aridity, continental sabkhas experience an evaporitic sedimentary environment, with ephemeral lakes formed within a desertic landscape. Thin layers of evaporite minerals accumulate on the surface after evaporation of the shallow-water body (Figure 8.11).

Continental sabkhas are formed by groundwater and nonmarine sedimentary processes, forming permanent or ephemeral lakes called playa lakes. These lakes are often desiccated during dry periods and can be found in various geomorphological settings, such as salt-crusted evaporitic mudflats, distal alluvial fans, sand flats, saline pans, and dune fields (Figure 8.12). Coastal sabkhas are observed near gulfs or seas, while continental sabkhas may be found near or below sea level in regions like the Death Valley in western North America, or in depressions in highlands, such as the salt flats in the Andean Altiplano in South America.

The continental sabkha environment includes a playa mudflat far from marine influence (Figure 8.11). This region is dry most of the time, but the capillary fringe and the water table may intersect the land surface at the lowest points, resulting in springs that form brine-saturated salt

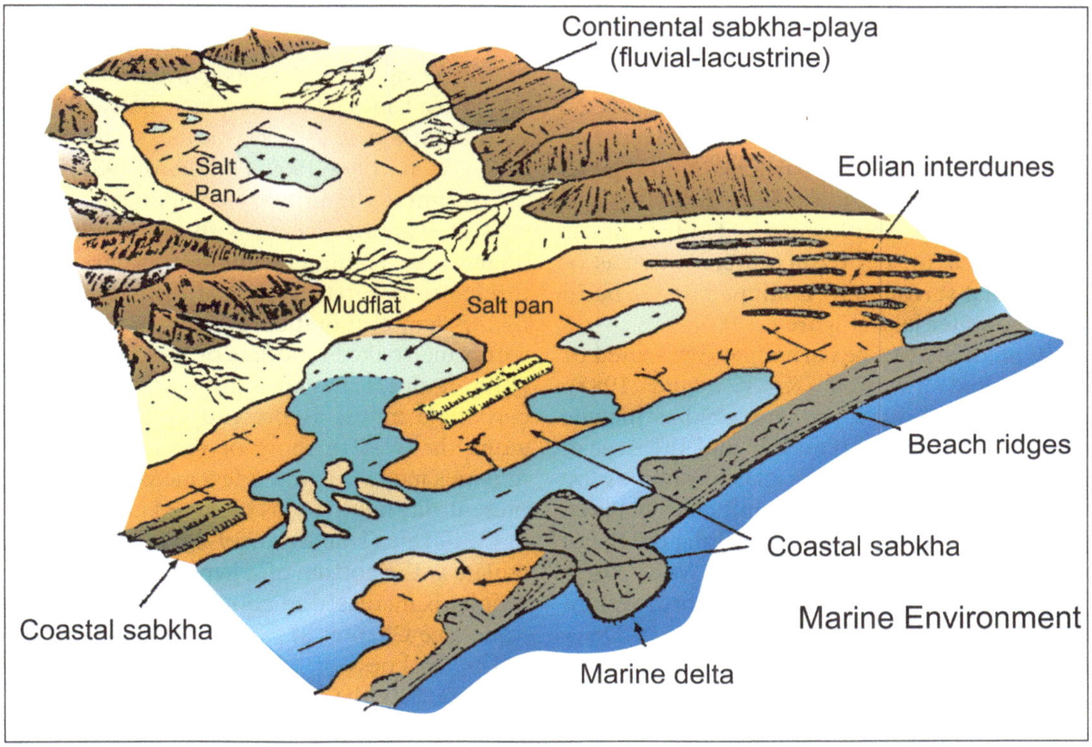

Figure 8.11 Schematic diagram showing the main continental, coastal, and shallow marine settings related to evaporite deposition. Dried-up sabkha-playa lakes surround the salt pans in endorheic basins (top left), whereas coastal sabkhas are formed adjacent to inland seas or gulfs. Carbonate and siliciclastic sabkha sedimentation is observed in interdune depressions.

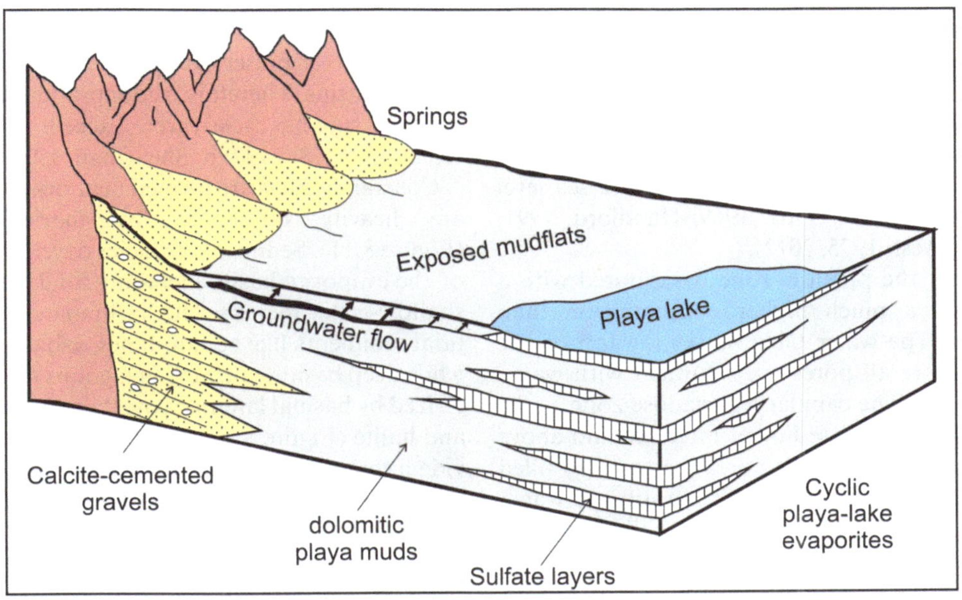

Figure 8.12 Playa lake setting with exposed mudflats adjacent to alluvial fans sourced from highlands. The proximal coarse-grained facies are characterized by calcite-cemented conglomerates that result from the erosion of uplifted mountain ranges. The mudflats in the evaporite basin are characterized by sulfate and chloride layers intercalated with dolomitic playa mudstones, forming cyclic playa-lake evaporite sequences. The playa lake is a permanent water body with hypersaline brines.

lakes (Figure 8.12). These lakes can be perennial (filled with water during the wet and dry seasons) or ephemeral, drying out many times in a century, especially during extreme climate aridity (Yechieli & Wood, 2002). Sediments in the basin margin may include coarse-grained, sand-prone alluvial fans, particularly in regions near tectonically uplifted blocks, such as the Basin and Range Province in the southwestern United States (Faulds et al., 1997, 2016). The basin center is often associated with a mudflat surrounding a salt pan on the basin floor of a shallow-water lake (Figure 8.12).

Sediments in mudflats can be carbonate-dominated or siliciclastic-dominated (Warren & Kendall, 1985). Examples of predominantly carbonate sabkhas include the southern coast of the Persian Gulf and the Trucial coast of the UAE (Figure 8.2). Siliciclastic sabkhas are found in the Sinai Peninsula, near the confluence between the Red Sea and the Gulf of Aqaba, and along the Saudi Arabian coast between Qatar and Kuwait (West, 2012). Fluvio-deltaic coastal sabkhas occur in the northwestern end of the Persian Gulf, where the Shatt al-Arab Delta occupies a large depression characterized by siliciclastic mudflats between Iraq and Kuwait (Saleh et al., 1999).

Coastal sabkhas are typically located in arid settings and occur along the margins of an ocean, an inland sea, or a marine gulf (Figure 8.11). The mudflats are highly influenced by sea-level variations, as they are periodically flooded by seawater that quickly evaporates due to climate aridity, as observed along the coastal flats adjacent to the Sahara Desert in North Africa and Arabia (Figure 8.2). The region that is above the higher sea level corresponds to the supratidal zone, whereas the region that is permanently underwater is the subtidal zone. Sabkha sediments are deposited in the intertidal zone, which locally forms a wide depression fringing the inland sea or gulf with a lagoon marked by local carbonate banks slightly above sea level (Evans, 1995; Evans et al., 1964, 1969; Handford, 1991; Shearman, 1978; West, 1975, 2012).

In arid sabkhas, the phreatic zone is saturated with a brine that shows a much higher concentration than normal seawater. The water table marks the top of the phreatic zone, where all pores are saturated with water, forming the aquifer. The capillary, or vadose zone, is the porous layer below the air-sediment interface and above the water table. In this zone, the pores are partially filled with air and saturated with a small quantity of water, which forms a thin film coating the solid particles (Handford, 1981a, 1981b; Warren, 1989, 2006, 2016).

Coastal sabkhas form primarily by the evaporation of seawater that impregnates the pores of carbonate or siliciclastic sediments in the vadose zone. The groundwater found a few centimeters to a few meters below the surface is enriched in several ions (mainly sulfates and chlorides). In the subsurface, the evaporite minerals are deposited in the capillary zone, filling the pore interstices of the rock formations that contain the aquifer. Sabkha evaporites thus grow by displacing the matrix as gypsum or anhydrite nodules are formed in a mudflat, which is usually composed of detrital carbonates, siliciclastic rocks, and biogenic material (Evans et al., 1964, 1969; Handford, 1981a, 1981b).

Gypsum is the primary component of sabkha evaporites in evaporite basins with saline lakes, lagoons, and inland seas. Halite precipitation occurs through various mechanisms. First, crystals of displacive halite may grow by capillary evaporation of supersaturated pore waters in soft brine-soaked mudflats or sabkha sediments. Second, halite hoppers may also grow as secondary minerals in carbonate muds, as presently observed on the surface along the western shore of the Dead Sea. Third, halite minerals may also grow a few meters below the surface of present-day mudflats, as observed in the Bristol Dry Lake in California (Handford, 1991). Fourth, halite may accumulate at the bottom of a hypersaline lake when the brine salinity is increased to about 10 times the salinity of seawater (see Chapter 2).

The thickness of nodular anhydrite in sabkha flats is controlled by the pore throat size of the host sediment. Sedimentological studies show that the narrower the pore throat, the thicker the capillary zone. In fine-grained silt and clay, the capillary zone can reach several meters, while in coarse-grained permeable sands, the capillary zone is only a few centimeters thick, resulting in negligible evaporite sediment accumulation. Most evaporite sediments above the top of the capillary zone are blown away or dissolved during rainfall or freshwater incursions and are rarely preserved in the stratigraphic column of ancient basins. Therefore, the supratidal evaporite unit in a single sabkha cycle rarely exceeds 1 m in thickness (Kendall, 1978a, 1978b; Shearman, 1978).

Coastal sabkhas are formed near open seas or gulfs and are heavily influenced by marine environments (Figure 8.11). Sedimentary facies develop in various parts of the evaporite basin, including tidal flats, lagoons, and shallow-water regions. The basin margin contains supratidal sediments like eolian dunes, sabkhas, and mudflats, while deep basin sequences in lagoons or gulfs are characterized by basinal laminites overlain by laminated gypsum and halite (Figure 8.12). Studies in the Persian Gulf provide a template for analyzing evaporitic environments in salt basins formed in the past (Alsharhan & Kendall, 2011; Kendall et al., 2014; Warren, 2016).

8.6.2. Geochemical Controls on Salt Precipitation from Interstitial Brines in Sabkhas

Field studies in the 1960s focused on the Holocene evaporite deposits along the Trucial Coast of the UAE, near Abu Dhabi. These studies revealed that minerals found in the supratidal flats of the sabkhas are controlled by the chemistry of interstitial brines that fill the pores of the Pleistocene sands and Quaternary carbonate rocks (Butler, 1969; Evans et al., 1969). Anhydrite is stable only in the capillary zone above the water table, where temperatures are higher than 60°C. Below the water table, temperatures are cooler and rock pores are filled with brine, making gypsum the only stable sulfate mineral (Warren, 2006, 2016).

Sedimentological investigations in the Persian Gulf recovered interstitial fluids from the carbonate rocks by squeezing sedimentary samples from different parts of the coastal plain. The formation age of the evaporite rocks was obtained using Carbon-14 methods. The radiocarbon method indicated that coastal sabkhas along the Trucial coast formed about 7,000 years before the present (Evans et al., 1969). The sea level reached an apparent high of about one meter above its present level, and the marine transgression peaked about 4,000 years BCE. Since then, the sea level has been falling, and the progradation of intertidal and supratidal sediments has taken place, forcing the land area to expand seaward (Alsharhan & Kendall, 2003, 2011; Evans et al., 1969).

The chemical analyses of the interstitial water involve the determination of the density and concentration of major ions, such as chlorides, sulfates, calcium, and magnesium. These analyses indicate that the brines are concentrated by the dissolution of pre-existing soluble salts in the subsurface and by the continued evaporation of the interstitial fluids. Several evaporite minerals are associated with supersaturated brines, such as gypsum, anhydrite, dolomite, and halite. These minerals are found within a narrow belt near the coastline, which is characterized by a gypsum crystal mush with a thickness that varies from a few centimeters to up to 40 cm (Alsharhan & Kendall, 2003, 2011; Evans et al., 1969).

Gypsum is progressively replaced landward by anhydrite, which is a secondary diagenetic mineral because it has not been observed to precipitate directly from concentrated seawater. Extensive dolomitization of the carbonate sediments is also observed toward the landward margin (McKenzie, 1981).

The brine and diagenetic evaporite minerals formed in the Abu Dhabi sabkha are related to flood recharge. The intertidal zones of the sabkha depressions are periodically flooded by Persian Gulf seawater. Due to evaporation in the arid climate, the chlorine concentration in the brine increases from 44 ‰ to 155‰ from the coastline toward the sabkha plain, whereas the sulfate concentration rises from 6 to 17 g/kg and then falls to less than 3 g/kg. The concentration of these minerals decreases with increasing depth, suggesting that surface evaporation is the main process controlling the brine concentration. When interstitial brines have chlorinity less than 145‰, gypsum is the stable calcium sulfate mineral, but with higher brine concentrations, anhydrite becomes the most common calcium sulfate (Butler, 1969).

The lagoon waters near Abu Dhabi Island have a moderately basic pH value of about 8.3, but interstitial water in the sediments along the sabkha plain shows more acidic pH values, ranging from 7.5 to 6 (Butler, 1969). Sediments deeper than about 2 m, which correspond to sandy mud layers and Pleistocene sands, display pH values smaller than 6. The increased acidity results from the decomposition of organic matter, which releases carbon dioxide and hydrogen sulfide. Most of the gas content in the pores of sabkha sediments corresponds to nitrogen (95%), whereas the quantity of free oxygen gas is only about 4.7% mol, compared with the 20% mol in the atmosphere (Butler, 1969).

8.7. THE PERSIAN GULF: THE PARADIGM FOR SABKHA EVAPORITES

The sabkha evaporites along the southern margin of the Persian Gulf are a well-studied depositional environment. The sediments are characterized by large volumes of nodular anhydrite and Holocene dolomite currently forming in the supratidal zone along the Gulf Coast (Alsharhan & Kendall, 2003, 2011; Kendall, 1978a, 1978b, 1984). Several sabkha environments along the southern margins of the Persian Gulf have been thoroughly described in the geological literature (Alsharhan & Kendall, 2003, 2011; Evans, 1995; Evans et al., 1964; Kendall, 1978a, 1978b; Kendall et al., 2014; Shearman, 1978; Warren & Kendall, 1985). Classical examples of this evaporitic environment include the Trucial Coast in the UAE, the Abu Dhabi sabkha, the Dukhan Sabkha in western Qatar, the sabkhas in the Sinai Peninsula, and the south Oman sabkha (Figure 8.2).

The Persian Gulf is one of the warmest seas in the world, with temperatures varying from 20°C in winter to more than 40°C in summer. Annual rainfall is extremely small, ranging from less than 1 mm to about 70 mm. Rain floodwaters can sometimes cover portions of the sabkha mudflats for a few months, dissolving the halite crusts present at the surface. During periods of sea-level rise, marine waters of the Persian Gulf may flood the inland sabkha depressions, resulting in the accumulation of evaporite crusts in the intertidal zone when the seawater evaporates (Goodall et al., 2000).

The Persian Gulf region corresponds to a foredeep encompassing a subrectangular basin trending NW-SE with a dimension of about 850 km in length and a maximum width of about 330 km (Figure 8.13). The Strait of Hormuz separates the slightly saline waters of the Persian Gulf (about 40,000 ppm) from the normal salinity waters of the Gulf of Oman and the Indian Ocean. Where coastal islands restrict the exchange of waters with the open Persian Gulf, as in the vicinity of Abu Dhabi Island

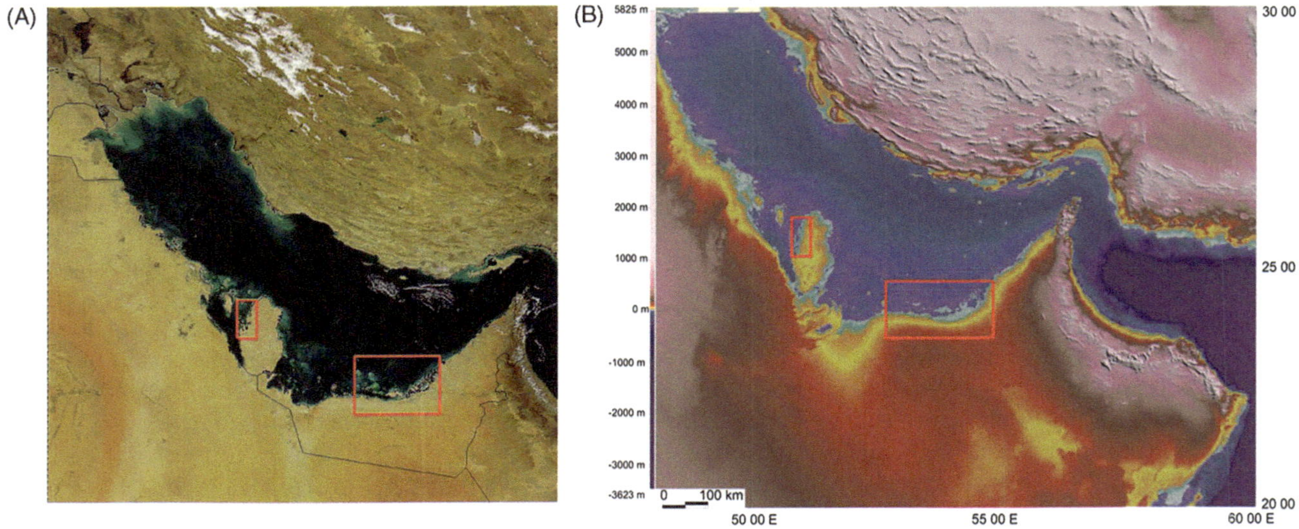

Figure 8.13 Satellite image (courtesy of NASA) of the Persian Gulf (A) and a topobathymetric map (B) showing the deeper bathymetry in the Gulf of Oman, east of the Hormuz Strait. The gulf extends from the Tigris and Euphrates delta plains to the Hormuz Strait in the east, which connects with the Indian Ocean via the Gulf of Oman. The red rectangles indicate the Dukhan Sabkha in western Qatar and the Trucial Coast in the UAE. The Sahara Desert with eolian dunes extends across Saudi Arabia toward Oman. In the satellite image, the desert dune fields are characterized by the yellowish colors and by the higher topography as compared with the coastal depression adjacent to the gulf, where the coastal sabkhas are well developed. The Oman Mountains and the Zagros Mountains in Iran are characterized by elevated topography, exceeding 4 km. The small red rectangle shows the sabkha plains in west Qatar, whereas the large red rectangle shows the coastal sabkha on the Trucial Coast.

on the UAE coast, lagoon salinities can vary from 54,000 to almost 70,000 ppm, nearly double the salinity of normal seawater (Kendall & Skipwith BT, 1969).

Geologically, the Persian Gulf region is of great interest to the salt sciences as it involves the formation of carbonate and evaporite sequences, making it a modern analog for shallow inland seas where ramp carbonates can accumulate in a shallow-water environment (Alsharhan & Kendall, 2003; Kendall & Harwood, 1996; Warren, 2006, 2016).

The Arabian Gulf is marked shallow-water depths, ranging from 40 m on the southern side to 100 m on the northern side. The euphotic zone can exceed 30 m in the axial parts but is less than 20 m in turbid waters near the Abu Dhabi coastline. The Gulf has relatively clear waters due to low terrigenous input, except for the Tigris-Euphrates fluvial system in Mesopotamia. The lack of fluvial systems in most of the central gulf results in the presence of tropical carbonate-secreting biota, forming carbonate islands along the Trucial coast (Figure 8.14).

The siliciclastic sediment inflow into the Persian Gulf is primarily longitudinal, coming from three main rivers in the Mesopotamian region (Tigris, Euphrates, and Karun) that form the Shatt Al-Arab delta plain between Kuwait and Iraq. The northeastern side of the gulf in Southern Iran is supplied with smaller amounts of sediment derived from the erosion of the Zagros Mountains, which has higher topography and smaller river systems that feed alluvial fans (Figure 8.13B).

The Persian Gulf is a unique region with modern continental sedimentary environments ranging from alluvial and fluvial systems near the Zagros Mountains in Iran to a foredeep basin, carbonate ramp, and sabkha environments in the Mesopotamian depression (Pirouz et al., 2011). The modern evaporite facies correspond to Neogene geological units in the subsurface, dating from the Miocene to the Quaternary, and include: (1) the evaporitic deposits of the Gachsaran Formation, interpreted as formed in a supratidal environment, sabkha, and lagoon; (2) the alluvial fan conglomeratic sediments associated with braided river gravel sheets (Bakhtyari Fm.); (3) the meandering fluvial system and coastal sedimentary wedges (Agha Jari Fm.); (4) the marine sediments deposited in a foredeep marine environment (Mishan Fm.); (5) the marl and carbonate deposits in a carbonate ramp (Guri Mb.); and (6) the modern sabkha environments of the supratidal region of the southern Persian Gulf, onshore of the UAE (Pirouz et al., 2011).

The Holocene sedimentation on the southern margin of the Persian Gulf is characterized by carbonates and

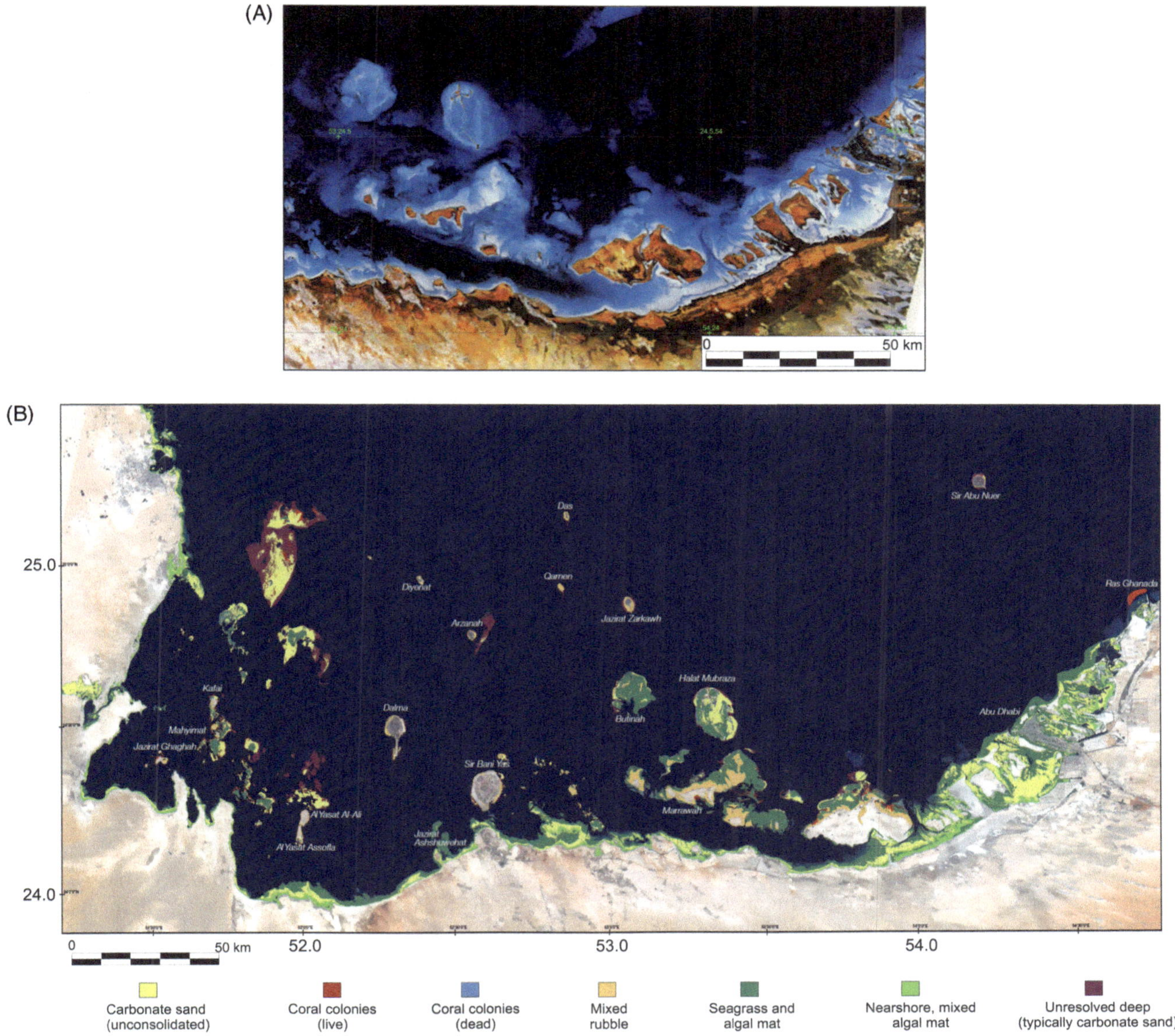

Figure 8.14 Satellite image (courtesy of NASA) of the Trucial Coast in the UAE (A). The image shows the transition from the eolian dune fields (orange to yellow color) to the carbonate sabkha flats (brown color), the lagoon, the carbonate banks, and submerged reefs in the shallow waters of the Arabian Gulf. The satellite image at the bottom (B) shows the eastern part of the Qatar Peninsula and the coastline of the UAE adjacent to the Khor-al-Bazam lagoon. The image displays the ecological facies map of the coral colonies associated with the transition from the sabkha flats, algal mats, and the Persian Gulf marine environment. Several types of coral reefs are recognized in the Great Pearl Bank, composed of living corals and oolite shoals.

evaporites that form a belt over an area of 500 km long and up to 60 km wide. Rather than depositing in the subaqueous marine environment of the gulf, the early studies from the 1960s indicated that most of the Holocene evaporites were found as thin sedimentary accumulations in the supratidal coastal plains (Evans, 2011).

The Persian Gulf is a marine carbonate system today, but around 20 ka BCE, when the sea level was 120 m below the present level, the whole Gulf was a desiccated basin with active eolian dune fields stretching from Arabia to Iran. The Pleistocene-Holocene transgression covered and reworked the dune sands, and these reworked eolian siliciclastics are now covered by the Holocene marine carbonates. Algal mats and several coral colonies are registered in the transition from the sabkha flats to the lagoon (Moati et al., 2008).

8.7.1. The Trucial Coast Sabkha

The coastal plains of the UAE, previously known as the Trucial Coast, are characterized by a belt of evaporite sediments associated with the sabkha environment (Figure 8.14). The Persian Gulf sabkha sediments along the Trucial Coast extend landward some 15 to 20 km from the extremely shallow waters of the Khor al Bazam coastal lagoon, across the intertidal zone where algal mats and mangroves occur as narrow strips (Kendall & Skipwith, 1969). The sediments in the sabkha flats are typically composed of carbonate sands and muds intercalated with evaporites (gypsum, anhydrite, and some halite) between the continental eolian dune fields and the marine environments of the Persian Gulf (Figure 8.14).

The depositional environments of the coastal sabkhas along the Trucial Coast include the sand desert (eolian environment), extensive mud and algal flats, and lagoonal environments at the transition to the marine environment, where several carbonate islands are present (Figure 8.14). The sabkha depositional system is characterized by (1) the subtidal environment, which in the Persian Gulf is covered by shallow marine waters; (2) the intertidal zone located along the Trucial Coast, which is periodically exposed and flooded; and (3) the supratidal zone that extends inland toward the desert dunes (Figure 8.14).

The coast of Abu Dhabi is locally protected from the open marine conditions of the Persian Gulf by several peninsulas, islands, shallow carbonate banks, and coral reefs (Figure 8.14). The east-west trending Great Pearl Bank is fringed by a biota of living corals and oolite shoals (Moati et al., 2008). The Khor-al-Bazam lagoon is characterized by a seawater salinity much higher than normal seawater, reaching 70 parts per thousand during dry periods with high evaporation rates. As the lagoon water evaporates, the lagoon becomes shallower and eventually is exposed at low tide, forming an intertidal flat covered by carbonate mud, gypsum sand, and algal mat (Kendall & Skipwith, 1969).

The Abu Dhabi sabkha was formed circa 7,000 years ago by wind erosion of desert dunes and the progradation of subtidal (lagoon), intertidal, and supratidal carbonate sediments that advanced toward the Persian Gulf (Evans et al., 1969). This extensive mudflat measures approximately 320 km in length and up to 24 km in width, extending along the Persian Gulf shoreline between the Qatar Peninsula and Ras Ghanada to the east (Figure 8.14).

The sedimentary successions deposited in the Abu Dhabi sabkha environments form progradational sequences with lithological facies that Warren and Kendall (1985) divided into three physiographic settings stacked one atop the other, prograding basinward and forming a tripartite, shallowing upward cycle. These units are known as the Sabkha Trilogy cycle, corresponding to subtidal, intertidal, and supratidal facies, each with specific sedimentary characteristics as shown in a schematic block diagram (Figure 8.15) and a generalized stratigraphic column (Figure 8.16).

The simplified stratigraphic column (Figure 8.16) illustrates the stacking of different facies, from subtidal carbonates and silts to intertidal carbonate sands and muds. The algal and microbial mats are overlain by supratidal gypsum sand (gypsarenite), anhydrite, and halite. Many of these lithological facies that are registered in modern sabkha depositional environments on the Trucial Coast correspond to sedimentary successions observed in cores from the subsurface of the Persian Gulf, onshore and offshore of the Arabian Peninsula, and in other ancient Mesozoic and Cenozoic basins worldwide (Warren, 2006, 2010, 2016).

The three main lithological facies associated with the sabkha cycle on the Trucial Coast are summarized as follows:

1. *Subtidal facies:*
   The subtidal facies correspond to the open marine sedimentation found seaward of carbonate barrier island systems near the coastline (Figure 8.14). An "offshore tidal bar" corresponds to a submarine embankment of sediments that lie approximately parallel to the shoreline. Tidal bars are often covered by the sea, but eventually, they may reach the water level and form islands or shoals that are not attached to the shores at either end. The open sea in the distal gulf is characterized by bioclastic and peloidal medium sand on the shoreface, grading into hemipelagic marls in deeper waters. Canyons and turbidite fans are observed basinward of the shelf, carrying clastic carbonate fragments to the deep basin (Figure 8.15). The carbonate banks in the Khor al Bazam lagoon are composed of Holocene and cemented Pleistocene oolite shoals, which are separated by channels that reach depths of about 7–10 m (Purser, 1973; Warren, 1989). Open marine sedimentation occurs on the seaward side of the northeast-trending barrier island system, forming carbonate mudstones and bioturbated carbonate rocks (packstones and wackestones). The landward side of the barrier island contains a shallow-water lagoon with organic-rich muds and microbial mats (Figure 8.16). In the intertidal zone, sabkha sediments are associated with evaporite deposits, mainly gypsum sand, anhydrite, and halite.
2. *Intertidal facies (lagoonal):*
   The intertidal zone can be divided into two facies: an upper and a lower intertidal facies. The lower intertidal facies grade into the lagoonal facies, which are the site of the accumulation of organic debris on the lagoon's floor, forming an algal mat. The lower intertidal facies are characterized by algal peat, a black

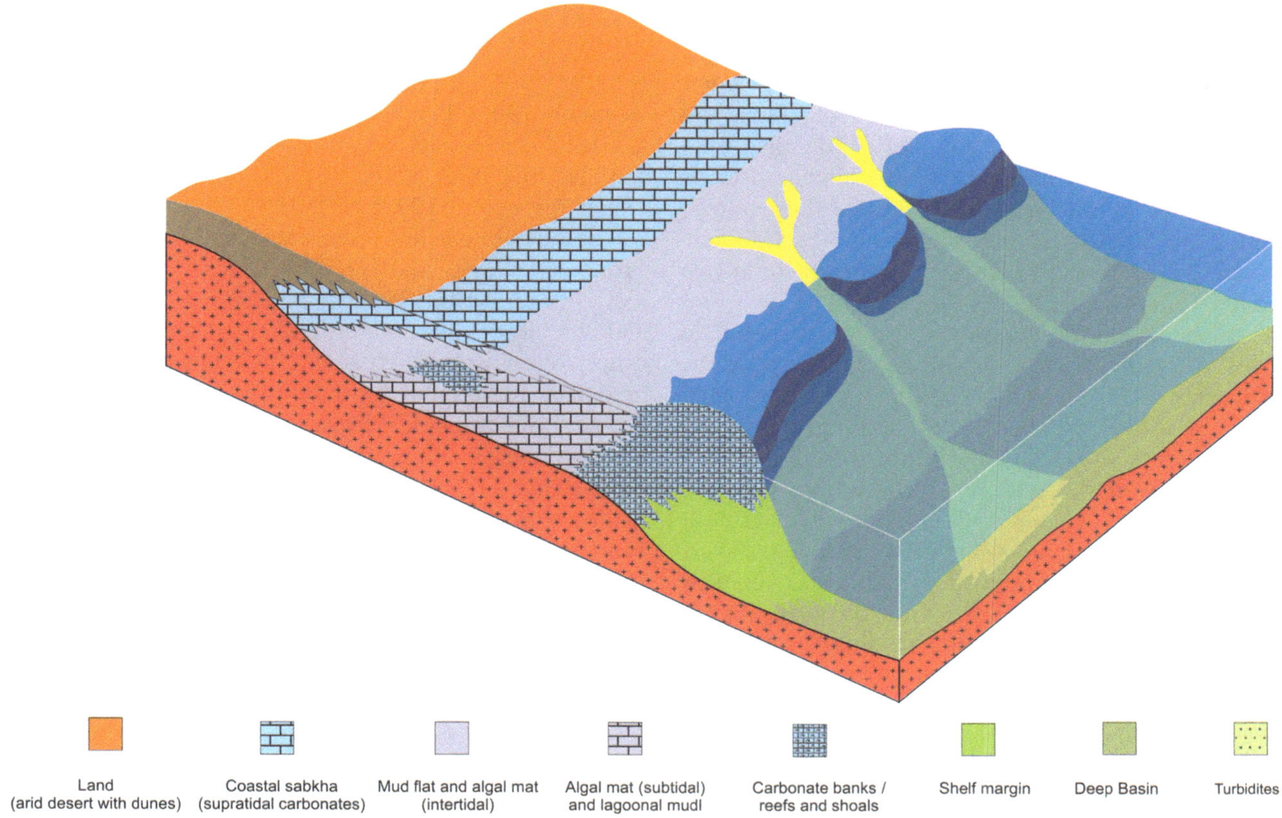

Figure 8.15 Block diagram showing the depositional environments of sand desert and coastal sabkha in the Trucial Coast of the UAE. The southern margin of the Arabian Gulf is characterized by eolian dunes, which toward the north pass into the supratidal carbonates of the coastal sabkha, lagoonal sediments (intertidal mudflats and algal mats), and carbonate banks and reefs in the shallow waters of the Arabian Gulf. Channels between the banks connect with the submarine canyons that transport sediments into the deep basin. Credit: Adapted from Silver and Todd (1969).

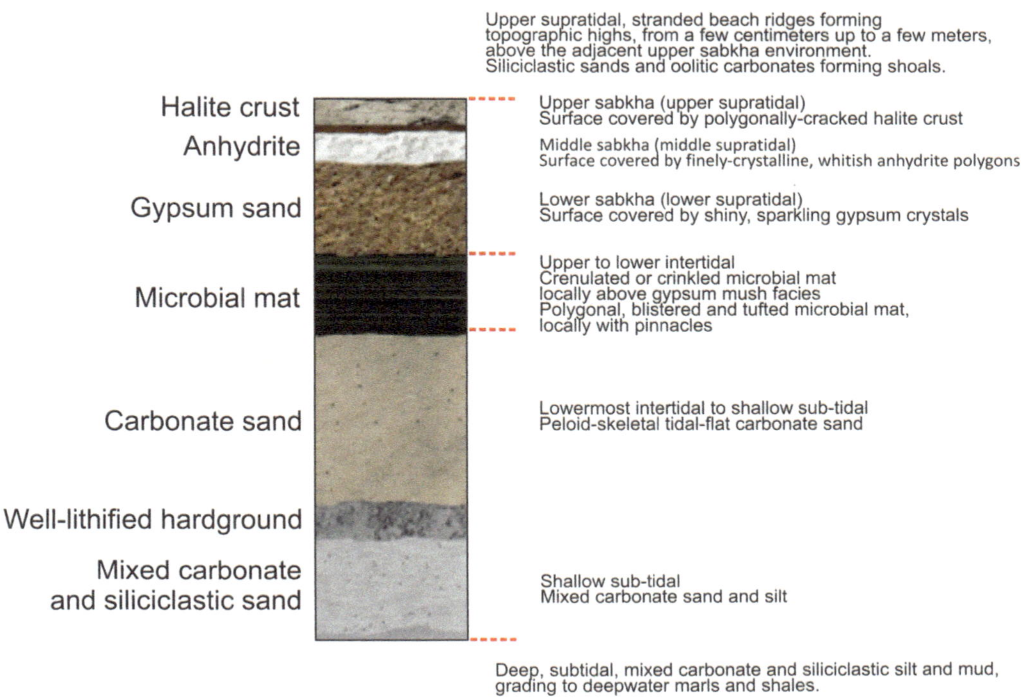

Figure 8.16 Simplified sedimentological column of sabkha facies (supratidal, intertidal, and subtidal) with main lithologies as observed in the Trucial Coast (UAE) and Abu Dhabi.

organic-rich, fine-grained sedimentary layer composed of bioturbated remnants of the algal mat and microbial mats (Figure 8.16). Below the algal mat, the sediments are composed of laminated mudstones up to 50 cm in thickness, intercalated with carbonate muds and sands composed of gray-colored aragonite layers and centimetric lenses of cyanobacteria-rich muds (Figure 8.16).

The upper intertidal facies include a laminated algal mat, forming an extensive belt of algal flats above the water level, extending for large swaths that may reach 1 to 2 km in width (Figure 8.14). These sediments are often exposed to subaerial conditions, with desiccation cracks formed on the surface of the evaporite crust. Locally displacive gypsum crystals and anhydrite nodules appear dispersed in millimetric lenticular layers (Figure 8.17). When the intertidal muds are covered by prograding supratidal layers, the Trucial Coast sabkhas are often dolomitized in the shallow subsurface (McKenzie, 1981). This sedimentary environment corresponds to one of the few examples in the world where the deposition of primary dolomites has been recorded in the Holocene.

The upper intertidal facies may also exhibit beach ridges that form linear belts with low relief, up to a few meters. These ridges are composed of skeletal grainstone, fragments of gastropods, bivalve shells, foraminifers, peloids, and ooids (Kendall, 1978; Kendall & Skipwith, 1969). Buried carbonate sand ridges have been studied by researchers based on shallow trenches that sampled the mid-sabkha setting (Warren & Kendall, 1985; West et al., 1979, 1983).

Coastal sabkhas are often associated with a complex of elongated spits and lagoons. A "spit" is an embankment of sediments that is still attached to the land at one end of a bay or cove and terminates in the open sea at the other side. During high tides, the connection with the land may be covered by water, and the spit is disconnected from the land bridge, forming an island at the lagoon's margin (Figure 8.13). In many sabkhas, the carbonate ridges reach a few meters above the surface sediments, resulting in the sequential development of ancient beaches along the lagoon's shoreline. Some sabkhas in Qatar are characterized by a series of parallel inactive ridges that indicate progradation, as the active beaches are located seaward (Billeaud et al., 2014; West, 2012).

3. *Supratidal facies:*

The Trucial Coast is characterized by a supratidal belt less than 10–15 km in width, where the evaporite sediments are formed by growing inside a matrix-rich sedimentary layer that is less than a meter thick. The evaporite deposits include gypsum sand, which is commonly covered by an evaporitic crust composed of anhydrite and halite (Figure 8.16).

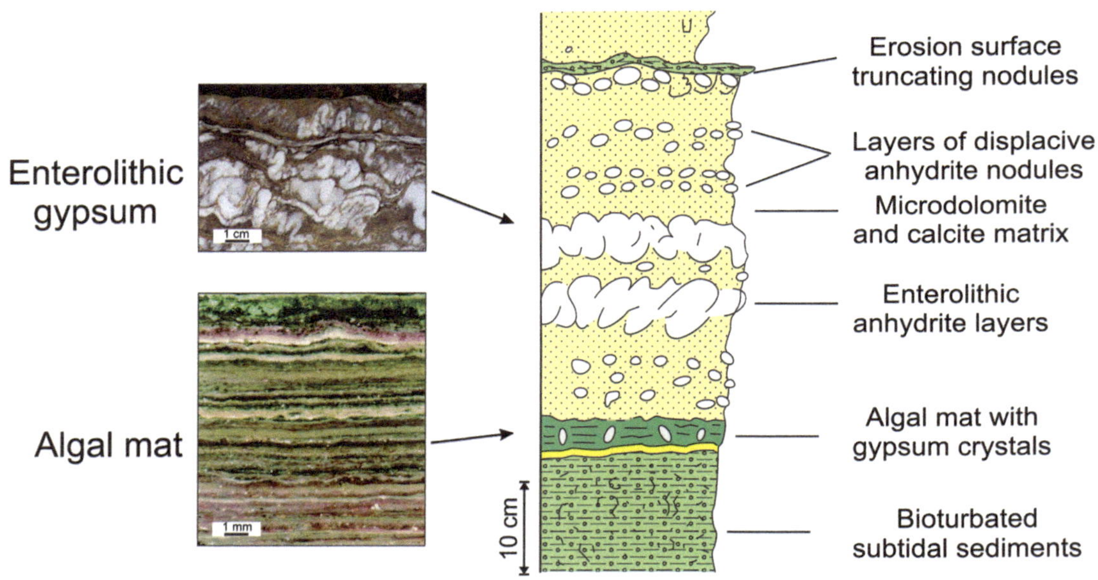

Figure 8.17 The stratigraphic column to the right illustrates the characteristic sedimentary features of coastal sabkha evaporites, with examples of enterolithic anhydrite/gypsum, and algal mat layers overlying bioturbated subtidal sediments. An unconformity separates the subtidal sediments from the intertidal and supratidal sediments. The upper sedimentary facies correspond to carbonate rocks with displacive anhydrite nodules and enterolithic gypsum and anhydrite layers. Below the unconformity, the sabkha flat plain is characterized by bioturbated subtidal sediments and dark gray to green, flat-laminated, algal mats of the intertidal zone. The images to the left show examples of these sedimentary facies. Credit: West (1975)/with permission of Elsevier.

The supratidal facies in the Persian Gulf are characterized by a belt of gypsum or anhydrite mush, which can extend up to 3 km in width and reach a thickness of about 30 cm (Kendall & Harwood, 1996; Kendall & Skipwith, 1969; Warren, 2006, 2010). Much of the gypsum grows within the capillary zone, forming a bedded unit of displacive gypsum crystals. During hot, dry periods, the gypsum layer loses some of its structural water and is altered to bassanite or anhydrite. However, in cooler rainy seasons, these minerals are rehydrated and transformed into gypsum again, leading to the formation of enterolithic gypsum layers (Figure 8.17).

The transformation of anhydrite into gypsum by hydration involves the addition of two water molecules to the crystal structure, resulting in a volume increase of about 63%. This expansion induces progressive fracturing of the rock matrix, resulting in the folding of the gypsum layers and forming enterolithic gypsum (Figure 8.17). The growth of anhydrite layers may also result in compressional structures, particularly ptygmatic and enterolithic folds. Locally, diapir-like structures and small-scale thrust faults are observed in desiccation cracks affected by contraction. Halite crusts are characterized by polygons with sizes that may reach tens of centimeters, forming concave-shaped plates that often expand laterally, resulting in uplifted margins a few centimeters above the ground, which are usually composed of an anhydrite pavement in the middle supratidal zone (Figure 8.16).

The middle supratidal zone corresponds to a belt of evaporitic sediments up to 2 km wide containing nodules of sulfate minerals of diagenetic origin. Nodular anhydrite lenses within a carbonate matrix sediment are commonly found in supratidal sequences overlying gypsiferous mud and algal mats (Figure 8.18). Below the sabkha water table, there is widespread precipitation of displacive and pore-filling gypsum minerals that form anhydrite nodules and veins in algal mats. The middle supratidal carbonate layer is often capped by a surface crust of ephemeral halite, which is locally associated with desiccation cracks and tepees (Warren, 1982a, 1982b, 1983).

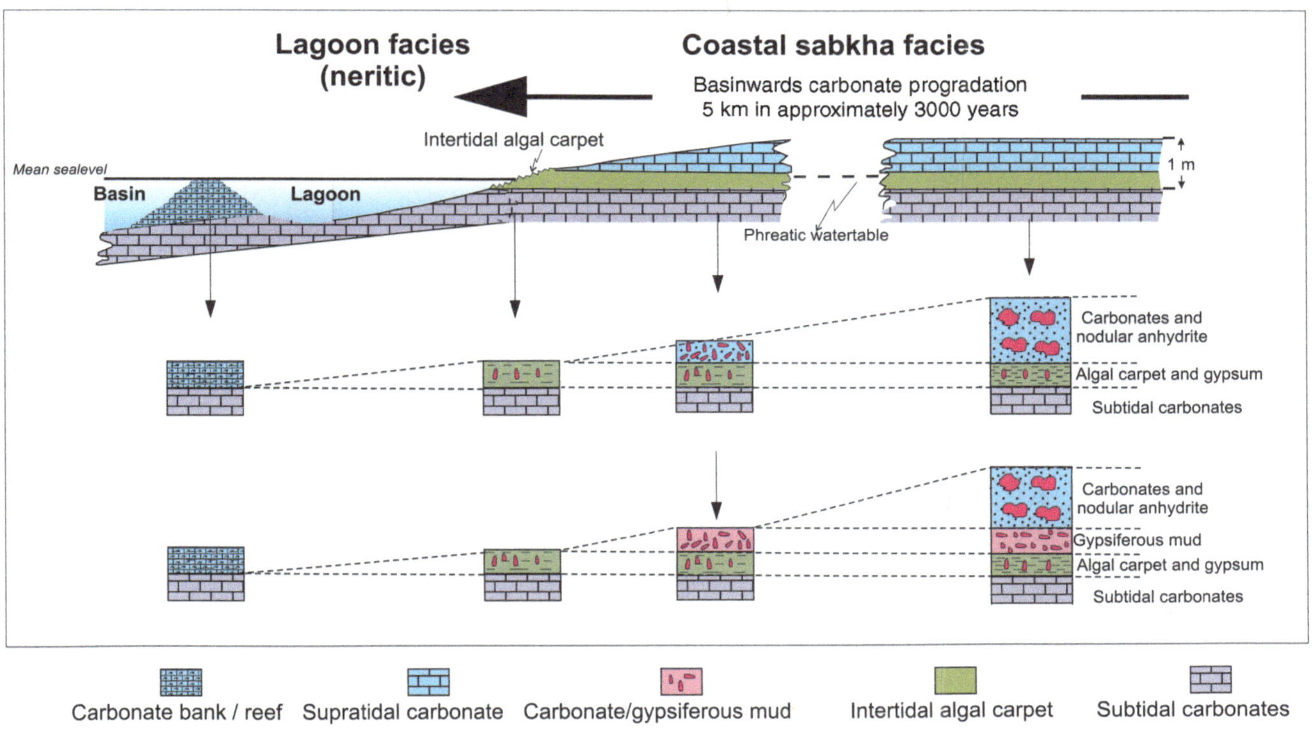

Figure 8.18 Trucial Coast classical sabkha and lagoon facies in an idealized transect extending from the offshore open-marine environment, where carbonate muds occur associated with skeletal sediments, including organic buildups forming carbonate islands, to the onshore sabkha facies. The deep basin is separated from the lagoon by belts of oolitic grainstone, forming a barrier with neritic facies. In the intertidal zone, algal mats form organic-rich carpets overlying subtidal carbonates. The supratidal coastal sabkha sequences change facies landward into desert sands. The carbonate sediments in the coastal plain of the Abu Dhabi sabkha show a progradation of about 13 km during the last 6,000 years, or 5 km in the past 3,000 years. The marine transgression reached a peak in sea level about 4,000 years BCE. Since then, the sabkha has been advancing basinward, resulting in regressive depositional facies where carbonate and evaporite sediments overlie subtidal carbonates, intertidal algal mats, and gypsiferous muds.

Toward the landward side of the lagoon, the surface anhydrite is capped by reworked siliciclastic and carbonate sands, or silty loess, which are blown across the sabkha surface. The desiccation cracks and anhydrite polygons appear truncated in cross-section, indicating eolian erosion or deflation affecting the sediments that have been uplifted slightly above the capillary zone and into the vadose zone by the continued subsurface precipitation of anhydrite (Figure 8.17). The erosional surface is often covered by siliciclastic sediments (quartz arenites) that originated from the eolian dunes in the sabkha's continental part (Figures 8.15 and 8.17).

An idealized transect extending from the coastal sabkha to the offshore open-marine environment illustrates the classical sabkha trilogy with the lagoon facies (Figure 8.18). In the open marine environment of the Persian Gulf, carbonate muds occur associated with skeletal sediments, including organic buildups or reefs. The supratidal sabkha sediments are characterized by the subaerial accumulation of carbonate mud and the growth of evaporite minerals in the interstices of the sediments that occur in the capillary zone, above the water table. The sulfate minerals are formed according to their position relative to the water table. Anhydrite is always found above the water table, whereas gypsum is formed in the phreatic zone, where the sediment pores are saturated with water. However, the gypsum is transformed into anhydrite when the sediments are buried to depths greater than about 1 km (Warren, 2006, 2016).

Extensive sedimentological research in the past 50 years has investigated both the surface distribution of the sabkha sediments and the subsurface by digging pits and drilling shallow sedimentological wells (Evans, 1995, 2011). The present sabkha surface reflects the progradation of the sedimentary facies during the last 4000 years, following the peak of marine transgression, and since then the sabkha has been advancing seaward, toward the deep gulf (Dean & Schreiber, 1978; Evans et al., 1969; Shearman, 1978).

This has resulted in regressive depositional facies (Figure 8.18), where carbonate and evaporite sediments overlie subtidal carbonates, intertidal algal mats, and gypsiferous muds. It is estimated that the carbonate sands advanced more than 5 km toward the Persian Gulf during the last 3,000 years. This progradation results in an average sabkha progradation of about 1.6 m/year (Dean & Schreiber, 1978; Evans et al., 1969; Shearman, 1978).

8.7.2. The Abu Dhabi Sabkha

Abu Dhabi's coastal and inland sabkhas (Figure 8.19) are situated in a hot, arid environment, with temperatures exceeding 50°C in the summer. Where ground temperatures exceed 40°C, the near-surface gypsum loses its

(A) (B)

Figure 8.19 Satellite image of the Persian Gulf showing the islands near Abu Dhabi and the coastal sabkha adjacent to the shoreline (A). Credit: NASA. The sabkha patches display different colors, and the beach ridges form elongated features near the lagoon. Channels between the carbonate banks connect the lagoon with canyons that transport sediments to the deep gulf. The salt flats may occur adjacent to the beach ridges, which may reach a few meters above the sabkha plains, which are locally covered by halite crusts with metric polygonal mudcracks (B). Credit: Patrick/Adobe Stock. The halite crusts form polygons that often expand laterally, resulting in uplifted margins a few centimeters above the ground. The beach ridges are a few meters high, providing an approximate scale.

compositional water and dehydrates to form anhydrite. The sabkhas are characterized by a narrow belt of dark microbial algal mats, up to 1 km in width, mainly composed of blue-green algae interlaminated with carbonate mud (Figure 8.17).

The intense evaporation and concentrated surficial pore waters in the carbonate sediments result in the precipitation of evaporites and authigenic minerals like aragonite, calcite, dolomite, celestite, and magnetite. Locally, the sediments are enriched with sulfates, and gypsum can make up more than 50% of the top meter of the sabkha sediments, forming nodular and contorted enterolithic gypsum. Gypsum can also form layers or crystals in the carbonate mud, growing with their long axis along a vertical plane. Halite hoppers can crystalize in the mud and cover the sabkha surface with a thick centimetric layer of gypsum and halite crusts (Figure 8.19B).

These flat-lying, featureless plains display no vegetation and are characterized by a salt crust at the surface. Halite crusts accumulate during some parts of the year, forming metric polygons with uplifted margins forming ridges (Figure 8.19B), similar to desiccation cracks in siliciclastic mudflats. Some playa lakes have a pervasively cracked surface, forming polygons ranging from a few centimeters to several meters in diameter (Kendall, 1984, 1992, 2010).

Based on the sedimentological analysis of the shallow sabkha trilogy, a marine transgression was interpreted to have formed the Abu Dhabi sabkha sediments over 7,000 years ago (Evans et al., 1969). The tidal range on the Trucial Coast is small (1–2 m), resulting in narrow belts of supratidal and intertidal sediments. The intertidal zone includes algal mats, which are associated with swamps where mangroves and cyanobacteria can accumulate in protected, low-energy environments adjacent to the lagoon (Figure 8.18). Total organic carbon in the intertidal microbial mats and mangrove paleosoils of the Abu Dhabi lagoon exhibits values of 2–8% (Alsharhan & Kendall, 2003). This high organic content is comparable to the Persian Gulf source rock intervals developed in ancient sabkha environments during the Mesozoic (see Chapter 12).

8.7.3. The Dukhan Sabkha in Western Qatar

The Dukhan Sabkha in western Qatar is a modern example of a sabkha environment in the Central Persian Gulf. The Qatar Peninsula is located to the east of the Gulf of Salwah, between Saudi Arabia and Bahrain (Figure 8.20), and it is primarily covered by carbonate rocks, while the southern margin of the Persian Gulf is mostly covered by siliciclastic sediments from the Sahara Desert (Ashour, 2013). The Dukhan Sabkha is covered by a compact crust of fine-grained deposits of sand, clay mud, and crystals of halite and gypsum, which reach a thickness of a few millimeters and locally may reach about 1 cm. Petrographic and scanning microscopy analyses of subsurface samples indicate that the sabkha sediments are mainly composed of fine prismatic to acicular gypsum crystals and halite (Al-Youssef, 2015).

The Qatar peninsula is associated with the Qatar Arch, a major structure that controls the occurrence of large hydrocarbon accumulations in this segment of the Persian Gulf (Figure 8.13). The Dukhan Sabkha is the largest coastal sabkha in the peninsula, covering an area of 73 km^2 and occupying a synclinal area between the Dukhan anticline in the west and the Qatar anticline in the east. The Dukhan Anticline and the Qatar Arch are deeper features related to Mesozoic to Cenozoic compressional tectonics involving the Arabian plate (Warren, 2006, 2016; West, 2012).

The sabkha's surface is extremely flat, with the center lying about one meter below sea level, leading to high groundwater inflow. Climate aridity results in high rates of evaporation and high brine concentrations. Chemical analyses of brines in sediments from about 30 cm below the ground to a depth of about 1 m indicate a predominance of chlorine (mean of about 62 ppt) and sodium (mean of about 39 ppt). The standard seawater concentration (in grams/liter) is about 19 (chlorine) and 11 (sodium), which indicates that the brine in the Dukhan Sabkha has been concentrated more than three times the normal seawater salinity (Al-Youssef, 2015).

Based on comparisons between the sabkhas of the Trucial Coast and Qatar, Billeaud et al. (2014) suggested a sedimentological model for the development of sabkha facies in this evaporitic setting (Figure 8.21). The supratidal sabkha environments include desert dunes and locally desiccated lakes that are filled with evaporites (gypsum and halite). The intertidal back barrier mud flat often displays microbial mats. The back barrier is characterized by tidal channels and alluvial plains with fine- to coarse-grained sediments (Billeaud et al., 2014).

The regressive sedimentation pattern in the sabkha is characterized by the progradation of the supratidal environments toward the deep basin, resulting in the progressive abandonment of beaches and spits, which became inactive in the Quaternary (Figure 8.21). The active beach and spit, made up of bioclastic coarse sand, may periodically isolate the sabkha environment by obstructing communication with the open sea (Persian Gulf). The open sea environment is characterized by bioclastic and peloidal medium sands that grade into hemipelagic marls in deeper waters. Tidal and submarine bars occur offshore, from the shoreline to the shelf break, while local canyons and turbidite fans occur basinward of the shelf break, reaching the deep basin (Figure 8.21).

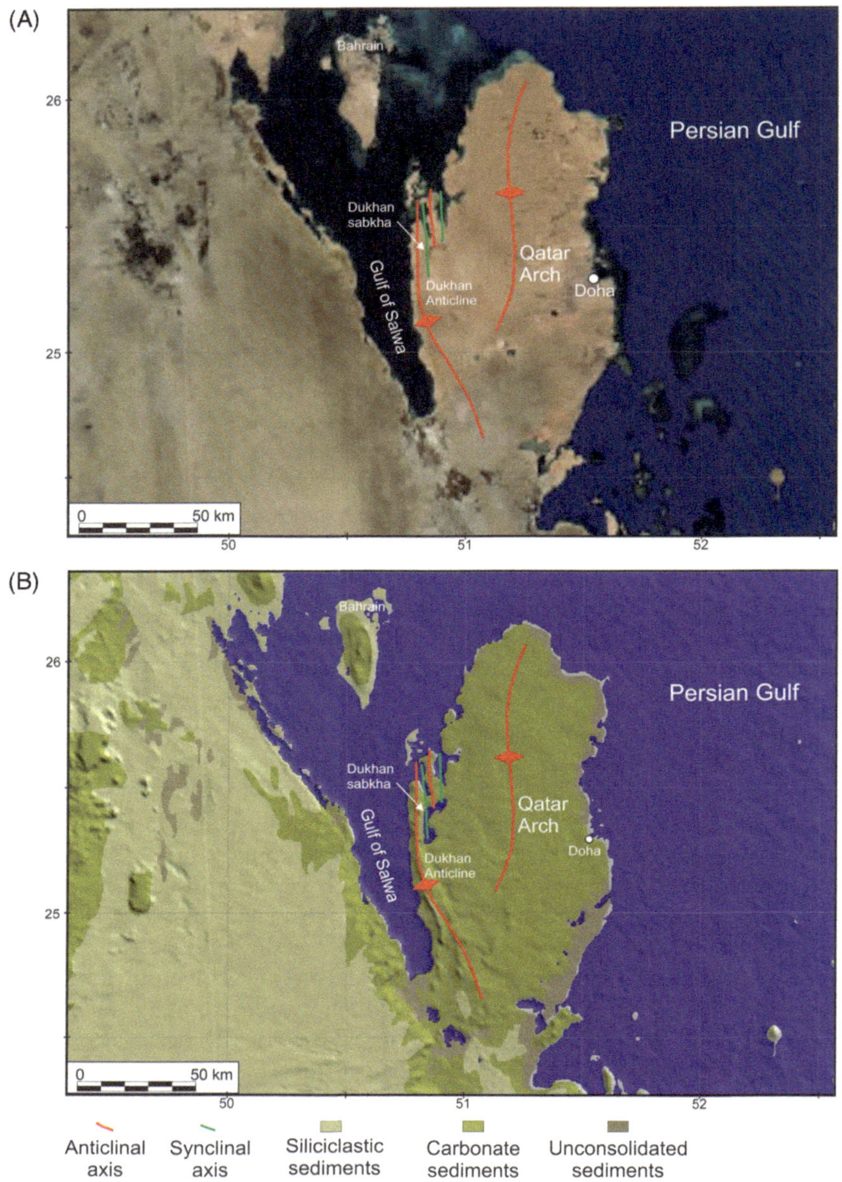

Figure 8.20 (A) Satellite image (courtesy of NASA) of the Persian Gulf margin between Saudi Arabia and Qatar, showing the Dukhan Sabkha on the western Qatar coast. (B) Simplified geological map. Credit: Hartmann and Moosdorf (2012)/with permission of John Wiley & Sons. Most of the southern margin of the Persian Gulf is covered by siliciclastic sediments derived from the Sahara Desert, but Bahrain and the Qatar Peninsula are mainly covered by carbonate rocks. The Dukhan anticline and the Qatar Arch are deeper features related to Mesozoic to Cenozoic compressional tectonics. The Dukhan Sabkha surface appears in the satellite image with greenish colors, and at present, its surface is about 1 m below sea level.

8.7.4. The El-Alamein Sabkha in Egypt

The El-Alamein sabkha is a depression in northern Egypt, located west of the Nile Delta and near the Mediterranean Sea (Figure 8.22). The sabkha is located to the west of a large gypsum-saturated lagoon near the Mediterranean Sea (Figure 8.23), and covers a depression with a low-relief, flat surface almost at sea level. The sabkha flats are separated from the sea by Pleistocene and Holocene beach ridges, which form sedimentary barriers. These NE-SW-trending ridges are composed of carbonate sediments (West et al., 1983).

The El-Alamein sabkha sediments are between the carbonate beach ridges (Figure 8.24) and correspond to mudflats occupying two elongated depressions (Figure 8.25). The sabkha surface is covered by brown

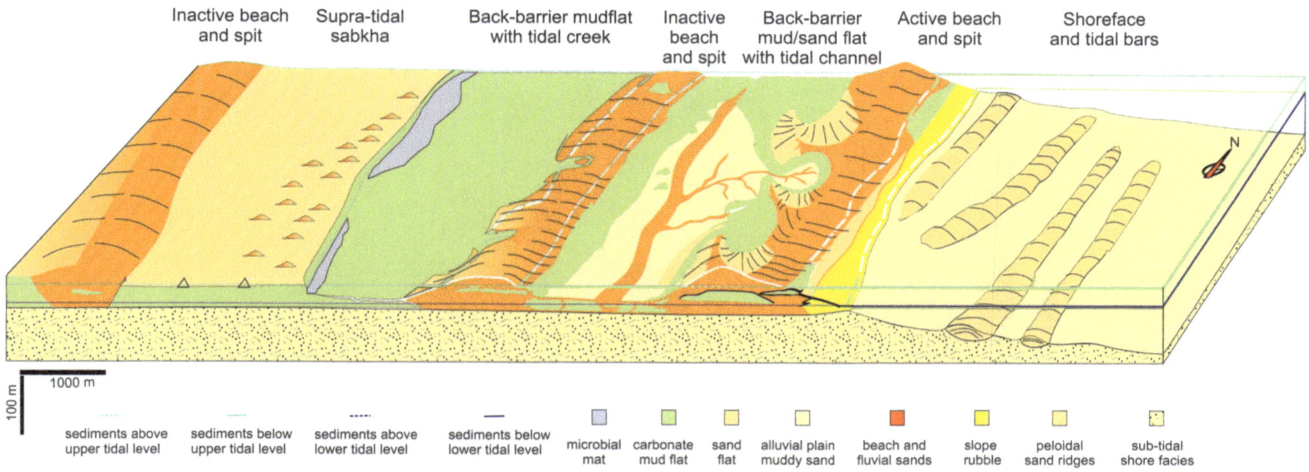

Figure 8.21 Schematic block diagram showing the sabkha model in the Trucial Coast and Qatar. The supratidal sabkha environment includes eolian dunes and locally desiccated lakes that may be filled with evaporites (gypsum and halite). The intertidal back-barrier mudflat is characterized by microbial mats that locally display halite crusts due to basin desiccation. Tidal channels and alluvial plains are characterized by fine- to coarse-grained sediments that occur in the back barrier. Progradation of the sabkha environment toward the deep basin resulted in the abandonment of beaches and spits, which became inactive. The active beach at the border of the gulf is constituted by bioclastic coarse sand that forms a barrier that may isolate the sabkha environment by obstructing communication with the open sea. Credit: Adapted from Billeaud et al. (2014).

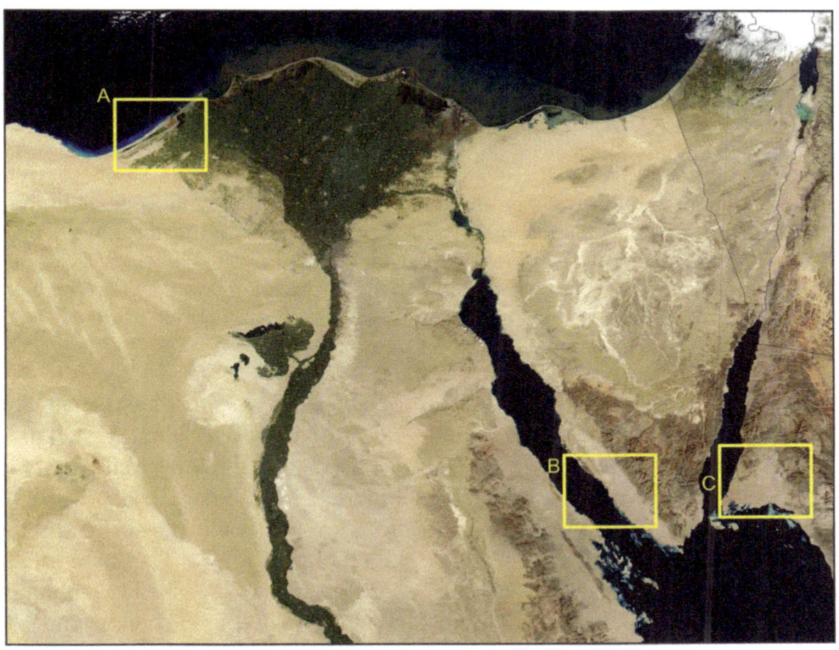

Figure 8.22 NASA satellite image (courtesy of NASA) with zoom rectangles showing the location of the El Alamein sabkha west of the Nile Delta in northern Egypt (A), the El Qa plain in the western Sinai Peninsula, adjacent to the Gulf of Suez (B), and the plains that locally form sabkha flats in the Midyan Basin in western Saudi Arabia, adjacent to the northern Red Sea, near the Gulf of Aqaba (C). Precambrian rocks outcrop at the Sinai Peninsula and onshore Saudi Arabia, at the northern sector of the Red Sea.

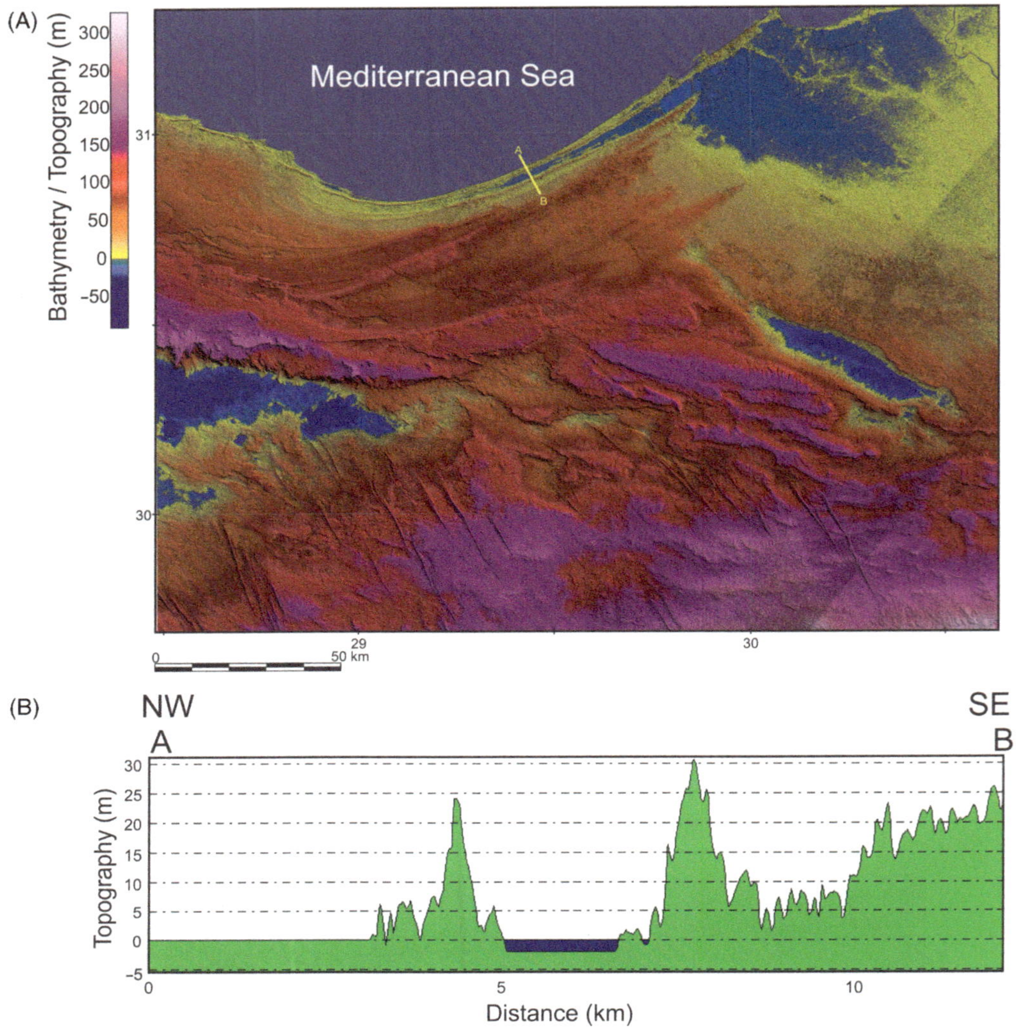

Figure 8.23 (A) Topographic map of the southern Mediterranean region near the Sabkha El Alamein in Northern Egypt, showing the gypsum-saturated lagoon near the Mediterranean Sea and the Pleistocene and Holocene beach ridges that form barriers reaching elevations of about 20–30 m above sea level; (B) Cross-section A-B showing topography and the occurrence of sabkha sediments in the First Depression between the Coastal Ridge, with an elevation of about 25 m. The Abu-Sir Ridge is marked by an elevation of about 30 m.

desert loess with nodular gypsum, while the underlying carbonate sediments are formed by oolitic grains (carbonate sand) and shell fragments (Figure 8.25).

The present-day sabkha sediments, with gypsum nodules dispersed in the carbonate matrix, form a thin surficial cover of only a few centimeters. A geological cross-section (profile A-B) across the First Depression shows the sabkha sediments between the Coastal Ridge (oolitic sands) and Abu Sir Ridge (carbonate rocks). The sedimentary facies are aligned parallel to the Mediterranean Sea coastline (Figure 8.25). The cross-section sedimentological profile (Figure 8.25) shows the flat topography between the ridges and the occurrence of sabkha sediments in the First Depression between the Coastal Ridge, with an elevation of about 25 m, and the Abu-Sir Ridge, with a maximum elevation of about 30 m (West et al., 1979).

Based on shallow wells and rectangular pits (e.g., Figure 8.26), West et al. (1979, 1983) identified five depositional sequences in the surficial sedimentation of the sabkha (Figure 8.25). The base of the sabkha sediments corresponds to gray shelly carbonate and silt at the bottom (I) and gypsiferous carbonate and silt at the top (II, III, IV).

The rectangular pits that were dug for the study are rather small and shallow, measuring about 1 m inside by 1 m in depth. The water table is located just below the nodular gypsum layer, 0.5–0.6 m below the surface (Figure 8.26A). The nodular gypsum layer occurs above the water table, forming a layer about 0.5 m below the

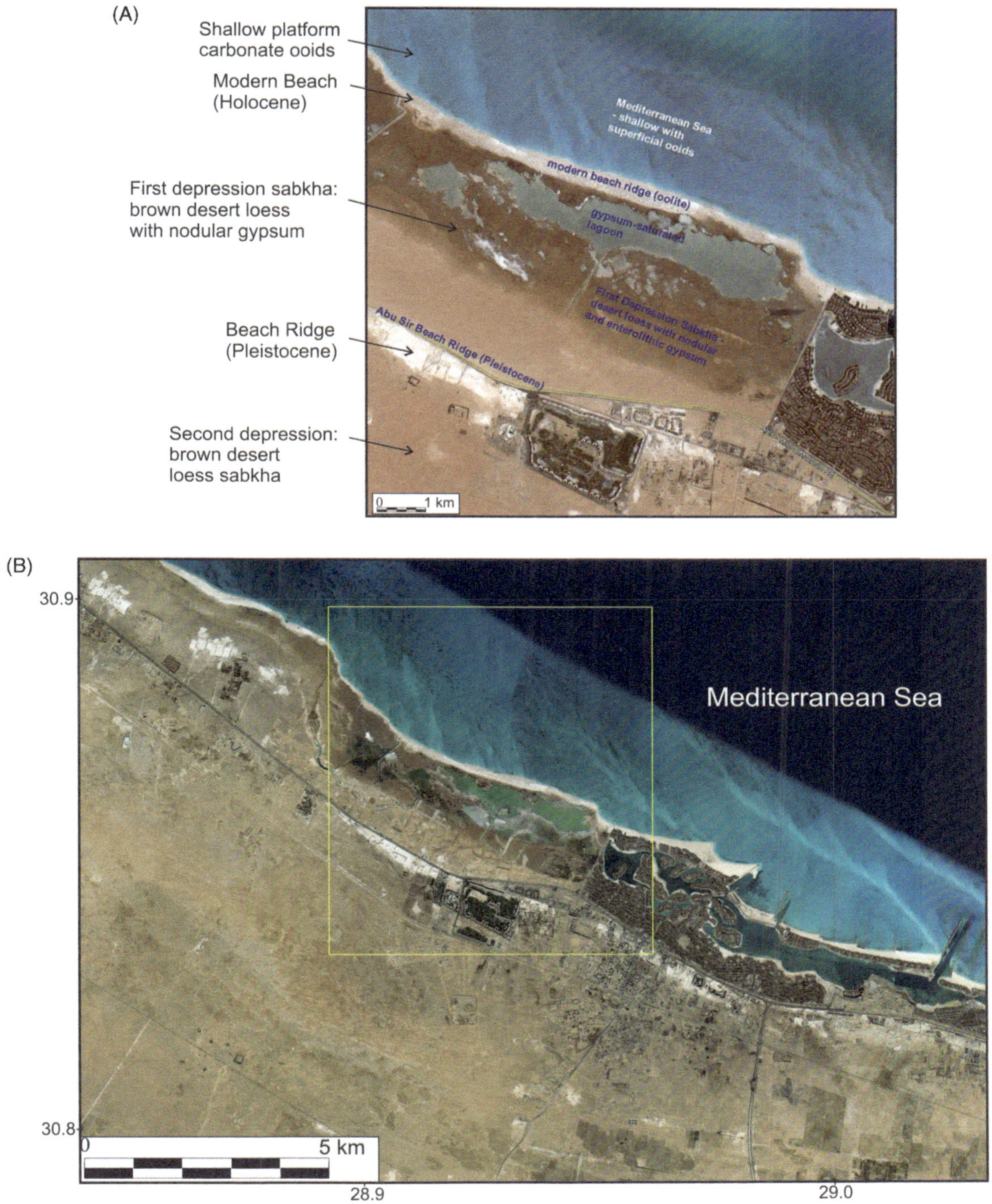

Figure 8.24 (A) Sabkha El Alamein in northern Egypt is a typical coastal sabkha with a gypsum-saturated lagoon near the Mediterranean Sea, between the Pleistocene and Holocene beach ridges. The yellow rectangle marks a detailed image of the depression, which is filled with sabkha sediments between the beach ridges. (B) The detail image shows the sabkha environment, characterized by brown desert loess with nodular gypsum between the carbonate sediments near the Mediterranean Sea coastline. Credit: Planet Labs, Inc./Wikimedia Commons/CC BY-SA 4.0.

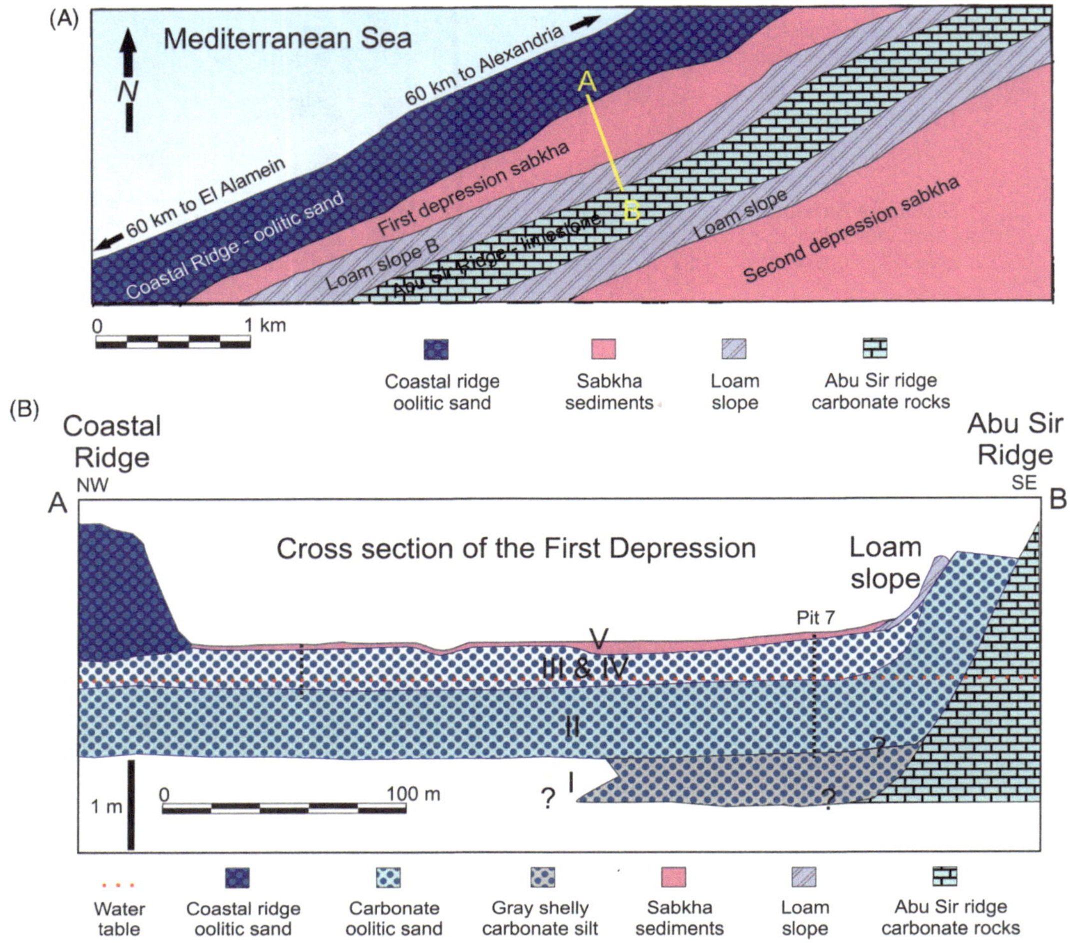

Figure 8.25 (A) Schematic geological diagram showing the Sabkha El Alamein in northern Egypt. The yellow line A-B shows the location of a topographic profile across the first depression sabkha (middle of line A-B). (B) Cross-section (profile A-B) across the First Depression showing the sabkha sediments between the Coastal Ridge (oolitic sands) and the Abu Sir Ridge (carbonate rocks). The sedimentation is characterized by five deposition sequences, with gray shelly carbonate and silt at the bottom (I) and gypsiferous carbonate and silt at the top (II, III, and IV). The present-day sabkha sediments with gypsum nodules form a thin surficial cover.

surface, and is characterized by centimetric nodules in a silty loess-carbonate matrix (Figure 8.26B). The interstitial brine above the nodular gypsum layer has a higher concentration than the sulfate saturation level. Gypsum minerals are deposited in the interstices of carbonate rocks, and enterolithic veins and nodules form a thick layer (up to tens of centimeters) above the nodular gypsum layer. At the bottom of the pit, below the vadose zone, the sulfate-rich brine was sampled in the phreatic zone (Figure 8.26B).

Chemical measurements of the interstitial brine indicate that salinity increases upward, reaching 300–400‰ near the surface, allowing the precipitation of sulfate and chloride minerals that form a thin gypsum-halite crust covering the surface (West et al., 1983). The sedimentological model for the development of the El Alamein sabkha (Figure 8.27) suggests that the coastal region was initially covered by the Mediterranean Sea. Carbonate buildups resulted in the isolation of a hypersaline lagoon that was desiccated after a beach ridge isolated the depression. The sedimentary barrier hindered communication with the Mediterranean Sea, resulting in the concentration of brines and evaporite deposition in the lagoon (West et al., 1983).

8.7.5. Sabkhas in the Sinai Peninsula and in the Western Red Sea

Another example of a modern sabkha environment is observed at the western margin of the Sinai Peninsula, near the Gulf of Suez, one of the hottest and driest places in the world (Figure 8.28). The carbonate components are less widespread in the northern Red Sea than in the

Figure 8.26 The top photograph (A) shows a pit in the Sabkha El Alamein in northern Egypt, showing enterolithic veins and nodules of gypsum in layers near the surface. The pit dimensions (length, width, and depth) measure about 1 m. The water table is located just below the nodular gypsum layer, 50–60 cm below the surface. The bottom photograph (B) shows a detailed image of a nodular gypsum layer just above the water table, forming a thin layer below the surface. The whitish gypsum is characterized by centimetric nodules with enterolithic structures, dispersed in silty loess and a carbonate matrix exhibiting a brownish color. Credit: West et al. (1979)/with permission of Geological Society of America.

Persian Gulf sabkhas, which are located along the Trucial Coast, near Abu Dhabi, far from the Tigris-Euphrates-Karun delta system, which transports mainly fine-grained siliciclastic sediments to the northern part of the Persian Gulf. The borders of the Sinai Peninsula have been uplifted during the continental rifting that formed the Gulf of Suez and Gulf of Aqaba, resulting in coarse siliciclastic sediments transported to the coastal plains and basins (Figure 8.28).

The Sinai Peninsula sabkhas consist of siliciclastic minerals derived from the erosion of Precambrian gneisses that outcrop on the highlands of the peninsula (Figure 8.28). The evaporite minerals in the sabkhas at the eastern margin of the Gulf of Suez correspond to gypsum crystals scattered in the interstices of the sandstone grains. The main component of the sabkha sedimentary matrix is quartz, so these evaporitic sediments have not undergone as much diagenesis as observed in the carbonate sabkhas of the Trucial Coast (Gavish, 1980).

The northern Red Sea is characterized by two gulfs around the Sinai Peninsula: the Gulf of Suez, which connects the Red Sea with the Mediterranean Sea, and the Gulf of Aqaba (or Elat), which separates Saudi Arabia and Jordan from Egypt (Figure 8.28). These gulfs differ significantly in their geologic history and tectonic development, causing extreme differences in bathymetry (see Chapter 1 in Mohriak, 2025). The Gulf of Aqaba is related to a transform fault zone controlling an extremely deep and narrow trough, while the Gulf of Suez is associated with extensional tectonics that formed a rift basin similar to the Red Sea, but its bathymetry is relatively shallow with an average water depth of about 70 m (Gavish, 1980). The coastal area in the eastern Persian Gulf is characterized by an extensive sand flat associated with the El Qa plain (Figure 8.28). Thick evaporite layers dated to the Middle Miocene occur in the central part of the Gulf of Suez (see Chapter 1 in Mohriak, 2025).

Similar to the Persian Gulf, most evaporites in the sabkha at the eastern margin of the Gulf of Suez (El Qa Plain) are formed in the supratidal zone (Figure 8.29). However, the sabkha sediments are characterized by a quartz-rich matrix with some evaporite minerals (sulfate and chloride) above the water table, distributed in sedimentary units with a thickness of less than 1 m (Gavish, 1980). The upper portions of the supratidal sediments in the coastal plain include siliciclastic and carbonate sediments with interstitial pores filled with high-salinity brines due to evaporation. In contrast to sediments formed in a more seaward position, the supratidal zone sedimentary facies show little or no dolomite. Aragonite and dolomite minerals increase in percentage toward the subtidal zone of the Gulf of Suez (Gavish, 1980).

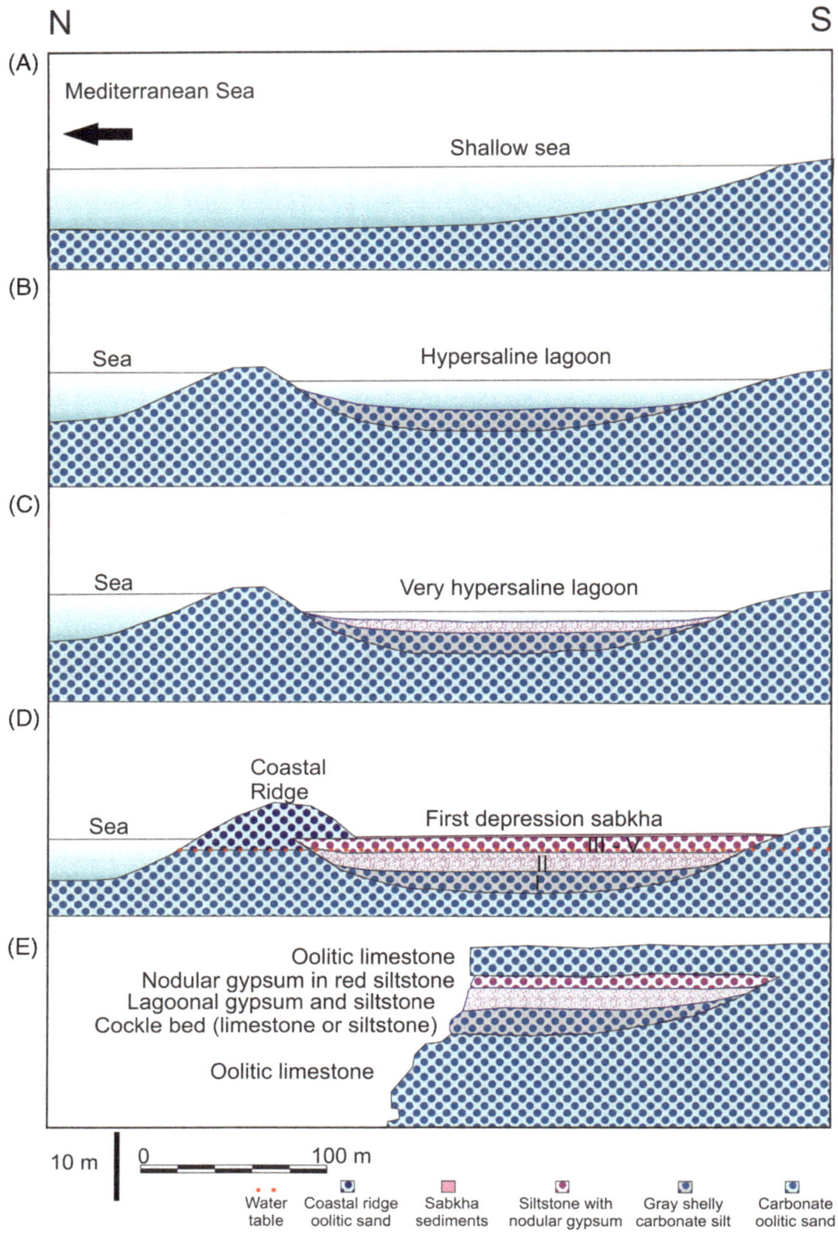

Figure 8.27 Schematic diagram showing the geological model for the development of the Sabkha El Alamein in northern Egypt. The sedimentary barrier associated with oolitic carbonate sands formed a coastal ridge that isolated the sabkha depression from the open sea. The lagoon became hypersaline and was filled with evaporites that precipitated when the brine concentration exceeded the gypsum solubility. Credit: Adapted from West et al. (1983).

8.7.6. The Bar Al Hikman Sabkha in Southeast Oman

The Bar Al Hikman sabkha flats are situated on the southeastern coast of Oman, south of the alluvial fan systems derived from the erosion of the Oman Mountains (Figure 8.30). This sabkha provides a modern analog for a desert dune field in a carbonate-evaporite system. The sabkha is located in a geographic setting that lies close to the Indian Ocean and extends from the Sahara Desert toward the coast and continental shelf (e.g., Homewood et al., 2007).

The sabkha flats in the Bar Al Hikman depression are located on the eastern edge of the Arabian Peninsula, where Masirah Island protects against strong Indian Ocean currents. Carbonate sediments are preserved from erosion and accumulate on land and in shallow

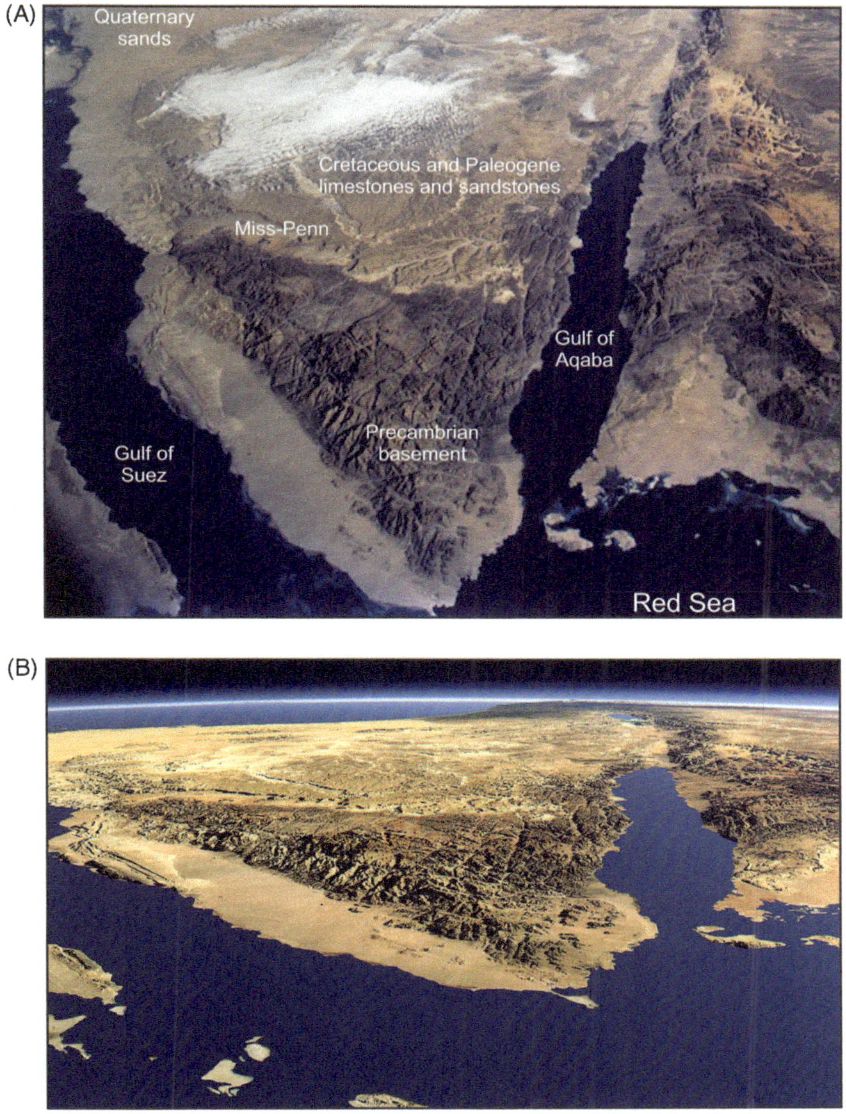

Figure 8.28 The top diagram (A) shows a NASA satellite image with the simplified geological setting of the Sinai Peninsula. The Precambrian igneous and metamorphic rocks outcropping at the southern tip are characterized by highlands. The center of the Sinai Peninsula is covered by Cretaceous and Paleogene sediments. The southwest part of the peninsula is characterized by a flat desertic region on the eastern side of the Gulf of Suez, corresponding to the El Qa Plain, adjacent to the Precambrian basement outcrops. The Midyan Basin in Saudi Arabia, near the confluence of the Gulf of Aqaba with the northern Red Sea, is characterized by a flat region where sabkha sediments are associated with gypsum outcrops. The oblique satellite image at the bottom (B) shows the confluence of the Gulf of Suez and the Gulf of Aqaba with the northern Red Sea. The modern sabkha depressions are characterized by a flat topography at the southern tip of the Sinai Peninsula. Credit: NASA/Wikimedia Commons/ Public domain.

waters, covering the shallow platform with unconsolidated carbonate sand grains. The Bar Al Hikman carbonate-evaporite system comprises lithologies deposited in both continental and coastal sabkhas. Sedimentary facies on land continue offshore as shallow marine carbonates, forming a carpet of rubble to fine-grained fragments derived from coral debris, reefal bodies, and abundant coarse-grained carbonate sediments that are constantly reworked by waves (Homewood et al., 2007).

In the onshore region, the sabkha is characterized by carbonate sand flats that extend from the siliciclastic dunes to the west, which display a yellowish color in the satellite image, and reach the extensive sabkha plain near

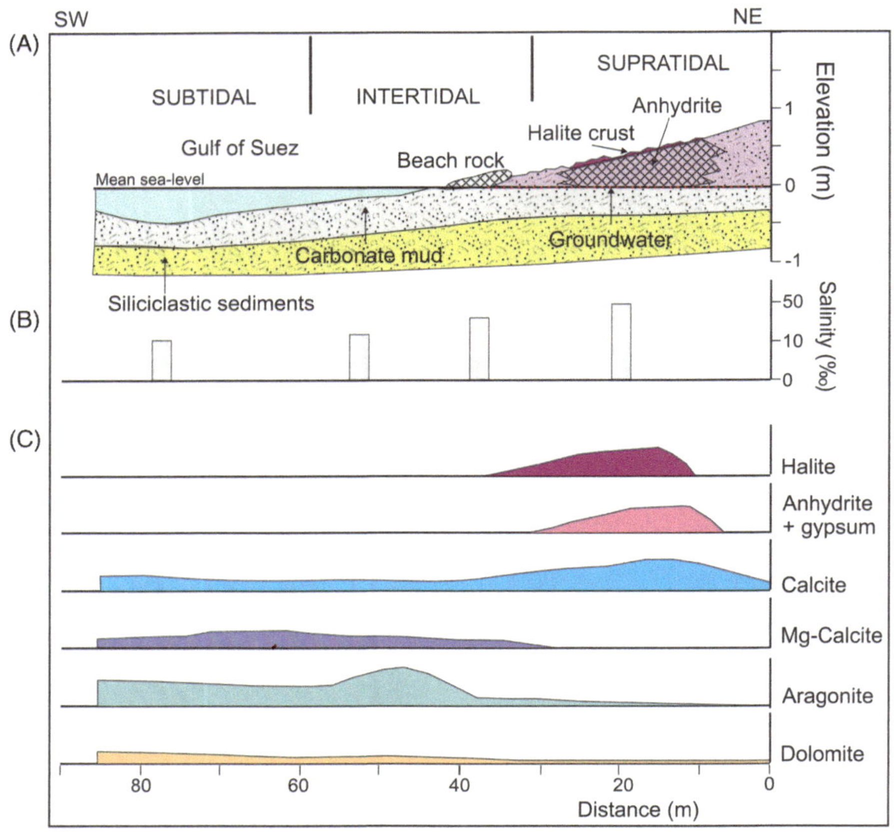

Figure 8.29 Schematic diagram showing the evaporitic mudflat at the western side of the Sinai Peninsula (El Qa plain). Halite and anhydrite minerals occur in the upper part of the surficial sediments adjacent to the Gulf of Suez, with elevations up to 1 m above sea level. The intertidal zone is characterized by a 50-cm variation around the sea level. The saline groundwater is evaporated in the interstitial pores of the sabkha sediments, which are characterized by a quartz-rich matrix. The brine concentration exceeded the gypsum and halite solubility, precipitating these minerals in the supratidal zone. The subtidal and intertidal sediments are mainly characterized by carbonate minerals (calcite and aragonite). Credit: Gavish (1980)/with permission of Elsevier.

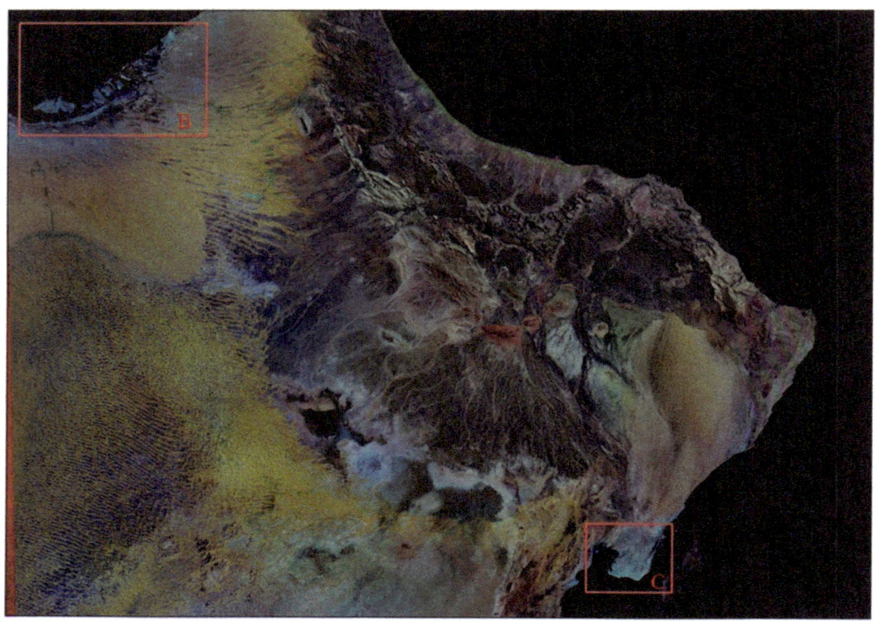

Figure 8.30 Satellite image showing the sabkha plains in the UAE (Trucial Coast, Abu Dhabi sabkha, rectangle B). The eastern Oman sabkha (Bar Al Hikman, rectangle C), is located south of the huge alluvial fan system derived from the erosion of the Oman Mountains, composed of mafic rocks that exhibit a dark color in the satellite image. The eolian dunes southwest of the Oman Mountains are characterized by yellow reflectivity. The sabkha sediments are marked by a bright bluish color reflectivity. Credit: Courtesy of NASA.

Figure 8.31 Sedimentary facies of the Bar Al Hikman sabkha on the southeastern coast of Oman. The photograph shows an upper intertidal microbial belt with desiccation polygons in the foreground. A sandy beach ridge and sparse black mangrove trees are seen in the background. Credit: University of Illinois at Chicago/CC BY 4.0.

the Indian Ocean coastline (Figure 8.30). Although locally there are halite and gypsum crusts on the sabkha's surface (Figure 8.31), most of the evaporite deposition takes place within the sediment column in the subsurface, with the vadose zone being up to 1 m thick. Overlying salt flats with halite crusts and intertidal microbial mats are recognized by their distinctive bluish color in satellite images (Figure 8.30).

The geomorphic features of the Bar Al Hikman sabkha include shallow tidal lagoons marked by intertidal shoals, channels, sand bars, and scattered salt ponds at abandoned lagoons or littoral depressions. Some depressions are partially filled by evaporites, appearing as white patches in satellite images (Mettraux et al., 2011).

Geological work in the offshore region of the Bar Al Hikman sabkha indicates that carbonate muds form a thin layer on the seafloor, below the tidal level. Spectacular shallow reefs lie in water depths of less than 5 m and are exposed during low tide levels. Locally, carbonate sand shoals are seen above sea level, forming barrier bars, small islands, and arcuate storm spits derived from coral rubble. Several submerged carbonate sand beaches, which are associated with tidal currents and sea-level fluctuations, are recognized in the offshore region (Homewood et al., 2007; Mettraux et al., 2011).

8.8. SYNTHESIS OF SALT DEPOSITION MODELS AND SABKHA ENVIRONMENTS

Evaporite deposits are found in various sedimentary environments related to the different plate tectonic stages of the Wilson cycle. The main depositional environments associated with ancient "Salt Giants" correspond to arid deserts, lakes, sabkhas, shorelines, lagoons, gulfs, shelves, continental slopes, and deep oceanic basins. Several climate and geological factors control salt deposition, particularly high global temperatures, high evaporation rates, and basin isolation from global oceans, resulting in brine concentration and salt precipitation. The depressions occur on land, forming lacustrine systems or playa lakes and, along continental margins, forming coastal sabkhas. Episodic marine incursions resulted in repeated flooding, replenishing the evaporated brine and forming several cycles of salt deposits.

Some geological intervals are characterized by widespread accumulation of evaporite sediments, indicating major episodes of global climate aridity, such as the Neoproterozoic, Late Permian, Late Triassic, Middle Jurassic, and Early Cretaceous. The last episodes of widespread salt deposition on Earth were recorded in the Red Sea and the Mediterranean Sea. During the Middle Miocene, the Red Sea formed a giant salt basin by closing the connection to the global oceans, which corresponds to a narrow strait separating the African and Arabian plates. The MSC in the Mediterranean Sea corresponds to the last major interval of salt deposition. It resulted from the convergence of the African, Iberian, European, Arabian, and Asian plates during the Late Miocene.

The relative plate movement of the Sinai and Arabian plates is responsible for the development of a major transform fault zone that developed rift valleys or pull-apart basins. The Dead Sea basin corresponds to a small continental rift basin that was connected to the Mediterranean Sea during the Pliocene via a small channel. Due to solar heating, seawater evaporated, and the brine was extremely concentrated, exceeding the solubility of salt minerals. The deep lake precipitated thick salt layers, predominantly composed of halite. The Dead Sea is the only modern analog for halite precipitation in deep water settings, but its use as an analog for the giant salt basins is questionable.

There is no modern analog for significant salt deposition in deep oceanic basins, continental platforms, gulfs, and inland seas that might be compared to the classical "Salt Giants." The only sedimentary environments where some salt is being deposited at present correspond to a few isolated lakes, lagoons connected to major seas, playa lakes and sabkha systems in desert regions, and elevated plateaus associated with plate convergence.

Therefore, in terms of salt basin development, the present is not the key to the past. The only sedimentary environments where some salt is being deposited at present correspond to a few isolated lakes, such as the Great Salt Lake in Utah; some lagoons connected to major seas, such as the Kara Bogaz Gol; the playa lakes and sabkha systems in desert regions, such as the Chotts in the Sahara

Desert; or the Basin and Range rifts in southwest North America. The mudflats and sabkhas along the northern coast of the Arabian Peninsula, adjacent to the southern margin of the Persian Gulf, have been considered a paradigm for the salt giants of the ancient past. In addition, there are important salt deposits on some elevated plateaus, such as the Qarhan salt flats in China or the Altiplano salt flats in the Andean Mountains, which are related to the shrinking of much larger lacustrine systems. Understanding the factors controlling salt precipitation, accumulation, and preservation is critical to interpreting the brine chemistry and sedimentary facies of the evaporite layers in the subsurface.

REFERENCES

Almalki, K. A., Betts, P. G., & Ailleres, L. (2015). The Red Sea—50 years of geological and geophysical research. *Earth-Science Reviews*, *147*, 109–140.

Alsharhan, A. S., & Kendall, C. G. St. C. (2003). Holocene coastal carbonates and evaporites of the southern Arabian Gulf and their ancient analogues. *Earth-Science Reviews*, *61*, 191–243.

Alsharhan, A. S., & Kendall, C. G. St. C. (2011). Introduction to Quaternary carbonate and evaporite sedimentary facies and their ancient analogues. In: C. G. St. C. Kendall & A. S. Alsharhan (Eds.), *Quaternary carbonate and evaporite sedimentary facies and their ancient analogues: A tribute to Douglas James Shearman* (Vol. *43*, pp. 1–10). International Association of Sedimentologists, Special Publication. Sussex, UK: Wiley-Blackwell.

Ashour, M. M. (2013). Sabkhas in Qatar Peninsula. *Landscape and Geodiversity*, *1*, 10–35.

Al-Youssef, M. (2015). Gypsum crystals formation and habits, Dukhan Sabkha, Qatar. *Journal of Earth Science and Climatic Change*, *6*(10), 321, 12 pp.

Babel, M., & Schreiber, B. C. (2014). Geochemistry of evaporites and evolution of seawater. *Treatise on geochemistry* (2nd ed., Chapter 9.17, pp. 483–560). Amsterdam: Elsevier.

Barbieri, R., Stivaletta, N., Marinangeli, L., & Ori, G. G. (2006). Microbial signatures in sabkha evaporite deposits of Chott el Gharsa (Tunisia) and their astrobiological implications. *Planetary and Space Science*, *54*(8), 726–736.

Baucon, A. (2010). Leonardo da Vinci, the founding father of Ichnology. *Palaios*, *25*(5/6), 361–367.

Billeaud, I., Caline, B., Livas, B., Tessier, B., Davaud, E., Frebourg, G., et al. (2014). The carbonate-evaporite lagoon of Al Dakhirah (Qatar): An example of a modern depositional model controlled by longshore transport. In I. P. Martini & H. R. Wanless (Eds.), *Sedimentary coastal zones from high to low latitudes: Similarities and differences* (Vol. *388*, pp. 561–587). Geological Society, London, Special Publications. https://doi.org/10.1144/SP388.7

Blatt, H., Middleton, G., & Murray, R. (1980). *Origin of sedimentary rocks* (2nd ed.). New Jersey, USA: Prentice-Hall, 634 pp.

Boggs, S., Jr. (2009). *Petrology of sedimentary rocks*. Cambridge, UK: Cambridge University Press, 600 pp.

Boggs, S., Jr. (2014). *Principles of Sedimentology and Stratigraphy*. (5th ed.). Essex, England: Pearson Education Limited, 660 pp.

Branson, E. B. (1915). Origin of thick gypsum and salt deposits. *Geological Society of America Bulletin*, *26*(1), 231–242.

Briere, P. R. (2000). Playa, playa lake, sabkha: Proposed definitions for old terms. *Journal of Arid Environments*, *45*, 1–7. doi:10.1006/jare.2000.0633

Bryant, R. G., Sellwood, B. W., Millington, A. C., & Drake, N. A. (1994). Marine-like potash evaporite formation on a continental playa: Case study from Chott el Djerid, southern Tunisia. *Sedimentary Geology*, *90*, 269–291.

Burke, K. (1975). Atlantic evaporites formed by evaporation of water spilled from Pacific, Tethyan and Southern Oceans. *Geology*, *3*, 613–616.

Butler, G. P. (1969). Modern evaporite deposition and geochemistry of coexisting brines, The Sabkha, Trucial Coast, Arabian Gulf. *Journal of Sedimentary Petrology*, *39*(1), 70–89.

Carlson, A. E. (2011). Ice sheets and sea level in Earth's past. *Nature Education Knowledge*, *3*(10), 3.

Chaboureau, A.-C., Donnadieu, Y., Sepulchre, P., Robin, C., Guillocheau, F., & Rohais, S. (2012). The Aptian evaporites of the South Atlantic: A climatic paradox? *Climate of the Past*, *8*, 1047–1058.

CIESM (2008). The Messinian salinity crisis from mega-deposits to microbiology—A consensus report. In F. Briand (Ed.), *CIESM Workshop Monographs* (Vol. *33*), Monaco, 168 pp.

Cita, M. B. (1983). The Messinian salinity crisis in the Mediterranean: A review. In H. Berckhemer & K. J. Hsü (Eds.), *Alpine mediterranean geodynamics, Geodynamics Series* (Vol. *7*, 113–140). Washington, DC: American Geophysical Union.

Clark, P. U., Dyke, A. S., Shakun, J. D., Carlson, A. E., Clark, J., Wohlfarth, B., et al. (2009). The Last Glacial Maximum. *Science*, *325*(5941), 710–714.

Davaud, E., Jedoui, Y., & Strasser, A. (1996). Holocene peritidal and evaporitic sedimentation in southern Tunisia. *17th IAS meeting of Sedimentology, Sfax, Fieldtrip guidebook*, pp. 71–82.

Dean, W. E., & Schreiber, B. C. (1978). *Marine Evaporites: Lecture Notes for Short Course No. 4*, Tulsa, Okla: Society of Economic Paleontologists and Mineralogists, 188 pp.

DeGolyer, E. (1925). Origin of North American salt domes. *American Association of Petroleum Geologists Bulletin*, *9*(5), 831–874.

Degroot, D., Anchukaitis, K. J., Tierney, J. E., Riede, F., Manica, A., Moesswilde, E., & Nicolas Gauthier, N. (2022). The history of climate and society: A review of the influence of climate change on the human past. *Environmental Research Letters*, *17*(10), 103001. doi 10.1088/1748-9326/ac8faa

Duan, Z., & Hu, W. (2001). The accumulation of potash in a continental basin: The example of the Qarhan Saline Lake, Qaidam Basin, *West China. European Journal of Mineralogy*, *13*(6), 1223–1233.

Essefi, E., & Tagorti, M. A. (2021). Geoeconomic interest of minerals assemblage of sebkha El Melah, southeastern Tunisia. *Water Practice and Technology*, *16*(2), 633–647. https://doi.org/10.2166/wpt.2021.00

Evans, G. (1995). The Arabian Gulf: A modern carbonate-evaporite factory; A review. *Cuadernos de Ceologia Ibérica, 19*, 61–96. https://revistas.ucm.es/index.php/CGIB/article/viewFile/CGIB9595110061A/2557

Evans, G. (2011). An historical review of the Quaternary sedimentology of the Gulf (Arabian/Persian Gulf) and its geological impact. In C. G. St. C. Kendall & A. S. Alsharhan (Eds.), *Quaternary Carbonate and evaporite sedimentary facies and their ancient analogues, A tribute to Douglas James Shearman* (Vol. *43*, pp. 11–44). International Association of Sedimentologists Special Publication. doi:10.1002/9781444392326.ch2

Evans, G., Kendall, C. G. S. C., & Skipwith, P. (1964). Origin of the coastal flats, the Sabkha, of the Trucial Coast, *Persian Gulf. Nature*, *202*(4934), 759–761.

Evans, G., Schmidt, V., Bush, P., & Nelson, H. (1969). Stratigraphy and geologic history of the sabkha, Abu Dhabi, Persian Gulf. *Sedimentology*, *12*, 145–159.

Faulds, J. E., Schreiber, B. C., Langenheim, V. E., Hinz, N. H., Shaw, T. H., Heizler, M. T., et al. (2016). Paleogeographic implications of late Miocene lacustrine and nonmarine evaporite deposits in the Lake Mead region: Immediate precursors to the Colorado River. *Geosphere*, *12*(3), 721–767. doi: 10.1130 /GES01143.1.

Faulds, J. E., Schreiber, B. C., Reynolds, S. J., Gonzalez, L. A., & Okaya, D. (1997). Origin and paleogeography of an immense, nonmarine Miocene salt deposit in the Basin and Range (Western USA). *Journal of Geology*, *105*, 19–36.

Friedman, G. M. (1972). Significance of Red Sea in problem of evaporites and basinal limestones. *The American Association of Petroleum Geologists Bulletin*, *56*(6), 1072–1086.

Gavish, E. (1980). Recent sabkhas marginal to the southern coasts of Sinai, Red Sea. In A. Nissenbaum (Ed.), *Hypersaline brines and evaporitic environments, Developments in Sedimentology* (Vol. *28*, pp. 233–251). Amsterdam: Elsevier.

Goodall, T. M., North, C. P., & Glennie, K. W. (2000). Surface and subsurface sedimentary structures produced by salt crusts. *Sedimentology*, *47*(1), 99–118. doi:10.1046/j.1365-3091.2000.00279.x

Gould, S. J. (1965). Is uniformitarianism necessary? *American Journal of Science*, *263*, 223–228.

Grabau, A. W. (1924). *Principles of stratigraphy*. New York, NY: Seiler, 1185 pp.

Hall, D. J. (1995). The continent-ocean boundary in the Gulf of Mexico and the Red Sea as an analog for early Gulf History: Salt basins underlain by oceanic crust. In C. J. Travis, H. Harrison, M. R. Hudec, B. C. Vendeville, F. J. Peel, & R. E. Perkins (Eds.), *Salt, sediment and hydrocarbons, GCSSEPM Foundation, 16th Annual Research Conference* (pp. 71–81).

Handford, C. R. (1981a). A process-sedimentary framework for characterizing recent and ancient sabkhas. *Sedimentary Geology*, *30*, 255–265.

Handford, C. R. (1981b). Coastal sabkha and salt pan deposition of the lower clear fork formation (Permian), *Texas. Journal of Sedimentary Research*, *51*(3), 761–778.

Handford, C. R. (1982). Sedimentology and evaporite genesis in a Holocene continental sabkha playa basin—Bristol Dry Lake, California. *Sedimentology*, *29*, 239–253.

Handford, C. R. (1991). Marginal marine halite: Sabkhas and salinas. In J. L. Melvin (Ed.), *Evaporites, petroleum and mineral resources, Developments in Sedimentology* (Vol. *50*, Chapter 1, pp. 1–66). Amsterdam: Elsevier.

Hardie, L. A. (1984). Evaporites: Marine or non-marine. *American Journal of Science*, *283*, 193–240.

Hardie, L. A. (1987). Dolomitization: A critical view of some current views. *Journal of Sedimentary Petrology*, *57*, 166–183.

Hardie, L. A. (1990). The roles of rifting and hydrothermal $CaCl_2$ brines in the origin of potash evaporites: An hypothesis. *American Journal of Science*, *290*, 43–106.

Hardie, L. A. (1991). On the significance of evaporites. *Annual Review Earth and Planetary Science*, *290*, 43–106.

Hardie, L. A., & Eugster, H. P. (1971). The depositional environment of marine evaporites: A case for shallow, clastic accumulation. *Sedimentology*, *16*, 187–220.

Hardie, L. A., & Lowenstein, T. K. (2004). Did the Mediterranean Sea dry out during the Miocene? A reassessment of the evaporite evidence from DSDP Legs 13 and 42A Cores. *Journal of Sedimentary Research*, *74*(4), 453–461. doi:10.1306/112003740453

Hartmann, J., & Moosdorf, N. (2012). The new global lithological map database GLiM: A representation of rock properties at the Earth surface. *Geochemistry, Geophysics, Geosystems*, 13, Q12004. doi: 10.1029/2012GC004370

Homewood, P., Vahrenkamp, V., Mettraux, M., Mattner, J., Vlaswinkel, B., Droste, H., & Kwarteng, A. (2007). Bar Al Hikman: A modern carbonate and outcrop analogue in Oman for Middle East Cretaceous fields. *First Break*, *25*, 55–61.

Hovland, M., Rueslåtten, H., & Johnsen, H. K. (2018a). Large salt accumulations as a consequence of hydrothermal processes associated with "Wilson cycles": A review, Part 1: Towards a new understanding. *Marine and Petroleum Geology*, *92*, 987–1009.

Hovland, M., Rueslåtten, H., & Johnsen, H. K. (2018b). Large salt accumulations as a consequence of hydrothermal processes associated with "Wilson cycles": A review, Part 2: Application of a new salt-forming model on selected cases. *Marine and Petroleum Geology*, *92*, 128–148.

Hovland, M., Rueslåtten, H. G., Johnsen, H. K., Kvamme, B., & Kuznetsova, T. (2006). Salt formation associated with sub-surface boiling and supercritical water. *Marine and Petroleum Geology*, *23*, 855–869.

Hsü, K. J. (1972). Origin of saline giants: A critical review after the discovery of the mediterranean evaporite. *Earth-Science Reviews*, *8*, 371–396.

Hsü, K. J. (1977). Tectonic evolution of the mediterranean basins. In A. E. M. Nairn, W. H. Kanes, & F. G. Stehli (Eds.), *The ocean basins and margins, Volume 4A, The eastern Mediterranean* (pp. 29–75). New York: Springer, Plenum Press.

Hsü, K. J., Cita, M. B., & Ryan, W. B. F. (1973). The origin of the Mediterranean evaporites. In Ryan, W. B. F., Hsü, K. J. et al. (Eds.), *Initial reports of the Dead Sea Drilling Project* (Vol. *13*, pp. 1203–1231). Washington: United States Government Printing Office.

Hsü, K. J., Montadert, L., Bernoulli, D., Cita, M. B., Erickson, A., Garrison, R. E., et al. (1977). History of the Mediterranean salinity crisis. *Nature*, *267*, 399–403.

Hsü, K. J., Montadert, L., Bernoulli, D., Cita, M. B., Erickson, A., Garrison, R. E., et al. (1978). History of the Mediterranean salinity crisis. In K. J. Hsü, L. Montadert, et al. (Eds.), *Initial reports of the Deep Sea Drilling Project* (Vol. *42*, Part 1, pp. 1053–1078), Washington: U.S. Government Printing Office. doi:10.2973/dsdp.proc.42-1.149.1978

Hsü, K. J., Stoffers, P., & Ross, D. A. (1978). Messinian evaporites from the Mediterranean and Red Seas. *Marine Geology*, *26*, 71–72.

Hutton, J. (1788, 1795). *Theory of the Earth*. Global Grey, 447 pp.

Jackson, M. P. A. (1995). Retrospective salt tectonics. In M. P. A. Jackson, D. G. Roberts, & S. Snelson (Eds.), *Salt tectonics: A global perspective. American Association of Petroleum Geologists Memoir* (Vol. *65*, pp. 1–28).

Jager, J., & Visser, C. (2017). Geology of the Groningen field—An overview. *Netherlands Journal of Geosciences (Geologie in Mijnbouw)*, *96*(5), s3–s15.

Jolivet, L., Augier, R., Robin, C., Suc, J.-P., & Rouchy, J. M. (2006). Lithospheric-scale geodynamic context of the Messinian salinity crisis. *Sedimentary Geology*, *188–189*, 9–33. doi:10.1016/j.sedgeo.2006.02.004

Kendall, A. C. (1978a). Facies Models 11. Continental and Supratidal (Sabkha) evaporites. In R.G. Walker (Ed.), *Facies Models* (Vol. *5*, No.2, pp. 66–78). Geoscience Canada.

Kendall, A. C. (1978b). Facies Models 12. Subaqueous evaporites. In R. G. Walker (Ed.), *Facies Models* (Vol. *5*, No. 3, pp. 124–139). Geoscience Canada.

Kendall, A. C. (1984). Evaporites. In R. G. Walker (Ed.), *Facies Models* (Vol. *1*, pp. 259–296). Geoscience Canada, Reprint Series.

Kendall, A. C. (1992). Evaporites. In R. G. Walker & N. P. James (Eds.), *Facies Models: Response to sea level change* (pp. 375–409). Geological Association of Canada.

Kendall, A. C. (2010). Marine evaporites. In N. P. James & R. W. Dalrymple (Eds.), *Facies Models* (Vol. *6*, pp. 505–539). Geological Association of Canada, Series Geotext.

Kendall, A. C., & Harwood, G. M. (1996). Marine evaporites: Arid shorelines and basins. In H. G. Reading (Ed.), *Sedimentary environments: Processes, facies and stratigraphy* (3rd ed., pp. 281–324). Oxford: Blackwell Scientific Publications.

Kendall, C. G. C., Alshrahan, A. S., & Marlow, L. (2014). Stratigraphy and depositional systems of the Southern Tethyan Region. In L. Marlow, C. Kendall, & L. Yose (Eds.), *Petroleum systems of the Tethyan region, American Association of Petroleum Geologists, Memoir* (Vol. *106*, pp. 29–57).

Kendall, C. G. St. C., & Skipwith, Bt., P. A. d'E. (1969). Geomorphology of a recent shallow-water carbonate province: Khor Al Bazam, Trucial Coast, Southwest Persian Gulf. *Geological Society of America Bulletin*, *80*(5), 865–892. https://doi.org/10.1130/0016-7606(1969)80[865:GOARSC]2.0.CO;2

Kendall, C. G. St. C., Weber, L. J., & Alsharhan, A. S. (2009). The Giant Oil Field Evaporite Association: A function of the Wilson cycle, climate, basin position and sea level. *AAPG Search and Discovery*, Article #40471 (2009).

Kinsman, D. J. J. (1969). Modes of formation, sedimentary associations, and diagnostic features of shallow-water and supratidal evaporites. *American Association of Petroleum Geologists Bulletin*, *53*(4), 830–840.

Kinsman, D. J. J. (1975). Salt floors to geosynclines. *Nature*, *255*, 375–378 (29 May 1975).

Kostick, D. S. (1993). The material flow of salt. *United States Department of the Interior, Bureau of Mines*, Information Circular/1993, IC 9343, 32 pp.

Last, W. M. (1989). Continental brines and evaporites of the northern Great Plains of Canada. *Sedimentary Geology*, *64*, 207–221.

Lugli, S. (2009). Evaporites. In Gornitz, V. (Eds.), *Encyclopedia of paleoclimatology and ancient environments, Encyclopedia of Earth Sciences Series* (pp. 321–325). Dordrecht: Springer. https://doi.org/10.1007/978-1-4020-4411-3_84

Lugli, S., Manzi, V., Roveri, M., & Schreiber, B. C. (2015). The deep record of the Messinian salinity crisis: Evidence of a non-desiccated Mediterranean Sea. *Palaeogeography, Palaeoclimatology, Palaeoecology*, *433*, 201–218. http://dx.doi.org/10.1016/j.palaeo.2015.05.017

Lyell, C. (1830). Principles of geology, being an attempt to explain the former changes of the Earth's surface, by reference to causes now in operation, *Murray*. London, 385 pp.

McKenzie, J. A. (1981). Holocene dolomitization of calcium carbonate sediments from the coastal Sabkhas of Abu Dhabi, U.A.E.: A stable Isotope study. *The Journal of Geology*, *89*(2), 185–198. https://www.jstor.org/stable/30078285

Meilijson, A., Steinberg, J., Hilgen, F., Bialik, O. M., Waldmann, N. D., & Makovsky, Y. (2018). Deep-basin evidence resolves a 50-year-old debate and demonstrates synchronous onset of Messinian evaporite deposition in a non-desiccated Mediterranean. *Geology*, *46*(3), 243–246.

Mettraux, M., Homewood, P. W., Kwarteng, A. Y., & Mattner, J. (2011). Coastal and continental sabkhas of Barr Al Hikman, Sultanate of Oman. In C. G. St. C. Kendall, & A. S. Alsharhan (Eds.), *Quaternary carbonate and evaporite sedimentary facies and their ancient analogues, A tribute to Douglas James Shearman* (Vol. *43*, pp. 183–204). International Association of Sedimentologists Special Publication.

Moati, M. A., Al Mansoori, A., Al Cibahy, A., Al Harthi, S., Foster, K., Purkis, S., et al. (2008). *Conservation and management plan for Abu Dhabi and Eastern Qatar Coral Reefs*, Supreme Council for the Environment and Natural Reserves (SCENR) of the State of Qatar. Environment Agency of Abu Dhabi Emirate (EAD), National Coral Reef Institute (NCRI), Emirates Wildlife Society in association with the World Wide Fund for Nature (EWS-WWF), 70 pp.

Mohriak, W. U. (2014). *Birth and development of continental margin basins: Analogies from the South Atlantic, North Atlantic, and the Red Sea*. AAPG Search & Discovery, Article #41502, Posted December 22, 2014.

Mohriak, W. (2025). *Salt in the Earth sciences: Basin analysis and salt tectonics*. Washington, DC: American Geophysical Union. doi:10.1002/9781394307067

Murray, G. E. (1966). Salt structures of Gulf of Mexico Basin—A review. *American Association of Petroleum Geologists Bulletin*, *50*(3), 439–475.

Nettleton, L. L. (1934). Fluid mechanics of salt domes. *American Association of Petroleum Geologists Bulletin*, *18*(9), 1175–1204.

Nettleton, L. L. (1943). Recent experimental and geophysical evidence of mechanics of salt dome formation. *American Association of Petroleum Geologists Bulletin*, *27*(1), 51–63.

Ochsenius, K. (1877). Die Bilding der Steinsalzlager und ihrer mutter augensalze. *C.E.M. Pfeffer, Halle*, 172 pp.

Ochsenius, K. (1888). On the formation of rock salt beds and mother liquor salts. *Philadelphia Academic Sciences, Proceedings*, Pl *2*. 181–187.

Perthuisot, J. P. (1971). Recent polyhalite from Sebkha el Melah (Tunisia). *Nature*, *232*, 186–187.

Perthuisot, J. P. (1981). Les vicissitudes conceptuelles sur la genèse d'un bassin salin: la sebkha El Melah (Tunisie). Comité Français d'Historie de la géologie, Première série,

Philip, D., Hughes, P. D., Gibbard, P. L., & Ehlers, J. (2013). Timing of glaciation during the last glacial cycle: Evaluating the concept of a global "Last Glacial Maximum" (LGM). *Earth-Science Reviews*, *125*, 171–198.

Pirouz, M., Simpson, G., Bahroudi, A., & Azhdari, A. (2011). Neogene sediments and modern depositional environments of the Zagros foreland basin system. *Geological Magazine*, *148*(5–6), 838–853. doi:10.1017/S0016756811000392

Puls, D. D., Jameson, J., Kozar, M., Al-Ansi, H., & LeBlanc, J. (2009). *The Dukhan Sabkha: A modern analog for the Arab C carbonate reservoir, Dukhan Field, Qatar. International Petroleum Technology Conference* (pp. 1–13), IPTC 13629.

Purser, B. H. (1973). Sedimentation around bathymetric highs in the southern Persian Gulf. In B. H. Purser (Ed.), *The Persian Gulf: Holocene carbonate sedimentation and diagenesis in a shallow epicontinental sea* (pp. 157–177). Berlin: Springer.

Renaut, R. W., & Gierlowski-Kordesch, E. H. (2010). Lakes. In N. P. James & R. W. Dalrymple (Eds.), *Facies Models* (pp. 541–575). Geological Association of Canada, Series Geotext, 6.

Romano, M. (2015). Reviewing the term uniformitarianism in modern Earth sciences. *Earth-Science Reviews*, *148*, 65–76. http://dx.doi.org/10.1016/j.earscirev.2015.05.010

Ruiz, F., Abad, M., Galán, E., González, I., Aguilá, I., Olía, M., et al. (2006). The present environmental scenario of El Melah Lagoon (NE Tunisia) and its evolution to a future sabkha. *Journal of African Earth Sciences*, *44*, 289–302.

Saleh, A., Al-Ruwaih, F., Al-Reda, A., & Gunatilaka, A. (1999). A reconnaissance study of a clastic coastal sabkha in Northern Kuwait, Arabian Gulf. *Journal of Arid Environments*, *43*(1), 1–19. doi:10.1006/jare.1999.0529

Schmalz, R. F. (1969). Deep-water evaporite deposition: A genetic model. *American Association of Petroleum Geologists Bulletin*, *53*(4), 798–823.

Schmalz, R. F. (1991). The Mediterranean salinity crisis: Alternative hypotheses. *Carbonates and Evaporites*, *6*(2), 121–126.

Schreiber, B. C. (1978). Environments of subaqueous gypsum deposition. In W. E. Dean, & B. C. Schreiber (Eds.), *Marine evaporites, Lecture Notes for Short Course* (Vol. *4*, pp. 43–73). Tulsa, Okla: Society of Economic Paleontologists and Mineralogists.

Schreiber, B. C. (1986). Arid shorelines and evaporites. In H. G. Reading (Ed.), *Sedimentary environments and facies* (pp. 189–228). Oxford, UK: Blackwell Scientific Publications.

Schreiber, B. C. (1988). *Evaporites and hydrocarbons*. New York: Columbia University Press, 475 pp.

Schreiber, B. C., & Friedman, G. M. (1976). Depositional environments of Sicily as determined from analysis of intercalated carbonates. *Sedimentology*, *23*, 255–270.

Schreiber, B. C., & Hsü, K. J. (1980). Evaporites. In G. D. Hobson (Ed.), *Developments in petroleum geology* (Vol. *2*, pp. 87–138). London: Applied Science Publishers.

Schreiber, B. C., & Marshak, S. (1981). Evaporites, carbonates and oil. In *Geology of passive continental margins, AAPG Education Course Note Series* (Vol. *19*, pp. 6.1–6.32).

Schreiber, B. C., Bąbel, M., & Lugli, S. (2007). Introduction and overview. In B. C. Schreiber, S. Lugli, & M. Bąbel (Eds.), *Evaporites through space and time* (Vol. *285*, pp. 1–13). Geological Society, London, Special Publication. doi: 10.1144/SP285.1

Scribano, V., Carbone, S., Manuella, F. C., Hovland, M., Rueslåtten, H., & Johnsen, H. K. (2017). Origin of salt giants in abyssal serpentinite systems. *International Journal of Earth Sciences*, 1–14.

Scribano, V., Carbone, S., & Manuella, F. C. (2018). Tracking the Serpentinite Feet of the Mediterranean Salt Giant. *Geosciences*, *8*, 352. doi:10.3390/geosciences8090352

Serra, O. (1984). *Fundamentals of Well-Log Interpretation. 1. The Acquisition of Logging Data. Developments in Petroleum Science* (Vol. *15A*). Amsterdam, The Netherlands: Elsevier Science Publishers B.V., 423 pp.

Serra, O. (1986). *Fundamentals of well-log interpretation. 2. The interpretation of logging data. Developments in Petroleum Science* (Vol. *15B*). Amsterdam, The Netherlands: Elsevier Science Publishers B.V., 679 pp.

Shea, J. H. (1982). Twelve fallacies of uniformitarianism. *Geology*, *10*(9), 455–460. doi:10.1130/0091-7613(1982)10 <455:tfou>2.0.co;2

Shearman, D. J. (1966). Origin of marine evaporites by diagenesis. *Transactions of the Institute of Mining and Metallurgy*, *B75*, 207–215.

Shearman, D. J. (1978). Evaporites of coastal sabkhas. In W. E. Dean & B. C. Schreiber (Eds.), *Marine evaporites, Lecture Notes for Short Course* (Vol. *4*, pp. 6–42). Tulsa, Okla: Society of Economic Paleontologists and Mineralogists.

Silver, B. A., & Todd, R. G. (1969). Permian cyclic strata, Northern Midland and Delaware Basins, West Texas and Southeastern New Mexico. *American Association of Petroleum Geologists Bulletin*, *53*(11), 2223–2251.

Sonnenfeld, P., & Perthuisot, J. -P. (1989). *Brines and evaporites. Short Course in Geology, Volume 3, 28th International Geological Congress*, American Geophysical Union, Washington, DC, 126 pp.

Soto, J. I., Flinch, J. F., & Tari, G. (2017). Permo-Triassic Basins and Tectonics in Europe, North Africa and the Atlantic Margins: A synthesis. In J. I. Soto, J. F. Flinch, & G. Tari (Eds.), *Permo-Triassic Salt Provinces of Europe, North Africa and the Atlantic Margins Tectonics and Hydrocarbon Potential* (pp. 3–41). Amsterdam: Elsevier. http://dx.doi.org/10.1016/B978-0-12-809417-4.00038-0

Stauble, A. J., & Milius, G. (1970). Geology of Groningen Gas Field, Netherlands. In M. T. Halbouty (Ed.), *Geology of Giant Petroleum Fields, AAPG Memoir* (Vol. *14*, pp. 359–369).

Talbot, C. J. (1998). Extrusions of Hormuz salt in Iran. In D. J. Blundell & A. C. Scott (Eds.), *Lyell: The past is the key to the present* (Vol. *143*, pp. 315–334). Geological Society, London, Special Publication.

Taylor, J. C. M. (1985). *The role of evaporites in hydrocarbon exploration*. Joint Association of Petroleum Exploration Courses (UK)—Japec Course No. 39.

Trusheim, F. (1960). Mechanism of salt migration in northern Germany. *American Association of Petroleum Geologists Bulletin*, *44*(9), 1519–1540.

Usiglio, J. (1849a). Analyse de l'eau de la Mediterraneé sur le côtes de France. *Annales de Chemie et de Physique, Ser. 3*, 27, 92–107.

Usiglio, J. (1849b). Etudes sur la composition de l'eau de la Mediterranee et sur l'exploitation des sel quy'elle conteint. *Annales de Chemie et de Physique, Ser. 3*, 27, 172–191.

Warren, J. K. (1982a). The hydrological setting, occurrence and significance of gypsum in late Quaternary salt lakes in South Australia. *Sedimentology*, *29*, 609–637.

Warren, J. K. (1982b). The hydrological significance of Holocene tepees, stromatolites, and boxwork limestones in coastal salinas in South Australia. *Journal of Sedimentary Petrology*, *52*, 1171–1201.

Warren, J. K. (1983). Tepees, modern (Southern Australia) and ancient (Permian-Texas and New Mexico)—A comparison. *Sedimentary Geology*, *34*, 1–19.

Warren, J. K. (1986). Shallow-water evaporitic environments and their source rock potential. *Journal of Sedimentary Petrology*, *56*(3), 442–454.

Warren, J. K. (1989). *Evaporite sedimentology: Importance in hydrocarbon accumulation*. New Jersey, USA: Prentice Hall Advanced Reference Series, 285 pp.

Warren, J. K. (1991). Sulfate dominated sea-marginal and platform settings: Sabkhas and salinas, mudflats and salterns. In J. L. Melvin (Ed.), *Evaporites, petroleum and mineral resources, Developments in Sedimentology* (Vol. *50*, Chapter 2, pp. 69–187). Amsterdam: Elsevier.

Warren, J. K. (1992). Evaporites and their importance to petroleum exploration. *IHRDC, GL 508 Petroleum Geology*, 234 pp.

Warren, J. K. (2000). Evaporites, brines and base metals: Low-temperature ore emplacement controlled by evaporite diagenesis. *Australian Journal of Earth Sciences*, *47*, 179–208.

Warren, J. K. (2006). *Evaporites: Sediments, resources and hydrocarbons*, 1035p, Berlin: Springer Verlag.

Warren, J. K. (2010). Evaporites through time: Tectonic, climatic and eustatic controls in marine and nonmarine deposits. *Earth-Science Reviews*, *98*, 217–268. doi:10.1016/j.earscirev.2009.11.004

Warren, J. K. (2012). Understanding hydrocarbons in an evaporite context. *AAPG Search and Discovery*, Article #50549 (2012).

Warren, J. K. (2014). *Evaporite sedimentology: Importance in hydrocarbon accumulation*. New Jersey, USA: Prentice Hall Advanced Reference Series, 285 p.

Warren, J. K. (2016). *Evaporites—A geological compendium*. Berlin: Springer, 1807 pp. ISBN 978-3-319-13511-3.

Warren, J. K. (2021). Evaporite deposits. In S. Elias, & D. Alderton (Eds.), *Encyclopedia of geology* (2nd ed., pp. 945–977). Amsterdam: Elsevier. https://doi.org/10.1016/B978-0-08-102908-4.00165-X

Warren, J. K., & Kendall, C. G. St. C. (1985). Comparison of sequences formed in marine sabkha (subaerial) and salina (subaqueous) settings-modern and ancient. *American Association of Petroleum Geologists Bulletin*, *69*(6), 1013–1023.

West, I. M. (1975). Evaporites and associated sediments of the basal Purbeck Formation (Upper Jurassic) of Dorset. *Proceedings of the Geologists' Association*, *86*(2), 205–225 (updated in 2012).

West, I. M. (2012). Qatar—Geology, Sabkhas, Evaporites and Desert environments. Internet Webpage. http://www.soton.ac.uk/~imw/Qatar-Sabkhas.htm

West, I. M., Ali, Y. A., & Hilmy, M. E. (1979). Primary gypsum nodules in a modern sabkha on the Mediterranean coast of Egypt. *Geology*, *7*, 354–358.

West, I. M., Ali, Y. A., & Hilmy, M. E. (1983). Facies associated with primary gypsum nodules of northern Egyptian sabkhas. *Sixth International Symposium on Salt* (Vol. *1*, pp. 171–183). Salt Institute.

Whitmarsh, R. B., Weser, O. E., Ross, D. A., Ali, S., Boudreaux, J. E., Coleman, R., et al. (1974). *Initial reports of the Deep Sea Drilling Project* (Vol. *23*). Washington (U.S. Government Printing Office), 1180 pp.

Wu, Z., Yin, H., Wang, X., Zhao, B., Zheng, J., Wang, X., & Wang, W. (2015). The structural styles and formation mechanism of salt structures in the Southern Precaspian Basin: Insights from seismic data and analog modeling. *Marine and Petroleum Geology*, *62*, 58–76. http://dx.doi.org/10.1016/j.marpetgeo.2015.01.010

Yechieli, Y., & Wood, W. W. (2002). Hydrogeologic processes in saline systems: Playas, sabkhas, and saline lakes. *Earth-Science Reviews*, *58*, 343–365.

9
Saline Lakes and Salt Deposition in Extensional Settings

ABSTRACTS

Salt deposition in extensional settings is primarily linked to lacustrine environments in rift basins, such as the hypersaline lakes in the East African Rift System, and some foreland basins associated with plate convergence, like the Tibetan Plateau in China or the salt flats in the Andean region of South America. Salt basins with significant halite and gypsum accumulation are often associated with hypersaline lakes near passive continental margins, like the northwest Australian margin or the Danakil Depression in the southern Red Sea. The Afar region, located at the triple junction of the East African rifts, the Red Sea, and the Gulf of Aden, also hosts modern halite deposits in lacustrine environments, such as Lake Asal in Djibouti.

Continental lacustrine evaporite deposits are found in various sedimentary environments, including sabkhas in North Africa and the Persian Gulf (see Chapter 8), as well as lacustrine systems and lagoons in arid and subtropical regions. In sabkha environments, modern evaporite deposition occurs in subaerial flats adjacent to inland seas or mountainous regions far from the oceans (see Chapter 10). The process starts by replacing the sedimentary matrix with carbonate and sulfate nodules, which grow and displace the previous mineral framework (see Chapter 8). Salt deposition in saline lakes mainly occurs in subaquatic environments. Hypersaline lagoons periodically dry out, and salinas form within permanent lakes when the brine is concentrated and saturated to the point of precipitating diverse evaporite minerals on the lake's bottom (see Chapter 4).

A salina corresponds to a subaqueous setting for salt deposition, with seawater influence when the basins are located in coastal sabkhas adjacent to marine gulfs and shallow inland seas. Lagoons are commonly fed by marine waters, such as the MacLeod Lake in Northwest Australia (Handford, 1991; Logan, 1987). There are many continental lakes with no seawater input, resulting in salinas with a suite of evaporite minerals typical of nonmarine minerals, such as the hypersaline lakes in the Basin and Range Province of the Southwestern United States (Renaut & Gierlowski-Kordesch, 2010; Renaut & Tiercelin, 1993; Renaut et al., 2012, 2021; Smoot & Lowenstein, 1991; Warren, 2006).

Carbonate, sulfate, and chloride minerals precipitate underwater at different brine concentrations, exceeding the salt solubility (Handford, 1991; Warren, 1989, 1992, 1997, 2000, 2006, 2010; Warren & Kendall, 1985). The salina evaporites are commonly formed in arid regions when the water body is isolated from marine or continental water inflow. The hypersaline brines in salinas are concentrated by solar evaporation, resulting in the precipitation of massive, homogeneous salt layers on the lake's bottom and margins (Warren, 2006, 2016).

Salterns are evaporite-filled seaways, gulfs, or lakes that accumulate salt in subaqueous environments and are exploited industrially in saltworks to produce valuable salt minerals, particularly halite and potassium-bearing salts. They are associated with large, deep lakes, inland seas, gulfs, continental platforms, and basinwide marine settings that precipitated vast amounts of halite and other evaporite minerals. Most authors interpret that there are no modern analogs for these settings. Evaporite deposition in the Holocene is mainly associated with small hypersaline lakes and coastal sabkhas (Kendall, 1992, 2010; Renaut & Gierlowski-Kordesch, 2010; Warren, 2006, 2016).

The sediments precipitated in hypersaline lakes are formed from increasingly concentrated brines, resulting in the accumulation of the most soluble minerals on top

Salt in the Earth Sciences: Evaporite Rocks and Salt Deposition, Geophysical Monograph 286, First Edition. Webster Mohriak.

DOI: 10.1002/9781119405245.ch09

of the evaporite cycle (see Chapter 4). These cycles range from a few centimeters to tens of meters in thickness. Salt layers are often capped by thin subaerial sabkha sequences or by a salt pan, corresponding to a desiccated salt basin covered by gypsum or halite crusts with no water in the system (see Chapter 8). Modern examples include Lake MacLeod in NW Australia (Logan, 1987) or the El Melah lagoon and associated sabkha environments in Tunisia (Alsharhan & Kendall, 1994; Ruiz et al., 2006).

Hypersaline lakes are shallow, highly saline water bodies that precipitate carbonate and evaporite minerals along the margins and in the basin center. They are found along coastal regions and in intracontinental settings far from marine influence, such as the Basin and Range Province of the southwestern United States or the Great Salt Lake in Utah (e.g., Carozzi, 1962; Nicoll, 2021; Spencer et al., 1984). These environments correspond to shallow-water lakes, filled with only a few meters of water. When desiccated, these lakes form sedimentary facies, forming a bull's-eye pattern (e.g., Hsü, 1972; Schmalz, 1969, 1970). The only modern deep lake with halite deposition is the Dead Sea, between Israel and Jordan, which is characterized by a deep northern lake and an extremely shallow saltern in the southern lake (see Chapter 2).

Evaporite facies distribution is marked by concentric belts with more soluble salt in the center, resulting from the evaporation of the entire water body in an endorheic basin. These evaporite facies include carbonate rocks fringing the lake margins and a belt around the central core, associated with the precipitation of gypsum, which may occur as loose crystals (gypsarenite) or massive gypsum (selenite). During periods of maximum aridity, when the brine concentration exceeds the mineral solubility, the lake's center precipitates soluble minerals (see Chapters 4 and 11).

Some lakes develop in divergent plate tectonic settings during the last stages of the rift phase that precedes oceanic crust development or during the early stages of oceanization. Examples include Lake Asal (also known as Lake Assal, or Asal Lake) in Djibouti and the Danakil Depression in Ethiopia. Both regions are characterized by volcanic barriers separating the water bodies from the open sea. Other lakes are remnants of larger inland seas, such as Kara Bogaz Gol in Turkmenistan, located adjacent to the Caspian Sea (see Chapter 10).

Five major types of continental saline lakes associated with plate divergence can be distinguished: (1) lakes in continental rifts far from oceanic waters; (2) lakes in arid desert regions; (3) lakes adjacent to passive continental margins; (4) lakes in active rifts near gulfs and incipient divergent margins; and (5) lakes that are remnants of larger inland seas or continental lacustrine systems, such as the Great Salt Lake.

This chapter addresses a few modern examples of the most important types of hypersaline lakes and presents a summary of the sedimentary evolution of the selected basins to highlight the geological controls on salt deposition.

9.1. ALLUVIAL FAN-EPHEMERAL SALINE LAKES IN CONTINENTAL RIFTS

The Basin and Range province, located in the southwestern United States, is characterized by a continental environment with alluvial fan and saline lake systems, which are often linked to active tectonic processes forming rifts or foreland basins (Figure 9.1). Several hypersaline lakes are observed in this region, which has been affected by various extensional, strike-slip, and contractional tectonic events during different geological periods (Faulds et al., 1997, 2016). Water inflow into depressions is primarily from groundwater seepage supplied from the adjacent uplifted mountain ranges and occasional storm runoff, as observed in the Death Valley of eastern California (Hooke, 1972; Knott et al., 2005).

Death Valley is a steep-sided depression with sabkhas and hypersaline lakes, with its floor 86 m below sea level and forming one of the driest spots in the southwestern United States, with an annual rainfall of less than 5 cm (2 in). Dry mudflats and salt pans are characterized by the deposition of various evaporite minerals, such as glauberite, gypsum, halite, and minor calcite, which have been accumulating for more than 10,000 years (10 ka) due to climatic changes following the demise of the last glaciation (Hooke, 1972; Knott et al., 2005). A perennial lake with a maximum water depth of around 90 m existed between 35 and 10 ka, as evidenced by mud deposits above the present saline pan surface (Li et al., 1996; Lowenstein et al., 1999). Since 10 ka, the paleolake has dried out, resulting in a mudflat with a saline pan covered by an evaporite crust (Figure 9.2). The Badwater salt pan, the largest modern lake in the Death Valley region (Figure 9.3), records several climate fluctuations during the Holocene.

A borehole drilled in the Death Valley salt pan penetrated subsurface sediments and recovered 185 m of evaporitic sediments, indicating several desiccation events followed by replenishment with water in wet periods, forming several evaporite cycles (Li et al., 1996). The sedimentary sequences in the subsurface correspond to a saline pan composed of different lithological facies (Li et al., 1996).

The most important lithological units in the subsurface are represented by interbedded halite, chaotic muddy halite, and siliciclastic deposits, predominantly composed of mud and silt. These sediments were deposited in a changing saline lacustrine system that involved different cycles of aridity and water inflow to the basin (Li et al., 1996; Lowenstein et al., 1999; Roberts et al., 1994).

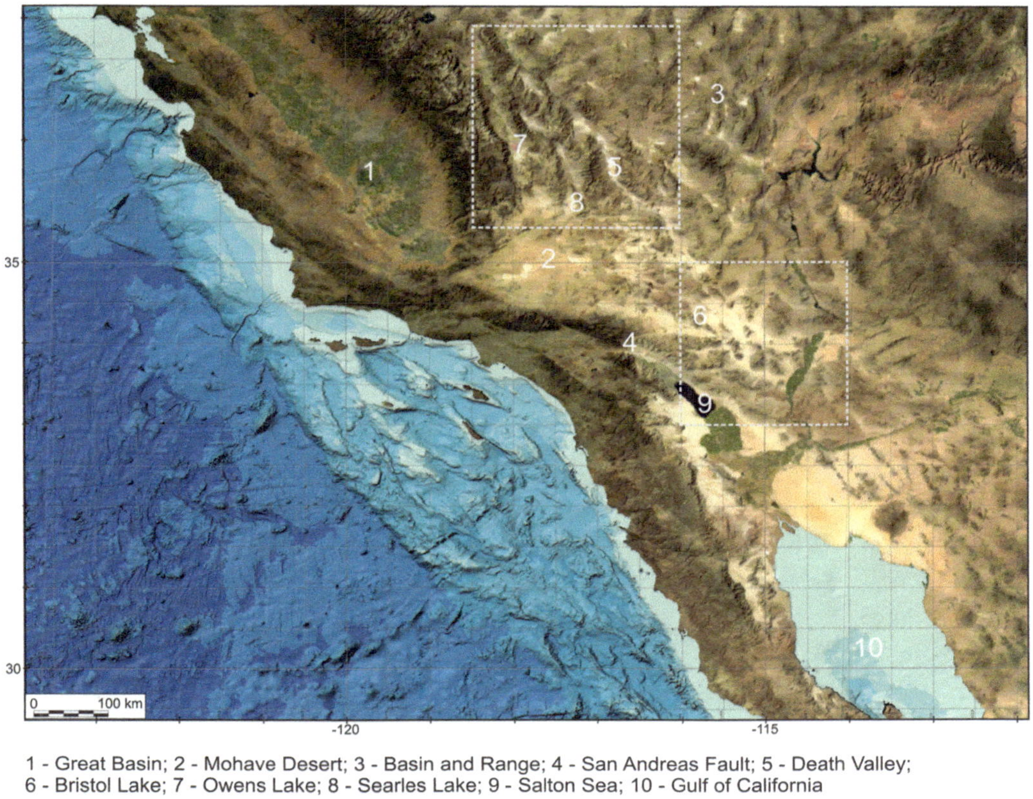

Figure 9.1 The satellite image depicts the southwestern region of the United States, including the Great Basin near the Pacific Coast and the Gulf of California. The Mohave Desert, north of the Gulf of California and south of the Great Basin, is characterized by the San Andreas transcurrent fault zone. The larger rectangle shows the region adjacent to Death Valley in California, while the smaller rectangle shows Bristol Lake, northeast of the Salton Sea, appearing as a dark blue lake in the satellite image. The smaller Owens and Searles lakes are located on the southern border of the large rectangle. Credit: With permission of The General Bathymetric Chart of the Oceans.

Radiometric dating by Uranium-series isochron methods of the drilling core sediments indicated that the borehole penetrated evaporitic mudstone sequences at the bottom, with an estimated age of circa 100 ka (Li et al., 1996). Some researchers (e.g., Lowenstein et al., 1999) have proposed that the interval 60 to 10 ka might be characterized by a generally wet period when Death Valley had a perennial lake with mud layers and carbonate sediments intercalated with minor halite deposition. The halite layers recovered from the drilling cores correlate with several hypersalinity events recorded in the basin (Li et al., 1996; Lowenstein et al., 1999). There is evidence of a major desiccation event in Death Valley between 60 and 35 ka, which resulted in evaporite deposits preserved in the subsurface as thick layers of bedded halite (Lie et al., 1996).

The core interval between 18 and 61 m, dated as 35–60 ka, corresponds to poorly bedded halite, indicating that during this interval, the hypersaline brines precipitated halite with some mud intercalations, suggesting a salt pan with an ephemeral lake that was repeatedly desiccated and flooded by dilute waters (Li et al., 1996). Overlying the halite layer, black lacustrine muds are observed between 35 and 10 ka, suggesting a deepwater environment and a wetter interval. However, the last 10 ka are characterized by an ephemeral lake with a salina depositing multiple intervals of halite layers (Li et al., 1996). These periods of salt deposition show that aridity increased during the Holocene. As a result, salt pans dried out during the hottest parts of the aridity events and formed crusts of efflorescent salt (Li et al., 1996).

9.2. EPHEMERAL SALINE LAKES IN ARID DESERTS

In lacustrine settings with permanent bodies of water, the sediments are mostly made up of siliciclastic sand and carbonate mud, with some intercalations of evaporite rocks. Significant evaporite deposits are formed when lakes are located in arid regions and isolated from marine influence or fluvial drainage. When the lake brines are

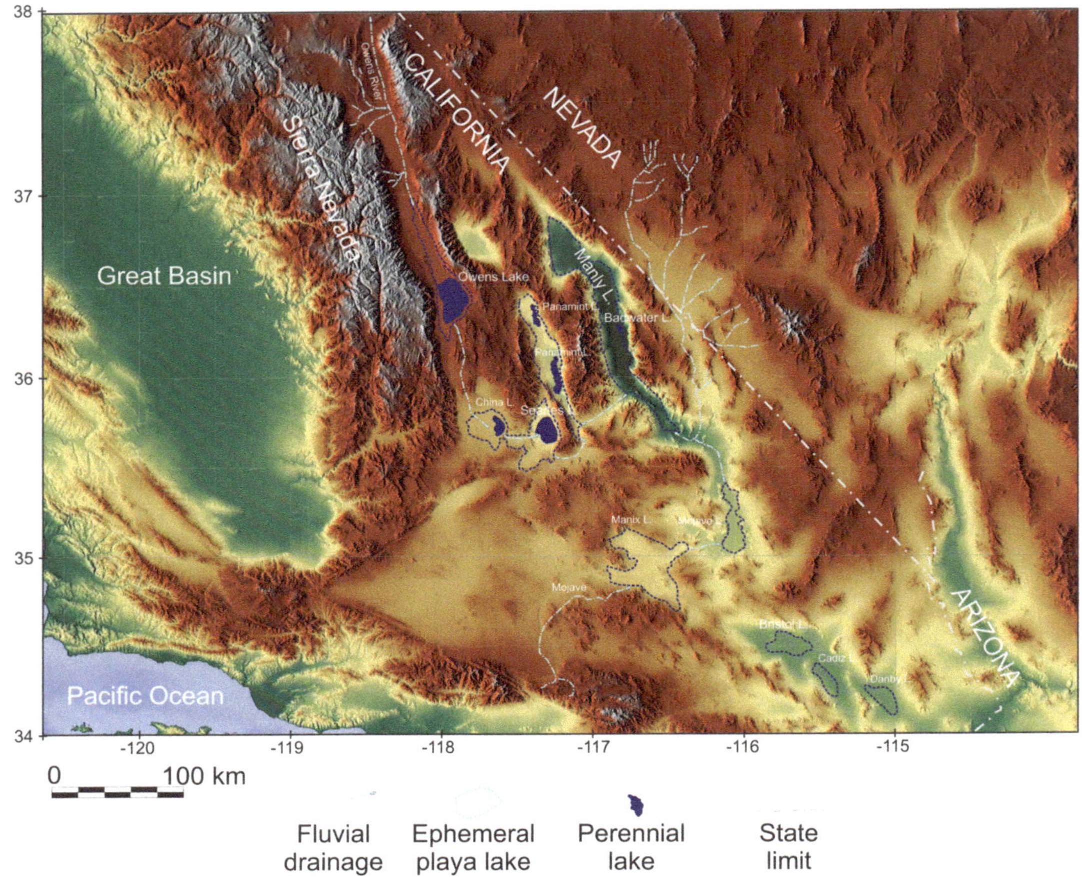

Figure 9.2 The shaded relief topographic map depicts the southwestern region of the United States, highlighting the Pleistocene-Holocene paleo-drainage systems. Lake Manly, a large paleolake, occupied the Death Valley rift valley depression around 12 ka. After the Wisconsin glaciation, the lake dried out, leaving the remnant Badwater Lake occupying the lowest part of the depression. Other smaller lakes in California include Owens, Searles, and Bristol. Most lakes eastward of the Mojave Desert were ephemeral playa lakes. Credit: Smith (1979)/with permission of USGS.

concentrated due to solar evaporation, relatively thick (more than 1 m) layers of evaporite minerals (gypsum and halite) may accumulate in the salinas. The bull's-eye model is observed in isolated, endorheic evaporitic basins or in basins with partial connection with an open sea or gulf, resulting in different facies distribution (Figure 9.4).

In a vertical sequence, ephemeral braided stream floodplain sequences are the dominant sedimentary facies. Pluvial, fluvial, and storm runoff predominantly control water inflow in these desert environments. Sporadic catastrophic storm events are often spaced several years or even decades apart (Renaut & Gierlowski-Kordesch, 2010; Warren, 2006, 2016).

Ephemeral saline lakes occur in many active rift settings worldwide, such as the Basin and Range province in the southwestern United States and the Danakil Depression in Ethiopia, south of the Red Sea (see Chapter 8). These arid regions with dune fields and mudflats are located in tectonic depressions that are periodically flooded by continental pluvial and fluvial waters, as in the Basin and Range Province (Figure 9.1), or by seawater in regions adjacent to gulfs or oceans, as in the Danakil Depression (e.g., Holwerda & Hutchinson, 1968; Tazieff et al., 1972). Some hypersaline lakes are characterized by active rifting, forming depressions below the global sea level, which are periodically flooded by seawater, as observed in Lake Asal in Djibouti (Gasse & Fontes, 1989; Jackson et al., 2000).

Lacustrine systems in the Basin and Range Province are associated with tectonic depressions flanked by uplifted rift flanks (see Chapter 11). The highlands bordering the depressions are characterized by alluvial fans developed in canyons that cut the mountain ranges, forming fluvial drainage floodplains and lacustrine systems that accumulate siliciclastic and evaporitic sediments on the lake's floor. These regions are characterized by mudflats and sabkha environments formed adjacent to continental deserts and dune fields (see Chapter 8). This is the

Figure 9.3 Top: NASA satellite images (natural color) show the Death Valley lacustrine system and evaporite deposition during wet (2005) and dry (2007) climate episodes. In 2005 (A), the desert valley received four times the average annual rainfall, resulting in ephemeral lakes filled with water, such as Badwater Lake. Halite crusts covered the depression during basin desiccation episodes in 2007 (B). Evaporite sediments initially displayed a whitish color but became dark gray over time. Bottom: Photographic images of the Badwater Basin, located in the central to southern Death Valley depression. The lower left (C) is a view toward the west (Panamint Mountains in the background). A salt pan covers the Badwater Basin, with a lake level 86 m below the global sea level, making it the lowest depression in North America. It is periodically flooded during periods of wet climate, as in the 2005 rainy season. Climate aridity resulted in rapid evaporation of water in the ephemeral lake. The lower right image (D) shows the dried-up salt pan covered with a halite-rich mud crust. The salt crust is characterized by polygons measuring tens of centimeters, sometimes reaching more than a meter in length. Credits: (A), (B) courtesy of NASA and (C), (D) courtesy of USGS.

case of the Mojave Desert (Figure 9.1), where several ephemeral lakes are periodically desiccated due to extreme aridity and the surrounding mountain ranges are dissected by canyons that carry coarse sediments that are deposited in alluvial fans, while the fine-grained sediments are deposited in the mudflats (see Chapter 11).

Examples of continental sabkhas associated with ephemeral saline lakes in arid regions include the Bristol Dry Lake and Owens Lake in California, which provide modern analogs for ancient saline lakes (Figure 9.1). Modern evaporitic lacustrine systems fed by marine water also exist along the continental margins of several continents. Lake MacLeod, located at the northwestern extremity of Australia, corresponds to a hypersaline lake that accumulated halite-rich evaporitic sedimentary sequences in the Holocene salinas. The Dallol Depression, located in the southwestern margin

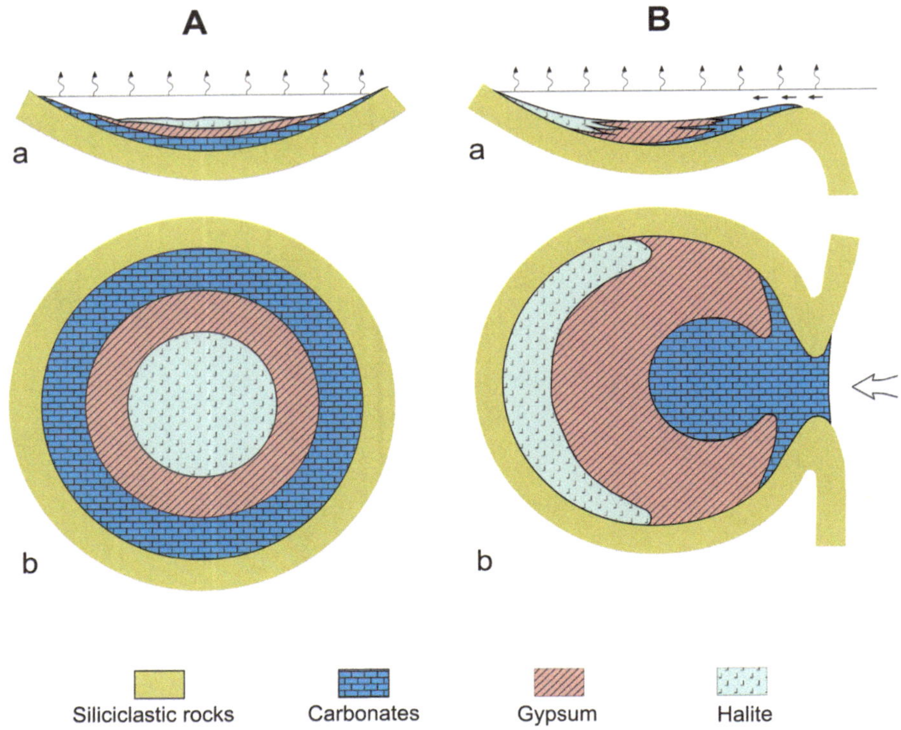

Figure 9.4 Schematic diagrams showing lacustrine evaporite facies in isolated basins with a bull's-eye pattern (A-a). The tear-drop pattern (B-a) is observed in basins with partial connection with an open sea or gulf. The more soluble salt deposits are located at the basin center, with carbonate rocks forming a circular belt at the basin periphery (A-b). The tear-drop pattern (B-b) is marked by an asymmetry in the evaporite facies. The soluble salt deposits are located far away from the channel where water flows into the basin, and the less soluble evaporites (carbonate rocks and gypsum) are located near the entrance. Credit: Hsu (1972)/with permission of Elsevier.

of the Red Sea, corresponds to a modern example of a coastal lagoon that was desiccated in the Holocene, due to extreme climate aridity and basin isolation (Bastow et al., 2018; Bonatti et al., 1971, 2016; Cavalazzi et al., 2019; López-García et al., 2020).

9.2.1. Ancestral Lacustrine Systems in the Basin and Range Province

One of the best-studied ancestral lacustrine systems with saline lakes and continental sabkha playa lakes in the world is the southwestern United States, where several modern ephemeral lakes occur in a large desert area extending from the Great Basin toward the Basin and Range Province (Figure 9.1). The uplifted mountain ranges flanking the Death Valley in California are formed by transpression along strike-slip faults, whereas the depressions are related to transtensional releasing bends in strike-slip fault zones (Sylvester, 1988). The largest of these pull-apart basins in Southern California is the Salton Sea, a rift basin formed by oblique extension across the San Andreas strike-slip fault zone (Figure 9.5). The dextral shear zone resulted in subsidence and the formation of a rift trough in the Neogene. The Salton Sea is the largest lake in California, corresponding to a shallow endorheic rift lake with slightly saline waters. At present, the Salton Sea occupies a large depression filled with deltaic, lacustrine, and fluvial sediments, resulting in a continental fluvial-lacustrine environment with no marine connection (Brothers et al., 2009).

The ancient Salton Sea formed a large lake or inland sea in the Late Miocene to Pleistocene, when a much larger depression hosted a fluvial-lacustrine system that periodically dried out, similar to the modern ephemeral lakes in the Death Valley region (Figure 9.3). The Salton Sink was a remnant depression previously occupied by a much larger lake, known as Lake Cahuilla, which had its maximum extension during the Pleistocene (Waters, 1983). The Cahuilla paleolake extended across the Imperial Valley toward the Colorado River in the south (Figure 9.5). In the Holocene, the lake repeatedly dried out, and by the first decade of the twentieth century, most of the surface was covered by mud and salt flats. In the Death Valley, the Badwater Basin is characterized by periodic flooding and desiccation events during the Holocene (Figure 9.3).

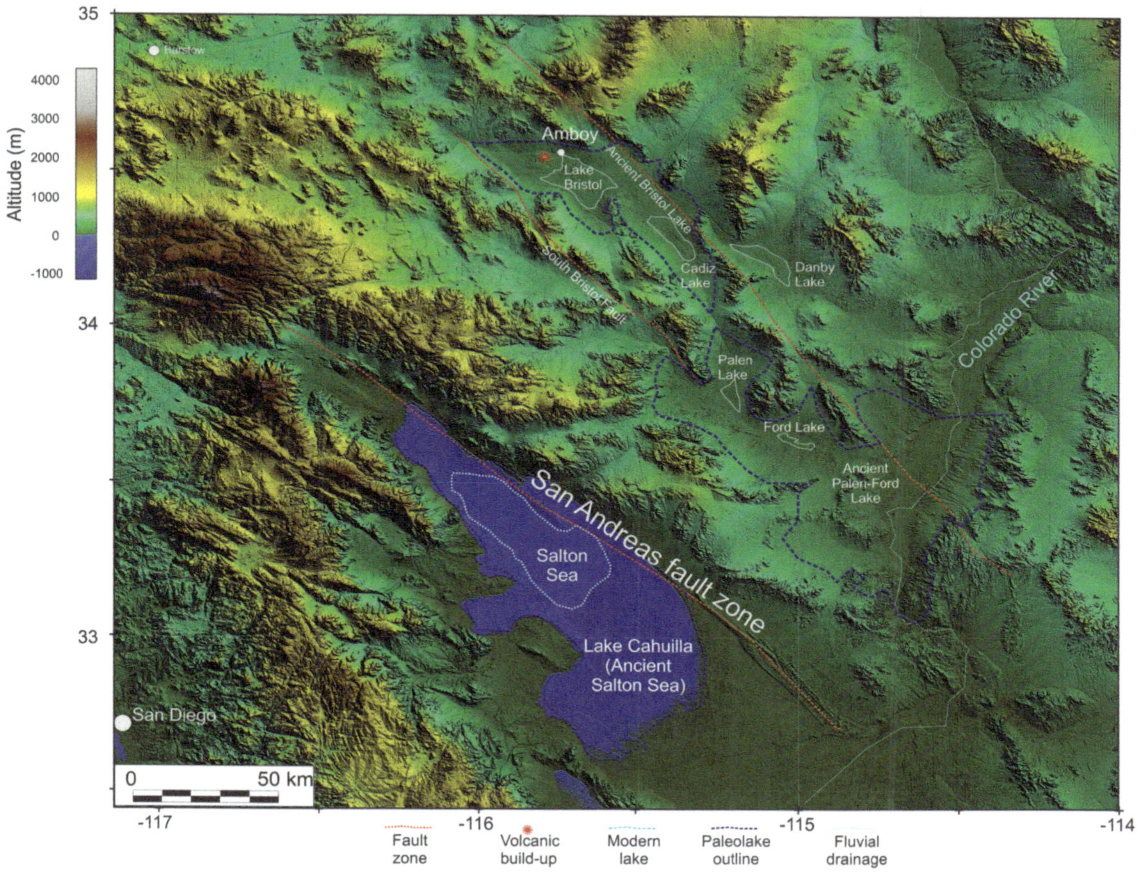

Figure 9.5 Regional topographic map of the Salton Sea and Bristol Lake. The ancient lacustrine systems north of the San Andreas fault zone (indicated by a thin red line) were much larger in the Pleistocene and probably linked with the Colorado River, which fed the ancient Palen-Ford lacustrine system in the south and the ancient Bristol Lake to the NW. The dark blue dashed line indicates the extent of the ancient Pleistocene lake (Lake Cahuilla), while the light blue dashed line represents the present-day extent of the lake waters, which have significantly decreased due to the arid climate in the region.

The present-day shallow lakes across the Mojave Desert (Figure 9.1) are remnants of ancient lacustrine systems, such as the Palen-Ford and Bristol Lakes in California (Figure 9.2). The Colorado River fed these lakes, forming perennial lakes in a fluvial-lacustrine system (Brothers et al., 2009). These paleolakes dried up in the Holocene, as suggested by scattered lacustrine sediments intercalated with evaporite rocks in several segments of the depression. The larger paleolake that preceded the Salton Sea corresponds to Lake Cahuilla, a Late Pleistocene lacustrine system that deposited several cycles of siliciclastic fluvial, deltaic, and lacustrine sequences intercalated with evaporitic sediments (Rockwell et al., 2022; Waters, 1983).

During the Pleistocene and Holocene, the Salton Sea paleolake periodically isolated, forming an endorheic basin. The brines were concentrated to the point of precipitation of several evaporitic minerals, resulting in thick beds of sodium sulfate minerals, such as thenardite ($Na_2(SO_4)$) and blödite ($Na_2Mg(SO_4)2{\cdot}4H_2O$), as indicated by the outcrops at the lake's margins (McKibben et al., 1988).

In the subsurface, drillcores of commercial geothermal boreholes drilled south of the Salton Sea shoreline indicated the presence of bedded evaporites (anhydrite layers, up to 0.6 m in thickness) intercalated with a shale-rich sequence (McKibben et al., 1988). The abnormally high geothermal gradient results from the northern continuation of spreading centers and transform fault zones in the Gulf of California. Magmatic activity in the region of the Salton Sea formed fumaroles, mud volcanoes, and other hydrothermal features, such as vents and brine pools. Some buildups formed by rhyolitic volcanic rocks were extruded during the Holocene (Adams et al., 2017).

The Salton Sea is a lake formed from a partially filled transtensional rift basin south of the San Andreas strike-slip fault zone (Figure 9.5). The present-day lake is 80 m below sea level, corresponding to the remnant water body of the larger ancient lake, Lake Cahuilla, which covered the depression during the Pleistocene (Rockwell et al., 2022; Waters, 1983). Lake Cahuilla was one of the largest lakes in southwest North America, forming a fluvial-lacustrine system south of the San Andreas fault zone, while a few other lacustrine systems were formed to the north of the transcurrent fault (Figure 9.5).

The Palaeo-Lake Bonneville, the largest lake developed during the Last Glacial Maximum in the United States, covered an area of about 50,000 km^2 between 23 and 19 ka (Nicoll, 2021). The modern Great Salt Lake (GSL) is the remnant of this megalake (e.g., Atwood et al., 2016; Spencer et al., 1984). The GSL is a shallow, hypersaline, perennial lake with environmental facies including siliciclastic, carbonate, and evaporite deposits (see Chapter 11).

The Cahuilla paleolake, a freshwater lake six times the size of the present Salton Sea, extended across the border between Mexico and California, from the Gulf of California to the mountain ranges associated with the San Andreas fault zone (Figure 9.5). The lake's length was about 160 km, its maximum width was over 50 km, and its water depth was around 90 m. In the early 2000s, the Salton Sea lake level dropped due to climate aridity, lower inflows, and increased water consumption in agricultural projects. In 2022, the Salton Sea's water surface was 70 m below global sea level, with a slightly higher sodium chloride content (56 g/L) than the Pacific Ocean's normal seawater concentration (35 g/L). The salt concentration in the region reflects the region's aridity, with evaporation being higher than precipitation.

The Salton Sea would be a dry pan today if the hydrology of the region was not managed by the U.S. government agencies during the first decade of the twentieth century. The modern Salton Sea was anthropogenically formed in 1905 by filling the Salton Sink with water diverted from the Colorado River, which covered the evaporitic mudflat with salt crusts. At that time, the Salton Sink was a large depression with a surface almost 80 m below sea level, and the evaporitic mudflat was covered by salt crusts. Much of the present-day salinity in the Salton Sea is due to the dissolution of the salt crust covering the lower parts of the depression.

9.2.2. Bristol Dry Lake

Bristol Dry Lake, located in the rift depressions north of the Imperial Valley in California, is an ephemeral lacustrine system that is only filled with water during wet periods. It is a continental playa sabkha isolated from marine influence, lying 300 km to the north of the Gulf of California at an altitude of approximately 190 m above sea level (Figure 9.5).

Most authors (e.g., Handford, 1982) interpret the Bristol Lake as a strictly continental lacustrine basin that never received any marine water inflow. However, McDougall and Martínez (2014), based on biostratigraphic analysis of foraminiferal assemblages, proposed marine incursions during deposition of the lower part of the Bouse Formation, which underlies the evaporite sequences. The salt layers are assigned to the upper parts of the Bouse Formation, which mainly correspond to sedimentary sequences deposited in hypersaline lacustrine environments. Late Miocene sediments are composed of carbonate and siliciclastic sediments (limestone, marls, and mudstones) deposited in the Blythe Basin, which extended from the north of the proto-Gulf of California to the paleo-Bristol Basin, located north of the San Andreas fault zone (McDougall & Martínez, 2014).

Geological maps of the Salton Sea region and other areas adjacent to the San Andreas fault zone indicate expressive volcanic activity in the depression (e.g., Adams et al., 2017). Seismological monitoring of the region around the Salton Trough (e.g., Lohman & McGuire, 2007) indicates some clusters of small and moderate earthquakes (magnitudes smaller than 5). However, many seismological centers recorded earthquakes in the Imperial Valley with magnitudes between 5 and 6.6 from the 1980s to the 2010s (Hauksson et al., 2017; Lohman & McGuire, 2007). Seismological activity is mainly recorded along the strike-slip fault zones, indicating an active rifting system associated with the San Andreas transform fault zone and other faults affecting the transtensional troughs formed north of the Gulf of California (Hauksson et al., 2017).

The lacustrine system involves Neogene-to-Quaternary sedimentation in rift basins that overlie Late Cretaceous sediments in the pre-rift sequences (Brothers et al., 2009). Volcanic rocks dated to the Pleistocene occur northwest of the Bristol Lake lacustrine system region, near the city of Amboy (Figure 9.6).

Recent basaltic lavas occupy an area of about 70 km^2 and cover the flat-lying alluvial valley, which forms the playas near Bristol Lake (Figure 9.6). Mineralogical analysis of some muds intercalated with the salt layers indicated a volcanic ash origin. Lowenstein and Risacher (2009) suggested that the inflow of hydrothermal waters associated with the Neogene magmatism enriched the brines that deposited the evaporite minerals in Bristol Lake with calcium and chlorine (Ca-Cl brines).

Bristol Dry Lake, with a surface area of 155 km^2, is continental sabkha playa lake that has dried up, forming a salt pan (Figure 9.6). The depression is filled with at least 300 m of interbedded terrigenous clastic rocks, mainly mudstones, intercalated with evaporite rocks such as gypsum, anhydrite, halite, and calcium chloride, a rare, highly soluble mineral with a composition of $CaCl_2(H_2O)x$,

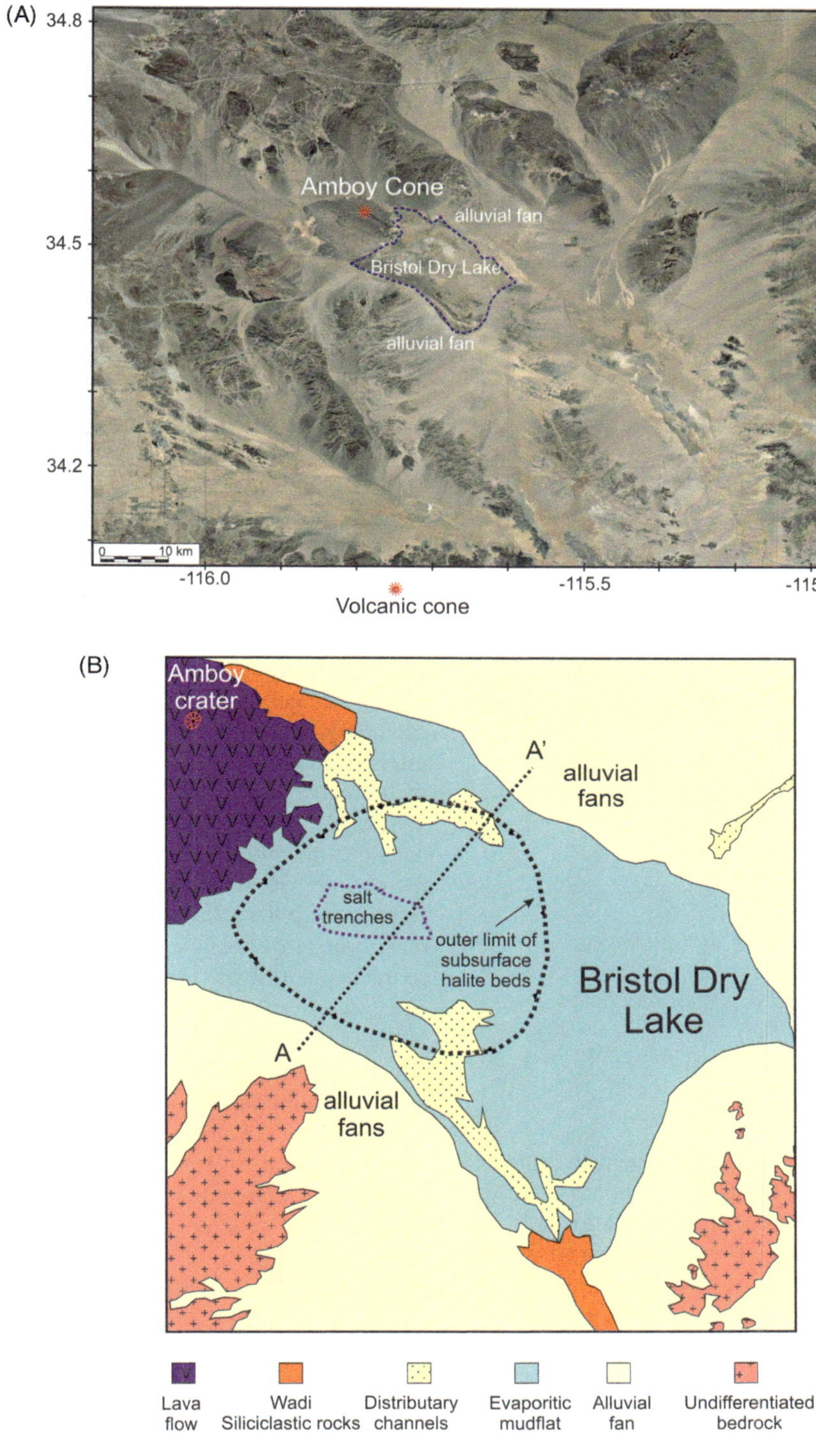

Figure 9.6 (A) Satellite image (natural color) showing Bristol Dry Lake, located south of Amboy. The salt flat is characterized by a halite crust on its surface and dark colors to the NW, indicating volcanic rocks surrounding the Amboy crater. (B) Simplified geological map of the Bristol Dry Lake, surrounded by alluvial fans made of Quaternary siliciclastic rocks. A thin black line marks the geological cross-section A-A′ across the central part of the salt pan. A dashed black line indicates the area of halite in the subsurface, as indicated by boreholes drilled near the salt trenches. Credit: Adapted from Handford (1982).

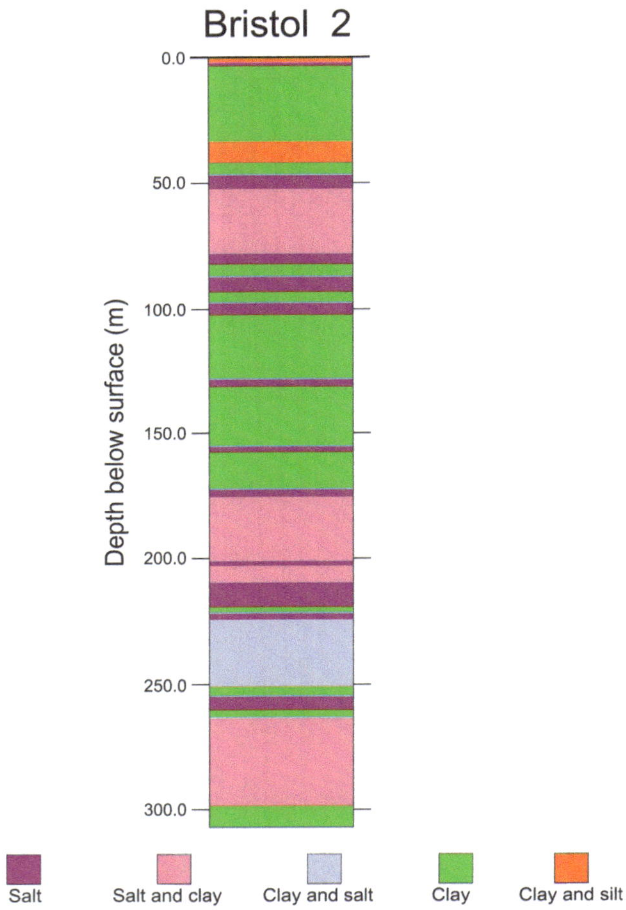

Figure 9.7 The exploratory well Bristol-2, located near salt trenches in Bristol Dry Lake, has a lithologic column of evaporitic rocks, with intercalations of siliciclastic sequences (clay, silt, and sand) and evaporite deposits, mainly halite and crystalline gypsum, from the surface to a depth of about 300 m. Credit: Adapted from Bassett et al. (1959).

where x is the number of water molecules in the mineral structure, which can vary from 0 to 6. The lake's evaporite facies fits the bull's-eye model of inland sabkhas (Figure 9.4). The basin center is characterized by more soluble salts that form layers or crusts on the surface, and the lake's margins are characterized by sulfate, carbonate, and siliciclastic rocks (Bassett et al., 1959; Handford, 1982).

In the 1950s, exploratory wells were drilled to investigate the evaporite deposits in the subsurface of the Bristol Dry Lake (e.g., Bassett et al., 1959). The results revealed abundant massive beds of halite intercalated with red and green clays, characterized by scattered halite crystals (Figure 9.7). The exploratory borehole Bristol-2, drilled south of the village of Amboy, penetrated over 300 m of sediments below the surface and showed siliciclastic sequences with several intercalations of evaporite minerals, mainly halite and gypsum (Figure 9.7). Amboy is famous for the Amboy Crater, a volcanic structure with a pyroclastic cone about 300 m in height. This volcano lastly erupted in the Upper Pleistocene, probably around 80 ka, but the tuffs in the basal Bouse Formation indicate Pliocene ages of about 4.8 Ma (Miller et al., 2014; Rosen et al., 2020).

Geological mapping and subsurface basin analysis studies of Bristol Dry Lake have identified four main lithofacies during evaporite deposition (e.g., Handford, 1982). These facies range from alluvial fans at the lake's borders to a salt pan across the center of the sabkha playa lake. The evaporite sediments form a bull's-eye pattern, with gypsum and anhydrite surrounding the basin depocenter with about 40% of halite (Handford, 1982; Rosen et al., 2020; Warren, 2006, 2016). The basin margins are marked by chaotic mudflat sediments, predominantly composed of silt and clay, interdigitated with the halite beds (Figure 9.8).

From the alluvial fans on the edge of the playa to the middle of the salt pan, the main rock types are alluvial fan clastic rocks with coarse bedding, playa edge sand flats, wadi sand and silt flats, evaporite units, salt-pan halite beds, and crusts on the surface (Hndford, 1982). Alluvial fan clastic rocks, playa-margin sand flats, and saline mud flats interfinger with other lithological facies. Evaporite units, including gypsum, anhydrite, halite, chaotic mud, and clay, form crusts on the surface. The hypersaline lake brines were periodically diluted by the inflow of fluvial and pluvial freshwater. However, the climate aridity, intense evaporation, and capillary discharge resulted in brine-level lowering (evaporative drawdown) and brine concentration, leading to the deposition of several evaporite sequences and evaporite cycles marked by the reprecipitation of halite. Salt-pan halite beds, up to 4 m thick, formed crusts on the surface due to subaqueous and intrasedimentary precipitation of evaporite minerals. Basin desiccation events can be seen up to the present day (Figure 9.9).

Salt exploration boreholes drilled in Bristol Dry Lake proved the occurrence of thick evaporite deposits in the center of the sabkha playa lake, indicating the paleoclimate and hydrology of the basin led to recurrent episodes of salt deposition (Rosen et al., 2020). Some beds reach a thickness of up to 4 m of pure halite, indicating long periods of extremely concentrated brine (Figure 9.7). The evaporite deposits were likely formed through a complex history of alternating subaqueous precipitation in salinas and intrasedimentary precipitation of evaporite minerals in the supratidal zone of a sabkha environment. The intense evaporation resulted in increasing brine concentration and lowering the lake level, leading to the capillary discharge of brines onto the depositional surface, forming evaporite crusts. The cyclicity observed in deep wells indicates that the deposition of these lacustrine evaporites is related to a playa lake environment

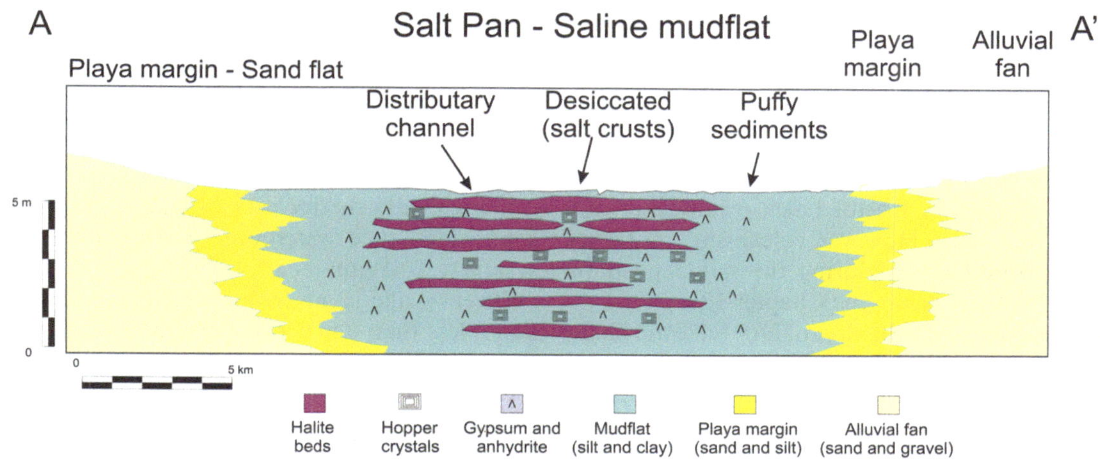

Figure 9.8 Schematic geological cross-section A-A′ (see location in Figure 9.7) shows the sedimentary facies distribution in Bristol Dry Lake. The lithological units vary from the siliciclastic rocks in the alluvial fans at the basin margins to the evaporite deposits at the center of the salt pan. Credit: Adapted from Smith (1979).

Figure 9.9 Photographs of Bristol Dry Lake. The top image (A) shows the salt crust formed on the playa lake surface. The bottom image (B) shows Bristol Lake after rainfall, with a thin water cover that will quickly evaporate in the extremely dry climate of the Mojave Desert. Credits: (A) High Plains Drifter/Wikimedia Commons/Public domain, (B) Adbar/Wikimedia Commons/CC BY-SA 3.0.

followed by the progradation of mudflats over the salt pan halite beds (Figure 9.7).

Some authors have suggested that the Bristol Dry Lake provides a modern analog model for the evaporitic sediments in the Permian Basin of the southwestern United States (e.g., Handford, 1982). The core descriptions of the exploratory wells in the Bristol Dry Lake reported chaotic mud halite that is almost indistinguishable from the evaporite deposits in the Permian Basin, located in Texas and dated as Late Permian (see Chapter 12). These ancient evaporite deposits are related to a sabkha flat flanking an extensive, shallow playa lake setting with a salt-pan complex in the center. Handford (1982) suggests that the lithological similarities indicate the deposition of Permian Basin evaporite sediments in periodically desiccated continental lakes. However, there are important differences that hinder using the Bristol Dry Lake as a modern analog for the Permian Basin.

First, Bristol Dry Lake is a continental inland salt pan with no indications of marine influence on the evaporite deposits. The Permian Basin, on the other hand, was fed by marine water (see Chapter 12). The marine connection was probably related to the periodic incursions from the Panthalassa Ocean to the west (Blakey, 2008, 2013).

Second, the Bristol Dry Lake is a very small basin related to extensional and transcurrent faults, while the Permian Basin was a much larger epicontinental sea developed on top of an earlier rift system. Subsequently, the region was affected by compressional tectonics that resulted in large foreland depressions that hosted a large inland sea with a carbonate-evaporite platform and deep water settings in the basin depocenter (see Chapter 12).

Third, some authors envisaged a possible enrichment of salt ions by volcanic activity near the lake (Lowenstein & Risacher, 2009). The Amboy Volcano and its associated lava flows were formed above a magma chamber, resulting in salt deposits rich in chlorides. The hydrothermal waters were enriched with Ca-Cl compounds and mixed with the brines in Bristol Lake, resulting in salt deposits rich in chlorides. However, volcanism is not considered an important factor during the salt deposition episode in the Permian Basin (see Chapter 12).

Fourth, the evaporites in the Bristol Dry Lake are characterized by a very shallow lacustrine environment that grades into a continental setting with an eolian reworking of alluvial fans and playa margin sediments, resulting in surface halite crusts, salt efflorescence, and even small barchan dunes. In contrast, the much wider and deeper sedimentary environments for evaporite deposition in the Permian Basin ranged from shallow to deep water. A regional high with a carbonate platform separated the depositional environment into two major depocenters, corresponding to the Delaware and Midland sub-basins (see Chapter 12). The Permian Basin hydrology and brine chemistry were highly influenced by marine water, contrasting with the continental brine in the Bristol Basin, which displays a more complex and varied evaporite mineral association typical of continental lakes (Smoot & Lowenstein, 1991).

9.2.3. Owens Lake

Several saline lakes and continental playa lakes are located to the east of the Sierra Nevada in Southern California, near the Mojave Desert (Figure 9.1). Owens Lake and other lakes in the region, such as Mono Lake in the north and Panamint Lake in the south, were part of much larger fluvial-lacustrine systems that extended across northern California and Nevada during the Pleistocene Epoch (Figure 9.2). The ancient Owens River might have flowed during episodes of wet climate, as the fluvial system was probably fed by meltwater from alpine glaciers in the Sierra Nevada (Smith, 1979).

The fluvial-lacustrine drainage extended along the rift valleys of the Basin and Range extensional province from about 3 million to 12,000 years ago. During episodes of wet climate, the ancient Owens River likely flowed as meltwater from the alpine glaciers in the Sierra Nevada, feeding the fluvial system (Smith, 1979). The lakes in Southern California might have reached depths of about 200 m with maximum areas of about 1,000 km^2 during the last stages of the glacial epoch (Smith, 1979).

As the global climate became warmer and drier during the demise of the Wisconsin period, the ice sheets in North America started to melt around 20,000 years ago. About 10,000 years ago, a dry climate dominated the southwestern region of the United States, and most of the valleys became arid mudflats with ephemeral drainage. The Holocene has been marked by extreme aridity, as indicated by the evaporite crusts covering the surface of most desiccated lakes, which support no vegetation (Smith, 1979). The climate aridity has continued from 10 ka to the present day, intercalating extremely dry periods with some short, wet seasons when some water was present in the hydrological system (e.g., Adams et al., 1999).

Owens Lake is an ephemeral, mostly dry lake (water depth less than 1 m), located in the Owens Valley region of Death Valley, surrounded by high mountains such as the Sierra Nevada to the west and the White and Inyo Mountains to the east (Figure 9.11). Unlike most lakes in the Basin and Range region, which have been dry for thousands of years in the Holocene, Owens Lake had significant water until the first decade of the twentieth century. Historical reports mention that it was a blue, saline lake covering hundreds of square kilometers near the border between California and Nevada. Reports from the nineteenth century described the waters as deep enough to support boat and steamer traffic.

However, in 1913, the Owens River was diverted to supply water to Los Angeles, located about 266 km to the south (Figure 9.10). The lake waters were already receding in the previous decades, not only due to climate or natural causes but also because of human activity, as many farmers exploited the lake and river waters for agricultural purposes, leading to total desiccation by 1926. The lake formed a playa lake covered by dust, and government measures were taken to control windblown dust, which affected the health of the inhabitants of nearby villages. Some of the water flow was restored from the fluvial system, and by the end of the twentieth century, the lake basin contained some water in shallow ponds (Hollett et al., 1991).

As seen in the satellite image (Figure 9.11A), the central part of Owens Lake is covered by a small brine pool that is fed by shallow groundwater, local springs, and seeps, which support minor ponds and wetlands in the playa mudflats. In the 21st century, Owens Lake is a large salt pan with a mixture of clay, sand, and various evaporite minerals, including halite, burkeite, mirabilite, thenardite, and trona. In wet seasons, the in-flowing water dissolves these minerals, which are then redeposited during dry periods when the shallow waters evaporate. The shallow pool is filled with a concentrated brine that is hostile to most forms of life. The natural color satellite images show gray and white materials within the central part of the salt pan, corresponding to the exposed sediments and salt crusts. Some shallow water ponds at the lake margins are colored pink to red due to the presence of halophilic bacteria and other microorganisms that have adapted to the extremely salty environment (Figure 9.11B).

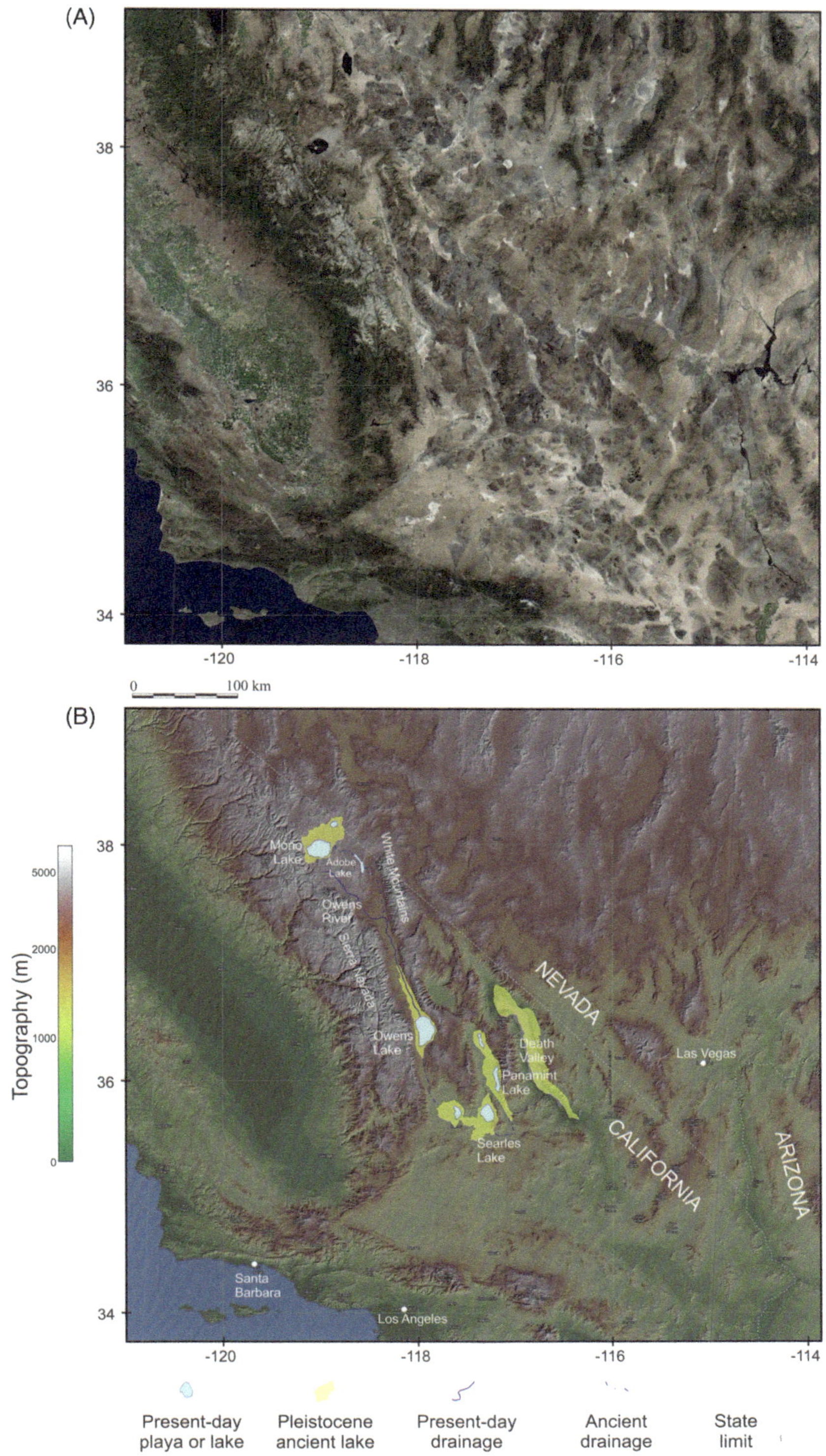

Figure 9.10 The NASA satellite image (A) and topographic map (B) depict the outline of ancient and modern lakes in southern California. The Sierra Nevada highlands separate the Great Basin from the Pleistocene continental deposits north of the Mojave Desert (shown by yellow colors). Credit: Smith (1979)/USGS/Public domain.

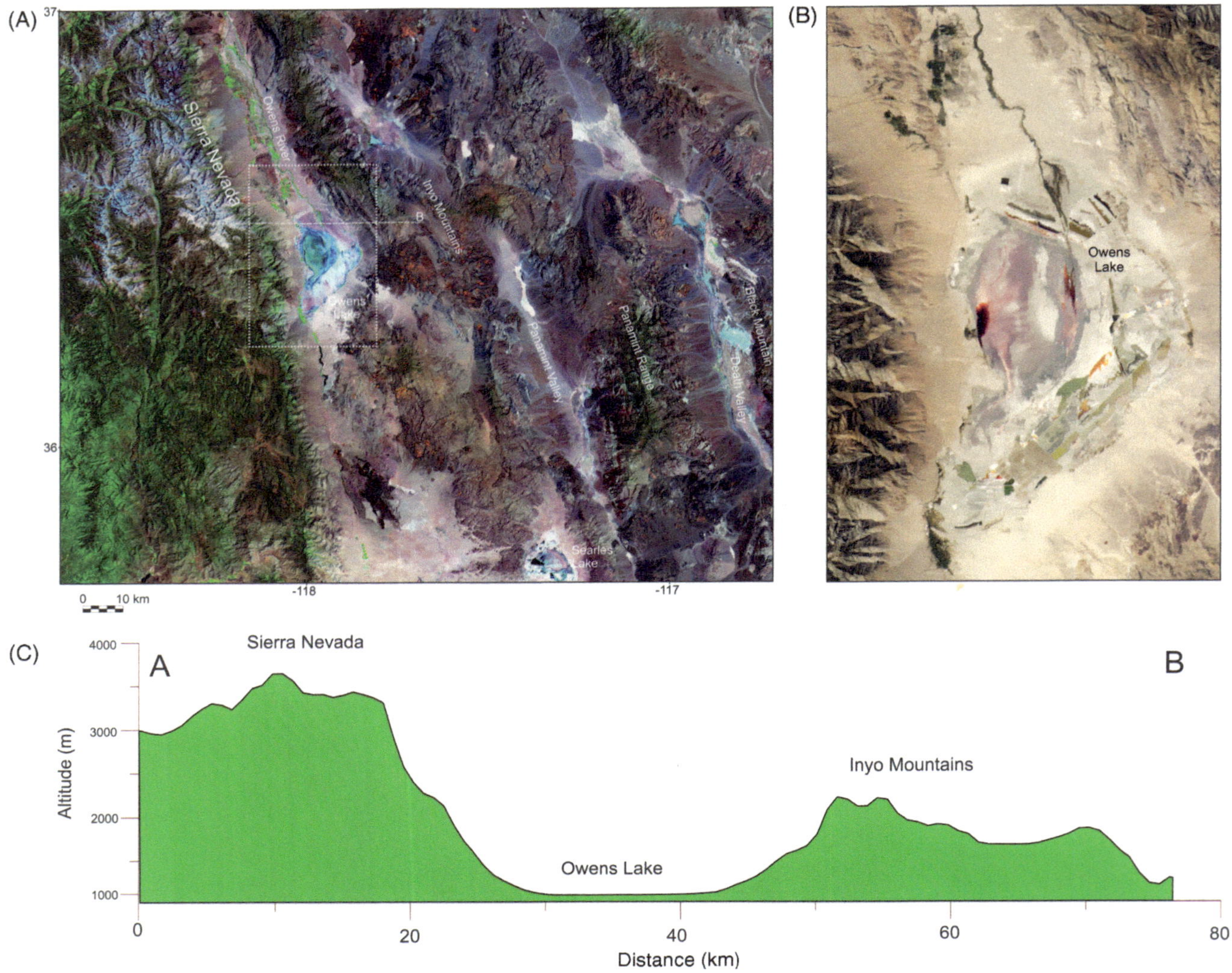

Figure 9.11 Satellite images of Owens Lake, east of the Sierra Nevada (A). The top right (B) image shows whitish-to-pinkish reflectivity in the salt flats and red coloration in shallow ponds due to halophilic bacteria. The lower diagram (C) displays a regional cross-section A-B (see location in Figure 9.12A) showing the topographic profile across the Owens River valley. Credit: Courtesy of NASA.

9.2.4. Searles Lake

Searles Lake, located in the Searles Valley, is an almost dry salt pan with a width of about 10 km and an area of about 100 km^2 (Figure 9.10). The small, residual lake is known for its abundance of rare chemical elements, including lithium, and evaporite minerals such as trona, hanksite, and halite. After the Wisconsin glacial period in the Pleistocene, Searles Lake was part of a chain of continental lakes fed by a downstream flow from the melting of snow from the Sierra Nevada to the west (Figure 9.2).

Successive layers of lacustrine sediments were deposited as lake levels fluctuated, preserving an important record of regional climate changes from about 130 ka to the present (Olson & Lowenstein, 2021; Smith, 1979). Mud deposited during wet periods and evaporite sediments precipitated in hypersaline lakes during dry periods. At its highest level, Searles Lake reached an area of about 1,000 km^2 and was >200 m deep, contrasting with the present-day area that is 1/10 of its original size, resulting in an almost dry salt pan (Olson & Lowenstein, 2021; Smith, 1979).

Some deep boreholes drilled in the salt pan penetrated sedimentary sequences to a depth of 929.6 m (Olson & Lowenstein, 2021). The sedimentological analysis of some cores indicates that during the interval 32 to 6 ka, Owens Lake was a perennial lake with a maximum depth of several meters and deposited siliciclastic sediments (mud and silt). However, during arid intervals, the intense evaporation resulted in the brine concentration and deposition of several bedded evaporites, including the Upper Salt (10 to 6 ka) and the Lower Salt (32 to 24 ka) layers, both reaching a maximum thickness of about 15 m. These evaporite sediments (mainly trona, halite, and borax) were cored in some boreholes at shallow depths (< 40 m). The evaporite beds are separated by a 3-m-thick mud layer that was penetrated at a depth of about 22 m (Olson & Lowenstein, 2021).

The mineral trona ($Na_3(HCO_3)(CO_3){\cdot}2H_2O$) constitutes most of the edge facies of the evaporite deposits that precipitated on the floor of ancient Searles Lake. Salt mining operations extract sodium- and potassium-rich minerals (primarily borax and salt) from the subsurface deposits (Figure 9.12). Spectacular columns formed by trona-rich sediments rise above the dried lake surface, reaching heights of about 45 m, constituting a natural

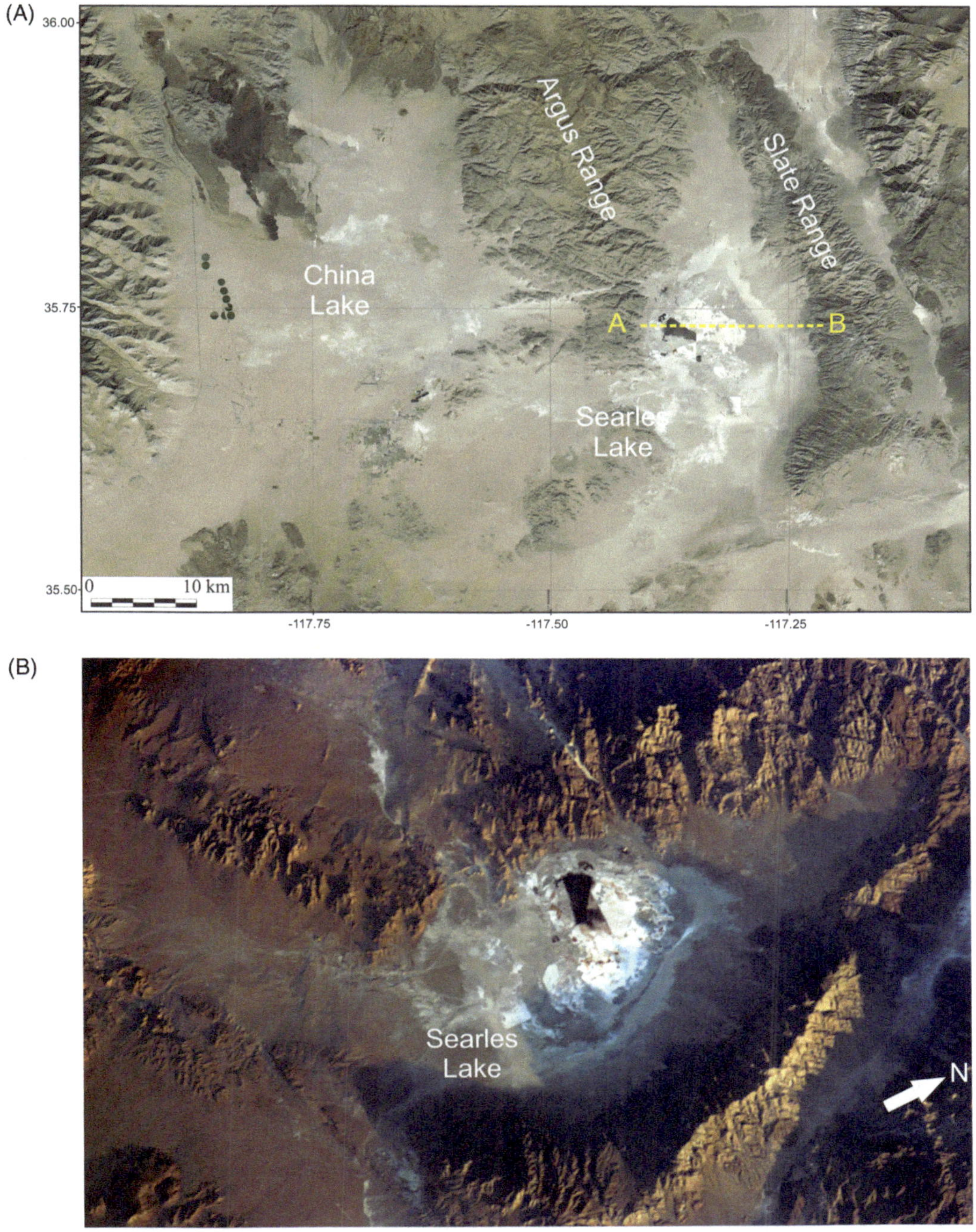

Figure 9.12 (A) NASA satellite image (natural colors) of Searles Lake showing the valley depression and playa flats between the Argus Mountains and Slate Mountains, corresponding to the continuation of the Panamint Mountains in California. A thin yellow line shows a schematic geological section A-B across the central part of the playa lake (see Figure 9.14). (B) Satellite image (enhanced colors) showing a residual shallow pond at the center of a playa salt flat. Credit: Courtesy of NASA.

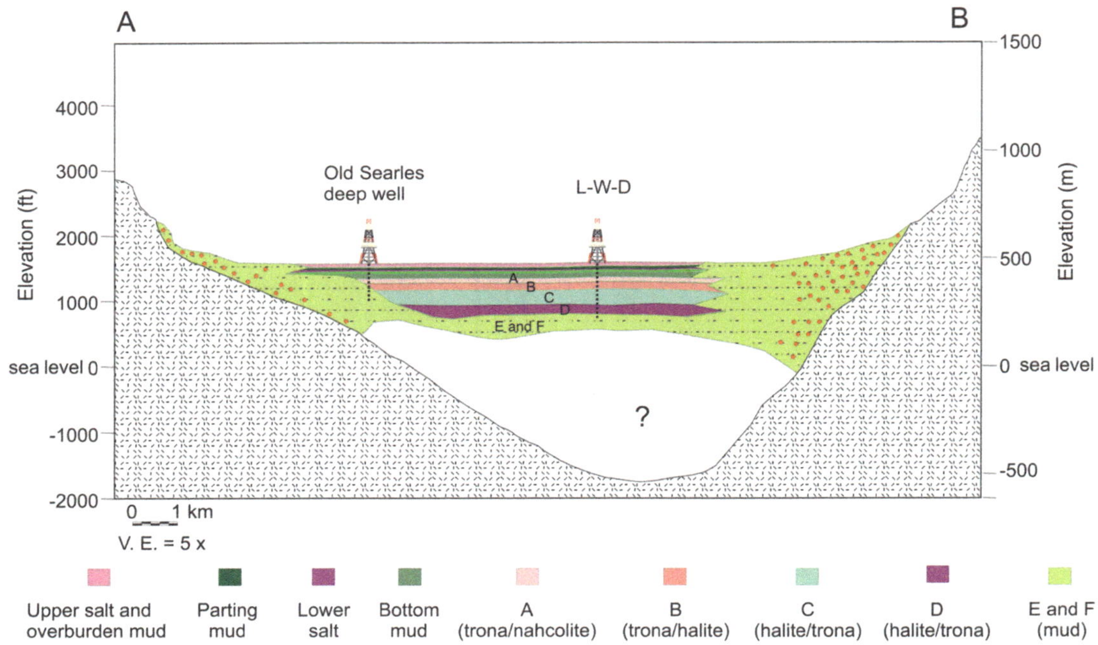

Figure 9.13 Schematic regional cross-section A-B across the valley between the Argus and Slate mountains (see Figure 9.12A for location). The Searles playa lake, located in the valley between the Argus and Slate mountains, is a salt flat at an elevation of around 500 m above sea level. The lake's distribution of evaporite sedimentary facies is shown in the cross-section. Early exploration wells penetrated only the margins of the evaporite basin. The deeper boreholes L-W-D, drilled by the salt industry in the central part of the playa, penetrated several evaporite sequences, which are mainly composed of halite and trona. Credit: Adapted from Smith (1979).

attraction that is protected by heritage conservation agencies. The carbonate tufa mounds are composed of calcite and aragonite. These minerals are associated with the hydrothermal springs at the bottom of the Pleistocene lake (Guo & Chafetz, 2012).

In the subsurface, the salt beds are separated by mud layers that were deposited during more humid climates, when the lake was deeper and anoxic conditions prevailed at the bottom (Figure 9.13). These siliciclastic fine-grained sediments contain interstitial brines that are pumped to the surface and evaporated to crystallize several minerals extracted in the saltworks, primarily borax, halite, and potassium-bearing minerals. The salt pan is characterized by a large proportion of mud, which comprises about two-thirds of the exposed area, and one-third corresponds to a hard salt crust (Smith, 1979).

The United States holds the largest trona deposits in the world, concentrated in the Green River Basin in Wyoming, with about 47 billion tons of soda ash resources from bedded trona and trona-halite sequences. Trona deposits constitute the main target for mining companies in Searles and Owens lakes, but the reserves are estimated at only 0.8 billion tons of soda ash (U.S. Geological Survey, 2020). Other minerals, such as halite and burkeite, are more abundant in the central or intermediate portions of the evaporite bodies but are not considered as valuable as sodium carbonate.

9.3. HYPERSALINE LAKES ADJACENT TO PASSIVE CONTINENTAL MARGINS

Hypersaline lakes adjacent to passive margins with salt deposits in the Holocene are rare. Most lakes in coastal regions adjacent to passive margins are well flushed and have fresh to slightly less than normal water salinity (Eugster & Hardie, 1978). Two of the largest water bodies on the southeastern Brazilian margin are Patos Lake in the southern region and Lagoa Feia Lake, at the western border of the Campos Basin in southeast Brazil. Both are shallow lakes that are highly influenced by fluvial freshwater. Lagoa Feia Lake, with a maximum depth of 2–3 m, is marked by freshwater fishing activity due to its fluvial water input from several rivers. Patos Lake is the largest lagoon in South America and is connected to the South Atlantic Ocean by a narrow channel. The salinity varies depending on pluvial and fluvial water input, but never exceeds normal seawater salinity.

Hypersaline lakes commonly occur in tectonically stable areas associated with local depressions in arid-zone

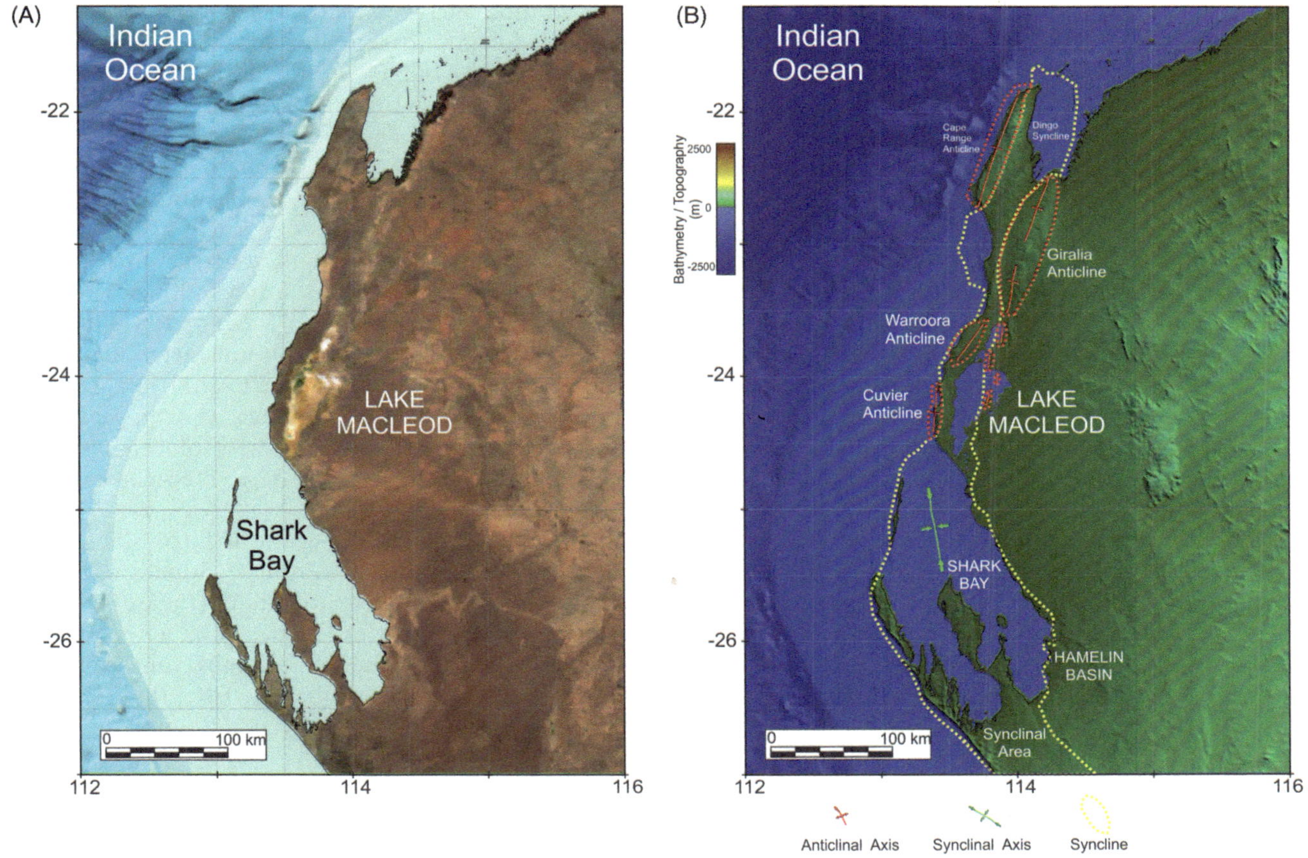

Figure 9.14 The left-side image (A) displays a natural-colored satellite image of Northwestern Australia's continental margin, while the right-side image (B) provides a topobathymetric map outlining the shorelines of Shark Bay and Lake MacLeod and the main structures in the basin.

subtropical high-pressure belts, such as the NW Australian margin and the southern provinces of central Australia. In central and southern Australia, there are inland depressions with sand flats that have been periodically inundated by fluvial waters during wet seasons, which are subsequently evaporated, precipitating evaporite minerals (Mernagh, 2013; Mernagh et al., 2016). Lake MacLeod in NW Australia is an example of hypersalinity in a modern lake adjacent to a passive continental margin. The lake is located in a very arid region with limited influence of fluvial or marine waters, and several salt deposition episodes are recorded in the Holocene.

9.3.1. Salterns in the Continental Margin of Australia

A significant example of a modern saltpan adjacent to a continental margin exists in NW Australia, north of the coastal city of Carnarvon. During the Quaternary, Lake MacLeod, the westernmost lake in the country, accumulated evaporites in tectonically controlled depressions (Mernagh, 2013; Whitney & Hengesh, 2015). This region is located south of the Cape Cuvier in the Gascoyne Platform (Figure 9.14). Saltworks in the region surrounding Lake MacLeod produce most of the salt consumed by the Australian population, making it the site of a modern salt industry.

Lake MacLeod is a 120-km-long depression in NW Australia, covering an area of over 2,000 km^2 (Figure 9.14A). It is the largest active marine-fed coastal salina in the world, known as the Lake MacLeod saltern (Logan, 1987). Marine water from the nearby Indian Ocean feeds the lake depression. The lake is characterized by a Holocene sedimentary infill of more than 10 m of bedded gypsum and halite intercalated with small amounts of carbonate (Handford, 1991; Handford et al., 1984; Logan, 1987).

The main components of the lacustrine sedimentary infill are halite and gypsarenite. Carbonate rocks form moat facies around the lake's rim due to aragonitic seepage. The older carbonate rocks (Egret limestone) outcrop near the lake's western border, forming a belt located to the east of the Indian Ocean shoreline (Logan, 1987). At present, the salina is mainly covered by a gypsum layer that was deposited in the Holocene. The lake has been almost totally desiccated for the past 5,000 years, and the lake level in the remaining ponds is about 3–4 m below the Indian Ocean sea level (Figure 9.14B).

The coastal barrier that separates Lake MacLeod from the Indian Ocean is known as the Quobba Barrier, or Quobba Horst. This sedimentary barrier is related to Neogene anticlinal structures, particularly the Cuvier and Waroora anticlines (Figure 9.14B). The sedimentary cover is characterized by Holocene dunes consisting of carbonate and siliciclastic sand-sized grains (Logan, 1987). There are several anticlinal structures in the subsurface. These structures are related to the Late Quaternary deformation that formed major anticlines, such as the Cape Range, Giralia, and Cuvier (Whitney & Hengesh, 2015). Both the Lake MacLeod and the Shark Bay depressions are located on synclinal structures along the western coast of Australia (Figure 9.14B).

Tectonic activity in the Western coastal region of Australia dates back to the Miocene and continued into the Quaternary, resulting in a subsidence of about 100 m for the coastal depressions (e.g., Mernagh, 2013). Several lithologic types, including eolian sands and thin marine sediments (carbonate sands), have filled these depressions. The Indian Ocean and Lake MacLeod are separated by the Qoobba Barrier, a permeable sand bank (Figure 9.15). The southern barrier is associated with a siliciclastic bank composed of Quaternary deltaic sediments, forming the Bejalin Barrier that outcrops as a 10-m-high sandy ridge (Figure 9.16). These sedimentary barriers allowed seawater to seep into the lake, discharging as springs along the shorelines and beneath the lakebed (Logan, 1987; Playford et al., 2013). The lake floor (several meters below sea level) was the main driver for the hydrological controls on the marine incursions and seawater seepage. During episodes of basin isolation in the Holocene, gypsum and halite filled some of the desiccated coastal lakes (Logan, 1987).

Shark Bay, located to the south of Lake MacLeod, is marked by NW-trending peninsulas within a wide shelf that comprises several bays and islands (Figure 9.14). The waters in Shark Bay are tinted green due to the presence of algae and some sediments. A shallow sand bank bars the southeastern part of Shark Bay from the rest of the

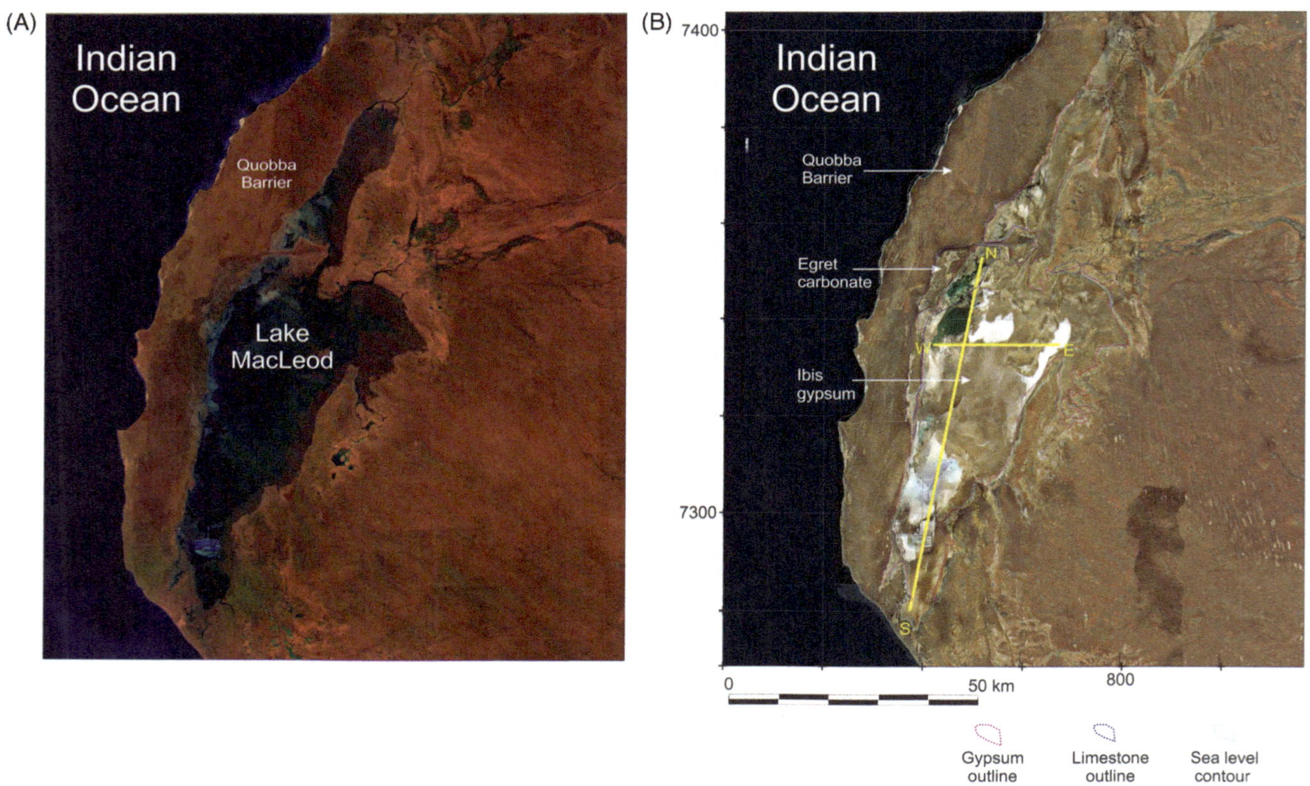

Figure 9.15 (A) Zoom of the satellite image showing Lake MacLeod in NW Australia. (B) NASA satellite image (natural colors) illustrating the geological setting of the lake. A sandy barrier is located to the west of the lake, resulting in a Quaternary depression that was filled with evaporite sediments. Thin yellow lines show regional cross-sections along the N-S and E-W directions. Credit: Photos by NASA/Public domain.

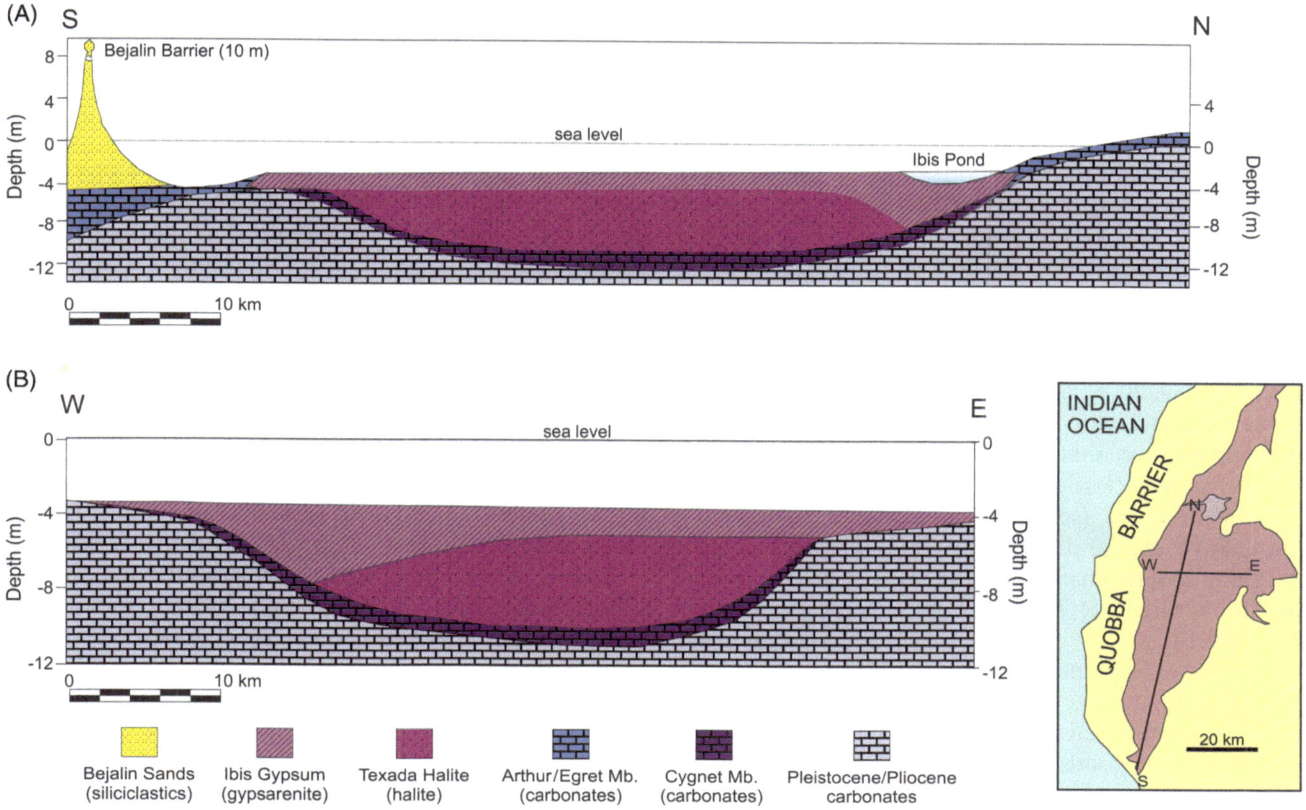

Figure 9.16 The Schematic diagrams show Lake MacLeod's N-S (A) and E-W (B) cross-sections across Lake MacLeod (see location in Figure 9.16). During the Quaternary, the depression was filled with evaporite sediments from Indian Ocean marine ingressions, such as Texada halite and Ibis gypsum. Credit: Adapted from Logan (1987).

bay, restricting water interchange with the Indian Ocean. This results in a hypersaline lagoon rimmed by intertidal flats and occupied by Hamelin Pool, which has a maximum water depth of about 11 m (Playford et al., 2013). Hamelin Pool is a classic example of a barred basin filled with a brine that is twice the salinity of normal seawater, resulting in the development of stromatolites and small bivalves that thrive in the hypersaline environment (Playford et al., 2013).

Satellite images showing Lake McLeod show a brown landscape desert region typical of the western provinces in Australia (Figure 9.15B). The Lake MacLeod depression appears as a greenish pond encircled by whitish patches that correspond to alluvial fans and playa lake mudflats (Figure 9.15B). Lake MacLeod appears as an almost dry salt flat with only a shallow pond in the north, exhibiting a greenish color reflectivity in the satellite image (Figure 9.15B). The large white to light brown areas in the Lake MacLeod depression correspond to salt flats, with light bluish colors indicating standing water or humid soil may still be present in the depression (Figure 9.15B).

The hydrologic balance in the lake system is delicate, with ephemeral water bodies covering parts of the salt flats during wet seasons (Figures 9.15A and 9.15B). The lowest topographic depression at the northern end corresponds to the Ibis Pond, which is currently covered by a water film with a depth of only 1 m (Logan, 1987). Given the extreme aridity of the region, the pond may desiccate in a few years. The carbonate rocks that immediately underlie the modern evaporite deposits correspond to the Cygnet Mb., a marine limestone bed less than a meter thick (Logan, 1987). These rocks were deposited in a lagoon during the peak of the Holocene transgression (9 to 6 ka) and covered the Pliocene-Pleistocene consolidated carbonates. The stratigraphic relationships between the underlying carbonate sequences and the Holocene evaporite deposits are presented in two regional cross-sections along the N-S and E-W directions (Figure 9.16).

The lagoonal carbonate sediments are characterized by intercalations of aragonite and gypsum minerals, with calibrated radiocarbon ages of 6.6 to 5.8 ka (Logan, 1987). The overlying evaporite layers (gypsum and halite) form the main Holocene fill of the lake, indicating a modern episode of basin restriction. The evaporite deposits are divided into two members: the Texada Halite and the Ibis Gypsite (Figure 9.16). The Texada Halite is characterized by a thickness of about 3–5 m of halite and minor gypsum intercalations, with maximum thickness at the southern and eastern parts of the MacLeod salt basin (Logan, 1987). Radiocarbon dating of associated carbonate beds yielded ages of 6.1 to 4.2 ka for these evaporite sediments. Lake MacLeod holds the thickest deposit of Holocene halite in Australia and constitutes one of the few examples worldwide of preserved thick halite deposits with salt accumulation persisting up to the Holocene (Logan, 1987; Warren, 1989, 2006, 2010, 2016).

The Texada Halite is overlain by the Ibis Gypsite, a sedimentary layer that shows a thickness of about 2–6 m of alternating clastic gypsarenite beds and massive gypsum deposits (selenite), intercalated with minor amounts of clay and aragonite mud. The Ibis Gypsite presently covers much of Lake MacLeod's floor and locally interdigitates with the Egret Carbonate Member (Figure 9.16).

The pre-salt Pliocene-Pleistocene carbonate rocks locally outcrop along the lake's western margin. Seepage from the Indian Ocean infiltrates the depression at rates that are high enough to form ponds where aragonite mud is deposited. These evaporite and carbonate rocks are overlain by the more recent coastal dune deposits of the Bejaling Sands, which form the main barrier to the west of the lake. These carbonate and sandstone banks and ridges separate the depression from the open sea (Figure 9.16).

Lake MacLeod's geological cross-sections show asymmetry in infilling evaporite sequences, including layers of gypsum and halite (Logan, 1987). The S-N section shows thickening of gypsarenites toward Ibis Pond, characterized by a massive, 5-m-thick gypsum layer. The W-E cross-section, extending from near the western barrier toward the eastern alluvial fans, shows syn-depositional thickening of gypsarenites toward the western side of the lake. The central and southern sectors of the lake are covered by the Texada massive halite unit, reaching a thickness of about 7 m (Figure 9.16).

9.3.2. Sedimentological Model of Evaporite Deposition in Lake Macleod

Figure 9.17 shows a schematic model illustrating the development of the Lake Macleod saltern. Lake Macleod corresponds to a hypersaline lake that was formed by the evaporation of inflowing seawater during the Pleistocene (e.g., Logan, 1987). The lake initially had a salinity close to normal seawater, depositing carbonate sediments (Figure 9.17A). Over time, the basin has been periodically isolated from the Indian Ocean, desiccated, and re-flooded by Indian Ocean seawater and freshwater from monsoonal rains (Figure 9.17B). Basin isolation by a carbonate barrier led to increased evaporation and brine concentration, leading to massive halite deposition (Figure 9.17C). In recent times, brine salinity has resulted in the deposition of gypsum layers (Logan, 1987).

Two sources of water fill the lake: saline water from springs at carbonate outcrops, and saline water derived from the Indian Ocean that flooded the basin during sea-level rises. The seawater initially had normal marine salinity when it entered Lake Macleod. Around 6,000 years before the present, the Indian Ocean (Australasian Sea) rose to a level that was within a meter or so above its present level. Since then, the sea has fallen to its present level. The marine incursions have supplied water to the lake by a combination of evaporative drawdown, brine reflux, and spillover (Logan, 1987).

Evaporite precipitation followed the brine salinity cycle. Limestone sediments accumulated on the lake's bottom during the early stages, creating an aquitard that reduced the volume of brine seeping out through the lake floor, increasing brine salinity with evaporation. Precipitation of gypsum across the lake floor created an even more effective aquitard than carbonate units. Brine salinity increased, leading to halite precipitation, particularly at the southern end of the lake (Logan, 1987).

Increased aridity in the Holocene reduced water in the hydrological system, causing the lake to become a dried-up playa lake with evaporites exposed at the surface. In modern times, surface water salinity has reduced to the point where gypsum has become the stable mineral phase. Today, gypsarenite is the main mineral phase of gypsum deposited on the lake floor, with carbonate layers making up less than 10% of the sediments (Logan, 1987; Playford et al., 2013).

Lake MacLeod in Australia exhibits a unique sequence of evaporite deposition in the Holocene, characterized by a halite-rich sequence. This is distinct from other salt lakes in Australia, which are mostly saline, filled with gypsum, and surrounded by boxwork limestones (Warren, 2016). The unique occurrence of thick halite layers in Lake MacLeod can be attributed to factors such as proximity to marine water inflow, arid environments with reduced meteoric water input, and higher brine concentrations due to the basin's isolation from the ocean (Logan, 1987).

Logan (1987) postulates that subaqueous halite accumulates in three main forms: detrital halite, crust halite on salt flats, and halite precipitation from suspension (also known as "rain from heaven"). Four mechanisms explain the thick layers of halite minerals at Lake MacLeod's bottom in the Holocene: 1) suspension from surface-grown hopper rafts sinking to the bottom; 2) small individual crystals form on the lake's floor; 3) surface brines precipitate at the air/water interface, sinking to the bottom without dissolution due to the high salt concentration; and 4) brine mixing within the water column as a result of brine mixing (Raup 1970, 1982).

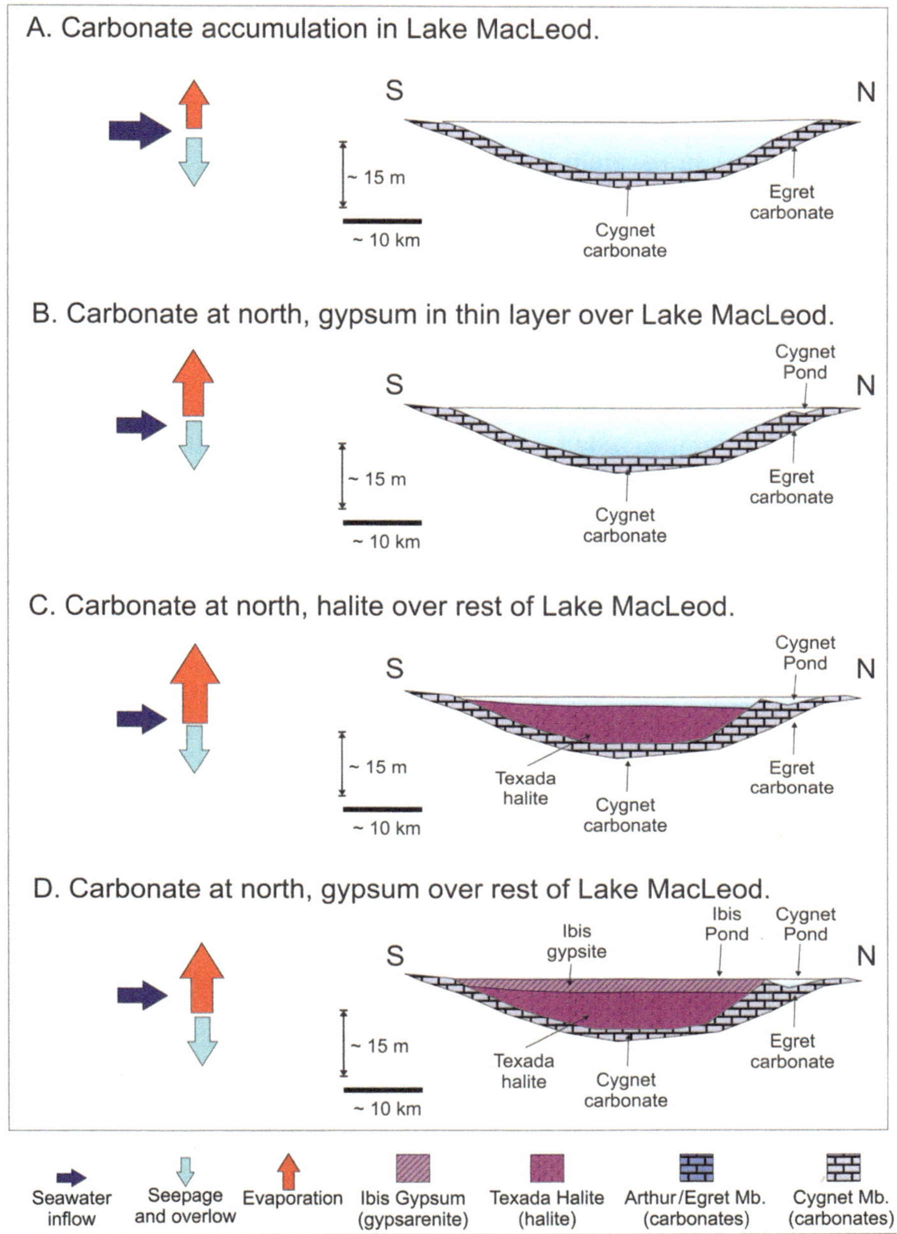

Figure 9.17 The schematic diagrams illustrate the sequential development of the evaporite deposition in Lake MacLeod. Carbonate deposition with some gypsum occurred during the early stages (diagrams A and B). Increased evaporation and brine concentration led to massive halite deposition (diagram C). In more recent times, the reduction in brine salinity has resulted in the deposition of gypsum layers (diagram D). Credit: Adapted from Alsharhan and Kendall (1994).

Australia's total salt production in 2019 was around 13 million metric tons, with Lake MacLeod having the largest salt accumulation in the country (U.S. Geological Survey, 2020). With an annual production of over 35,000 metric tons, the lake has the potential to produce 3 million metric tons annually. The saturated brine in the lake, such as Ibis Pond, is 10 times saltier than normal seawater and is close to the point of halite precipitation. The Lake MacLeod saltworks are more profitable than most salterns worldwide due to the high salt concentration in subsurface, deep aquifers. A floating dredge is used to extract gypsum at the saltworks, while the salt brine is pumped from the aquifers and evaporated in artificial ponds and crystallizers. When the hypersaline brines reach the sodium chloride saturation point, they precipitate as almost pure halite, which can be harvested. The residual brine contains ions from more soluble potassium salts, including sylvinite, used in the chemical and agricultural industries.

9.4. EPHEMERAL TO PERENNIAL SALINE LAKES IN ACTIVE RIFTS AND INCIPIENT DIVERGENT MARGINS

Source rocks in saline to hypersaline lacustrine environments are essential components for the prolific petroleum systems in the South Atlantic (see Chapter 3 in Mohriak, 2025). These lakes were formed during the rift event that separated South America and West Africa in the Early Cretaceous. The East African Rift System (EARS) is considered the best analog for South Atlantic lacustrine systems (e.g., Olsen, 1997; Tiercelin et al., 1994). However, salt deposition in some basins of the South Atlantic postdates the source rock intervals in rift troughs, which are associated with brackish to saline water. However, there is evidence for source rock accumulation in small hypersaline lakes in northeastern Brazil (e.g., Martins et al., 2019).

Modern hypersaline lakes with Holocene evaporite deposits exhibit characteristics analogous to those of ancient salt lakes. Some lakes in the East African Rift System are located far from any seawater input, but exhibit significant evaporite deposition, notably sodium carbonate (trona), such as Lake Natron. Lakes with marine influence, such as the Danakil Depression in Ethiopia and the Gulf of Tadjoura in Djibouti, record the deposition of thick salt layers (halite) in the Holocene.

9.4.1. Saline Lakes in the East African Rift System

Perennial saline lakes and salt pans in the East African Rift System (EARS) are modern examples of continental lacustrine evaporite basins in divergent rifts that are not influenced by marine waters. These lakes occupy the lowest topographic depressions in closed basins formed by extensional faults. Lake Magadi and Lake Natron in Kenya are examples of hyper saline lakes within the Eastern Branch of the EARS, resulting in hydrologic sumps (Warren, 2006, 2016; Owen et al., 2019; Getenet et al., 2022). The water surface in these lakes is just 600 m above sea level, while mountain ranges bordering the rift troughs may attain altitudes of more than 2,000 m (Figure 9.18). These elevated regions are likely associated with a pre-rift domal structure caused by mantle plumes, such as in the Kenya Dome in East Africa (Bonini et al., 2005). The long-wavelength relief indicates an

Figure 9.18 The shaded topographic map shows the eastern branch of the EARS between Kenya and Tanzania. Lake Natron and Lake Magadi's saline waters filled the Quaternary depression between the highlands that border the rift zone. The region features several volcanic edifices, including Mt. Kilimanjaro in Tanzania, the highest mountain in Africa, and the Gelai Mount and other large volcanoes to the south, between Lake Natron and Eyasi and Manyara. Credit: With permission of The General Bathymetric Chart of the Oceans.

uplift related to deep mantle processes, but the local relief variations adjacent to the lakes are associated with active faulting and magmatism within the rift depression or as off-axis volcanoes (Corti, 2009; Corti et al., 2015).

Located in the northern part of the Afar Triangle, the Danakil Horst exhibits less elevated topography compared to the Ethiopian Escarpment, which is closer to the Danakil Rift depocenter (Bastow et al., 2018). These differences in elevation result from extensional rift mechanisms, with the flexural rift shoulder developed on the opposite side of the rift master fault (Bott, 1995). The Danakil salt flats are located between the Ethiopian highlands to the west and Danakil Horst to the east, adjacent to the southern Red Sea. The salt flats are characterized by thick deposits of evaporite rocks precipitated by the evaporation and concentration of marine seawater sourced by the Red Sea (Bastow et al., 2018; Bonatti et al., 1971; Holwerda & Hutchinson, 1968).

The EARS was formed by continental extension and normal faults during the Neogene, resulting in elongated rift valleys. Some regions within syn-rift valleys are occupied by deepwater lacustrine systems, resulting in freshwater lakes like Lake Tanganyka (Chorowicz, 2005; Tiercelin et al., 1994). Some lakes in the EARS have hydrological and chemical characteristics of saline to hypersaline water. Rapid evaporation in the arid climate shallows the lakes, leading to brine concentration and salt deposition on the lakes' margins and depocenters (e.g., Renaut & Gierlowski-Kordesch, 2010; Warren, 2016).

The present-day environment and tectonic setting of EARS deep lakes may correspond to Mesozoic rift basins developed during the Gondwana breakup (see Chapter 3 in Mohriak, 2025). Some regions formed deep lacustrine rift basins that never accumulated evaporite sediments, such as the Recôncavo Basin in northeast Brazil (e.g., Figueiredo et al., 1994). Other rift basins are characterized by lacustrine systems that formed thick salt deposits during the first marine incursions in the Central and South Atlantic basins (Katz, 2001; Kukla et al., 2018; Sabato Ceraldi & Green, 2016).

Several lakes in the EARS exhibit organic-rich lacustrine sediments deposited in anoxic environments. For example, the lagoonal sediments around Lake Natron are rich in organic material, with muds along the lake edges displaying a total organic carbon content (TOC) of up to 10% (e.g., Demaison & Moore, 1980; Katz, 2001). However, these marginal deposits are unlikely to produce petroleum-source rocks due to rapid oxidation in subaerial settings. Large, deep, stratified, and anoxic lakes, such as Lake Tanganyika, are more prone to accumulating and preserving the richest lacustrine source rocks (e.g., Demaison & Moore, 1980; Huc et al., 1990; Tiercelin et al., 1994).

The Gregory Rift Valley is the main rift in the Eastern Branch of the EARS, extending from Mozambique and Malawi to Ethiopia and Djibouti. The northern part of the rift branch reaches the Afar Triangle, a region affected by magmatism and close to continental breakup (Chorowicz, 2005; Corti, 2009). Several lakes, including Lake Tanganyika, are present within the rift depressions. Lake Tanganyika, located in the western branch of the rift system, corresponds to one of the deepest lakes in the world, with a maximum water column of about 1470 m, and is characterized by anoxic zones at depths below circa 100 m (Huc et al., 1990; Tiercelin et al., 1994).

Some lakes, located at the eastern branch of the EARS, near the border between Tanzania and Kenya (Figure 9.18), are characterized by hypersaline brines, such as Lake Natron and Lake Magadi (Eugster, 1980; Owen et al., 2019). Unlike the western branch, the eastern branch of the EARS and the Afar region have been affected by intense volcanic activity in the Holocene (Beyene & Abdelsalam, 2005; Getenet et al., 2022; Macgregor, 2015). During the Early Holocene, Lake Natron and Lake Magadi were connected, forming a larger paleolake within the rift system. However, they are now separated by a volcanic barrier at an elevation of 635 m, rising tens of meters above the present-day lake shores, with a water level of about 600 m above sea level (Figure 9.18).

Lake Natron is a shallow lake (water depth less than 1 m), located in a depression that extends some 55 km in an N-S direction and reaches a width of about 24 km in the E-W direction. The lake's total surface area is about 950 km^2 (Figure 9.19). Lake Natron is a soda lake with extremely alkaline waters, with an average pH of 12. The lake is considered the most caustic lake in the world, with extremely toxic waters for most organisms (Owen et al., 2019; Vincens et al., 1986; Vincens & Casanova, 1987). The salt flats on the lake's surface are mainly composed of sodium carbonate, known as the mineral trona ($Na_2CO_3.10H_2O$). Satellite images show some residual lagoons filled with saline water and extremely shallow ponds or moats adjacent to the desiccated basin, which appear as dark blue colors at the southern and northern parts of the lake (Figure 9.19).

The extreme toxicity in the ponds around the salt flats (known as moats) results in a few species adapted to this harsh environment, such as the Spiricum algae, which feed on the halophilic cyanobacteria known as spirulina. The Spiricum algae are part of the food chain of many birds that flock around the lake. The water is unsuitable for mammals, and the lake's remoteness has repelled larger animals that feed on birds. This has turned Lake Natron into an ecological niche internationally famous for attracting pink flamingos that use the lake's shores as breeding grounds. The lake is also famous for its bright red colors imprinted on the lake's surface by the proliferation of halophilic organisms in the hypersaline brines (Scoon, 2018).

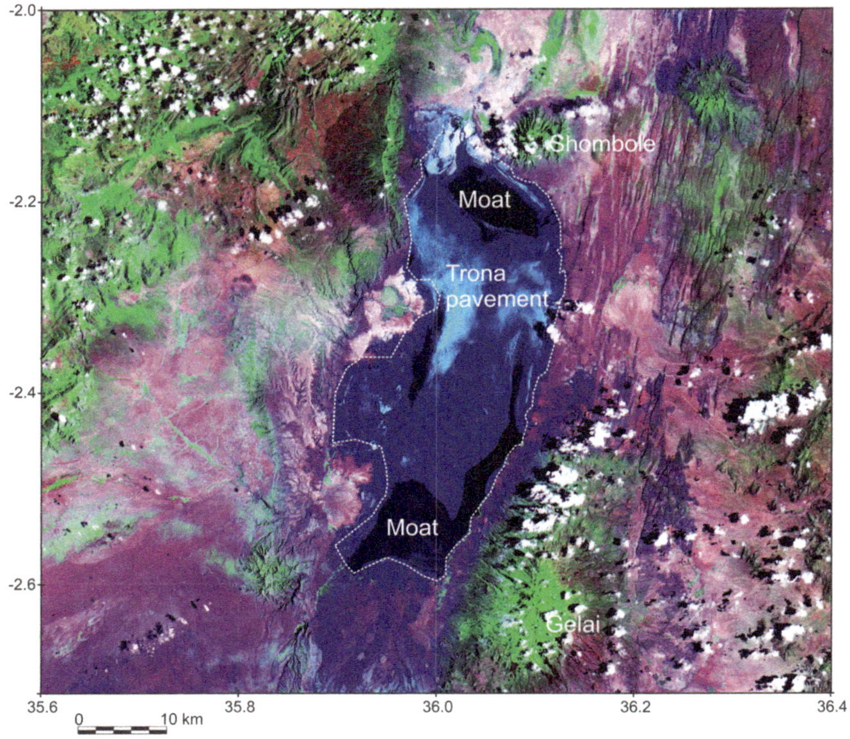

Figure 9.19 The Landsat satellite image (false colors) depicts volcanic rocks in the southern part of the EARS near Lake Natron. The lake is characterized by a depression with a trona salt pan and residual saline lagoons. Several volcanic structures, including Gelai Mount and Shombole Volcano, occur in the region. The volcanic basement east of the depression is characterized by N-S rift faults. Credit: Courtesy of NASA.

The lake occupies a low-lying region bordered by the Ethiopian Western Escarpment, a major uplifted region constructed by the piling up of Pliocene trachytic lava flows and tephra (Chorowicz, 2005). The eastern flexural border of the basin is covered by volcanic rocks less affected by extensional normal faults. The hydrographic closure of this endorheic basin is provided by extensional faults that affected the basement rocks and formed uplifted rift shoulders (Figure 9.18).

This region is marked by several tall volcanic structures, including Mount Kilimanjaro, the highest mountain in Africa, located to the east of the rift system (Figure 9.18). The giant Gelai volcano is located near the southeastern margin of Lake Natron. Southwest of the lake, the symmetric volcanic cone of Mt. Oldoinyo Lengai rises more than 2 km above the valley floor. This volcano, one of the most active Holocene volcanoes within the Gregory Rift, is famous for the mixture of alkaline silicate and sodium carbonate minerals in the magma that periodically erupts at the surface as low-temperature lava flows (e.g., Scoon, 2018). The rift depression continues south of Tanzania, with Lake Eyesi and Lake Manyara being large lakes filled with less saline waters and lacking evaporite deposits (Warren, 2006, 2016).

During the wet season, Lake Natron has a deeper water column than Lake Magadi, reaching depths of about 1–3 m. Small rivers feed the lake, but most of the water comes from a swamp area that forms shallow lagoons. Lake Natron is one of the most cited saline-alkaline lakes in geological literature that focuses on evaporitic sedimentary environments (e.g., Scoon, 2018). In addition, both Lake Magadi and Lake Natron are cited in anthropological works as the birthplace or cradle for the first humanoids (e.g., Deocampo et al., 2022; Owen et al., 2018).

There are dozens of lakes within the depressions of the EARS, but only Lake Magadi and Lake Natron are characterized by the deposition of bedded evaporites in modern times (Deocampo et al., 2022; Eugster, 1982, 1986). These lakes are located near the equator, and the evaporation rate in the rift valley is much higher than precipitation. The negative hydrological balance leads to a concentration of chemical components that are presently depositing along the lake's shores, forming a trona pavement (Figure 9.19).

Lake Magadi is a large ephemeral continental saline lake located in a depression near a large master fault to the west of the main rift system. At an elevation of 579 m, the lake surface of this asymmetric graben structure corresponds to the lowest point in the depression. Other lakes in the EARS show lower lake surfaces, such as Lake Turkana (360 m) and Lake Abhe (243 m) in Ethiopia-Djibout. The lowest lake surface in Africa is Lake Asal in Djibouti, a small rift depression periodically flooded by Gulf of Aden seawater (Gasse & Fontes, 1989).

Lake Magadi is considered a typical example of a large ephemeral continental saline lake with alkaline sodium carbonate minerals forming a well-developed salt pan, lagoonal mudflats, and dry mudflats, some of which correspond to old lake bottoms. Ephemeral stream complexes and small alluvial fans are recorded in this sedimentary setting, but steep shorelines characterized by extensional faults associated with earthquake activity prevent the formation of wide mudflats, or sabkhas (Figure 9.20).

Currently, the Magadi Basin exhibits several recognized sedimentary facies, including: (1) alluvial fans; (2) ephemeral stream floodplains; (3) spring deposits; (4) delta; (5) dry sandflat and mudflat; (6) saline mudflats; and (7) salt pan.

In the subsurface, exploratory drilling has revealed thick layers of lacustrine sediments overlying trachyte flows, indicating two earlier phases of sediment accumulation: the High Magadi beds and the ancient Lake Oloronga beds, which are associated with lacustrine sediments deposited at least 800 ka (Eugster, 1980). These ancient lakes were characterized by less saline or freshwaters, and the lake sediments were mainly volcanic sand, silt, and clay, with chert beds indicating high silica content (Eugster, 1980).

Jones et al. (1977) collected data from brines obtained from boreholes as deep as 297 m and characterized two distinct brine bodies below the present-day lake. The shallow brines near the surface have a high concentration of dissolved solids (260 g/kg on average). The brines in the deeper lacustrine sediments or fractured lavas exhibit only half the concentration of the superficial brines. Jones et al. (1977) suggested that meteoric waters infiltrate from areas with higher rainfall and bring several ions to the groundwater (Figure 9.21). Groundwater and seeps migrate through the sediments and concentrate in shallow lakes, which are evaporated due to solar heat, precipitating evaporite sediments (mainly sodium carbonate) on the depression's lowest points and forming the trona pavement (Warren, 2006, 2016).

The perennial lake system in a relatively arid environment requires continuous freshwater input from rain, rivers, and spring seeps. Ephemeral streams and hydrothermal springs provide a constant influx of fresh and saline water into the EARS saline lacustrine systems (Handford, 1991). The springs are characterized by thermal waters enriched in sodium and bicarbonate ions

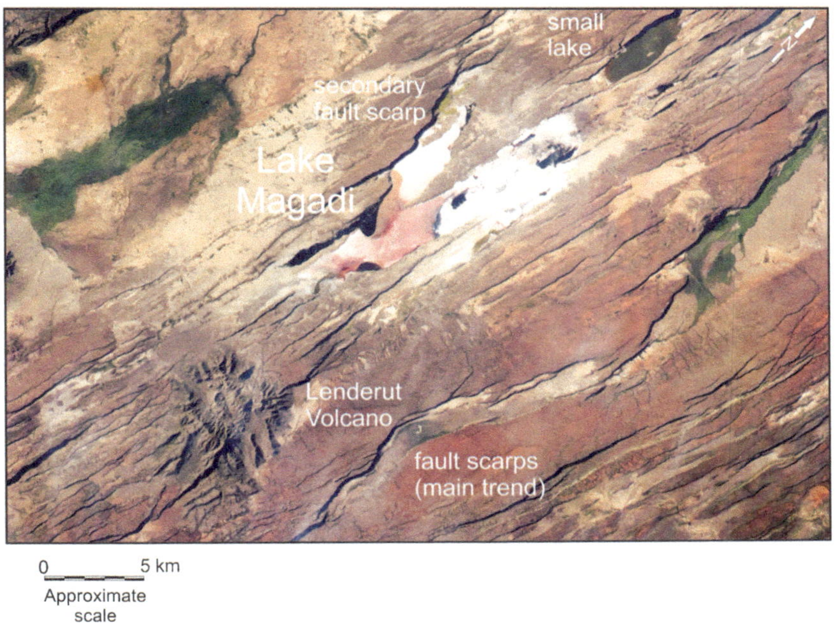

Figure 9.20 The NASA satellite image (oblique view) shows the East African rift system. Lake Magadi's salt pan and evaporite sediments, primarily soda carbonate, form a trona pavement with varying reflectivity. Trachytic volcanic rocks are associated with various volcanoes, and extensional faults affect lava flows. The Lenderut Volcano, with a radial dyke pattern, is affected by a major normal fault that bifurcates toward Lake Magadi, forming a relay ramp. Lake Natron is located south of the Lenderut Volcano within a rift valley depression marked by large fault scarps. Credit: Courtesy of NASA.

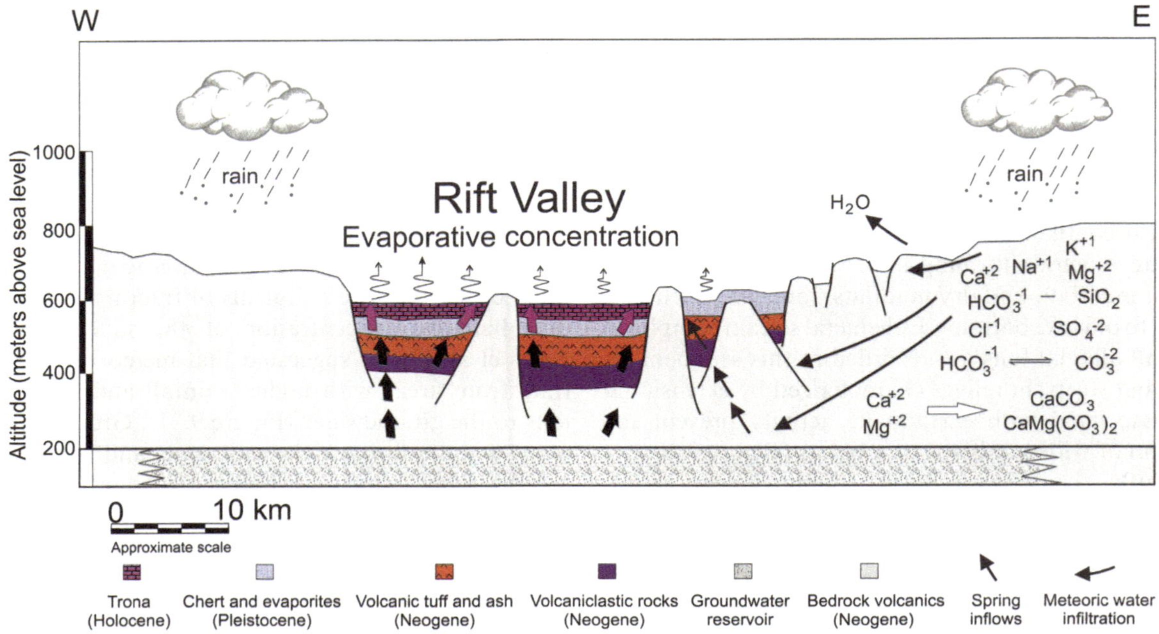

Figure 9.21 Schematic geological cross-section showing a rift valley with lacustrine systems depositing evaporite sediments on the lowest points of the depression. Meteoric waters infiltrate areas with higher rainfall, bringing ions into the depression. Solar evaporation causes brine concentration and migration through sediments, causing sodium carbonate precipitation on the lake's floor when nonmarine brine reaches saturation. Desiccation causes salt crusts to cover the lake's floor, forming a trona pavement. Credit: Warren (2006)/with permission of Springer Nature.

because of their location in volcanically active terrains (Figure 9.21). Some springs form lagoons with different hydrology around the lake's edge, while others are responsible for the only permanent surface water input into the hydrological cycle, forming perennial ponds that do not dry up even during times of increased aridity (Warren, 2006, 2016).

Evaporite sediments are initially formed in a subaqueous environment, forming salinas. When the lake is desiccated, the evaporite deposits become subaerial, forming a salt pan with an evaporite pavement in the central part of the lake. The extensive salt flats appear with a white-to-pink tint in satellite images (Figure 9.22). Moats correspond to depressions with bathymetry that may reach several meters and exhibit a deep blue reflectivity on satellite images (Figure 9.22).

In the rainy seasons (March-April), there is enough runoff to fill the lake with fresh water, which dissolves part of the previously deposited trona crusts surrounding the central pan. Lake Magadi is a very shallow body of water, often desiccated in dry seasons due to intense evaporation, precipitating evaporite minerals, particularly trona ($NaHCO_3$. Na_2CO_3.$2H_2O$), which accumulate on the lake's bottom. This results in an annual cyclicity of salt deposition, dissolution, and re-deposition of the evaporite layers (Warren, 2006, 2016).

The central part of Lake Magadi is often subaerially exposed, and most of the surface is covered by evaporites (trona and other salts), which appear as a white sheet in satellite images (Figure 9.22). Floodplains associated with the drainage system locally accumulate fluvial sediments when the rivers dry up. When the lake is flooded again during wet periods, another cycle of lake replenishment begins. Trona dissolves and is subsequently reprecipitated or transported as detrital sediments that accumulate on top of previous layers. The result is a thick trona layer that may reach a thickness of one to a few tens of meters, covering areas that may reach almost 100 km^2 (Vincens & Casanova, 1987).

Sodium carbonate minerals form layers that can be of commercial value in some hypersaline lakes. The Kenyan salt industry has been exploiting evaporite resources in Lake Magadi for decades, and plans are underway to exploit sodium carbonate minerals in Lake Natron, located in northern Tanzania (Kadigi et al., 2014).

9.4.2. Saline Lakes in the Afar Region of Northeastern Africa

The Afar region is a triangular-shaped area in northeastern Africa, located north of the EARS and southwest of the Red Sea (Figure 9.23). It is home to the Afar Volcanic

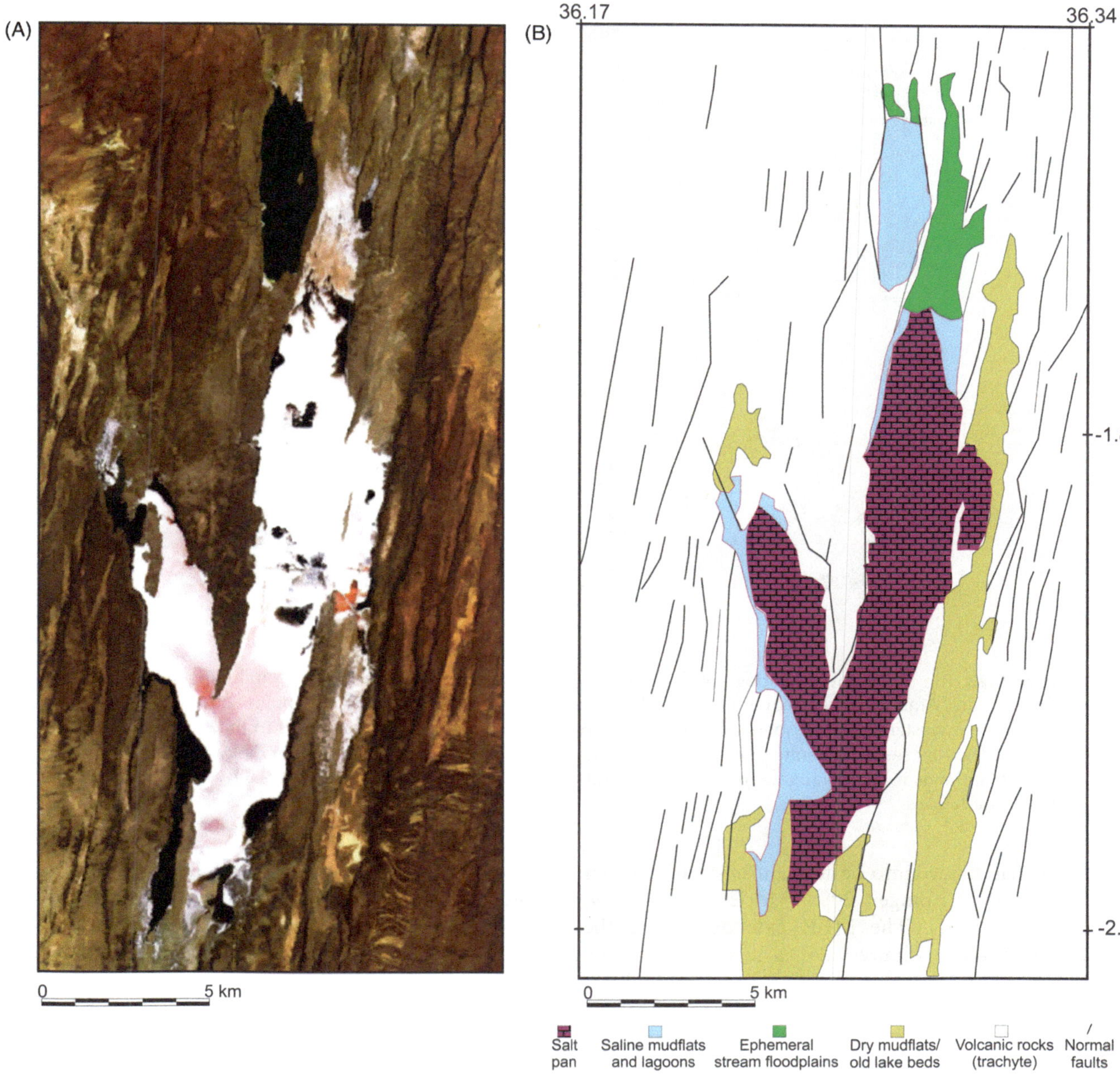

Figure 9.22 (A) Zoom of a NASA satellite image (true colors) showing Lake Magadi salt pan with whitish-to-pinkish reflectivity and volcanic rocks in a rift depression with brown colors. Credit: Courtesy of NASA. (B) Simplified sedimentary facies map of Lake Magadi salt pan. Most evaporite sediments correspond to trona (sodium carbonate). Credit: Adapted from Eugster (1980).

Province, one of the most important Large Igneous Provinces (LIPs) in the world, with widespread volcanic rock outcrops dating back to the Neogene (Barberi & Varet, 1977; Coffin & Eldhom, 1994; Ernst, 2014; Stab et al., 2016; Tazieff et al., 1972). The Afar Triangle spans approximately 600 km in length and 300 km in width, covering an area of around 200,000 km^2 (Figure 9.23).

The Afar province is situated at the triple junction involving the Nubian, Somalian, and Arabian plates, with three rift systems meeting in the region. The integration of topobathymetric maps (e.g., GEBCO, 2019) and geological maps (e.g., Chorlton, 2007) indicate that two branches of the triple junction correspond to depressions marked by oceanic rifts with active spreading centers: the southern Red Sea and the western Gulf of Aden (Bosworth et al., 2005). The third arm of the triple junction is associated with the Main Ethiopian Rift (MER), corresponding to the northern extension of the

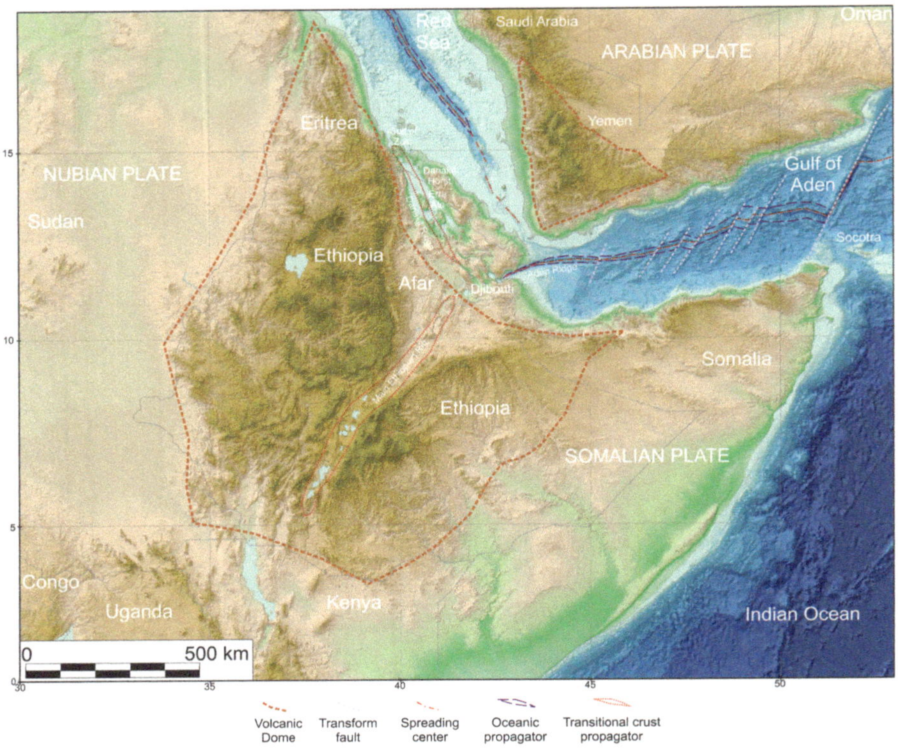

Figure 9.23 The shaded-relief topographic map shows the northern African margin, including the Nubian, Somalian, and Arabian plates, that are separated by the Red Sea and Gulf of Aden oceanic rifts. A triple junction, including the western Gulf of Aden axial ridge, Danakil Ridge, and the MER, is located in the Afar region near Djibouti. The Afar region is centered in the Ethiopian Dome. Neogene basalts extend from the Afar Triangle to the Yemen Highlands in the conjugate margin of the southern Red Sea. Credit: Stab et al. (2016)/with permission of John Wiley & Sons.

EARS, which is divided into western and eastern branches (Chorowicz, 2005; Ebinger, 2005; Morley et al., 1999; Tazieff et al., 1972). The MER extends toward the Danakil Depression (Figure 9.24), a region marked by recent volcanic and seismic activity (Bastow et al., 2018; Bonini et al., 2005).

According to several researchers, the Afar lithosphere corresponds to a more advanced stage in the development of oceanic crust than the EARS (e.g., Ebinger, 2005). Some authors postulate that the Afar rift system formed on attenuated continental crust intruded by magmatic centers. The depression was covered by flood basalts during a prolonged rift stage that started in the Neogene and has been active up to the present (Beyene & Abdelsalam, 2005). Other authors postulate that the magnetic stripes and other geophysical data in the Danakil depression indicate a transitional stage or proto-oceanic crust underlying the rifts. Mohr (1989) and Bridges et al. (2012) propose a highly intruded rifted crust and a newly formed oceanic crust with embryonic spreading centers developed during the Pliocene-Pleistocene. According to this hypothesis, the Danakil Ridge in the subsurface probably extends to the Afar region near Lake Asal in Djibouti (Figure 9.24).

Hutchinson and Engels (1970) proposed that rift processes in the southern Red Sea were active in the Upper Miocene, with spreading centers dying out near the Strait of Bab el Mandeb as the magmatic and tectonic activities were shifted westward during the Pliocene-Pleistocene interval (Bastow et al., 2018; Beyene & Abdelsalam, 2005). This is indicated by major faults and earthquake activity in the western part of the Danakil Depression in Ethiopia, characterized by a Quaternary salt basin overlying the rift deposits (Figure 9.25). The Danakil Depression is marked by a magmatic complex with large active volcanoes, such as the Erta Ale (Barberi & Varet, 1977; Moore et al., 2019).

The Afar Province volcanic rocks span from northern Africa to the southwestern margin of Arabia, forming the Ethiopian Plateau and the volcanic complexes in the depression. The Danakil Depression, located in the northern part of the Afar Triangle, is characterized by Quaternary siliciclastic, carbonate, and evaporite sediments overlying the upper layers of the Neogene rift trough (Figure 9.25). The volcano-sedimentary sequences in the Danakil rift basin include the Neogene basalt and the

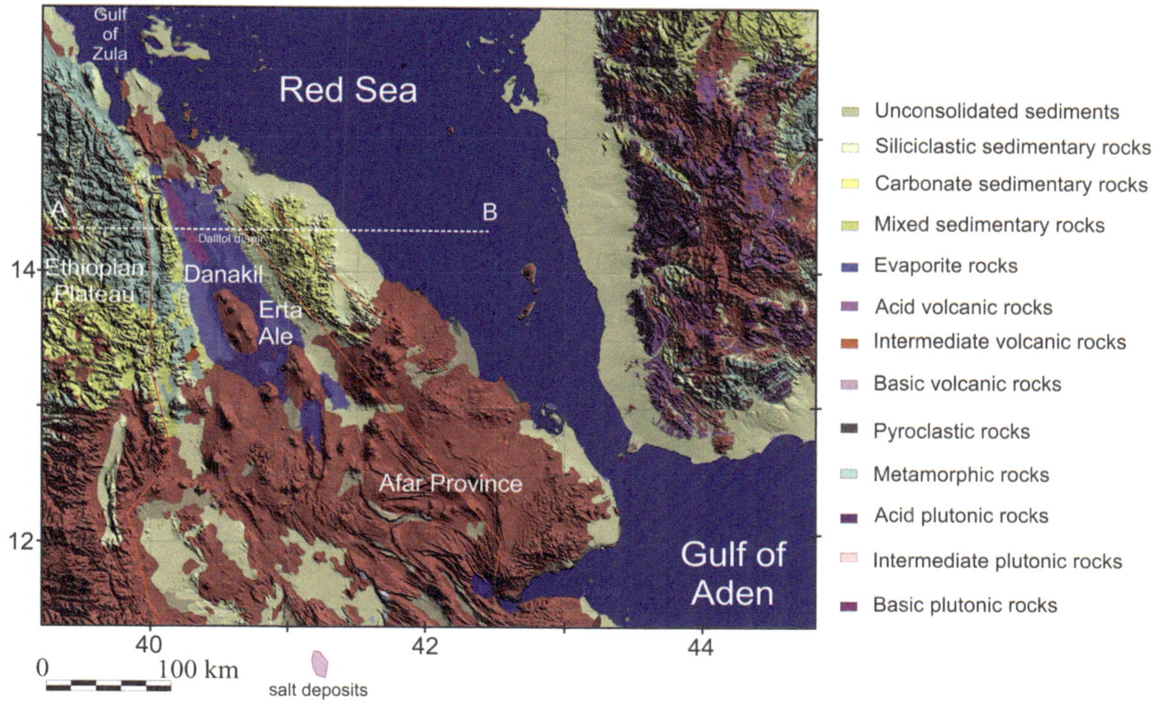

Figure 9.24 Zoom image of the simplified lithological units draped on the digital elevation model of the Afar volcanic province and the Danakil Depression. The Afar volcanic province and the Danakil Depression are centered at the intersection of the southern Red Sea, the Gulf of Aden, and the East African rift system. The light blue colors indicate the presence of evaporite sediments covering the Dallol depression. One of the largest magmatic centers in the Afar region is the Erta Ale Volcanic Complex, characterized by an ellipsoidal structure trending NW-SE. The Danakil salt flat is 120 m below global sea level, and the volcanic structures near the Gulf of Zula in Eritrea likely formed a barrier to isolate the depression from the southern Red Sea. During the Quaternary, repeated marine ingressions formed a lagoon that periodically desiccated, resulting in the precipitation of salt layers. Credit: Hartmann and Moosdorf (2012)/with permission of John Wiley & Sons.

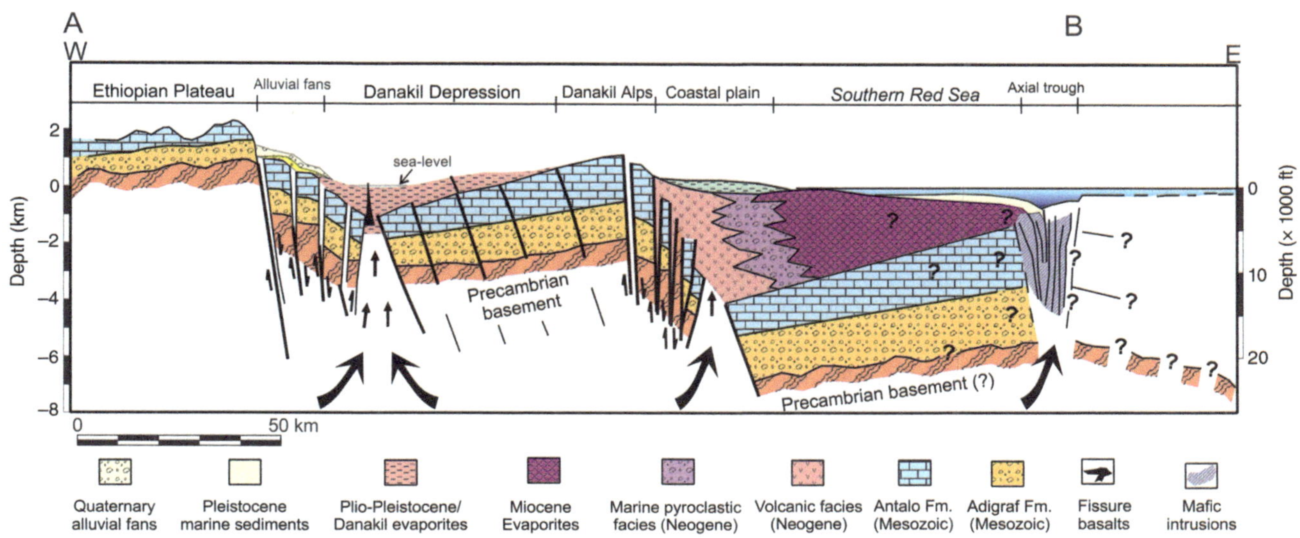

Figure 9.25 The schematic geological cross-section, extending from the Ethiopian Plateau to the southern Red Sea axial trough, shows the Danakil Depression in the Afar region. The Danakil Horst separates the evaporite-filled Quaternary depression from the coastal plains bordering the Red Sea continental margin. The Red Sea coastal plain sediments overlie volcanic rocks and possibly Mesozoic pre-rift sediments above the Precambrian basement. The offshore basin in the southern Red Sea is characterized by massive Middle Miocene salt deposits in the deep-water region, forming a wedge that pinches out toward the axial trough with the active spreading center.

Cenozoic to Mesozoic pre-rift sedimentary sequences that outcrop in the Ethiopian highlands and in the Danakil Horst that separates the depression from the Red Sea (Abbate et al., 2015; Bastow et al., 2018; Beyene & Abdelsalam, 2005; Corti et al., 2015; Holwerda & Hutchinson, 1968; Manighetti et al., 1998).

Four main physiographic provinces are identified in the Afar region: (1) the Ethiopian Plateau highlands that border the rift depressions; (2) the salt plains and volcanic fields in the Danakil Depression; (3) the Danakil Horst, also known as the Danakil Alps, located east of the Danakil Depression; and (4) the Red Sea coastal plains and continental margins offshore Eritrea and Yemen. In the Yemeni conjugate margin, the volcanic rocks of the Yemeni Plateau are located on land, eastward of the coastal plains adjacent to the Red Sea (Figure 9.24).

The widespread occurrence of volcanic rocks forming igneous massifs and basaltic lava flows in the Afar Province indicates a mantle plume influence and proximity to the oceanic spreading center in the southern Red Sea (Bosworth et al., 2005; Corti, 2009; Tazieff et al., 1972; Tesfaye et al., 2003). Most of the volcanic and earthquake activity in the central part of the Afar region follows a direction trending NNW, sub-parallel to the axial trough in the southern Red Sea (e.g., Illsley-Kemp et al., 2018).

The Erta Ale Volcanic Complex is located north of the Ethiopian Dome, at the center of the Danakil Depression, forming magmatic centers surrounded by Quaternary sedimentary deposits (Figure 9.23). The igneous massifs exhibit an NW trend sub-parallel to the axis of the southern Red Sea (see Chapter 1 in Mohriak, 2025). North of the Erta Ale Volcanic Complex, the Danakil Depression is filled by volcanic basaltic rocks, forming igneous massifs and extrusive lava flows, and evaporite rocks, which appear as white patches in the satellite image (Figure 9.24).

The Ethiopian Plateau, westward of the Danakil Depression, is characterized by high elevations with a mean altitude of around 2 km above sea level, but locally reaching up to 4 km in some ranges (Figures 9.23–9.25). The plateau is uplifted with Precambrian basement rocks exposed and locally covered by Mesozoic siliciclastic and carbonate rocks (Bonini et al., 2005). The plateau was strongly affected by magmatic activity and extensional tectonics during the Neogene (Corti, 2009; Corti et al., 2015; Ebinger, 2005; Hutchinson & Engels, 1970; Momenzadeh, 1990). The plateau uplift probably resulted from the impact of the Afar plume in the Neogene, forming magmatic additions to the crust, and flexural uplift of the rift border by extensional faults (Bosworth et al., 2005; Stab et al., 2016).

At the eastern edge of the Ethiopian Plateau, there is a steep, rugged escarpment some 50 km wide that is dissected by fluvial systems with alluvial fans that advance toward the depression (Figure 9.25). This feature, known as the Ethiopian Escarpment, was formed by a system of basement-involved, eastward-dipping faults developed in the Neogene (Abbate et al., 2015; Bosworth et al., 2005; Stab et al., 2016). The downfaulted blocks developed an elevated rift shoulder with active alluvial fans that advanced toward the Danakil Depression during the Quaternary. The rift border fault corresponds to the western boundary of the Danakil Depression (Figure 9.24).

The Danakil Depression corresponds to an asymmetric rift trough characterized by a sedimentary depocenter developed from the Miocene to Recent (Bastow et al., 2018; Corti et al., 2015; Hutchinson & Engels, 1970; Momenzadeh, 1990). The topographic elevation of more than 2 km in the Ethiopian Plateau rapidly decreases to elevations below sea level in the Afar depression by normal faults that dip to the east (Abbate et al., 2015; Corti et al., 2015; Bastow et al., 2018). The sediments in the rift basin accumulated thick layers of siliciclastic sediments interdigitated with evaporite deposits that precipitated in a lagoon that covered the depression during the Plio-Pleistocene and Quaternary (Figure 9.25).

The Danakil Depression exhibits a general N-S direction, with a length of about 185 km and a maximum width of about 70 km in the south (Figure 9.24). The surface of the Danakil Depression corresponds to a salt flat at an altitude of 125 m below sea level (Bastow et al., 2018; Bosworth et al., 2005; Tazieff et al., 1972). This desertic, hyperarid, and hostile environment occupies an almost uninhabited area measuring about 100,000 km^2, located north of the volcanic rocks that outcrop in the Afar triple junction and Erta Ale Volcanic Complex (Figure 9.24). The depression is one of the hottest and lowest places on Earth's surface, with mean annual temperatures above 35°C and maximum temperatures exceeding 50°C in the summer. Annual rainfall in the depression is typically less than 200 mm each year, as reported by Fazzini et al. (2015). Some perennial and ephemeral lakes exhibit a surface that is more than 100 m below the global sea level, such as Lake Assale, Lake Bakili, and Lake Afrera, located north of the Erta Ale Volcanic Complex (Figure 9.24).

Most of the Danakil Depression surface is covered by Holocene mud and salt flats (Figure 9.25). However, the Neogene extensional tectonics developed local horst structures in the subsurface. The depression extends toward the Danakil Horst, also known as the Danakil Alps, a large block reaching elevations up to 730 m (Figure 9.25), with a sedimentary cover that dips to the west due to the flexural uplift of the basement (Bastow et al., 2018; Hutchinson & Engels, 1970; Momenzadeh, 1990).

The eastern flank of the structure is characterized by the outcrop of pre-rift Mesozoic sedimentary rocks that overlie the Precambrian metamorphic basement

(Figure 9.25). The Danakil Horst separates the depression from the southwestern Red Sea coastal plain. The structural high is downfaulted by a system of normal faults that dip toward the Red Sea, forming an uplifted border adjacent to the asymmetric Red Sea rift structure in the offshore region (Figure 9.25). The Red Sea basin is characterized by Neogene sediments overlying syn-rift and pre-rift sedimentary sequences, including an extremely thick salt layer deposited during the Middle to Late Miocene (see Chapter 1 in Mohriak, 2025).

9.4.3. The Danakil Depression

The magmatic centers in the Danakil Depression formed large volcanic structures surrounded by Neogene to Quaternary sedimentary sequences, including thick Quaternary evaporite deposits (Barberi & Varet, 1977; Hutchinson & Engels, 1970; López-García et al., 2020; Tazieff et al., 1972). The depression is home to several active volcanoes, including the famous Erta Ale and Gada Ale, which are located between Assale (or Karum) Lake and Bakili Lake (Figure 9.24). The Erta Ale volcano, located at the northern part of a magmatic center, is one of the most active shield volcanoes in the world, with a spectacular permanent lake of molten lava.

The Danakil Depression is characterized by a Quaternary carbonate-to-evaporite mudflat surrounding the NW-trending Erta Ale volcanic complex, with several magmatic centers forming volcanic edifices (Figure 9.24). The northern part of the salt flats, located adjacent to Lake Assale (north of the Erta Ale volcanic complex), is much narrower, reaching a width of about 10 km. The deepest evaporite-filled troughs are located to the west, near the Ethiopian Plateau downfaulted blocks (Bastow et al., 2018; Hutchinson & Engels, 1970, 1972). Some alluvial fans from the Ethiopian highlands and the Danakil Horst have advanced toward the center of the Danakil Depression (López-Garcia et al., 2020).

The hypersaline lacustrine systems in the Danakil Depression constitute modern examples of ephemeral to perennial salt lakes on incipient volcanic margins. These lakes were formed in a tectonic setting of plate divergence, leading to the inception of transitional to oceanic crust through the development of magmatic centers. This transitional stage is characterized by the widespread magmatism recorded in the Danakil Depression, from the northern Dallol plains to the southern Afar province, near Lake Asal in Djibouti (Figure 9.24). Geological maps indicate the widespread occurrence of basaltic volcanic rocks in the region extending from Afar to the Danakil Depression (e.g., Hagos et al., 2016; Hartmann & Moosdorf, 2012)

The Danakil Depression was connected to the southern Red Sea in the Pleistocene, forming a lagoon that extended toward the Erta Ale magmatic centers (Almalki et al., 2015; Bonatti et al., 1971). During the Quaternary, stratovolcanoes formed a volcanic barrier near the Gulf of Zula, isolating the depression from the open marine environment (Bonatti et al., 1971; Mitchell et al., 1992; Talbot, 2008).

The Quaternary lagoon was filled by transient marine incursions of Red Sea waters, which periodically flooded the basin, depositing limestones and siliciclastic rocks (Almalki et al., 2015; Bonatti et al., 1971). Carbon and oxygen isotope studies indicate that coral debris was deposited under normal open marine seawater around 54 ka, indicating effective communication between the lagoon and the Red Sea. Subsequently, the embayment was isolated from the Red Sea by a volcanic barrier, as indicated by the volcanic construction at the northern boundary of the depression, south of the Gulf of Zula (Figure 9.24). As the lagoon was evaporated, the carbonate rocks were recrystallized, and the brine salinity increased, depositing salt layers (Bonatti et al., 1971).

The lagoon was desiccated around 32 ka, with widespread halite deposits forming a salt pan at the northern part of the depression (Figure 9.24). Locally, the salt thickness probably exceeds 3,000 m (Holwerda & Hutchinson, 1968). Evaporite cycles in the subsurface indicate partial or total desiccation episodes during the Holocene. Dry periods resulted in the deposition of gypsum and halite, which alternated with siliciclastic and carbonate deposits during marine transgressions and wet periods (Almalki et al., 2015; Bonatti et al., 1971).

9.4.4. The Dallol Depression Salt Flats

The Dallol Depression, located north of the Erta Ale volcanic complex, is characterized by the Dallol salt flats, a Quaternary salt pan floored by siliciclastic, carbonate, and evaporite deposits (Figure 9.26). The sediments on the surface are characterized by an alternation of mudstone and carbonate rocks deposited in submarine lacustrine settings, covered by gypsum and halite crusts, which are associated with desiccation events in the Quaternary (Bonatti et al., 1971; Cavalazzi et al., 2019; Hutchinson & Engels, 1970; Holwerda & Hutchinson, 1968). Evaporite sequences are also registered in the subsurface, including halite, anhydrite, and thin layers of more soluble minerals (carnallite, sylvinite, and bischofite), particularly in the northern part of the depression (Bastow et al., 2018; López-Garcia et al., 2020; Pedley et al., 2016).

The evaporite sediments in the Dallol Depression were precipitated in a lagoon connected to the Red Sea before the uplift of the Danakil Horst to the east. The basin isolation is related to the development of a volcanic barrier that interrupted the lagoon's communication with the

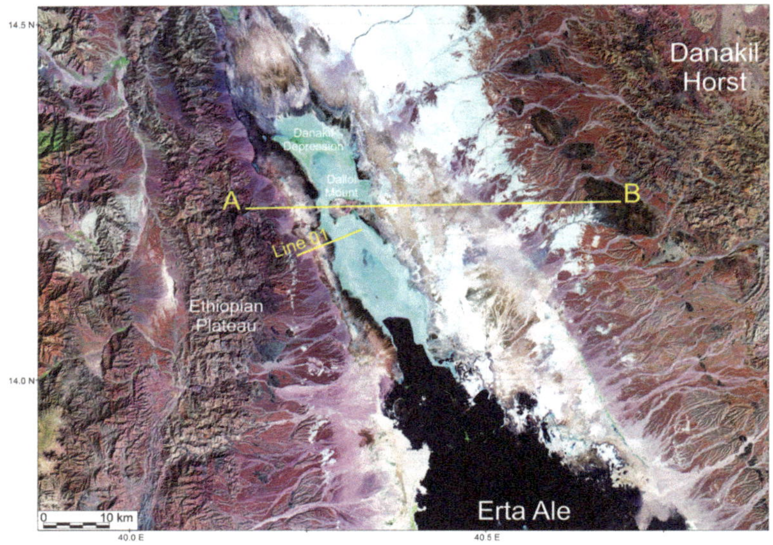

Figure 9.26 The satellite image (false colors) depicts the Dallol salt flats in bright blue, north of the volcanic complex near the Assale (Karum) and Bakili lakes, which exhibit black reflectivity. The Dallol Mount is associated with an elliptic structure with an E-W trend, rising 40–60 m above the salt flats. A thin yellow line A-B crosses the Danakil Depression, with a crater near the summit of the structures. Seismic line 01 is located 4 km south of the Dallol Mount, extending from the alluvial fans in the Ethiopian Escarpment to the Danakil Depression. Credit: Image by NASA.

Southern Red Sea in the Quaternary. The piling up of basaltic lavas formed a volcanic barrier in the region of the Gulf of Zula in Eritrea, north of the Dallol depression, isolating the lagoon from the southern Red Sea (Figure 9.24). The arid climate and water evaporation resulted in a hypersaline lake, and some researchers postulate that the Dallol salt flats corresponded to a shallow salina until at least 32 ka, when the water was evaporated and halite was deposited, forming a salt pan (Bonatti et al., 1971; Carniel et al., 2010).

The total desiccation of the paleolake resulted in a dry playa lake that has been dry since the end of the last glaciation. However, in the early Holocene (circa 12 to 5 ka), some episodic intervals of a more wet climate are registered, causing lake levels in the Danakil Depression to rise substantially. During these episodes, lake levels in the Danakil Depression rose substantially, as indicated by paleolake shorelines in the Lake Afrera region (Bonatti et al., 2016).

The salt flats in the Dallol region are characterized by extensive gypsum plain and halite crust, likely derived from remobilized evaporites dissolved from deeper salt levels (Bonatti et al., 2016). In the subsurface, the Quaternary salt layers are mainly composed of halite, mixed with a few thin beds of potash minerals (sylvite). Evaporites in the Dallol salt flats have been exploited since the beginning of the last century, particularly chlorides like sylvite (KCl). Sulfate layers with economic importance, such as kainite ($MgSO_4.KCl.3H_2O$), occur at relatively shallow depths in the western footwall blocks (Pedley et al., 2016).

The Dallol salt flats have been the subject of extensive research, with borehole drilling and seismic profiles revealing a minimum thickness of over 2 km of Quaternary evaporites in the deeper troughs of the rift depocenter (Figure 9.27). According to several researchers (e.g., Bastow et al., 2018; Bonatti et al., 1971; Franzson et al., 2015; Hutchinson & Engels, 1970, 1972; Pedley et al., 2016; Talbot, 2008), the evaporite deposits overlie clastic units affected by rift-related normal faults. The evaporite layers can be divided into three main salt sequences: Upper Rock Salt, Middle Rock Salt (also known as the Houston Formation), and Lower Rock Salt (Pedley et al., 2016).

The Upper Rock Salt is mainly composed of halite, with gypsum and anhydrite layers interbedded with clays. This very thick sequence was drilled in the depocenter located east of the rift border fault (Figure 9.27) by the exploratory well DK-10-04B, located about 4 km southward of the Dallol Mount crater. The well penetrated 975 m of halite (reflectors A, B, and C) with two thin intervals of bittern salts at the base (Holwerda & Hutchinson, 1968). The Middle or Intermediate evaporite sequence, associated with the Houston Formation, consists of anhydrite with clay and halite layers, rich in potash minerals such as kainite, carnallite, and sylvite. The Lower Rock Salt Formation consists of halite with coarse anhydrite and clay intercalations (Bastow et al., 2018; Cavalazzi et al., 2019; Pedley et al., 2016).

The Houston Formation comprises various lithological units, including the sylvinite layer, upper carnallite, bischofite, lower carnallite, and kainite members

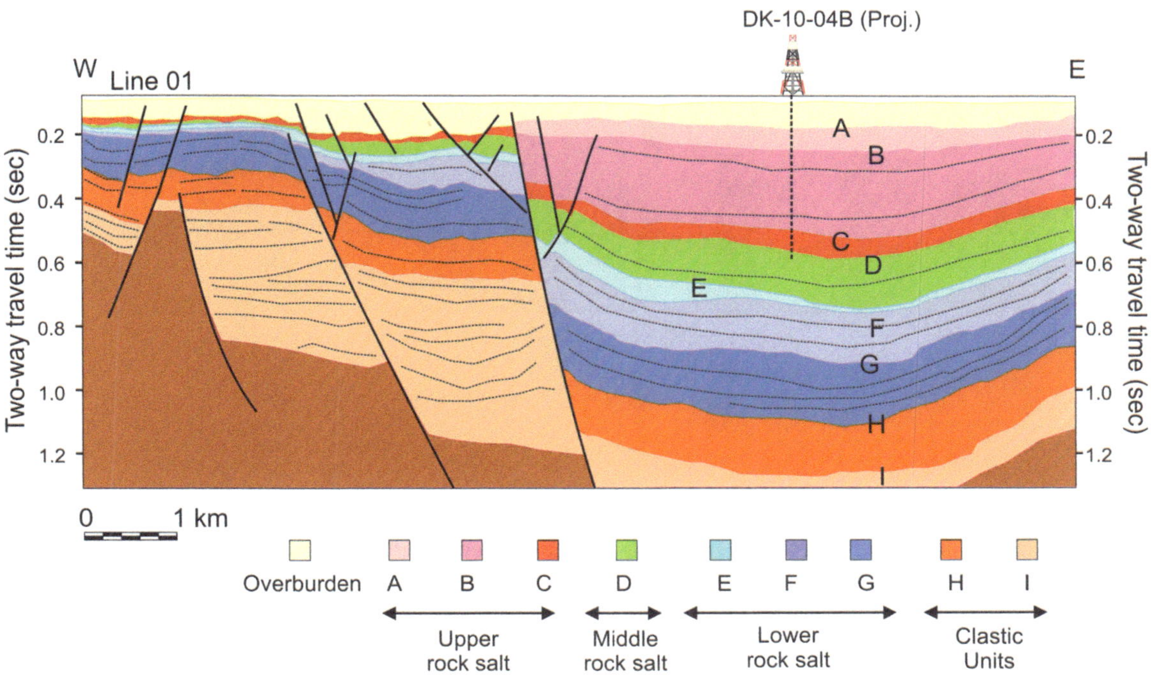

Figure 9.27 The geoseismic profile in the Dallol salt flat (Line 01; see location in Figure 9.26) reveals the Upper, Middle, and Lower rock salt units in the Dallol Depression. A borehole drilled to 975 m penetrated the upper rock salt units and the middle rock salt. The Middle Rock Salt unit is associated with various lithological types of the Houston Formation, including sylvite ores. Lower evaporite sequences likely extend at least 2 km below the surface. Credit: Adapted from Bastow et al. (2018).

(Pedley et al., 2016). The upper sequence is separated from the Middle Rock Salt Sequence by a marker bed corresponding to the seismic reflector D (Figure 9.27). The sylvinite layer is the main potassium ore, forming productive zones with thicknesses ranging from 15 to 40 m. The Lower Rock Salt Sequence has been penetrated in the footwall block, where the underlying clastic units are shallower (Figure 9.27).

Potash salt deposits in the Dallol salt flats cannot be explained by the solar evaporation of normal Red Sea seawater. Geochemical analysis suggests that these evaporites are the result of ion enrichment associated with nearby hot springs and rift-related igneous intrusions (Bastow et al., 2018; Cavalazzi et al., 2019; López-García et al., 2020). Hydrothermal fluids carrying $CaCl_2$ brines mixed with overlying brines precipitated primary $MgSO_4$-poor potash evaporites and formed halite-carnallite deposits. Sylvite layers are likely formed as secondary minerals, selectively leaching extremely soluble $MgCl_2$ from carnallite (Holwerda & Hutchinson, 1968).

9.4.5. The Dallol Salt Dome

The northern part of the Danakil Depression is a lowland salt pan with a surface about 100 m below sea level, formed by the evaporation of the hypersaline lagoon in the Holocene. The central part of the Dallol salt flats is characterized by an elliptically-shaped topographic anomaly forming a low-relief swell with a diameter of about 2 km, known as Dallol Mount or Dallol Volcano, according to some authors (e.g., Cavalazzi et al., 2019; Warren, 2016). This structure stands out with a top circa 40–60 m above the flat salt plains, which is almost 120 m below the global sea level (Figure 9.26). The flanks of the dome are gently dipping from a central vent or collapsed crater on its top, forming a structure that is remarkably similar to a volcano, but the flanks are composed of salt layers instead of lava flows (Figure 9.28).

The landscape around the Dallol Dome is marked by tall salt columns that have been eroded along canyons that cut across the uplifted sediments (Figure 9.26). The strata on the top of the exposed dome are characterized by anhydrite layers that serve as a protective cover and prevent the dissolution of the underlying halite (López-Garcia et al., 2020). The central part of the circular structure corresponds to a collapse crater resulting from salt dissolution or explosive volcanism (Holwerda & Hutchinson, 1968).

The only igneous rocks that outcrop in the northern Danakil Depression near the Dallol dome correspond to the lava flows from the Gada Ale shield volcano. This structure is the northernmost cone of a series of volcanic craters associated with the magmatic complex that extends to the Erta Ale shield volcano, located about

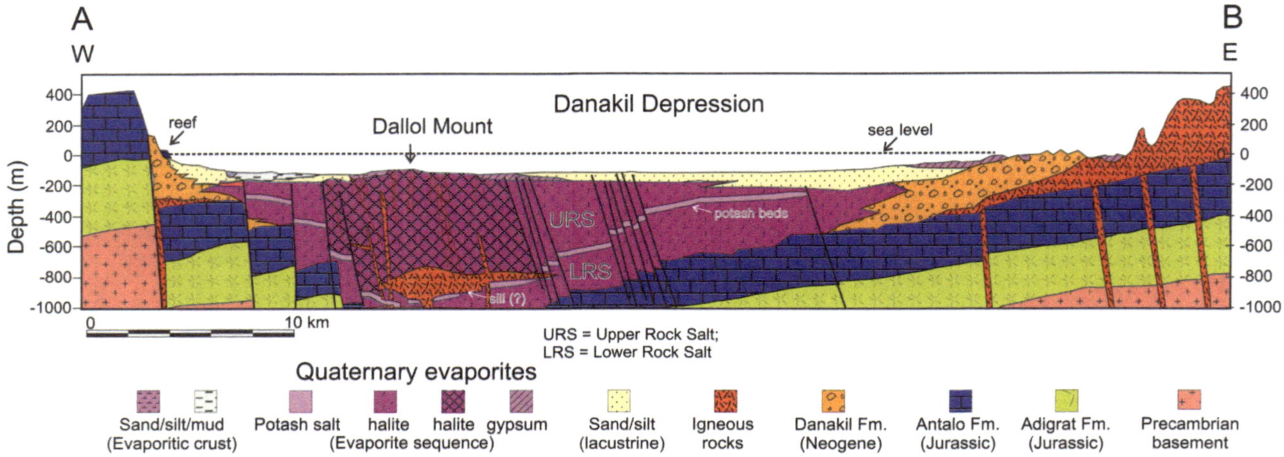

Figure 9.28 Schematic geological cross-section from the western Dallol Depression toward the Danakil Horst (see location in Figure 9.26). The fan deltas from the western flank of the Danakil Horst advanced toward the Danakil depression, covering the salt flats and reaching the Dallol Mount near its summit. Credit: López-Garcia et al. (2020)/Frontiers Media S.A./CC BY 4.0.

80 km to the south of Lake Assale (Figure 9.24). These basaltic volcanic rocks occupy large areas south of the Danakil Depression and extend toward the Afar triple junction, which involves the southern Red Sea, the western Gulf of Aden, and the MER (Figures 9.23 and 9.24). In the region of the Erta Ale volcanic complex, Quaternary evaporite layers onlap the igneous structures and are locally interdigitated with the lava flows (Holwerda & Hutchinson, 1968).

Geological maps of the Dallol dome indicate that only evaporite rocks occur at the flanks of the central depression, and the lake at the crater is filled with a salt pan, not volcanic rocks (Cavalazzi et al., 2019; Warren, 2006, 2016). However, hydrothermal vents are observed nearby and suggest the presence of magma chambers in the subsurface at upper crustal levels (Carniel et al., 2010; López-Garcia et al., 2020).

An explosive event at Dallol in 2011 was probably associated with hydrothermal vents and hot springs (Global Volcanism Program, 2013). According to historical reports, the crater was formed by a major phreatic eruption in 1926. Other hydrothermal features observed in the Dallol Mount crater include the spectacular brightly colored hot springs that are associated with diverse features such as geysers, fumaroles, and hot water pools with boiling, supersaturated, extremely acidic brines (pH ~ 0) that formed sulfur deposits around degassing vents (López-Garcia et al., 2020).

The Dallol Mount (Figure 9.28) is interpreted above an igneous sill in the subsurface, which was probably fed by a magma chamber in the deep crust (López-Garcia et al., 2020). Hydrothermal vents, geysers, fumaroles, and melted evaporites are probably products of the thermal anomaly associated with igneous masses at depth. The eastern flank of the asymmetric rift structure is characterized by volcanic rocks exposed at the western flank of the Danakil Horst, which overlie the Mesozoic sedimentary sequences (Figure 9.28).

López-Garcia et al. (2020) analyzed the Dallol waters from several pools, obtaining a very high total dissolved solids (up to 525 g/kg). These waters can be grouped into three chemical suites, all of which contain high levels of chloride ions. X-ray diffraction of evaporite samples from hydrothermal deposits within the crater indicated the major constituents of most samples are halite, sulfur, calcite, sodalite, and hematite, with minor levels of silica. The salt crusts are brown, yellow, or bluish due to metallic oxides and potassium or fluorine impurities. Hydrothermalism is common in the Dallol Depression, with some pools in the Black Mountain hot spring characterized by $MgCl_2$-saturated brine rising to the surface at 130°C (e.g., Holwerda & Hutchinson, 1968). Melted bischofite flows are locally observed, indicating high temperatures in the subsurface (López-Garcia et al., 2020).

Indirect evidence for the presence of magma chambers underlying the Dallol dome suggests that an along-rift dike intrusion affected the subsurface of the dome, possibly fed by an unidentified magma chamber deep in the crust. Nobile et al. (2012) reported the occurrence of an earthquake with a magnitude of 5.5 near Dallol Mount in 2004. The active magma reservoir is estimated to be about 2.4 km below the surface. As there are no lava flows on the surface, the magmatic system of Dallol is still in the proto-volcanic stage, with an extrusive/intrusive ratio of zero. In other regions of the Danakil Depression, the volcanic rocks on the surface indicate more advanced stages of magmatism, as in the region of the Gada Ale volcano (López-Garcia et al., 2020).

López-Garcia et al. (2020) postulated that the geological history of the dome can be attributed to a deep magmatic intrusion associated with basaltic igneous rocks, similar to the composition of the Erta Ale volcanic complex. The intrusion occurred later than 6 ka, resulting in an uplift of lacustrine deposits and developing the Dallol Mount with a halite core. The melting of evaporite minerals with low melting points, such as bischofite, carnallite, and halite, results in evaporite flows at the surface (López-Garcia et al., 2020).

The eastern parts of the Dallol salt flats extend toward the gently west-sloping flank of the Danakil Horst (Figure 9.28). The gypsum layers deposited closer to the Danakil Horst are intercalated with red clastic sedimentary rocks outcropping to the east of the salt pan (Figure 9.26). These Holocene sequences are overlain by thin reefal limestones (Bonatti et al., 1971). The evaporite rocks that outcrop on the surface are underlain and overlain by basaltic lava flows, which result from modern eruptions associated with fissures connected with deeper magma reservoirs (Holwerda & Hutchinson, 1968; López-Garcia et al., 2020).

The Dallol salt flats are home to evaporitic sediments linked to active volcanism, specifically around the Erta Ale volcanic complex (Bastow et al., 2018; Bonatti et al., 2015; Cavalazzi et al., 2019; Corti et al., 2015). Magmatic activity includes lava flows and hydrothermal vents, indicating subsurface hot magma chambers and hot fluids. Igneous rock intrusions, emplaced at depths greater than about 2 km, are believed to be the heat source for the geothermal system. Spectacular salt outcrops with unique geomorphologies and mineral deposits are produced by supersaturated hydrothermal water and brine (Figure 9.29). These features result from complex interactions between climate and hydrological systems. Part of the pluvial waters accumulated during wet periods are infiltrated into the subsurface, where the magmatic system involves active and inactive hydrothermal alteration of the bedrock. The rapid evaporation of surface waters, associated with hydrothermal springs, leads to the recrystallization of evaporite minerals at the surface, resulting in multicolored chemical precipitates of iron chlorides and hydroxides (Figure 9.28).

Geothermal anomalies in the Dallol Mount crater have created various geomorphologic features, including karst topography, salt columns, hydrothermal vents, fumaroles, and geysers (Figure 9.29). These features include salt pillars, chimneys, and circular structures indicative of gas escape (vents). Some sulfuric hot springs have formed extremely acidic pools. The larger ponds can reach tens of

(A)

(B)

Figure 9.29 (A) Salt columns in the Dallol salt mountain, formed by evaporite and siliciclastic sediments deposited in the Quaternary desiccated lagoon. (B) Sulfur-rich hydrothermal springs in the Dallol salt flat, with colorful chemical precipitates associated with iron chlorides and hydroxides. Credit: Courtesy of A. Salvin (2019).

meters in diameter and display an upward flow of groundwater hypersaline hot fluids. The brines in some Dallol ponds are characterized by temperatures above the boiling point of water (up to 112°C). These fluids exhibit hyperacidity, with an extremely low pH value (less than 1), and extremely high salinity, with values of total dissolved solids exceeding 350 g/kg, corresponding to ten times the normal seawater salinity (see Chapter 2). The brine is enriched in several ions, including sodium, chlorine, potassium, magnesium, sulfur, iron, and aluminum (Cavalazzi et al., 2019; Franzson et al., 2015). As hot brine cools, various combinations of evaporite minerals are precipitated, including halite, carnallite, and bischofite. Some ponds display multicolored salt crusts due to the presence of chemical elements like sulfur and iron (Figure 9.29B). Between 2011 and 2012, episodic effusions of hypersaline hot fluids, including molten bischofite flows, have been reported in the region near the Black Mountain (Franzson et al., 2015; López-Garcia et al., 2020).

Astrobiological studies suggest that arid environments in the Dallol Mount region produce extremely acidic and hypersaline fluids, providing an extraordinary analog for similar extraterrestrial environments, such as the habitats hypothesized to occur on Mars' surface (Cavalazzi et al., 2019). Chapter 9 in Mohriak (2025) provides an overview of extreme Earth environments that serve as analogs for extraterrestrial environments observed on some planets in the solar system.

9.4.6. Lake Asal

The Gulf of Aden, also known as the Gulf of Tadjoura or the Gulf of Djibouti (Figure 9.30A), is a small rift-controlled depression that marks an active oceanic propagator advancing from the Gulf of Aden toward the Afar region (Dauteuil et al., 2001; Leroy et al., 2012; Manighetti et al., 1998). Geophysical methods, ongoing seismicity, and magmatic activity indicate that incipient oceanic spreading structures affect the thinned continental or transitional crust in the western extremity of the Gulf of Aden (Bridges et al., 2012; Illsley-Kemp et al., 2018). The landward extension of the oceanic propagator is characterized by extensive lava flows and shallow hypersaline lakes in the rift depression, such as Lake Asal (e.g., Jackson et al., 2000).

Lake Asal is separated from the Gulf of Djibouti by a volcanic ridge measuring circa 10 km (Figure 9.30A). The Kadda Soma volcanic buildup is located within a depression that trends NW-SE, which is called the Asal-Ghoubbat al-Kharab rift. The volcanic barrier separates the hypersaline waters in the lake from the normal seawater in the Gulf of Aden (Figure 9.31). At present, there is no connection between the ocean and the lake waters.

The volcanic rocks that surround the lake correspond to basalts that erupted during the Quaternary. The tectonic interpretation of the system of faults appearing on the surface, integrated with radiometric dating of the basalt layers on land and magnetic anomalies offshore, suggests that the incipient spreading center is propagating from the Gulf of Aden to the Afar region, with a spreading rate of about 30–60 mm per year (Barberi & Varet, 1977).

Lake Asal is a small water body with about 52 km^2 of area and is the lowest point on land in the African continent, with the lake level at about –155 m at present. However, the lake's surface will probably become lower due to the high evaporation rates in the Afar region. The lake presents a bathymetric profile with an average water depth of 7–8 m, reaching a maximum water depth of about 30 m in the lake center. The extreme aridity in the region caused rapid evaporation and concentration of the seawater, resulting in the hypersaline brine. Lake Asal's waters are the second most saline in the world, with a brine concentration more than 10 times the salinity in the Gulf of Aden (see Chapter 2). The brine concentration and salt precipitation are related to the sporadic communication with the Gulf of Tadjoura during the Holocene, with a steady flow of seawater through highly fractured basalts. The brine exhibits a salinity of about 35% and thus deposits gypsum and halite (Gasse & Fontes, 1989; Gasse & Stieltjes, 1973).

The Asal lakeshores are characterized by halite deposits accumulated at the desiccated margins (Figure 9.31). The stacked halite crusts cover much of the saline pan facies at the northern part of the lake. The evaporites form a thick (20–80 m) layer extending for a wide belt (about 60 km) at the fringe of the lake, overlying Quaternary basalts (Gasse & Fontes, 1989; Gasse & Stieltjes, 1973).

The evaporite facies vary from halite crusts overlying the gypsum layers to lacustrine sediments (Figure 9.31). These halite beds have been exploited by inhabitants of the region since prehistoric times, as it holds the world's largest lacustrine salt reserves. Salt extraction activity in Lake Asal is marked by prehistoric scenes of camel caravans coming to the lakeshore to be loaded with halite slabs manually cut from the desiccated lakebed (see Chapter 1).

Lake Asal, despite being small compared to giant salt basins like the South Atlantic or the Red Sea, is a unique example of an evaporitic basin with a salina below sea level fed by a nearby gulf during oceanic spreading (Jackson et al., 2000). The lake provides a modern analog for drawdown of marine seepage through volcanic barriers, and is considered a prototype of salt accumulation in rift basins affected by intense volcanic activity. While some hypersaline lakes are separated from their connection with the sea by sedimentary barriers, such as Lake MacLeod in Australia, Lake Asal was isolated from the Gulf of Aden by a volcanic barrier that was built up in the Quaternary (Gasse & Fontes, 1989; Gasse & Stieltjes, 1973).

Lake Asal's geological setting corresponds to a hypersaline lake at the edge of an oceanic crust basin and near regions with active rifting and magmatism associated with basalt extrusions. This geological setting is similar to the South Atlantic salt basins, where lacustrine sedimentation and evaporite deposition were associated with active volcanism during continental breakup and the early stages of plate divergence (Hutchinson & Engels, 1970; Jackson et al., 2000). Salt deposition in Central South Atlantic basins is believed to result from the isolation of saline lakes and gulfs from the oceanic basins further south by volcanic barriers near the Florianópolis Fracture Zone (e.g., Kukla et al., 2018; Mohriak & Fainstein, 2012; Szatmari & Milani, 2016).

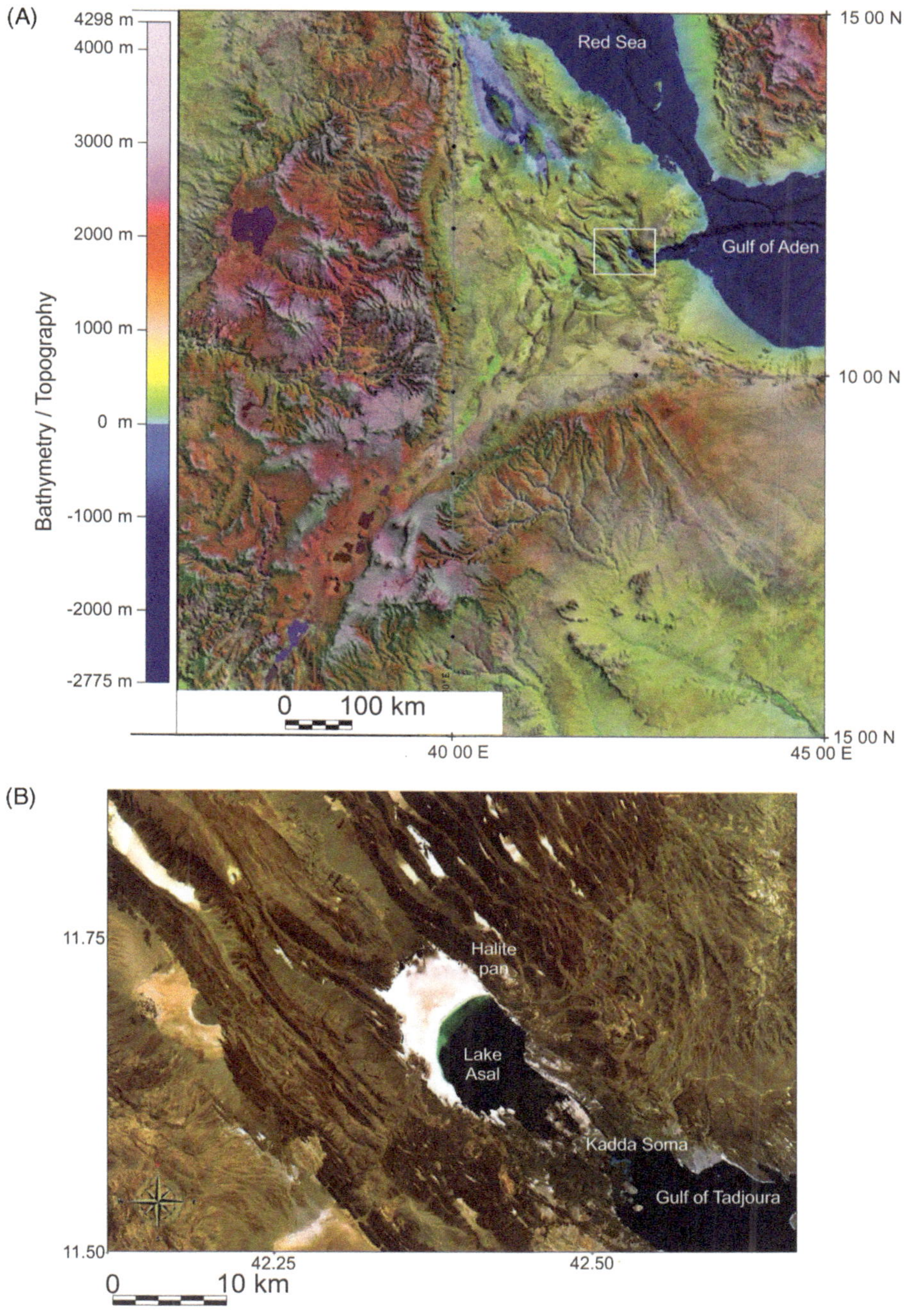

Figure 9.30 (A) Satellite image draped on the digital elevation model of the Afar region. The Bab El Mandeb Strait connects the southern Red Sea with the western Gulf of Aden, and it is marked by an E-W bathymetric trough that corresponds to an oceanic propagator. The salt flats in the Danakil Depression show light-bluish reflectivity, while the Erta Ale Volcanic Complex has dark blue reflectivity and corresponds to an ellipsoidal magmatic structure with an NW trend. The thin white line rectangle highlights the location of Lake Asal at the western extremity of the Gulf of Tadjoura. (B) NASA satellite image (true colors) showing a detailed zoom of the volcanic field (Kadda Soma) that separates Lake Asal from the Gulf of Tadjoura. White reflectivity patches near the hypersaline lake margins correspond to salt flats associated with halite crusts. Credit: Courtesy of NASA.

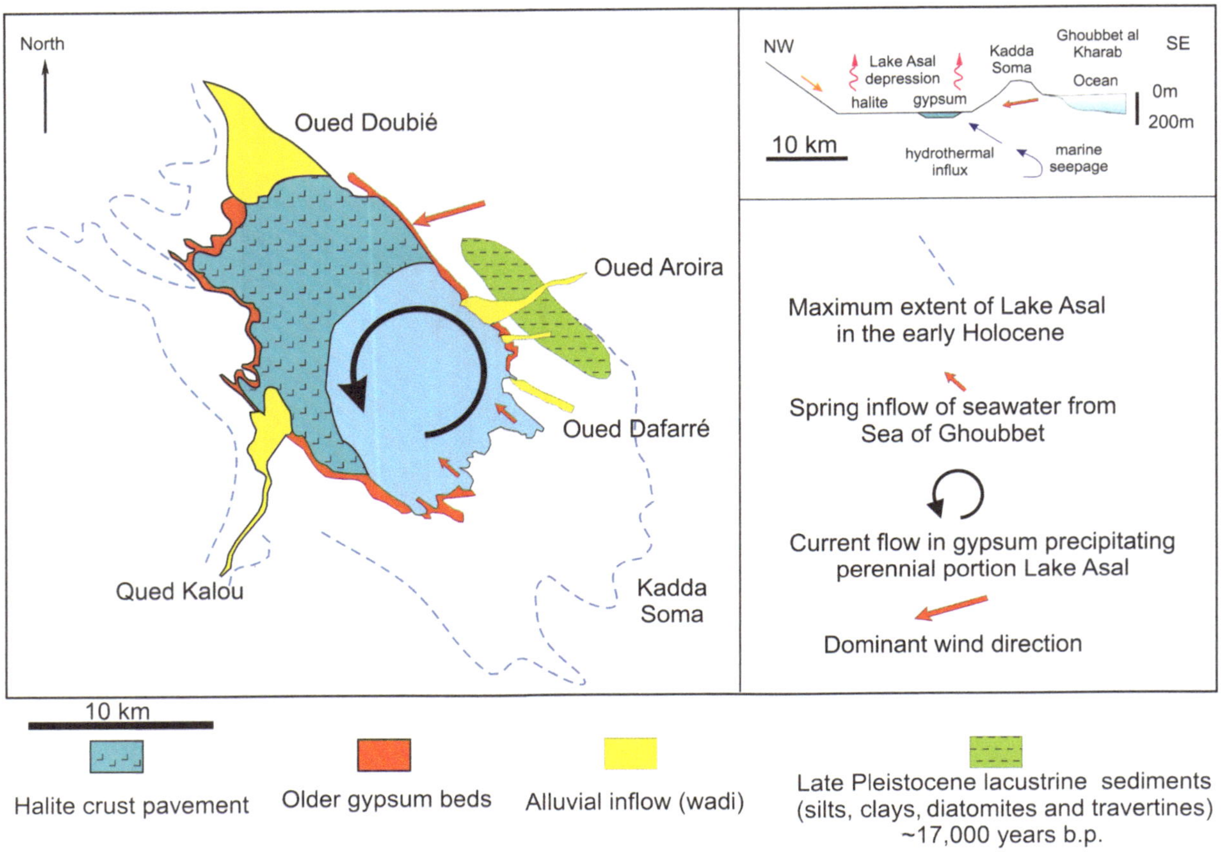

Figure 9.31 The simplified geological map of Lake Asal shows the outline of evaporite deposits, forming a halite crust pavement on the northern shore. The halite pan overlies the older gypsum beds that outcrop at the margins of the evaporite basin. The simplified cross-section at the upper right panel shows the Kadda Soma volcanic buildup. The structure constituted a barrier that restricted communication between the hypersaline lake and the Gulf of Aden normal seawater during the Quaternary. Credit: Adapted from Gasse and Stieltjes (1973).

REFERENCES

Abbate, E., Bruni, P., & Sagri, M. (2015). Geology of Ethiopia: A review and geomorphological perspectives. In P. Billi (Ed.), *Landscapes and landforms of Ethiopia, World Geomorphological Landscapes* (pp. 33–64). Dordrecht, The Netherlands: Springer.

Adams, J., Maslin, M., & Thomas, E. (1999). Sudden climate transitions during the Quaternary. *Progress in Physical Geography*, *23*(1), 1–36.

Adams, P. M., Lynch, D. K., Buckland, K. N., Johnson, P. D., & Tratt, D. M. (2017). Sulfate mineralogy of fumaroles in the Salton Sea Geothermal Field, Imperial County, California. *Journal of Volcanology and Geothermal Research*, *347*, 15–43.

Almalki, K. A., Betts, P. G., & Ailleres, L. (2015). The Red Sea—50 years of geological and geophysical research. *Earth-Science Reviews*, *147*, 109–140.

Alsharhan, A. S., & Kendall, C. G. St. C. (1994). Depositional setting of the Upper Jurassic Hith Anhydrite of the Arabian Gulf: An analog to holocene evaporites of the United Arab Emirates and Lake MacLeod of Western Australia. *American Association of Petroleum Geologists Bulletin*, *78*(7), 1075–1096.

Atwood, G., Wambeam, T. J., & Anderson, N. J. (2016). The present as a key to the past: Paleoshoreline correlation insights from Great Salt Lake. In C. G. Oviatt & J. F. Shrode (Eds.), *Lake Bonneville: A scientific update, Developments in Earth Surface Processes* (Vol. *20*, 1–27). Amsterdam: Elsevier. doi:10.1016/b978-0-444-63590-7.00001-9

Barberi, F., & Varet, J. (1977). Volcanism of Afar: Small-scale plate tectonics implications. *Geological Society of America Bulletin*, *88*, 1251–1266.

Bassett, A. M., Kupfer, D. H., & Barstow, F. C. (1959). Core logs from Bristol Cadiz, and Danby Dry Lakes, San Bernardino County, *California. USGS Geological Survey Bulletin 1045-D*, 138 pp.

Bastow, I. D., Booth, A. D., Corti, G., Keir, D., Magee, C., Jackson, C. A.-L., et al. (2018). The development of late-stage continental breakup: Seismic reflection and borehole evidence from the Danakil Depression, Ethiopia. *Tectonics*, *37*, 2848–2862. https://doi.org/10.1029/2017TC004798

Beyene, A., & Abdelsalam, M. G. (2005). Tectonics of the Afar depression: A review and synthesis. *Journal of African Earth Sciences*, *41*(1–2), 41–59.

Blakey, R. C. (2008). Pennsylvanian-Jurassic sedimentary basins of the Colorado Plateau and Southern Rocky Mountains. In A. D. Miall (Ed.), *Sedimentary basins of the World, Vol. 5, The Sedimentary Basins of the United States and Canada* (pp. 245–296). Amsterdam: Elsevier.

Blakey, R. C. (2013). Using Paleogeographic maps to portray Phanerozoic Geologic and Paleotectonic history of Western North America. *AAPG Search and Discovery*, Article #30267 (2013).

Bonatti, E., Emiliani, C., Ostlund, G., & Rydell, H. (1971). Final desiccation of the Afar Rift, *Ethiopia, Science*, *172*(3982), 468–469. doi:10.1126/science.172.3982.468

Bonatti, E., Gasperini, E., Vigliotti, L., Lupi, L., Vaselli, O., Polonia, A., & Gasperini, L. (2016). Lake Afrera, a structural depression in the Northern Afar Rift (Red Sea). *Heliyon*, *3*, e00301, 1–15. doi: 10.1016/j.heliyon.2017.

Bonini, M., Corti, G., Innocenti, F., Manetti, P., Mazzarini, F., Abebe, T., & Pecskay, Z. (2005). Evolution of the main Ethiopian rift in the frame of Afar and Kenya rifts propagation. *Tectonics*, *24*(1), TC1007. https://doi.org/10.1029/2004TC001680

Bosworth, W., Huchon, P., & McClay, K. (2005). The Red Sea and Gulf of Aden basins. *Journal of African Earth Sciences*, *43*, 334–378.

Bott, M. H. P. (1995). Mechanisms of rifting: Geodynamic modeling of continental rift systems. In K. H. Olsen (Ed.), *Continental rifts: Evolution, structure, tectonics, Developments in Geotectonics* (Vol. *25*, pp. 27–43). Amsterdam: Elsevier.

Bridges, D. L., Mickus, K., Gao, S. S., Abdelsalam, M. G., & Alemu, A. (2012). Magnetic stripes of a transitional continental rift in Afar. *Geology*, *40*, 203–206. doi:10.1130/G32697.1

Brothers, D. S., Driscoll, N. W., Kent, G. M., Harding, A. J., Babcock, J. M., & Baskin, R. L. (2009). Tectonic evolution of the Salton Sea inferred from seismic reflection data. *Nature Geoscience*, *2*(8), 581–584.

Carniel, R., Munoz, J. E., & Jones, J. (2010). A geophysical multi-parametric analysis of hydrothermal activity at Dallol, Ethiopia. *Journal of African Earth Sciences*, *58*, 812–819.

Carozzi, A. V. (1962). Observations on algal biostromes in the Great Salt Lake, *Utah. The Journal of Geology*, *70*(2), 246–252. doi:10.1086/626814

Cavalazzi, B., Barbieri, R., Gómez, F., Capaccioni, B., Olsson-Francis, K., Pondrelli, M., et al. (2019). The Dallol Geothermal Area, Northern Afar (Ethiopia)—An exceptional planetary field analog on Earth. *Astrobiology*, *19*(4), 553–578.

Chorlton, L. B. (2007). Generalized geology of the world: Bedrock domains and major faults in GIS format: A small-scale world geology map with an extended geological attribute database. *Geological Survey of Canada, Open File 5529*, 2007, 48 pages.

Chorowicz, J. (2005). The East African Rift system. *Journal of African Earth Sciences*, *43*, 379–410.

Coffin, M. F., & Eldholm, O. (1994). Large igneous provinces: Crustal structure, dimensions, and external consequences. *Reviews of Geophysics*, *32*(1), 1–36. doi:10.1029/93RG02508

Corti, G. (2009). Continental rift evolution: From rift initiation to incipient break-up in the Main Ethiopian Rift, East Africa. *Earth-Science Reviews*, *96*, 1–53.

Corti, G., Bastow, I. D., Keir, D., Pagli, C., & Baker, E. (2015). Rift-related morphology of the Afar depression. In: P. Billi (Ed.), *Landscapes and landforms of Ethiopia, World Geomorphological Landscapes* (Chapter 15, pp. 251–274). Dordrecht, The Netherlands: Springer. doi 10.1007/978-94-017-8026-1_15.

Dauteuil, O., Huchon, P., Quemeneur, F., & Souriot, T. (2001). Propagation of an oblique spreading centre: The western Gulf of Aden. *Tectonophysics*, *332*, 423–442.

Demaison, G. J., & Moore, G. T. (1980). Anoxic environments and oil source bed genesis. *American Association of Petroleum Geologists Bulletin*, *64*(8), 1179–1209. https://doi.org/10.1306/2F91945E-16CE-11D7-8645000102C1865D

Deocampo, D. M., Owen, R. B., Lowenstein, T. K., Renaut, R. W., Rabideaux. N. M., Billingsley, A., et al. (2022). Orbital control of Pleistocene Euxinia in Lake Magadi, *Kenya. Geology*, *50*(1), 42–47. https://doi.org/10.1130/G49140.1

Ebinger, C. (2005). Continental break-up: The East African perspective. *Astronomy & Geophysics*, *46*(2), 2.16–2.21. https://doi.org/10.1111/j.1468-4004.2005.46216.x

Ernst, R. E. (2014). *Large igneous provinces*. Cambridge, UK: Cambridge University Press, 653 pp.

Eugster, H. P. (1980). Lake Magadi, Kenya, and its precursors. In A. Nissenbaum (Ed.), *Hypersaline brines and evaporitic environments, Developments in Sedimentology* (Vol. *28*, pp. 195–232). Amsterdam: Elsevier. doi:10.1016/s0070-4571(08)70239-5

Eugster, H. P. (1982). Climatic significance of lake and evaporite deposits. In *Climate in Earth history: Studies in Geophysics* (pp. 105–111). National Research Council, Washington, DC: The National Academies Press.

Eugster, H. P. (1986). Lake Magadi, Kenya: A model for rift valley hydrochemistry and sedimentation? In Frostick, L. E. et al. (Eds.), *Sedimentation in the African rifts* (Vol. *25*, pp. 177–189). Geological Society, London, Special Publications.

Eugster, H. P., & Hardie, L. A. (1978). Saline lakes. In A. Lerman (Ed.), *Lakes, Chemistry, Geology, Physics* (pp. 237–293). Berlin, Germany: Springer-Verlag.

Faulds, J. E., Schreiber, B. C., Reynolds, S. J., Gonzalez, L. A., & Okaya, D. (1997). Origin and paleogeography of an immense, nonmarine Miocene salt deposit in the Basin and Range (Western USA). *Journal of Geology*, *105*, 19–36.

Faulds, J. E., Schreiber, B. C., Langenheim, V. E., Hinz, N. H., Shaw, T. H., Heizler, M. T., et al. (2016). Paleogeographic implications of late Miocene lacustrine and nonmarine evaporite deposits in the Lake Mead region: Immediate precursors to the Colorado River. *Geosphere*, *12*(3), 721–767. doi: 10.1130/GES01143.1.

Fazzini, M., Bisci, C., & Billi, P. (2015). The climate of Ethiopia. In P. Billi (Ed.), *Landscapes and Landforms of Ethiopia, World Geomorphological Landscapes* (pp. 65–87). Dordrecht, The Netherlands: Springer.

Figueiredo, A. M. F., Braga, J. A. E., Zabalaga, J. C., Oliveira, J. J., Aguiar, G. A., Silva, O. B., Mato, L. F., Daniel, L. M. F., and Magnavita, L. P. (1994). Reconcavo Basin, Brazil: a prolific intracontinental rift basin. In: Landon, S.M., (Ed.), *Interior Rift Basins, AAPG Memoir, 59*, pp. 157–203.

Franzson, H., Helgadóttir, H. M., & Óskarsson, F. (2015). *Surface exploration and first conceptual model of the Dallol Geothermal area, Northern Afar, Ethiopia, Proceedings* (pp. 1–11). World Geothermal Congress 2015, Melbourne, Australia.

Gasse, F., & Fontes, J. -C. (1989). Palaeoenvironments and palaeohydrology of a tropical closed lake (Lake Asal, Djibouti) since 10,000 yr B.P. *Palaeogeography, Palaeoclimatology, Palaeoecology*, *69*, 67–102.

Gasse, F., & Stieltjes, L. (1973). Les sédiments du Quaternaire récent du lac Asal (Afar central territoire français des Afars et des Issas). *Bulletin du bureau de recherches géologiques et minières: géologie générale*, n° spécial, *Paris*, *4*, 229–251.

GEBCO (2019). General bathymetric chart of the oceans. *British Oceanographic Data Centre*, U.K. website www.gebco.net.

Getenet, M., García-Ruiz, J. M., Otálora, F., Emmerling, F., Al-Sabbagh, D., & Verdugo-Escamilla, C. (2022). A comprehensive methodology for monitoring evaporitic mineral precipitation and hydrochemical evolution of saline lakes: The case of Lake Magadi Soda Brine (East African Rift Valley, Kenya). *Crystal Growth and Design*, *22*(4), 2307–2317. https://doi.org/10.1021/acs.cgd.1c01391

Global Volcanism Program (2013). Report on Dallol (Ethiopia). In R. Wunderman (Ed.), *Bulletin of the Global Volcanism Network* (Vol. *38*, No. 5). Smithsonian Institution. https://doi.org/10.5479/si.GVP.BGVN201305-221041.

Guo, X., & Chafetz, H. S. (2012). Large tufa mounds, Searles Lake, California. *Sedimentology*, *59*, 1509–1535. doi: 10.1111/j.1365-3091.2011.01315.x.

Hagos, M., Koeberl, C., & van Wyk de Vries, B. (2016). The Quaternary volcanic rocks of the northern Afar Depression (northern Ethiopia): Perspectives on petrology, geochemistry, and tectonics. *Journal of African Earth Sciences*, *117*, 29–47. doi:10.1016/j.jafrearsci.2015.11.022

Handford, C. R. (1982). Sedimentology and evaporite genesis in a Holocene continental sabkha playa basin—Bristol Dry Lake, California. *Sedimentology*, *29*, 239–253.

Handford, C. R. (1991). Marginal marine halite: Sabkhas and salinas. In J. L. Melvin (Ed.), *Evaporites, petroleum and mineral resources, Developments in Sedimentology* (Vol. *50*, Chapter 1, 1–66). Amsterdam: Elsevier.

Handford, C. R., Kendall, A. C., Prezbindowski, D. R., Dunham, J. B., & Logan, B. W. (1984). Salina margin tepees, pisoliths and aragonite cements, Lake MacLeod, Western Australia: Their significance in interpreting ancient analogs. *Geology*, *12*, 523–527.

Hartmann, J., & Moosdorf, N. (2012). The new global lithological map database GLiM: A representation of rock properties at the Earth surface. *Geochemistry, Geophysics, Geosystems*, *13*, Q12004. doi:10.1029/2012GC004370

Hauksson, E., Meier, M.-A., Ross, Z. E., & Jones, L. M. (2017). Evolution of seismicity near the southernmost terminus of the San Andreas Fault: Implications of recent earthquake clusters for earthquake risk in southern California. *Geophysical Research Letters*, *44*, 1293–1301. https://doi.org/10.1002/2016GL072026.

Hollett, K. J., Danskin, W. R., McCaffrey, W. F., & Walti, C. L. (1991). Geology and water resources of Owens valley, California. *United States Geological Survey Water-Supply Paper 2370-B*, 77 pp.

Holwerda, J. G., & Hutchinson, R. W. (1968). Potash-bearing evaporites in the Danakil area, *Ethiopia. Economic Geology*, *63*(2), 124–150.

Hooke, R. L. (1972). Geomorphic evidence for Late-Wisconsin and Holocene tectonic deformation, Death Valley, *California. Geological Society of America Bulletin*, *83*(7), 2073–2098.

Hsü, K. J. (1972). Origin of saline giants: A critical review after the discovery of the Mediterranean evaporite. *Earth-Science Reviews*, *8*, 371–396.

Huc, A. Y., Le Fournier, J., Vandenbroucke, M., & Bessereau, G. (1990). Northern Lake Tanganyika; An example of organic sedimentation in an anoxic rift lake. In B. J. Katz (Ed.), *Lacustrine basin exploration; Case studies and modern analogs, AAPG Memoir* (Vol. *50*, pp. 169–185).

Hutchinson R. W., & Engels G. G. (1970). Tectonic significance of regional geology and evaporite lithofacies in northeastern Ethiopia. *Philosophical Transactions of the Royal Society of London. Series A*, *Mathematical and Physical Sciences*, *267*, 1181, 313–329.

Hutchinson, R. W., & Engels, G. G. (1972). Tectonic evolution in the southern Red Sea and its possible significance to older rifted continental margins. *Geological Society of America Bulletin*, *83*, 2989–3002.

Illsley-Kemp, F., Keir, D., Bull, J. M., Gernon, T. M., Ebinger, C., Ayele, A., et al. (2018). Seismicity during continental breakup in the Red Sea rift of Northern Afar. *Journal of Geophysical Research, Solid Earth*, *123*, 2345–2362. https://doi.org/10.1002/2017JB014902

Jackson, M. P. A., Cramez, C., & Fonck, J. M. (2000). Role of subaerial volcanic rocks and mantle plumes in creation of South Atlantic margins: Implications for salt tectonics and source rocks. *Marine and Petroleum Geology*, *17*, 477–498.

Jones, B. F., Eugster, H. P., & Rettig, S. L. (1977). Hydrochemistry of the Lake Magadi Basin, Kenya. *Geochimica et Cosmochimica Acta*, *41*, 53–72.

Kadigi, R. M. J., Mwathe, K., Dutton, A., Kashaigili, J., & Kilima, F. (2014). Soda Ash Mining in Lake Natron: A reap or ruin for Tanzania? *Journal of Environmental Conservation Research*, *2*(2), 37–49.

Katz, B. J. (2001). Lacustrine basin hydrocarbon exploration—Current thoughts. *Journal of Paleolimnology*, *26*, 161–179. https://doi.org/10.1023/A:1011173805661

Kendall, A. C. (1992). Evaporites. In R. G. Walker & N. P. James (Eds.), *Facies Models: Response to sea level change* (pp. 375–409). Geological Association of Canada.

Kendall, A. C. (2010). Marine evaporites. In N. P. James & R. W. Dalrymple (Eds.), *Facies Models* (pp. 505–539). Geological Association of Canada IV, Series GEOtext.

Knott, J. R., Sarna-Wojcicki, A. M., Machette, M. N., & Klinger, R. E. (2005). Upper Neogene stratigraphy and tectonics of Death Valley—A review. *Earth-Science Reviews*, *73*(1–4), 245–270.

Kukla, P. A., Strozyk, F., & Mohriak, W. U. (2018). South Atlantic salt basins—Witnesses of complex passive margin evolution. *Gondwana Research*, *53*, 41–57.

Leroy, S., Razin, P., Autin, J., Bache, F., Acremont, E., Watremez, L., et al. (2012). From rifting to oceanic spreading in the Gulf of Aden: A synthesis. *Arabian Journal of Geosciences*, *5*, 859–901. doi 10.1007/s12517-011-0475-4.

Li, J., Lowenstein, T. K., Brown, C. B., Ku, T.-L., & Luo, S. (1996). A 100 ka record of water tables and paleoclimates from salt cores, Death Valley, California, *Palaeogeography, Palaeoclimatology, Palaeoecology*, *123*, 179–203.

Logan, B. W. (1987). *The MacLeod evaporite basin, Western Australia: Holocene environments, sediments and geological evolution, American Association of Petroleum Geologists Memoir* (Vol. *44*), 140 pp.

Lohman, R. B., & McGuire, J. J. (2007). Earthquake swarms driven by aseismic creep in the Salton Trough, *California. Journal of Geophysical Research*, *112*, B04405. doi:10.1029/2006JB004596

López-García, J. M., Moreira, D., Benzerara, K., Grunewald, O., & López-García, P. (2020). Origin and evolution of the Halo-Volcanic Complex of Dallol: Proto-Volcanism in Northern Afar (Ethiopia). *Frontiers in Earth Science*, *7*, 351. https://doi.org/10.3389/feart.2019.00351

Lowenstein, T. K. (1988). Origin of depositional cycles in a Permian "saline giant": The Salado (McNutt zone) evaporites of New Mexico and Texas. *Geological Society of America Bulletin*, *100*, 592–608.

Lowenstein, T. K., & Risacher, F. (2009). Closed basin brine evolution and the influence of Ca-Cl inflow waters: Death Valley and Bristol Dry Lake California, Qaidam Basin, China, and Salar de Atacama, Chile. *Aquatic Geochemistry*, *15*, 71–94.

Lowenstein, T. K., Li, J., Brown, C. B., Roberts, S. M., Ku, T. L., Luo, S., & Yang, W. (1999). 200 k. y. paleoclimate record from Death Valley salt core. *Geology*, *27*, 3–6.

Macgregor, D. (2015). History of the development of the East African Rift System: A series of interpreted maps through time. *Journal of African Earth Sciences*, *101*, 232–252. http://dx.doi.org/10.1016/j.jafrearsci.2014.09.016

Manighetti, I., Tapponnier, P., Gillot, P. Y., Jacques, E., Courtillot, V., Armijo, R., et al. (1998). Propagation of rifting along the Arabia–Somalia plate boundary: Into Afar. *Journal of Geophysical Research*, *103*(B3), 4947–4974.

Martins, G. S., Mohriak, W. U., & Destro, N. (2019). Synrift evaporite deposition and structural characterization of the onshore Alagoas subbasin. *Interpretation*, *7*(4), SH19–SH31.

McDougall, K., & Martínez, A. Y. M. (2014). Evidence for a marine incursion along the lower Colorado River corridor. *Geosphere*, *10*(5), 842–869. doi:10.1130/GES00975.1;

McKibben, M. A., Williams, A. E., & Okubo, S. (1988). Metamorphosed Plio-Pleistocene evaporites and the origins of hypersaline brines in the Salton Sea geothermal system, California: Fluid inclusion evidence. *Geochimica et Cosmochimica Acta*, *52*(5), 1047–1056.

Mernagh, T. P. (2013). A review of Australian salt lakes and assessment of their potential for strategic resource. *Geoscience Australia*, Record 2013/39, Canberra.

Mernagh, T. P., Bastrakov, E. N., Jaireth, S., de Caritat, P., English, P. M., & Clarke, J. D. A. (2016). A review of Australian salt lakes and associated mineral systems. *Australian Journal of Earth Sciences*, *63*(2), 131–157. doi:10.1080/08120099.2016.1149517

Miller, D. M., Reynolds, R. E., Bright, J. E., & Starrat, S. W. (2014). Bouse Formation in the Bristol basin near Amboy, California, USA. *Geosphere*, *10*(3), 462–475.

Mitchell, D. J. W., Allen, R. B., Salama, W., & Abouzakm, A. (1992). Tectonostratigraphic framework and hydrocarbon potential of the Red Sea. *Journal of Petroleum Geology*, *15*(2), 187–210.

Mohr, P. (1989). Nature of the crust under Afar: New igneous, not thinned continental. *Tectonophysics*, *167*, 1–11.

Mohriak, W. (2025). *Salt in the Earth sciences: Basin analysis and salt tectonics*. Washington, DC: American Geophysical Union. doi:10.1002/9781394307067

Mohriak, W. U., & Fainstein, R. (2012). Phanerozoic regional geology of the Eastern Brazilian margin. In D. G. Roberts & A. W. Bally (Eds.), *Phanerozoic passive margins, cratonic basins and global tectonic maps* (Vol. *1C*, Chapter 7, pp. 223–282). Amsterdam: Elsevier B.V. doi: 10.1016/B978-0-444-56357-6.00006-8

Momenzadeh, M. (1990). Saline deposits and alkaline magmatism: A genetic model. *Journal of Petroleum Geology*, *13*(3), 341–356.

Moore, C., Wright, T., Hooper, A., & Biggs, J. (2019). The 2017 eruption of Erta 'Ale Volcano, Ethiopia: Insights into the shallow axial plumbing system of an incipient mid-ocean ridge. *Geochemistry, Geophysics, Geosystems*, *20*(12), 5727–5743. https://1029/2019GC008692

Morley, C. K., Ngenoh, D. K., & Ego, J. K. (1999). Introduction to the East African Rift System. In C. K. Morley (Ed.), *Geoscience of rift systems—Evolution of East Africa. AAPG Studies in Geology* (Vol. *44*, pp. 1–18).

Nicoll, K. (2021). What's new about the Old Bonneville Basin? Fresh insights about the modern Limnogeology of Great Salt Lake. In M. R. Rosen, D. Finkelstein, L. Park-Boush, & S. Pla-Pueyo (Eds.), *Limnogeology: Progress, challenges and opportunities: A tribute to Beth Gierlowski-Kordesch, Syntheses in Limnogeology* (No. *2*, pp. 273–294). Springer Nature AG. https://doi.org/10.1007/978-3-030-66576-0_9

Nobile, A., Pagli, C., Keir, D., Wright, T. J., Ayele, A., Ruch, J., & Acocella, V. (2012). Dike-fault interaction during the 2004 Dallol intrusion at the northern edge of the Erta Ale Ridge (Afar, Ethiopia). *Geophysical Research Letters*, *39*(19), 2–7. doi: 10.1029/2012GL053152

Olsen, P. E. (1997). Stratigraphic record of the Early Mesozoic breakup of Pangea in the Laurasia-Gondwana rift system. *Annual Review of Earth Planetary Sciences*, *25*, 337–401.

Olson, K. J., & Lowenstein, T. K. (2021). Searles Lake evaporite sequences: Indicators of late Pleistocene/Holocene lake temperatures, brine evolution, and pCO2. *GSA Bulletin*, *133*(11–12), 2319–2334. https://doi.org/10.1130/B35857.1

Owen, R. B., Muiruri, V. M., Lowenstein, T. K., Renaut, R. W., Rabideaux, N., Luo, S., et al. (2018). Progressive aridification in East Africa over the last half million years and implications for human evolution. *PNAS*, *115*(44), 11174–11179. https://doi.org/10.1073/pnas.1801357115

Owen, R. B., Renaut, R. W., Muiruri, V. M., Rabideaux, N. M., Lowenstein, T. K., McNulty, E. P., et al. (2019). Quaternary history of the Lake Magadi Basin, southern Kenya Rift: Tectonic and climatic controls. *Palaeogeography, Palaeoclimatology, Palaeoecology*, *518*, 97–118.

Pedley, A., Neubert, J., & van der Kauw, S. (2016). Potash deposits in Africa. *Episodes*, *39*(2), 447–457. doi: 10.18814/epiiugs/2016/v39i2/95787

Playford, P. E., Cockbain, A. E., Berry, P. F., Roberts, A. P., Haines, P. W., & Brooke, B. P. (2013). The geology of Shark Bay. *Geological Survey of Western Australia, Bulletin 146*, 281 pp.

Raup, O. B. (1970). Brine mixing: an additional mechanism for formation of basin evaporites. *American Association of Petroleum Geologists Bulletin*, *54*(12), 2246–2259.

Raup, O. B. (1982). Gypsum precipitation by mixing seawater brines. *American Association of Petroleum Geologists Bulletin*, *66*(3), 363–367.

Renaut, R. W., & Gierlowski-Kordesch, E. H. (2010). Lakes. In N. P. James (Ed.), *Facies Models* (Vol. *4*, pp. 541–575). Geological Association of Canada, St. John's, Newfoundland.

Renaut, R. W., & Tiercelin, J.-J. (1993). Lake Bogoria, Kenya: Soda, hot springs and about a million flamingoes. *Geology Today*, *9*(2), 56–61. doi:10.1111/j.1365-2451.1993.tb00981.x

Renaut, R. W., Owen, R. B., Jones, B., Tiercelin, J.-J., Tarits, C., Ego, J. K., & Konhauser, K. O. (2012). Impact of lake-level changes on the formation of thermogene travertine in continental rifts: Evidence from Lake Bogoria, *Kenya Rift Valley. Sedimentology*, *60*(2), 428–468. https://doi.org/10.1111/j.1365-3091.2012.01347.x

Renaut, R. W., Owen, R. B., Lowenstein, T. K., De Cort, G., Mcnulty, E., Scott, J. J., & Mbuthia, A. (2021). The role of hydrothermal fluids in sedimentation in saline alkaline lakes: Evidence from Nasikie Engida, *Kenya Rift Valley. Sedimentology*, *68*(1), 108–134. https://doi.org/10.1111/sed.12778

Roberts, S. M., Spencer, R. J., & Lowenstein, T. K. (1994). Late Pleistocene saline lacustrine sediments, Badwater Basin, Death Valley, California. In A. J. Lomando, B. C. Schreiber, & P. M. Harris (Eds.), *Lacustrine reservoirs and depositional systems* (pp. 61–103), SEPM Society for Sedimentary Geology, Core Workshop 19.

Rockwell, T. K., Meltzner, A. J., Haaker, E. C., & Madugo, D. (2022). The late Holocene history of Lake Cahuilla: Two thousand years of repeated fillings within the Salton Trough, Imperial Valley, *California. Quaternary Science Reviews*, *282*, 107456 doi: 10.1016/j.quascirev.2022.107456

Rosen, M. R., Stillings, L. L., Kane, T., Campbell, K., Vitale, M., & Spanjers, R. (2020). Li and Ca enrichment in the Bristol dry lake brine compared to brines from Cadiz and Danby dry lakes, Barstow-Bristol Trough, California, USA. *Minerals*, *10*(3), 284. https://doi.org/10.3390/min10030284

Ruiz, F., Abad, M., Galán, E., González, I., Aguilá, I., Olía, M., et al. (2006). The present environmental scenario of El Melah Lagoon (NE Tunisia) and its evolution to a future sabkha. *Journal of African Earth Sciences*, *44*, 289–302.

Sabato Ceraldi, T., & Green, D. (2016). Evolution of the South Atlantic lacustrine deposits in response to Early Cretaceous rifting, subsidence and lake hydrology. In T. Sabato Ceraldi, R. A. Hodgkinson, & G. Backe (Eds.), *Petroleum Geoscience of the West Africa Margin* (Vol. *438*, pp. 77–98). Geological Society, London, Special Publications.

Schmalz, R. F. (1969). Deep-water evaporite deposition: A genetic model. *American Association of Petroleum Geologists Bulletin*, *53*(4), 798–823.

Schmalz, R. F. (1970). Environment of marine evaporite deposition. *Mineral Industry*, *35*(8), 1–7.

Scoon, R. N. (2018). Lake Natron and the Oldoinyo Lengai Volcano. In R. N. Scoon (Ed.), *Geology of National Parks of Central/Southern Kenya and Northern Tanzania* (pp. 193–206). Switzerland: Springer Nature. doi:10.1007/978-3-319-73785-0_17

Smith, G. I. (1979). Subsurface stratigraphy and geochemistry of Late Quaternary evaporites, Searles Lake, California. *USGS Professional Paper 1043*, 130 pp.

Smoot, J. P., & Lowenstein, T. K. (1991). Depositional environments of non-marine evaporites. In J. L. Melvin (Ed.), *Evaporites, petroleum and mineral resources. Developments in Sedimentology* (Vol. *50*, Chapter 3, pp. 189–347). Amsterdam: Elsevier.

Spencer, R. J., Baedecker, M. J., Eugster, H. P., Forester, R. M., Goldhaber, M. B., Jones, B. F., et al. (1984). Great Salt Lake, and precursors, Utah: The last 30,000 years. *Contributions to Mineralogy and Petrology*, *86*(4), 321–334. doi:10.1007/bf01187137

Stab, M., Bellahsen, N., Pik, R., Quidelleur, X., Ayalew, D., & Leroy, S. (2016). Modes of rifting in magma-rich settings: Tectono-magmatic evolution of Central Afar. *Tectonics*, *35*(1), 2–38. doi:10.1002/2015TC003893

Sylvester, A. G. (1988). Strike-slip faults. *Geological Society of America Bulletin*, *100*, 1666–1703.

Szatmari, P., & Milani, E. J. (2016). Tectonic control of the oil-rich large igneous-carbonate-salt province of the South Atlantic rift. *Marine and Petroleum Geology*, *76*, 567–596.

Talbot, C. J. (2008). Hydrothermal salt—But how much? Discussion. *Marine and Petroleum Geology*, *25*, 191–202.

Tazieff, H., Varet, J., Barberi, F., & Giglia, G. (1972). Tectonic significance of the Afar (or Danakil) depression. *Nature*, *235*(5334), 144–147.

Tesfaye, S., Harding, D. J., & Kusky, T. M. (2003). Early continental breakup boundary and migration of the Afar triple junction, Ethiopia. *Geological Society of America Bulletin*, *115*, 1053–1067.

Tiercelin, J.-J., Cohen, A. S., Soreghan, M. J., & Lezzar, K.-E. (1994). Pleistocene-modern deposits of the Lake Tanganyika rift basin, East Africa: A modern analog for lacustrine source rocks and reservoirs. *SEPM Special Publications, AAPG Denver Meeting*, pp. 37–59.

U.S. Geological Survey (2020). Mineral commodity summaries 2020. *U.S. Geological Survey*, 200 pp. https://doi.org/10.3133/mcs2020

Vincens, A., & Casanova, J. (1987). Modern background of Natron-Magadi basin (Tanzania-Kenya): Physiography, climate, hydrology and vegetation—Contexte actuel du bassin Natron-Magadi (Tanzanie-Kenya): Physiographie, climat, hydrologie et vegetation. *Sciences Geologiques, Bulletin*, *40*, 9–21.

Vincens, A., Casanova, J., & Tiercelin, J. J. (1986). Palaeolimnology of lake Bogoria (Kenya) during the 4500 BP high lacustrine phase. In L. E. Frostick et al. (Eds.),

Sedimentation in the African rifts (Vol. *25*, pp. 323–333). Geological Society Special Publication. https://doi.org/10.1144/GSL.SP.1986.025.01.26

Warren, J. K. (1989). *Evaporite Sedimentology: Importance in hydrocarbon accumulation*. New Jersey, USA: Prentice Hall Advanced Reference Series, 285 pp.

Warren, J. K. (1992). *Evaporites and their importance to petroleum exploration*, IHRDC, GL 508 Petroleum Geology, *234* pp.

Warren, J. K. (1997). Evaporites, brines and base metals: fluids, flow and "the evaporite that was." *Australian Journal of Earth Sciences*, *44*, 149–183.

Warren, J. K. (2000). Dolomite: Occurrence, evolution and economically important associations. *Earth-Science Reviews*, *52*, 1–81.

Warren, J. K. (2006). *Evaporites: Sediments, resources and hydrocarbons*. Berlin: Springer Verlag, 1035p.

Warren, J. K. (2010). Evaporites through time: Tectonic, climatic and eustatic controls in marine and nonmarine deposits. *Earth and Science Review*, *98*, 217–268.

Warren, J. K. (2016). *Evaporites—A geological compendium*. Berlin: Springer, 1807 pp. ISBN 978-3-319-13511-3.

Warren, J. K., & Kendall, C. G. St. C. (1985). Comparison of sequences formed in marine sabkha (subaerial) and salina (subaqueous) settings—Modern and ancient. *American Association of Petroleum Geologists Bulletin*, *69*(6), 1013–1023.

Waters, M. R. (1983). Late Holocene lacustrine chronology and archaeology of ancient Lake Cahuilla, *California. Quaternary Research*, *19*(3), 373–387. doi:10.1016/0033-5894(83)90042-X

Whitney, B. B., & Hengesh, J. V. (2015). Geomorphological evidence for late Quaternary tectonic deformation of the Cape Region, coastal west central Australia. *Geomorphology*, *241*, 160–174.

10

Saline Lakes in Convergent Settings

ABSTRACT

Currently, there are several hypersaline lakes with halite accumulation in various locations worldwide, such as Baja California, the Great Salt Lake in the United States, Kara-Bogaz Gol in Turkmenistan, the Andean salt flats in South America, and the Qaidam Basin in China. These lakes are significant sources of potash and lithium. Both convergent and divergent plate tectonic settings often link hypersaline lakes to petroleum source rocks.

This chapter focuses on the characteristics of evaporite basins formed in convergent plate tectonic settings, which are often linked to oceanic plate subduction and continental block collision. The largest modern hypersaline lakes with evaporite deposits are found in elevated plateaus, mountain ranges, and foreland basins, such as the Great Salt Lake (GSL) in the United States (Figure 10.1), the Andes Altiplano in South America (Figure 10.2), and the Tibetan Plateau in China (Figure 10.3). These Quaternary hypersaline lakes are remnants of larger lakes that have been shrinking due to intense evaporation in the rain shadow zone of the elevated mountain ranges formed by plate tectonics. However, some lagoons are located in depressions near the coast, as the Baja California region of western Mexico (Figure 10.1).

The topobathymetric maps of these regions (based on Amante & Eakins, 2009) include some simplified tectonic elements based on the global maps by Bally et al. (2012). The tectonic setting involves the subduction of oceanic lithosphere (Pacific, Juan de Fuca, and Nazca plates) under continental lithosphere (North and South American plates), forming fold-thrust belts and foreland basin systems. The collision of Indian and Asian continental blocks (Figure 10.3) resulted in the uplift of the Tibetan Plateau in the Cenozoic and the deformation of sedimentary basins by contraction and strike-slip faults (e.g., Aitchison et al., 2007; Wang et al., 2012, 2017; Wu et al., 2014). The Indian plate is moving toward the northeast, as evidenced by a diffuse zone of earthquakes (Bally et al., 2012).

The overall plate tectonic setting of sedimentary basins in western North America is related to the subduction of the Juan de Fuca and Pacific Plates and transcurrent movements involving the boundary with the North American Plate (Bally et al., 2012). The Pacific Plate motion is toward the northwest (NW), and the boundary with the North American Plate is marked by subduction zones and large transform fault zones, such as the San Andreas fault, located south of the Great Basin (Figure 10.1).

The lakes in the southwestern United States are primarily linked to extensional and transtensional tectonics, with depositional complexes in rift basins involving alluvial fans, fluvial floodplains, and lacustrine systems with shallow water. Several ephemeral lakes occur in transtensional depressions along the San Andreas fault zone in California (see Chapter 9).

In the Basin and Range Province, there are many examples of perennial and ephemeral saline lakes formed on terranes previously associated with convergent settings subsequently affected by extensional tectonics (Constenius, 1996). Some of the largest salt flats on Earth correspond to remnants of larger lakes or inland seas formed on foreland depressions, such as the GSL in Utah (Atwood et al., 2016; Chen & Maloof, 2017).

The hypersaline lakes in the Andean Altiplano are associated with the Nazca Plate subduction, which resulted in fold-thrust belts, volcanic activity, and elevated plateaus in the upper plate, corresponding to the western

Salt in the Earth Sciences: Evaporite Rocks and Salt Deposition, Geophysical Monograph 286, First Edition. Webster Mohriak.

DOI: 10.1002/9781119405245.ch10

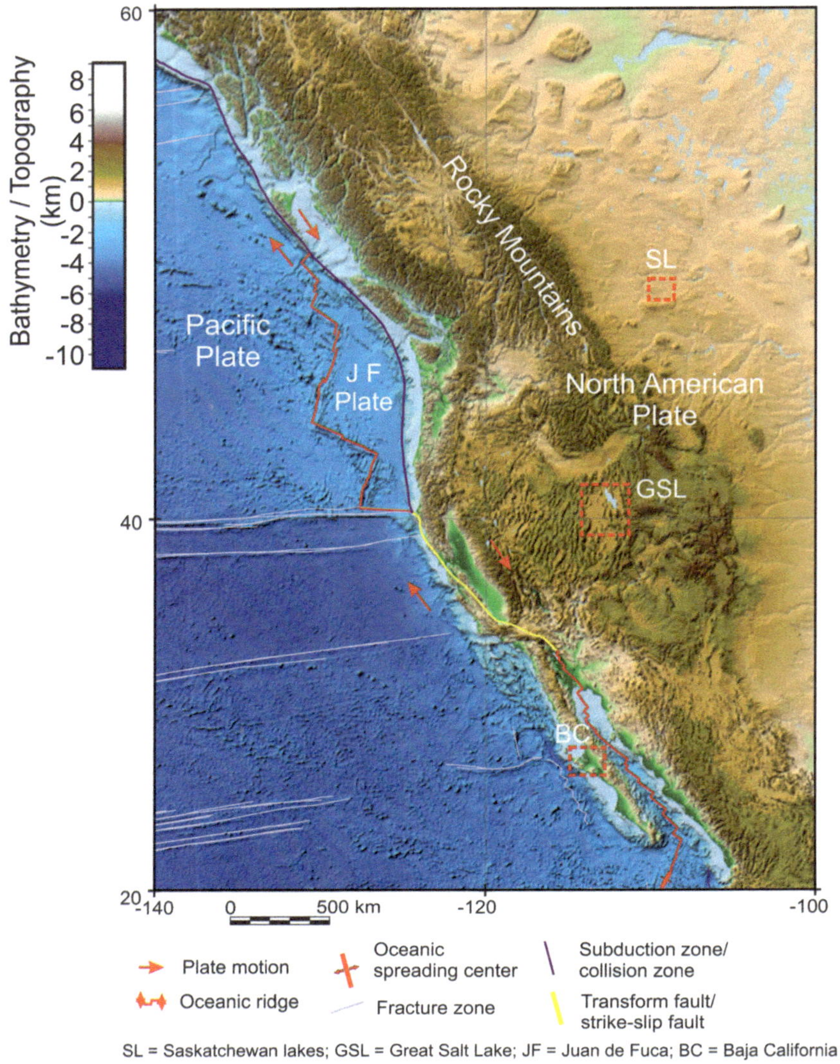

Figure 10.1 The topobathymetric map depicts the subduction zones of the Pacific Plate and Juan de Fuca Plates in northwest North America. The Gulf of California separates the Baja California Peninsula from the Mexican mainland, with a spreading center established in the Neogene. The thin, dashed-red rectangle in Baja California (BC) shows the approximate location of the Laguna Ojo de Liebre, adjacent to the Pacific Ocean coastline. The thin-dashed-red rectangle in the United States shows the GSL in Utah. The GSL in Utah is a remnant of a large lacustrine system from the Pleistocene. In the Holocene (7 to 6 ka), an abrupt fall of lake level led to the development of a large salt flat with carbonate rocks and evaporite crusts. The thin, dashed-red rectangle in Canada shows the Saskatchewan Lakes (SL), located in the foreland of the Rocky Mountains. After the last glaciation, evaporite deposits were formed by the evaporation of continental brines in shallow lakes during arid intervals. Credit: Bally et al. (2012)/with permission of Elsevier.

South America landmass (Figure 10.2). The absolute motion of the Nazca plate is toward the southeast, resulting in a major inflection of the subduction zone and Andean ranges, changing from the N-NE trend in Chile to the N-NW trend in northern South America (Bally et al., 2012). The Altiplano is located near this inflection, which is marked by an expansion of the fold-thrust belt width, reaching circa 700 km in the Central Andes, where the highest average elevations are observed. In the emblematic Altiplano Plateau, several salt flats occupy the depressions, such as Salar de Uyuni (Figure 10.2). South of the Altiplano, Salar de Atacama (also known as Atacama Salar or Atacama Depression) is the largest salt pan in Chile. The depression is located between mountain ranges to the west and volcanic terranes to the east, adjacent to the Western Cordillera. It is characterized by a

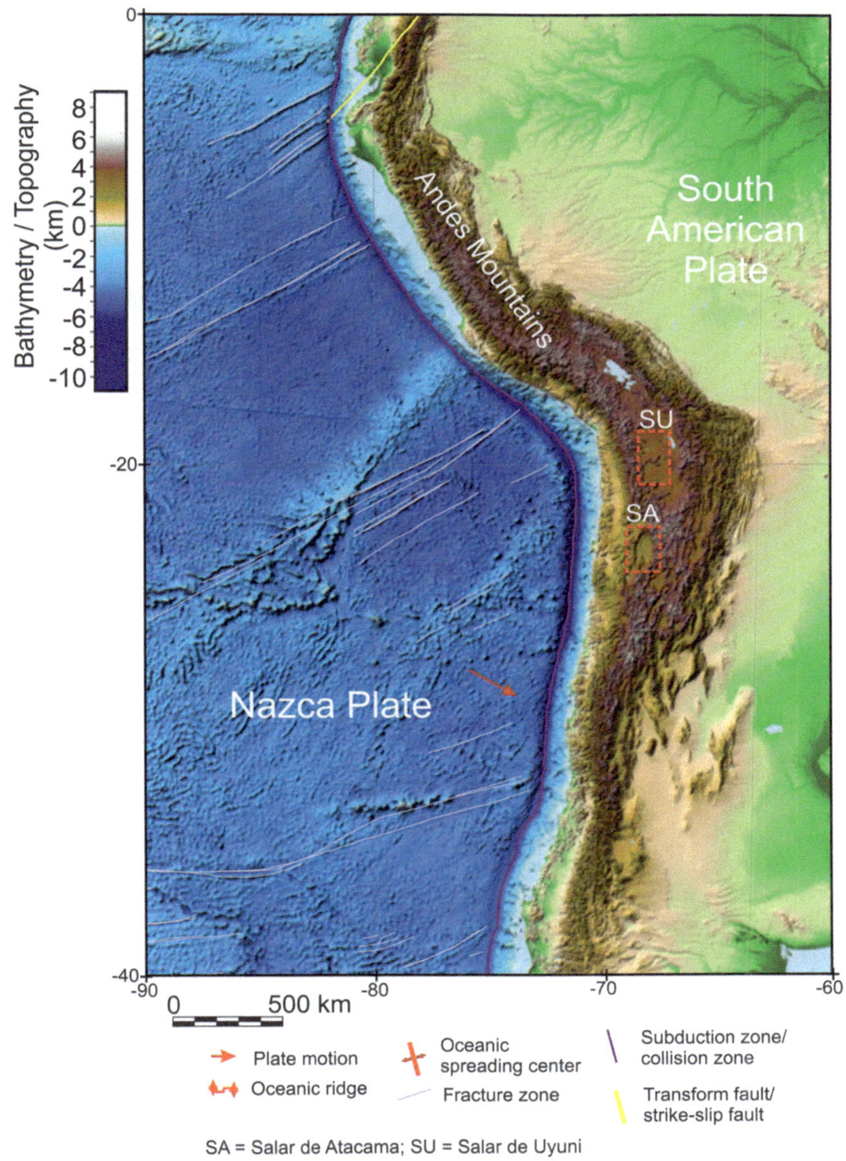

Figure 10.2 The topobathymetric map depicts the subduction zone of the Nazca Plate under South America, extending from Colombia to Chile. The Andean Mountains are highlighted in Bolivia and Chile, with the Salar de Uyuni in Bolivia and the Salar de Atacama in Chile. The Atacama depression hosts the Salar de Atacama, highlighted by a dashed-red rectangle. The depression is located south of the Altiplano and east of the Precordillera (Domeyko Range), with volcanic fields near the Western Cordillera. Credit: Bally et al. (2012)/with permission of Elsevier.

lower altitude than the Altiplano. Both the Uyuni and Atacama depressions exhibit a salt crust overlying thick halite deposits precipitated in a hypersaline lake during the Pleistocene and Quaternary (Lowenstein et al., 2003).

Quaternary to recent salt accumulations are linked to lacustrine environments near major inland seas. Some lakes in Asia have been shrinking due to plate reorganization in convergent settings, such as Kara-Bogaz Lake in Turkmenistan. The lake is fed by brackish waters from the Caspian Sea, which is surrounded by orogenic belts reflecting the closure of the ancient Tethys Sea in the Cenozoic. Kara-Bogaz Lake was affected by a major desiccation event in the past century due to basin isolation from the Caspian Sea. This lake represents a unique modern analog for basin desiccation events affecting shallow-water inland seas adjacent to larger and deeper seas (Figure 10.3).

In northern South America, important evaporitic deposits are associated with convergent plate tectonic

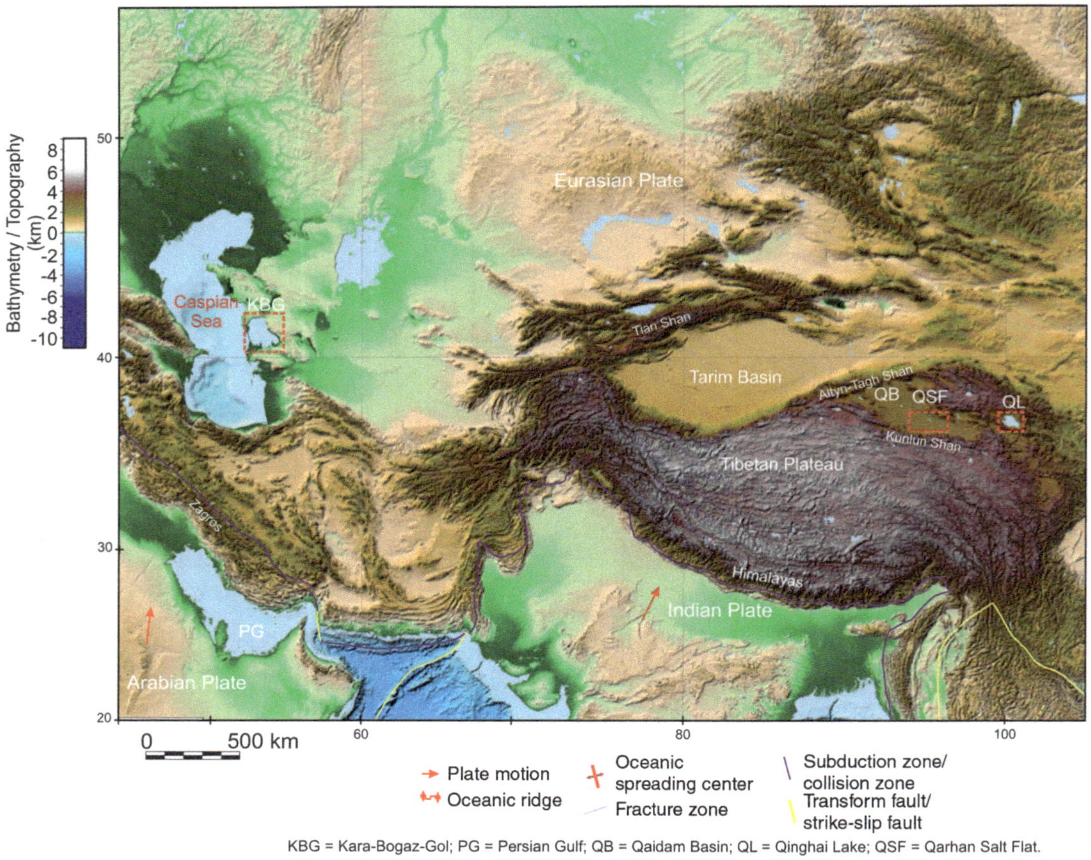

Figure 10.3 The topobathymetric map shows the eastern part of the Alpine Orogeny, extending from the Zagros Orogen south of the Caspian Sea to the Tibetan Plateau in China. Kara-Bogaz-Gol (KBG) is a shallow lake with continental brines, formed adjacent to the Caspian Sea. The Persian Gulf (PG) is characterized by several episodes of evaporite deposition in the Mesozoic and Cenozoic, associated with the convergence between the Arabian and Eurasian plates. The Tien Shan limits the Tarim Basin to the north and the Altyn-Tagh Shan limits it to the southeast, while the Kunlun Shan borders the Qaidam Basin to the south. The Qinghai Lake is a large brackish water lake eastward of the Qaidam Basin. The Qarhan salt flats host several hypersaline lakes in the eastern part of the Qaidam Basin, north of the Kunlun Shan. Credit: Bally et al. (2012)/with permission of Elsevier.

settings. The active subduction of the Nazca oceanic lithosphere under the South American continental lithosphere resulted in magmatic arcs and compressional tectonics, forming the Andes Mountains (e.g., Charrier et al., 2007; Martínez et al., 2018). Lacustrine systems on the Andean highlands are characterized by Quaternary salt deposits with no marine influence, suggesting that the evaporites are related to brine enrichment by volcanic activity (e.g., Hovland et al., 2018a, 2018b). The Andean Altiplano is marked by several salt flats that correspond to ancient saline lakes that dried up, precipitating evaporite layers (e.g., Alonso et al., 1991).

This chapter discusses some of the most important salt flats in the region, particularly Salar de Uyuni in Bolivia and Salar de Atacama in Chile. These salt basins are characterized by exceptional geological and economic importance due to intensive salt mining and the extraordinary possibilities of exploiting lithium-rich brines (see Chapter 9 in Mohriak, 2025).

Hypersaline lakes, such as those in the Tibetan Plateau, are often linked to continent-continent plate convergence (Figure 10.3). The Qaidam Basin, located in the northern part of the Tibetan Plateau, is the largest intramontane, endorheic, and nonmarine salt basin in the region. The strike-slip basins on the Tibetan Plateau were formed by differential motions between the Indian and Asian plates (Bally et al., 2012). The Qaidam Basin is desertic and cold, and the central basin floor is 2.7 km above sea level (a.s.l.). The basin is characterized by large delta plains associated with the surrounding mountain ranges that exhibit altitudes of more than 5 km (e.g., Sun et al., 2020).

The Qaidam Basin depocenter is predominantly covered by dry salt flats, with many shallow, hypersaline lakes, such as Qarham Lake in the eastern part of the basin. The lacustrine system in the Qarhan salt flat is characterized by CaCl brines deficient in $MgSO_4$. These lakes have been used as modern analogs for the deposition of potassium-rich minerals in nonmarine basins (e.g., Lowenstein et al., 1989). Several complex mineralogical associations are recognized in the lacustrine deposits. Qarhan Salt Lake (QSL) hosts the largest potash brine deposit in the world, and the lake also produces lithium and boron salt resources (Du et al., 2018; Li et al., 2021; Stober et al., 2023).

These modern hypersaline basins in the Tibetan Plateau, mostly associated with the convergence of tectonic plates and the collision of continental blocks, provide useful analogs for ancient foreland basins formed in the geological past. For example, the Permian Basin, located in the Southwest United States, resulted from plate tectonic rearrangements following the Marathon-Ouachita Orogen during the Paleozoic (see Chapter 12). Most giant potash deposits worldwide are associated with marine basins, while only a few are associated with nonmarine lacustrine environments. The largest potassium deposits in China are hosted in nonmarine lacustrine basins on the high-altitude Tibetan Plateau. The Andean Altiplano salt flats are characterized by the occurrence of salt deposits and lithium-rich brines in subsurface reservoirs. These hypersaline basins are important economic assets for Bolivia, Chile, and Argentina due to their large mineral resources in the subsurface (U.S. Geological Survey, 2023).

10.1. PERENNIAL SALINE LAKES IN BAJA CALIFORNIA

A modern example of salt deposition in a saline lake associated with convergent margins is observed in western Mexico (Figure 10.4). The Baja California Peninsula, located between the Pacific Ocean and the Mexican mainland, corresponds to a block of continental crust limited by the subducting oceanic crust in the west and the nascent spreading centers in the Gulf of California (Umhoefer, 2011; Umhoefer et al., 2007). The back-arc and pull-apart basins in the Gulf of California are related to the oblique rifting of the continental crust and strike-slip motions involving the Pacific and North American Plates. The dextral movement along the San Andreas fault has resulted in the development of small transtensional basins adjacent to transform fault zones. Several basins are recognized in the region, from the Salton Sea onshore, at the border between the United States and Mexico, to the southern part of the gulf, eastward of the Baja California Peninsula (see Chapter 9).

The Laguna Ojo de Liebre is a large hypersaline lagoon formed between the Pacific Ocean coastline and the magmatic arc volcanic rocks in central to eastern Baja California (Figure 10.4). The lagoon is characterized by shallow brine pans separated from the ocean by a low sandy barrier that is breached by small channels and extensive salt flats (Phleger & Ewing, 1962). The magmatic arc, located east of the lagoon, is related to the subduction of the Pacific Plate under the North American Plate (Figure 10.1).

Large volcanic fields occur in the coastal region of the Baja California Peninsula and the Gulf of California (Umhoefer, 2011). The western part of the gulf is characterized by an elevated topographic relief. The center part of the gulf is characterized by deepwater and young spreading centers offset by large transform fault zones formed in the Neogene (Figure 10.4A). Several volcanic domes occur adjacent to the Pacific Ocean and in the vicinity of the gulf (Figure 10.4B). The San Borja Volcanic Field is characterized by widespread basaltic to andesitic lava flows, ignimbrite buildups, and cinder cones (Varley, 2012).

Laguna Ojo de Liebre, the most important coastal depression in Baja California, is connected to the Pacific Ocean through two main channels (Figure 10.4B). The lagoon was filled with marine sediments during Quaternary transgressions, indicating a continuous connection with the ocean (Phleger & Ewing, 1962; Pierre et al., 1984). High evaporation rates in the region result in concentrated brines and evaporite minerals precipitating at the lake's bottom and margins, forming an extensive salt flat north of the lagoon. Carbonate rocks, mainly composed of aragonite and calcite ($CaCO_3$), accumulate in brine pans, while evaporite sediments form in more restricted parts of the lagoon, including gypsum ($CaSO_4.2H_2O$) and halite (NaCl) deposits in salt flats (Phleger, 1969). The sedimentary context of these Quaternary evaporites is predominantly related to a siliciclastic coastal sabkha with large eolian dunes, except for carbonate rocks that outcrop at the southern border of the lagoon (Pierre et al., 1984).

The supratidal flats surrounding Laguna Ojo de Liebre extend up to 15 km into the mainland and have been transformed into artificial saline ponds since the late 1950s. The Guerrero Negro saltworks in Baja California are one of the largest salt-making facilities on Earth. In 2022, China was the largest salt mine producer in the world, with 64 million tons, followed by India (45 million tons) and the United States (42 million tons), according to the U.S. Geological Survey (2023).

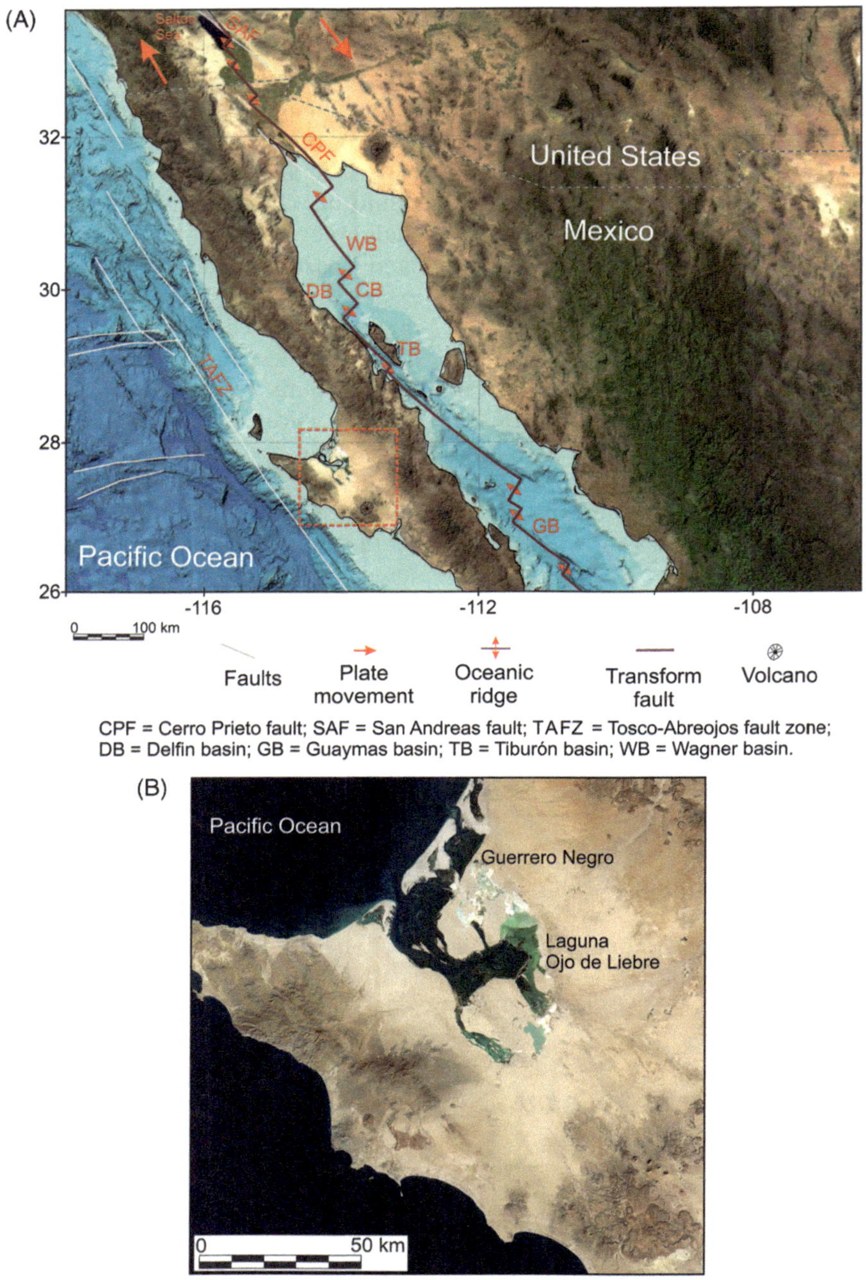

Figure 10.4 (A) Satellite image of the Baja California Peninsula between the Pacific Ocean and the Mexican mainland. The Baja California Peninsula is characterized by high topographic elevations to the west of the spreading centers and the transform fault zones in the gulf. (B) NASA satellite image showing the western, central part of the Baja California Peninsula (highlighted by a red rectangle in the previous image). The Laguna Ojo de Liebre is a large hypersaline lagoon located near the Pacific Ocean continental margin. The lava flows and ignimbrites display a dark brown reflectivity in the satellite image. Credit: (A) Adapted from Umhoefer (2011) and (B) NASA/Public domain.

10.2. PERENNIAL SALINE LAKES NORTH OF THE BASIN AND RANGE

Perennial salt lakes are large, shallow lakes surrounded by salt flats or small, deep lakes that lack extensive sabkhas. In the Death Valley of California, some small lakes are flanked by steep slopes in rift-related mountain ranges, with alluvial fans and fluvial systems associated with ephemeral playa lakes (see Chapter 9). The substrate of these perennial saline lakes commonly corresponds to well-consolidated ancient lake-bottom sedimentary deposits. These lacustrine sediments accumulated during previous wet periods and are not directly associated with the depositional settings of the active sabkhas.

The sabkha sedimentary facies occur as belts fringing the perennial salt lakes and are commonly composed of unconsolidated carbonate sand, mudflats, and salt crusts (see Chapter 8).

Evaporite sediments may accumulate in sabkha plains adjacent to perennial saline lakes during periods of maximum aridity. During these desiccation episodes, the water table falls abruptly, and evaporite minerals precipitate in the pores of the previously deposited subaqueous sediments. The supratidal evaporites become subaerial during sea level fall, as observed in the arid environments of the Persian Gulf (see Chapter 8).

Falling lake levels due to climate aridity result in important saline lakes, as observed in some regions affected by ancient orogenies in the southwestern United States, such as the large lakes in Utah, near the Basin and Range Province, located on the Wasatch Culmination region (Figure 10.1). The regions bordering the GSL in Utah are marked by thin-skinned thrust faults related to basement-involved contraction (Costenius, 1996). Several sedimentary basins were formed during the collapse and westward gravity spreading of the thickened crust, resulting in a system of half-grabens controlled by extensional faults associated with the Basin and Range Province (Constenius, 1996).

Ancient lacustrine environments in the western part of the Rocky Mountains developed on a substratum involving a foreland basin and rift depressions. These structures resulted from extensional faults associated with the gravitational collapse of the Cordilleran Orogen fold-thrust belt in the Late Cretaceous to Paleogene (Constenius, 1996). The rift depocenters formed between the middle Eocene and early Miocene, as indicated by exploratory boreholes and seismic interpretation (Mohapatra & Johnson, 1998).

A later stage of crustal extension and magmatism developed between the Middle Miocene and Quaternary, forming a network of half-grabens and magmatism that resulted in the present-day geomorphology in the northern part of the Basin and Range province (Constenius, 1996). The GSL, the largest hypersaline lake in the Western Hemisphere, is the fourth-largest terminal lake in the world. The lacustrine sediments occupy one of the lowest depressions in the extensional fault-block systems of Southwestern United States. The basin architecture developed from the interaction of the Wasatch Mountains and the reactivated thrust fault systems inherited from the Sevier Orogeny (Figure 10.5).

The lake's surface fluctuates at an altitude of about 1280 m a.s.l., but the water level has been steadily declining in the past decades. The GSL water volume declined to its lowest value in 2016, reaching a water volume of less than half the record high level observed in 1986 (Wurtsbaugh et al., 2017). At present, the salinity of the GSL varies from 50 to 270 g/kg, much above the normal seawater salinity of 35 g/kg (see Chapter 2).

The evaporite sediments accumulated on the lake's bottom and around its periphery include carbonate, sulfate, and chloride minerals, particularly aragonite muds, gypsum, mirabilite, and halite. The white patches in satellite images (Figure 10.5) correspond to widespread evaporite deposits associated with the desiccated, ancient lake that existed in the Pleistocene, known as Lake Bonneville.

The GSL Desert is a remnant of the much larger and deeper lake that covered much of western Utah, eastern Nevada, and southern Idaho during the Pleistocene, as a consequence of glacier melting circa 30 ka (e.g., Lindsay et al., 2020; Spencer et al., 1984). The highest lake level, about 1552 m a.s.l., was attained around 15 ka, as indicated by the paleoshorelines and evaporite crusts (Atwood et al., 2016). The Wisconsin glaciation waned during the transition from the Pleistocene to the Holocene, resulting in increased aridity and evaporation, leading to the drying up of paleolakes and hypersaline lakes. Abundant lacustrine deposits (carbonate muds and evaporite sediments) currently form salt flats surrounding the remnant GSL in the region once occupied by the much larger Lake Bonneville during the Pleistocene (Atwood et al., 2016; Lindsay et al., 2020).

The Bonneville Salt Flats (Figure 10.5) correspond to the desiccated lakebed relicts of this ancient lake, which was much larger and deeper. Lake Bonneville probably reached depths of more than 300 m, whereas the present-day lake has a water depth of only 10–15 m (Benson et al., 2011). During the initial phase of lake desiccation in the Holocene, carbonate muds formed biostromes composed of blue-green colonial algae, resulting in microbialite buildups and reefal mounds at the interface between the land and the lake brine (Carozzi, 1962; Bouton et al., 2016). As the lake level has fallen since the last glacial epoch, the brines have concentrated, precipitating sulfates (gypsum) and chlorides (halite) during intervals of increased brine concentration, developing sabkha flats along the lake's margins. These gypsum and halite crusts outcrop along the lake's margins, exhibiting white reflectivity in satellite images (Figure 10.5).

The GSL in Utah is divided into two halves by a railroad causeway, the Southern Pacific Railroad causeway, which was built in 1904 and rebuilt in 1959. This creates a barrier that restricts water circulation between the two halves, resulting in different colors and hues (Figure 10.5). The northern sector is saltier than the southern sector, with a salinity of 27%, eight times the salinity of seawater. The lake brines are extremely concentrated and precipitate halite in isolated ponds. Almost no organisms can survive in this hostile environment except for some algae and halophile sulfur bacteria that proliferate in the hypersaline brine and tint the water reddish (Figure 10.5). Freshwaters enter the lake from an active drainage system in the southern part of the depression, near Salt Lake City. This southern sector appears dark blue in the satellite

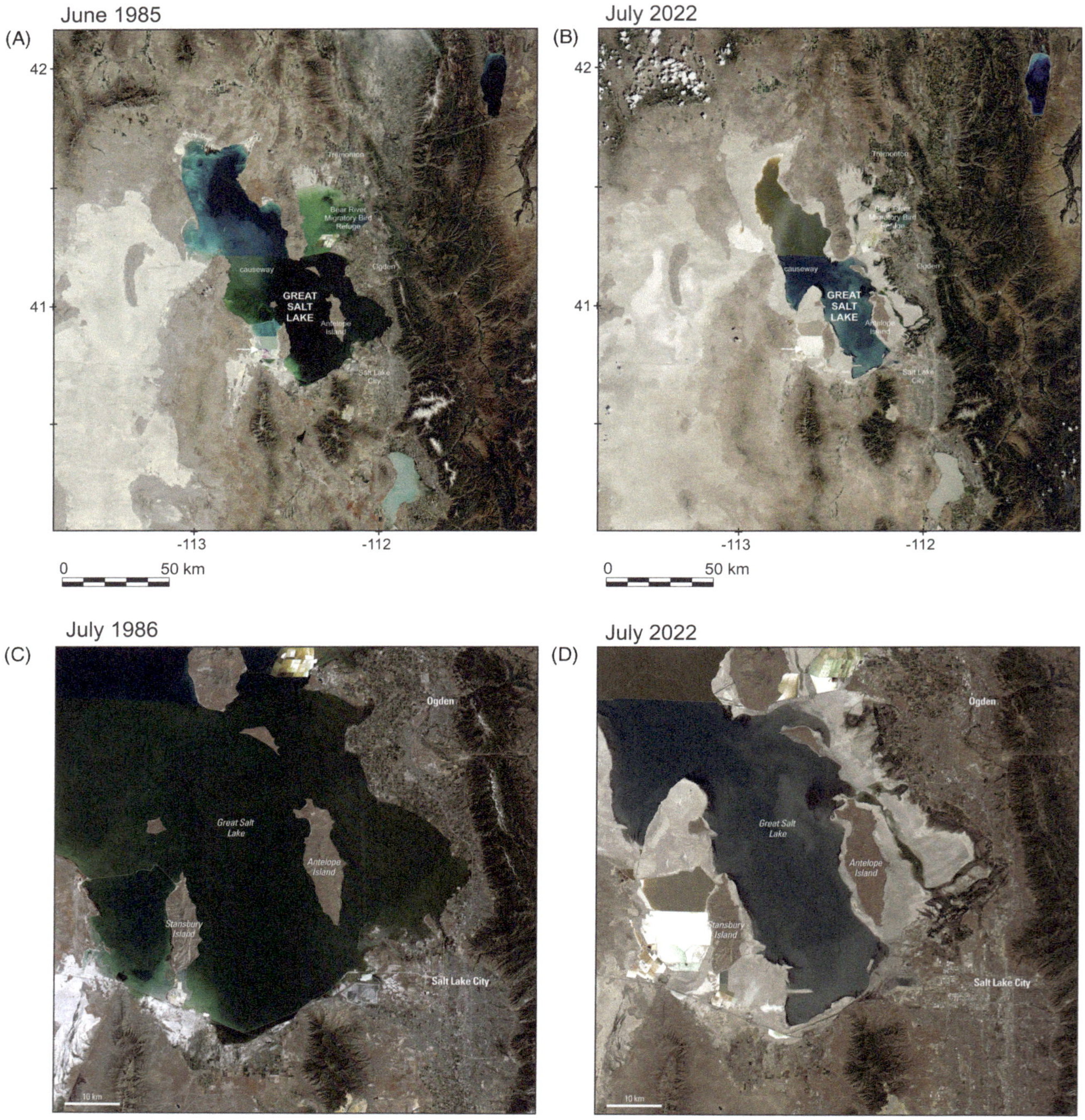

Figure 10.5 Top: comparison of lake level and geomorphology of the GSL in June 1985 (A) and July 2022 (B). In 1985, the lake was at 1284 m above sea level (a.s.l), resulting in a high level. In 2022, the lake level reached 1277 m a.s.l., exposing large areas on mudflats adjacent to the remnant lake. Evaporites precipitated in the northern part of the lake, with a higher salinity brine. The southern part was less concentrated due to river water inflow. The railroad causeway built in 1959 splits the lake into two arms that display different colors in satellite images. This difference in reflectivity is due to the halophilic bacteria that adapted to the more concentrated brine in the northern basin. Bottom: detailed views of Landsat images showing the southern part of the GSL at a record high level in July 1986 (C) and at a historic low elevation in July 2022 (D). Credit: NASA.

image, indicating a reduced salinity of around 5%, close to seawater with 3.5% (see Chapter 2).

Although water levels in the GSL fluctuate by year, the lake level has been generally declining for decades due to intensive use by agriculture and human consumption (Wurtsbaugh et al., 2017). The highest recorded lake level was in 1986, when heavy rains resulted in a mean water elevation of 1284 m. Since the late 1980s, the lake surface has dropped about 7 m, reaching a new record low of 1277 m in 2021 (Utah Water Science Center, 2021). The comparison of the satellite images obtained with an interval of a few decades illustrates the rapid variations in lake geomorphology following the drought that affected the southwestern region of the United States since the late 1980s (Figure 10.5).

The GSL, the largest saline lake in the United States and eighth-largest globally, exhibits large salt flats at its southwest border, adjacent to the Basin and Range arid regions (Figures 10.5C and 10.5D). Saltworks exploit the brines by solar evaporation, producing magnesium, salt for deicing roads, and potassium for fertilizer. However, the United States has a fraction of the world's magnesium compound reserves. China was the leading producer in 2022, with mine production estimated at 17 million tons (U.S. Geological Survey, 2023). Primary magnesium metal has been recovered from brines from the GSL, but U.S. production has been withheld to avoid disclosing industry-proprietary data. China was responsible for over 90% of the total primary production of magnesium metal in 2022 (U.S. Geological Survey, 2023).

10.2.1. The Great Salt Lake Carbonate and Evaporite Sediments

The GSL is situated at the northeastern edge of the Basin and Range Province, eastward of the Wasatch Front (Figure 10.6) The Wasatch Front is a tectonically active fault, as indicated by earthquake activity and seismic data (e.g., Colman et al., 2002; Liberty et al., 2021). In the 1970s and 1980s, the petroleum industry acquired 2-D reflection seismic lines in the GSL region and drilled exploratory wells, finding heavy oil intervals in fractured Pliocene volcanic rocks (Bortz, 1987).

The GSL overlies the Cenozoic rift basins that are separated by an NE-trending high from Carrington Island toward the Promontory Range on the eastern side of the lake. The structural high flanking the rift trough is composed of lower and middle Paleozoic rocks (Figure 10.6B). Regional seismic profiles from the 1970s and 1980s revealed that the sedimentary troughs in this segment of the Basin and Range Province are mainly filled with Miocene, Pliocene, and Quaternary sediments and volcanic rocks, with maximum depth in the sedimentary depocenters estimated around 4–5 km (Mohapatra & Johnson, 1998).

The GSL margins are characterized by carbonate mounds and reef-like microbialite structures developed on algal mats covering the shallow-water platforms (Lindsay et al., 2020). These structures also occur in the lake bottom, covering about 20% of the lake, forming one of the most extensive occurrences of microbialite rocks in the world, rivaling Shark Bay in Australia (Baskin et al., 2017; Jahnert & Collins, 2013).

The lake shore on Antelope Island corresponds to a structural high at the footwall of a normal fault in the southern part of the GSL (Mohapatra & Johnson, 1998). The lake sediments along the shores include conglomeratic quartzite grains and ooid sands composed of carbonate (aragonite) and fecal pellets. Desiccation polygons and mudcracks suggest subaerial exposure of the mud and microbialite in previously submerged playas (Baskin et al., 2017). Laminated microbialite and cauliflower stromatolite structures occur up to 3 m above the lake surface, indicating a higher paleoshoreline elevation (Chidsey et al., 2015; Lindsay et al., 2020; Oviatt & Shroder, Jr., 2016; Oviatt et al., 2021). The submerged microbial and subaerial sedimentary structures, recently covered by lake water, indicate rapid fluctuations in lake level in the Holocene. In deeper areas, the ooid sands grade into a finely laminated quartz-rich clay sediment with 30% dolomite and calcite (Bouton et al., 2016). The deepest parts of the lake are characterized by carbonate mud. However, Colman et al. (2002) reported the occurrence of deeper microbialite sediments circa 9 m below the lake level (1271 m a.s.l.), indicating a rapid rise of the lake level in the Holocene.

The radiocarbon age of microbialites indicates a variation for paleolake shorelines from a mean of 1280 m a.s.l. to a maximum water level of about 1550 m a.s.l. during episodes of high lake levels, corresponding to the Bonneville shoreline circa 15 ka (e.g., Atwood et al., 2016; Chen & Maloof, 2017). The Lake Bonneville paleoshoreline extends far west and south of the remnant modern lake (Figure 10.6). The lake geomorphology has changed dramatically in the past few decades, from a maximum lake level in 1985–1986 to a minimum lake level in 2022 (Figures 10.6C and 10.6D).

The GSL experienced a sudden drop in lake level from 15 to 12 ka, leading to the formation of desiccation polygons between 7 and 6 ka (Bouton et al., 2016). This period coincided with the Holocene water level drop in many lakes across the United States, resulting in an increase in salt concentration and the deposition of a layer of mirabilite (e.g., Spencer et al., 1984). High-resolution seismic data revealed an evaporite layer at a depth of about 2 m below the lake bottom (Colman et al., 2002). The mirabilite layer was cored and revealed the evaporite deposition was associated with a layer of volcanic fragments (Mazama tephra), indicating a basin desiccation event circa 6.7 ka (Colman et al., 2002).

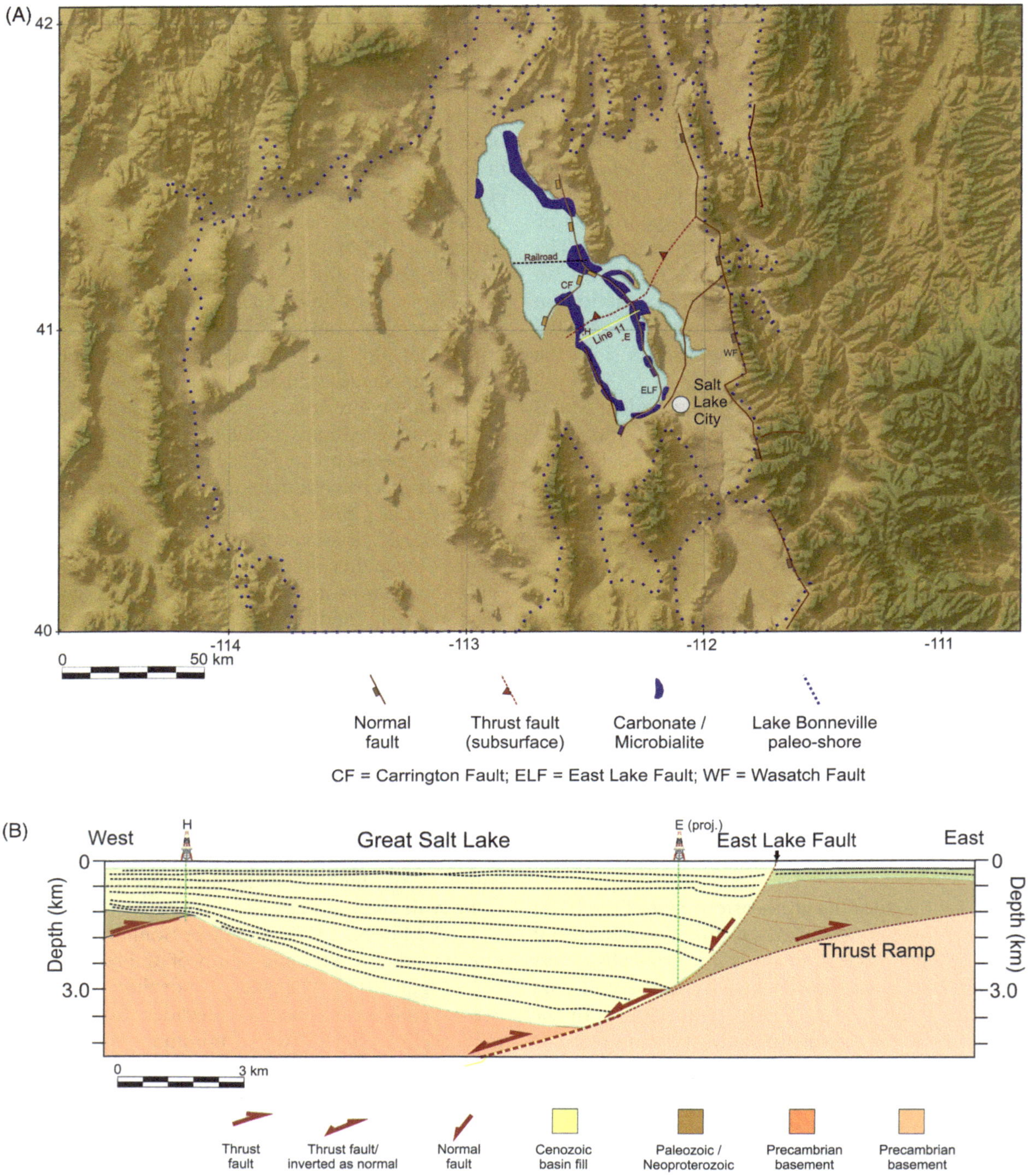

Figure 10.6 (A) The topographic relief map depicts the GSL and the paleoshoreline of the Pleistocene-Holocene Lake Bonneville at its highest level around 15 ka. The salt flats exhibit a smooth texture and are characterized by carbonate microbialite rocks. A thin yellow line shows the location of Line 11, a petroleum industry seismic line crossing a half-graben NW of Salt Lake City. (B) Schematic interpretation of a regional seismic line in the southern arm of the GSL (Line 11). The seismic profile extends from Carrington Island to Antelope Island, near borehole H. At the western parts of the rift basin, Paleozoic to Neoproterozoic rocks in the footwall of the East Lake Fault outcrop or are overlain by Holocene sediments. Borehole E projects along the eastern segment of the profile, near the basin depocenter. At the western and eastern extremities of the seismic profile, researchers observe microbialite buildups and carbonate reefs near the surface. Credits: (A) Colman et al. (2002)/with permission of Elsevier and (B) adapted from Mohapatra and Johnson (1998).

Mirabilite layers with thicknesses of up to a few meters were also recorded in the northern arm of the GSL (Spencer et al., 1984). High-resolution seismic profiles in the offshore region of the southern part of the lake, near Antelope Island, revealed normal faults with offsets exceeding a few meters in the Holocene sediments and locally affecting the lake bottom (Colman et al., 2002). The shallow portions of the sedimentary sequences are affected by synthetic and antithetic normal faults, likely linked with listric extensional faults detaching on pre-existing thrust ramps deeper in the basin (Figure 10.6B). The extensional faults controlling the Neogene rift basins in the GSL region probably correspond to a system of imbricate listric normal faults formed by extensional reactivation of the Sevier-age Willard thrust in the subsurface (Mohapatra & Johnson, 1998).

High-resolution seismic data (e.g., Colman et al., 2002) reveal several normal faults displacing upper sedimentary layers and the lake floor by several meters. Sediments typically reach a maximum thickness of 13 m adjacent to the main active fault (East Lake Fault). Recent drilling has shown that local grabens are filled with Holocene sediments that reach a thickness of over 20 m above the Bonneville Lake horizon. These layers correspond to a maximum lake level at around 15 ka, when the lake expanded due to the demise of the last glacial epoch and melting of glaciers, with the northern shorelines reaching Idaho and the western shorelines reaching Nevada (e.g., Oviatt et al., 2021; Thompson et al., 2016). The minimum lake level was estimated at around 6.7 ka, resulting in the precipitation of mirabilite interbedded with calcareous muds (Colman et al., 2002; Spencer et al., 1984).

High-resolution seismic profiles show mound structures on the basin floor, interpreted as bioherms, similar to the stromatolite structures around the lake's shores. Some mounds are located adjacent to faults, suggesting that the buildups formed by enhanced carbonate precipitation due to basinal fluids migrating upward and reaching the depositional surface through fault planes or vertical conduits (Colman et al., 2002). The microbialite constructions are observed both above the lake level and in the shallow portions of the lake (Baskin et al., 2017; Lindsay et al., 2020).

Shallow seismic data and cores reveal the presence of listric normal faults controlling an abrupt fall in the lake's upper sedimentary layers. These disturbances were induced by modern tectonic activity, as indicated by the large number of earthquakes affecting the GSL region (Liberty et al., 2021).

Climate change is a significant factor controlling microbial deposits in the Great Southern Lake. Lake level fluctuations are associated with seasonal variations, with water volume increasing in spring and summer due to snowmelt and rainfall. Water volumes decrease in late summer and autumn due to extensive evaporation (Atwood et al., 2016; Bouton et al., 2016). However, these annual fluctuations do not explain the main trends of lake-level variation and microbial deposition, which are controlled by longer-term climatic factors and lake hydrology (Lindsay et al., 2020). Some bioherms are aligned for tens of meters, possibly following fault zones affecting lake sediments (Colman et al., 2002).

The GSL has limited siliciclastic input, limited groundwater recharge, and insignificant hydrothermal input. The GSL waters are at saturation or supersaturated with calcium carbonate, forming microbialite deposits (e.g., Baskin et al., 2017; Bouton et al., 2016). The main water contributions are from fluvial waters and glacier melting, direct pluvial precipitation, and hydrothermal and groundwater influxes (e.g., Jones et al., 2009). However, hydrothermal fluids and groundwater account for 18% of the total ionic influx into the lake (Bouton et al., 2016). The ion-rich fluid circulation through fractures might enhance the mineralization of the microbial structures (Lindsay et al., 2020).

10.3. PERENNIAL TO EPHEMERAL SALINE LAKES IN ASIA: THE KARA-BOGAZ-GOL

Modern examples of perennial and ephemeral saline lake systems have been described in the Caspian Sea region of Asia, particularly in Turkmenistan (Figure 10.7). Kara-Bogaz-Gol is a large lagoon located on the eastern side of the Caspian Sea. The lake is about 160 km long and 140 km wide, covering an area of approximately 20,000 km^2 (Kosarev et al., 2009; Varushchenko et al., 2000). The lake is an extension of the much larger Caspian Sea, from which it is separated by a narrow strip of land with a channel that intermittently communicates the two water bodies. Kara-Bogaz-Gol is a shallow lagoon, with maximum water depths of less than 10 m, and the brine is oversaturated with halite, precipitating salt during desiccation episodes. The lagoon is considered by several researchers as an important modern analog for salt deposition in ancient sedimentary basins (Hsü, 1972; Ochsenius, 1877, 1888; Schreiber, 1978; Warren, 2006).

The Caspian Sea, the world's largest saltwater body by volume (see Chapter 2), is an inland sea located in Asia, south of Russia and Kazakhstan, and extends to Iran in the south. It is a brackish water lake with a salinity of 1.2%, one-third of normal seawater. However, from north to south, the Caspian Sea exhibits marked differences in salinity and water depth. The water is saltiest in the south and freshest in the north, with freshwater input provided by large rivers like the Volga and the Ural (Figure 10.7). The most important contribution to the water balance comes from runoff associated with the Volga River, which is responsible for almost 80% of the total fluvial water input (Kosarev & Kostianoy, 2005).

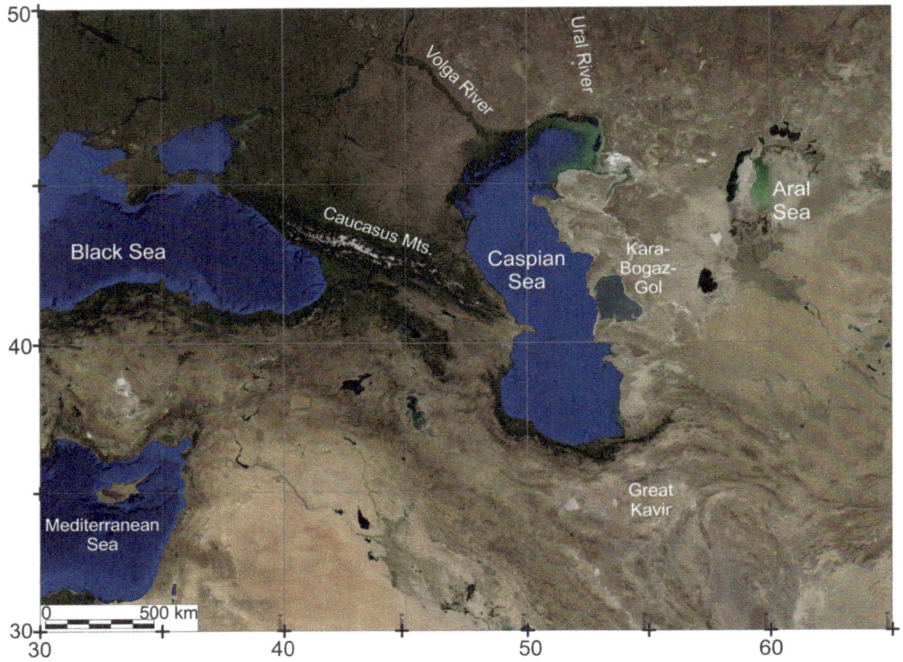

Figure 10.7 NASA's natural-color satellite image of Asia shows the eastern Mediterranean Sea, the Black Sea, the Caspian Sea, and the Aral Sea. An NW-trending mountain ridge separates the Black Sea from the Caspian Sea, extending to the Absheron Peninsula in Azerbaijan, near the city of Baku. Kara-Bogaz-Gol, located immediately to the east of the central Caspian Sea, is a shallow lake marked by green reflectivity in the satellite image, similar to the northern sector of the Caspian Sea. A narrow land bridge with a channel connects the Caspian Sea and Kara-Bogaz-Gol, which are filled with water with different salt concentrations. The Caspian Sea is filled with brackish waters, while Kara Bogaz brines are saltier and precipitate evaporites. The Aral Sea appears almost dried up in the desert region, while the Great Kavir Desert in northern Iran displays whitish patches indicating salt flats. North of the Black Sea, the more vegetated regions appear greenish. Credit: Courtesy of NASA.

The water cascades through a channel in the elongated barrier that separates the Caspian Sea and the lagoon. The modest freshwater inflow controls the high salinity in the lagoon, making up only 5% of the Caspian Sea water cycle. There are important differences in water depth along this large inland sea. Most of the northern part of the Caspian Sea is very shallow, with a depth of about 10 m or less, which constitutes an environmental hazard for seismic acquisition and drilling activities conducted by petroleum exploration companies. The southern part of the Caspian Sea is much deeper, with bathymetries measured in hundreds of meters, with some areas exceeding 1,000 m. As the Caspian Sea is located in the middle of an arid region, the evaporation rate is very high, and the sea surface is actually below the mean global sea level. According to Chen et al. (2017), the mean yearly evaporation rate in the past two decades (1995–2015) was 116 cm, exceeding precipitation (75 cm) and surface run-off fluvial input (35 cm). Consequently, the Caspian Sea water level dropped by a mean rate of 6.7 cm/year, resulting in a water level fall of almost 1.5 m. If this trend continues into the next decades, the Caspian Sea water surface might reach the historically low levels of the late 1970s, with catastrophic consequences for Kara-Bogaz-Gol (Chen et al., 2017).

The water flow into the Kara-Bogaz-Gol was reduced from 22.5 km^3/year in 1930 to 5.4 km^3/year in 1979 due to irrigation projects in the former Soviet Union (Kurilenko et al., 1988). The Caspian Sea water level dropped steadily during the twentieth century, from –25 to about –29 m in the late 1970s (Figure 10.8). Some drastic actions were undertaken in the 1980s and 1990s to reverse this trend. In 1980, a dam was built to control the water flow into Kara-Bogaz-Gol. However, the drastic reduction of water flow into the lake led to almost complete basin desiccation. In 1992, the dam was imploded, restoring communication with the Caspian Sea. The Kara-Bogaz basin was rapidly refilled with Caspian Sea water, dissolving some of the previous evaporite deposits. In the last decade of the past century, the lake level returned to its previous level, with the surface about 4 m below the Caspian Sea level (Kosarev et al., 2009).

The Caspian Sea's surface level has fallen abruptly since 1995–2015, likely due to global temperature and climate change (Chen et al., 2017). This affected the water flowing into Kara Bogaz Lake, which has a mean water

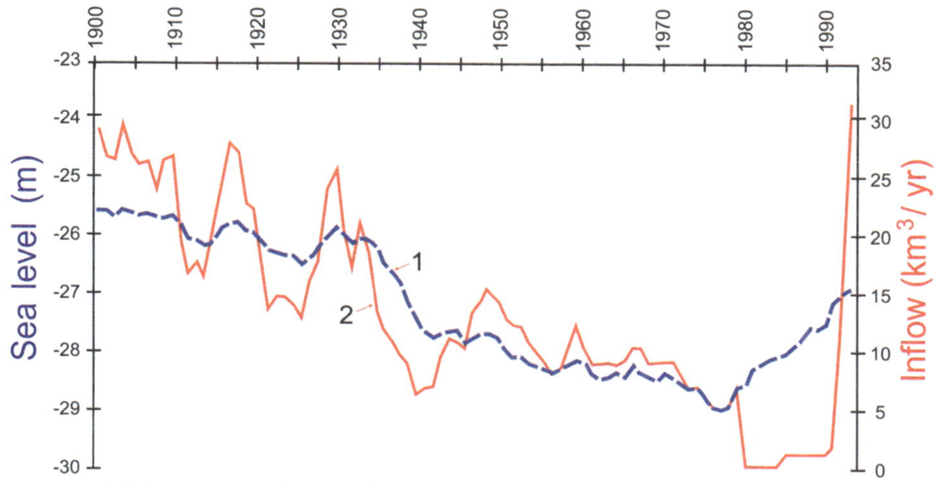

Figure 10.8 The Kara-Bogaz-Gol lake experienced a steady decline in sea level from 1900 to 1975, with the lake level occasionally falling below the Caspian Sea. Rapid sea level falls in the 1930s and 1940s reduced water inflow, leading to a minimum sea level of –28 m in the 1970s. The Soviet government implemented measures to prevent increased salinization, including the construction of a dam in 1980, which reduced water inflow in the 1980s. The lake nearly desiccated in 1985, but the dam was destroyed in 1992 by the Turkmenistan government, allowing water inflow to the lagoon through a narrow channel. In the present century, the lake level has returned to its original level, about 4 m below the Caspian Sea surface. Credit: Varushchenko et al. (2000)/with permission of John Wiley & Sons.

level of around –31 m. The Caspian Sea waters flow eastward, feeding the lake with less saline, nutrient-rich, and darker waters. These waters cascade toward the lagoon as a giant waterfall, forming a delta-like feature (Dickerson, 2000; Varushchenko et al., 2000).

In the nineteenth century, many scientists studied the narrow barrier that separated the Kara-Bogaz-Gol lake from the open sea. This barrier isolated the Caspian Sea from the lagoon and was only trespassed during sea level rises, resulting in the periodic communication of the two water masses (Ochsenius, 1877, 1878). Ochsenius' classical bar theory proposes that evaporite basins form by the episodic isolation of restricted lakes from larger seas. In the Kara-Bogaz-Gol, the barrier is a sedimentary ridge crossed by a narrow channel that connects the Caspian Sea with the lagoon. Closing the strait stops the water flow and isolates the basin. Solar evaporation causes brine concentration, leading to chemical compound precipitation according to the mineral solubility (see review by Hsü, 1972).

Satellite images show that the Kara-Bogaz-Gol lake is fed by turbulent, nutrient-rich, dark waters from the Caspian Sea, forming a stream of tongue-shaped deltaic sediments. The dark color is due to cyanobacteria and dissolved organic matter in the Caspian seawater. The fresh, nutrient-rich water disappears when mixed with the hypersaline waters in the lagoon, which have a higher concentration of sodium chloride. Kara-Bogaz-Gol brine is one of the world's saltiest, with a concentration of ten times seawater salinity (see Chapter 2). The evaporite minerals precipitating at Kara-Bogaz-Gol reveal a continental lacustrine basin affinity, not a marine bay with ocean water evaporites (Kosarev et al., 2009).

Brine concentration has fluctuated since the nineteenth century due to the decreasing sea level from 1900 to the late 1970s (Figure 10.4). An artificial dam built by the Soviet Union in 1980 barred the lagoon's entrance and caused an abrupt decrease in water flow in the 1980s, leading to basin desiccation. The basin precipitated several minerals, mainly chlorides and sulfates, while more soluble salts remained dissolved in the brine (Kosarev et al., 2009). The change in hydrology was not caused by nature but by anthropogenic intervention in the fluvial system, such as the deviation of the Volga River for agricultural purposes in the 1960s and 1970s (Figure 10.8).

10.3.1. The Desiccation of Kara-Bogaz-Gol Lagoon in Modern Times

The Kara-Bogaz-Gol lagoon has experienced significant changes in water salinity due to intense evaporation. The brine concentration in the lagoon is high, reaching 300–350 g/L of salt (35%), precipitating several evaporite minerals in ponds around the lagoon. Kara-Bogaz-Gol is the only salina in the world that produces mirabilite ($Na_2(SO_4)\cdot 10H_2O$), halite (NaCl), epsomite

($MgSO_4 \cdot 7H_2O$), glauberite ($Na_2Ca(SO_4)_2$), and a variety of blödite ($Na_2Mg(SO_4)2{\cdot}4H_2O$) known as astrakhanite (e.g., Kosarev et al., 2009). Large amounts of sulfate-rich evaporite minerals have accumulated on the bottom of the lagoon, which have been exploited by salt companies operating in the region since the early decades of the twentieth century. The salt-bearing deposits in the subsurface (down to a depth of about 30 m below the ground) belong to the sulfate-sodium type (Kurilenko et al., 1988).

Changes in the water salinity of Kara-Bogaz-Gol lagoon have been observed for a long time, as the climate aridity in the region has been steadily increasing since the last century (Figure 10.8). Most hydrological events related to water inflow into the lake are cyclic and result from natural causes, mainly associated with sea level fluctuations in the Caspian Sea. However, the dramatic events that led to the basin desiccation in the 1980s were caused by human manipulation of the hydrological cycle, particularly the deviation of the Volga River to agricultural projects and the construction of a dam in 1980.

In the mid-1970s, the Soviet Union's central government was determined to irrigate the lowlands to the north of the Caspian Sea, aiming to implement a strategic plan for increasing agricultural production on Russian farms and cotton plantations. A political decision was taken to deviate the waters from the Volga River, which drains much of the lowlands in Russia and runs off large volumes of water into the northern Caspian Sea. This led to a net reduction of water flow into the Caspian, resulting in an acceleration of the sea level (Varushchenko et al., 2000).

As the sea level in the Caspian Sea was dropping, the water level in the Kara-Bogaz-Gol also decreased. When the water level reached the sill at the barrier that separated the two water bodies, the flow of water from the Caspian to the Kara-Bogaz was interrupted, dropping sharply to about 5 km^3/year in the mid-1970s. With this decrease in water supply, the already low level of Kara-Bogaz fell sharply, and there was an abrupt increase in the brine concentration. As a result, salt started to precipitate in the lake and along the margins, accumulating in the northern sector of the lagoon (Figure 10.9A).

In 1979, the Soviet Union considered that the reduced flow of water across the channel between the Caspian Sea and Kara-Bogaz-Gol was one of the reasons for the drop in the lake level. The central government determined the construction of a dam to completely obstruct the flow of water between the inland sea and the lagoon, aiming to better preserve the Caspian Sea water level. This dam was completed in 1980, isolating the Caspian from the lake. However, instead of regulating the sea level in the Caspian Sea, this dam further increased the speed of lake-level fall in the Kara-Bogaz. Within a few years, the extreme concentration of the brines resulted in widespread desertification of the lowlands in the lagoon. The catastrophic event led to a near-complete desiccation of the lake in the mid-1980s, and evaporites precipitated throughout the basin. A satellite image from 1987 (Figure 10.9B) shows that the isolation of the lake waters by the artificial dam resulted in a barren salt flat across most of the lagoon. The closure resulted in the transformation of the lagoon into a salt pan, causing an environmental disaster

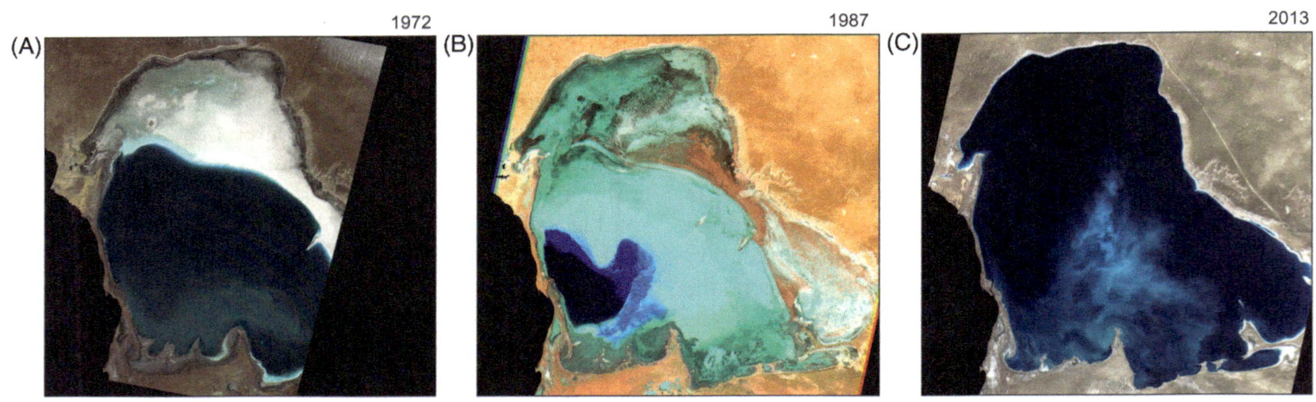

Figure 10.9 NASA satellite images show the desiccation and re-flooding of Kara-Bogaz-Gol since the 1980 dam's construction and its removal in 1992. The left-side image (A) was acquired in 1972 and shows evaporite sediments on the northern lake's shore, which is marked by whitish and bluish reflectivity. The center image (B) shows a satellite image (enhanced colors) from 1987, when the lagoon was almost desiccated. A dam built in 1980 obstructed the channel across the sedimentary ridge, restricting the basin and causing the precipitation of evaporite sediments. The light blue reflectivity indicates salt deposits, mainly gypsum and halite. The darker blue color at the lake entrance near the central Caspian Sea indicates less concentrated brine. The right-side image (C) shows the dark blue waters of Kara-Bogaz-Gol in 2013, when the lagoon was refilled by Caspian Sea water and partially dissolved the previously deposited evaporite minerals. Credit: Courtesy of NASA (2017).

comparable to the Aral Sea. The Aral Sea disaster transformed the fourth-largest lake in the world into a dried-out remnant water body with a volume of only 10% of the previous volume and an increase in salinity from 10 to more than 100 g/L (Micklin, 2007).

During the 1980s, salt dust was blown over the land adjacent to the Kara-Bogaz-Gol, polluting the environment and causing health problems in the population that lived in the coastal villages around the lake. This catastrophic event affected fishing activities and salt extraction. Aiming to reverse the situation, the Soviet government decided to reduce the diversion of the Volga waters into the Caspian Sea because the artificial barrier reduced the water inflow 10 times faster than Soviet hydrologists had predicted. The canal across the sand ridge was partially opened in 1984, and the dam was imploded and destroyed in 1992 (Kosarev & Kostianoy, 2005). These measures resulted in restoring the natural connection between the sea and the lagoon. The rising waters of the Caspian Sea spread across the barrier and refilled the lagoon in the 1990s (Figure 10.8). In the 21st century, the Kara-Bogaz lagoon has returned to its original situation in terms of lake surface and water volume (Figure 10.9C). After the fall of the Soviet Union, the Turkmenistan government adopted several actions to remedy the effects on the lake's biodiversity, aiming to restore the fishing and salt industries to their normal conditions.

10.3.2. The Re-Flooding of the Kara-Bogaz-Gol and Implications for Evaporite Cycles

During the first two decades of the present century, the Kara-Bogaz-Gol returned to its average lake surface at –28 m, and the Caspian Sea was about 1–2 m above this level, with water level at –27 to –26 m (Kosarev et al., 2009). The lake has experienced rapid changes in water inflow over the past decades. The lake's surface shifted from a perennial lake to a desiccated salt flat, exposing the lake bottom to subaerial erosion in the 1980s. Communication with the Caspian Sea resumed in 1992, causing the lake level to rise. By 1996, the lake level was restored to its original equilibrium. Rapid flooding events created a lake full of brackish water, leading to the dissolution of soluble evaporites deposited on the lake's shores and in the basin center (Figure 10.9C).

These changes in water salinity caused extreme stresses on biological systems, as some organisms could not adapt to extreme brine concentrations. A comparison between the high and low lake levels and the resulting paleogeographic situations (Figure 10.10) shows the dramatic effects on evaporite deposition in the lagoon.

The evaporite deposition in the lagoon was significantly affected by sea level variations in the Caspian Sea. The brine chemistry in hypersaline lakes controls the precipitation of different evaporite minerals based on their solubility (see Chapter 4). In Kara-Bogaz-Gol, the brine concentration changed rapidly, resulting in the sequential precipitation of calcite, aragonite, dolomite, gypsum, and mirabilite, followed by halite, astrakhanite, and epsomite (Kurilenko et al., 1988).

Evaporite deposits in the Kara-Bogaz-Gol basin are more typical of basins fed by continental water than marine sources. Many continental evaporite basins worldwide record similar short-term facies variations (Schreiber & Hsu, 1980; Warren, 2006, 2016). These compositional changes are illustrated in a simplified evaporite facies map for the Kara-Bogaz-Gol in 1920 and 1950 (Figure 10.11).

During dry periods, different minerals were deposited on the lake's bottom, which was originally covered by glauberite and gypsum. During low sea levels in the Caspian Sea, brines in the Kara Bogaz were concentrated, replacing these minerals with halite, a more soluble mineral. Desiccation events in the mid-1980s led to the formation of halite crusts in subaerial regions, which are characterized by desiccation polygons, typical of sabkha flats (see Chapter 8). Small ponds with increased salt concentrations deposited more soluble minerals, including potassium and magnesium salts like carnallite ($KMgCl_3 \cdot 6H_2O$) and bischofite ($MgCl_2 .\ 6H_2O$). During flooding events, more soluble minerals in the lake were dissolved, resulting in crusts formed by carbonate (calcite, aragonite) and sulfate (gypsum) minerals (Figure 10.11). The map distribution of sedimentary facies in the evaporite deposits on salt flats exhibits an approximately circular pattern symmetrical to the basin's edges, corresponding to the bull's-eye evaporitic sedimentation model (Hsu, 1972; Sonnenfeld, 1979, 1989, 1992; Warren, 1989).

The evaporite facies map of the Kara-Bogaz-Gol lagoon shows calcium and sodium sulfate (gypsum, mirabilite) on the margins and chloride facies (halite) in the central portions of the basin. The northern part of the lake is floored by subaqueous epsomite and halite, while the southwestern part is floored by glauberite (sodium and calcium sulfate). The exposed sabkha flats exhibit different evaporite crusts, with the southern margins dominated by gypsum and halite crusts in the northern and eastern salt flats (Figure 10.11).

The recent dramatic desiccation episode in the Kara-Bogaz-Gol lagoon can be compared to catastrophic events in the great inland seas that were desiccated and re-flooded in the ancient past, such as the Mediterranean Sea. However, these giant salt basins exhibit many differences in scale, chemistry, and paleogeography compared with the desiccation episode in Kara-Bogaz-Gol. The Mediterranean Sea, a giant salt basin formed during the Late Miocene, was fed by seawater, and the pre-salt sediments were deposited in a deepwater marine setting (see Chapter 5 in Mohriak, 2025). In the Kara-Bogaz-Gol

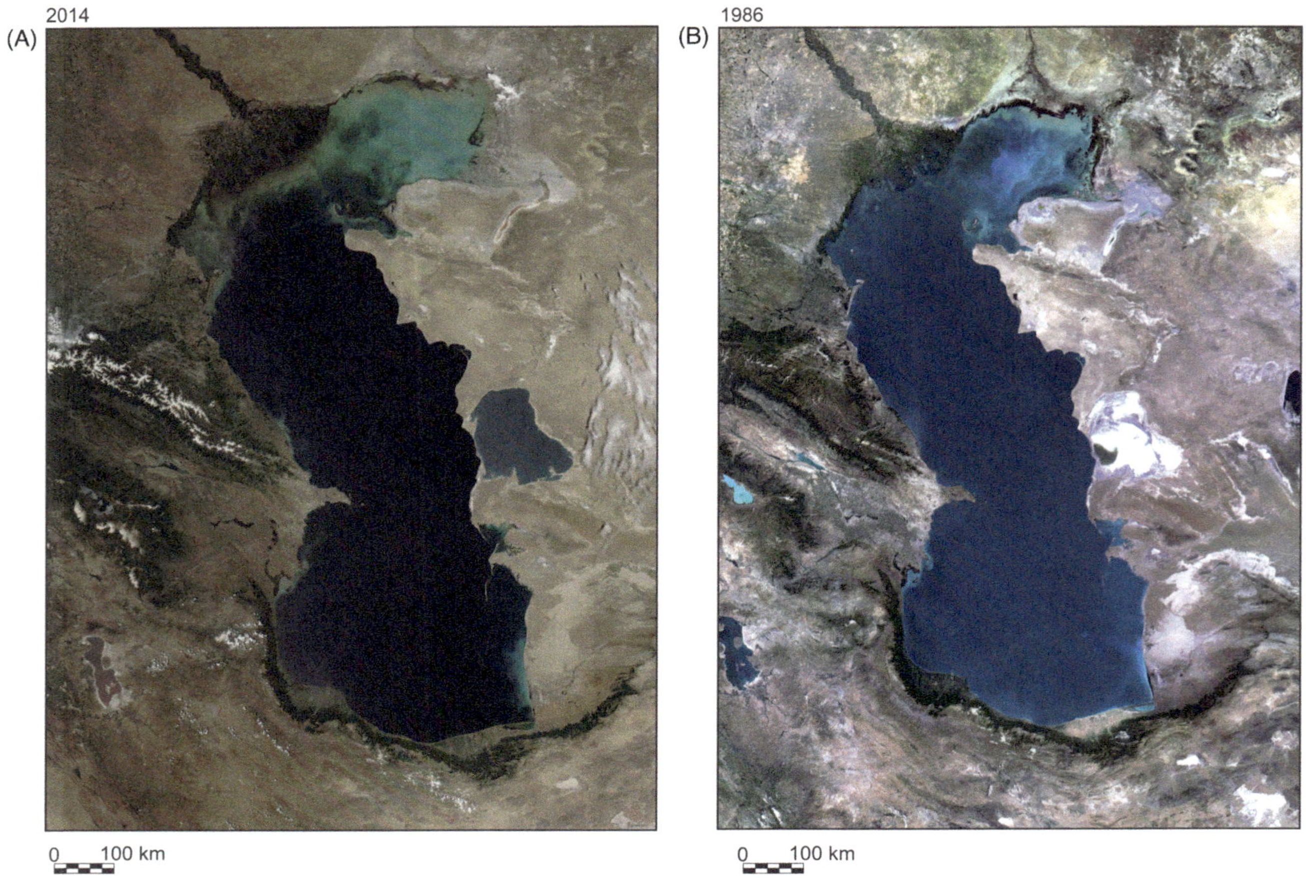

Figure 10.10 Comparison of NASA satellite images illustrating the evaporite cycles in Kara-Bogaz-Gol. In 1986 (B), the lake experienced a low level, resulting in the exposure of most evaporite deposits on the dried-up lake surface. This was followed by a high lake level in 2014 (A). The basin desiccation event occurred due to the Soviet Union's construction of a dam in 1980, reducing water inflow and causing rapid evaporation of the lagoon's water. Kara-Bogaz-Gol brine concentrated and precipitated evaporite minerals on the lagoon's floor and margins. In the 1980s, only a small remnant body of water was present near the Caspian Sea channel. After the dam was removed in 1992, the basin was refilled with Caspian Sea water. Credit: Courtesy of NASA (2017).

case history, the basin was a shallow-water continental lake throughout the desiccation and flooding events. Most basinwide salt deposition models assume that the inland gulfs and lakes were periodically flooded and partially or completely dried up when the barriers and their crossing channels failed to communicate with the global oceans, as in the South Atlantic salt basin (see Chapter 3 in Mohriak, 2025).

Around the Late Aptian (115 Ma), a barrier south of the Santos Basin separated the hypersaline gulf in the Central South Atlantic, between Brazil and Angola, isolating the lacustrine systems from the open sea (Davison, 2007; Kukla et al., 2018). Geological processes, mainly related to tectonic and magmatic activity and sea level fluctuations, periodically obstructed the channels of the transient barriers, isolating the hypersaline lagoons and causing an increase in brine salinity. This led to the precipitation of large amounts of evaporite minerals along the salt basins in the conjugate margins (e.g., Sonnenfeld, 1992; Szatmari et al., 2021).

There are some analogies between the South Atlantic salt basins and the Caspian Sea and Kara-Bogaz-Gol lagoon systems, but major differences are present. First, the barrier in the South Atlantic is considered a volcanic barrier, formed by a magmatic episode that affected large areas in the basin, not a sedimentary barrier that shifts periodically with variations in currents and sediment input. Second, the hypersaline gulf of the South Atlantic was more than 1,000 km long, surpassing the size of the Kara-Bogaz-Gol by one order of magnitude. Third, the sea level in the Caspian Sea varied by only a few meters during the desiccation and flooding events recorded in the twentieth century, while the lake level in the Kara Bogaz was only tens of meters below the global sea level.

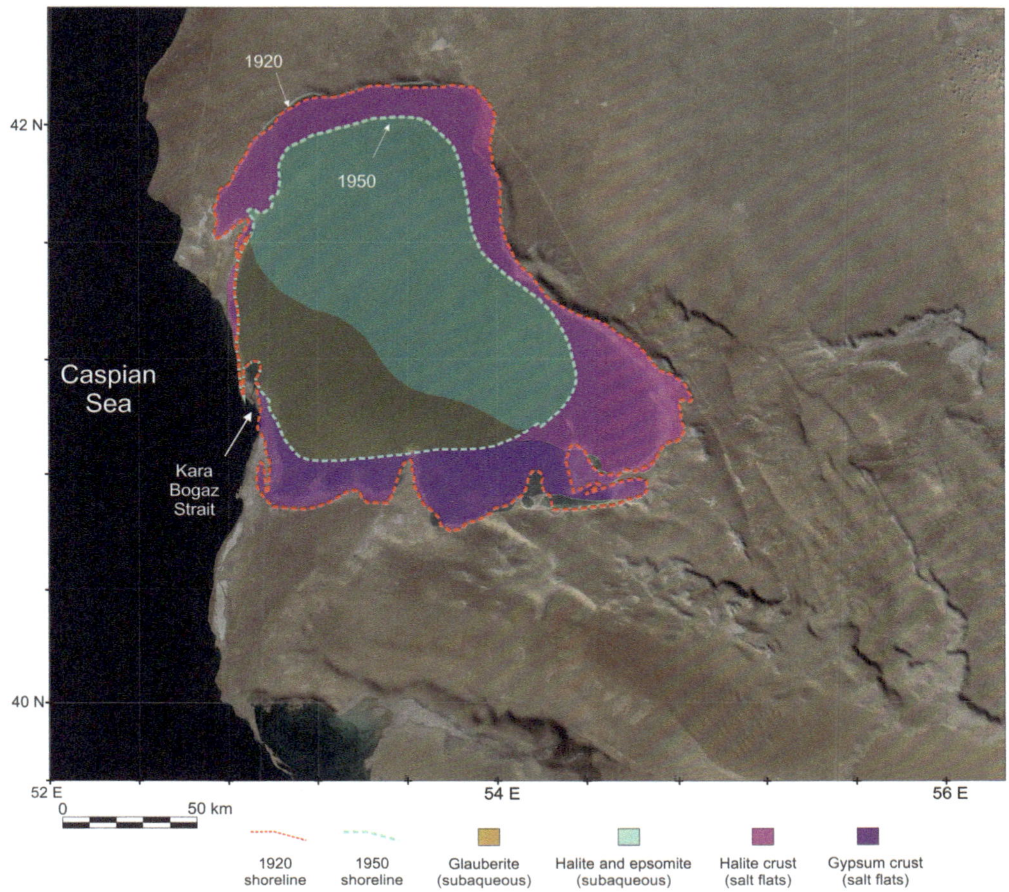

Figure 10.11 The diagram depicts the distribution of Kara-Bogaz-Gol evaporite facies in 1920 and 1950. In 1950, halite crusts formed salt flats at the lake's northern and eastern segments, while the southern margin was mainly gypsum flats. In the 1950s, subaqueous halite and epsomite covered the central lake. In the late 1980s, the basin desiccated due to anthropogenic factors, and the salt flats were exposed subaerially. Credit: Image courtesy Jeff Schmaltz/NASA/Public domain.

Although some authors postulate that the South Atlantic was only a few meters below the global sea level (Burke & Sengor, 1988), most authors interpret that the sea level in the South Atlantic Gulf during the Late Aptian was hundreds of meters below the average global sea level, resulting in accommodation space to allow the deposition of thick salt layers (Szatmari & Milani, 2016).

Kara-Bogaz-Gol is a unique case study in the modern world where a large basin undergoes a complete cycle of desiccation and refilling. Many giant salt basins record the catastrophic processes related to periodic seawater inflow and abrupt evaporation. This is akin to the South Atlantic, which formed a hypersaline gulf in the Late Aptian, or the Mediterranean Sea desiccation event due to the Messinian salinity crisis. Kara-Bogaz-Gol is considered a modern analog for salt deposition in sedimentary basins with intermittent connection with an open sea and periodic isolation, resulting in brine concentration through solar evaporation and salt precipitation (e.g., Hsü, 1972). These dry-out events are likely associated with climate aridity and decreased water input due to reduced run-off or the lack of communication with larger water bodies (Chaboureau et al., 2012, 2013). These processes are analogous to the ones observed in the Kara-Bogaz-Gol in the past century. However, in most of the ancient giant salt basins, the salt deposits accumulated from the evaporation of normal oceanic water instead of continental, brackish brine, and the catastrophic events leading to basin desiccation were measured in time scales of thousands of years, not years or decades (e.g., Warren, 2006, 2016).

10.4. THE ANDEAN SALT FLATS

The Andean region in western South America is home to several modern salt basins related to compressional settings and plate convergence. The Andean Altiplano is

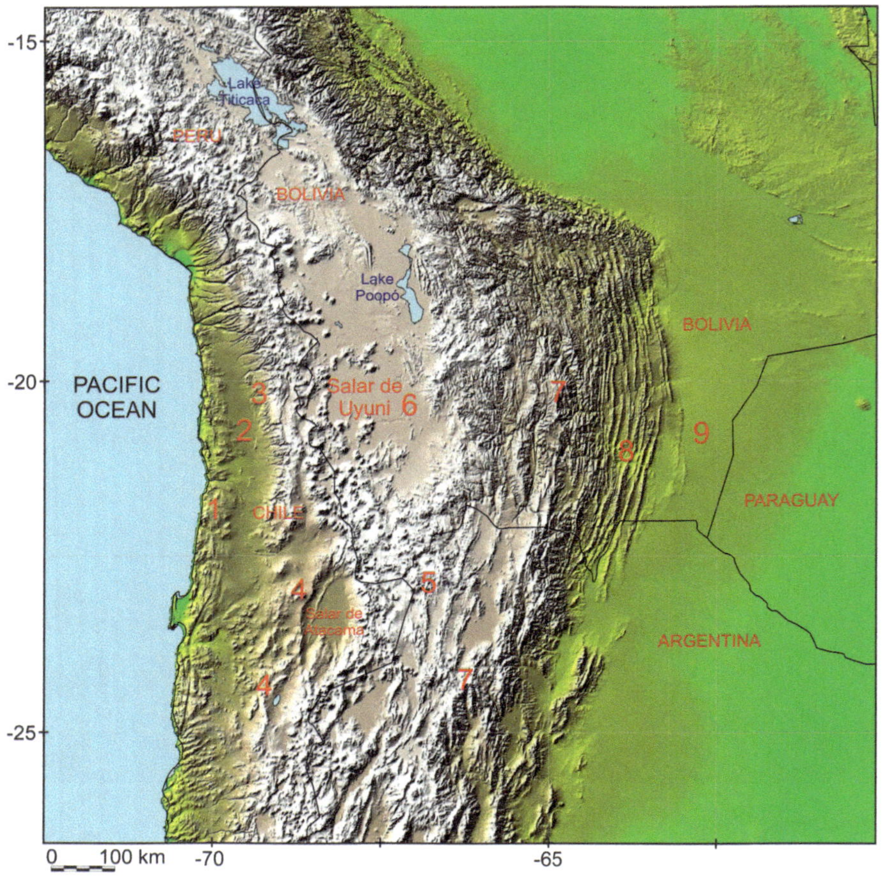

1 = Coast Ranges; 2 = Central Valley/Atacama Desert; 3 = Pediplanes;
4 = Precordillera/Domeyko Range; 5 = Western Cordillera; 6 = Altiplano;
7 = Eastern Cordillera; 8 = Fold-thrust belt; 9 = Foreland basins.

Figure 10.12 The NASA Shuttle Radar shaded-relief digital elevation model of the Central Andes identifies the main geomorphologic provinces, including Lake Titicaca and Lake Popoó, with altitudes of 3,808 and 3,718 m, respectively. Salar de Uyuni in Bolivia has a mean altitude of 3,653 m, while Coipasa Salar, located slightly to the north, has a mean altitude of 3,656 m. South of the Altiplano, Salar de Atacama is located to the east of the Domeyko Range (Precordillera) in Chile, while salt flats overlie a depression at 2,300 m. The Central Andean fold-thrust belt (identified by the number 8) is located between the Eastern Cordillera and the foreland basins in Argentina, Bolivia, and Peru. Credit: NASA/Wikimedia Commons/Public domain.

marked by tall volcanoes and numerous salt flats, forming spectacular geological features and the largest expanses of modern evaporite sediments on Earth (Figure 10.12).

The Andean Altiplano salt flats are located between the western and eastern branches of the Andes Mountains. The salt flats are characterized by large thicknesses and volumes of salt minerals that accumulate in intermontane basins or high-relief arid deserts (Alonso, 2006; Alonso et al., 1991). Most of the modern evaporite deposits are concentrated in the salt flats of the Central Andes Altiplano (Figure 10.12). The tectonic and volcanic activities in the highlands of the Central Andes are associated with global climate evolution (Armijo et al., 2015).

The Andean region has several geomorphological provinces, from low-relief ranges near the coast to the highlands in the Cordillera. The Central Andean magmatic arc resulted from several phases of compressional tectonics and different episodes of magmatic activity. Volcanic terranes correspond to the basement of many of the depressions in the Andean region, where salt layers have been deposited since the Neogene (e.g., Ericksen, 1993; Stern, 2004).

The Altiplano, located between the western and eastern branches of the Andes Mountains, corresponds to a highland with elevations around 3–4 km, forming a plateau extending from Bolivia to Argentina (Figure 10.12). Large lakes, such as Lake Titicaca and Lake Poppó, occur at altitudes reaching almost 4 km. Salar de Uyuni, the largest salt flat in the world, is located in the Bolivian Altiplano at an altitude of 3,656 m. Most of the salt flats and salt pans correspond to dried-up lakes located in depressions occupied by perennial lakes in the Neogene. Paleolake Pauca constituted a larger lacustrine environment in the Altiplano when the lake surface was about 3,760 m (Nunnery et al., 2018). Due to evaporation, the ancient lake formed several small lakes and desiccated salt flats. The contemporary hydrological system, with a lake's surface below 3,700 m, resulted in Lake Poopó (Figure 10.12). The Uyuni and Coipasa salt flats are characterized by many isolated, ephemeral lakes that are remnants of the larger paleolake (Nunnery et al., 2018).

The Andean region is home to numerous salt-encrusted pans or playas, known as salars (also spelled salares), a Spanish word that means desiccated lakes covered with salt deposits. The largest salt flats and perennial lakes in the Central Andes are located within the Altiplano (Figure 10.13). The comparison between the topobathymetric map (based on GEBCO, 2019) and the satellite image of the Central Andes (Figure 10.13) shows a deflection of the subduction zone and the Andean chains from NW-SE to almost N-S south of latitude 20° S. Salar de Uyuni in Bolivia, at latitude 20° S, occupies an area of 10,582 km^2. Salar de Atacama in Chile, at latitude 23° S, is located at lower altitude and stretches across an area of almost 3,000 km^2 (Figure 10.14).

Salars are genetically related to ancient, deep lakes that dried up during periods of increased aridity. Endorheic drainage brings water and ions to these depressions, which are characterized by shallow lakes that are dry most of the time but are sporadically filled with water that quickly evaporates in the extremely dry climate. Titicaca Lake, between Peru and Bolivia, is a permanent freshwater mass located at an altitude of 3,808 m a.s.l. In natural-color satellite images, the lake appears dark blue (Figure 10.13).

The Eastern Cordillera exhibits elevations that exceed 5 km, and the fold-thrust belts east of the Andean highlands display a more subdued topography, forming a salient associated with thrust faults with east and west vergences (Figure 10.12). The foreland basins in the forested Amazon region are characterized by even lower altitudes, forming flatlands covered by Quaternary drainage systems. Some basins exhibit major low-angle thrust faults detaching on evaporite layers, such as the Huallaga Basin in Peru, where Permian salt deposits deep in the subsurface have been involved in contractional structures (Borderie et al., 2019).

The Pediplain and the Precordillera are characterized by moderate elevations, whereas the Western Cordillera (the present magmatic arc) exhibits elevations up to 6 km (Risacher & Fritz, 2009). The Central Valley in Chile (also called the Central Depression or Longitudinal Valley) is located between the Coast Ranges and the Pediplain. The Coast Ranges are characterized by moderate elevations ranging from about 1 up to 3 km. The Central Depression is filled by alluvial fan sediments derived from erosion of the Precordillera and Western Cordillera (Figure 10.15).

The simplified E-W geological cross-section in the Atacama salt flat (Figure 10.15) shows the different landforms from the coastal plains to the elevated ranges in the Central Andes (Risacher & Fritz, 2009). Variations in topography exert a profound influence over the climate. The regions at the highest altitudes (> 5 km) record 100–400 mm of rain precipitation per year, whereas the Central Valley depressions record 5–100 mm/year. The driest provinces are in the rain shadow of coastal ranges. Salar de Atacama, with less than 5 mm of rain per year, is the driest place on Earth (Figure 10.12).

10.4.1. Origin of the Evaporites in the Andean Salars

Salars in the Central Andes are formed within depressions that receive their constituents from weathering volcanic rocks associated with magmatic arcs (Rettig et al., 1980; Lowenstein et al., 2003). These igneous and felsic rocks are formed by partial melting of the mantle wedge overlying the subducting plate at great depths (Stern, 2004), whereas the more felsic rocks are likely formed by partial melting of the crust (Coldwell et al., 2011). Rivers flow from the Altiplano highlands to flat intermontane depressions, discharging the ionic content in endorheic basins (Figure 10.15). Most rivers do not run toward the Pacific Ocean due to the topographic barrier resulting from uplifted coastal mountain ranges. However, the salt flats on the Altiplano are not regarded as truly terminal lakes or closed basins, as the drainage leaks the salty water into groundwater. Consequently, some of the incoming salts are transported to other places by subsurface flow (Risacher et al., 2003).

Sedimentary processes leading to salt deposition have been in operation since the mid-Miocene, when the uplift of the Andes Mountains resulted in mountain ranges near the coastline (Rohrmann et al., 2016). These mountains created an obstacle that hindered the transfer of humidity from the Pacific Ocean to the interior of the continent, creating a rain shadow (Alonso et al., 1991). Due to its location between two mountain ranges that block moisture from both sides, the west side of the Andes Mountains is marked by extreme aridity, with Salar de Atacama being the driest desert on Earth. The Andes Mountains in the east block the humid air of the Amazon region, and the Chilean Coastal Ranges block moisture from the Pacific Ocean (Figure 10.14).

Drainage systems in the Andean salt flats are characterized by rivers that flow toward depressions adjacent to the Precordillera ranges, forming lakes that have been drying up since the late Neogene. The brine concentration

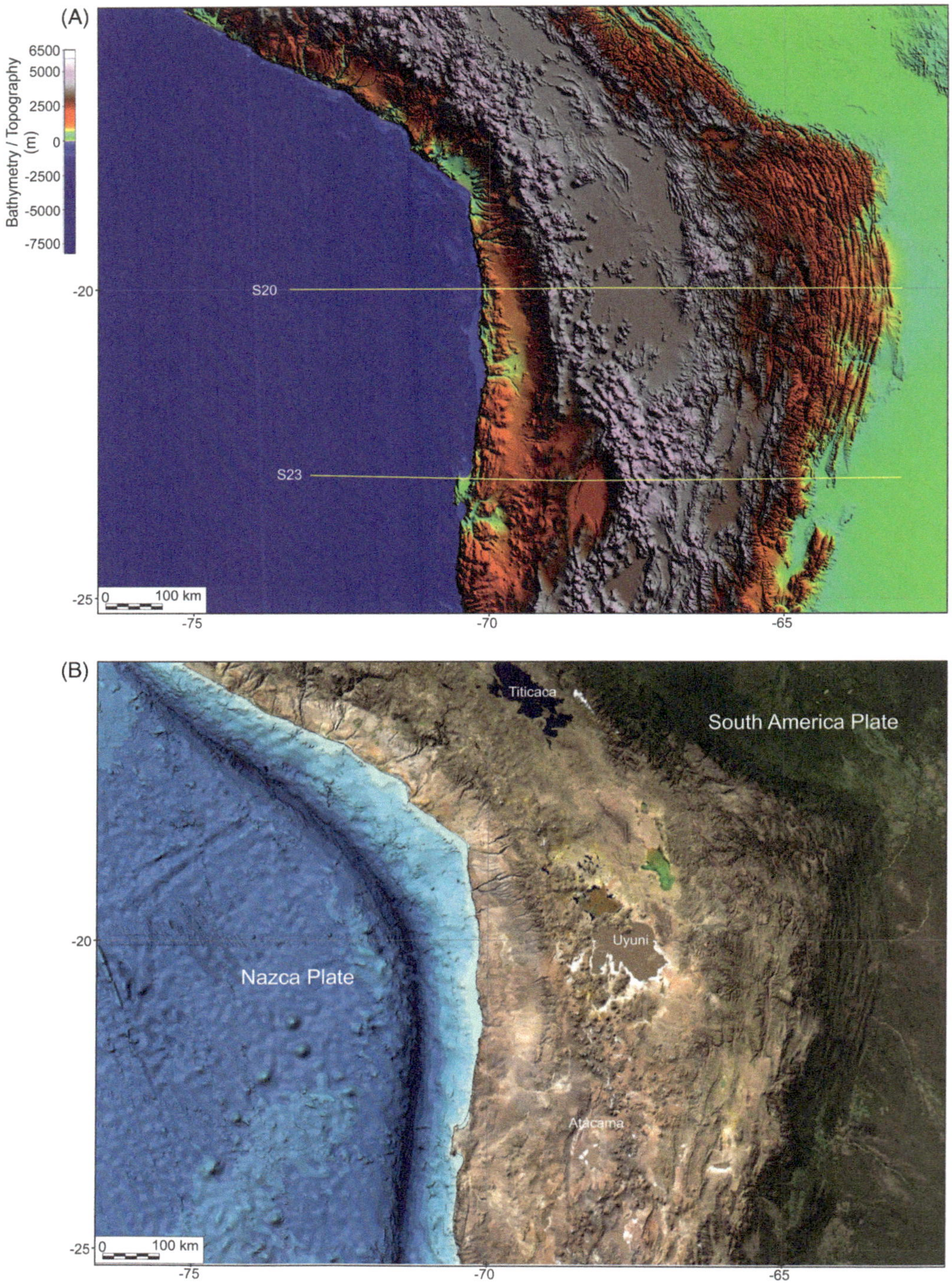

Figure 10.13 A topobathymetric map (A) and satellite image (B) depict the Central Andes region in northern South America. The locations of the regional topographic cross-sections at 20° and 23° S are shown by thin, yellow lines. The subduction zone of the Nazca plate under the South American plate forms a deep trench ocean-ward, resulting in a magmatic arc within the Andean Mountains. This has led to the formation of numerous active volcanoes. The Altiplano is a plateau surrounded by volcanic complexes and a tectonically elevated highland to the east, corresponding to the Andean fold-thrust belt. The high-altitude depression is associated with endorheic, internally drained basins with several lakes and salt flats formed in the Neogene.

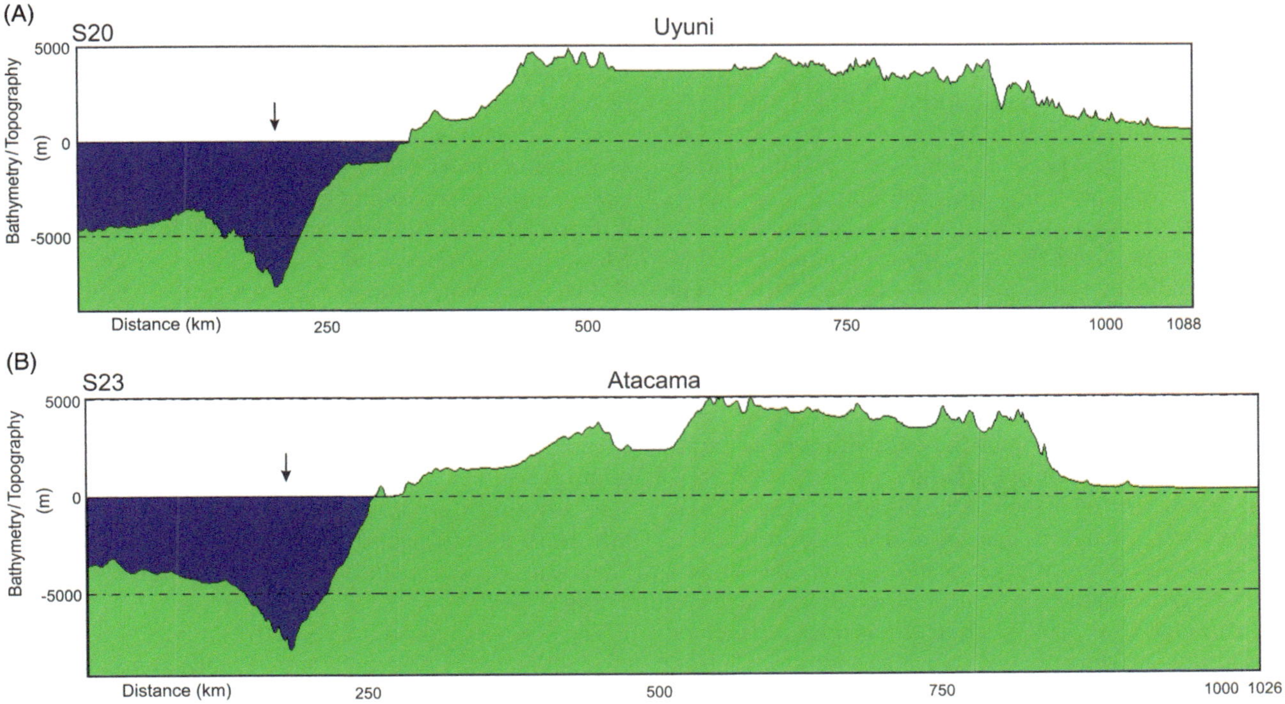

Figure 10.14 The topographic profiles at 20° S and 23° S reveal significant altitude differences between the Pacific Ocean trench and salt flats in Salar de Uyuni (A) and Salar de Atacama (B). A subduction zone, marked by a black arrow, corresponds to the boundary between the Nazca and South American plates. The Uyuni salt flat is 3,653 m high, while the Atacama salt flat is around 2,300 m high. These high-altitude depressions are associated with endorheic basins receiving fluvial waters that bring several ions from the leaching of the basement volcanic rocks.

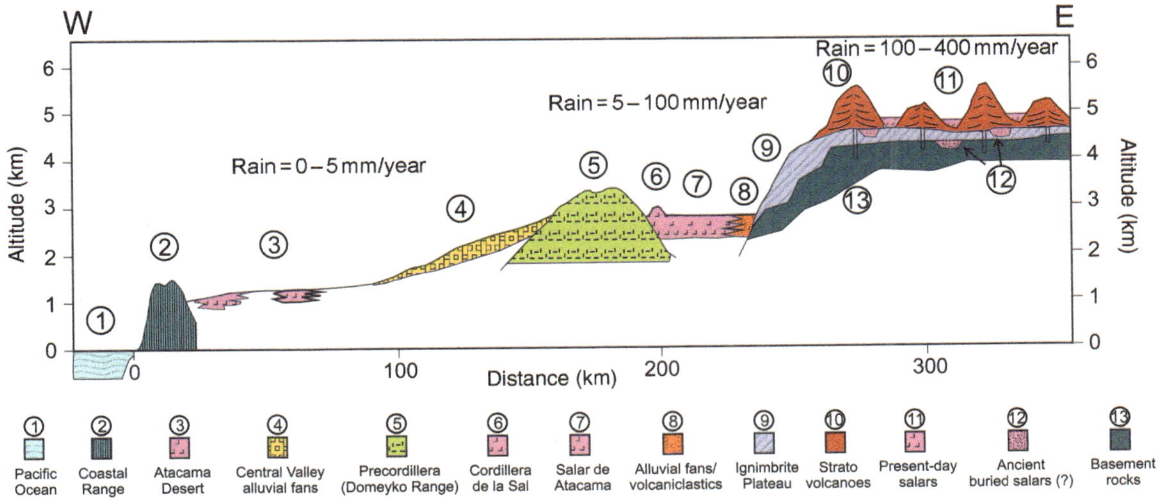

Figure 10.15 A schematic cross-section depicts the geomorphology of the western margin of South America, extending from the Pacific Ocean and Coast Ranges to the Central Valley, Precordillera, Salar de Atacama, and Western Cordillera highlands, with varying rainfall levels in different sectors of the profile. Credit: Risacher and Fritz (2009)/with permission of Springer Nature.

resulted in salt precipitation, forming saline pans that accumulated large thicknesses of evaporite sequences in some depressions, such as Salar de Atacama (Figure 10.12).

Active volcanoes in the Central Andes magmatic belt have extruded felsic to intermediate lavas (rhyolites and andesites) since the Miocene, with several WNW-trending volcanic ridges cutting the Altiplano and Puna (Stern, 2004). Many of the salars are characterized by geothermal vents and thermal springs around the edges of the paleolakes (Hovland et al., 2018a, 2018b). The volcanic complexes are covered by ignimbrites, a pyroclastic rock formed by density flows during explosive volcanic outbursts and composed of hardened tuff and rock fragments within a matrix composed of glass shards. The chemical composition of these ignimbrites is characterized by a high content of alkaline elements.

The weathering of intermediate intrusive and extrusive igneous rocks, including dacites, andesites, and volcanic tuffs, releases chemical compounds into the surface waters, resulting in the enrichment of underground brines (e.g., Risacher et al., 2003). These fluids not only include several evaporite-forming ions, such as sodium, chlorine, and sulfur, but also display anomalously elevated contents of lithium and boron. The alkaline elements are sequestered in the interstitial brines that fill the pore space of the siliciclastic sedimentary layers that fill the depressions. Most of the salars precipitate thick evaporite layers, predominantly composed of gypsum, halite, and borax. The Central Andean salt flats contain the largest reserves of boron ores in the Southern Hemisphere, which are associated with hydroboracite ($CaMg[B_3O_4(OH)_3]_2 \cdot 3H_2O$) and ulexite ($NaCaB_5O_6(OH)_6 \cdot 5H_2O$). In some shallow evaporite basins, hydroboracite accumulates in the center of the lake and gypsum ($CaSO_4.2H_2O$) at the lake margins (Alonso et al., 1991; Ortí and Alonso, 2000).

Salars in the Andes are characterized by salt crusts, gypsum, and halite, with a karstic topography due to dissolved salt layers. Subsurface sediments are siliciclastic rocks (mudstones) intercalated with evaporite sequences. Deeper in the sedimentary sequences, the basins include ancient salt deposits, which are recorded in specific geological intervals, including the evaporite-bearing Tonel Fm., which unconformably overlies the Late Carboniferous–Early Permian Agua Dulce Fm. (Charrier et al., 2007).

Some diapiric structures are interpreted in seismic profiles in the Andes, particularly in the Eastern Cordillera in Colombia, north of the Central Andes (Parravano et al., 2015). Salt diapirs in the Altiplano are indicated by elongated anticlines or circular structures, characterized by outcrops with evaporite minerals in the caprock (Risacher & Fritz, 2000). Halite is commonly dissolved, and gypsum is the only mineral present in the outcrops. One of the elongated structures is the Cordillera de La Sal, near the Atacama salt flats (Alonso et al., 1991).

Salt mines in the Andean region date back to prehistoric times (see Chapter 1). However, these evaporite deposits have only been exploited using modern mining technologies since the last century. Figure 10.16 shows a simplified map with the identification of a large number of salars in the Andean region, based on compilations by several authors (including Ericksen, 1981, 1993; Risacher & Fritz, 2000, 2009; Risacher et al., 2003; Warren, 2006). Most of the Andean salars are located in Peru, Chile, Bolivia, and Argentina, particularly in the Altiplano between Bolivia and Chile (Figure 10.12).

The Andean salt flats yield several evaporite minerals, such as chloride, nitrate, borate, and iodide salts. More recently, large brine extraction and processing plants have been in operation in some salars, particularly for lithium mining (see Chapter 9 in Mohriak, 2025). These plants take advantage of the much larger scale of the Andean salt flats as compared with their European, North American, African, and Asian counterparts (Cabello, 2022; Crangle, 2018).

The U.S. Geological Survey (2023) reported that boron production in the Andean region reached almost 1 million metric tons in 2022, primarily from Chile, Bolivia, Peru, and Argentina. The largest ulexite deposits are in Chile's Salar de Suirire, with reserves estimated at 1.5 million metric tons (Crangle, 2018). Turkey produced the largest amount of refined borates worldwide, with 1.7 million metric tons in 2022, and the country hosts the largest boron reserves in the world, with 1,200 million metric tons of boron (U.S. Geological Survey, 2023).

Lithium appears in the Earth's upper crust with a very small abundance (only 0.002%, or 20 ppm, according to McLennan, 2001). However, lithium is present in many minerals, particularly pegmatitic granitic igneous rocks and some clays. Extracting lithium from these rocks is expensive due to crushing and treating the mineral powder with acid and heat. The most economical method for recovering lithium from the subsurface is based on its occurrence as dissolved ions in formation waters filling sedimentary rocks. Bradley et al. (2017) estimate that lithium resources worldwide exceed 39 million tons. The South American salt flats will probably dominate the world market for lithium salt mining in the future, as there are immense resources in the subsurface of the salt flats in Bolivia, Chile, and Argentina (e.g., Cabello, 2022).

The average lithium content in Salar de Atacama brine is about 1400 ppm (Bradley et al., 2017). Mining companies employ simple techniques for obtaining the metal from the subsurface, pumping brine to surface ponds, and allowing the salty water to evaporate in the arid climate to precipitate lithium minerals, especially zabuyelite ($Li_2(CO_3)$), a lithium carbonate that was discovered at Lake Zabuye in the Tibetan Plateau in China. It is estimated that the lithium reserves in the evaporite crusts in

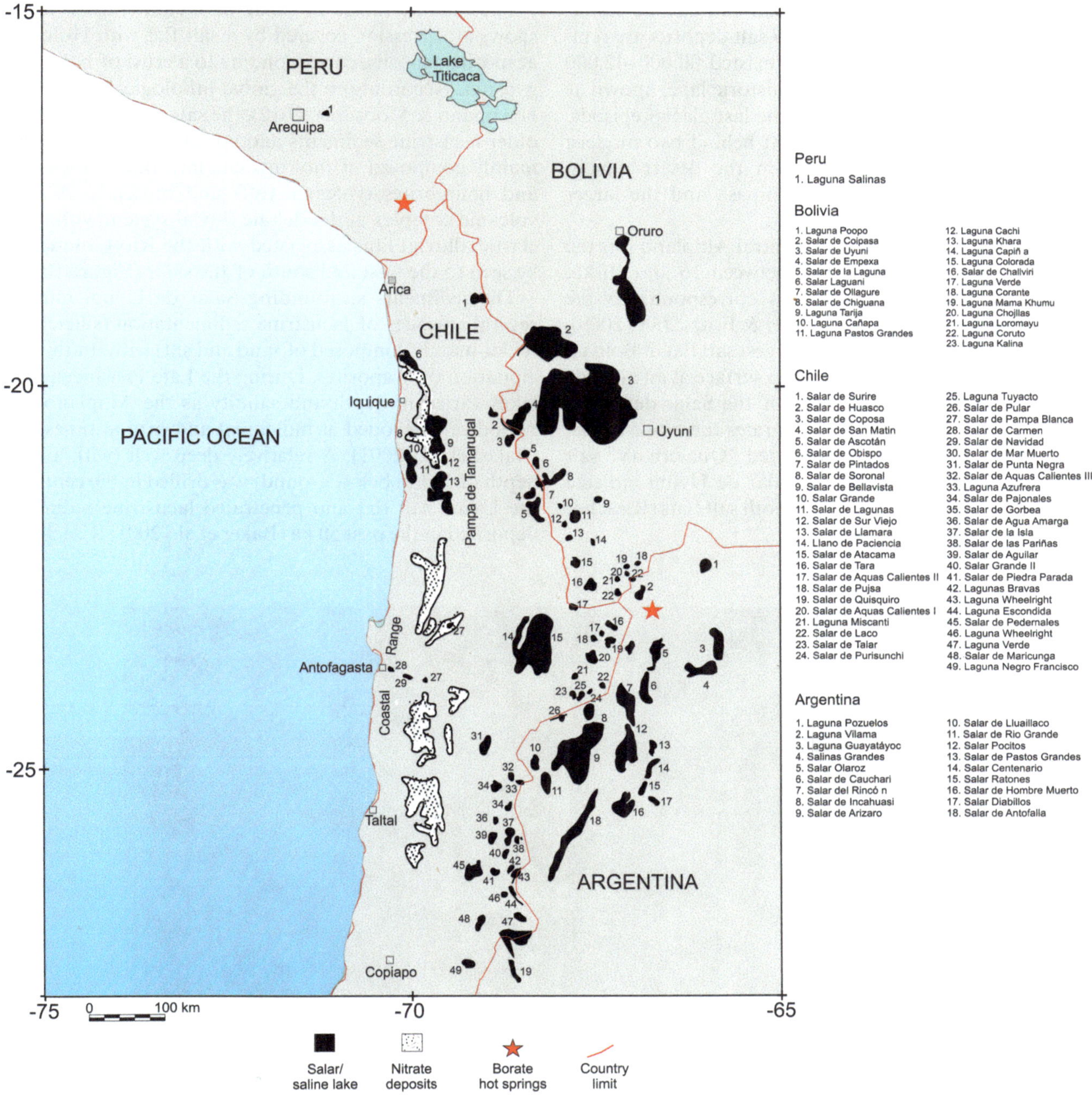

Figure 10.16 The simplified map of the Central Andes features salt flats (salars) in Peru, Bolivia, Chile, and Argentina. Lake Titicaca, a perennial lake, is located between Peru and Bolivia. The ephemeral saline lakes, overlain by desiccated mudflats and salt pans covered by evaporite crusts, receive some water during episodic rains. Salar de Atacama is located eastward of the Coastal Ranges in the Antofagasta region of Chile. Borate hot springs (red stars) are located in northern Argentina, near the border with Chile and Bolivia.

the Andean salars, which may reach a thickness of tens of meters in some places, presently correspond to 50–70% of the world's reserves. Chile (9.3 million tons) and Australia (6.2 million tons) hold the largest reserves of lithium in the world (U.S. Geological Survey, 2023).

10.4.2. Salar de Uyuni, Bolivia

Salar de Uyuni in Bolivia is the world's largest salt flat, covering 10,582 km^2 in the Andean Altiplano (Figure 10.12). It is located on the highest plateau where salt deposits are seen today, at an altitude of about 3,653 m (Figure 10.17). The Uyuni salt deposits are remnants of a larger depression that existed 30,000–42,000 years ago (30–42 ka). A giant prehistoric lake, known as Lake Minchin, was formed after the last glacial episode. The ancient lake dried out and left behind two modern lakes and two major salt pans in the desert region, corresponding to the Salar de Coipasa and the larger Salar de Uyuni (Figure 10.16).

The last salt lake covering the central Altiplano was the Paleolake Tauna, which existed between 16 and 10 ka, covering the area that nowadays corresponds to the Coipasa and Uyuni salars (Risacher & Fritz, 2000, 2009). The Coipasa Salar is the second-largest salt flat in Bolivia, with an area of 2,218 km^2 and a top surface at an altitude of 3,656 m, whereas the surface of the Salar de Uyuni is 3 m lower. A sill at 3,657 m separates these two salars. Exploratory wells that penetrated Quaternary salt sequences in the subsurface of Salar de Uyuni indicate that thick layers of halite underlie both salt flats (Risacher & Fritz, 2000).

The satellite image of Salar de Uyuni (Figure 10.17) shows a depression covered by a salt flat with Holocene evaporite deposits, corresponding to a crust of halite and gypsum. According to the global lithological maps (e.g., Hartmann & Moosdorf, 2012), the salar is surrounded by older lacustrine sediments and volcanic rocks, which are mainly composed of intermediate lava flows (andesites) and ignimbrites (volcanic tuffs and breccias). A large volcanic complex and a deltaic fluvial system with siliciclastic alluvial fans associated with the Rio Grande can be seen to the west and south of the salar (Figure 10.17).

The sediments surrounding Salar de Uyuni indicate several episodes of lacustrine sedimentation (siliciclastic rocks, mainly composed of mud and silt) without the precipitation of evaporites. During the Late Pleistocene, the lakes varied in depth and salinity as the Altiplano was periodically flooded at millennial and orbital timescales (Baker et al., 2001). A relatively deep well (with a total depth of 121 m below ground) was drilled in the center of the Uyuni salt flat and penetrated lacustrine sediments deposited in the past 50 ka (Baker et al., 2001; Risacher &

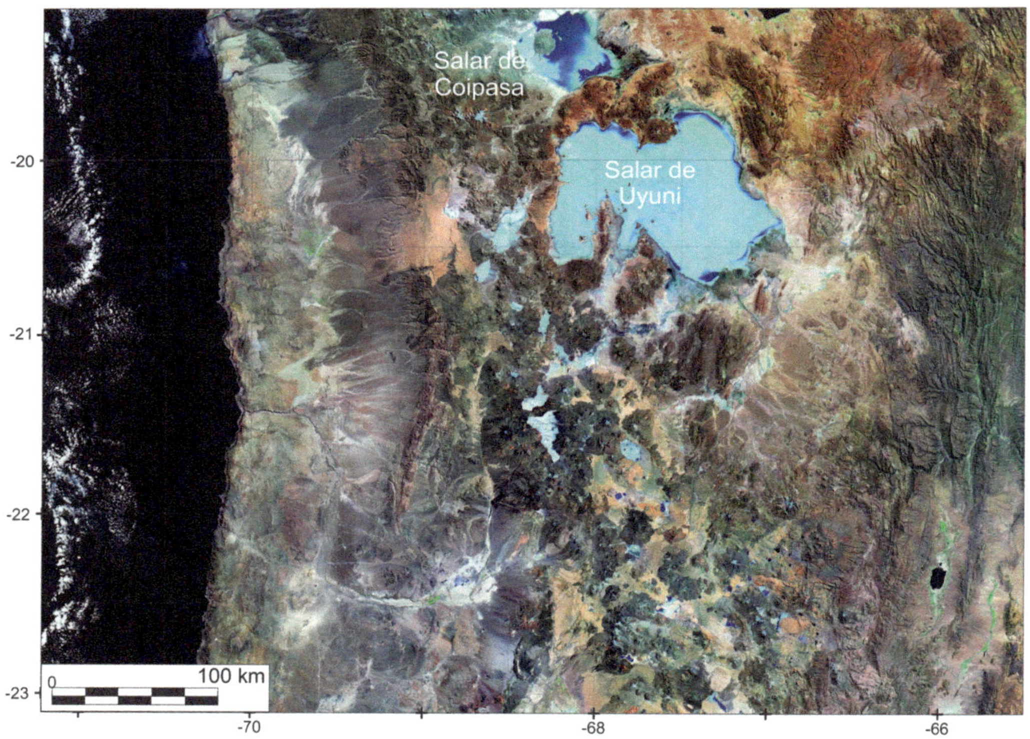

Figure 10.17 The NASA Landsat satellite image (false colors) of Salar de Uyuni in Bolivia's Altiplano shows salt flats at an altitude of 3,653 m. The evaporite deposits exhibit light blue reflectivity, and the shallow lagoons are dark blue, forming moats bordering the salt pans. The volcanic rocks are characterized by a rugose texture and appear dark brown to black. Some volcanoes with a round crater on top of the mountain mark the large volcanic complex located southwest of Salar de Uyuni. The drainage systems in the southern sector form an endorheic basin receiving fluvial waters. Salar de Coipasa, located slightly to the north, has a large perennial lake. Credit: Courtesy of NASA.

Fritz, 2000). The subsurface lithologies penetrated by the borehole record several intervals of salt accumulation intercalated with mud layers, pointing to wet and dry episodes in the Quaternary. The mud intervals typically contain abundant diatoms, indicating deposition in perennial lakes with shallow to deepwater and variable salinities. The freshening episodes are characterized by carbonate sedimentation associated with microbial bioherms (microbialites).

Dry episodes are characterized by salt deposition, mainly consisting of halite and thin layers of gypsum. The halite textures indicate paleoenvironments ranging from perennial hypersaline lakes to ephemeral, annually desiccated saline pans. The salt layers, predominantly composed of layered halite, are thicker at the base of the well, reaching 20 m in thickness (Baker et al., 2001).

Chemical analyses of the salt crust and halite layers penetrated by the deep well revealed that the interface between the surficial redistributed halite and the primary halite is characterized by a high peak in bromide content a few meters below the surface, reaching values of up to 5–10 mg/kg (Risacher & Fritz, 2000). However, the bromide content in the salt minerals ranges from 1 to 10 mg/kg, a much lower value than observed in halite deposited in a marine environment, which is typically above 70 mg/kg (see Chapter 7).

The bromide content in the intracrustal brines recovered from the deep well shows a maximum value at a shallow depth of 20 m, reaching almost 140 mg/L (Risacher & Fritz, 2000). The deep well shows high Br contents in brines to a depth of about 40 m, reaching a minimum value of about 50 mg/L at a depth of 50 m. A mean value of about 60 mg/L is observed for the 60–120 m interval, where thick halite layers are observed. The low values of bromide in the salt crusts penetrated by the well, as compared with the abundance of halite minerals, indicate that the origin of the evaporite crusts is related to the leaching of previously deposited salt deposits (Risacher & Fritz, 2000).

On the surface of Salar de Uyuni, gypsum forms eolian dunes that are known as lunettes, due to their moon-white appearance. These sulfate salt flats are connected to the deflation of ancient salt lake deposits. They are made up of beds of gypsum particles with sand-sized grains that gathered on the floor of short-lived salt lakes. These lacustrine deposits were desiccated and then subjected to deflation in desert areas, forming gypsum dunes, as observed in several other playa lakes in the world. Some examples are the Bonneville Salt Flats in Utah, the White Sands and the Salt Flat Playa in New Mexico, and the Laguna Ojo de Liebre in Baja California (Warren, 2006, 2016).

The Salar de Uyuni is fed by the Rio Grande's fresh waters, which enter the depression at its southwestern part (Figure 10.17). The river has built a large delta, bringing waters enriched with dissolved ions from the weathering and leaching of volcanic rocks in the highlands. The halite crust in the salar intercalates with fluvial sediments, such as silt and clay. Local conglomeratic sand lenses incised the muddy, fine-grained lagoonal sediments, indicating the presence of fluvial channels that underlie both the deltaic and evaporitic sequences (Risacher & Fritz, 2000).

The deep well drilled in Salar de Uyuni indicates that lacustrine and evaporitic environments have accumulated since the Pleistocene, with wet periods forming deep lakes and thicker halite layers during dry periods. The region where hypersaline brine mixed with less saline water resulted in the precipitation of less soluble minerals, such as calcite, gypsum, and ulexite. The mixing of waters with different salinities is commonly associated with the genesis of the largest borax deposits in Bolivia, which accumulated at the interface between fresh and more saline waters at the edges of an ancient salt pan. The ulexite interval in the Rio Grande de Lipez delta prism, located at the southern margin of Salar de Uyuni, has an average thickness of only 10 cm but spreads across over 100 km^2. The high boron concentrations in the Rio Grande drainage system suggest the leaching of the volcanic terranes at the southern end of Salar de Uyuni (Rettig et al., 1980). According to the U.S. Geological Survey (2023), boron production in Bolivia in 2019 was 210,000 metric tons, or about one-half of the total boron production in Chile.

10.4.3. Salar de Atacama, Chile

Salar de Atacama is a desert located at a moderate altitude of 2,300 m on the western flank of the high Central Andes Mountains in Chile. It spans almost 3,000 km^2 (Figures 10.12 and 10.13). The Central Valley depression stretches along a coastal belt in northern Chile, east of the Coastal Ranges in the Antofagasta region (Figure 10.14). This intermontane depression (Figure 10.18) is considered a hyperarid region, as the present-day mean rainfall is about 50 mm/year and is nil in some years, as rains in this desert may occur once every 10 or 20 years. Salar de Atacama is located within the lowest depression at the center of the Atacama Desert in Chile, near the border with Argentina and Bolivia (Figure 10.14). It is considered the third-largest saline pan in the world, surpassing only Uyuni in Bolivia and Salinas Grandes in Argentina. The Atacama Desert is considered the driest place on Earth and has become one of the most visited natural attractions in Chile (e.g., Rech et al., 2002, 2019).

Two barriers prevent moisture from reaching the desert: the Coastal Ranges near the Pacific Ocean and the obstacle formed by the more elevated Central Andean

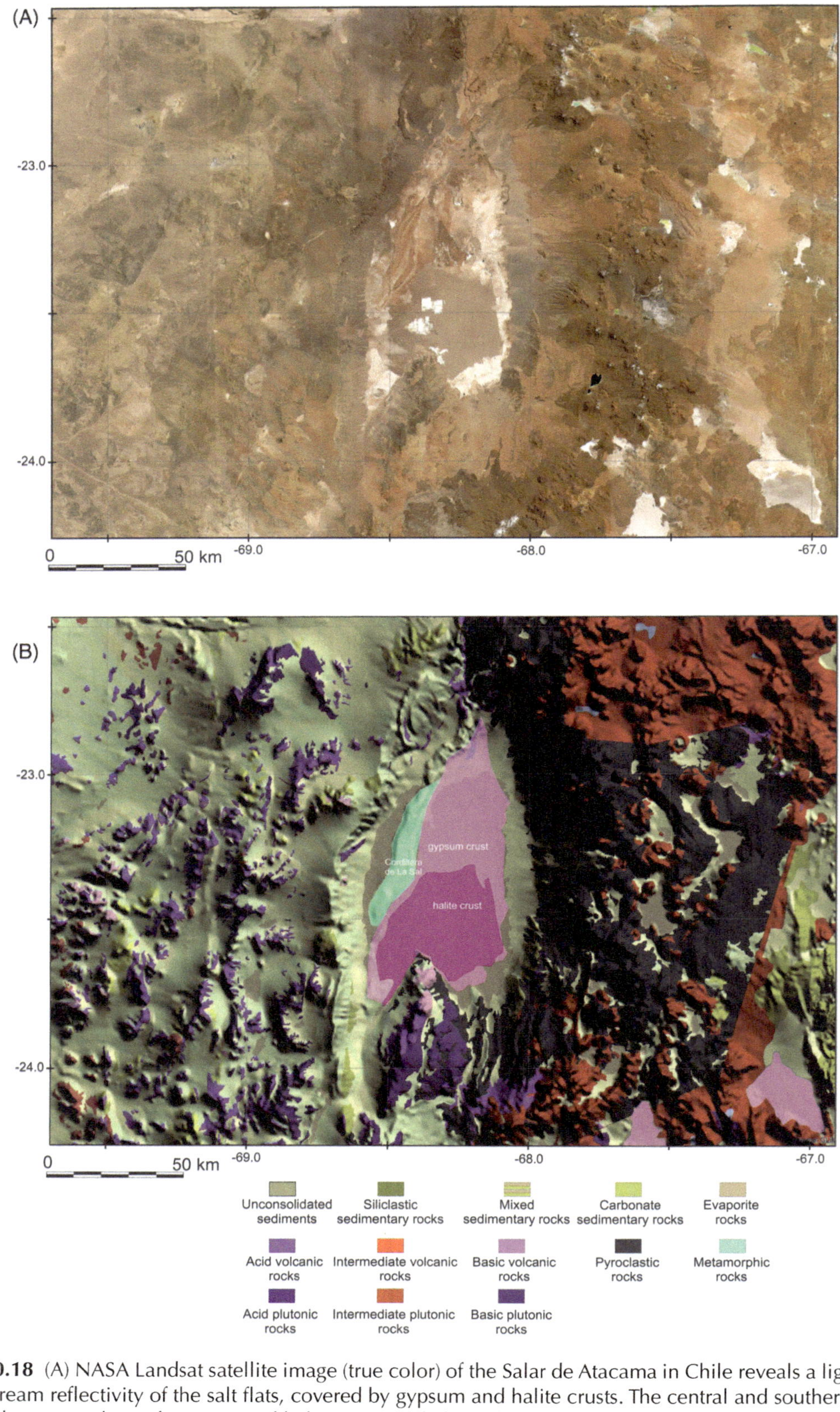

Figure 10.18 (A) NASA Landsat satellite image (true color) of the Salar de Atacama in Chile reveals a light brown to light cream reflectivity of the salt flats, covered by gypsum and halite crusts. The central and southern parts of the salar have a nucleus of gypsum and halite, surrounded by siliciclastic sediments. Alluvial fans and volcanic rocks have dark brown to black reflectivity. The image also shows several volcanic cones and subcircular craters. (B) A simplified geological map highlights the outlines of evaporite facies using semitransparent pink and magenta colors. Eastward of the Cordillera de la Sal, the halite crust in the nucleus of the Atacama salt flat is surrounded by the gypsum pavement. Credit: Hartmann and Moosdorf (2012)/with permission of John Wiley & Sons.

branch to the east, which is a barrier to the advection of moisture in the forested Amazon region (see Figure 10.15). There is no permanent fluvial drainage across the depression, resulting in an endorheic basin. The brines are concentrated by evaporation, precipitating evaporites that accumulate on the salt flats, which appear light brown on natural-color satellite images (Figure 10.18).

Salar de Atacama was formed in a north-south trending depression bounded by elevated flanks to the east and west, occupying a basin about 150 km long and 60–90 km wide. It is bounded to the west by the Western Cordillera and by Quaternary volcanic rocks close to the Licancabur Volcano at the eastern boundary. The present-day salt flat overlies a deep sedimentary basin that includes Paleozoic, Mesozoic, and Cenozoic sediments, resulting in a foreland basin that is more than 9 km in total thickness (Charrier et al., 2007).

The Atacama salt flats contain over 1,000 m of Pliocene to Quaternary evaporite sediments intercalated with Neogene siliciclastic rocks (Jordan et al., 2002; Rech et al., 2002, 2019; Warren, 2006, 2016). These sediments formed under conditions that alternated between a hypersaline lake and a dry salt pan. Compressional tectonics have affected the foreland basin since the mid-Cretaceous (Arriagada et al., 2006).

The salt flat sediments in the Atacama Depression are overlain and interdigitated with alluvial fans derived from the erosion of the surrounding mountains. The alluvial fans are formed by siliciclastic and volcanic rocks that locally develop some soil on the desert surface. On satellite images, the volcanic rocks surrounding the salt flat appear grey, tan, brown, and black (Figure 10.18). The volcanic facies are characterized by a rugose texture in the images, contrasting with the smooth texture of the halite crust. Several volcanic cones are visible in the satellite images, some of which display subcircular craters. The geological map shows a schematic outline of the salt flat evaporites (Figure 10.18). The halite crusts in the Atacama salt flats are surrounded by the gypsum pavements located eastward of the Cordillera de La Sal. Ignimbrites and extrusive intermediate volcanic rocks (based on Hartmann & Moosdorf, 2012) outcrop to the northeast of the salt flat (Figure 10.18).

The Atacama salt flats are covered by a crust of evaporite minerals, primarily gypsum and halite, characterized by different reflectivity in satellite images (Figure 10.18). The salt flat is composed of mixed siliciclastic, carbonate, and sulfate sediments, covering about 40% of the salar surface. The central nucleus, with an area of almost 1,800 km^2, is formed by almost pure halite (e.g., Rech et al., 2002, 2019). The salt pan is fringed by a belt of clay and carbonate sediments, which exhibit whitish reflectivity in satellite images (Figure 10.19). These sediments are locally covered by surficial evaporite crusts, as sporadic

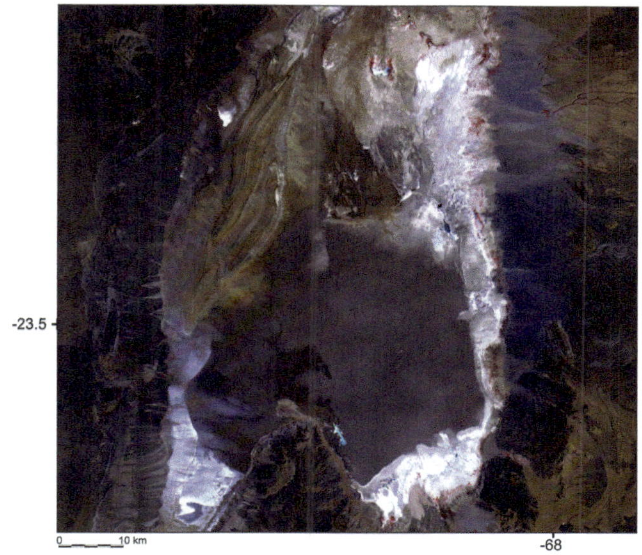

Figure 10.19 A NASA satellite image depicts the halite nucleus in Salar de Atacama, surrounded by a white gypsum pavement. The Cordillera de la Sal, NW of the salar, has a sigmoidal shape indicating strike-slip movement. The Cordillera Domeyko, located west of the salt flat, appears dark brown. The east-facing triangular-shaped feature to the east of the Cordillera de la Sal is the delta system associated with the San Pedro River, which runs from the northern part of the salar and terminates in the lagoon at the center. South of the salar, a triangular-shaped feature, known as Cordon de Lila, separates the gypsum pavements. Cordon de Lila is an uplifted block with exposed Mesozoic and Paleozoic basement rocks. Credit: Courtesy of NASA.

water from rain or groundwater dissolves the more soluble minerals. The water table rises and falls according to hydrological infiltration, dissolving salt and leaving a thin veneer of fine-grained sand and silt. Several small saline ponds with salt crusts are breeding grounds for flamingos adapted to the arid environment.

The most conspicuous feature of Salar de Atacama is the sedimentary ridge known as the Cordillera de la Sal, which extends in an NE-SW direction to the west of the salt flat (Figure 10.19). The Cordillera de la Sal is characterized by a sigmoidal shape, indicating strike-slip movements controlling its geomorphology (Figure 10.19). Simplified geological maps (e.g., Chorlton, 2007) indicate that the structure is cored by Oligocene to Miocene sediments (Figure 10.20). Although forming an outstanding structural high in satellite images and geological maps, the Cordillera de la Sal displays rather modest elevations, as on average the structure rises 200 m above the level of the surrounding Quaternary playa sediments. The structure displays an N-S length of about 80 km, adjacent to the Cordillera Domeyko, an anticlinal structure affected by thrust faults forming an uplifted block (Figure 10.20).

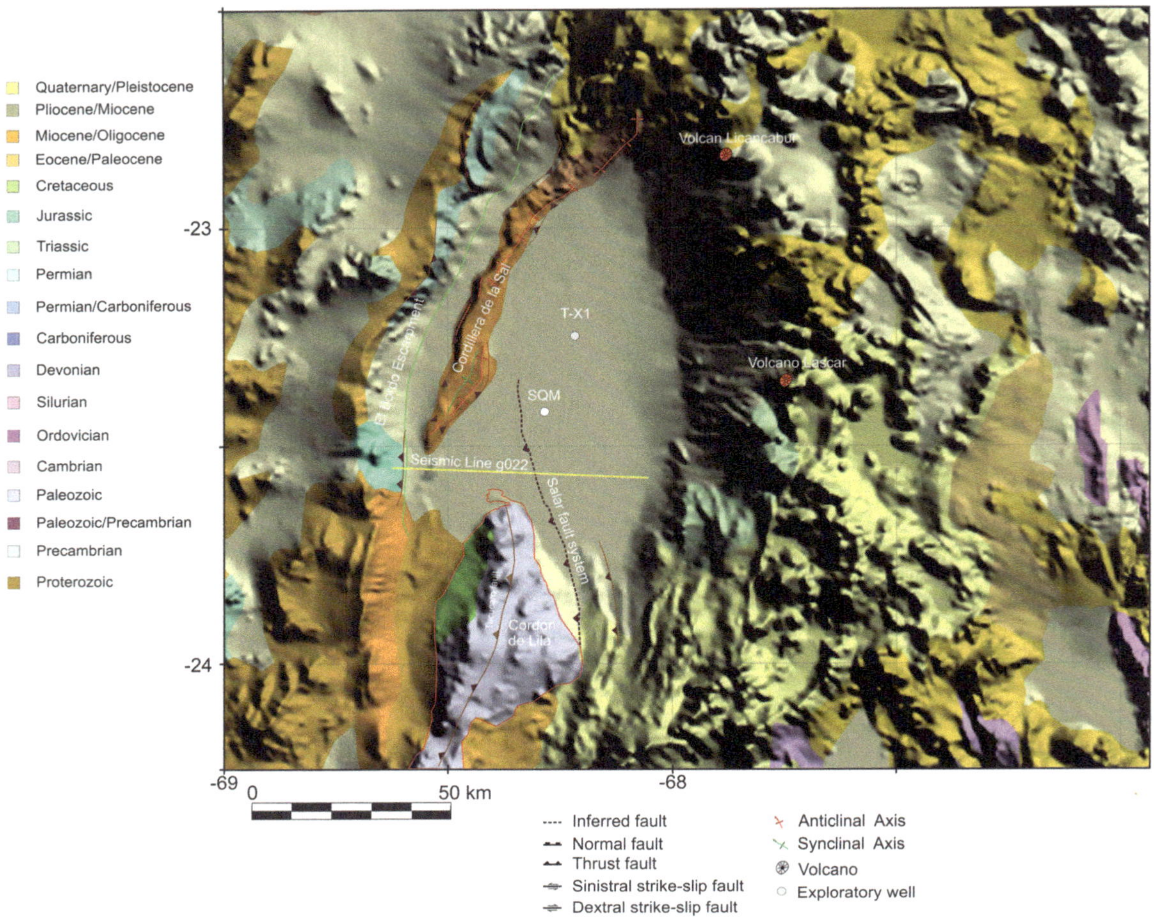

Figure 10.20 The simplified geological map of the Atacama Depression shows the Cordillera de la Sal, characterized by Oligocene to Miocene sediments, and the Cordillera Domeyko, a major NE-trending structural high cored by Mesozoic sediments. The deep petroleum exploration well Toconao-X1 (T-X1) and the shallow exploratory SQM wells were drilled eastward of Cordillera de la Sal, in the northern part of the depression. The seismic line g022 crosses the southern part of the depression along an E-W direction. The seismic profile is located north of the Cordon de Lila, a basement horst where Mesozoic and Paleozoic rocks outcrop at the surface. Neogene volcanic rocks occur eastward of Salar de Atacama, toward the Western Cordillera. The Licancabur Volcano, located northeast of the depression, reaches an altitude of almost 6 km. Credit: Rubilar et al. (2018)/with permission of Elsevier.

This NE-trending structural high is characterized by outcrops of Paleozoic to Mesozoic rocks (Wilkes & Gorler, 1994), indicating an upthrust involving the metamorphic basement (Figure 10.21).

The E-W seismic profile in the southern sector of Salar de Atacama (Figure 10.21) indicates that the Cordillera de la Sal corresponds to an intrabasinal uplift that exposes strongly folded and faulted Neogene sedimentary sequences intercalated with Late Miocene to Pliocene ignimbrites (Pananont et al., 2004). Transpression has strongly folded and faulted the Cretaceous to Paleogene sedimentary layers in the northern Pre-Andean Depression, inverting their originally extensional structures (Flint et al., 1993; Lin et al., 2016; Rubilar et al., 2018). Regional seismic profiles indicate that the Cordillera de la Sal overlies Mesozoic to Cenozoic anticlinal structures in the subsurface, which are probably associated with halokinesis and compressional tectonics (Figure 10.21). Rubilar et al. (2018) interpreted that the Cordillera de la Sal formed within a half-graben structure controlled by a normal fault dipping to the east, with indications of Neogene compressional tectonics inverting the rift basin depocenter, located to the east of the Cordillera Domeyko (Figure 10.21).

The pre-Eocene strata in the Atacama Basin have been strongly folded during a compressional orogenic

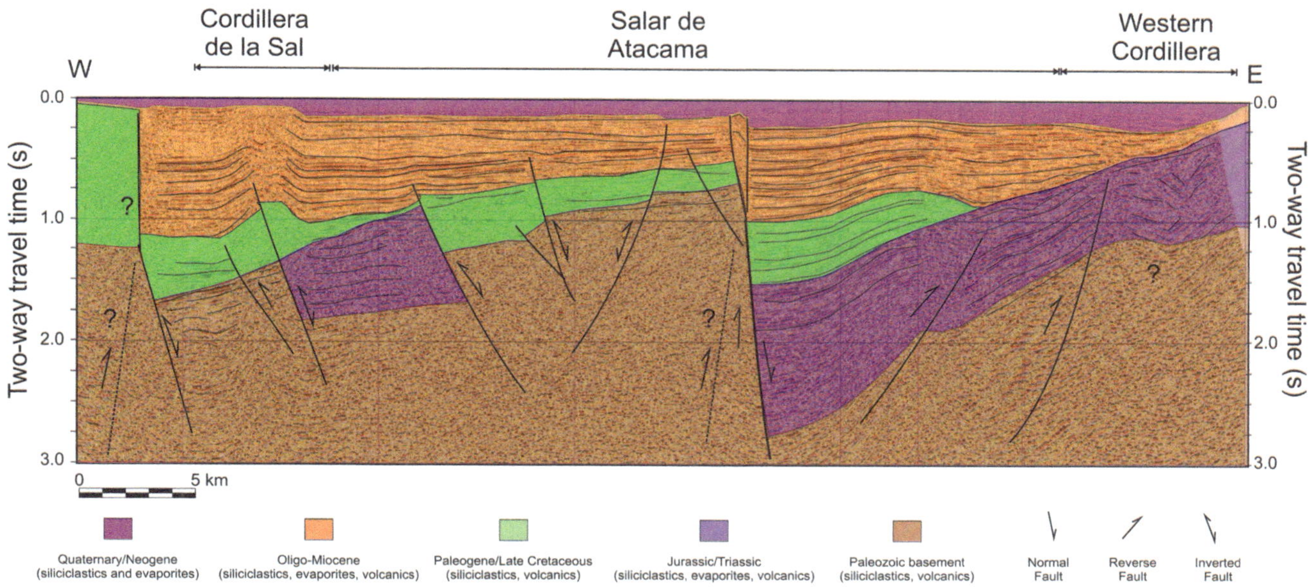

Figure 10.21 Cross-section showing a schematic line interpretation of a seismic profile (line g022) extending from the southern part of the Cordillera de la Sal across the Atacama Basin. The Cordillera Domeyko, located on the western side of the profile, corresponds to an upthrusted block. The profile shows Quaternary to Miocene sequences that include thick evaporite deposits filling the depression. Due to poor seismic imaging and borehole control, the profile infers Paleogene and older sequences in the rift blocks. Extensional normal faults were developed in the Mesozoic but were inverted in the Neogene, resulting in reverse faults with an eastward vergence. The Cordillera de la Sal is characterized by diapiric salt structures below the Quaternary salt flats. The Salar Fault System is interpreted at the center of the profile, but most faults slightly affect the Quaternary, flat-lying sedimentary layers. Eastward of Salar de Atacama, Neogene volcanic rocks outcrop in the Western Cordillera. Credit: Rubilar et al. (2018)/with permission of Elsevier.

event known as the Incaic Phase, with the climax dated as late Eocene (Hammerschmidt et al., 1992; Pananont et al., 2004). The Oligo-Miocene sequence, up to 1,300 m thick, is composed of coarse detrital and evaporitic sediments included in the Tambores Formation, forming a depocenter east of the Cordillera Domeyko (Figure 10.21). Radiometric dating indicates that the ignimbrites and volcanic rocks to the east of the Atacama Depression are from the Miocene period, corresponding to a volcanic event associated with the Quechua phase of magmatic activity in the Central Andes (Wilkes & Gorles, 1994).

The Oligocene sediments that outcrop in the Cordillera de la Sal are mainly composed of continental red beds (San Pedro Formation, Oligo-Miocene), which locally reach a thickness of more than 3,000 m and are covered by Miocene ignimbrites. According to Rubilar et al. (2018), these sediments were deposited in a playa environment that was periodically desiccated and accumulated large amounts of evaporite minerals (gypsum, anhydrite, glauberite, and especially halite). The origin of the sigmoidal-shaped structure, which shows a complicated structural pattern in outcrops, is probably the result of a combination of extensional, compressional, and sinistral strike-slip movements.

The Cordillera de la Sal is bounded by a steep reverse fault at its eastern margin, which displays an eastward vergence, and by back thrust faults to the west, forming a "pop-up" or positive flower structure (Figure 10.21). This uplifted zone divided the Atacama Depression into two depocenters that overlie the graben structure in the subsurface (Figure 10.21). The seismic interpretation suggests an anticlinal structure with a halite core in a diapiric structure, which probably acted as the main source of the Quaternary salt (Rubilar et al., 2018). As the outcropping evaporites were intensively cannibalized by weathering and erosion, the resulting halite and gypsum were incorporated into the paleolake and resedimented in the modern Atacama depression (Figure 10.21).

10.4.4. Subsurface Structures in Salar de Atacama

Based on field geology and seismic interpretation, several authors (e.g., Cembrano et al., 2007; Mpodozis et al., 2005; Muñoz et al., 2002; Pananont et al., 2004; Reutter et al., 2006) suggest that the steep dip observed in reverse faults at the western margin of Salar de Atacama originated as normal faults during the Late Jurassic-Early Cretaceous development of back-arc basins. Compressional events associated with the Incaic orogenetic phase in the Late Cretaceous to Paleogene inverted these normal faults, superseding the extensional regime (Amilibia et al., 2008; Cembrano et al., 2007; Martínez et al., 2018; Reutter et al., 2006; Rubilar et al., 2018).

During the Mesozoic, the magmatic arc was located near the present coastline, but in the Neogene, it shifted east, forming igneous massifs along the Western Cordillera (Stern, 2004). The Atacama Depression is currently in the forearc of volcanic terranes formed during the Mesozoic. A regional seismic profile extending from the southern extension of the Cordillera de la Sal to the eastern part of the salar (Figure 10.21) shows that large extensional faults have been inverted in the Neogene, resulting in compressional structures with a vergence to the east (Muñoz et al., 2002; Rubilar et al., 2018).

The Cordón de Lila, a triangular-shaped ridge at the southern part of Salar de Atacama, is an upthrusted block forming a basement high exposing Pre-Cretaceous basement rocks (Figure 10.20). The pre-Andean Lila Igneous and Sedimentary Complex formed a structural high that interacted with the sedimentary cover during the Paleozoic, Mesozoic, and Cenozoic phases of the basin evolution (Arriagada et al., 2006; Muñoz et al., 2002). At a regional scale, the Cordón de Lila High can be interpreted as an asymmetric, north-plunging anticline bordered to the east by one of the basement faults of the "Salar Fault System," which does not penetrate the Quaternary sediments (Jordan et al., 2002, 2007), suggesting a large fault-propagation fold.

Toward the northern part of the Atacama Depression, the thickness of the halite layers increases in the Neogene sequences. This salt level controlled the transfer of the basement thrust to subhorizontal detachment levels in the Pliocene sequences, deforming the sequences through systems of detachment folds (Muñoz et al., 2002). Seismic data suggest that the Neogene salt layers in the northern part of the Cordillera de la Sal have deformed and remobilized, forming small salt diapirs. These structures were locally folded and breached the overlying stratigraphic sequences by active diapirism (Pananont et al., 2004; Rubilar et al., 2018).

Seismic profiles collected across the central part of Salar de Atacama revealed a large fault system that extends for at least 30 km along an N-NW direction across the basin (Flint et al., 1993; Jordan et al., 2002, 2007). The interpretation of vintage seismic data (e.g., Figure 10.22) indicates that the total thickness of the halite layer might reach 1,500 m in the halite nucleus at the central part of the salar. The basin depocenter is crossed by a major fault zone, which is seen intersecting the upper stratigraphic sequences in the southern part of the salar (Jordan et al., 2002, 2007; Rubilar et al., 2018). This fault can be interpreted as a reverse fault with an eastward vergence and a down-to-the-east offset of

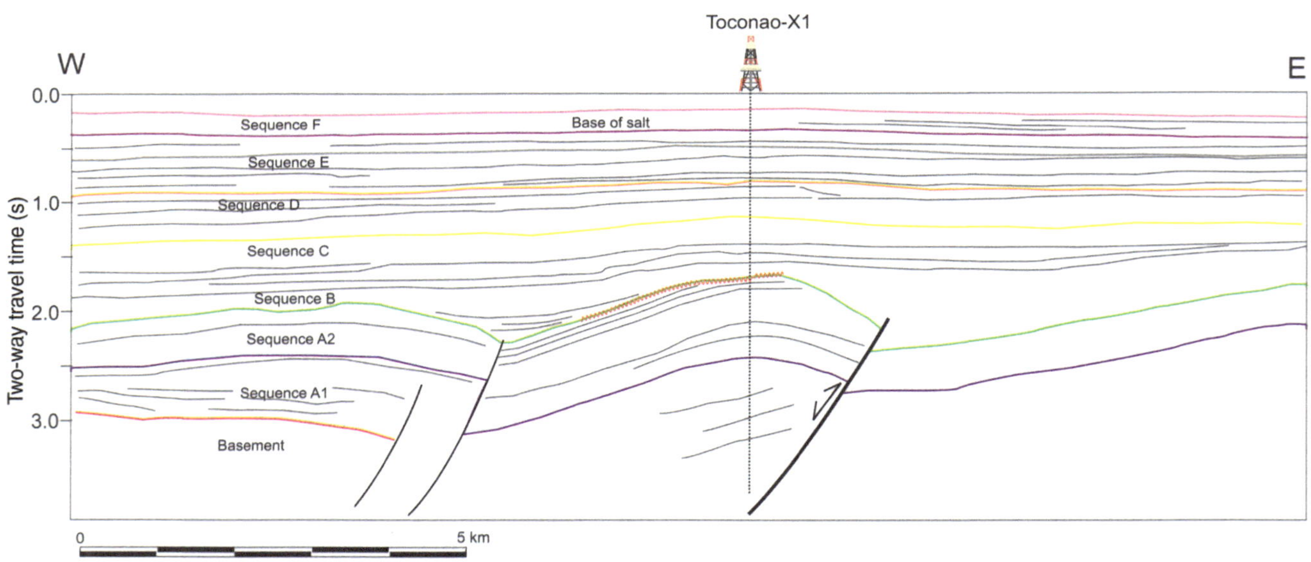

Figure 10.22 A vintage seismic profile near the Toconao-X1 deep exploratory borehole in the northern part of Salar de Atacama reveals an anticlinal structure associated with a reverse fault with an eastward vergence. Extensional fault blocks involving Sequence A (Upper Cretaceous) have been inverted and onlapped by Sequence B (Paleogene), forming an anticlinal structure associated with a reverse fault with an eastward vergence. The Neogene sequences (C, D, and E) show minor compressional features. Minor thinning above the anticlinal structure drilled by the exploratory well characterizes the evaporite layers of Sequence F, which span the Pliocene to Pleistocene. Credit: Adapted from Muñoz et al. (2002).

about 900 m (Figure 10.22). Most of the fault movement is dated as Pliocene, with only 200 m of the fault offset dated as Quaternary (Jordan et al., 2002, 2007).

Although rates of fault displacement throughout the Quaternary have been estimated to be about 0.1–2.0 mm/year (Jordan et al., 2002), these large displacement rates and large offsets do not show any geomorphological disturbance affecting the salt flat surface (Figure 10.22). Seismic data clearly show the faults, but there is no evidence of a modern topographic scarp in the halite nucleus. This suggests blind faults, with deformation in the subsurface but lacking topographic expression on the present-day flat surface (Figure 10.22).

Thrust faults with an eastward vergence have impacted the Atacama Depression, forming anticlinal structures in the Cordillera de Domeyko, Cordillera de la Sal, Cordon de Lila, and Tilocalar Hill, located to the east of the salt flat (e.g., Bascuñán et al., 2019; Lin et al., 2016). One example of a blind thrust fault is observed eastward of the Cordillera de la Sal, at the center of the salt flat (Figure 10.20).

The structural high drilled by the borehole Toconao-X1 is associated with a fault controlling a depocenter with a thickening of Paleogene strata (Figure 10.22). This is one of the few exploratory boreholes that drilled the Salar de Atacama salt deposits and penetrated the Mesozoic sediments (Muñoz et al., 2002). López Steinmetz et al. (2020) postulate three main mechanisms for contractional deformation and block uplift in the region of the Atacama Basin: (1) propagation of east-vergent basement-involved reverse faults and thrusts; (2) tectonic inversion of pre-existing normal faults formed during extensional events; and (3) activation of shallow, thin-skinned thrust faults.

Following this interpretation, some conclusions can be derived from the structural analysis of the Atacama Depression. The basin is limited by a system of uplifted blocks to the west (Domeyko Ranges) and east (Altiplano basement blocks). First, the Domeyko Cordillera was formed by a series of east-vergent basement-involved thrust faults that deformed the Mesozoic and Paleogene sequences. Second, the Cordillera de La Sal was formed by thin-skinned deformation with thrust faults showing a double vergence and involving mainly Neogene strata. Third, some reverse faults mainly affected the Late Cretaceous to Oligo-Miocene sequences in the eastern part of the Atacama Basin, with minimal disturbance of the Plio-Pleistocene sediments and Quaternary evaporite deposits (López Steinmetz et al., 2020).

10.4.5. Stratigraphic Sequences in Salar de Atacama

The deep exploratory borehole Toconao-X1 (T-X1; see location in Figure 10.20) was drilled in the central part of the Salar de Atacama Basin, eastward of the Cordillera de la Sal (Arriagada et al., 2006). Analysis of seismic reflection data and drill cores from this exploratory well (Figure 10.23) revealed several stratigraphic successions correlated to the eastern flank of the Domeyko Cordillera (Charrier et al., 2007; Muñoz et al., 2002). The Domeyko Cordillera consists of elongated (30 to more than 100 km in length) and wide (up to 10 km) basement highs to the west of the Cordillera de la Sal (Figure 10.20). The upthrusted blocks form the core of a major anticlinal structure, bounded by high-angle reverse faults (Cembrano et al., 2007). Salar de Atacama is characterized by a Neogene depocenter located to the east of this major structure (Figure 10.21). Outcrops of Mesozoic rocks in the upthrusted block indicate basement-involved compressional tectonics affecting Cenozoic sediments. The main thrust fault exhibits an eastward vergence toward the Cordillera de la Sal (Figure 10.20). In the Atacama Basin, Quaternary evaporites cover the complex structures associated with the deformed Cenozoic and Mesozoic sedimentary sequences, as recorded by field mapping and seismic datasets (Figure 10.21).

The T-X1 borehole was drilled to a depth of about 5,500 m and penetrated pre-Cretaceous Bordo and Peine formations, which were interpreted as basement rocks (Figure 10.23). These meta-sediments correspond to Mesozoic (Triassic) and Paleozoic (Permian to Carboniferous) rocks included in a rift block that was inverted in the Late Cretaceous (Muñoz et al., 2002). The overlying successions correspond to the lower part of the Late Cretaceous to Paleogene sedimentary sequences, mainly limestones of the Yacoraite and Tonel Formations (Muñoz et al., 2002). These sequences can be correlated with carbonate rocks in the Salta rift basin in Argentina and marine sediments from the Lomas Negras Fm., which outcrop in the Altiplano, near the Chile-Argentina border (Charrier et al., 2007).

The Late Cretaceous to Miocene sedimentary deposits drilled by the deep borehole (sequences A to E, Figure 10.23) are interpreted as formed in extensional basins that were subsequently inverted during the Andean orogenies (Figure 10.22). The thick and coarse Late Cretaceous to Paleogene terrigenous deposits of the Purilactis Fm. outcrop in the Domeyko Range region (Figure 10.20). Compressional tectonics associated with a foreland setting has affected this segment of the Andes since the mid-Cretaceous, leading to the deposition of Miocene and Plio-Pleistocene siliciclastic and evaporitic sediments (Arriagada et al., 2006; Bascuñán et al., 2019; López Steinmetz et al., 2020; Mpdozis et al., 2005).

The deep exploratory borehole T-X1 revealed that the Atacama sedimentary depression accumulated thick evaporite sediments (halite) in the Quaternary (more than 1,000 m of evaporitic sediments, corresponding to the upper part of the sequence F; Figure 10.23). Additionally,

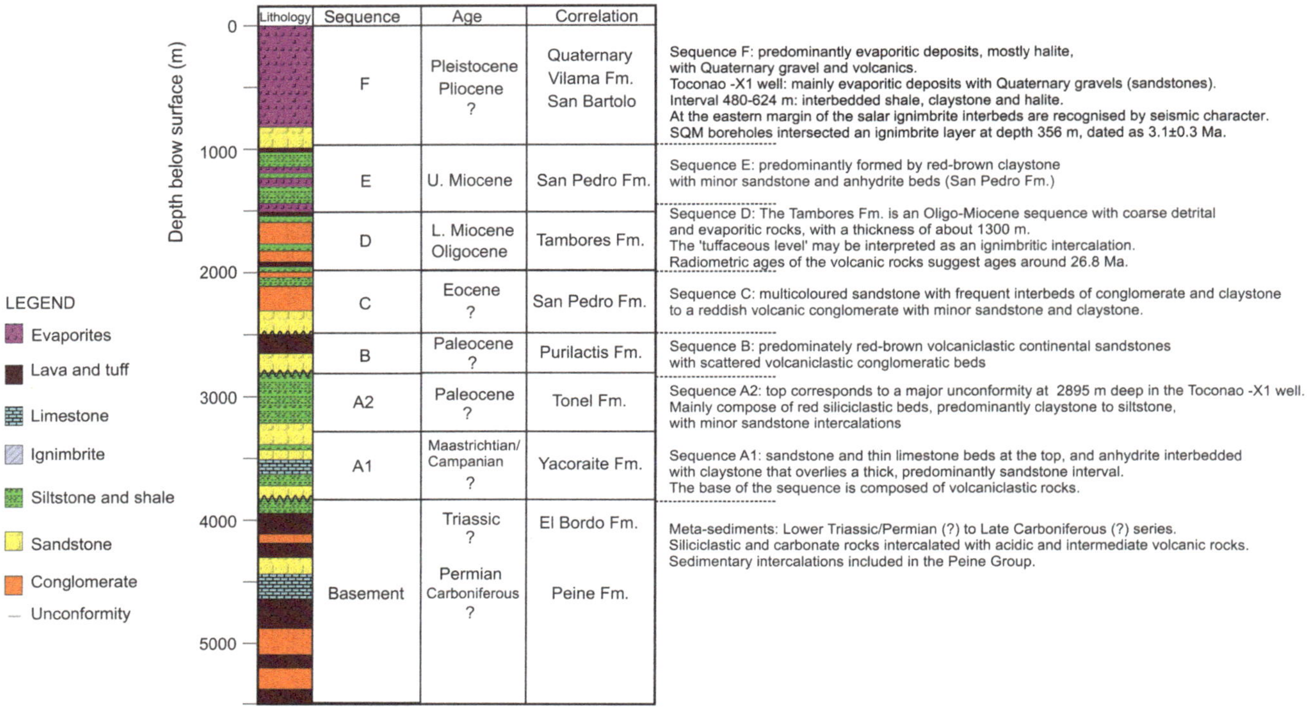

Figure 10.23 The stratigraphic column shows a simplified interpretation of the lithostratigraphic sequences drilled by the deep exploratory well Toconao-X1 in the northern part of the Atacama Basin (see Figure 10.25). Sequence F is a thick evaporitic sequence with siliciclastic and volcanic intercalations, interpreted as Pliocene to Pleistocene. Sequences E, D, C, and B are siliciclastic and carbonate rocks with intercalations of volcanic layers. Sequence A occurs below a major unconformity and includes Paleogene to Late Cretaceous sediments. The pre-Cretaceous strata correspond to slightly metamorphic rocks interpreted as the basement of the Atacama depression, including volcanic series associated with Mesozoic and Paleozoic magmatic events recorded in the Cordon de Lila region. Credit: Adapted from Muñoz (2002) and Cembrano et al. (2007).

the Neogene sequences also contain evaporites, such as the Miocene salt deposits of the San Pedro Fm., which include several halite layers. These sediments were deposited in the region of the Cordillera de La Sal, east of the Domenyko Range, and close to the active faults of the Atacama Depression fault system (Figure 10.20). Some faults formed topographic scarps, resulting in the growth of sedimentary strata in the downthrown block, but some faults did not affect the depositional surface (Jordan et al., 2002; López Steinmetz et al., 2020; Muñoz et al., 2002).

Several wells drilled in the Salar de Atacama aimed at investigating the potential to extract and generate lithium (Li) and potassium (K) products from the subsurface brines (Sociedad Química y Minera de Chile, 2022). The exploratory borehole SQM No. 2005 (Figure 10.20) corresponds to a shallow well that was drilled in the halite nucleus at the center of Salar de Atacama in 1986 to determine the subsurface stratigraphy of the basin (Bobst et al., 2001). The well cores recovered 100 m of evaporitic sediments in the upper parts of the salt flat, sampling sedimentary sequences deposited from about 100 ka to the Holocene. The main lithologies corresponded to halite (87% of the cores), siliciclastic lacustrine rocks (10%), and sulfate minerals (3% of the cores). Paleoclimate studies indicated that deposition was dominated by extremely arid conditions similar to today, with subaerial halite forming in a dry salt pan (Bobst et al. 2001).

The SQM well recorded two significant wetter intervals when sedimentation occurred in saline mudflats adjacent to a perennial saline lake. The first interval was dated as 13–20 ka, and the second one, marked at a depth of 48 m (about 61 ka), corresponded to a saline mudflat or gypsum flat. The lower part of the core (between depths of 62 and 100 m) was dominated by subaerial halite, forming a thick crust. These lithologies indicate that evaporite cycles in the Central Andean region were controlled by climatic events that intercalated periods of permanent and ephemeral lakes. During the extremely dry periods, the salina desiccated and accumulated thick halite layers, forming evaporite crusts (Bobst et al., 2001).

Salar de Atacama is presently exploited for its abundant salt minerals, which occur in the subsurface to a depth of more than 1,000 m, as indicated by the T-X1 well (Figure 10.23). The outcrops in the Cordillera de la Sal indicate that halite layers also occur in the Oligo-Miocene São Pedro Formation (Wilkes & Gorles, 1994). Trace element analysis of halite samples from outcrops shows very low levels of bromide, which is more consistent with continental sources for the salt deposits (Wilkes & Gorler, 1994). The bromide isotopes indicate that arid environments resulted in the development of continental sabkhas with no marine influence. These playa lakes have been effectively accumulating halite minerals from the Neogene to the Quaternary and Holocene (Bobst et al., 2001).

Since the 1990s, companies have built modern evaporation ponds in the Atacama Depression to exploit the subsurface brines, which hold significant potential for lithium production (Godfrey & Álvarez-Amado, 2020; see Chapter 9 in Mohriak, 2025). Currently, Australia's mine production of lithium is the largest in the world, with 61,000 metric tons extracted from igneous minerals like spodumene (U.S. Geological Survey, 2023). Chile's mine production is second, with 39,000 metric tons. The Andean countries have the largest proven reserves, accounting for over 50% of global reserves. The revised reserves include 21 million tons in Bolivia, 20 million tons in Argentina, and 11 million tons in Chile (U.S. Geological Survey, 2023).

10.5. SALT LAKES IN CHINA

The Qaidam (also spelled Chaidam and Tsaidam) Basin in the Tibetan Plateau is a well-studied example of brine concentration and salt deposition in continental settings affected by compressional tectonics (Duan & Hu, 2001; Eugster & Hardie, 1978; Gao et al., 2019; Guo et al., 2018; Lowenstein & Risacher, 2009; Lowenstein et al., 1989). Salt lakes in China formed within highland depressions related to the continent-continent plate convergence involving the Indian and Asian plates in the Cenozoic (Bally et al., 2012; Evenick, 2021; Wu et al., 2014). These elevated plateaus are associated with the Alpine-Himalayan orogenies, as observed in mountainous regions that extend from southern Europe, Turkey, and Iran, forming a deformation zone north of the Mediterranean Sea and Arabia (see Chapter 3).

The Tibetan Plateau is a high-elevated region associated with the uplift of crustal blocks by basement-involved reverse and strike-slip faults (Wang et al., 2012). The Tibetan Plateau is surrounded by several mountain ranges that formed during the Cenozoic, such as the Kunlun Shan, the Karakoram Range, the Altyn-Tagh (also spelled Altun) Mountains, and the Tien Shan mountains north of the Tarim Basin (Figure 10.24). The highlands correspond to structurally positive features reaching more than 6 km in altitude. The uplifted blocks are associated with strike-slip, thrust, and reverse faults resulting from the convergence between the Indian and Eurasian plates (Wang et al., 2006). The Altyn-Tagh fault zone corresponds to a large system of mainly left-lateral strike-slip faults forming shear zones that separate the Tarim and Qaidam basins, with the main phase of tectonic activity starting in the Eocene (Cheng et al., 2014, 2016; Wang et al., 2006). The southern margin of the basin is marked by the Kunlun Shan, with altitudes exceeding 5 km (Qinmin & Coward, 1990; Wu et al., 2014).

The Qaidam depression is associated with crustal adjustments during the collision of the Indian and Eurasian plates during the Mesozoic and Cenozoic. The basin architecture is controlled by transcurrent and normal faults that formed rift basins in the Mesozoic. Sedimentary sequences were inverted in the Paleogene due to compressional events that extended up to the Neogene (Pei et al., 2017; Qinmin & Coward, 1990).

The Tarim and Qaidam basins are depressions where fluvial and lacustrine systems have developed since the Neogene. The high-elevated regions north of the Tibetan Plateau host over 400 inland salt lakes, covering a total area exceeding 8,500 km^2. Most Quaternary to Holocene salt lakes in China are found in the Qaidam Basin, which means "salt marsh" in Mongolian. Some lakes are slightly saline, like Lake Qinghai, while others contain hypersaline brines, like Dabussun Lake in the Qaidam Basin, where the hypersalinity (>300,000 ppm) exceeds the solubility of halite (Zheng & Li, 2009).

The Qaidam Basin lies southeast of the Tarim Basin, stretching 800 km from east to west and extending for almost 350 km from north to south (Figure 10.24). The southern boundary of the basin is marked by the NW-SE-trending Kunlun Shan mountain range. The NW portion of the Qaidam Basin is a true desert, similar to the desiccated Tarim Basin (Figure 10.24).

The Qaidam depression is a large, endorheic intermontane basin with an average elevation between 2,400 and 3,000 m (Figure 10.24). It covers an area of approximately 120,000 km^2, with one-fourth covered by saline lakes and playas. The drainage basin covers a much larger area of about 270,000 km^2, surrounded by elevated plateaus and mountainous regions (Figure 10.24). The most important hypersaline lakes in the Qaidam depression are located in the Qarhan (also spelled Charhan) salt flat (Kezao & Bowler, 1986). These playa flats are covered by gypsum and halite crusts formed subaerially at high altitudes. However, evaporite deposits also occur in the subsurface of the perennial lakes associated with the Holocene lacustrine systems. The brines responsible for the salt deposits were concentrated by the evaporation of nonmarine waters and by ion enrichment derived from volcanic intrusions or deep hydrothermal sources in the crust (Spencer et al., 1990).

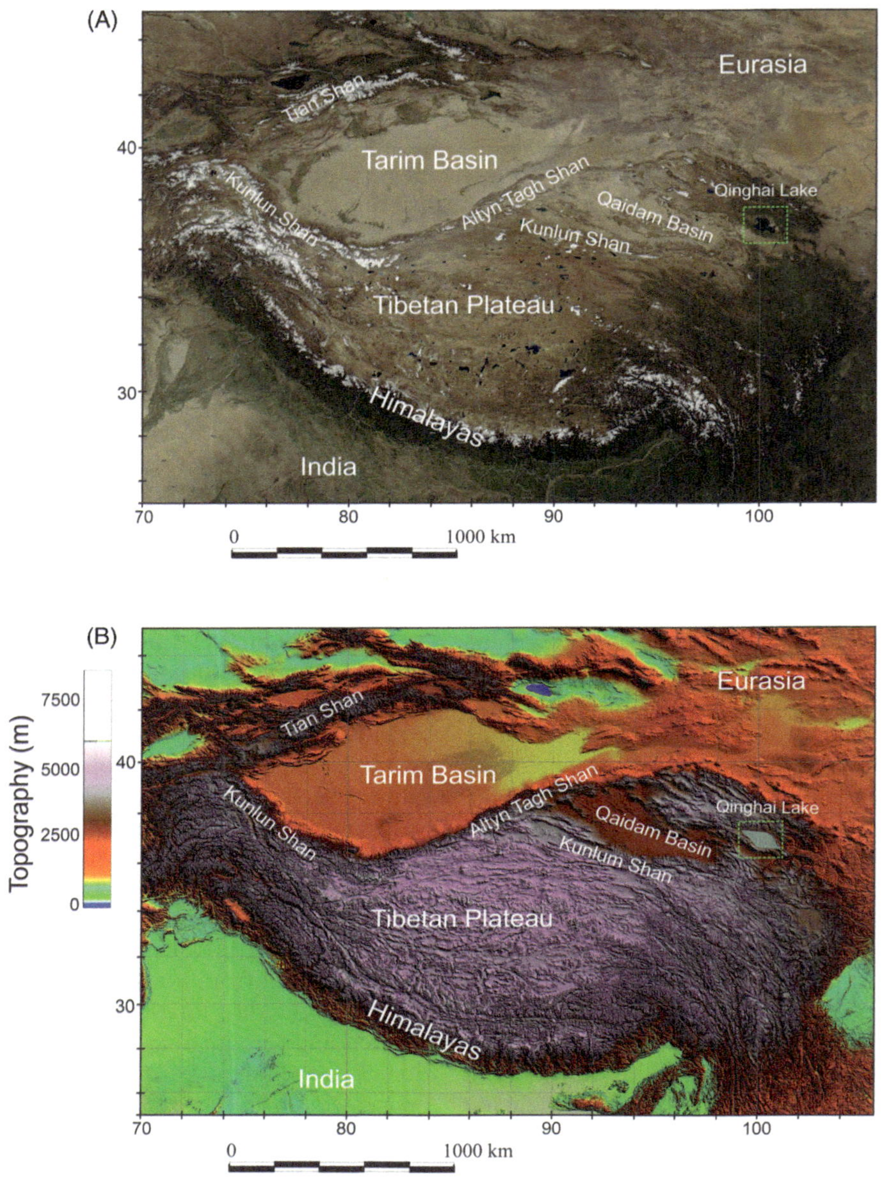

Figure 10.24 (A) A NASA satellite image centered on the Tibetan Plateau and the Tarim and Qaidam basins in NW China shows the Himalayas mountain ranges formed by the collision of the Indian and Eurasian plates. (B) The topographic map of the Tibetan Plateau shows the Tarim and Qaidam basins separated by the Altyn-Tagh Shan. Qinghai Lake, located eastward of the Qaidam Basin, is a brackish to slightly saline lake in a depression surrounded by highlands. Hypersaline lakes in the Qaidam Basin are located north of the Kunlun Shan, a mountain range reaching altitudes of over 5 km. Credit: NASA/Wikimedia Commons/Public domain.

Hypersaline lakes are formed in arid, hot, low-elevation deserts, but most saline lakes in China occur in highly elevated, cold, and extremely dry regions. The Qaidam Basin is considered hyperarid due to its five times the accepted index of evapotranspiration/average annual precipitation (e.g., UNCOD, 1978). The region's cold and dry climate only occasionally fills shallow lakes with water. Some regions in the Qaidam Basin are so arid that they are considered analogs for extraterrestrial environments, such as the Dalangtan salt flats (Anglés & Li, 2017; Cheng et al., 2017; Xiao et al., 2017).

The Qaidam depression contains the largest brine potash deposit in the world, mainly located in the Qarhan salt marshes. These evaporite resources hold the greatest reserves of lithium, magnesium, potassium, and sodium that have been found anywhere in China. The lakes are mainly exploited for diverse mineral deposits, including halite, mirabilite, boron, potash, and lithium minerals (Duan & Huo, 2001; Du et al., 2018). The total salt reserves in some hypersaline lakes in the Qaidam Basin are estimated at around 100 billion tons of sodium chloride and potash salts, with most production concentrated in Lake Dabussun (also spelled Dabusun or Dabuxun) in the Qarhan salt flat, north of Kunlun Shan (Figure 10.24). The Qarhan salt flats constitute the main source of potash and other chemical compounds produced in China, including the large Qinghai potash fertilizer plant (Casas et al., 1992; Chen & Bowler, 1986).

The Qaidam Basin is the only oil and gas province within the Tibetan Plateau of NW China and China's fourth-largest onshore gas-producing region, surpassed only by the Sichuan, Ordos, and Tarim basins (Sun et al., 2020; Wu et al., 2014). Qinghai Province is the most important petroleum region in the NW part of the basin, where several producing oil fields have been discovered in the past decades. These hydrocarbon traps are characterized by Neogene reservoirs associated with anticlinal structures in the subsurface of the salt flats (Suotang, 2016; Suotang et al., 2015, 2016).

Figure 10.25 The satellite image (natural colors) depicts Lake Qinghai in NW China, surrounded by highlands north of the Tibetan Plateau. Fluvial systems around the lake have developed deltaic environments at the western and northern lakeshores. Credit: NASA/Wikimedia Commons/Public domain.

10.5.1. Qinghai Lake

Qinghai Lake, also known as Kokonor in Mongolia, is China's largest inland salt lake, located to the east of the Qaidam Basin (Figure 10.24). Qinghai Lake (which means "blue lake" in Chinese) is known for its clear, transparent waters, which are blue to green in satellite images (Figure 10.25). Nestled on Quaternary sedimentary deposits in the eastern segment of the Qaidam depression, it is surrounded by eolian sands and an arid climate under monsoons. Research projects have investigated Holocene paleoclimate changes in Asia by drilling deep boreholes in the lake sediments (Zhang et al., 2011).

Qinghai Lake is nestled within a depression on the northeastern Tibetan Plateau, where elevations vary mainly from 2,500 to 4,500 m a.s.l. Its surface is currently 3,193 m a.s.l., and its area is estimated at 4,300 km^2. The lake has an average depth of 19 m and a maximum depth of 28 m. It is a large, deep, brackish water lake with a salinity of approximately 14 parts per thousand, or about 12.5 g/L of sodium chloride (Zhang et al., 2011), corresponding to one-third of the normal seawater salinity. The lake's slightly salty waters are alkaline, with a pH ranging from 8.9 to 9.5 (Jin et al., 2016).

Climatic studies show that the lake level in the early Holocene (12 ka) was about 20 m lower than today (Chen et al., 2016). The highest lake level occurred between 5.5 and 4 million years ago, at the early to middle Pliocene, when the lake level was about 10 m higher than today. At the beginning of the Holocene, Qinghai Lake was isolated from its main source of water, the Yellow River. Although the lake's surface has been falling steadily and the lakeshores have been shrinking, it still constitutes a relatively deep lake and a famous tourist attraction in China (Figure 10.25). Qinghai Lake is a modern analog of a permanent lake that may become an ephemeral water body due to natural or anthropogenic factors.

10.5.2. Neogene Salt Deposits in the Qaidam Basin

The Tibetan Plateau is a high-altitude arid steppe with local depressions that form brackish to hypersaline lakes in several large endorheic basins (Figure 10.26). The plateau is surrounded by high mountain ranges and hosts a large number of lakes, including Qinghai Lake and several hypersaline lakes in the Qarhan salt flats (Zheng et al., 2016). The Tarim Basin, located to the NW of the Tibetan Plateau, has developed desert conditions since the Pleistocene, with eolian dunes occupying most of the depositional surface (Figure 10.26). Most lakes fed by the Tarim River have desiccated in the Holocene, resulting in a desolate landscape with no active lacustrine systems (Figure 10.24).

During the Late Miocene to the Pliocene, a freshwater mega-lake system existed in the Qaidam Depression, forming a lacustrine complex covering areas of as much

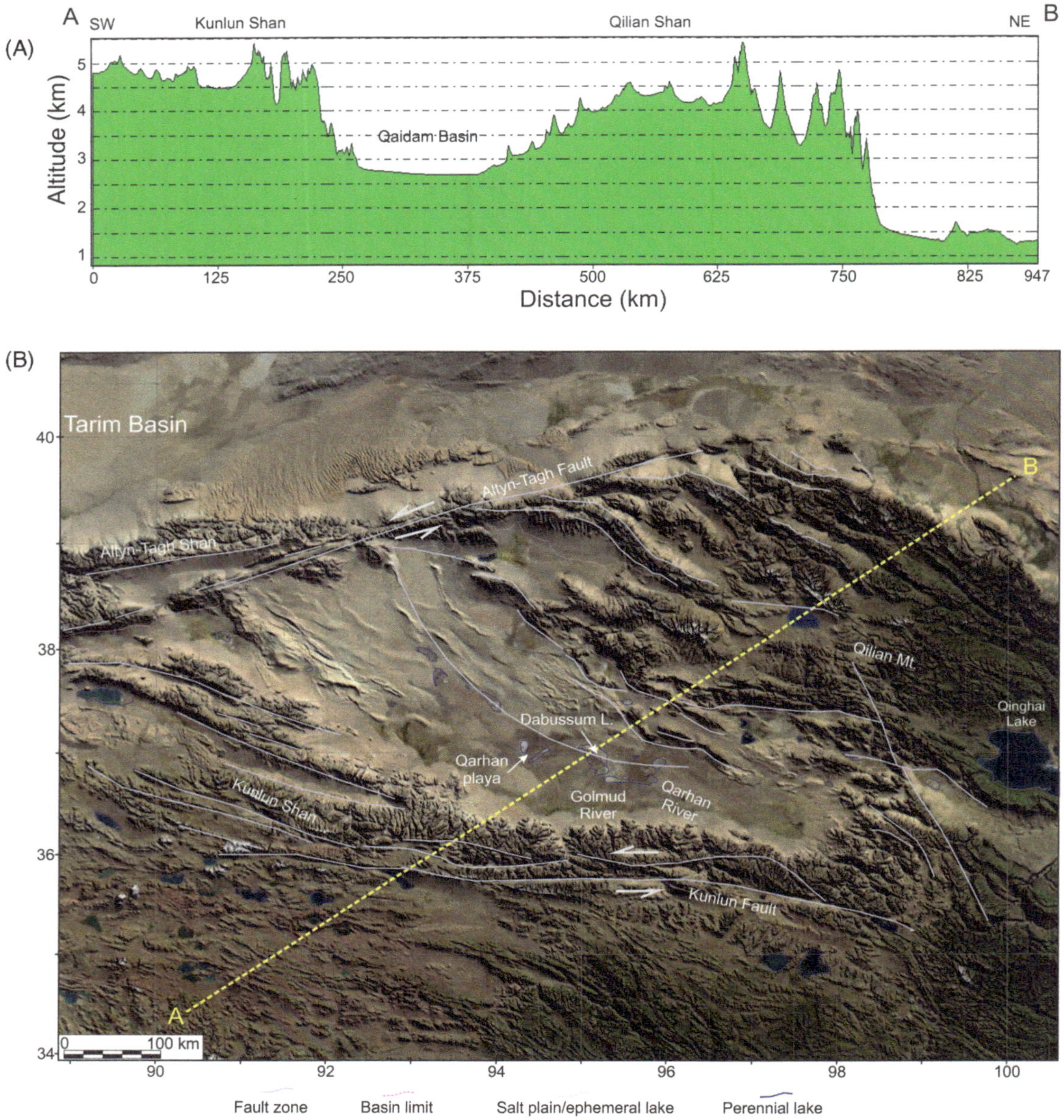

Figure 10.26 (A) Topographic profile A-B (SW-NE) from the highlands in the Kunlun Shan to the Qilian Shan mountain ranges. The Qaidam Basin is situated in a depression with an altitude of around 2.7 km a.s.l. The basin is surrounded by the Kunlun Shan highlands and Qilian Shan mountain ranges. Dabussun Lake is located at the lowest point of the Qarhan Depression, in the eastern part of the basin. (B) A satellite image of the Qaidam Basin shows several lakes in the Qarhan salt flats north of Kunlun Shan. The Altyn-Tagh mountain ranges, located north of the basin, are associated with a left-lateral strike-slip shear zone that separates the Qaidam from the Tarim basins. Qinghai Lake is located southeast of the Qilian Shan deformation zone. The lacustrine systems in the Qaidam Basin are located in the southern sector of the basin, forming the Qarhan salt flats. A topographic profile (Line A-B) is shown by a thin yellow line extending from the East Kunlun fault zone toward the Qilian mountain ranges. Credit: Photo by NASA/Public domain.

as 100,000 km^2, while the surrounding regions experienced an increasingly arid climate (Zhenhan et al., 2008). The onset of Pleistocene glaciations resulted in the shrinking of the permanent lakes, and several ephemeral lakes were almost desiccated. After the demise of the last glacial episode around 20 ka, meltwater filled the depressions with freshwater, forming large lakes, like Qinghai Lake (Zhenhan et al., 2008). During the Holocene, the hyperarid conditions in the Qaidam depression resulted in some remnant saline ponds in the deepest parts of the basin. Several lakes were dried out, forming large salt flats in the Qarhan Playa, from the western sector of the basin to the depression north of the Golmud River alluvial fan (Figure 10.26).

The Qaidam Basin is the largest and thickest intramontane sedimentary basin in the northern part of the Tibetan Plateau, corresponding to the deepest intracontinental basin in China (Cheng et al., 2018). The Qarhan salt flats are one of the largest playas in the world, extending from the Altyn-Tagh Mountains in the NW to the Kunlun Shan in the south (Figure 10.27). These mountain belts are related to sinistral strike-slip faults (Qinmin & Coward, 1990; Suotang et al., 2015). Quaternary sediments in the Qaidam Basin overlie a thick sedimentary basin with a total sedimentary thickness exceeding 15 km (Cheng et al., 2018; Pang et al., 2004; Qinmin & Coward, 1990). The deeper parts of the Qaidam Basin are characterized by remnants of Paleozoic to Mesozoic sequences (mainly Jurassic to Triassic strata) that are quite thin, with a maximum thickness of about 2 km. The Cenozoic sequences in the subsurface are extremely thick, reaching a maximum thickness of about 16 km in some depocenters (Kezao & Bowler, 1986; Pang et al., 2004; Pei et al., 2017). The Qarhan salt flats are characterized by evaporite crusts and remnant lakes covering a large expansion in the center of the depression (Figure 10.27).

The Qaidam depression is characterized by the presence of Neogene siliciclastic, carbonate, and mixed evaporite rocks overlying Cenozoic depocenters (e.g., Hartmann & Moosdorf, 2012). Metamorphic and igneous rocks are found at the southern edge of the depression, related to NW-trending uplifts forming the Kunlun mountain ranges. These basement rocks are related to NW-trending uplifts forming the Kunlun mountain ranges (Figure 10.26). Paleozoic and Precambrian rocks are also observed in outcrops along E-W transcurrent faults in the highlands at the northern edge of the basin. The Altyn-Tagh mountain ranges are associated with active left-lateral transcurrent fault zones (Figure 10.26). The northeastern border of the depression is marked by an NW-SE trending fault zone, which extends toward the region of Qinghai Lake (Figure 10.27).

The development of the Qaidam Basin in the Cenozoic has been controlled by NE-SW compression, derived from the uplift of the Tibetan Plateau as India approached Asia (Figure 10.3). The Indian Plate began to impact ancient Tibetan continental margins in the Paleogene, resulting in the left-lateral transcurrent fault in the Altyn-Tagh mountains and a major strike-slip dextral fault system near the Qaidam River in the center of the basin, near the Qarhan salt plains (Spencer et al., 1990).

In the Neogene, the Qaidam Basin was completely separated from the ancient Tethys Ocean and became a typical intermontane basin due to the fast uplift of the Qinghai-Tibet Plateau. This led to climatic effects that drastically affected the environment, leading to vegetation changes from forest steppes to desert steppes and sedimentary environments changing from fluvial-lacustrine to desert settings. During the Eocene, some of the lakes started to dry up, precipitating bedded halite layers (Guo, P. et al., 2017).

The Neogene sediments are mainly affected by compressional tectonics at the margins of the depression, with the vergence of major thrust faults directed toward the basin center (Cooper, 2007; Guo, J. et al., 2017). A schematic cross-section in the Qaidam Basin (Figure 10.27A) illustrates the sedimentary successions and the main structures. The Mesozoic rift sequences are observed in the depocenter in the northeastern sector, whereas thick Neogene sequences are widespread, covering the basin center (Figure 10.27A). The structural framework of the depression is characterized by a concentration of reverse and strike-slip faults at the periphery of the basin, but some transcurrent and reverse faults also affect Meso-Cenozoic sediments in the center of the depression (Cooper, 2007; Pei et al., 2017; Sun et al., 2020; Wang et al., 2012). In the basin depocenter, the Oligocene to Triassic sedimentary successions are less affected by compressional tectonics (Figure 10.27).

The Neogene lacustrine deposits in the Qaidam Basin are very thick (up to about 6 km), with Pliocene siliciclastic sequences reaching more than 4 km in thickness (Kezao & Bowler, 1986). A scientific borehole in the western Qaidam Basin recorded several levels of evaporite deposits intercalated with Pliocene to Quaternary muds, silts, and marls (Fang et al., 2016; Stober et al., 2023). The ages of the lacustrine sequences range from circa 4 Ma (Pliocene) to 5–25 ka (Holocene), indicating different episodes of salt deposition in diverse segments of the depression (Fang et al., 2016; Kezao & Bowler, 1986; Stober et al., 2023).

Guo et al. (2018) analyzed the carbonate and chloride contents of brine recovered from several boreholes in the basin, revealing that during the early Miocene, a large palaeolake was established in the Qaidam Basin, which migrated southeastward from the Late Miocene to the Pliocene. Increased evaporation in the Pleistocene was associated with global and regional climate changes, leading to the precipitation of evaporite sediments (Guo et al., 2018). During the Quaternary, the brine reached a chloride content exceeding 100,000 mg/kg, resulting in the precipitation of gypsum and some halite in hypersaline depocenters (Guo et al., 2018).

The Qarhan salt flats are remnants of a larger salt lake that existed during the Pleistocene (~2 Ma) and dried up in the Upper Pleistocene to Holocene (<50 ka), forming a large number of saline pans throughout the Qaidam Basin (Figure 10.26). The paleolake, which covered at least 25,000 km^2, had a surface of 50–60 m above the present-day lake level and alternated periods of fresh and saline waters. Around 30–25 ka, the lake dried up,

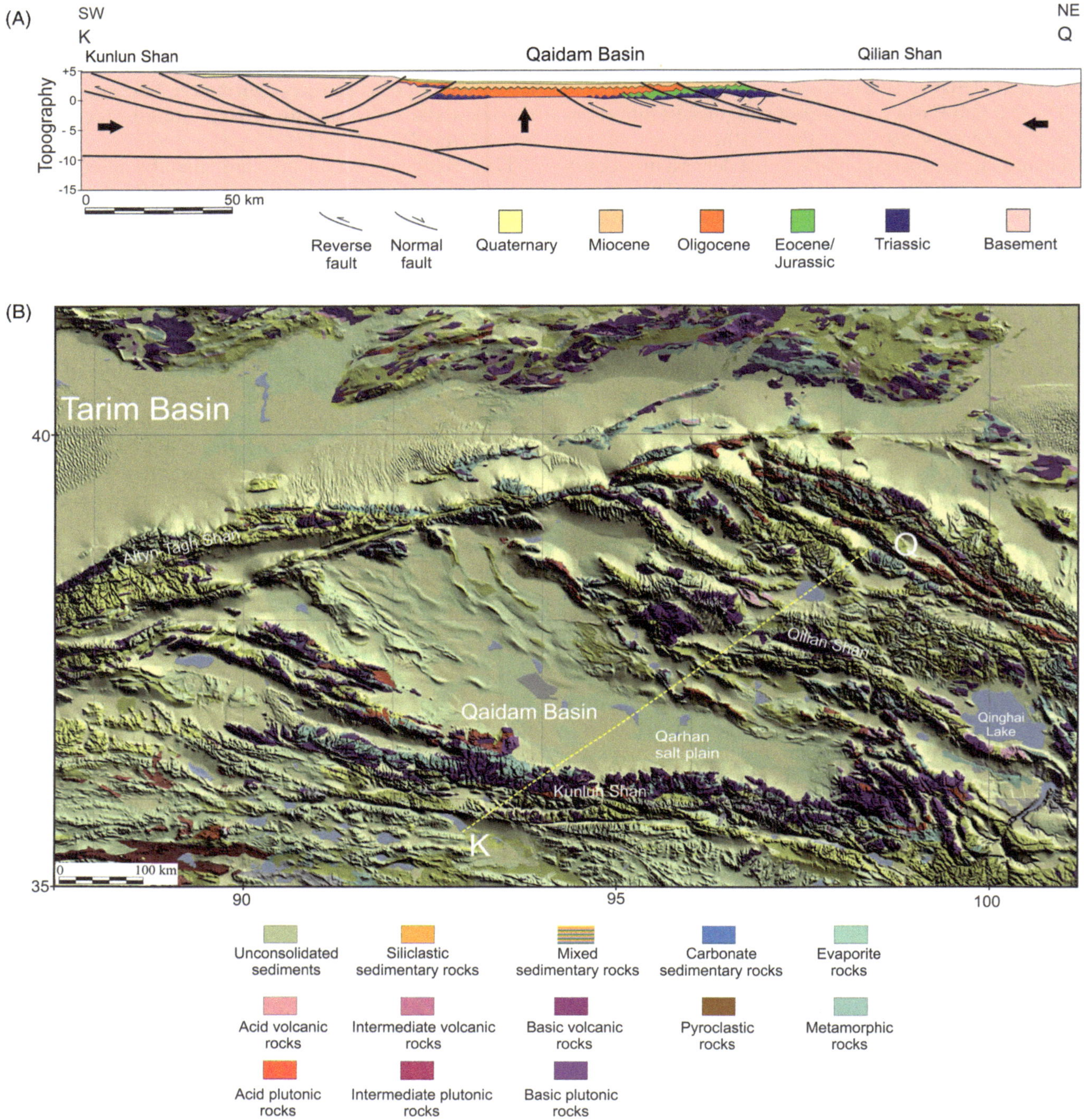

Figure 10.27 (A) Schematic geological cross-section in the Qaidam Basin showing the inversion of the Mesozoic basin by compressional tectonics. Large thrust faults occur at the basin's borders, exhibiting a vergence directed toward the basin center. In the central depression, the Neogene sedimentary sequences are made up of siliciclastic lacustrine sediments and some Quaternary evaporite deposits. (B) Map showing the main geological units draped on the topographic map of the eastern Tarim and Qaidam basins. The Kunlun mountain ranges are characterized by NW-SE trending faults affecting volcanic rocks that outcrop to the south of the Qaidam Basin. The transcurrent fault movements at the northern and southern edges of the basin are linked to compressional structures associated with plate convergence, such as folds, thrusts, and reverse faults. The Altyn-Tagh is a left-lateral strike-slip fault that forms a shear zone separating the Tarim and Qaidam basins. The Qarhan salt plain is located in the central to eastern part of the depression. Lake Qinghai, located in an NW-SE trending fold belt eastward of the Qaidam Basin, is surrounded by highlands associated with uplifted basement rocks. Credits: (A) Adapted from Cooper (2007) and (B) Hartmann and Moosdorf (2012)/with permission of John Wiley & Sons.

forming isolated salt pans, which were cut off from sufficient fresh water in the Qarhan salt plains, resulting in several evaporite cycles (Kezao & Bowler, 1986).

The salt-bearing sequences in the subsurface may reach 1,000 m in total thickness, with evaporite sequences (sulfate and chloride) adding to a total thickness of around 200 m, mainly corresponding to halite (Kezao & Bowler, 1986). The uppermost subsurface salt layers formed principally in an environment of dry and cold climate, corresponding to the late Wurm glaciation, lasting from 115 ka to circa 12 ka. The widespread evaporite precipitation episode in the Qarhan salt flat (~ 25 ka) coincided with the demise of the glaciation (Zheng et al., 2000).

The evolution of lake salinity in the Qaidam Basin was controlled by climatic factors, including humid climates between 40 and 25 ka, drier climates around 25 ka, drying up between 9 and 15 ka, and the formation of new saline lakes over ancient playa flats in the last 9 ka (Kezao & Bowler, 1986), resulting in the modern Qarhan salt flats with the associated lacustrine systems (Figure 10.28).

10.5.3. Geomorphology of the Qaidam Basin

The Qaidam Basin is an endorheic basin within a depression fed by interior drainage, characterized by a cold, arid environment with a mean annual precipitation of 100 mm and a mean annual potential evapotranspiration of 1,528 mm, typical of the elevated plateaus across Asia. Most rivers discharge into shallow salt lakes and saline swamps that are distributed around

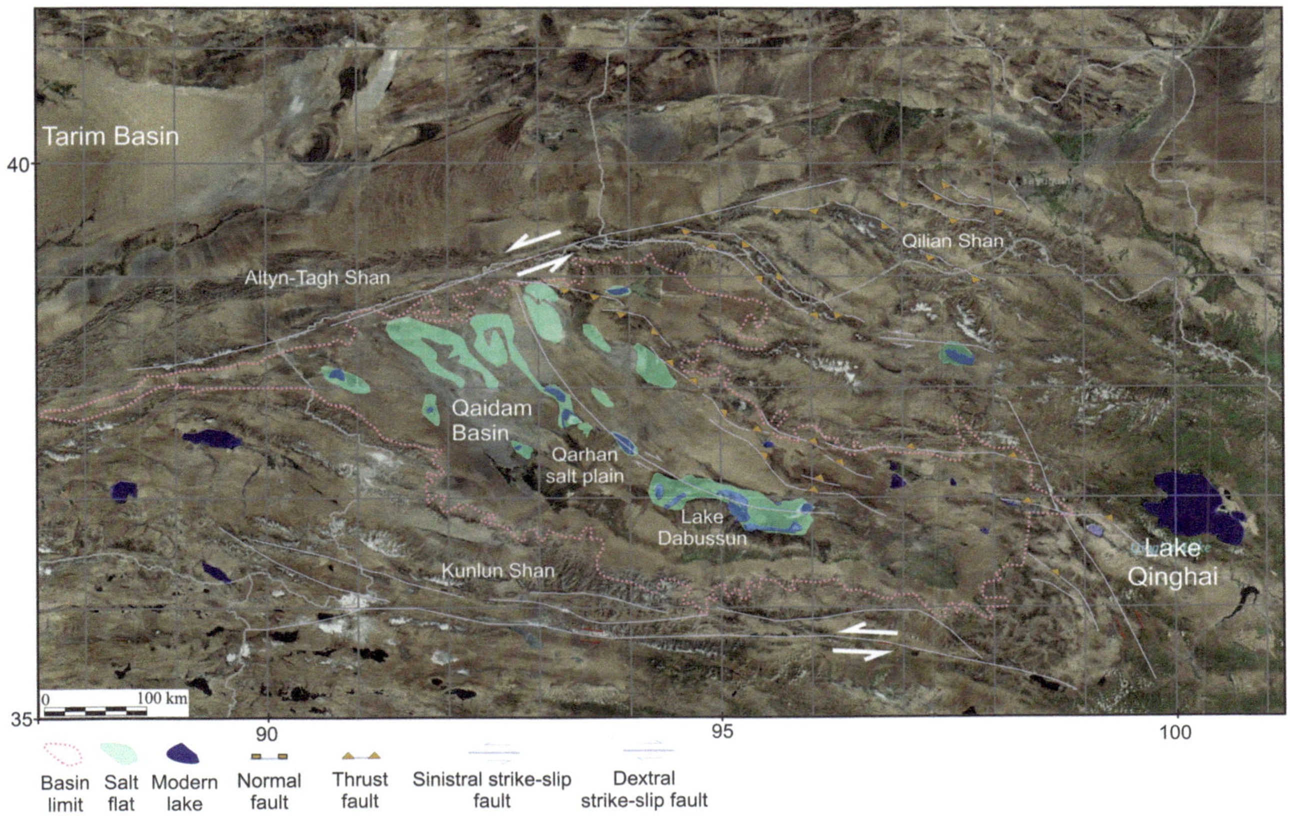

Figure 10.28 The satellite image (true colors) depicts the Qaidam Basin depression and its surrounding highlands, which are characterized by strike-slip, thrust, and reverse faults. Vegetated areas are greenish, while ephemeral lakes and desiccated salt flats are light blue and white. Deeper lakes, like Lake Qinghai, are dark blue. Light gray lines show the basin limits, while brick red lines depict the most important structures. The E-W trending fault zone in the Kunlun Mountains marks the southern border, while the northern edge is marked by SW-NE strike-slip fault zones with left-lateral motion. The NW sector of the Qaidam Basin is characterized by desertic salt plains, while the eastern sector, located north of the Golmud delta, is a depression containing salt marshes, salt pans, and hypersaline lakes. North of Dabussun Lake, a dextral strike-slip fault with an NW trend is mapped in the subsurface. Hydrothermal waters developed a karst topography and formed several springs in this region. Credit: Photo by NASA/Public domain.

the basin's central area. The Golmud River is part of the drainage system, forming a deltaic plain north of the eastern Kunlun Shan. The Qarhan River and other rivers feed the lacustrine system in the depression. There are over forty shallow perennial and ephemeral saline lakes and playa lakes in the Qaidam Basin. These lakes are associated with shallow groundwater brines and evaporite layers in the subsurface. In addition, several dry saline pans are located in the western part of the basin and the Qarhan salt flats, north of the Golmud Delta (Figure 10.26).

The two largest lakes in the Qarhan salt flats are Dabussun and Taijinier, which represent the contracted remnants of an ancient, much larger lacustrine system that stretched across the depression during the Pliocene (Kezao & Bowler, 1986). The desiccating Qarhan Salt Marsh is one of the largest inland saline pans in the world, covering an area of 6,000 km^2 and stretching 160 km from east to west and 20-40 km from south to north (Fang et al., 2016; Herb et al., 2015; Spencer et al., 1990; Wang et al., 2017). The playa lakes are subjected to intense evaporation, resulting in ephemeral hypersaline lakes with subaerial salt crusts on the margins. Concentrated brine precipitated halite, potash, and borate minerals at the bottom of the playa lakes. The evaporite deposits are mixed with Pliocene to Recent siliciclastic sedimentary sequences (Chen & Bowler, 1986; Lowenstein et al., 1989).

Dabussun Lake, located in the southeastern part of the Qaidam Basin, is the largest modern salt lake in the Qarhan salt flats (Figures 10.29). Its surface area varies from 184 to 334 km^2, with a water surface at an altitude of around 2,675 m. Dabussun is a rather shallow lake, with water depths of only a few meters (Casas et al., 1992; Schubel & Lowenstein, 1997). This lake is a famous tourist attraction due to its greenish water and extremely concentrated brine (> 300,000 ppm NaCl), which exceeds the halite solubility. Most of the salt flats in the depression correspond to a dry saline pan that forms a belt of evaporite crust north of the remnant water body. Spectacular masses of salt crystals are clustered along the lake margins. The salt deposits are exploited commercially by saltworks near Qarhan and used by chemical and agricultural industries across China.

Dabussun Lake is notable for its high level of lithium salts, which appear mixed with halite and more soluble salts of potassium and magnesium (Lowenstein et al., 1989). During the Pleistocene, the hypersaline waters evaporated and desiccated the shallow ephemeral lake, precipitating many evaporite minerals and rare metals, including lithium (Kezao & Bowler, 1986). Dabussun Lake holds special a interest in evaporite studies due to its

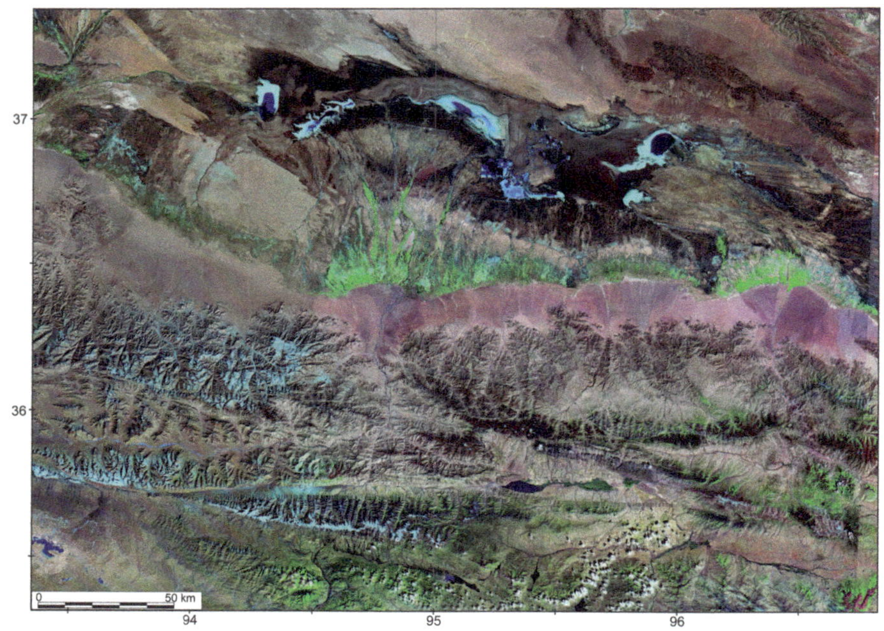

Figure 10.29 The NASA Landsat satellite image (false colors) depicts the Qaidam Basin depression, the Qarhan salt plain, and the modern fluvial-lacustrine systems. The surface is mostly covered by siliciclastic rocks (loess), but salt flats with evaporites (gypsum and halite) correspond to salt pans. Dabussun Lake is located in the center of the Qarhan salt flat, north of the Golmud River delta plain. Vegetated areas in the delta plain are greenish, desiccated salt flats are light blue, and deep lakes exhibit a dark blue reflectivity. Credit: Courtesy of NASA.

unique high-altitude lacustrine environment formed in a tectonic setting of plate convergence. It is one of the few natural lacustrine environments in the world that record the natural precipitation of significant quantities of potash salts (Lowenstein et al., 1989).

10.5.4. Salt Deposition in the Qaidam Salt Basin

The Qaidam Basin is a nonmarine basin filled with more than 16 km of Jurassic and younger sediments. It includes the Qarhan salt flats, which extend outside present salt marshes and contain several remnant lakes (Figure 10.29). Dabussun Lake is the largest and most active saltern with modern evaporite deposition in the basin. The present-day depositional setting of the Qarhan salt plain includes surface playas with hypersaline lakes with extremely concentrated, high-density brine (over 1.3 g/c). The brine in the subsurface is highly concentrated, with a total dissolved solids (TDS) generally exceeding 300 g/L (e.g., Stober et al., 2023).

There is evidence of multiple salt precipitation episodes in the Qaidam Basin. In the Shizigou area, located south of the Altyn-Tagh mountain ranges, bedded halite layers occur intercalated with lacustrine carbonate mud (Guo, P. et al., 2017). In the Late Eocene, the Shizigou depocenter was isolated from the larger, brackish Qaidam paleolake, resulting in evaporation and brine concentration during several periods of basin isolation (Guo, P. et al., 2017). During humid climate stages, the saltern was intermittently connected with the main paleolake, resulting in the alternation of deepwater carbonate muds and evaporite sequences. The nonmarine evaporite deposits are associated with an anoxic environment that accumulated Eocene source rocks. The evaporite cycles resulted in the alternation of source rocks with thick-bedded halite layers (Qiang et al., 2002).

The large hypersaline lakes of the Qarhan salt flats record evaporite layers that precipitated in the Neogene and Quaternary (Lowenstein et al., 1989). In the Holocene, the remnant lake is characterized by K-rich brines, including shallow brines at the lake bottom and intercrystalline brines in the upper sedimentary layers. Boreholes have revealed deep-seated sedimentary layers with pores filled with highly concentrated brine in the subsurface (Zhang et al., 2019).

The dense brines are enriched with calcium and chlorine ions sourced from hydrothermal springs and seeps that flow into the basin along deep-seated faults located in the northern karstified region of the Qarhan salt flat (Lowenstein & Risacher, 2009). The hyperconcentrated brines precipitate most of the sodium chloride originally dissolved in the lake waters, forming a residue that corresponds to a magnesium chloride solution derived from carnallite ($MgKCl.6H_2O$). The brines descend in a plume, precipitating carnallites on top of the evaporite layers that accumulate on the lake's bottom (Spencer et al., 1990). Fieldwork and petrographic analysis of the carnallite layers in the Qarhan salt flat suggest that they represent modern analogs of ancient salt deposits. The Qarhan evaporites have been interpreted as modern analogs of the Permian Salado Fm. in New Mexico and the Devonian Prairie Fm. in Saskatchewan (Casas et al., 1992).

Qarhan salt flats contain the largest amount of nonmarine evaporite deposits in China. During the Pliocene and Quaternary, hypersaline lacustrine environments precipitated evaporite layers dominated by halite and potash-bearing minerals. At present, extensive salt deposits (dominated by halite and potash-bearing minerals, mainly sylvite (KCl) and carnallite ($KMgCl_3 \cdot 6H_2O$) are exploited by the solar-driven concentration of brine water in saltworks around the lake margins. A large potash extraction industry has been established near the city of Golmud, at the southern border of Lake Dabussun (Figure 10.29). The Qaidam Basin is an important source of lithium produced from the formation waters of deep subsurface petroleum reservoir rocks, but some lakes also produce lithium from brine (e.g., Ma et al., 2021; Yu et al., 2013).

10.5.5. Geochemistry and the Origin of Brines in the Qaidam Basin

The Qaidam Basin hosts significant quantities of potash salts in the large lakes of the Qarhan salt flat, located north of the Golmud delta plain (Casas et al., 1992; Schubel & Lowenstein, 1997; Spencer et al., 1990). The lacustrine systems occupy a depression extending along an E-W direction, forming several lakes, including Dabussun Lake (Figure 10.29). Dabussun Lake provides a unique case history to understand the origin of brines responsible for precipitating soluble evaporites in continental lacustrine systems in compressional settings. This modern analog may apply to other ancient nonmarine $MgSO_4$-deficient salt basins related to the convergence of continental blocks and the development of high plateaus with closed lacustrine systems influenced by hydrothermal fluids (Lowenstein et al., 1989).

Despite extensive study, there are still points of dissension among researchers. The origin of Qarhan brines has been attributed to several alternative hypotheses, including eastward migration of mega-paleolakes, dissolution of Neogene halite layers, chemical weathering of igneous rocks, paleo-river capture of underground formation waters, and the mixing of modern river water and CaCl water from hydrothermal springs (e.g., Li et al., 2020; Stober et al., 2023). The Holocene evaporites deposited in the Qaidam Basin demonstrate that extensive nonmarine potash deposits may exist in continental settings if the brine is associated with two end-member sources, one related to hydrothermal springs and the other to fluvial waters that infiltrate the subsurface. These brines are mixed, reach the lakes via deep faults, and provide the necessary ions to precipitate the soluble potash salts (Lowenstein & Risacher, 2009; Lowenstein et al., 1989).

The Qarhan salt flats contain many evaporite minerals, such as borates, sulfates, and chlorides, which occur as layered halite-carnallite sequences along the shores of the perennial saline lakes. These deposits have been drilled by exploratory boreholes that penetrated the Quaternary to Paleogene sedimentary sequences, including Late Eocene bedded halite intercalated with lacustrine carbonate mud (Guo, P. et al., 2017).

The meteoric waters in the Qaidam Basin are associated with several perennial rivers that flow onto the Qarhan salt plain. The larger rivers, such as the Qaidam River in the north and the Golmud River in the south, form an extensive delta plain north of the Kunlun Mountains (Casas et al., 1992; Spencer et al., 1990). The Golmud River is the main source of water for Dabussun Lake, the largest hypersaline lake in the depression. Several ephemeral playa lakes occur along the dry stream beds of small rivers running along the Qarhan salt plain during wet seasons (Figure 10.28). These lakes and their salt pans are fed by these ephemeral streams or by meteoric groundwater, which flows a few meters below the stream channels. The TDS in the Qarhan fluvial waters (> 400 ppm) is four times larger than the average TDS content (105 ppm) of average world river water (Spencer et al., 1990). Recent measurements by Zhang et al. (2019) calculated the TDS of intercrystalline brines in the Qaidam depression sediments and obtained extremely high values, averaging 359.4 g/L, which is about 10 times higher than the TDS in normal seawater (35.0 g/L). Although these fluvial waters are enriched in Na^+ and HCO_3^-, the chemical composition is rather similar to that of the average world river waters, which are characterized by $HCO_3^- > Cl^-, SO_4^{2-}$ (see Chapter 4, Figure 4.5).

The high salinity values in brine from the Qarhan salt flats results from the dissolution of ephemeral evaporites and brine concentration by solar evaporation in the extremely arid environment. Compared to the global river water contents, fluvial waters entering the Qarhan salt plain display a low content of HCO_3^- and Ca^{2+}. However, the K^+, Ca^{2+}, Mg^{2+}, and Cl- contents of hydrothermal fluids in springs and seeps are very high (Figure 10.30). This is likely due to the precipitation of $CaCO_3$, which forms carbonate minerals at the base of the evaporite layers. However, the range of K^+ concentrations measured in rivers in the Qaidam depression is 2.7–122.2 mg/L (average 19.5 mg/L). This value is much higher than the world's average river water, which is generally around 3.9 mg/L (Zhang et al., 2019). The origin of the K^+ ions in the fluvial waters is related to the high percentage of K-bearing minerals in the granitic rocks that outcrop in the Kunlun and Altyn mountains, with an average weight percentage that can be as high as 3.5% (Zhang et al., 2019). These granitic rocks have been upthrusted during the Cenozoic, leading to the weathering and leaching of silicate minerals and abnormal enrichment of the fluvial waters that feed the lacustrine systems (Figure 10.31).

The amount of water that is brought into the lakes by the groundwater flow is much less than the river inflow, which is geochemically characterized by a typically fluvial water composition (Na^+ - HCO_3^- - SO_4^{2-}). Mixing of these waters results in a Na^+ - Mg^{+2} - K^+ - Cl^- - SO_4^{2-} water. Due to evaporation, brines in modern lakes far from oceans (such as Dabussun Lake) are enriched in ions that precipitate several evaporite minerals typical of nonmarine sources. The Qarhan salt flats are characterized by a distinct distribution of evaporite deposits according to chemical composition. Sulfates (mainly gypsum) occur in the western part of the depression, and chlorides and potash-bearing minerals in the eastern part. The more soluble potash-bearing minerals, such as carnallite, bischofite, and tachyhydrite, precipitate in the eastern part of the basin (Figure 10.31).

Figure 10.31 shows a schematic diagram illustrating the migration of evaporite facies from the Western to the Middle and Eastern Qaidam basins, as proposed by Zhang et al. (2019). The Chahansilatu playa in the west part of the basin is characterized by a salt flat mainly composed of sulfate minerals, with the mean surface at an altitude of 2,710 m (Figure 10.31). The Yiliping and Taijinaier playas are transitional-type salt flats, with

Ions	Na^+	Cl^-	Mg^{2+}	K^+	Ca^{2+}	SO_4^{2-}	HCO_3^-
Rivers	0.00232	0.00158	0.00102	0.0010	0.00107	0.00102	0.00358
Springs	2.786	5.394	0.944	0.309	0.231	0.0005	0.0014
Brine	0.883	7.471	3.257	0.462	0.022	0.060	0.041

Figure 10.30 Ion content (values in molalities) of the inflow waters to the Qarhan salt plain from fluvial, springs, and lake brine.

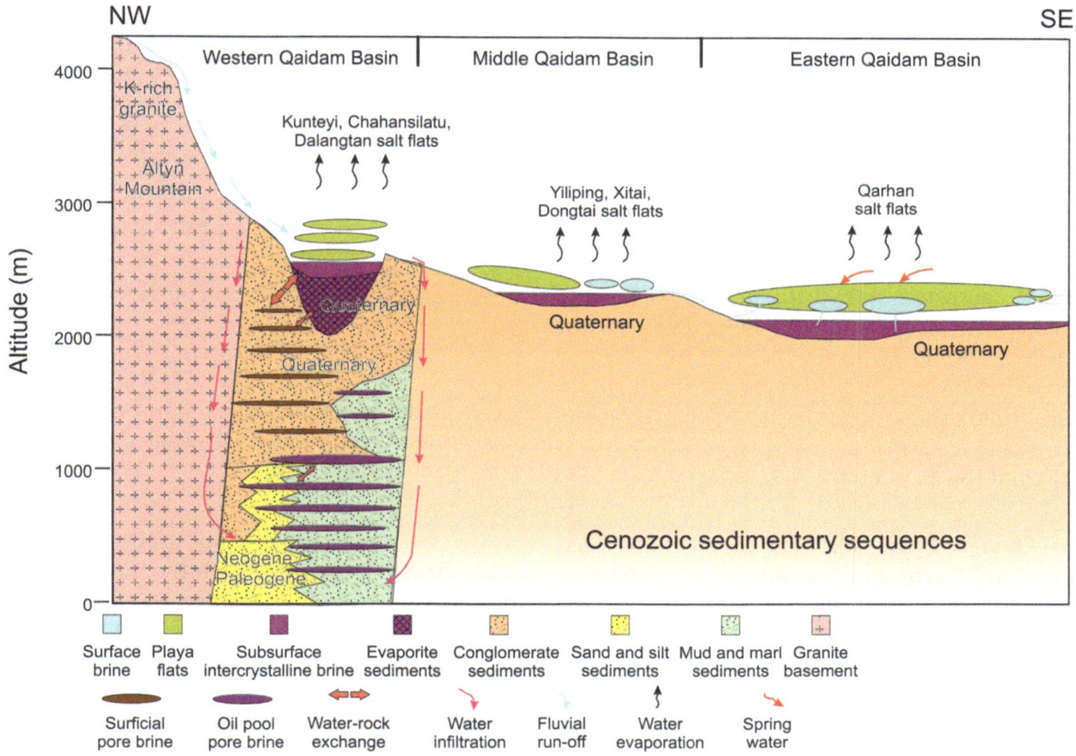

Figure 10.31 Distribution and origin of different types of K-rich brine in the Qaidam Basin. The granitic basement rocks outcrop in the elevated mountain ranges. Weathering and leaching of the K-rich minerals supply K^+ ions to fluvial waters, which in turn feed the lacustrine systems. The Eastern Qaidam Basin is characterized by a playa salt flat with a large distribution of subsurface intercrystalline brines enriched by spring waters. The western part of the basin is associated with the $ClSO_4$ field, with a large proportion of sulfate minerals. The eastern part is characterized by ions in the CaCl field, with a large proportion of chloride minerals in the Qarhan salt flats. Credit: Zhang et al. (2019)/MDPI/CC BY 4.0.

sulfates and chloride minerals precipitating at an altitude of 2,683 m. The Qarhan salt flat is characterized by a chloride salt flat surrounding Dabussun Lake, with a water level presently at an altitude of 2,675 m, one of the lowest in the Qaidam Basin (Kezao & Bolwer, 1986).

The migration of stratigraphic sequences from the northwestern parts of the basin to the Qarhan depression in the southeast is influenced by variable compressional stresses during the Neogene (Suotang et al., 2016). This tectonosedimentary interpretation is based on geomorphic evidence of migration of paleoshorelines and mapping of basin depocenters. The evaporite deposits are characterized by a younging of the depocenters, from the western to the eastern depressions (Kezao & Bowler, 1986; Cheng et al., 2018).

Compared to normal seawater, the Qarhan salt flats exhibit a marked $CaCl_2$ enrichment in concentrated brines that deposit highly soluble minerals (e.g., Möller et al., 2007; Spencer, 1987; Spencer et al., 1990). This puzzle has been investigated by several researchers (e.g., Lowenstein & Risacher, 2009; Stober et al., 2023). Three main hypotheses for CaCl brines in the tectonically active closed basins have been proposed: (1) deep circulating hydrothermal fluids formed by magmatic activity; (2) hydrothermal springs originated in continental rifts and strike-slip systems; and (3) formation waters from deep reservoirs within sedimentary basins, including basinal brines and oil-field brines (Lowenstein & Risacher, 2009).

Groundwater brines in the Qarhan salt flats have a distinct chemical composition compared to normal fluvial or seawater. These brines are rich in $CaCl_2$ ions, which probably reached the surficial layers and formed a series of springs and seeps (e.g., Lowenstein et al., 1989). Although Ca^{2+} precipitates as carbonate minerals due to its low solubility, the brines in the Qarhan salt flats have an abnormally high concentration of calcium ions compared to river water (Figure 10.30). The enrichment in Ca^{2+}- Cl^- ions results in brine with a geochemical fingerprint similar to hydrothermal fluids in seeps, formation waters from oil field brine, and formation waters from deep reservoirs (Spencer et al., 1990). Some authors (e.g., Casas et al., 1992; Lowenstein & Risacher, 2009; Spencer et al., 1990) suggest that the

$CaCl_2$ -enriched waters in the Qaidam Basin are associated with deep-seated dextral strike-slip faults at the northern edge of Dabussun Lake, which is marked by a well-developed karst zone (Figure 10.28). These hydrothermal fluids mixed with the brine that filled paleolakes formed on the Qarhan depression during the Quaternary, as indicated by the analysis of selected water samples from rivers, springs, and lake brine (Figure 10.30).

The sequential precipitation of the evaporitic minerals includes calcite, gypsum, anhydrite, halite, polyhalite ($K_2Ca_2Mg(SO_4)_4{\cdot}2H_2O$), epsomite ($MgSO_4{\cdot}$ $7H_2O$), hexahydrite ($Mg(SO_4){\cdot}6H_2O$), kieserite ($Mg(SO_4){\cdot}H_2O$), carnallite, and bischofite at the upper part of the cycles (Spencer et al., 1990). Other lakes in the Qarhan salt flats do not follow the seawater evaporation sequence because they lack $MgSO_4$ salts (polyhalite, epsomite, hexahydrite, and kieserite).

The eastern Qaidam Basin is home to depressions filled with Paleogene to Quaternary sedimentary sequences (Figure 10.31). The oil field brines in reservoir rocks are enriched in K^+ and Li^+ (Li et al., 2021; Qiang et al., 2002). The shallower evaporite sequences in Quaternary lakes have K-rich layers forming continuous beds or K-rich minerals disseminated in the halitic mud matrix. Intercrystalline pore fluids in subsurface reservoirs also contain a high content of K^+ (Spencer et al., 1990). The Qarhan salt flats in the eastern sector of the Qaidam Basin display surface brines with high TDS content, mainly fed by fluvial systems and incorporating ions from hydrothermal springs and dissolution of previous evaporite layers. High evaporation rates result in brine concentration and precipitation of soluble evaporite minerals on the lake's bottom and playas. Contrasting with other large potassium-rich salt basins, the hypersaline lakes in the Qaidam Basin formed brine without any influence of marine seawater.

10.5.6. Strontium Isotopic Analysis and Sources the Qaidam Basin Brines

Fan et al. (2018) studied the geochemical factors that influenced the formation of evaporite minerals in the Qaidam Basin, including hydrothermal, diagenetic, and a mix of both processes. The western playas have intercrystalline brines enriched in $SO_4^{\,2-}$ and are classified as $CaSO_4$ brines, contrasting with the Qarhan depression at the eastern part of the basin (Figure 10.31), where brines are CaCl type (Fan et al., 2018; Lowenstein & Risacher, 2009).

Based on Sr isotope ratios of river waters, intercrystalline brines, spring waters, and minerals such as halite, gypsum, and carbonate shells from Qarhan playa lakes, researchers suggest different sources for brine origin in the Qaidam Basin (e.g., Fan et al., 2018; Stober et al., 2023). Fluvial waters from the Qarhan River and Golmud River delta plain have the lowest $^{87}Sr/^{86}Sr$ values in the basin, ranging from 0.71060 to 0.71110 (Fan et al., 2018), contrasting with the more radiogenic values for CaCl spring inflow waters. Normal seawater is characterized by an $^{87}Sr/^{86}Sr$ value of 0.70920, whereas the spring waters yielded values of 0.71713, 0.71711, and 0.71158 (Figure 10.32).

Source	$^{87}Sr/^{86}Sr$
Seawater	0.70920
Golmud River	0.71060 to 0.71110
Surface intercrystalline brine	0.71119 to 0.71196
Oil field formation water	0.71121 to 0.71220
Spring water	0.71713, 0.71711, 0.71158
Halite and gypsum layers	0.71112 to 0.71216

Figure 10.32 Strontium isotope measurements in the Qaidam Basin waters compared with seawater. Credit: Fan et al. (2018)/ with permission of Elsevier.

The Golmud River waters have higher $^{87}Sr/^{86}Sr$ values than seawater but lower than spring waters. The strontium isotope ratios in river waters are similar to moraine deposits from weathered bedrock in the Kunlun Mountains, where basement rocks outcrop in the highlands south of the basin (Figure 10.27). These similar isotope ratios indicate an origin associated with glacial meltwater, which probably leached the potassium feldspar in granitic rocks (Figure 10.31).

Halite minerals in the subsurface were cored by some boreholes for isotope measurements, yielding $^{87}Sr/^{86}Sr$ values intermediate between fluvial and spring water, indicating a mixture of the two. The upper and lower salt layers have Sr isotope ratios ranging from 0.71112 to 0.71216. Formation waters in subsurface reservoirs of oil fields overlap with these values, ranging from 0.71121 to 0.71220 (Figure 10.32). Fan et al. (2018) noted an increase in $^{87}Sr/^{86}Sr$ values from west to east in CaCl inflow waters entering QSL. The same pattern is observed in the $^{87}Sr/^{86}Sr$ values measured in halite and formation waters of subsurface reservoirs, indicating that the present-day processes have been in operation for the past 50 ka.

10.5.7. The Qaidam Basin as a Modern Analog for Ancient Potash Deposits

The Qaidam salt flats are a prime example of salt deposition in nonmarine continental settings, far from marine water sources (Li et al., 2020; Lowenstein et al., 1989; Smoot & Lowenstein, 1991). The brine concentration leads to the precipitation of halite and potash salt minerals through the evaporation of brines sourced from springs and seeps developed along thrust faults in the northern part of the depression (Casas et al., 1992; Guo, J. et al., 2017; Spencer et al., 1990).

Although the brines are nonmarine, the Qaidam salt deposits may serve as analogs for ancient salt basins with marine-influenced brines, such as the Sergipe Basin in Brazil, the Zechstein Basin in Europe, the Permian Basin in the southwest United States, the Green River Basin in the central United States, and the Alberta Basin in Canada (Smoot & Lowenstein, 1991; Spencer et al., 1990). Several of these basins have recorded an SO_4 deficiency in the evaporite mineral associations (Stober et al., 2023; Warren, 2006, 2016). Some potassium-bearing minerals are economically exploited in some salt basins, particularly sylvite and polyhalite (see Chapter 4). However, most of the giant K-rich salt deposits are associated with marine settings, such as the Devonian Prairie evaporites in Canada (see Chapter 8 in Mohriak, 2025) and the late Permian Zechstein evaporites in northern central Europe (see Chapter 2 in Mohriak, 2025).

Most giant K-rich salt deposits in the world are associated with marine settings. Potash salts precipitated in ancient basins, such as the Oligocene Rhine Graben in Europe, the Permian Salado Fm. in the southwestern United States, and the Devonian Prairie evaporites in northern North America, were associated with syndepositional brines that were compositionally distinct from modern seawater (Lowenstein & Spencer, 1990). The sulfate deficiency observed in the Qaidam Basin and other ancient salt basins has been attributed to several factors, including biological reduction by bacteria feeding on sulfate minerals and modification of the syndepositional, parent marine brine by nonmarine waters rich in calcium chloride from hydrothermal springs (Lowenstein & Risacher, 2009; Lowenstein et al., 1989; Sonnenfeld, 1989, 1992; Stober et al., 2023; Warren, 2006, 2016).

10.6. EVAPORITES IN THE WESTERN CANADA CONTINENTAL LAKES

The Great Plains of western Canada and the northern United States are sedimentary provinces with diverse hydrological systems and a wide range of water chemistry in the modern brine. The lacustrine system in the Western Canada Sedimentary Basin is associated with the foreland of the Cordilleran Orogen (Figure 10.1), while the substratum of these lakes corresponds to sediments derived from the Pleistocene continental glaciation (Fenton et al., 1994). Most sediments are of glacial origin, covering about 70% of the area, while continental lacustrine environments occupy about 20% of the surface (Fenton et al., 1994).

Last (1989) observed that the northern Great Plains contain more than 3 million lakes, but most of these water bodies are brackish (1,000–5,000 mg/L) or slightly saline (> 5,000 mg/L). The water chemistry of the more saline lakes is dominated by the ions Na^+ and SO_4^{2-}. About 99% of the saline lakes are small (less than 100 km^2) and very shallow (less than 3 m deep). The lake level in these lakes is mainly controlled by seasonal variations of temperature and precipitation, and most lakes are intermittently filled with water derived from glacial melt. These ephemeral playa lakes are filled with water during the spring, resulting from snowmelt and rain, but dry up during the late summer, exposing the lake bottom and accumulating dry white salts that look like snow, forming a thin crust (Last & Ginn, 2005).

The basic facies patterns commonly observed in the modern playa lakes of western Canada include the outer shoreline/nearshore complex, comprised of colluvium, mudflat/sand flat, and beach facies. The sediments include detrital material derived from the adjacent glacial deposits, grading basinward into a salt pan complex where evaporite minerals are precipitated depending on the chemical composition of the lake water. The mudflats are commonly colonized by algae and cyanobacterial mats, which change basinward into deepwater basinal facies composed of carbonate and siliciclastic muds (Last & Ginn, 2005).

The organic-rich sediments are laminated and may host biogenic carbonate buildups. Geothermal springs affect the saline lakes by creating the platform and pinnacle accumulations of salt minerals. These structures are mainly formed in winter, when the colder temperature causes rapid and massive carbonate and salt precipitation, particularly the dissolved Mg and Na sulfates (Last & Ginn, 2005). Some saline spring brines exhibit a chemical composition dominated by Na and Cl ions, markedly different from the majority of the Great Plains hypersaline lakes, which are dominated by sulfate-rich brine. This anomalous Cl-rich brine probably reflects the dissolution of the deeply buried Devonian halite deposits by groundwater, as suggested by the pinch-out of the evaporite layers in the subsurface (see Chapter 8 in Mohriak, 2025).

Many evaporite minerals precipitated during the dry-up episodes following the last glaciation, particularly carbonates (aragonite, calcite) and sulfates (mirabilite, thenardite, epsomite, and gypsum). As a result, these evaporites form crusts and hardgrounds that cover salt flats next to playa lakes, massive and bedded salt layers, spring deposits (such as travertine and tufa), and sabkha-related evaporite minerals that form in the pores of the subsurface sediments or groundwater-related accumulations (Last & Ginn, 2005).

The source of the ions in the salt lakes of the Great Plains is still a subject of considerable debate, but there is a geographic coincidence between the salt lakes and the subsurface solution edge of the Devonian Prairie

Evaporite Fm., eastward of the Cordilleran Deformation Front, from Alberta to Saskatchewan (Spencer, 1987; Grobe, 2000). Many researchers assume that the source of the ions that form the evaporite minerals accumulating in modern lakes is the dissolution of the Paleozoic layers by subsurface groundwater flow; the ionic ratios observed in the lake waters are not indicative of this mechanism. The most abundant anions in more than 95% of modern lakes correspond to sulfates and carbonates (Last, 1989). The Devonian evaporite deposits include halite and K-rich soluble evaporites, such as sylvite (KCl). These Paleozoic evaporite minerals precipitated in a marine environment, forming a depocenter in the foredeep of the Rocky Mountain fold-thrust belt, from Alberta to Saskatchewan (see Chapter 8 in Mohriak, 2025).

Crater Lake, located in Saskatchewan, corresponds to a small depression characterized by a circular shape with a diameter of about 244 m (Christiansen, 1971). However, Crater Lake was not formed by a meteor impact but by the removal of the Prairie evaporites in the subsurface by groundwater. Shallow seismic data and drilling have revealed that the lake overlies a collapse chimney with a top less than 1 km below the surface (Gendzwill & Hajnal, 1971).

The evaporite deposits in the northern Great Plains have considerable economic importance, as more than 20 different salt lakes in the region produce anhydrous sodium sulfate (salt cake). From a high of approximately 700,000 metric tons in 1973, the region now produces almost half of the total North American demand for sodium sulfate. The hypersaline Ingebright Lake hosts the largest sodium sulfate extraction plant in North America. The evaporite deposits at the lake's bottom exhibit an average thickness of around 6–7 m but some layers locally exceed 40 m (Last & Ginn, 2005).

The sodium sulfate industry in the Great Plains relies on three primary resources: dissolved sodium sulfate in lake water, hydrated sodium sulfate mineral mirabilite, and bedded salts composed mainly of mirabilite and thenardite (Last & Ginn, 2005). Most sodium and magnesium sulfate production in hypersaline lakes in the Great Plains comes from brines pumped from the basin into the holding ponds of salt plants, similar to the lithium production in the Andean region (see Chapter 9 in Mohriak, 2025).

10.7. SOURCE ROCKS AND CARBONATE RESERVOIRS IN HYPERSALINE LAKES

Evaporitic lacustrine environments in both extensional and compressional settings are marked by the deposition of carbonate and siliciclastic rocks with petroleum source rock potential (Warren, 1986, 1989, 2006, 2016). These rocks are typically deposited during the early stages of brine concentration or after the main episode of salt precipitation (Kirkland & Evans, 1981; Sonnenfeld, 1985). For instance, some hydrocarbon discoveries in the western part of the Qaidam Basin are associated with the Eocene source rocks deposited in lacustrine, deepwater environments that accumulated carbonate muds with high values of total organic carbon (Qiang et al., 2002; Pang et al., 2004). The Late Eocene paleo-Qaidam Lake precipitated thick layers of bedded halite intercalated with these source rocks, indicating abrupt changes in the lake level and brine concentration (Guo, P. et al., 2017).

In the Tarim Basin, Cambrian source rocks were deposited in a shallow-water evaporitic platform, as evidenced by organic-rich claystone interbedded with bedded evaporite and carbonate layers (Wei et al., 2021). In the Persian Gulf, Triassic, Jurassic, and Cretaceous source rocks are intercalated with sediments from normal marine to hypersaline environments in some foredeep depocenters (see Chapter 12).

The Early Cretaceous lacustrine to transitional sedimentary sequences in the South Atlantic divergent margins are characterized by organic-rich sedimentary sequences that preceded the widespread salt deposition in the Late Aptian (Evans, 1978; Evans & Kirkland, 1988; Hite & Anders, 1991). In northeastern Brazil, lacustrine source rocks were deposited in the Sergipe Basin during the anoxic events associated with the Late Aptian evaporitic intervals (Mello et al., 1993; Souza Junior et al., 2013).

Modern lacustrine environments with source rock potential have been discussed by several authors (e.g., Warren, 1986; Hite & Anders, 1991). Many deep lakes display high total organic carbon contents (TOCs) in basinal sediments. For example, muds from Lake Tanganyika and Lake Kivu in the East African Rift System exhibit TOCs of 11% and 15%, respectively (Sonnenfeld, 1985; Warren, 2006, 2016). Some shallow coastal saline lakes adjacent to the subduction zone in Western North America are characterized by very high organic content in anoxic intervals, such as Laguna Guerrero in Baja California, with TOCs ranging from 9% to 29% in some layers (Warren, 1986).

Hypersaline lakes located close to inland seas and gulfs present favorable conditions for source rock accumulation due to the drastic environmental changes preceding basinwide evaporite deposition. Abundant planktonic and benthonic algal growth occurs at the end of the evaporite cycle, when surface waters are freshened by marine transgression. During times of basin restriction and brine concentration, the lake becomes stratified, and anoxic conditions prevail at the lake's bottom. Few species can adapt to hypersaline brines, but those that survive commonly proliferate due to the absence of predators (Kirkland & Evans, 1981).

Sedimentological and geochemical studies of hypersaline lakes, lagoons, and sabkhas indicate that source rocks formed during anoxic intervals are associated with cyclic basin restrictions and climatic changes. The high organic content of black shales is commonly associated with wetter intervals when fresher water flows into the concentrated brine and forms a halocline, with the denser, more saline layers at the bottom of the lake or lagoon. This stratification inhibited the free circulation of oxygen on the lake's bottom layers, enhancing the preservation potential of organic-rich sedimentary successions (Hussain & Warren, 1991). Some of the best source rocks in the world are associated with saline environments in hostile and arid settings, such as the algal mats in the mudflats and lagoons of the Persian Gulf (see Chapter 8).

Hypersaline environments where salinity exceeds the solubility of halite can also deposit source rocks. The Kara-Bogaz-Gol hypersaline lake is characterized by a scarcity of living organisms due to extreme brine concentrations, resulting in an inhospitable environment for most species. However, some halophilic organisms that have adapted to the harsh environment face few competitors and proliferate. These life forms are highly stressed organisms restricted to a few species (algae and crustaceans). Some algae exist in large quantities in some hypersaline environments, but die massively when the brine concentration is increased at the apex of the evaporitic cycle. Due to the extremely high organic productivity, flocks of algae are commonly seen floating on the surface of some hypersaline lakes (Kosarev & Kostianoy, 2005). Some birds take advantage of this fact, such as the flamingo flocks that proliferate around the Kara-Bogaz-Gol, the saline lakes in the Andean Altiplano, and Lake Natron in Africa (see Chapter 9).

10.7.1. Carbonate Rocks and Microbialite Deposits in Evaporitic Environments

In lacustrine environments, siliciclastic and carbonate sediments are the most common lithologies. These sediments form before evaporite minerals were widely deposited during episodes of basin desiccation. There are several stratigraphic intervals in lacustrine basins where carbonate rocks predominate over siliciclastic contributions, such as the pre-salt microbialite sequences that form extremely prolific reservoir rocks in the South Atlantic (see Chapter 3 in Mohriak, 2025).

In the Mediterranean Sea, the Messinian Event is characterized by carbonate sediments (Calcare di Base) that preceded the main phase of salt deposition (Manzi et al., 2011; Schreiber & Hsu, 1980). When sealed by the Messinian salt, some pre-salt carbonate reservoirs forming reefs constitute an important play in the Levantine Basin (see Chapter 5 in Mohriak, 2025).

Alkaline earth carbonates and evaporite minerals typically accumulate in perennial and ephemeral lakes following a bull's eye pattern, with a facies variation depending on mineral solubility (see Chapter 2). The occurrence of carbonate reservoirs associated with evaporite basins has been discussed by many researchers (e.g., Sarg et al., 2001). Commonly, there are carbonate outliers well out into the basin center, where evaporite minerals normally predominate following total basin desiccation (see Chapter 8). This facies variation is due to spring mounds flowing with fresher water, resulting in a local reduction of the salt concentration. The result is the development of carbonate buildups surrounded by evaporite deposits near the channels where unsaturated water enters the salina. Tall carbonate columns in the lake center are locally observed in the East African Rift System and the GSL in Utah. Carbonate sediments are commonly precipitated along the lakeshores as spring mounds, forming travertine rocks, algal mounds, stromatolites, and oolite shoals (Carozzi, 1962; Schreiber & Hsu, 1980; Warren, 1989, 2014).

The genesis of some pre-salt carbonate deposits is usually associated with two end-member models involving organic and inorganic components (e.g., Dupraz et al., 2009). Stromatolites represent buildups formed by organic activity, as observed in the GSL in Utah (e.g., Chidsey et al., 2015). Shark Bay in Western Australia is characterized by stromatolite growth in shallow waters, forming carbonate deposits that have been considered modern analogs for ancient carbonate sequences interpreted as microbialite reservoir rocks in the South Atlantic (e.g., Jahnert & Collins, 2013).

Reefal buildups in ancient sedimentary basins are often linked to chemical precipitates with abiotic origins (Wright & Barnett, 2015, 2017). In the East African Rift System (EARS), hot springs at Lake Manyara and Lake Natron have created travertine deposits intercalated with algae buildups, providing modern analogs for ancient sedimentary basins (Scoon, 2018). In Lake Natron, the mixing of $CaHCO_3$ waters with alkaline brine has formed a pisolite pavement remarkably similar to oolitic limestones (Eugster, 1980a, 1980b). Mono Lake in California exhibits spring tufas forming columns, pinnacles, and needles that reach a height of up to 4 m, resembling reefs. Many basins accumulate inorganic carbonate rocks and abundant microbial precipitates, forming extensive platforms and ramps on lacustrine and marine shorelines. Several basins worldwide have recorded microbialite deposits with organic and inorganic affinity, spanning from the Precambrian to the Cenozoic (see review by Grotzinger & Knoll, 1995).

10.7.2. Pre-Salt Microbialite Petroleum Reservoir Rocks

Sedimentary basins with hypersaline lakes, gulfs, and inland seas have deposited a mixture of carbonate and evaporite sequences associated with microbialite rocks (e.g., Moore & Wade, 2013; Sarg, 2001). Microbialites are lacustrine carbonate rocks formed by organic accumulation (laminated structures, or stromatolites), spring precipitates (travertine), and soil encrustations in salt flats surrounding perennial water bodies in evaporitic lacustrine systems. During periodic floods, these sediments are eroded, dissolved, reworked, and carried into the center of the lake as detrital carbonate and evaporite clastic rocks, which may form a rock known as tufa and are associated with clastic evaporites, particularly gypsarenites. Accumulations of fragments of carbonate shells known as coquinas are frequently observed in lacustrine environments, resulting in lithological units that constitute prolific petroleum reservoirs in several South Atlantic rift basins (Muniz & Bosence, 2018; Thompson et al., 2015).

Several examples of microbialite carbonate rocks are observed in modern saline lakes, both in convergent and divergent plate tectonic settings (e.g., Kara-Bogaz-Gol, GSL, Natron Lake in the East African Rift System, etc.) and also interpreted in many ancient salt basins (e.g., Chidsey et al., 2015; Pace et al., 2016; Seard et al., 2013; Słowakiewicz et al., 2013). Microbialite rocks constitute the most important petroleum reservoirs in several giant petroleum accumulations discovered in the present century, particularly in the northern Caspian Sea and South Atlantic continental margins (e.g., Mohriak, 2015). Carbonate reefs and microbialite reservoirs are recorded in many salt basins in compressional settings, such as the Northern Caspian Sea, where the Kashagan oil field in Kazakhstan's territorial waters constitutes the largest worldwide hydrocarbon discovery in the 21st century. The Kashagan giant oil accumulation is associated with pre-salt Devonian to Carboniferous carbonate reservoirs forming a rimmed platform, sealed by thick Lower Permian (Kungurian) evaporite layers that have been deformed by halokinesis and presently form tall salt diapirs (Kartmagambetova et al., 2011; Mohriak, 2015).

The deepwater pre-salt carbonate sequences in the South Atlantic were formed during the late rift stage. There are two main types of reservoir rocks recognized in these basins: shell fragments (coquinas) within the rift blocks, and microbialite rocks formed at the transition from the sag basin to the evaporite deposition event. Both rock types commonly overlie structural highs in the basin (Thompson et al., 2015; Wright & Barnett, 2017; Wright & Rodriguez, 2018). Some of the most important oil fields that have been recently discovered in the Brazilian margin are related to microbialite carbonate reservoir rocks forming reefal buildups overlain by thick salt layers (Mohriak, 2015; Muniz & Bosence, 2015; Szatmari & Milani, 2016; Wright & Rodriguez, 2018).

Farias et al. (2019) proposed that the calcium content in the pre-salt carbonate rocks of the Santos Basin was sourced by hydrothermal brines that infiltrated into the paleolake and mixed with the continental and seawater that fed the lake intermittently through the volcanic barrier at the southern part of the basin. According to the model, the high geothermal gradient due to the Moho uplift favored the generation of $CaCl_2$ brines in the hydrological system with an already elevated Ca^{2+} ion and low sulfate concentration. This high concentration of calcium ions resulted in the precipitation of carbonate rocks in an internally drained lacustrine system (Farias et al., 2019).

In recent years, several large gas accumulations have been discovered in the Levantine Basin of the Eastern Mediterranean Sea. The main reservoir rocks correspond to pre-salt Miocene carbonate rocks (microbialites), deposited on top of structural highs, such as the Eratosthenes Seamount (see Chapter 5 in Mohriak, 2025).

10.8. SUMMARY OF SALINE LAKES IN CONVERGENT PLATE TECTONIC SETTINGS

Figure 10.33 provides a synthetic global overview of Quaternary saline lakes, focusing on their geographic locations and plate tectonic settings. Some basins are associated with oceanic-continent subduction zones, like the Andean region in South America, or continent-continent collision zones, like the Asian salt basins in the Alpine-Himalayan orogenies. Evaporite sediments can be found on salt flats at high altitudes or in ephemeral lakes with concentrated brines from solar evaporation. Salt ions can be derived from the dissolution of older salt layers, the weathering of evaporitic and magmatic rocks, or deep hydrothermal sources in the crust. The selected examples of saline lakes in convergent plate tectonic settings include the Ojo de Liebre Lagoon in Baja California, the GSL in Utah, Salar de Uyuni in Bolivia, Salar de Atacama in Chile, Kara-Bogaz-Gol in Turkmenistan, and the hypersaline lakes in the Qarhan salt flats of the eastern Qaidam Basin in China (Figure 10.33).

Many lakes are located on high-elevation plateaus, such as the Andean salt flats and hypersaline lakes on the Tibetan Plateau, while others occur at relatively lower altitudes, such as the GSL. The Ojo de Liebre lagoon is connected with the Pacific Ocean, and the water surface is at sea level. The Kara-Bogaz-Gol is located north of the suture zone between the Arabian and the Eurasian plates, east of the Caspian Sea, and both the inland sea and the lagoon exhibit water levels below the global sea level (Figure 10.3).

These Quaternary lakes are associated with distinct plate tectonic settings, such as the Baja California Peninsula, which is separated from the North American Plate by the Neogene oceanic spreading centers developed in the back-arc of the Pacific Plate subduction zone

Basin/ Lake	Tectonic setting	Area (km^2)	Lake level (m)	Evaporite minerals	Remarks
Baja California (Ojo de Liebre lagoon)	Subduction of the Pacific Plate under North America	40	0	Gypsum, halite.	Biggest commercial saltworks plant in the world
Great Salt Lake	Foreland basin, collapse of the Cordilleran Orogen	120	1280	Carbonate rocks, mirabilite.	Holocene evaporite precipitation (6-7 ka)
Salar de Uyuni	Subduction of the Nazca Plate under South America	10582	3653	Gypsum, halite. Lithium-rich brine.	Largest salt flat on Earth
Salar de Atacama	Subduction of the Nazca Plate under South America	3000	2300	Gypsum, halite. Lithium-rich brine.	Driest desert on Earth
Kara-Bogaz-Gol	Convergence between the Arabian and Eurasian plates	20000	−28	Gypsum, halite, carnallite.	High-salinity lagoon (35%); desiccation episode in the 1980s
Qarhan salt flats (Dabussun Lake)	Convergence between the Indian and Eurasian plates	6000	2675	Gypsum, halite, carnallite, sylvite.	High-salinity lake (> 30%); most important salt basin in China

Figure 10.33 Overview of the selected Quaternary salt basins that have been formed in convergent plate tectonic settings. Credit: Warren (2006)/with permission of Springer Nature.

(Figure 10.1). The Ojo de Liebre Lagoon is sourced with normal seawater from the Pacific Ocean but becomes slightly concentrated due to climate aridity. The artificial salt ponds on land were built by the Mexican salt industry in the second half of the twentieth century. The salt ponds are located near the city of Guerrero Negro, in a desert region surrounding the lagoon, constituting the biggest industrial saltworks in the world (Figure 10.1).

The GSL, located in the northern part of the Basin and Range Province, is the largest saline lake in the Western Hemisphere and is a remnant of Bonneville Lake, a paleolake formed by accumulating pluvial, fluvial, and meltwater from the last glaciation episode that affected the northern hemisphere during the Quaternary. The water salinity ranges from 5% to 27%, precipitating evaporite sediments on the desert flats surrounding the depression. The predominant sediments accumulating on the lake's bottom are carbonate rocks, including microbialite buildups, but an abrupt lake level fall circa 7 to 6 ka resulted in the precipitation of sulfate rocks, particularly mirabilite (Figure 10.33). Artificial salt ponds built by the salt industry on the lake margins produce several magnesium-rich evaporite minerals. Saltworks in the Qarhan salt plain in China (Figure 10.28) are responsible for the largest accumulations of nonmarine evaporite resources in the world, including halite, sylvite, carnallite, and lithium ores. Salar de Uyuni and Salar de Atacama are salt flats in the Andean highlands of South America, located near the Andes Mountains (Figure 10.2). These salt flats are remnants of older lakes that were desiccated during the Quaternary. Salar de Uyuni is the largest salt flat on Earth and can be seen in satellite images taken from space. Salar de Atacama is the driest desert on Earth, with less than 5 mm of rain per year. Both regions exhibit gypsum and halite crusts outcropping on the dried-out margins of ancient lakes. In the subsurface, some salt deposits formed evaporite layers that were locally involved in compressional deformation associated with the Andean orogenies. Both salt flats are characterized by lithium-rich brines in subsurface reservoirs, resulting in extremely large reserves. The salt industry is building modern facilities to extract lithium from the brine and export the production as lithium carbonate. Some technological advances are based on lithium batteries for electric vehicles, computers, and cell phones. The large lithium reserves in the Andean salt flats constitute an important economic asset for South American countries. These reserves are the largest in the world and will be increasingly exploited during the next decades (e.g., U.S. Geological Survey, 2023).

Kara-Bogaz-Gol in Turkmenistan is a large hypersaline lagoon formed in a convergent tectonic setting involving the Arabian and Eurasian plates (Figure 10.3). The lagoon communicates with the brackish-water Caspian Sea, a remnant of the Tethys Ocean. Anthropogenic factors interrupted the intermittent water flow from the Caspian Sea to the lagoon in the 1980s, resulting in almost total basin desiccation. In the late 1990s, the channel communicating the two water masses was reestablished, and the lagoon exhibits one of the highest salt concentrations worldwide, precipitating gypsum, halite, carnallite, and more soluble minerals (Figure 10.33).

The Qaidam Basin is characterized by highlands with cold deserts and shallow, hypersaline lakes (Figure 10.33). The Qarhan salt flats host many salt-producing lakes, including Lake Dabussun. Dabussun's lacustrine system is characterized by high concentrations of potassium-rich minerals, precipitating gypsum, halite, carnallite, and sylvite. The Qaidam Basin also produces hydrocarbons from Paleogene to Neogene reservoir rocks.

In summary, Quaternary salt lakes formed in tectonic settings associated with plate convergence mainly precipitated evaporites from nonmarine, continental brines. Some basins formed on highlands associated with plate convergence, subduction, and volcanism, while others formed close to sea level. Some lakes are connected to a nearby ocean, such as the Ojo de Liebre lagoon. Many of the selected examples correspond to continental lakes far from any marine sources, such as the GSL, the lakes on the Tibetan Plateau, and the Andean salt flats. Some researchers consider these Quaternary lakes as possible analogs for giant salt basins of the ancient past, such as the Permian Basin in the United States, the Persian Gulf between Arabia and Iran, or the Mediterranean Sea during the Miocene. However, these modern analogs display sizes that are only a fraction of their ancient counterparts, and the salt thickness in Quaternary analogs is much less than in some geological intervals and localities associated with basinwide salt deposition.

The Quaternary salt basins are characterized by extremely shallow basins and shallow water, contrasting with conceptual models developed for some classical salt basins. There is no modern analog for the precipitation of extremely thick halite layers formed during episodes of plate convergence and climate aridity.

REFERENCES

Aitchison, J. C., Ali, J. R., & Davis, A. M. (2007). When and where did India and Asia collide? *Journal of Geophysical Research*, *112*(B5), B05423, 1–19.

Alonso, R. N. (2006). Ambientes evaporíticos continentales de Argentina. *INSUGEO, Serie Correlacion Geologica*, *21*, 155–170.

Alonso, R. N., Jordan, T. E., Tabbutt, K. T., & Vandervoort, D. S. (1991). Giant evaporite belts of the Neogene central Andes. *Geology*, *19*, 401–404.

Amante, C., & Eakins, B. W. (2009). ETOPO1 1 arc-minute global relief model: Procedures, data sources and analysis. NOAA Technical Memorandum NESDIS NGDC-24. *National Geophysical Data Center Marine Geology and Geophysics Division*, Boulder, Colorado, March 2009, 19 pp.

Amilibia, A., Sàbat, F., McClay, K. R., Muñoz, J. A., Roca, E., & Chong, G. (2008). The role of inherited tectono-sedimentary architecture in the development of the central Andean mountain belt: Insights from the Cordillera de Domeyko. *Journal of Structural Geology*, *30*, 1520–1539.

Anglés, A., & Li, Y. (2017). The western Qaidam Basin as a potential Martian environmental analogue: An overview. *Journal of Geophysical Research Planets*, *122*(5), 856–888. doi:10.1002/2017JE005293

Armijo, R., Lacassin, R., Coudurier-Curveur, A., & Carrizo D. (2015). Coupled tectonic evolution of Andean orogeny and global climate. *Earth-Science Reviews*, *143*, 1–35. http://dx.doi.org/10.1016/j.earscirev.2015.01.005

Arriagada, C., Cobbold, P. R., & Roperch, P. (2006). Salar de Atacama basin: A record of compressional tectonics in the central Andes since the mid-Cretaceous. *Tectonics*, *25*, TC1008, 19 p.

Atwood, G., Wambeam, T. J., & Anderson, N. J. (2016). The present as a key to the past: Paleoshoreline correlation insights from Great Salt Lake. In C. G. Oviatt, & J. F. Shroder Jr. (Eds.), *Lake Bonneville: A scientific update, Developments in Earth Surface Processes* (Vol. 20, pp. 1–27). Amsterdam: Elsevier. doi:10.1016/b978-0-444-63590-7.00001-9

Baker, P. A., Rigsby, C. A., Seltzer, G. O., Fritz, S. C., Lowenstein, T. K., Bacher, N. P., & Veliz, C. (2001). Tropical climate changes at millennial and orbital timescales on the Bolivian Altiplano. *Nature*, *409*, 698–701.

Bally, A. W., Roberts, D. G., Sawyer, D., & Sinkewich, A. (2012). Tectonic and basin maps of the world. In D. G. Roberts & A. W. Bally (Eds.), *Regional geology and tectonics: Phanerozoic passive margins, cratonic basins and global tectonic maps* (Vol. 1C, pp. 973–1151). Amsterdam: Elsevier Science Ltd.

Bascuñán, S., Maksymowicz, A., Martínez, F., Becerra, J., César Arriagada, C., & Deckart, K. (2019). Geometry and late Mesozoic-Cenozoic evolution of the Salar de Atacama Basin (22°30′-24°30′S) in the northern Central Andes: New constraints from geophysical, geochronological and field data. *Tectonophysics*, *759*, 58–78. https://doi.org/10.1016/j.tecto.2019.04.008

Baskin, R. L, Wright, V. P., Driscoll, N., & Kent, G. (2017). Great Salt Lake, Utah: A natural laboratory for assessing Geobody dimensions, Reservoir Architectures, and Microbialite Controls. *AAPG Search and Discovery*, Article #51430 (*2017*).

Benson, L. V., Lund, S. P., Smoot, J. P., Rhode, D. E., Spencer, R. J., Verosub, K. L., et al. (2011). The rise and fall of Lake Bonneville between 45 and 10.5 ka. *Quaternary International*, *235*, 57–69.

Bobst, A. L., Lowenstein, T. K., Jordan, T. E., Godfrey, L. V., Ku, T. L., & Luo, S. D. (2001). A 106 ka paleoclimate record from drill core of the Salar de Atacama, northern Chile. *Palaeogeography Palaeoclimatology Palaeoecology*, *173*(1–2), 21–42.

Borderie, S., Vendeville, B. C., Graveleau, F., Witta, C., Dubois, P., Baby, P., & Calderon, Y. (2019). Analogue modeling of large-transport thrust faults in evaporites-floored basins: Example of the Chazuta Thrust in the Huallaga Basin, Peru. *Journal of Structural Geology*, *123*, 1–17.

Bortz, L. C. (1987). Heavy-oil deposit, Great Salt Lake, Utah. In R. F. Meyer (Ed.), Exploration for heavy crude oil and natural bitumen. *AAPG Studies in Geology*, *25*, 555–563.

Bouton, A., Vennin, E., Boulle, J., Pace, A., Bourillot, R., Thomazzo, C., et al. (2016). Linking the distribution of microbial deposits from the Great Salt Lake (Utah, USA) to tectonic and climatic processes. *Biogeosciences*, *13*, 5511–5526. doi: 10.5194/bg-13-5511-2016

Burke, K., & Sengor, A. M. C. (1988). Ten metre global sea-level change associated with South Atlantic Aptian salt deposition. *Marine Geology*, *83*, 309–312.

Cabello, J. (2022). Reserves, resources and lithium exploration in the salt flats of northern Chile. *Andean Geology*, *49*(2), 297–306. doi: https://dx.doi.org/10.5027/andgeoV49n2-3444

Carozzi, A. V. (1962). Observations on Algal Biostromes in the Great Salt Lake, Utah. *The Journal of Geology*, *70*(2), 246–252. doi:10.1086/626814

Casas, E., Lowenstein, T. K., Spencer, R. J., & Penghi, Z. (1992). Carnallite mineralization in the nonmarine, Qaidam Basin, China: Evidence for the early diagenetic origin of potash evaporites. *Journal of Sedimentary Petrology*, *62*(5), 881–898.

Cembrano, J., Lavenu, A., Yañez, G., Riquelme, R., García, M., González, G., & Hérail, G. (2007). Neotectonics. In T. Moreno & W. Gibbons (Eds.), *The geology of Chile* (pp. 231–261). Geological Society of London.

Chaboureau, A.-C., Donnadieu, Y., Sepulchre, P., Robin, C., Guillocheau, F., & Rohais, S. (2012). The Aptian evaporites of the South Atlantic: A climatic paradox? *Climate of the Past*, *8*, 1047–1058.

Chaboureau, A. C., Guillocheau, F., Robin, C., Rohais, S., Moulin, M., & Aslanian, D. (2013). Paleogeographic evolution of the central segment of the South Atlantic during Early Cretaceous times: Paleotopographic and geodynamic implications. *Tectonophysics*, *604*, 191–223.

Charrier, R., Pinto, L., & Rodrígues, M. P. (2007). Tectonostratigraphic evolution of the Andean Orogen in Chile. In T. Moreno & W. Gibbons (Eds.), *The geology of Chile* (pp. 21–114). Geological Society of London.

Chen, C. Y., & Maloof, A. (2017). Revisiting the deformed high shoreline of Lake Bonneville. *Quaternary Science Reviews*, *159*, 169–189. https://doi.org/10.1016/j.quascirev.2016.12.019

Chen, F., Wu, D., Chen, J., Zhou, A., Yu, J., Shen, J., et al. (2016). Holocene moisture and East Asian summer monsoon evolution in the northeastern Tibetan Plateau recorded by Lake Qinghai and its environs: A review of conflicting proxies. *Quaternary Science Reviews*, *154*, 111–129.

Chen, J. L., Pekker, T. Wilson, C. R., Tapley, B. D., Kostianoy, A. G., Cretaux, J.-F., & Safarov, E. S. (2017). Long-term Caspian Sea level change. *Geophysical Research Letters*, *44*, 6993–7001. doi:10.1002/2017GL073958

Chen, K. Z., & Bowler, J. M. (1986). Late Pleistocene evolution of Salt Lakes in the Qaidam Basin, Qinghai Province, China. *Palaeogeography, Palaeoclimatology, Palaeoecology*, *54*, 87–104.

Cheng, F., Garzione, C., Jolivet, M., Guo, Z., Zhang, D., & Zhang, C. (2018). A new sediment accumulation model of Cenozoic depositional ages from the Qaidam basin, Tibetan Plateau. *Journal of Geophysical Research: Earth Surface*, *123*, 3101–3121. https://doi.org/10.1029/2018JF004645

Cheng, F., Jolivet, M., Fu, S., Zhang, Q., Guan, S., Yu, X., & Guo, Z. (2014). Northward growth of the Qimen Tagh Range: A new model accounting for the Late Neogene strike-slip deformation of the SW Qaidam Basin. *Tectonophysics*, *632*, 32–47.

Cheng, F., Jolivet, M., Fu, S., Zhang, C., Zhang, Q., & Guo, Z. (2016). Large-scale displacement along the Altyn Tagh Fault (North Tibet) since its Eocene initiation: Insight from detrital zircon U–Pb geochronology and subsurface data. *Tectonophysics*, *677–678*, 261–279.

Cheng, Z., Xiao, L. Wang, H., Yang, H., Li, J., Huang, T., et al. (2017). Bacterial and archaeal lipids recovered from subsurface evaporites of Dalangtan Playa on the Tibetan Plateau and their astrobiological implications. *Astrobiology*, *17*(11), 1–11.

Chidsey, T. C., Jr., Vanden Berg, M. C., & Eby, D. E. (2015). Petrography and characterization of microbial carbonates and associated facies from modern Great Salt Lake and Uinta Basin's Eocene Green River Formation in Utah, USA. In D. W. J. Bosence, K. A. Gibbons, D. P. Le Heron, W. A. Morgan, T. Pritchard, & B. A. Vining (Eds.), *Microbial carbonates in space and time: Implications for global exploration and production* (Vol. 418, pp. 261–286). Geological Society, London, Special Publications. http://dx.doi.org/10.1144/SP418.6

Chorlton, L. B. (2007). Generalized geology of the world: Bedrock domains and major faults in GIS format: A small-scale world geology map with an extended geological attribute database. *Geological Survey of Canada, Open File 5529*, 48 pp. doi: 10.4095/223767.154/3192695

Christiansen, E. A. (1971). Geology of the Crater Lake collapse structure in Southeastern Saskatchewan. *Canadian Journal of Earth Sciences*, *8*, 1505–1513. doi: https://doi.org/10.1139/e71-140

Coldwell, B., Clemens, J., & Petford, N. (2011). Deep crustal melting in the Peruvian Andes: Felsic magma generation during delamination and uplift. *Lithos*, *125*(1–2), 272–286. doi:10.1016/j.lithos.2011.02.011

Colman, S. M., Kelts, K. R., & Dinter, D. A. (2002). Depositional history and neotectonics in Great Salt Lake, Utah, from high-resolution seismic stratigraphy. *Sedimentary Geology*, *148*, 61–78.

Constenius, K. N. (1996). Late Paleogene extensional collapse of the Cordilleran foreland fold and thrust belt. *Geological Society of America Bulletin*, *108*(1), 20–39. doi: 10.1130/0016-7606(1996)108<0020:LPECOT>2.3.CO;2

Cooper, M. (2007). Structural style and hydrocarbon prospectivity in fold and thrust belts: A global review. In A. C. Ries, R. W. H. Butler, & R. H. Graham (Eds.), *Deformation of the continental crust: The legacy of Mike Coward* (Vol. 272, pp. 447–472). Geological Society, London, Special Publications.

Crangle, R. D. (2018). Boron. In *2016 Minerals Yearbook* (pp. 13.1–13.7). U. S. Geological Survey.

Davison, I. (2007). Geology and tectonics of the South Atlantic Brazilian salt basins. In A. C. Ries, R. W. H. Butler, & R. H. Graham (Eds.), *Deformation of the continental crust: The legacy of Mike Coward* (Vol. 272, pp. 345–359). Geological Society, London, Special Publications.

Dickerson, P. W. (2000). A Caspian chronicle: Sea-level fluctuations between 1982 and 1997. In K. P. Lulla & L. V. Dessinov (Eds.), *Dynamic Earth environments: Remote sensing observations from Shuttle-Mir missions* (pp. 145–148, 266, 280–281). New York: John Wiley & Sons.

Du, Y., Fan, Q., Gao, D., Wei, H., Shan, F., Li, B., et al. (2018). Evaluation of boron isotopes in halite as an indicator of the salinity of Qarhan paleolake water in the eastern Qaidam Basin, western China. *Geoscience Frontiers*, *10*, 253–262.

Duan, Z., & Hu, W. (2001). The accumulation of potash in a continental basin: The example of the Qarhan Saline Lake, Qaidam Basin, *West China. European Journal of Mineralogy*, *13*(6), 1223–1233.

Dupraz, C., Reid, R. P., Braissant, O., Decho, A. W., Norman, R. S., & Visscher, P. T. (2009). Processes of carbonate precipitation in modern microbial mats. *Earth-Science Reviews*, *96*, 141–162.

Ericksen, G. E. (1981). Geology and origin of the Chilean nitrate deposits. *USGS Professional Paper*, *1188*, 37pp.

Ericksen, G. E. (1993). Upper Tertiary and Quaternary continental saline deposits in the central Andean region. *Geological Association of Canada Special Paper*, *40*, 89–102.

Eugster, H. P. (1980a). Geochemistry of evaporitic lacustrine deposits. *Annual Review of Earth and Planetary Sciences*, *8*(1), 35–63. doi:10.1146/annurev.ea.08.050180.000343

Eugster, H. P. (1980b). Lake Magadi, Kenya, and its precursors. In A. Nissenbaum (Ed.), *Hypersaline brines and evaporitic environments, Developments in Sedimentology* (Vol. 28, pp. 195–232). Elsevier: Amsterdam. doi:10.1016/s0070-4571(08)70239-5

Eugster, H. P., & Hardie, L. A. (1978). Saline lakes. In A. Lerman (Ed.), *Lakes, Chemistry, Geology, Physics* (pp. 237–293). Amsterdam: Springer-Verlag.

Evans, R. (1978). Origin and significance of evaporites in basins around Atlantic margins. *American Association of Petroleum Geologists Bulletin*, *62*(2), 223–234.

Evans, R., & Kirkland, D. W. (1988). Evaporitic environments as a source of petroleum. In B. C. Schreiber (Ed.), *Evaporites and hydrocarbons* (pp. 256–299). New York: Columbia University Press.

Evenick, J. C. (2021). Glimpses into Earth's history using a revised global sedimentary basin map. *Earth-Science Reviews*, *215*, 103564. https://doi.org/10.1016/j.earscirev.2021.103564

Fan, Q. S., Lowenstein, T. K., Wei, H. C., Yuan, Q., Qin, Z. J., Shan, F. S., & Ma, H. Z. (2018). Sr isotope and major ion compositional evidence for formation of Qarhan Salt Lake, western China. *Chemical Geology*, *497*,128–145. https://doi.org/10.1016/j.chemgeo.2018.09.001

Fang, X., Li, M., Wang, Z., Wang, J., Li, J., Liu, X., & Zan, J. (2016). Oscillation of mineral compositions in Core SG-1b, western Qaidam Basin, *NE Tibetan Plateau. Nature Scientific Reports*, *6*, 32848, 1–7.

Farias, F., Szatmari, P., Bahniuk, A., & França, A. B. (2019). Evaporitic carbonates in the pre-salt of Santos Basin—Genesis and tectonic implications. *Marine and Petroleum Geology*, *105*, 251–272.

Fenton, M. M., Schreiner, B. T., Nielsen, E., & Pawlowicz, J. G. (1994). Chapter 26—Quaternary geology of the western plains. In G. D. Mossop & I. Shetsen (Eds.), *Geological Atlas of the Western Canada Sedimentary Basin* (pp. 413–420). Canadian Society of Petroleum Geologists and Alberta Research Council.

Flint, S., Turner, P., Jolley, E. J., & Hartley, A. J. (1993). Extensional tectonics in convergent margin basins: An example from the Salar de Atacama, Chilean Andes. *Geological Society of America Bulletin*, *105*, 603–617.

Gao, C., Yu, J., Min, X., Cheng, A., Hong, R., Zhang, L., & Li, T. (2019). The sedimentary evolution of Da Qaidam Salt Lake in Qaidam Basin, northern Tibetan Plateau: implications for hydro-climate change and the formation of pinnoite deposit. *Environmental Earth Sciences*, *78*, 463, 15 pp.

GEBCO (2019). *General Bathymetric Chart of the Oceans*, British Oceanographic Data Centre, U.K., website www.gebco.net.

Gendzwill, D. J., & Hajnal, Z. (1971). Seismic investigation of the Crater Lake Collapse Structure in Southeastern Saskatchewan. *Canadian Journal of Earth Sciences*, *8*(12), 1514–1524. https://doi.org/10.1139/e71-141

Godfrey, L., & Álvarez-Amado, F. (2020). Volcanic and saline lithium inputs to the Salar de Atacama. *Minerals*, *10*(2), 201. https://doi.org/10.3390/min10020201

Grobe, M. (2000). Distribution and thickness of salt within the Devonian Elk Point Group, Western Canada sedimentary basin. Alberta Energy and Utilities Board, Alberta Geological Survey, *Earth Sciences Report 2000-02*, 12 pp + Maps.

Grotzinger, J. P., & Knoll, A. H. (1995). Anomalous carbonate precipitates: Is the Precambrian the key to the Permian? *PALAIOS*, *10*(6), 578–596. doi:10.2307/3515096

Guo, J., Wei, X., Long, G., Wang, B., Fan, H., & Xu, S. (2017). Three-dimensional structural model of the Qaidam basin: Implications for crustal shortening and growth of the northeast Tibet. *Open Geosciences*, *9*, 174–185.

Guo, P., Liu, C., Huang, L., Wang, P., Wang, K., Yuan, H., et al. (2017). Genesis of the late Eocene bedded halite in the Qaidam Basin and its implication for paleoclimate in East Asia. *Palaeogeography, Palaeoclimatology, Palaeoecology*, *487*, 364–380.

Guo, P., Liu, C., Huang, L., Yu, M., Wang, P., & Zhang, G. (2018). Palaeohydrological evolution of the late Cenozoic saline lake in the Qaidam Basin, NE Tibetan Plateau: Tectonic vs. climatic control. *Global and Planetary Change*, *165*, 44–61.

Hammerschmidt, K., Döbel, R., & Friedrichsen, H. (1992). Implication of ^{40}Ar / ^{39}Ar dating of early Tertiary volcanic rocks from the north Chilean Precordillera. *Tectonophysics*, *202*, 55–81.

Hartmann, J., & Moosdorf, N. (2012). The new global lithological map database GLiM: A representation of rock properties at the Earth surface. *Geochemistry, Geophysics, Geosystems*, *13*, Q12004. doi:10.1029/2012GC004370

Herb, C., Appel, E., Voigt, S., Koutsodendris, A., Pross, J., Zhang, W., & Fang, X. (2015). Orbitally tuned age model for the late Pliocene–Pleistocene lacustrine succession of drill core SG-1 from the western Qaidam Basin (NE Tibetan Plateau). *Geophysical Journal International*, *200*, 35–51.

Hite, R. J., & Anders, D. E. (1991). Petroleum and evaporites. In J. L. Melvin (Ed.), *Evaporites, petroleum and mineral resources, Developments in Sedimentology* (Vol. 50, Chapter 4, pp. 349–411). Amsterdam: Elsevier. doi:10.1016/s0070-4571(08)70262-0

Hovland, M., Rueslåtten, H., & Johnsen, H. K. (2018a). Large salt accumulations as a consequence of hydrothermal processes associated with "Wilson cycles": A review, Part 1:

Towards a new understanding. *Marine and Petroleum Geology*, *92*, 987–1009.

Hovland, M., Rueslåtten, H., & Johnsen, H. K. (2018b). Large salt accumulations as a consequence of hydrothermal processes associated with "Wilson cycles": A review, Part 2: Application of a new salt-forming model on selected cases. *Marine and Petroleum Geology*, *92*, 128–148.

Hsü, K. J. (1972). Origin of saline giants: A critical review after the discovery of the Mediterranean evaporite. *Earth-Science Reviews*, *8*, 371–396.

Hussain, M., & Warren, J. K. (1991). Source rock potential of shallow-water evaporites: An investigation in Holocene–Pleistocene Salt Flat sabkah (playa), west Texas-New Mexico. *Carbonates evaporites*, *6*, 217–224. https://doi.org/10.1007/BF03174424

Jahnert, R. J., & Collins, L. B. (2013). Controls on microbial activity and tidal flat evolution in Shark Bay, Western Australia. *Sedimentology*, *60*, 1071–1099. doi: 10.1111/sed.12023

Jin, C., Li, S., Zhang, H., Liu, J., Chen, W., & Jiang, Y. (2016). Hydro-Chemical processes in Lake Qinghai throughout climate warming: In situ investigations of the largest lake in China. *Natural Science*, *8*, 574–590.

Jones, B. F., Naftz, D. L., Spencer, R. J., & Oviatt, C. G. (2009). Geochemical Evolution of Great Salt Lake, Utah, USA. *Aquatic Geochemistry*, *15*(1–2), 95–121. doi:10.1007/s10498-008-9047-y

Jordan, T. E., Mpodozis, C., Munoz, N., Blanco, N., Pananont, P., & Gardeweg, M. (2007). Cenozoic subsurface stratigraphy and structure of the Salar de Atacama Basin, northern Chile. *Journal of South American Earth Sciences*, *23*, 122–146.

Jordan, T. E., Muñoz, N., Hein, M., Lowenstein, T., Godfrey, L., & Yu, J. (2002). Active faulting and folding without topographic expression in an evaporite basin, *Chile. Geological Society of America Bulletin*, *114*(11), 1406–1421. doi:10.1130/0016-7606.

Kartmagambetova, Z., Daulbayev, B., Galiyeva, G., Gaz, I., & Ibragimov, F. (2011). Kashagan field geological description. *First EAGE Caspian Region Workshop, European Association of Geoscientists & Engineers, Conference Proceedings*, *11780*, cp-260-00006. doi: https://doi.org/10.3997/2214-4609.20144348

Kezao, C., & Bowler, J. M. (1986). Late Pleistocene evolution of salt lakes in the Qaidam Basin, Qinghai Province, China. *Palaeogeography, Palaeoclimatology, Palaeoecology*, *54*, 87–104.

Kirkland, D. W., & Evans, R. (1981). Source rock potential of evaporitic environment. *American Association of Petroleum Geologists Bulletin*, *65*(1), 181–190.

Kosarev, A. N., & Kostianoy, A. G. (2005). Kara-Bogaz-Gol Bay. In Kostianoy, A. G., & Kosarev, A. N. (Eds.), *The Caspian sea environment: The handbook of Environmental Chemistry* (Vol. 5P, pp. 211–221). Berlin, Heidelberg: Springer. https://doi.org/10.1007/698_5_011

Kosarev, A. N., Kostianoy, A. G., & Zonn, I. S. (2009). Kara-Bogaz-Gol Bay: Physical and chemical evolution. *Aquatic Geochemistry*, *15*(1–2), 223–236. https://doi.org/10.1007/s10498-008-9054-z

Kukla, P. A., Strozyk, F., & Mohriak, W. U. (2018). South Atlantic salt basins—Witnesses of complex passive margin evolution. *Gondwana Research*, *53*, 41–57.

Kurilenko, V. V., Ruday, I. G., & Shvarts, A. A. (1988). The origin and commercial exploitation of subsurface brines in the northern half of Kara-Bogaz-Gol. *International Geology Review*, *30*(11), 1238–1245. doi: 10.1080/00206818809466105

Last, W. M. (1989). Continental brines and evaporites of the northern Great Plains of Canada. *Sedimentary Geology*, *64*, 207–221.

Last, W. M., & Ginn, F. M. (2005). Saline systems of the Great Plains of western Canada: An overview of the limnogeology and paleolimnology. *Aquatic Biosystems*, *1*, 10 (2005). https://doi.org/10.1186/1746-1448-1-10

Li, J., Chen, F., Ling, Z., & Li, T. (2021). Lithium sources in oilfield waters from the Qaidam Basin, Tibetan Plateau: Geochemical and Li isotopic evidence. *Ore Geology Reviews*, *139*(A), 104481. https://doi.org/10.1016/j.oregeorev.2021.104481

Li, Q., Fan, Q., Wei, H., Qin, Z., Zhang, X., Du, Y., & Shan, F. (2020). Sulfur isotope constraints on the formation of $MgSO_4$-deficient evaporites in the Qarhan salt Lake, western China. *Journal of Asian Earth Sciences*, *189*, 104160. https://doi.org/10.1016/j.jseaes.2019.104160

Liberty, L. M., St. Clair, J., & McKean, A. P. (2021). A broad, distributed active fault zone lies beneath Salt Lake City, Utah. *The Seismic Record*, *1*, 35–45. doi: 10.1785/0320210009.

Lin, Y. -S., Chuang, Y.-R., Shyu, J. B. H., González, G., Shen, C.-C. Lo, C.-H., & Liou, Y.-H. (2016). Structural characteristics of an active fold-and-thrust system in the southeastern Atacama Basin, northern Chile. *Tectonophysics*, *685*, 44–59.

Lindsay, M. R., Dunham, E. C., & Boyd, E. S. (2020). Microbialites of Great Salt Lake. In B. Baxter & J. Butler (Eds.), *Great Salt Lake Biology* (pp. 87–118). Switzerland: Springer Nature. https://doi.org/10.1007/978-3-030-40352-2_4

López Steinmetz, R. L., Salvi, S., Sarchi, C., Santamans, C., & López Steinmetz, L. C. (2020). Lithium and Brine Geochemistry in the Salars of the Southern Puna, *Andean Plateau of Argentina. Economic Geology*, *115*(5), 1079–1096. doi:10.5382/econgeo.4754

Lowenstein, T. K., & Risacher, F. (2009). Closed basin brine evolution and the influence of Ca–Cl inflow waters: Death Valley and Bristol Dry Lake California, Qaidam Basin, China, and Salar de Atacama, Chile. *Aquatic Geochemistry*, *15*, 71–94.

Lowenstein, T. K., & Spencer, R. J. (1990). Syndepositional origin of potash evaporites: petrographic and fluid inclusion evidence. *American Journal of Science*, *290*, 1–42.

Lowenstein, T. K., Hein, M. C., Bobst, A. L., Jordan, T. E., Ku, T.-L., & Luo, S. (2003). An assessment of stratigraphic completeness in climate-sensitive closed-basin lake sediments: Salar de Atacama, *Chile. Journal of Sedimentary Research*, *73*(1), 91–104.

Lowenstein, T. K., Spencer, R. J., & Zhang, P. (1989). Origin of ancient potash evaporites; clues from the modern Qaidam Basin, western China. *Science*, *245*(4922), 1090–1092. DOI: 10.1126/science.245.4922.1090

Ma, Z., Han, F., Chen, T., Yi, L., Lu, X., Chen, F., Liu, X., & Yuan, W. (2021). The Forming Age and the Evolution Process of the Brine Lithium Deposits in the Qaidam Basin Based on Geochronology and Mineral Composition. *Frontiers Earth Science*, *9*, 702223. https://doi.org/10.3389/feart.2021.702223.

Manzi, V., Lugli, S., Roveri, M., Schreiber, B. C., & Gennari, R. (2011). The Messinian "Calcare di Base" (Sicily, Italy) revisited. *Geological Society of America Bulletin*, *123*(1/2), 347–370.

Martínez, F., López, C., Bascuñan, S., & Arriagada, C. (2018). Tectonic interaction between Mesozoic to Cenozoic extensional and contractional structures in the Preandean depression (23°–25°S): Geologic implications for the Central Andes. *Tectonophysics*, *744*, 333–349.

McLennan, S. M. (2001). Relationships between the trace element composition of sedimentary rocks and upper continental crust. *Geochemistry Geophysics Geosystems*, *2*(4), 2000GC000109, 24 p. http://dx.doi.org/10.1029/2000GC000109.

Mello, M. R., Koutsoukos, E. A. M., Santos Neto, E. V., & Telles A. C. S., Jr. (1993). Geochemical and micropaleontological characterization of lacustrine and marine hypersaline environments from Brazilian sedimentary basins. In B. J. Katz, & L. M. Pratt (Eds.), *Source Rocks in a sequence stratigraphic framework, Studies in Geology* (Vol. 37, pp. 17–34). American Association of Petroleum Geologists.

Micklin, P. (2007). The Aral sea disaster. *Annual Review of Earth and Planetary Sciences*, *35*, 47–72. https://doi.org/10.1146/annurev.earth.35.031306.140120

Mohapatra, G. K., & Johnson, R. A. (1998). Localization of listric faults at thrust fault ramps beneath the Great Salt Lake basin, Utah: Evidence from seismic imaging and finite element modeling. *Journal of Geophysical Research*, *103*(B5), 10047–10063.

Mohriak, W. U. (2015). Pre-Salt carbonate reservoirs in the South Atlantic and Worldwide Analogs. AAPG Search and Discovery, Article #51086 (2015). Posted April 6, 2015.

Mohriak, W. (2025). *Salt in the Earth sciences: Basin analysis and salt tectonics*. Washington, DC: American Geophysical Union. doi:10.1002/9781394307067

Möller, P., Lüders, V., & De Lucia, M. (2007). Formation of Rotliegend Ca-Cl brines in the North German Basin compared to analogues in the geological record. *Chemical Geology*, *459*, 32–42. https://doi.org/10.1016/j.chemgeo.2017.04.001

Moore, C. H., & Wade, W. J. (2013). The basic nature of carbonate sediments and sedimentation. In C. H. Moore and W. J. Wade (Eds.), *Carbonate ereservoirs—Porosity and diagenesis in a sequence stratigraphic framework* (pp. 3–21). doi:10.1016/b978-0-444-53831-4.00001-x

Mpodozis, C., Arriagada, C., Basso, M., Roperch, P., Cobbold, P., & Reich, M. (2005). Late Mesozoic to Paleogene stratigraphy of the Salar de Atacama Basin, Antofagasta, Northern Chile: Implications for the tectonic evolution of the Central Andes. *Tectonophysics*, *399*, 125–154.

Muniz, M. C., & Bosence, D. W. J. (2015). Pre-salt microbialites from the Campos Basin (offshore Brazil): Image log facies, facies model and cyclicity in lacustrine carbonates. In D. W. J. Bosence, K. A. Gibbons, D. P. Le Heron, W. A. Morgan, T. Pritchard, & B. A. Vining, *Microbial carbonates in space and time: Implications for global exploration and production* (Vol. 418, pp. 221–242). Geological Society, London, Special Publications.

Muniz, M. C., & Bosence, D. W. J. (2018). Lacustrine carbonate platforms: Facies, cycles, and tectonosedimentary models for the presalt Lagoa Feia Group (Lower Cretaceous), Campos Basin, Brazil. *American Association of Petroleum Geologists Bulletin*, *102*(12), 2569–2597.

Muñoz, N., Charrier, R., & Jordan, T. (2002). Interactions between basement and cover during the evolution of the Salar de Atacama Basin, northern Chile. *Revista Geológica de Chile*, *29*(1), 55–80.

Nunnery, J. A., Fritz, S. C., Baker, P. A., & Salenbien, W. (2018). Lake-level variability in Salar de Coipasa, Bolivia during the past ~40,000 yr. *Quaternary Research*, *91*(2), 881–891. https://doi.org/10.1017/qua.2018.108

Ochsenius, K. (1877). Die Bilding der Steinsalzlager und ihrer mutter augensalze. *C.E.M. Pfeffer*, Halle, 172 pp.

Ochsenius, K. (1888). On the formation of rock salt beds and mother liquor salts. *Philadelphia Academy of Science, Proceedings*, *Pl 2*, 181–187.

Ortí, F., & Alonso, R. N. (2000). Gypsum-Hydroboracite Association in the Sijes formation (Miocene, NW Argentina): Implications for the genesis of Mg-Bearing Borates. *Journal of Sedimentary Research*, *70*(3), 664–681. https://doi.org/10.1306/2DC4092F-0E47-11D7-8643000102C1865D

Oviatt, C. G., & Shroder Jr., J. F. (2016). Introduction. In C. G. Oviatt, and J. F. Shroder Jr. (Eds.), *Lake Bonneville: A scientific update. Developments in Earth Surface Processes* (Vol. 20, pp. xxv–xxxvi). Amsterdam: Elsevier.

Oviatt, C. G., Atwood, G., & Thompson, R. S. (2021). History of Great Salt Lake, Utah, USA: Since the termination of Lake Bonneville. In M. R. Rosen, D. B. Finkelstein, L. Park Boush, & S. Pla-Pueyo (Eds.), *Limnogeology: Progress, challenges and opportunities. syntheses in Limnogeology* (pp. 233–271). Switzerland: Springer Nature. https://doi.org/10.1007/978-3-030-66576-0_8

Pace, A., Bourillot, R., Bouton, A., Vennin, E., Galaup, S., Bundeleva, I., et al. (2016). Microbial and diagenetic steps leading to the mineralisation of Great Salt Lake microbialites. *Scientific Reports*, *6*, 31495. doi: 10.1038/srep31495 (2016).

Pananont, P., Mpodozis, C., Blanco, N., Jordan, T. E., & Brown, L. D. (2004). Cenozoic evolution of the northwestern Salar de Atacama Basin, northern Chile. *Tectonics*, *23*(6), TC6007. doi: 10.1029/2003TC001595

Pang, X. Q., Li, Y. X., & Jiang, Z. X. (2004). Key geological controls on migration and accumulation for hydrocarbons derived from mature source rocks in Qaidam Basin. *Journal of Petroleum Science and Engineering*, *41*, 79–95.

Parravano, V., Teixell, A., & Mora, A. (2015). Influence of salt in the tectonic development of the frontal thrust belt of the eastern Cordillera (Guatiquía area, Colombian Andes). *Interpretation*, *3*(4), SAA17–SAA27. http://dx.doi.org/10.1190/INT-2015-0011

Pei, Y., Paton, D. A., Wu, K., & Xie, L. (2017). Subsurface structural interpretation by applying trishear algorithm: An example from the Lenghu5 fold-and-thrust belt, Qaidam Basin, Northern Tibetan Plateau. *Journal of Asian Earth Sciences*, *143*, 343–353.

Phleger, F. B. (1969). A modern evaporite deposit in Mexico. *American Association of Petroleum Geologists Bulletin*, *53*(4), 824–829.

Phleger, F. B., & Ewing, G. C. (1962). Sedimentology and oceanography of coastal lagoons in Baja California, *Mexico. Geological Society of America Bulletin*, *73*(2), 145–182.

Pierre, C., Ortlieb, L., & Person, A. (1984). Supratidal evaporitic dolomite at Ojo de Liebre lagoon: mineralogical and isotopic arguments for primary crystallization. *Journal of Sedimentary Petrology*, *54*(4), 1049–1061.

Qiang, J., Ming, Z., Zhen, L., XianZhi, G., Dehua, P., & Lamei, L. (2002). Geology and geochemistry of source rocks in the Qaidam Basin, NW China. *Journal of Petroleum Geology*, *25*(2), 219–238. doi:10.1111/j.1747-5457.2002.tb00005.x

Qinmin, W., & Coward, M. P. (1990). The Chaidam Basin, NW China; Formation and hydrocarbon potential. *Journal of Petroleum Geology*, *13*, 93–112. https://doi.org/10.1111/j.1747-5457.1990.tb00255.x

Rech, J. A., Quade, J., & Betancourt, J. (2002). Late Quaternary paleohydrology of the central Atacama Desert (lat 22°–24°S), Chile. *Geological Society of America Bulletin*, *114*, 334–348.

Rech, J. A., Currie, B. S., Jordan, T. E., Riquelme, R., Lehmann, S. B., Kirk-Lawlor, N. E., et al. (2019). Massive middle Miocene gypsic paleosols in the Atacama Desert and the formation of the Central Andean rain-shadow. *Earth and Planetary Science Letters*, *506*, 184–194.

Rettig, S. L., Jones, B. F., & Risacher, F. (1980). Geochemical evolution of brines in the Salar of Uyuni, Bolivia. *Chemical Geology*, *30*, 57–79.

Reutter, K. J., Charrier, R., Gotze, H. J., Schurr, B., Wigger, P., Scheuber, E., et al. (2006). The Salar de Atacama Basin: A subsiding block within the western edge of the Altiplano-Puna Plateau. In O. Oncken et al. (Eds.), *The Andes Active Subduction Orogeny* (pp. 303–325). Berlin Heidelberg: Springer.

Risacher, F., & Fritz, B. (2000). Bromine geochemistry of salar de Uyuni and deeper salt crusts, Central Altiplano, Bolivia. *Chemical Geology*, *167*, 373–392.

Risacher, F., & Fritz, B. (2009). Origin of salts and brine evolution of Bolivian and Chilean salars, *Aquatic Geochemistry*, *15*(1–2), 123–157.

Risacher, F., Alonso, B., & Salazar, C. (2003). The origin of brines and salts in Chilean salars: A hydrochemical review. *Earth-Science Reviews*, *63*, 249–293.

Rohrmann, A., Sachse, D., Mulch, A., Pingel, H., Tofelde, S., Alonso, R. N., & Strecker, M. R. (2016). Miocene orographic uplift forces rapid hydrological change in the southern central Andes. *Scientific Reports*, *6*, 35678, 1–7.

Rubilar, J., Martínez, F., Arriagada, C., Becerra, J., & Bascuñan, S. (2018). Structure of the Cordillera de la Sal: A key tectonic element for the Oligocene-Neogene evolution of the Salar de Atacama basin, Central Andes, northern Chile. *Journal of South American Earth Sciences*, *87*, 200–210. https://doi.org/10.1016/j.jsames.2017.11.013

Sarg, J. F. (2001). The sequence stratigraphy, sedimentology, and economic importance of evaporite-carbonate transitions: A review. *Sedimentary Geology*, *140*, 9–42.

Schreiber, B. C. (1978). Environments of subaqueous gypsum deposition. In W. E. Dean & B. C. Schreiber (Eds.), *Marine evaporites, Society of Economic Paleontologists and Mineralogists, Lecture Notes for Short Course* (Vol. 4, pp. 43–73). Tulsa, Okla.

Schreiber, B. C., & Hsu, K. J. (1980). Evaporites. In G. D. Hobson (Ed.), *Developments in Petroleum geology* (Vol. 2, pp. 87–138). London: Applied Science Publishers.

Schubel, K. A., & Lowenstein, T. K. (1997). Criteria for the recognition of shallow-perennial-saline-lake halites based on Recent sediments from the Qaidam Basin, Western China. *Journal of Sedimentary Research*, *67*(1), 74–87.

Scoon, R. N. (2018). Lake Natron and the Oldoinyo Lengai Volcano. In *Geology of National Parks of Central/Southern Kenya and Northern Tanzania* (pp. 193–206). Switzerland: Springer Nature. doi:10.1007/978-3-319-73785-0_17

Seard, C., Camoin, G., Rouchy, J.-M., & Virgone, A. (2013). Composition, structure and evolution of a lacustrine carbonate margin dominated by microbialites: Case study from the Green River formation (Eocene; Wyoming, USA). *Palaeogeography, Palaeoclimatology, Palaeoecology*, *381–382*, 128–144.

Słowakiewicz, M., Tucker, M. E., Pancost, R. D., Perri, E., & Mawson, M. (2013). Upper Permian (Zechstein) microbialites: Supratidal through deep subtidal deposition, source rock, and reservoir potential. *American Association of Petroleum Geologists Bulletin*, *97*(11), 1921–1936.

Smoot, J. P., & Lowenstein, T. K. (1991). Depositional environments of non-marine evaporites. In J. L. Melvin (Ed.), *Evaporites, petroleum and mineral resources, Developments in Sedimentology* (Vol. 50, Chapter 3, pp. 189–347). Amsterdam: Elsevier.

Sociedad Química y Minera de Chile (2022). Technical Report Summary – Operation Report Salar de Atacama. 251 pp. https://minedocs.com/22/Salar-de-Atacama-TR-042022.pdf

Sonnenfeld, P. (1979). Brines and evaporites. *Geoscience Canada*, *6*(2), 83–90.

Sonnenfeld, P. (1985). Evaporites as oil and gas source rocks. *Journal of Petroleum Geology*, *8*(3), 253–271. doi:10.1111/j.1747-5457.1985.tb00270.x

Sonnenfeld, P. (1989). Part I. Genesis of evaporites. In P. Sonnenfeld & J. -P. Perthuisot (Eds.), *Brines and evaporites, Short Course in Geology: Volume 3*, 1–64. *28th International Geological Congress*, American Geophysical Union, Washington, DC.

Sonnenfeld, P. (1992). Genesis of marine evaporites—A summation. *Geologica Carpathica*, *43*(5), 259–274.

Sousa Júnior, G. R., Santos, A. L. S., Lima, S. G., Lopes, J. A. D., Reis, F. A. M., Santos Neto, E. V., & Chang, H. K. (2013). Evidence for euphotic zone anoxia during the deposition of Aptian source rocks based on aryl isoprenoids in petroleum, Sergipe–Alagoas Basin, northeastern Brazil. *Organic Geochemistry*, *63*, 94–104. https://doi.org/10.1016/j.orggeochem.2013.07.009

Spencer, R. J. (1987). Origin of Ca-Cl brines in Devonian formations, western Canada sedimentary basin. *Applied Geochemistry*, *2*, 373–384.

Spencer, R. J., Baedecker, M. J., Eugster, H. P., Forester, R. M., Goldhaber, M. B., Jones, B. F., et al. (1984). Great Salt Lake, and precursors, Utah: The last 30,000 years. *Contributions to Mineralogy and Petrology*, *86*(4), 321–334. doi:10.1007/bf01187137

Spencer, R. J., Lowenstein, T. K., Casas, E., & Pengxi, Z. (1990). Origin of potash salts and brines in the Qaidam Basin, China. In Spencer, R. J., & I-Ming Chou (Eds.), *Fluid-Mineral interactions: A tribute to H. P. Eugster* (No.2, pp. 395–408). The Geochemical Society, Special Publication.

Stern, C. R. (2004). Active Andean volcanism: Its geologic and tectonic setting. *Revista Geológica de Chile*, *31*(2), 161–206.

Stober, I., Zhong, J., & Buche, K. (2023). From freshwater inflows to salt lakes and salt deposits in the Qaidam Basin, *W China. Swiss Journal of Geosciences*, *116*, 5. https://doi.org/10.1186/s00015-023-00433-4

Sun, Y., Shaw, J. H., Guan, S., Ma, D., & Ma, X. (2020). Development and growth of basement-involved structural wedges in the northwestern Qaidam Basin, China. *American Association of Petroleum Geologists Bulletin*, *104*(5), 1091–1113. doi: 10.1306/11111918076

Suotang, F. (2016). Potential oil and gas exploration areas in Qaidam Basin. *China Petroleum Exploration*, *21*(5), 1–11.

Suotang, F., Dade, M., Yan, C., Guoqink, Z., & Kunyu, W. (2016). New advance of petroleum and gas exploration in Qaidam Basin. *Acta Petrolei Sinica*, *37*(S1), 1–10. doi: 10.7623/syzb2016S1001

Suotang, F., Dade, M., Zhaojie, G., & Feng, C. (2015). Strike-slip superimposed Qaidam Basin and its control on oil and gas accumulation, *NW China. Petroleum Exploration and Development*, *42*(6), 778–789.

Szatmari, P., & Milani, E. J. (2016). Tectonic control of the oil-rich large igneous-carbonate-salt province of the South Atlantic rift. *Marine and Petroleum Geology*, *76*, 567–596.

Szatmari, P., Moré de Lima, C., Fontaneta, G., de Melo Lima, N., Zambonato, E., Menezes, M. R., et al. (2021). Petrography, geochemistry and origin of South Atlantic evaporites: The Brazilian side. *Marine and Petroleum Geology*, *127*, 104805. https://doi.org/10.1016/j.marpetgeo.2020.104805

Thompson, D. L., Stilwell, J. D., & Hall, M. (2015). Lacustrine carbonate reservoirs from Early Cretaceous rift lakes of Western Gondwana: Pre-Salt coquinas of Brazil and West Africa. *Gondwana Research*, *28*, 26–51.

Thompson, R. S., Oviatt, C. G., Honke, J. S., & McGeehin, J. P. (2016). Late quaternary changes in lakes, vegetation, and climate in the Bonneville basin reconstructed from sediment cores from Great Salt Lake. In C. G. Oviatt & J. F. Shroder, Jr. (Eds.), *Developments in Earth Surface Processes* (Vol. 20, pp. 221–291). doi:10.1016/b978-0-444-63590-7.00011-1

Umhoefer, P. J. (2011). Why did the Southern Gulf of California rupture so rapidly?—Oblique divergence across hot, weak lithosphere along a tectonically active margin. *GSA Today*, *21*(11), 4–10. doi: 10.1130/G133A.1

Umhoefer, P. J., Schwenicke, T., Del Margo, M. T., Ruiz-Geraldo, G., Ingle, J. C., Jr., & McIntosh, W. (2007). Transtensional fault-termination basins: An important basin type illustrated by the Pliocene San Jose Island basin and related basins in the southern Gulf of California, Mexico. *Basin Research*, *19*, 297–322.

UNCOD (1978). *United Nations Conference on Desertification, 29 August–9 September 1977. Round-Up, Plan of Action and Resolutions*. New York: United Nations, 43 pp. http://www.ciesin.org/docs/002-478/002-478.html

U.S. Geological Survey (2023). Mineral commodity summaries. *U.S. Geological Survey*, 210 pp. https://pubs.er.usgs.gov/publication/mcs2023

Utah Water Science Center (2021). Great salt lake elevations. Utah Water Science Center, *Science*, website. https://www.usgs.gov/centers/utah-water-science-center/science/great-salt-lake-elevations

Varley, N. (2012). Volcanoes of Mexico. In R. A. Meyers (Ed.), *Encyclopedia of sustainability science and technology*. New York, NY: Springer. https://doi.org/10.1007/978-1-4419-0851-3_477

Varushchenko, A. N., Lukyanova, S. A., Solovieva, G. D., Kosarev, A. N., & Kurayev, A. V. (2000). Evolution of the Gulf of Kara-Bogaz-Gol in the past century. In K. P. Lulla & L. V. Dessinov (Eds.), *Dynamic Earth environments: Remote sensing observations from Shuttle-Mir missions* (pp. 201–210). New York: John Wiley & Sons.

Wang, E., Xu, F.-Y., Zhou, J.-X., Wan, J., & Burchfiel, B. C. (2006). Eastward migration of the Qaidam basin and its implications for Cenozoic evolution of the Altyn Tagh fault and associated river systems. *Geological Society of America Bulletin*, *118*(3/4), 349–365.

Wang, W., Zheng, W., Zhang, P., Li, Q., Kirby, E., Yuan, D., et al. (2017). Expansion of the Tibetan Plateau during the Neogene. *Nature Communications*, *8*, 15887, 1–12. doi: 10.1038/ncomms15887.

Wang, Y., Zheng, J., Zhang, W., Li, S., Liu, X., Yang, X., & Liu, Y. (2012). Cenozoic uplift of the Tibetan Plateau: Evidence from the tectonic–sedimentary evolution of the western Qaidam Basin. *Geoscience Frontiers*, *3*(2), 175–187.

Warren, J. K. (1986). Shallow-water evaporitic environments and their source rock potential. *Journal of Sedimentary Petrology*, *56*(3), 442–454.

Warren, J. K. (1989). *Evaporite sedimentology: Importance in hydrocarbon accumulation*, New Jersey, USA: Prentice-Hall Advanced Reference Series, 285 pp.

Warren, J. K. (2006). *Evaporites: Sediments, resources and hydrocarbons*. Berlin: Springer Verlag, 1035 pp.

Warren, J. K. (2012). Understanding hydrocarbons in an evaporite context. *AAPG Search and Discovery*, Article #50549 (*2012*).

Warren, J. K. (2014). *Evaporite sedimentology: Importance in hydrocarbon accumulation*, New Jersey, USA: Prentice-Hall Advanced Reference Series, 285 p.

Warren, J. K. (2016). *Evaporites—A geological compendium*. Berlin: Springer, 1807 pp. ISBN 978-3-319-13511-3.

Wei, M., Bao, Z., Munnecke, A., Liu, W., Harrison, G. W. M., Zhang, H., et al. (2021). Paleoenvironment of the Lower–Middle Cambrian Evaporite Series in the Tarim Basin and its impact on the organic matter enrichment of shallow water source rocks. *Minerals*, *11*, 659. https://doi.org/10.3390/min11070659

Wilkes, E., & Gorler, K. (1994). Sedimentary and structural evolution of the Salar de Atacama depression. In K.-J. Reutter, E. Scheuber, & P. J. Wigger (Eds.), *Tectonics of the Southern Central Andes: Structure and evolution of an active continental margin* (pp. 171–188). Berlin: Springer-Verlag.

Wright, P., & Rodriguez, K. (2018). Reinterpreting the South Atlantic Pre-Salt "Microbialite" reservoirs: Petrographic, isotopic and seismic evidence for a shallow evaporitic lake depositional model. *First Break*, *36*(5), 71–77.

Wright, V. P., & Barnett, A. J. (2015). An abiotic model for the development of textures in some South Atlantic Early Cretaceous lacustrine carbonates. In D. W. J. Bosence, K. A. Gibbons, D. P. Le Heron, W. A. Morgan, T. Pritchard, & B. A. Vining (Eds.), *Microbial carbonates in space and time: Implications for Global Exploration and Production* (Vol. 418, pp. 209–219). Geological Society, London, Special Publications.

Wright, V. P., & Barnett, A. (2017). Critically evaluating the current depositional models for the Pre-Salt Barra Velha formation, *Offshore Brazil. AAPG Search and Discovery*, Article #51439 (*2017*), Posted November 20, 2017.

Wu, L., Xiao, A., Ma, D., Li, H., Xu, B., Shen, Y., & Mao, L. (2014). Cenozoic fault systems in southwest Qaidam Basin, northeastern Tibetan Plateau: Geometry, temporal development, and significance for hydrocarbon accumulation. *American Association of Petroleum Geologists Bulletin*, *98*(6), 1213–1234.

Wurtsbaugh, W. A., Miller, C., Null, S. E., DeRose, R. J., Wilcock, P., Hahnenberger, M., et al. (2017). Decline of the world's saline lakes. *Nature Geoscience*, *10*(11), 816–821. doi:10.1038/ngeo3052

Xiao, L., Wang, J., Dang, Y., Cheng, Z., Huang, T., Zhao, J., et al. (2017). A new terrestrial analogue site for Mars research: The Qaidam Basin, Tibetan Plateau (NW China). *Earth-Science Reviews*, *164*, 84–101.

Yu, J., Gao, C., Cheng, A., Liu, Y., Zhang, L., & He, X. (2013). Geomorphic, hydroclimatic and hydrothermal controls on the formation of lithium brine deposits in the Qaidam Basin, northern Tibetan Plateau, China. *Ore Geology Reviews*, *50*, 171–183. http://dx.doi.org/10.1016/j.oregeorev.2012.11.001

Zhang, G., Xie, H., Duan, S., Tian, M., & Yi, D. (2011). Water level variation of Lake Qinghai from satellite and in situ measurements under climate change. *Journal of Applied Remote Sensing*, *5*, 053532.1–053532.15.

Zhang, X., Fan, Li, Q., Du, Y., Qin, Z., Wei, H., & Shan, F. (2019). The source, distribution, and sedimentary pattern of K-rich brines in the Qaidam Basin, *Western China. Minerals*, *9*(11), 655. https://doi.org/10.3390/min9110655

Zheng, M., & Liu, X. (2009). Hydrochemistry of salt lakes of the Qinghai-Tibet Plateau, *China. Aquatic Geochemistry*, *15*(1), 293–320. doi:10.1007/s10498-008-9055-y

Zheng, M., Meng, Y., & Wei, L. (2000). Evidence of the Pan-Lake stage in the period of 40-28 ka B.P. on the Qinghai-Tibet Plateau. *Acta Geologica Sinica*, *74*(2), 266–272.

Zheng, M., Zhang, Y., Liu, X., Qi, W., Kong, F., Nie, Z., et al. (2016). Progress and prospects of Salt Lake Research in China. *Acta Geologica Sinica*, *90*(4), 1195–1235. https://doi.org/10.1111/1755-6724.12767

Zhenhan, W., Barosh, P. J., Zhonghai, W., Daogong, H., Xun, Z., & Peisheng, Y. (2008). Vast early Miocene lakes of the central Tibetan Plateau. *Geological Society of America Bulletin*, *120*(9/10), 1326–1337.

11

Evaporite Deposition in Lacustrine and Shallow Marine Carbonate Platform Environments

ABSTRACT

Genetic models for evaporite deposition in lacustrine and shallow marine carbonate platforms are based on modern sedimentary environments in rift basins like the Death Valley in California and ancient lacustrine environments like the Rhine Graben in Central Europe. The Dead Sea, a transtensional sedimentary basin with a deep lake, is the only modern example of halite accumulation in a deepwater lake. Ancient carbonate platforms with evaporite deposition in lagoons or inland seas are recorded in several Paleozoic intracratonic basins in North America and South America.

This chapter presents the sedimentary models for the development of evaporite sequences in lacustrine and shallow marine carbonate platform environments. Some classical examples of Quaternary to Holocene rift basins in extensional, convergent, and strike-slip settings are discussed as modern analogs for evaporite deposition in hypersaline lakes. The chapter presents a summary of the hydrological characteristics of modern and ancient evaporite basins with continental fluvial-lacustrine environments, sabkhas, and shallow marine carbonate platforms.

The selected examples of Cenozoic, Mesozoic, and Paleozoic evaporitic basins are discussed based on the observations and concepts established by applying the uniformitarianism principle (see Chapter 8). However, the small size of the modern continental lakes and marginal sabkha carbonate platforms is several orders of magnitude less than the ancient lacustrine and carbonate platform salt basins (Figure 11.1).

Quaternary to Holocene salt deposits occur in limited locations globally, including lacustrine environments, sabkha flats, shallow marine carbonate platforms, and nonmarine continental basins. These salt accumulations are smaller in size and thickness compared to their ancient counterparts (Figure 11.1). Evaporite sediments formed in continental lakes and shallow carbonate platforms differ in hydrology and lithological associations. Hypersaline lakes are associated with fluvial sources, surface runoff, and hydrothermal sources, while lacustrine sediments are influenced by climate factors that control the brine salinity (Hardie et al., 1978; Smoot & Lowenstein, 1991; Zeng et al., 2012). Evaporating lakes are characterized by periodic desiccation events that expose the lake's bottom, forming subaerial evaporite crusts (Figure 11.2). In shallow carbonate platforms, evaporite deposits are mainly associated with concentrated brine derived from seawater entering the isolated basin by marine ingressions across a permeable barrier or a shallow sill (Hsü, 1972; Shearman, 1978; Warren, 1991, 2006, 2010, 2016). The evaporitic basin is associated with a sabkha environment that merges with a shallow carbonate platform and a deep basin, where salterns may accumulate evaporite sediments in a submarine environment (Figure 11.3).

The evaporite sediments on continental platforms obey the chemical precipitation sequence following the concentration of normal marine seawater that is concentrated by solar evaporation. The basin brine is dominated by the ions Na^+ and Cl^-, with lesser amounts of SO_4^{2-}, Mg^{2+}, Ca^{2+}, K^+, CO_3^{2-}, and HCO_3^- (Bąbel & Schreiber, 2014; Hardie, 1984, 1991).

Evaporite minerals precipitate when the brine concentration reaches the mineral solubility limit (see Chapter 2). The evaporite sequences are normally associated with a suite of minerals that range from carbonate

Salt in the Earth Sciences: Evaporite Rocks and Salt Deposition, Geophysical Monograph 286, First Edition. Webster Mohriak.

DOI: 10.1002/9781119405245.ch11

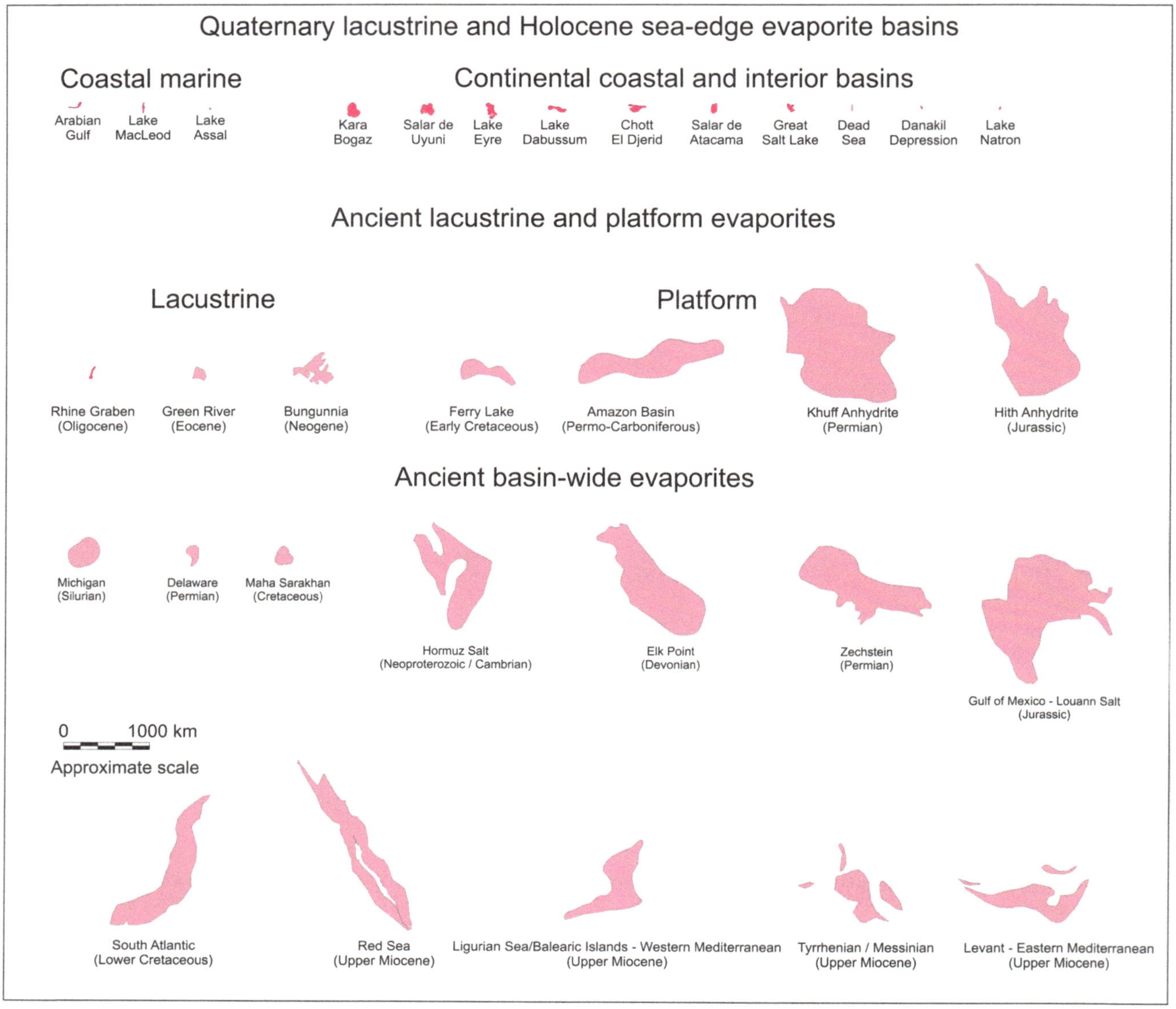

Figure 11.1 Comparison of Quaternary lacustrine and Holocene sea-edge evaporite basins with simplified outlines of ancient salt basins at the same scale. Quaternary lacustrine salt basins form interior basins in coastal and desert regions with no oceanic connection. Lacustrine and platform environments near open oceans or seaways are related to ancient continental and platform carbonate-evaporite basins. Rift basins (e.g., Hormuz salt), foreland basins (Elk Point, Zechstein), divergent margins (Gulf of Mexico, South Atlantic, Red Sea), and convergent oceanic basins (Mediterranean Sea) record basinwide evaporites.

(calcite and aragonite) and sulfate (gypsum and anhydrite), to chloride (halite, carnallite, sylvite, tachyhydrite, etc.). Evaporite minerals deposited in continental lacustrine environments are more diverse than their counterparts precipitated in normal marine environments. The mineralogy of the evaporite sequences depends on several factors that affect the hydrology of the continental lakes, including (1) the ionic content of the surface runoff, fluvial water, and hydrothermal discharge feeding the lake; (2) the type of lithology constituting the substratum of the elevated regions adjacent to the lake margins; (3) the weathering and leaching processes, dissolving the most mobile chemical compounds, which are transferred to the underground aquifers; (4) the chemical composition of the sedimentary layers in the subsurface and the groundwater; (5) the geological events affecting the surface runoff, such as volcanic events with pyroclastic material; and (6) the biological activity in the lacustrine

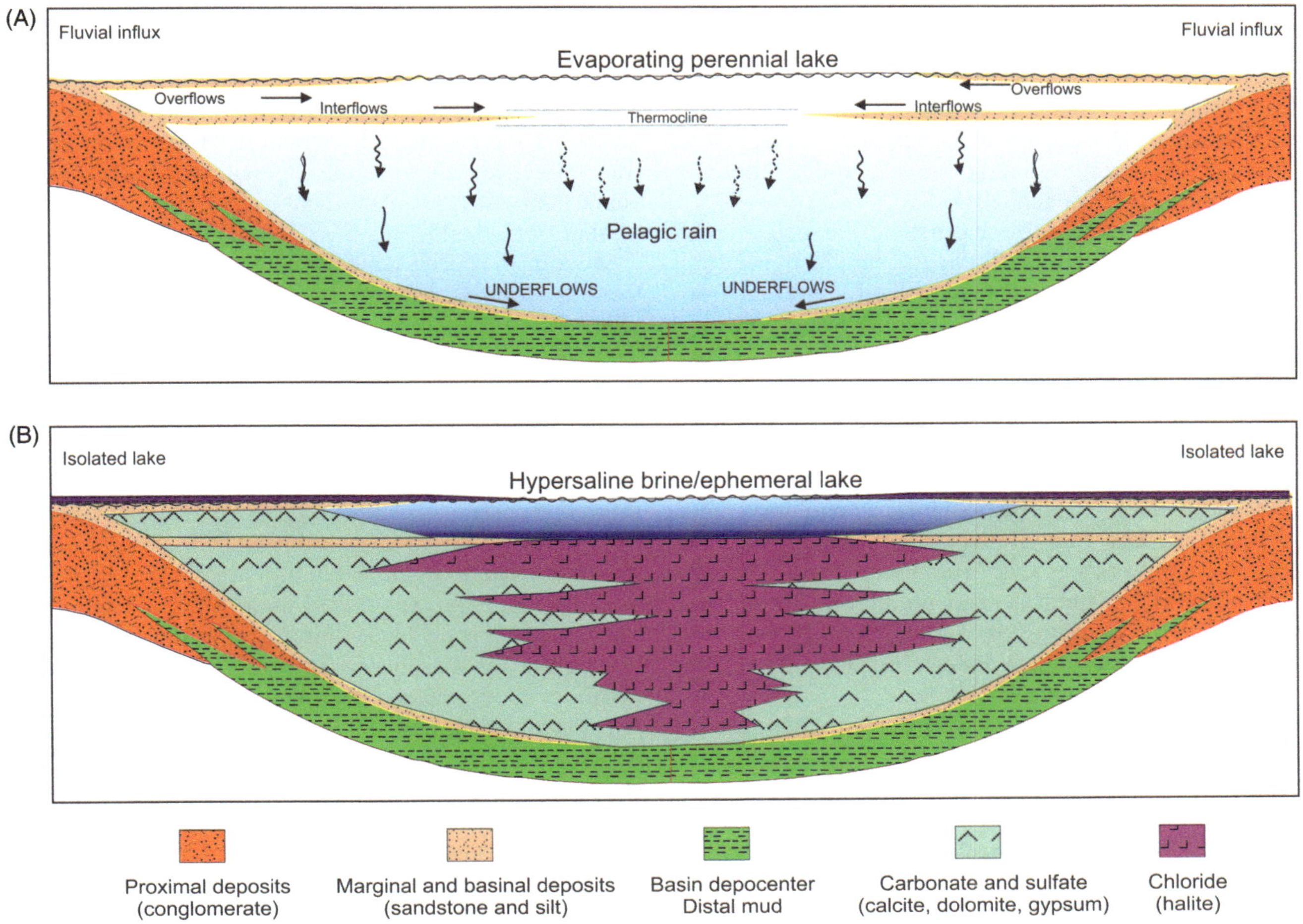

Figure 11.2 Schematic diagrams illustrate the deposition of evaporite sediments in lacustrine basins subjected to desiccation. (A) shows an evaporating perennial lake with fluvial input and pelagic rain of siliciclastic, carbonate, and evaporite sediments. (B) shows an ephemeral lake with salt deposition when the hypersaline brine is almost completely evaporated, forming a remnant brine pool. Credit: Galloway and Hobday (1983)/with permission of Springer Nature.

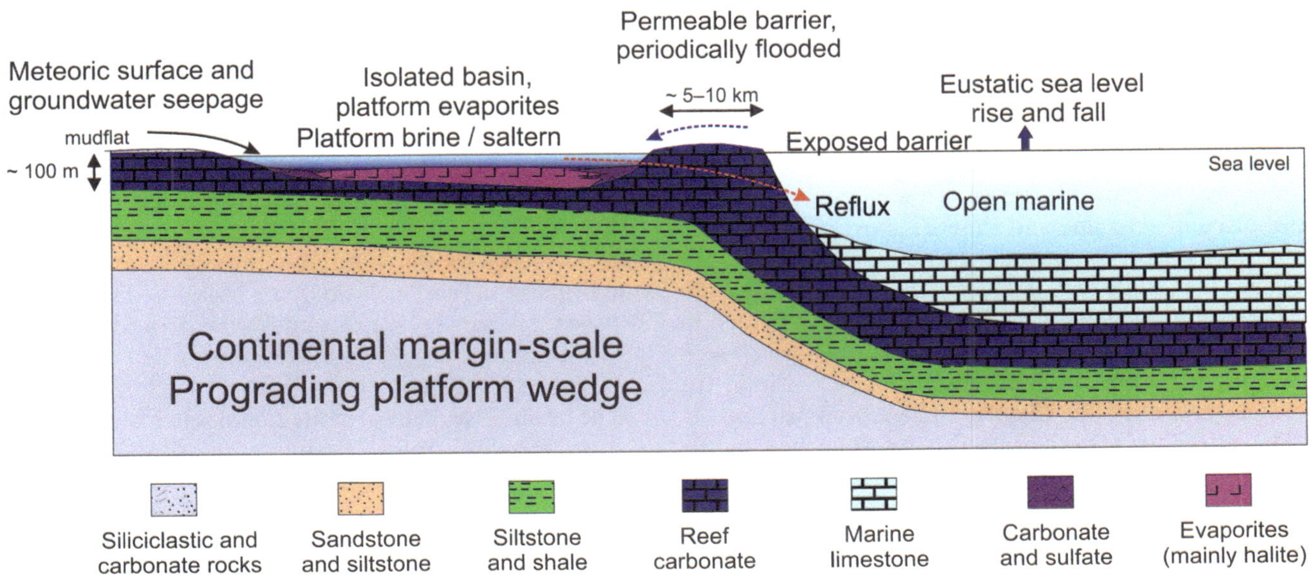

Figure 11.3 The diagram illustrates a genetic model for evaporite sedimentation in a closed basin platform, where a permeable barrier separates the platform from the open marine environment. Lagoons and carbonate platforms are filled with groundwater seepage, meteoric water runoff, and marine ingressions. Solar evaporation results in the concentration and precipitation of evaporite sediments in a subaqueous environment, forming salterns. Sea level rise and fall control the flooding and desiccation episodes. The different evaporite facies include carbonate, sulfate, and chloride deposits. Hypersaline brine reflux through the permeable barrier affects the ocean side, but no evaporites precipitate in the open sea, except for clastic evaporite layers formed by reworking and resedimentation of platform evaporite deposits. Credit: Warren (2016)/with permission of Springer Nature.

environments (Deocampo & Jones, 2014; Eugster & Hardie, 1978; Schnurrenberger et al., 2003; Smoot & Lowenstein, 1991; Warren, 2018).

Many different minerals are genetically associated with a continental environment (lakes and sabkhas), resulting in a suite of minerals that are distinctly different from evaporite deposits sourced from marine sources (Smoot & Lowenstein, 1991). Hypersaline continental lakes are characterized by sodium carbonate and sulfate, such as natron, trona, burkeite, and mirabilite (see Chapter 4). The distinction between hydrological types is not straightforward and depends on the original brine and the abundance of common evaporite minerals accumulated on the basin's floor. For example, some continental lakes in Central Turkey, such as Lake Tuz, Lake Seyfe, and Sivas Basin, exhibit both marine and nonmarine sources for the Quaternary evaporite deposits (Çemen et al., 1999; Gürbüz & Kazanci, 2014; Palmer et al., 2004; Pichat et al., 2018; Önalgil et al., 2015).

Isotope analysis of modern evaporite deposits in several Turkish lakes indicates the recycling from older sequences of sedimentation in the Paleocene to the Miocene (e.g., Palmer et al., 2004). Facies distribution maps of continental and marine-derived salt deposits in the Anatolia region of Central Turkey show a belt of gypsum-rich sediments in lacustrine basins closer to marine input from the Mediterranean Sea during the Messinian (Palmer et al., 2004; Pichat et al., 2018).

Lake Tuz, the second-largest lake in Turkey, forms an evaporitic basin that overlies a Late Cretaceous to Tertiary rift basin characterized by continental sedimentation associated with alluvial, fluvial, and lacustrine environments (e.g., Aydin et al., 2020). The Tuz Gölü Basin is the largest and deepest inner basin among the Central Anatolian Cenozoic basins, with over 5k Neogene sediments (Çemen et al., 1999; Galamay et al., 2023). The endorheic basin is currently isolated from the Mediterranean Sea and precipitates evaporite minerals on the lake's margins, forming a crust of halite, gypsum, and kieserite above a muddy layer with other evaporite minerals, including gypsum, magnesite, and thenardite (e.g., Kiliç & Kilic, 2010). In the subsurface, thick, shallow marine sediments are recorded in the Neogene, including evaporite (gypsum) deposits, which are associated with the Messinian Salinity Crisis in the Mediterranean Sea (see Chapter 5 in Mohriak, 2025).

The overlying Quaternary lacustrine evaporite deposits in Lake Tuz is derived from continental sources (Gürbüz & Kazanci, 2014; Palmer et al., 2004). Geochemical analysis of sulfate evaporite minerals from the southeastern part of the Tuz Gölü Basin indicates that the brines that precipitated the salt layers are mainly of marine origin. The inflow of seawater into the sedimentary basin occurred due to Messinian transgressions. The isotopic composition of fluid inclusions in evaporite minerals suggests that the chemical composition of the basin brines was influenced by volcanic emanations, hydrothermal vents, meteoric (pluvial), fluvial (river), and groundwater (Galamay et al., 2023).

Modern evaporite deposits in Lake Seyfe, located in Central Turkey, are associated with the reworking of the Oligo-Pliocene continental evaporites, interpreted as the source for the evaporite layers presently accumulating on the lake's bottom (Önalgil et al., 2015). The Sivas Basin, located in northeast Turkey, is marked by Quaternary evaporite deposits derived from the reworking of the thick salt layers formed during the Late Eocene marine ingressions. The Paleogene evaporite basin was fed from marine sources associated with the closure of the southern branch of the Neotethys Sea, as the Anatolian and Arabian plates subducted under the Eurasian plate. There is isotopic evidence that the Quaternary evaporites result from the dissolution and recycling of the Eocene marine evaporite layers (Pichat et al., 2018).

The shortening and uplift of the continental crust in the Anatolian plate during the Neogene resulted in the shallowing of the marine environment, forming several evaporite basins in the arid to semiarid climate. Subsequently, the Messinian Salinity Crisis resulted in widespread evaporite deposition in the Mediterranean Sea region during the transition to the Pliocene. Modern evaporite deposits in the Anatolian region are sourced from the reworking and resedimentation of different sources (Çemen et al., 1999; Gürbüz & Kazanci, 2014; Önalgil et al., 2015; Palmer et al., 2004; Pichat et al., 2018). More advanced chemical analysis involving stable isotope measurements (e.g., strontium, sulfur, and oxygen) is required to distinguish between the environments of deposition and brine sources of evaporites associated with hypersaline lakes (e.g., Deocampo & Jones, 2014).

Aiming at distinguishing different hydrological types, some researchers have integrated sedimentological, mineralogical, and geochemical data in several peripheral basins adjacent to the Mediterranean Sea. These techniques, including the isotopic composition of sulfates and bromine in fluid inclusions, are useful to differentiate between the primary sources and brine derived from the dissolution and recycling of older evaporite layers (e.g., Taberner et al., 2000).

The bromine content in halite samples has been used as a parameter to distinguish between the continental and marine origins of many evaporite deposits worldwide, including the Dead Sea and the Neogene basins in the Mediterranean Sea region. For instance, the Cenozoic lacustrine sediments in northern Turkey and central Spain are characterized by the precipitation of several evaporite minerals that indicate continental sources, with no influence of marine water from the Mediterranean Sea (Smoot & Lowenstein, 1991).

The isotope measurements and environmental geochemical markers in several Neogene basins indicate that some deposits were affected by marine ingressions, whereas others are genetically linked to continental water inflow only (e.g., Palmer et al., 2004; Taberner et al., 2000). Based on the strontium and sulfur isotope composition of the halite minerals, several researchers have discussed the sedimentological and hydrological interpretation of the evaporite deposits in the Jordan Valley and the Dead Sea (e.g., Raab et al., 1997, 2000).

Some hypersaline lakes are separated from the main body of normal marine seawater by volcanic barriers, sedimentary barriers, and coastal depressions, such as Lake Assal in Djibouti, which is intermittently connected with the Gulf of Aden. Other shallow hypersaline lakes are separated from the inland seas or adjacent oceans by sedimentary barriers, such as the Kara-Bogaz in Turkmenistan, which is fed by the Caspian Sea. The McLeod Lake in northwest Australia is an example of a coastal depression adjacent to an oceanic region. The lagoon was intermittently connected to the Indian Ocean in the Holocene, accumulating thick successions of halite and gypsum (see Chapter 9). The hypersaline lakes and salt flats in the Tibetan Plateau and the Andean Altiplano are remnants of much larger lacustrine systems developed at the end of the last glacial age on Earth (see Chapter 10).

This chapter presents selected examples of hypersaline lakes within rift basins, continental sabkhas, marginal depressions, and continental platforms. The evaporite sequences in ancient lacustrine environments and carbonate platforms are analyzed within a plate tectonics context, including information obtained from the subsurface, as most of these basins have been drilled by the salt and petroleum industries. The genesis of the ancient evaporite basins has been historically studied by applying direct geological methods, including field descriptions of scattered outcrops in orogenic belts (see Chapter 3). However, the giant saline basins have been mostly analyzed by indirect methods, particularly the interpretation of reflection seismic data (see Chapter 5).

The interpretation of seismic horizons in the subsurface is often based on stratigraphic data from drilling exploratory boreholes. Ancient evaporite basins show significant intervals of salt accumulation with geological characteristics similar to modern sedimentary environments (Schreiber, 1988; Schreiber & Hsü, 1980; Sonnenfeld & Perthuisot, 1989; Warren, 1989, 1991, 2010, 2016).

Seismic data and results from exploratory boreholes are used to present a comprehensive overview of salt deposition models associated with the Rhine Graben in Central Europe. This chapter provides an overview of the geology of carbonate platforms with evaporite deposition in the Paleozoic and Mesozoic foreland and intracratonic basins of North America. It also discusses Cretaceous carbonate-evaporite platform deposits in the northeastern Gulf of Mexico and onshore Florida. The Permo-Carboniferous Amazon salt basin in Brazil is discussed based on seismic data and results from exploratory wells. Finally, some geochemical characteristics of the Early Cretaceous lacustrine deposits associated with prolific source rocks in the South Atlantic rift basins are summarized.

11.1. SALT DEPOSITION IN ACTIVE CONTINENTAL RIFTS

The study of evaporite deposition in sabkha and coastal marine environments has long been a topic of interest in geological literature (see Chapter 8). The search for analogs involving lacustrine salt deposits in active continental rifts has provided numerous examples of small ephemeral hypersaline lakes where thick halite deposits have accumulated in the Holocene (see Chapter 9 and Chapter 10). Most published studies have focused on Pleistocene-Holocene examples in desert areas of Australia, northern Africa, and the southwestern United States. Examples of modern salt basins formed in lacustrine environments related to extensional tectonics in the East African Rift System include Lake Natron, located at the border between Tanzania and Kenya (see Chapter 9). Several examples of hypersaline lakes associated with strike-slip faults are reported in the Death Valley of California, such as the Searles and Badwater lakes (Ku et al., 1998; Smoot & Lowenstein, 1991).

The Basin and Range and the Baja California region are marked by sedimentary basins with shallow water evaporite deposits accumulated in continental environments (playa lakes and sabkhas), shallow marine environments (platforms), and deep marine basins associated with transform fault zones, such as the Gulf of California of southwestern United States and Mexico (see Chapter 9). This region constitutes a classical example of a deep marine pull-apart basin that accumulated thick evaporite sequences in the Miocene (Miller & Lizarralde, 2013).

Marine-fed evaporite basins forming lagoons and coastal sabkhas correspond to onshore depressions connected to an open sea that periodically overflows or seeps into the basin depocenter, as the Afar region adjacent to the Red Sea, the Trucial Coast plain in the Persian Gulf, and the MacLeod Lake in NW Australia. Excessive evaporation results in lowering the water surface in the coastal depression, eventually reaching altitudes below the global sea level, as the Danakil Depression in the Afar region, or Lake Assal in Djibouti (see Chapter 9).

Marine waters flow into lacustrine basins during the sea-level rise, replacing the original pore fluids with more concentrated brines after evaporation, resulting in a lowering of the lake level. This replenishment mechanism is known as an evaporative drawdown. Most modern shallow-water evaporite deposits are forming in coastal lakes or salinas located in depressions close to or even below the sea level (Warren & Kendall, 1985).

Subaqueous evaporite environments can be divided into two types depending on the water chemistry that feeds the basin. The first hydrological end-member corresponds to a nonmarine setting, with the lacustrine environments exclusively supplied by continental waters. The second type corresponds to marine waters filling coastal lakes, gulfs, inland seas, and oceanic embayments (Handford, 1991). Continental lacustrine environments that have no marine input depend exclusively on meteoric groundwaters, surface runoff, and springs or seeps to renew the water supply to the evaporite basin. These continental hypersaline lakes are often observed in regions affected by active continental rifting, such as the California lakes in the Basin and Range Province. In addition, there are hypersaline lakes in elevated regions affected by plate convergence, such as the Andean highlands or the Tibetan Plateau in China (see Chapter 10).

Modern examples of salt deposition in continental lacustrine environments associated with extensional, convergent, and transform plate boundaries are found in the Death Valley in California, the Tularosa Basin in New Mexico, and Lake Tuz in Central Turkey. The Kara-Bogaz Gol, one of the saltiest lakes in the world, is a classic example of a salt basin with evaporite precipitation in the Holocene (see Chapter 10).

Rift basins with evaporite deposits are dominantly associated with closed hydrology lakes, with the water level below the lake sill or barrier. The evaporite layers accumulate in underfilled basins, with the lake level controlled by the climate. The hypersaline brine precipitates evaporite minerals according to the salinity (see Chapter 2). Most modern sedimentary environments accumulating evaporite deposits are related to desert regions where continental lakes were formed by active extensional or compressional tectonics. These lakes are periodically desiccated by evaporation due to the extremely arid climates, as in the Death Valley of California and the Tibetan Plateau in China (see Chapters 9 and 10). Three main lithological facies are recognized in the lacustrine environments: clastic, biogenic, and chemical, forming siliciclastic, carbonate, and evaporite layers (Bohacs et al., 2003).

11.1.1. Facies Distribution in Rift Basin Lacustrine Systems

The modern continental hypersaline lakes in desert regions (for example, the Saharan Desert in northern Africa, or the active rift basins in the Basin and Range Province of the southwestern United States) have developed sedimentary facies varying from alluvial fans, deltaic aprons, evaporitic mudflats, sabkhas, and salinas. These lakes are characterized by three main lithological associations.

1. Alluvial fans are associated with active rift border faults located adjacent to highlands at the uplifted flank of the rift basin. Sediments are transported across canyons by rivers that flow during flash floods. As the sedimentary debris enters the depression, the flow velocity is reduced and the sediments are deposited in fan deltas and ephemeral fluvial systems (Blair & McPherson, 1994). The process results in the deposition of conglomerates and poorly sorted sandstones within fault-bounded troughs. Evaporitic mudflats in continental sabkhas and adjacent salinas develop in subaerial to subaqueous settings, such as intermontane playas in the Basin and Range Province of the southwest United States, salt lakes in the East African rift valleys, and saline valleys at the southern end of the Dead Sea transform fault (DSTF) zone (see Chapter 9).

2. Ephemeral stream floodplains and deltaic aprons are formed within desert regions where intermittent drainage with a fluvial-lacustrine system is observed during wet seasons. These systems link with the delta and lacustrine systems in the depression. Ephemeral hypersaline lakes are found in the central part of depressions, while ephemeral floodplains are found in cold, arid settings associated with uplifted regions in compressional provinces. The groundwater flow system is the source of the brines that flow from the elevated regions to the depression, where the water table is extremely shallow, a few centimeters below the ground. In Central Turkey, Lake Tuz is marked by an uplifted eastern margin of the rift depression, where an alluvial fan system links with the lacustrine environment in the basin depocenter. Due to climate aridity, the evaporite sediments that precipitated on the bottom of ephemeral lakes are exposed subaerially when the basin dries out, forming a gypsum crust (Aydin et al., 2020; Kiliç & Kilic, 2010).

3. Perennial stream floodplains are characterized by perennial saline lakes, where brine is concentrated by evaporation, reaching the point of halite saturation. These environments are often associated with larger and deeper paleolakes and extensive paleo drainage systems, such as the Great Salt Lake in Utah See Chapter 10).

Modern analogs of lacustrine systems in rift basins with Holocene evaporite deposits include the Death Valley in California and the Tularosa Basin in New Mexico.

11.1.2. The Death Valley in California

Death Valley in eastern California is a modern example of a fluvial-lacustrine system developed within an active rift in a desert region Figure 11.4. The rift basin's deep

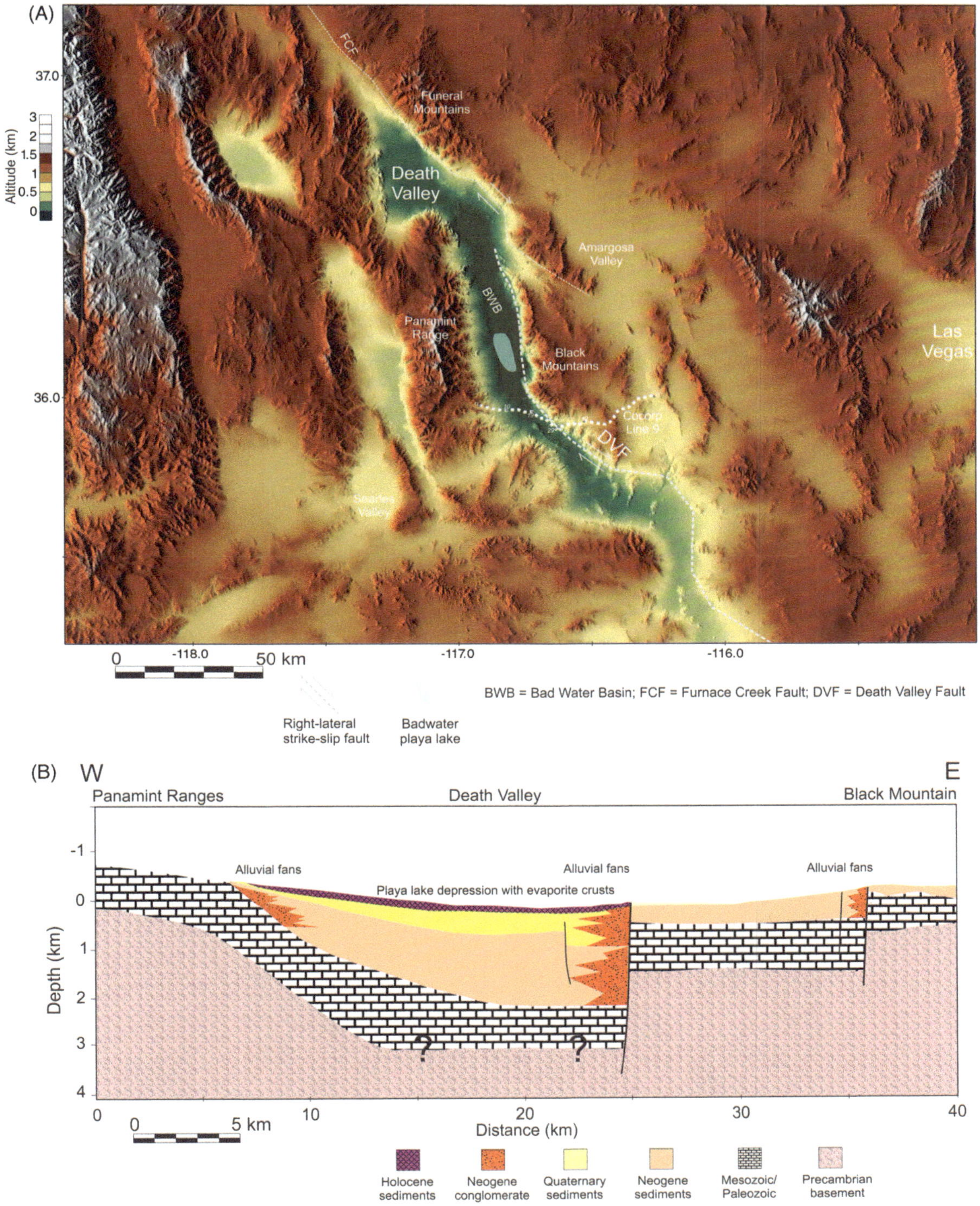

Figure 11.4 (A) Digital relief map of the southwestern region of the United States, showing the Death Valley depression in eastern California, a pull-apart basin with siliciclastic deposits and mudflats with evaporite deposits. Badwater Lake, located in the central part of the depression, shows the water level at 86 m below the global sea level. The Consortium for Continental Reflection Profiling (COCORP) regional seismic line is shown by a dashed white line extending from the Panamint to the Black Mountain Ranges. (B) Schematic interpretation of the regional seismic line acquired by COCORP across the Death Valley (Line 9). The Death Valley fault zone controls the main rift trough at the eastern side of the depression, resulting in alluvial fans with conglomerate and coarse sands. The depression is characterized by a Holocene mudflat with evaporite deposition and ephemeral playa lakes. Credit: (B) Adapted from Geist and Brocher (1987).

structure was investigated by a regional seismic line acquired by the Consortium for Continental Reflection Profiling (COCORP) near the basin depocenter (Geist & Brocher, 1987). During the Pleistocene ice age, which ended roughly 10–12 ka, the Sierra Nevada and other mountain ranges were wetter and hosted large glaciers. The demise of the ice age resulted in the melting of the ice, and Death Valley was filled with a huge and deep lake that occupied the pull-apart basin (Lowenstein et al., 1999). The ancient lake, known as Glacial Lake Manly, extended almost 240 km in the N-S direction and reached a maximum depth of about 200 m (Hooke, 1999; Ku et al., 1998). The evaporation of lake water in the last few thousand years was due to climate aridity and lack of fluvial drainage in the region, resulting in periodic basin desiccation and precipitation of evaporite sediments, which are intercalated with the lacustrine and fluvial sediments in the mudflats.

Active rift border faults formed mountain ranges, such as the Panamint Mountains to the west and the Black Mountains to the east of Death Valley (Figure 11.4). The uplifted flanks have been dissected by canyons that formed alluvial fans during episodic rainfalls and flash floods, accumulating sediments at the mouth of the fluvial drainage. The coarse sedimentary facies in the alluvial fans grade into the mudflats and evaporite deposits in the lowest points of the depression (Figure 11.5).

The ephemeral lakes in the depression, such as Badwater Lake, exhibit a shallow-water body during wet periods, as in the 2005 rainy season (Figure 11.4). Two years later, the satellite image shows the climate aridity resulted in rapid evaporation of the water as the lake was almost desiccated, precipitating some evaporitic sediments and forming a halite crust on the surface (Figure 11.4).

11.1.3. White Sands in New Mexico

Similar to the Great Salt Lake of Utah, the White Sands region in the Tularosa Basin of southern New Mexico is an endorheic basin with no active drainage system and surface water exit from the depression (Figure 11.5). It is a topographically closed, partly drained basin in a desert region with a water table of several meters beneath the land surface, and an aquifer not at its flow capacity (Hibbs, 2022; Holliday et al., 2019;).

The White Sands National Park is situated in a depression in the southern part of the Tularosa Basin, near Mexico's border, between the San Andres and Sacramento mountains (Figure 11.5). The basin is filled with Paleozoic, Mesozoic, and Cenozoic sediments within rift troughs (Figure 11.6). These rift blocks are associated with the Rio Grande rift system, which developed in the southwestern part of the United States during the

Figure 11.5 The NASA Landsat image (false colors) shows the Tularosa rift basin in New Mexico. The White Sands National Park, located between the San Andres and Sacramento mountain ranges, displays light blue reflectivity in gypsum deposits and eolian gypsum dunes. Quaternary lava flows exhibit a dark brown color in the region north of the White Sands gypsum deposits. Credit: NASA/Public domain.

Neogene, with the earliest syn-rift deposits dating circa 32 Mabp (Chapín, 1979). The main rift trough is marked by Neogene to Quaternary lava flows, forming spectacular outcrops to the north of the White Sands National Park (Figure 11.6). Paleozoic deposits in the pre-rift sequences were deposited in a foreland basin within the Pangea supercontinent, with marine sediments accumulating in a shallow sea during the Permian, including some evaporite deposits. The gypsum sand grains in the White Sands dune field were derived from the reworking of evaporite minerals that precipitated at the bottom of a shallow large hypersaline lake or inland sea that covered the southwestern region of North America in the Permian (about 280–250 Ma). The Permian Basin, located in the southwestern United States, is a larger area encompassing the depocenters of the Delaware and Midland subbasins (see Chapter 12). The Permian Sea extended to the northern region of Texas and New Mexico, forming a shallow inland sea that was periodically evaporated, concentrating the marine brine and accumulating thick sequences of carbonate, gypsum, and halite (Blakey, 2013; Blakey & Ranney, 2018; Handford & Bassett, 1982).

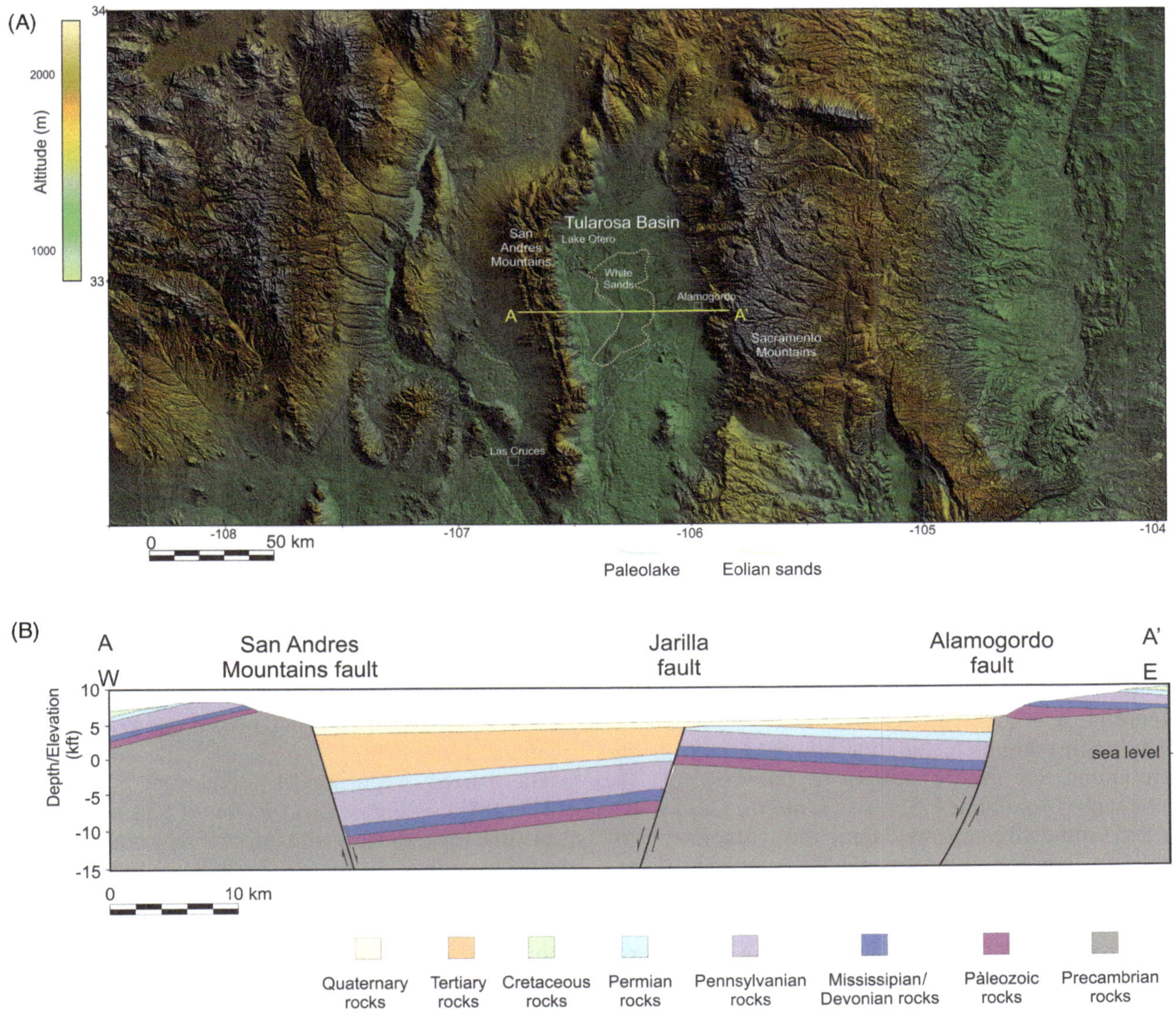

Figure 11.6 (A) Simplified geological map of the Tularosa Basin in New Mexico, showing the maximum extent of the Otero paleo-lake between the San Andres and Sacramento mountain ranges. The yellow line shows the location of the regional geological cross-section A-A′ across the White Sands. (B) Schematic geological cross-section A-A′ illustrating the rift architecture of the Tularosa Basin in New Mexico. Permian evaporite sediments outcrop on the uplifted rift flanks, both along the San Andres and Sacramento mountain ranges. The San Andres and Alamogordo normal fault zones control the Neogene rift trough. Credit: Love et al. (2020)/with permission of Elsevier.

The Gondwana breakup in the Mesozoic led to the receding of the Permian Sea and sedimentary basins accumulating Mesozoic and Cenozoic sediments in continental environments. The southwestern U.S. region experienced compressional tectonics, culminating with the Laramide Orogeny during the Late Cretaceous. The mountain-building process was associated with a thrust-and-fold belt resulting from the subduction of the Pacific Plate under the North American plate, uplifting the Rocky Mountains and forming highlands near the border with Mexico (Dickinson, 2002). The regional uplift developed a broad arch that collapsed in the Cenozoic due to gravity and extensional tectonics, resulting in the Basin and Range extensional province. During the Neogene, extensional forces developed a system of rift basins, forming structural highs and lows in the Basin and Range Province, locally accompanied by magmatism (Putirka & Platt, 2012).

The Tularosa Basin in southern New Mexico is a down-dropped block at the southern segment of the Rio Grande Rift (Figure 11.6). The intermontane basin is approximately 250 km long and 100 km wide, covering an area of 13,700 km^2 (Love et al., 2020). The elongated graben is controlled by north-south trending faults, with the uplifted rift flanks corresponding to the Sacramento and San Andres mountains (Figure 11.5).

Older sedimentary sequences, dated from the Paleozoic, Mesozoic, and Paleogene, correspond to the pre-rift sequence in the basin. The Neogene deposits accumulated mainly during the Late Miocene, about 10 Mabp. The sedimentary sequences were formed in a fluvial-lacustrine environment, with rock fragments derived from the erosion and transport of the adjacent highlands. Permian siliciclastic, carbonate, and evaporitic sedimentary rocks outcrop in the uplifted flanks of the rift basin, constituting the main source of clasts and chemical compounds recycled and accumulated in the Tularosa Basin (Figure 11.6).

The basin margins include the alluvial fans from the surrounding San Andres ranges to the west and the Sacramento Mountains to the east. Until the Quaternary, the Rio Grande River flowed along the southern edge of the basin, but the sediments that accumulated downstream blocked the connection with the sea. The isolated basin was occupied by Otero Lake during the Pleistocene and Quaternary, with hypersaline brines covering a large part of the depression. The last glacial age saw a paleolake much larger than the remnant lakes in the southwestern part of the basin. After the last glacial age (about 11 ka), climate aridity led to the evaporation of water in the depression. Gypsum layers at the central part of the basin formed a playa that has been deflated by strong winds. The modern depression is characterized by a saline mudflat known as Alkali Flat, which hosts a small ephemeral playa occupied by Lake Lucero, a small remnant lake formed at the foot of the San Andres mountains (Figure 11.6).

The erosion, dissolution, transportation, and reworking of gypsum grains formed the White Sands, an extensive sea of white dunes composed of sand-sized sulfate grains reworked by the wind into minute flakes (Love et al., 2020; Szynkiewicz et al., 2010). The gypsum sand grains in the White Sands National Monument derive from the Holocene lake sediments and modern gypsum crystals that precipitated on the surface of the Alkali Flat. The strong winds predominantly blow from the southwest, forming dunes that migrate to the northeast. The deflated area to the west is the source of the gypsum that has accumulated in the White Sands dune field (Figure 11.5). The thickness of the gypsum sand deposits increases from the Alkali Flat and Lake Lucero to the northeast region, where the largest dunes rise more than 30 m above the playa. Several dune types are recognized in the White Sands region, from the dome, barchanoid, barchan, and parabolic dunes (Love et al., 2020). The White Sands is the largest gypsum dune field in the world and is considered one of the best terrestrial analogs for the desert dunes observed on Mars' surface (see Chapter 9 in Mohriak, 2025).

11.1.4. Facies Distribution of Isolated Lacustrine Basins

The evaporation of the water in isolated lakes within rift depressions results in brine concentration and precipitation of evaporite minerals following a facies distribution pattern known as the bull-eye model (Figure 11.7). This symmetric pattern is formed when an endorheic, isolated basin dries out by excessive evaporation. The more soluble salts are deposited at the center of the basin, as the more concentrated brines form residual ponds during the last stages of desiccation (Handford, 1991; Hsü, 1972; Schmalz, 1969, 1970).

The barred basin model, developed by Ochsenius (1877, 1888), is the classical model of salt deposition in lacustrine environments with intermittent connections with the open sea (e.g., Perthuisot, 1989; Schreiber, 1986, 1988; Taylor, 1985; Warren, 2006, 2014, 2016). The facies distribution of evaporite minerals results in different configurations, indicating either water inflow or gradual brine concentration by evaporation of endorheic basins. Basins that are isolated and dry out completely have a bull's eye pattern with more soluble minerals at the center of the depression (Figure 11.7). Basins partially connected to seaways have an asymmetric facies distribution with less soluble evaporite mineral facies closer to the normal marine water input (such as gypsum deposits). The periodic water

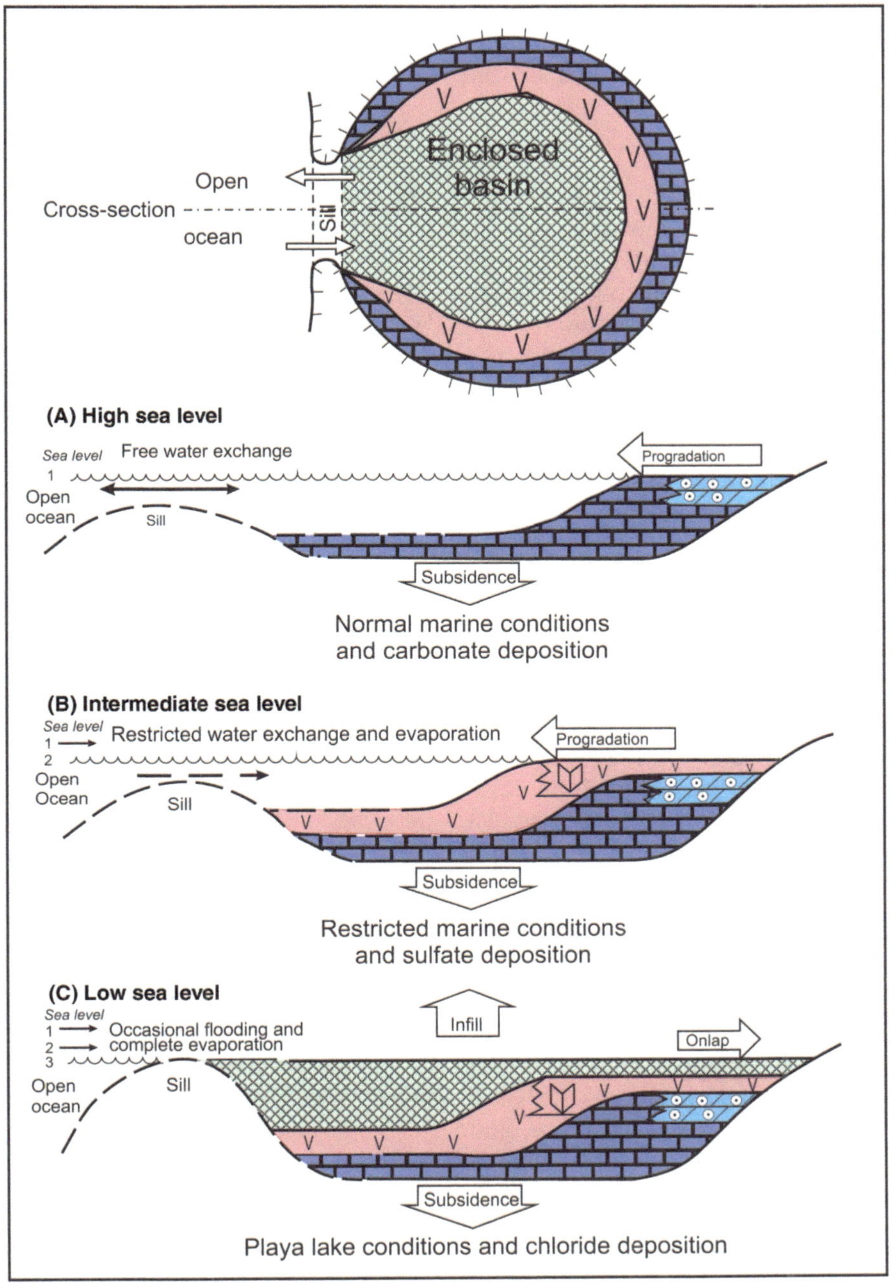

Figure 11.7 The barred basin model illustrates evaporite cycles formed by carbonate deposition during high sea levels when lake water salinity is close to normal seawater (A). As sea levels fall, brines concentrate, depositing sulfate minerals like gypsum (B). During basin isolation and low sea levels, brines concentrate for chloride deposition, precipitating halite and more soluble minerals and forming massive salt layers that fill the basin (C). Credit: Adapted from Taylor (1985).

inflow in barred basins results in less concentrated basin brines near the entrance channels, where carbonates and sulfates are deposited (Figure 11.7).

Evaporite cycles are formed with the precipitation of different mineral types depending on the brine concentration. The early phases of evaporite deposition are characterized by carbonate deposition during periods of high sea level, when the salinity of the lake water is close to normal seawater (Figure 11.7-A). When the sea level falls and the brines are concentrated, sulfate minerals are precipitated, forming gypsum and anhydrite (Figure 11.7-B). During periods of basin isolation, when the sea level is below the barrier, the brines are concentrated to the point of chloride precipitation. This results in the deposition of massive halite layers that fill the basin (Figure 11.7-C).

11.1.5. Marginal Sabkhas and Lagoons

Quaternary evaporite depositional environments in Northern Africa and Arabia are associated with continental and marginal sabkhas, which are located in playa lakes on land or coastal depressions adjacent to oceanic seaways or gulfs (e.g., Handford, 1991). Lacustrine evaporite accumulations are observed in the Sahara Desert and other arid regions bordering the Mediterranean Sea and the Caspian Sea (see Chapter 3). These inland seas and continental sabkhas accumulate carbonate, sulfate, and chloride minerals that precipitate following their chemical solubility (see Chapter 4). Until the 1950s, there were no well-documented modern evaporite analogs that could be compared to ancient giant salt basins. Evaporite basins were recognized in Europe (Zechstein and Mediterranean salt basins) and the United States (Gulf of Mexico coastal regions), but sedimentological models were based on sparse descriptions of modern environments, particularly in the desert areas of the Sahara Desert or along the United Arab Emirates (UAE) coast (see Chapter 8). The evaporite sabkha deposits along the southern margin of the Arabian Gulf, along the Trucial Coast of the UAE, and the marginal depressions in the Qatar Persian Gulf have been thoroughly described in seminal papers, constituting the paradigm for equivalent sedimentary sequences in ancient basins (Alsharhan & Kendall, 2003, 2011).

The sedimentological studies along the UAE coasts were supported by teamwork research conducted by oil companies searching for evaporite depositional models in the prolific sedimentary basins in Arabia, Europe, and the Americas. The main interest of petroleum companies working with carbonate reservoirs was to refine depositional models based on modern analogs and apply them to ancient reservoirs in a predictive way to identify evaporitic lithological facies. There are dozens of published works with descriptions of genetic models for carbonate reservoirs associated with ancient evaporite sequences formed in marine and continental sabkhas. These studies covered a wide spectrum of basins with different ages, like the Ordovician Red River Fm. in the Williston Basin, the Permian Zechstein Basin in Europe, the Permian Lower Clear Fork Fm. in Texas, the Jurassic evaporite-carbonate sequences from the Gulf of Mexico, and the Miocene carbonate-evaporite sequences in the Gulf of Suez, Red Sea, and peripheral basins around the Mediterranean Sea (e.g., Kendall, 1978, 1992, 2010; Kendall, & Harwood, 1996).

11.1.6. Facies Distribution of Evaporite—Carbonate Platforms

Sedimentary environments in the transition from sabkha evaporites to shallow carbonate platforms, subtidal, lagoon, and distal basins are described by several authors using outcrops and subsurface information from petroleum exploration datasets (e.g., Kendall, 1978, 1992, 2010; Warren, 1989, 2006, 2014, 2016). Classical examples include the Permian Basin in the southwest United States (Ewing, 2017, 2019; Hills, 1972; Hovorka et al., 2007; King, 1942), Mesozoic carbonate platforms in the Persian Gulf foreland basin (e.g., Alsharhan & Kendall, 2003, 2011), and Devonian carbonate platform—evaporite sequences in the Elk Point Basin of western Canada (Garret, 1996; Liu et al., 2018; Meijer Drees, 1994). These basins are characterized by evaporite deposits in supratidal (sabkha) and subaqueous (salina) lagoonal environments at the edge of the carbonate platform (Figure 11.8).

The proximal lagoon is a shallow, low-energy zone on the landward side of the epicontinental sea, characterized by brine concentration and hypersalinity during evaporite deposition episodes. Lagoonal evaporites are associated with mudflat evaporites that pass landward into a zone of carbonate and siliciclastic sediments interdigitated with eolian dunes in desert environments. Basinward, lagoonal evaporites are commonly limited by a reefal buildup forming a barrier (Figure 11.8).

The slope intermediate zone is a narrow, high-energy belt affected by waves and sediment reworking, characterized by sandstone and grainstone shoals, reef mounds, and patch reefs forming buildups that rim the lagoon depression. The back-reef includes packstone, wackestone, and stromatolites, whereas the forereef is characterized by slumps and debris derived from the erosion of the carbonate shelf edge, with rudstones, floatstones, and bindstones. The shelf edge region, located adjacent to the reefal buildup, is characterized by a skeletal sand apron with grainstones and packstones (e.g., Lucia, 2007).

Carbonate buildups near the shelf edge can form islands, which are reefal buildups with a core of bindstones and rudstones (Figure 11.8). The erosion of these platforms results in slumps and debris, which develop sigmoidal wedges that prograde basinward. The distal margin corresponds to a wide low-energy zone on the open-marine side of the platform, resulting in relatively deepwater settings. Sedimentary successions are characterized by low-energy, basinal facies (bioturbated mudstones, marls, and wackestones) deposited below the wave base (e.g., Taylor, 1985).

Most ancient carbonate platforms transition to subaerial environments with redbeds, such as the Permian Basin (e.g., Handford, 1991). The continental eolian environments and sabkhas formed in coastal regions are adjacent to the lagoon, where carbonate and evaporite

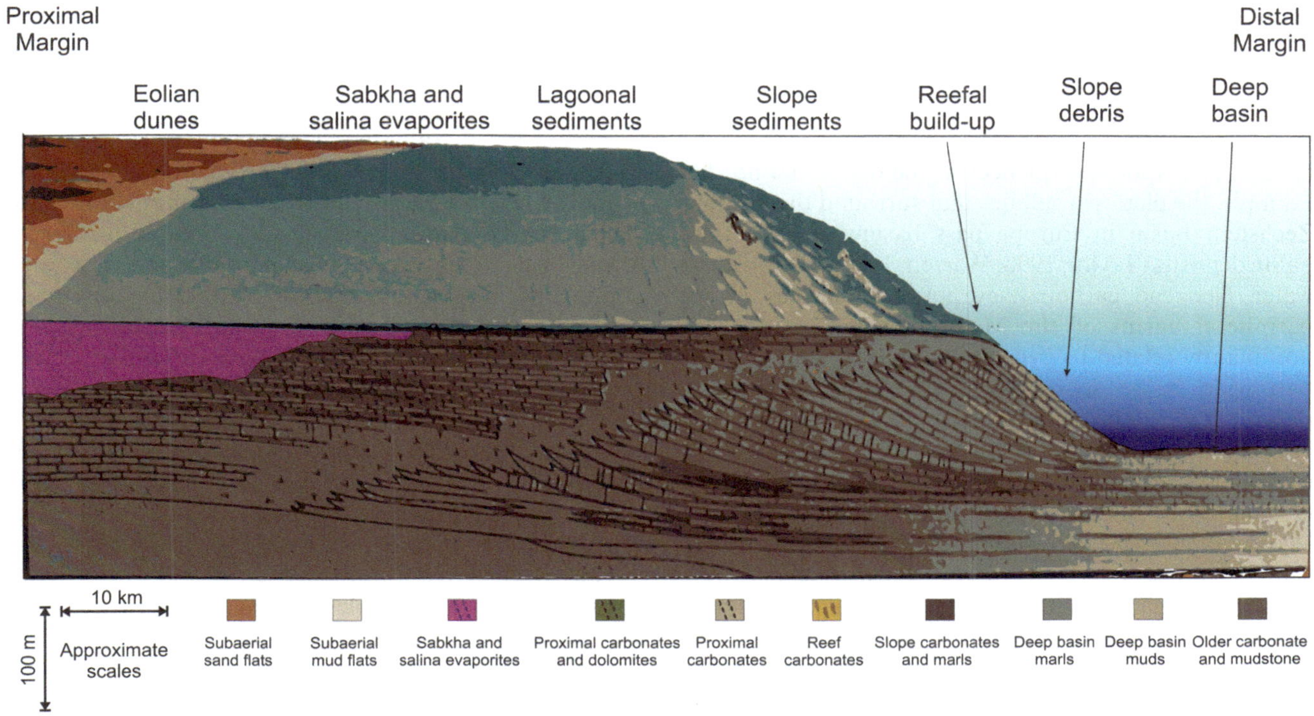

Figure 11.8 The schematic 3D diagram (oblique plan view on top, vertical cross-section below) shows the transition from continental sabkha evaporites to lagoonal, carbonate platform, slope, and deep basin environments where the salinity is close to normal seawater. The evaporite-carbonate platform is characterized by reefs at the shelf break, separating the basin depocenter from the slope and open sea. The lagoon is isolated from the open sea by the carbonate barrier, concentrating brines into a salina that is filled with evaporite sequences (gypsum and halite). The erosion of carbonate banks that rim the platform results in debris, forming sigmoidal wedges that prograde basinward, while deepwater, low-energy sedimentary facies accumulate in the distal margin.

sequences are limited by a reef buildup at the shelf edge (Figure 11.8). The Permian Basin in the southwestern United States is rimmed by reef constructions at the edge of the carbonate platform. The Capitan Reef constitutes a classical example of a carbonate buildup separating the basin depocenter from the sabkha environment landwards (Harris & Kerans, 2016). In other basins, the main reservoir rocks in oil and gas fields are eolian sandstones intercalated with evaporite sequences, such as the Amazon Basin in Northern Brazil. The Permo-Carboniferous evaporites mainly correspond to anhydrite and halite layers, but also include soluble salts, such as carnallite and sylvinite (Mohriak et al., 2012).

Carbonate platforms can be divided into two main types: a shallow water province near the coast, with a gentle bathymetric increase toward the deep basin, forming a ramp where carbonates and evaporites are deposited, or a shallow water environment marked by a steeper shelf edge rimmed by a carbonate bank or reef at the transition to the deep basin (Figure 11.8). Modern-day examples of this setting are observed in the Persian Gulf margins, particularly in Abu Dhabi and the United Arab Emirates (e.g., Kendall & Skipwith, 1969). In ancient basins, these zones might have extended for hundreds of kilometers on land, forming a wide shallow depression (often only a few centimeters deep) characterized by limited water circulation, as interpreted for the lagoonal environments of the Permian Basin in the southwest United States. This restriction resulted in occasional stagnation and anoxic conditions in the back-reef or lagoon, leading to the deposition of organic-rich laminated sediments. In arid settings, the proximal margin was frequently covered by hypersaline brines that deposited thin layers of platform evaporites or formed gypsum and halite crusts. Open-marine carbonate rocks with source rock potential are rare in this environment, as the salinity was too high for most carbonate-secreting organisms, except for cyanobacteria, which in some basins formed thinly laminated algal mats (Schreiber & Marshak, 1981).

The geological literature on the carbonate platform genetic model for salt deposition in ancient basins is extremely large, with dozens of case histories based on detailed sedimentological, petrographic, and well-log interpretations from sedimentary basins dated from Neoproterozoic to Neogene (e.g., Kendall, 1978, 1992, 2010; Warren, 2006, 2010, 2016). Shallow-water carbonate platform evaporites are often associated with basinwide evaporite sequences in the distal margin. For example, the platform sulfates that surround the Permian Zechstein Basin in Europe pass basinward into thick halite deposits (Taylor, 1998; Warren, 1991). The shallow-water Neogene carbonate-evaporite sequences in the peripheral basins of the Mediterranean Sea are predominantly related to gypsum deposits, with equivalent basinwide halite sequences in the deepwater region (Roveri et al., 2008a, 2008b).

The distinction between the genetic models involving carbonate platform evaporites and deep basin evaporites is tenuous and elusive. Several salt basins are characterized by tabular bedded evaporite sequences that do not exhibit salt diapirs, which is considered a typical geomorphological characteristic of the evaporite—carbonate model in sabkha environments. In some basins, the bedded salt layers are affected by the incipient development of salt structures such as salt pillows. For example, in vintage petroleum exploration reports for the Michigan Basin, published in the late 1960s, some dome-like features are reported (e.g., Ells, 1967).

The debate concerning the water depth of the evaporite deposits in carbonate platforms is also subject to different interpretations by different authors. There is sedimentological evidence for shallow-water environments during the deposition of the Salina Group salt layers in the Michigan and Appalachian basins. However, more soluble minerals (halite and sylvite) precipitated in the basin depocenter, which might have attained depths of a few tens of meters based on palinspastic reconstructions (e.g., Dellwig & Evans, 1969; Mesolella, 1978). Some major evaporite basins worldwide include stratigraphic units interpreted as evaporitic sequences deposited in shallow water carbonate platforms, such as the Amazon Basin in Northern Brazil. Although salt diapirs are not observed in the Amazon Basin, the seismic profiles suggest the presence of salt mounds that locally resemble turtle structures, a feature that is characteristic of halokinetic processes indicating salt flow (e.g., Mohriak et al., 2012).

11.2. ANCIENT RIFT-RELATED LACUSTRINE SALT BASINS

Lacustrine evaporites are found in numerous ancient lacustrine systems worldwide, such as the Ebro Basin in Spain (Urgeles et al., 2011) and the Potosi area in the Andes (Rouchy et al., 1993). These evaporite deposits were primarily sulfate (gypsum) layers intercalated with siliciclastic and carbonate sequences. Other lacustrine basins in extensional settings precipitated both sulfate and chloride minerals, such as the Sergipe Basin in Northeast Brazil (Mohriak et al., 2012).

Ancient evaporite basins fed mainly by marine waters formed evaporite salterns, accumulating thick sequences of gypsum and halite in submarine environments. These evaporite layers were laid down in various environments, such as marginal areas along shallow oceans, inland seas formed near open oceans, oceanic gulfs that separated continental masses, and shallow platforms along divergent continental margins. The evaporite sequences are associated with distinct geomorphological types, such as ramps and continental shelves. Ramps are characterized by a smooth increase in water depth from the coastline to the deepwater region. Continental shelves are characterized by a shallow water region with a steep slope where the bathymetry increases rapidly at the shelf edge (Warren, 1989, 2006, 2016).

The Green River Fm. in the Uinta and Green River basins is a classic example of a lacustrine environment with evaporite accumulation in the Eocene, with salt deposition almost concomitant with the formation of kerogen-rich shales (Katz, 1995). This lake is related to a large hydrographic system that spread across Colorado, Utah, and Wyoming during the Paleogene, covering an area of nearly 100,000 km^2 (Eugster, 1982). The lake margin is characterized by a carbonate platform dominated by microbialites (Seard et al., 2013). The lake margin is characterized by a carbonate platform dominated by microbialites, and the sediments have an aggregate evaporite thickness of about 3 km in the basin depocenter and 1 km in the northern part of the lake, in Wyoming. Carbonates compose the most important evaporite minerals in the Green River Basin, especially trona sequences ($NAHCO_3 \cdot Na_2CO_3 \cdot 2H_2O$), which occur in vast amounts and are exploited economically. Other minerals such as halite (NaCl) occur mixed with trona, as observed in the southern part of the basin, where the salt layer reaches a maximum thickness of about 6 m (Eugster & Hardie, 1975, 1978).

Several examples of saline lacustrine basins are associated with organic-rich source rocks, such as the oil shales in the Uinta Basin (Eugster, 1982). The Eocene lake levels fluctuated from brackish to saline, but several episodes of basin desiccation occurred when the lake dried up (Eugster & Hardie, 1975, 1978). The sedimentary model for the accumulation of evaporite and kerogen-rich sediments is climate-dependent, with wet periods resulting in large and shallow lakes depositing organic-rich sediments (oil shale), while dry periods resulted in lowered surface levels and increased brine concentration. Eventually, the hypersaline lake was desiccated and formed a salt pan surrounded by a salt flat, particularly the northern region, near the Green River, north of the border between Utah and Wyoming. The lake surface lowered due to evaporation and the lake shrank drastically, increasing the brine concentration and depositing trona minerals, followed by thin halite layers (Eugster, 1982).

11.2.1. The Cenozoic Rhine Graben Salt Basin

The Rhine Graben, located between Germany and France, is part of the European Cenozoic Rift System, which transects the Variscan Orogen, Permo-Carboniferous rift troughs, and Late Permian to Mesozoic thermal sag sedimentary basins in Central Europe (Figure 11.9-A). The basin was formed in the Paleogene (Middle Eocene to Early Oligocene) due to an E-W extensional regime that affected pre-rift sediments (Hinsken et al., 2007; Rotstein, Schaming, & Rousse, 2005; Ziegler & Dézes, 2005). The Rhine Graben is bordered by two uplifted blocks, forming massifs cored by Variscan crystalline rocks surrounded by Triassic sedimentary sequences. The Vosges and the Black Forest basement massifs border the graben to the west and east, reaching modest elevations, up to almost 1,500 m (Figure 11.9-B).

The base of the syn-rift sequence is associated with a major regional unconformity, marking the absence of most Late Jurassic, Cretaceous, and Paleocene sedimentary sequences (Hinsken et al., 2007). The pre-rift sequences correspond to Jurassic, Triassic, and Permian sediments associated with a previous tectonic event characterized by widespread subsidence and sedimentation in the Central European Basin (Berger et al., 2005).

The main rift phase occurred during the Middle Eocene to Early Oligocene and is associated with fluvial-lacustrine environments (Figure 11.10). The rifting of the continental crust was related to regional WNW-ESE extension of the lithosphere, resulting in extensional faults that formed deep troughs. These depressions locally resulted in large lakes between the uplifted flanks of the rift. The rift shoulders have been deeply eroded and basement rocks outcrop in the Vosges in France and the Black Forest in Germany (Asch, 2005; Ziegler & Dézes, 2005).

During the Late Eocene, a major acceleration of subsidence developed widespread syn-rift sedimentation across the basin, with mixed lacustrine to brackish environments (Figure 11.10). Some authors suggest that marine ingressions might have entered the southern and central parts of the Rhine Graben during this phase, along seaways from the south, across the Jura thrust and fold belt (Berger et al., 2005).

During the Neogene, the Rhine Graben formed several subbasins, with depositional environments frequently changing between terrestrial (deserts and mudflats), fluvial, lacustrine, brackish, and evaporitic lagoons that formed subaqueous salterns. The main salt deposition episodes are recorded in the Late Eocene to Early Oligocene (Figure 11.10).

Evaporitic lagoons precipitated salt layers during an active phase of extensional tectonics, resulting in syn-rift deposits. The Potash Basin, located in the central and southern Upper Rhine Graben, is associated with a sedimentary depocenter that accumulated over 1,800 m of evaporites, including potash minerals (Figure 11.11).

The Dannemarie subbasin, located in the southern part of the Upper Rhine Graben, is characterized by Eocene-Oligocene syn-rift sedimentary sequences composed of siliciclastic, carbonate, and sulfate (gypsum) sediments (Figure 11.11-A). The rift geometry is marked by a normal fault in the western border, forming alluvial fans that grade into an evaporitic sabkha mudflat with gypsum deposits in the depocenter (Rotstein, Schaming, & Rousse, 2005).

The syn-rift evaporite sequences belong to the Salt Fm., constituting the main sedimentary successions that fill the Upper Rhine Graben (Hinsken et al., 2007). The southern margin of the Upper Rhine Graben is characterized by abrupt changes in sedimentary facies, including conglomerate rocks in alluvial fans and siliciclastic sequences composed of marl and shales in the mudflats at the basin depocenter. Some subordinate evaporite intercalations include sulfate (gypsum and anhydrite) and halite. The Potash Basin depocenter contains thick evaporite sequences, including gypsum, halite, and more soluble minerals, represented by thin lenses of potassium-bearing minerals like sylvite (Figure 11.11-B).

In the Miocene, an NW-SE compressional phase resulted in the inversion of previously extensional structures and the development of strike-slip faults and local uplifts associated with transpression (Rotstein, Behrmann, et al., 2005). The Middle Miocene igneous rocks of the Kaiserstuhl Volcano, which was active between 19 and 16 Ma, rest unconformably on faulted and deeply truncated Late Oligocene sediments (Braunger et al., 2018; Wimmenauer, 1974). This phase corresponds to the Alpine fore-bulge, a compressional phase in the Late Miocene to Pliocene, which resulted in regional uplift and erosion of the Neogene sequences in the Vosges and Black Forest regions (Figure 11.10).

11.2.2. Seismic Interpretation of the Rhine Graben

The Upper Rhine Graben is characterized by thick evaporite layers forming the Late Eocene-Early Oligocene depositional sequences, accumulated in the sedimentary trough between the Vosges and the Black Forest (Figure 11.11-A). The lacustrine facies correspond to coarse-grained siliciclastic rocks (conglomerates and sandstones) that grade into a mud flat and salt lake.

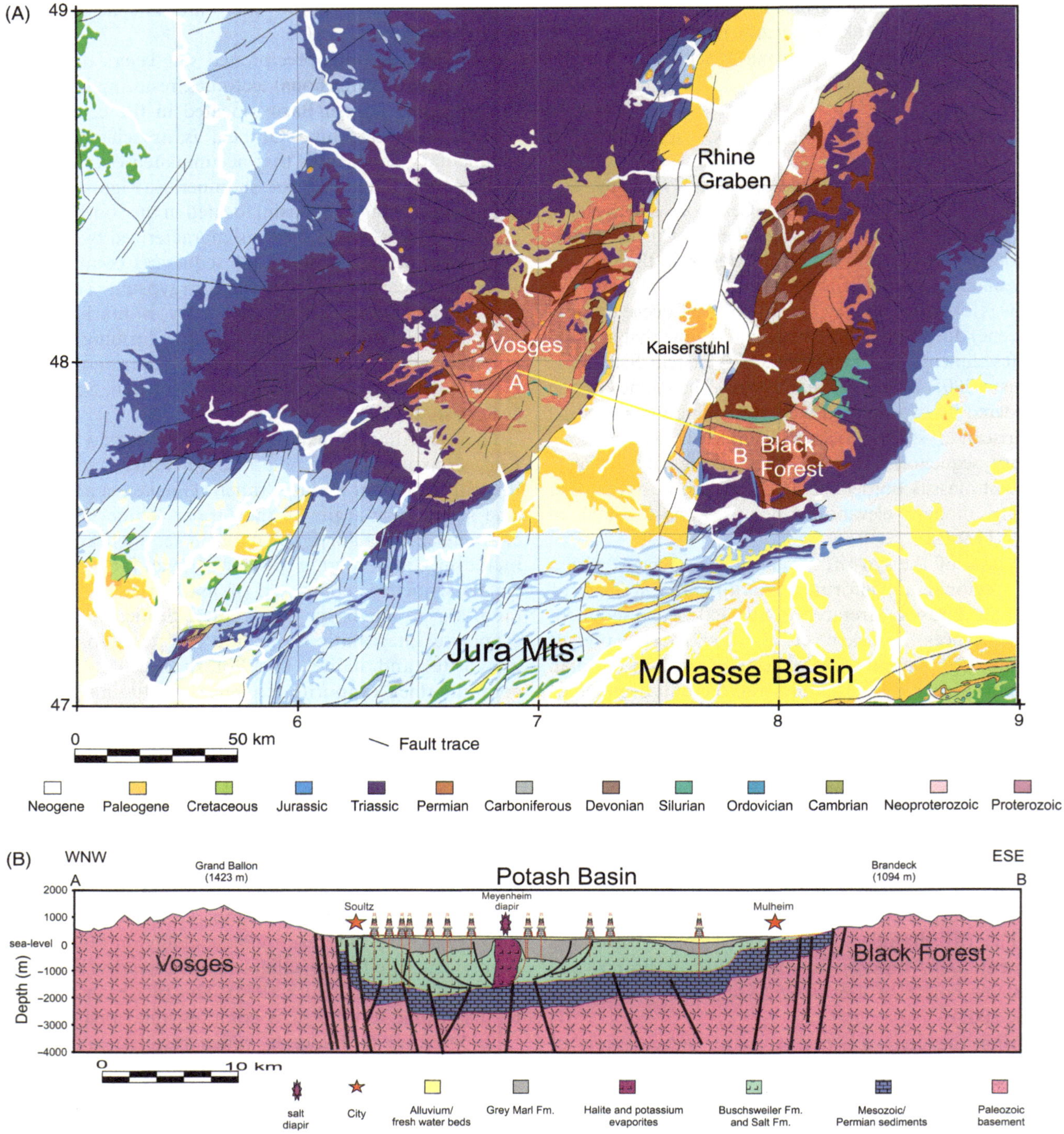

Figure 11.9 (A) The simplified geological map shows the Upper Rhine Graben, located north of the Jura Mountains in the Alpine Chain, and the Molasse Basin, formed in the Neogene. The Kaiserstuhl is a volcanic structure within the Rhine Graben depression. A thin yellow line marks the location of the geological cross-section A-B, extending from the Vosges to the Black Forest. (B) Geological cross-section (A-B) across the Rhine Graben, from the Vosges in France to the Black Forest in Germany. The Eocene Buschsweler Fm. is characterized by alluvial fans near the rift border fault, grading into gypsum and salt toward the basin depocenter. The Potash Basin has thick layers of evaporite that were formed in a lacustrine environment during the Eocene. These layers contain sulfates (anhydrite) and chlorides (halite and potash salts). The salt diapir in the basin depocenter is mainly composed of halite. Credits: (A) Asch (2005)/with permission of BGR and (B) adapted from Hinsken et al. (2007).

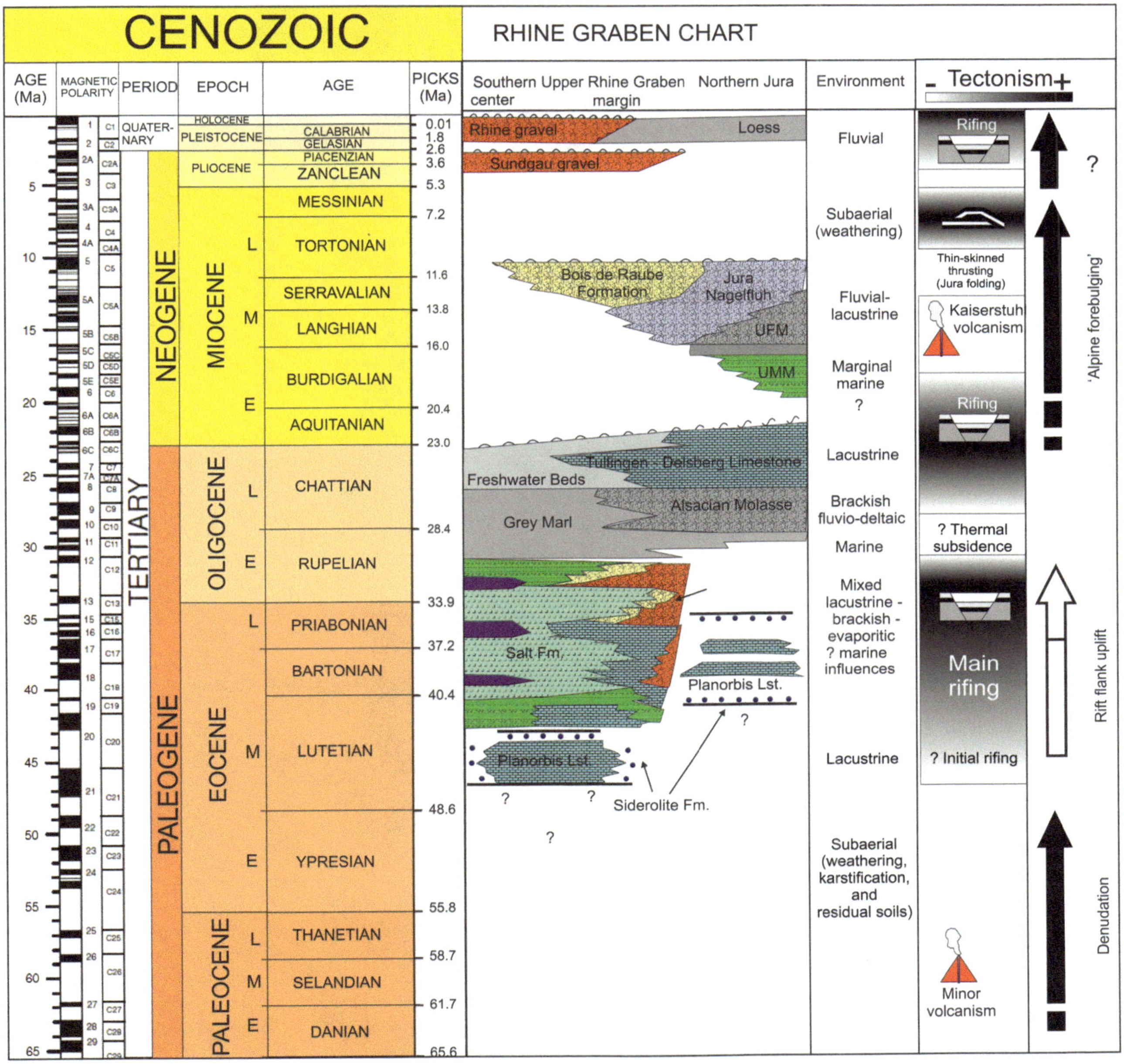

Figure 11.10 A simplified tectonostratigraphic chart of the Rhine Graben shows the main lithological units and tectonic events. The Paleogene evaporites (Salt Fm.) were deposited in a lacustrine environment during the Late Eocene, concomitant with the main rifting event in the basin.

The main syn-rift stratigraphic units overlie the pre-rift Mesozoic sequences in the basin depocenter (Figure 11.11-B). The results from exploratory boreholes indicate a major unconformity separating the Paleogene sediments from the Jurassic pre-rift strata (Figure 11.10).

Seismic stratigraphy of the Upper Rhine Graben has been established through the integration of 2D seismic profiles acquired by petroleum companies in the late 1980s (e.g., Veslud et al., 2005). These profiles were calibrated by drilling results from exploratory boreholes, such as the Bellemagny-1 (Figure 11.12). The borehole was drilled at the southern margin of the graben, close to the exploratory well Brechaumont, which projects onto the N-S stratigraphic section (Figure 11-B). The borehole Bellemagny-1 penetrated the rift section with the Série Grise (Gray Marls) dated as Upper Ruppelian (~ 30–31 Ma). The Salt Fm. (also called Saliferous Fm.) includes three sequences, the upper, middle, and lower Saliferous Fm., characterized by marls and saliferous limestones (Figure 11.12). Boreholes in the Potash Basin register

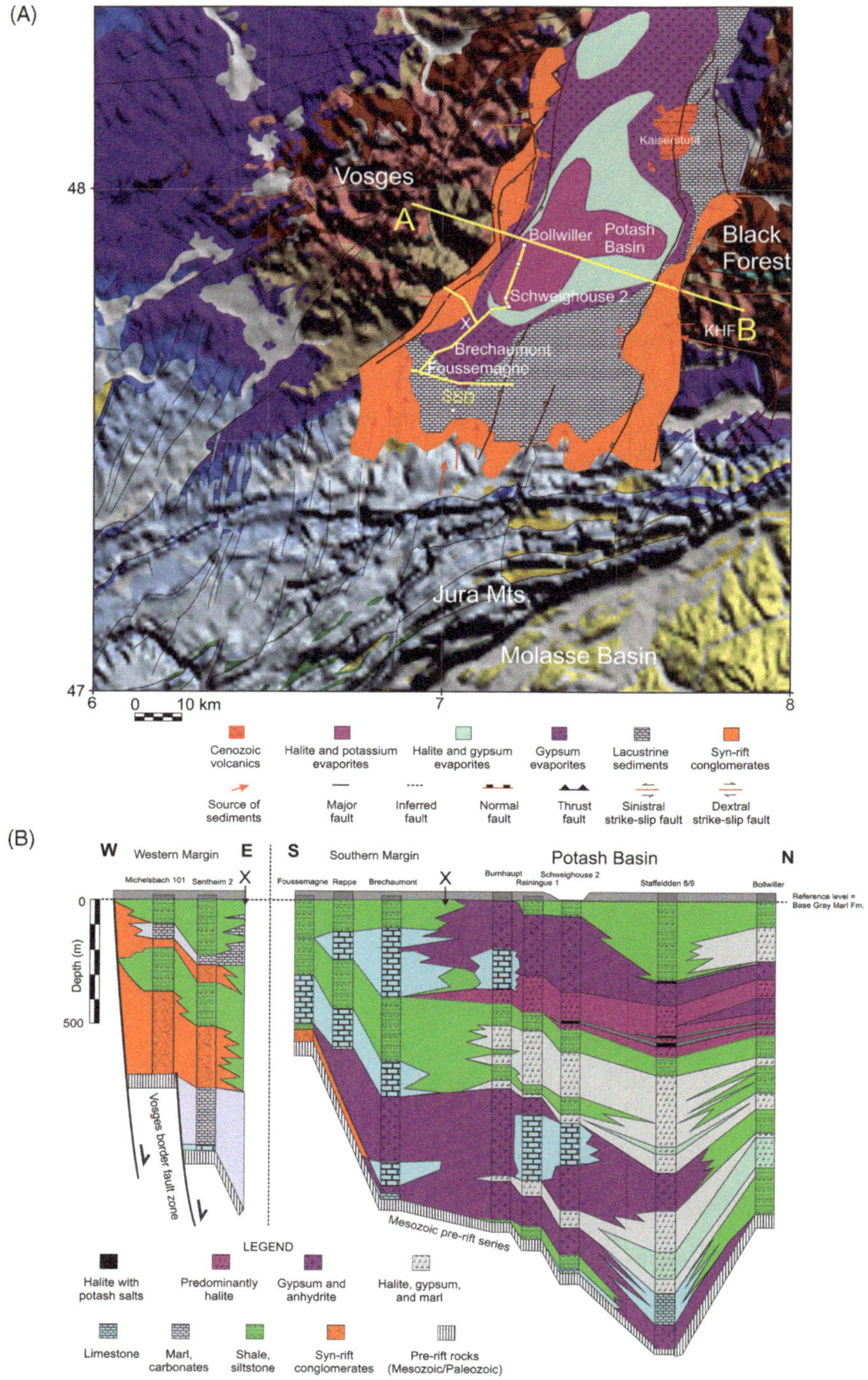

Figure 11.11 (A) The geological map of the southern Rhine Graben is draped on present-day topography. The schematic evaporite facies map of the Eocene Salt Fm. is overlain on the Rhine Graben outline. The east-west regional cross-section A-B from the Vosges to the Black Forest extends from the rift border faults to the basin depocenter, where Paleogene evaporites were deposited in a continental lacustrine environment. The Potash Basin includes halite and potash minerals in the center, with gypsum and siliciclastic rocks toward the rift margins. The stratigraphic sections S-N and W-E run from the southern part of the basin to the Botswiller well in the Potash Basin. The Dannemarie subbasin is characterized by alluvial fans and gypsiferous mudflats south of the Potash Basin. The stratigraphic sections intersect at (X) between the wells Brechaumont and Schweighouse-2. The E-W seismic profile in the southern part of the basin (SSD, yellow line) is located near the exploratory well Foussemagne, slightly to the south of the gypsum evaporite facies. The longer stratigraphic section (S-N) extends from the southern margin (Dannemarie subbasin) to the Potash Basin in the Upper Rhine Graben depocenter. (B) Stratigraphic sections (W-E and S-N) along exploratory boreholes in the Rhine Graben. The sections were horizontalized at the base of the Gray Marl Fm. (Early Oligocene) to illustrate the stratigraphic relationships of Eocene sequences. The short W-E section in the western margin extends from the Vosges border fault in France to the basin center, illustrating conglomerate facies in alluvial fans. The longer stratigraphic section extends from the southern margin to the Potash Basin in the Upper Rhine Graben depocenter. The Potash Basin is characterized by thick evaporite layers and potash salts in the Saliferous Fm., as indicated by the exploratory boreholes located in the northern segment of the profile. Credit: Hinsken et al. (2007)/with permission of Springer Nature.

Bellemagny 1 borehole

Topographic surface: 370 m above sea-level (asl)

Stratigraphy	Age of the base	Lithology	Depth
			0 = 370 m asl
Upper Rupelian	31 Ma	'Série Grise' consisting of: 'Grey Marl'; 'Foraminifera Marl'; 'Fish Marl'; 'Meletta Marl'; 'Cyrenea Marl'	294 m
Middle Rupelian	~ 32 Ma	Upper Saliferous Fm. (marls and saliferous limestones)	489 m
Lower Rupelian to Priabonian	34 Ma	Middle Saliferous Fm. (marls and saliferous limestones)	1034 m
	35 Ma	Lower Saliferous Fm. (marls and saliferous limestones)	1504 m
Jurassic	~ 160 Ma ~ 200 Ma	Marls and limestones	2310 m
Triassic		Upper Triassic (gypsiferous marls) Middle Triassic (carbonates) Lower Triassic (sandstones)	2741 m

Borehole bottom: 1.45 s TWT (2741 m)

Base Tertiary Unconformity — Upper Oligocene (siliciclastic rocks) — Oligocene / Eocene (Salt Fm.) — Jurassic (carbonates) — Triassic (carbonates, siliciclastics)

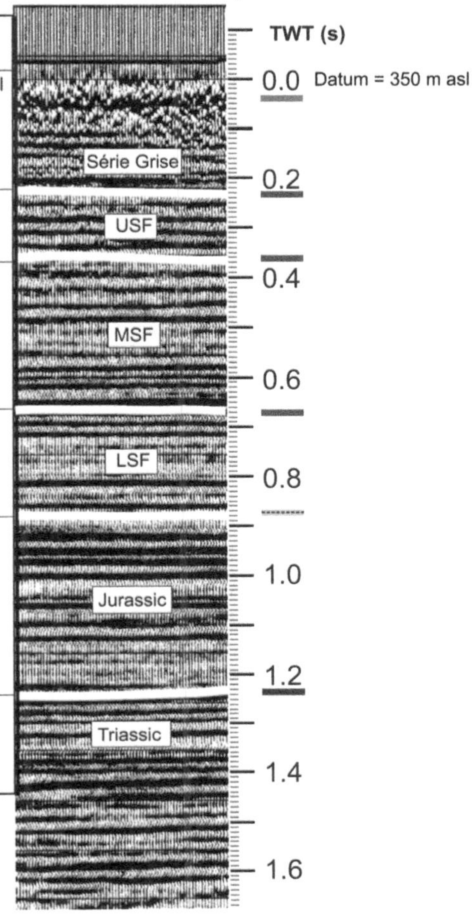

Figure 11.12 A simplified seismogram of the exploratory well Bellemagny, located north of well Brechaumont, shows that the borehole mainly penetrated the Saliferous Formation (Early Oligocene–Late Eocene). The borehole drilled Jurassic sequences and bottomed at Triassic siliciclastic sediments. The seismogram indicates that the basement is deeper than 1.6 s TWT.

thick layers of gypsum and halite, with subordinate lenses of sylvite (Figure 11.11-B).

The rift sequences in the southwesternmost depocenter of the Upper Rhine Graben are well-studied in the Dannemarie subbasin by seismic datasets calibrated by exploratory boreholes (Figure 11.11-A). The correlation of seismic data with stratigraphic markers in the Bellemagny-1 borehole indicates that pre-rift Jurassic carbonate rocks and Triassic marls and carbonates were penetrated (Figure 11.12). The borehole Bellemagny-1 was drilled to a total depth of 2741 m, bottoming at Triassic sandstones (Veslud et al., 2005). A major unconformity separates the Jurassic pre-rift layers from the overlying Eocene strata, resulting in a stratigraphic gap of about 125 Ma (Figure 11.12). The unconformity is associated with the denudation that affected the Rhine Graben and removed the Lower Eocene, Paleocene, and Cretaceous strata (Figure 11.10).

The pre-rift Mesozoic sediments are primarily composed of Jurassic marl and limestone, overlying Triassic gypsiferous marls, carbonates, and sandstones (Figure 11.12). Above the unconformity, the Paleogene sediments are mainly composed of saliferous limestones, marls, and evaporite sequences. The Saliferous Fm. is divided into lower, middle, and upper units, while the Gray Marl Fm. is dated as Early to Late Oligocene (Figure 11.10). The Eocene—Oligocene saliferous sequences form the main syn-rift deposits in the Upper Rhine Graben, as recorded in the Bellemagny-1 borehole (Figure 11.12).

The Dannemarie subbasin is bounded to the west by the Vosges fault zone, with pre-rift Triassic and Jurassic sediments forming a ramp toward the basin center

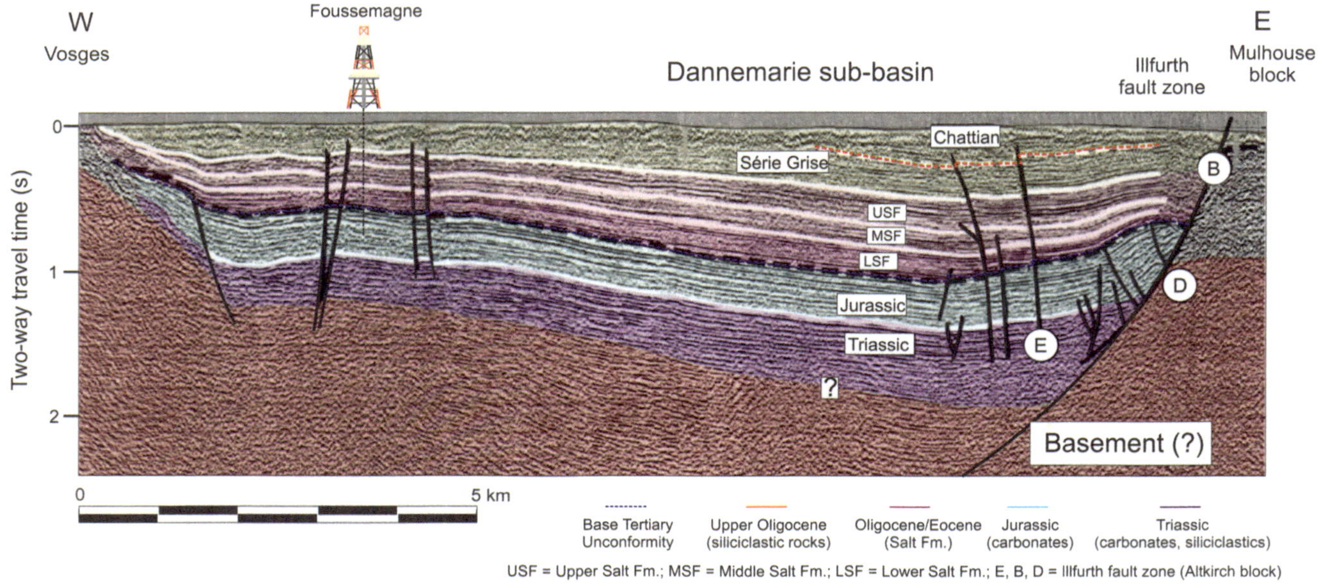

Figure 11.13 Simplified interpretation of the E-W seismic section SSD in the Dannemarie subbasin, extending from the Vosges Massif fault zone in the west to the Mulhouse block (Altkirch Horst) in the east. The profile is located near the Foussemagne well, which penetrated the rift and pre-rift sediments near the western border fault. The Dannemarie subbasin is characterized by carbonate rocks with minor evaporite intercalations in the Salt Formation. The carbonate-evaporite deposits overlie Jurassic and Triassic sequences that occur below the Base of Tertiary Unconformity. The evaporitic sediments extend across the Dannemarie subbasin, reaching the Mulhouse Horst border fault in the east. Credit: Veslud et al. (2005)/with permission of GeoScienceWorld.

(Figure 11.13). The eastern border is characterized by the Illfurth fault zone that limits the Mulhouse block, a major horst structure in the footwall of the rift master fault (Veslud et al., 2005). The opposite western margin is characterized by a flexural border with post-rift inversion in the Vosges Massif (Figure 11.13). The southern border of the Upper Rhine Graben is associated with the Jura fold belt, characterized by tectonic activity between 11 and 3 Ma, as indicated by sedimentary layers that onlap the folded strata. The Miocene–Pliocene sedimentary sequences that onlap the pre-rift strata suggest synchronous rifting and subsidence associated with the distal effects of Alpine collision (Veslud et al., 2005).

The seismic data at the southern segment of the graben does not show structural features associated with diapiric salt tectonics. The Paleogene sequences correspond to sub-horizontal sedimentary layers that formed bedded salt deposits within the rift trough (Figure 11.13). The Foussemagne well penetrated a sedimentary sequence predominantly composed of siliciclastic and carbonate rocks at the southern margin of the graben (Figure 11.11-Band 11.13).

The Bellemagny-1 borehole (Figure 11.11-B) shows that evaporite sediments in the southern margin of the Rhine Graben are primarily thin sulfate layers (gypsum and anhydrite) intercalated with carbonate rocks (Veslud et al., 2005). Sedimentary facies maps show that halite and soluble evaporites are concentrated in the Potash Basin, north of the Dannemarie subbasin (Figure 11.11-B). The evaporite facies distribution can be correlated with the simplified model of lacustrine salt deposition in an isolated basin with a bull's eye pattern (Figure 11.7). The Salt Fm. in the Potash Basin is thick and predominantly composed of halite with intercalations of anhydrite and marls (Figure 11.11-B).

The regional cross-section from the Vosges to the Black Forest indicates that isolated salt diapirs occur in the Upper Rhine Graben depocenter. These structures are restricted to the Lower Salt unit and occur only in the depocenter of the Potash Basin (Figure 11.11-A). The Manheim salt diapir is a large and tall structure intruding into the Gray Marl Fm., dated as Upper Oligocene. The top of the salt diapir is covered by thin sequences of Quaternary sediments (Figure 11.9). The lack of diapirism within the Upper Saliferous Fm. (Middle Rupelian) suggests that the Neogene, post-salt sediments did not accumulate halite in considerable thickness. In addition, these evaporitic deposits were not covered by an overburden thick enough that could cause some diapirism by differential loading (Figure 11.13).

The Upper Rhine Graben halite layers are generally believed to be associated with a continental lacustrine environment with nonmarine waters, although some authors suggest a possible linkage with a marine source (e.g., Berger et al., 2005). Hinsken et al. (2007) suggested that some Tertiary salt may have originated from the dissolution of Triassic evaporites. This interpretation is based on the presence of Triassic salt-bearing strata within the rift border conglomerates, which are dated as Paleogene and correlate laterally with the Salt Fm. (Figure 11.11-B).

The Triassic debris suggests a Paleogene uplift of graben shoulders, exposing older evaporite deposits to weathering, erosion, transportation, and dissolution. Moreover, the sulfate isotopic determinations in gypsum minerals indicated a Triassic, rather than Paleogene, marine origin for the evaporites (Fontes et al., 1991). The analysis of isotope data supports a genetic model of evaporite deposition in a continental lacustrine environment, within a closed hydrological system, precluding any marine connection with the Alpine Sea, the North Sea, or the Paris Basin (Fontes et al., 1991).

11.2.3. Petroleum and Salt Exploration in the Rhine Graben

The Upper Rhine Graben is the oldest known oil province in Western Europe, with tar exploitation dating back to the medieval ages. Böcker et al. (2016) reported that tar exploitation dates to the medieval ages, as sandstones impregnated with oil were collected in the Alsace region of France during the 15th century.

The most important petroleum accumulation in the region is the Pechelbronn field in France, which produced about 3 million metric tons until the late 2000s (Böcker et al., 2016). In Germany, the Landau oil field is the most important oil accumulation. The field was discovered in 1955 and reached a peak of oil production in the early 1960s, with declining production since the 1970s (Reinhold et al., 2016). Most of the productive zones are found in shallow (500–1000 m depth) Oligocene reservoirs within the Pechelbronn Formation. The most productive zones are found adjacent to normal faults, suggesting that extensional fractures play a major role in oil migration and entrapment during the Neogene (Reinhold et al., 2016).

Although some authors postulate that oil accumulations in the Rhine Graben region are related to organic-rich marls deposited in Paleogene lacustrine environments (e.g., Hofmann et al., 1993), there is still some discussion on the geochemical biomarkers associated with the possible source rocks. Böcker et al. (2016) identified two potential source rocks in the region: Middle Jurassic Toarcian shale (Type I kerogen), and the Oligocene Pechelbronn Formation (type III kerogen).

At the base of the Middle Eocene sedimentary successions, just below the Salt Fm. succession, there is a prominent sapropelic coal layer composed of bituminous shale with exceptionally high TOC values of 20%–32 % (analyzed from ditch sample cuttings) and hydrogen index (HI) values up to 640–760 mg HC/g TOC. These values indicate an extremely rich Paleogene source rock with potential for oil generation. However, the highly variable lateral distribution in terms of thickness and source rock facies results in rather small oil accumulations throughout the basin (Böcker et al., 2016; Hofmann et al., 1993). The Rhine Graben oil accumulations are related to anoxic environments formed in continental lacustrine systems preceding the evaporite deposition, similar to most South Atlantic rift basins (Evans & Kirkland, 1988).

There have been over 50 hydrocarbon discoveries in the Upper Rhine Graben, but only a few are considered commercial and still produce some oil since the end of the last century. However, a geothermal well drilled in 2003 penetrated Triassic Buntsandstein reservoirs saturated with oil and discovered the Römerberg oil field. This serendipitous discovery sparked some new interest in oil exploration in a basin that was considered senile (Reinhold et al., 2016). The delineation of the oil discovery resulted in the largest onshore oil field ever found in Germany, with hydrocarbon reserves estimated at more than 60 MM barrels of oil in place. Nonetheless, this is a tiny oil accumulation when compared to the giant oil fields discovered in other salt basins worldwide (e.g., Böcker et al., 2016).

Salt deposits have been mined for potassium ores in the Potash Basin of the Rhine Graben for over a century. These deposits are mainly located in the Alsace region of France, in the Mulhouse Potash Salt area (Figure 11.11-A). The sylvite layers are found within the Oligocene halite sequences in the Upper Saliferous Zone of the evaporite succession (Figure 11.11-B). As exploitation of sylvite ores in small mines declined by the end of the last century, production was interrupted at the beginning of the 21st century. At present, some of the mines have been converted into salt museums, constituting a famous tourist attraction in the region.

11.3. NEOGENE SALT BASINS IN TRANSTENSIONAL RIFTS: THE DEAD SEA

The northern Arabian Peninsula (Figure 11.14) is home to two of the world's largest salt basins: the Red Sea, where Middle to Late Miocene salt accumulated above rift blocks, and the Mediterranean Sea, where Upper Miocene (Messinian) evaporites accumulated on top of the Middle Miocene marine sediments (e.g., Bosworth et al., 2005). The Levantine Basin in the eastern Mediterranean is marked by pre-Messinian sedimentary successions that were deposited in deepwater environments associated with the remnants of the Tethys Sea (Hsü, 1972, 1977; Hsü & Bernoulli, 1978).

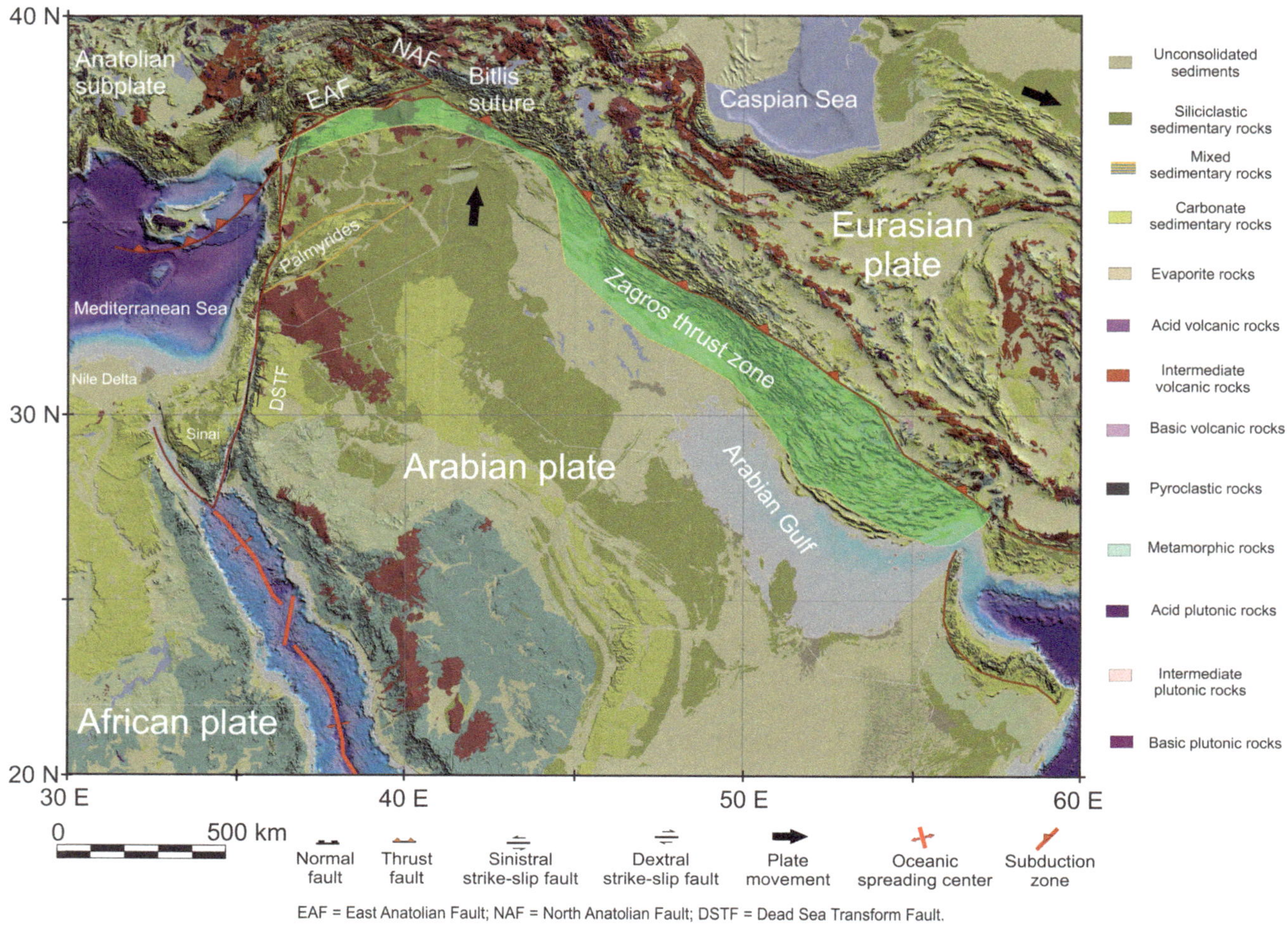

Figure 11.14 The map shows the geological map of the African, Arabian, and Eurasian plates draped on topography. The Red Sea spreading center is connected to the Dead Sea transform fault zone through a left-lateral transcurrent fault extending from the Gulf of Aqaba to the Arava Valley, reaching the Palmyrides and East Anatolian fault zone in the Taurus Mountains. Turkey's highlands are associated with the North Anatolian fault zone and the Bitlis suture zone. The Zagros thrust and fold belt deformation zones are highlighted by a transparent green color. The Gulf of Suez and Gulf of Aqaba correspond to Neogene rift zones bordering the Sinai Peninsula. Transtensional movements along the Dead Sea transform fault formed the Dead Sea, a pull-apart basin associated with a terminal lake. The transform fault zone, east of the Sinai Peninsula, links the Gulf of Aqaba to the Red Sea spreading center, forming a triple junction with the Gulf of Suez rift.

Some authors suggest that the Upper Miocene interval (Messinian) was marked by salt deposition both in the Red Sea and the Mediterranean Sea (e.g., Hsü et al., 1978). However, most researchers interpret that salt precipitation in the two salt basins was diachronous (e.g., Taviani, 1985), with the Red Sea evaporites predating the widespread Messinian Salinity Crisis in the Mediterranean Sea region (e.g., Bosworth et al., 2005).

The Dead Sea basin is a deep rift basin associated with a tectonically active transform fault related to the sinistral movement between the Arabian and Eurasian plates (Figure 11.14). The Dead Sea Transform Fault (DSTF) is a NE-trending fault zone that links the spreading center in the northern Red Sea to the Gulf of Aqaba deep trough, forming tectonic depressions, including the Arava Valley and the present-day Dead Sea depression (e.g., Brew et al., 2001; Bosworth et al., 2005). The DSTF began in the Late Miocene, after the opening of the Northern Red Sea by oceanic spreading centers, around 11 Ma (Reilinger & McClusky, 2011).

The DSTF is characterized by a series of N-NE strike-slip faults, pull-apart basins, and compressional uplifts (Figure 11.14). Some pull-apart basins are associated with modern lakes, such as the Dead Sea and the Sea of Galilee, which are characterized by a water surface below the global sea level, as indicated by topobathymetric maps (e.g., GEBCO, 2022). The fault zone extends further north along the Jordan River, connecting with deformation zones in the Palmyrides fold belt in Syria and the Taurus Mountains in Central Turkey (e.g., Brew et al., 2001; Bosworth et al., 2005). Transpressional zones are observed in the Taurus Mountains, a collision zone in Central Turkey (Anatolia). The East Anatolian and North Anatolian fault zones mark the northwestern portion of the Zagros fold-thrust belt, which is associated with the interaction between the Arabian and Eurasian plates (Figure 11.14).

The highlands in Turkey extend to Iran, forming a fold-thrust belt associated with the Zagros Orogeny, which developed mainly in the Cenozoic (e.g., Agard et al., 2011). The collision between Arabian and Eurasian plates formed the Bitlis suture zone in eastern Anatolia (Figure 11.14). Deformation continues along the NW-SE trending Zagros thrust and fold belt, extending from the Anatolian block to the Gulf of Hormuz at the southeastern extremity of the Arabian Gulf (Figure 11.14). Tectonic activity continues at a rate of 7 to 10 mm/year along transcurrent fault zones. The Arabian Plate is moving northward toward the Bitlis suture zone, as indicated by the Holocene earthquake and volcanic activity (Bosworth et al., 2005, 2019; Stern & Johnson, 2010).

The Dead Sea rift valley's origin has been discussed by several authors (e.g., Al-Zoubi & ten Brink, 2001; Al-Zoubi et al., 2002; Garfunkel & Ben-Avraham, 1996; Mart, 1991; Noda, 2013). There are two main hypotheses for its formation: one suggests that the DSTF marks the beginning of continental breakup and corresponds to a true extensional rift zone, similar to the Eastern African Rift System or the Red Sea Rift (Mart, 1991; Mart et al., 2005). Another hypothesis suggests that the Dead Sea depression is associated with a transtensional basin formed within a continental plate due to a "step-over" discontinuity along the N-NE sinistral strike-slip fault zones associated with the Dead Sea Transform (Ben-Avraham et al., 2008, 2010; Lazar et al., 2006).

The Dead Sea Transform Fault (DSTF) zone is one of the most seismically active regions in the Middle East since prehistoric times. A devastating earthquake in 1078 AD occurred in the Arava Valley, affecting the entire east Mediterranean region (Zilberman et al., 2005). Jericho, one of the world's oldest cities, is located within the valley and has been heavily affected by earthquake hazards since at least 70 ka before the present (Ben-Avraham et al., 1997). The Dead Sea lake margins display outcrops of Holocene lacustrine sediments, as in the Amiaz Plain located to the west of Mt. Sedom. Contorted sedimentary layers, known as seismites, provide evidence of earthquake activity affecting salt deposits and the overlying lacustrine sequences (Lu, Waldmann, Alsop, & Marco, 2017). Geologists from Israel have correlated these seismites to ancient earthquakes, known since biblical times (Alsop & Marco, 2011; Ben-Avraham et al., 2005; Kagan et al., 2018).

The Dead Sea Lake is currently part of an endorheic basin with no connection to the Red Sea or the Mediterranean Sea. The separation from the Red Sea is marked by a valley north of the Gulf of Aqaba, forming a land bridge about 177 km long associated with the depressions along the DSTF (Figure 11.15). The separation from the Mediterranean Sea is characterized by the uplifted regions west of the lake, such as the Lebanon Mountains. In some places, a strip of land with a width of less than 80 km constitutes the geomorphological barrier that has isolated the basin since the Pliocene. Several authors postulated that this barrier was breached in the Late Miocene/Lower Pliocene by canyons along valleys that linked with the marine waters from the Levantine Basin in the eastern Mediterranean Sea (Zak & Bentor, 1972; Zak, 1974).

Geophysical interpretation and physical models (using silicone and sand to represent salt and sediments, respectively) indicate that the Dead Sea lacustrine system was implanted on a pull-apart basin related to left-lateral strike-slip movements (Bowman, 2019; Neev & Hall, 1979; Smit, Brun, Cloetingh, & Ben-Avraham, 2008; Smit, Brun, Fort, et al., 2008; Smit et al., 2010). The pull-apart basin was formed around 17–18 Ma, due to the transcurrent, sinistral movements involving the Arabian and African plates (ten Brink & Ben-Avraham, 1989). The modern depression is an elongated valley extending from the Gulf of Aqaba to the Central Arava Valley, connecting the Dead Sea and the Sea of Galilee. The present-day Dead Sea Lake measures 80 km in length by 20 km in width, with a lake area of about 810 km^2 (Figure 11.15).

The Dead Sea basin has been subdivided into two subbasins, the northern and southern lakes, separated by a shallow sill that corresponds to the Lisan Peninsula, an almost buried salt diapir that appears on satellite images with white reflectivity (see Chapter 3). However, the southern lake has desiccated and only artificial saltworks are present. The Jordan River is the main source of freshwater for the depression. In the past century, the Dead Sea lake has been drying up, and only the northern part corresponds to a perennial hypersaline lake with a maximum water column of around 300 m (see Chapter 2).

The Dead Sea depression resulted in a lacustrine environment with a water surface much lower than the Mediterranean and Red Sea, which are at global sea level. At present, there is no communication between the Dead Sea and these larger water bodies. Due to the tectonic barriers, the Dead Sea has not been fed by marine waters from either source since the Pleistocene, despite the hydraulic gradient of more than 400 m (Lazar, 2019; Lazar et al., 2006).

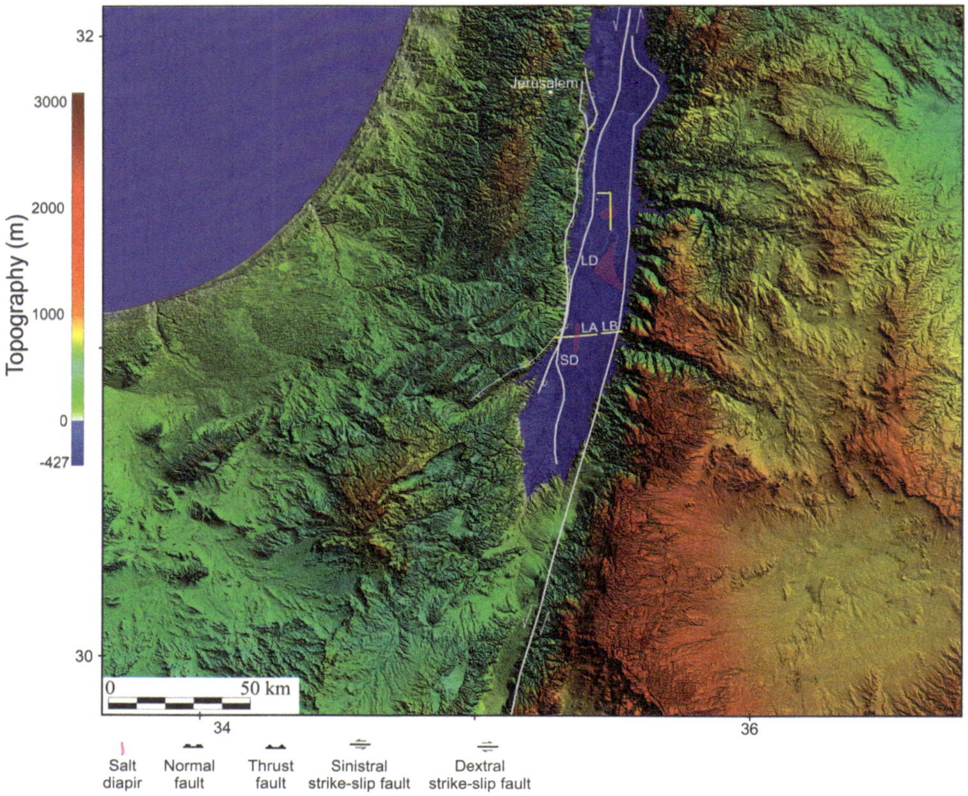

Figure 11.15 The high-resolution topographic map of the Arabian Peninsula near the Dead Sea transform fault shows a higher topography eastward of the sinistral transcurrent fault. Seismic Line A crosses Sedom diapir (SD) near the western shore of the southern lake on the Israel side. The continuation of the transect on the Jordan side corresponds to seismic line B. Lisan diapir (LD) is located between the northern and southern Dead Sea lakes. The present-day Dead Sea lake level is around −430 m, and both Sedom and Lisan diapirs are above the lake level, indicating salt is currently rising from source layers. The bluish colors on land correspond to altimetry below global sea level, forming a depression extending from the southern basin toward the Sea of Galilee in the north.

The Dead Sea northern lake is the only modern example of a highly concentrated brine(circa 300,000 ppm of NaCl, with a brine density of about 1.23 g/cc) accumulating halite and bedded salts on a deep lake floor (e.g., Warren, 2016). The southern basin was extremely shallow, with a mean water depth of 1 to 2 m (Stein, 2001; Torfstein, 2008), and currently has been replaced by artificial ponds. Periodically, parts of the southern lake sediments were exposed at the surface, forming a playa with a residual, extremely shallow saline pan (see Chapter 2).

11.3.1. Basin Subsidence and Lake-level Variations in the Dead Sea

The subsidence in the Dead Sea rift basin is controlled by active normal and transcurrent faults that formed a depression that is 80 km long and 15 to 17 km wide but is very deep, reaching 6–12 km deep in local depocenters, predominantly filled with Neogene sediments (Ben-Avraham, 1997). The rift sediments overlie Paleogene and Mesozoic pre-rift rocks that occur below and outside of the rift basin, outcropping on the uplifted flanks of the Dead Sea depression (Garfunkel & Ben-Avraham, 1996). The Holocene section corresponds to sediments associated with the Lisan paleo-lake, reaching a thickness of about 100 m in some places, indicating active subsidence.

Several water bodies occupied the tectonic depressions along the DSTF during the Miocene to Holocene interval. The remnants of the sediments deposited within the paleolakes are observed along the margins of the present-day Dead Sea Lake, as in the Amiaz Plain, where the elevated terraces from paleo-lake shores are locally cut by canyons (Stein, 2001). During the Pliocene, the Dead Sea region was occupied by the Sedom Lake, which probably received water input from the Mediterranean Sea, which was concentrated by solar evaporation, resulting in the Sedom-Dead Sea brine that precipitated thick layers of halite (Zak, 1974, 1997).

Some authors have postulated that during the Pliocene or slightly earlier, water from the Mediterranean Sea entered the rift valley through northern Israel, flowing south and mixing with freshwaters from the previous lakes. The sediments deposited during the influence of marine waters are associated with the Hazeva Fm. (Stein, 2001). When the connection with the Mediterranean Sea was interrupted by the end of the Miocene, the lake began to evaporate and a thick layer of evaporite minerals, predominantly halite, was deposited on the lake's bottom, forming a layer that was up to about 2 km (Warren, 2016; Zak & Bentor, 1972). These evaporites are included in the Sedom Fm., which nowadays have been deformed by salt tectonics, forming several halokinetic structures such as the Lisan and Sedom salt diapirs (Figures 11.15 and 11.16).

After the main phase of salt deposition, the Dead Sea basin was consecutively occupied by a sequence of terminal lakes that correspond to Lake Amora, Lake Samra, and Lake Lisan (Stein, 2001; Torfstein, 2017). These limnological variations resulted in the deposition of sedimentary layers that correspond to the three stratigraphic divisions of the Pleistocene, which correspond to the Amora, Samra, and Lisan formations. The Holocene lake (the Dead Sea) is associated with the Zeelim Fm. (Stein, 2001). These fluvial to lacustrine sediments include siliciclastic and carbonate rocks, mainly silt, marl, and chalk. These deposits are intercalated with thin lacustrine evaporite layers, forming cyclic laminations similar to varves (Ben Dor et al., 2019).

The present-day Dead Sea Lake is the remnant of a much larger lake (Lake Lisan) that extended across the valley from the Late Pliocene to the Quaternary. In the Pleistocene, the lake level was much higher (~ 180 m) than at present (~ 430 m). Thus, the lake extended to a much larger area, forming the Quaternary Lisan Fm., which can be seen today outcropping in terraces 200 m above the lake level, as in the Amiaz Plain (see Chapter 2).

These sediments are characterized by laminar dark-light carbonate laminae and lesser amounts of sulfate (gypsum)—carbonate couplets. Deposition of the Lisan Fm. is estimated to have started at some 60 ka and ended at about 18 ka (Coianiz et al., 2019; Stein, 2001, 2014). The paleo-lake has shrunk in size and depth since the last glacial episode on Earth, which ended about 40–20 ka. During the glaciation, the lake level fell from about −160 m to the mean altitude of −400 m, which has been registered throughout the Holocene (see Chapter 2).

The maximum extent of Lake Lisan (maximum flooding surface) corresponds to a maximum elevation of the lake level, estimated at ~ 150 m below mean sea level (bmsl), reaching around 27 ka (Coianiz et al., 2019). The lake has shrunk in size and depth, with its maximum extent reaching around 27 ka. Around 14 ka, the lake level dropped abruptly to 465 m bmsl, probably the lowest late Pleistocene level. The lake recovered the water column quickly until another extreme climatic event, around 11 ka, resulted in the water level falling again, depositing a thick layer of salt (Stein et al., 2010). This massive salt deposition event coincided with the Neolithic Revolution when humankind quickly spread around the Arabian Peninsula (Lu, Waldmann, Nadel, & Marco, 2017).

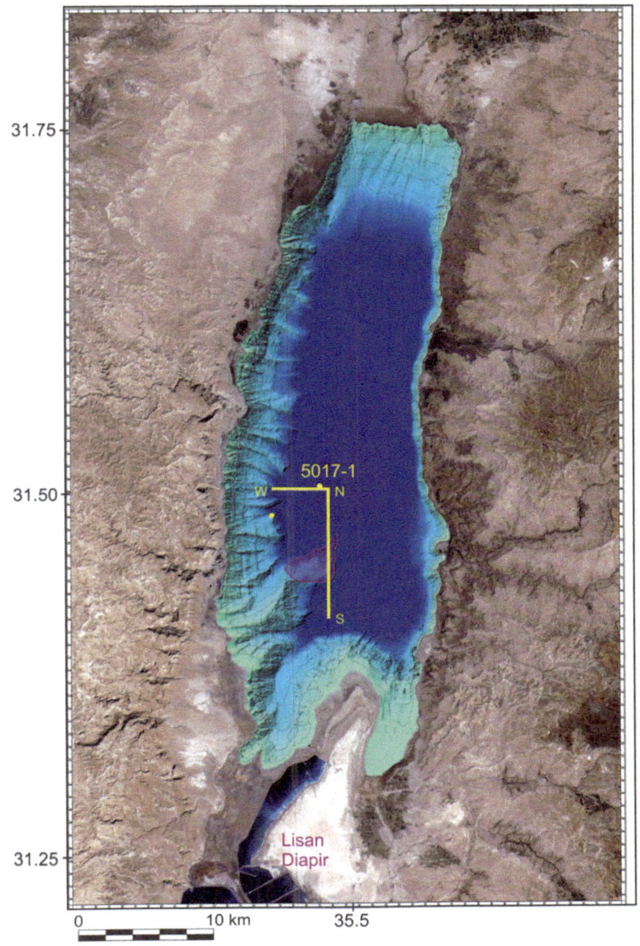

Figure 11.16 The high-resolution bathymetric map of the northern Dead Sea shows the location of borehole 5017-1 that cored the Holocene to Pleistocene sedimentary successions. The thin yellow line marks the location of the seismic line with N-S and E-W segments. The Lisan Diapir is a triangular-shaped white area separating northern and southern lakes. The Ein-Gedi salt diapir is characterized by a bathymetric high north of the Lisan structure.

During most of the Holocene, the lake stabilized at around 400 m bmsl, but in recent decades, the lake level has been falling rapidly. At present (2022), the water surface of the Dead Sea is 433 m bmsl, making this area the deepest water surface on continental crust worldwide. The maximum water depth in the northern lake is located near the region where the scientific well 5017-1 was drilled (Figure 11.16).

The Dead Sea's northern lake loses about 1 m of water every year by evaporation, and in a few centuries, it could completely dry up. The water body is disappearing fast, and if nothing is done to compensate for the water lost due to climate aridity and lack of rain, only the valley depression will remain. However, the rate of lake-level fall will probably decrease when high salt concentrations are attained. Extremely high brine concentrations lead to a chemical compensation mechanism, resulting in the decrease of the evaporation processes. The lake is resilient and has gone through several cycles of shrinking and expanding in the past 225 ka (Coianiz et al., 2019, 2020).

11.3.2. Holocene Sedimentary Deposits in the Dead Sea Lake

The Dead Sea is a deep hypersaline lake formed in a transform plate boundary, with salt deposits associated with Pliocene to Pleistocene lakes along the Dead Sea transform fault zone. These salt deposits have been extensively studied as a paradigm of deep lacustrine systems formed in transtensional settings (e.g., Coianiz et al., 2019; Smit, Brun, Cloetingh, & Ben-Avraham, 2008; Smit, Brun, Fort, et al., 2008). Most sediments in the Dead Sea are siliciclastic fine-grained sediments (clay and silt), with carbonate precipitation in the form of mineral aragonite occurring throughout the water surface in the central part of the northern Dead Sea Lake, and the fine particles settle to the bottom of the lake forming light-colored aragonite-rich lamina, characterized as a microcrystalline carbonate ooze. Intrasediment gypsum occurs as individual crystal disks that formed at the air-water interface and settled down to the bottom, composing about 6% of the deep basinal muddy sediments.

Deepwater accumulation of halite is recorded only in the southern part of the northern basin, with small cubes of salt crystals making only a tiny proportion of the mud, usually only a few percent to less than 10% of the bottom sediments. Detrital clay and marl make up most of the deepwater sediments. The only modern analog available for deepwater evaporite deposition is characterized by a notorious lack of massive evaporite sediments in the deep basin. In the past decades, following the Dead Sea overturn in the late 1970s, halitic mud and some crystalline masses of pure halite have formed periodically, but no massive precipitation of halite layers has been recorded covering the whole basin (e.g., Warren, 2016).

There is evidence for subaqueous salt diapirs in the northern Dead Sea lake, north of the Lisan salt diapir (Figure 11.16 and 11.17). Vintage seismic data (e.g., Neev & Hall, 1979) indicated a high-relief positive structure east of Ein-Gedi, at the western shores of the lake. The feature was interpreted as a Pliocene salt diapir adjacent to a structural low that formed a Pleistocene sedimentary depocenter. This led to the International Continental Scientific Drilling Program (ICDP) decision to drill two deep boreholes in the northern lake in 2010–2011, which aimed at investigating the sedimentary section in the subsurface (Neugebauer et al., 2014; Stein, 2012; Stein et al., 2011).

The deepwater scientific borehole 5017-1 (Figure 11.16) was drilled near the deepest part of the lake. The borehole was located about 10 km to the northeast of the village of Ein-Gedi, in water depths of 300 m, eastward of a large salt diapir that forms a protuberance at the lake floor (Figures 11.17). The borehole penetrated the sedimentary deposits to a maximum depth of 455 m below the lake floor (blf) and reached a total depth of about 1175 m bmsl.

Borehole 5017-1 was drilled near the E-W segment of a high-resolution seismic line, revealing detailed stratigraphic sequences in the depression adjacent to the salt diapir (Figure 11.17). The borehole penetrated a lithological column associated with the Zeelim, Lisan, Samra, and Amora formations, sampling the depositional history of the Dead Sea for the past 220 ka (Neugebauer et al., 2014; Ben Dor et al., 2019; Coianiz et al., 2019, 2020). The lithologies recovered from the boreholes are composed of siliciclastic, carbonate, and evaporite rocks, corresponding to deepwater lacustrine sediments characterized by alternations of carbonate mud (rich in aragonite) and thin layers of evaporites (Figure 11.18). Halite is more abundant in the deepwater site, while the shallower sites are dominated by laminated marls beneath the thin Holocene sedimentary section (Coianiz et al., 2019, 2020; Neugebauer et al., 2014).

The sedimentary records obtained from the cores cover the Zeelim (Holocene) and Lisan and Samra formations (Pleistocene). The Lisan Fm. is associated with the last glacial episode, while the Samra Fm. is related to the last interglacial episode, between ~135 ka and 75 ka (Stein, 2001; Waldmann et al., 2009). The well bottomed at the Amora Formation, which corresponds to the penultimate glacial episode (Torfstein, 2017).

During the Holocene, several evaporite intervals are recognized in the Zeelim Fm., forming thinly laminated halite intervals intercalated with detrital sediments (Figure 11.18). Evaporites also occur in the interglacial Samra Fm., forming sequences dominated by halite and salty mud, with intercalations of gypsum and aragonite-rich layers. The electric logs of the borehole indicate several stratigraphic sequences formed during the Quaternary, which are associated with low-stand, transgressive, and high-stand systems tracts (Coianiz et al., 2019, 2020).

The shallower sedimentary successions of the Dead Sea northern lake, penetrated by the DSDDP borehole

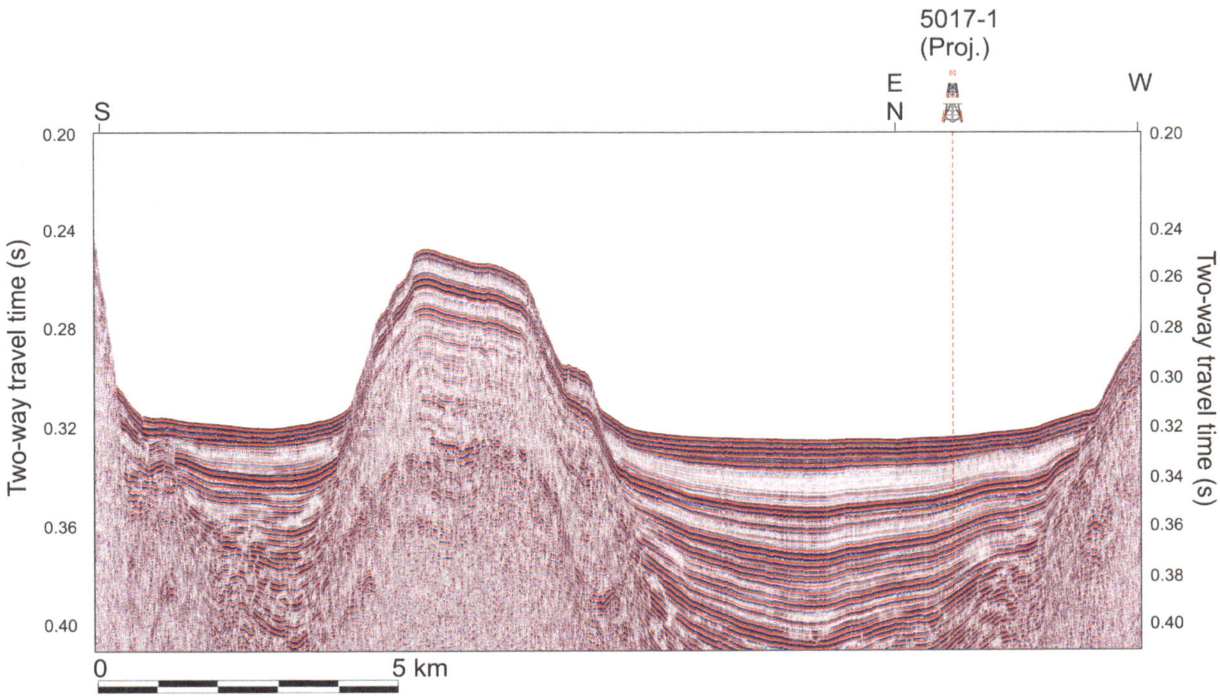

Figure 11.17 A high-resolution seismic line shows a subaqueous active salt diapir east of Ein-Gedi on the Dead Sea's western shores. The diapir is characterized by tilted sedimentary layers and rises 50 m above the lake floor. Well 5017-1 was drilled near the E-W segment of the seismic line. The borehole penetrated the Zeelim, Lisan, Samra, and Amora formations. The total depth of the well, 455 m below the lake water level, is deeper than the seismic data, which shows only the Zeelim and Lisan formations.

1017-1, are predominantly associated with marls, not evaporites. About 20% of the sedimentary record deposited in the last 220,000 years is composed of halite (Neugebauer et al., 2014).

Older borehole cores taken below water depths of 40 m at the southern end of the northern lake penetrated hard halite layers below 10 to 80 cm of laminated carbonate sediment (Neev & Emery, 1967). The origin of this compacted halite is not clear. It might correspond to older evaporite deposits or Recent layers deposited a few centuries ago, during a previous overturn event that preceded the lake overturn recorded in 1979 (see Chapter 2). Since this overturn, halite has been presently precipitating during some intervals of the year (Kiro et al. 2016; Steinhorn, 1983; Sirota et al. 2016, 2017; Warren, 2006).

The scarcity of bedded salt layers during the past centuries, from about 1650 to 1979, indicates that during modern times the Dead Sea waters were stratified and almost completely separated in terms of temperature and salinity, resulting in a meromictic stage. The absence of evaporites at the bottom of the lake for the past five centuries is surprising, as brine salinity has been close to halite saturation in the past millennia. However, only relics of halite minerals are observed in the deepwater sediments that have accumulated recently on the lake's bottom, contrasting with the dozens of halite layers that are registered in the well 5017-1 (Figure 11.18).

These thin salt intervals indicate that the Dead Sea is probably the only deep hypersaline lake in the world that deposited rhythmically repeating sequences of finely laminated sediments, including siliciclastic (silt and mud), carbonate (marl and aragonite-rich sediments), and evaporite layers, represented by cycles with aragonite, gypsum, and halite (Figure 11.18). Recent studies report that the brine concentration and evaporite minerals presently precipitating in different parts of the northern Dead Sea lake are characterized by temperature and sedimentary facies that are seasonally controlled (Kiro et al., 2016; Sirota et al., 2016, 2017). Ben Dor et al. (2019) suggested that the rhythmites in the well 5017-1 were deposited following the annual cycle of temperature changes, and thus might correspond to varves. During the Holocene period (last 11 ka), the Dead Sea witnessed 13 major climatic intervals ranging from monsoonal to arid. The dry episodes are characterized by increased salinity, forming an aggregate thickness of at least 85 m of evaporite sediments that accumulated on the northern lake's bottom (Charrach, 2018).

Currently, carnallite, sylvite, and other more soluble minerals are only produced artificially in the salt works along the shores of the southern lake. Potassium minerals,

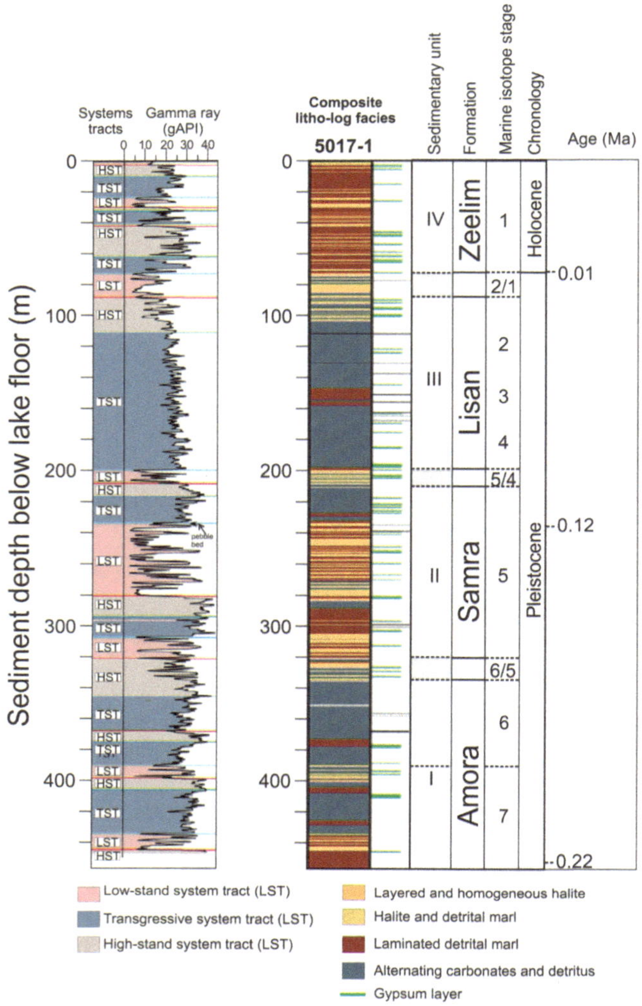

Figure 11.18 Detailed lithostratigraphic column of well 5017-1 drilled by the Dead Sea Drilling Project. The borehole penetrated sedimentary units dated from the Holocene to the Pleistocene. The borehole's gamma-ray log reveals low-stand, high-stand, and transgressive system tracts. The borehole was drilled to a depth of 455 m below the lake water level, revealing strata estimated to be around 220 ka. Holocene deposits are mainly laminated detrital marl with gypsum and halite intercalations.

such as sylvite, have not been registered in the Dead Sea. The only lakes in the world where significant amounts of K-rich evaporites are presently precipitating naturally are the high-altitude lakes in the Qarhan salt flats in China (see Chapter 10). Some authors have interpreted that during the Pliocene, when the Sedom salt precipitated in the Dead Sea, potash salt minerals might have accumulated at the end of the evaporite cycles (García-Veigas et al., 2009).

The southern desiccated lake in the Dead Sea corresponds to extremely shallow saltworks that produce several salts, including bromine, magnesium, and potash salt. Sodium chloride is not economically exploited in the area due to transportation distances and costs. The Dead Sea Lake is an extraordinary economic asset for Israel, with salt industry companies responsible for 10% of all potassium chloride production worldwide (U.S. Geological Survey, 2021).

11.3.3. Salt Diapirs in the Dead Sea

The Early Pliocene salt deposits in the Dead Sea formed tens of evaporite cycles, forming a total thickness of over 2,000 m (García-Veigas et al., 2009). The climate ranged from humid to extremely arid, forming various siliciclastic and evaporitic minerals such as muds, marls, gypsum, halite, and possibly more soluble salts like carnallite. The brine concentration in the ancient Sedom Lake was related to the interplay between water inflow and evaporation. Low inflow and high evaporation rates resulted in increasing the brine concentration and caused the precipitation of multiple beds of soluble salts, such as carnallite and bischofite. The aridity in the Late Miocene-Early Pliocene was much greater than the present-day Dead Sea climate, as only halite has been precipitated naturally from the concentrated brine.

The thick salt layers in the Sedom Fm. record a period of widespread evaporite precipitation that affected the Dead Sea in the Pliocene, indicating a salinity crisis in the basin. This episode has been correlated to the Mediterranean and the Red Sea salinity crises (e.g., Charrach, 2018). However, even during this period of extreme aridity, there were intervals of more humid climate with significant rainfall in the hydrological system, resulting in flooding of the basin. The entering of freshwater into the system likely resulted in the dissolution of carnallite layers and the precipitation of sylvite and polyhalite as secondary minerals (Charrach, 2018).

A few salt diapirs rising to the surface are observed in the Dead Sea basin, such as the submerged Ein-Gedi salt diapir, which forms a protuberance of circa 50 m above the lake floor (Figure 11.17). The large Lisan Diapir, located in the peninsula that separates the southern and northern lakes of the Dead Sea basin, is characterized by a cap rock that can be seen just above the lake water level (Figure 11.16).

The Sedom salt diapir is a 4 km tall structure located at the western edge of the southern Dead Sea subbasin (Figure 11.19). It is controlled by a strike-slip longitudinal fault parallel to the main rift trough and intersected by the NW-SE trending Amazyahu cross fault (Alsop et al., 2012, 2014). Some authors interpret these faults as related to salt tectonics or halokinesis, forming a low-angle listric fault (Ben-Avraham, 1997; Larsen et al., 2002;

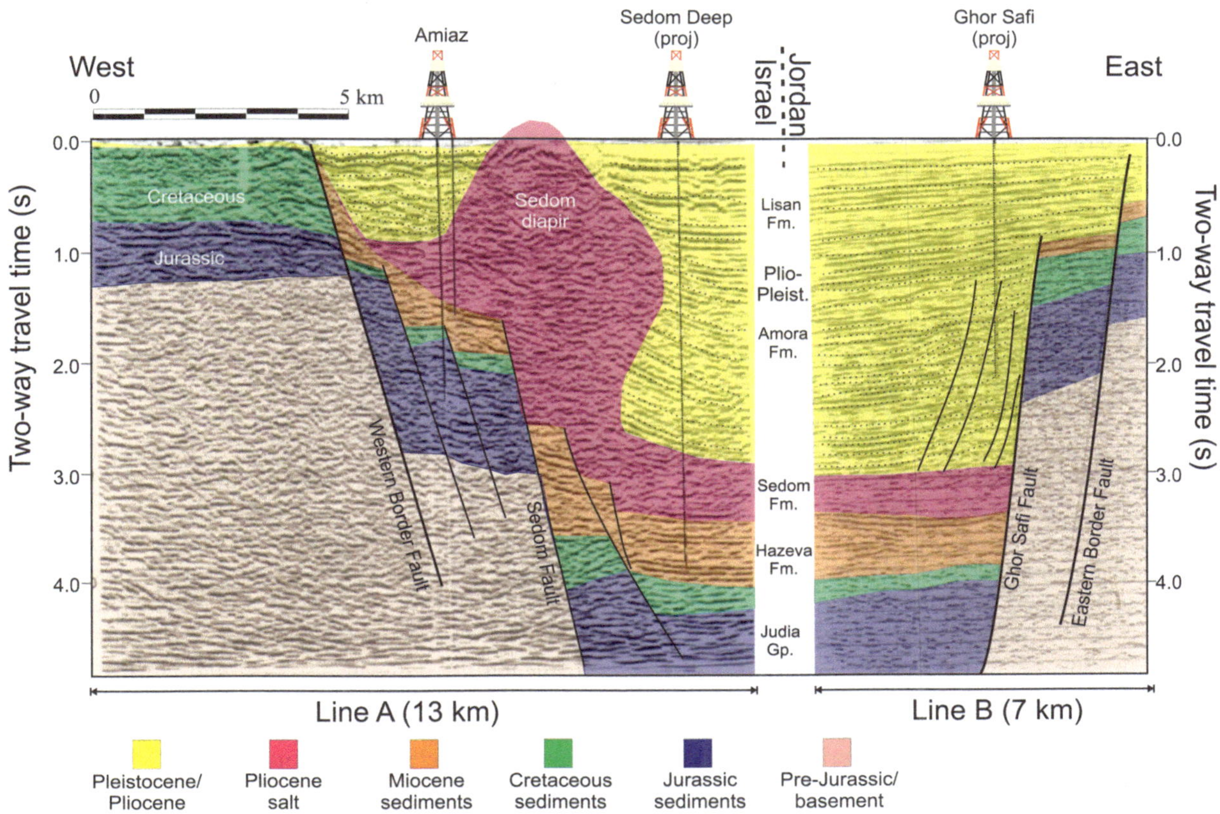

Figure 11.19 Composite seismic profile across the Sedom diapir in the southern Dead Sea, from Israel to Jordan. Line A (West) is located on the Israeli side, while Line B (East) is located on the Jordanian side (see Figure DS-2). The seismic line reveals an asymmetric salt distribution with thicker Pleistocene to Pliocene sedimentary layers toward the eastern border fault system in Jordan. The Miocene rift includes Hazeva Fm. and Sedom Fm., while the Pliocene and Pleistocene sediments include the Amora and Lisan formations.

ten Brink & Ben-Avraham, 1989). As seen in seismic profiles in some segments of the Dead Sea basin, the detached listric faults form large roll-over structures involving the Pleistocene to Holocene sedimentary layers (e.g., Smit, Brun, Fort, et al., 2008). Other authors, on the contrary, have suggested that these faults correspond to basement-involved faults associated with the ongoing rifting process (Ginzburg et al., 2006).

The Sedom salt diapir outcrops near the southern lake edge, forming the famous Mt. Sedom, which has been thoroughly studied by several researchers (e.g., Charrach, 2018; Weinberger et al., 2006; Zak & Freund, 1980). Mt. Sedom is an elongated salt ridge with recumbent folds as the flanks are almost vertical (Alsop et al., 2015, 2016). At the surface, the diapir is about 11 km long and rises about 200 m above the lake level, but its top is still below the world's sea level. It is considered one of the most intriguing salt structures in the world, with salt columns outcropping at its margin and caves carved in stone, which have caught the attention of many prehistoric cultures. Some features with anthropomorphic shapes (such as Lot's wife) are linked to ancient tales. Mt. Sedom is associated with the biblical stories describing the destruction of Sodom and Gomorrah (see Chapter 1 and Chapter 3).

11.4. MARGINAL MARINE EVAPORITE SEDIMENTATION IN FORELAND BASINS

Marginal marine evaporites are associated with several carbonate platforms in ancient foreland basins, including the Permian Basin in the southwest United States and the Persian Gulf in the Middle East. These hydrocarbon provinces, which developed during periods of plate convergence, are related to sabkhas, lagoons, and shallow to deep inland seas connected to a larger ocean. The paleo-shelf-edges in these basins are marked by carbonate buildups or reefs (Figure 11.8).

The Permian Basin, a major oil-producing region in the United States, has been extensively studied since the early 20th century as a paradigm for evaporite sequences extending from shallow carbonate platforms to deep water environments. The basin's sequence stratigraphy has been extensively analyzed due to excellent 3D outcrop exposures at the basin margins (Harris & Kerans, 2016). The Permian Basin is a highly prolific petroleum province that has been exploited since the last century (Dutton et al., 2005; Ewing, 2017, 2019; Robinson, 1988; Wright, 2011).

The Delaware and Midland subbasins host the prolific Permian age carbonate and siliciclastic reservoirs, which span the time interval between 250 and 300 Ma (Ewing, 2017; Hills, 1972; Robinson, 1988). Other sedimentary provinces in North America are associated with Paleozoic and Mesozoic inland seas, which have accumulated carbonate rocks and evaporites during their geological evolution. Salt structures in these basins have been used as potential storage sites for radioactive waste materials (e.g., Dean & Johnson, 1989; Johnson & Gonzales, 1978; Perry et al., 2014).

Ancient marginal marine environments formed large salt deposits in North America, unlike the relatively small hypersaline lakes in desert regions. Evaporite sedimentation was recorded in various sedimentary provinces throughout geological time, from Paleozoic intracratonic depressions (e.g., Williston Basin) to foreland troughs in the northeastern region (e.g., the Appalachian Basin). During the breakup of Pangea in the Mesozoic, several rifted continental margin basins were formed in southern North America, such as the Gulf of Mexico and the Eastern Atlantic divergent margin (see Chapter 4 in Mohriak, 2025).

11.4.1. Ancient Carbonate Platforms with Evaporite Deposits

Ancient basins with carbonate platform-evaporite sequences were formed in two major paleogeographic settings: (1) continental margins with deserts, sabkhas, lagoons, and shallow seas; and (2) basinwide evaporite deposition associated with epeiric seas with deep water evaporites in the basin depocenter (e.g., Handford, 1991; Kendall, 1978, 1992, 2010; Warren, 1989, 1991, 2006, 2016) The carbonate-evaporite model (Figure 11.8) is characterized by a carbonate bank forming a barrier that intermittently connected the restricted basin with an open ocean with normal marine sedimentation. The sedimentological characteristics of these models are summarized below.

Platform evaporites correspond to bedded salt deposits formed by stratiform layers with a thickness of 1 to 10 m, rarely forming a cycle with more than 50 m in total salt thickness. These geological units may spread laterally across the subsiding region, covering thousands of square kilometers, as observed in several intracratonic basins (e.g., Michigan Basin, Figure 11.20). Sedimentary facies related to the proximal margins include subaerial deserts, mudflats, coastal sabkhas, lagoons, and shallow water marine environments in the continental shelf (Figure 11.8). The lithologies include siliciclastic rocks intercalated with carbonate and evaporite rocks. The evaporite sequences in lagoons and shallow inland seas are composed of alternations of carbonate, sulfate, and chloride minerals deposited according to the brine salinity, which varies from the marginal lagoon to the basin depocenter (Figure 11.8).

Proximal coastal zones include mudflats with evidence of subaerial exposure, as indicated by mud cracks, erosional surfaces, and karst structures. Modern and ancient examples of platform evaporitic mudflats appear interlayered with eolian or fluvial sediments (Figure 11.8). Subaerial components include ephemeral salt pans with evaporite crusts intercalated with shallow-water deposits. Displacive evaporites include nodular, enterolithic gypsum, or anhydrite. Platform evaporites commonly form isolated pods of halite in salinas, which pass upward and seaward into time-equivalent open-marine marls and mudstones (Figure 11.8). Modern counterparts of this marginal marine depositional setting can be found in the sabkhas of the Persian Gulf (Alsharhan & Kendall, 2003; Handford, 1991). Examples of ancient platform evaporites include Carboniferous evaporites in the Amazon Basin in northern Brazil and the Cretaceous Ferry Lake Anhydrite in the Gulf Coast.

Basin restriction by sedimentary and tectonic barriers resulted in anoxic environments in the basin depocenters of several Paleozoic intracratonic basins marked by extensive shallow carbonate platforms. The Silurian hot shales in North Africa and Arabia present anomalously rich organic carbon intervals, which are responsible for more than 80% of the oil accumulations discovered in some petroliferous basins (Luning et al., 2000; Ziegler et al., 1977).

11.5. EVAPORITE CARBONATE PLATFORMS IN NORTH AMERICA

Figure 11.20 shows a simplified map of salt deposits in onshore basins across the United States. The map illustrates the areal extent of diverse geological units, including carbonate and evaporite sequences from the Paleozoic, Mesozoic, and Cenozoic eras (e.g., Eia-U. S. Energy Information Administration, 2016; Johnson & Gonzales, 1978; Perry et al., 2014).

Paleozoic-Mesozoic evaporite basins on land are characterized by bedded salt deposits in carbonate platforms formed in convergent and divergent margins (Warren, 20 06, 2010, 2014, 2016, 2018). The main bedded salt deposits in the southwestern United States are associated with the Permian Basin, which extends across Kansas, Texas, and New Mexico. Several other basins include

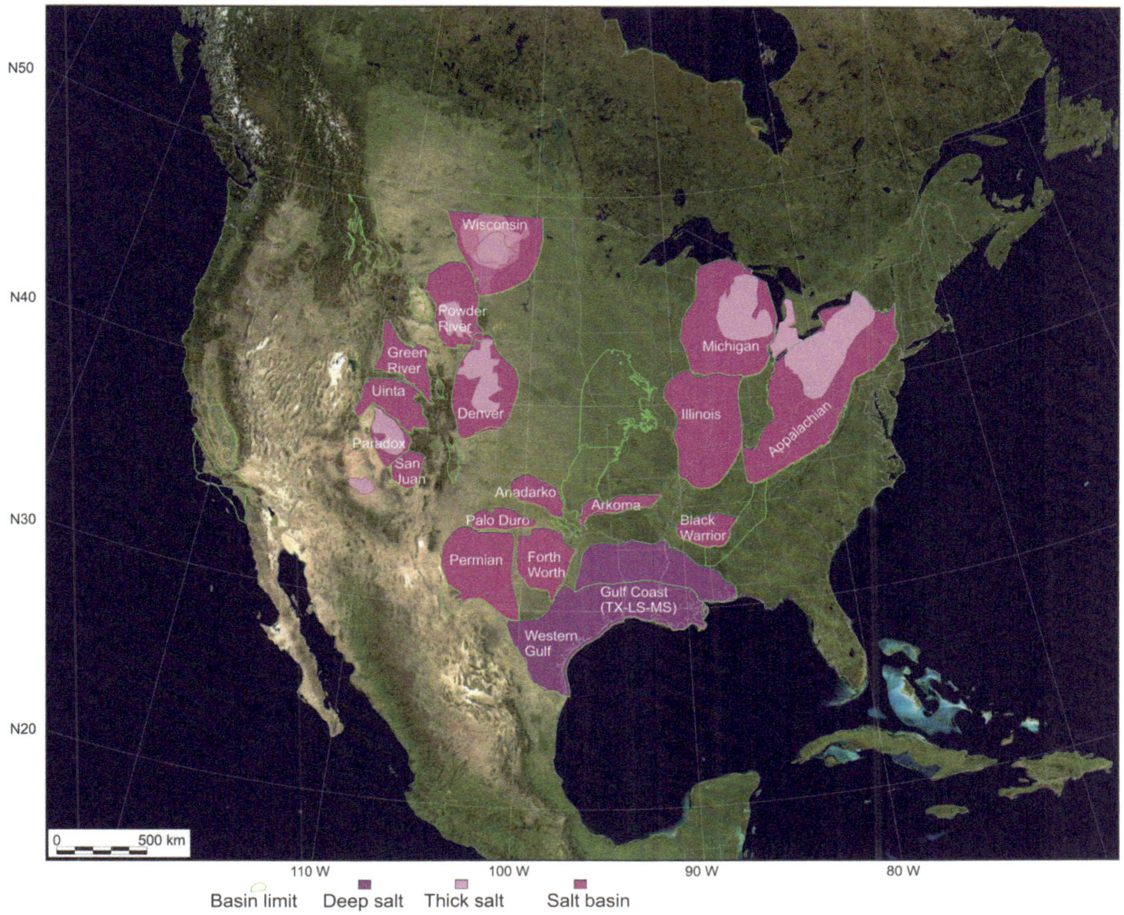

Figure 11.20 The satellite image of the United States shows the outline of major sedimentary basins with green lines. The simplified outlines of major salt provinces with bedded salt, such as in the Permian Basin, are shown in magenta, while the Gulf Coast is depicted in purple. Light pink is used to represent the thick-bedded salt deposits in depocenters. The Williston Basin at the border with Canada has different ages of evaporite layers, while the Gulf Coast has thick and deep Jurassic evaporite layers extending from the coastal plain to the deep-water region of the Gulf of Mexico. Credit: Photo by NASA/Public domain.

evaporite sediments, such as the Appalachian Basin (Silurian salt) and the Gulf of Mexico (Jurassic salt).

Autochthonous salt deposits are typically deep and hidden from view by overlying strata, as most basins are covered by Cenozoic sediments (Figure 11.20). For instance, evaporite deposits in the Williston Basin are restricted to the subsurface. In some provinces, evaporitic units are shallow, like in the Michigan and Permian basins, where the salt layers are locally covered by less than 1000 m of sediments (Perry et al., 2014).

In the Gulf of Mexico, most Jurassic salt is overlain by thick post-salt sequences, but locally salt diapirs rise to near-surface levels. These shallow bedded salt layers and diapiric structures have been considered potential waste material repositories or hydrocarbon storage sites (Johnson & Gonzales, 1978; Seni et al., 1985). Salt layers occur in half of the conterminous United States, covering a wide geologic time period, with evaporite sequences ranging from Silurian to Neogene. The largest and most important geological units with evaporite deposits in the onshore part of the United States are presented in Figure 11.21.

The largest salt provinces in the United States are along the Gulf Coast, with the Louann salt extending both onshore and offshore of the Gulf of Mexico (Figure 11.21). Jurassic salt layers occur in the continental shelf and deep waters, forming a giant salt basin that reaches the Mexican territory in the western part of the Gulf and the Yucatan Peninsula in the east (see Chapter 4 in Mohriak, 2025).

Several basins in the United States are characterized by evaporite sedimentation in sabkha environments, particularly during the Permian, when desertic conditions prevailed in the western part of the country (e.g., Benison & Goldstein, 2000). Permian redbeds extend across much

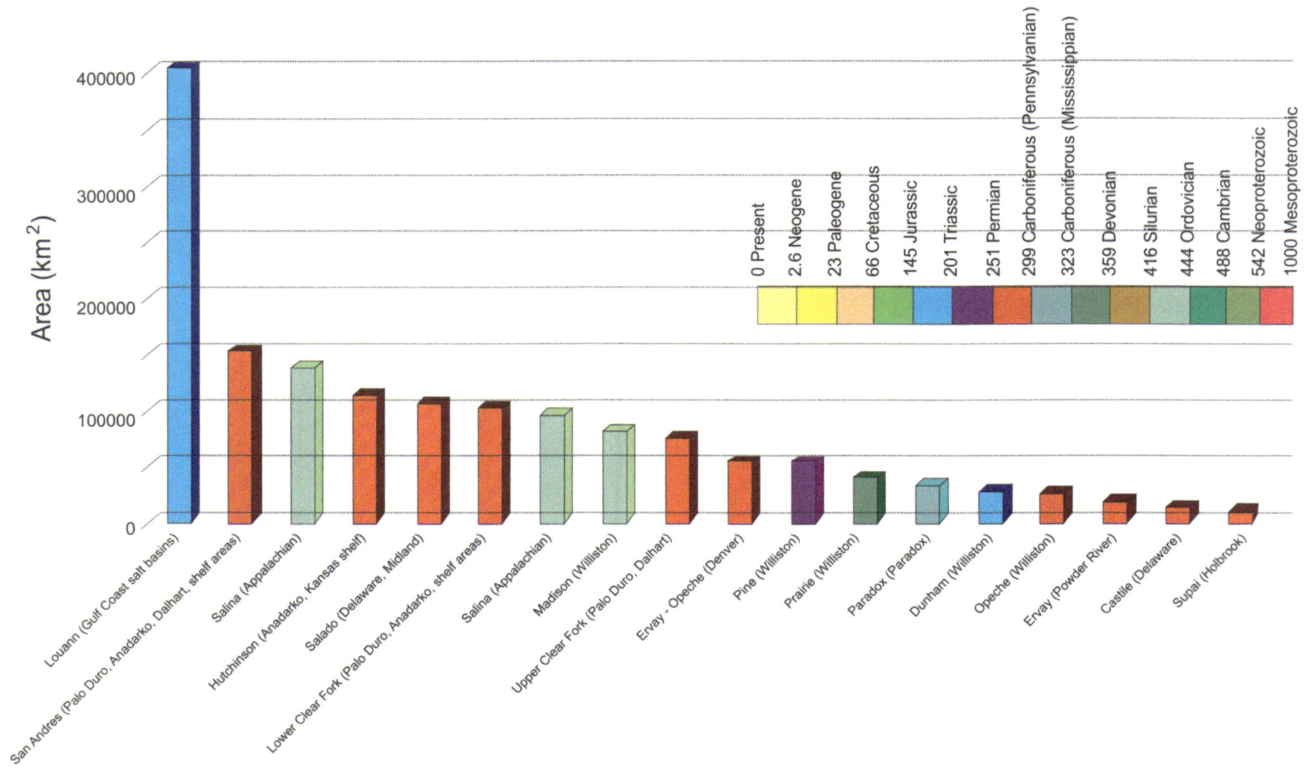

Figure 11.21 The diagram shows the onshore sedimentary basins in the United States with diverse geological units and evaporite sequences. The 3D column height indicates the area extent of subsurface geological units, while the vertical column color indicates the depositional interval age of the evaporite layer, with approximate ages in million years.

of the central to western parts of the country, with bedded evaporite layers concentrated in the Permian Basin Figure 11.20). Evaporite sediments are also recorded in other geological intervals, such as the Silurian (Appalachian Basin) and Cretaceous (Ferry Lake Anhydrite in the onshore part of the Gulf of Mexico).

Figure 11.22 shows a synthesis of the geological units associated with salt deposits in the sedimentary basins onshore of the United States. Paleozoic salt sequences are recorded in many stratigraphic levels, particularly the Silurian, Devonian, Carboniferous, and Permian (Figure 11.21).

Permian bedded salt sequences are found in several onshore basins, such as the greater Permian Basin in Texas, New Mexico, and Oklahoma. In the Permian Basin, these deposits are associated with the Castile and Salado geological units. These sediments correspond to marginal marine evaporite sequences with bedded anhydrite and halite overlying the Permian platform carbonate rocks (e.g., Scholle, 2020; Scholle et al., 2007).

Jurassic salt is mainly found in the southern part of the United States along the Gulf Coast margin. The Middle Jurassic Louann salt layers formed large diapirs in the Mesozoic and Cenozoic, particularly in the offshore region. Major salt structures have been delineated from the onshore part of the Gulf, extending from the western part at the border with Mexico to several subbasins in south Texas, Louisiana, Mississippi, and Florida (Figure 11.20).

11.5.1. Salt Deposition in the North American Paleozoic Basins

Orogenic events have been occurring in the southwest United States since the Late Precambrian, when the North American plate was the site of major tectonic events that led to the rupture of the early mega-continent (Blakey, 2008, 2013; Scotese, 2001). The rupture of the Rodinia landmass formed several Neoproterozoic rifts and continental margins adjacent to large oceans. The continental masses were separated by oceanic regions in the Early Paleozoic, but started to approach again during the Late Paleozoic, closing the intervening oceans by plate convergence and forming suture zones.

Before the Permian, the western region of North America (Western Canada, the United States, and northern Mexico)

Basin	Geological unit	Age
Gulf Coast salt basins	Louann	Jurassic
Palo Duro, Anadarko, Dalhart, shelf areas	San Andres	Permian
Appalachian	Salina	Silurian
Anadarko, Kansas shelf	Hutchinson	Permian
Midland, Delaware	Salado	Permian
Palo Duro, Anadarko, shelf areas	Lower Clear Fork	Permian
Michigan	Salina	Silurian
Williston	Madison	Mississippian
Palo Duro, Dalhart	Upper Clear Fork	Permian
Denver	Ervay-Opeche-San Andres equivalents	Permian
Williston	Pine	Permo-Triassic
Williston	Prairie (within US)	Devonian
Paradox	Paradox	Pennsylvanian
Williston	Dunham	Jurassic
Williston	Opeche	Permian
Powder River	Ervay	Permian
Delaware	Castile	Permian
Holbrook	Supai	Permian

Figure 11.22 Sedimentary basins in the United States have significant evaporite deposits, mostly associated with marginal marine environments. These deposits are found in several stratigraphic units, including the Silurian Salina Fm. in Michigan and Appalachian basins, the Devonian Prairie Fm. in the Williston Basin, the Carboniferous Paradox Group in the Paradox Basin, and several Permian units in smaller basins.

was located within the Laurentia Plate, with an active continental margin in the Panthalassa Ocean, west of the landmass (Figure 11.23). During most of the Paleozoic, a subduction zone with an eastward vergence was formed as the Panthalassa Ocean subducted at the western margin of Laurentia and Gondwana. Several exotic terranes were accreted and amalgamated to this large continental landmass, such as the Wrangellia terrane. Wrangellia was accreted to the North American Cordillera and presently forms most of Vancouver Island (Blakey & Ranney, 2018).

The southern margin of Laurentia was separated from the northern margin of Gondwana by the Rheic Ocean around 370 Ma (Nance et al., 2012). The Iapetus Ocean was closed by the collision of North America and Africa during the Upper Carboniferous to Permian, forming the Alleghanian-Appalachian orogeny (Figure 11.23). This event also affected Laurentia's southern margin until the Permian when the Marathon-Ouachita orogeny closed the Rheic inland sea. According to Hatcher (2010), this event lasted from the Late Mississippian until the Late Permian (340 to 250 Ma). By the Permian, a single supercontinent known as Pangea was formed by the coalescence of several cratonic blocks separated by fold belts and suture zones (Nance et al., 2012; Torsvik & Cocks, 2012, 2017).

The Paleozoic collisional event resulted in the amalgamation of most landmasses in the Pangea supercontinent. The closure of these oceanic regions led to shallow marine environments and the formation of widespread salt deposits in North America, particularly near the border between the United States and Canada (Figure 11.20). These desiccation episodes are recorded in several stratigraphic intervals with salt precipitation, from Silurian to Permian (Figure 11.22).

11.5.2. Carbonate-Evaporite Platforms in the Paleozoic Intracratonic and Foreland Basins

Most of the Paleozoic intracratonic basins in North America accumulated evaporite successions in the lower parts of the sedimentary depressions, forming bedded salt sequences. Silurian evaporites occur in the depocenter and borders of the intracratonic Michigan Basin (Harrison & Voice, 2017). Silurian salt is also recorded in the depocenter of foreland basins affected by compressional tectonics, such as the Appalachian Basin

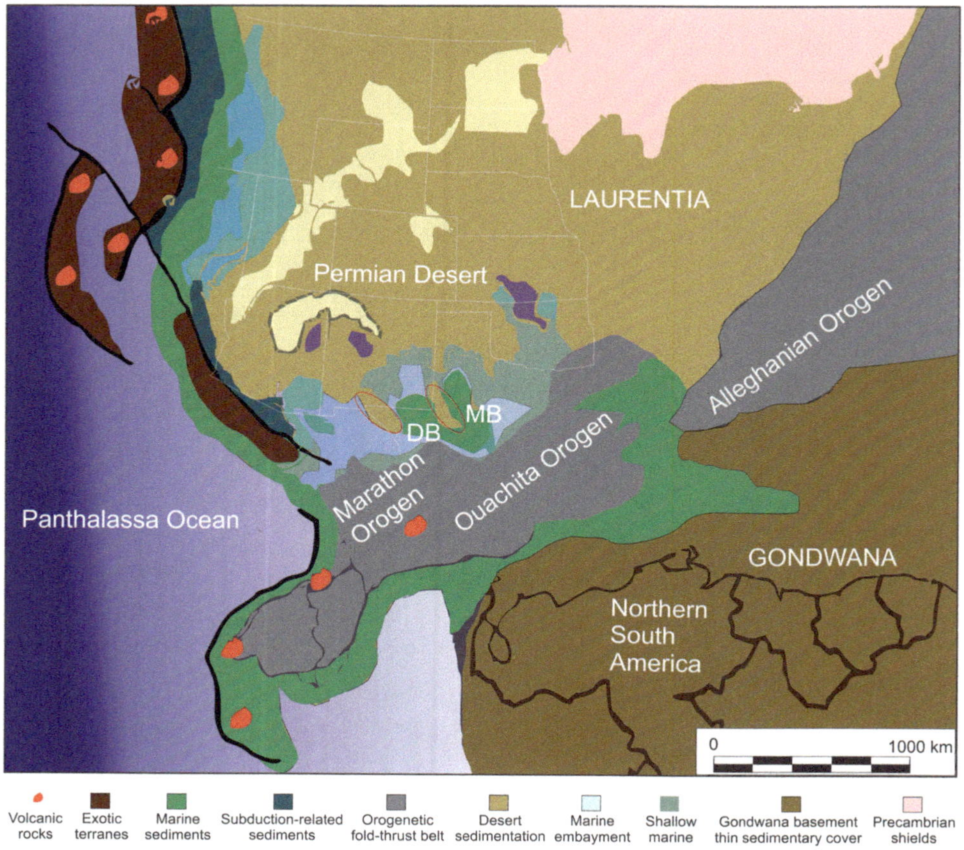

Figure 11.23 A simplified paleogeographic map shows the reconstruction of the western North American landmass during the Permian. The Panthalassa Ocean was subducting at the western margin, leading to the accretion of exotic terranes. The seaway between southern North America and northern South America was closing, resulting in the Marathon-Ouachita orogenic belts. The eastern North American and northwestern African plates were converging, resulting in the Alleghanian-Appalachian Orogen. Sedimentary depocenters, such as the Permian and Anadarko basins in north Texas and Oklahoma, are related to flexural depressions in the continental crust. The thin white dashed lines on the Permian Desert mark the approximate location of the U.S. southwestern state limits. The dark brown lines south of the Ouachita Orogen mark the approximate country limits of northwestern South America.

(Ettensohn, 2008). Younger deposits are recorded in the Williston, Paradox, and Permian basins.

Four principal regions in the United States have thick evaporite sequences associated with carbonate platforms in the Paleozoic intracratonic and foreland basins (Figure 11.22):

- Michigan and Appalachian Basins—Silurian salt
- Williston Basin—Devonian salt
- Paradox Basin—Carboniferous salt
- Permian Basin of New Mexico/Texas to Kansas—Permian salt

The intracratonic and foreland basins in the northern United States are characterized by Silurian evaporite successions linked to the Salina Fm., a significant source of salt production during the Industrial Revolution (see Chapter 1). This region was the hub for halite and sylvite production in the 19th century, using the old technique of evaporating hypersaline brine from deep subsurface reservoirs.

The evaporite layers in the Paleozoic intracratonic basins are geographically distributed, with a maximum thickness near the basin depocenter, according to the bull's eye model (Figure 11.7). The sedimentary layers are characterized by equivalent stratigraphic intervals with thinner layers of less soluble evaporites extending toward the borders (Dellwig & Evans, 1969; Horton, 2017). This model is exemplified by the Michigan and Appalachian sedimentary basins, where Paleozoic evaporites exhibit halite and more soluble minerals in the depocenter, while the edges are characterized by anhydrite layers, carbonate platforms, and reefal buildups (11.24).

Middle to Upper Silurian evaporites in the Michigan Basin are crucial for petroleum systems in the pinnacle reef play (e.g., Catacosinos et al., 1991; Droste & Shaver, 1977; Miller & Dupal, 1987). These large stromatoporoid barrier reefs, known as bioherms, are composed of massive carbonate buildups that surround the edges of the Michigan Basin Clustered along a 16–24 km wide belt, they reach thicknesses of 150–200 m (Figure 11.25). The basinal equivalent of these carbonate reefs is a thin laminated carbonate sequence with a thickness of about 18 to 24 m (Rine et al., 2017). The evaporite layers are composed of finely banded halite and dolomite-anhydrite couplets, with the laminae displaying thicknesses of less than 10 cm (Dellwig & Evans, 1969; Harrison & Voice, 2017; Sonnenfeld & Al-Aasm, 1991).

The Salina Group's laminated evaporites have been attributed to deepwater origin (e.g., Droste & Shaver, 1977), while others suggest a shallow-water origin, possibly linked to lagoons and sabkhas (e.g., Nurmi & Friedman, 1977). Cercone (1988) points out that shallow seas must have existed during the deposition of the thickest evaporite sequences in the Michigan Basin (corresponding to the basal A-1 salt layer) because of the following evidence: (1) the presence of vertically oriented gypsum pseudomorphs; (2) chevron-zoned halite crystals; (3) nodular anhydrite; (4) calcium borate nodules; (5) algal stromatolites; (6) diagenetic processes (dissolution, dolomitization, and recrystallization) indicating subaerial exposure and weathering of reef carbonates within the basin; and (7) reduced deposition of sulfate and magnesium minerals in the basin center.

The Michigan Basin's water level may have been lowered by evaporative drawdown, resulting in a basin below global sea level due to isolation and rapid evaporation under an arid climate, as proposed for the Mediterranean Sea in the Late Miocene by several authors (e.g., Hsü, 1972, 1977; Hsü et al., 1978). Large reefal carbonate buildups in the Silurian rim of the basin created a barrier, hindering communication between normal Silurian seas and hypersaline brines. Evaporite sequences covered the basin floor and reefs during marine ingression and basin desiccation, resulting in dozens of cycles in the Salina evaporite sequences (Cercone, 1988).

Figure 11.25 shows a schematic cross-section illustrating the stratigraphic relationship between the Niagaran platform carbonates, the shelf-edge reef, and the lowest evaporite cycles (A1, A2, and B) at the northwest part of the basin. The younger salt oversteps the older layers, extending from the depocenter to the back-reef region (Figure 11.25).

The Silurian mother salt layers in the Michigan Basin include several cycles of evaporite deposition, represented by intercalations of carbonate, sulfate, and chloride minerals, indicating that the brines reached saturations that allowed the precipitation of halite and sylvite (Figure 11.26). These evaporite minerals were deposited in a restricted marine basin that was partially connected to the open ocean and was periodically desiccated during the Silurian, but evaporite deposits are also recorded in the Devonian (Harrison & Voice, 2017; Sonnenfeld & Al-Aasm, 1991).

Intracratonic basins in the northern U.S., such as the Williston Basin, record multiple episodes of salt deposition (Figure 11.29). The Paleozoic salt deposits are extensively exploited in Canada, forming one of the largest salt provinces in the world, which accounts for a large percentage of potash salt production in North America (mainly sylvinite). These deposits form evaporite sequences intercalated with carbonate and siliciclastic sequences. Canada is the leading country in global potash salt production, with 14,000 million tons of K_2O, accounting for one-third of the global production capacity (U. S. Geological Survey, 2021).

The salt depth also varies from basin to basin in North America, with the deepest salt layers occurring in the Gulf Coast and Appalachian basins. The Permian Basin is characterized by a rather shallow top of the salt layers, which are often covered by less than 1 km of sediments (Perry et al., 2014; Scholle et al., 2007).

The thickness of individual salt bodies in these basins is small, with thin halite lenses often intercalated with different lithologies, including carbonate and sulfate sequences. Most geological reports that display salt isopach maps (e.g., Johnson & Gonzales, 1978) correspond to the total summed thickness of several salt layers included within a particular geologic formation. This is not the total thickness of the formation because the geological units containing halite layers typically contain interbedded carbonate, sulfate, and siliciclastic beds The evaporite cycles reflect the alternation of shallow marine and near-shore depositional environments associated with marine transgressions and regressions.

The Williston Basin is a sedimentary province with the largest number of different episodes of salt deposition in the United States (Figure 11.24). This intracratonic basin is characterized by a depocenter with a total thickness of sediments reaching a maximum column of about 4,900 m, and the stratigraphic sequences span the interval from Cambrian to Recent (Catacosinos et al., 1991). The Williston Basin is mainly characterized by the accumulation of Devonian salt, which is related to the Prairie Fm., corresponding to the equivalent geological unit in the Elk Point Basin in Canada. However, the Williston Basin records multiple events of salt deposition, including the Carboniferous, Permian, Permo-Triassic, and Jurassic (Figure 11.26 and Figure 11.27).

The salt formations in the basin are almost flat-lying and vary in depth of occurrence from about 1,000 m near the borders to more than 3000 m in the depocenter, where

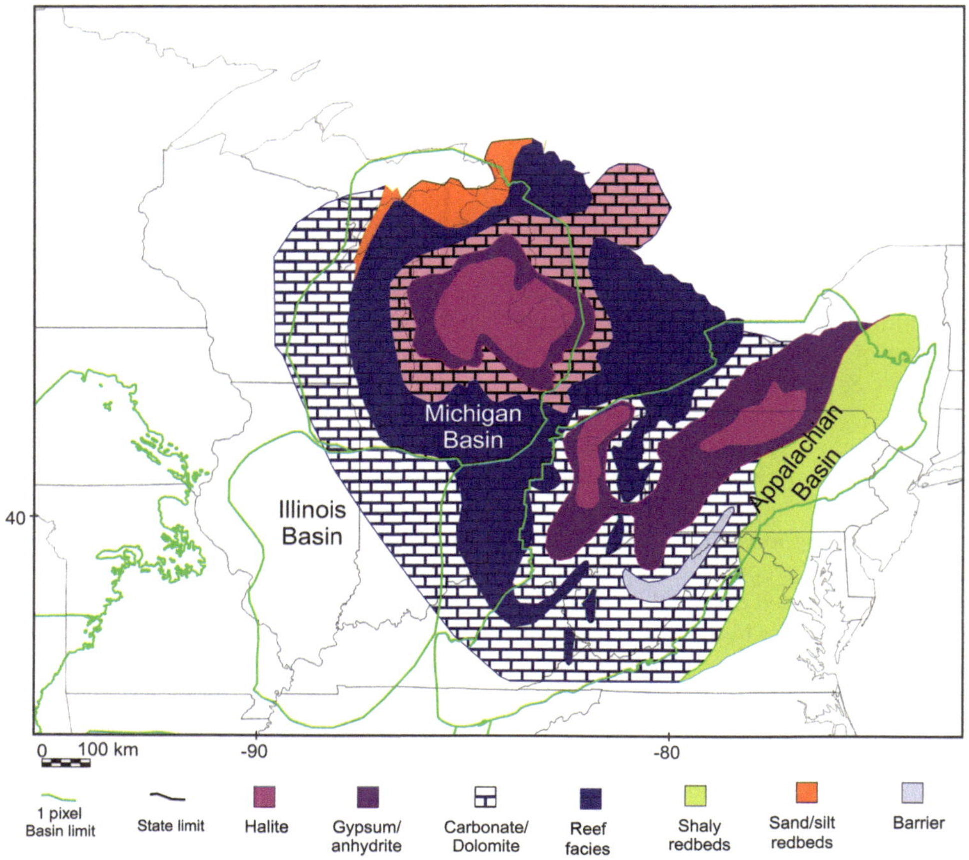

Figure 11.24 The schematic map shows the carbonate and evaporite facies in the Michigan and Appalachian basins during the deposition of Salina evaporites. Carbonate and dolomite lithologies dominate at the basin borders, while sulfate and chloride minerals are found in the central parts. A belt of reef carbonates surrounds the basin depocenter, forming buildups near the shelf edge. The interruption of reef construction at the NE portion of the Michigan Basin suggests a Silurian entrance of seawater. The dolomites associated with the Salina evaporites gradually change into limestones toward the west and south, indicating a transition to environments with normal marine salinity.

most of the soluble salt deposits are accumulated (Dellwig & Evans, 1969). The Devonian Prairie Fm., the oldest and lowest evaporite unit, is characterized by a combined salt thickness of about 130 m in the American portion of the basin. The top of this salt layer varies from 1,300 m at the borders to more than 3,600 m at the basin depocenter (Johnson & Gonzalez, 1978; Perry et al., 2014).

Marginal marine evaporites from the Permian period are crucial for petroleum systems in the most prolific basins in the southwest U.S., including Delaware, Midland, Palo Duro, Anadarko, and Denver. These bedded evaporites were deposited during the Late Paleozoic (Permian), but the resultant combined thickness of the halite layers was not large enough to form salt diapirs. Thick mobile evaporite layers forming large salt diapirs are only observed in a few onshore basins in the southwest region of the U.S., such as the Paradox Basin (Carboniferous salt) and particularly the Gulf Coast salt basin. In the onshore region of the Gulf of Mexico, from Texas to Louisiana and Mississippi, Jurassic salt accumulated in several depocenters (see Chapter 4 in Mohriak, 2025).

11.5.3. The Mesozoic Carbonate Platforms in the Northeastern Gulf of Mexico

The Gulf of Mexico exhibits multiple phases of evaporite deposition throughout its stratigraphic column, with the most significant salt deposits being the Middle Jurassic Louann salt basin in the Gulf Coast plain and offshore regions (Salvador, 1987; Snedden & Galloway, 2019). The northeastern margin of the Gulf of Mexico, in the Florida Peninsula, is characterized by the deposition of some Cretaceous evaporites in carbonate platforms (Figure 11.28).

The onshore Eastern Texas salt basin has large thicknesses of Louann evaporites (predominantly halite),

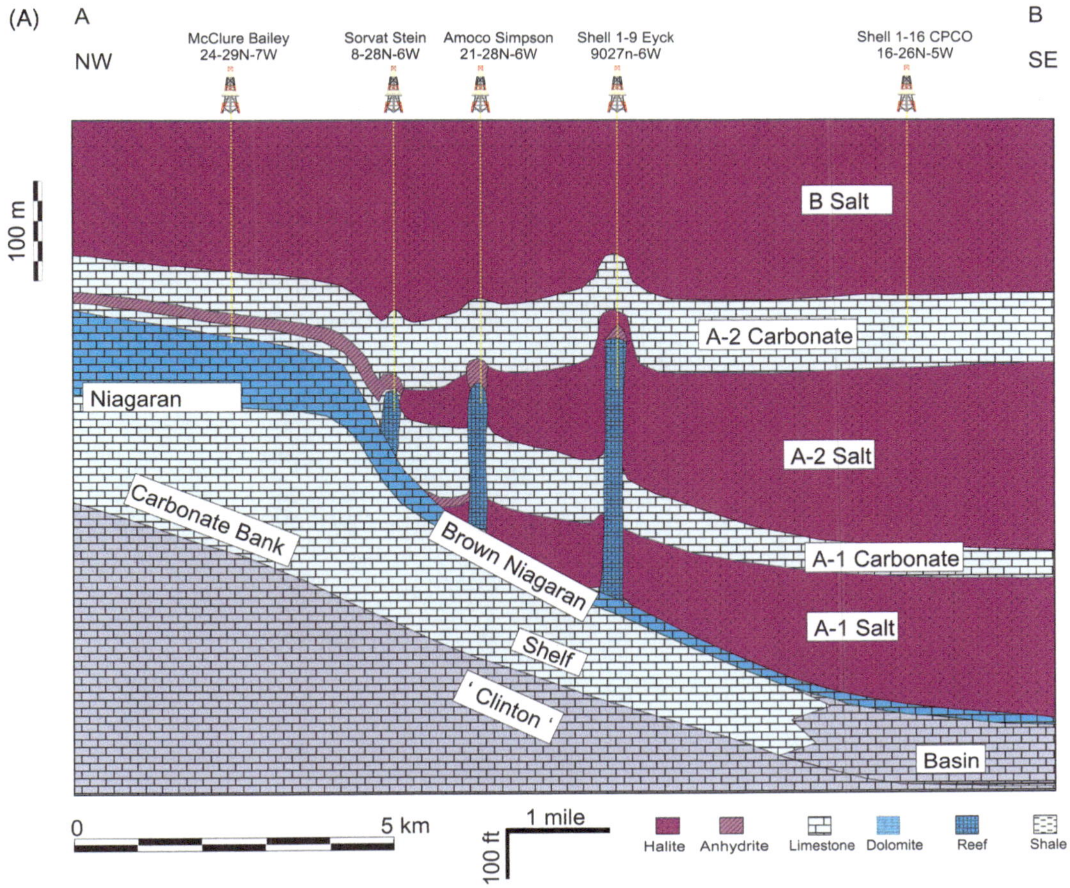

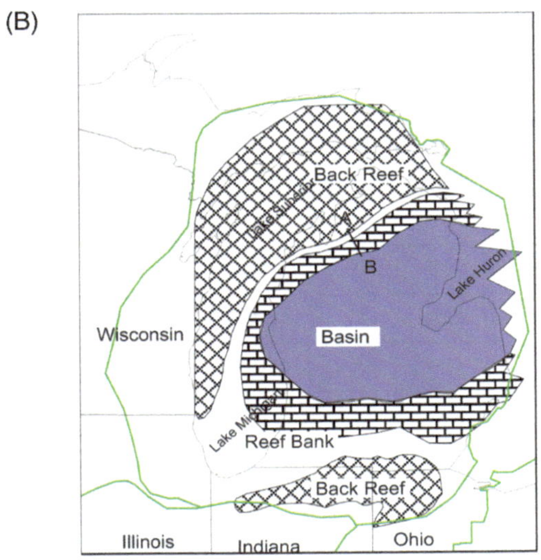

Figure 11.25 (A) The schematic cross-section illustrates the transition from shallow-water carbonate platforms to deepwater regions in the central part of the Michigan Basin during the Silurian. (B) The simplified inset map shows the carbonate platform, the fringing reefs, and the basin depocenter. The geological profile (location is shown by line A-B in the inset) illustrates the Niagaran back-reef carbonates deepening rapidly basinward of the shelf edge. Tall reef pinnacles (bioherms) are potential petroleum exploration targets. The Salina evaporites are divided into several sequences, with younger salts overstepping older ones. In the geological profile, the B salt extends to the northwest (landward) of the A-1 and A-2 halite sequences. Credit: Adapted from Miller and Dupal (1987).

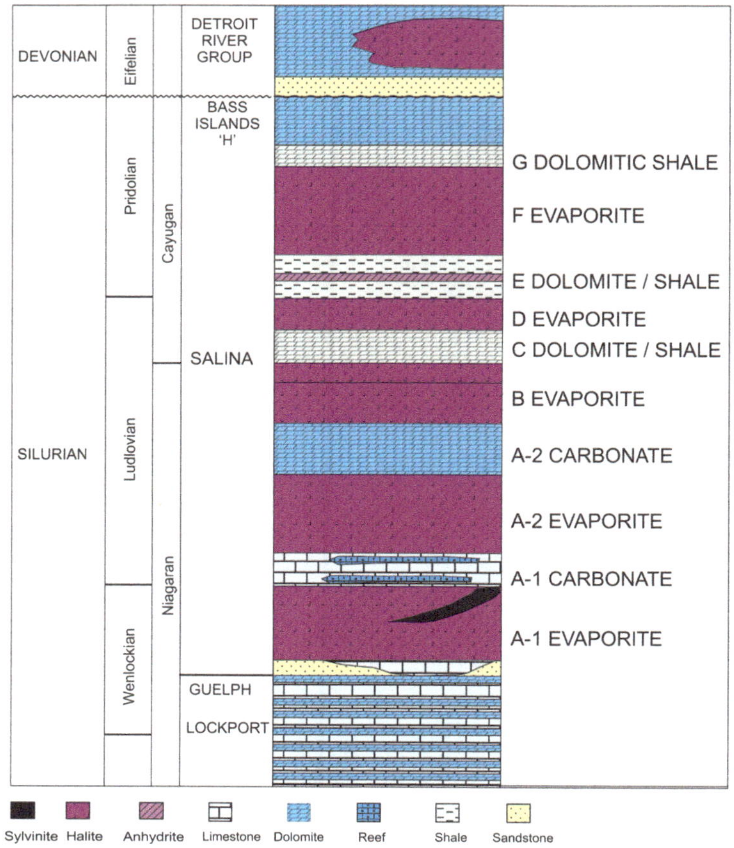

Figure 11.26 Stratigraphic chart of the Michigan Basin's Silurian and Devonian evaporite sequences. The more soluble salt minerals, primarily sylvinite, occur at the top of the A-1 evaporite sequence near the base of the Salina Group. Credit: Sonnenfeld and Al-Aasm (1991)/with permission of Geological Society of America.

Age of evaporites	Geological formation	Salt layer thickness (m)	Top of salt layer (m)	Base of salt layer (m)	Area of occurrence
Jurassic	Dunham	40	1400	2300	eastern margin and basin center
Permo-Triassic	Pine	100	1200	2300	southern margin and basin center
Permian	Opeche	50	1600	2300	border and basin center
Mississippian	Madison	100	1100	3000	border and basin center
Devonian	Prairie (in the U.S.)	130	1700	3600	border and basin center

Figure 11.27 Age of evaporite sequences in the Williston Basin and the geological characteristics of individual salt layers. Credit: Adapted from Perry et al. (2014).

resulting in tall salt diapirs that reach the surface (Seni et al., 1985; Snedden et al., 2019). After the deposition of the Louann salt in the Middle Jurassic, the Late Jurassic was marked by the development of a carbonate platform with evaporitic and siliciclastic sediments in a rapidly deepening depositional environment. The Late Jurassic platform carbonate sequences developed reefal buildups, forming high-energy peloid and ooid grainstones, which are classical exploratory targets for the petroleum industry. The sabkha evaporites-carbonate platform in the Early Cretaceous resulted in the accumulation of the Ferry Lake Anhydrite in the coastal sabkha and lagoon

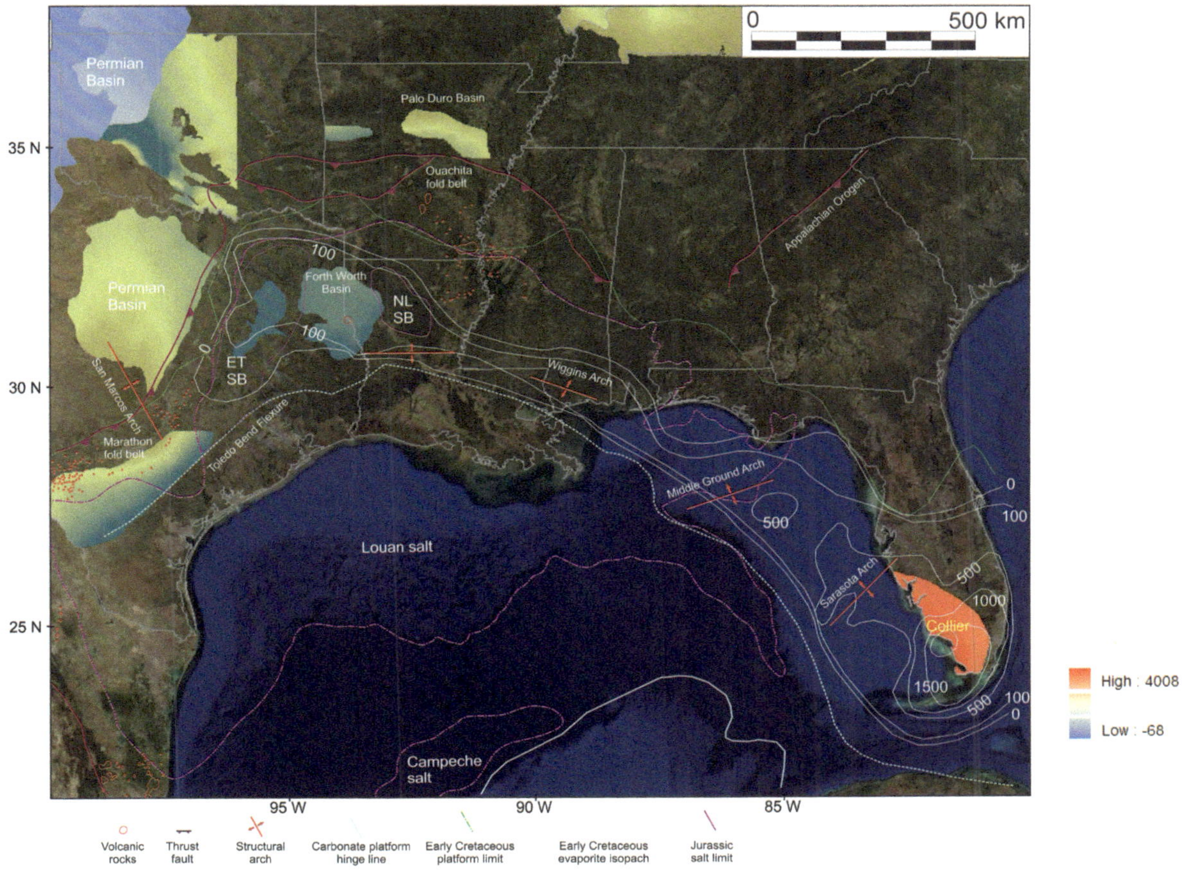

Figure 11.28 The simplified salt isopach map of Early Cretaceous evaporite layers in the northern Gulf of Mexico shows the distribution of the Ferry Lake Anhydrite in Texas and Louisiana and the Punta Gorda Fm. in Florida. The contour lines vary from 100 m in the depocenters of the onshore East Texas and North Louisiana basins to more than 1,500 m in southern Florida. The Cretaceous carbonate-evaporite deposits form a belt extending over 2,000 km from Mexico's border to the Wiggins Arch region on land. The satellite image also shows the Permian salt deposits in the Permian and Forth Worth basins in northern Texas. Thin dashed lines with magenta color show the proximal and distal limits of the Jurassic Louann and Campeche salt basins in the Gulf of Mexico. The Toledo Bend flexure, marked by a thin dashed line with a turquoise color, corresponds to a basement high associated with a hinge line controlling the Jurassic to Cretaceous platform carbonates. The Toledo Bend flexure also controls the abrupt deepening of the base of the autochthonous Louann salt. Credit: Photo by NASA/Public domain.

during periods of sea-level fall and basin isolation, particularly in the Early Albian (Phelps et al., 2014). The carbonate platforms were established by the Valanginian, and extended into the Cenomanian when marine shales were deposited throughout the basin (Ewing & Galloway, 2019; Handford & Baria, 2007). The Ferry Lake Anhydrite forms tabular layers in the interior salt basins, with thicknesses up to 100 m (Figure 11.28). The Punta Gorda Fm. salt isopach reaches about 2,000 m near the exploratory well Collier (Figure 11.28).

The genetic models that have been proposed for the Early Cretaceous carbonate-evaporite sequences in Gulf Coast environments are similar to the models advocated for the Persian Gulf Trucial Coast (see Chapter 8). However, exploratory well composite logs and 3D seismic profiles in the Gulf Coast show a faster increase in water depth from the shelf to the deep basin (e.g., Ewing & Galloway, 2019; Phelps et al., 2014). Shelf-edge reefs in the northern Gulf Coast were stacked forming different carbonate buildups from Barremian to Cenomanian, as the Sligo and Stuart City reefs (Figure 11.29).

Perry et al. (2014) mapped the depth of Permian salt layers in the southwestern United States, revealing shallow deposits in West Texas and deeper Jurassic Louann salt in Gulf Coast interior basins. The Toledo Bend flexure marked an abrupt increase in depth to the

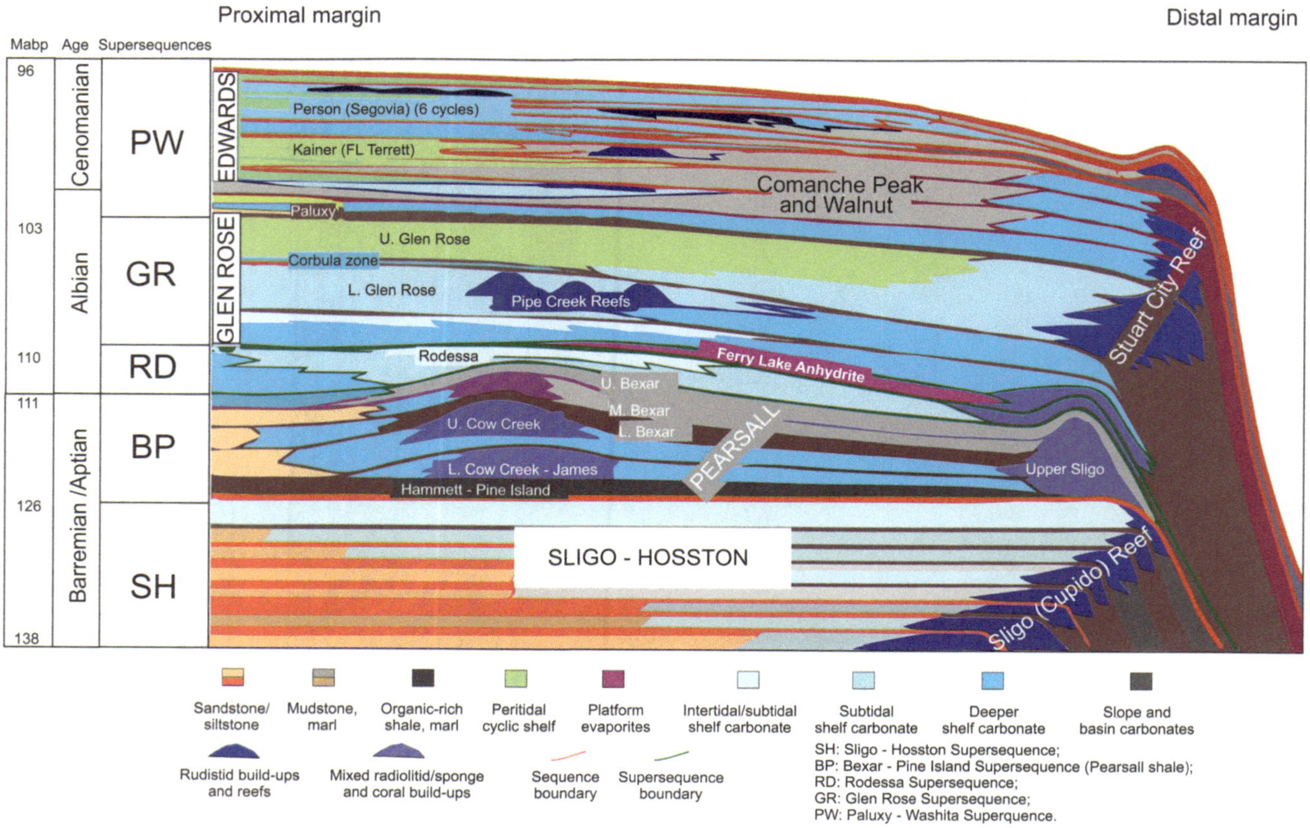

Figure 11.29 A schematic stratigraphic section from the northern Gulf of Mexico's proximal to distal margin during the Early Cretaceous shows supersequences characterized by carbonate-evaporite successions. Carbonate buildups occur at the shelf-to-deep basin transition. At the base of the Glen Rose supersequence, Ferry Lake Anhydrite is characterized by evaporitic deposits. Reefal buildups such as Upper Sligo and Stuart City are characterized by carbonate rocks with good reservoir properties.

autochthonous Jurassic salt layer. Basinward of the hinge line, the depth to the base of the salt horizon is greater than 4 km in the Gulf Coast plain and deepens toward the continental shelf and slope (Figure 11.28).

The stratigraphic sequences in the northern Gulf of Mexico are associated with several cycles of deposition from the platform to the deep basin. Regional unconformities separating sedimentary packages define the supersequences, which are identified by seismic horizons representing timelines (Ewing & Galloway, 2019; Galloway & Hobday, 1983). Each of these stratigraphic supersequences can be characterized by a paleo shelf break with reefal buildups, with a rapid deepening of the depositional environment basinward. For example, the platform edge in the Barremian to Apian is marked by the Sligo carbonate buildup, whereas the Stuart City reef marks the shelf edge in the Albian (Figure 11.29).

The Early Cretaceous evaporite layers are part of a carbonate-evaporite sequence that accumulated mainly sulfate minerals in sabkha environments in the northern Gulf of Mexico. In the subsurface, the evaporite deposits extend along a curved belt from Texas to Florida, north of the Toledo Bend flexure (Phelps et al., 2014). Episodes of sea-level fall and basin restriction in the Early Albian resulted in the deposition of evaporite layers, predominantly composed of sulfate minerals (gypsum) intercalated with carbonate rocks. This carbonate-evaporite sequence is characterized by thicknesses ranging from 10 m to over 100 m in the onshore region of the East Texas Salt Basin and North Louisiana Salt Basin. The depth of occurrence varies from a few hundred meters, near the updip limit of the Cretaceous anhydrite deposits, to more than 4,000 m in northern Texas and Louisiana (Dean & Johnson, 1989; Maher, 1971). In Texas, the Albian evaporites, mainly composed of anhydrite intercalated with carbonate layers, are assigned to the Ferry Lake Fm., included in the Glen Rose supersequence (Figure 11.29).

The carbonate buildups at the edges of the Early Cretaceous lagoon constitute exploratory targets for the petroleum industry (Sarg, 2001; Snedden &

Galloway, 2019). The Fairway Field in East Texas is characterized by a porous carbonate reservoir rock dated as Aptian, corresponding to the James Limestone. The reservoirs are sealed by the overlying Ferry Anhydrite rocks, which are characterized by gypsum nodules that are vertically aligned (Loucks & Longman, 1982). The evaporite sequences in the Fairway oil field are genetically related to a sabkha environment extending to a lagoon and a shallow shelf rimmed by a carbonate bank. The evaporite layers correspond to supratidal and subtidal sediments deposited in a broad, shallow, and hypersaline lagoon that was up to 260 km wide and precipitated evaporites in local depressions, forming salterns (Warren, 1991).

11.6. THE PALEOZOIC CARBONATE-EVAPORITE PLATFORM OF THE AMAZON BASIN

The Parnaíba Basin and the Amazon Basin are two major Paleozoic basins in Northern Brazil, characterized by evaporite deposits. The Parnaíba Basin is mainly composed of gypsum and anhydrite layers in continental sabkhas and salinas, while the Amazon Basin is a large intracratonic basin with Permo-Carboniferous bedded salt deposits with thick layers of anhydrite and halite. The Amazon Basin is divided into three subbasins: Solimões (Upper Amazon), Middle Amazon, and Lower Amazon basins (Figure 11.30). These basins

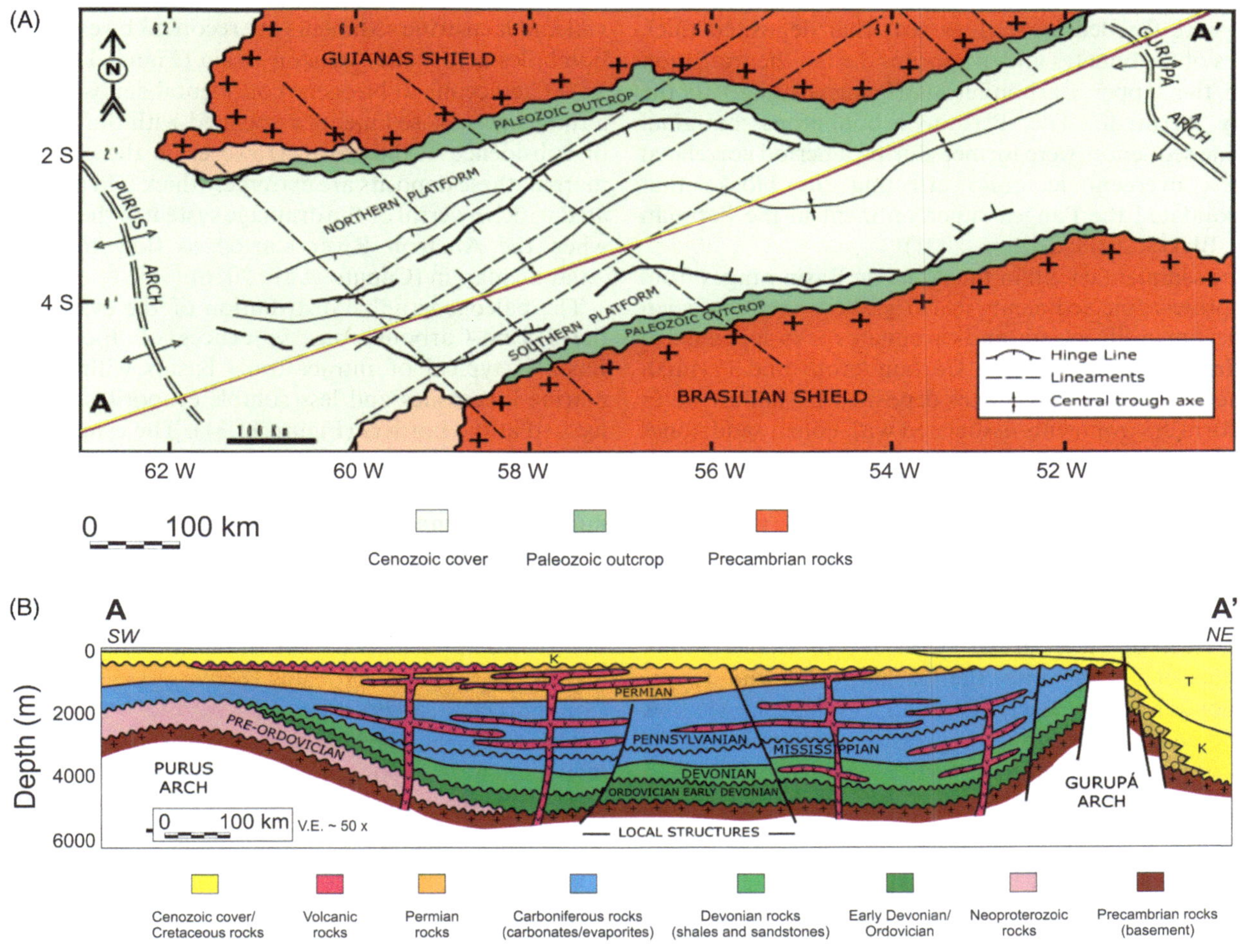

Figure 11.30 (A) Regional geological map showing the Amazon Basin in northern Brazil. The intracratonic Paleozoic basin is located between the Guianas Shield and the Brazilian Shield and is divided into the Upper Amazon basin and the Amazonas basin, separated by the Purus Arch. The location of a regional profile (A–A′) extending from the Purus Arch in the west to the Gurupá Arch in the east is shown by a thin yellow line. (B) Regional geological cross-section A–A′ extending from the Purus Arch to the Gurupá Arch in the east. The depocenter of the Amazon Basin is rather shallow, with a total sediment thickness of less than 6 km. Igneous rocks dated around 200 Ma heavily intruded the Paleozoic stratigraphic sequences (Devonian, Permo-Carboniferous, and Permian). Several evaporite cycles, alternating carbonate rocks with anhydrite and halite, are recorded in the Permo-Carboniferous sequences.

cover an area of about 1 million km^2 and display different depths to the basement (Costa & Wanderley Filho, 2008). Seismic interpretation and borehole control indicate the Solimões Basin is shallower than the Amazon Basin (Caputo & Soares, 2016). The Amazon Basin corresponds to a Paleozoic depression located to the east of the Purus Arch, a Middle Proterozoic rift basin that was inverted and formed a structural high separating the Solimões) from the Middle and Lower Amazon basins (Caputo & Soares, 2016; Costa & Wanderley Filho, 2008).

The Amazon Basin has been the focus of petroleum exploration since the 1950s, with several gas discoveries associated with eolian sandstones as the main reservoir rocks. These sediments indicate desert environments that were flooded by shallow seas that deposited thick layers of carbonates and evaporites during the transition from the Upper Carboniferous (Pennsylvanian) to the Early Permian. The Permo-Carboniferous marginal marine sequences were formed during a period correlated with convergent tectonics affecting the blocks that consolidated the Pangea supercontinent in the Permian (e.g., Blakey, 2008; Scotese, 2001).

The sedimentary fill in the Amazon Basin ranges from Neoproterozoic sediments in rift grabens to Ordovician to Devonian siliciclastic and carbonate rocks deposited in marine environments. The Carboniferous and Permian sequences include carbonates deposited in shallow water platform environments associated with eolian sandstones and evaporites. The evaporite deposits were formed following a period of glaciation that deposited diamictites in the Devonian.

Continental environments predominated in the basin before the Permo-Carboniferous evaporitic interval. Desertic environments with eolian sandstones are the main reservoir rocks in the hydrocarbon accumulations. Marine incursions into the basin during the Upper Carboniferous to Early Permian resulted in shallow inland seas affected by compressional tectonics in the Mesozoic. The Cretaceous-Tertiary sequences overlie a major unconformity separating the Permian deposits from the younger sequences (Figure 11.30-B).

A regional cross-section extending from the western to the eastern parts of the Amazon Basin shows a marked thickening of the Tertiary sequences eastward of the Gurupá Arch, near the continental margin (Figure 11.30-B). This arch separates the Lower Amazon Basin (with Paleozoic successions) from the Marajó rift at the continental margin (Caputo & Soares, 2016; Costa & Wanderley Filho, 2008). The North Brazilian continental margin is characterized by thick accumulations of Mesozoic and Cenozoic sediments that formed Cretaceous and Tertiary depocenters in the offshore region (Figure 11.30-A). These rifts developed during the Early Cretaceous, as the central part of the South Atlantic equatorial margin was affected by large transform fault zones (Mohriak et al., 2012).

The northern Brazilian region was affected by widespread volcanic events during the Late Triassic and Early Jurassic, related to the Central Atlantic Magmatic Province (CAMP), which affected large areas of North America and northwest Africa, with radiometric ages around 200 Ma (Marzoli et al., 1999; Rezende et al., 2021). Igneous intrusive and extrusive rocks are recorded in the Amazon Basin as sills intruding into Paleozoic strata (Figure 11.30-B). A major regional angular unconformity post-dates a period of compressional tectonics associated with the widespread magmatic event (e.g., Rezende et al., 2021). Above this unconformity, some fluvial and lacustrine sediments are recorded by exploratory boreholes across the Amazon Basin (Figure 11.30). The Late Cretaceous to Neogene continental deposits formed a thin veneer of sediments associated with the last phase of subsidence in the basin. However, in the continental margin, these deposits are extremely thick and are associated with a reversal of the drainage system in the Neogene, when the Amazon River started to flow toward the Atlantic margin (Caputo et al., 2016).

The paleogeographic distribution of the evaporites in the Permo-Carboniferous sequences of the Amazon Basin is typical of intracratonic basins with marginal marine limestones and less soluble evaporites (gypsum) located at the borders (Figure 11.31). The evaporite successions are composed of rhythmically bedded carbonate, sulfate, and chloride minerals deposited in a shallow sea during the Pennsylvanian (about 300 Ma). The basin depocenter is characterized by the accumulation of halite and more soluble salts in restricted lagoons during desiccation episodes. During this episode of salt accumulation, the basin was partially cut off from the open ocean, with marine ingressions probably coming from the west (Szatmari et al., 1979).

The Permo-Carboniferous carbonate-evaporite sequences were deposited in a shallow sea that was periodically desiccated, forming several evaporite cycles that can be correlated regionally. The salt layers form the main regional seal for hydrocarbon accumulations in the Upper Carboniferous sandstone reservoirs (Costa & Wanderley Filho, 2008; Szatmari et al., 1979).

Seismic profiles in the Amazon Basin indicate that igneous sills and volcanic rocks, dated around 200 Ma by K-Ar and Ar-Ar methods, are commonly found intercalated with the Permo-Carboniferous evaporite layers (Figure 11.32). These sills also occur within the Devonian to Carboniferous siliciclastic sequences and above the carbonate-evaporite layers. Similar to other Paleozoic salt basins worldwide, the igneous intrusions have thermally affected the pre-evaporitic source rocks,

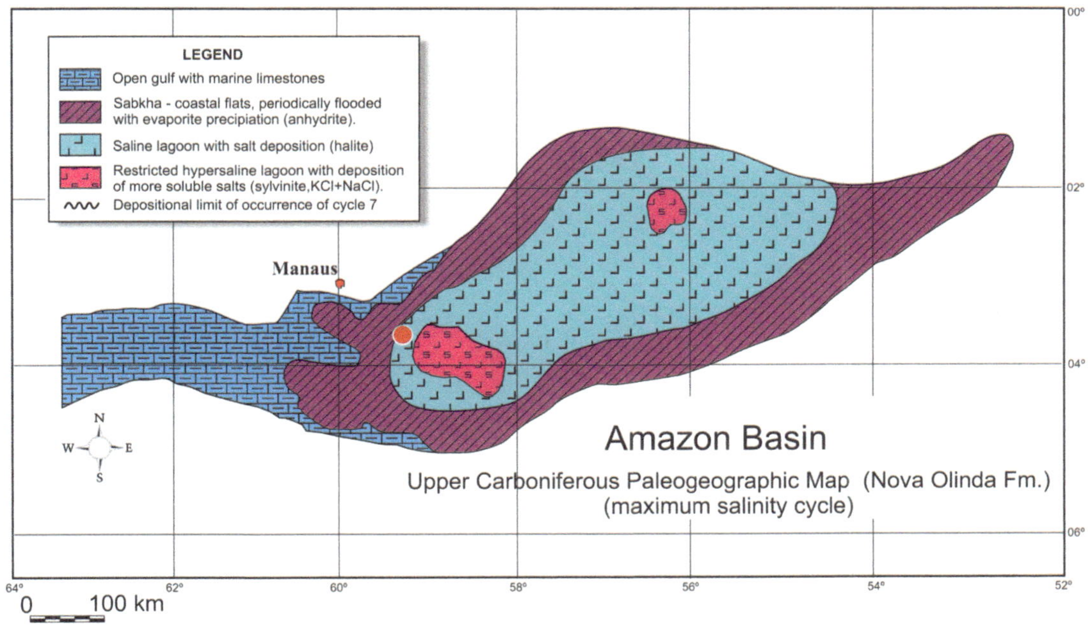

Figure 11.31 The simplified map shows the paleogeographic distribution of evaporites in the Permo-Carboniferous sequences of the Amazon Basin in northern Brazil. The western part of the evaporite basin is characterized by marine limestones. The basin depocenter is characterized by the accumulation of halite and more soluble salts in restricted lagoons during desiccation episodes in the Upper Carboniferous (Pennsylvanian). Credit: Szatmari et al. (1979)/with permission of GeoScienceWorld.

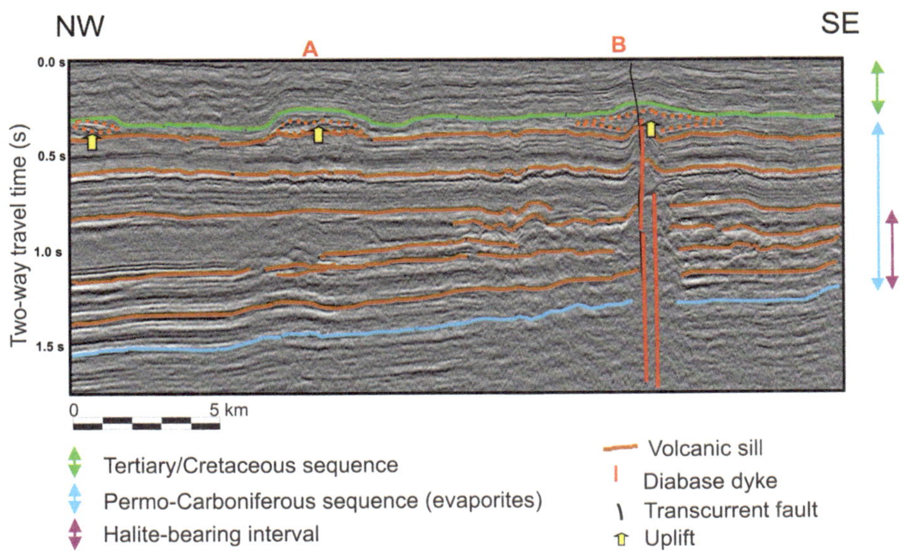

Figure 11.32 The seismic profile in the Amazon Basin reveals the layer-cake stratigraphy of the Permo-Carboniferous sequence, heavily intruded by volcanic sills related to the Central Atlantic Magmatic Province. The Cretaceous reflector (green color) is deformed by igneous intrusions. Some uplifts (points A and B) have vertical diabase dikes cutting evaporite layers and forming Triassic-age laccoliths below the Cretaceous unconformity. These features are likely associated with compactional effects and transcurrent faulting, as indicated in point B. A magenta arrow indicates the interval with halite deposits at the base of the Permo-Carboniferous carbonate-evaporite sequences. The green arrow corresponds to the Cretaceous-Tertiary sedimentary sequences.

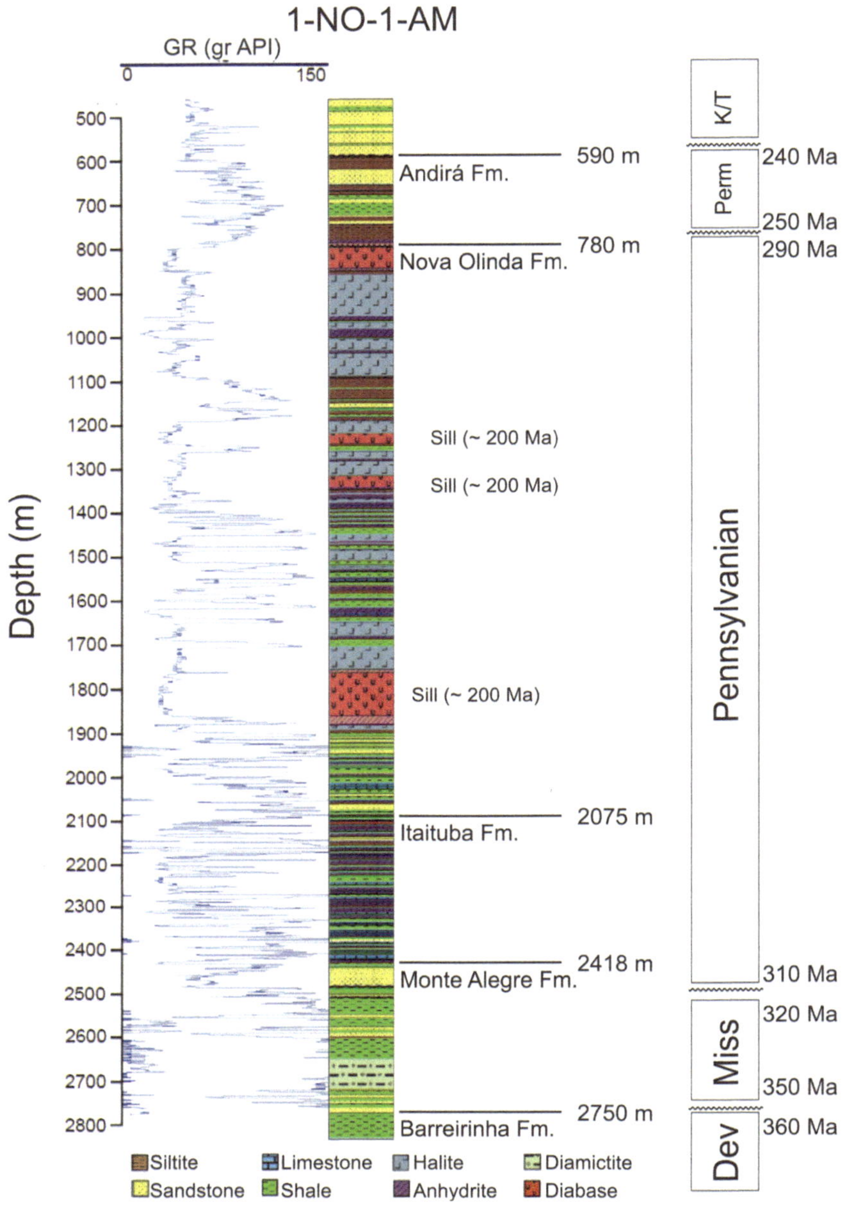

Figure 11.33 Composite log of the Nova Olinda exploratory well in the Amazon Basin with Gamma Ray readings and stratigraphic tops. The rock column displays the evaporite layers that form a thick Pennsylvanian sequence on top of the Mississippian siliciclastic rocks. Evaporite layers composed of anhydrite and halite intercalate with the platform carbonates.

the evaporite layers, and the post-evaporitic sequences, resulting in atypical petroleum systems (Heimdal et al., 2019; Mosmann et al., 1986; Rezende et al., 2021; Warren, 1989, 2016).

Figure 11.33 shows a composite log illustrating the main lithological types and stratigraphic units of the Amazon Basin (Nova Olinda region). The Pennsylvanian carbonate-evaporite sequences (780–2075 m) are associated with the Nova Olinda Fm., which includes shallow marine carbonate rocks intercalated with anhydrite and halite, but locally more soluble minerals (e.g., sylvinite) formed in the depocenters (Figure 11.31).

The main source rocks in the basin are the Devonian shales (Barreirinha Fm.), characterized by marine shales with high gamma-ray values. The carbonate-evaporite sequences were deposited between 310 and 290 Ma, with intrusive sills and volcanic layers exhibiting radiometric ages of about 200 Ma (Mazoli et al., 1999; Rezende et al., 2021). The Andirá Fm. (Permian) is overlain by less than 600 m of Cretaceous and Tertiary sediments, indicating a major erosional episode after the Triassic magmatic episode (Figure 11.33). The petroleum systems in the Amazon Basin are considered atypical or nonconventional. This is due to the immature Devonian source rocks that are heavily intruded by igneous rocks, resulting in gas generation with few oil accumulations (Mosmann et al., 1986). Despite the discovery of oil and gas fields in the Solimões and Amazonas basins in the past two decades, most hydrocarbon accumulations remain as potential reserves.

11.7. PETROLEUM SYSTEMS ASSOCIATED WITH LACUSTRINE SEQUENCES

Sedimentary sequences associated with continental lacustrine systems are widely recognized as prolific source rocks in many basins formed since the Neoproterozoic to the Cenozoic (e.g., Bohacs et al., 2000; Busson, 1991; Katz, 1995, 2001). These organic-rich sediments are found in hypersaline lakes and freshwater to brackish water lakes, such as the South Atlantic salt basins off Brazil and West Africa (see Chapter 3 in Mohriak, 2025).

Lake basins are categorized into overfilled, balanced-fill, and underfilled basins (Figure 11.34). Underfilled basins have high evaporation rates, leading to basin desiccation and precipitation of evaporite layers that seal the pre-salt source rocks and reservoirs (Bohacs et al., 2000). These basins exhibit closed hydrology, chemical stratification, a low groundwater table, and a high concentration of evaporite ions. When brine concentration exceeds mineral solubility, evaporite minerals precipitate on the lake's margins and bottom, including carbonates (aragonite, calcite, dolomite, and trona), sulfates (gypsum), and chlorides (halite).

Organic production within continental rift lakes is primarily controlled by the photosynthetic fixation of CO_2 by organisms such as algae and aquatic plants. Primary organic productivity in lacustrine systems is influenced by various factors such as water chemistry, climate, solar input, wind, precipitation and evaporation, temperature, and lake stratification. The presence of source rock depends on primary organic production, destruction by scavengers, and dilution by mixing with nonorganic material (Bohacs et al., 2000).

Most lacustrine basins are dominated by an aggradational shoreline architecture, but during dry climate episodes, the water body shrinks, and remnant lakes are concentrated at the lowest points of the rift depression (Figure 11.34). Some modern basins are characterized by ephemeral lakes resulting in evaporitic muds interspersed

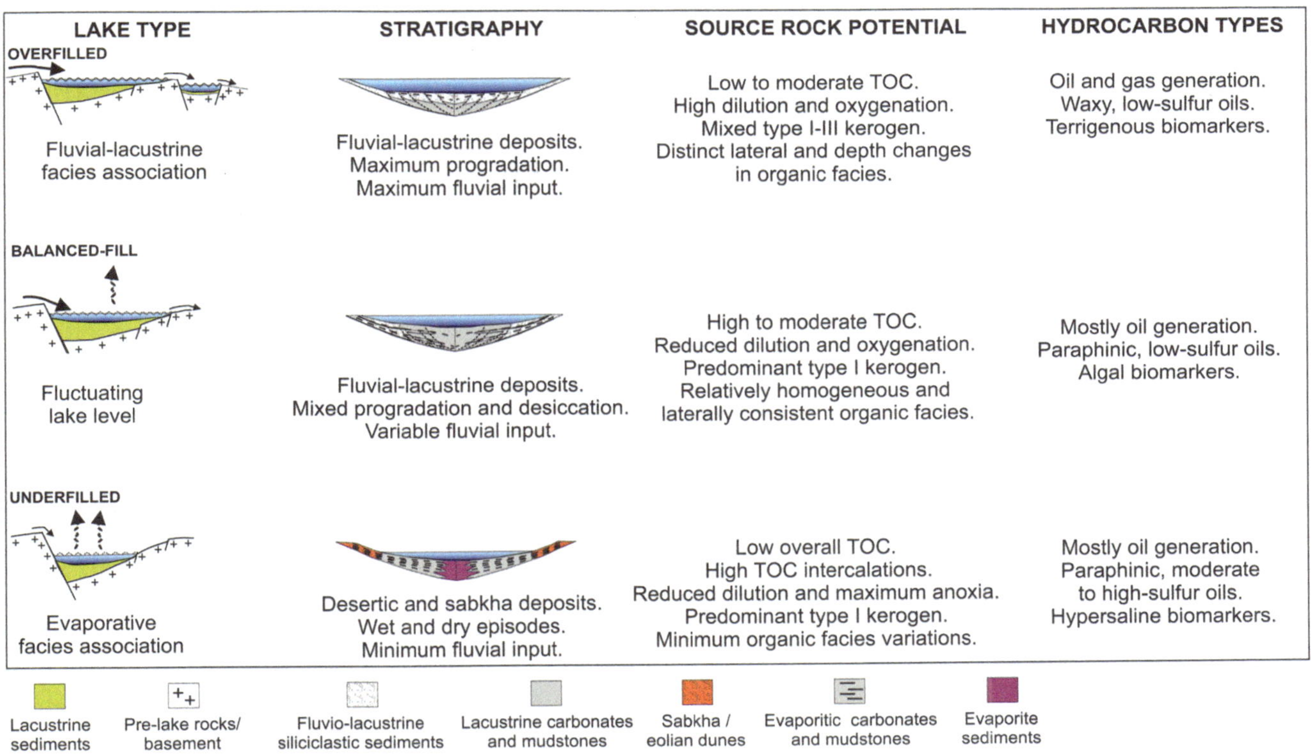

Figure 11.34 The schematic diagram depicts the main lake types based on accommodation space, hydrology, stratigraphic characteristics, source rock potential, and typical hydrocarbon types. Credit: Bohacs et al. (2000)/ with permission of GeoScienceWorld.

with playa lake evaporite crusts when the basin is desiccated, while others are characterized by deepwater environments, with reworking and resedimentation of clastic and chemical evaporites, as observed in the Dead Sea (see Chapter 3).

Hydrocarbon generation in lacustrine environments depends on the type of organic matter produced, its preservation through time, and adequate temperature during burial. Overfilled lakes are more likely to contain larger amounts of terrestrial organic matter by fluvial input, making them more gas-prone than balanced-fill and underfilled lakes. Some underfilled lake basins are characterized by extreme evaporative concentration and a low influx of siliciclastic sediments, leading to thermal and chemical stratification that preserves sulfur-rich kerogens. Oils derived from these hypersaline environments may exhibit concentrations of up to 12% of sulfur (Bohacs et al., 2000).

Geochemical analyses of rock samples from petroliferous basins include the determination of biomarkers, corresponding to organic molecules or molecular fossils derived from organisms whose skeleton was composed of carbon. These markers provide useful information about the depositional environment, thermal maturity, migration pathways, and hydrocarbon alteration by several factors, such as biodegradation and the influx of freshwater into reservoir rocks.

The interpretation of the depositional paleoenvironments, including the lake hydrology and brine salinity, is supported by the determination of key biomarkers, such as tricyclic terpanes β-carotene, gammacerane, and steranes (Bohacs et al., 2000; Katz, 2001). Many authors have reported the presence of gammacerane and β-carotene in highly saline lacustrine anoxic environments (e.g., Mello et al., 1994; Mohriak et al., 2021; Peters & Moldowan, 1993).

Some giant saline basins in the world, such as the South Atlantic salt basins offshore Brazil and West Africa, are characterized by a rift phase with lacustrine sedimentation preceding the sag basin sedimentation that culminated with the deposition of the widespread Late Aptian evaporite rocks. The sedimentation of the rift sequences started in the Neocomian with fluvio-lacustrine deposits locally overlying a volcanic basement. During the Barremian, the rift climax is marked by the deposition of coquinas on structural highs (see Chapter 3 in Mohriak, 2025). The electric log and lithologies of exploratory boreholes in the Campos Basin, offshore Brazil, show rich source rocks in the lacustrine sequences underlying the Aptian coquinas (Figure 11.35).

The RJS-492 exploratory well (Figure 11.35) drilled through the Late Aptian evaporite sequences (Retiro Fm.), the sag basin siliciclastic and carbonate rocks (Macabu Fm.), and the syn-rift deposits (Coqueiros Fm.). The Lagoa Feia Group includes the lacustrine and transitional deposits in the Campos Basin, pointing to a

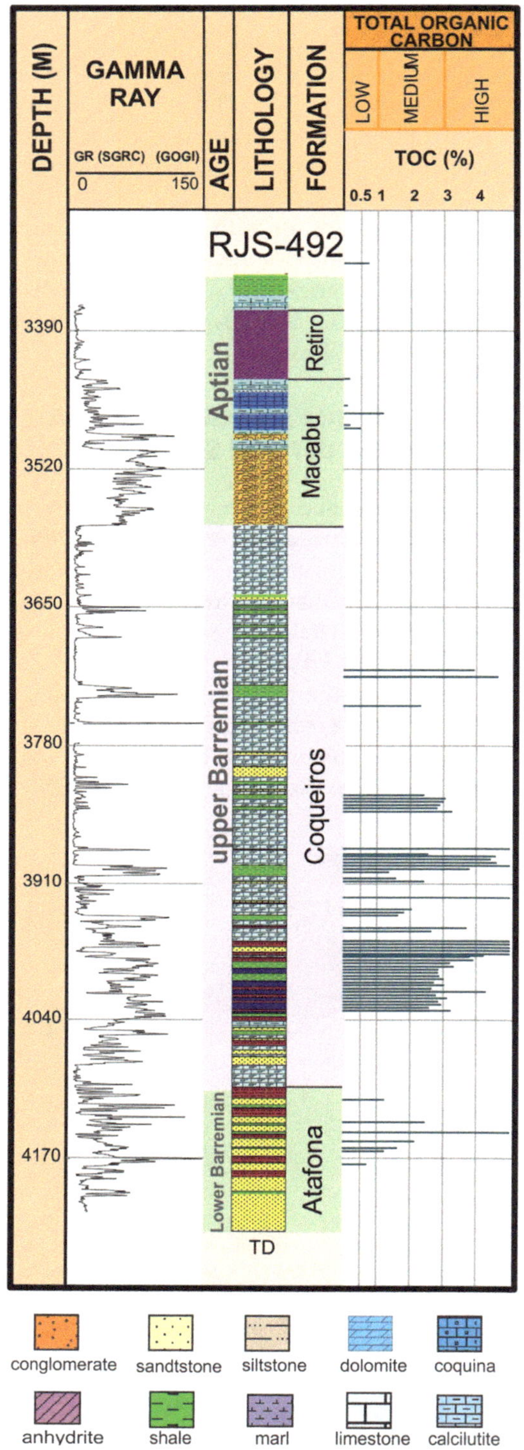

Figure 11.35 The exploratory well RJS-492 was drilled in the Campos Basin, offshore Brazil. The composite log shows the Late Aptian evaporites (Retiro Fm.), the sag basin sediments (Macabu Fm.), and the rift sequences (Coqueiros Fm. and Atafona Fm.). The main source rock intervals are associated with the middle part of the Barremian lacustrine sequence, corresponding to type I kerogen with total organic carbon values locally exceeding 5%. Credit: Mohriak et al. (2021)/with permission of AAPG.

balanced-fill to an underfilled lake. Paleontological analysis of ostracod assemblages suggests the Barremian lacustrine facies are related to initially freshwater hydrology that evolved to a brackish and hypersaline brine in the Late Aptian (Mohriak et al. (2021)/with permission of AAPG.).

Geochemical analyses of source rocks in the Campos Basin reveal a type I kerogen associated with calcareous black shales and marls. These sediments were deposited in a saline-alkaline lacustrine environment, with TOC up to 9% and low sulfur content (< 0.3%). High organic productivity in the Lagoa Feia Group lacustrine sediments is related to algae and bacteria, with a more anoxic, saline, and denser bottom, suggesting density and chemical stratification (Mello et al., 1994; Mohriak et al., 2021; Trindade et al., 1995).

The Bucomazi Fm. in the Lower Congo Basin is characterized by bituminous shales accumulated in a lacustrine environment during the Neocomian-Barremian syn-rift sequence (Burwood et al., 1995). The depositional environment varied from deep to shallow alkaline lakes, with anoxic bottom waters. A lake contraction phase during the Late Aptian resulted in brine concentration and evaporite sediment accumulation in saltpans and ephemeral lakes. The presence of gammacerane suggests hypersaline waters, followed by marine transgressions that developed shallow carbonate platforms under more oxygenated conditions and periodic fluvial inputs (Burwood et al., 1995).

The Neocomian Kissenda shales are pre-salt, lacustrine freshwater source rocks in the Interior Basin and N'Komi Rift system in northern Gabon, with TOC values ranging from 0.5% to 2.2% (Kuo, 1994). The biomarkers obtained from the Early Cretaceous shales indicate an oligotrophic lacustrine system primarily containing bacteria and cyanobacteria, with minimal input of terrestrial higher plants, resulting in waxy source rocks. As lake hydrology evolved, the overlying Melania shales became less waxy and more effective source rocks for oil generation. The Barremian shales contain 0.8%–4.8% TOC and are mainly composed of type I and type II kerogens (Kuo, 1994).

The Green River Fm. was deposited in hypersaline, alkaline lake basins in the western United States, resulting from the Laramide Orogeny. Two main paleolakes are interpreted in the region between Wyoming, Colorado, and Utah: Lake Gosiute in the north and Lake Uinta in the south (Eugster & Hardie, 1975, 1978). Eocene evaporite sequences are intercalated with dolomitic marls, forming several intervals of lacustrine sediments associated with the Green River Fm., including the Tipton, Wilkes Peak, and Laney Shale members (Eugster & Hardie, 1975; Katz, 1995). The middle member (Wilkes Peak) corresponds to an underfilled lake type, whereas the upper and lower units correspond to balanced-fill lakes (Figure 11.36).

The Wilkes Peak paleo-lake was characterized by a stratified water mass, with deeper parts of the water column occupied by a highly concentrated brine rich in

Geological unit Basin	Hydrologic type	Rift age	Salt deposition	Petroleum systems
Kissenda Fm. N'Komi Gabon	Overfilled	Neocomian/ Barremian	Late Aptian	Mean TOC varying from 0.5 to 2.2%; Kerogen type I and II Fresh-water lacustrine system evolving to the Melania lake in the Barremian, with TOC varying from 0.8 to 4.8%..
Bucomazi Fm. Lower Congo, Cabinda	Overfilled	Neocomian/ Barremian	Late Aptian	Mean TOC varying from 1.3 to 7.9%; Kerogen type I and II. Presence of gammacerane biomarker.
Lagoa Feia Gp., Coqueiros Fm. Campos Basin, Brazil	Balanced-fill	Neocomian/ Barremian	Late Aptian	Source rocks deposited under anoxic conditions in brackish to saline alkaline waters, with increasing upward salinity. Calcareous black shales rich in bacterial and algal-derived type I kerogen (up to 9% TOC).
Green River Fm. Green River Basin, USA	Balanced-fill	Paleogene foreland basin	Eocene	Marls and dolomitic mudstones with oil shales. Mean TOC about 3%.
Lagoa Feia Gp. Coqueiros/Macabu Fm. Campos Basin, Brazil	Underfilled	Barremian/ Aptian	Late Aptian	Diagnostic biomarker features include low concentrations of steranes, presence of B-carotene and gammacerane, indicating hypersalinity.
Wilkes Peak Mb. Green River Basin, USA	Underfilled	Paleogene	Eocene	Marls and dolomitic mudstones intercalated with trona and halite. Type I kerogen in the oil shales.
Dabussun Lake Qaidam Basin, China	Underfilled	Neogene	Neogene	Marls and dolomitic mudstones intercalated with gypsum, halite, and K-rich evaporite minerals.

Figure 11.36 Hydrologic types and geological characteristics of lacustrine sequences in various evaporite basins, which are crucial components in petroleum systems. Credit: Bohacs et al. (2000)/with permission of GeoScienceWorld.

sodium carbonate, resulting in the precipitation of trona and halite (Eugster & Hardie, 1975, 1978). These sequences are intercalated with marl and dolomite deposited in shallow lakes. During the early Eocene, the lake water level rose above the structural high that separated the two water masses, forming one large lake. During the Middle Eocene (about 44 Mabp), Lake Gosiute was filled by volcaniclastic sediments, mainly volcanic tuff, resulting in a playa mudflat (Eugster & Hardie, 1975).

Evaporite deposition is a critical element in the petroleum systems of many sedimentary basins. Shallow-water evaporites formed in sabkha environments and carbonate-evaporite platforms may act as seals for hydrocarbon traps involving the pre-salt reservoir rocks. Examples of carbonate-evaporite platform petroleum systems are recorded in various regions worldwide. Selected case histories include the Silurian reefs of the Michigan Basin, the San Andres Fm. and the Guadalupian carbonate plays in the Permian Basin of West Texas and New Mexico, the Permian Zechstein Group in Europe, the Upper Permian Khuff Fm. in the Arabian Gulf, and the Cretaceous Ferry Lake Anhydrite in the East Texas Basin (e.g., Kendall et al., 2009; Warren, 2006, 2010, 2016).

REFERENCES

Agard, P., Omrani, J., Jolivet, L., Whitechurch, H., Vrielynck, B., Sparkman, W., et al. (2011). Zagros orogeny: A subduction-dominated process. *Geological Magazine, 148*(5–6), 692–725, Cambridge University Press. doi: 10.1017/S001675681100046X

Al-Zoubi, A., & ten Brink, U. S. (2001). Salt diapirs in the Dead Sea basin and their relationship to Quaternary extensional tectonics. *Marine and Petroleum Geology*, *18*, 779–797.

Al-Zoubi, A., Shulman, H., & Ben-Avraham, Z. (2002). Seismic reflection profiles across the southern Dead Sea basin. *Tectonophysics*, *346*, 61–69.

Alsharhan, A. S., & Kendall, C. G. St. C. (2003). Holocene coastal carbonates and evaporites of the southern Arabian Gulf and their ancient analogues. *Earth-Science Reviews*, *61*, 191–243.

Alsharhan, A. S., & Kendall, C. G. St. C. (2011). Introduction to Quaternary carbonate and evaporite sedimentary facies and their ancient analogues. In C. G. St. C. Kendall & A. S. Alsharhan (Eds.), *Quaternary carbonate and evaporite sedimentary facies and their ancient analogues: A tribute to Douglas James Shearman* (Vol. 43, pp. 1–104). International Association of Sedimentologists, Special Publication, Sussex, UK: Wiley-Blackwell, 496 pp.

Alsop, G. I., & Marco, S. (2011). Soft-sediment deformation within seismogenic slumps of the Dead Sea Basin. *Journal of Structural Geology*, *33*, 433–457.

Alsop, G. I., Archer, S. G., Hartley, A. J., Grant, N.T., & Hodgkinson, R. (2012). *Salt tectonics, sediments and prospectivity: An introduction* (Vol. 363, pp. 1–6). In G. I. Alsop, S. G. Archer, A. J. Hartley, N. T. Grant, & R. Hodgkinson (Eds.). Geological Society, London, Special Publications. http://dx.doi.org/10.1144/SP363.1

Alsop, G. I., Weinberger, R., Levi, T., & Marco, S. (2015). Deformation within an exposed salt wall: recumbent folding and extrusion of evaporites in the Dead Sea Basin. *Journal of Structural Geology*, *70*, 95–118.

Alsop, G. I., Weinberger, R., Levi, T., & Marco, S. (2016). Cycles of passive versus active diapirism recorded along an exposed salt wall. *Journal of Structural Geology*, *84*, 47–67.

Alsop, G. I., Weinberger, R., Marco, S., & Levi, T. (2014). Fault and fracture patterns around a strike-slip influenced salt wall. *Journal of Structural Geology*, *106*, 103–124.

Asch, K. (2005). *IGME 5000: 1:5 Million International Geological Map of Europe and Adjacent Areas*. BGR (Hannover).

Aydin, F., Erlat, E., & Türkeş, M. (2020). Impact of climate variability on the surface of Lake Tuz (Turkey), 1985–2016. *Regional Environmental Change*, *20*, 68. doi:10.1007/s10113-020-01656-z

Bąbel, M., & Schreiber, B. C. (2014). 9.17—Geochemistry of evaporites and evolution of seawater. In Holland, H. D., & Turekian, K. K. (Eds.), *Treatise on geochemistry* (2nd ed., Vol. 9, pp. 483–560). Amsterdam: Elsevier.

Ben-Avraham, Z. (1997). Geophysical framework of the Dead Sea: Structure and tectonic. In T. M. Niemi, Z. Ben-Avraham, & J. R. Gat (Eds.), *The Dead Sea: The lake and its settings, Oxford Monographs on Geology and Geophysics* (Vol. 36, pp. 22–35). New York: Oxford University Press.

Ben-Avraham Z., Lazar M., Schattner U., & Marco, S. (2005). The Dead Sea fault and its effect on civilization. In F. Wenzel (Ed.), *Perspectives in modern seismology*, *Lecture Notes in Earth Sciences* (Vol. 105, pp. 145–167). Berlin, Heidelberg: Springer-Verlag

Ben-Avraham, Z., Garfunkel, Z., & Lazar, M. (2008). Geology and evolution of the southern Dead Sea fault with emphasis on subsurface structure. *Annual Review of Earth and Planetary Sciences*, *36*, 357–387.

Ben-Avraham, Z., Lyakhovsky, V., & Schubert, G. (2010). Drop-down formation of deep basins along the Dead Sea and other strike-slip fault systems. *Geophysical Journal International*, *181*, 185–197. doi: 10.1111/j.1365-246X.2010.04525.x

Ben Dor, Y., Neugebauer, I., Enzel, Y., Schwab, M. J., Tjallingii, R., Erel, Y., & Brauer, A. (2019). Varves of the Dead Sea sedimentary record. *Quaternary Science Reviews*, *215*, 173–184.

Benison, K. C., & Goldstein, R. H. (2000). Sedimentology of ancient saline pans: An example from the Permian Opeche shale, Williston Basin, North Dakota, U.S.A. *Journal of Sedimentary Research*, *70*(1), 159–169.

Berger, J.-P., Reichenbacher, B., Becker, D., Grimm, M., Grimm, K., Picot, L., et al. (2005). Paleogeography of the Upper Rhine Graben (URG) and the Swiss Molasse Basin (SMB) from Eocene to Pliocene. *International Journal of Earth Sciences (Geol Rundsch)*, *94*, 697–710.

Blair, T. C., & McPherson, J. G. (1994). Alluvial Fans and their natural distinction from rivers based on morphology, hydraulic processes, sedimentary processes, and facies assemblages. *SEPM Journal of Sedimentary Research*, *64A*, 450–489. doi:10.1306/d4267dde-2b26-11d7-8648000102c1865d

Blakey, R. C. (2008). Gondwana paleogeography from assembly to breakup—A 500 m.y. odyssey. In C. R. Fielding, T. D. Frank, & J. L. Isbell (Eds.), *Resolving the late Paleozoic ice age in time and space* (Vol. 441, pp. 1–28). The Geological Society of America Special Paper.

Blakey, R. C. (2013). Using Paleogeographic maps to portray Phanerozoic Geologic and Paleotectonic history of Western North America. *AAPG Search and Discovery*, Article #30267 *(2013)*.

Blakey, R. C., & W. D. Ranney (2018). *Ancient landscapes of western North America: A geologic history with paleogeographic maps*. Switzerland: Springer International Publishing, 228 pp.

Böcker, J., Littke, R., & Forster, A. (2016). An overview on source rocks and the petroleum system of the central Upper Rhine Graben. *International Journal of Earth Sciences*, *106*(2), 707–742.

Bohacs, K. M., Carroll, A. R., & Neal, J. E. (2003). Lessons from large lake systems—Thresholds, nonlinearity, and strange attractors. In Chan, M. A., & Archer, A. W. (Eds.), *Extreme depositional environments: Mega end members in geologic time* (Vol. 370, pp. 75–90). Boulder, CO: Geological Society of America Special Paper.

Bohacs, K. M., Carroll, A. R., Neal, J. E., & Mankiewicz, P. J. (2000). Lake-basin type, source potential, and hydrocarbon character: An integrated-sequence-stratigraphic–geochemical framework. In E. H. Gierlowski-Kordesch & K. R. Kelts (Eds.), *Lake basins through space and time* (Vol. 46, pp. 3–34). AAPG Studies in Geology.

Bosworth, W., Huchon, P., & McClay, K. (2005). The Red Sea and Gulf of Aden basins. *Journal of African Earth Sciences*, *43*, 334–378.

Bosworth, W., Taviani, M., & Rasul, N. M. A. (2019). Neotectonics of the Red Sea, Gulf of Suez and Gulf of Aqaba. In N. M. A. Rasul & I. C. F. Stewart (Eds.), *Geological setting, Palaeoenvironment and Archaeology of the Red Sea* (pp. 11–35). Switzerland: Springer Nature.

Bowman, D. (2019). The regional approach: Alluvial Fans along the Dead Sea-Arava Rift Valley. In Bowman, D. (Ed.), *Principles of alluvial fan morphology* (pp. 135–151). Dordrecht: Springer. https://doi.org/10.1007/978-94-024-1558-2_19

Braunger, S., Marks, M. A. W., Walter, B. F., Neubauer, R., Reich, R., Wenzel, T., et al. (2018). The Petrology of the Kaiserstuhl Volcanic Complex, SW Germany: The importance of metasomatized and oxidized lithospheric mantle for carbonatite generation. *Journal of Petrology*, *59*(9), 1731–1762.

Brew, G., Barazangi, M., Al-Maleh, A. K., & Sawaf, T. (2001). Tectonic and geologic evolution of Syria. *GeoArabia*, *6*(4), 573–616.

Burwood, R., De Witte, S. M., Mycke, B., & Paulet, J. (1995). Petroleum geochemical characterisation of the Lower Congo Coastal Basin Bucomazi Formation. In B. J. Katz (Ed.), *Petroleum source rocks, Casebooks in Earth Sciences* (pp. 235–263). Berlin, Heidelberg: Springer.

Busson, G. (1991). Relationship between different types of evaporitic deposits, and the occurrence of organic-rich layers (potential source-rocks), *Carbonates and Evaporites*, *6*(2), 177–192. doi:10.1007/bf03174442

Caputo, M. V., & Soares, E. A. A. (2016). Eustatic and tectonic change effects in the reversion of the transcontinental Amazon River drainage system. *Brazilian Journal of Geology*, *46*(2), 301–328.

Catacosinos, P. A., Harrison, W. B., III, & Daniels, P. A., Jr. (1991). Structure, stratigraphy, and petroleum geology of the Michigan Basin. In M. W. Leighton, D. R. Kolata, D. F. Oltz, & J. J. Eidel (Eds.), *Interior cratonic basins, American Association of Petroleum Geologists, Memoir* (Vol. 51, pp. 561–601).

Çemen, I., Goncuoglu, M. C., & Dirik, K. (1999). Structural evolution of the Tuzgolu Basin in Central Anatolia, Turkey. *The Journal of Geology*, *107*, 693–706.

Cercone, K. R. (1988). Evaporative sea-level drawdown in the Silurian Michigan basin. *Geology*, *16*, 387–390.

Chapin, C. E. (1979). Evolution of the Rio Grande rift—A summary. In R. E. Riecker (Ed.), *Rio Grande rift: Tectonics and magmatism* (Vol. 14, pp. 1–7). American Geophysical Union, Special Publications,.

Charrach, J. (2018). The Sdom evaporite formation in Israel and its relationship with the Messinian Salinity Crisis. *Carbonates and Evaporites*, *33*, 727–766. https://doi.org/10.1007/s13146-017-0410-1

Coianiz, L., Bialik, O. M., Ben-Avraham, Z., & Lazar, M. (2019). Late Quaternary lacustrine deposits of the Dead Sea basin: High resolution sequence stratigraphy from downhole logging data. *Quaternary Science Reviews*, *210*, 175–189.

Coianiz, L., Schattner, U., Lang, G., Ben-Avraham, Z., & Lazar, M. (2020). Between plate and salt tectonics—New stratigraphic constraints on the architecture and timing of the Dead Sea basin during the Late Quaternary. *Basin Research*, *32*(4), 636–651. https://doi.org/10.1111/bre.12387

Costa, A. R. A., & Wanderley Filho, J. R. (2008). Os evaporitos e halocinese na Amazônia. In W. Mohriak, P. Szatmari, & S. M. C. Anjos (Eds.), *Sal: Geologia e Tectônica* (pp. 208–219). São Paulo, Brazil: Editora Beca.

Dean, W. E., & Johnson, K. S. (1989). Anhydrite deposits of the United States and characteristics of anhydrite important for storage of radioactive wastes. *U.S. Geological Survey*, Bulletin 1794, 132 pp.

Dellwig, L. F., & Evans, R. (1969). Depositional processes in Salina salt of Michigan, Ohio, and New York. *American Association of Petroleum Geologists Bulletin*, *53*(4), 949–956.

Deocampo, D. M., & Jones, B. F. (2014). Geochemistry of Saline Lakes. In Holland, H. D., & Turekian, K. K. (Eds.), *Treatise on geochemistry* (2nd ed., Chapter 7.13, pp. 437–469). Elsevier. http://dx.doi.org/10.1016/B978-0-08-095975-7.00515-5

Dickinson, W. R. (2002). The basin and range province as a composite extensional domain. *International Geology Review*, *44*(1), 1–38. doi: 10.2747/0020-6814.44.1.1

Droste, J. B., & Shaver, R. H. (1977). Synchronization of deposition: Silurian reef-bearing rocks on Wabash platform with cyclic evaporites of Michigan basin. In J. H. Fisher (Ed.), *Reefs and evaporites—Concepts and depositional models, Studies in Geology* (Vol. 5, pp. 93–109). American Association of Petroleum Geologists.

Dutton, S. P., Kim, E. M., Broadhead, R. F., Raatz, W. D., Breton, C. L., Ruppel, S. C., & Kerans, C. (2005). Play analysis and leading-edge oil-reservoir development methods in the Permian basin: Increased recovery through advanced technologies. *American Association of Petroleum Geologists Bulletin*, *89*(5), 553–576.

Eia-U. S. Energy Information Administration (2016). *Maps: Oil and Gas Exploration, Resources and Production*. Sedimentary basin boundaries in Lower 48 States. https://www.eia.gov/maps/maps.htm

Ells, G. D. (1967). Michigan's Silurian oil and gas pools, State of Michigan Department of Conservation. *Geological Survey Division, Report of Investigation 2*, 49 pp.

Ettensohn, F. R. (2008). The Appalachian foreland basin in eastern United States. In A. D. Miall (Ed.), *Sedimentary Basins of the World, Vol. 5, The Sedimentary Basins of the United States and Canada* (pp. 105–179). Amsterdam: Elsevier.

Eugster, H. P. (1982). Climatic significance of lake and evaporite deposits. In *Climate in earth history: Studies in Geophysics* (pp. 105–111). National Research Council, Washington, DC: The National Academies Press.

Eugster, H. P., & Hardie, L. A. (1975). Sedimentation in an ancient playa-lake complex: The Wilkins Peak Member of the Green River Formation of Wyoming. *Geological Society of America Bulletin*, *86*, 319–334.

Eugster, H. P., & Hardie, L. A. (1978). Saline lakes. In A. Lerman (Ed.), *Lakes, Chemistry, Geology, Physics* (pp. 237–293). Berlin, Heidelberg: Springer-Verlag.

Evans, R., & Kirkland, D. W. (1988). Evaporitic environments as a source of petroleum. In B. C. Schreiber (Ed.), *Evaporites and hydrocarbons* (pp. 256–299). New York: Columbia University Press.

Ewing, T. E. (2017). The Permian Basin—Intracratonic subsidence triggered by ARM compression, making an American "Superbasin". *AAPG Search and Discovery*, Article #10988 *(2017)*.

Ewing, T. E. (2019). Tectonics and Subsidence in the West Texas (Permian) Basin: A Model for Complex Intracratonic Basin Development, *AAPG Search and Discovery* Article #30606 *(2019)*

Ewing, T. E., & Galloway, W. E. (2019). Evolution of the Northern Gulf of Mexico sedimentary basin. In A. D. Miall (Ed.), *The sedimentary basins of the United States and Canada* (pp. 627–694). Amsterdam: Elsevier.

Fontes, J. C., Filly, A., Gaudant, J., & Duringer, P. (1991). Origine continentale des evaporites paleogenes de Haute Alsace; arguments paleoecologiques, sedimentologiques et isotopiques, *Bulletin de la Société Géologique de France*, *162*, 725–737. doi: https://doi.org/10.2113/gssgfbull.162.4.725

Galamay, A. R., Karakaya, M. Ç., Bukowski, K., Karakaya, N., & Yaremchuk, Y. (2023). Geochemistry of Brine and Paleoclimate reconstruction during sedimentation of Messinian Salt in the Tuz Gölü Basin (Türkiye): Insights from the study of fluid inclusions. *Minerals*, *13*, 171. https://doi.org/10.3390/min13020171

Galloway, W. E., & Hobday, D. K. (1983). *Terrigenous clastic depositional systems—Applications to Petroleum, Coal, and Uranium exploration* (1st ed.). New York, USA: Springer-Verlag. 423 pp.

García-Veigas, J., Rosell, L., Zak, I., Playà, E., Ayora, C., & Starinsky, A. (2009). Evidence of potash salt formation in the Pliocene Sedom Lagoon (Dead Sea Rift, Israel). *Chemical Geology*, *265*, 499–511.

Garfunkel, Z., & Ben-Avraham, Z. (1996). The structure of the Dead Sea basin. *Tectonophysics*, *226*, 155–176.

Garret, D. E. (1996). Potash deposits. In D. E. Garret (Ed.), *Potash deposits, processing, properties and uses* (pp. 81–211). London, UK: Chapman & Hall.

GEBCO (2022). General Bathymetric Chart of the Oceans, British Oceanographic Data Centre, U.K., website www.gebco.net (Grid. doi:10.5285/e0f0bb80-ab44-2739-e053-6c86abc0289c).

Geist, E. L., & Brocher, T. M. (1987). Geometry and subsurface lithology of southern Death Valley basin, California, based on refraction analysis of multichannel seismic data. *Geology*, *15*(12), 1159–1162.

Ginzburg, A., Reshef, M., Ben-Avraham, Z., & Schattner, U. (2007). The style of transverse faulting in the Dead Sea basin from seismic reflection data: The Amazyahu fault. *Israel Journal of Earth Sciences*, *55*, 129–139. doi: 10.1560/IJES_55_3_129.

Gürbüz, A., & Kazanci, N. (2014). Facies characteristics and control mechanisms of Quaternary deposits in the Lake Tuz Basin. *Bulletin of the Mineral Research and Exploration*, *149*, 1–18.

Handford, C. R. (1991). Marginal marine halite: sabkhas and salinas. In J. L. Melvin (Ed.), *Evaporites, petroleum and mineral resources, Developments in Sedimentology* (Vol. 50, Chapter 1, pp. 1–66). Amsterdam: Elsevier.

Handford, C. R., & Baria, L. (2007). Geometry and seismic geomorphology of carbonate shoreface clinoforms, Jurassic Smackover Formation, north Louisiana. In R. J. Davies, H. W. Posamentier, L. J. Wood, & J. A. Cartwright (Eds.), *Seismic geomorphology: Applications to hydrocarbon exploration and production* (Vol. 277, pp. 171–185). London: Geological Society, Special Publications.

Handford, C. R., & Bassett, R. L. (1982). Permian facies sequences and evaporite depositional styles, Texas Panhandle. In C. R. Handford, R. G. Loucks, & G. R. Davies (Eds.), *Depositional and diagenetic spectra of evaporites—A Core Workshop* (Vol. 3, pp. 210–237). SEPM Society for Sedimentary Geology.

Hardie, L. A. (1984). Evaporites: Marine or non-marine. *American Journal of Science*, *283*, 193–240.

Hardie, L. A. (1991). On the significance of evaporites. *Annual Review Earth and Planetary Science*, *290*, 43–106.

Hardie, L. A., Smoot, J. P., & Eugster, H. P. (1978). Saline lakes and their deposits: A sedimentological approach. In A. Matter & M. E. Tucker (Eds.), *Modern and ancient lake sediments. International Association of Sedimentologists Special Publication*, *2*, 7–41.

Harris, P. M., & Kerans, C. (2016). The role of the Guadalupe Mountains and Capitan research in the evolution of carbonate conceptual models and paradigms, *AAPG Search and Discovery*, Article #60052 *(2016)*

Harrison, W. B., III, & Voice, P. J. (2017). Evaporite facies of the Michigan Basin. In G. M. Grammer, W. B. Harrison, III, & D. A. Barnes (Eds.), *Paleozoic stratigraphy and resources of the Michigan Basin* (Vol. 531, pp. 197–216). Geological Society of America Special Paper. https://doi.org/10.1130/2017.2531(10).

Hatcher, R. D. (2010). The Appalachian orogen: A brief summary. In R. P. Tollo, M. J. Bartholomew, J. P. Hibbard, & P. M. Karabinos (Eds.), *From Rodinia to Pangea: The Lithotectonic record of the Appalachian Region, Geological Society of America, Memoir* (Vol. 206, pp. 1–19).

Heimdal, T. H., Callegaro, S., Svensen, H. H., Jones, M. T., Pereira, E., & Planke, S. (2019). Evidence for magma–evaporite interactions during the emplacement of the Central Atlantic Magmatic Province (CAMP) in Brazil. *Earth and Planetary Science Letters*, *506*, 476–492.

Hibbs, B. J. (2022). Commentary and review of modern environmental problems linked to historic flow capacity in arid groundwater basins. *Geosciences*, *12*, 124. https://doi.org/10.3390/geosciences12030124

Hills, J. M. (1972). Late Paleozoic sedimentation in West Texas Permian Basin, *American Association of Petroleum Geologists Bulletin*, *56*(12), 2303–2322.

Hinsken, S., Ustaszewski, K., & Wetzel, A. (2007). Graben width controlling syn-rift sedimentation: the Palaeogene southern Upper Rhine Graben as an example. *International Journal of Earth Sciences*, *96*(6), 979–1002. doi:10.1007/s00531-006-0162-y

Hofmann, P., Huc, A. Y., Carpentier, B., Schaeffer, P., Albrecht, P., Kelly, B., et al. (1993). Organic matter of the Mulhouse Basin, France: A synthesis. *Organic Geochemistry*, *20*(8), 1105–1123. doi:10.1016/0146-6380(93)90003-t

Holliday, V. T., Harvey, A., Cuba, M. T., & Weber, A. M. (2019). Paleoindians, paleolakes and paleoplayas: Landscape geoarchaeology of the Tularosa Basin, New Mexico. *Geomorphology*, *331*, 92–106. https://doi.org/10.1016/j.geomorph.2018.08.012

Hooke, R. L. (1999). Lake Manly(?) shorelines in the eastern Mojave Desert, California. *Quaternary Research*, *52*, 328–336.

Horton, J. D. (2017). The State Geologic Map Compilation (SGMC) geodatabase of the conterminous United States (ver. 1.1, August 2017), U.S. Geological Survey data release, https://doi.org/10.5066/F7WH2N65.

Hovorka, S. D., Holt, R. M., & Powers, D. W. (2007). Depth indicators in Permian Basin evaporites. In B. C. Schreiber, S. Lugli, & M. Babel (Eds.), *Evaporites through space and time* (Vol. 285, pp. 335–364). London: Geological Society, Special Publications.

Hsü, K. J. (1972). Origin of saline giants: A critical review after the discovery of the Mediterranean evaporite. *Earth-Science Reviews*, *8*, 371–396.

Hsü, K. J. (1977). Tectonic evolution of the Mediterranean Basins. In A. E. M. Nairn, W. H. Kanes, & F. G. Stehli (Eds.), *The ocean basins and margins: The Eastern Mediterranean* (Vol. 4A, pp. 29–75). New York: Springer, Plenum Press.

Hsü, K. J., & Bernoulli, D. (1978). Genesis of the Tethys and the Mediterranean. In K. J. Hsü, L. Montadert, et al. (Eds.), *Initial Reports of the Deep Sea Drilling Project* (Vol. 42, Part 1, pp. 943–949). Washington: U.S. Government Printing Office. doi:10.2973/dsdp.proc.42-1.149.1978

Hsü, K. J., Montadert, L., Bernoulli, D., Cita, M. B., Erickson, A., Garrison, R. E., et al. (1978). History of the Mediterranean salinity crisis. In K. J. Hsü, L. Montadert, et al. (Eds.), *Initial Reports of the Deep Sea Drilling Project* (Vol. 42, Part 1, pp. 1053–1078). Washington: U.S. Government Printing Office. doi:10.2973/dsdp.proc.42-1.149.1978

Johnson, K., & S. Gonzales (1978). Salt deposits in the United States and regional geologic characteristics important for storage of radioactive waste, prepared for *Union Carbide Corp., Nuclear Division, Oak Ridge National Laboratory*, Office of Waste Isolation, Y/OWI/SUB-7414/1, 188 p.

Kagan, E., Stein, M., & Marco, S. (2018). Integrated paleoseismic chronology of the last glacial Lake Lisan: From lake margin seismites to deep-lake mass transport deposits. *Journal of Geophysical Research, Solid Earth*, *123*, 2806–2824. https://doi.org/10.1002/2017JB014117

Katz, B. J. (1995). The Green River Shale: An Eocene carbonate lacustrine source rock. In Katz, B. J. (Ed.), *Petroleum source rocks* (pp. 309–324). Berlin: Springer-Verlag.

Katz, B. J. (2001). Lacustrine basin hydrocarbon exploration—Current thoughts. *Journal of Paleolimnology*, *26*, 161–179. https://doi.org/10.1023/A:1011173805661

Kendall, A. C. (1978). Facies Models 12. Subaqueous Evaporites. In R. G. Walker (Ed.), *Facies Models, Geoscience Canada*, *5*(3), 124–139.

Kendall, A. C. (1992). Evaporites. In R. G. Walker, & N. P. James (Eds.), *Facies Models: Response to sea level change* (pp. 375–409). Geological Association of Canada.

Kendall, A. C. (2010). Marine evaporites. In N. P. James & R. W. Dalrymple (Eds.), *Facies Models* (pp. 505–539). Geological Association of Canada IV, Series GEOtext.

Kendall, A. C., & Harwood, G. M. (1996). Marine evaporites: arid shorelines and basins. In H. G. Reading (Ed.), *Sedimentary environments: Processes, facies and stratigraphy* (3rd ed., pp. 281–324). Oxford: Blackwell Scientific Publications.

Kendall, C. G. St. C., & Skipwith BT., P. A. d'E. (1969). Geomorphology of a recent shallow-water carbonate province: Khor Al Bazam, Trucial Coast, Southwest Persian Gulf. *Geological Society of America Bulletin*, *80*(5), 865–892. https://doi.org/10.1130/0016-7606(1969)80[865:GOARSC]2.0.CO;2

Kendall, C. G. St. C., Weber, L. J., & Alsharhan, A. S. (2009). The Giant Oil Field Evaporite Association: A function of the Wilson Cycle, climate, basin position and sea level. *AAPG Search and Discovery*, Article #40471 (*2009*).

Kiliç, O., & Kilic, A. M. (2010). Salt crust mineralogy and geochemical evolution of the Salt Lake (Tuz Gölü), *Turkey. Scientific Research and Essays*, *5*(11), 1317–1324.

King, P. B. (1942). Permian of west Texas and southeastern New Mexico. *American Association of Petroleum Geologists Bulletin*, *26*(4), 535–763

Kiro, Y., Goldstein, S. L., Lazar, B., & Stein, M. (2016). Environmental implications of salt facies in the Dead Sea. *Geological Society of American Bulletin*, *128*(5/6), 824–841. doi: 10.1130/B31357.1.

Ku, T.-L., Luo, S., Lowenstein, T. K., & Li, J. (1998). U-series chronology of lacustrine deposits in Death Valley, California. *Quaternary Research*, *50*, 261–275.

Kuo, L.-C. (1994). Lower Cretaceous lacustrine source rocks in northern Gabon: effect of organic facies and thermal maturity on crude oil quality. *Organic Geochemistry*, *22*(2), 257–273.

Larsen, B. D., Ben-Avraham, Z., & Shulman, H. (2002). Fault and salt tectonics in the southern Dead Sea basin. *Tectonophysics*, *346*, 71–90.

Lazar, M. (2019). Tectonic segmentation of the Dead Sea Fault System: A review of geophysical evidence. In João C. Duarte (Ed.), *Transform plate boundaries and fracture zones* (pp. 417–436). Amsterdam: Elsevier.

Lazar, M., Ben-Avraham, Z., & Schattner, U. (2006). Formation of sequential basins along a strike–slip fault—Geophysical observations from the Dead Sea basin. *Tectonophysics*, *421*, 53–69.

Liu, C., Wang, L., Yan, M., Zhao, Y., Cao, Y., Fang, X., et al. (2018). The Mesozoic-Cenozoic tectonic settings, paleogeography and evaporitic sedimentation of Tethyan blocks within China: Implications for potash formation. *Ore Geology Reviews*, *102*, 406–425.

Loucks, R. G., & Longman, M. W. (1982). Lower Cretaceous Ferry Lake Anhydrite, Fairway Field, East Texas: Product of shallow-subtidal deposition. In C. R. Handford, R. G. Loucks, & G. R. Davies (Eds.), *Depositional and diagenetic spectra of evaporites—A core workshop* (Vol. 3, pp. 130–173). SEPM Society for Sedimentary Geology.

Love, D. W., Allen, B. D., Scholle, P. A., & Bustos, D. (2020). White Sands National Park. In P. A. Scholle, D. S. Ulmer-Scholle, S. M. Cather, & S. A. Kelley (Eds.), *The Geology of Southern New Mexico's parks, monuments, and public lands, New Mexico Bureau of Geology and Mineral Resources* (pp. 221–231).

Lowenstein, T. K., Li, J., Brown, C. B., Roberts, S. M., Ku, T. L., Luo, S., & Yang, W. (1999). 200 k. y. paleoclimate record from Death Valley salt core. *Geology*, *27*, 3–6.

Lu, Y., Waldmann, N., Alsop, G. I., & Marco, S. (2017). Interpreting soft sediment deformation and mass transport deposits as seismites in the Dead Sea depocenter. *Journal of Geophysical Research: Solid Earth*, *122*, 8305–8325.

Lu, Y., Waldmann, N., Nadel, D., & Marco, S. (2017). Increased sedimentation following the Neolithic Revolution in the Southern Levant. *Global and Planetary Change*, *152*, 199–208.

Lucia, F. J. (2007). *Carbonate reservoir characterization—An integrated approach* (2nd ed.). Berlin: Springer, 336 pp.

Luning, S., Craig, J., Loydell, D. K., Storch, P., & Fitches, B. (2000). Lower Silurian "hot shales" in North Africa and Arabia: Regional distribution and depositional model. *Earth-Science Reviews*, *49*, 121–200.

Maher, J. C. (1971). Geological framework and petroleum potential of the atlantic coastal plain and continental shelf. *Geological Survey Professional Paper 659*, 98 pp.

Mart, Y. (1991). The Dead Sea Rift: from continental rift to incipient ocean, *Tectonophysics*, *197*, 155–179.

Mart, Y., Ryan, W. B. F., & Lunin, O. V. (2005). Review of the tectonics of the Levant Rift system: the structural significance of oblique continental breakup. *Tectonophysics*, *395*, 209–232.

Marzoli, A., Renne, P. R., Piccirillo, E. M., Ernesto, M., Bellieni, G., & De Min, A. (1999). Extensive 200 Ma-year-old continental flood basalts of the Central Atlantic Magmatic Province. *Science*, *284*, 616–618.

Meijer Drees, N. (1994). Devonian Elk Point Group of the Western Canada Sedimentary Basin. In G. Mossop & I. Shetsen (Eds.), *Geological Atlas of the Western Canada Sedimentary Basin* (pp. 129–147). Canadian Society of Petroleum Geologists/Alberta Research Council, Calgary, Alberta.

Mello, M. R., Mohriak, W. U., Koutsoukos, E. A. M., & Bacoccoli, G. (1994). Selected Petroleum Systems in Brazil. In L. B. Magoon & W. G. Dow (Eds.), *The petroleum system-from source to trap, AAPG Memoir* (Vol. 60, pp. 499–512).

Mesolella, K. J. (1978). Paleogeography of some Silurian and Devonian reef trends, Central Appalachian basin. *American Association of Petroleum Geologists Bulletin*, *62*(9), 1607–1644.

Miller, D. E., & Dupal, I. (1987). Reef delineation by multiple offset borehole seismic data collection: a case study. *AAPG Studies in Geology*, *27*(1), 110–115.

Miller, N. C., & Lizarralde, D. (2013). Thick evaporites and early rifting in Guaymas Basin, *Gulf of California. Geology*, *41*(2), 283–286.

Mohriak, W. (2025). *Salt in the Earth sciences: Basin analysis and salt tectonics*. Washington, DC: American Geophysical Union. doi:10.1002/9781394307067

Mohriak, W. U., Szatmari, P., & Anjos, S. (2012). Salt: geology and tectonics of selected Brazilian basins in their global context. In G. I. Alsop, S. G. Archer, A. J. Hartley, N. T. Grant, & R. Hodgkinson (Eds.), *Salt tectonics, sediments and prospectivity* (Vol. 363, pp. 131–158). Geological Society, London, Special Publications.

Mohriak, W. U., Gordon, A. C., & Mello, M. R. (2021). Origin and petroleum system of the Cabo Frio High Between the Santos and Campos basins: Reviewed integration of structural and paleogeographic reconstruction with the oil and gas systems. In M. R. Mello, P. O. Yilmaz, & B. J. Katz (Eds.), *The supergiant Lower Cretaceous pre-salt petroleum system of the Santos Basin, Brazil, American Association of Petroleum Geologists, Memoir* (Vol. 124, pp. 273–324).

Mosmann, R., Falkenhein, F. U. H., Goncalves, A., & Nepomuceno Filho, F. (1986). Oil and Gas Potential of the Amazon Paleozoic Basins. In M. T. Halbouty (Ed.), *Future petroleum provinces of the world, American Association of Petroleum Geologists, Memoir* (Vol. 40, pp. 207–241).

Nance, R. D., Gutiérrez-Alonso, G., Keppie, J. D., Linnemann, U., Murphy, J. B., Quesada, C., et al. (2012). A brief history of the Rheic Ocean. *Geoscience Frontiers*, *3*(2), 125–135.

Neev, D., & Emery, K. 0. (1967). The Dead Sea. *Geological Survey of Israel Bulletin*, *41*, 147 pp.

Neev, D., & Hall, J. K. (1979). Geophysical investigations in the Dead Sea. *Sedimentary Geology*, *23*, 209–238. https://doi.org/10.1016/0037-0738(79)90015-0

Neugebauer, I., Brauer, A., Schwab, M. J., Waldmann, N. D., Enzel, Y., Kitagawa, H. A., et al. (2014). Lithology of the long sediment record recovered by the ICDP Dead Sea Deep Drilling Project (DSDDP). *Quaternary Science Reviews*, *102*, 149–165.

Noda, A. (2013). Strike-slip basin – Its configuration and sedimentary facies, in Y. Itoh (Ed.), *Mechanism of sedimentary basin formation – multidisciplinary approach on active plate margins* (pp. 27–57). London, UK: IntechOpen. DOI: 10.5772/56593. https://www.intechopen.com/chapters/45342

Nurmi, R. D., & Friedman, G. M. (1977). Sedimentology and depositional environments of basin-center evaporites, lower Salina Group (Upper Silurian), Michigan Basin, in J. H. Fisher (Ed.), *Reefs and evaporites—Concepts and depositional models, Studies in Geology* (Vol. 5, pp. 23–52). American Association of Petroleum Geologists.

Ochsenius, K. (1877). Die Bilding der Steinsalzlager und ihrer mutter augensalze, *C.E.M. Pfeffer, Halle*, 172 pp.

Ochsenius, K. (1888). On the formation of rock salt beds and mother liquor salts. *Philadelphia Academic Sciences, Proceedings, Pl 2*, 181–187.

Önalgil, N., Kadir, S., Külah, T., Eren, M., & Gürel, A. (2015). Mineralogy, geochemistry and genesis of the modern sediments of Seyfe Lake, Kırsehir, central Anatolia, Turkey. *Journal of African Earth Sciences*, *102*, 116–130. http://dx.doi.org/10.1016/j.jafrearsci.2014.10.020

Palmer, M. R., Helvacı, C., & Fallick, A. (2004). Sulphur, sulphate oxygen and strontium isotope composition of Cenozoic Turkish evaporites. *Chemical Geology*, *209*, 341–356. doi:10.1016/j.chemgeo.2004.06.027

Perry, F. V., Kelley, R. E., Dobson, P. F., & Houseworth, J. E. (2014). Regional geology: A GIS Database for Alternative Host Rocks and Potential Siting Guidelines, *U.S. Department of Energy*, FCRD-UFD-2014-000068, 168 pp.

Perthuisot, J. P. (1989). Part II. Recent Evaporites. In P. Sonnenfeld & J.-P. Perthuisot (Eds.), *Brines and evaporites, Short Course in Geology* (Vol. 3, pp. 65–118), 28th International Geological Congress. Washington, DC: American Geophysical Union.

Phelps, R. M., Kerans, C. Loucks, R. G., et al. (2014). Oceanographic and eustatic control of carbonate platform evolution and sequence stratigraphy on the Cretaceous (Valanginian–Campanian) passive margin, northern Gulf of Mexico. *Sedimentology*, *61*, 461–496, doi:10.1111/sed.12062.

Pichat, A., Hoareau, G., Callot, J.-P., Legeay, E., Kavak, K. S., Révillon, S., et al. (2018). Evidence of multiple evaporite recycling processes in a salt-tectonic context, Sivas Basin, Turkey. *Terra Nova*, *30*(1), 40–49. doi:10.1111/ter.12306 10.1111/ter.12306

Putirka, K., & Platt, B. (2012). Basin and Range volcanism as a passive response to extensional tectonics. *Geosphere*, *8*(6), 1274–1285. https://doi.org/10.1130/GES00803.1

Raab, M., Friedman, G. M., Spiro, B., Starinsky, A., & Zak, I. (1997). The geological history of Messinian (Upper Miocene) evaporites in the Central Jordan Valley (Israel) and how strontium and sulfur isotopes relate to their origin. *Carbonates Evaporites*, *12*, 296–324. https://doi.org/10.1007/BF03175424

Raab, M., Friedman, G. M., Spiro, B., Starinsky, A., & Zak, I. (2000). The geological history of Pliocene-Pleistocene evaporites in Mount Sedom (Israel) and how strontium and sulfur isotopes relate to their origin. *Carbonates Evaporites*, *15*, 93–114. https://doi.org/10.1007/BF03175817

Reijs, J., & McClay, K. (1998). Salar Grande pull-apart basin, Atacama Fault System, northern Chile. In R. E. Holdsworth, R. A. Strachan, & J. F. Dewey (Eds.), *Continental transpressional and transtensional tectonics* (Vol. 135, pp. 127–141). London: Geological Society, Special Publications.

Reilinger, R., & McClusky, S. (2011). Nubia–Arabia–Eurasia plate motions and the dynamics of Mediterranean and Middle East tectonics. *Geophysical Journal International*, *186*(3), 971–979. https://doi.org/10.1111/j.1365-246X.2011.05133.x

Reinhold, C., Schwarz, M., Bruss, D., Heesbeen, B., Perner, M., & Suana, M. (2016). The Northern Upper Rhine Graben-Re-dawn of a mature petroleum province? *Swiss Bulletin für angewandte Geologie*, *21*/2, S.35–56.

Rezende, G. L., Martins, C. M., Nogueira, A. C. R., Domingos, F. G., & Ribeiro-Filho, N. (2021). Evidence for the Central Atlantic magmatic province (CAMP) in Precambrian and Phanerozoic sedimentary basins of the southern Amazonian Craton, Brazil. *Journal of South American Earth Sciences*, *108*, 103216.

Rine, M. J., Kaczmarek, S. E., & Garrett, J. D. (2017). A new sequence stratigraphic model for the Silurian A-1 Carbonate (Ruff Formation) of the Michigan Basin. In G. M. Grammer, W. B. Harrison, III, & D. A. Barnes (Eds.), *Paleozoic stratigraphy and resources of the Michigan Basin* (Vol. 531, pp. 175–195). Geological Society of America Special Paper.

Robinson, K. (1988). Petroleum Geology and hydrocarbon plays of the Permian Basin petroleum province, West Texas and Southeast New Mexico. *United States Department of the Interior, U.S. Geological Survey*, Open-File Report 88-450Z, 53 p.

Rotstein, Y., Behrmann, J. H., Lutz, M., Wirsing, G., & Lutz, A. (2005). Tectonic implications of transpression and transtension: Upper Rhine Graben. *Tectonics*, *24*(6), 1–12. doi:10.1029/2005tc001797

Rotstein, Y., Schaming, M., & Rousse, S. (2005). Tertiary tectonics of the Dannemarie Basin, upper Rhine graben, and regional implications. *International Journal of Earth Sciences (Geol Rundsch)*, *94*, 669–679. doi 10.1007/s00531-005-0473-4

Roveri, M., Lugli, S., Manzi, V., & Schreiber, B. C. (2008a). The Messinian Sicilian stratigraphy revisited: New insights for the Messinian salinity crisis. *Terra Nova*, *20*, 483–488. doi:10.1111/j.1365-3121.2008.00842.x

Roveri, M., Lugli, S., Manzi, V., & Schreiber, B. C. (2008b). The Messinian salinity crisis: A sequence-stratigraphic approach. *GeoActa*, *7984*, 169–190.

Salvador, A. (1987). Late Triassic-Jurassic paleogeography and origin of Gulf of Mexico Basin. *American Association of Petroleum Geologists Bulletin*, *71*(4), 419–451.

Sarg, J. F. (2001). The sequence stratigraphy, sedimentology, and economic importance of evaporite-carbonate transitions: A review. *Sedimentary Geology*, *140*, 9–42.

Schmalz, R. F. (1969). Deep-water evaporite deposition: A genetic model. *American Association of Petroleum Geologists Bulletin*, *53*(4), 798–823.

Schmalz, R. F. (1970). Environment of marine evaporite deposition. *Mineral Industries*, *35*(8), 1–7.

Schnurrenberger, D., Russell, J., & Kelts, K. (2003). Classification of lacustrine sediments based on sedimentary components. *Journal of Paleolimnology*, *29*, 141–154. https://doi.org/10.1023/A:1023270324800

Scholle, P. (2020). *Introduction to the Geology of the Permian reef complex, Guadalupe and Delaware Mountains, New Mexico & West Texas*. New Mexico Bureau of Geology and Mineral Resources. https://geoinfo.nmt.edu/tour/federal/parks/PermianReef/home.html

Scholle, P. A., Goldstein, R. H., & Ulmer-Scholle, D. S. (2007). *Classic upper Paleozoic reefs and bioherms of west Texas and New Mexico*. Open file Report 504, New Mexico Bureau of Geology and Mineral Resources, Socorro, New Mexico, 174 pp. https://geoinfo.nmt.edu/publications/openfile/downloads/500-599/504/ofr_504.pdf

Schreiber, B. C. (1986). Arid shorelines and evaporites. In H. G. Reading (Ed.), *Sedimentary environments and facies* (pp. 189–228). Oxford: Blackwell Scientific Publications.

Schreiber, B. C. (1988). *Evaporites and hydrocarbons*. New York: Columbia University Press, 475 pp.

Schreiber, B. C., & Friedman, G. M. (1976). Depositional environments of Sicily as determined from analysis of intercalated carbonates, *Sedimentology*, *23*, 255–270.

Schreiber, B. C., & Hsü, K. J. (1980). Evaporites. In G. D. Hobson (Ed.), *Developments in petroleum geology* (Vol. 2, pp. 87–138). London: Applied Science Publishers.

Schreiber, B. C., & Marshak, S. (1981). Evaporites, carbonates and oil. In *Geology of passive continental margins, AAPG Education Course Note Series* (Vol. 19, pp. 6.1–6.32).

Scotese, C. R. (2001). *Atlas of Earth History, Volume 1, Paleogeography*, PALEOMAP Project, Arlington, Texas, 52 pp.

Seard, C., Camoin, G., Rouchy, J.-M., and Virgone, A. (2013). Composition, structure and evolution of a lacustrine carbonate margin dominated by microbialites: Case study from the Green River formation (Eocene; Wyoming, USA). *Palaeogeography, Palaeoclimatology, Palaeoecology*, *381–382*, 128–144.

Seni, S. J., Mullican, W. F., III, & Hamlin, H. S. (1985). Texas salt domes: Natural resources, storage caverns, and extraction technology. *Report for Texas Department of Water Resource*. The University of Texas at Austin Bureau of Economic Geology, 161 pp.

Shearman, D. J. (1978). Evaporites of coastal sabkhas. In W. E. Dean & B. C. Schreiber (Ed.), *Marine evaporites, Lecture Notes for Short Course* (Vol. 4, pp. 6–42). Society of Economic Paleontologists and Mineralogists, Tulsa, Okla.

Sirota, I., Arnon, A., & Lensky, N. G. (2016). Seasonal variations of halite saturation in the Dead Sea, *Water Resources Research*, *52*, 7151–7162, doi:10.1002/2016WR018974.

Sirota, I., Enzel, Y., & Lensky, N. G. (2017). Temperature seasonality control on modern halite layers in the Dead Sea: In situ observations, *Geological Society of America Bulletin*, *129*(9/10),1181–1194. doi: 10.1130/B31661.1.

Smit, J., Brun, J. P., Cloetingh, S., & Ben-Avraham, Z. (2008). Pull-apart basin formation and development in narrow transform zones with application to the Dead Sea basin. *Tectonics*, *27*, TC6018. doi:10.1029/2007TC002119.

Smit, J., Brun, J.-P., Cloetingh S., & Ben-Avraham Z. (2010). The rift-like structure and asymmetry of the Dead Sea Fault, *Earth and Planetary Science Letters*, *290*, 74–82. doi:10.1016/j.epsl.2009.11.060

Smit, J., Brun, J. P., Fort, X., Cloetingh, S., & Ben-Avraham, Z. (2008). Salt tectonics in pull-apart basins with application to the Dead Sea Basin. *Tectonophysics*, *449*, 1–16, doi:10.1016/j.tecto.2007.12.004

Smoot, J. P., & Lowenstein, T. K. (1991). Depositional environments of non-marine evaporites. In J. L. Melvin (Ed.), *Evaporites, petroleum and mineral resources, Developments in Sedimentology* (Vol. 50, Chapter 3, pp. 189–347). Amsterdam: Elsevier.

Snedden, J. W., & Galloway, W. E. (2019). *The Gulf of Mexico Sedimentary Basin: Depositional evolution and petroleum applications*. Cambridge: Cambridge University Press, 326 pp.

Snedden, J. W., Norton, I. O., Hudec, M. R., & Peel, F. (2019). Paleogeographic Reconstruction of the Louann Salt Basin, 37th Annual GCSSEPM Foundation Perkins-Rosen Research Conference. *Salt Tectonics, Associated Processes, and Exploration Potential: Revisited: 1989–2019* (pp. 24–27). Abstracts.

Sonnenfeld, P., & Al-Aasm, I. (1991). The Salina evaporites in the Michigan Basin. In P. A. Catacosinos & Paul A. Daniels, Jr. (Eds.), *Early sedimentary evolution of the Michigan Basin* (Vol. 256, pp. 139–154). Geological Society of America, Special Publications.

Sonnenfeld, P., & Perthuisot, J.-P. (1989). *Brines and evaporites, Short Course in Geology, Volume 3, 28th International Geological Congress*. Washington, DC: American Geophysical Union, 126 pp.

Stein, M. (2001). The sedimentary and geochemical record of Neogene-Quaternary water bodies in the Dead Sea basin-inferences for the regional paleoclimatic history. *Journal of Paleolimnology*, *26*(3), 271–282.

Stein, M. (2012). The ICDP Dead Sea Deep Drilling Project. Report GSI/23/2012, Ministry of Energy and Water Resources, *Geological Survey of Israel*, 13 pp.

Stein, M. (2014). The evolution of Neogene-Quaternary water-bodies in the Dead Sea Rift Valley, in Dead Sea Transform fault system: Reviews. In Z. Garfunkel, Z. Ben-Avraham, and E. Kagan (Eds.), *Modern Approaches in Solid Earth Sciences* (Vol. 6, pp. 279–316). Dordrecht: Springer. doi 10.1007/978-94-017-8872-4_10

Stein, M., Ben-Avraham, Z., & Goldstein, S. L. (2011). Dead Sea cores: A window into past climate and seismicity. *EOS*, *92*(49), 453–454.

Stein, M., Torfstein, A., Gavrieli, I., & Yechieli, I. (2010). Abrupt aridities and salt deposition in the post-glacial Dead Sea and their North Atlantic connection. *Quaternary Science Reviews*, *29*(3), 567–575.

Steinhorn, I. (1983). In situ salt precipitation at the Dead Sea. *Limnology Oceanography*, *28*(3), 580–583.

Stern, R. J., & Johnson, P. (2010). Continental lithosphere of the Arabian Plate: A geologic, petrologic, and geophysical synthesis. *Earth-Science Reviews*, *101*, 29–67.

Szatmari, P., Carvalho, R. S., & Simões, I. A. (1979). A comparison of evaporite facies in the late Paleozoic Amazon and the Middle Cretaceous South Atlantic salt basins. *Economic Geology*, *74*(2), 432–447.

Szynkiewicz, A., Moore, C. H., Glamoclija, M., Bustos, D., & Pratt, L. M. (2010). Origin of coarsely crystalline gypsum domes in a saline playa environment at the White Sands National Monument, New Mexico. *Journal of Geophysical Research*, *115*, F02021. doi:10.1029/2009JF001592

Taberner, C., Cendón, D. I., Pueyo, J. J., & Ayora, C. (2000). The use of environmental markers to distinguish marine vs. continental deposition and to quantify the significance of recycling in evaporite basins. *Sedimentary Geology*, *137*, 213–240.

Taviani, M. (1985). Red Sea Miocene evaporites: A discussion about their possible relationships with the Mediterranean ones. *CIESM*, *29*(2), 125–126.

Taylor, J. C. M. (1985). *The role of evaporites in hydrocarbon exploration. Joint Association of Petroleum Exploration Courses (UK)*. Japec Course No. 39.

Taylor, J. C. M. (1998). Upper Permian–Zechstein. In K. W. Glennie (Ed.), *Petroleum geology of the North Sea: Basic concepts and recent advances* (4th ed., pp. 174–211). Oxford, UK: Blackwell Publishing Limited.

ten Brink, U. S., & Ben-Avraham, Z. (1989). The anatomy of a pull-apart basin: Seismic reflection observations of the Dead Sea Basin. *Tectonics*, *8*(2), 333–350.

Torfstein, A. (2008). Brine–freshwater interplay and effects on the evolution of saline lakes: The Dead Sea Rift terminal lakes. *Geological Survey of Israel*, Report GSI /20/08, 290 pp.

Torfstein, A. (2017). The Amora formation, Dead Sea Basin. In Y. Enzel & O. Bar (Eds.), *Quaternary of the levant-environments, climate change, and humans* (pp. 91–98). Cambridge, UK: Cambridge University Press. doi: https://doi.org/10.1017/9781316106754.010

Torsvik, T. H., & Cocks, L. R. M. (2012). From Wegener until now: The development of our understanding of Earth's Phanerozoic evolution. *Geologica Belgica*, *15*(3), 181–192.

Torsvik, T. H., & Cocks, L. R. M. (2017). *Earth history and palaeogeography*. Cambridge, UK: Cambridge University Press, 317 pp.

Trindade, L. A. F., Dias, J. L., & Mello, M. R. (1995). Sedimentological and geochemical characterization of the Lagoa Feia formation, Rift Phase of the Campos Basin, Brazil. In B. J. Katz (Ed.), *Petroleum source rocks, Casebooks in Earth Sciences* (pp. 149–165). Berlin, Heidelberg: Springer.

Urgeles, R., Camerlenghi, A., Garcia-Castellanos, D., De Mol, B., Garces, M., Verges, J., et al. (2011). New constraints on the Messinian sealevel drawdown from 3D seismic data of the Ebro margin, western Mediterranean. *Basin Research*, *23*, 123–145, 10.1111/j.1365-2117.2010.00477.x

U.S. Geological Survey (2021). Mineral commodity summaries 2020. *U.S. Geological Survey*, 200 p. https://doi.org/10.3133/mcs2020

Veslud, C. L. C., Bourgeois, O., Diraison, M., & Ford, M. (2005). 3D stratigraphic and structural synthesis of the Dannemarie basin (Upper Rhine Graben). *Bulletin de la Societe Geologique de France*, *176*(5), 433–442.

Waldmann, N., Stein, M., Ariztegui, D., & Starinsky, A. (2009). Stratigraphy, depositional environments and level reconstruction of the last interglacial Lake Samra in the Dead Sea basin. *Quaternary Research*, *72*(01), 1–15. doi:10.1016/j.yqres.2009.03.005

Warren, J. K. (1989). *Evaporite sedimentology: Importance in hydrocarbon accumulation*. New Jersey, USA: Prentice Hall Advanced Reference Series, 285 p.

Warren, J. K. (1991). Sulfate dominated sea-marginal and platform settings. In J. L. Melvin (Ed.), *Evaporites, petroleum and mineral resources, Developments in Sedimentology* (Vol. 50, Chapter 2, pp. 477–533). Amsterdam: Elsevier.

Warren, J. K. (2006). *Evaporites: Sediments, resources and hydrocarbons, 1035p*. Berlin: Springer Verlag.

Warren, J. K. (2010). Evaporites through time: Tectonic, climatic and eustatic controls in marine and nonmarine deposits. *Earth and Science Review*, *98*, 217–268.

Warren, J. K. (2014). *Evaporite sedimentology: Importance in hydrocarbon accumulation*. New Jersey, USA: Prentice Hall Advanced Reference Series, 285 p.

Warren, J. K. (2016). *Evaporites—A geological compendium*. Berlin: Springer, 1807 pp. ISBN 978-3-319-13511-3.

Warren, J. K. (2018). Evaporites. In W. M. White, W. H. Casey, B. Marty, & H. Yurimoto (Eds.), *Encyclopedia of geochemistry* (pp. 464–471). Switzerland: Springer.

Warren, J. K., & Kendall, C. G. St. C. (1985). Comparison of sequences formed in marine sabkha (subaerial) and salina (subaqueous) settings—Modern and ancient. *American Association of Petroleum Geologists Bulletin*, *69*(6), 1013–1023.

Weinberger, R., Begin, Z. B., Waldmann, N., Gardosh, M., Baer, G., Frumkin, A., & Wdowinski, S. (2006). Quaternary rise of the Sedom diapir, Dead Sea basin. In Y. Enzel, A. Agnon, & M. Stein (Eds.), *New frontiers in Dead Sea paleoenvironmental research* (Vol. 401, pp. 33–51). Geological Society of America Special Paper. doi: 10.1130/2006.2401(03).

Wimmenauer, W. (1974). The alkaline provinces of Central Europe and France. In H. Sörensen (Ed.), *The Alkaline Rocks* (pp. 238–271). New York: Wiley Intersciences.

Wright, W. R. (2011). Pennsylvanian paleodepositional evolution of the greater Permian Basin, Texas and New Mexico: Depositional systems and hydrocarbon reservoir analysis. *American Association of Petroleum Geologists Bulletin*, *95*(9), 1525–1555.

Zak, I. (1974). Sedimentology and bromine geochemistry of marine and continental evaporites in the Dead Sea Basin. In *Fourth International Symposium on Salt* (pp. 349–361). Northern Ohio Geological Society.

Zak, I. (1997). Evolution of the Dead Sea brines. In T. M. Niemi, Z. Ben-Avraham, & J. R. Gat (Eds.), *The Dead Sea: The lake and its setting* (pp. 133–144). Oxford, UK: Oxford University Press.

Zak, I., & Bentor, Y. K. (1972). Some new data on the salt deposits of the Dead Sea area, Israel. In G. Richter-Bernburg (Ed.), *Geology of saline deposits, Proceedings of the Hanover Symposium*, 15–21 May 1968, International Union of Geological Sciences, United Nations Educational, Scientific and Cultural Organization, *Earth Sciences*, *7*, 137–146.

Zak, I., & Freund, R. (1980). Strain measurements in eastern marginal shear zone of Mount Sedom salt diapir, Israel. *American Association of Petroleum Geologists Bulletin*, *64*(4), 568–581.

Zeng, H., Zhu, X., Zhu, R., & Zhang, Q. (2012). Guidelines for seismic sedimentologic study in non-marine postrift basins. *Petroleum Exploration and Development*, *39*(3), 295–304.

Ziegler, A. M., Hansen, K. S., Johnson, M. E., Kelly, M. A., Scotese, C. R., & Van der Voo, R. (1977). Silurian continental distributions, paleogeography, climatology, and biogeography. *Tectonophysics*, *40*, 13–51.

Ziegler, P. A., & Dézes, P. (2005). Evolution of the lithosphere in the area of the Rhine rift system. *International Journal of Earth Sciences (Geol Rundsch)*, *94*, 594–614.

Zilberman, E., Amit, R., Porat, N., Enzel, Y., & Avner, U. (2005). Surface ruptures induced by the devastating 1068 AD earthquake in the southern Arava valley, Dead Sea Rift, Israel. *Tectonophysics*, *408*, 79–99.

12

Genetic Models for Basinwide Evaporites

ABSTRACT

Genetic models for basinwide evaporite sequences are based on observations of giant salt basins like the Mediterranean Sea, where the Messinian salinity crisis led to halite accumulation during the Upper Miocene. Researchers have proposed various genetic models for basinwide evaporites, such as salt deposition in shallow-water environments of shallow basins, shallow-water environments of deep basins, deepwater environments of deep basins, and composite models. Important basins with basinwide evaporites include the Permian Basin in the southwest United States and the carbonate platform evaporite basins in the Persian Gulf. The Gulf of California is also known for its Late Miocene evaporite layers that extend from the continental shelf to the continent-oceanic crust transition in the deepwater region.

Basinwide subaqueous evaporites are the largest and thickest salt basins on Earth, deposited in diverse sedimentary environments. These giant salt accumulations are found in sabkhas, lagoons, continental platforms, and deep basins, with bathymetry ranging from tens to hundreds of meters. The Jurassic salt deposits in the Gulf of Mexico, the Late Aptian salt in the South Atlantic Ocean, the Upper Miocene salt in the Red Sea, and the Messinian salt in the Mediterranean Sea are basinwide evaporite deposits mainly located in the deepwater region. Chapters 1 to 5 in Mohriak (2025) will discuss in detail how these giant evaporite provinces formed diapiric salt structures that locally overlay the oceanic crust.

This chapter focuses on basinwide evaporite deposits that form bedded salt in sedimentary basins associated with convergent and strike-slip settings, such as the Permian Basin in the southwestern United States, the Late Paleozoic to Mesozoic evaporite sequences in the Persian Gulf, and the Neogene evaporite deposits in the Gulf of California. The geological analysis of these selected salt basins includes a short description of the different stages of their tectonic evolution.

The Zagros region is currently associated with a convergent setting involving the subduction of the Arabian plate under Eurasia. During the Neoproterozoic, the Hormuz salt was deposited in an extensional phase, followed by a divergent margin configuration throughout most of the Paleozoic and Mesozoic. The Guaymas salt basin in the central segment of the Gulf of California is associated with a transtensional setting that developed pull-apart rift basins in the Upper Miocene. The Gulf of California illustrates the development of basinwide evaporite deposits on a transform margin with marine environments from the rift edge to the boundary between continental and oceanic crust.

Until the beginning of the 21st century, no reliable descriptions of Quaternary sedimentary models for basinwide subaqueous evaporite deposits were available in the geological literature (Kendall, 1978, 1992, 2010; Warren, 1992, 2006, 2010, 2016). Since the 1970s, several genetic models have been proposed for giant salt basins worldwide. One of these models is the desiccation model of deep oceanic settings, initially proposed for the Mediterranean Sea (e.g., Hsü, 1972, 1977; Hsü et al., 1973, 1977, 1978). Several researchers who have thoroughly analyzed the original evidence presented by the model proponents have recently challenged the desiccation model, which has been extensively applied as a paradigm for the South Atlantic and Gulf of Mexico salt basins (see Chapter 5 in Mohriak, 2025).

Subaqueous marine environments deposited basinwide evaporite sediments in giant salt basins of the ancient past, forming relatively pure, massive units composed of

Salt in the Earth Sciences: Evaporite Rocks and Salt Deposition, Geophysical Monograph 286, First Edition. Webster Mohriak.

DOI: 10.1002/9781119405245.ch12

thick halite sequences with minor intercalations of carbonate, sulfate, and more soluble chloride minerals, such as carnallite and sylvite (Jackson et al., 2015, 2019; Warren, 2006, 2010, 2016). The evaporite cycles are distinguished by lithological variations and identified by the analysis of electric logs and ditch samples, as these rocks are rarely cored during drilling activities in petroleum exploration projects. Drilling activities in the South Atlantic margins primarily focus on evaluating pre-salt carbonate reservoirs, which are estimated to contain around 60% of all reservoir rocks worldwide and are more complex and heterogeneous than their siliciclastic counterparts (Burchette, 2012).

Basinwide evaporites are formed in a spectrum of different environments, from the widespread deposition of gypsum layers in continental sabkhas, lagoons, and carbonate platforms to thick halite layers deposited in deep marine settings. Carbonate-evaporite deposits on shallow platforms are characterized by thick layers of evaporite minerals that reflect brine saturation. The regional occurrence of sulfate deposits in some basins indicates that the brine concentration remained well below the point of depositing halite. This is the case of the Upper Permian Khuff Fm. in the Arabian Peninsula or the Upper Jurassic Hith Anhydrite in the Persian Gulf (Ehrenberg et al., 2007; Warren, 1991).

The evaporite facies associated with basinwide salt deposits are related to different settings (Figure 12.1). The sedimentary environments include (1) supratidal sabkhas to subtidal lagoons, such as the proximal, subaerial facies of the Jurassic Arab Fm. in central Abu Dhabi (Grötsch et al., 2003); (2) bedded evaporites in shallow carbonate platforms, such as the proximal, subaqueous facies of the Jurassic Arab Fm. in central Abu Dhabi (Grötsch et al., 2003); and (3) deep basins that accumulated bedded carbonate, sulfate, and chloride minerals, such as the carbonate platform and deep basin of the Permian Basin in the southwestern United States (Scholle, 2020; Scholle et al., 2007), and the resedimented clastic evaporite deposits, such as the gypsarenites in the Mediterranean Sea (e.g., Manzi et al., 2005; Roveri et al., 2006, 2019).

Basinwide carbonate-evaporite platforms of ancient salt basins did not accumulate huge thicknesses of mobile salt. Both the Permian Basin in the United States and the Permian and Jurassic evaporites in the Arabian Peninsula are characterized by bedded salt, mainly gypsum, with subordinate halite intercalated with carbonate rocks (Figure 12.1). These events reflect repeated marine incursions into the basin, forming several cycles according to the brine concentration (see Chapter 8).

Although the bedded evaporite sequences in the Permian Basin include halite deposits in the Castile and Salado formations, these layers have not been mobilized significantly and have not formed diapiric structures. The evaporite sequences are characterized by relatively undeformed stratification, although locally there are folds and the development of incipient pillows, indicating some salt flowage (Kendall, 1978, 1984, 1992; Taylor, 1985; Warren, 1991).

Some genetic models for the Mediterranean Sea that salt basin involves erosion, reworking, transport, and redeposition of clastic evaporite sequences in deep water settings. Manzi et al. (2005) discussed the correlation of shallow to deepwater facies in the Adriatic foredeep, indicating that they reworked and deposited the distal margin salt as clastic evaporite sequences. Lugli et al. (2015) reexamined the borehole cores from the Upper Evaporites in the deepwater region of the Mediterranean Sea and observed these rocks consist mainly of clastic evaporites deposited subaqueously, with no evidence of basin desiccation. Deepwater deposits near the platform edge (often rimmed by a carbonate reef) are commonly characterized by slope deposits, resulting in debris and talus that indicate deformation by gravity gliding and sliding. Submarine reworking of the lagoon and platform evaporites results in resedimented evaporites that are remobilized basinward (Figure 12.1).

Remobilized evaporite deposits are interpreted as a wedge of reworked carbonates and clastic evaporites forming slumps and turbidites in carbonate ramps (Figure 12.1). Outcrops and cores of the Zechstein evaporites of the North Sea, the Messinian salt basins of the Mediterranean Sea, and the Permian Delaware Basin in the western United States provide evidence of salt redeposition and recrystallization in deep marine settings (Kendall, 1978, 1992, 2010; Manzi et al., 2005, 2012, 2018, 2020; Taylor, 1984, 1985, 1998; Warren, 2016). Laterally continuous, fine-grained, laminated evaporites in the center of giant salt basins form stratified sequences of different evaporite minerals, suggesting subaqueous sedimentation in a deep basin either as autochthonous deposits or as resedimented allochthonous deposits (Schmalz, 1969; Warren, 2016).

Some evaporite basins with basinwide salt deposits show abrupt changes in depositional setting, intercalating deepwater evaporite facies with shallow water saltern and mudflat units during extreme aridity. The desiccation model suggests these events resulted in the evaporative drawdown and isolation of salt pans in the central part of the depression. Sedimentary structures such as mud cracks and tepees indicate subaerial to shallow water depth during basinwide salt deposition (Hardie & Eugster, 1970, 1971).

12.1. DEPOSITIONAL MODELS FOR BASINWIDE SUBAQUEOUS EVAPORITES

Ancient evaporite basins with thick sequences of evaporites are often interpreted using three basic models involving water depth during salt deposition.

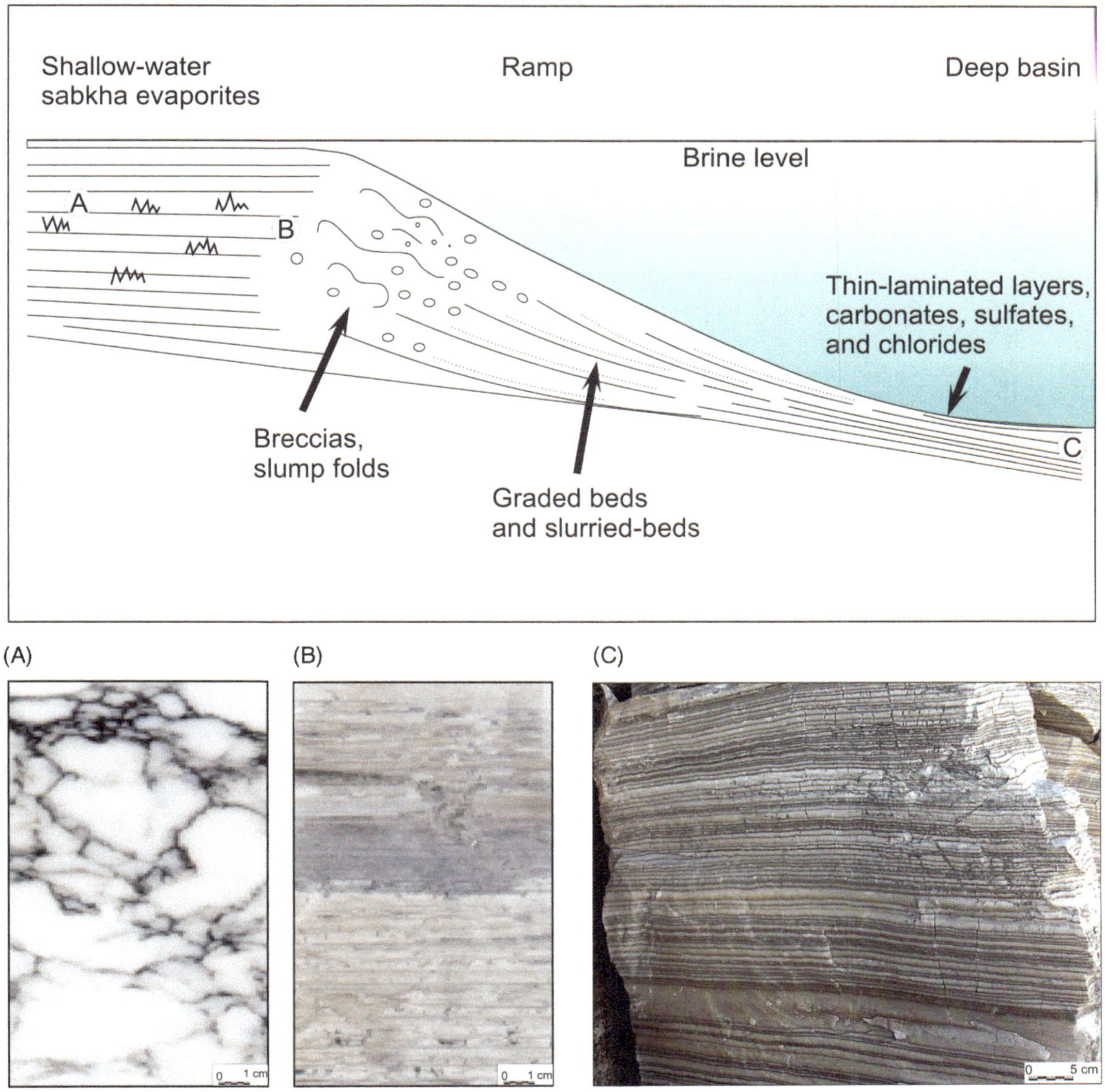

Figure 12.1 Top: The schematic diagram shows depositional environments associated with basinwide carbonate platform-evaporite sequences. Bottom: Lithological examples of basinwide carbonate-evaporite sequences formed at different water depths. The pictures of the different outcrops show evaporite samples from different environments at different scales, such as supratidal sabkha (A), subtidal lagoonal (B), and subtidal deep basin (C). The supratidal chicken-wire anhydrite nodules (A) are associated with subaerial sabkha environments, such as the proximal facies of the Arab Fm. in central Abu Dhabi. Carbonate mud is deposited in shallow water lagoons, such as the Arab Fm., corresponding to subtidal lagoonal sediments (B). The thin-laminated bedded evaporites in the distal margin (C) correspond to couplets of carbonate and sulfate minerals, mainly composed of calcite and gypsum, as observed in the Permian Basin of the southwestern United States. Credits: (A),(B) Grötsch et al. (2003)/ with permission of GeoScienceWorld and (C) adapted from Scholle (2020).

The paleobathymetry in ancient salt basins corresponds to the height of the water column that filled the depression (Kendall, 1984, 1992; Sonnenfeld & Perthuisot, 1989; Warren, 1989). The classical barred basin model was proposed by Ochsenius in the late 19th century, and was commonly used to interpret the development of giant salt basins (see Chapter 8). This seminal model was subsequently modified by several researchers that suggested variations of the original concepts (see reviews by Hsü, 1972; Perthuisot, 1989).

The formations of basinwide salt deposits are associated with some end-member models (Figure 12.2): (A) shallow-basin-shallow water; (B) deep basin-deep water; (C) deep basin-shallow water; and (D) composite models

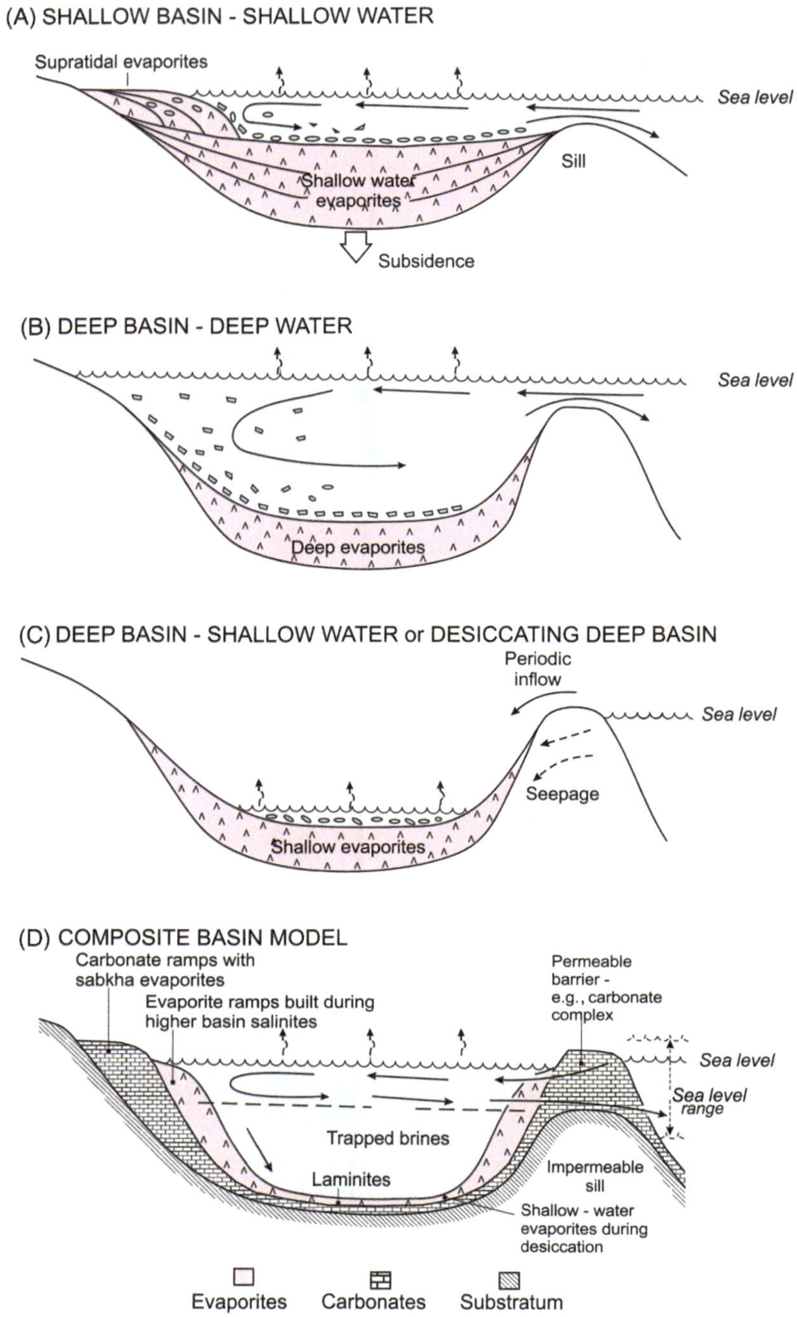

Figure 12.2 The schematic diagrams illustrate the four different schemes for basinwide salt deposition. Model (A) involves shallow sedimentary basins with shallow water, periodically connected to the open sea by a small gap at the entrance; model (B) features deep sedimentary basins with deep water and a barrier or sill that periodically communicates with the open sea; model (C) features deep sedimentary basins with shallow water formed below the global sea level during total basin isolation; and model (D) combines a deep sedimentary basin with a permeable barrier with intermittent communication with the open sea. Credit: Adapted from Taylor (1985).

involving a permeable barrier allowing communication between the barred basin and the open sea (e.g., Taylor, 1985). These simplified conceptual models represent a wide spectrum of basin types that change rapidly during salt deposition events, as evaporites precipitate in an extremely short geological time. These models have implications for other geological features in salt basins, such as deep erosional canyon incisions observed in the Mediterranean Sea during the Messinian (see Chapter 5 in Mohriak, 2025). Some authors suggest a deep basin or shallow water model for the Mediterranean salt deposits, indicating subaerial erosion and fluvial systems (e.g., Madof et al., 2019; Mascle & Mascle, 2019; Roveri & Manzi, 2006; Roveri et al., 2008), while others propose that these erosional features formed in a subaqueous environment due to submarine channels and turbidity currents (e.g., Andreetto et al., 2021; Christeleit et al., 2015; CIESM, 2008; Roveri et al., 2014).

12.1.1. Shallow Water-Shallow Basin Model (Model A)

The shallow water-shallow basin model assumes an isolated, shallow sedimentary basin or lagoon connected to the sea through a narrow channel (Figure 12.2-A). The lagoon evaporates and dries out under an arid climate with little freshwater input from fluvial or pluvial systems. The salinity increases and brine concentration reaches the saturation point of different evaporite minerals, which accumulate at the bottom of the lake or its margins (Perthuisot, 1989).

The evaporite sequences are dominated by shallow water sedimentation, leading to almost dry salt flats. Salt layers are deposited as shoaling saltern and mudflat cycles in a shallow basin, with water columns less than a few meters. The evaporite accumulation rate keeps pace with the subsidence rate, resulting in salt deposits in a continuously shallow environment. Repeated marine incursions result in the accumulation of thick sequences of evaporites stacked on the lake's floor, with sedimentary structures indicating subaerial to shallow water environments throughout the evaporation cycle (Kendall & Harwood, 1996; Kinsman, 1969).

The basin subsidence in the shallow water-shallow basin model is slow enough to maintain restrictions above the wave base and a reduced water column in the basin throughout the evaporite cycles (Figure 12.2). Van den Belt and De Boer (2007) postulated that gradual subsidence mechanisms in the shallow water depositional model result from an isostatic response of changing brine density (1.2 g/cc) to salt mineral density (> 2.2 g/cc). Sedimentary environments usually start with continental, lacustrine, and fluvial deposits overlain by carbonate and evaporite rocks. In passive continental margins, the marine invasion follows the rift phase and oceanic crust formation, as envisaged by several authors for the South Atlantic, the Gulf of Mexico, and the Red Sea (Kukla et al., 2018; Warren, 2016). The lagoonal evaporite successions may pass up section and basinward into marine evaporites or basinal carbonate and siliciclastic successions, as characterized in the Permian Basin (see Chapter 11, Figure 11.8).

12.1.2. Deepwater-Deep Basin Model (Model B)

The deepwater-deep basin model (Figure 12.2-B) assumes that a partially isolated sedimentary basin forms a perennial lake or inland sea separated from the global ocean by a barrier or "sill." Seawater from the nearby ocean flows into the basin as a perennial surface inflow during high sea level intervals, interrupted when sea levels fall below the barrier, resulting in brine concentration and salt precipitation (Sonnenfeld & Perthuisot, 1989; Taylor, 1985). Evaporite deposits in the basin center are formed in a relatively deepwater setting, reaching tens of meters between the coastline and the shelfbreak in rimmed platforms and hundreds of meters in the distal margin or basin depocenter (Schmalz, 1969; Taylor, 1985).

The basin center usually starts as a deepwater environment, with anoxic deposits formed when salinity increases to the saturation point of precipitating carbonate minerals. Laminated deepwater normal salinity carbonates and siliciclastic rocks underlie the basinal evaporites in many basins, forming organic-rich layers. Millimeter-laminated fine-grained salts (mainly halite and gypsum) dominate the central areas of some large evaporite basins, while reworked platform carbonates and evaporites deposited as slumps, debris, and turbidites dominate the basin slopes (Kendall, 1978; Lugli et al., 2010, 2013, 2015; Warren, 1991). The deepwater-deep basin model is characterized by a platform with saltern evaporites that may accumulate thick sequences during marine transgressions (see Chapter 11, Figure 11.8).

Depending on the duration of the evaporitic conditions, the evaporite successions may fill the basin to a level equivalent to the original water depth attained before the basin started to evaporate. The evaporite fill starts as deepwater laminated and resedimented salts, passing up-section into shallow-water salterns, mudflats, or continental playas, indicating an overall regressional episode (Handford, 1981a, 1991; Hardie, 1991; Sonnenfeld & Perthuisot, 1989).

The Delaware Basin in southwest Texas is an ancient example of this basin type, with bedded evaporites and some halite layers. The basin depocenter was filled and shallowed into the Salado Formation, which is mainly composed of anhydrite and halite minerals (Ewing, 2017; Hovorka et al., 2007). Another example of this type of basin with deepwater evaporites at the basin center is the Permian Zechstein salt basin in Europe. The basinwide

Zechstein evaporite deposits are overlain by Triassic sedimentary sequences involving shallow marine carbonates and continental sandstones, indicating a shallowing of the depositional settings (Glennie, 1998; Taylor, 1998).

12.1.3. Shallow Water-Deep Basin Model (Model C)

The shallow water-deep basin model (Figure 12.2-C) is a concept based on observations of salt basins that accumulate evaporite sediments in depressions below the global sea level, such as the Dead Sea (see Chapter 9). Infiltrating seawater from the adjacent oceanic basin compensates for the evaporative drawdown in these basins (Ryan, 2008; Schreiber & Hsü, 1980; Warren, 1989). The water feeds the depression through marginal seeps and occasional waterfalls through channels that link the depression with a larger water body (Kendall, 1992, 2010).

Deepwater marine sediments often underlie the evaporite sequences, deposited before the basin became isolated from the continuous supply of global seawater. As the brine level is maintained below the nearby water masses due to evaporative drawdown, deepwater sediments often overlay the evaporite layers, which are immediately deposited when the connection with the global sea is reset and open-marine waters reenter the basin. The Mediterranean Sea is considered the best example of an oceanic basin filled by Messinian evaporites in a short time interval and refilled with oceanic waters in the Upper Miocene (Hsü, 1972, 1977; Hsü et al., 1973; Schreiber & Hsü, 1980).

12.1.4. Deepwater-Deep Basin Composite Model (Model D)

Taylor (1985) proposed a composite basin model that incorporated a carbonate barrier at the shelf edge of a deep basin to impede the communication of the evaporitic lagoon with the oceanic basin with normal salinity (Figure 12.2-D). The model assumed that evaporite deposits accumulated in deep basins near a shelf edge barrier made of permeable sedimentary rocks or volcanic constructions. The evaporite-carbonate platforms formed evaporite cycles during basin dry-out episodes when communication with the open sea was greatly restricted by the barrier. This model is likely applicable to evaporite deposits in the Permian Basin in the southwestern United States (Castille and Salado Fm.) and the Early Cretaceous evaporite deposits in the northern Gulf of Mexico (Ferry Lake Anhydrite). This model suggests basinwide accumulation of hundreds of meters of carbonate, anhydrite, and halite sequences along a belt that stretched for hundreds of kilometers in the back reef (see Chapter 11, Figure 11.4).

The evaporating basin in this model is partially cut off from the main body of ocean water by a permeable barrier, usually a carbonate bank, reef buildup, or volcanic rocks (Figure 12.2-D). This allows intermittent flow as seawater seeps through the barrier. The hypothesis requires that the output of water in the inland sea or gulf is greater than the input of freshwater by rivers, rainfall, or reflux from the sea. Unlike the "narrow channel model" described in Model A, when the water in the isolated basin evaporates, it is free to drop below the adjacent sea level, resulting in a depression filled with air instead of water. This process produces a pressure differential between the two sides of the permeable barrier. The hydrodynamic laws ensure that water flows in just one direction, from the region with higher pressure (the main body of water in the adjacent ocean) into the isolated basin in the depression (Sonnenfeld & Perthuisot, 1989).

Salt deposition episodes in Holocene lagoons occur when carbonate banks or siliciclastic islands above sea level form a semipermeable barrier, allowing intermittent water communication between the gulf and the lagoon, such as the MacLeod Lake in Australia or the Sabkha El Melah in Tunisia (see Chapter 8).

12.1.5. Basinwide Salt Sedimentation in Subaerial and Subaqueous Environments

The isostatic principle plays a crucial role in the genetic models for basinwide salt deposition in sedimentary basins. This principle suggests that an initially shallow basin filled rapidly with salt deposits will deepen through isostatic adjustments, resulting in increased water depth and salt thickness (Montaron & Tapponier, 2009; Van den Belt & de Boer, 2007). The subsidence results in additional depositional space, eliminating the need for desiccation and refilling the basin with gigantic waterfalls, as proposed by Hsü (1972) for the Mediterranean Basin (Lugli et al., 2015).

Many researchers advocate a trilogy of supratidal, intertidal, and subtidal lithological facies for the modern and ancient sabkha evaporites (see Chapter 8). The distinction between depositional environments and paleowater depths can be qualitatively inferred from the analysis of sedimentary structures. The following fingerprints suggest a shallow water or sabkha environment for bedded evaporites: laminated sapropelic shales with gypsum and anhydrite lenses, mud cracks, tepee structures, corrosion planes, dissolution features, stromatolites, nodular anhydrite with chicken-wire texture, enterolithic folding in anhydrite and polyhalite zones, selenite crystals, and chevron and displacive hopper crystals in halite zones. These sedimentary structures are suggestive of a very shallow evaporite basin, with a water column ranging from 0.5 to 2 m (Handford, 1991; Schreiber, 1986; Schreiber et al., 1982; Warren, 2014).

Subaqueous evaporites have been described in the literature with a wide range of stratigraphic ages and water depths. Salterns are recognized in several regions and ages, such as the Holocene MacLeod Lake in NW Australia. Shallow platform sulfates with distal facies of basinal chlorides are observed in the sabkha, lagoon, carbonate ramp, and deep basin of the Palo Duro and Permian basins in the southwest United States.

The basinal evaporites of ancient saline giants, such as the Permian Zechstein Basin, the Aptian South Atlantic, or the Messinian Mediterranean Sea, have been considered by several authors to have formed in subaqueous deepwater environments, as indicated by the massive deposits of stratified evaporites (Christeleit et al., 2015; Kukla et al., 2018; Warren, 2006, 2016).

The pure subaqueous evaporite deposits, especially chemically precipitated gypsum and halite, are subject to early recrystallization, flowage, and intense burial diagenesis, and the remaining evaporitic units display only indistinct relics of the original texture. This contrasts with the less intense burial overprinting observed in sediments from sabkha sequences, where the carbonate or siliciclastic matrix tends to be more rigid and less mobile than the sulfate or chloride minerals (Handford, 1981a; Schreiber, 1978; Schreiber & El Tabakh, 2000; Schreiber et al., 1982).

During the Messinian salinity crisis (MSC), shallow-water evaporite sediments were deposited in depressions at the edges of the Mediterranean platform, forming perched basins separated from the deep ocean realm (Miramontes et al., 2023; Riding et al., 1998; Schreiber & Hsü, 1980). Some authors advocate that the Mediterranean Sea peripheral basins recorded subaerial sabkha and salt pan sequences during the MSC, formed simultaneously with the massive subaqueous gypsum and halite deposits in the deeper parts of the basin (e.g., Hsü et al., 1973, 1977, 1978).

A sabkha origin for the basinwide sulfate deposits in the Mediterranean Sea and other basins such as the Permian Basin in the southwest United States is often envisaged based on a simplistic interpretation of the presence of nodular gypsum or anhydrite in borehole cores (see Chapter 5 in Mohriak, 2025). Detailed petrographic analysis of these cores indicates that the gypsum rocks have undergone deep diagenetic changes related to (1) dehydration to anhydrite because of burial; and (2) subsequent re-hydration to microcrystalline gypsum during exhumation and contact with phreatic waters (Schreiber et al., 1982; Sonnenfeld & Perthuisot, 1989). Sometimes, the original facies are still visible, but when the diagenesis is strong, the original texture can be obliterated and only a diagenetic gypsum facies, known as saccharoidal, chicken-wire, or alabastrine, is visible (Hardie et al., 1985; Holliday, 1970; Jarzyna et al., 2022). During the MSC, some researchers have interpreted these facies as forming in a sabkha environment, leading to arguments for a supratidal paleo-water depth (e.g., Garrison et al., 1978; Gvirtzman & Buchbinder, 1978).

However, several researchers have challenged the original hypothesis of sabkha deposits and the shallow water or deep basin model proposal for the MSC, highlighting several interpretation problems related to the original descriptions of the DSDP borehole cores and the possible unrealistic reconstruction of the paleo-water depths (e.g., Hardie & Lowenstein, 2004; Lugli et al., 2015). In some cases, the evaporite rocks in the DSDP cores are diagenetic components of evaporitic cement that filled the interstices of porous carbonates in evaporitic sequences. In other cases, the evaporites are recrystallized deposits from the reworking of previous evaporite deposits (Hardie & Lowenstein, 2004; Manzi et al., 2005; Roveri & Manzi, 2006). Lugli et al. (2013, 2015) suggest that the anhydrite nodules could be a diagenetic product resulting from clastic subaqueous gypsum layers, while the original salt deposits might have formed in deeper environments. Many studies disagree with the sabkha environments even in the peripheral basins surrounding the Mediterranean Sea. The gypsum layers in these basins might correspond to clastic evaporite deposits that formed underwater as reworked gypsarenites (Hardie & Lowenstein, 2004; Manzi et al., 2005; Roveri et al., 2014).

The proper understanding of the sedimentary environments associated with thick evaporite layers accumulated in giant salt basins requires the integration of geophysical datasets (seismic profiles) and geological datasets (well logs and cores). Some classical giant salt basins, such as the Red Sea, North-Central Europe, the South Atlantic, the Gulf of Mexico, and the Mediterranean Sea, present additional difficulties, as the analysis of the bathymetry during salt deposition requires removing the deformation that affected the original geometry.

12.1.6. Salt Deposition and Global Sea Level Fluctuations

Sea level variations can effectively concentrate brine and deposit evaporites in restricted lagoons and adjacent sabkhas. Global sea level fluctuations control two main types of evaporite deposits observed in ancient basins: subaerial sabkha and subaqueous salt deposits (e.g., Blatt et al., 1980; Shearman, 1978).

Figure 12.3 shows a schematic drawing of a lagoon on a carbonate platform separated from the deep basin by a carbonate barrier or sill. Carbonate banks or islands also occur at the lagoon edge near the tidal flats, as observed in the Persian Gulf (Evans, 2011). To isolate the lagoon from the open sea and form thick evaporitic units, the crest of the carbonate reef should be positioned at or just

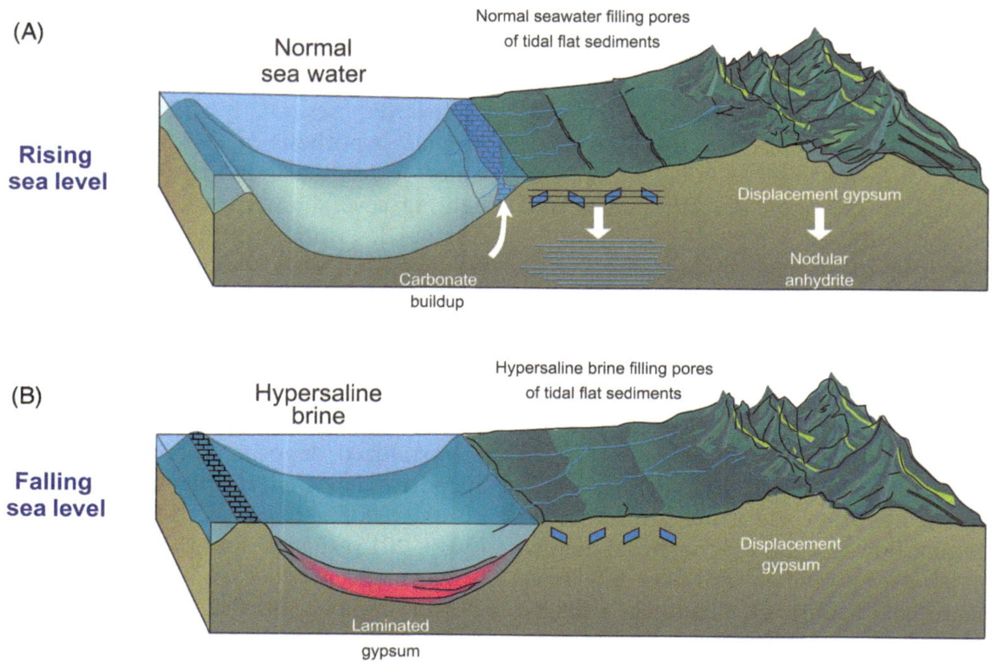

Figure 12.3 The block diagrams depict a barred basin model with a carbonate bank or reef separating the lagoon from the open ocean during the episodes of rising and falling sea levels. Carbonate buildups and islands mark the lagoon margin, near the tidal flats. (A) During sea-level rise, marine waters flood the coastal sabkha environment, forming gypsum crystals that displace the matrix. During basin subsidence, the displacement gypsum transforms into nodular anhydrite layers. (B) The carbonate bank may isolate the lagoon during sea-level fall, causing brine concentration and gypsum precipitation in subaqueous settings of a shallow lagoon, gulf, or deep inland sea. Credit: Adapted from Blatt et al. (1980).

above sea level (Figure 12.3). To restrict the free exchange of marine and lagoon waters, it is necessary to periodically close any channels that connect the seawater with the lagoon. If the channel or strait is wide and deep enough to allow permanent communication between water bodies, the brine is not concentrated enough to precipitate evaporites, either subaqueously (salterns) or in the subaerial mudflats and shallow lakes.

If the sill constitutes a shallow and narrow barrier to isolate the basin, two mechanisms for salt precipitation are envisaged. During episodes of sea level rise, the carbonate bank is below the wave base and does not obstruct the communication between the open sea and the lagoon. Marine waters enter the lagoon and flood the coastal plain and sabkhas, accumulating only carbonate sediments (Figure 12.3A). The evaporation and brine concentration precipitate gypsum in the pore interstices of the supratidal carbonate sands, displacing the matrix and transforming into nodular anhydrite during burial (Figure 12.3A).

If sea-level fall occurs, the carbonate rim at the outer edge of the lagoon becomes subaerial, forming an island a few meters above sea level, isolating the lagoon from the gulf or seaway (Figure 12.3B). The brines are concentrated, and gypsum layers are precipitated in bathymetric lows within the lagoon. Locally, the depocenters may include more soluble minerals, as seen in the Punta Gorda Formation in the South Florida salt basin. Exploratory wells in some depocenters have proven the presence of halite, although the Punta Gorda evaporites are generally represented by anhydrite sequences deposited on an Early Cretaceous carbonate platform (see Chapter 11).

The water column in basins with thick salt deposits is challenging to interpret quantitatively due to the absence of sedimentary structures and evaporite minerals being affected by diagenetic alterations (Handford, 1981a; Warren, 1989, 1991, 2006, 2016). The MSC in the Mediterranean Sea has been estimated to correspond to a sea level drawdown of 1.5 to 2 km, according to several basinwide evaporite models (e.g., Ryan, 2008).

This model can also be applied to the northern Gulf of Mexico, where the Early Cretaceous hinge line separates the carbonate platform from the deep basin and is associated with reefal buildups at the shelf edge (see Chapter 11, Figure 11.4). A stacked mudflat sequence with intercalated sabkha or salt pan sequences may occur simultaneously with subaqueous gypsum or halite deposition in deeper parts of evaporite basins, as recorded in the Permian Basin (Handford, 1981a, 1981b, 1991).

In a critical review of ancient and modern halite sedimentation, Warren (2010) considered that the brine depth for halite deposition in most salinas was less than 5 m. Quantitative analysis of subsidence and paleobathymetry estimates for the Mediterranean Sea indicate that the basin margin evaporites were formed in shallow water, whereas the evaporites in the basin center were more probably deposited in deepwater settings (Christeleit et al., 2015; CIESM, 2008; Manzi et al., 2005, 2018, 2020; Meilijson et al., 2019; Ryan, 2008).

12.1.7. Sedimentation Rates and Water Level Variations in Salt Basins

The sedimentation rate of Messinian evaporites in the Mediterranean Sea depocenter has been calculated by Meilijson et al. (2018), who recognized evaporitic cycles from seismic data and used a stratigraphic framework that has been challenged by some authors. Based on multiple borehole analyses that constrained the duration of the salt in a shorter time window, Manzi et al. (2018) calculated a salt accumulation rate similar to that in modern settings. Furthermore, other researchers (e.g., Manzi et al., 2012) have described the smallest scale cyclicity in gypsum and halite laminites, recognizing annual and pluriannual periodicities.

The only reliable calculations of sea-level drop during the deposition of Messinian salt in the Mediterranean Sea are recorded in basins near the Betic Cordillera, northeast of Almeria (e.g., Clauzon et al., 2015; Riding et al., 1998, 1999). Assuming the pre-MSC carbonate shelf sediments were deposited near sea level, a sea level drop of around 200 m has been estimated for the deposition of the Messinian salt in the Sorbas Basin depocenter (Riding et al., 1998; Roveri et al., 2008, 2019). In the Levant basin, a sea level drop was estimated at around 800 m by Druckman et al. (1995), based on the assumption that the Mavquim anhydrite was deposited in a very shallow environment. According to the facies description by Lugli et al. (2013), the anhydrite is a diagenetic product of clastic subaqueous deposits that could have been deposited in deeper environments. These estimations are based on hard rock evidence and are less speculative than other methods based on seismic interpretation (e.g., Ryan, 2008).

Braga et al. (2006) reported that the evaporite sediments in the Sorbas Basin in southeast Spain correspond mainly to gypsum, as no halite is recorded in the outcrops. These evaporite beds correspond to the Yesares Member and are separated from the underlying normal marine sediments (Abad Member, mainly composed of marl) by an erosional surface. The evaporite deposits in the Sorbas Basin allow the reconstruction of the maximum paleo-water depth of a basin that hosted the deposition of the primary lower gypsum (PLG) deposits during Stage 1 of the MSC. Some authors have proposed that the carbonate reefs at the margins of the peripheral basins are related to the PLG (e.g., Martinez del Olmo, 2011; Riding et al., 1999). Assuming a very shallow (0–15 m) paleo-water depth for the top of the carbonate reef drilled by the Torrevieja Marino-1 well, the PLG unit was deposited at around 200 m below, indicating a relatively deep basin (Martinez del Olmo, 2011).

The original relief difference between the top of the last pre-evaporitic reef, corresponding to a carbonate buildup known as the Fringing Reef Unit (located at the northern margin of the Sorbas Basin), and the first gypsum bed of the Yesares Member (located in the basin depocenter), has been estimated at around 240 m (Riding et al., 1998, 1999). However, the nature of the base of the PLG in the Sorba Basin has been heatedly debated by some authors (e.g., Fortuin et al., 2000; Riding et al., 1999, 2000) because of the rapid closure of the PLG against the basin edges, which is marked by a carbonate buildup (Riding et al., 1999). Manzi et al. (2013) interpreted that the base of the PLG in the Sorbas Basin is conformable and proposed a refined age of 5.971 Ma for the onset of the MSC, agreeing reasonably well with the 5.9 Ma age for the base of the unconformity proposed by Riding et al. (1999). Roveri et al. (2020) suggested the thickness of the Yesares Member (150 m) corresponds to the filling of a deep basin that accumulated gypsum in progressively shallower conditions. The depositional rate of the gypsum layers is very high, as some beds reach a thickness of up to 40 m. The gypsum beds directly onlap the flanks of the fringing reef (e.g., Braga et al., 2006; Riding et al., 1999, 2000).

The evaporite layers of the Yesares Member gradually change upwards into the siliciclastic layers of the Sorbas Member by progressive thickening of the terrigenous beds and concomitant thinning of the gypsum layers, until they disappear at the top of the succession (Clauzon et al., 2015). This observation contradicts the genetic models that assume the drawdown of the Mediterranean Sea resulted in the erosion of the peripheral basins (Clauzon et al., 1996). Riding et al. (1998, 1999) postulated that the evaporite deposits in the Sorbas Basin formed after the Mediterranean drawdown, as there was a gradual increase in water level following the evaporite deposition in the Messinian. This hypothesis admits that the gypsum layers formed during the final refilling of the Western Mediterranean Sea, as there is a gradual increase in water level following the evaporite deposition in the Messinian (Braga et al., 2006).

In the Gulf of Mexico, the most accepted interpretation of the depositional settings of the Middle Jurassic Louann salt assumes evaporite deposition near sea level on the landward side of the basin, where it is associated with anhydrite deposits. At the distal margin, near the continent-oceanic boundary, the basin was at oceanic abyssal depths shortly after salt deposition (e.g., Hudec & Peel, 2019). Most researchers correlate the proximal gypsum layers to the distal halite layers, which results in a huge difference in level at the base of the evaporites. This problem has been discussed by several authors who have questioned the physical connection of the peripheral basin salt deposits in the Mediterranean Sea with the halite deposits in the basin center (e.g., Roveri & Manzi, 2006). As shown in Figure 12.3, evaporite sequences might form both during periods of high sea level (in sabkha flats) and low sea level, precipitating evaporites in sabkha flats and also in subaqueous settings. However, these deposits might not be physically or temporally connected, as they formed during different stages of salt deposition (see Chapter 5 in Mohriak, 2025).

12.2. SALT DEPOSITION STAGES IN DEEPWATER BASINS

The deposition of salt in deepwater settings has been a topic of debate among researchers, with no consensus yet reached, even in well-studied basins like the Mediterranean Sea. A classic paper by Schmalz (1969) discussed several stages of the evolution of salt basins developed within a deepwater setting. The genetic model includes the pre-evaporitic phase and several stages of evaporite deposition in deep lakes, gulfs, and inland seas (Figure 12.4).

The pre-evaporitic phase is characterized by evaporation rates slightly exceeding water input in the basin by runoff and precipitation, with salinity close to seawater (1.02–1.03 g/cc) and well-oxygenated waters throughout the water column. Normal marine benthos and nekton proliferate in this environment, as seen in the Calcare di Base in the Mediterranean Sea, which precedes the deposition of the massive halite beds at the top of the Lower Evaporites (Lugli et al., 2010, 2013; Manzi et al., 2011, 2013, 2018).

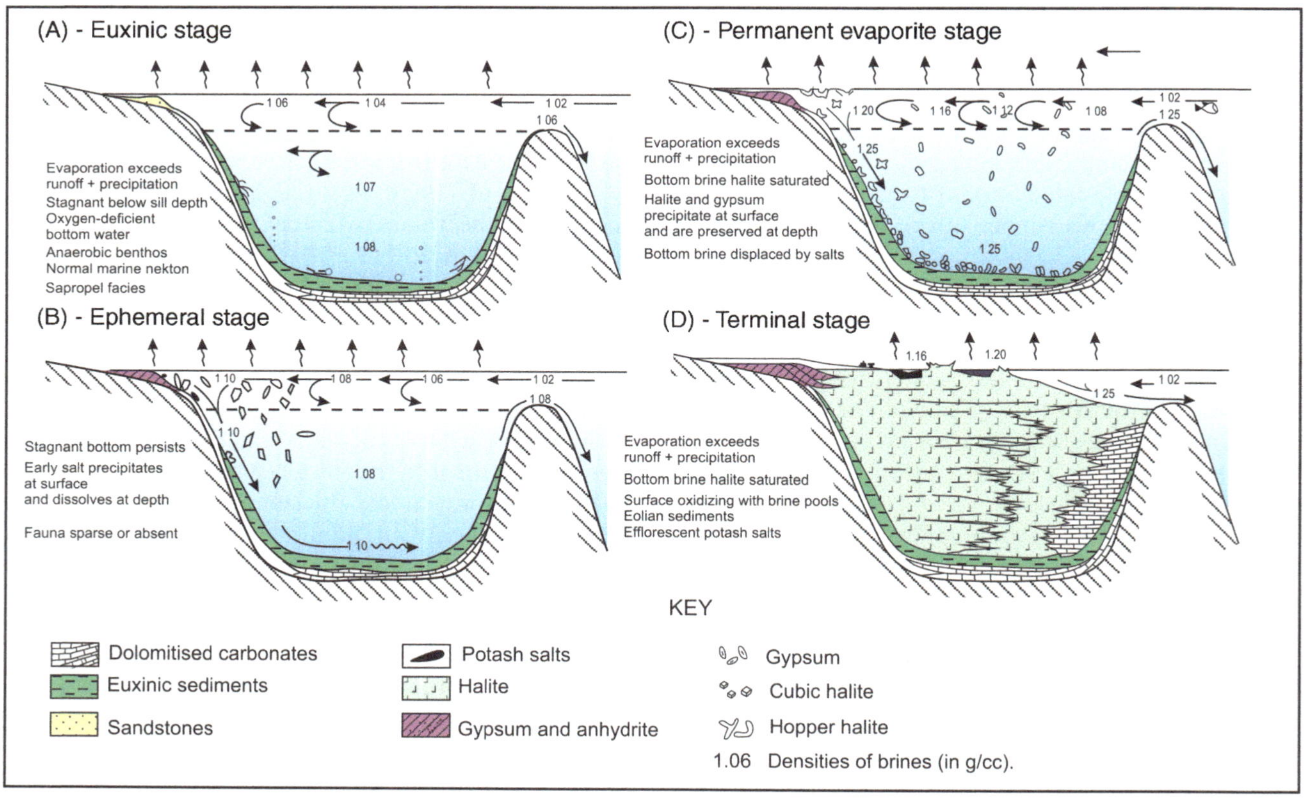

Figure 12.4 The schematic diagrams illustrate a deep basin evaporite sedimentation model involving different stages of brine concentration. The euxinic stage (A) results in anoxic conditions at the water body's bottom. The ephemeral stage (B) forms carbonate and sulfate minerals at the air–water interface but dissolves before forming thick accumulations on the lake's bottom. The permanent evaporite stage (C) concentrates brine, precipitating evaporites that accumulate at the bottom. The terminal stage (D) fills the depositional space with massive evaporite deposits. Credit: Adapted from Schmalz (1969) and Taylor (1985).

Hypersaline water bodies that form evaporite deposits are characterized by several stages of brine concentration and salt precipitation (e.g., Schmalz, 1969; Sonnenfeld & Perthuisot, 1989; Taylor, 1985). The evaporite cycles in the deepwater basin model (Figure 12.4) include the following stages:

A. Euxinic stage: During this phase, evaporation largely exceeds the water input in the basin. The salinity is increased (1.02–1.08 g/cc at the bottom of the lake), and the waters become stagnant below the sill depth. This results in deficient oxygen in the lake, allowing only anaerobic benthos to survive. Sapropel facies are deposited at the bottom, and normal marine nekton can still proliferate in the upper parts of the water column.
B. Ephemeral stage: Salinity increases substantially (from 1.02 to 1.10 g/cc at the bottom of the lake), resulting in stagnant water with higher salt concentrations. Early salt crystals precipitate at the air–water interface but dissolve when falling to the lake bottom, resulting in sparse fauna and only some types of bacteria and organisms that can adapt to this environment.
C. Permanent evaporite stage: Evaporation largely exceeds water input and salinity increases (1.02–1.25 g/cc at the bottom of the lake). The bottom brine is saturated in NaCl, resulting in a hypolimnion with density stratification. Gypsum and halite crystals formed at the air–water interface can accumulate at the bottom, displacing the bottom brine and reducing the water column.
D. Terminal evaporite stage: During this phase, the evaporation largely exceeds the water input, resulting in the complete drying out of the basin, leaving only residual shallow pools. The surface of the salt flats is under oxidizing conditions, with brine pools with extreme concentration and high density (up to 1.40 g/cc), depositing more soluble salts (carnallite and sylvite). The salt pans are surrounded by eolian sediments, mudflats, salt flats, and efflorescent salt crusts. Communication with the open sea is established only at the sill, with a remnant water body overlying the salt deposits. Carbonate and gypsum minerals precipitate closer to the water entrance, controlling the lithological facies through the salinity of the water.

In the terminal stage of small basins, Stage (D) represents a complete infill of the basin by evaporite sediments, while interconnected deep basins can remain in the previous stage of permanent salt deposition (Stage C) when ultradeep brines are highly concentrated in large and deep basins.

12.3. CHARACTERISTICS OF BASINWIDE EVAPORITES IN SHALLOW AND DEEPWATER SETTINGS

Basinwide evaporites are thick (often > 50 m) and sometimes reach several hundred to a few thousand meters in thickness. The sequences are composed of layered evaporite minerals, predominantly halite and gypsum, and mainly occur in inland seas, gulfs, and continental margin settings (Kendall, 1978, 1984, 1992, 2010; Warren, 2006, 2016). They are recorded in the stratigraphy of some large rift basins, such as the Neoproterozoic basins in the Arabian Peninsula (Hormuz salt) or the oceanic regions of the Mediterranean Sea. On divergent continental margins, widespread salt layers overlie continental rift blocks, extending from the continental platform to the shelfbreak, slope, and deep basin at the transition to the oceanic crust. Examples of ancient passive margins with basinwide evaporites include divergent margins, such as the Red Sea, the South Atlantic, and the Gulf of Mexico. Several ancient foreland basins developed basinwide evaporite layers in convergent settings, such as the Devonian Prairie evaporite in the Elk Point Basin of Western Canada or the Permian Zechstein salt basin in Central Europe. One of the largest salt provinces in the world, the Mediterranean Sea, is also included in this category (see Chapter 5 in Mohriak, 2025).

Sabkha models developed from observations in the Persian Gulf are not adequate to explain the presence of massive salt in the giant salt basins of the geological past (Warren & Kendall, 1985). Subaerial evaporites in sabkha environments are characterized by gypsum and nodular anhydrite; massive continuous salt layers are not observed. Most giant salt basins exhibit sabkha evaporite deposits in the proximal margins. In most cases, the distinction between subaerial and shallow-water evaporite sediments from deepwater evaporite sequences is elusive. Subaqueous evaporite units are usually composed of relatively pure evaporite lithologies (sulfates and chlorides, depending on the brine salinity) and tend to be thicker than sabkha evaporites. On carbonate platforms, the evaporite cycles show at least a few meters of thickness, as observed in the Permian Basin (Scholle, 2020). In some ancient basins, the stratified evaporite sequences reach a thickness of thousands of meters, as in the Mediterranean Sea (Meilijson et al., 2018). The electric log response to subaqueous evaporites is characterized by lower-frequency readings and a blocky appearance associated with the constant physical properties of the evaporite unit.

The genetic mechanisms for the deposition of evaporite sequences in deepwater settings are poorly understood because there are no modern analogs on Earth for the deposition of thick halite layers on continental margins or inland seas. Salt precipitation is an extremely fast geological process, and the depositional space necessary to accumulate a substantial thickness of evaporite sediments at the bottom of a deep basin should predate the salt deposition. A dense brine can bring ions and crystals to deepwater settings without the complete saturation of the water column. A deep basin with a large water column (> 500 m) requires the saturation of the bottom brines to equilibrium with the gypsum and/or halite precipitation to preserve the evaporite beds from dissolution (Figure 12.4).

Several researchers have proposed that a dense brine cascading from the shallow shelf may spill over the shelf edge and form a near-bottom gravity current or an intermediate-depth intrusion, bringing ions and crystals into the deepwater setting without the complete saturation of the water column (see a global review by Ivanov et al., 2004; Roveri et al., 2014). Modern gulfs do not exhibit bottom water saturation with common evaporite ions, even in regions where the basins are almost totally enclosed by arid desert lands, such as in the Red Sea or the Gulf of California (Kendall, 2010; Warren, 2010, 2016).

The best approximation to a deepwater setting with modern salt accumulation is the Dead Sea between Israel and Jordan, where evaporite sediments (predominantly halite) precipitate from highly concentrated brines that fill a 300 m deep lake (see Chapter 11). At present, the whole water column is saturated to the point of halite precipitation. However, the Dead Sea is a small transtensional rift basin with a width of only 20–30 km, and the volume of dissolved salt necessary to maintain the hypersaline lacustrine system precipitating halite is small compared to large ancient salt basins. The Dead Sea is not an appropriate analog for ancient deep marine evaporite basins that extended for hundreds of kilometers from the platform, slopes, and deep basins, such as the Zechstein, the Gulf of Mexico, or the South Atlantic. During the Holocene, most of the sedimentary sequences in the Dead Sea were carbonate muds and siliciclastic rocks, with evaporite minerals forming only thin centimetric lenses (see Chapter 11).

Lithological cross-plots based on different electric log properties show a higher percentage of halite and gypsum in subaqueous evaporite deposits when compared to sabkha sequences. Supratidal sabkha evaporites are characterized by sulfate-dominated facies and a larger proportion of matrix minerals, particularly calcite and quartz (Kendall, 1984; Schreiber, 1986; Warren, 1986, 19 89, 2006, 2016). Subaqueous evaporites are commonly composed of monomineralic layers, indicating a homogeneous brine concentration, both horizontally and vertically. Most massive layers exhibit small-scale cyclicity, commonly characterized by layers of evaporitic minerals separated by other deposits (Manzi et al., 2012). The central part of the Mediterranean Sea has been drilled by several boreholes that penetrated the pre-Messinian deposits. These deep boreholes confirmed the seismically transparent layers are composed of homogeneous, almost pure halite, while the reflective intrasalt layers correspond to bundles of thin claystone and diatomite layers (Feng et al., 2016; Meilijson et al., 2018, 2019).

Based on well-log interpretation, the lithologies of the evaporites in the distal parts of the South Atlantic, the North Sea, and the Gulf of Mexico salt basins are estimated to be composed of more than 90% halite (Jackson et al., 2019; Warren, 2016). In the Permian Basin of the southwest United States, the evaporite facies in evaporite cycles can be correlated for large distances, forming varves with intercalations of halite, anhydrite, and carbonate layers (Figure 12.1). Similarly, some sylvite peaks are recorded in gamma-ray logs of the Devonian Prairie Evaporites in the Elk Point Basin in Western Canada, implying evaporite sequences with maximum desiccation at the top of the cycles (Grobe, 2000; Meijer Drees, 1994).

Evaporite cycles can be interpreted through the lithological associations in evaporite sequences, as different minerals are precipitated when brine concentration is changed (see Chapter 4). The simplest way to reduce brine concentration at the bottom of a large saline lake or epicontinental seas and gulfs from a halite concentration back to lower salinities is to input new water into the system and change the overall salinity. This results from mixing the water column with fresh water that comes from rain (pluvial input), rivers (fluvial input), the melting of glaciers, or marine seepage across a barrier. The fresher, less saline water that enters a lake or gulf is less dense than the gypsum-saturated bottom brines and will always float atop the denser brines (Schmalz, 1969). During the permanent evaporitic stage, the brine density reaches about 1.25 g/cc at the lake's bottom (Figure 12.4-C).

Halite precipitation in deepwater environments has been registered in relatively small lakes or gulfs that went through overturning processes in modern times, such as the Dead Sea (see Chapter 2). However, in basins hundreds to thousands of kilometers in area and hundreds of meters deep, it is almost physically impossible to overturn the waters on a short-term time scale to precipitate the salt from the bottom to the top of the basin, as assumed for the terminal stage of deepwater deposition (Figure 12.4-C). The alternation of chloride and sulfate layers with carbonate sequences in the Permian Basin of the southwest United States indicates that the brine salinity strongly fluctuated in the deepwater depositional environment (Figure 12.1).

Although recognized in outcrops since the early geological descriptions of salt mines, layered evaporites have only recently been recognized in giant salt basins such as the Mediterranean Basin or the South Atlantic. These layered evaporites appear as laminated sequences intercalating halite and anhydrite, forming laminae that display thicknesses of only a few centimeters, which could only be formed in short time intervals, resembling varves (example, Figure 12.1-C). These evaporite sequences are interpreted to have been formed in basinal positions on platform margins, located away from the sediment sources (Kirkland, 2003).

Some authors have attributed the cyclicity in the Dead Sea salt deposition to annual climatic changes, which affected the mineralogy, sedimentation rate, and amount of organic material that may be incorporated into the evaporite sequences (e.g., Ben Dor et al., 2019). Other authors (e.g., Kirkland, 2003) suggest a millennial time scale (up to 3 ka) to explain the varves in the Castile Fm. evaporite sediments of the Permian Basin. Extremely high cyclicity has been recognized in the Mediterranean Sea Messinian evaporites. Some periodicity peaks vary from a few to tens of years, indicating climatic oscillations associated with solar and solar-lunar tidal cycles (Manzi et al., 2012).

Brine mixing within the water column is also considered another potential mechanism for the formation of evaporite minerals at various water depths (Raup, 1970, 1982). The brine-mixing mechanism involves the influx of normal seawater that mixes with more concentrated brines derived from the dissolution of underlying evaporites. The dense brine is ponded in the deeper parts of the lake, resulting in the precipitation of evaporite minerals as fine-grained masses that grow on the bottom of the lake, forming an evaporitic sedimentary layer without additional evaporation. Several authors (e.g., Jenyon, 1986) interpreted that the gypsum minerals presently precipitating on the floor of the Mediterranean Sea, at depths of 3.3 km, where an anoxic environment prevails, were formed by brine mixing.

The brine-mixing hypothesis assumes that the evaporites were not formed by the evaporative concentration of the basin waters but by the recycling of previous deposits in a subaqueous environment. For example, the Persian Gulf is characterized by the Neoproterozoic Hormuz salt diapirs, which originally formed a basinwide evaporite sequence in several rift basins in the Saudi Arabian plate (Alsharhan & Nairn, 2003). These evaporites were deformed in the Paleozoic and Mesozoic, resulting in salt diapirs with tops close to the sea floor, where dissolution resulted from contact with unsaturated water. Multiple salt deposition intervals are recorded in the Persian Gulf throughout the stratigraphic column, up to the Neogene. In several basins worldwide (e.g., the Gulf of Mexico), there are examples of the dissolution and re-precipitation of ancient salt layers forming brine ponds in the deepwater region (see Chapter 7 in Mohriak, 2025).

12.4. BASINWIDE EVAPORITE DEPOSITS IN ANCIENT MARINE SALT BASINS

Basinwide evaporites are mostly found in marine depositional environments, such as sabkhas, shallow embayments, deepwater gulfs, inland seas, and oceanic basins. Giant salt basins of the geological past are associated with sedimentary sequences that correspond to thick, basinwide evaporitic units that usually reach thicknesses of hundreds to thousands of meters (Allen & Beaumont, 2016; Jackson & Vendeville, 1994). Several salt basins record evaporite deposition during an active rift phase, such as the Neoproterozoic Hormuz Salt in Iran (e.g., Alsharhan & Nairn, 2003; Husseini & Husseini, 1990). Other basins record the main phase of salt precipitation after the rift phase, such as the South Atlantic salt basins (e.g., Kukla et al., 2018).

Basinwide evaporites commonly display textural evidence of many different depositional settings, from shallow to deep waters, including the typical geomorphology of a platform, slope, and deep basin (Figure 12.1). The basin geomorphology is associated with different salt thicknesses and environments of deposition, including isolated lagoons, gulfs, inland seas, and marine or continental settings with homogeneous brine, as expressed by the long-distance correlation of the evaporite beds. For example, when evaporite deposition prevailed in the Pliocene Dead Sea margins due to brine concentration, the whole basin was evaporitic, accumulating halite deposits on the lake's bottom. This indicates that the hypolimnion brines were concentrated to the point of gypsum or halite deposition, from the surface to the bottom of the water body (Zak, 1997).

The Appalachian Basin and the Permian Basin in the United States are classical examples of foreland basins with convergent margins where evaporite deposition in the proximal or peripheral part of the basin was associated with shallow lagoons, evaporite, and carbonate platforms, while the deepwater distal margins were characterized by predominantly siliciclastic sedimentation (Sarg, 2001). Other basins, such as the Gulf of Mexico, the South Atlantic, and the Red Sea, are characterized by basinwide salt deposition, with massive salt layers from the slope to the continent-oceanic crust boundary. Oceanic basins such as the Mediterranean Sea deposited basinwide evaporites during the MSC, during the African–Eurasian plate convergence since the Mesozoic, and in the Neogene, when the convergent plates entered the subduction zones (see Chapter 5 in Mohriak, 2025).

Several evaporite basins are characterized by multiple events of salt deposition, such as the Gulf of Mexico's basinwide evaporite deposition during the Middle Jurassic, followed by carbonate platform sediments in the Late Jurassic. Locally, some younger evaporite sequences were deposited in sabkha environments, as suggested for the Early Cretaceous Ferry Lake Anhydrite in South Florida (see Chapter 4 in Mohriak, 2025).

The Hormuz salt in the Middle East was deposited as basinwide evaporite layers in the Neoproterozoic, during a rift phase that developed a passive margin during the Paleozoic and Mesozoic. The margin was covered by an intra-shelf, basinwide platform-evaporite sequence in the

Late Permian-Early Triassic and the Late Jurassic. The Late Permian Khuff Fm. evaporites and the Late Jurassic Hith Anhydrite display thick layers of anhydrite and halite intercalated with shallow marine carbonates (Alsharhan, 2006; Vaslet et al., 2005; Warren, 2010, 2012). Several other cycles of evaporite deposition are recorded in the Persian Gulf region, including the Neogene evaporite sequences deposited on shallow marine carbonate platforms. The southeastern region of the Persian Gulf is characterized by thick salt layers (mainly halite and anhydrite) forming sequences between the Asmari and Gachsaran carbonates, suggesting a Miocene evaporitic event (Sajadi et al., 2016).

12.5. BASINWIDE EVAPORITES AND SOURCE ROCK DEPOSITION

Schmalz's (1969) model of deepwater salt sedimentation emphasized the occurrence of sapropels at the bottom of saline lakes during the euxinic phase (Figure 12.4-A). Woolnough (1937) discussed the barred basin model, which describes the depositional environment in deep basins as favorable for the accumulation and preservation of organic material. Many other authors have analyzed the anoxic environments associated with the early phases of evaporite deposition (e.g., Friedman, 1982; Hite & Anders, 1991). However, most researchers have emphasized the lack of modern analogs for the association of evaporite environments with source rock deposition. Anoxic environments associated with rich source rocks were widespread in large barred basins of the ancient past, both in divergent and convergent plate tectonic settings (e.g., Huerta & Harry, 2012; Warren, 2010, 2012).

The Late Jurassic marine sediments of the Persian Gulf region are associated with a carbonate-evaporite basin that resulted in very rich oil source rocks (Stoneley, 1987). The Arab carbonate-evaporite sequences in the Persian Gulf include some of the richest source rocks in the Upper Jurassic worldwide, which constitute a key element for petroleum systems responsible for giant hydrocarbon accumulations in the Middle East (Hite & Anders, 1991; Schreiber, 1988). One possible explanation for the enrichment of organic matter in these barred basins is their paleogeomorphology. The substratum of the evaporite sequences was marked by the maturity of peneplanation. The denudation of the previous topographic relief resulted in wide expanses of shallow water regions where carbonates and evaporites were deposited in periods of increased aridity, overlying black shales that might constitute the source rocks for oil. This corresponds to the euxinic stage preceding the brine concentration (Figure 12.4-A).

Several authors (e.g., Busson, 1991; Kirkland & Evans, 1981; and Warren, 1989, 2006, 2016) have summarized the four major depositional settings of source rock deposition associated with evaporite sediments, including (a) basinwide marine source rocks intercalated with evaporites or formed by stratification of the water column in deep marine basins with stagnant anoxic waters in the hypolimnion; (b) carbonate platforms with local lows where source rocks accumulated during periods of anoxia; (c) rift lakes where source rocks were deposited preceding the first marine incursions; and (d) mini-basins in evaporite basins where depressions resulted from salt flow under local sedimentary depocenters.

The euxinic stage is characterized by water stratification and stagnant bottom conditions, leading to the deposition of sapropels and organic-rich material (Schmalz, 1969). Two main situations are envisaged for episodes of anoxia leading to source rock deposition and basinwide salt deposition: (1) deep basins where salt concentration was not high enough to precipitate evaporite minerals, but dense brines formed in lagoons or evaporitic platforms; and (2) basins where the aridity resulted in evaporative drawdown and the brines were concentrated beyond mineral solubility accumulated on the deep basin's floor (Figure 12.4-C).

Source rock deposition is a crucial aspect of petroleum systems, particularly in carbonate-evaporite shallow-water platforms overlying salt deposits. Extremely organic-rich source rocks were deposited when most continents had extensive epeiric seas and were located in low latitudes (Warren, 2010). This is the case of the Devonian, Carboniferous, and Permian shales in the Permian Basin of the southwest United States (Busson, 1991).

Prolific source rocks are found in basins floored by thick salt layers, such as the Late Jurassic Smackover Fm. in the Gulf of Mexico (Mancini et al., 2003). These depocenters form below the wave base, restricting water circulation and corresponding to deepwater environments where organic material accumulates during anoxic intervals. The close relationship between carbonate rocks on the platform and buildups at the shelf edge favors a petroleum system with carbonate banks surrounded by source rocks. The proximity of the source rocks to the sabkha and desert sediments resulted in a prolific play in the northern Gulf of Mexico, associated with the Norphlet eolian sand reservoirs (Godo, 2017, 2019).

The Mediterranean Sea exhibits one possible example of the euxinic stage in a giant salt basin. Roveri et al. (2016) assessed the source rock potential of the deep water, non-evaporitic sedimentary successions that were deposited in the basin depocenter during the MSC and formed the Stage 1 gypsum deposits, which are ascribed to the Lower Evaporites in the peripheral basins (see Chapter 5 in Mohriak, 2025). In the deep basin, this interval is characterized by the euxinic shales and dolostones deposited below the Stage 2 evaporite sediments (Resedimented Lower Gypsum, RLG). The stage 1 evaporites in the perched basins are associated with a shallow-water–shallow basin model, whereas the coeval euxinic shales in the basin depocenter represent a deepwater–deep basin model (Figure 12.4-A).

In the South Atlantic divergent margins, Late Aptian salt layers overlie a sag basin with prolific carbonate reservoirs composed of microbialites. These rocks are associated with a deep lacustrine environment where rich source rocks were deposited in the Barremian (Dickson et al., 2019; Schiefelbein et al., 1999). The source-reservoir-seal system is extremely favorable for petroleum systems, resulting in giant hydrocarbon accumulations in the Santos Basin (Carminatti et al., 2008, 2009; Mello et al., 2021). Evaporite layers form a perfect seal for reefal buildups adjacent to pre-salt depressions, with source rocks ideally located in nearby depocenters. This model also applies to other prolific salt basins, such as the Northern Caspian Sea, where a Paleozoic petroleum systems involve the carbonate reservoirs and the evaporite layers as seals, resulting in several giant hydrocarbon accumulations (Mohriak, 2015).

The last major setting for source rock deposition associated with evaporite basins is related to local depressions or mini-basins above autochthonous or allochthonous salt. Salt flowage controls the development of depocenters between rising salt diapirs, resulting in mini-basins where source rocks and recycled evaporites accumulate within the depressions. Salt dissolution from nearby shallow salt diapirs results in hyper-concentrated, dense brine masses that flow toward the lowest depressions on the seafloor, forming stratified pools underlying the seawater with normal salinity. For example, the Orca Basin, a deepwater mini-basin located in the northern part of the Gulf of Mexico slope, is characterized by deepwater brine pools at the bottom of the ocean with salinities close to halite saturation (Pilcher & Blumstein, 2007). Anoxic intervals with very high contents of organic material are registered in these mini-basins. However, these pods are commonly encapsulated in allochthonous salt masses. The source rock intervals in these mini-basins are often disconnected regionally, contributing only a small percentage of oil to the petroleum systems (Pilcher & Blumstein, 2007).

12.6. GEOLOGICAL EVOLUTION OF THE PERMIAN BASIN, SOUTHWEST UNITED STATES

The western part of the United States is characterized by Permian sedimentary deposits, which are associated with desert, transitional, and marine environments. These deposits are found in several sedimentary basins, including the Permian Basin in NW Texas and SE New Mexico (Ewing, 2017, 2019; Hills, 1972; Scholle, 2020; Scholle et al., 2007; Silver & Todd, 1969; Ward et al., 1986; Wright, 2011). The sedimentary rocks consist of siliciclastic redbeds, carbonate rocks, and evaporites, forming a belt extending from the Mexican to Canadian borders (Benison & Goldstein, 2000). The Permian evaporite sequences accumulated within the depocenters of Paleozoic basins, and the Permian redbeds are related to arid conditions that formed a widespread continental cover throughout the western United States (see Chapter 11, Figure 11.20).

Permian sedimentary rocks overlie older sequences in the Paradox Basin, which accumulated halite deposits in a foreland setting during the Late Carboniferous (Middle Pennsylvanian, 306–315 Ma). The Paradox Basin was occupied by eolian sediments and redbeds in the Permian and Triassic (Trudgill, 2011). This indicates that during the Late Permian (about 250–260 Ma), the North American plate was located near the equator and covered by eolian dune fields (Blakey, 2008a, 2008b).

The Permian Basin corresponds to a large asymmetric structural depression of the Precambrian basement at the southern margin of the North American craton. The sedimentary basin boundaries form an ellipsoidal region, covering an area of about 300,000 km^2, including the Delaware and Midland subbasins, which were separated by the Central Basin Platform (Figure 12.5). The depression was primarily filled by Paleozoic sediments, and the Permian strata comprise about 50% of the sedimentary column (Ewing, 2017; Hartman & Woodard, 1971).

Paleogeographic reconstructions for the Late Paleozoic continental landmasses indicate that the Panthalassa Ocean was located to the west of the Permian Basin (see Chapter 11, Figure 11.22). The western margin of Pangea during the Late Paleozoic was characterized by the Panthalassa subduction zone to the west of the Diablo Platform and the fold-thrust belt to the south of the Permian Basin, separating Laurentia and Gondwana (Blakey, 2013; Scotese, 2001).

The greater Permian Basin encompasses two subbasins. The Delaware subbasin straddles the border between New Mexico and West Texas, eastward of the Sacramento Mountains, and the Midland subbasin is located in West Texas, north of the Llano Uplift and Ozona Arch (Figure 12.5). The depocenters of these subbasins are separated by a regional high known as the Central Basin Platform, which corresponds to an uplifted basement horst probably associated with reverse faults (Hartman & Woodard, 1971; Hills, 1972, 1984; Yang & Dorobek, 1995). The depocenter of the Delaware basin is filled with sediments that reach a combined thickness of about 8 km whereas the Midland basin attains a maximum thickness of about 5 km at the deepest parts (Ewing, 2017; Robinson, 1988).

The Delaware and Midland basins were characterized by carbonate banks, which developed reefal constructions

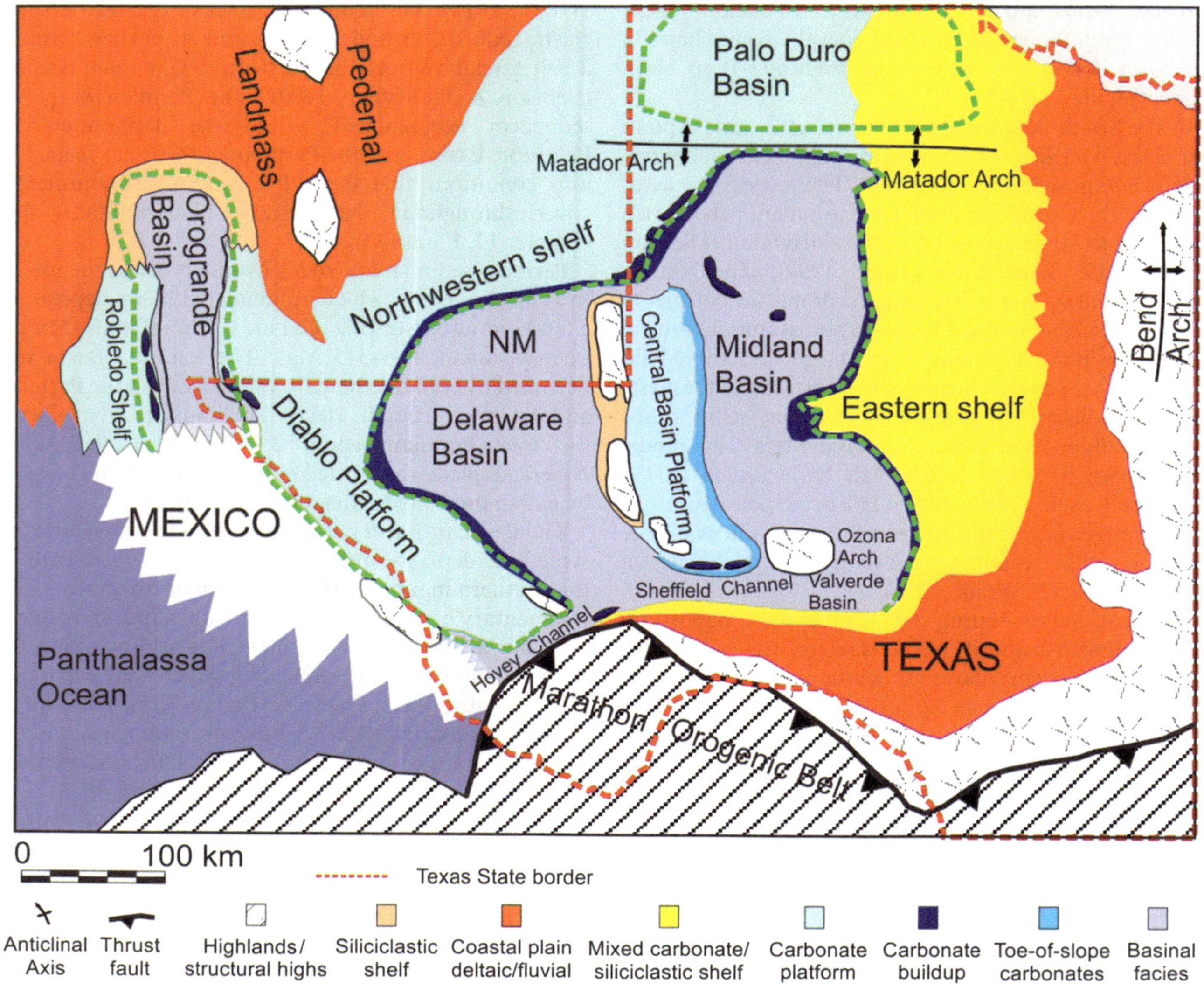

Figure 12.5 The simplified reconstruction map of the Permian Basin paleogeography during the Permian shows a branch of the Panthalassa Ocean west of the Diablo Platform and Northwestern Shelf. Shallow carbonate rocks encircled the Delaware Basin during the Late Permian. The southern border was marked by the Marathon-Ouachita orogenic belt, corresponding to a fold-thrust belt with a northward vergence. Credit: Adapted from Wright (2011) and Blomquist (2016).

such as the Capitan Reef, northeast of the Central Basin Platform. The Pedernal Uplift was characterized by siliciclastic rocks that formed a large desert northward of the carbonate platform. The Eastern Shelf was marked by siliciclastic rocks formed westward of the Bend Arch, resulting in highlands with alluvial fans that advanced toward the coastal plain, which was characterized by fluvial to deltaic environments. The Eastern Shelf corresponded to a mixed carbonate-siliciclastic shelf that extended from the Matador Arch in the north to the Marathon Orogenic Belt (Figure 12.5).

As the basins strongly subsided during the Permian, some major depressions were formed as foreland basin troughs related to the Marathon Orogenic Front. This fold-thrust belt straddled the region that presently corresponds to the border between the United States and Mexico (Figure 12.5). Minor sedimentation was received from the lowlands that surrounded the basin to the north, resulting in a greater bathymetry being developed at the basin depocenter. The oceanic waters communicated with the almost encroached inland sea through gaps or channels at the southwestern portion of the carbonate reef that rimmed the depression (Figure 12.5). The Hovey Channel was formed at the junction of the Diablo Platform with the Marathon Orogenic Belt and the Eastern Shelf (Figure 12.5). This channel probably constituted the main entrance for the marine waters that filled the basin (Blomquist, 2016; Kirkland et al., 2000; Wright, 2011).

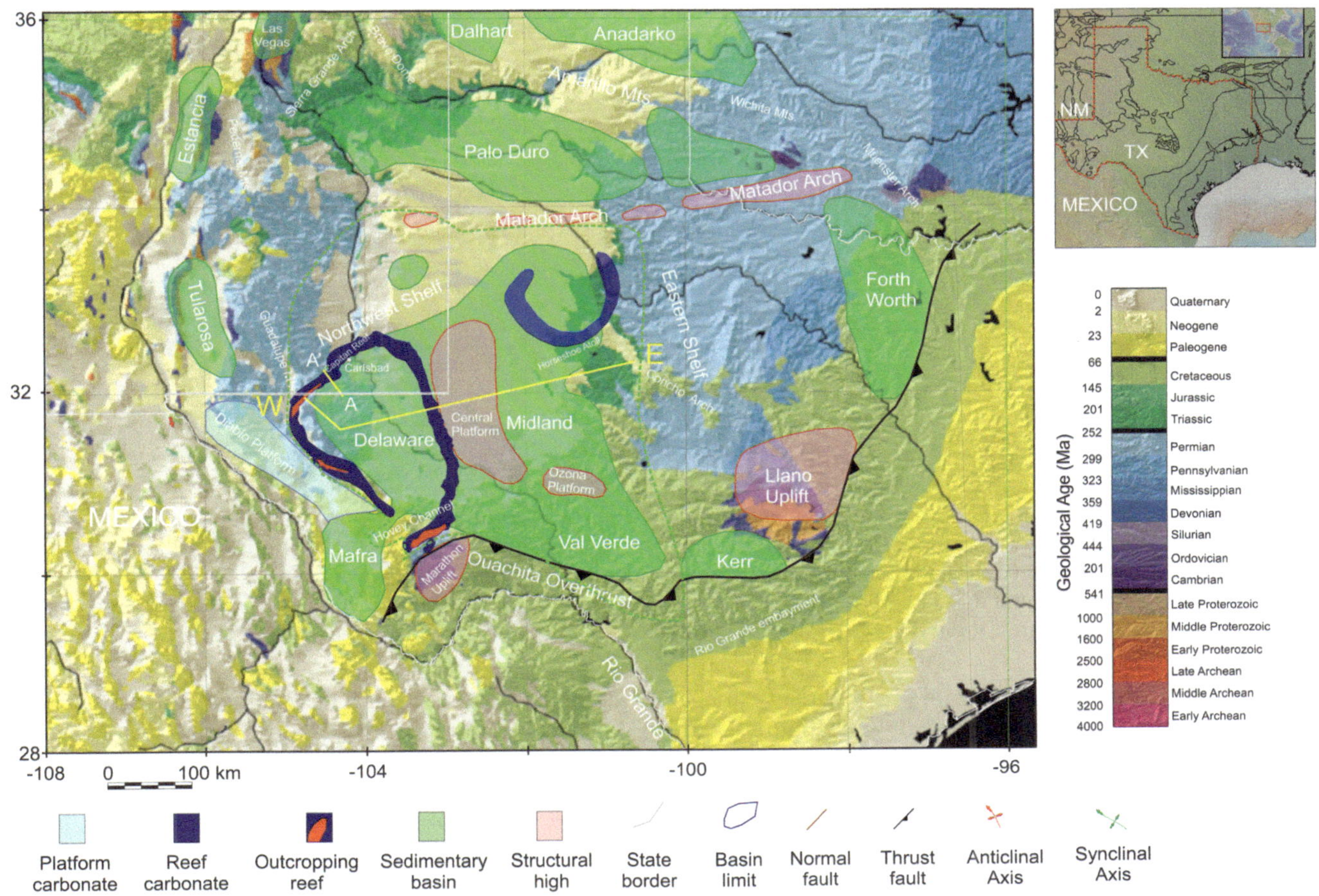

Figure 12.6 The geological map of the southwestern United States shows bedrock units and bedrock age draped on the topographic map. The Permian deposits are highlighted in blue, while the Texas state borders are shown in the upper right inset map. The Permian Basin depocenters are surrounded to the west and east by the Permian to Carboniferous sediments. The Delaware Basin features carbonate reefal structures in dark blue, with red colors indicating the segments where reefal carbonate rocks outcrop at the surface. The regional profile W-E runs from the Capitan Reef in the Delaware Basin, crossing the Central Platform and reaching the Midland Basin in the east. The shorter stratigraphic section A'-A crosses the Capitan Reef in the northern part of the Delaware Basin. Credit: Vigil et al. (2000)/USGS.

The southern and northern margins of the Permian Basin were covered by desert and sabkha environments, while shallow carbonate platforms surrounded the depression from the Palo Duro Basin in the north to the Northwestern Shelf and Diablo Platforms in the west (Figure 12.5). The transition to the Palo Duro Basin is associated with the E-W trending Matador Arch, which runs from the carbonate platform to a mixed carbonate-siliciclastic shelf in the east. The Eastern Shelf was formed by a mixed platform that graded into a desert landscape adjacent to the highlands to the east and south (Figure 12.5).

During the Permian, the depression formed a large lagoon that deposited carbonate-evaporite sediments and was rimmed by a reef structure at the shelf edge. The circular ring around the basin depocenter extended from the border between New Mexico and Texas to the Marathon Uplift in the south, adjacent to the Ouachita Overthrust (Figure 12.6). The deep basin in the central depression was starved of clastic sediments and deposited mainly evaporite sediments in the Upper Permian (Ewing, 2017; King, 1942).

The sedimentary environments included sabkhas, carbonate platforms, and a deep basin. Northward of the Delaware Basin depocenter, a shallow-water platform was established landward of the reef. The depositional setting involved the transition from lagoonal siliciclastic and carbonate sediments to sabkha tidal flats and continental deserts with redbeds and eolian sandstones (Presley, 1987; Ward et al., 1986). The basin depocenter was characterized by varves and resedimented clastic evaporites. The environments are schematically represented by the schematic cross-section previously discussed in Chapter 11 (see Figure 11.4).

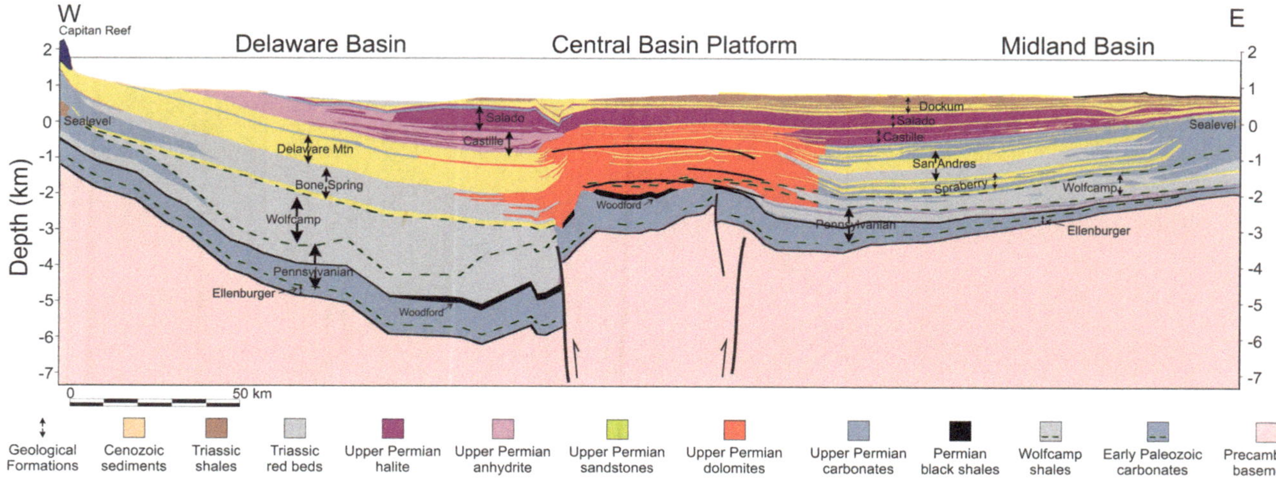

Figure 12.7 The geological cross-section extends from the Capitan Reef in the west to the Midland Basin in the east. The profile shows the Upper Permian dolomites on the basement high at the Central Basin Platform, where a horst structure was controlled by reverse faults. The Castile and Salado evaporite layers overlie the basin depocenters and the Central Basin Platform. The Capitan Reef is a major carbonate buildup fringing the northern edge of the Delaware Basin. Credit: Adapted from Scanlon et al. (2017) and Matchus and Jones (1984).

The simplified geological map of the greater Permian Basin and adjacent sedimentary troughs (Figure 12.6) shows a foreland depression related to the Marathon-Ouachita overthrust, which extends to the east toward the Llano Uplift (Ewing, 2017; Vigil et al., 2000). The fold-thrust belt is related to the closure of the ocean that separated the southwest part of the North American plate (Laurentia) from the northwest part of the South American plate (Gondwana) in the Carboniferous and Permian (Blakey, 2008a, 2008b; Scotese, 2001).

The regional schematic geological cross-section across the Permian Basin (Figure 12.7) extends from the Guadalupe Mountains in southeast New Mexico, crossing the Central Basin Platform, which corresponds to dolomitized carbonate rocks that overlie a basement high associated with reverse faults (Ewing, 2017). This cross-section is based on exploratory wells drilled in West Texas, illustrating the main structural style and stratigraphic units in the Early Paleozoic successions of the Delaware and Midland subbasins. The Delaware subbasin's sedimentary layers are deeper in the depocenter, forming a trough west of the central high that separates from the Midland subbasin (Matchus & Jones, 1984).

The Central Basin Platform is characterized by thick dolomitized sequences of Permian carbonate rocks stacked on the basement floor (Figure 12.7). Dolomitization processes are believed to be related to the infiltration of meteoric waters that dissolved the Upper Permian evaporites and became enriched with magnesium (e.g., Engle et al., 2016, 2020; Saller & Stueber, 2018; Scanlon et al., 2017). At present, meteoric waters enter the basin in the highlands to the west, where Paleozoic rocks outcrop in the Sacramento and Guadalupe mountains and descend into the San Andres Fm., which underlies the Upper Permian evaporites in the northern shelf in the Midland Basin and the dolomites in the Central Basin Platform (Figure 12.7). However, in the outcrop areas around the Capitan Reef, no evaporites are preserved in the reef and fore-reef sections, and only partial remnants of evaporite minerals are recorded in the back-reef settings, as evidenced by pseudomorphs of carbonate minerals after halite and gypsum (e.g., Scholle et al., 1992, 2007).

12.6.1. Tectono-sedimentary Evolution and Stratigraphy of the Permian Basin

The tectono-sedimentary evolution of the Permian Basin is characterized by four phases (e.g., Ewing, 2017, 2019): (1) pre-Permian clastic marine and carbonate deposition, following the Neoproterozoic breakup of Rodinia and the development of continental margins in the Early Paleozoic; (2) Late Carboniferous to Permian marine foreland basin associated with a continental collision involving Laurentia and Gondwana; (3) Permian basin filling under desert, sabkha, and marine conditions, resulting in the deposition of evaporite sediments in different environments; and (4) post-Permian sedimentation under shallow marine to continental environments. A simplified tectonostratigraphic chart of the Permian Basin, based on geochronological divisions proposed by Gradstein et al. (2012) and Walker et al. (2018), presents a synopsis of the main lithological types and tectonic events associated with the geological evolution of the basin (Figure 12.8). The chart also includes the relative contribution of several stratigraphic sequences to the total oil production in the

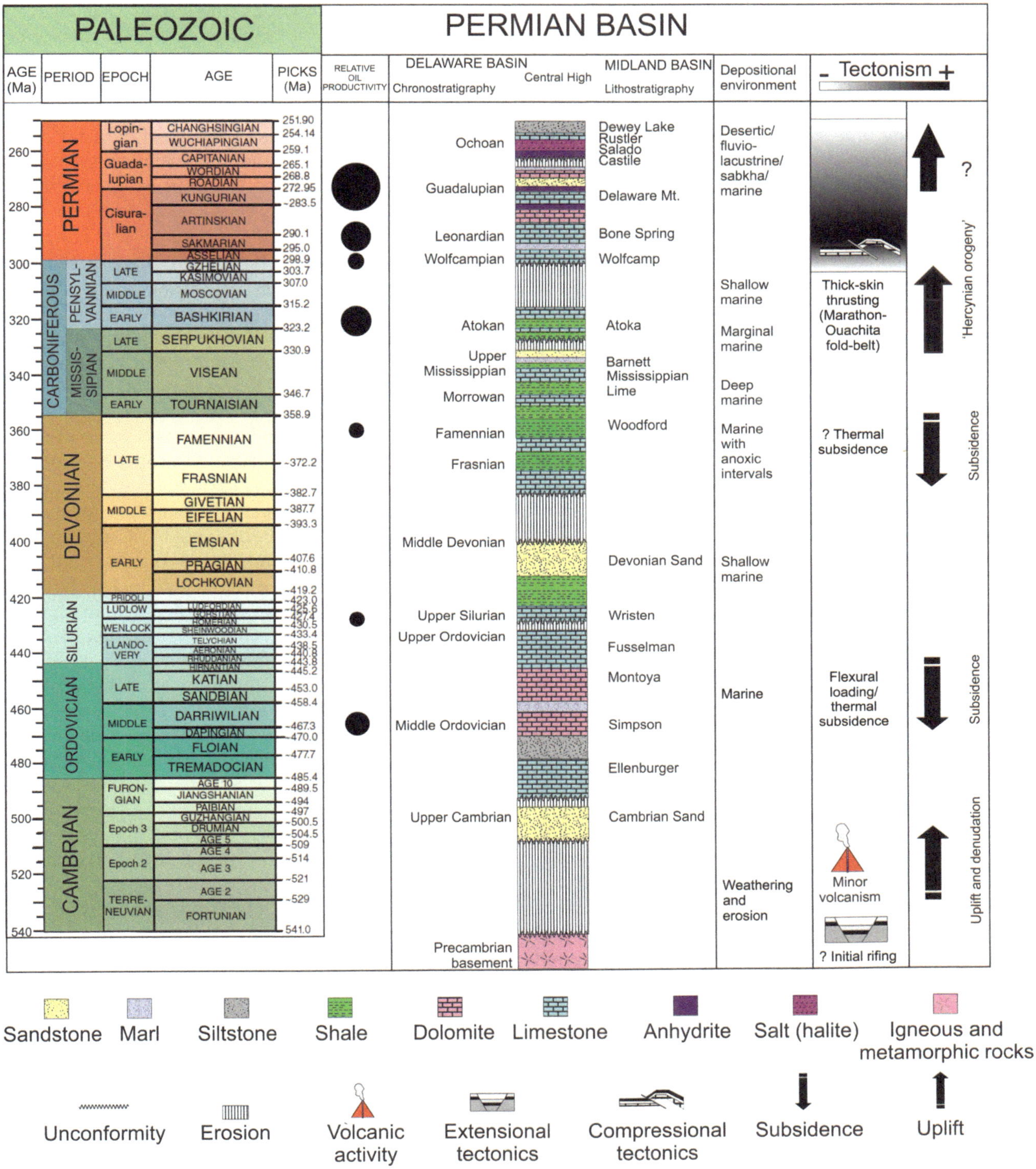

Figure 12.8 The tectonostratigraphic chart in the Permian Basin shows the main geological units and tectonic events. The Upper Permian carbonate sequences are associated with the Guadalupian Series and overlie evaporite sequences of the Ochoan Series. The relative amounts of oil production are shown by the size of the black circles, indicating that most of the conventional oil production is associated with the Upper Permian (Guadalupian) reservoirs. Marine sedimentation predominated in the Paleozoic, including anoxic events in the Late Devonian, resulting in organic-rich marine source rocks. The Upper Permian is marked by a transitional environment with sabkha and lagoonal evaporites, with anhydrite and halite salt deposits in the Castile and Salado Fm. (Ochoan Epoch).

Permian Basin. The Upper Permian Guadalupian Series is responsible for the largest percentage of petroleum production in the basin (Figure 12.8).

The Early Paleozoic sediments that underlie the Permian sequences are related to a larger marine seaway developed during the subduction of the Panthalassa Ocean under the North American plate (Figure 12.5). The inland sea occupied a large NW-SE trending downwarp that accumulated carbonate sediments from the Early Paleozoic to Early Carboniferous (approximately 541 to 323 Ma). This sedimentary province is known as the Tobosa Basin, a broad intracratonic basin with shallow marine environments (Adams, 1965). This Paleozoic shallow sea occupied the region that now corresponds to west Texas and eastern New Mexico and is limited to the south by the Marathon Mountains at the border with Mexico (Figure 12.5).

The Tobosa Basin has accumulated important stratigraphic sequences drilled by exploratory wells aimed at deep targets in the Permian Basin. Sedimentation in the basin started with the deposition of Cambrian sandstone and was followed by the deposition of Ellenburger carbonate rocks in the Ordovician (Figure 12.8). Shales and carbonate rocks were deposited in the Silurian and Devonian when the southwestern portion of Laurentia was covered by an inland sea. The basin was connected to the Panthalassa Ocean to the west and bordered by the Rheic Ocean to the east and south (Figure 12.5).

The deepest exploratory targets in the Permian Basin include the Ellenburger carbonates and Simpson dolomites (Cambrian to Ordovician). The meteoric waters that descended from the Sacramento Mountains and infiltrated through the Upper Permian Castile and Salado formations dolomitized the Early Paleozoic carbonates (Ellenburger Fm.), which constitute an important reservoir in the Permian Basin (Lee & Friedman, 1987).

During the upper part of the Devonian period, anoxic intervals deposited black fossiliferous shales in the central part of the Permian Basin. These black shales correspond to distal facies, forming low-energy sedimentary layers deposited in a marine environment. These Devonian shales are one of the main source rocks for numerous oil accumulations in the Delaware and Midland subbasins. The Silurian to Devonian black shales present thin intercalations of tight sands that locally constitute hydrocarbon reservoirs with a small percentage in the total basin production (Figure 12.8). This is the case of the Woodford shale, dated as Late Devonian to Early Carboniferous (Mississippian). This rich source rock displays high contents of total organic carbon (TOC), up to about 10% in weight (Infante-Paez et al., 2016). Other source rock intervals are observed in the Mississippian Barnett Fm. shales, which are one of the main targets for shale gas exploitation activities in the Fort Worth Basin of Texas (Ewing, 2017).

During the Carboniferous, the seaway that separated Laurentia and Western Gondwana became smaller and shallower due to plate convergence. In the Late Carboniferous, the proto-South American continent (part of Western Gondwana) collided with the proto-North American continent (part of Laurasia), developing major thrust faults and uplifts affecting the region. This tectonic event formed the Pangea supercontinent, resulting in orogenic belts between the southern North America and northern South America plates (Figure 12.5). The most important fold-thrust belt affecting the Early Paleozoic to Carboniferous sedimentary sequences is the Marathon–Ouachita–Alleghanian Orogen (Yang & Dorobek, 1995). The Marathon orogenic belt resulted in the splitting apart of the region previously occupied by the Tobosa Basin. Several new sedimentary depocenters were created and formed independent sag basins separated by regional uplifts and platforms (Figure 12.6).

The plate tectonic compressional processes resulted in buckling of the crust, leading to major erosion associated with the Hercynian orogenic event that formed the highlands between Laurentia and Gondwana (see Figure 11.22). The foreland depression was marked by the development of fault-controlled sedimentary troughs, where increased subsidence was observed in the Late Carboniferous to Early Permian. The Paradox Basin was one of these foreland troughs and accumulated Late Carboniferous salt in the depocenter (see Chapter 3).

The Wolfcampian fine-grained carbonate and siliciclastic rocks were deposited in the Permian Basin above the Hercynian unconformity (Figure 12.8). The Middle and Upper Permian sequences include marine carbonates and evaporite sequences (Benison & Goldstein, 2000; Ewing, 2017). The carbonate-evaporite deposits in the Permian Basin are correlated to equivalent lithostratigraphic units registered in the Val Verde, Kerr, Fort Worth, Palo Duro, Anadarko, Denver, Powder River, and Williston basins (Figure 12.6). The most important of these subbasins are the Delaware and Midland, which share similar tectono-sedimentary characteristics and exhibit equivalent stratigraphic sequences (Figure 12.7). These depocenters correspond to structural and topographic lows that were intermittently connected with the marine waters from the Panthalassa Ocean during most of the Permian (Blakey, 2008a, 2008b; Wright, 2011).

The sedimentary environments during the Late Permian varied according to paleogeography and water depths in the depressions (Figure 12.5). The Central Basin Platform remained at a higher elevation, resulting in less sedimentation on the structural high, which was mainly covered by shallow-water carbonate sediments (dolomites). The Delaware Basin exhibits a deep depocenter to the west of the high (Figure 12.7), forming deepwater environments that accumulated terrigenous clastic rocks in the Permian, including deep marine turbidites (Figure 12.8).

The Delaware Basin is marked by a topographic profile with increasing altitudes to the west of the central high (Figure 12.9). The Sacramento Mountains are an uplifted block located to the west of the basin. This western segment of the Delaware basin is characterized by the outcrop of Paleozoic rocks due to the Laramide tectonic event (late Cretaceous to Paleogene) and the development of the Basin and Range province in the Tertiary (Blakey & Ranney, 2018).

A regional E-W geological cross-section extending from the Sacramento Mountains, crossing the northern part of the Delaware Basin, the Central Platform, and the Midland Basin, illustrates the characteristics of these basins (Figure 12.9). The topography gradually decreases from the Sacramento Mountains to the Central Platform, which currently has an average altitude of nearly 1 km. This creates a plateau covered by flat-lying Upper Permian evaporites and Triassic redbeds, extending eastward toward the Midland Basin. Upper Permian dolomites constitute the core of the Central Basin Platform, which separates the carbonate platforms in the west from the more siliciclastic sequences in the east (Figure 12.9).

The Paleozoic subbasins display depocenters forming asymmetric structures that are in the Delaware Basin and Midland Basin separated by the Central Basin Platform. The maximum depth of the sedimentary sequences east of the Central Platform corresponds to a depth of −6,000 m, forming a trough with Early Paleozoic sediments overlying the Precambrian basement (Figure 12.9).

The sedimentary sequences in the Delaware Basin rise toward the west, while in the Midland Basin, they rise eastward. The Permian lithologies are characterized by a higher input of siliciclastic rocks, with the Bend Arch highlands and alluvial fans advancing toward the Eastern Shelf (Figure 12.5).

During the Permian, thick subaqueous deltaic systems progressively filled the basin with progradational wedges (Figure 12.10). Subsequently, the deltaic system was replaced by shallow marine and continental floodplains. The Middle Permian (Guadalupian) evaporite facies was associated with gypsum and halite layers that accumulated in sabkha to lacustrine environments, resulting in the San Andres and Yates formations. The thick evaporite deposits during the Upper Permian form cycles that include carbonates, sulfates, and chlorides. The Capitan Reef is the carbonate buildup at the shelf edge of this epoch, isolating the sabkha from the deep basin (Figure 12.10).

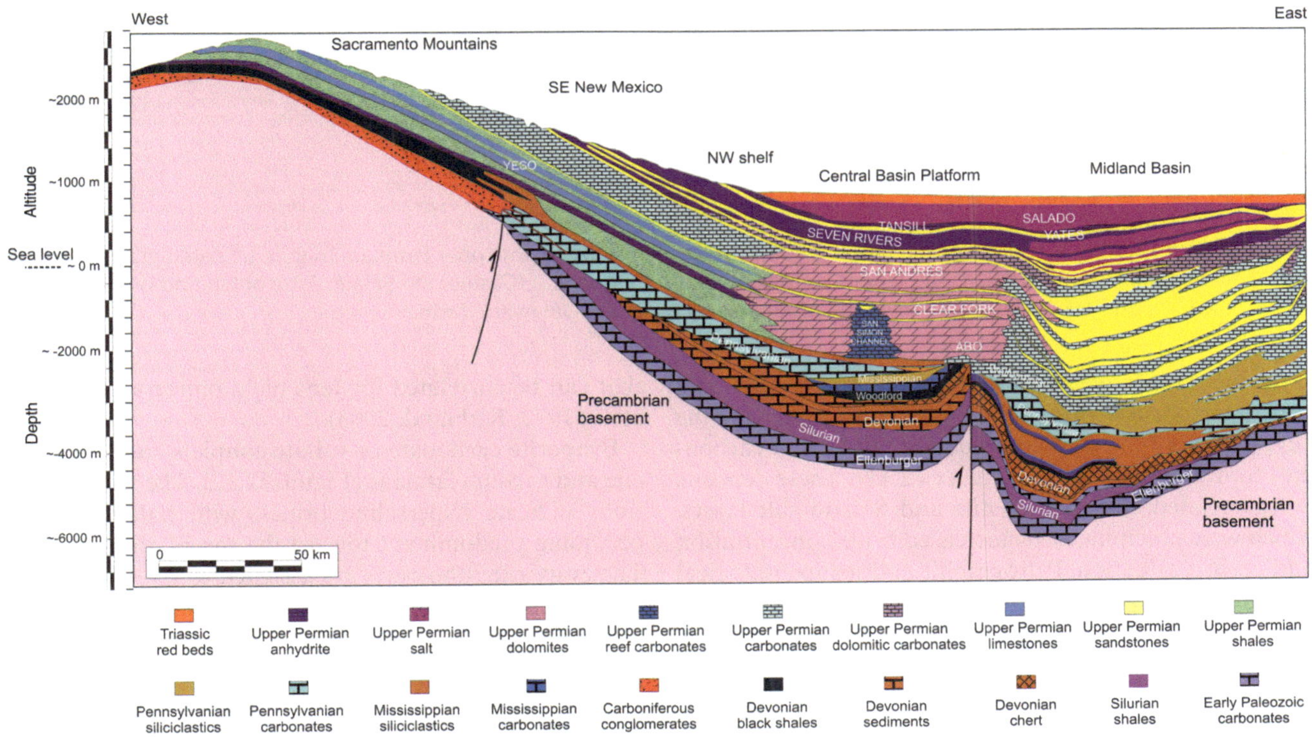

Figure 12.9 The schematic geological cross-section (W-E) in the Permian Basin extends from the Sacramento Mountains in New Mexico to the Midland Basin in northwest Texas. The vertical scale above sea level (altitude) is vertically exaggerated to show the topography of the western margin of the Delaware Basin in southeast New Mexico. The Central Basin Platform, a basement high associated with reverse faults, is dolomitized by meteoric waters infiltrating from the highlands to the west.

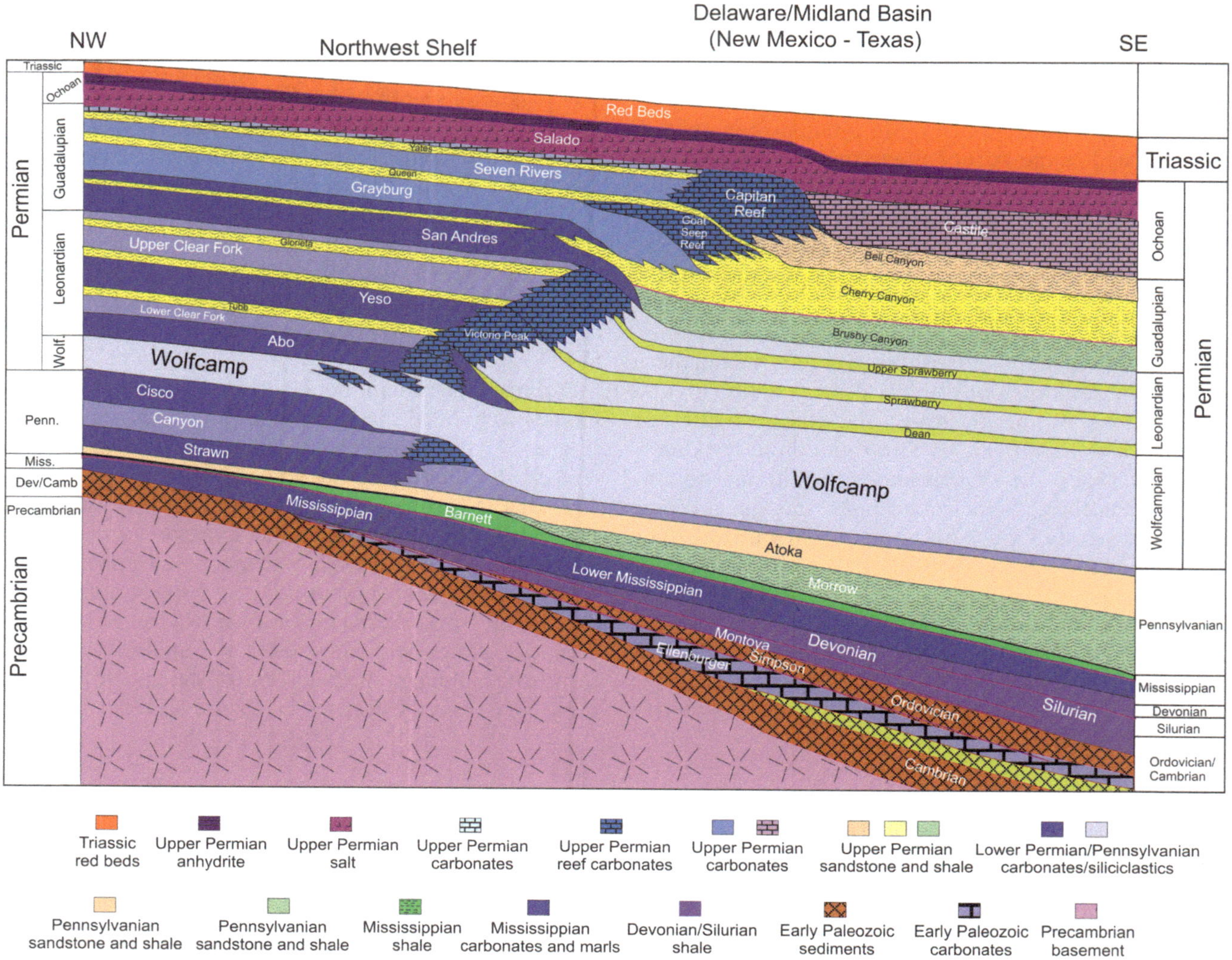

Figure 12.10 The regional stratigraphic profile shows the Paleozoic to Upper Permian-Triassic lithostratigraphic units in the Permian Basin. The Upper Permian evaporite sequences (Castile and Salado Fm.) onlap and overlie the Capitan Reef carbonate rocks in the northwest sector of the Delaware Basin.

In the Upper Permian (Ochoan Epoch), marginal marine evaporites covered the basinal area, overlying the Seven Rivers Fm., reef construction, and distal carbonates from the Castile Fm. (Figure 12.11). These deposits are associated with the Castile and Salado salt layers, which were precipitated by increased brine concentration and basin desiccation. Petrographic and sedimentological research suggests that the evaporite layers were formed in shallow to deep marine or deep hypersaline lacustrine settings, with maximum water depths reaching 500 m (Kirkland et al., 2000; Scholle et al., 1992).

The Castile Fm. corresponds to the first stage of evaporite sedimentation. It is made up of a thick sequence of millimeter-scale beds of carbonate, sulfate, and organic-rich black mud. Annual climate variations control the formation of varves, which are composed of thin laminae that can be correlated for tens of kilometers (Anderson et al., 1972; Kirkland, 2003).

Evaporite carbonate or sulfate couplets consist of calcite and gypsum laminae (Figure 12.12). The facies variations indicate hypersaline brines, with halite deposits becoming predominant toward the top of the sequence. The evaporite lithologies are assigned to the Castile Fm. and the Salado Fm., forming Upper Permian (Ochoan Epoch) evaporite sequences with a maximum thickness of about 600 m each. The total evaporite sedimentary column is composed of 5% carbonate mud, 45% gypsum and anhydrite, and 50% halite, but it is estimated that about half of the halite has been dissolved since the Late Permian (Anderson, 1981). The salt layers were finally covered by a thin layer of continental deposits, expressed by Triassic red beds (Figure 12.10).

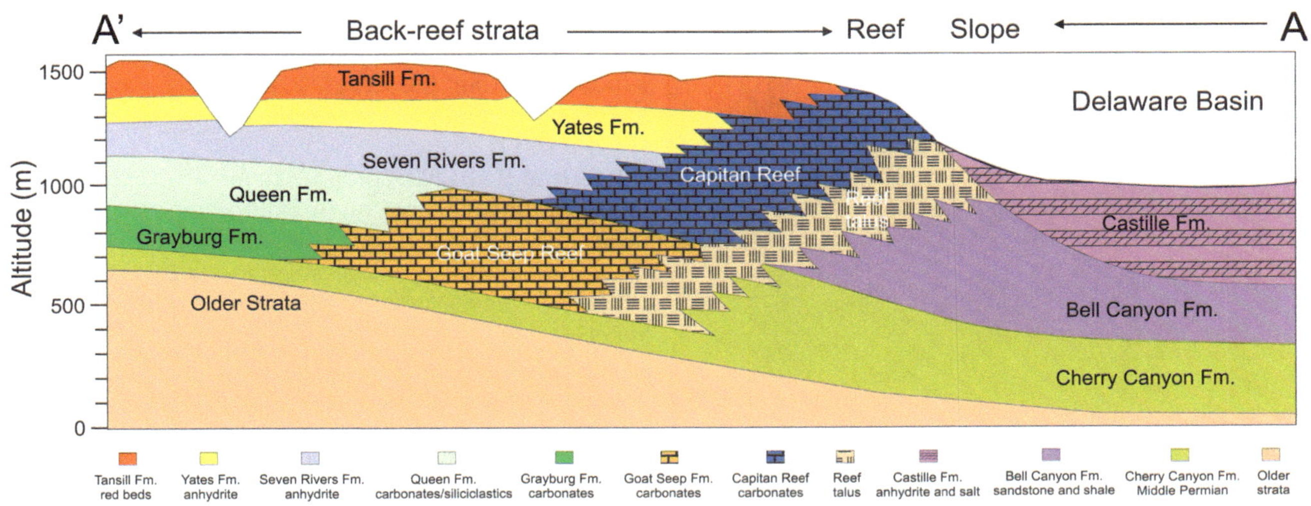

Figure 12.11 The geological cross-section in the northern Delaware Basin shows a paleoreconstruction of Upper Permian stratigraphy before the deposition of Salado Salt. The sedimentary facies include a deepwater basin with siliciclastic rocks and evaporites, an apron of allochthonous carbonates forming a reef talus, a carbonate shelf margin with a reefal buildup, and a carbonate mudflat with some evaporite intercalations. The Capitan Reef, formed at the transition from the back-reef strata in the lagoon and the slope adjacent to the continental margin. During the Late Permian, the deep basin accumulated gypsum and anhydrite (Castile Formation). Basinwide evaporites, associated with the Salado Fm., covered these successions during the Ochoan Epoch.

12.6.2. The Carbonate Reefs Surrounding the Permian Basin

The Permian Basin is characterized by carbonate buildups formed by rudist colonies, which fringed the shelf edge at the transition from the lagoon to the deep basin (Figure 12.6). Landwards of the reef, sedimentary facies include carbonate platform sediments, such as those on the Diablo Platform or the Northwest Shelf. The eastern depression, corresponding to the Midland Basin, is characterized by a more siliciclastic input, forming a mixed carbonate siliciclastic carbonate shelf that changed into coastal plain deltaic and fluvial systems (Figure 12.5).

The Delaware Basin is marked by a spectacular reef construction that extends along the northern margin for over 500 km (Handford & Bassett, 1982; Scholle, 2020; Scholle et al., 1992, 2007; Ward et al., 1986). This Late Permian carbonate buildup now outcrops in the Guadalupe Mountains in northwest Texas and southeast New Mexico, forming the Capitan Reef (Figure 12.6). The Capitan Reef is composed of different carbonate rocks derived from sponges, algae, and microbial organisms, which separated the shallow-water region with sabkha carbonate and evaporite sediments from the deepwater deposits in the basin depocenter (Figure 12.1). Carbonate debris and slumps accumulated directly basinward from the reef front, forming a carbonate talus, indicating an unstable slope (Figure 12.11). Many caves are presently seen in the carbonate rocks, constituting a tourist attraction (Palmer & Palmer, 2012).

Early sedimentological studies that analyzed the Guadalupe Mountains and the Capitan Reef in the past century played a key role in the development of conceptual models and paradigms for sequence stratigraphy (e.g., King, 1942). These investigations focused on the petrographic characterization of the carbonate buildups at the transition from the continental evaporitic lagoon to the deep marine environments. The high-relief carbonate outcrops and deep marine sedimentary successions are exposed in canyons, providing a three-dimensional visualization of the stratigraphic sequences (Harris & Kerans, 2016; Scholle, 2020; Scholle et al., 1992, 2007). The reefal constructions constituted barriers that separated the shallow sabkha from the depression (Figure 12.11).

Evaporite deposition began with thin beds of aragonitic mud intercalated with centimetric layers of gypsum that accumulated on the basin floor. The basinal evaporites formed the characteristically varved Castile salt deposits, represented by thousands of couplets of carbonate (aragonite and calcite) and sulfate (gypsum) layers, which were followed by the sedimentation of couplets of gypsum and halite (Anderson et al., 1972). The Castile Fm. hand specimen samples and the cores from exploratory boreholes are composed of dark-colored organic-rich calcite laminae that alternate with lighter-colored gypsum-anhydrite laminae. Locally, these laminated sediments appear folded and convoluted, indicating syn-depositional

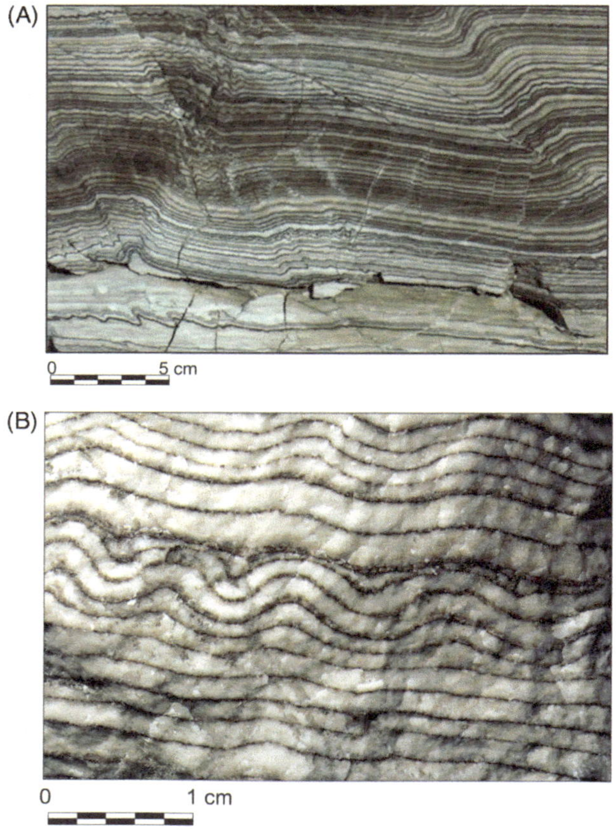

Figure 12.12 (A) Castile Fm. outcrop in the Delaware Basin features gypsum and calcite couplets, with centimetric-scale compressional structures formed during and after evaporite layers' deposition. (B) A detailed image of the sulfate-carbonate couplets shows seasonal variation in brine salinity, with varves deposited during summer and winter. Nondeposition, erosion, or dissolution features affect the contorted layers. The Castile Fm. evaporites were covered by basinwide evaporites associated with the Salado Fm., indicating an increasing brine concentration in the basin depocenter. Parts of the Salado Fm. salt deposits have been dissolved by meteoric waters. Credit: Courtesy of Peter Scholle (2017).

deformation by gravitational processes (slides and slumps) and minor folding and faulting (Figure 12.12).

12.6.3. Paleogeographic Reconstructions of the Permian Basin

The Upper Permian Castile evaporites were deposited in a closed basin in deepwater settings, with seawater entering the basin primarily through seepage along the Capitan Fm. carbonate bank, which constituted one of several barriers at the basin's borders (Figure 12.13). The carbonate bank rimmed the region from the northwest shelf to the southern parts of the basin, where channels might have communicated with the open ocean to the west (Figure 12.5). In the northern part of the Delaware Basin, the Capitan Reef carbonates separated the basin depocenter from the exposed shelf with sabkha and lagoon sedimentation in the carbonate-evaporite platform (Figure 12.14).

Kirkland et al. (2000) proposed a modified closed-basin drawdown model based on Sr isotope analysis and sedimentological observations, suggesting marine-derived brine enrichment by evaporation and partial reflux of denser brines to the ocean. The cyclicity observed in the varves of the Castile and Salado Fm. (Figure 12.12) can be correlated for several kilometers. This long-distance correlation suggests that the periodic influx of seawater into the deep basin persisted over thousands of years. This rhythmical water influx could not be obtained via a long-lasting channel or strait separating the Panthalassa Ocean from the Castile basin (Kirkland et al., 2000). The source of the water that infiltrated into the basin probably came from intermittent seepage through the porous carbonate rock debris that constituted the fore-reef talus of the Capitan Reef barrier (Figure 12.13-B). A comparison with the composite model of basinwide salt deposition seems appropriate (Figure 12.2-D).

The reflux of highly concentrated brine was effective through the reef and fore-reef debris around the entire circumference of the Delaware Basin (Kirkland et al., 2000). Communication with the open ocean was established through a connection that might correspond to the Hovey channel, a narrow inlet that linked the ocean with the southern part of the Castile evaporite basin during the Permian (Figure 12.5).

Assuming a paleogeographic reconstruction involving a closed basin, with the depocenter surrounded by the carbonate reefs (Figure 12.13-A), the Castile brine surface might have been only slightly below sea level during the basin desiccation and deposition of the Salado Salt (Figure 12.11). During sea level rises, seepage of normal marine seawater into the basin (which might have been hundreds of meters deep) would reduce the brine concentration from the chloride to the sulfate field of chemical precipitation, forming the couplets of halite and anhydrite (Figure 12.12).

This interpretation is based on detailed studies of outcrops and subsurface analysis, as several boreholes have been drilled by academic institutions and the petroleum industry (Ward et al., 1986). The Castile evaporite basin is a well-studied paradigm of inland seas that deposits evaporite sediments in deepwater settings (Figure 12.6 and Figure 12.13). Sedimentologic and petrographic analysis distinguishes several facies from the deep basin to the onshore mudflats, including (1) a deepwater basin with lime mudstones; (2) an apron of allochthonous reworked carbonates; (3) a carbonate bank rimming the shelf margin, forming a prominent reef (Capitan Reef); (4) carbonate

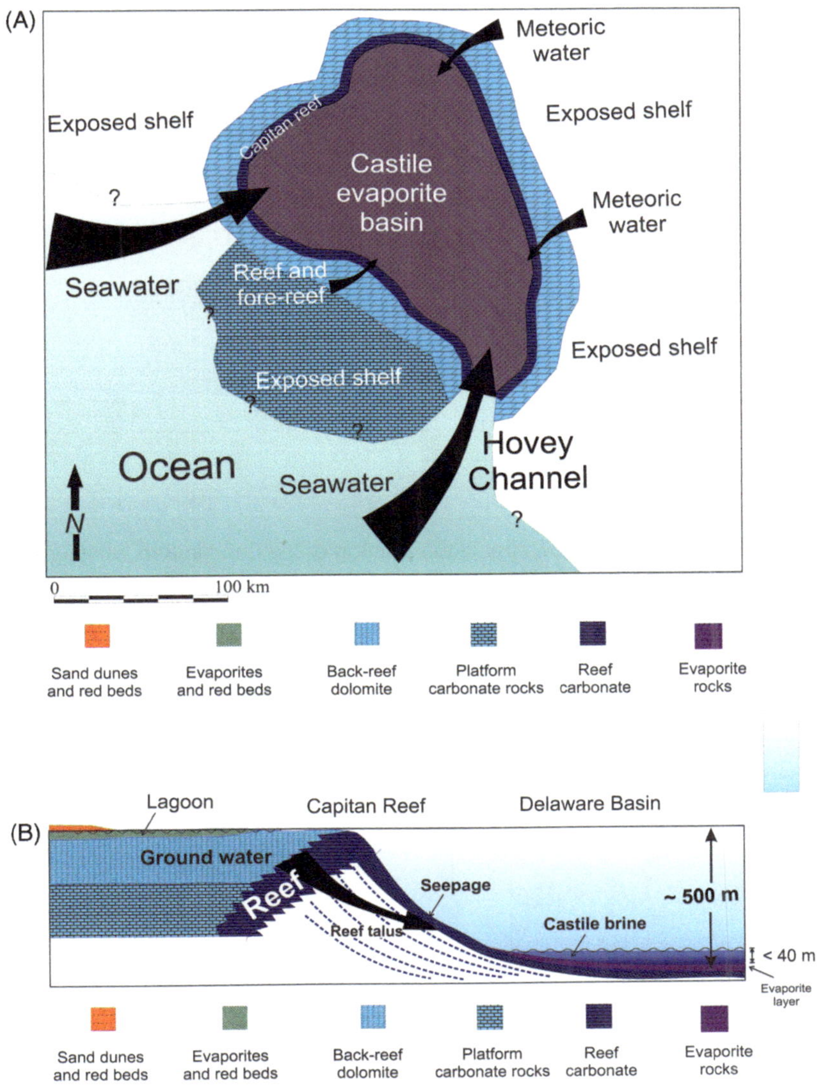

Figure 12.13 (A) The simplified paleogeographic reconstruction of the Permian Basin during the Upper Permian period shows the exposed shelf with arid deserts in the north (New Mexico), while the Delaware Basin is bordered by sabkha and carbonate rocks. The hypersaline basin had intermittent connections with the open sea, with meteoric water inflow mainly from the north and east. (B) The simplified N-S geological cross-section from New Mexico to the Delaware Basin depocenter in Texas illustrates the sabkha flats, lagoon, and Capitan Reef that fringed the Delaware Basin at the lagoon's edge. The basin depocenter was characterized by a deep basin with water depths estimated at 0.5 km and a brine level much lower than the global sea level. Less than 40 m of bedded salt deposits accumulated on the lake's bottom, indicating a starved basin. Credit: Adapted from Kirkland et al. (2000).

sand flats, composed of limestone and dolomites; (5) carbonate barrier islands, composed of dolomitic pisolitic grainstones; (6) a restricted lagoon, with dolomitic mudstones and wackestones; and (7) coastal playas, supratidal mudflats, and salt flats (Sabkhas), with intercalations of limestone, mudstone, gypsum, and anhydrite layers.

The northern part of the Delaware Basin is characterized by an extensive salt flat along a broad lowland region adjacent to continental deserts that extend to the Palo Duro Basin in northern Texas (Figure 12.6). The Castile brines were concentrated in local basins due to extreme climate aridity during the Ochoan Epoch, precipitating thick layers of halite and other soluble minerals (Figure 12.8).

The paleogeographic interpretation proposed by Handford (1981b) for Late Permian sequences in the Palo Duro Basin corresponds to a salt pan surrounded by marginal salt flats in a sabkha environment extending southward toward the Permian Basin (Figure 12.15).

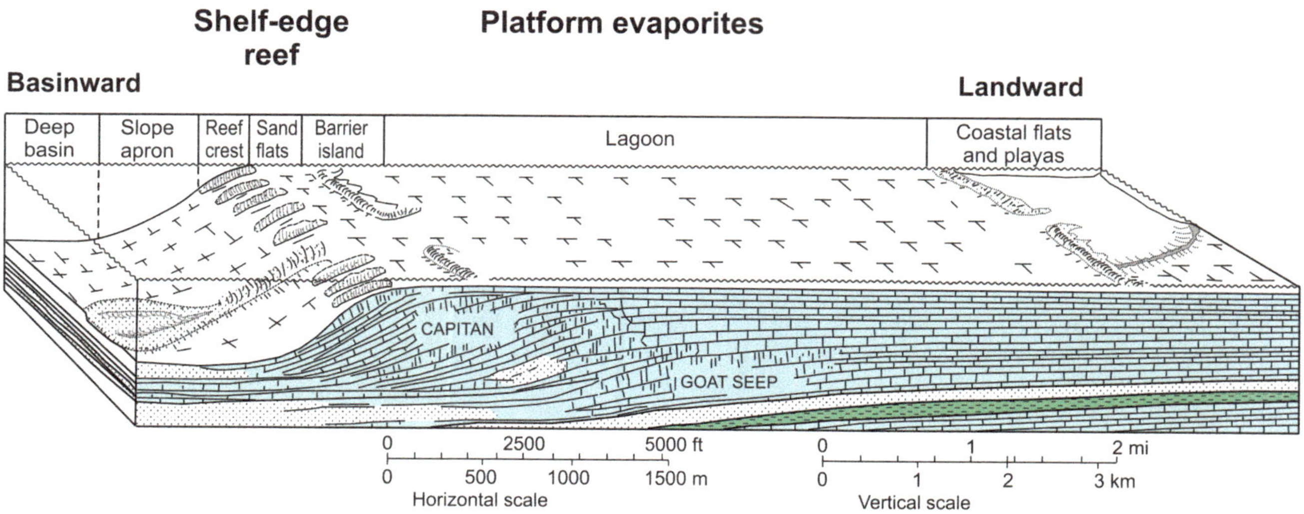

Figure 12.14 A geological 3D diagram shows the reconstruction of the Capitan Reef fringing the shelf edge of the Delaware Basin in the Permian. The diagram illustrates the Sabkha flats in the north and the deep basin in the south. The lagoon environment changed landward to coastal flats and playas. The sedimentary environments include a barrier reef at the shelf edge, carbonate sand, subtidal mudflats, and supratidal mudflats. Basinward, slumped sediments and carbonate debris occur at the slope, while deepwater sediments deposit on the distal basin floor, forming turbidite fans and reworked clastic evaporites. Credit: Adapted from Ward et al. (1986) and King (1942).

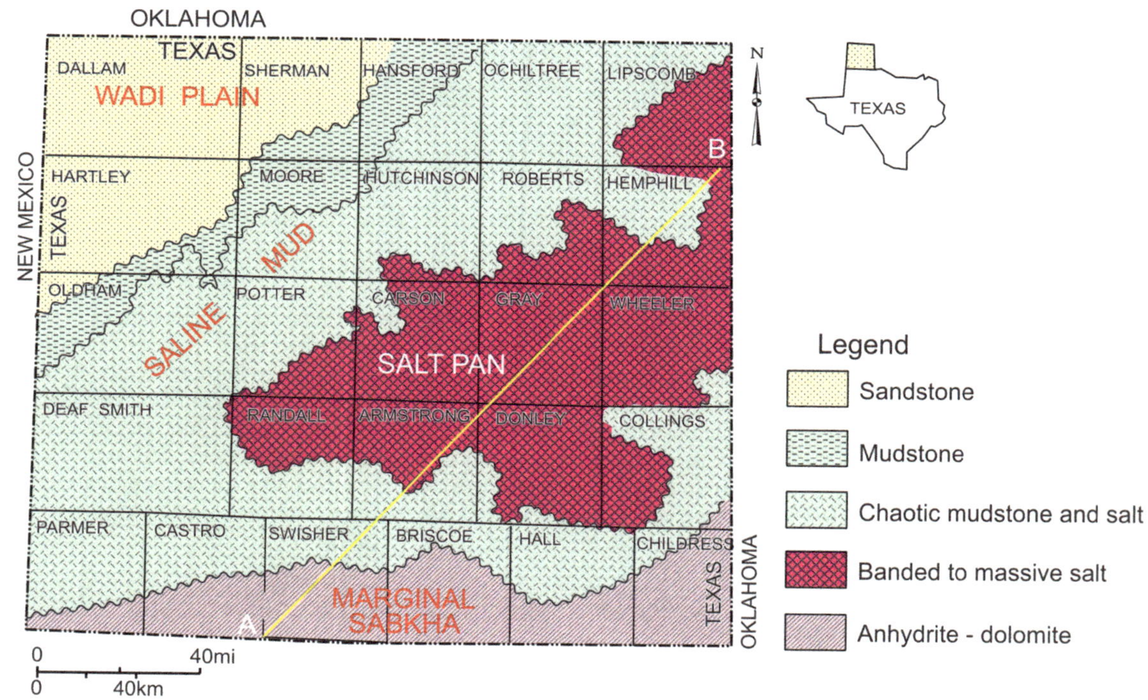

Figure 12.15 The simplified lithofacies map shows the interpretation of the evaporite cycles associated with the Lower Clear Fork Fm. in the Palo Duro Basin, north of the Permian Basin in northwest Texas. Marginal sabkha sediments, such as dolomite and anhydrite, occur south of the Amarillo Uplift, which separates the Palo Duro from the Anadarko Basin in Oklahoma. The geological cross-section A-B extends north of the Matador Arch, which is associated with reverse faults involving the Upper Permian evaporite sequences. Credit: Adapted from Handford (1981).

The genetic model assumes that eolian sandstones formed continental deserts near the border between Texas and New Mexico. The laminated red siltstones and mudstones were deposited in a desert-alluvial plain and dry mudflat setting that graded into the saline mudflat that surrounded a salt pan. Deeper lagoons formed in the central part of the depression, depositing chaotic mudstones intercalated with evaporitic sediments. During the Lower Permian, changes in brine concentration resulted in the precipitation of bedded halite and laminated anhydrite-dolomite couplets in the Delaware Basin depocenter (Handford, 1981b).

The Lower Clear Fork Fm. is characterized by an evaporite facies variation from the Matador Arch toward the Anadarko Basin, where carbonates and sulfates (dolomite and anhydrite) change into chloride (halite) from south to north (Figure 12.15), indicating a saltern in the region of the Amarillo Uplift in Gray County (Handford, 1981a).

Johnson and Gonzales (1978) proposed that the evaporite facies variations involving the Late Permian salt sequences in the Palo Duro Basin were controlled by a structural high likely associated with a reverse fault with a vergence toward the Anadarko Basin (Figure 12.16). The Anadarko Basin in Oklahoma is characterized by subhorizontal evaporite beds with a larger percentage of halite in the anticlinal structure region in Gray County (Figure 12.15).

The lowermost evaporite layer corresponds to the Hutchinson Salt, with a depositional center to the north in the Anadarko Basin (Handford, 1981b). The Salado Salt is characterized by a depositional center to the south, toward the Permian Basin. As the salt layers approach the surface, they are affected by salt dissolution when affected by canyons that cut deep into the subsurface, as observed in the Palo Duro Canyon. Salt dissolution is also characterized in the Anadarko Basin, particularly in the subsurface of the region near the Cimarron River in Woodward County (Figure 12.16).

12.6.4. Genetic Models for the Salt Deposition in the Permian Basin

Handford (1981b) proposed a two-stage development for the Permian Basin, with marine water infilling an arid depression during Stage 1. This stage was followed by rapid subsidence, leading to aggradation and the accumulation of thick sequences of subaqueous, bottom-nucleated evaporites in an extensive brine pan behind the marginal sabkha (Stage 2). The sabkha environment changed southward into a marine basin or inland sea environment formed in the Permian Basin depocenter, where marine limestones and marls were deposited in deepwater settings (Figure 12.5).

The Castile brine in the Permian Basin has been considered by several researchers to be derived from oceanic seawater (e.g., King, 1947; Kirkland, 2003). However, Kirkland et al. (2000) proposed a closed-basin drawdown model involving a nonmarine to mixed marine brine as the source for the thick Upper Permian evaporite sequences in the basin. This model assumes a drop of the lake level by as much as 500 m, forming a major sink for groundwater, which would probably feed the basin from the surrounding Capitan Reef complex. Assuming this model, a large fraction of the concentrated brines in the hypersaline lake originated from recycling from older Permian evaporites, which were deposited on the surrounding flat, arid, and generally dry carbonate shelf located north of the reef (Figure 12.14).

King (1947) interpreted the Castile Fm. evaporites in the Delaware Basin as deposited in a restricted marine embayment. Following this model, the inland shallow sea was partially connected with the open sea by a channel located at the southwestern part of the basin (Figure 12.5). However, Kirkland et al. (2000) proposed several other connections with the open sea in the western part of the Delaware Basin, adjacent to the Capitan Reef, such as the Hovey channel at the southern part of the basin (Figure 12.13).

King (1947) postulated that the marine-derived brine level was hundreds of meters deeper than the global sea level, forming a denser and more concentrated layer below the average wave base. According to the model, reflux or submarine currents through the barrier resulted in a continuous flow of denser brine toward the open sea. The barrier sporadically isolated the basin from the open sea, resulting in basin desiccation (King, 1947).

Lucia (1972) challenged this evaporitic model, proposing that the Delaware Basin might be associated with an inland sea or a hypersaline lake with no channel connecting it to the open sea. Intermittent communication might occur through reflux through the carbonate barrier, particularly the Capitan buildup that encircled the basin in the Upper Permian. The brine level during periods of extreme aridity might have been drawn hundreds of meters below the global sea level, exposing the Capitan Reef carbonate rocks and their basinal reef debris. This drop in brine level might reach levels much lower than the groundwater table, and thus the Delaware Basin would be a major sink for the groundwater flow, involving marine and meteoric waters (Figure 12.13). A mixed brine with hybrid waters resulted in a higher salinity concentration, favoring the precipitation of sulfates rather than chlorides and forming varves (Figure 12.12-B).

12.6.5. Hydrocarbon and Mineral Resources in the Permian Basin

Although salt is an important component of the petroleum systems in the Permian Basin, it does not form large diapiric salt structures with hydrocarbon traps. The Castile and Salado evaporite layers are bedded salt (halite) intercalated with sulfates (gypsum and anhydrite)

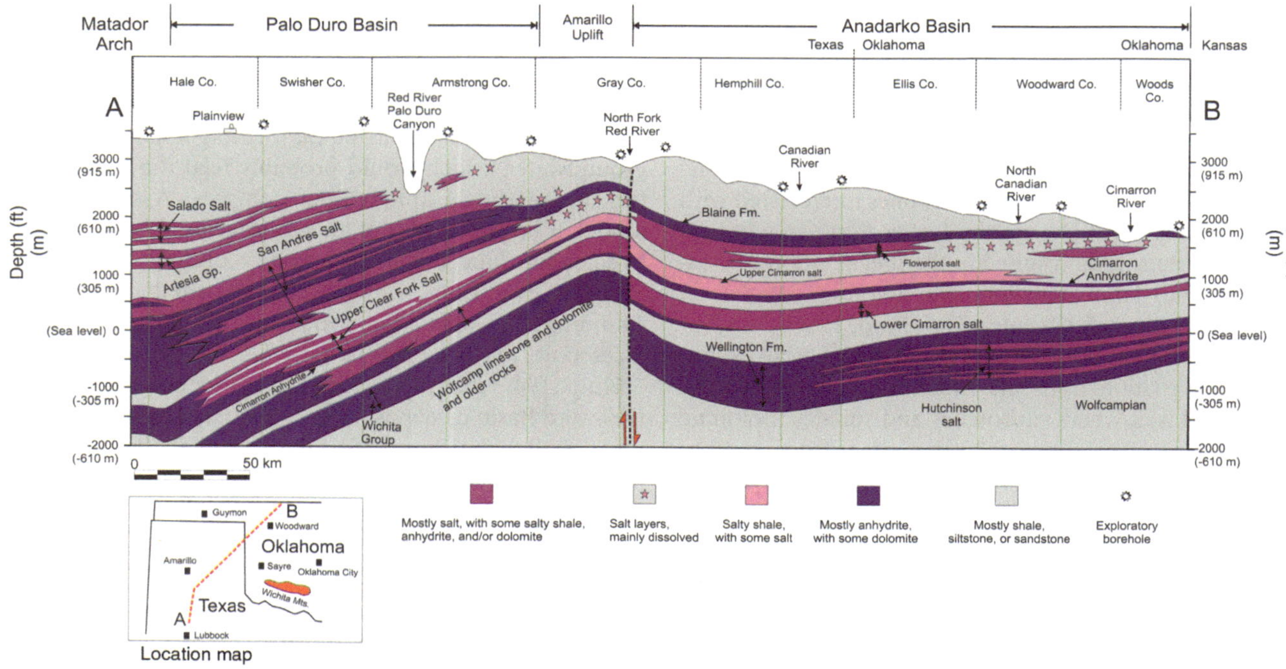

Figure 12.16 The regional cross-section from the Matador Arch to the Anadarko Basin corresponds to sections A-B in the paleogeographic map of the previous figure, crossing the salt pan north of the marginal sabkha. Upper Permian evaporite deposits in the Palo Duro Basin are affected by thrust faults with a northern vergence, resulting in an anticlinal axis in the Amarillo Uplift. Salt dissolution is observed in the Palo Duro canyon incision and above the Amarillo Uplift, as indicated by the star symbol. Extensive salt dissolution is interpreted in the subsurface of the Anadarko Basin. Credit: Adapted from Handford (1981).

and carbonates (calcite and dolomite), which overlie the Middle to Late Permian (Guadalupian Epoch) carbonates and siliciclastic rocks, which correspond to the main reservoirs in the basin (Figure 12.8). The main source rocks in the Permian Basin are organic-rich calcareous shale and shaly limestone, deposited within restricted lagoons in a carbonate shelf environment developed during the Lower Permian (Figure 12.14). Geochemical analyses indicate that the most important source rocks are shales from the Wolfcamp Fm., with TOC content ranging from less than 2.0% to about 8.0% (Ward et al., 1986).

The Permian Basin is one of the most drilled and prolific hydrocarbon provinces in the world, with more than half a million exploratory wells. The oil production in the basin dates back almost a century, with estimates suggesting that the basin has had cumulative production of more than 40 billion barrels of oil since 1921 when the first exploration well was drilled and hydrocarbons were discovered at shallow depths (760 m) in carbonate reservoirs of the San Andres Fm. (Ward et al., 1986; Robichaud & Robichaud, 2019). The daily production reached 2 million barrels of oil per day in the late 2010s (Engle et al., 2016, 2020). Hydrocarbons have been produced from different reservoirs, ranging from the Early Paleozoic to the Upper Permian (Ewing, 2017; Saller & Stueber, 2018; Stueber et al., 1998).

Most of the production is related to reservoirs associated with the Upper Permian carbonate rocks, particularly the San Andres Formation (Guadalupian). When dolomitized, the carbonate rocks from the San Andres Formation constitute important hydrocarbon reservoirs, responsible for more than 50% of all production from the Permian Basin. The thick Upper Permian bedded salt layers in the subsurface provide an excellent regional seal for hydrocarbon accumulations in the underlying dolomite, carbonate, and siliciclastic reservoirs (Warren & Kendall, 1985).

The Permian Basin is also responsible for a large production percentage of evaporite ores in the United States, particularly bedded sylvite deposits associated with the Upper Permian Salado Fm. evaporites. The total salt (halite) production in the United States in 2020 was about 39 million metric tons (U.S. Geological Survey, 2021). Most of the production comes from salt mines in the northern states (Kansas, Michigan, New York, and Ohio), including brine salt produced by the evaporation of the formation water pumped from the subsurface reservoirs, and only a small percentage is produced in Texas.

When the soluble evaporite layers of the Salado Fm. occur near the surface, they are exploited in salt mines, as in the region near the city of Carlsbad in southeastern New Mexico. The Carlsbad Mining District concentrates most of the potash deposits produced from the Permian Basin, either by underground mines or by deep-well solution mines, contributing to New Mexico's role as the number one producer of potash salt in the United States.

12.7. THE CARBONATE PLATFORM EVAPORITES IN THE ARABIAN GULF

The Arabian Peninsula (Figure 12.17) is linked to several Neoproterozoic extensional intracratonic rift basins, such as the Rub' Al Khali Basin in Saudi Arabia, the Hormuz salt basin in the Persian Gulf, the Zagros region in Iran, and the Ara salt basins in Oman (Al-Johi & Al-Laboun, 2015; Osterloff et al., 2004; Stewart, 2016; Ziegler, 2001). During the Permian, a carbonate-evaporite platform was established across the northern part of the margin, from Saudi Arabia to the Persian Gulf (Figure 12.18).

During the late rift phase (Neoproterozic–Early Cambrian), the Hormuz salt layers were deposited in pull-apart rift basins located between Mesopotamia and the Qatar Arch (Western Gulf) and in the region between the Zagros suture zone and the Oman ophiolite (Southern Gulf). The salt basin depocenters recorded in Saudi Arabia, Iraq, Iran, and Oman are separated by regional highs and arches (Stewart, 2016; Thomas et al., 2015).

The rifted crust of the Arabian Peninsula evolved to form passive continental margin platforms in the Paleozoic. During the Mesozoic, the passive margin basins developed oceanic crust associated with the Neotethys Ocean. The passive margin extended from Turkey to Iran in the north and from Iraq to northern Arabia and Oman in the south (Alsharhan & Nairn, 1997, 2003; Osterloff et al., 2004). During the Cretaceous and Cenozoic, the convergence between the Arabian and Eurasian plates resulted in subduction zones

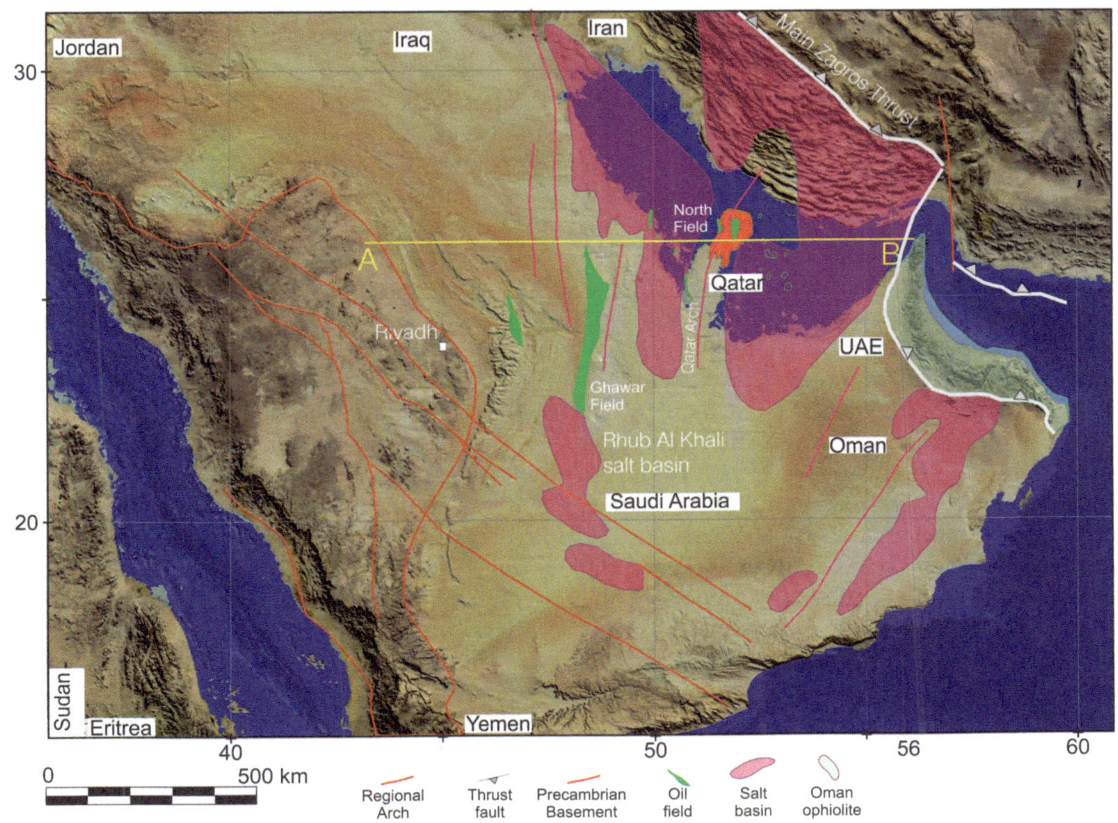

Figure 12.17 The satellite image of the Arabian Gulf shows the Zagros deformation zone, the Zagros Main Thrust Fault, and the outline of Precambrian basement rocks in western Saudi Arabia. The Ghawar oil field in Saudi Arabia corresponds to an elongated N-S structure, and the North Field in Qatar overlies a major basement arch. A thin yellow line shows the location of the regional east-west cross-section corresponding to profile A-B, extending from Precambrian outcrops to the Hormuz Strait.

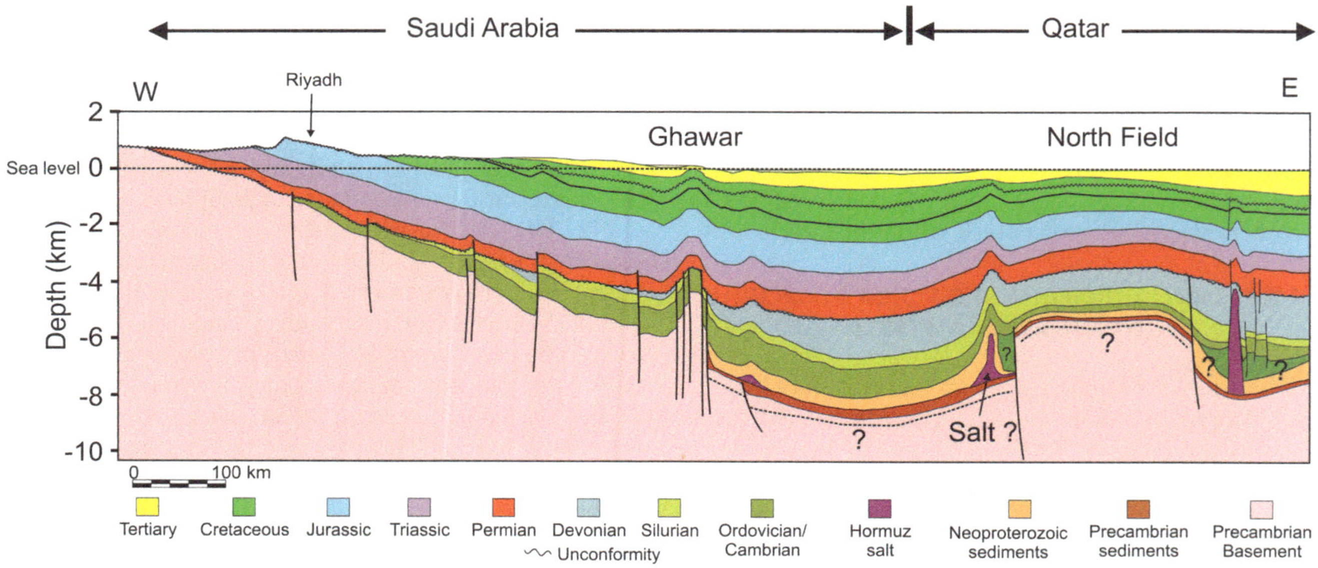

Figure 12.18 The Ghawar oil field and the North Field gas accumulation are visible in a regional cross-section from western Saudi Arabia to eastern Qatar. The Neoproterozoic–Early Paleozoic Hormuz salt formed salt diapirs in the Persian Gulf, while a carbonate-evaporite platform developed in the Permian. The Cenozoic deformation of the Hormuz salt led to outcrops in the Zagros region. The Permian and Jurassic sequences also record evaporite deposits, which form basinwide carbonate-evaporite sequences. Credit: Adapted from Afifi (2005).

and the formation of foreland basins (Agard et al., 2011; Alsharhan & Nairn, 2003; Konert et al., 2001).

The Hormuz salt basin in the southern Gulf and the Oman salt basins have been partially subducted along the Zagros suture zones (Figure 12.17). The Hormuz salt basin in the western Gulf (Mesopotamia region) has been only mildly deformed by the plate convergence (Thomas et al., 2015). In the Zagros fold-thrust belt of central Iran, several salt structures reach the surface, resulting in spectacular outcrops of the Hormuz salt and younger evaporite sequences extruded as salt glaciers (see Chapter 3). A major unconformity, associated with the Hercynian Orogeny, separates the Permian sequences from the underlying Carboniferous, Devonian, and older sections (Faqira et al., 2009; Konert et al., 2001; Ziegler, 2001).

Following the Gondwana breakup in the Triassic, the margins developed a wide, shallow, epeiric sea in the Jurassic, when oceanic spreading centers were formed between the Arabian and Eurasian plates, resulting in the Neotethys (Beydoun, 1991). A subduction zone involving the Jurassic oceanic crust was developed in the Neotethys margin during the Late Jurassic and Early Cretaceous. The subduction of the Arabian plate under Eurasia and Arabia consumed the younger and older oceanic crusts, resulting in continental block collisions in the Cenozoic (Agard et al., 2011; Faqira et al., 2009; Ziegler, 2001).

The Hormuz evaporite sequences were deposited in the Neoproterozoic and formed extensional salt diapirs in the Paleozoic (Asl et al., 2019). During the Cretaceous and Cenozoic, the diapiric salt structures were deformed by compressional tectonics. The salt diapirs are seismically imaged in the subsurface of the Persian Gulf (Mesopotamian Basin) and locally outcrop, forming some islands near the southern Iranian coastline. On land, some salt structures are exposed due to contraction, forming salt glaciers in the Zagros fold-thrust in Iran (Agard et al., 2011; Glennie, 2010; Hassanpour et al., 2020).

In the Neogene, the closure of the Neothethys Ocean and the collision of the continental blocks involving Arabia and Eurasia resulted in the Zagros fold-thrust belt, one of the most seismically active mountain ranges in the world, with intense earthquake activity (Nissen et al., 2011). The Mesopotamian and Persian Gulf depressions correspond to a foreland basin associated with the deformation belt extending from Turkey, Iraq, and Iran, reaching the Strait of Hormuz at the eastern extremity (Figure 12.17).

The Arabian Gulf is characterized by diverse episodes of salt deposition. The oldest salt deposits are related to the basinwide evaporites associated with the initial rift phase of the basin, which resulted in the accumulation of the Neoproterozoic Hormuz salt. The early to late Paleozoic post-salt interval is characterized by shallow marine evaporitic sedimentation above a rifted continental crust. Oceanward, at the transition to the oceanic crust of the ancient Tethys Ocean, the northeastern Arabian passive margin formed a thick sedimentary prism with a diapiric salt province associated with the Hormuz evaporites (Alsharhan & Nairn, 2003; Hassanpour et al., 2020; Konert et al., 2001; Stampfli et al., 1991).

Following the development of the Tethys margin, which resulted in a late Paleozoic phase of thermal subsidence, the Permian sediments were deposited across a large shallow-water carbonate-evaporite platform at the southern margin of the Permian Tethys Ocean, also known as Permotethys (Stampfli et al., 1991). The Permotethys E-W spreading centers formed a V-shaped oceanic crust segment, which separated the Arabian block in the south from the Armenia-Iran-Afghanistan-Lhasa blocks in the north.

The Late Permian carbonate-evaporite sequences include the Khuff Anhydrite, which is characterized by thick layers of sulfate deposits intercalated with carbonate rocks. The breakup of Gondwana during the Late Triassic-Early Jurassic resulted in the formation of spreading centers that separated the African, Arabian, and Eurasian landmasses as the Neotethys oceanic propagator advanced toward the Alps (Stampfli et al., 1991; Ziegler, 2001). The passive margin carbonate-evaporite platform is associated with the Jurassic Hith Anhydrite deposits on the continental shelf of the northern Arabian–Persian Gulf margin.

The subduction zone formed between the Arabian plate and the Armenian–Iranian block in the Cretaceous resulted in the development of an accretionary prism and magmatism in the upper plate. The Arabian plate was subducted under the Eurasian plate in the Cenozoic, closing the remaining embayment of the Tethys Ocean (Glennie, 2010; Kendall et al., 2009; Ziegler, 2001). The collision between the Arabian plate and the Anatolian and Iranian blocks in the Neogene resulted in the suture zones that trend NW-SE, following the trend of the thrust faults (Alsharhan, 2014).

The satellite image of the Arabian plate between the Red Sea and the Zagros deformation zone (Figure 12.17) shows the divergent plate boundary separating the African and Arabian plates, corresponding to the Red Sea spreading axis developed in the Neogene (see Chapter 1 in Mohriak, 2025). The convergent plate boundary involving the Arabian and Eurasian plates is marked by a fold-thrust belt between the Persian Gulf and the Main Zagros Thrust Fault (MZT). The MZT corresponds to a deformation zone with crushed rocks near the suture between the Arabian and Iranian blocks. The Zagros deformation zone is located to the northeast of the Mesopotamian Basin and the Persian Gulf. The fold structures are characterized by anticlinal and synclinal axes trending predominantly along the NW-SE and E-W directions. Most of the thrust faults exhibit a vergence to the southwest (Figure 12.17).

The schematic regional cross-section in the Arabian Peninsula, extending from Saudi Arabia to Qatar (Figure 12.18) illustrates the structural style of the margin and the main stratigraphic units from the Neoproterozoic to the Neogene. Both the Ghawar and North Field giant hydrocarbon accumulations are located on structural highs (Figure 12.18). The western part of the basin is characterized by sedimentary layers forming a homoclinal structure. The strata overlying the Precambrian outcrops in the west are characterized by a regional dip toward the basin depocenter, north of the Qatar Peninsula. Compressional structures have affected the Hormuz salt structures and the Mesozoic and Cenozoic stratigraphic sequences, resulting in major anticlinal structures that constitute exploratory targets (Alsharhan, 2014).

12.7.1. Permian Marginal Marine Carbonate–Evaporite Platform

The Khuff Fm. evaporites are primarily anhydrite layers deposited during the Late Permian (around 250–270 Ma), following a denudation episode associated with the Hercynian Orogeny (Konert et al., 2001). The shallow platform sedimentation continued until the early Triassic when a major rifting episode affected the entire region. The Permian Tethys Ocean formed a V-shaped oceanic basin, extending from the NW African and Arabian mainlands in the south to the Eurasian continental block in the north (Scotese, 2001; Stampfl et al., 1991).

The shallow carbonate platform extended northward from the Precambrian outcrops in central Saudi Arabia, covering the southern continental margin of the Permian Tethys Ocean (Konert et al., 2001). The continental margins along the northern Arabian plate accumulated thick sequences of Permian carbonate and evaporite rocks in a shallow platform rimmed by carbonate banks at the shelf edge. The Khuff Fm. reaches a maximum thickness of about 1,000 m in the United Arab Emirates (Alsharhan, 2006) and is composed of several different lithologies, mainly dolomite, limestone, and dolomitic limestone, with subordinate anhydrite and minor halite (Ehrenberg et al., 2007; Vaslet et al., 2005).

The sedimentological studies and petrophysical characteristics of the Middle Permian–Early Triassic Khuff Fm. carbonate-evaporite rocks indicate four distinct depositional settings: (1) supratidal (sabkha); (2) lagoon; (3) carbonate banks, shoals, and reefs; and (4) shallow continental shelf. Most of the Khuff Fm. carbonate-evaporite deposits are composed of several cycles of carbonate-sulfate minerals, indicating an upward shallowing after each marine transgression, resulting in carbonate sequences capped by anhydrite deposits (Ehrenberg et al., 2007; Vaslet et al., 2005).

During the Late Permian, the Khuff Fm. formed an extensive carbonate platform extending from central Saudi Arabia to Qatar and Abu Dhabi in the east, rimmed by a carbonate reef buildup near the shelf edge (Figure 12.19). Northward, the basin deepened toward a subduction zone with a trench where deep marine carbonate rocks (calcilutite and marls) accumulated along an NW-SE trough (Agard et al., 2011; Beydoun, 1991). Siliciclastic sedimentation (eolian dunes, redbeds, and mudflats) predominated in the Central Saudi Arabia region, north of the Precambrian outcrops. Carbonate and evaporite facies formed shallow-water sequences that covered the region of the present-day Persian Gulf, while the region near the Zagros Main Thrust was characterized by argillaceous deep marine carbonates (Figure 12.19).

The exploratory boreholes that drilled the Permian sequences indicate an increase in the siliciclastic facies toward the west and toward the base of the sequence, especially in Saudi Arabia. The evaporite/siliciclastic ratio increases toward the top of the sequence, which straddles the boundary between the Permian and Triassic and is marked by a low sea-level curve (Alsharhan, 2006). The Middle Anhydrite layer separates the Upper Khuff (248–260 Ma) and Lower Khuff Fm. (260–269 Ma), which overlie Early Permian siliciclastic rocks (Figure 12.19). These carbonate sequences are characterized by several porous reservoir units intercalated with dense, tight, and non-reservoir units (Alsharhan, 2006; Dasgupta et al., 2001).

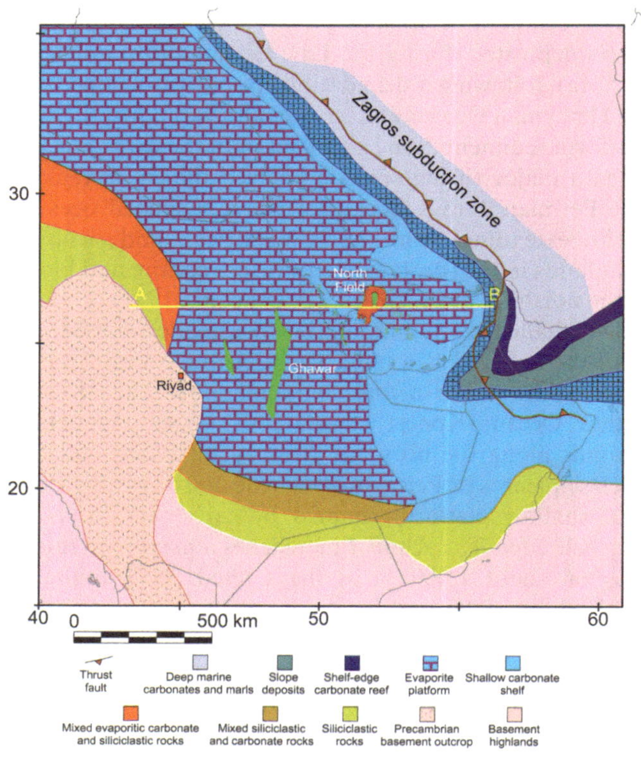

Figure 12.19 The paleogeographic reconstruction of the Arabian Plate during the Permian reveals giant hydrocarbon accumulations, such as Ghawar and North Field, on the evaporite carbonate platform. Shelf-edge reefal buildups are located near the Zagros thrust fault, where the continental margin was subducted under the Eurasian Plate. The Zagros subduction zone formed on the carbonate platform's distal margin, oceanward of lagoonal evaporite deposits and shelf-edge carbonate reefs. The thrust faults were formed within a zone of deep-marine carbonates and marls, exhibiting southwest vergence.

The isopach maps of the Khuff Fm. indicate that the evaporite layers thicken from about 260 m in southwestern Saudi Arabia to more than 915 m in the central Rub 'al Khali Basin, attaining large thicknesses in Oman (1.2 km) and Iran (1.5 km). This carbonate-sulfate cycle resulted in a prolific petroleum play associated with Lower Paleozoic (Silurian) source rocks and Permian reservoir rocks capped by salt layers. This play is responsible for several giant hydrocarbon accumulations in Saudi Arabia, Bahrain, Qatar, Iran, and the United Arab Emirates, including the Ghawar oil field and the North Field in Qatar (Alsharhan, 2014; Faqira et al., 2009). A schematic stratigraphic section showing the carbonate-evaporite facies variation in the Khuff Fm. (Figure 12.20) illustrates the development of reefal buildups at the shelf edge. The transition to the deep basin was very abrupt, as the shelf break and slope were located very close to the trench zone in the subduction zone (Figure 12.19).

The Khuff Fm. depositional sequences represent a classical basinwide carbonate-evaporite system developed on a shallow platform. The stratigraphic units are subdivided based on sequence stratigraphy, electric log interpretation, and the petrophysical properties of the carbonate reservoirs. The lithostratigraphic units correspond to two main groups, separated by the Middle Anhydrite, which is dated around 255 Ma (Alsharhan, 2006). The lower lithological group corresponds to the Khuff A and B reservoir units, whereas the upper group is associated with the Khuff C and D units. The Khuff carbonate lithologies are divided into KS1 to KS4 above the Middle Anhydrite and KS5 to KS7 below the Middle Anhydrite (Alsharhan, 2006). These intricate subdivisions are important in the characterization of the permeable reservoir rock intervals in the restricted evaporitic carbonate shelf facies developed in the northern part of the Arabian plate during the Permian (Dasgupta et al., 2001; Warren, 2012, 2014).

12.7.2. Jurassic Marginal Marine Carbonate–Evaporite Platform

The Jurassic platform carbonate–evaporite facies maps of the Arabian Gulf region reveal a shallow carbonate shelf in the south with three main intra-shelf depressions (Figure 12.21). These depressions acted as intra-shelf

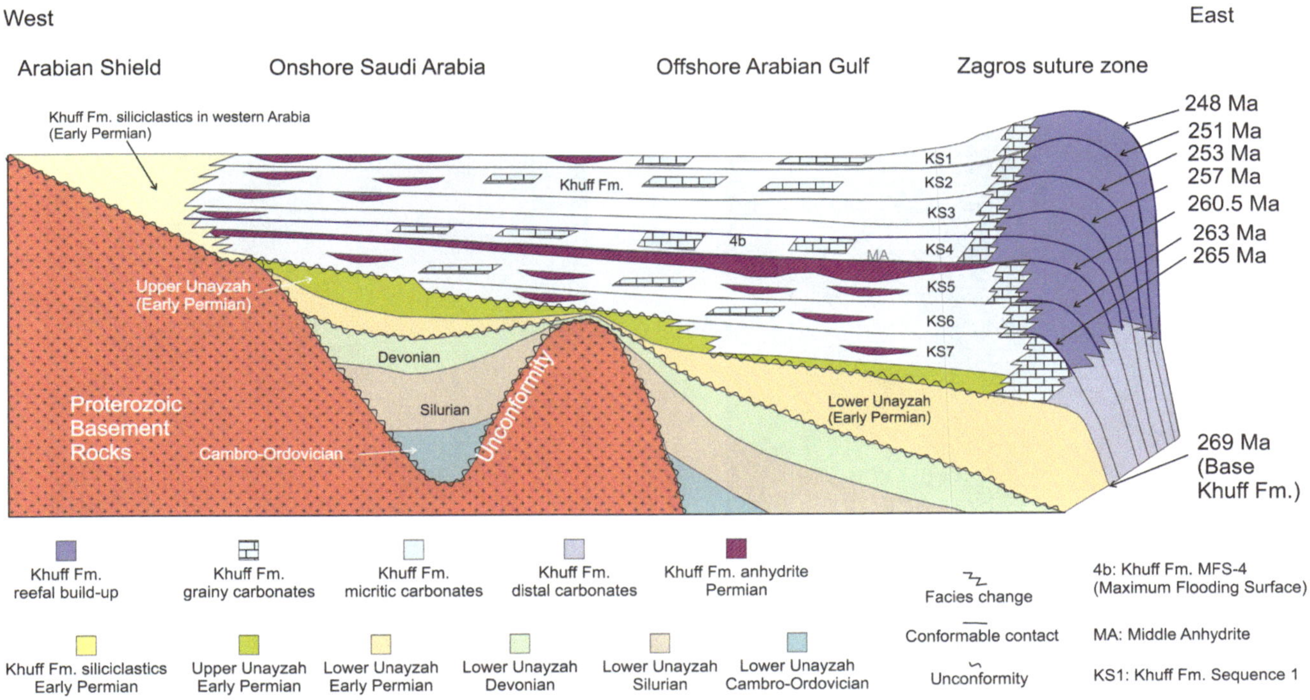

Figure 12.20 The stratigraphic section of the Arabian Gulf shows the distribution of Permian carbonate-evaporite facies from onshore Saudi Arabia to the Zagros suture zone. The carbonate buildup at the edge of the continental shelf is close to the subduction trench. The Middle Anhydrite (MA) divides the stratigraphic sequences into Upper (KS1 – KS4) and Lower (KS5-KS7) Lower Khuff Fm. units. Credit: Adapted from Alsharhan (2006).

depressions that accumulated marine shales during highly anoxic intervals in the Jurassic. The Kimmeridgian shales are the main source rocks for hydrocarbon accumulations associated with the Upper Jurassic to Early Cretaceous petroleum systems. Most oil fields are located in the southern part of the gulf, including the super-giant Ghawar oil field in Saudi Arabia and the Dunkhan oil field in Qatar, which is located to the southwest of the super-giant gas accumulation in the North Field (Figure 12.21).

The shallow carbonate platform was characterized by basinwide deposition of sulfate evaporites during sea-level fall episodes, such as the Hith Anhydrite, deposited between the Kimmeridgian and the Tithonian (circa 150 Ma). During the Upper Jurassic, the sulfate layers were deposited in a shallow marginal basin that covered most of the present-day region adjacent to the Persian Gulf (Figure 12.22). The anhydrite layers form a regional top seal for the Arab-D reservoirs in many oil fields in the Arabian region (Alsharhan & Whittle, 1995; Grötsch et al., 2003; Swart et al., 2005).

The Late Jurassic (Tithonian) salt basins in the Persian Gulf area show two depocenters: the Gotnia salt basin in the northern region and the Rub Al-Khali salt basin in the southern region (Figure 12.22). The northern one (Gotnia salt basin) was affected by compressional tectonics associated with the Zagros orogeny during the Cenozoic (Alsharhan & Whittle, 1995). The southern one (Rub Al-Khali salt basin) is located in central Saudi Arabia, to the east of the Precambrian basement high (Stewart, 2016). The Rub Al-Khali salt basin is characterized by anhydrite deposits overlying the Jurassic evaporitic carbonate platform east of the Arabian Shield Precambrian outcrops. The Gotnia salt basin corresponds to a saltern located in the Mesopotamian region, where halite predominates in the evaporite cycles (Figure 12.22).

The Jurassic Arab Fm. has better reservoir characteristics compared to the Permian–Triassic Khuff Fm., with an average porosity of 12%–26% (Ehrenberg et al., 2007). The shallower Arab reservoirs have greater porosity and permeability due to the greater burial and porosity loss by chemical compaction, diagenesis, and cementation. During the Permian, the world's oceans precipitated carbonate rocks with the mineral aragonite, which is less stable than the mineral calcite that predominated in the Jurassic seas. The calcite in the Jurassic carbonates resulted in more stable mineralogical components, which are less affected by diagenesis. Consequently, the Arab carbonate reservoirs are characterized by better preservation of primary depositional pore types, exhibiting a greater original porosity associated with coarse crystalline dolomite fabrics. Moreover, the Permian carbonates are often affected by anhydrite

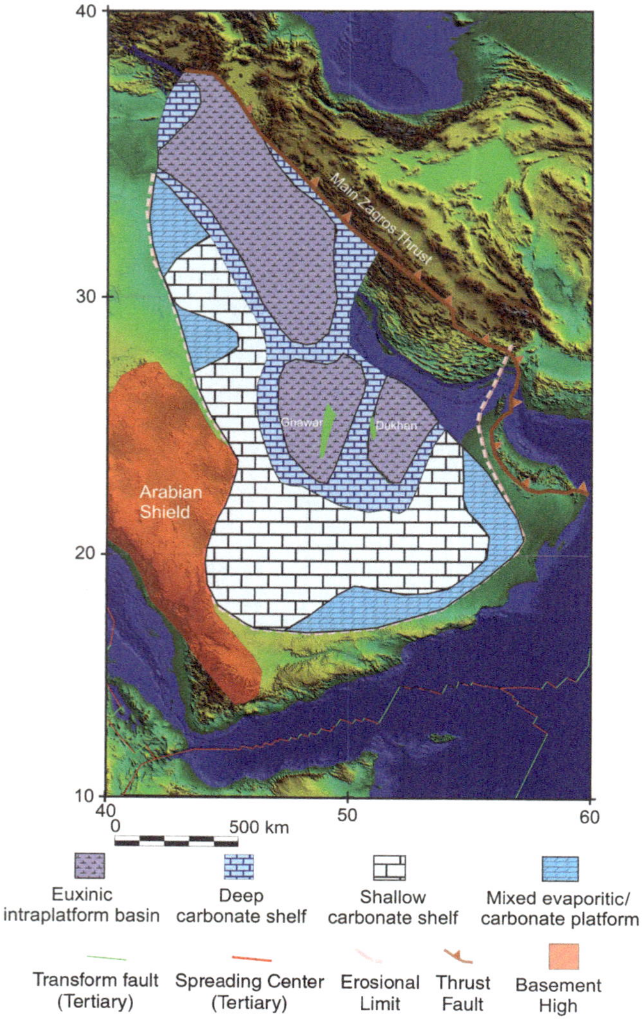

Figure 12.21 The three main Jurassic intra-shelf basins in the Arabian Gulf area, including the Ghawar oil field in Saudi Arabia and the Dunkhan oil field in Qatar, are characterized by regional facies distribution and giant hydrocarbon accumulations. During the Jurassic period, basinwide anhydrite layers formed on the shallow carbonate shelf. The Hith Anhydrite covered the Jurassic basin when the sea level dropped, creating the top seal for the prolific Arab-D carbonate reservoirs in many oil fields in the Middle East. Credit: Adapted from Alsharhan and Kendall (1986).

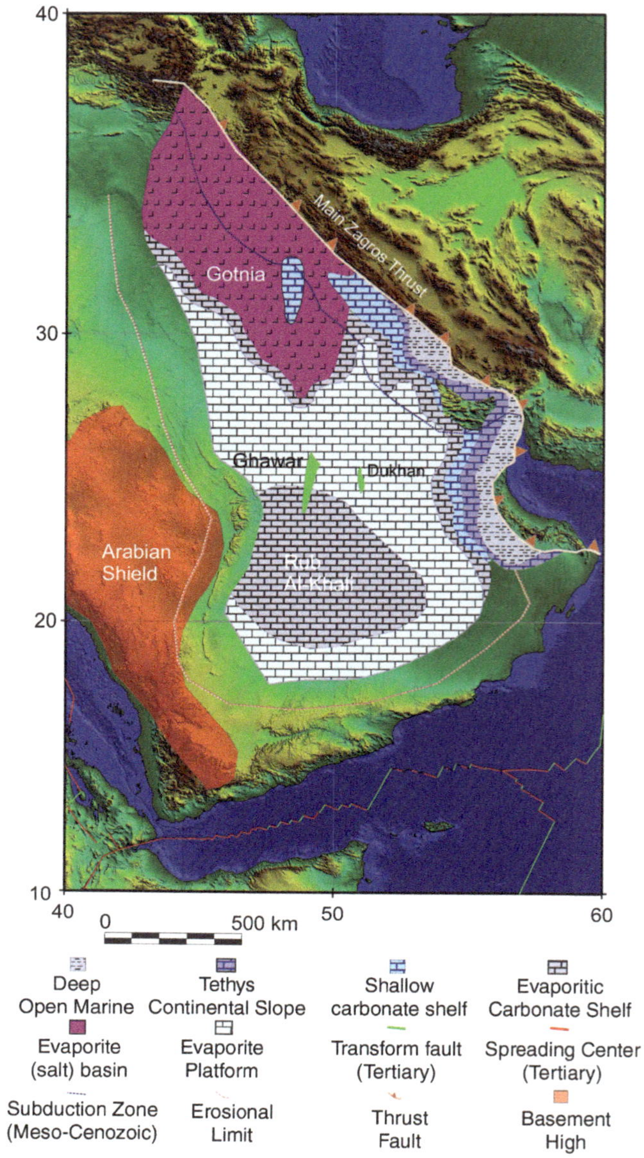

Figure 12.22 The two main Late Jurassic salt basins in the Arabian Gulf area, Gotnia and Rub Al-Khali, are located in the northern extremity of the Gulf. Gotnia is a saltern with halite dominating in evaporite cycles, while Rub Al-Khali has the giant Ghawar field in Saudi Arabia and the Dunkhan oil field in Qatar. Compressional tectonics affected the northern part of the Gotnia salt basin during the Neogene. The southwestern limit of the Jurassic evaporite platform is erosional due to the uplift of the Arabian Shield Precambrian basement.

plugging the original porosity, resulting in denser, tighter intervals with reduced production capacity (Ehrenberg et al., 2007).

The Upper Jurassic Arab carbonate–evaporite sequences include the most prolific reservoir rocks in the Persian Gulf, known as the Arab carbonates. There are four carbonate cycles in the Arab Fm. (A-D), capped by the Hith Anhydrite. The oldest cycle (D), ascribed to the Late Kimmeridgian, is associated with bioclastic packstones and grainstones, developing oolitic ridges at the top of the sequences. The stratigraphic distribution of the Arab Fm. carbonate sequences displays a general younging toward the east (Figure 12.23).

The evaporite cycles represent shoaling up from a marine environment, with bioclastic lime mudstone at the base,

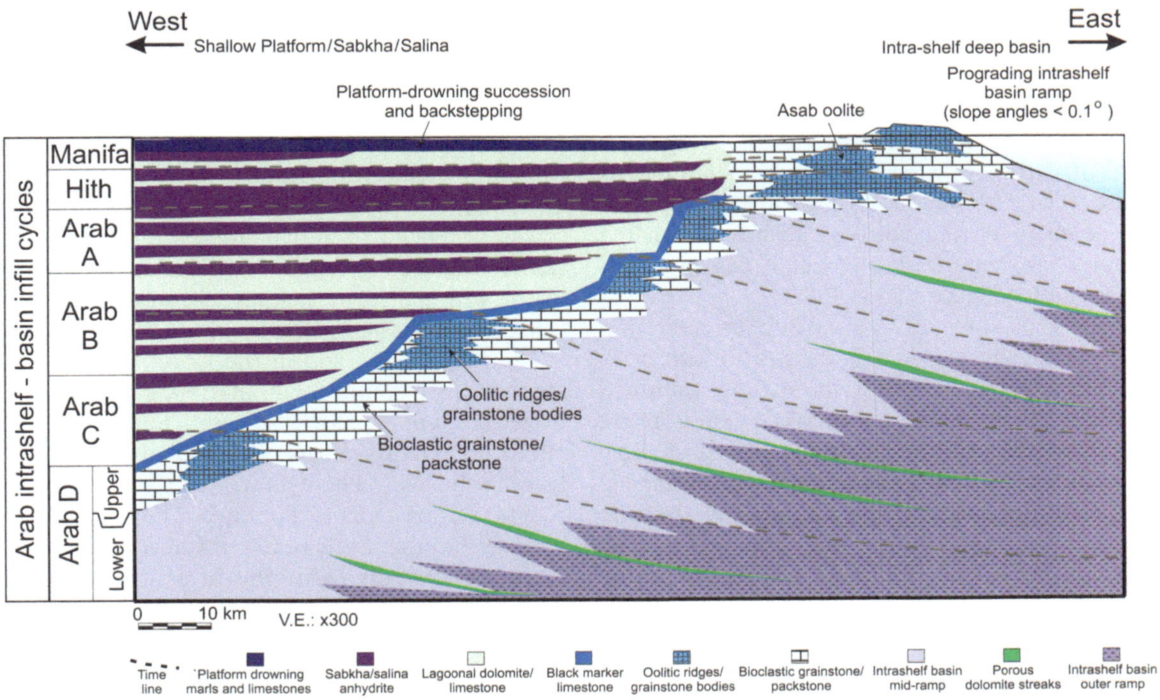

Figure 12.23 The schematic stratigraphic section shows the general cycle of the Late Jurassic Arab carbonate–evaporite sequences forming a series of marine deposits, characterized by bioclastic lime mudstone, calcarenite, and time-transgressive reefal constructions. The lower unit is a bioclastic lime mudstone, while the middle unit is a bioclastic grainstone. The sequences are capped by an upper dolomite unit, which separates the carbonates from lagoonal sediments with anhydrite intercalations. The upper unit grades into massive to nodular anhydrite layers, forming the Hith Anhydrite at the Kimmeridgian to Tithonian transition. These sequences were deposited in a sabkha environment, resulting in thick units of supratidal sabkha evaporites and subaqueous salterns. Credit: Adapted from Grötsch et al. (2003).

evolving into a restricted marine environment with tidal flat dolomites. The top of the cycles developed an evaporitic subaqueous environment (saltern), commonly marked by gypsum deposition. The Tithonian evaporites, formed at the top of the sequence, are mainly characterized by sulfate deposits (originally gypsum) that presently correspond to the Hith Anhydrite (Al-Hussein, 1997; Alsharhan & Whittle, 1995). The massive nodular anhydrite layers included in the Hith Fm. are widely distributed across the carbonate–evaporite platform, extending from the Mesopotamia Basin to the Strait of Hormuz (Figure 12.22).

The Late Jurassic carbonate platform was characterized by depressions that, with total isolation, were desiccated and accumulated halite layers in salterns (Al-Hussein, 1997). These halite deposits in local basin depocenters are ascribed to the Gotnia Evaporite Basin in the Mesopotamian Basin, located north of the Persian Gulf and slightly to the south of the Zagros subduction zone. The large saltern associated with the Rub Al-Khali Evaporite Basin was located in the southwestern and central parts of Saudi Arabia (Figure 12.22).

Regional facies maps calibrated by detailed analysis of cores and electric logs show that the carbonate reservoirs in the Jurassic Arab Fm. are characterized by a shoreline progradation toward the east and northeast (Alsharhan & Whittle, 1995). The cycles culminated with high-energy banks of oolitic and bioclastic grainstones, which climb up in the stratigraphy along the eastward direction (Figure 12.23). The structural highs to the east-northeast are marked by high-energy oolitic/bioclastic grainstone banks, such as the Asab oolite (Grötsch et al., 2003). These porous carbonate banks are capped by a thick unit of subaqueous anhydrite and halite deposited in the salterns. This sulfate layer (Hith Anhydrite) extended across the basin and formed a major seal covering the main hydrocarbon reservoirs in the largest accumulations of the Arabian Gulf, such as the Ghawar oil field in Saudi Arabia and the North Field in Qatar (Figure 12.19).

Based on observations of Jurassic carbonate outcrops and cores from oil fields, several authors (e.g., Marchionda et al., 2018; Warren, 1989) suggest that the Arab Fm. evaporite facies were controlled by paleo-water depths varying from subtidal to supratidal. According to this interpretation, the evaporite rocks located on the apex of structural highs that constitute the major hydrocarbon accumulations in the Persian Gulf were deposited in a subaerial sabkha environment, whereas the evaporites that formed off-structure were deposited in subaqueous environments. The carbonate reservoirs display a paleobathymetric control on the lithological facies, with the shoal complex characterized by reefal constructions exposed subaerially (Figure 12.23 - PG-7).

This implies that some of the carbonate reefs that locally form petroleum reservoirs in the subsurface correspond to the islands that existed in the shallow sea north of the Arabian Peninsula during Jurassic times. These islands formed from sea-level fluctuations in a shallow marine carbonate platform influenced by salt movements. Salt flow toward anticlinal structures led to the development of porous carbonates, such as grainstones and packstones, associated with reefal buildups at the paleo-lagoon border (Marchionda et al., 2018).

The interpretation of carbonate islands is based on analogies with the southern Red Sea, where some carbonate islands developed shallow-water and subaerial carbonate sediments in the Quaternary (e.g., Farasan Islands). These subaerial, high-porosity carbonate banks are related to their development above the rising salt diapirs (Almalki et al., 2015).

Modern analogs from the Red Sea and the southern margin of the Persian Gulf indicate that anticlinal structures associated with halokinetic structures influenced the development of Jurassic islands in the Arabian Gulf, affecting the shallow marine platform. During the Jurassic, the flow of the underlying basinal evaporites, sourced from Neoproterozoic Hormuz Fm., caused the syndepositional shoaling of the seafloor. At present, some islands along the Arabian Gulf are related to salt structures fed by Hormuz salt, such as the southern Iranian coast or Abu Dhabi coast (see Chapter 8). Sir Bani Yas Island, located in the United Arab Emirates (UAE) margin, is characterized by carbonate banks with corals on top of a large salt structure, with carbonate mud sedimentation seaward along the carbonate ramp (e.g., Read, 1985).

12.7.3. Hydrocarbon Resources in the Persian Gulf Carbonate–Evaporite Platform

The Arabian Gulf oil province is home to numerous giant hydrocarbon accumulations, including reservoirs related to carbonate–evaporite sequences. The UAE has five petroleum systems, divided into gas and oil accumulations. The petroleum systems associated with the gas accumulations include (1) Lower Silurian source rocks and Upper Permian Khuff reservoirs and (2) Upper Jurassic source rocks and Lower Cretaceous reservoirs. The petroleum systems associated with oil accumulations include the following source rock and reservoir components: (1) Upper Jurassic/Upper Jurassic, (2) Upper Jurassic/Lower Cretaceous/Lower Cretaceous, and (3) Middle Cretaceous/Middle to Upper Cretaceous/Cenozoic (Yin et al., 2018). Most hydrocarbon accumulations are related to anticlinal structures with carbonate reservoirs sealed by Mesozoic to Cenozoic platform evaporites, with anhydrite-rich layers constituting a common trap seal (Alsharhan, 2014).

The largest gas field in the world that is not associated with petroleum accumulation is the North Field in Qatar (Figure 12.18). The gas field was discovered in 1971 and extends across the Arabian Gulf as far as the Iranian border. The Qatari part of this giant field covers an area of 6,000 km^2. The total recoverable reserves in the anticlinal structure are estimated to exceed 1,000 Tcfg, resulting in a field life that might extend for two centuries at planned production rates (Alsharhan, 2014).

The overlying Upper Jurassic carbonates of the Arabian Gulf are ascribed to the Arab Formation, a carbonate–evaporite sequence that includes carbonate reservoir rocks associated with some of the most prolific oil fields in the world. The Ghawar Field in Saudi Arabia holds most of the oil in the Jurassic carbonate reservoirs. Several other oil fields with equivalent reservoirs are registered in the Persian Gulf, such as the Berri, Damman, and Qatif oil fields (Alsharhan, 2014; Alsharhan & Kendall, 1986, 1994, 2003, 2011; Alsharhan & Whittle, 1995; Lindsay et al., 2006). The source rock kitchens across the Arabian Gulf are distributed in several stratigraphic sequences responsible for the supergiant hydrocarbon accumulations in this extraordinarily prolific oil province. The Upper Jurassic mudstone and marlstone source rocks are characterized by a total organic content varying from 1.0% to 8.4%. The hydrocarbon generation potential ranges from 20 to 50 mg/g, and the kerogen type belongs to Type I sapropel. The source rocks are not very thick (ranging from 50 to 300 m), but exhibit excellent oil-generation properties (Alsharhan, 2014; Swart et al., 2005).

Integrated geological and geophysical datasets led to the discovery of the giant Ghawar oil field in 1948 (e.g., Afifi, 2005). The hydrocarbon trap is related to an anticlinal structure involving marginal marine carbonate and evaporite rocks (Afifi, 2005; Alsharhan, 2014; Lindsay et al., 2006). The dolomitization of the Arab-D reservoir in the Ghawar field is associated with hypersaline brines derived from the overlying anhydrite-halite beds. This diagenetic process probably occurred shortly after the Jurassic salt deposition, as indicated by petrographic and geochemical analysis (Swart et al., 2005).

12.8. CONTINENTAL TO MARINE EVAORITES IN THE GULF OF CALIFORNIA RIFT BASINS

The Gulf of California shares some similarities with the Dead Sea, as it corresponds to a rift basin associated with strike-slip faulting related to a major transform fault zone. Both basins were formed by transcurrent movements that resulted in rhomboid-shaped pull-apart basins that were filled with Neogene clastic sediments and evaporites. However, the Dead Sea Basin corresponds to a depression filled by sediments deposited on a continental crust substratum (see Chapter 11). On the other hand, the Gulf of California has developed spreading centers and incipient divergent margins floored by an oceanic crust emplaced in the last few million years (Miller & Lizarralde, 2013).

The origin of the evaporites in the Dead Sea is controversial, with some authors suggesting that the halite deposits were formed by marine water incursions from the Mediterranean Sea in the Pliocene (Zak, 1997). However, recent works challenge this interpretation, postulating that salt minerals were precipitated by the evaporation of nonmarine brines (see Chapter 11).

The Gulf of California evaporites filled depressions below the global sea level and are associated with marine incursions from the Pacific Ocean, which sourced the continental rift basins with marine ingressions in the Upper Miocene (Figure 12.24). The nature of salt deposits in the deepwater region of the Gulf of California is open to debate. The evaporites are indicated by indirect methods, mainly geophysical interpretation, but the salt masses at the continental-oceanic crust transition have not been directly sampled by drilling. The DSDP investigated the Gulf of California in the late 1970s, but the drilling of the scientific boreholes in the Guaymas Basin was mainly aimed at characterizing the nature of the basement. These wells targeted the volcanic basement rocks in the axial trough, where evaporites are not present (Moore et al., 1982). The Gulf of California salt basin, known as the East Guaymas Evaporites (EGE), is located on the North American plate, eastward of the spreading center that separates Baja California from the mainland (Figure 12.24).

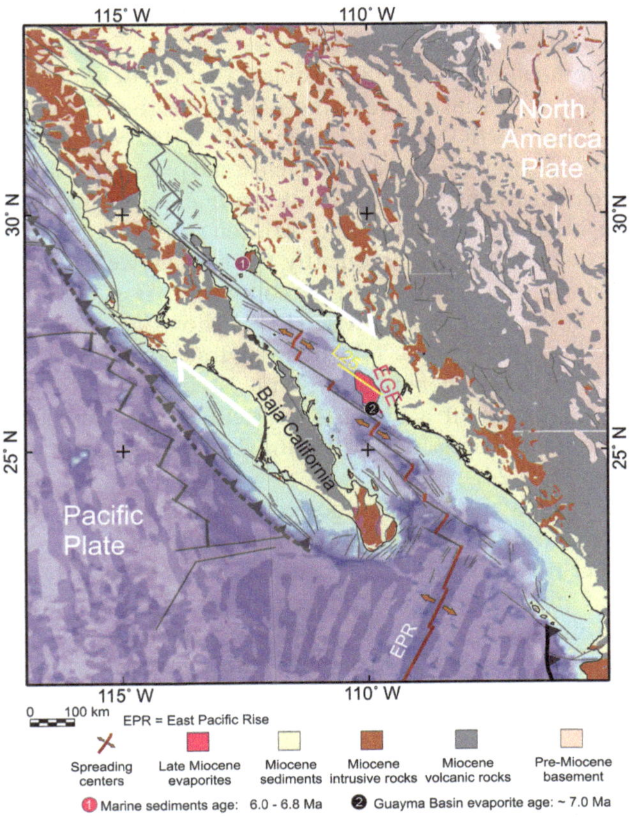

Figure 12.24 The simplified geological map depicts the dextral movement between the Pacific and North American plates along the San Andreas fault. Transform faults in the Gulf of California connect the Salton Sea trough with the East Pacific Rise, southeast of Baja California. The East Guaymas Evaporites (EGE) are a thick salt mass interpreted by 2D seismic lines at the North American plate near the southern part of the Gulf of California. Evaporite sediments are also observed in the central portion of the Gulf of California, west of the brown circle on the Baja California Peninsula. Joining the purple circle (1) in the middle of the Gulf of California with the black circle (2) south of the Guaymas Basin reconstructs the plate movement, revealing that these evaporites are conjugate to the EGE. A thin yellow line crosses the EGE, showing the location of the seismic profile L25. Credit: Adapted from Miller and Lizarralde (2013).

Salt deposits are known in the region of Baja California since the first Spanish settlements in the late 17th century identified halite in the sedimentary layers outcropping in the Isla del Carmen, located on the western margin of the gulf, near the city of Loreto. Kirkland et al. (1966) reported that in 1717, one of the Catholic Church missionaries that settled in the region wrote a letter to the King of Spain asking for permission to explore the salinas in the region of Carmen Island because "there might be enough salt there to supply the whole world." In modern times, some salt is still extracted by small-scale industry projects on the island, which produce salt by pumping brine from the subsurface and evaporating the water in the salt pans that cover the salt flats. However, the most important salt industry in Mexico is located on the western margin of Baja California, in the Guerrero Negro region, adjacent to the Pacific Ocean (see Chapter 10).

Dorsey and Umhoefer (2012) interpret the boundary between the Pacific and North American plates as corresponding to a deformation belt with a width of about 50–60 km. Transcurrent movement by strike-slip faults resulted in oblique rifting and the development of pull-apart sedimentary basins on land and across the Gulf of California (see Chapter 9).

Umhoefer et al. (2018) postulate that the development of the rift basins between the Baja California Peninsula and the North American mainland started with strike-slip faults that developed transtensional pull-apart basins. The rift troughs were filled with continental sediments in the Neogene, overlying Late Miocene volcanic rocks in the distal margin. Subsequently, these basins were flooded by marine incursions from the southern part of the Gulf, where the East Pacific Rise (EPR) shows an NE trend and is offset by major NW-trending transform faults that resulted in the opening of the Gulf. Miller and Lizarralde (2013) proposed a channel adjacent to the mainland, forming a connection with the Pacific Ocean waters (Figure 12.24).

The Guaymas Basin is located in the offshore region at the central part of the gulf, slightly to the north of Isla del Carmen (Figure 12.24). The basin was formed during the Upper Miocene and Pliocene by strike-slip faults and transtensional zones. The southern extremity of the gulf was close to the EPR spreading center, while the northern part was characterized by continental crust with lacustrine basins affected by extensional processes and magmatism (Bennett et al., 2015; Miller & Lizarralde, 2013). Thick sequences of evaporites were deposited in the Guaymas Basin during periods of basin restriction and extreme climate aridity. The barrier that isolated the evaporating basins from the open ocean was a topographic sill related to strike-slip faults. Extensional processes resulted in the breaching of the transtensional fault zones, leading to continental breakup in a short geological time interval. This process led to the creation of oceanic basins floored by a post-Miocene volcanic crust (Umhoefer, 2011).

Miller and Lizarralde (2013) presented seismic evidence of thick salt deposition in the deepwater region of the Guaymas Basin. Although not previously documented in the deepwater region of the Gulf of California, the salt masses appear to be concentrated at the distal margin, eastward of the present-day spreading center (Figure 12.25). These massive evaporites are possibly correlated with well-known gypsum layers that outcrop on the Baja California Peninsula, near Santa Rosalia, located to the northwest of the seismic profiles (Miller & Lizarralde, 2013). The Santa Rosalia Basin is characterized by the gypsum layers that are well exposed on the nearby island of San Marcos. This island is renowned for exhibiting a continuous outcrop of gypsum that extends for several kilometers, constituting one of the largest gypsum mines in the world. When the plates are reconstructed to their pre-breakup stage, the EGE is positioned close to the gypsum occurrences in Santa Rosalia and Isla San Marcos (Miller & Lizarralde, 2013).

The seismic profile in the Guaymas Basin (Figure 12.25) suggests a salt mass with a folded top and a thickness of

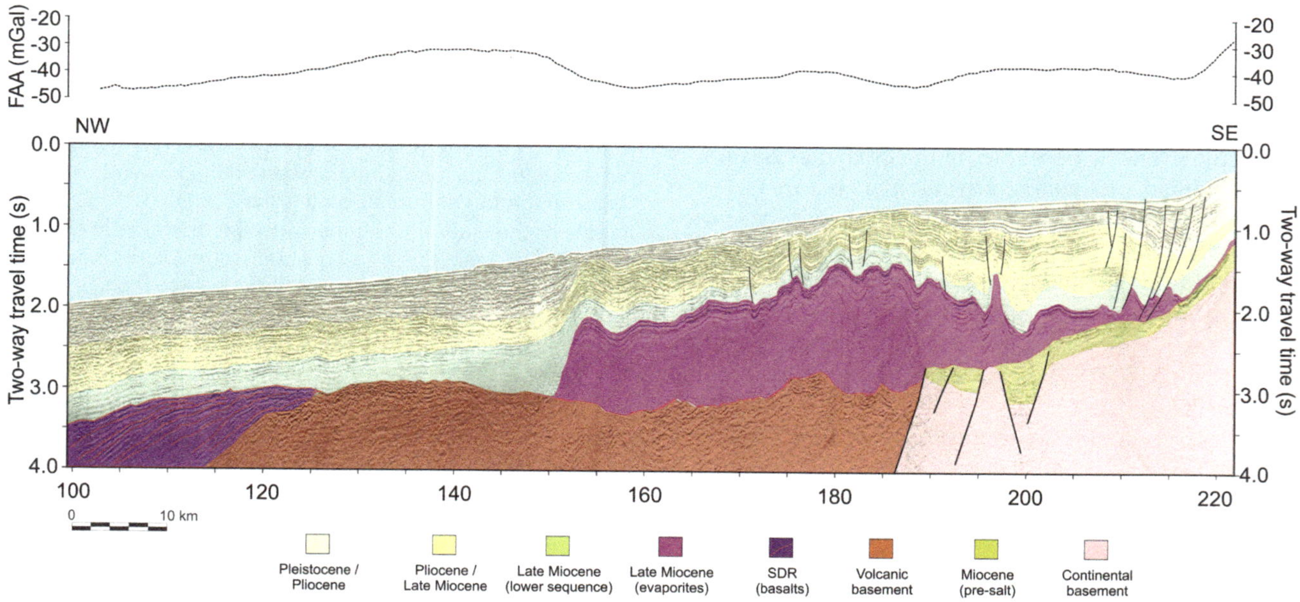

Figure 12.25 The seismic profile L25 shows that the salt mass extends from the continental crust to a region with a volcanic basement in the deepwater area of the Guaymas Basin, located in the central part of the Gulf of California. The transitional crust is located near seaward-dipping reflectors, marking the transition from the continental to the oceanic crust. The Miocene and Pliocene sedimentary sequences are characterized by extensional and compressional salt tectonic provinces forming depocenters from the platform toward the deep basin. Credit: Adapted from Miller and Lizarralde (2013).

about 2 km, pinching out toward the Mexican mainland. The thick evaporite mass is limited oceanward by wedges of seaward-dipping reflectors that exhibit a sigmoidal shape in seismic profiles. Seaward-dipping reflectors (SDRs) are indicative of strong magmatic activity during the inception of the oceanic crust (Jackson et al., 2000). These volcanic features are located about 100 km from the present-day spreading ridge. The gentle folding of the salt layers also affected part of the overburden, indicating Neogene compressional stresses. The salt distribution indicates that salt flowed downslope toward the oceanic crust, resulting in extensional structures upslope and salt withdrawal near the shelf edge, with growth sequences associated with listric faults (Figure 12.25).

12.8.1. Magmatism and the Age of Salt Deposition in the Guaymas Basin

The volcanic rocks of central Baja California trace the transition from calc-alkaline arc magmatism to syn-rift magmatism and volcanism, which occurred around 24–12 Ma. These plutonic rocks were followed by syn-rift magmatism and volcanism associated with the opening of the Gulf of California oceanic basin (Busby et al., 2020). The western margin of the Guaymas basin corresponds to the most magmatically active segment of the Gulf of California oceanic rift. Seafloor spreading began around 6 Ma and continued during the development of transtensional pull-apart basins, controlled by transform fault zones (Figure 12.24).

Volcanic and hypabyssal rocks on Santa Rosalia Island include volcanic sequences (lava flows, ash-flow tuffs, and ignimbrites) and hypabyssal intrusions (plugs, dikes, and peperites), yielding Ar-Ar ages ranging from 13.8 to 8.8 Ma, followed by andesitic lavas contemporaneous with the onset of seafloor spreading in the Guaymas basin, at ca. 6 Ma (Busby et al., 2020). The rift transitional volcanic rocks underlying the Santa Rosalia sedimentary basin have isotopic ages of about 9 Ma, providing a maximum age for the basin fill that underlies the salt layers (Figure 12.25).

The age of salt sequences in the Guaymas Basin is correlated with gypsum occurrences in Santa Rosalia and Isla San Marcus, forming massive beds of up to 70 m in thickness (Busby et al., 2020). Paleontological analysis of the underlying limestones and shales dated these deposits to the Late Miocene. Miller and Lizarralde (2013) suggest that salt deposition can be positioned at circa 7 Ma, close to the MSC in the Mediterranean Sea (see Chapter 5 in Mohriak, 2025), based on absolute radiometric ages provided by volcanic rocks above the gypsum.

Magmatic centers are imaged near the axial trough of the rift basin (Figure 12.25). The volcanic basement reflector is characterized by a typically rugose seismic aspect, suggesting extrusive igneous rocks (basalts). Hydrothermal vents and volcanic features, such as plugs and chimneys, have also been interpreted in other seismic profiles in the Guaymas Basin (e.g., Ondréas et al., 2018). These magmatic features are similar to the structures observed in the axial troughs of other young oceanic rifts, such as the Red Sea (see Chapter 1 in Mohriak, 2025).

12.8.2. Reconstruction of Basin Evolution and Salt Deposition in the Gulf of California

Geodetic studies (e.g., Plattner et al., 2007) indicate that a narrow zone between the Baja California Microplate (BCM) and the North American Plate (NAM) mainland confines over 90% of modern-day Pacific-North America (PAC-NAM) plate motion. This plate boundary concentrates most of the deformation between the mainland (western Mexico) and the Baja California Peninsula. This region experienced deformation during the Upper Miocene, with continental strike-slip faults and linked pull-apart basins. The oceanic crust was restricted to the region around the EPR in the south, while a continental depression extended for hundreds of kilometers toward the Salton Trough, sourced by fluvial waters from the Colorado River drainage at the border between the United States and Mexico. These geodynamic settings are similar to the depression north of the Gulf of Aqaba, which extends from the northern Red Sea to the Dead Sea along the Dead Sea Transform Fault Zone (see Chapter 1 in Mohriak, 2025).

During most of the Miocene, sediments related to a fluvial-lacustrine system filled the rift troughs in pull-apart basins (Umhoefer et al., 2018). During the Tortonian (about 8 Ma), the Gulf of California had marine sedimentation restricted to the embayment at its southern extremity, where the EPR approached the North American Plate limit (Figure 12.26-A).

By 7 Ma. (Messinian), marine incursions from the Pacific Ocean breached across the strike-slip faults in the south and flooded the Pescadera and Farallon basins (Figure 12.26-B). A transtensional fault zone extending for about 75–100 km separated the Farallon and the Guaymas basins but was only breached between 7 and 6.5 Ma when marine waters invaded the Guaymas Basin, which formed a depression below sea level. Intermittent marine incursions and high evaporation rates led to the deposition of a thick sequence of evaporites in the Guaymas Basin (EGE, Figure 12.24).

North of the Guaymas Basin, marine sediments dated between 6.8 and 6.0 Ma are recorded in the Isla del Tiburón, located close to the eastern margin of the Gulf of California (point 2 in Figure 12.24). Between 6.5 and 6.3 Ma, the strike-slip faults and transtensional zones between the Guaymas and Tiburón basins ceased to

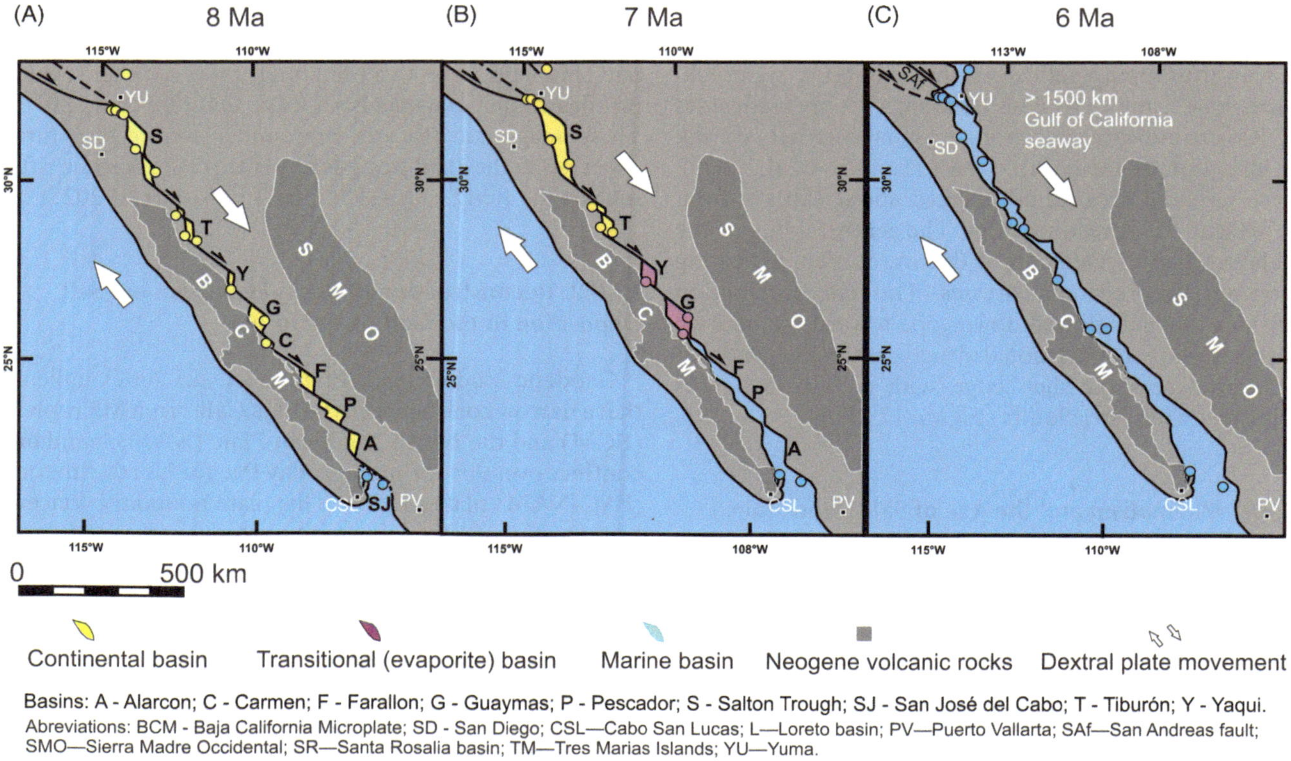

Figure 12.26 The simplified drawing shows the plate reconstruction and movements between the Pacific Plate-Baja California microplate relative to the North American plate. Transtensional motions formed rhombic-shaped pull-apart basins along the San Andreas-Gulf of California transform fault zone at 8 Ma. These basins were filled with continental lacustrine sediments. At 7 Ma, the Guaymas basin was filled with evaporites, while the Salton Trough was filled with continental sediments. The southernmost basins (Farallon, Pescador, and Alarcon) were filled with marine sediments. The Gulf of California opened around 6 Ma, flooding all basins and stopping evaporite sedimentation in the Guaymas Basin. Credit: Adapted from Umhoefer et al. (2018).

function as a barrier for marine incursions, and the northern part of the Gulf of California oblique rift was flooded with marine waters, ending the episode of salt precipitation (Figure 12.26-C).

By 6.0 Ma (Messinian), embryonic spreading centers started to rupture the continental crust, forming oceanic ridges with magmatic centers that split the salt in the Guaymas Basin. The seafloor spreading in the Guaymas basin resulted in basin isolation for 3–4 million years (Dorsey & Umhoefer, 2012; Umhoefer et al., 2018). During this interval, the regions to the north and south were still in the rifting stage, forming a series of pull-apart and transtensional basins, as registered in the Salton Trough (Figures 12.24). Hydrothermal vents are presently observed both in the axial trough and in the regions adjacent to the oceanic rift spreading center. The modern shallow seismicity indicates an active, ongoing process of seafloor spreading (Plattner et al., 2007).

The Guaymas Basin is characterized by pods of salt precipitation in the center of the Gulf of California, forming thick evaporite deposits that are geographically and chronologically associated with the incipient development of the divergent margins. The salt deposits in the Guaymas Basin are genetically related to the first marine incursions from a nearby ocean that provided a marine source for the brine that filled the depression in a short geological interval. The salt layers were formed between the continental rift phase and the establishment of active spreading centers that eventually split the salt masses by emplacement of the oceanic crust. After the depression was breached by the ridges and transform-fault corridors that connected to the global ocean, the hydrology in the gulf reflected the concentration of normal seawater, terminating the deposition of evaporite sediments (Figure 12.26).

The Guaymas Basin is a small but useful analog for the development of basinwide evaporite deposits in strike-slip and extensional settings, providing insights for the tectono-sedimentary evolution of much larger basins such as the Red Sea, the South Atlantic, and the Gulf of Mexico.

REFERENCES

Adams, J. E. (1965). Stratigraphic-tectonic development of Delaware Basin. *American Association of Petroleum Geologists Bulletin*, *49*(11), 2140–2148.

Afifi, A. M. (2005). Ghawar: The Anatomy of the World's Largest Oil Field, *AAPG Search and Discovery*, Article #20026.

Agard, P., Omrani, J., Jolivet, L., Whitechurch, H., Vrielynck, B., Sparkman, W., et al. (2011). Zagros orogeny: A subduction-dominated process. *Geological Magazine*, *148*(5–6), 692–725. Cambridge University Press. doi: 10.1017/S001675681100046X

Al-Husseini, M. I. (1997). Jurassic sequence stratigraphy of the Western and Southern Arabian Gulf. *GeoArabia*, *2*(4), 361–382.

Al-Johi, A., & Al-Laboun, A. (2015). Neoproterozoic (infracambrian)—Paleozoic Total Petroleum Systems of Arabian Peninsula; An Overview. *Arabian Journal of Earth Sciences*, *2*(1), 1–21.

Allen, J., & Beaumont, C. (2016). Continental margin syn-rift salt tectonics at intermediate width margins. *Basin Research*, *28*, 598–633. doi: 10.1111/bre.12123

Almalki, K. A., Ailleres, L., Betts, P. G., & Bantan, R. A. (2015). Evidence for and relationship between recent distributed extension and halokinesis in the Farasan Islands, southern Red Sea, Saudi Arabia. *Arabian Journal of Geosciences*, *8*, 8753–8766.

Alsharhan, A. S. (2006). Sedimentological character and hydrocarbon parameters of the Middle Permian to Early Triassic Khuff Formation, *United Arab Emirates. GeoArabia*, *11*(3), 121–158.

Alsharhan, A. S. (2014). Petroleum systems in the Middle East. In H. R. Rollinson, M. P. Searle, I. A. Abbasi, A. Al-Lazki, & M. H. Al Kindi (Eds.), *Tectonic evolution of the Oman Mountains* (Vol. 392, pp. 361–408). Geological Society, London, Special Publications.

Alsharhan, A. S., & Kendall, C. G. St. C. (1986). Precambrian to Jurassic rocks of Arabian Gulf and Adjacent areas: Their facies, depositional setting, and hydrocarbon habitat. *American Association of Petroleum Geologists Bulletin*, *70*(8), 977–1002.

Alsharhan, A. S., & Kendall, C. G. St. C. (1994). Depositional Setting of the Upper Jurassic Hith Anhydrite of the Arabian Gulf: An analog to Holocene evaporites of the United Arab Emirates and Lake MacLeod of Western Australia. *American Association of Petroleum Geologists Bulletin*, *78*(7), 1075–1096.

Alsharhan, A. S., & Kendall, C. G. St. C. (2003). Holocene coastal carbonates and evaporites of the southern Arabian Gulf and their ancient analogues. *Earth-Science Reviews*, *61*, 191–243.

Alsharhan, A. S., & Kendall, C. G. St. C. (2011). Introduction to Quaternary carbonate and evaporite sedimentary facies and their ancient analogues. In C. G. St. C. Kendall & A. S. Alsharhan (Eds.), *Quaternary carbonate and evaporite sedimentary facies and their ancient analogues: A tribute to Douglas James Shearman* (Vol. 43, pp. 1–10). International Association of Sedimentologists, Special Publication. Sussex, UK: Wiley-Blackwell.

Alsharhan, A. S., & Nairn, A. E. M. (1997). *Sedimentary Basins and Petroleum Geology of the Middle East* (pp. 651–736). Amsterdam: Elsevier.

Alsharhan, A. S., & Nairn, A. E. M. (2003). The geological history and structural elements of the Middle East. In A. S. Alsharhan & A. E. M. Nairn (Eds.), *Sedimentary basins and petroleum geology of the Middle East* (pp. 15–63). Amsterdam: Elsevier.

Alsharhan, A. S., & Whittle, G. L. (1995). Carbonate-evaporite sequences of the Late Jurassic, Southern and Southwestern Arabian Gulf. *American Association of Petroleum Geologists Bulletin*, *79*(11), 1608–1630.

Anderson, R. Y. (1981). *Deep-seated salt dissolution in the Delaware Basin, Texas and New Mexico* (Vol. 10, pp. 133–145). New Mexico Geological Society, Special Publication.

Anderson, R. Y., Dean, W. E., Jr., Kirkland, D. W., & Snider, H. I. (1972). Permian Castile varved evaporite sequence, West Texas and New Mexico. *Geological Society of America Bulletin*, *83*, 59–86.

Andreetto, F., Alois, G., Raad, F., Heida, H., Flecker, R., Agiadi, K., et al. (2021). Freshening of the Mediterranean Salt Giant: controversies and certainties around the terminal (Upper Gypsum and Lago-Mare) phases of the Messinian Salinity Crisis. *Earth-Science Reviews*, *216*, 103577. https://doi.org/10.1016/j.earscirev.2021.103577

Asl, M. E., Faghih, A., Mukherjee, S., & Soleimany, B. (2019). Style and timing of salt movement in the Persian Gulf basin, offshore Iran: Insights from halokinetic sequences adjacent to the Tonb-e-Bozorg salt diapir. *Journal of Structural Geology*, *122*, 116–132.

Ben Dor, Y., Neugebauer, I., Enzel, Y., Schwab, M. J., Tjallingii, R., Erel, Y., & Brauer, A. (2019). Varves of the Dead Sea sedimentary record. *Quaternary Science Reviews*, *215*, 173–184.

Benison, K. C., & Goldstein, R. H. (2000). Sedimentology of ancient saline pans: An example from the Permian Opeche shale, Williston Basin, North Dakota, U.S.A. *Journal of Sedimentary Research*, *70*(1), 159–169.

Bennett, S. E. K., Oskin, M. E., Dorsey, R. J., Iriondo, A., & Kunk, M. J. (2015). Stratigraphy and structural development of the southwest Isla Tiburón marine basin: Implications for latest Miocene tectonic opening and flooding of the northern Gulf of California. *Geosphere*, GES01153-1.

Beydoun, Z. R. (1991). Arabian plate hydrocarbon geology and potential—A plate tectonic approach. *American Association of Petroleum Geologists, Studies in Geology*, *33*, 77 pp.

Blakey, R. C. (2008a). Pennsylvanian-Jurassic sedimentary basins of the Colorado Plateau and Southern Rocky Mountains. In A. D. Miall (Ed.), *Sedimentary Basins of the World, Vol. 5, The Sedimentary Basins of the United States and Canada* (pp. 245–296). Amsterdam: Elsevier.

Blakey, R. C. (2008b). Gondwana paleogeography from assembly to breakup—A 500 m.y. odyssey. In C. R. Fielding, T. D. Frank, & J. L. Isbell (Eds.), *Resolving the late Paleozoic ice age in time and space* (Vol. 441, pp. 1–28). The Geological Society of America Special Paper.

Blakey, R. C. (2013). Using Paleogeographic maps to portray Phanerozoic Geologic and Paleotectonic History of Western

North America, *AAPG Search and Discovery*, Article #30267 (*2013*).

Blakey, R. C., & Ranney, W. D. (2018). *Ancient Landscapes of Western North America: A Geologic History with Paleogeographic Maps*, Springer International Publishing, 228 pp.

Blatt, H., Middleton, G., & Murray, R. (1980). *Origin of sedimentary rocks* (2nd ed.). New Jersey, USA: Prentice-Hall, 782 pp.

Blomquist, P. K. (2016). Wolfcamp Horizontal Play, Midland Basin, West Texas, *AAPG Search and Discovery*, Article #10890(*2016*).

Braga, J. C., Martín, J. M., Riding, R., Aguirre, J., Sánchez-Almazo, I. M., & Dinarès-Turell, J. (2006). Testing models for the Messinian Salinity Crisis: The Messinian record in Almería, SE Spain. *Sedimentary Geology*, *189–190*, 131–154.

Burchette, T. P. (2012). Carbonate rocks and petroleum reservoirs: a geological perspective from the industry. In Garland, J., Neilson, J. E., Laubach, S. E., & Whidden, K. J. (Eds), *Advances in carbonate exploration and reservoir analysis* (Vol. 370, pp. 17–37). Geological Society, London, Special Publications. http://dx.doi.org/10.1144/SP370.14

Busby, C., Graettinger, A., López Martínez, M., Sarah Medynski, S., Niemi, T., Andrews, C., et al. (2020). Volcanic record of the arc-to-rift transition onshore of the Guaymas basin in the Santa Rosalía area, Gulf of California, *Baja California. Geosphere*, *16*(4), 1012–1041. https://doi.org/10.1130/GES02094.1

Busson, G. (1991). Relationship between different types of evaporitic deposits, and the occurrence of organic-rich layers (potential source-rocks). *Carbonates and Evaporites*, *6*(2), 177–192.

Carminatti, M., Dias, J. L., & Wolf, B. (2009). From turbidites to carbonates: Breaking paradigms in deep waters. *Offshore Technology Conference, Abstract* Volume, *OTC-20124*, 7 pp.

Carminatti, M., Wolff, B., & Gamboa, L. A. P. (2008). New exploratory frontiers in Brazil. *19th World Petroleum Congress*, Madri, Spain, WPC Proceedings, Abstracts CD, 11pp.

Christeleit, E. C., Brandon, M. T., & Zhuang, G. (2015). Evidence for deep-water deposition of abyssal Mediterranean evaporites during the Messinian salinity crisis. *Earth and Planetary Science Letters*, *427*, 226–235.

CIESM (2008). The Messinian Salinity Crisis from mega-deposits to microbiology—A consensus report. N° 33 in *CIESM Workshop Monographs* [F. Briand, Ed.], 168 pages, Monaco.

Clauzon, G., Suc, J.-P., Gautier, F., Berger, A., & Loutre, M.-F. (1996). Alternate interpretation of the Messinian salinity crisis: Controversy resolved? *Geology*, *24*(4), 363–366.

Clauzon, G., Suc, J.-P., Couto, D. D., Jouannic, G., Melinte-Dobrinescu, M. C., Jolivet, L., et al. (2015). New insights on the Sorbas Basin (SE Spain): The onshore reference of the Messinian Salinity Crisis. *Marine and Petroleum Geology*, *66*(Part 1), 71–100. https://doi.org/10.1016/j.marpetgeo.2015.02.016

Dasgupta, S. N., Hong, M. R., & Al-Jallal, I. A. (2001). Reservoir characterization of Permian Khuff-C carbonate in the supergiant Ghawar Field of Saudi Arabia. *The Leading Edge*, *20*(7), 706–717. https://doi.org/10.1190/1.1487282

Dickson, W., Schiefelbein, C., & Odegard, M. (2019). Defining a supergiant petroleum system in Brazil's Santos Basin with multidisciplinary methods: One template for exploration success. *Interpretation*, *7*(4), SH133–SH152.

Dorsey, R. J., & Umhoefer, P. J. (2012). Influence of sediment input and plate-motion obliquity on basin development along an active oblique divergent plate boundary: Gulf of California and Salton Trough. In C. J. Busby & A. Azor (Eds.), *Tectonics of sedimentary basins: Recent advances* (pp. 209–225). Oxford, UK: Wiley-Blackwell Publishing.

Druckman, Y., Buchbinder, B., Martinotti, G. M., Simon Tov, R., & Aharon, P. (1995). The buried Afiq Canyon (eastern Mediterranean, Israel): A case study of a Tertiary submarine canyon exposed in Late Messinian times. *Marine Geology*, *123*, 167–185. doi:10.1016/0025-3227(94)00127-7.

Ehrenberg, S. N., Nadeau, P. H., & Aqrawi, A. A. M. (2007). A comparison of Khuff and Arab reservoir potential throughout the Middle East. *American Association of Petroleum Geologists Bulletin*, *91*(3), 275–286.

Engle, M. A., Doolan, C. A., Pitman, J. A., Varonka, M. S., Chenault, J., Orem, W. H., et al. (2020). Origin and geochemistry of formation waters from the lower Eagle Ford Group, Gulf Coast Basin, south central Texas. *Chemical Geology*, *550*, 119754. https://doi.org/10.1016/j.chemgeo.2020.119754

Engle, M. A., Reyes, F. R., Varonka, M. S., Orem, W. H., Ma, L., Ianno, A. J., et al. (2016). Geochemistry of formation waters from the Wolfcamp and "Cline" shales: Insights into brine origin, reservoir connectivity, and fluid flow in the Permian Basin, USA. *Chemical Geology*, *425*, 76–92. http://dx.doi.org/10.1016/j.chemgeo.2016.01.025

Evans, G. (2011). An historical review of the Quaternary sedimentology of the Gulf (Arabian/Persian Gulf) and its geological impact. In C. G. St. C. Kendall & A. S. Alsharhan (Eds.), *Quaternary carbonate and evaporite sedimentary facies and their ancient analogues, A tribute to Douglas James Shearman* (Vol. 43, pp. 11–44). International Association of Sedimentologists Special Publication. doi:10.1002/9781444392326.ch2

Ewing, T. E. (2017). The Permian Basin—Intracratonic subsidence triggered by ARM compression, making an American "Superbasin", *AAPG Search and Discovery*, Article #10988 (*2017*).

Ewing, T. E., & Galloway, W. E. (2019). Evolution of the Northern Gulf of Mexico Sedimentary Basin. In A. D. Miall (Ed.), *The sedimentary basins of the United States and Canada* (pp. 627–694). Amsterdam: Elsevier.

Faqira, M., Rademakers, M., & Afifi, A. M. (2009). New insights into the Hercynian Orogeny, and their implications for the Paleozoic Hydrocarbon System in the Arabian Plate. *GeoArabia*, *14*(3), 199–228.

Feng, Y. E., Yankelzon, A., Steinberg, J., & Reshef, M. (2016). Lithology and characteristics of the Messinian evaporite sequence of the deep Levant Basin, eastern Mediterranean. *Marine Geology*, *376*, 118–131.

Friedman, G. M. (1982). Evaporites as source rock for petroleum. In: C. R. Handford, R. G. Loucks, & G. R. Davies (Eds.), *Depositional and diagenetic spectra of evaporites—A core workshop* (Vol. 3, pp. 385–395). SEPM Society for Sedimentary Geology.

Garrison, R. E., Schreiber, B. C., Bernoulli, D., Fabriciius, F. H., Kidd, R. B., & Mélières, F. (1978). Sedimentary petrology and structures of Messinian evaporitic sediments in the

Mediterranean Sea. In Hsü, K. J., et al. (Eds.), *Initial Reports of the Deep Sea Drilling Project* (Vol. 42, Part 1, pp. 571–612). Washington, DC: U.S. Government Printing Office.

Glennie, K. W. (1998). *Petroleum geology of the North Sea: Basic concepts and recent advances* (4th ed.), 656 pages. Oxford, UK: Blackwell Publishing Limited.

Glennie, K. W. (2010). Structural and stratigraphic evolution of Abu Dhabi in the context of Arabia. *Arabian Journal of Geosciences*, *3*(4), 331–349. https://doi.org/10.1007/s12517-010-0194-2

Godo, T. (2017). The Appomattox field: Norphlet aeolian sand dune reservoirs in the deep-water Gulf of Mexico. In R. K. Merrill & C. A. Sternbach (Eds.), *Giant fields of the decade 2000–2010, American Association of Petroleum Geologists Memoir* (Vol. 113, pp. 29–54).

Godo, T. J. (2019). The Smackover-Norphlet petroleum system, deepwater Gulf of Mexico: Oil fields, oil shows, and dry holes, *GCAGS Journal*, *8*, 104–152.

Gradstein, F. M., Ogg, J. G., Schmitz, M. D., & Ogg, G. (2012). *The Geologic Time Scale 2012*. Amsterdam: Elsevier, 1144 pp. https://doi.org/10.1016/B978-0-444-59425-9.00004-4

Grobe, M. (2000). Distribution and thickness of salt within the Devonian Elk Point Group, Western Canada sedimentary basin. Alberta Energy and Utilities Board, Alberta Geological Survey, *Earth Sciences Report 2000-02*, 12 pp + Maps.

Grötsch, J., Suwaina, O., Ajlani, G., Taher, A., El-Khassawneh, R., Lokier, S., et al. (2003). The Arab Formation in central Abu Dhabi: 3-D reservoir architecture and static and dynamic modeling, *GeoArabia*, *8*(1), 47–86.

Gvirtzman, G., & Buchbinder, B. (1978). The Late Tertiary of the Coastal Plain and Continental Shelf of Israel and its bearing on the history of the eastern Mediterranean. In K. J. Hsü, L. Montadert, et al. (Eds.), *Initial reports of the Deep Sea Drilling Project* (Vol. 42, Part 2, pp. 1195–1222). Washington: U.S. Government Printing Office. doi:10.2973/dsdp.proc.42-2.160.1978

Handford, C. R. (1981a). A process-sedimentary framework for characterizing recent and ancient sabkhas. *Sedimentary Geology*, *30*, 255–265.

Handford, C. R. (1981b). Coastal sabkha and salt pan deposition of the lower Clear Fork Formation (Permian), *Texas. Journal of Sedimentary Research*, *51*(3), 761–778.

Handford, C. R. (1991). Marginal marine halite: Sabkhas and salinas. In J. L. Melvin (Ed.), *Evaporites, petroleum and mineral resources, Developments in Sedimentology* (Vol. 50, Chapter 1, pp. 1–66). Amsterdam: Elsevier.

Handford, C. R., & Bassett, R. L. (1982). Permian facies sequences and evaporite depositional styles, Texas Panhandle. In C. R. Handford, R. G. Loucks, & G. R. Davies (Eds.), *Depositional and diagenetic spectra of evaporites—A core workshop* (Vol. 3, pp. 210–237). SEPM Society for Sedimentary Geology.

Hardie, L. A. (1991). On the significance of evaporites. *Annual Review Earth and Planetary Science*, *290*, 43–106.

Hardie, L. A., & Eugster, H. P. (1970). *The evolution of closed basin brines* (Vol. 3, pp. 273–290). Mineralogical Society of America, Special Paper.

Hardie, L. A., & Eugster, H. P. (1971). The depositional environment of marine evaporites: A case for shallow, clastic accumulation. *Sedimentology*, *16*, 187–220.

Hardie, L. A., & Lowenstein, T. K. (2004). Did the Mediterranean Sea dry out during the miocene? A reassessment of the evaporite evidence from DSDP Legs 13 and 42A Cores. *Journal of Sedimentary Research*, *74*(4), 453–461. doi:10.1306/112003740453

Hardie, L. A., Lowenstein, T. K., & Spencer, R. J. (1985). The problem of distinguishing between primary and secondary features in evaporites. In B. C. Schreiber & H. L. Harner (Eds.), *Sixth International Symposium on Salt* (Vol. 1, pp. 11–39). Alexandia, VA: The Salt Institute.

Harris, P. M., & Kerans, C. (2016). The role of the Guadalupe Mountains and Capitan Research in the evolution of carbonate conceptual models and paradigms, *AAPG Search and Discovery*, Article #60052 (*2016*).

Hartman, J. K., & Woodard, L. R. (1971). Future petroleum resources in post-Mississippian strata of North Central, and West Texas and eastern New Mexico. In I. H. Cram (Ed.), *Future petroleum provinces of the United States their geology and potential, American Association of Petroleum Geologists Memoir*, *15*(2), 752–803.

Hassanpour, J., Yassaghi, A., Muñoz, J. A., & Jahani, S. (2020). Salt tectonics in a double salt-source layer setting (Eastern Persian Gulf, Iran): Insights from interpretation of seismic profiles and sequential cross-section restoration. *Basin Research*. doi:10.1111/bre.12459

Hills, J. M. (1972). Late Paleozoic sedimentation in West Texas Permian Basin. *American Association of Petroleum Geologists Bulletin*, *56*(12), 2303–2322.

Hills, J. M. (1984). Sedimentation, tectonism, and hydrocarbon generation in Delaware Basin, West Texas and southeastern New Mexico. *American Association of Petroleum Geologists Bulletin*, *68*(3), 250–267.

Hite, R. J., & Anders, D. E. (1991). Petroleum and evaporites. In J. L. Melvin (Ed.), *Evaporites, petroleum and mineral resources, Developments in Sedimentology* (Vol. 50, Chapter 4, pp. 349–411). Amsterdam: Elsevier. doi:10.1016/s0070-4571(08)70262-0

Holliday, D. W. (1970). The petrology of secondary gypsum rocks: A review. *Journal of Sedimentary Petrology*, *40*(2), 734–744.

Hovorka, S. D., Holt, R. M., & Powers, D. W. (2007). Depth indicators in Permian Basin evaporites. In B. C. Schreiber, S. Lugli, & M. Babel (Eds.), *Evaporites through space and time* (Vol. 285, pp. 335–364). 6.

Hsü, K. J. (1972). Origin of saline giants: A critical review after the discovery of the Mediterranean evaporite. *Earth-Science Reviews*, *8*, 371–396.

Hsü, K. J. (1977). Tectonic evolution of the Mediterranean Basins. In A. E. M. Nairn, W. H. Kanes, & F. G. Stehli (Eds.), *The ocean basins and margins: The eastern Mediterranean* (Vol. 4A, pp. 29–75). New York: Springer, Plenum Press.

Hsü, K. J., Cita, M. B., & Ryan, W. B. F. (1973). The origin of the Mediterranean evaporites. In W. B. F. Ryan, K. J. Hsü, et al. (Eds.), *Initial reports of the Dead Sea Drilling Project* (Vol. 13, pp. 1203–1231). Washington: United States Government Printing Office.

Hsü, K. J., Montadert, L., Bernoulli, D., Cita, M. B., Erickson, A., Garrison, R. E., et al. (1977). History of the Mediterranean salinity crisis. *Nature*, *267*, 399–403.

Hsü, K. J., Montadert, L., Bernoulli, D., Cita, M. B., Erickson, A., Garrison, R. E., et al. (1978). History of the Mediterranean salinity crisis. In K. J. Hsü, L. Montadert, et al. (Eds.), *Initial reports of the Deep Sea Drilling Project* (Vol. 42, Part 1, pp. 1053–1078). Washington: U.S. Government Printing Office. doi:10.2973/dsdp.proc.42-1.149.1978

Hudec, M. R., & Peel, F. J. (2019). A critical review of models for deposition of the Louann salt, and implications for Gulf of Mexico evolution, *37th Annual GCSSEPM Foundation Perkins-Rosen Research Conference, Salt Tectonics, Associated Processes, and Exploration Potential: Revisited: 1989–2019* (pp. 222–238).

Huerta, A. D., & Harry, D. L. (2012). Wilson cycles, tectonic inheritance, and rifting of the North American Gulf of Mexico continental margin. *Geosphere*, *8*(2), 374–385. https://doi.org/10.1130/GES00725.1

Husseini, M. I., & Husseini, S. I. (1990). Origin of the Infra-Cambrian salt basin of the Middle East. In J. Brooks (Ed.), *Classic Petroleum Provinces* (Vol. 50, pp. 279–292). Geological Society, London, Special Publication.

Infante-Paez, L., Cardona, L.-F., McCullough, B., & Slatt, R. (2016). Seismic analysis of paleotopography and stratigraphic controls on total organic carbon: Rich sweet spot distribution in the Woodford Shale, Oklahoma, USA. *Interpretation*, *5*(1), T33–T47.

Ivanov, V. V., Shapiro, G. I., Huthnance, J. M., Aleynik, D. L., & Golovin, P. N. (2004). Cascades of dense water around the world ocean. *Progress in Oceanography*, *60*, 47–98.

Jackson, C. A. L., Jackson, M. P. A., Hudec, M. R., & Rodriguez, C. R. (2015). Enigmatic structures within salt walls of the Santos Basin—Part 1: Geometry and kinematics using 3D seismic reflection and well data. *Journal of Structural Geology*, *75*, 135–162. doi: 10.1016/j.jsg.2015.01.010.

Jackson, C. A. -L., Elliott, G. M., Royce-Rogers, E., Gawthorpe R. L., & Aas, T. E. (2019). Salt thickness and composition influence rift structural style, northern North Sea, offshore Norway. *Basin Research*, *31*(3), 514–538. doi: 10.1111/bre.12332

Jackson, M. P. A., & Vendeville, B. C. (1994). Regional extension as a geologic trigger for diapirism. *Geological Society of America Bulletin*, *106*, 57–73.

Jackson, M. P. A., Cramez, C., & Fonck, J. M. (2000). Role of subaerial volcanic rocks and mantle plumes in creation of South Atlantic margins: implications for salt tectonics and source rocks. *Marine and Petroleum Geology*, *17*, 477–498.

Jarzyna, A., Bąbel, M., Ługowski, D., & Vladi, F. (2022). Petrographic Record and Conditions of Expansive Hydration of Anhydrite in the Recent Weathering Zone at the Abandoned Dingwall Gypsum Quarry, Nova Scotia, *Canada. Minerals*, *12*(1), 58. https://doi.org/10.3390/min12010058.

Jenyon, M. K. (1986). *Salt Tectonics*. Amsterdam: Elsevier Applied Science Publishers Ltd., 191 pp.

Johnson, K., & Gonzales, S. (1978). *Salt deposits in the United States and regional geologic characteristics important for storage of radioactive waste*. Prepared for Union Carbide Corp., Nuclear Division, Oak Ridge National Laboratory, Office of Waste Isolation, Y/OWI/SUB-7414/1, 188 p.

Kendall, A. C. (1978). Facies Models 12. Subaqueous evaporites. In R. G. Walker (Ed.), *Facies Models, Geoscience Canada*, *5*(3), 124–139.

Kendall, A. C. (1984). Evaporites. In R. G. Walker (Ed.), *Facies Models, Geoscience Canada, Reprint Series* (Vol. 1, pp. 259–296).

Kendall, A. C. (1992). Evaporites. In R. G. Walker & N. P. James (Eds.), *Facies Models: Response to sea level change* (pp. 375–409). Geological Association of Canada.

Kendall, A. C. (2010). Marine evaporites. In N. P. James & R. W. Dalrymple (Eds.), *Facies Models, Geological Association of Canada IV* (pp. 505–539). Series GEOtext.

Kendall, A. C., & Harwood, G. M. (1996). Marine evaporites: arid shorelines and basins. In H. G. Reading (Ed.), *Sedimentary environments: Processes, facies and stratigraphy* (3rd ed., pp. 281–324). Oxford: Blackwell Scientific Publications.

Kendall, C. G. St. C., Weber, L. J., & Alsharhan, A. S. (2009). *The Giant Oil Field Evaporite Association: A function of the Wilson Cycle, climate, basin position and sea level, AAPG Search and Discovery, Article #40471 (2009)*.

King, P. B. (1942). Permian of west Texas and southeastern New Mexico. *American Association of Petroleum Geologists Bulletin*, *26*(4), 535–763.

King, R. H. (1947). Sedimentation in the Permian Castile sea. *American Association of Petroleum Geologists Bulletin*, *31*(3), 470–477.

Kinsman, D. J. J. (1969). Modes of formation, sedimentary associations, and diagnostic features of shallow-water and supratidal evaporites. *American Association of Petroleum Geologists Bulletin*, *53*(4), 830–840.

Kirkland, D. W. (2003). An explanation for the Varves of the Castile evaporites (Upper Permian), Texas and New Mexico, USA. *Sedimentology*, *50*, 899–920. doi: 10.1046/j.1365-3091.2003.00588.x

Kirkland, D. W., & Evans, R. (1981). Source rock potential of evaporitic environment. *American Association of Petroleum Geologists Bulletin*, *65*(1), 181–190.

Kirkland, D. W., Bradbury, J. P., & Dean, W. E., Jr. (1966). Origin of Carmen Island salt deposit, Baja California, *Mexico. The Journal of Geology*, *74*(6), 932–938.

Kirkland, D. W., Denison, R. E., & Dean, W. E. (2000). Parent brine of the castile evaporites (Upper Permian), Texas and New Mexico, *Journal of Sedimentary Research*, *70*(3), 749–761.

Konert, G., Afifi, A. M., Al-Hajri, S. A., & Droste, H. J. (2001). Paleozoic stratigraphy and hydrocarbon habitat of the Arabian Plate. *GeoArabia*, *6*(3), 407–442.

Kukla, P. A., Strozyk, F., & Mohriak, W. U. (2018). South Atlantic salt basins—Witnesses of complex passive margin evolution. *Gondwana Research*, *53*, 41–57.

Lee, Y. I., & Friedman, G. M. (1987). Deep-burial dolomitization in the Ordovician Ellenburger Group carbonates, West Texas and Southeastern New Mexico. *Journal of Sedimentary Petrology*, *57*(3), 544–557.

Lindsay, R. F., Cantrell, D. L., Hughes, G. W., Keith, T. H., Mueller, H. W., III, & Russell, S. D. (2006). Ghawar Arab-D reservoir: Widespread porosity in shoaling-upward carbonate cycles, Saudi Arabia. In P. M. Harris & L. J. Weber (Eds.), *Giant hydrocarbon reservoirs of the world: From rocks to reservoir characterization and modeling, AAPG Memoir 88/SEPM Special Publication* (pp. 97–137).

Lucia, F. J. (1972). Recognition of evaporite-carbonate shoreline sedimentation. In J. K. Rigby & W. K. Hamblin (Eds.), *Recognition of ancient sedimentary environments* (Vol. 16, pp. 160–191). SEPM Special Publication.

Lugli, S., Gennari, R., Gvirtzman, Z., Manzi, V., Roveri, M., & Schreiber, B. C. (2013). Evidence of clastic evaporites in the canyons of the Levant Basin (Israel): Implications for the Messinian salinity crisis. *Journal of Sedimentary Research*, *83*, 942–954. doi: 10.2110/jsr.2013.72

Lugli, S., Manzi, V., Roveri, M., & Schreiber, B. C. (2010). The Primary Lower Gypsum in the Mediterranean: A new facies interpretation for the first stage of the Messinian salinity crisis. *Palaeogeography, Palaeoclimatology, Palaeoecology*, *297*, 83–99. doi:10.1016/j.palaeo.2010.07.017

Lugli, S., Manzi, V., Roveri, M., & Schreiber, B. C. (2015). The deep record of the Messinian salinity crisis: Evidence of a non-desiccated Mediterranean Sea. *Palaeogeography, Palaeoclimatology, Palaeoecology*, *433*, 201–218.

Madof, A. S., Bertoni, C., & Lofi, J. (2019). Discovery of vast fluvial deposits provides evidence for drawdown during the late Miocene Messinian salinity crisis. *Geology*, *47*(2), 171–174. doi:10.1130/g45873.1

Mancini, E. A., Parcell, W. C., Puckett, T. M., & Benson, D. J. (2003). Upper Jurassic (Oxfordian) Smackover carbonate petroleum system characterization and modeling, Mississippi Interior Salt Basin area, northeastern Gulf of Mexico, USA. *Carbonates and Evaporites*, *18*(2), 125–150. doi:10.1007/bf03176234

Manzi, M., Lugli, S., Lucchi, F. R., & Roveri, M. (2005). Deep-water clastic evaporites deposition in the Messinian Adriatic foredeep (northern Apennines, Italy): Did the Mediterranean ever dry out? *Sedimentology*, *52*, 875–902. doi: 10.1111/j.1365-3091.2005.00722.x

Manzi, V., Gennari, R., Lugli, S., Roveri, M., Scafetta, N., & Schreiber, B. C. (2012). High-frequency cyclicity in the Mediterranean Messinian evaporites: Evidence for solar–lunar climate forcing. *Journal of Sedimentary Research*, *82*, 991–1005. doi:10.2110/jsr.2012.81.

Manzi, V., Lugli, S., Roveri, M., Schreiber, B. C., & Gennari, R. (2011). The Messinian "Calcare di Base" (Sicily, Italy) revisited. *Geological Society of America Bulletin*, *123*(1/2), 347–370.

Manzi, V., Gennari, R., Hilgen, F., Krijgsman, W., Lugli, S., Roveri, M., & Sierro, F. J. (2013). Age refinement of the Messinian salinity crisis onset in the Mediterranean. *Terra Nova*, *25*, 315–322. doi:10.1111/ter.12038.

Manzi, V., Gennari, R., Lugli, S., Persico, D., |Reghizz, M., Roveri, M., et al. (2018). The onset of the Messinian salinity crisis in the deep Eastern Mediterranean basin. *Terra Nova*, *30*(3), 189–198. https://doi.org/10.1111/ter.12325

Manzi, V., Argnani, A., Corcagnani, A., Lugli, S., & Roveri, M. (2020). The Messinian salinity crisis in the Adriatic foredeep: Evolution of the largest evaporitic marginal basin in the Mediterranean. *Marine and Petroleum Geology*, *115*, 104288. https://doi.org/10.1016/j.marpetgeo.2020.104288

Marchionda, E., Deschamps, R., Cobianchi, M., Nader, F. H., Di Giulio, A., Morad, D. J., et al. (2018). Field-scale depositional evolution of the Upper Jurassic Arab Formation (onshore Abu Dhabi, UAE). *Marine and Petroleum Geology*, *89*(2), 350–369.

Martinez del Olmo, W. (2011). El arrecife messiniense del sondeo Torrevieja Marino C-1 desde las líneas sísmicas (SE. de España). *Revista Sociedad Geologica De Espana*, *24*(3–4), 173–185.

Mascle, G., & Mascle, J. (2019). The Messinian salinity legacy: 50 years later. *Mediterranean Geoscience Reviews* (pp. 1–11). https://doi.org/10.1007/s42990-019-0002-5

Matchus, E. J., & Jones, T. S. (1984). East-West cross section through Permian Basin of West Texas. *West Texas Geological Society, Publication No. 84*, p. 79.

Meijer Drees, N. (1994). Devonian Elk Point Group of the Western Canada sedimentary basin. In G. Mossop & I. Shetsen (Eds.), *Geological Atlas of the Western Canada Sedimentary Basin* (pp. 129–147). Canadian Society of Petroleum Geologists/Alberta Research Council, Calgary, Alberta.

Meilijson, A., Hilgen, F., Sepúlveda, J., Steinberg, J., Fairbank, V., Flecker, R., et al. (2019). Chronology with a pinch of salt: Integrated stratigraphy of Messinian evaporites in the deep Eastern Mediterranean reveals long-lasting halite deposition during Atlantic connectivity. *Earth-Science Reviews*, *194*, 374–398.

Meilijson, A., Steinberg, J., Hilgen, F., Bialik, O. M., Waldmann, N. D., & Makovsky, Y. (2018). Deep-basin evidence resolves a 50-year-old debate and demonstrates synchronous onset of Messinian evaporite deposition in a non-desiccated Mediterranean. *Geology*, *46*(3), 243–246.

Mello, M. R., Rostirolla, S. P., Elias, V., Ferreira, M. Ahulalli, A. P., Becker, S., et al. (2021). Lacustrine source rocks and oil systems present in the lower cretaceous pre-salt sections of the Santos Basin, Brazil. In M. R. Mello, P. O. Yilmaz, & B. J. Katz (Eds.), *The supergiant Lower Cretaceous pre-salt petroleum system of the Santos Basin, Brazil, American Association of Petroleum Geologists, Memoir* (Vol. 124, pp. 35–76).

Miller, N. C., & Lizarralde, D. (2013). Thick evaporites and early rifting in Guaymas Basin, Gulf of California. *Geology*, *41*(2), 283–286.

Miramontes, E., Déverchère, J., Pellegrini, C., & Chiarella, D. (2023). Chapter 2—Mediterranean Sea evolution and present-day physiography. In K. Schroeder & J. Chiggiato (Eds.), *Oceanography of the Mediterranean Sea: An introductory guide* (pp. 13–39). Amsterdam: Elsevier.

Mohriak, W. U. (2015). Pre-Salt carbonate reservoirs in the South Atlantic and World-wide Analogs, AAPG Search and Discovery, Article #51086 (2015). Posted April 6, 2015.

Mohriak, W. (2025). *Salt in the Earth sciences: Basin analysis and salt tectonics.* Washington, DC: American Geophysical Union. doi:10.1002/9781394307067

Montaron, B., & Tapponnier, P. (2009). A quantitative model for salt deposition in actively spreading basins, *AAPG Search and Discovery*, Article #30117 (*2010*)

Moore, D. G., Curray, J. R., Blakeslee, J., Platt, L. W., Stout, L. N., Aguayo, J. E., & Vacquier, V. (1982). Objectives of drilling on young passive continental margins; application to the Gulf of California, *Initial Reports of the Deep Sea Drilling Project*, *64*(1), 27–33. doi:10.2973/dsdp.proc.64.102.1982

Nissen, E., Tatar, M., Jackson, J. A., & Allen, M. B. (2011). New views on earthquake faulting in the Zagros fold-and-thrust belt of Iran. *Geophysical Journal International*, *186*, 928–944. doi: 10.1111/j.1365-246X.2011.5119.x

Ondréas, H., Scalabrin, C., Fouquet, Y., & Godfroy, A. (2018). Recent high-resolution mapping of Guaymas hydrothermal fields (Southern Trough). *BSGF—Earth Sciences Bulletin*, *189*, 6.

Osterloff, P., Al-Harthy, A., Penney, R., Spaak, P., Williams, G., Al-Zadjali, F., et al. (2004). Depositional sequences of the Gharif and Khuff formations, subsurface Interior Oman. In Al-Husseini, M. I. (Ed.), *Carboniferous, Permian and Early Triassic Arabian Stratigraphy* (Vol. 3, pp. 83–147). GeoArabia Special Publication.

Palmer, M. V., & Palmer, A. N. (2012). Petrographic and isotopic evidence for late-stage processes in sulfuric acid caves of the Guadalupe Mountains, New Mexico, USA. *International Journal of Speleology*, *41*(2), 231–250.

Perthuisot, J. P. (1989). Part II. Recent evaporites. In P. Sonnenfeld and J.-P. Perthuisot, *Brines and evaporites, Short Course in Geology* (Vol. 3, pp. 65–118). 28th International Geological Congress. Washington, DC: American Geophysical Union.

Pilcher, R. S., & Blumstein, R. D. (2007). Brine volume and salt dissolution rates in Orca Basin northeast Gulf of Mexico. *American Association of Petroleum Geologists Bulletin*, *91*(6), 823–833. doi:10.1306/12180606049.

Plattner, C., Malservisi, R., Dixon, T. H., LaFemina, P., Sella, G. F., Fletcher J. M., & Suarez-Vidal, F. (2007). New constraints on relative motion between the Pacific plate and Baja California microplate (México) from GPS measurements. *Geophysical Journal International*, *170*, 1373–1380.

Presley, M. W. (1987). Evolution of Permian evaporite basin in Texas Panhandle. *American Association of Petroleum Geologists Bulletin*, *71*(2), 167–190.

Raup, O. B. (1970). Brine mixing: an additional mechanism for formation of basin evaporites. *American Association of Petroleum Geologists Bulletin*, *54*(12), 2246–2259.

Raup, O. B. (1982). Gypsum precipitation by mixing seawater brines. *American Association of Petroleum Geologists Bulletin*, *66*(3), 363–367.

Read, J. F. (1985). Carbonate platform facies models. *American Association of Petroleum Geologists Bulletin*, *69*, 1–21.

Riding, R., Braga, J. C., Martín, J. M., & Sánchez–Almazo, I. M. (1998). Mediterranean Messinian Salinity Crisis: Constraints from a coeval marginal basin, Sorbas, southeastern Spain. *Marine Geology*, *146*, 1–20.

Riding, R., Braga, J. C., & Martín, J. M. (1999). Late Miocene Mediterranean desiccation: topography and significance of the Salinity Crisis erosion surface on-land in southeast Spain. *Sedimentary Geology*, *123*, 1–7. https://doi.org/10.1016/S0037-0738(98)00115-8

Riding, R., Braga, J. C., & Martín, J. M. (2000). Discussion: Late Miocene Mediterranean desiccation: Topography and significance of the "Salinity Crisis" erosion surface on-land in southeast Spain: Reply. *Sedimentary Geology*, *133*, 175–184.

Robichaud, S. R., & Robichaud, N. W. (2019). Mineral Resource–Lithium–Recovery from naturally occurring Permian Basin Waste Water. *AAPG Search and Discovery Article #42475 (2019)*, 27 pp.

Robinson, K. (1988). Petroleum Geology and hydrocarbon plays of the Permian Basin petroleum province, West Texas and Southeast New Mexico. *United States Department of the Interior, U.S. Geological Survey*, Open-File Report 88-450Z, 53 p.

Roveri, M., & Manzi, V. (2006). The Messinian salinity crisis: Looking for a new paradigm? *Palaeogeography, Palaeoclimatology, Palaeoecology*, *238*(1–4), 386–398. http://dx.doi.org/10.1016/j.palaeo.2006.03.036

Roveri, M., Lugli, S., Manzi, V., & Schreiber, B. C. (2008). The Messinian salinity crisis: A sequence-stratigraphic approach. *GeoActa*, *7984*, 169–190.

Roveri, M., Manzi, V., Bergamasco, A., Falcieri, F. M., Gennari, R., Lugli, S., & Schreiber, B. C. (2014). Dense shelf water cascading and Messinian canyons: A new scenario for the Mediterranean salinity crisis. *American Journal of Science*, *314*, 751–784. doi 10.2475/05.2014.03

Roveri, M., Gennari, R., Lugli, S., Manzi, V., Minelli, N., Reghizzi, M., et al. (2016). The Messinian salinity crisis: Open problems and possible implications for Mediterranean petroleum systems. *Petroleum Geoscience*, *22*, 283–290. doi:10.1144/petgeo2015-089

Roveri, M., Gennari, R., Ligi, M., Lugli, S., Manzi, V., & Reghizzi, M. (2019). The synthetic seismic expression of the Messinian salinity crisis from onshore records: Implications for shallow- to deep-water correlations. *Basin Research*, *31*, 1121–1152. https://doi.org/10.1111/bre.12361

Roveri, M., Lugli, S., Manzi, V., Reghizzi, M., & Rossi, F. P. (2020). Stratigraphic relationships between shallow-water carbonates and primary gypsum: Insights from the Messinian succession of the Sorbas Basin (Betic Cordillera, Southern Spain). *Sedimentary Geology*, *105678*. doi:10.1016/j.sedgeo.2020.105678

Ryan, W. B. F. (2008). Modeling the magnitude and timing of evaporite drawdown during the Messinian salinity crisis. *Stratigraphy*, *5*(3–4), 227–243.

Sajadi, S. H., Baghbani, D., Daneshian, J., & Moezabad, M. K. (2016). Stratigraphy of Oligocene–Miocene salt deposits in the SE Persian Gulf. *Carbonates Evaporites*, *31*, 277–288.

Saller, A. H., & Stueber, A. M. (2018). Evolution of formation waters in the Permian Basin, USA: Late Permian evaporated seawater to Neogene meteoric water. *American Association of Petroleum Geologists Bulletin*, *102*(3), 401–428.

Sarg, J. F. (2001). The sequence stratigraphy, sedimentology, and economic importance of evaporite-carbonate transitions: A review. *Sedimentary Geology*, *140*, 9–42.

Scanlon, B. R., Reedy, R. C., Male, F., & Walsh, M. (2017). Water Issues related to transitioning from conventional to unconventional oil production in the Permian Basin. *Environmental Science Technology*, *51*(18), 10903–10912. doi: 10.1021/acs.est.7b02185.

Schiefelbein, C. F., Zumgerge, J. E., Cameron, N. R., & Brown, S. W. (1999). Petroleum systems in the South Atlantic margins. In N. R. Cameron, R. H. Bate, & V. S. Clure (Eds.), *The oil and gas habitats of the South Atlantic* (Vol. 153, pp. 169–179). Geological Society, London, Special Publications.

Schmalz, R. F. (1969). Deep-water evaporite deposition: A genetic model, *American Association of Petroleum Geologists Bulletin*, *53*(4), 798–823.

Scholle, P. (2020). Introduction to the Geology of the Permian Reef Complex, Guadalupe and Delaware Mountains, New Mexico & West Texas, New Mexico Bureau of Geology and Mineral Resources. https://geoinfo.nmt.edu/tour/federal/parks/PermianReef/home.html

Scholle, P. A., Ulmer, D. S., & Melim, L. A. (1992). Late-stage calcites in the Permian Capitan Formation and its equivalents,

Delaware Basin margin, West Texas and New Mexico: Evidence for replacement of precursor evaporites. *Sedimentology*, *39*(2), 207–234. doi:10.1111/j.1365-3091.1992.tb01035.x

Scholle, P. A., Goldstein, R. H., & Ulmer-Scholle, D. S. (2007). *Classic upper Paleozoic reefs and bioherms of west Texas and New Mexico*. Open file Report 504, New Mexico Bureau of Geology and Mineral Resources, Socorro, New Mexico, 174 pp. https://geoinfo.nmt.edu/publications/openfile/downloads/500-599/504/ofr_504.pdf

Schreiber, B. C. (1978). Environments of subaqueous gypsum deposition. In W. E. Dean & B. C. Schreiber (Eds.), *Marine evaporites, Lecture Notes for Short Course* (Vol. 4, pp. 43–73). Society of Economic Paleontologists and Mineralogists, Tulsa, Okla.

Schreiber, B. C. (1986). Arid shorelines and evaporites. In: H. G. Reading (Ed.), *Sedimentary environments and facies* (pp. 189–228). Oxford, UK: Blackwell Scientific Publications.

Schreiber, B. C. (1988). *Evaporites and hydrocarbons*. New York: Columbia University Press, 475 pp.

Schreiber, B. C., & Friedman, G. M. (1976). Depositional environments of Sicily as determined from analysis of intercalated carbonates. *Sedimentology*, *23*, 255–270.

Schreiber, B. C., & Hsü, K. J. (1980). Evaporites. In G. D. Hobson (Ed.), *Developments in petroleum geology* (Vol. 2, pp. 87–138). London: Applied Science Publishers.

Schreiber, B. C., & El Tabakh, M. (2000). Deposition and early alteration of evaporites. *Sedimentology*, *47*(1), 215–238. https://doi.org/10.1046/j.1365-3091.2000.00002.x

Schreiber, B. C., Roth, M. S., & Helman, M. L. (1982). Recognition of primary facies characteristics of evaporites and the differentiation of these forms from diagenetic overprints. In C. R. Handford, R. G. Loucks, & G. R. Davies (Eds.), *Depositional and diagenetic spectra of evaporites—A core workshop* (Vol. 3, pp. 1–32). SEPM Society for Sedimentary Geology.

Scotese, C. R. (2001). *Atlas of Earth History, Volume 1, Paleogeography*. PALEOMAP Project, Arlington, Texas, 52 pp.

Shearman, D. J. (1978). Evaporites of coastal sabkhas. In W. E. Dean & B. C. Schreiber (Ed.), *Marine evaporites, Lecture Notes for Short Course* (Vol. 4, pp. 6–42). Society of Economic Paleontologists and Mineralogists, Tulsa, Okla.

Silver, B. A., & Todd, R. G. (1969). Permian cyclic strata, Northern Midland and Delaware Basins, West Texas and Southeastern New Mexico. *American Association of Petroleum Geologists Bulletin*, *53*(11), 2223–2251.

Sonnenfeld, P., & Perthuisot, J. -P. (1989). *Brines and evaporites, Short Course in Geology* (Vol. 3), 28th International Geological Congress, Washington, DC: American Geophysical Union, 126 pp.

Stampfli, G., Marcoux, J., & Baud, A. (1991). Tethyan margins in space and time. *Palaeogeography, Palaeoclimatology, Palaeoecology*, *87*, 373–409.

Stewart, S. A. (2016). Structural geology of the Rub' Al-Khali Basin, Saudi Arabia. *Tectonics*, *35*, 2417–2438.

Stoneley, R. (1987). A review of petroleum source rocks in parts of the Middle East. In J. Brooks & A. J. Fleet (Eds.), *Marine petroleum source rocks* (Vol. 26, pp. 263–269). Geological Society Special Publication.

Stueber, A. M., Saller, A. H., & Ishida, H. (1998). Origin, migration, and mixing of brines in the Permian Basin: Geochemical evidence from the Eastern Central Basin Platform, Texas. *American Association of Petroleum Geologists Bulletin*, *82*(9), 1652–1672.

Swart, P. K., Cantrell, D. L., Westphal, H., Handford, C. R., & Kendall, C. G. (2005). Origin of dolomite in the Arab-D reservoir from the Ghawar Field, Saudi Arabia: Evidence from petrographic and geochemical constraints. *Journal of Sedimentary Research*, *75*(3), 476–491.

Taylor, J. C. M. (1984). Late Permian Zechstein. In K. W. Glennie, (Ed.), *Introduction to the petroleum geology of the North Sea* (Chapter 4, pp. 61–101). London: Blackwell Scientific Publishers.

Taylor, J. C. M. (1985). *The role of evaporites in hydrocarbon exploration*. Joint Association of Petroleum Exploration Courses (UK), Japec Course No. 39.

Taylor, J. C. M. (1998). Upper Permian–Zechstein. In K. W. Glennie (Ed.), *Petroleum Geology of the North Sea: Basic concepts and recent advances* (4th ed., pp. 174–211). Oxford, UK: Blackwell Publishing Limited.

Thomas, R. J., Ellison R. A., Goodenough, K. M., Roberts, N. M. W., & Allen, P. A. (2015). Salt domes of the UAE and Oman: Probing eastern Arabia. *Precambrian Research*, *256*, 1–16.

Trudgill, B. (2011). Evolution of salt structures in the northern Paradox Basin: Controls on evaporite deposition, salt wall growth and supra-salt stratigraphic architecture. *Basin Research*, *23*, 208–238.

Umhoefer, P. J. (2011). Why did the Southern Gulf of California rupture so rapidly?—Oblique divergence across hot, weak lithosphere along a tectonically active margin. *GSA Today*, *21*(11), 4–10. doi: 10.1130/G133A.1

Umhoefer, P. J., Darin, M. H., Bennett, S. E. K., Skinner, L. A., Dorsey, R. J., & Oskin, M. L. E. (2018). Breaching of strike-slip faults and successive flooding of pull-apart basins to form the Gulf of California seaway from ca. 8–6 Ma. *Geology*, *46*(8), 695–698.

U.S. Geological Survey (2021). Mineral commodity summaries 2021. *U.S. Geological Survey*, 200 p. https://pubs.usgs.gov/periodicals/mcs2021/mcs2021.pdf

Van den Belt, F. J. G., & De Boer, P. L. (2007). A shallow-basin model for "saline giants" based on isostasy-driven subsidence. In: G. Nichols, E. Williams, and C. Paola (Eds.), *Sedimentary processes, environments and basins: A tribute to Peter Friend* (Vol. 38, pp. 241–252). International Association of Sedimentologists Special Publication. doi:10.1002/9781444304411.ch11

Vaslet, D., Le Nindre, Y.-M., Vachard, D., Broutin, J., Crasquin-Soleau, S., Berthelin, M., et al. (2005). The Permian-Triassic Khuff formation of central Saudi Arabia. *GeoArabia*, *10*(4), 77–134. doi:10.2113/geoarabia100477

Vigil, J. F., Pike, R. J., & Howell, D. G. (2000). *A tapestry of time and terrain, U.S. Geological Survey Geologic Investigations Series 2720*. https://pubs.usgs.gov/imap/i2720/

Walker, J. D., Geissman, J. W., Bowring, S. A., & Babcock, L. E. (Compilers) (2018). *Geologic Time Scale*, v. *5.0*, Geological Society of America, https://doi.org/10.1130/2018.CTS005R3C.

Ward, R. F., Kendall, C. G. St. C., & Harris, P. M. (1986). Late Permian (Guadalupian) facies and their association with hydrocarbons. *American Association of Petroleum Geologists Bulletin*, *70*(3), 239–262.

Warren, J. K. (1986). Shallow-water evaporitic environments and their source rock potential. *Journal of Sedimentary Petrology*, *56*(3), 442–454.

Warren, J. K. (1989). *Evaporite Sedimentology: Importance in hydrocarbon accumulation*. New Jersey, USA: Prentice-Hall Advanced Reference Series, 285 p.

Warren, J. K. (1991). Sulfate dominated sea-marginal and platform settings. In J. L. Melvin, *Evaporites, petroleum and mineral resources, Developments in Sedimentology* (Vol. 50, Chapter 2, pp. 477–533). Amsterdam: Elsevier.

Warren, J. K. (1992). Evaporites and their importance to petroleum exploration. *IHRDC, GL 508 Petroleum Geology,* 234 pp.

Warren, J. K. (2006). *Evaporites: Sediments, resources and hydrocarbons, 1035p*. Berlin: Springer Verlag.

Warren, J. K. (2010). Evaporites through time: Tectonic, climatic and eustatic controls in marine and nonmarine deposits. *Earth-Science Reviews*, *98*, 217–268. doi:10.1016/j.earscirev.2009.11.004

Warren, J. K. (2012). Understanding hydrocarbons in an evaporite context, *AAPG Search and Discovery*, Article #50549 *(2012)*.

Warren, J. K. (2014). *Evaporite sedimentology: importance in hydrocarbon accumulation*. New Jersey, USA: Prentice-Hall Advanced Reference Series, 285 p.

Warren, J. K. (2016). *Evaporites—A geological compendium*. Berlin: Springer, 1807 pp. ISBN 978-3-319-13511-3.

Warren, J. K., & Kendall, C. G. St. C. (1985). Comparison of sequences formed in marine sabkha (subaerial) and salina (subaqueous) settings—Modern and ancient. *American Association of Petroleum Geologists Bulletin*, *69*(6), 1013–1023.

Woolnough, W. G. (1937). Sedimentation in barred basins, and source rocks of oil. *American Association of Petroleum Geologists Bulletin*, *21*(9), 1101–1157.

Wright, W. R. (2011). Pennsylvanian paleodepositional evolution of the greater Permian Basin, Texas and New Mexico: Depositional systems and hydrocarbon reservoir analysis. *American Association of Petroleum Geologists Bulletin*, *95*(9), 1525–1555.

Yang, K.-M., & Dorobek, S. (1995). The Permian Basin of West Texas and New Mexico: Tectonic history of a "composite" foreland basin and its effects on stratigraphic development. In *Stratigraphic Evolution of Foreland Basins* (*No 52*, pp. 149–174). SEPM Special Publication.

Yin, S.-Z., Bai, G.-P., Gao, L., & Xu, C.-Y. (2018). A facies and palaeogeography-based approach for analysis of petroleum systems in United Arab Emirates. *Journal of Palaeogeography*, *7*(2), 168–178.

Zak, I. (1997). Evolution of the Dead Sea brines. In T. M., Niemi, Z. Ben-Avraham, & J. R. Gat (Eds.), *The Dead Sea: The lake and its setting* (pp. 133–144). Oxford, UK: Oxford University Press.

Ziegler, M. A. (2001). Late Permian to Holocene Paleofacies evolution of the Arabian Plate and its hydrocarbon occurrences. *GeoArabia*, *6*(3), 445–504.

INDEX

Note: *Italic* page numbers refer to *figures* and **Bold** page numbers reference to **tables**.

Salt in the Earth Sciences: Evaporite Rocks and Salt Deposition, Geophysical Monograph 286, First Edition. Webster Mohriak.

DOI: 10.1002/9781119405245.index

Printed in the USA/Agawam, MA
September 16, 2026

910661.205